VOLUME EIGHTY

VITAMINS AND HORMONES

Insulin and IGFs

VITAMINS AND HORMONES

VOLUME EIGHTY

VITAMINS AND HORMONES

Insulin and IGFs

Editor-in-Chief

GERALD LITWACK

Chair of Basic Sciences and
Interim Associate Dean for Research
Commonwealth Medical College
Scranton, Pennsylvania

Former Professor and Chair
Department of Biochemistry and Molecular Pharmacology
Thomas Jefferson University Medical College
Philadelphia, Pennsylvania

Former Visiting Scholar
Department of Biological Chemistry
David Geffen School of Medicine at UCLA
Los Angeles, California

ELSEVIER

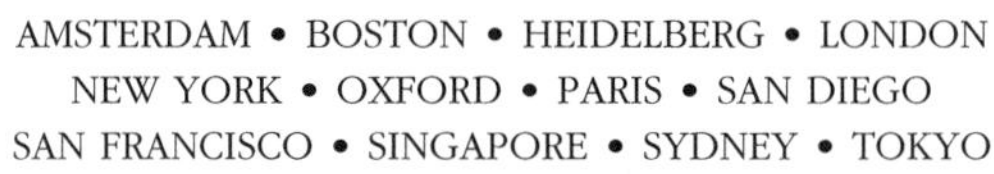

AMSTERDAM • BOSTON • HEIDELBERG • LONDON
NEW YORK • OXFORD • PARIS • SAN DIEGO
SAN FRANCISCO • SINGAPORE • SYDNEY • TOKYO

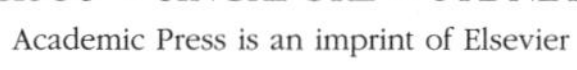

Academic Press is an imprint of Elsevier

Cover photo credit:
Litwack, G. Human Biochemistry and Disease (2008), pp. 131–188.

Academic Press is an imprint of Elsevier
32 Jamestown Road, London, NW1 7BY, UK
Radarweg 29, PO Box 211, 1000 AE Amsterdam, The Netherlands
Linacre House, Jordan Hill, Oxford OX2 8DP, UK
30 Corporate Drive, Suite 400, Burlington, MA 01803, USA
525 B Street, Suite 1900, San Diego, CA 92101-4495, USA

First edition 2009

ISBN: 978-0-12-374408-1
ISSN: 0083-6729

For information on all Elsevier Academic Press publications
visit our website at elsevierdirect.com

Printed and bound in USA
09 10 11 12 10 9 8 7 6 5 4 3 2 1

Contents

Contributors

Rebecca A. Bachmann
Department of Cell and Developmental Biology, University of Illinois at Urbana-Champaign, 601 S. Goodwin Ave. B107, Urbana, IL 61801

Jonathan S. Bogan
Section of Endocrinology and Metabolism, Department of Internal Medicine and Department of Cell Biology, Yale University School of Medicine, New Haven, Connecticut 06520-8020

Michael Bordonaro
The Commonwealth Medical College, Department of Basic Sciences, Scranton, Pennsylvania 18510

Sigalit Boura-Halfon
Department of Molecular Cell Biology, Weizmann Institute of Science, Rehovot 76100, Israel

J. Brown
Cancer Research UK Receptor Structure Research Group, Division of Structural Biology, Wellcome Trust Centre for Human Genetics, University of Oxford, Roosevelt Drive, Headington, Oxford OX3 7BN, United Kingdom

Maria Francesca Cassarino
Department of Internal Medicine, Endocrinology Unit, University of Catania, P.O. Garibaldi Nesima, Via Palermo 636, 95122 Catania, Italy

Jie Chen
Department of Cell and Developmental Biology, University of Illinois at Urbana-Champaign, 601 S. Goodwin Ave. B107, Urbana, IL 61801

Ashok Kumar Dilly
Departments of Ophthalmology and Cell Biology, and Dean A. McGee Eye Institute, University of Oklahoma Health Sciences Center, 608 Stanton L. Young Boulevard, Oklahoma City, Oklahoma 73104

Hesham M. El-Shewy
Department of Medicine, Medical University of South Carolina, Charleston, South Carolina 29425

Craig A. Eyster
Department of Biochemistry and Molecular Biology, Oklahoma University Health Sciences Center, Oklahoma City, Oklahoma 73126

WuQiang Fan
Department of Medicine and Bioregulatory Science, Graduate School of Medical Science, Kyushu University, Maidashi 3-1-1, Higashi-ku, Fukuoka 812-8582, Japan

B. E. Forbes
School of Molecular and Biomedical Science, The University of Adelaide, Adelaide 5005, Australia

Francesco Frasca
Department of Internal Medicine, Endocrinology Unit, University of Catania, P.O. Garibaldi Nesima, Via Palermo 636, 95122 Catania, Italy

Marco Genua
Department of Internal Medicine, Endocrinology Unit, University of Catania, P.O. Garibaldi Nesima, Via Palermo 636, 95122 Catania, Italy

Takahiko Hara
Stem Cell Project Group, The Tokyo Metropolitan Institute of Medical Science, Tokyo Metropolitan Organization for Medical Research, 3-18-22 Honkomagome, Bunkyo-ku, Tokyo 113-8613, Japan

June Chunqiu Hou
Department of Pharmacological Sciences, Stony Brook University, Stony Brook, NY 11794

Raymond E. Hulse
Committee on Neurobiology, The University of Chicago, Chicago, Illinois 60637

Olof Idevall-Hagren
Department of Medical Cell Biology, Uppsala University, Biomedical Centre, SE-751 23 Uppsala, Sweden

Ivana Ivanovic
Departments of Ophthalmology and Cell Biology, and Dean A. McGee Eye Institute, University of Oklahoma Health Sciences Center, 608 Stanton L. Young Boulevard, Oklahoma City, Oklahoma 73104

Maja Jensen
Receptor Systems Biology Laboratory, Hagedorn Research Institute, Niels Steensensvej 6, 2820 Gentofte, Denmark

E. Y. Jones
Cancer Research UK Receptor Structure Research Group, Division of Structural Biology, Wellcome Trust Centre for Human Genetics, University of Oxford, Roosevelt Drive, Headington, Oxford OX3 7BN, United Kingdom

Jeong-Ho Kim
Department of Cell and Developmental Biology, University of Illinois at Urbana-Champaign, 601 S. Goodwin Ave. B107, Urbana, IL 61801

Su-Jin Kim
Department of Cellular and Physiological Sciences, The Diabetes Research Group, Life Sciences Institute, University of British Columbia, 2350 Health Sciences Mall, Vancouver, BC, Canada V6T 1Z3

Feng Liu
Departments of Pharmacology and Biochemistry, The University of Texas Health Science Center, San Antonio, Texas 78229

Louis M. Luttrell
Department of Biochemistry and Molecular Biology, Medical University of South Carolina, Charleston, South Carolina 29425, and Research Service of the Ralph H. Johnson Veterans Affairs Medical Center, Charleston, South Carolina 29401, and Department of Medicine, Medical University of South Carolina, Charleston, South Carolina 29425

Geert A. Martens
Diabetes Research Center, Brussels Free University-VUB, Laarbeeklaan 103, 1090 Brussels, Belgium

Cynthia Corley Mastick
Department of Biochemistry and Molecular Biology, University of Nevada School of Medicine, Reno, Nevada 89557

Christopher H. S. McIntosh
Department of Cellular and Physiological Sciences, The Diabetes Research Group, Life Sciences Institute, University of British Columbia, 2350 Health Sciences Mall, Vancouver, BC, Canada V6T 1Z3

Joseph L. Messina
Veterans Affairs Medical Center, Birmingham, Alabama 35233, and Department of Pathology, Division of Molecular and Cellular Pathology, 1670 University Boulevard, University of Alabama at Birmingham, Birmingham, Alabama 35294-0019

Rosa Linda Messina
Department of Internal Medicine, Endocrinology Unit, University of Catania, P.O. Garibaldi Nesima, Via Palermo 636, 95122 Catania, Italy

Pierre De Meyts
Receptor Systems Biology Laboratory, Hagedorn Research Institute, Niels Steensensvej 6, 2820 Gentofte, Denmark

Le Min
Division of Endocrinology, Diabetes and Hypertension, Brigham and women's hospital, 221 Longwood Ave, Boston, MA 02115

Joseph M. Muretta
Department of Biochemistry, Molecular Biology, and Biophysics, University of Minnesota, Minneapolis, Minnesota 55455

Yuki Nakayama
Stem Cell Project Group, The Tokyo Metropolitan Institute of Medical Science, Tokyo Metropolitan Organization for Medical Research, 3-18-22 Honkomagome, Bunkyo-ku, Tokyo 113-8613, Japan, and Priority Organization for Innovation and Excellence, Kumamoto University, 2-39-1 Kurokami, Kumamoto City, Kumamoto, 860–8555, Japan

Lilia Noriega
Departamento de Fisiología de la Nutrición, Instituto Nacional de Ciencias Médicas y Nutrición Vasco de Quiroga No 15, México DF 14000, Mexico

Ann Louise Olson
Department of Biochemistry and Molecular Biology, Oklahoma University Health Sciences Center, Oklahoma City, Oklahoma 73126

Giuseppe Pandini
Department of Internal Medicine, Endocrinology Unit, University of Catania, P.O. Garibaldi Nesima, Via Palermo 636, 95122 Catania, Italy

Jeffrey E. Pessin
Departments of Medicine and Molecular Pharmacology, Albert Einstein College of Medicine, Bronx, NY 10461

Daniel Pipeleers
Diabetes Research Center, Brussels Free University-VUB, Laarbeeklaan 103, 1090 Brussels, Belgium

Raju V. S. Rajala
Departments of Ophthalmology and Cell Biology, and Dean A. McGee Eye Institute, University of Oklahoma Health Sciences Center, 608 Stanton L. Young Boulevard, Oklahoma City, Oklahoma 73104

Luis A. Ralat
Ben-May Department for Cancer Research, The University of Chicago, Chicago, Illinois 60637

Bradley R. Rubin
Section of Endocrinology and Metabolism, Department of Internal Medicine and Department of Cell Biology, Yale University School of Medicine, New Haven, Connecticut 06520-8020

Alexei Y. Savinov
Burnham Institute for Medical Research, Inflammatory and Infectious Disease Center, La Jolla, CA 92037

Frances Separovic
School of Chemistry, University of Melbourne, Victoria 3010, Australia

Fazel Shabanpoor
School of Chemistry, and Howard Florey Institute, University of Melbourne, Victoria 3010, Australia

Alex Y. Strongin
Burnham Institute for Medical Research, Inflammatory and Infectious Disease Center, La Jolla, CA 92037

Tian-Qiang Sun
Rigel Pharmaceuticals, Inc., South San Francisco, California 94080

Xiao Jian Sun
Department of Medicine, The University of Chicago, Chicago, Illinois 60637

Wei-Jen Tang
Committee on Neurobiology, and Ben-May Department for Cancer Research, The University of Chicago, Chicago, Illinois 60637

Anders Tengholm
Department of Medical Cell Biology, Uppsala University, Biomedical Centre, SE-751 23 Uppsala, Sweden

Nimbe Torres
Departamento de Fisiología de la Nutrición, Instituto Nacional de Ciencias Médicas y Nutrición Vasco de Quiroga No 15, México DF 14000, Mexico

Armando R. Tovar
Departamento de Fisiología de la Nutrición, Instituto Nacional de Ciencias Médicas y Nutrición Vasco de Quiroga No 15, México DF 14000, Mexico

John D. Wade
School of Chemistry, and Howard Florey Institute, University of Melbourne, Victoria 3010, Australia

Michael A. Weiss
Department of Biochemistry, Case Western Reserve University, Cleveland, Ohio 44106

Scott Widenmaier
Department of Cellular and Physiological Sciences, The Diabetes Research Group, Life Sciences Institute, University of British Columbia, 2350 Health Sciences Mall, Vancouver, BC, Canada V6T 1Z3

Jie Xu
Department of Medicine, Division of Endocrinology, Diabetes, and Metabolism, University of Alabama at Birmingham, Birmingham, Alabama 35294-0019

Toshihiko Yanase
Department of Medicine and Bioregulatory Science, Graduate School of Medical Science, Kyushu University, Maidashi 3-1-1, Higashi-ku, Fukuoka 812-8582, Japan

Jiandi Zhang
Rigel Pharmaceuticals, Inc., South San Francisco, California 94080

Yehiel Zick
Department of Molecular Cell Biology, Weizmann Institute of Science, Rehovot 76100, Israel

Preface

Diabetes, especially Type 2 diabetes, has reached epidemic proportions in the United States. Because of the greater emphasis on solving the problems associated with this disease, it is critical to bring forward an up-to-date review of the basic science aspects. Accordingly, this volume focuses on insulin as the pivotal hormone, the understanding of which is essential for fundamental laboratory approaches to the clinical problem. Although the subject of diabetes occurs in this treatise, the emphasis is upon the basic science of the hormone, insulin, and also IGFs, themselves. We need to know about the three-dimensional structure, receptors, signaling, secretion mechanisms, regulation and relationships to other factors in the body. These considerations and others are highlighted in this book.

The newer structural aspects of insulin are topics at the beginning: F. Shabanpoor, F. Separovic and J.D. Wade start off with "The human insulin superfamily of polypeptide hormones" followed by "The structure and function of insulin: decoding the TR transition" by M.A. Weiss. Signaling from the insulin receptor and related subjects come next. "Molecular mechanisms of differential intracellular signaling from the insulin receptor" is reviewed by M. Jensen and P. De Meyts. M. Genua, G. Pandini, M.F. Cassarino, R.L. Messina and F. Frasca offer "C-ABL and insulin receptor signaling". "CXCL14 and insulin action" is reported by T. Hara and Y. Nakayama. J. Xu and J.L. Messina write on "Crosstalk between growth hormone and insulin signaling".

GLUT4, a key glucose transporter, is the topic of the following. "Intracellular retention and insulin-stimulated mobilization of GLUT4 glucose transporters", by B.R. Rubin and J.S. Bogan and "Compartmentalization and regulation of insulin signaling to GLUT4 by the cytoskeleton" by C.A. Eyster and A.L. Olson.

A number of subjects are addressed in the papers that follow. "Nutrient modulation of insulin secretion" is reviewed by N. Torres, L. Noriega and A.R. Tovar. J.M Muretta reports "How insulin regulates glucose transport in adipocytes". "Spatio-temporal dynamics of phosphatidylinositol-3,4,5-trisphosphate signaling" is the work of A. Tengholm and O. Idevall-Hagren. S. Boura-Halfon and Y. Zick introduce "Serine kinases of insulin receptor substrate proteins". X.J. Sun and F. Liu report "Phosphorylation of IRS-proteins: Yin-Yang regulation of insulin signaling". "IRS-2 and its involvement in diabetes and aging" is reviewed by J. Zhang and T.-Q. Sun. "Glucose-dependent insulinotropic polypeptide (gastric inhibitory polypeptide; GIP)" by C.H.S. McIntosh, S. Widenmaier and S.-J. Kim and "Insulin granule biogenesis, trafficking and exocytosis" by J.C. Hou and J.E. Pessin, follow. The next subject is "Glucose, regulator of survival and phenotype of pancreatic beta cells" by G.A. Martens and D. Pipeleers. Several offerings follow: "Matrix metalloproteinases, T cell homing and *beta*-cell mass in type 1 diabetes" by A.Y. Savinov and A.Y. Strongin; "Role of Wnt signaling in the development of type 2 diabetes" by M. Bordonaro; "Retinal insulin receptor signaling in hyperosmotic stress" by R.V.S. Rajala, I. Ivanovic and A.K. Dilly; "Interleukin-6 and insulin resistance" by J.-H. Kim, R.A. Bachmann and J. Chen; "Structure, function and regulation of insulin-degrading enzyme" by R.E. Hulse, L.A. Ralat and W.-J. Tang; "Modification of androgen receptor function by IGF-1 signaling: implications in the mechanism of refractory prostate carcinoma" by T. Yanase and W.-Q. Fan; "Insulin-like growth factor-2/mannose-6 phosphate receptors" by H.M. El-Shewy and L.M. Luttrell, and lastly, "Interaction of IGF-II with the IGF2R/cation-independent mannose-6-phosphate receptor: mechanism and biological outcomes" by B.J. Jones and B.E. Forbes.

The structure on the cover is a cartoon of the insulin receptor taken from *Human Biochemistry and Disease*, Academic Press/Elsevier, 2008.

Renske van Dijk and Tari Paschall Broderick helped with the early phases and then Gayle Luque and, more recently, Lisa Tickner and Janice Hackenberg of Academic Press/Elsevier facilitated this publication.

Gerald Litwack
October 3, 2008

CHAPTER ONE

The Human Insulin Superfamily of Polypeptide Hormones

Fazel Shabanpoor,[*,†] Frances Separovic,[†] *and* John D. Wade[*,†]

Contents

Abstract

The identification in the 1950s of insulin, an essential carbohydrate regulatory hormone, as consisting of not one but two peptide chains linked by three disulfide bonds in a distinctive pattern was a milestone in peptide chemistry. When it was later found that relaxin also possessed a similar overall structure, the term 'insulin superfamily' was coined. Use of methods of conventional protein chemistry followed by recombinant DNA and more recently bioinformatics has led to the recognition that insulin is the precursor to a large protein superfamily that extends beyond the human. Insulin-like peptides are found not only in vertebrates such as mammals, birds, reptiles, amphibians but also in the invertebrates such as chordates, molluscs and insects. All superfamily members share the distinctive insulin structural motif. In the human, there

* Howard Florey Institute, University of Melbourne, Victoria 3010, Australia
† School of Chemistry, University of Melbourne, Victoria 3010, Australia

Vitamins and Hormones, Volume 80
ISSN 0083-6729, DOI: 10.1016/S0083-6729(08)00601-8

exists ten members of the superfamily, each of which are expressed on the ribosome as a single-chain pre-prohormone that undergoes proteolytic processing to produce eight double-chain mature proteins and two single-chain forms. The six cysteine residues that form the three insulin disulfide cross-links – one intramolecular within the A-chain and two intermolecular between that A- and B-chains – are absolutely conserved across all members of the superfamily. They are responsible for imparting a similar overall tertiary structure. The human insulin superfamily members have each evolved to assume remarkably distinctive biological functions ranging from glucose homeostasis to neuroendocrine actions. That such diversity is contained within a modestly sized superfamily is testament to efficiency of the insulin structural motif as an evolutionary template.

I. Introduction

The insulin superfamily of peptide hormones in the human consists of insulin, insulin-like growth factors (IGF) I and II and seven members of the relaxin-like peptide family which includes gene 1 (H1) relaxin, gene 2 (H2) relaxin, gene 3 (H3) relaxin, insulin-like peptide 3 (INSL3, also called Leydig insulin-like peptide and relaxin-like factor (RLF)), insulin-like peptide 4 (INSL4, also known as placentin and early placenta insulin-like (EPIL)), insulin-like peptide 5 (INSL5), insulin-like peptide 6 (INSL6) (Adham *et al.*, 1993; Bathgate *et al.*, 2002; Chassin *et al.*, 1995; Conklin *et al.*, 1999; Hudson *et al.*, 1983, 1984; Lok *et al.*, 2000). The grouping of the relaxin family of peptide hormones within the insulin superfamily is due to their structural similarities and also the fact that they have derived from the ancestral insulin gene early in vertebrate evolution (Hsu, 2003; Wilkinson *et al.*, 2005b). Insulin was the first member of this superfamily to be discovered by Banting and Best (1922) and a few years later relaxin was discovered by Hisaw (1926). The most recent member of this superfamily was H3 relaxin which was discovered by the Howard Florey Institute in Melbourne in 2002. Genomic data and phylogenetic analysis have shown the presence of homologs of the members of insulin superfamily in other species such as apes, mouse, frogs and fish (Wilkinson *et al.*, 2005b).

Each expressed as an immature pre-prohormone with four functional segments: (1) an N-terminal signal peptide for secretion, (2) a conserved B chain, (3) a non-conserved C peptide, and (4) a C-terminal A chain. The co-translational loss of N-terminal signal peptide results in the conversion of pre-prohormone to prohormone. Subsequent formation of two disulphide bonds between A- and B-chains and one disulphide bond within A-chain followed by proteolytic cleavage of the C peptide (with the exception of

IGFs I and II which are single chain peptides) results in a mature, active heterodimeric A-B peptide (Hsu *et al.*, 2005; Wilkinson *et al.*, 2005b) (Fig. 1.1). The primary structures of each member has been either determined or predicted (Fig. 1.2). There is very little homology between members with only the six cysteines that make up the three disulfide bonds and a single glycine residue within the B-chain being absolutely conserved. Despite this low homology, the members each adopt a similar tertiary structure (Fig. 1.3). The tertiary structures of insulin (Blundell *et al.*, 1971a,bc), IGF-1 (Brzozowski *et al.*, 2002; Cooke *et al.*, 1991; Sato *et al.*, 1993; Siwanowicz *et al.*, 2005; Vajdos *et al.*, 2001; Zeslawski *et al.*, 2001), IGF-2 (Terasawa *et al.*, 1994; Torres *et al.*, 1995), H2 relaxin (Eigenbrot *et al.*, 1991) and INSL3 (Rosengren *et al.*, 2006) have been resolved either by X-ray crystallography or NMR spectroscopy. The tertiary structure has yet to be elucidated for other members of this superfamily.

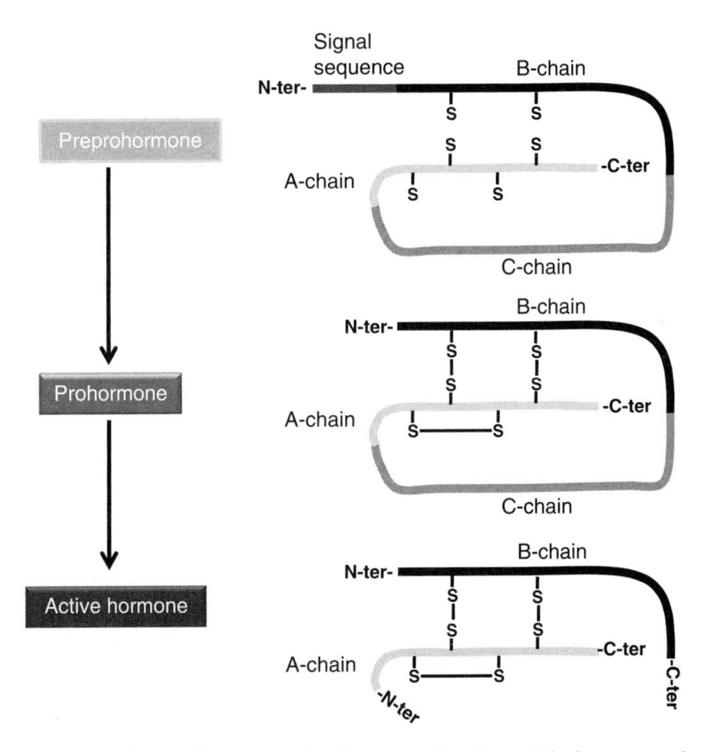

Figure 1.1 Expression and proteolytic cleavage of polypeptide hormones in insulin superfamily. Each is expressed as a pre-prohormone. The cleavage of N-terminal signal sequence followed by formation of three disulfide bonds gives rise to a prohormone which is further processed to a mature hormone as a result of excision of connecting C-chain. The exceptions are IGF-1 and IGF-2 for which the active form of these peptides is the prohormone having the C-chain.

A-Chain

```
Insulin                G I V E Q C C T S I C S L Y Q L E N Y C N
Relaxin-1        R P Y V A L F E K C C L I G C T K R S L A K Y C
Relaxin-2        Z L Y S A L A N K C C H V G C T K R S L A R F C
IGF-1          ~ A P Q T G I V D E C C F R S C D L R R L E M Y C A ~
IGF-2          ~ R R S R G I V E E C C F R S C D L A L L E T L C A ~
INSL3            A A A T N P A R Y C C L S G C T Q Q D L L T L C P Y
INSL4          R S G R H R F D P F C C E V I C D D G T S V K L C
INSL5                Q D L Q T L C C T D G C S M T D L S A L C
INSL6                  G Y S E K C C L T G C T K E E L S I A C
Relaxin-3        D V L A G L S S S C C K W G C S K S E I S S L C
```

B-chain

```
Insulin              F V N Q H L C G S H L V E A L Y L V C G E R G F F Y T P K A
Relaxin-1        K W K D D V I K L C G R E L V R A Q I A I C G M S T W S
Relaxin-2      D S W M E E V I K L C G R E L V R A Q I A I C G M S T W S
IGF-1                ~ G P E T L C G A E L V D A L Q F V C G D R G F Y F N K P ~
IGF-2                ~ P S E T L C G G E L V D T L Q F V C G D R G F Y F S R P ~
INSL3            P T P E M R E K L C G H H F V R A L V R V C G G P R W S T E A
INSL4            Z S L A A E L R G C G P R F G K H L L S Y C P M P E K T F T T T P
INSL5                S K E S V R L C G L E Y I R T V I Y I C A S S R W
INSL6            S D I S S A R K L C G R Y L V K E I E K L C G H A N W S F R
Relaxin-3        R A A P Y G V R L C G R E F I R A V I F T C G G R W
```

Figure 1.2 Primary structure of each member of insulin superfamily that has been either determined or predicted. Each has very low sequence homology except for the six cysteines that make up the three disulfide bonds and a singe glycine residue within the B-chain.

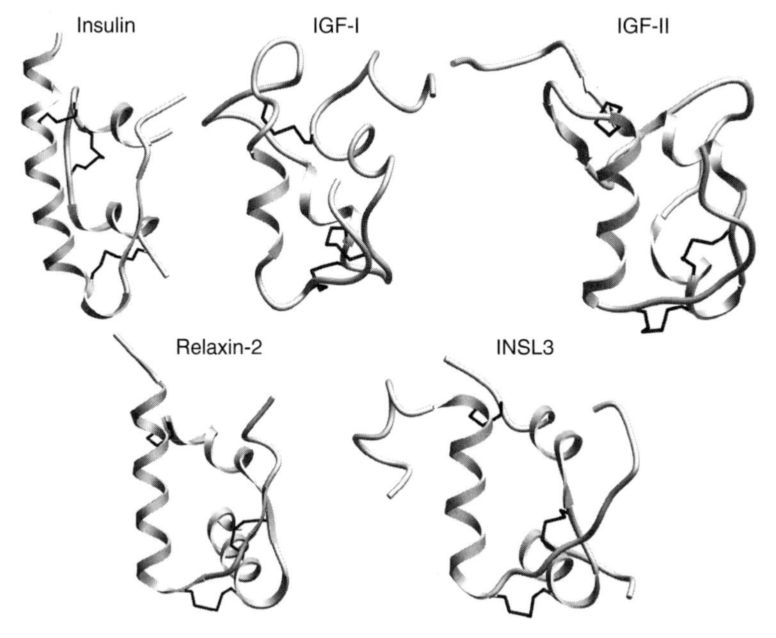

Figure 1.3 Tertiary structures of peptide hormones in the insulin superfamily that have been determined either by NMR-spectroscopy or X-ray crystallography. These peptides adopt a similar core structure especially in the regions confined by disulfide bonds.

Despite the high degree of structural similarity between the members of this superfamily, they bind and activate two different types of receptors. Insulin and IGFs-1 and 2 bind and activate cell surface tyrosine kinase receptors (reviewed in Nakae *et al.*, 2001). On the other hand, the peptide hormones within relaxin-like peptide family bind and activate G-protein coupled-receptors (Bogatcheva *et al.*, 2003; Hsu *et al.*, 2002; Kumagai *et al.*, 2002; Sudo *et al.*, 2003).

Among the peptides in insulin superfamily, insulin, IGF-1, IGF-2, H2 relaxin and INSL3 have been more extensively studied compared to other members. In this review, each member of this superfamily is introduced together with an overview of available structural information and known biological function.

A. Insulin

Insulin is the main hormone controlling intermediary metabolism. Its most obvious acute effect is to lower blood glucose. The main factor controlling the synthesis and secretion of insulin is the blood glucose concentration (Beardsall *et al.*, 2008). Other stimuli to insulin release include fatty acids, amino acids (particularly arginine and leucine) and gastrointestinal hormones. Insulin decreases blood glucose by increasing glucose uptake into muscle and fat cells via GLUT-4 (an insulin-sensitive glucose transporter present in muscle and fat cells), increasing glycogen and fatty acids synthesis, DNA replication and protein synthesis, decreasing proteolysis, lipolysis, gluconeogenesis and glycogen breakdown (Beardsall *et al.*, 2008).

The polypeptide hormone insulin was discovered in 1922 by Banting and Best the latter of whom subsequently shared the Noble Prize in Medicine in 1923 (Banting and Best, 1922). In their landmark experiment, they tied a string around the pancreatic duct of several dogs. A number of weeks later, they found that the dogs' pancreas had each shrunken to about one-third of its normal size. Removal of the remaining pancreatic tissue was followed by its dicing. A small amount was injected into the vein of a depancreatized or diabetic dogs. The animals became more active and their blood sugar levels were significantly reduced (Rosenfeld, 2002). The active component was named insulin.

In 1955, insulin was the first peptide to have its primary amino acid sequence determined (Ryle *et al.*, 1955). Several years later, it became the first hormone to have its tertiary structure determined by X-ray crystallography (Fig. 1.3) (Blundell *et al.*, 1971a,c). By 1975, scientists had isolated and sequenced insulin from all classes of vertebrates including hagfish (Chan and Steiner, 2000). Insulin is synthesised by clusters of pancreatic endocrine cells, β-cells of islet of Langerhans as a pre-prohormone which is processed into a biologically active polypeptide (Cheatham and Kahn, 1995) before secretion into the blood stream. Insulin circulates in the

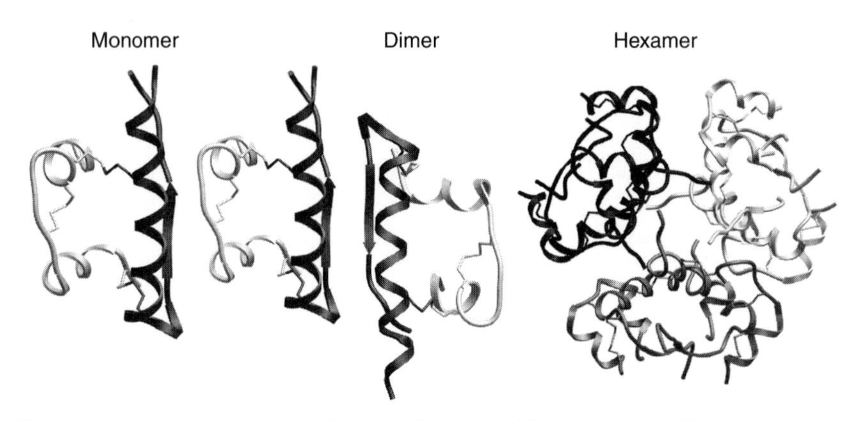

Figure 1.4 Tertiary structure of insulin determined by X-ray crystallography. Insulin forms dimers at micromolar concentrations which can then assemble into hexamers in the presence of zinc ions.

blood as a monomer form consisting of two peptide chains, A-chain and B-chain of 21 and 30 amino acids (in man) respectively. This polypeptide is held together with three disulfide bonds, two inter-chain disulfide bonds and a third intra-A-chain disulfide bond (reviewed in De Meyts, 2004). It has been demonstrated that insulin can dimerise at micromolar concentrations and, in the presence of zinc ions, it can associate into hexamers (Fig. 1.4) (De Meyts, 2004).

Insulin exerts its effect through a receptor tyrosine kinase (RTK). The first study that reported the presence of an insulin receptor (IR) was by Cuastrecasas in 1971 using ^{125}I-insulin to label a protein in the plasma membrane on insulin-responsive cells (Cuastrecasas, 1971). IR was first reported to belong to RTK family in 1982 (Kasuga *et al.*, 1982). IR is a large transmembrane glycoprotein complex (Sparrow *et al.*, 2007) that, has been shown, using SDS gel electrophoresis to be a homodimer consisting of two α- and two β-subunits (Fig. 1.5) covalently linked by three disulfide bridges (Hedo *et al.*, 1981; Massague *et al.*, 1981; Siegel *et al.*, 1981). The two α-subunits are entirely extracellular and contain two different binding sites (referred to as sites 1 and 2) which bind insulin asymmetrically, cross-linking the constituent monomer (De Meyts, 2004; De Meyts and Whittaker, 2002; De Meyts *et al.*, 2004; McKern *et al.*, 2006). The two β-subunits have a transmembrane segment and an intracellular domain contains a tyrosine kinase catalytic domain flanked by two regulatory regions (a juxtamembrane region and the C-tail) (De Meyts, 2004).

The two monomers of IR are covalently linked in the absence of ligand unlike other RTKs which dimerise or oligomerise upon ligand binding (De Meyts, 2004). IR has two isoforms (A and B) which have different affinity for insulin and it does not only bind insulin, it can also bind IGF-1 and 2, although with very low affinity (De Meyts, 2004; Jones and

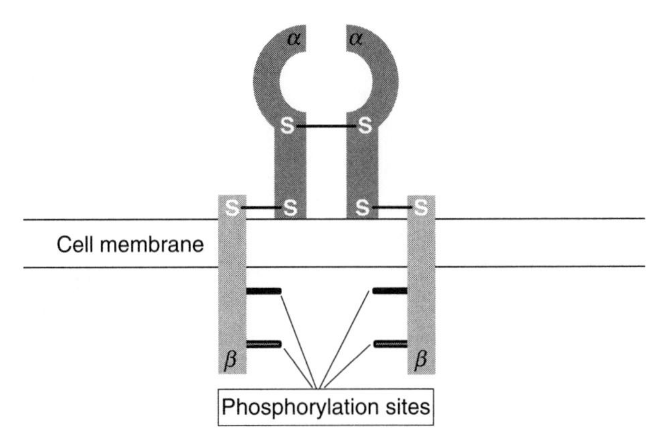

Figure 1.5 The insulin receptor (IR), a large transmembrane glycoprotein complex consisting of two α-subunits which are entirely extracellular. Each carries an insulin-binding site and two transmembrane β-subunits with phosphorylation sites and tyrosine kinase activity.

Clemmons, 1995). This indicates the structural similarity between insulin and IGFs and it is also not surprising that their receptors are classified together in a subfamily of tyrosine kinase receptors.

B. IGF-1

In 1963, a polypeptide was identified in human serum with insulin-like activities which were not neutralized by anti-insulin antibodies, so was called non-suppressible insulin-like activity, NSILA (Froesch *et al.*, 1963). IGF-1 was also known as 'sulfation factor' in 1970s and 'somatomedin C' in 1980s (Daughaday *et al.*, 1987; Salmon and Burkhalter, 1997). It was originally isolated from a Cohn fraction of human serum (Rinderknecht and Humbel, 1976b).

IGF-1 is a single chain peptide consists of 70 amino acids in four domains, B, C, A and D. The A- and B-domains are structural homologs of the insulin A- and B-chains with 50% sequence similarity and domain C is the analogous to the connecting C-peptide in proinsulin, and the D domain in not found in insulin (Humbel, 1990; Pavelic *et al.*, 2007).

The three dimensional structure of IGF-1 has been determined by both NMR spectroscopy (Cooke *et al.*, 1991; Sato *et al.*, 1993) and X-ray crystallography method (Brzozowski *et al.*, 2002; Siwanowicz *et al.*, 2005; Vajdos *et al.*, 2001; Zeslawski *et al.*, 2001). The B- and A-domains are similar to those observed in the crystal structure of insulin (Baker *et al.*, 1988; Bentley *et al.*, 1976). It has three α-helices one in the B-domain similar to that of insulin B-chain and two within the A- domain equivalent of insulin A-chain helices. The single chain IGF-1 is folded into a tertiary structure which is held together by three intramolecular disulfide bridges that are in the same disposition as in insulin (Denley *et al.*, 2005).

Pituitary growth hormone (GH) as well as insulin stimulates the biosynthesis of IGF-1 in the liver and in the other organs and tissues which circulates at high concentration in blood (Humbel, 1990). The IGF-1 synthesis is higher during postnatal development (Pollak, 2004). During puberty, the serum level of IGF-1 increases by 2–3 fold in both male and female and this is due to the increase in the pulsatile secretion of GH (Mauras *et al.*, 1987) and increase in sex steroid (Harris *et al.*, 1985). IGF-1 act as endocrine hormone via the blood and as paracrine and autocrine growth factor locally (Humbel, 1990).

In contrast to insulin that freely circulates in the blood stream, the IGFs circulate in complexes with IGF-binding proteins (IGFBPs). The level of free IGF-1 in the plasma is controlled by a family of six high affinity IGF binding proteins (IGFBPs-1 to -6) which are proteins of about 30 kDa (Hwa *et al.*, 1999). The IGFBPs regulate the interaction of IGF-1 with its receptor IGF-1R by binding 99% of the circulating IGF (Hwa *et al.*, 1999). In human about 80% of circulating IGF-1 is carried by IGFBP-3/acid labile subunit complex (Lewitt *et al.*, 1994). All IGFBPs inhibit IGF action by sequestering IGFs and some of IGFBPs (IGFBP-1, -3 and -5) can also potentiate IGF action (Pavelic *et al.*, 2007). A fine balance of IGF level is required for normal growth and this is achieved by the availability of free IGF which is regulated by IGFBPs. In tissues, IGF is released from its complex with IGBPs by either proteolysis of IGFBPs or binding of IGBPs to extracellular matrix (Baxter, 2000; Clemmons, 2001). The abnormal changes in the level of circulating IGF-1 can lead to growth abnormalities such acromegaly due to overproduction of GH which increases the level of IGF-1. On the other hand, a low level of IGF-1 resulting from an inactive GH receptor leads to Laron dwarfism (Baumann, 2002; Laron, 2004).

The polypeptide hormone IGF-1 binds to a number of cell surface receptors such as IGF receptor type 1 and 2 (IGF-1R, IGF-2R) and IR with different affinities. This also explains that there is not only structural similarity between the IGF ligands and insulin but there is also significant similarity between the IGF-1R and the IR which results in cross-talk between the two systems (Nakae *et al.*, 2001). IGF-1R can pair with either isoforms of IR (IR-A and IR-B) to form functional hybrid receptors which their role in cellular response is not clear (reviewed in Denley *et al.*, 2005).

IGF-1 binding to the extracellular α-subunit leads to conformational change resulting in tyrosine phosphorylation of the intracellular β-subunits, which causes an increase in the intrinsic kinase activity of the receptor which brings about various cellular responses (Rubin *et al.*, 1983). The cellular responses are the results of stimulation of multiple intracellular signalling pathways which regulate cell proliferation and survival (Pavelic *et al.*, 2007). The key downstream signalling pathways include mitogen activated protein kinase (MAPK), extracellular signal regulated kinase (ERK) and phophatidylinositide-3-kinase (PI3-K)/Akt-1 (Baserga, 1999; Yu and Rohan, 2000). Various studies such alanine scanning mutagenesis, antibody studies and receptor chimeras have shown that IGF-1 binds to its receptor using several residues located on the opposite surface of the molecule (Denley *et al.*, 2005). Activation of IGF-1R by IGF-1 binding leads to a cascade of signalling networks which results in various cellular responses, such as cell proliferation, differentiation, migration and protection from apoptosis (Pavelic *et al.*, 2007).

C. IGF-2

After the discovery of NSILA in 1963, it was then sequenced in 1976 and found to be two distinct proteins which were originally named NSILA-1 and NSILA-2 and then renamed to IGF-1 and IGF-2 based on their structural resemblance to proinsulin (Rinderknecht and Humbel, 1976a,b).

IGF-2 is a single-chain polypeptide with 67 amino acids, the primary source of which is reported to be the liver. The tertiary structure of IGF-2 has been determined by NMR spectroscopy (Terasawa *et al.*, 1994; Torres *et al.*, 1995) and is very similar to that of IGF-1. It, too, has four domains (A–D); the B and A domains of IGF-2 have 50% sequence similarity to the B- and A-chains of insulin (Rinderknecht and Humbel, 1976a). The three intramolecular disulfide bonds are in analogous positions to those in pro-insulin (Rinderknecht and Humbel, 1978).

IGF-2 is usually paternally expressed in the foetus and placenta and is crucial for placental development and foetal growth (Constancia *et al.*, 2002). The synthesis of IGF-2 is growth hormone-independent and it is mainly expressed during foetal development (Pavelic *et al.*, 2007). The IGF II gene (*Igf2*) is transcribed from four different promoters (P1–P4), of which only the P1 promoter is expressed biallelically. In contrast, the other promoters contains CpG dinucleotides which are targets of methylation and hence are subjected to imprinting such that only one allele is expressed (DeChiara *et al.*, 1991; Tang *et al.*, 2006). Various forms of IGF-2 are expressed with different molecular weights; however, the most active form which binds IGF receptors is 7.5 kDa (Kiess *et al.*, 1994).

Like IGF-1, IGF-2 also binds to a number of cell surface receptors with different affinities. It binds to type 2 IGF receptor (IGF-2R) and IR-A with high affinity, with intermediate affinity to IR-B, and with low affinity to IGF-1R. It can also bind to a heterodimer of IR-A/IGF-1R (reviewed in Chao and D'Amore, 2008). IGF-2R is also known as cation-independent mannose-6-phosphate receptor (IGF-2R/M-6-P) which is structurally and functionally different from IGF-1R. Unlike IGF-1R, IGF-2R is a monomeric membrane spanning glycoprotein with a 15 repeating extracellular domains (containing one binding site for IGF-2 and two binding sites for

M-6-P), a 23 amino acid transmembrane domain and a 163 amino acid intracellular region (reviewed in Denley *et al.*, 2005; Pavelic *et al.*, 2007). IGF-2R binds IGF-1 and IGF-2 with low and high affinities respectively and it does not bind insulin. IGF-2R does not have intracellular signalling domain unlike IGF-1R, therefore, it has no intrinsic signalling transduction capability and its main role is to bind and internalize IGF-2 which is then degraded and in this way it clears the IGF-2 from the circulation (Denley *et al.*, 2005; Pavelic *et al.*, 2007). IGF-2R also binds other ligands such as protein containing M-6-P, which is critical in mediating cell proliferation, (Scott and Firth, 2004) and lysosomal enzymes (Kornfeld, 1992). Similar to IGF-1, IGF-2 also binds to IGFBPs, preferentially IGFBP2 and 6 (Clemmons, 1997).

IGF-2 has a key physiological role in muscle and bone development and is crucial for placental development and foetal growth (reviewed in Chao and D'Amore, 2008). There is evidence that IGF-2 may be involved in tumour growth and progression as it has been shown to increase tumour cell proliferation in cell culture (reviewed in Pavelic *et al.*, 2007).

II. Relaxin Peptide Hormone Subfamily

Relaxin was first identified in 1926 by Frederick Hisaw as a substance in the serum of pregnant female guinea pig which causes the relaxation of pelvic ligaments and influence the female reproductive tract (reviewed in Samuel *et al.*, 2007). This substance was isolated from a crude extract of sow corpora lutea (Fevold *et al.*, 1930) and it was shown to have the characteristics of a peptide and to cause the relaxation of the interpubic ligaments and was hence named relaxin.

The first human relaxin gene was identified in the early 1980s and named H1 relaxin (Hudson *et al.*, 1983). One year later a second relaxin (H2) gene was found (Hudson *et al.*, 1984). Both genes are located on the short arm of chromosome 9 and were thought to have arisen from a common ancestor gene (Crawford *et al.*, 1984). A third relaxin (H3) gene was recently discovered using relaxin-1 sequence to screen the human genome data base for relaxin-related peptides (Bathgate *et al.*, 2002).

A. Relaxin-1

Relaxin-1 gene is only found in higher primates (Klonisch *et al.*, 2001). The product of this gene in human is unknown and native H1 relaxin peptide has yet to be actually isolated. The mRNA expression of H1 relaxin has been detected in human deciduas, prostate gland and placenta trophoblast (Hansell *et al.*, 1991). However, its functional significance remains unknown.

B. Relaxin-2

Unlike the relaxin-1 gene, which has been found only in humans and higher primates, the relaxin-2 gene has been found in most species (review in Halls *et al.*, 2007). The product of human gene-2 relaxin (H2 relaxin) is the functional orthologue of relaxin gene-1 product from non-primate species (H1 relaxin) which share greater than 90% identity (Park *et al.*, 2005) and they are the result of gene duplication in the ancestor of higher primates (reviewed in Scott *et al.*, 2003). For the rest of this review the use of term 'relaxin' will refer to H2 relaxin.

The phylogenetic analysis of relaxin has shown high sequence variability among closely related species and surprising similarities have been also observed in some distantly related species such pigs and whales (reviewed in Wilkinson *et al.*, 2005b). The presence of a hormone with 'relaxin-like' properties have been reported in invertebrates such as protozoa (*T. pyriformis*), ascidians (*H. momus*) and tunicates (*C. intestinalis*) (reviewed in Wilkinson *et al.*, 2005b). There is a great deal of variation in relaxins between as well as within species which range from 40–60% (Schwabe and Büllesbach, 1994) and despite the differences, there are also similarities in relaxin among species specially the conserved residues such as cysteines and glycines (Schwabe and Büllesbach, 1994).

Relaxin is the most extensively studied and well characterised member of relaxin family. Like insulin, relaxin is also expressed as an immature pre-prohormone having a signal sequence, and B-, C- and A-chains which is then processed into an active hormone by cleavage of signal sequence, formation of three disulfide bonds and excision of C-chain (Adham *et al.*, 1993; Hsu, 2003). The tertiary structure of relaxin-2 (Fig. 1.3) has been determined by X-ray crystallography (Eigenbrot *et al.*, 1991). The A chain (24 amino acids) consists of two alpha helices connected via a short loop. The B-chain (30 amino acids) consists of an alpha helix and a strand. As predicted from earlier studies, these two chains are connected via two inter-chain and one intra-A-chain disulphide bond. The interaction of relaxin with its receptors has been shown to be dependent on the presence of a motif, (Arg^{13}-X-X-X-Arg^{17}-X-X-Ile^{20}), in the mid-region of the B-chain helix (Büllesbach and Schwabe, 2000; Büllesbach *et al.*, 1992). The two positively charged arginine residues are of great importance to binding of relaxin to its receptor as their substitution with citrulline, a neutral analogue of arginine, inactivated relaxin (Büllesbach *et al.*, 1992). These residues collectively form a binding surface which interacts with a binding pocket on the receptor.

Due to the structural similarity between relaxin, insulin and IGFs, it was long thought that the receptors for the relaxin would, like the IR, belong to the RTK family. Recently, relaxin has been shown to bind, and activate two previously orphan G-protein couple receptors through a G_s-cAMP dependent pathway (Hsu *et al.*, 2002). These two receptors were originally named as LGR7 (leucine-rich G-protein coupled receptor 7) and LGR8. They were recently renamed based upon an International Union of Pharmacology recommendation as relaxin family peptide (RXFP) receptors 1 and 2, respectively (Bathgate *et al.*, 2006a). They belong to a subfamily of glycoprotein receptors known LGR (leucine-rich repeat-containing G protein-coupled receptor) (Fig. 1.6A; Ascoli *et al.*, 2002). There are three subtypes (A, B and C) of LGRs and studies have suggested that LGRs might have originated during the early evolution of metazoans (Hsu *et al.*, 2005). Type A LGRs include receptors for follicle stimulating (FS) and luteinizing hormones (LHs) and thyroid-stimulating hormone (TSHR). At present type B LGRs remain orphan with only three members, LGR4-6. RXFP1 and RXF2 are type C LGRs and they are the only two members of this subtype (Hsu, 2003; Hsu *et al.*, 2000, 2002). The RXFP receptors have 60% sequence identity and a common domain structure of these receptors is the presence of an N-terminal low density lipoprotein (LDL) region, which is not found in other LGR family members (Hsu, 2003; Hsu *et al.*, 2000, 2002). These are further distinguished from other glycoprotein hormone receptors by this unique low-density lipoprotein receptor-like cysteine-rich motif (low-density lipoprotein class A module (LDLa)) which is located near the N-terminus of the extracellular region. The LDLa domain has been postulated to be involved in receptor activation but not in primary ligand

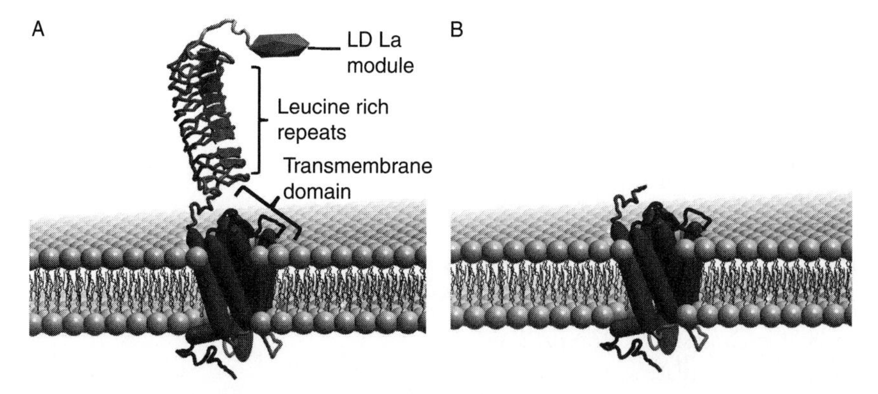

Figure 1.6 Receptors for relaxin family peptides. (A) Receptor for human relaxin-2 (RXFP1 or LGR7) and INSL3 (RXFP2 or LGR8) with large leucine rich repeats ectodomain which is the primary ligand binding site and classical seven transmembrane helices. These are coupled to intracellular G-protein receptor for signalling. A low-density lipoprotein class A module (LDLa) is located near the N-terminus of the extracellular region. This domain has been postulated to be involved in receptor activation. (B) A classic G-protein coupled receptor with seven transmembrane helices, receptor for human relaxin-3 (RXFP3) and INSL5 (RXFP4 also known as GPCR142 or GPR100).

binding (Adham and Agoulnik, 2004; Hsu, 2003). Recent structural and functional studies on mutated full-length RXFP1 has shown that a hydrophobic surface within the N-terminal region of the LDLa module is essential for activation of the RXFP1 receptor signalling cascade in response to relaxin stimulation (Hopkins *et al.*, 2007). Recent studies using truncated and chimeric receptors have revealed two binding sites on RXFPs 1 and 2. The leucine rich repeats within the ectodomain has been shown to be the primary, high affinity binding site and a secondary, low affinity site within the transmembrane region (Halls *et al.*, 2005). Using Northern blot analysis the RXFP1 transcript inhuman has been identified in various tissues such uterus, ovary, adrenal gland, prostate, skin, testis, brain, kidney, liver, thyroid, salivary gland and heart (Hsu *et al.*, 2002; Luna *et al.*, 2004; Mazella *et al.*, 2004). Relaxin-2 binds to and activates both RXFP1 and RXFP2;however, it has a lower affinity for RXFP2 (Hsu, 2003; Hsu *et al.*, 2002, 2003; Sudo *et al.*, 2003). RXFP1 and RXFP2 have high sequence similarity with greater than 90% in human and rodent orthologs and there is a high degree of conservation from fish to mammals (reviewed in Halls *et al.*, 2007). They have been shown to have emerged before the divergence of fish and they were 'acquired' as RXFP receptor during mammalian evolution (reviewed in Halls *et al.*, 2007).

With the identification of relaxin and its receptor in various tissues in human, relaxin is emerging as a pleiotropic hormone which displays a wide variety of biological effects. Relaxin was initially regarded as a hormone of pregnancy based on its action in the female reproductive tract during parturition (Sherwood, 2004). It is produced during pregnancy in the corpus luteum (Hudson *et al.*, 1984) and it is the major circulating form of relaxin. Additionally, relaxin expression has also been detected in placenta and prostate as well as in small quantity within the male reproductive tract which is released into seminal fluid (Samuel *et al.*, 2007). Apart from its expression by reproductive organs, relaxin expression has also been detected in several other non-reproductive organs such as brain, heart and kidney (Gunnersen *et al.*, 1995).

The biological actions of relaxin on female reproductive system have been extensively studied particularly in rodents. The circulating level of relaxin changes at different times during pregnancy in different species. In rodent, circulating relaxin (secreted from corpus luteum) is absent during early pregnancy and first detectable in the serum at midpregnancy, and reaches its maximum level before labour (Sherwood and Rutherford, 1981) with similar pattern observed in sows and dogs (Steinetz *et al.*, 1987). In human females, relaxin is detectable after conception and reaches its maximum level in the first trimester after which the plasma level of relaxin is reduced and remains stable throughout the remainder of pregnancy (Bell *et al.*, 1987; Eddie *et al.*, 1986). Relaxin promotes growth and cervical softening in several species and thereby enables litter delivery as the relaxin gene knockout female mice showed abnormal parturition and reduced level of foetal delivery and survival (reviewed in Samuel *et al.*, 2007). Relaxin soften the connective tissue of the interpubic ligament by decreasing collagen accumulation, and increasing collagen metabolism (Samuel *et al.*, 1998; Zhao *et al.*, 1999b) which leads to softening and dilation of birth canal. Studies from knockout mice have shown that relaxin homozygous knockout female mice has prolonged labour and some were unable to deliver their litter normally (Zhao *et al.*, 1999b). Relaxin is not only required for cervical softening, it is also required for the growth of mammary gland and nipples (Hansell *et al.*, 1991; Sherwood, 2004) as the study of mice deficient for relaxin showed impaired nipple development during late pregnancy and were unable to feed their pups (Zhao *et al.*, 1999a). Relaxin has been shown to be involved during the early stage of follicle development in human (reviewed in Park *et al.*, 2005). It has been suggested that relaxin might be involved in embryo implantation as relaxin is able to interact with its receptor on endometrial stromal cell and causes differentiation of endometrial cells and vascularisation which is an essential process for embryo implantation (reviewed in Park *et al.*, 2005).

The physiological roles of relaxin are not limited to reproductive system. Increasing evidence from recent studies has shown that relaxin also plays important roles in cardiovascular, pulmonary and renal systems. Other functions of relaxin are inhibition of inflammation, wound healing, fluid balance and body homeostasis (reviewed in Samuel *et al.*, 2007). Relaxin is currently regarded as one of the most pleiotropic hormone ever known. Most of the protective of function of relaxin is due to its anti-fibrotic effects as recombinant relaxin has been shown to inhibit collagen over expression *in vitro*, by inhibiting TGF-β1-stimulated collagen synthesis (Unemori and Amento, 1990). Relaxin, in combination with interferon-gamma (IFN-γ), has been shown to decrease collagen expression in scleroderma cells *in vitro* (Unemori *et al.*, 1992). Relaxin can also reduce dermal scarring which is thought to be due to the ability of relaxin to regulate fibrillin-2 expression (reviewed in Samuel *et al.*, 2007). Relaxin has been shown to prevent lung-injury-induced collagen accumulation, reduce glomerular and interstitial fibrosis and reverse established cardiac fibrosis in relaxin null mice *in vivo* and hence restore lung, kidney and heart functions (reviewed in Samuel *et al.*, 2007). Relaxin is able to inhibit histamine release by inflammatory cells and also prevent their influx into injured organs (reviewed in Samuel *et al.*, 2007). Relaxin also promotes wound healing possibly by increasing the expression of the vascular endothelial growth factors (VEGF) which stimulate angiogenesis (Koos *et al.*, 2005; Palejwala *et al.*, 2002). In addition to antifibrotic effects of relaxin, it has been shown that relaxin can induce vasodilation in heart, liver and kidney by increasing nitric oxide (NO) production, via stimulation of NO synthase and endothelin B (ETB) receptor (reviewed in Samuel *et al.*, 2007).

Despite the many physiological roles of relaxin, it has also been implicated in having role in certain cancers such breast and prostate as relaxin stimulate the *in vitro* invasiveness of breast and endometrial cancer cell lines which may be due to the angiogenesis effect of relaxin. Therefore, relaxin antagonists can be potential treatment for these types of cancers as a recent study has shown that an analogue of relaxin exhibited antagonistic properties and impaired tumor growth progression *in vivo* (Silvertown *et al.*, 2007).

Relaxin has a number of functions in reproductive and non-reproductive organs which have been reported to be a potential therapeutic agent in many processes such as allergic responses, cardiac protection, fibrosis, wound healing and cancer (reviewed in Samuel *et al.*, 2007). Relaxin is currently in clinical trial for congestive heart failure (CHF) in Germany. The clinical trial of recombinant relaxin in human subjects for treatments of cervical ripening, scleroderma, fibromyalgia and orthodontics has not shown any effectiveness of relaxin use in clinics.

C. Relaxin-3

The search for relaxin-related peptides in the human genome led to the unexpected identification of a third human relaxin (H3 relaxin) (Bathgate *et al.*, 2002). Phylogenetic analysis of RXFPs has shown that relaxin-3 is highly conserved among a range of species from fish to mammals and is likely to be the ancestral relaxin which might have emerged prior to the divergence of fish (Wilkinson *et al.*, 2005b).

Relaxin-3, also known as insulin-like peptide 7 (INSL7), is a peptide with two chains (A and B) held together with three disulfide bonds similar to relaxin-2. Relaxin-3 is primarily expressed in the brain of human, mouse and rat (Bathgate *et al.*, 2002; Burazin *et al.*, 2002). Relaxin-3 gene transcript has been identified in tissues such as spleen, thymus and testis in human (Bathgate *et al.*, 2002). The distribution of relaxin-3 in brain has been shown by immunohistochemistry to be in neurons of the nucleus incertus (NI) (Ma *et al.*, 2007).

The relaxin-3 receptor was first identified in human cortical cDNA library and originally named Somatostatin and angiotensin-like peptide receptor (SALPR) because it had 35% sequence homology with somatostatin type-5 receptor (SSTR5) and 31% sequence homology with angiotensin receptor type-1 (AT1) (Matsumoto *et al.*, 2000). Recent studies have been shown that relaxin-3 is the endogenous ligand for the orphan G-protein coupled receptor SALPR (also known as GPCR135) (Boels and Schaller, 2003; Liu *et al.*, 2003b; Matsumoto *et al.*, 2000). Recently, this receptor has been renamed as relaxin family peptide receptor-3 (RXFP3) (Bathgate *et al.*, 2006a). Unlike RXF1 and RXFP2, RXFP3 (Fig. 1.6B) lacks a large ectodomain and it also has a short amino and carboxy-terminal tails compared to RXFP1 and RXFP2 (Halls *et al.*, 2007). The mRNA expression of RXFP3 has been detected in human brain, adrenal gland, thymus and testis (Chen *et al.*, 2005; Liu *et al.*, 2003b). Recently, it has been shown that relaxin-3 can also activate another G-protein couple receptor known as GPCR142 (RXFP4) (Fig. 1.6B) (Liu *et al.*, 2003a). The mRNA expression of this receptor has been detected in a variety of human tissues such as brain, kidney, testis, thymus, placenta, prostate thyroid, colon and salivary gland (Liu *et al.*, 2003a). RXFP4 was first discovered by searching the human genomic data bank using RXFP3 gene sequence and these two receptors have 43% identity in human (Boels and Schaller, 2003; Liu *et al.*, 2003b). Relaxin-3 binds RXFP1 with high affinity and it also has a low binding affinity at RXFP2 (Bathgate *et al.*, 2006b).

The physiological role of relaxin-3 remains to be fully elucidated. However, based on the distribution of relaxin-3 and its receptor RXFP3, it has been suggested that relaxin-3 may be important in higher-order processes such stress and cognition (Ma *et al.*, 2007). Bolus-administered intracerebroventricular (ICV) relaxin-3 to rats has been shown to increase food intake and therefore it has been suggested a possible role of relaxin-3 in energy homeostasis and appetite regulation (Hida *et al.*, 2006; McGowan *et al.*, 2006). It has also been suggested that relaxin-3 antagonists may be potential anti-obesity agents (Hida *et al.*, 2006).

D. INSL3

The peptide hormone, INSL3, was discovered in the early 1990s as a result of differential cloning of testis specific genes in pig (Adham *et al.*, 1993) and mice (Pusch *et al.*, 1996). It was originally designated Ley-IL (Leydig cell insulin-like peptide) because its mRNA was initially found in Leydig cells of the testis (Burkhardt *et al.*, 1994). It has also been referred to as RLF (relaxin-like-factor) due to its relaxin-like activity in mouse interpubic ligament bioassay (Bullesbach and Schwabe, 1995).

INSL3 is a 6-kDa novel circulating hormone that is expressed exclusively in high levels by the Leydig cells of adult testes in a differentiation-dependent manner (Ivell and Bathgate, 2002; Ivell and Einspanier, 2002). It acts as a marker for fully differentiated adult-type Leydig cells (Ivell and Bathgate, 2002). In the female, INSL3 is expressed in the ovarian follicle and in the corpus luteum, but at lower levels compared to that of male (Roche *et al.*, 1996; Tashima *et al.*, 1995).

INSL3 is a polypeptide that consists of two chains, a 26 amino acid A- and 31 amino acid B-chain, that are held together by the three insulin-like disulfide bonds. The tertiary structure of human INSL3 has been recently resolved using NMR spectroscopy (Fig. 1.3) (Rosengren *et al.*, 2006). It, and indeed all other members of the insulin superfamily for which structures are known, adopts a very similar insulin core structure with differences being mostly around the termini. The N- and C-termini of INSL3

A-chain are structurally disordered compared to relaxin-2, in which the termini of A-chain is more defined (Rosengren *et al.*, 2006). Another important structural distinction between relaxin-2 and INSL3 is the number of helical turns in their B-chain. The helical segment of relaxin B-chain consists of 15 residues whereas in INSL3 the B-chain helix is shorter by 3 residues.

Based on the primary binding motif (Arg^{12}-X-X-X-Arg^{16}-X-X-Ile^{19}, where X is any amino acid) of relaxin-2, it was thought that the binding of INSL3 to its receptor is dependent on residues in the helical region of the B-chain. The residues at equivalent positions in INSL3 are His^{12}, Arg^{16} and Val^{19}. Recent structure-activity studies by our group using single Ala substitution has shown that substituting Arg^{16} and Val^{19} with Ala significantly reduced the receptor binding affinity of these analogues (Rosengren *et al.*, 2006). On the other hand, multi-Ala-substitution showed that His^{12} and Arg^{20} has a strong synergistic effect with Arg^{16} suggesting that His^{12} and Arg^{20} may be involved in the initial step of receptor recognition that drive the electrostatic interaction between the basic residues of the peptide and acidic residues on the receptor (Rosengren *et al.*, 2006). Apart from these residues, Trp^{B27} toward the C-terminus of the B-chain has been shown to be crucial for binding of INSL3 to its receptor as the mutation or deletion of Trp^{B27} leads to loss of receptor binding affinity (Büllesbach and Schwabe, 1999; Rosengren *et al.*, 2006).

The mechanism of receptor activation by INSL3 has recently been reported to be independent of the amino acid side chains and is a function of certain peptide bonds at the N-terminus of the A-chain (Büllesbach and Schwabe, 2007). In this study, the authors have shown that progressive truncation of A-chain from N-terminus diminishes the signalling activity of LGR8 but single residue mutation of Arg^{A8} and Tyr^{A9} to alanine does not perturb the signalling activity (Büllesbach and Schwabe, 2007). In order to further explore the involvement of backbone amide bond in the signalling activity of INSL3, the Tyr^{A9} has been replaced to D-Pro and had been shown to have no impact of receptor binding while severely retarding the signalling activity. Further replacement of Tyr^{A9} with *N*-methyl alanine has also been shown to have similar effect as that of D-Pro replacement (Büllesbach and Schwabe, 2007).

Like relaxin-2 receptor, INSL3 receptor is also a leucine-rich repeat-containing G protein-coupled receptor (LGR8) which recently has been renamed as RXFP2 (Bathgate *et al.*, 2006a). The RXFP2 (also known as GREAT in rodent: G-protein coupled receptor affecting testicular descent) expression has been observed in several organs and tissues, such as the gubernaculum, testis, brain, skeletal muscle, uterus, peripheral blood cell, thyroid and bone marrow (Gorlov *et al.*, 2002; Hsu *et al.*, 2002; Overbeek *et al.*, 2001). Relaxin-2 binds to and activates both RXFP1 and RXFP2; however, it has a lower affinity for RXFP2 (Hsu, 2003; Hsu *et al.*, 2002). In contrast, INSL3 has been shown to bind exclusively to and activate

RXFP2 to induce an intracellular increase in cAMP, both in transfected cells overexpressing cloned RXFP2 receptor and in gubernacular and prostate cells with naturally expressed receptors (Overbeek *et al.*, 2001).

A direct interaction between INSL3 and RXFP2 receptor has also been demonstrated through different studies such as receptor binding studies, changes in cAMP production, and ligand-receptor cross-linking studies (Claasz *et al.*, 2002; Hsu *et al.*, 2002; Kumagai *et al.*, 2002). In order to define the primary INSL3 binding site in LRR of RXFP2, our group has used alanine scanning to show that RXFP2 Asp-227 is crucial for binding INSL3 Arg-B16 whereas RXFP2 Phe-131 and Gln-133 are involved in INSL3 Trp-B27 binding (Scott *et al.*, 2007).

Like relaxin-2, INSL3 has both reproductive and non-reproductive roles. The identification of the former roles has been facilitated by analysis of INSL3 knock out mice, in which testicular descent during development was shown to be severely retarded due to loss of INSL3-mediated gubernaculum growth (i.e. they show bilateral cryptorchidism (Greek: hidden-gonad)) (Nef and Parada, 1999; Spiess *et al.*, 1999; Zimmermann *et al.*, 1999). The cryptorchid phenotype, with the testis being retained in the abdomen, results in disrupted spermatogenesis and infertility. INSL3 has been suggested to be involved in female fertility as female homozygous INSL3$^{-/-}$ mice display impaired fertility associated with an abnormal estrous cycle (Nef and Parada, 1999). Phenotypically, male mice deficient for RXFP2 have been shown to be similar to that of male INSL3 deficient mice in that both exhibit bilateral cryptorchidism and have impaired spermatogenesis (Bogatcheva *et al.*, 2003; Feng *et al.*, 2004; Foresta and Ferlin, 2004). On the other hand, female RXFP2 receptor null mice remain fertile (Nef and Parada, 1999; Nef *et al.*, 2000).

Recently, it has also been shown that INSL3 acts as a paracrine factor in mediating gonadotropin actions. LH, released by anterior pituitary gland, stimulates Leydig insulin-like 3 (INSL3) transcripts in ovarian theca and testicular Leydig cells. INSL3, in turn, binds to RXFP2 receptor expressed in germ cells to activate the inhibitory G protein thus leading to decreases in cAMP production. This leads to initiation of meiotic progression of arrested oocytes in preovulatory follicles *in vitro* and *in vivo* and suppresses male germ cell apoptosis *in vivo* (Kawamura *et al.*, 2004).

The expression of INSL3 has been found to be up-regulated in hyperplastic and neoplastic human thyrocytes suggesting that it may play a role in human thyroid carcinoma. The expression of human INSL3 has also been detected in neoplastic Leydig cells and mammary epithelial cells which also suggests that it may be involved in tumour biology (Klonisch *et al.*, 1999, 2005).

Although INSL3 plays role in gonadal and other physiological processes, the intracellar signalling pathways involved are still unclear. Most of the signalling studies have thus far focused on cAMP pathways. The level of cAMP has been shown to either increase or decrease based on coupling of RXFP2 to different G protein. INSL3 stimulation of HEK cells expressing recombinant RXFP2 and gubernaculum cells (which endogenously express RXFP2) causes increase in cAMP as a result of G_s-mediated activation of adenylate cyclase in these cells (Kumagai *et al.*, 2002). RXFP2 couples to G_i/G_o proteins in testicular germ cells and oocytes which upon stimulation by INSL3 cause a decrease in cAMP that is prevented by pertussis toxin (Kawamura *et al.*, 2004).

INSL3 plays important role in foetal development and in fertility regulation and it has shown that both agonists and antagonists of the peptide will have significant clinical promise for use in the reproductive system in both males and females. Recently, a group of men receiving a male contraceptive (a combination of testosterone and progesterone which acts as inhibitor of pituitary gonadotropins) showed that most men are azoospermic or severely oligospermic ($\leq$1 million sperm/ml). However, 10–20% of men have persistent sperm production despite profound gonadotropin suppression. This was found to be associated with a higher serum INSL3 concentration which prevents complete suppression of spermatogenesis in those men on hormonal contraceptive regimens (Amory *et al.*, 2007). This finding suggests that INSL3 antagonists could have potential for development as male contraceptives. Recent studies by our group, on developing short peptide-based INSL3 antagonists using INSL3 B-chain, has led to the identification of short cyclic analogues of INSL3 B-chain which we have shown to be antagonists *in vitro* (Del Borgo *et al.*, 2006; Shabanpoor *et al.*, 2007). The administration of one of these analogues into the testes of rats resulted in a substantial decrease in testis weight probably due to the inhibition of germ cell survival (Del Borgo *et al.*, 2006).

E. INSL4

INSL4 was identified by screening a human cytotrophoblast-subtracted cDNA library of first-trimester human placenta (Chassin *et al.*, 1995). It was initially named placentin or EPIL peptide because it was first discovered in placenta. It is expressed more strongly in the differentiated syncytiotrophoblast than in cytotrophoblast cells (Laurent *et al.*, 1998). It has also been detected in amniotic fluid and maternal serum during normal pregnancy (Mock *et al.*, 1999), and in higher levels during pregnancies with trisomy (Mock *et al.*, 2000). INSL4 expression is highly restricted in adult tissues (Bruni *et al.*, 2007). The human INSL4 gene consists of 2 exons and 1 introns (Chassin *et al.*, 1995) and is clustered along with relaxin-1, 2 and INSL6 genes on chromosome 9 and it has been suggested that they have arisen though local gene duplication (Sankoff, 2001). The actual length of INSL4, if it is expressed, remains unknown. However, it has about 44% homology with human relaxins and only 15% homology with insulin (Chassin *et al.*, 1995). There is evidence that suggests INSL4 is probably the only insulin-like growth factor gene which is only expressed in higher primates and there is an insertion of human endogenous retrovirus element into the human INSL4 gene promoter which has been suggested to have occurred after the divergence of New World and Old World monkeys (Bieche *et al.*, 2003).

Due to its high expression during early gestation, it has been suggested that INSL4 might be important in foetal and placental growth and development (Millar *et al.*, 2005). The mRNA of INSL4 has been identified in the perichondrium of the four limbs, vertebra, ribs and interbone ligaments which has been suggested an involvement of INSL4 in regulation of bone formation (Laurent *et al.*, 1998).

Unlike insulin and relaxins which are fully processed from pre-prohormone to a two chain mature peptides, INSL4 has been shown to have none of the three dimensional feature of insulin and relaxin and lack hydrophobic core and helical structure at physiological pH (Büllesbach and Schwabe, 2001). Immunoblot and gel electrophoresis analyses of histidine-tagged recombinant placentin and also the presence of only dibasic recognition sites for putative enzyme cleavage between the C and A domain indicated that INSL4 is composed of two chains of apparent molecular masses of 4 and 13 kDa (Koman *et al.*, 1996). Synthetic INSL4 fails to interact with RXFP1 and its splice variant as well as RXFP2 (Lin *et al.*, 2004). Further, the tertiary structure of this hormone has not yet been determined nor has a receptor been identified. Very little is known about the biological function of INSL4 peptide although it has been shown to be able to stimulate tyrosine phosphorylation, DNA synthesis and human chorionic gonadotropin (HCG) production which may suggests that INSL4 may be involve in cellular proliferation and differentiation during placental development (Koman *et al.*, 1996).

F. INSL5

INSL5 is a *bona fide* member of insulin superfamily that was identified by searching EST databases for the presence of the conserved insulin B-chain cysteine motif (Conklin *et al.*, 1999). INSL5, a polypeptide of 139 amino acids, is characterised by a signal peptide, a B-chain, a connecting C-chain and an A-chain which is a signature motif of peptides in insulin superfamily (Conklin *et al.*, 1999). Human INSL5 gene has been localised to chromosome 1 and it has been shown to have 48%, 40% and 34% identity to shark relaxin, human relaxin and INSL3 respectively (Conklin *et al.*, 1999). The expression of human INSL5, using northern blot analysis, has been detected in rectal, colon, and uterine tissue (Conklin *et al.*, 1999). A more recent study using quantitative RT-PCR has shown the mRNA expression of INSL5 in human brain, although in low level, and the pituitary which expresses INSL5 at higher level (Liu *et al.*, 2005b). In mouse, the INSL5 expression has been detected in colon (highest level), thymus and to lower extent in testis (Conklin *et al.*, 1999). The tertiary structure of INSL5 is yet to be resolved. INSL5 binds and activate a previously orphan G-protein coupled receptor, RXFP4, also known as (GPCR142 or GPR100). RXFP4 gene was discovered by searching the human genomic data base (Genebank) using the RXFP3 sequence (reviewed in Halls *et al.*, 2007). Human RXFP4 shares 43% sequence identity with RXFP3 and it was thought that they might have similar ligands (Halls *et al.*, 2007). RXFP4 is highly conserved among human, monkey, cow and pig and it is a non-functional pseudogenes in rats and dogs die to the lack of open reading frames (Wilkinson *et al.*, 2005a). Human RXFP4 gene has been localised on chromosome 1 at 1q21.2-1q21.3 and it shares 74% sequence identity to the mouse RXFP4 (Boels and Schaller, 2003). Recombinant human INSL5 has been shown to activate RXFP4 with high affinity but it does not activate human RXFP1-3 (Liu *et al.*, 2005b). INSL5 and RXFP4 have similar tissue expression profile and the fact that both INSL5 and RXFP4 genes are pseudogenes in rats suggest that INSL5 is an endogenous ligand for RXFP4 (Liu *et al.*, 2005a). A clear functional role for INSL5/RXFP4 system has not been determined yet but the expression of INSL5 in mouse brain and its ability to increase intracellular calcium in hypothalamic neurons has been suggested INSL5 to have a possible neuroendocrine function (Dun *et al.*, 2006).

G. INSL6

INSL6 is the latest member of insulin superfamily which has been identified as a result of searching EST databases for proteins containing the insulin family B-chain cysteine motif (Lok *et al.*, 2000). INSL6 sequences in rat and human encodes polypeptides of 213 and 188 amino acids respectively, which are in three motifs of B-chain, C-chain and A-chain which the signature of peptides in insulin superfamily (Lok *et al.*, 2000). INSL6 has been demonstrated be linked by disulfide bonds and it is processed from pre-prohormone to mature hormone in a similar manner to other members of insulin superfamily (Lu *et al.*, 2006). Human INSL6 shares 43% sequence identity to human relaxin-2 in the B-chain and A-chain regions and radiation hybrid mapping has localised INSL6 locus on chromosome 9p24 near INLS4 and testis-determining loci (Lok *et al.*, 2000). Northern blot analysis has shown that INSL6 is expressed at high levels in the human testis, and in situ hybridization analysis has shown that INSL6 is specifically expressed within seminiferous tubules in spermatocytes and round spermatids (Lok *et al.*, 2000). Recent study has reported that the meiotic and postmeiotic germ cells of the testis are the principal sites of expression of INSL6 (Lu *et al.*, 2006). In mouse, INSL6 is expressed at highest level in testes and its expression has been detected in other tissues such as kidney, small bowel, heart, brain and thymus, albeit at lower level (Kasik *et al.*, 2000). In mouse brain, immunoreactive-INSL6 (irINSL6) has been detected in two main areas, caudal hypothalamus and brainstem (Brailoiu *et al.*, 2005). The tertiary structure of INSL6 remains to be elucidated and the biological function(s) of this hormone is currently unknown; however, its specific expression in the germ cells of the human testis indicates a possible role of INSL6 in sperm development and fertilization (Lok *et al.*, 2000; Lu *et al.*, 2006).

III. Concluding Remarks

The human insulin superfamily of peptides has thus far provided an extraordinary insight into protein evolution, structure and function which remains incomplete today as research continues to elucidate the roles of the most recently discovered members. That insulin-like peptides bearing the insulin structural motif is present in all metazoa, both vertebrates and invertebrates, is testament to its effectiveness as a template for eliciting specific biological function. The results of ongoing studies on the human insulin superfamily combined with the emergence of further genomic data derived from other metazoan insulin superfamilies will ultimately lead to a full understanding of the evolution of this fascinating class of proteins.

Acknowledgments

The author is grateful to Professor Geoffrey Tregear for critical appraisal of the manuscript. Reported studies undertaken in the authors' laboratory were supported by NHMRC of Australia Project grants (#350245, 350284, 50899 and 509048).

References

Adham, I. M., and Agoulnik, A. I. (2004). Insulin-like 3 signalling in testicular descent. *Int. J. Androl.* **27,** 257–265.

Adham, I. M., Burkhardt, E., Benahmed, M., and Engel, W. (1993). Cloning of a cDNA for a novel insulin-like peptide of the testicular Leydig cells. *J. Biol. Chem.* **268,** 26668–26672.

Amory, J. K., Page, S. T., Anawalt, B. D., Coviello, A. D., Matsumoto, A. M., and Bremner, W. J. (2007). Elevated end-of-treatment serum INSL3 is associated with failure to completely suppress spermatogenesis in men receiving male hormonal contraception. *J. Androl.* **28,** 548–554.

Ascoli, M., Fanelli, F., and Segaloff, D. L. (2002). The lutropin/choriogonadotropin receptor, a 2002 perspective. *Endocr. Rev.* **23,** 141–174.

Baker, E. N., Blundell, T. L., Cutfield, J. F., Cutfield, S. M., Dodson, E. J., Dodson, G. G., Hodgkin, D. M., Hubbard, R. E., Isaacs, N. W., Reynolds, C. D., Sakabe, S., Sakabe, N., *et al.* (1988). The structure of 2Zn pig insulin crystals at 1.5 A resolution. *Philos. Trans. R. Soc. Lond. B Biol. Sci.* **319,** 369–456.

Banting, F., and Best, C. (1922). The internal secretion of the pancreas. *J. Lab. Clin. Med.* **7,** 251–266.

Baserga, R. (1999). The IGF-I receptor in cancer research. *Exp. Cell Res.* **253,** 1–6.

Bathgate, R. A., Samuel, C. S., Burazin, T. C., Layfield, S., Claasz, A. A., Reytomas, I. G., Dawson, N. F., Zhao, C., Bond, C., Summers, R. J., Parry, L. J., Wade, J. D., *et al.* (2002). Human relaxin gene 3 (H3) and the equivalent mouse relaxin (M3) gene. Novel members of the relaxin peptide family. *J. Biol. Chem.* **277,** 1148–1157.

Bathgate, R. A., Ivell, R., Sanborn, B. M., Sherwood, O. D., and Summers, R. J. (2006a). International Union of Pharmacology LVII: Recommendations for the nomenclature of receptors for relaxin family peptides. *Pharmacol. Rev.* **58,** 7–31.

Bathgate, R. A., Lin, F., Hanson, N. F., Otvos, L. Jr., Guidolin, A., Giannakis, C., Bastiras, S., Layfield, S. L., Ferraro, T., Ma, S., Zhao, C., Gundlach, A. L., *et al.* (2006b). Relaxin-3: Improved synthesis strategy and demonstration of its high-affinity interaction with the relaxin receptor LGR7 both *in vitro* and *in vivo*. *Biochemistry* **45,** 1043–1053.

Baumann, G. (2002). Genetic characterization of growth hormone deficiency and resistance: Implications for treatment with recombinant growth hormone. *Am. J. Pharmacogenomics* **2,** 93–111.

Baxter, R. C. (2000). Insulin-like growth factor (IGF)-binding proteins: Interactions with IGFs and intrinsic bioactivities. *Am. J. Physiol. Endocrinol. Metab.* **278,** E967–E976.

Beardsall, K., Diderholm, B. M., and Dunger, D. B. (2008). Insulin and carbohydrate metabolism. *Best Pract. Res. Clin. Endocrinol. Metab.* **22,** 41–55.

Bell, R. J., Eddie, L. W., Lester, A. R., Wood, E. C., Johnston, P. D., and Niall, H. D. (1987). Relaxin in human pregnancy serum measured with an homologous radioimmunoassay. *Obstet. Gynecol.* **69,** 585–589.

Bentley, G., Dodson, E., Dodson, G., Hodgkin, D., and Mercola, D. (1976). Structure of insulin in 4-zinc insulin. *Nature* **261,** 166–168.

Bieche, I., Laurent, A., Laurendeau, I., Duret, L., Giovangrandi, Y., Frendo, J. L., Olivi, M., Fausser, J. L., Evain-Brion, D., and Vidaud, M. (2003). Placenta-specific INSL4 expression is mediated by a human endogenous retrovirus element. *Biol. Reprod.* **68,** 1422–1429.

Blundell, T. L., Cutfield, J. F., Cutfield, S. M., Dodson, E. J., Dodson, G. G., Hodgkin, D. C., Mercola, D. A., and Vijayan, M. (1971a). Atomic positions in rhombohedral 2-zinc insulin crystals. *Nature* **231,** 506–511.

Blundell, T. L., Cutfield, J. F., Dodson, G. G., Dodson, E., Hodgkin, D. C., and Mercola, D. (1971b). The structure and biology of insulin. *Biochem. J.* **125,** 50P–51P.

Blundell, T. L., Dodson, G. G., Dodson, E., Hodgkin, D. C., and Vijayan, M. (1971c). X-ray analysis and the structure of insulin. *Recent Prog. Horm. Res.* **27,** 1–40.

Boels, K., and Schaller, H. C. (2003). Identification and characterisation of GPR100 as a novel human G-protein-coupled bradykinin receptor. *Br. J. Pharmacol.* **140,** 932–938.

Bogatcheva, N. V., Truong, A., Feng, S., Engel, W., Adham, I. M., and Agoulnik, A. I. (2003). GREAT/LGR8 is the only receptor for insulin-like 3 peptide. *Mol. Endocrinol.* **17,** 2639–2646.

Brailoiu, G. C., Dun, S. L., Yin, D., Yang, J., Chang, J. K., and Dun, N. J. (2005). Insulin-like 6 immunoreactivity in the mouse brain and testis. *Brain Res.* **1040,** 187–190.

Bruni, L., Luisi, S., Ferretti, C., Janneau, J. L., Quadrifoglio, M., Richon, S., Dangles-Marie, V., Bellet, D., and Petraglia, F. (2007). Changes in the maternal serum concentration of proearly placenta insulin-like growth factor peptides in normal vs abnormal pregnancy. *Am. J. Obstet. Gynecol.* **197,** 606 e1–e4.

Brzozowski, A. M., Dodson, E. J., Dodson, G. G., Murshudov, G. N., Verma, C., Turkenburg, J. P., de Bree, F. M., and Dauter, Z. (2002). Structural origins of the functional divergence of human insulin-like growth factor-I and insulin. *Biochemistry* **41,** 9389–9397.

Büllesbach, E. E., and Schwabe, C. (1995). A novel Leydig cell cDNA-derived protein is a relaxin-like factor. *J. Biol. Chem.* **270,** 16011–16015.
Büllesbach, E. E., and Schwabe, C. (1999). Tryptophan B27 in the relaxin-like factor (RLF) is crucial for RLF receptor-binding. *Biochemistry* **38,** 3073–3078.
Büllesbach, E. E., and Schwabe, C. (2000). The relaxin receptor-binding site geometry suggests a novel gripping mode of interaction. *J. Biol. Chem.* **275,** 35276–35280.
Büllesbach, E. E., and Schwabe, C. (2001). Synthesis and conformational analysis of the insulin-like 4 gene product. *J. Pept. Res.* **57,** 77–83.
Büllesbach, E. E., and Schwabe, C. (2007). Structure of the transmembrane signal initiation site of the relaxin-like factor (RLF/INSL3). *Biochemistry* **46,** 9722–9727.
Büllesbach, E. E., Yang, S., and Schwabe, C. (1992). The receptor-binding site of human relaxin II. A dual prong-binding mechanism. *J. Biol. Chem.* **267,** 22957–22960.
Burazin, T. C., Bathgate, R. A., Macris, M., Layfield, S., Gundlach, A. L., and Tregear, G. W. (2002). Restricted, but abundant, expression of the novel rat gene-3 (R3) relaxin in the dorsal tegmental region of brain. *J. Neurochem.* **82,** 1553–1557.
Burkhardt, E., Adham, I. M., Hobohm, U., Murphy, D., Sander, C., and Engel, W. (1994). A human cDNA coding for the Leydig insulin-like peptide (Ley I-L). *Hum. Genet.* **94,** 91–94.
Chan, S. J., and Steiner, D. J. (2000). Insulin through the ages: Phylogeny of a growth promoting and metabolic regulatory hormone. *Am. Zool.* **40,** 213–222.
Chao, W., and D'Amore, P. A. (2008). IGF2: Epigenetic regulation and role in development and disease. *Cytokine Growth Factor Rev.* **19,** 111–120.
Chassin, D., Laurent, A., Janneau, J. L., Berger, R., and Bellet, D. (1995). Cloning of a new member of the insulin gene superfamily (INSL4) expressed in human placenta. *Genomics* **29,** 465–470.
Cheatham, B., and Kahn, C. R. (1995). Insulin action and the insulin signaling network. *Endocr. Rev.* **16,** 117–142.
Chen, J., Kuei, C., Sutton, S. W., Bonaventure, P., Nepomuceno, D., Eriste, E., Sillard, R., Lovenberg, T. W., and Liu, C. (2005). Pharmacological characterization of relaxin-3/INSL7 receptors GPCR135 and GPCR142 from different mammalian species. *J. Pharmacol. Exp. Ther.* **312,** 83–95.
Claasz, A. A., Bond, C. P., Bathgate, R. A., Otvos, L., Dawson, N. F., Summers, R. J., Tregear, G. W., and Wade, J. D. (2002). Relaxin-like bioactivity of ovine Insulin 3 (INSL3) analogues. *Eur. J. Biochem.* **269,** 6287–6293.
Clemmons, D. R. (1997). Insulin-like growth factor binding proteins and their role in controlling IGF actions. *Cytokine Growth Factor Rev.* **8,** 45–62.
Clemmons, D. R. (2001). Use of mutagenesis to probe IGF-binding protein structure/function relationships. *Endocr. Rev.* **22,** 800–817.
Conklin, D., Lofton-Day, C. E., Haldeman, B. A., Ching, A., Whitmore, T. E., Lok, S., and Jaspers, S. (1999). Identification of INSL5, a new member of the insulin superfamily. *Genomics* **60,** 50–56.
Constancia, M., Hemberger, M., Hughes, J., Dean, W., Ferguson-Smith, A., Fundele, R., Stewart, F., Kelsey, G., Fowden, A., Sibley, C., and Reik, W. (2002). Placental-specific IGF-II is a major modulator of placental and fetal growth. *Nature* **417,** 945–948.
Cooke, R. M., Harvey, T. S., and Campbell, I. D. (1991). Solution structure of human insulin-like growth factor 1: A nuclear magnetic resonance and restrained molecular dynamics study. *Biochemistry* **30,** 5484–5491.
Crawford, R. J., Hudson, P., Shine, J., Niall, H. D., Eddy, R. L., and Shows, T. B. (1984). Two human relaxin genes are on chromosome 9. *EMBO J.* **3,** 2341–2345.
Cuastrecasas, P. (1971). Insulin-receptor interactions in adipose tissue cells: Direct measurment and properties. *Proc. Natl. Acad. Sci. USA* **68,** 1264–1268.
Daughaday, W. H., Hall, K., Salmon, W. D. Jr., Van den Brande, J. L., and Van Wyk, J. J. (1987). On the nomenclature of the somatomedins and insulin-like growth factors. *Endocrinology* **121,** 1911–1912.

De Meyts, P. (2004). Insulin and its receptor: Structure, function and evolution. *Bioessays* **26,** 1351–1362.
De Meyts, P., and Whittaker, J. (2002). Structural biology of insulin and IGF1 receptors: Implications for drug design. *Nat. Rev. Drug Discov.* **1,** 769–783.
De Meyts, P., Palsgaard, J., Sajid, W., Theede, A. M., and Aladdin, H. (2004). Structural biology of insulin and IGF-1 receptors. *Novartis Found Symp.* **262,** 160–171; discussion 171–176, 265–268.
DeChiara, T. M., Robertson, E. J., and Efstratiadis, A. (1991). Parental imprinting of the mouse insulin-like growth factor II gene. *Cell* **64,** 849–859.
Del Borgo, M. P., Hughes, R. A., Bathgate, R. A., Lin, F., Kawamura, K., and Wade, J. D. (2006). Analogs of insulin-like peptide 3 (INSL3) B-chain are LGR8 antagonists *in vitro* and *in vivo*. *J. Biol. Chem.* **281,** 13068–13074.
Denley, A., Cosgrove, L. J., Booker, G. W., Wallace, J. C., and Forbes, B. E. (2005). Molecular interactions of the IGF system. *Cytokine Growth Factor Rev.* **16,** 421–439.
Dun, S. L., Brailoiu, E., Wang, Y., Brailoiu, G. C., Liu-Chen, L. Y., Yang, J., Chang, J. K., and Dun, N. J. (2006). Insulin-like peptide 5: Expression in the mouse brain and mobilization of calcium. *Endocrinology* **147,** 3243–3248.
Eddie, L. W., Bell, R. J., Lester, A., Geier, M., Bennett, G., Johnston, P. D., and Niall, H. D. (1986). Radioimmunoassay of relaxin in pregnancy with an analogue of human relaxin. *Lancet* **1,** 1344–1346.
Eigenbrot, C., Randal, M., Quan, C., Burnier, J., O'Connell, L., Rinderknecht, E., and Kossiakoff, A. A. (1991). X-ray structure of human relaxin at 1.5 A. Comparison to insulin and implications for receptor binding determinants. *J. Mol. Biol.* **221,** 15–21.
Feng, S., Cortessis, V. K., Hwang, A., Hardy, B., Koh, C. J., Bogatcheva, N. V., and Agoulnik, A. I. (2004). Mutation analysis of INSL3 and GREAT/LGR8 genes in familial cryptorchidism. *Urology* **64,** 1032–1036.
Fevold, H., Hisaw, F. L., and Meyer, R. K. (1930). The relaxative hormone of the corpus luteum. Its purification and concentration. *J. Am. Chem. Soc.* **52,** 3340–3348.
Foresta, C., and Ferlin, A. (2004). Role of INSL3 and LGR8 in cryptorchidism and testicular functions. *Reprod. Biomed Online* **9,** 294–298.
Froesch, E. R., Buergi, H., Ramseier, E. B., Bally, P., and Labhart, A. (1963). Antibody-suppressible and nonsuppressible insulin-like activities in human serum and their physiologic significance. An insulin assay with adipose tissue of increased precision and specificity. *J. Clin. Invest.* **42,** 1816–1834.
Gorlov, I. P., Kamat, A., Bogatcheva, N. V., Jones, E., Lamb, D. J., Truong, A., Bishop, C. E., McElreavey, K., and Agoulnik, A. I. (2002). Mutations of the GREAT gene cause cryptorchidism. *Hum. Mol. Genet.* **11,** 2309–2318.
Gunnersen, J. M., Crawford, R. J., and Tregear, G. W. (1995). Expression of the relaxin gene in rat tissues. *Mol. Cell. Endocrinol.* **110,** 55–64.
Halls, M. L., Bond, C. P., Sudo, S., Kumagai, J., Ferraro, T., Layfield, S., Bathgate, R. A., and Summers, R. J. (2005). Multiple binding sites revealed by interaction of relaxin family peptides with native and chimeric relaxin family peptide receptors 1 and 2 (LGR7 and LGR8). *J. Pharmacol. Exp. Ther.* **313,** 677–687.
Halls, M. L., van der Westhuizen, E. T., Bathgate, R. A., and Summers, R. J. (2007). Relaxin family peptide receptors–former orphans reunite with their parent ligands to activate multiple signalling pathways. *Br. J. Pharmacol.* **150,** 677–691.
Hansell, D. J., Bryant-Greenwood, G. D., and Greenwood, F. C. (1991). Expression of the human relaxin H1 gene in the decidua, trophoblast, and prostate. *J. Clin. Endocrinol. Metab.* **72,** 899–904.
Harris, D. A., Van Vliet, G., Egli, C. A., Grumbach, M. M., Kaplan, S. L., Styne, D. M., and Vainsel, M. (1985). Somatomedin-C in normal puberty and in true precocious puberty before and after treatment with a potent luteinizing hormone-releasing hormone agonist. *J. Clin. Endocrinol. Metab.* **61,** 152–159.

Hedo, J. A., Kasuga, M., Van Obberghen, E., Roth, J., and Kahn, C. R. (1981). Direct demonstration of glycosylation of insulin receptor subunits by biosynthetic and external labeling: Evidence for heterogeneity. *Proc. Natl. Acad. Sci. USA* **78,** 4791–4795.

Hida, T., Takahashi, E., Shikata, K., Hirohashi, T., Sawai, T., Seiki, T., Tanaka, H., Kawai, T., Ito, O., Arai, T., Yokoi, A., Hirakawa, T., *et al.* (2006). Chronic intracerebroventricular administration of relaxin-3 increases body weight in rats. *J. Recept. Signal Transduct. Res.* **26,** 147–158.

Hisaw, F. (1926). Experimental relaxation of the pubic ligament of the guinea pig. *Proc. Soc. Exp. Biol. Med.* **23,** 661–663.

Hopkins, E. J., Layfield, S., Ferraro, T., Bathgate, R. A., and Gooley, P. R. (2007). The NMR solution structure of the relaxin (RXFP1) receptor lipoprotein receptor class A module and identification of key residues in the N-terminal region of the module that mediate receptor activation. *J. Biol. Chem.* **282,** 4172–4184.

Hsu, S. Y. (2003). New insights into the evolution of the relaxin-LGR signaling system. *Trends Endocrinol. Metab.* **14,** 303–309.

Hsu, S. Y., Kudo, M., Chen, T., Nakabayashi, K., Bhalla, A., van der Spek, P. J., van Duin, M., and Hsueh, A. J. (2000). The three subfamilies of leucine-rich repeat-containing G protein-coupled receptors (LGR): Identification of LGR6 and LGR7 and the signaling mechanism for LGR7. *Mol. Endocrinol.* **14,** 1257–1271.

Hsu, S. Y., Nakabayashi, K., Nishi, S., Kumagai, J., Kudo, M., Sherwood, O. D., and Hsueh, A. J. (2002). Activation of orphan receptors by the hormone relaxin. *Science* **295,** 671–674.

Hsu, S. Y., Nakabayashi, K., Nishi, S., Kumagai, J., Kudo, M., Bathgate, R. A., Sherwood, O. D., and Hsueh, A. J. (2003). Relaxin signaling in reproductive tissues. *Mol. Cell. Endocrinol.* **202,** 165–170.

Hsu, S. Y., Semyonov, J., Park, J. I., and Chang, C. L. (2005). Evolution of the signaling system in relaxin-family peptides. *Ann. N Y Acad. Sci.* **1041,** 520–529.

Hudson, P., Haley, J., John, M., Cronk, M., Crawford, R., Haralambidis, J., Tregear, G., Shine, J., and Niall, H. (1983). Structure of a genomic clone encoding biologically active human relaxin. *Nature* **301,** 628–631.

Hudson, P., John, M., Crawford, R., Haralambidis, J., Scanlon, D., Gorman, J., Tregear, G., Shine, J., and Niall, H. (1984). Relaxin gene expression in human ovaries and the predicted structure of a human preprorelaxin by analysis of cDNA clones. *EMBO J.* **3,** 2333–2339.

Humbel, R. E. (1990). Insulin-like growth factors I and II. *Eur. J. Biochem.* **190,** 445–462.

Hwa, V., Oh, Y., and Rosenfeld, R. G. (1999). The insulin-like growth factor-binding protein (IGFBP) superfamily. *Endocr. Rev.* **20,** 761–787.

Ivell, R., and Bathgate, R. A. (2002). Reproductive biology of the relaxin-like factor (RLF/ INSL3). *Biol. Reprod.* **67,** 699–705.

Ivell, R., and Einspanier, A. (2002). Relaxin peptides are new global players. *Trends Endocrinol. Metab.* **13,** 343–348.

Jones, J., and Clemmons, D. (1995). Insulin-like growth factors and their binding proteins: Biological actions. *Endocr. Rev.* **16,** 3–34.

Kasik, J., Muglia, L., Stephan, D. A., and Menon, R. K. (2000). Identification, chromosomal mapping, and partial characterization of mouse Ins16: A new member of the insulin family. *Endocrinology* **141,** 458–461.

Kasuga, M., Karlsson, F. A., and Kahn, C. R. (1982). Insulin stimulates the phosphorylation of the 95,000-dalton subunit of its own receptor. *Science* **215,** 185–187.

Kawamura, K., Kumagai, J., Sudo, S., Chun, S. Y., Pisarska, M., Morita, H., Toppari, J., Fu, P., Wade, J. D., Bathgate, R. A., and Hsueh, A. J. (2004). Paracrine regulation of mammalian oocyte maturation and male germ cell survival. *Proc. Natl. Acad. Sci USA* **101,** 7323–7328.

Kiess, W., Yang, Y., Kessler, U., and Hoeflich, A. (1994). Insulin-like growth factor II (IGF-II) and the IGF-II/mannose-6-phosphate receptor: The myth continues. *Horm. Res.* **41**(Suppl 2), 66–73.

Klonisch, T., Ivell, R., Balvers, M., Kliesch, S., Fischer, B., Bergmann, M., and Steger, K. (1999). Expression of relaxin-like factor is down-regulated in human testicular Leydig cell neoplasia. *Mol. Hum. Reprod.* **5,** 104–108.

Klonisch, T., Froehlich, C., Tetens, F., Fischer, B., and Hombach-Klonisch, S. (2001). Molecular remodeling of members of the relaxin family during primate evolution. *Mol. Biol. Evol.* **18,** 393–403.

Klonisch, T., Mustafa, T., Bialek, J., Radestock, Y., Holzhausen, H. J., Dralle, H., Hoang-Vu, C., and Hombach-Klonisch, S. (2005). Human medullary thyroid carcinoma: A source and potential target for relaxin-like hormones. *Ann. N Y Acad. Sci.* **1041,** 449–461.

Koman, A., Cazaubon, S., Couraud, P.-O., Ullrich, A., and Strosberg, A. D. (1996). Molecular characterization and *in vitro* biological activity of placentin, a new member of the insulin gene family. *J. Biol. Chem.* **271,** 20238–20241.

Koos, R. D., Kazi, A. A., Roberson, M. S., and Jones, J. M. (2005). New insight into the transcriptional regulation of vascular endothelial growth factor expression in the endometrium by estrogen and relaxin. *Ann. N Y Acad. Sci.* **1041,** 233–247.

Kornfeld, S. (1992). Structure and function of the mannose 6-phosphate/insulinlike growth factor II receptors. *Annu. Rev. Biochem.* **61,** 307–330.

Kumagai, J., Hsu, S. Y., Matsumi, H., Roh, J. S., Fu, P., Wade, J. D., Bathgate, R. A., and Hsueh, A. J. (2002). INSL3/Leydig insulin-like peptide activates the LGR8 receptor important in testis descent. *J. Biol. Chem.* **277,** 31283–31286.

Laron, Z. (2004). Laron syndrome (primary growth hormone resistance or insensitivity): The personal experience 1958–2003. *J. Clin. Endocrinol. Metab.* **89,** 1031–1044.

Laurent, A., Rouillac, C., Delezoide, A. L., Giovangrandi, Y., Vekemans, M., Bellet, D., Abitbol, M., and Vidaud, M. (1998). Insulin-like 4 (INSL4) gene expression in human embryonic and trophoblastic tissues. *Mol. Reprod. Dev.* **51,** 123–129.

Lewitt, M. S., Saunders, H., Phuyal, J. L., and Baxter, R. C. (1994). Complex formation by human insulin-like growth factor-binding protein-3 and human acid-labile subunit in growth hormone-deficient rats. *Endocrinology* **134,** 2404–2409.

Lin, F., Otvos, L. Jr., Kumagai, J., Tregear, G. W., Bathgate, R. A., and Wade, J. D. (2004). Synthetic human insulin 4 does not activate the G-protein-coupled receptors LGR7 or LGR8. *J. Pept. Sci.* **10,** 257–264.

Liu, C., Chen, J., Sutton, S., Roland, B., Kuei, C., Farmer, N., Sillard, R., and Lovenberg, T. W. (2003a). Identification of relaxin-3/INSL7 as a ligand for GPCR142. *J. Biol. Chem.* **278,** 50765–50770.

Liu, C., Eriste, E., Sutton, S., Chen, J., Roland, B., Kuei, C., Farmer, N., Jornvall, H., Sillard, R., and Lovenberg, T. W. (2003b). Identification of relaxin-3/INSL7 as an endogenous ligand for the orphan G-protein-coupled receptor GPCR135. *J. Biol. Chem.* **278,** 50754–50764.

Liu, C., Chen, J., Kuei, C., Sutton, S., Nepomuceno, D., Bonaventure, P., and Lovenberg, T. W. (2005a). Relaxin-3/insulin-like peptide 5 chimeric peptide, a selective ligand for G protein-coupled receptor (GPCR)135 and GPCR142 over leucine-rich repeat-containing G protein-coupled receptor 7. *Mol. Pharmacol.* **67,** 231–240.

Liu, C., Kuei, C., Sutton, S., Chen, J., Bonaventure, P., Wu, J., Nepomuceno, D., Kamme, F., Tran, D. T., Zhu, J., Wilkinson, T., Bathgate, R., *et al.* (2005b). INSL5 is a high affinity specific agonist for GPCR142 (GPR100). *J. Biol. Chem.* **280,** 292–300.

Lok, S., Johnston, D. S., Conklin, D., Lofton-Day, C. E., Adams, R. L., Jelmberg, A. C., Whitmore, T. E., Schrader, S., Griswold, M. D., and Jaspers, S. R. (2000). Identification of INSL6, a new member of the insulin family that is expressed in the testis of the human and rat. *Biol. Reprod.* **62,** 1593–1599.

Lu, C., Walker, W. H., Sun, J., Weisz, O. A., Gibbs, R. B., Witchel, S. F., Sperling, M. A., and Menon, R. K. (2006). Insulin-like peptide 6: Characterization of secretory status and posttranslational modifications. *Endocrinology*. **147,** 5611–5623.

Luna, J. J., Riesewijk, A., Horcajadas, J. A., Van Os Rd, R., Dominguez, F., Mosselman, S., Pellicer, A., and Simon, C. (2004). Gene expression pattern and immunoreactive protein localization of LGR7 receptor in human endometrium throughout the menstrual cycle. *Mol. Hum. Reprod.* **10,** 85–90.

Ma, S., Bonaventure, P., Ferraro, T., Shen, P. J., Burazin, T. C., Bathgate, R. A., Liu, C., Tregear, G. W., Sutton, S. W., and Gundlach, A. L. (2007). Relaxin-3 in GABA projection neurons of nucleus incertus suggests widespread influence on forebrain circuits via G-protein-coupled receptor-135 in the rat. *Neuroscience* **144,** 165–190.

Massague, J., Pilch, P. F., and Czech, M. P. (1981). A unique proteolytic cleavage site on the beta subunit of the insulin receptor. *J. Biol. Chem.* **256,** 3182–3190.

Matsumoto, M., Kamohara, M., Sugimoto, T., Hidaka, K., Takasaki, J., Saito, T., Okada, M., Yamaguchi, T., and Furuichi, K. (2000). The novel G-protein coupled receptor SALPR shares sequence similarity with somatostatin and angiotensin receptors. *Gene* **248,** 183–189.

Mauras, N. E., Blizzard, R., Link, K., Johnson, M. L., Rogol, A. D., and Veldhuis, J. D. (1987). Augmentation of growth hormone secretion during puberty: Evidence for a pulse amplitude modulated phenomenon. *J. Clin. Endocrinol. Metab.* **64,** 596–601.

Mazella, J., Tang, M., and Tseng, L. (2004). Disparate effects of relaxin and TGFbeta1: Relaxin increases, but TGFbeta1 inhibits, the relaxin receptor and the production of IGFBP-1 in human endometrial stromal/decidual cells. *Hum. Reprod.* **19,** 1513–1518.

McGowan, B. M., Stanley, S. A., Smith, K. L., Minnion, J. S., Donovan, J., Thompson, E. L., Patterson, M., Connolly, M. M., Abbott, C. R., Small, C. J., Gardiner, J. V., Ghatei, M. A., *et al.* (2006). Effects of acute and chronic relaxin-3 on food intake and energy expenditure in rats. *Regul. Pept.* **136,** 72–77.

McKern, N. M., Lawrence, M. C., Streltsov, V. A., Lou, M. Z., Adams, T. E., Lovrecz, G. O., Elleman, T. C., Richards, K. M., Bentley, J. D., Pilling, P. A., Hoyne, P. A., Cartledge, K. A., *et al.* (2006). Structure of the insulin receptor ectodomain reveals a folded-over conformation. *Nature* **443,** 218–221.

Millar, L., Streiner, N., Webster, L., Yamamoto, S., Okabe, R., Kawamata, T., Shimoda, J., Bullesbach, E., Schwabe, C., and Bryant-Greenwood, G. (2005). Early placental insulin-like protein (INSL4 or EPIL) in placental and fetal membrane growth. *Biol. Reprod.* **73,** 695–702.

Mock, P., Frydman, R., Bellet, D., Diawara, D. A., Lavaissiere, L., Troalen, F., and Bidart, J. M. (1999). Pro-EPIL forms are present in amniotic fluid and maternal serum during normal pregnancy. *J. Clin. Endocrinol. Metab.* **84,** 2253–2256.

Mock, P., Frydman, R., Bellet, D., Chassin, D., Bischof, P., Campana, A., and Bidart, J. M. (2000). Expression of pro-EPIL peptides encoded by the insulin-like 4 (INSL4) gene in chromosomally abnormal pregnancies. *J. Clin. Endocrinol. Metab.* **85,** 3941–3944.

Nakae, J., Kido, Y., and Accili, D. (2001). Distinct and overlapping functions of insulin and IGF-I receptors. *Endocr. Rev.* **22,** 818–835.

Nef, S., and Parada, L. F. (1999). Cryptorchidism in mice mutant for Insl3. *Nat. Genet.* **22,** 295–299.

Nef, S., Shipman, T., and Parada, L. F. (2000). A molecular basis for estrogen-induced cryptorchidism. *Dev. Biol.* **224,** 354–361.

Overbeek, P. A., Gorlov, I. P., Sutherland, R. W., Houston, J. B., Harrison, W. R., Boettger-Tong, H. L., Bishop, C. E., and Agoulnik, A. I. (2001). A transgenic insertion causing cryptorchidism in mice. *Genes* **30,** 26–35.

Palejwala, S., Tseng, L., Wojtczuk, A., Weiss, G., and Goldsmith, L. T. (2002). Relaxin gene and protein expression and its regulation of procollagenase and vascular endothelial growth factor in human endometrial cells. *Biol. Reprod.* **66,** 1743–1748.

Park, J. I., Chang, C. L., and Hsu, S. Y. (2005). New Insights into biological roles of relaxin and relaxin-related peptides. *Rev. Endocr. Metab. Disord.* **6,** 291–296.
Pavelic, J., Matijevic, T., and Knezevic, J. (2007). Biological & physiological aspects of action of insulin-like growth factor peptide family. *Indian J. Med. Res.* **125,** 511–522.
Pollak, M. N. (2004). Insulin-like growth factors and neoplasia. *Novartis Found Symp.* **262,** 84–98discussion 98–107, 265–268.
Pusch, W., Balvers, M., and Ivell, R. (1996). Molecular cloning and expression of the relaxin-like factor from the mouse testis. *Endocrinology* **137,** 3009–3013.
Rinderknecht, E., and Humbel, R. E. (1976a). Amino-terminal sequences of two polypeptides from human serum with nonsuppressible insulin-like and cell-growth-promoting activities: Evidence for structural homology with insulin B chain. *Proc. Natl. Acad. Sci. USA* **73,** 4379–4381.
Rinderknecht, E., and Humbel, R. E. (1976b). Polypeptides with nonsuppressible insulin-like and cell-growth promoting activities in human serum: Isolation, chemical characterization, and some biological properties of forms I and II. *Proc. Natl. Acad. Sci. USA* **73,** 2365–2369.
Rinderknecht, E., and Humbel, R. E. (1978). The amino acid sequence of human insulin-like growth factor I and its structural homology with proinsulin. *J. Biol. Chem.* **253,** 2769–2776.
Roche, P. J., Butkus, A., Wintour, E. M., and Tregear, G. (1996). Structure and expression of Leydig insulin-like peptide mRNA in the sheep. *Mol. Cell. Endocrinol.* **121,** 171–177.
Rosenfeld, L. (2002). Insulin: Discovery and controversy. *Clin. Chem.* **48,** 2270–2288.
Rosengren, K. J., Zhang, S., Lin, F., Daly, N. L., Scott, D. J., Hughes, R. A., Bathgate, R. A., Craik, D. J., and Wade, J. D. (2006). Solution structure and characterization of the LGR8 receptor binding surface of insulin-like peptide 3. *J. Biol. Chem.* **281,** 28287–28295.
Rubin, J. B., Shia, M. A., and Pilch, P. F. (1983). Stimulation of tyrosine-specific phosphorylation in vitro by insulin-like growth factor I. *Nature* **305,** 438–440.
Ryle, A. P., Sanger, F., Smith, L. F., and Kitai, R. (1955). The disulphide bonds of insulin. *Biochem. J.* **60,** 541–556.
Salmon, W. D. Jr., and Burkhalter, V. J. (1997). Stimulation of sulfate and thymidine incorporation into hypophysectomized rat cartilage by growth hormone and insulin-like growth factor-I *in vitro*: The somatomedin hypothesis revisited. *J. Lab. Clin. Med.* **129,** 430–438.
Samuel, C. S., Coghlan, J. P., and Bateman, J. F. (1998). Effects of relaxin, pregnancy and parturition on collagen metabolism in the rat pubic symphysis. *J. Endocrinol.* **159,** 117–125.
Samuel, C. S., Hewitson, T. D., Unemori, E. N., and Tang, M. L. (2007). Drugs of the future: The hormone relaxin. *Cell. Mol. Life Sci.* **64,** 1539–1557.
Sankoff, D. (2001). Gene and genome duplication. *Curr. Opin. Genet. Dev.* **11,** 681–684.
Sato, A., Nishimura, S., Ohkubo, T., Kyogoku, Y., Koyama, S., Kobayashi, M., Yasuda, T., and Kobayashi, Y. (1993). Three-dimensional structure of human insulin-like growth factor-I (IGF-I) determined by 1H-NMR and distance geometry. *Int. J. Pept. Protein Res.* **41,** 433–440.
Schwabe, C., and Büllesbach, E. E. (1994). Relaxin: Structures, functions, promises, and nonevolution. *FASEB J.* **8,** 1152–1160.
Scott, C. D., and Firth, S. M. (2004). The role of the M6P/IGF-II receptor in cancer: Tumor suppression or garbage disposal? *Horm. Metab. Res.* **36,** 261–271.
Scott, D. J., Wilkinson, T., Tregear, G. W., and Bathgate, R. A. (2003). The relaxin family and their novel G-protein coupled receptors. *Lett. Pept. Sci.* **10,** 393–400.
Scott, D. J., Wilkinson, T., Zhang, S., Ferraro, T., Wade, J., Tregear, G. W., and Bathgate, R. A. (2007). Defining the Lgr8 residues involved in binding Insl3. *Mol. Endocrinol.* **21,** 1699–1712.

Shabanpoor, F., Bathgate, R. A., Hossain, M. A., Giannakis, E., Wade, J. D., and Hughes, R. A. (2007). Design, synthesis and pharmacological evaluation of cyclic mimetics of the insulin-like peptide 3 (INSL3) B-chain. *J. Pept. Sci.* **13,** 113–120.

Sherwood, O. D. (2004). Relaxin's physiological roles and other diverse actions. *Endocr. Rev.* **25,** 205–234.

Sherwood, O. D., and Rutherford, J. E. (1981). Relaxin immunoactivity levels in ovarian extracts obtained from rats during various reproductive states and from adult cycling pigs. *Endocrinology* **108,** 1171–1177.

Siegel, T. W., Ganguly, S., Jacobs, S., Rosen, O. M., and Rubin, C. S. (1981). Purification and properties of the human placental insulin receptor. *J. Biol. Chem.* **256,** 9266–9273.

Silvertown, J. D., Symes, J. C., Neschadim, A., Nonaka, T., Kao, J. C., Summerlee, A. J., and Medin, J. A. (2007). Analog of H2 relaxin exhibits antagonistic properties and impairs prostate tumor growth. *FASEB J.* **21,** 754–765.

Siwanowicz, I., Popowicz, G. M., Wisniewska, M., Huber, R., Kuenkele, K. P., Lang, K., Engh, R. A., and Holak, T. A. (2005). Structural basis for the regulation of insulin-like growth factors by IGF binding proteins. *Structure* **13,** 155–167.

Sparrow, L. G., Gorman, J. J., Strike, P. M., Robinson, C. P., McKern, N. M., Epa, V. C., and Ward, C. W. (2007). The location and characterisation of the O-linked glycans of the human insulin receptor. *Proteins* **66,** 261–265.

Spiess, A. N., Balvers, M., Tena-Sempere, M., Huhtaniemi, I., Parry, L., and Ivell, R. (1999). Structure and expression of the rat relaxin-like factor (RLF) gene. *Mol. Reprod. Dev.* **54,** 319–325.

Steinetz, B. G., Goldsmith, L. T., and Lust, G. (1987). Plasma relaxin levels in pregnant and lactating dogs. *Biol. Reprod.* **37,** 719–725.

Sudo, S., Kumagai, J., Nishi, S., Layfield, S., Ferraro, T., Bathgate, R. A., and Hsueh, A. J. (2003). H3 relaxin is a specific ligand for LGR7 and activates the receptor by interacting with both the ectodomain and the exoloop 2. *J. Biol. Chem.* **278,** 7855–7862.

Tang, S. H., Yang, D. H., Huang, W., Zhou, M., Zhou, H. K., Lu, X. H., and Ye, G. (2006). Differential promoter usage for insulin-like growth factor-II gene in Chinese hepatocellular carcinoma with hepatitis B virus infection. *Cancer Detect. Prev.* **30,** 192–203.

Tashima, L. S., Hieber, A. D., Greenwood, F. C., and Bryant-Greenwood, G. D. (1995). The human Leydig insulin-like (hLEY I-L) gene is expressed in the corpus luteum and trophoblast. *J. Clin. Endocrinol. Metab.* **80,** 707–710.

Terasawa, H., Kohda, D., Hatanaka, H., Nagata, K., Higashihashi, N., Fujiwara, H., Sakano, K., and Inagaki, F. (1994). Solution structure of human insulin-like growth factor II; recognition sites for receptors and binding proteins. *EMBO J.* **13,** 5590–5597.

Torres, A. M., Forbes, B. E., Aplin, S. E., Wallace, J. C., Francis, G. L., and Norton, R. S. (1995). Solution structure of human insulin-like growth factor II. Relationship to receptor and binding protein interactions. *J. Mol. Biol.* **248,** 385–401.

Unemori, E. N., and Amento, E. P. (1990). Relaxin modulates synthesis and secretion of procollagenase and collagen by human dermal fibroblasts. *J. Biol. Chem.* **265,** 10681–10685.

Unemori, E. N., Bauer, E. A., and Amento, E. P. (1992). Relaxin alone and in conjunction with interferon-gamma decreases collagen synthesis by cultured human scleroderma fibroblasts. *J. Invest. Dermatol.* **99,** 337–342.

Vajdos, F. F., Ultsch, M., Schaffer, M. L., Deshayes, K. D., Liu, J., Skelton, N. J., and de Vos, A. M. (2001). Crystal structure of human insulin-like growth factor-1: Detergent binding inhibits binding protein interactions. *Biochemistry* **40,** 11022–11029.

Wilkinson, T. N., Speed, T. P., Tregear, G. W., and Bathgate, R. A. (2005a). Coevolution of the relaxin-like peptides and their receptors. *Ann. N Y Acad. Sci.* **1041,** 534–539.

Wilkinson, T. N., Speed, T. P., Tregear, G. W., and Bathgate, R. A. (2005b). Evolution of the relaxin-like peptide family. *BMC Evol. Biol.* **5,** 14.

Yu, H., and Rohan, T. (2000). Role of the insulin-like growth factor family in cancer development and progression. *J. Natl. Cancer Inst.* **92,** 1472–1489.
Zeslawski, W., Beisel, H. G., Kamionka, M., Kalus, W., Engh, R. A., Huber, R., Lang, K., and Holak, T. A. (2001). The interaction of insulin-like growth factor-I with the N-terminal domain of IGFBP-5. *EMBO J.* **20,** 3638–3644.
Zhao, L., Roche, P. J., Gunnersen, J. M., Hammond, V. E., Tregear, G. W., Wintour, E. M., and Beck, F. (1999a). Mice without a functional relaxin gene are unable to deliver milk to their pups. *Endocrinology* **140,** 445–453.
Zhao, L., Roche, P. J., Gunnersen, J. M., Hammond, V. E., Tregear, G. W., Wintour, E. M., and Beck, F. (1999b). Mice without a functional relaxin gene are unable to deliver milk to their pups. *Endocrinology* **140,** 445–453.
Zimmermann, S., Steding, G., Emmen, J. M., Brinkmann, A. O., Nayernia, K., Holstein, A. F., Engel, W., and Adham, I. M. (1999). Targeted disruption of the Insl3 gene causes bilateral cryptorchidism. *Mol. Endocrinol.* **13,** 681–691.

CHAPTER TWO

The Structure and Function of Insulin: Decoding the TR Transition

Michael A. Weiss

Contents

Abstract

Crystal structures of insulin are remarkable for a long-range reorganization among three families of hexamers (designated T_6, $T_3R^f_3$, and R_6). Although these structures are well characterized at atomic resolution, the biological implications of the TR transition remain the subject of speculation. Recent studies indicate that such allostery reflects a structural switch between distinct folding-competent and active conformations. Stereospecific modulation of this switch by corresponding D- and L-amino-acid substitutions yields reciprocal effects on protein stability and receptor-binding activity. Naturally occurring human mutations at the site of conformational change impair the folding of proinsulin and cause permanent neonatal-onset diabetes mellitus. The repertoire of classical structures thus foreshadows the conformational lifecycle of insulin *in vivo*. By highlighting the richness of information provided by protein crystallography—even in a biological realm far removed from conditions of crystallization—these findings validate the prescient insights of the late D. C. Hodgkin. Future studies of the receptor-bound structure of insulin may enable design of novel agonists for the treatment of diabetes mellitus. © 2009 Elsevier Inc.

Department of Biochemistry, Case Western Reserve University, Cleveland, Ohio 44106

Vitamins and Hormones, Volume 80
ISSN 0083-6729, DOI: 10.1016/S0083-6729(08)00602-X

I. Introduction

Insulin is a small globular protein containing two chains, A (21 residues) and B (30 residues) (Fig. 2.1A). Stored in the β cell as a Zn^{2+}-stabilized hexamer, the hormone dissociates in the bloodstream to function as a Zn^{2+}-free monomer (Fig. 2.1B). The structure of the free hormone has been well characterized by X-ray crystallography and NMR spectroscopy (Baker *et al.*, 1988; Hua *et al.*, 1996b; Olsen *et al.*, 1996). Key receptor-binding contacts (Baker *et al.*, 1988; Blundell *et al.*, 1972; Liang *et al.*, 1994; Pullen *et al.*, 1976) have been defined by rare naturally occurring mutations (positions A3, B24, and B25; red in Fig. 2.1A) associated with impaired activity and diabetes mellitus (Shoelson *et al.*, 1983). Classical structure–function relationships

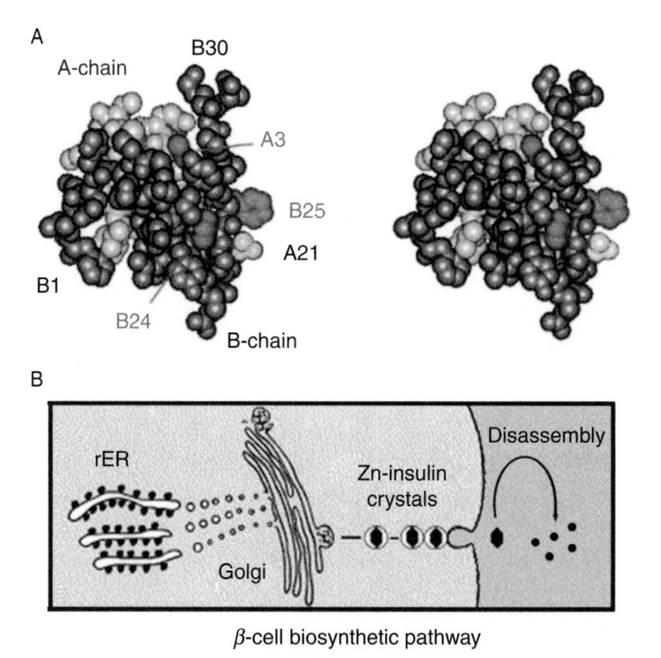

Figure 2.1 Globular structure of an insulin monomer and pathway of insulin biosynthesis. (A) Space-filling model of an insulin monomer highlighting sites of classical diabetes-associated mutations (red): $Val^{A3} \rightarrow Leu$, $Phe^{B24} \rightarrow Ser$, and $Phe^{B25} \rightarrow Leu$ (Shoelson *et al.*, 1983). The A- and B-chains are otherwise shown in light and dark gray, respectively. Atomic coordinates were obtained from Protein Databank entry 4INS (2-Zn molecule 1). (B) Nascent proinsulin folds as a monomer in ER wherein zinc-ion concentration is low; in Golgi apparatus zinc-stabilized proinsulin hexamer assembles. The prohormone is processed by cleavage of connecting peptide in post-Golgi vescicles to yield mature insulin. Zinc-insulin crystals are observed in secretory granules. Insulin hexamers dissociate in the bloodstream to release active zinc-free monomers.

were originally inferred from patterns of sequence conservation (Baker *et al.*, 1988) and subsequently tested through systematic studies of insulin analogs (Baker *et al.*, 1988; Hu *et al.*, 1993; Huang *et al.*, 2004; Kitagawa *et al.*, 1984; Liang *et al.*, 1994; Mirmira and Tager, 1989; Mirmira *et al.*, 1991; Nakagawa and Tager, 1986, 1992; Nakagawa *et al.*, 2000; Pullen *et al.*, 1976; Xu *et al.*, 2002b). Despite an abundance of such data and complementary advances in characterization of the insulin receptor (IR) (Lou *et al.*, 2006; McKern *et al.*, 2006), a molecular understanding of how insulin binds to its receptor has remained elusive (Ward *et al.*, 2008).

A rigorous foundation for studies of insulin folding, assembly, and dynamics has been provided by an extensive database of crystal structures (Adams *et al.*, 1969; Bentley *et al.*, 1976; Brader and Dunn, 1991; Ciszak *et al.*, 1995; Derewenda *et al.*, 1989). Such studies have spanned more than four decades and have played a central role in the history of structural biology. The structure of porcine insulin as a zinc-stabilized hexamer was first determined by the late Dorothy C. Hodgkin and coworkers in the 2-Zn form (Adams *et al.*, 1969). This celebrated structure, stabilized by two axial zinc ions, is now designated T_6. A second crystal form, known as 4-Zn insulin, was obtained in high concentrations of chloride ion by Schlichtkrull (1958). Now designated $T_3R^f_3$, the structure of this variant zinc hexamer was described by the Hodgkin laboratory (Bentley *et al.*, 1976) and extended to the corresponding structure of human insulin (Smith *et al.*, 1984). These two rhombohedral crystal forms exhibit extensive structural differences, designated the TR transition. (This nomenclature was adopted in honor of Max Perutz but is unrelated to the cooperative mechanism of oxygen binding to hemoglobin.) In 1989 this transition was observed in complete form by Dodson and coworkers in an R_6 phenol-stabilized zinc insulin hexamer (Derewenda *et al.*, 1989). The TR transition can be observed in crystals (Bentley *et al.*, 1978) and monitored in solution by spectroscopy (Roy *et al.*, 1989; Thomas and Wollmer, 1989).

The three families of insulin hexamers (T_6, $T_3R^f_3$, and R_6) are shown in Fig. 2.2A (top row). In the T-state conformation, the A-chain contains an N-terminal α-helix (residues A1–A8) followed by a noncanonical turn, second helix (A12–A18), and C-terminal segment (A19–A21); the B-chain contains an N-terminal segment (residues B1–B6), type II′ β-turn (B7–10), central α-helix (B9–B19), type I β-turn (B20–B23), and C-terminal β-strand (B24–B28), extended by less well-ordered terminal residues B29 and B30. In the R-state, the N-terminal portion of the B-chain participates in a single long α-helix. The resulting B1–B19 α-helix (or B3–B19 in the frayed R^f state) projects from the globular core of the protomer to make extensive hexamer contacts, including formation of specific phenol-binding pockets at symmetry-related trimer interfaces (Derewenda *et al.*, 1989). This change in the secondary structure of the B-chain is highlighted in Fig. 2.2A: the T-specific conformation of the B1–B8 is shown in green, the R-specific

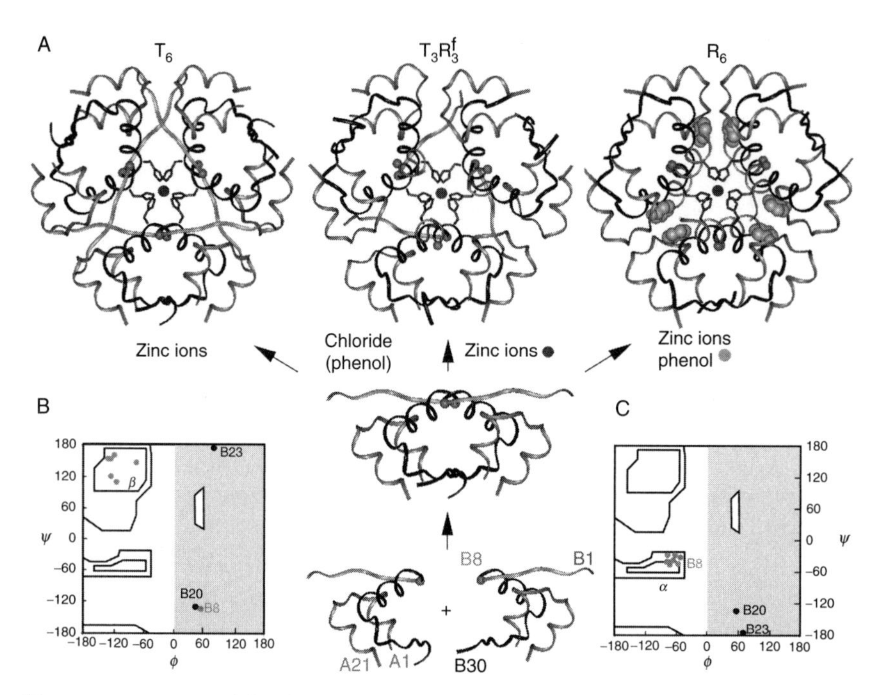

Figure 2.2 Structural families of insulin hexamers. (A) Left to right, ribbon models of T_6, $T_3R_3^f$, and R_6 zinc insulin hexamers. Central panel indicates pathway of insulin assembly and regulators of conformational reorganization (arrows). The C_α positions of Gly^{B8} are indicated by red balls; the variable secondary structures of the N-terminal segment of the B-chain (residues B1–B7) are shown highlighted in green (extended in T-state) or powder blue (α-helical in R-state). The B-chain is otherwise shown in black, and A-chain in gray. The side chains of His^{B10} in the metal-ion binding sites are shown in dark blue; zinc ions in magenta; and phenol in burnt amber. (B and C) The TR transition is associated with a conformational change of Gly^{B8} from right to left in the Ramachandran plot. Main-chain dihedral conformations of the three glycines in the B-chain (residues B8, B20, and B23) and residues B2–B8 in a representative T-state (B) or R-state (C) protomer. The conformation of Gly^{B8} is indicated by red; black circles indicate Gly^{B20} and Gly^{B23}. Residues B2–B7 are shown in green (in β-region in T-state) or blue (within α-helical island in R-state). (See Color Insert.)

conformation in powder blue. Interconversion among the three families of hexamers (T_6, $T_3R_3^f$, and R_6) is regulated by ionic strength (Bentley *et al.*, 1976; Blundell *et al.*, 1971) and the binding of small cyclic alcohols (Derewenda *et al.*, 1989; Krebs *et al.*, 2005) (arrows in Fig. 2.2). Crystal structures of zinc-free insulin dimers and monomeric fragments exhibit T-family features (Bi *et al.*, 1983, 1984; Whittingham *et al.*, 2006). Together, this set of structures has established a pathway of insulin assembly (central panel of Fig. 2.2).

Structural variation among crystal structures of insulin has provided an important model of conformational change and protein flexibility (Fig. 2.3)

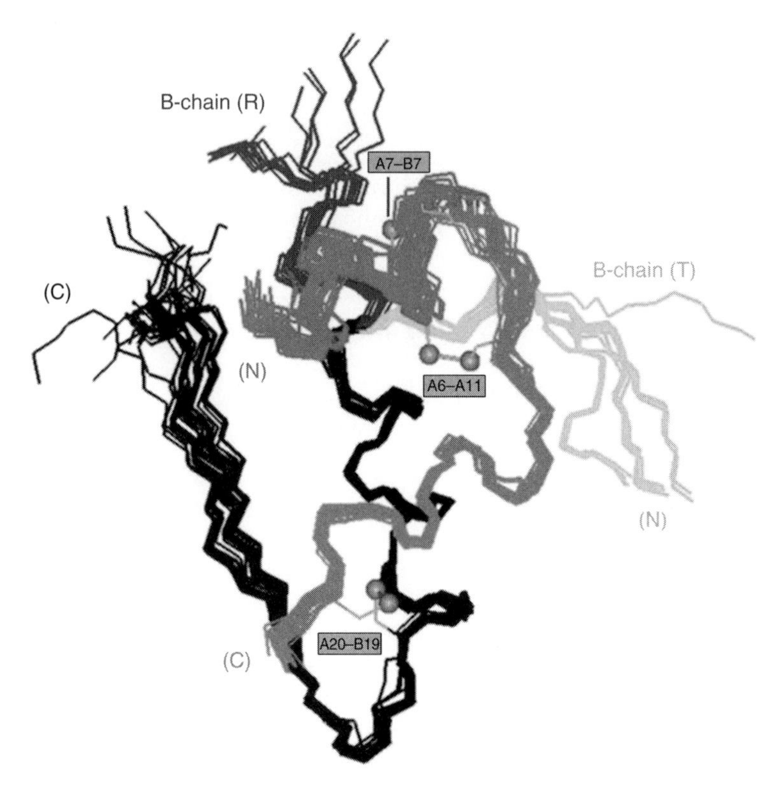

Figure 2.3 Structural variation among crystallographic protomers. Superposition of crystallographic protomers (15 T states and 15 R states). The structures were aligned according the main-chain atoms of residues B9–B24 and A12–A21. The A1–A8 α-helix in each protomer is shown in red; the variable secondary structure of the N-terminal segment of the B-chain is shown in green (extended in T state; right) or blue (extended α-helix in R state; left). Structures were obtained from the following entries in the Protein Data Bank: (T states), 4INS, 1APH, 1BPH, 1CPH, 1DPH, 1TRZ, 1TYL, 1TYM, 2INS, 1ZNI, 1LPH, 1G7A, 1MSO; (R states), 1EV6, 1ZNJ, 1TRZ, 1ZNI, 1LPH. (See Color Insert.)

(Chothia *et al.*, 1983). Yet the specific biological relevance of the TR transition, if any, has remained a matter of speculation. On the one hand, NMR studies have established that an insulin monomer in solution resembles the T-state (Hua *et al.*, 1991, 1996a; Olsen *et al.*, 1996). On the other hand, a variety of evidence suggests that the hormone undergoes a conformational change on receptor binding (Derewenda *et al.*, 1991; Hua *et al.*, 1991). In this chapter, we explore the possible biological relevance of the TR transition. Evidence is reviewed that the choreography of conformational changes in the crystalline state, in fact, identifies a functional switch (Hua *et al.*, 2006b; Nakagawa *et al.*, 2005; Wan *et al.*, 2008). We envisage that this switch operates to enable both the productive folding of proinsulin in the endoplasmic reticulum (ER) of pancreatic β-cells and binding of the mature hormone to the IR. Remarkably, naturally occurring mutations in the human insulin gene have recently been identified that lead to impairment of this switch and in turn to permanent neonatal-onset diabetes mellitus (Stoy *et al.*, 2007). These findings suggest that the lessons of protein structures can extend far beyond the immediate conditions of crystallization to inform a broad range of biological phenomena.

II. Structure–Activity Relationships

A. Chiral mutagenesis of insulin

The crux of the TR transition occurs at the junction between the N-terminal segment of the B-chain and its central α-helix. In the T-state, this junction forms a type II′ β-turn comprising residues B7–B10. The local structure of this turn requires a glycine at position B8 (arrow in Fig. 2.4A) whose main-chain (ϕ, ψ) dihedral angles lie on the right-hand side of the Ramachandran plane ($\phi > 0$; red circle in Fig. 2.2B). This conformation is ordinarily "forbidden" to L-amino acids. By contrast, in the R-state Gly^{B8} adopts a negative ϕ angle as expected within the interior of an α-helix (Fig. 2.2C). In either protomer, Gly^{B8} lies on the protein surface adjacent to Cys^{B7} (part of the canonical cystine A7–B7; this disulfide bridge provides a link between the junctional B-chain element and the A-chain (Fig. 2.4C)). The contrasting structural environments of Gly^{B8} in the T- and R-states are shown in Fig. 2.4B (red circles). Glycine is invariant at this site among mammalian insulins, insulin-like growth factors, and relaxin-like polypeptides (Blundell and Humbel, 1980).

In a recent series of studies "chiral mutagenesis"—comparison of corresponding D- and L amino-acid substitutions—has been employed to probe the functional importance of Gly^{B8} (Nakagawa *et al.*, 2005). The essential idea is that D or L substitutions impose a reciprocal bias on the setting of the B8 ϕ angle and may hence modulate the conformational equilibrium between T- and R states. Respective sites of L and D substituents are indicated in blue (the pro-L H_α of Gly^{B8}) or magenta (pro-D H_α) (Fig. 2.4D). Substitution of Gly^{B8} by L-amino acids was observed to impair the folding of a single-chain insulin precursor (Guo *et al.*, 2005) and to impede disulfide pairing in insulin chain combination (Nakagawa *et al.*, 2005). Substitution of this turn-specific "D-glycine" by a D-amino acid by contrast was observed to augment the stability of the native T-like fold. Remarkably, whereas unstable L-analogs can be highly active, D-analogs exhibit reduced binding to the IR. The extent of impairment caused by D-amino-acid substitutions at B8 (between 10^2- and 10^3-fold) is more marked than is usually observed on single amino-acid substitutions at other sites.

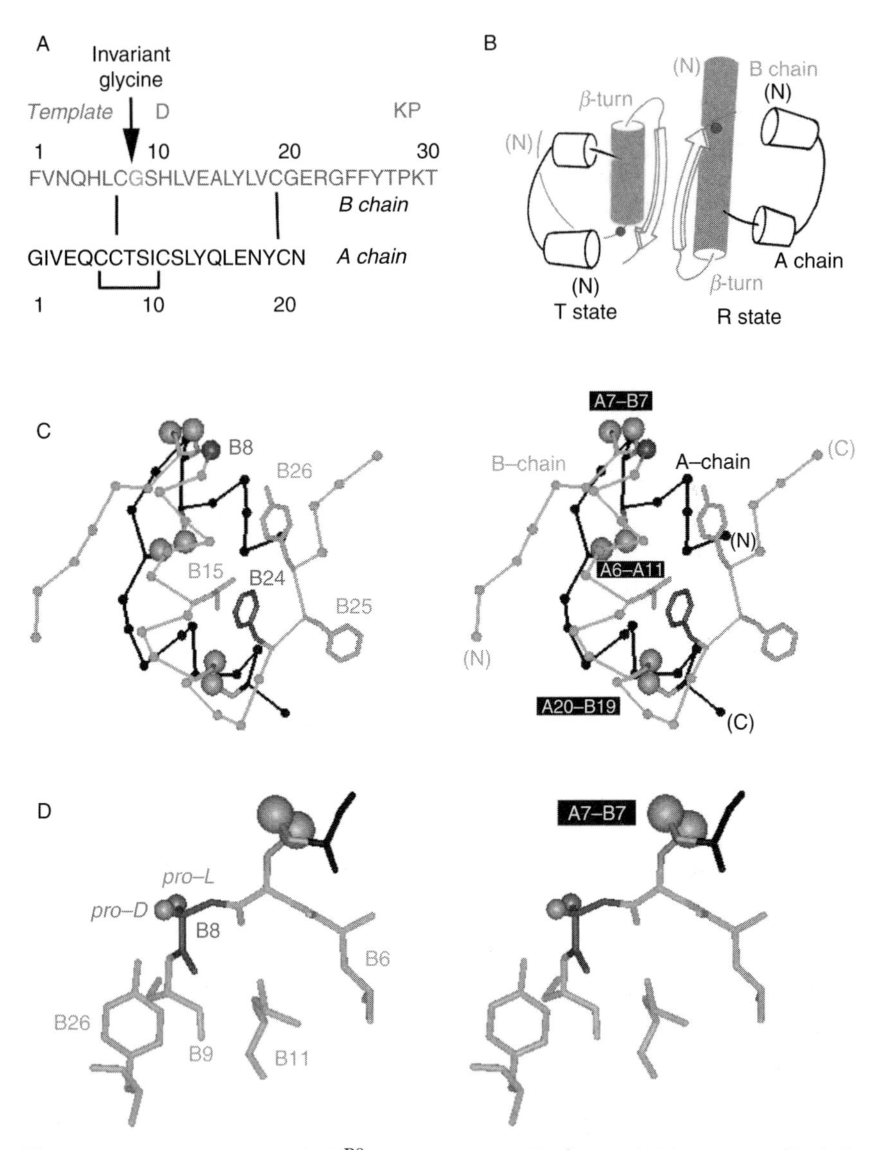

Figure 2.4 Conformation of Gly^{B8} in a T-state-specific β-turn. (A) Sequence of B chain (top) and A chain (bottom); arrow indicates invariant Gly^{B8} (red). Shown above B chain in magenta are the three substitutions in the monomeric DKP template. (B) Cylinder models of TR dimer based on crystal structure of zinc insulin hexamers (PDB ID: 1TRZ). The T state is at left and R state at right. B-chain α-helices are shown in green; the α-carbons of Gly^{B8} are shown as red circles. Three families of hexamers have been characterized, designated T_6, $T_3R_3^f$, and R_6. The R-state conformation has only been observed within hexamers. (C) Structure of insulin T state (stereo pair) showing positions of selected side chains (labeled at left) relative to Gly^{B8} C_α (red) and disulfide bridges (gold; labeled at right). The B chain is shown in green, and A chain in black. (D) Structure of T-state-specific B7–B10 β-turn (stereo pair). Main chain of Gly^{B8} is shown in red; its pro-L and pro-D H_α atoms are highlighted in blue and magenta, respectively. This figure is reprinted from Hua *et al.* (2006b). (See Color Insert.)

Such loss of activity is associated with impairment of the TR transition in crystallographic and NMR studies of hexameric assemblies of D-Ala^{B8}-insulin (Nakagawa *et al.*, 2005). Crystal structures of zinc insulin hexamers containing representative D- or L amino-acid substitutions at B8 nonetheless demonstrate that either chirality can be accommodated within such assemblies (Weiss and Dodson, unpublished results).

The structural basis of reciprocal stereospecific effects of B8 substitutions was investigated through comparative studies of diastereomeric analogs containing either D- or L-Ser^{B8} within an engineered T-like monomer (Hua *et al.*, 2006b). (Such engineering is generally required to avoid dimerization and higher-order protein assembly (Brange *et al.*, 1988), which would otherwise limit the feasibility of NMR analysis (Weiss *et al.*, 1991)). Although the NMR-derived structures of the analogs each resemble wild-type insulin, the unstable but active L-Ser^{B8} variant exhibits greater dynamic flexibility (Fig. 2.5). Enhanced flexibility or imprecision in the ensemble of L-specific structures is associated with an attenuated far-ultraviolet circular dichroism (CD) spectrum (Hua *et al.*, 2006b). Since the CD spectrum of insulin primarily reflects its α-helical subdomain and since the B8 site of substitution in a T-like monomer is peripheral to this subdomain, the attenuated CD spectrum of the L-Ser^{B8} analog indicates a global change in protein dynamics. The CD spectrum of the D-Ser^{B8} analog is by contrast essentially identical to that of the parent T-like monomer (Hua *et al.*, 2006b). Studies of amide proton exchange in D_2O suggest that D substitutions damp

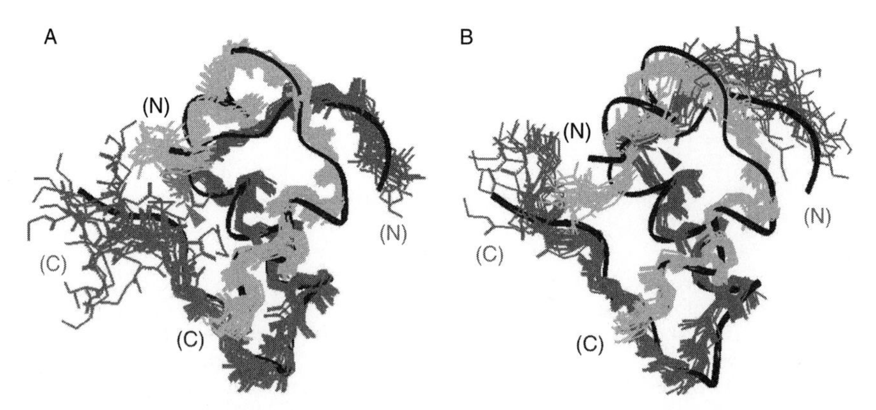

Figure 2.5 NMR-derived solution structures of D-Ser^{B8} and L-Ser^{B8} analogs of an engineered insulin monomer. Front and back views of (A) D-Ser^{B8}-DKP-insulin and (B) L-Ser^{B8}-DKP-insulin. In each case the A chain is shown in gray, and B chain in blue. The D- and L-^{SerB8} side chains are shown in green and purple, respectively (arrowheads). Ribbons indicate mean structure of DKP-insulin. This figure is reprinted from Hua *et al.* (2006b). (See Color Insert.)

local conformational fluctuations near the site of substitution (Hua and Weiss, unpublished results).

Together, the above studies highlighted the dual importance of Gly^{B8}. On the one hand, a T-like conformation (with positive ϕ angle) is required for initial disulfide pairing and for thermodynamic and dynamic stability of the T-like monomer once folding is achieved. On the other hand, stereospecific enhancement of such stability by D-amino-acid substitutions impedes receptor binding. Further, whereas L-amino-acid substitutions at B8 impede disulfide pairing and impair the stability of the T-like monomer, the high activity of such unstable analogs suggests that a local R-like conformation (with negative ϕ angle) is required for biological activity. We, therefore, suggest that Gly^{B8} recapitulates local aspects of the TR transition to function a switch between folding-competent and active conformations.

B. Uncoupling activity from allostery

Interpretation of crystallographic studies of insulin is limited by its state of assembly. Whereas crystal structures feature zinc-free dimers or zinc-stabilized hexamers, the hormone circulates in the bloodstream and binds to target cells as a monomer (Fig. 2.1B). Local modulation of the B8 dihedral angle by stereospecific substitutions (discussed earlier) does not address whether *global* aspects of the TR transition may pertain to the insulin monomer. As a further test of the relationship between the TR transition and biological activity, a recent study has exploited a species variant at position B5 (Wan *et al.*, 2008). In human insulin as in other eutherian mammals residue B5 is conserved as histidine (Fig. 2.6A). Attention was focused on its substitution by Arg (as observed in some hystricomorph mammals, birds, and reptiles) because of the distinctive structural environments of His^{B5} in wild-type insulin hexamers. In the T-state the imidazole ring packs within a solvated crevice at the edge of the protein surface (Fig. 2.6C; Baker *et al.*, 1988) whereas in the R-state the side chain packs at an internal interface between dimers. Although this interface is in part solvated, NMR studies of the pK_a of His^{B5} suggested that an R-specific interface would be destabilized by the uncompensated positive charge of an Arg^{B5} substituent. In accord with this prediction, spectroscopic results demonstrated that substitution of His^{B5} by Arg impedes the TR transition in solution. Further, crystals of Arg^{B5}-insulin grown under conditions leading to stabilization of wild-type R_6 hexamers (i.e., in the presence of phenolic ligands) were observed to contain only T_6 hexamers. The crystal structure of Arg^{B5}-insulin as a variant T_6 hexamer, determined at a resolution of 1.3 Å, was found to be essentially identical to that of the wild-type T_6 hexamer (Fig. 2.7A and B), including analogous interactions by the wild-type variant B5 side chain (His^{B5} and Arg^{B5}) (Wan *et al.*, 2008).

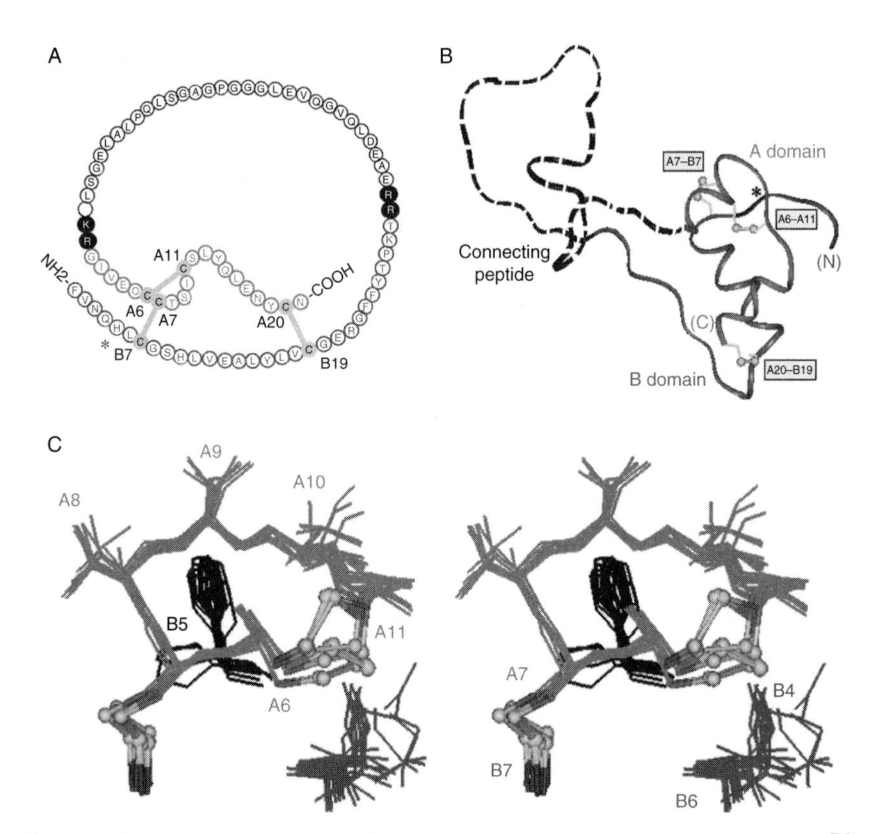

Figure 2.6 Structure of proinsulin and T-state-specific environment of His^{B5}. (A) Sequence of human proinsulin: insulin moiety is shown in red (A chain) and blue (B chain). The connecting region is shown in black: flanking dibasic cleavage sites (filled circles) and C-peptide (open circles). (B) Structural model of insulin-like moiety and disordered connecting peptide (dashed line). Cystines are labeled in yellow boxes. (C) Structural environment of His^{B5} within A-chain-related crevice. Structures are drawn from T-state coordinates given by PDB code in legend to Figure 2.3. This figure is reprinted from Hua *et al.* (2006a). (See Color Insert.)

The corresponding packing of His^{B5} and Arg^{B5} in similar inter-chain pockets is illustrated in Fig. 2.7C and D, respectively. Because Arg^{B5}-human insulin binds well to the IR (with affinity ca. 40% relative to wild-type human insulin), these results collectively demonstrate that competence to undergo the TR transition in a *hexamer* is not required for the biological activity of an insulin monomer. These findings are restricted to zinc insulin hexamers and so do not address whether competence of an insulin *monomer* to adopt an R-like conformation is required for high-affinity receptor binding. Whereas interfacial substitutions such as Arg^{B5} may be regarded as extrinsic probes of protein allostery, D and L substitutions at B8 modulate the intrinsic conformational repertoire of the monomer in relation to both the TR transition and receptor binding.

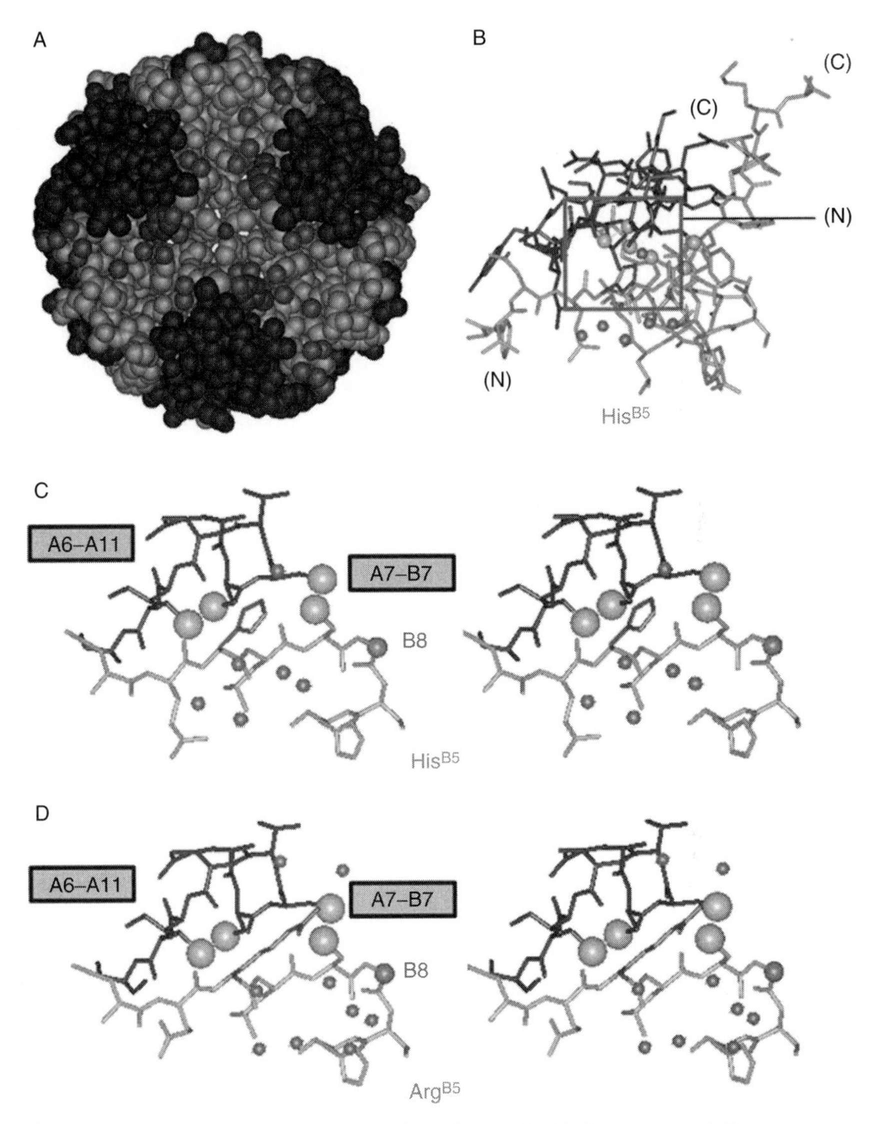

Figure 2.7 Structural environment of residue B5 in T_6 hexamer and T-state protomers. (A) Spacing filling model of T_6 hexamer showing side chain of His^{B5} (red) lying along protein surface near bound water molecules (blue). (B) Stick model of single T-state protomer with B-chain in black and A-chain in gray. Red box encloses environment of His^{B5} (red) in inter-chain crevice near A7–B7 and A6–A11 disulfide bridges (sulfur atoms shown in gold). (C) Expansion of boxed region in panel B providing stereo view of packing of His^{B5}. (D) Corresponding stereo view of Arg^{B5} (red) in crystallographic T-state protomer. Analogous side-chain NH functions of His^{B5} and Arg^{B5} are near the main-chain of the A-chain. Cystine A7–B7 in each case lies on the protein surface whereas cystine A6–A11 packs within the core of the protomer. Bound water molecules near respective B5-related crevices are shown as blue spheres. This figure is reprinted from Wan *et al.* (2008). (See Color Insert.)

III. Implications for the Genetics of Diabetes Mellitus

Insulin chain combination, developed more than four decades ago by Katsoyannis and colleagues (with independent and important contributions from the laboratories of Brandenburg and Zahn in Aachen, Germany, and from the laboratories of Cao and Gong in Shanghai, China) has provided a robust synthetic route to the preparation of insulin analogs (for review, see Katsoyannis, 1966). Despite its general success, chain combination fails in the case of selected substitutions in the A- or B-chains (Hu *et al.*, 1993; Hua *et al.*, 2002). Examples of a block to disulfide pairing are provided by substitutions at positions B5 and B8. Very low yields were encountered on substitution of His^{B5} by Ala or Met (Hua *et al.*, 2006a; Wan *et al.*, 2008) and on substitution of Gly^{B8} by any L-amino acid (Hua *et al.*, 2006b; Nakagawa *et al.*, 2005). The origin of these synthetic blocks is not well understood. Because efficient syntheses of unstable insulin analogs have previously been reported (Hua *et al.*, 2002; Weiss *et al.*, 2000; Xu *et al.*, 2002a), we have hypothesized that unfavorable substitutions at B5 or B8 imposes a kinetic barrier to disulfide pairing. It is possible that in a reaction intermediate a productive orientation of Cys^{B7} (and in turn its alignment with Cys^{A7}) requires (i) nascent interactions between the B5 side chain and the A-chain (resembling those observed in the native T-state; Figs. 2.6C and 2.7C) and (ii) a positive dihedral angle at B8 (as in the native T-state or as stabilized by D-amino-acid substitutions; Fig. 2.1B). The crystal structure of Arg^{B5}-insulin (discussed earlier) may thus have implications beyond the TR transition: His^{B5} and Arg^{B5} both contain nitrogenous side chains able to form specific hydrogen bonds to the main-chain of the A-chain (Fig. 2.7C and D). The biological relevance of these chemical findings is supported by their extension to corresponding studies of the biosynthetic expression of proinsulin variants and single-chain insulin analogs in eukaryotic cells (Guo *et al.*, 2005; Hua *et al.*, 2006a).

Recent advances in human genetics have identified a large collection of clinical mutations in the insulin gene causing permanent neonatal-onset diabetes mellitus (Colombo *et al.*, 2008; Edghill *et al.*, 2008; Molven *et al.*, 2008; Polak *et al.*, 2008; Stoy *et al.*, 2007). The majority of such mutations add or remove a cysteine from proinsulin, leading to an odd number of thiol groups and hence protein misfolding in the ER. Although the crystal structure of proinsulin has not been determined, a variety of spectroscopic evidence indicates that it consists of a folded insulin-like domain and unfolded connecting peptide (Fig. 2.6B). Misfolding of proinsulin in the ER presumably activates the unfolded protein response (UPR), leading to chronic ER stress and apoptosis of β-cells. Such a mechanism would rationalize the onset of apparent type 1 diabetes mellitus prior to maturation of an immune system capable of mounting an autoimmune attack (Stoy *et al.*, 2007). This hypothesis is supported by the observation of a human mutation associated with neonatal diabetes mellitus that recapitulates a classical diabetes-associated mutation in the mouse (the Akita mouse; Araki *et al.*, 2003; Oyadomari *et al.*, 2002; Ron, 2002) in which Cys^{A7} is substituted by Tyr. Affected patients are heterozygous, implying that misfolding of the variant proinsulin perturbs the folding and trafficking of the wild-type polypeptide. A subset of mutations may lead to less severe folding defects, leading to the onset of diabetes mellitus later in life (Edghill *et al.*, 2008; Molven *et al.*, 2008).

Remarkably, among patients with permanent neonatal-onset diabetes mellitus, residues B5 and B8 provide "hot spots" for mutations not involving cysteine directly. The clinical database includes (L) Ser^{B5}, providing an unusual example in which physical studies of a mutant protein preceded and anticipated a clinical phenotype (Guo *et al.*, 2005; Hua *et al.*, 2006b; Nakagawa *et al.*, 2005). This clinical correlation strongly suggests that insulin chain combination provides an informative peptide model for the physiological folding of nascent proinsulin in the ER of human β-cells as originally envisaged by Katsoyannis (1966). We propose that the conservation of His^{B5} and the invariance of Gly^{B8} are enjoined by the requirement to ensure the foldability of proinsulin—and conversely to avoid the severe pathological consequences of proteotoxity following its misfolding (Araki *et al.*, 2003; Liu *et al.*, 2007). This perspective thus highlights the achirality of glycine as a uniquely "ambidextrous" residue strategically placed at a site of conformational change.

IV. Concluding Remarks

Structural studies of insulin analogs strongly suggest that the classical insulin T-state represents an inactive conformation of the hormone. By identifying a critical hinge point at Gly^{B8}, studies of mirror-image D- and L substitutions in insulin have provided evidence for induced fit on receptor binding. The receptor-bound conformation of insulin may in part resemble the crystallographic R-state, but these processes may readily be uncoupled. We suggest that the classical allosteric reorganization of zinc insulin hexamers, so elegantly characterized by the late D. C. Hodgkin and colleagues, identifies sites of flexibility utilized by the insulin monomer on binding to its receptor. Remarkably, structural relationships observed in the inactive T-state are nonetheless of key biological importance: analogous interactions are required—presumably within the nascent structure of oxidative folding intermediates (Hua *et al.*, 2001, 2002; Weiss *et al.*, 2000)—for native disulfide pairing. The evolution of insulin sequences has thus been constrained by the dual and interlocking requirements of function and foldability. An understanding of structure–function relationships in insulin is of both basic and translational interest. Whereas analysis of ER-directed disulfide pairing is central to the emerging genetics of diabetes mellitus (Stoy *et al.*, 2007), characterization of the receptor-bound structure of insulin promises to enable design of novel agonists for the treatment of diabetes mellitus.

ACKNOWLEDGMENTS

The author thanks Profs. P. Arvan, G. G. Dodson, Q. Hua, P. Katsoyannis, D. F. Steiner, and J. Whittaker for helpful discussion and is grateful to members of his laboratory for all aspects of the work cited. Figure 2.2 was designed with the assistance of G. G. Dodson; the present studies have been greatly aided by his collaboration and deep historical understanding of insulin crystallography. Studies of insulin analogs at CWRU were supported in part by grants from the National Institutes of Health (DK40949 and DK069764) and American Diabetes Association. This chapter represents a contribution from the Cleveland Center for Membrane and Structural Biology.

REFERENCES

Adams, M. J., Blundell, T. L., Dodson, E. J., Dodson, G. G., Vijayan, M., Baker, E. N., Hardine, M. M., Hodgkin, D. C., Rimer, B., and Sheet, S. (1969). Structure of rhombohedral 2 zinc insulin crystals. *Nature* **224,** 491–495.

Araki, E., Oyadomari, S., and Mori, M. (2003). Endoplasmic reticulum stress and diabetes mellitus. *Intern. Med.* **42,** 7–14.

Baker, E. N., Blundell, T. L., Cutfield, J. F., Cutfield, S. M., Dodson, E. J., Dodson, G. G., Hodgkin, D. M., Hubbard, R. E., Isaacs, N. W., and Reynolds, C. D. (1988). The structure of 2Zn pig insulin crystals at 1.5 Å resolution. *Philos. Trans. R. Soc. Lond. B Biol. Sci.* **319,** 369–456.

Bentley, G., Dodson, E., Dodson, G., Hodgkin, D., and Mercola, D. (1976). Structure of insulin in 4-zinc insulin. *Nature* **261,** 166–168.

Bentley, G., Dodson, G., and Lewitova, A. (1978). Rhombohedral insulin crystal transformation. *J. Mol. Biol.* **126,** 871–875.

Bi, R. C., Cutfield, S. M., Dodson, E. J., Dodson, G. G., Giorgino, F., Reynolds, C. D., and Tolley, S. P. (1983). Molecular-replacement studies on crystal forms of despentapeptide insulin. *Acta Crystallogr. B* **39,** 90–98.

Bi, R. C., Dauter, Z., Dodson, E., Dodson, G., Giordano, F., and Reynolds, C. (1984). Insulin structure as a modified and monomeric molecule. *Biopolymers* **23,** 391–395.

Blundell, T. L., Cutfield, J. F., Cutfield, S. M., Dodson, E. J., Dodson, G. G., Hodgkin, D. C., Mercola, D. A., and Vijayan, M. (1971). Atomic positions in rhombohedral 2-zinc insulin crystals. *Nature* **231,** 506–511.

Blundell, T. L., Dodson, G. G., Hodgkin, D. C., and Mercola, D. A. (1972). Insulin: The structure in the crystal and its reflection in chemistry and biology. *Adv. Protein Chem.* **26,** 279–402.

Blundell, T. L., and Humbel, R. E. (1980). Hormone families: Pancreatic hormones and homologous growth factors. *Nature* **287,** 781–787.

Brader, M. L., and Dunn, M. F. (1991). Insulin hexamers: New conformations and applications. *Trends Biochem. Sci.* **16,** 341–345.

Brange, J., Ribel, U., Hansen, J. F., Dodson, G., Hansen, M. T., Havelund, S., Melberg, S. G., Norris, F., Norris, K., and Snel, L. (1988). Monomeric insulins obtained by protein engineering and their medical implications. *Nature* **333,** 679–682.

Chothia, C., Lesk, A. M., Dodson, G. G., and Hodgkin, D. C. (1983). Transmission of conformational change in insulin. *Nature* **302,** 500–505.

Ciszak, E., Beals, J. M., Frank, B. H., Baker, J. C., Carter, N. D., and Smith, G. D. (1995). Role of C-terminal B-chain residues in insulin assembly: The structure of hexameric LysB28ProB29-human insulin. *Structure (Lond.)* **3,** 615–622.

Colombo, C., Porzio, O., Liu, M., Massa, O., Vasta, M., Salardi, S., Beccaria, L., Monciotti, C., Toni, S., Pedersen, O., Hansen, T., and Federici, L., and Early Onset Diabetes Study Group of the Italian Society of Pediatric Endocrinology and Diabetes (2008). Seven mutations in the human insulin gene linked to permanent neonatal/infancy-onset diabetes mellitus. *J. Clin. Invest.* **118,** 2148–2156.

Derewenda, U., Derewenda, Z., Dodson, E. J., Dodson, G. G., Bing, X., and Markussen, J. (1991). X-ray analysis of the single chain B29-A1 peptide-linked insulin molecule. A completely inactive analogue. *J. Mol. Biol.* **220,** 425–433.

Derewenda, U., Derewenda, Z., Dodson, E. J., Dodson, G. G., Reynolds, C. D., Smith, G. D., Sparks, C., and Swenson, D. (1989). Phenol stabilizes more helix in a new symmetrical zinc insulin hexamer. *Nature* **338,** 594–596.

Edghill, E. L., Flanagan, S. E., Patch, A. M., Boustred, C., Parrish, A., Shields, B., Shepherd, M. H., Hussain, K., Kapoor, R. R., Malecki, M., MacDonald, M. J., and Stoy, J., and Neonatal Diabetes International Collaborative Group (2008). Insulin mutation screening in 1044 patients with diabetes: Mutations in the INS gene are a common cause of neonatal diabetes but a rare cause of diabetes diagnosed in childhood or adulthood. *Diabetes* **57,** 1034–1042.

Guo, Z. Y., Zhang, Z., Jia, X. Y., Tang, Y. H., and Feng, Y. M. (2005). Mutational analysis of the absolutely conserved B8Gly: Consequence on foldability and activity of insulin. *Acta Biochim. Biophys. Sin. (Shanghai)* **10,** 673–679.

Hu, S. Q., Burke, G. T., Schwartz, G. P., Ferderigos, N., Ross, J. B., and Katsoyannis, P. G. (1993). Steric requirements at position B12 for high biological activity in insulin. *Biochemistry* **32,** 2631–2635.

Hua, Q. X., Chu, Y. C., Jia, W., Phillips, N. B., Wang, R. Y., Katsoyannis, P. G., and Weiss, M. A. (2002). Mechanism of insulin chain combination. Asymmetric roles of A-chain α-helices in disulfide pairing. *J. Biol. Chem.* **277,** 43443–43453.

Hua, Q. X., Hu, S. Q., Frank, B. H., Jia, W., Chu, Y. C., Wang, S. H., Burke, G. T., Katsoyannis, P. G., and Weiss, M. A. (1996a). Mapping the functional surface of insulin by design: Structure and function of a novel A-chain analogue. *J. Mol. Biol.* **264,** 390–403.

Hua, Q. X., Liu, M., Hu, S. Q., Jia, W., Arvan, P., and Weiss, M. A. (2006a). A conserved histidine in insulin is required for the foldability of human proinsulin. Structure and function of an Ala^{B5} analog. *J. Biol. Chem.* **281,** 24889–24899.

Hua, Q. X., Nakagawa, S. H., Hu, S. Q., Jia, W., Wang, S., and Weiss, M. A. (2006b). Toward the active conformation of insulin. Stereospecific modulation of a structural switch in the B chain. *J. Biol. Chem.* **281,** 24900–24909.

Hua, Q. X., Nakagawa, S. H., Jia, W., Hu, S. Q., Chu, Y. C., Katsoyannis, P. G., and Weiss, M. A. (2001). Hierarchical protein folding: Asymmetric unfolding of an insulin analogue lacking the A7–B7 interchain disulfide bridge. *Biochemistry* **40,** 12299–12311.

Hua, Q. X., Narhi, L., Jia, W., Arakawa, T., Rosenfeld, R., Hawkins, N., Miller, J. A., and Weiss, M. A. (1996b). Native and non-native structure in a protein-folding intermediate:

Spectroscopic studies of partially reduced IGF-I and an engineered alanine model. *J. Mol. Biol.* **259,** 297–313.

Hua, Q. X., Shoelson, S. E., Kochoyan, M., and Weiss, M. A. (1991). Receptor binding redefined by a structural switch in a mutant human insulin. *Nature* **354,** 238–241.

Huang, K., Xu, B., Hu, S. Q., Chu, Y. C., Hua, Q. X., Qu, Y., Li, B., Wang, S., Wang, R. Y., Nakagawa, S. H., Theede, A. M., Whittaker, J., *et al.* (2004). How insulin binds: The B-chain α-helix contacts the L1 β-helix of the insulin receptor. *J. Mol. Biol.* **341,** 529–550.

Katsoyannis, P. G. (1966). Synthesis of insulin. *Science* **154,** 1509–1514.

Kitagawa, K., Ogawa, H., Burke, G. T., Chanley, J. D., and Katsoyannis, P. G. (1984). Critical role of the A2 amino acid residue in the biological activity of insulin: [2-glycine-A]- and [2-alanine-A]insulins. *Biochemistry* **23,** 1405–1413.

Krebs, M. R., Bromley, E. H., and Donald, A. M. (2005). The binding of thioflavin-T to amyloid fibrils: Localisation and implications. *J. Struct. Biol.* **149,** 30–37.

Liang, D. C., Chang, W. R., and Wan, Z. L. (1994). A proposed interaction model of the insulin molecule with its receptor. *Biophys. Chem.* **50,** 63–71.

Liu, M., Hodish, I., Rhodes, C. J., and Arvan, P. (2007). Proinsulin maturation, misfolding, and proteotoxicity. *Proc. Natl. Acad. Sci. USA* **104,** 15841–15846.

Lou, M., Garrett, T. P., McKern, N. M., Hoyne, P. A., Epa, V. C., Bentley, J. D., Lovrecz, G. O., Cosgrove, L. J., Frenkel, M. J., and Ward, C. W. (2006). The first three domains of the insulin receptor differ structurally from the insulin-like growth factor 1 receptor in the regions governing ligand specificity. *Proc. Natl. Acad. Sci. USA* **103,** 12429–12434.

McKern, N. M., Lawrence, M. C., Streltsov, V. A., Lou, M. Z., Adams, T. E., Lovrecz, G. O., Elleman, T. C., Richards, K. M., Bentley, J. D., Pilling, P. A., Hoyne, P. A., Cartledge, K. A., *et al.* (2006). Structure of the insulin receptor ectodomain reveals a folded-over conformation. *Nature* **443,** 218–221.

Mirmira, R. G., Nakagawa, S. H., and Tager, H. S. (1991). Importance of the character and configuration of residues B24, B25, and B26 in insulin-receptor interactions. *J. Biol. Chem.* **266,** 1428–1436.

Mirmira, R. G., and Tager, H. S. (1989). Role of the phenylalanine B24 side chain in directing insulin interaction with its receptor: Importance of main chain conformation. *J. Biol. Chem.* **264,** 6349–6354.

Molven, A., Ringdal, M., Nordbo, A. M., Raeder, H., Stoy, J., Lipkind, G. M., Steiner, D. F., Philipson, L. H., Bergmann, I., Aarskog, D., Undlien, D. E., Joner, G., *et al.* (2008). Mutations in the insulin gene can cause MODY and autoantibody-negative type 1 diabetes. *Diabetes* Epub ahead of print.

Nakagawa, S. H., and Tager, H. S. (1986). Role of the phenylalanine B25 side chain in directing insulin interaction with its receptor. Steric and conformational effects. *J. Biol. Chem.* **261,** 7332–7341.

Nakagawa, S. H., and Tager, H. S. (1992). Importance of aliphatic side-chain structure at positions 2 and 3 of the insulin A chain in insulin-receptor interactions. *Biochemistry* **31,** 3204–3214.

Nakagawa, S. H., Tager, H. S., and Steiner, D. F. (2000). Mutational analysis of invariant valine B12 in insulin: Implications for receptor binding. *Biochemistry* **39,** 15826–15835.

Nakagawa, S. H., Zhao, M., Hua, Q. X., Hu, S. Q., Wan, Z. L., Jia, W., and Weiss, M. A. (2005). Chiral mutagenesis of insulin. Foldability and function are inversely regulated by a stereospecific switch in the B chain. *Biochemistry* **44,** 4984–4999.

Olsen, H. B., Ludvigsen, S., and Kaarsholm, N. C. (1996). Solution structure of an engineered insulin monomer at neutral pH. *Biochemistry* **35,** 8836–8845.

Oyadomari, S., Koizumi, A., Takeda, K., Gotoh, T., Akira, S., Araki, E., and Mori, M. (2002). Targeted disruption of the Chop gene delays endoplasmic reticulum stress-mediated diabetes. *J. Clin. Invest.* **109,** 525–532.

Polak, M., Dechaume, A., Cave, H., Nimri, R., Crosnier, H., Sulmont, V., de Kerdanet, M., Scharfmann, R., Lebenthal, Y., Froguel, P., and Vaxillaire, M. (2008). Heterozygous missense mutations in the insulin gene are linked to permanent diabetes appearing in the neonatal period or in early-infancy: A report from the French ND Study Group. *Diabetes* **57,** 1115–1119.

Pullen, R. A., Lindsay, D. G., Wood, S. P., Tickle, I. J., Blundell, T. L., Wollmer, A., Krail, G., Brandenburg, D., Zahn, H., Gliemann, J., and Gammeltoft, S. (1976). Receptor-binding region of insulin. *Nature* **259,** 369–373.

Ron, D. (2002). Proteotoxicity in the endoplasmic reticulum: Lessons from the Akita diabetic mouse. *J. Clin. Invest.* **109,** 443–445.

Roy, M., Brader, M. L., Lee, R. W., Kaarsholm, N. C., Hansen, J. F., and Dunn, M. F. (1989). Spectroscopic signatures of the T to R conformational transition in the insulin hexamer. *J. Biol. Chem.* **264,** 19081–19085.

Schlichtkrull, J. (1958). Insulin Crystals: Chemical and Biological Studies on Insulin Crystals and Insulin Zinc Suspensions, p. 139. Kobenhavns universitet, Copenhagen, Munksgaard.

Shoelson, S., Haneda, M., Blix, P., Nanjo, A., Sanke, T., Inouye, K., Steiner, D., Rubenstein, A., and Tager, H. (1983). Three mutant insulins in man. *Nature* **302,** 540–543.

Smith, G. D., Swenson, D. C., Dodson, E. J., Dodson, G. G., and Reynolds, C. D. (1984). Structural stability in the 4-zinc human insulin hexamer. *Proc. Natl. Acad. Sci. USA* **81,** 7093–7097.

Stoy, J., Edghill, E. L., Flanagan, S. E., Ye, H., Paz, V. P., Pluzhnikov, A., Below, J. E., Hayes, M. G., Cox, N. J., Lipkind, G. M., Lipton, R. B., Greeley, S. A., *et al.* (2007). Insulin gene mutations as a cause of permanent neonatal diabetes. *Proc. Natl. Acad. Sci. USA* **104,** 15040–15044.

Thomas, B., and Wollmer, A. (1989). Cobalt probing of structural alternatives for insulin in solution. *Biol. Chem. Hoppe Seyler* **370,** 1235–1244.

Wan, Z., Huang, K., Whittaker, J., and Weiss, M. A. (2008). The structure of a mutant insulin uncouples receptor binding from protein allostery. An electrostatic block to the TR transition. *J. Biol. Chem.* **283,** 21198–21210.

Ward, C., Lawrence, M. C., Streltsov, V., Garrett, T., McKern, N., Lou, M. Z., Lovrecz, G., and Adams, T. (2008). Structural insights into ligand-induced activation of the insulin receptor. *Acta Physiol. (Oxf.)* **192,** 3–9.

Weiss, M. A., Hua, Q. X., Jia, W., Chu, Y. C., Wang, R. Y., and Katsoyannis, P. G. (2000). Hierarchiacal protein "un-design": Insulin's intrachain disulfide bridge tethers a recognition α-helix. *Biochemistry* **39,** 15429–15440.

Weiss, M. A., Hua, Q. X., Lynch, C. S., Frank, B. H., and Shoelson, S. E. (1991). Heteronuclear 2D NMR studies of an engineered insulin monomer: Assignment and characterization of the receptor-binding surface by selective ^{2}H and ^{13}C labeling with application to protein design. *Biochemistry* **30,** 7373–7389.

Whittingham, J. L., Youshang, Z., Zakova, L., Dodson, E. J., Turkenburg, J. P., Brange, J., and Dodson, G. G. (2006). I_{222} crystal form of despentapeptide (B26-B30) insulin provides new insights into the properties of monomeric insulin. *Acta Crystallogr. D Biol. Crystallogr.* **62,** 505–511.

Xu, B., Hua, Q. X., Nakagawa, S. H., Jia, W., Chu, Y. C., Katsoyannis, P. G., and Weiss, M. A. (2002a). A cavity-forming mutation in insulin induces segmental unfolding of a surrounding α-helix. *Protein Sci.* **11,** 104–116.

Xu, B., Hua, Q. X., Nakagawa, S. H., Jia, W., Chu, Y. C., Katsoyannis, P. G., and Weiss, M. A. (2002b). Chiral mutagensis of insulin's hidden receptor-binding surface: Structure of an *allo*-Ile^{A2} analogue. *J. Mol. Biol.* **316,** 435–441.

CHAPTER THREE

Molecular Mechanisms of Differential Intracellular Signaling From the Insulin Receptor

Maja Jensen *and* Pierre De Meyts

Contents

Abstract

Binding of insulin to the insulin receptor (IR) leads to a cascade of intracellular signaling events, which regulate multiple biological processes such as glucose and lipid metabolism, gene expression, protein synthesis, and cell growth, division, and survival. However, the exact mechanism of how the insulin-IR interaction produces its own specific pattern of regulated cellular functions is not yet fully understood. Insulin analogs, anti-IR antibodies as well as synthetic insulin mimetic peptides that target the two insulin-binding regions of the IR,

Receptor Systems Biology Laboratory, Hagedorn Research Institute,
Niels Steensensvej 6, 2820 Gentofte, Denmark

Vitamins and Hormones, Volume 80
ISSN 0083-6729, DOI: 10.1016/S0083-6729(08)00603-1

have been used to study the relationship between different aspects of receptor binding and function as well as providing new insights into the structure and function of the IR. This review focuses on the current knowledge of activation of the IR and how activation of the IR by different ligands initiates different cellular responses. Investigation of differential activation of the IR may provide clues to the molecular mechanisms of how the insulin-receptor interaction controls the specificity of the downstream signaling response. Differences in the kinetics of ligand-interaction with the IR, the magnitude of the signal as well as its subcelllar location all play important roles in determining/eliciting the different biological responses. Additional studies are nevertheless required to dissect the precise molecular mechanisms leading to the differential signaling from the IR.

I. Overview

Activation of the insulin receptor (IR) affects multiple biological processes such as glucose and lipid uptake/metabolism, gene expression, protein synthesis, and cell growth/division/survival (for a recent review, see Taniguchi *et al.*, 2006). The diverse effects of IR activation are mediated through a multicomponent signaling complex that assembles upon binding of insulin. Activation of the IR leads to phosphorylation of several intracellular protein substrates. These include IRS1, IRS2, and Shc. This initiates the two major signaling cascades: the MAPK pathway that includes activation of ERK1 and ERK2 leading to gene expression, protein translation, and cell growth and the PI3K pathway that includes activation of Akt leading to GLUT4 translocation, activation of glycogen synthase, and other enzymes/proteins necessary for the acute metabolic effects of insulin. Thus, the insulin signal is mediated through a complex network of signaling processes through multiple signal transduction pathways. These signaling pathways are also activated by a variety of other hormones and growth factors and the exact mechanism of how the insulin-IR interaction produces its own specific patterns of regulated cellular functions is not yet fully understood. The phosphorylation pattern of the IR, post-translational modifications of the IR as well as the cellular location of the IR have all been shown to be factors that modulate the cellular response induced by IR activation. Insulin analogs, anti-IR antibodies as well as synthetic insulin mimetic peptides that target the two insulin-binding regions on the IR, have been shown to mediate different cellular responses through the IR. The molecular mechanisms responsible for how different ligands activating the same receptor can initiate different biological responses in the same cell are still not completely known. This review will focus on the activation of the IR and the current knowledge of how activation of the IR with different ligands results in differences in the cellular response.

II. Insulin and the IR

A. Organization of the IR

The IR is a member of the receptor tyrosine kinase superfamily. Apart from the IR itself, the IR family includes the insulin-like growth factor-1 receptor (IGF-1R) and the IR related receptor (IRR) (Ebina *et al.*, 1985; Seino *et al.*, 1989). The IR is initially synthesized as a single-chain prepro-receptor. A 30-residue signal peptide directs it to the endoplasmic reticulum (ER) where it is cotranslationally cleaved off (Ullrich *et al.*, 1985). In the ER the precursor is glycosylated, folded, and dimerized under the assistance of the chaperones calnexin and calreticulin (Bass *et al.*, 1998) before its transport to the Golgi network. Once in the Golgi, the receptor is processed to yield the mature $\alpha_2\beta_2$-receptor dimer and subsequently transported to the plasma membrane. The IR exists in two isoforms that either contain (isoform B, IR-B) or lack (isoform A, IR-A) a 12-residue sequence (717–729) due to alternative splicing of exon 11 (Seino and Bell, 1989) of the IR gene. The modular structure and overall organization of the IR is shown in Fig. 3.1. The IR and the IGF-1R have a covalent dimeric $\alpha_2\beta_2$-structure. This is in contrast to the rest of the receptor tyrosine kinase family, which are single transmembrane polypeptides whose dimerization or oligomerization is induced or stabilized upon ligand binding (Ward *et al.*, 2008). The α-subunit consists of two successive homologous globular domains, L1 and L2 which are separated by a cysteine-rich (CR) region. These are followed by three fibronectin type III domains (FnIII-1, FnIII-2, and FnIII-3). The middle FnIII domain (FnIII-2) is split by an insert domain (ID) of ~120 amino acid residues that contains the IR proreceptor cleavage site and three disulfide bridges that link the α-subunits. The β-subunit spans the plasma membrane, and is linked to the α-subunit via disulfide bridges as well as noncovalent interactions. The intracellular part of the β-subunit contains the tyrosine kinase catalytic domain flanked by two regulatory regions–a juxtamembrane region that is involved in receptor internalization and docking of the insulin-receptor substrate (IRS) 1–4 and Shc and a carboxy-terminal tail that contains two phosphotyrosine-binding sites (Youngren, 2007).

B. Insulin binding to the IR

Binding of insulin to its receptor is complex and does not simply follow the law of mass action but exhibit negative cooperativity in binding, shown by curvilinear Scatchard plots and labeled ligand dissociation acceleration by cold ligand in an infinite dilution procedure. (reviewed in De Meyts and Whittaker, 2002). De Meyts *et al.* (1994) and Schäffer (1994) independently proposed binding models for insulin, integrating the available data

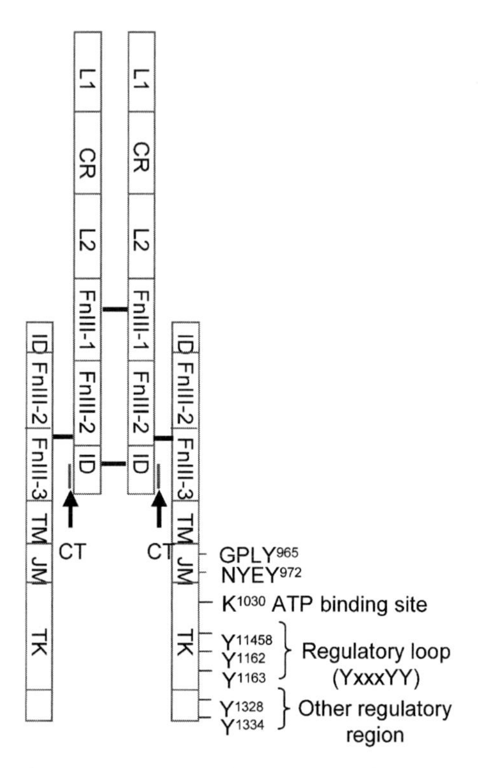

Figure 3.1 The modular organization of the IR. This figure shows a cartoon of the $\alpha_2\beta_2$ structure of the IR outlining the protein modules. L1 and L2, large domains 1 and 2 (leucine-rich repeats); CR, Cys-rich domain; FnIII-1, FnIII-2, FnIII-3, fibronectin type III domains; TM, transmembrane domain; JM, juxtamembrane domain; TK, tyrosine-kinase domain. ID, the insert domain, is the large ~120 amino acid insert in the CC′loop of the second FnIII domain. It contains the protease cleavage site that enables the IR prereceptor monomer to be cleaved into two chains, the N-terminal α chain and the C-terminal β chain. CT is the C terminal 16 residues of the IR α-chain involved in binding insulin. The corresponding region of the IGF-1R is essential for IGF-1 and IGF-2 binding. The black lines represent the disulfide bonds linking the different subunits. The lysine residue that binds ATP, the three phosphorylation sites in the regulatory loop and the four positively charged residues that mediate phosphorylation site selection are indicated (the amino acids are numbered based on the IR isoform B).

from structural and kinetic evidence. These models suggested that insulin acts as a bivalent ligand featuring two binding sites on the opposite faces of the molecule that contact two different binding domains located on each of the receptor α-subunits. De Meyts also proposed that the two α-subunits in each receptor are arranged in an antiparallel fashion, allowing alternative crosslinking of both pairs of binding sites (Fig. 3.2). This provides a plausible explanation for the kinetic behavior of insulin's binding to the receptor. The binding model has so far not been invalidated by any experimental

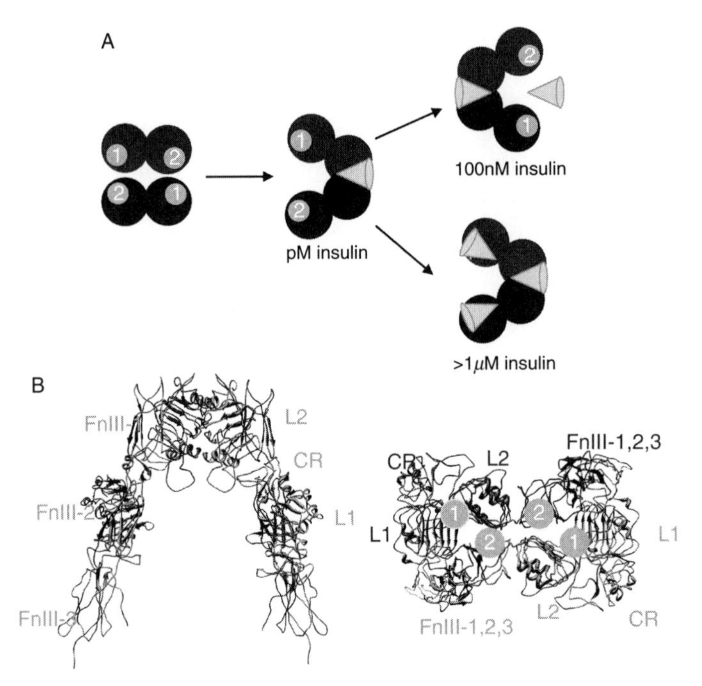

Figure 3.2 Insulin binding to the two binding sites on the IR. Panel A shows the proposed symmetrical model of bivalent-crosslinking insulin-binding mechanism (De Meyts *et al.*, 1994). This model suggests that insulin acts as a bivalent ligand featuring two binding sites on the opposite faces of the molecule that contact two different binding epitopes located on each of the receptor α-subunits. The two α-subunits in each receptor are arranged in an antiparallel fashion, allowing alternative crosslinking of both pairs of binding sites. At low concentrations one insulin molecule binds to a binding site pair (one site on each receptor half) thereby cross-linking the half-receptors (creating a high affinity site). This leaves the second pair of binding sites in a low affinity state. At higher concentrations of insulin (in the range 1–100 nM), insulin binds and crosslink the alternative pair of binding sites. This changes the conformation of the receptor and accelerates the dissociation of the already bound insulin molecule, since it is now monovalently bound. However, at very high insulin concentration (>1 μM), two insulin molecules will occupy the alternative binding sites simultaneously. This will prevent their crosslinking, stabilize the binding of the first bound insulin molecule, and thereby abolish the accelerated dissociation (De Meyts and Whittaker, 2002; De Meyts *et al.*, 1994). Panel B shows the three-dimensional structure of the ectodomain of the IR dimer determined by X-ray crystallography (McKern *et al.*, 2006). The view perpendicular to the ectodomain twofold axis is shown as well as the view along the twofold axis looking towards the cell membrane. One monomer is shown in green and the other in blue. The Individual domains are indicated as well as the insulin binding sites 1 and 2 on the receptor (gray circles). Modeling was carried out using Chimera (pdb file: 2DTG). (See Color Insert.)

data and the anti-parallel receptor construction has been supported by the recent crystal structure of the IR described below (Fig. 3.2B) (McKern *et al.*, 2006).

The three-dimensional structure of the insulin molecule was solved forty years ago (for review see Baker *et al.*, 1988; Kaarsholm and Ludvigsen, 1995). Most of the insulin residues involved in receptor binding have been mapped revealing two surfaces on the insulin molecule which bind to the IR; site 1 (the so-called "classical" surface) and site 2. The site 1 binding surface was the first to be described three decades ago and overlaps partly with the dimer forming surface. It includes the residues GlyA1, IleA2, ValA3, GlnA5, TyrA19, AsA21, ValB12, TyrB16, GlyB23, PheB24, PheB25, and TyrB26, many of which are evolutionarily conserved (De-Meyts, 2004; De Meyts *et al.*, 1978; Huang *et al.*, 2004, 2007; Kristensen *et al.*, 1997; Pullen *et al.*, 1976). The residues B23–26 have furthermore been shown to be essential for the negative cooperativity observed in insulin binding (De Meyts *et al.*, 1978). The site 2 binding surface has been mapped by alanine scanning mutagenesis in our laboratory (De Meyts, 2004; Jensen, Master thesis, University of Copenhagen, 1999) and involves the residues: SerA12, LeuA13, GluA17, His B10, GluB13, and LeuB17, with A13 and B17 making the major contributions.

Photoaffinity labeling, alanine scanning mutagenesis and 3-D-modelling data suggest that the site 1 binding surface of insulin binds to a site on the receptor (site 1) consisting of the L1 (Williams *et al.*, 1995) and CT domain (Kristensen *et al.*, 2002; Kurose *et al.*, 1994; Mynarcik *et al.*, 1996, 1997). This is further supported by the finding that a L1-CR-CT construct is able to bind insulin with nanomolar affinity (Kristensen *et al.*, 1998). The second IR binding site (site 2) is suggested to be located within the FnIII-1 and FnIII-2 domains (Benyoucef *et al.*, 2007; Hao *et al.*, 2006; Lawrence *et al.*, 2007) and binds the site 2 binding surface on insulin. Residues involved site 2 binding have now been mapped by alanine scanning mutagenesis (Whittaker, *et al.*, 2008). Mathematical modeling of insulin binding to the IR has now been done in our laboratory by Kiselyov *et al.* (2009) which suggests that site 1 has a K_d of about 6 nM and site 2 a K_d of about 400 nM.

The crystal structure of the ectodomain of the IR (McKern *et al.*, 2006) has recently been published (Fig. 3.2B). This structure was made in the absence of bound insulin but in the presence of a synthetic insulin mimetic peptide and two Fab fragments (83–7 and 83–14) (Soos *et al.*, 1986b). The peptide was however not visible in the structure. The crystal structure of the whole IR ectodomain reveals that each IR ectodomain monomer has an "inverted V layout" with respect to the cell membrane (McKern *et al.*, 2006). One leg of the "V" consists of the L1-CR-L2 fragment and the other of the three FnIII domains. In the structure the ID is mostly disordered. The exact arrangement of the IR monomer and CT peptide (that is a part of the ID and has been shown to play a role in binding of insulin (Kristensen *et al.*, 2002)) is therefore less certain, but Ward *et al.* (2007) have suggested two possible arrangements. The first suggestion is that the

CT peptide and the L1 domain that lies adjacent are part of the same IR monomer. The second suggestion is that the CT peptide is provided by the ID of the second IR monomer instead. The latter suggestion has been experimentally confirmed by Chan *et al.* (2007) using truncated IR receptors (mIR.Fn0-Ex10 construct described by Brandt *et al.* (2001)) that contain complementary mutations in each α subunit. The recent solving of the X-ray structure of the IR and the implications of this for the understanding of insulin induced IR activation has been extensively reviewed elsewhere (Ward *et al.*, 2008).

C. Activation of the IR

Binding of insulin to the IR leads to phosphorylation of several sites on the IR. In total seven tyrosine residues in the intracellular β-subunit have been identified as targets for autophosphorylation (Youngren, 2007). Two tyrosines, Tyr965 and Tyr972, are located in the juxtamembrane region and three tyrosines, Tyr1158, Tyr1162, and Tyr1163 are located in the activation loop of the kinase domain. The last two tyrosines, Tyr1328 and tyr1334, are located close to the C terminus. The illustration in Fig. 3.1 shows where the autophosphorylation sites are located in the β-subunit. The currently accepted model is one in which the cross-linking of the IR by insulin leads to a conformational change bringing the intracellular kinase domains close together, resulting in autophosphorylation of tyrosine residues on the β-subunit. Both cis- and trans-autophosphorylation have been shown. The tyrosines in the juxtamembrane domain have been shown to be phosphorylated by the phosphotransferase activity of the same subunit (*cis*-phosphorylated), whereas the tyrosines in the activation loop are trans-phosphorylated by the opposing β-subunit following ligand binding (Cann and Kohanski, 1997). This phosphorylation leads to the movement of the activation loop away from blocking the kinase catalytic site, which gives free access to ATP and substrates such as IRS and Shc (Fig. 3.3) (Hubbard and Miller, 2007).

III. Modulation of IR Activity

A. Modulation of signaling by different phosphorylation patterns on the IR

Activation of the IR tyrosine kinase and the phosphorylation of cellular substrates like IRS and Shc are essential for insulin signal transduction. The level and specific sites of autophosphorylation may regulate distinct but

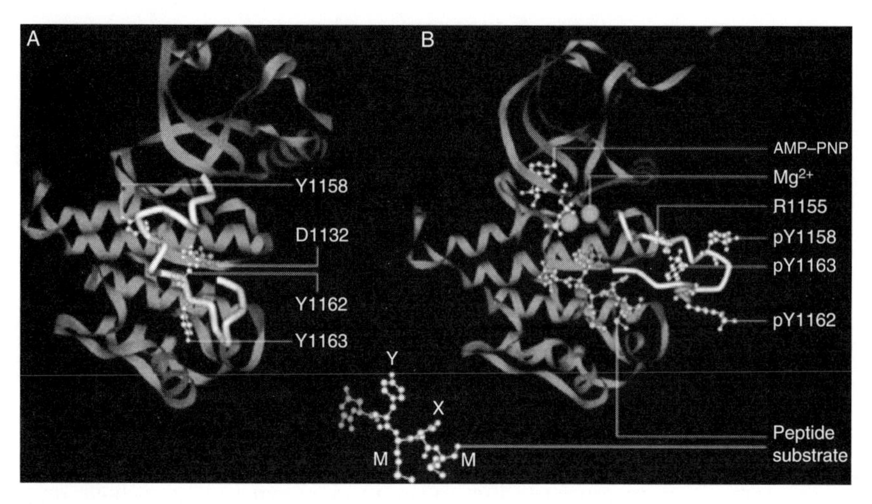

Figure 3.3 Activation of the tyrosine kinase. Panel a shows the inactive IR tyrosine kinase (Hubbard *et al.*, 1994) and panel b the activated kinase (Hubbard, 1997) with a ATP analogue (AMP–PNP), peptide substrate, and magnesium (Mg^{2+}) bound. This figure illustrates the autoinhibition mechanism: The tyrosine Tyr1158 (that is located in the activation loop) is bound in the active site and hydrogen-bonded to a conserved Asp1132 residue in the catalytic loop (panel A) and thereby competes with protein substrates (before autophosphorylation). In the activated state (panel B), the activation loop is tris-phosphorylated (on Tyr1158,Tyr1162, and Tyr1163) and moves out of the active site. Tyr1163 becomes hydrogen-bonded to a conserved Arg1155 residue in the beginning of the activation loop, which stabilizes the repositioned loop. Also shown is the peptide substrate, with the YMXM motif. AMP–PNP, adenylyl imidodiphosphate. The amino acids are numbered based on IR isoform B. From (De Meyts and Whittaker, 2002). (See Color Insert.)

functionally coordinated aspects of IR functions as described below and summarized in Fig. 3.4.

The three tyrosine residues (Tyr1158, Tyr1162, and Tyr1163) in the regulatory loop are regarded as the autophosphorylation sites primarily responsible for activation of substrate phosphorylation. Mutation at one, two or three of these tyrosine residues increasingly reduces insulin-stimulated kinase activity and results in a corresponding loss of biological activity (Wilden *et al.*, 1992). In normal cells only a small fraction of IRs get simultaneously phosphorylated on these three tyrosines (White *et al.*, 1988a). So whereas triple phosphorylation of the kinase domain is essential for full kinase activation and receptor internalization, it is not mandatory for signaling metabolic or growth effects. This suggests that IR signaling may be modulated by the degree of autophosphorylation. There is yet no clear evidence for the individual roles of these three phosphorylation sites. However, some specificity has been demonstrated when mutating these sites in terms of differential effects on glycogen synthesis and DNA synthesis

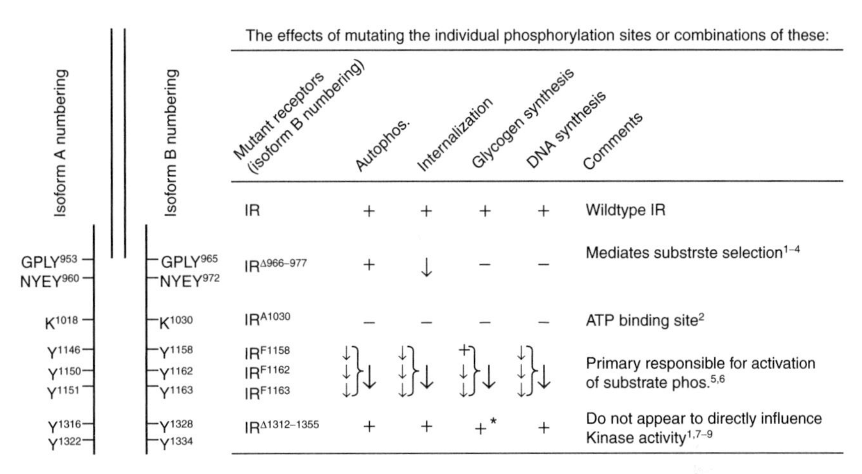

Figure 3.4 Locations and effects of mutating the phosphorylation sites on the IR β-subunit. Seven tyrosine residues in the intracellular β-subunit have been identified as targets for autophosphorylation (Youngren, 2007). The tyrosines Tyr965 and Tyr972, are located in the juxtamembrane region, Tyr1158, Tyr1162, and Tyr1163 in the activation loop of the kinase domain, and Tyr1328 and Tyr 1334, are located close to the C terminus. Moreover the lysine residue that binds ATP is also shown. Mutating the tyrosines in the different regions lead to differential effects on the biological activity. The effects on autophosphorylation, internalization, glycogen synthesis, and DNA synthesis are indicated as well as the references for the respective studies. Mutating the three tyrosines in the regulatory loop individually as well as in combination (triple mutant) has been done and is also indicated as some specificity has been demonstrated in terms of differential effects of the individual tyrosines on glycogen synthesis and DNA synthesis. *Contradicting results have been obtained as some studies show a reduced ability of this mutant receptor to stimulate glycogen synthesis. The IR exists in two isoforms A and B (representing alternate splice variants of a single gene without and with exon 11), this results in a difference of 12 amino acids in the C-terminal region of the α-subunit between the two isoforms. Therefore, the numbering of the amino acids is dependent on the isoform and we have for clarity numbered the amino acids based on both isoform A and B as both isoforms have been used in the references. In the text the numbering of the amino acids is based on the B isoform. See text for more details. 1, Backer *et al.* (1992); 2, White *et al.* (1988a); 3, Backer *et al.* (1990); 4, Kaburagi *et al.* (1993); 5, Wilden *et al.* (1990); 6, Wilden *et al.* (1992); 7, Myers *et al.* (1991); 8, Debant *et al.* (1988); 9, Maegawa *et al.* (1988).

(Wilden *et al.*, 1990, 1992). Further studies are needed to investigate how the differences between metabolic and growth responses are dependent on the extent of kinase activation.

The autophosphorylation of the tyrosines in the C terminus (Tyr1328 and Tyr1334) does not appear to play a role in kinase activation or internalization (Backer *et al.*, 1992; Maegawa *et al.*, 1988). Contradictory results have been obtained with regard to the role of C terminus for the biological response. Some studies show that the C terminus plays a minimal role in stimulating glycogen synthesis and DNA synthesis (Myers *et al.*, 1991) whereas others show that deleting the C terminus leads to low stimulation of glycogen synthesis but normal stimulation of thymidine incorporation (Debant *et al.*, 1988; Maegawa *et al.*, 1988), so the biological role of the C terminus is not clear. The tyrosines 965 and 972 in the juxtamembrane region may mediate substrate selection (Chaika *et al.*, 1999; Gustafson *et al.*, 1995; Kaburagi *et al.*, 1995; Sawka-Verhelle *et al.*, 1997b). The tyrosine 972, which exists in an NPXY972-motif (Feener *et al.*, 1993), presumably engages the phosphotyrosine binding (PTB) domains in the IRS1, Shc, and STAT5B. In contrast IRS2 is not dependent on Tyr972 for phosphorylation (Chaika *et al.*, 1999) but likely interacts with the Tyr1158, Tyr1162, and Tyr1163 in the activation loop instead. In this way blocking phosphorylation of Tyr972 would most likely only impact IRS1 but not IRS2 mediated signaling. Furthermore, a region unique to IRS2, termed the kinase regulatory-loop binding (KRLB) region, which interacts with the IR tyrosine kinase has been identified using yeast two hybrid studies (Sawka-Verhelle *et al.*, 1996, 1997a). The interaction is dependent upon phosphorylation of the kinase activation loop (He *et al.*, 1996; Sawka-Verhelle *et al.*, 1996) and a recent study indicates that the KRLB region function to regulate IRS2 activity by limiting tyrosine phosphorylation of IRS2 (Wu *et al.*, 2008). In addition, the juxtamembrane region also contains several serine phosphorylation sites that have been shown to be involved in regulation of IR activity (Youngren, 2007).

Overall, it seems like the individual phosphorylation sites as well as the different regions in the β-subunit of the IR contribute to the specificity of signaling either due to their influence on overall kinase activity (White *et al.*, 1988b) or through direct interaction with individual substrates (Sun *et al.*, 1991). However, further studies are needed to clarify whether specific phosphorylation patterns of the IR correspond with specific biological effects.

B. The effects of post-translational modifications of the IR

The activity of the IR is tightly regulated through several mechanisms. Tyrosine phosphatases are one class of regulatory proteins and reduce the IR activity by dephosphorylating important tyrosine residues on the IR. Other proteins reduce the IR activity by sterically blocking its interaction with the IRS proteins, or by modifying its kinase activity. Suppressor of cytokine signaling-1 (SOCS1) and SOCS3, growth-factor-receptor-bound protein 10 (Grb10) and plasma-cell-membrane glycoprotein-1 (PC1) are examples of such proteins (Ueki *et al.*, 2004). The ability and mechanisms of how these proteins can modulate IR activity have recently been thoroughly reviewed (Youngren, 2007) and are only briefly mentioned here.

The most studied of the tyrosine phosphatases is PTP1B. PTP1B interacts directly with the IR after insulin stimulation and is dependent on the residues Tyr1158, Tyr1162, and Tyr1163 on the IR (Bandyopadhyay *et al.*, 1997).

SOCS3 binds to phosphorylated tyrosine 972, whereas SOCS1 interacts with the phosphorylated form of the C-terminus containing the residues Tyr1158, Tyr1162, and Tyr1163 (Emanuelli *et al.*, 2001; Mooney *et al.*, 2001) and thereby blocks interaction between the IR and IRS. The effect of Grb10 and the highly homologous Grb14 on insulin has recently been reviewed in detail (Holt and Siddle, 2005). Grb10 binds phosphotyrosine residues on the IR including residues in the SH2 domain (Holt and Siddle, 2005) and seemingly inhibits the kinase activity (Langlais *et al.*, 2004; Mori *et al.*, 2005; Mounier *et al.*, 2001; Stein *et al.*, 2001). Studies suggest that PC1 interacts directly with the IR (residues 485–599). This region of the IR is required for the conformational change of the IR β-subunits that permits autophosphorylation and binding of PC1 may therefore block autophosphorylation in this way (Maddux and Goldfine, 2000). However, further studies are necessary to understand the precise mechanism.

Another mechanism of negative regulation is serine/threonine phosphorylation and O-linked glycosylation (Ahn *et al.*, 1993; Al-Hasani *et al.*, 1997; Bossenmaier *et al.*, 1997, 2000; Chin *et al.*, 1993; Coba *et al.*, 2004; Koshio *et al.*, 1989; Liu and Roth, 1994; Schmitz-Peiffer, 2002; Strack *et al.*, 2000) (reviewed in Schmitz-Peiffer, 2002). Many serine residues have been shown to be phosphorylated by PKC (protein kinase C). However, the effect of the individual phosphorylation sites is still unclear. Recently both N-linked and O-linked glycosylation of the IR has been characterized (Sparrow *et al.*, 2007, 2008). The IR is heavily N-linked glycosylated (Sparrow *et al.*, 2008). N-linked glycosylation has been shown to play a role in folding and processing of the IR, including the formation of the dimer and the cleavage of the IR protein into the α- and β-chains. N-linked glycosylation has also been shown to be involved in the transport of the newly synthesized IR to the cell surface. Recently a study has suggested a mechanism for metabolic regulation of mitogenic signaling through interaction with cell surface galectin-3 that links the number of N-glycan sites in receptor with the metabolic flux through the hexosamine pathway that regulates the degree of N-glycan branching (Lau *et al.*, 2007). The role of N-glycosylation of the IR has been reviewed (Adams *et al.*, 2000; Bass *et al.*, 1998) but determination of the extent to which N-glycosylation of the IR contributes to modulation of the IR activity needs further study. O-glycosylation by adding a GlcNAc moiety to the hydroxyl groups of serine/threonine residues may play a role in modulation of IR activity but experimental evidence in support thereof is still lacking (Sparrow *et al.*, 2007).

C. The role of cellular location of the IR

Another mechanism of negative regulation of IR activity is the down-regulation of the IR at the protein level by ligand-stimulated internalization and degradation (Gavin *et al.*, 1974). When the activated insulin–IR complex is internalized, insulin dissociates rapidly from its receptor and gets degraded, resulting in signal termination. The ligand free IR is recycled back to the surface (Carpentier *et al.*, 1986). Within endosomes, ligand bound tyrosine kinase receptors remain activated and maintain their ability to phosphorylate downstream intracellular effectors (Bergeron *et al.*, 1995; Grimes *et al.*, 1996; Sorkin *et al.*, 1993). Signaling molecules like Grb2, Shc, and ERK/MAPK, which are the molecules mainly responsible for the mitogenic response of insulin, have been shown to colocalize on endosomal membranes (Di Guglielmo *et al.*, 1994; Pol *et al.*, 1998; Rizzo *et al.*, 2000; Sorkin *et al.*, 2000). Therefore the mitogenic effect of insulin may result from IR signaling from endosomes. In agreement with this are studies where IR internalization was inhibited by either reduced temperature (Biener *et al.*, 1996), mutation of the receptor (Hamer *et al.*, 2002) or by the expression of a dominant-interfering mutant of dynamin (Ceresa *et al.*, 1998). These studies showed that inhibition of IR internalization inhibited insulin-stimulated Shc tyrosine phosphorylation (Biener *et al.*, 1996; Ceresa *et al.*, 1998; Hamer *et al.*, 2002) as well as activation of the mitogen-activated protein kinases ERK1 and ERK2 (Ceresa *et al.*, 1998; Hamer *et al.*, 2002). Furthermore, the inhibition of IGF-1R internalization by low temperature also been shown to inhibit Shc phosphorylation (Chow *et al.*, 1998). These studies all support the concept that internalization of IR as well as IGF-1R appears to be required for phosphorylation and activation of the Shc/MAPK pathway and indicates that signaling from the endosomes play an important role in the mitogenic response of insulin. Additionally, these studies also showed that inhibition of IR internalization had no effect on either receptor autophosphorylation, IRS1 tyrosine phosphorylation, or Akt kinase phosphorylation and activation (Biener *et al.*, 1996; Ceresa *et al.*, 1998; Hamer *et al.*, 2002), suggesting that these can be fully activated by the IR at the cell surface. This suggests that the metabolic response of insulin can be fully activated from IRs on the cell surface. In support of this, a study by Uhles *et al.* (2007) demonstrated that in beta cells, insulin activates the glucokinase gene from the plasma membrane (and via a PI3K/Akt pathway), while the c-*fos* gene is activated from early endosomes, and via an Shc/ERK pathway. In this way it appears that a link exists between receptor internalization and the type of signaling pathways that get initiated. Studies with EGF support this, as activation of the EGFR leads to greater and prolonged EGFR tyrosine phosphorylation in rat liver endosomal membranes (compared to EGFRs located at the plasma membrane) which is correlated with elevated Shc phosphorylation and Shc/Grb2 association at this intracellular compartment (Di Guglielmo *et al.*, 1994). Full activation of MAPK is also dependent on EGFR internalization (Vieira *et al.*, 1996). Internalization possibly provides access for the activated receptor to substrates that may not reside at the plasma membrane/cytosol interface. This has been shown for the tumor necrosis factor-α (TNF-α) receptor (TNFR).

Internalization of the TNFR in human T-cells stimulates an acidic sphingomyelinase present only in the endosomes which leads to activation of the transcription factor NF-kB (Wiegmann *et al.*, 1994).

A potential role of plasma membrane domains in regulation of insulin signaling has also been suggested. Disruption of caveolae via cholesterol depletion has been shown to inhibit insulin-induced Akt activation and glycogen synthesis in hepatocytes whereas the mitogenic ERK/MAPK signaling pathway was not inhibited (Parpal *et al.*, 2001). Furthermore, isoform specific IR signaling has been shown to involve different plasma membrane domains (Uhles *et al.*, 2003). However, the precise role of microdomains and caveolae in insulin signaling and specificity is still unclear and need further study (reviewed in Ishikawa *et al.*, 2005 and Saltiel and Pessin, 2003).

IV. Differential Activation of the IR

Activation of the IR with different ligands can lead to different biological effects most likely due to differences in receptor interaction, internalization rates, and phosphorylation patterns of the IR. Studies investigating the effect of activating the IR with insulin and IGF-2 (Frasca *et al.*, 1999; Morrione *et al.*, 1997; Pandini *et al.*, 2003; Sciacca *et al.*, 2002) found that insulin and IGF-2 bind with a similar affinity to the IR-A but whereas insulin led primarily to metabolic effects (measured by glucose uptake), IGF-2 led primarily to mitogenic effects (measured by ^{3}H-thymidine assays). These differences in the biological effects were associated with differential recruitment and activation of intracellular substrates as well as selective changes in gene expression. The details of the signal generated by the insulin-receptor interaction, which mediates the signaling and biological responses in the cell, remain unclear at the molecular level. However, insulin analogs, anti-IR antibodies, and synthetic insulin mimetic peptides that target the two insulin-binding regions on the IR, have been used to study the relationship between different aspects of receptor binding and functions as well as providing new insights into the structure and function of the IR.

A. Insulin analogs can initiate different cellular reponses than insulin through the IR

Differences in the effect on signaling pathways, thymidine incorporation, glycogen synthesis, glucose uptake, etc., between stimulation with insulin and insulin analogs have been shown in several studies (De Meyts and Shymko, 2000; Hansen *et al.*, 1996; Kurtzhals *et al.*, 2000). Insulin analogs with improved therapeutic properties have also been developed for the treatment of diabetes (Siddiqui, 2007). Some of the analogs showed an increased mitogenicity which has been a major concern related to the long-term use of insulin analogs (Kurtzhals *et al.*, 2000). The increased mitogenicity could either be due to increased affinity for the IGF-1R or a more mitogenic response through the IR. However it is not completely clear what determines the metabolic/mitogenic potency ratio.

Previous work showed that insulin analogues which dissociate very slowly from the IR induce a sustained activation of the IR and sustained phosphorylation of Shc (Hansen *et al.*, 1996). This correlated with an increased mitogenic activity, indicating that an increased duration of the insulin signal at the receptor level results in a shift towards a more mitogenic insulin response (De Meyts and Shymko, 2000; Hansen *et al.*, 1996; Ish-Shalom *et al.*, 1997; Kurtzhals *et al.*, 2000). This suggests that the mitogenic activity can be mediated via a sustained activation of the IR. Other studies show a correlation between the internalization of the analogs and the activation of the mitogenic pathway (MAPK pathway) (Drejer, 1992; Foti *et al.*, 2004; Rakatzi *et al.*, 2003). Rakatzi *et al.* found that high internalization rates correlated with higher Shc phosphorylation and ERK activation whereas a low internalization rate correlated with marginal activation of ERK in K5 myoblasts (Rakatzi *et al.*, 2003). Furthermore, sustained receptor binding decreases endosomal insulin degradation indicating that the analogs are protected against degradation when the insulin–IR complex is intact. This results in enhanced signaling from the endosomes (Bevan *et al.*, 1996, 2000). Other studies have shown that activation of MAPK also correlates to internalization of EGFR (Vieira *et al.*, 1996) and G protein-coupled receptors (Pierce *et al.*, 2000). This suggests that the mitogenic properties of insulin analogs results from a combination of the initial receptor interactions, internalization, and phosphorylation of the IR.

Increased interaction with the IGF1 receptor may also contribute to the higher mitogenicity of some insulin analogs (Kurtzhals *et al.*, 2000).

B. Anti-IR antibodies and insulin mimetic peptides can initiate different cellular responses than insulin through the IR

Besides insulin analogs, anti-IR antibodies and synthetic insulin mimetic peptides that target the two insulin-binding regions on the IR have also been used to study the relationship between different aspects of receptor binding and functions as well as providing new insights into the structure and function of the IR.

Several antibodies as well as the corresponding Fab fragments have been studied that can mimic a wide range of insulin effects (Brunetti *et al.*, 1989; Gu *et al.*, 1988; O'Brien *et al.*, 1986; Soos *et al.*, 1986ab; Taylor *et al.*, 1987; Wang *et al.*, 1988). They bind to several different epitopes on the IR (Soos *et al.*, 1986a). The binding and kinetic effects of several monoclonal anti-IR antibodies have been studied in detail both on purified soluble receptors as well as on whole cells. Some of these accelerated the dissociation (negative cooperativity) of prebound ^{125}I-insulin whereas others increased the binding (positive cooperativity) of ^{125}I-insulin. The presence of positive cooperativity suggests the existence of multiple conformational states of IR which may be induced or stabilized by binding of antibodies or insulin. It has also been suggested that the insulin—receptor complex can shift reversibly between interconvertible states by site–site interactions, which depend on the nature of the ligand occupying neighboring sites (Gu *et al.*, 1988; Wang *et al.*, 1988). The effects on IR phosphorylation, glucose uptake, thymidine incorporation (i.e., DNA synthesis) of several of these antibodies and corresponding Fabs have been studied (Brunetti *et al.*, 1989; O'Brien *et al.*, 1986; Soos *et al.*, 1986a; Taylor *et al.*, 1987). Some of the anti-IR antibodies studied have inhibitory effects and other stimulatory effects.

The finding that many different antibodies stimulated autophosphorylation suggest that kinase activation might depend on cross-linking of IRs by bivalent antibodies, rather than reaction at particular sensitive epitopes. This is further supported by the fact that many of the Fab fragments retain their binding ability to the IR but do not retain full activity in biological assays and furthermore antagonize the stimulatory effect of bivalent antibody when both are added together (Brunetti *et al.*, 1989; O'Brien *et al.*, 1986; Soos *et al.*, 1986a; Taylor *et al.*, 1987). As some Fabs stimulated partial activity in the biological assays it appears that with some of the antibodies a part of the activity is due to an epitope-dependent stimulation of autophosphorylation and the rest is due to a more general effect of cross-linking. Interestingly some studies find some monoclonal antibodies to be able to mimic insulin effects in cells but without significantly activating the IR kinase (Sung, 1992). This questions if a conformation state or aggregation state of the IR induced by either autophosphorylation or cross-linking of the IRs may be more important than kinase activation. However this needs further studies.

Synthetic insulin mimetic peptides that target the two insulin-binding sites on the IR have been generated which resulted in high affinity ligands for the IR (Pillutla *et al.*, 2002). Site 1 peptides act as agonists on the tyrosine kinase while site 2 peptides act as antagonists. This could indicate that while the site 1 interaction possibly is central to receptor activation, the site 2 interaction could be more responsible for affinity and selectivity. Whether or not these insulin mimetic peptides actually bind to the same sites as insulin on the IR is not known, but they are located close enough to be able to compete with insulin for binding (Pillutla *et al.*, 2002). Recently, homodimers and heterodimers of the site 1 and 2 mimetics have been generated to create molecules that activate the IR with high potency and specificity—possible through a mechanism similar to the binding of insulin (Schäffer *et al.*, 2003). We have recently shown that one of these synthetic insulin mimetic peptides that target the two insulin-binding sites on the IR led to significant differences on internalization, signaling response, gene expression, and cell proliferation compared to insulin. This peptide was able to differentially activate the IR and initiate a different biological response through the IR than insulin which is most likely due to the differences in receptor interaction, low phosphorylation of the IR and low degree of IR internalization compared to insulin stimulation (Jensen *et al.*, 2007, 2008). In this instance we obtained a good metabolic potency but low mitogenic potency. Such discriminating ligands would have a favorable profile for the development of new drugs for diabetes care. Besides activating the IR this peptide also enhanced insulin binding to the IR as seen earlier with the anti-IR antibodies described above. This opens up the idea that it may be possible to modulate IR action by activating the IR to different degrees (i.e., degree of IR phosphorylation). It is likely that downstream signaling pathways differ in the degree of receptor activation required for a full effect as the Akt signaling pathway seem to be fully activated at low levels of receptor activation, while full activation of the MAPK pathway require a higher level of receptor activation (Jensen *et al.*, 2007). Gene-regulatory events may in a similar fashion be differentiated dependent on the level of activation of the signaling pathways. Thus the different biological responses stimulated by different ligands could reflect the different phosphorylation degree of the IR (including a different dependence of the downstream signaling pathways on the degree of IR activation) as well as the location (cell surface vs. endosomal compartment) of the IR which influences which signaling molecules will be available for phosphorylation.

These findings raise the possibility of making improved drugs with specific insulin actions by either making noninternalizing mimetics or mimetics which induce specific phosphorylation patterns on the IR. The idea of using altered trafficking of ligands to enhance activity has already been shown to be feasible. Reddy *et al.* increased the mitogenicity of an EGF mutant by lowering the receptor binding affinity resulting in reduction of ligand depletion and receptor regulation. This EGF mutant was a more potent mitogenic stimulus for fibroblasts than natural EGF or transforming growth factor alpha because of its altered trafficking properties even though it had a lower receptor binding affinity (Reddy *et al.*, 1996). In the case of the IR, altering the trafficking of the IR could maybe enable us to enhance specific actions (metabolic responses) while minimizing others (mitogenic responses) by keeping the ligand–receptor complex at the cell surface (Jensen *et al.*, 2007). Hence, these studies suggest that optimization of the binding properties of a new ligand should include consideration of the processes from the initial binding event to the final cellular response.

The studies described above with the insulin analogs, anti-IR antibodies and insulin mimetic peptides all suggest that IR action is critically dependent on not only the concentration of the ligand but also its nature. It remains unclear what the precise mechanism underlying the differences in the signaling and biological responses of insulin, antibodies and peptides are, but the described studies suggest that a difference in the initial receptor interactions including internalization and phosphorylation of the IR are key factors in determining the subsequent cellular response.

V. Conclusions/Final Words

The details of the signal generated by the insulin–receptor interaction, which mediates the signaling and biological responses in the cell, remain unclear at the molecular level, although some aspects are beginning to emerge. The insulin–IR interaction might control the specificity of downstream signaling by a combination of several mechanisms. As such, the mitogenic and metabolic properties of IR action may result form a complex series of initial receptor/ligand interactions involving binding affinity and dissociation kinetics, internalization, the half-life of the ligand–IR complex and a differential phosphorylation pattern of the IR (Fig. 3.5). Future experiments will hopefully shed light on the precise molecular mechanisms leading to the differential signaling at the level of the IR.

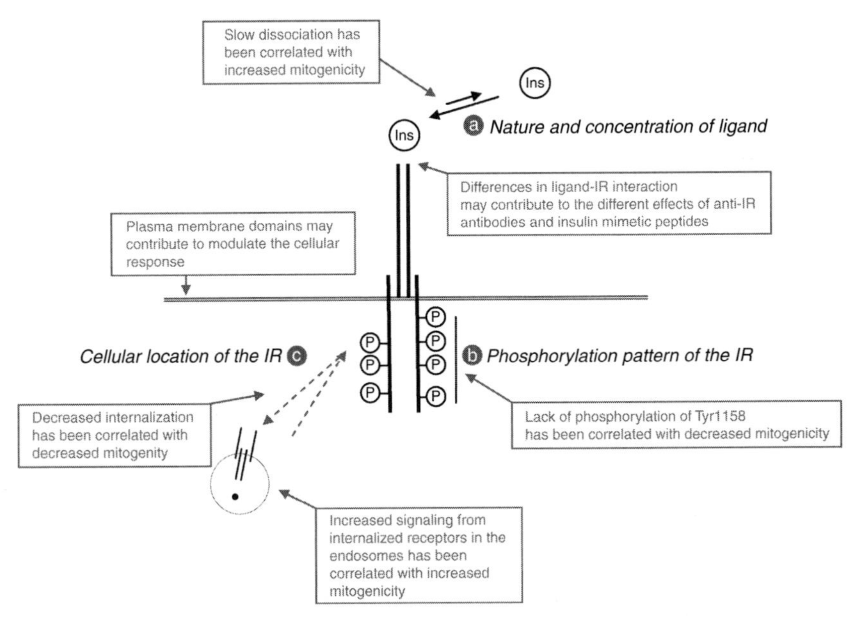

Figure 3.5 Modulation of a mitogenic versus metabolic response through the IR. The cellular response of IR activation have been shown to dependent on several factors: (a) differences in the nature and concentration of ligands (Section IV), (b) the phosphorylation pattern of the IR (Section IIIA–B), and (c) the cellular location of the activated IR (Section IIIC). See text for explanations.

REFERENCES

Adams, T. E., Epa, V. C., Garrett, T. P., and Ward, C. W. (2000). Structure and function of the type 1 insulin-like growth factor receptor. *Cell Mol. Life Sci.* **57,** 1050–1093.

Ahn, J., Donner, D. B., and Rosen, O. M. (1993). Interaction of the human insulin receptor tyrosine kinase from the baculovirus expression system with protein kinase C in a cell-free system. *J. Biol. Chem.* **268,** 7571–7576.

Al-Hasani, H., Eisermann, B., Tennagels, N., Magg, C., Passlack, W., Koenen, M., Müller-Wieland, D., Meyer, H. E., and Klein, H. W. (1997). Identification of Ser-1275 and Ser-1309 as autophosphorylation sites of the insulin receptor. *FEBS Lett.* **400,** 65–70.

Backer, J. M., Kahn, C. R., Cahill, D. A., Ullrich, A., and White, M. F. (1990). Receptor-mediated internalization of insulin requires a 12-amino acid sequence in the juxtamembrane region of the insulin receptor beta-subunit. *J. Biol. Chem.* **265,** 16450–16454.

Backer, J. M., Schroeder, G. G., Kahn, C. R., Myers, M. G., Wilden, P. A., Cahill, D. A., and White, M. F. (1992). Insulin stimulation of phosphatidylinositol 3-kinase activity maps to insulin receptor regions required for endogenous substrate phosphorylation. *J. Biol. Chem.* **267,** 1367–1374.

Baker, E. N., Blundell, T. L., Cutfield, J. F., Cutfield, S. M., Dodson, E. J., Dodson, G. G., Hodgkin, D. M., Hubbard, R. E., Isaacs, N. W., Reynolds, C. D., *et al.* (1988). The structure of 2Zn pig insulin crystals at 1.5 A resolution. *Philos. Trans. R Soc. Lond. B Biol. Sci.* **319,** 369–456.

Bandyopadhyay, D., Kusari, A., Kenner, K. A., Liu, F., Chernoff, J., Gustafson, T. A., and Kusari, J. (1997). Protein-tyrosine phosphatase 1B complexes with the insulin receptor *in vivo* and is tyrosine-phosphorylated in the presence of insulin. *J. Biol. Chem.* **272,** 1639–1645.

Bass, J., Chiu, G., Argon, Y., and Steiner, D. F. (1998). Folding of insulin receptor monomers is facilitated by the molecular chaperones calnexin and calreticulin and impaired by rapid dimerization. *J. Cell Biol.* **141,** 637–646.

Benyoucef, S., Surinya, K. H., Hadaschik, D., and Siddle, K. (2007). Characterization of insulin/IGF hybrid receptors: Contributions of the insulin receptor L2 and Fn1 domains and the alternatively spliced exon 11 sequence to ligand binding and receptor activation. *Biochem. J.* **403,** 603–613.

Bergeron, J. J., Di Guglielmo, G. M., Baass, P. C., Authier, F., and Posner, B. I. (1995). Endosomes, receptor tyrosine kinase internalization and signal transduction. *Biosci. Rep.* **15,** 411–418.

Bevan, A. P., Drake, P. G., Bergeron, J. J. M., and Posner, B. I. (1996). Intracellular signal transduction: The role of endosomes. *Trends Endocrinol. Metab.* **7,** 13–21.

Bevan, A. P., Seabright, P. J., Tikerpae, J., Posner, B. I., Smith, G. D., and Siddle, K. (2000). The role of insulin dissociation from its endosomal receptor in insulin degradation. *Mol. Cell. Endocrinol.* **164,** 145–157.

Biener, Y., Feinstein, R., Mayak, M., Kaburagi, Y., Kadowaki, T., and Zick, Y. (1996). Annexin II is a novel player in insulin signal transduction. Possible association between annexin II phosphorylation and insulin receptor internalization. *J. Biol. Chem.* **271,** 29489–29496.

Bossenmaier, B., Mosthaf, L., Mischak, H., Ullrich, A., and Häring, H. U. (1997). Protein kinase C isoforms beta 1 and beta 2 inhibit the tyrosine kinase activity of the insulin receptor. *Diabetologia* **40,** 863–866.

Bossenmaier, B., Strack, V., Stoyanov, B., Krützfeldt, J., Beck, A., Lehmann, R., Kellerer, M., Klein, H., Ullrich, A., Lammers, R., and Häring, H. U. (2000). Serine residues 1177/78/82 of the insulin receptor are required for substrate phosphorylation but not autophosphorylation. *Diabetes* **49,** 889–895.

Brandt, J., Andersen, A. S., and Kristensen, C. (2001). Dimeric Fragment of the Insulin Receptor α-Subunit Binds Insulin with Full Holoreceptor Affinity. *J. Biol. Chem.* **276,** 12378–12384.

Brunetti, A., Maddux, B. A., Wong, K. Y., Hofmann, C., Whittaker, J., Sung, C., and Goldfine, I. D. (1989). Monoclonal antibodies to the human insulin receptor mimic a spectrum of biological effects in transfected 3T3/HIR fibroblasts without activating receptor kinase. *Biochem. Biophys. Res. Commun.* **165,** 212–218.

Cann, A. D., and Kohanski, R. A. (1997). Cis-autophosphorylation of juxtamembrane tyrosines in the insulin receptor kinase domain. *Biochemistry* **36,** 7681–7689.

Carpentier, J. L., Gazzano, H., Van Obberghen, E., Fehlmann, M., Freychet, P., and Orci, L. (1986). Intracellular pathway followed by the insulin receptor covalently coupled to 125I-photoreactive insulin during internalization and recycling. *J. Cell Biol.* **102,** 989–996.

Ceresa, B. P., Kao, A. W., Santeler, S. R., and Pessin, J. E. (1998). Inhibition of clathrin-mediated endocytosis selectively attenuates specific insulin receptor signal transduction pathways. *Mol. Cell Biol.* **18,** 3862–3870.

Chaika, O. V., Chaika, N., Volle, D. J., Hayashi, H., Ebina, Y., Wang, L. M., Pierce, J. H., and Lewis, R. E. (1999). Mutation of tyrosine 960 within the insulin receptor juxtamembrane domain impairs glucose transport but does not inhibit ligand-mediated phosphorylation of insulin receptor substrate-2 in 3T3-L1 adipocytes. *J. Biol. Chem.* **274,** 12075–12080.

Chan, S. J., Nakagawa, S., and Steiner, D. F. (2007). Complementation analysis demonstrates that insulin cross- links both alpha subunits in a truncated insulin receptor dimer. *J. Biol. Chem.* **282,** 13754–13758.

Chin, J. E., Dickens, M., Tavare, J. M., and Roth, R. A. (1993). Overexpression of protein kinase C isoenzymes alpha, beta I, gamma, and epsilon in cells overexpressing the insulin receptor. Effects on receptor phosphorylation and signaling. *J. Biol. Chem.* **268,** 6338–6347.

Chow, J. C., Condorelli, G., and Smith, R. J. (1998). Insulin-like growth factor-I receptor internalization regulates signaling via the Shc/mitogen-activated protein kinase pathway, but not the insulin receptor substrate-1 pathway. *J. Biol. Chem.* **273,** 4672–4680.

Coba, M. P., Muñoz, M. C., Dominici, F. P., Toblli, J. E., Peña, C., Bartke, A., and Turyn, D. (2004). Increased in vivo phosphorylation of insulin receptor at serine 994 in the liver of obese insulin-resistant Zucker rats. *J. Endocrinol.* **182,** 433–444.

De-Meyts, P. (2004). Insulin and its receptor: Structure, function and evolution. *Bioessays* **26,** 1351–1362.

De Meyts, P., and Shymko, R. M. (2000). Timing-dependent modulation of insulin mitogenic versus metabolic signalling. *Novartis Found. Symp.* **227,** 46–57.

De Meyts, P., and Whittaker, J. (2002). Structural biology of insulin and IGF1 receptors: Implications for drug design. *Nat. Rev. Drug Discov.* **1,** 769–783.

De Meyts, P., Van-Obberghen, E., and Roth, J. (1978). Mapping of the residues responsible for the negative cooperativity of the receptor-binding region of insulin. *Nature* **273,** 504–509.

De Meyts, P., Wallach, B., Christoffersen, C. T., Ursø, B., Gronskov, K., Latus, L. J., Yakushiji, F., Ilondo, M. M., and Shymko, R. M. (1994). The insulin-like growth factor-I receptor. Structure, ligand-binding mechanism and signal transduction. *Horm. Res.* **42,** 152–169.

Debant, A., Clauser, E., Ponzio, G., Filloux, C., Auzan, C., Contreres, J. O., and Rossi, B. (1988). Replacement of insulin receptor tyrosine residues 1162 and 1163 does not alter the mitogenic effect of the hormone. *Proc. Natl. Acad. Sci. USA* **85,** 8032–8036.

Di Guglielmo, G. M., Baass, P. C., Ou, W. J., Posner, B. I., and Bergeron, J. J. (1994). Compartmentalization of SHC, GRB2 and mSOS, and hyperphosphorylation of Raf-1 by EGF but not insulin in liver parenchyma. *EMBO J.* **13,** 4269–4277.

Drejer, K. (1992). The bioactivity of insulin analogues from in vitro receptor binding to *in vivo* glucose uptake. *Diabetes Metab. Rev.* **8,** 259–285.

Ebina, Y., Ellis, L., Jarnagin, K., Edery, M., Graf, L., Clauser, E., Ou, J. H., Masiarz, F., Kan, Y. W., Goldfine, I. D., *et al.* (1985). The human insulin receptor cDNA: The structural basis for hormone-activated transmembrane signalling. *Cell* **40,** 747–758.

Emanuelli, B., Peraldi, P., Filloux, C., Chavey, C., Freidinger, K., Hilton, D. J., Hotamisligil, G. S., and Van-Obberghen, E. (2001). SOCS-3 inhibits insulin signaling and is up-regulated in response to tumor necrosis factor-alpha in the adipose tissue of obese mice. *J. Biol. Chem.* **276,** 47944–47949.

Feener, E. P., Backer, J. M., King, G. L., Wilden, P. A., Sun, X. J., Kahn, C. R., and White, M. F. (1993). Insulin stimulates serine and tyrosine phosphorylation in the juxtamembrane region of the insulin receptor. *J. Biol. Chem.* **268,** 11256–11264.

Foti, M., Moukil, M. A., Dudognon, P., and Carpentier, J. L. (2004). Insulin and IGF-1 receptor trafficking and signalling. *Novartis Found. Symp.* **262,** 125–141.

Frasca, F., Pandini, G., Scalia, P., Sciacca, L., Mineo, R., Costantino, A., Goldfine, I. D., Belfiore, A., and Vigneri, R. (1999). Insulin receptor isoform A, a newly recognized, high-affinity insulin-like growth factor II receptor in fetal and cancer cells. *Mol. Cell Biol.* **19,** 3278–3288.

Gavin, J. R., Roth, J., Neville, D. M., De-Meyts, P., and Buell, D. N. (1974). Insulin-dependent regulation of insulin receptor concentrations: A direct demonstration in cell culture. *Proc. Natl. Acad. Sci. USA* **71,** 84–88.

Grimes, M. L., Zhou, J., Beattie, E. C., Yuen, E. C., Hall, D. E., Valletta, J. S., Topp, K. S., LaVail, J. H., Bunnett, N. W., and Mobley, W. C. (1996). Endocytosis of activated TrkA: Evidence that nerve growth factor induces formation of signaling endosomes. *J. Neurosci.* **16,** 7950–7964.

Gu, J. L., Goldfine, I. D., Forsayeth, J. R., and De-Meyts, P. (1988). Reversal of insulin-induced negative cooperativity by monoclonal antibodies that stabilize the slowly dissociating ("Ksuper") state of the insulin receptor. *Biochem. Biophys. Res. Commun.* **150,** 694–701.

Gustafson, T. A., He, W., Craparo, A., Schaub, C. D., and O'Neill, T. J. (1995). Phosphotyrosine-dependent interaction of SHC and insulin receptor substrate 1 with the NPEY motif of the insulin receptor via a novel non-SH2 domain. *Mol. Cell Biol* **15,** 2500–2508.

Hamer, I., Foti, M., Emkey, R., Cordier-Bussat, M., Philippe, J., De Meyts, P., Maeder, C., Kahn, C. R., and Carpentier, J. L. (2002). An arginine to cysteine(252) mutation in insulin receptors from a patient with severe insulin resistance inhibits receptor internalisation but preserves signalling events. *Diabetologia* **45,** 657–667.

Hansen, B. F., Danielsen, G. M., Drejer, K., Sørensen, A. R., Wiberg, F. C., Klein, H. H., and Lundemose, A. G. (1996). Sustained signalling from the insulin receptor after stimulation with insulin analogues exhibiting increased mitogenic potency. *Biochem. J.* **315**(Pt 1), 271–279.

Hao, C. L., Whittaker, L., and Whittaker, J. (2006). Characterization of a second ligand binding site of the insulin receptor. *Biochem. Biophys. Res. Commun.* **347,** 334–339.

He, W., Craparo, A., Zhu, Y., O'Neill, T. J., Wang, L. M., Pierce, J. H., and Gustafson, T. A. (1996). Interaction of insulin receptor substrate-2 (IRS-2) with the insulin and insulin-like growth factor I receptors. Evidence for two distinct phosphotyrosine-dependent interaction domains within IRS-2. *J. Biol. Chem.* **271,** 11641–11645.

Holt, L. J., and Siddle, K. (2005). Grb10 and Grb14: Enigmatic regulators of insulin action - And more? *Biochem. J.* **388,** 393–406.

Huang, K., Xu, B., Hu, S. Q., Chu, Y. C., Hua, Q. X., Qu, Y., Li, B., Wang, S., Wang, R. Y., Nakagawa, S. H., Theede, A. M., Whittaker, J., *et al.* (2004). How insulin binds: The B-chain alpha-helix contacts the L1 beta-helix of the insulin receptor. *J. Mol. Biol.* **341,** 529–550.

Huang, K., Chan, S. J., Hua, Q. X., Chu, Y. C., Wang, R. Y., Klaproth, B., Jia, W. H., Whittaker, J., De-Meyts, P., Nakagawa, S. H., Steiner, D. F., Katsoyannis, P. G., *et al.* (2007). The A-chain of insulin contacts the insert domain of the insulin receptor: Photo-cross-linking and mutagenesis of a diabetes-related crevice. *J. Biol. Chem.* **282,** 35337–35349.

Hubbard, S. R. (1997). Crystal structure of the activated insulin receptor tyrosine kinase in complex with peptide substrate and ATP analog. *EMBO J.* **16,** 5572–5581.

Hubbard, S. R., and Miller, W. T. (2007). Receptor tyrosine kinases: Mechanisms of activation and signaling. *Curr. Opin. Cell Biol.* **19,** 117–123.

Hubbard, S. R., Wei, L., Ellis, L., and Hendrickson, W. A. (1994). Crystal structure of the tyrosine kinase domain of the human insulin receptor. *Nature* **372,** 746–754.

Ish-Shalom, D., Christoffersen, C. T., Vorwerk, P., Sacerdoti-Sierra, N., Shymko, R. M., Naor, D., and De Meyts, P. (1997). Mitogenic properties of insulin and insulin analogues mediated by the insulin receptor. *Diabetologia* **2**(Suppl 40), S25–S31.

Ishikawa, Y., Otsu, K., and Oshikawa, J. (2005). Caveolin; Different roles for insulin signal? *Cell Signal* **17,** 1175–1182.

Jensen, M., Hansen, B., De Meyts, P., Schäffer, L., and Ursø, B. (2007). Activation of the insulin receptor by insulin and a synthetic peptide leads to divergent metabolic and mitogenic signaling and responses. *J. Biol. Chem.* **282,** 35179–35186.

Jensen, M., Palsgaard, J., Borup, R., De Meyts, P., and Schäffer, L. (2008). Activation of the insulin receptor (IR) and a synthetic peptide has different effects on gene expression in IR-transfected L6 myoblasts. *Biochem J*.

Kaarsholm, N. C., and Ludvigsen, S. (1995). The high resolution solution structure of the insulin monomer determined by NMR. *Receptor* **5,** 1–8.

Kaburagi, Y., Momomura, K., Yamamoto-Honda, R., Tobe, K., Tamori, Y., Sakura, H., Akanuma, Y., Yazaki, Y., and Kadowaki, T. (1993). Site-directed mutagenesis of the juxtamembrane domain of the human insulin receptor. J. Biol. Chem. **268,** 16610–16622.

Kaburagi, Y., Yamamoto-Honda, R., Tobe, K., Ueki, K., Yachi, M., Akanuma, Y., Stephens, R. M., Kaplan, D., Yazaki, Y., and Kadowaki, T. (1995). The role of the NPXY motif in the insulin receptor in tyrosine phosphorylation of insulin receptor substrate-1 and Shc. *Endocrinology* **136,** 3437–3443.

Kiselyov, V. V., Versteyhe, S., Gauguin, L., and De Meyts, P. (2009). Harmonic oscillator model of the insulin and IGF-I receptors allosteric binding and activation. *Mol. Sys. Bio.*, in press.

Koshio, O., Akanuma, Y., and Kasuga, M. (1989). Identification of a phosphorylation site of the rat insulin receptor catalyzed by protein kinase C in an intact cell. *FEBS Lett.* **254,** 22–24.

Kristensen, C., Kjeldsen, T., Wiberg, F. C., Schäffer, L., Hach, M., Havelund, S., Bass, J., Steiner, D. F., and Andersen, A. S. (1997). Alanine scanning mutagenesis of insulin. *J. Biol. Chem.* **272,** 12978–12983.

Kristensen, C., Wiberg, F. C., Schäffer, L., and Andersen, A. S. (1998). Expression and characterization of a 70-kDa fragment of the insulin receptor that binds insulin. Minimizing ligand binding domain of the insulin receptor. *J. Biol. Chem.* **273,** 17780–17786.

Kristensen, C., Andersen, A. S., Østergaard, S., Hansen, P. H., and Brandt, J. (2002). Functional reconstitution of insulin receptor binding site from non-binding receptor fragments. *J. Biol. Chem.* **277,** 18340–18345.

Kurose, T., Pashmforoush, M., Yoshimasa, Y., Carroll, R., Schwartz, G. P., Burke, G. T., Katsoyannis, P. G., and Steiner, D. F. (1994). Cross-linking of a B25 azidophenylalanine insulin derivative to the carboxyl-terminal region of the α-subunit of the insulin receptor. Identification of a new insulin-binding domain in the insulin receptor. *J. Biol. Chem.* **269,** 29190–29197.

Kurtzhals, P., Schäffer, L., Sørensen, A., Kristensen, C., Jonassen, I., Schmid, C., and Trub, T. (2000). Correlations of receptor binding and metabolic and mitogenic potencies of insulin analogs designed for clinical use. *Diabetes* **49,** 999–991005.

Langlais, P., Dong, L. Q., Ramos, F. J., Hu, D., Li, Y., Quon, M. J., and Liu, F. (2004). Negative regulation of insulin-stimulated mitogen-activated protein kinase signaling by Grb10. *Mol. Endocrinol.* **18,** 350–358.

Lau, K. S., Partridge, E. A., Grigorian, A., Silvescu, C. I., and Reinhold, V. N. (2007). Complex N-glycan Number and degree of Branching Cooperate to Regulate Cell Proliferation and Differentiation. *Cell* **129,** 123–134.

Lawrence, M. C., McKern, N. M., and Ward, C. W. (2007). Insulin receptor structure and its implications for the IGF-1 receptor. *Curr. Opin. Struct. Biol.* **17,** 699–705.

Liu, F., and Roth, R. A. (1994). Identification of serines-1035/1037 in the kinase domain of the insulin receptor as protein kinase C alpha mediated phosphorylation sites. *FEBS Lett.* **352,** 389–392.

Maddux, B. A., and Goldfine, I. D. (2000). Membrane glycoprotein PC-1 inhibition of insulin receptor function occurs via direct interaction with the receptor alpha-subunit. *Diabetes* **49,** 13–19.

Maegawa, H., McClain, D. A., Freidenberg, G., Olefsky, J. M., Napier, M., Lipari, T., Dull, T. J., Lee, J., and Ullrich, A. (1988). Properties of a human insulin receptor with a COOH-terminal truncation. II. Truncated receptors have normal kinase activity but are defective in signaling metabolic effects. *J. Biol. Chem.* **263,** 8912–8917.

McKern, N. M., Lawrence, M. C., Streltsov, V. A., Lou, M. Z., Adams, T. E., Lovrecz, G. O., Elleman, T. C., Richards, K. M., Bentley, J. D., Pilling, P. A., Hoyne, P. A., Cartledge, K. A., *et al.* (2006). Structure of the insulin receptor ectodomain reveals a folded-over conformation. *Nature* **443,** 218–221.

Mooney, R. A., Senn, J., Cameron, S., Inamdar, N., Boivin, L. M., Shang, Y., and Furlanetto, R. W. (2001). Suppressors of cytokine signaling-1 and -6 associate with and inhibit the insulin receptor. A potential mechanism for cytokine-mediated insulin resistance. *J. Biol. Chem.* **276,** 25889–25893.

Mori, K., Giovannone, B., and Smith, R. J. (2005). Distinct Grb10 domain requirements for effects on glucose uptake and insulin signaling. *Mol. Cell. Endocrinol.* **230,** 39–50.

Morrione, A., Valentinis, B., Xu, S. Q., Yumet, G., Louvi, A., Efstratiadis, A., and Baserga, R. (1997). Insulin-like growth factor II stimulates cell proliferation through the insulin receptor. *Proc. Natl. Acad. Sci. USA* **94,** 3777–3782.

Mounier, C., Lavoie, L., Dumas, V., Mohammad-Ali, K., Wu, J., Nantel, A., Bergeron, J. J., Thomas, D. Y., and Posner, B. I. (2001). Specific inhibition by hGRB10zeta of insulin-induced glycogen synthase activation: Evidence for a novel signaling pathway. *Mol. Cell. Endocrinol.* **173,** 15–27.

Myers, M. G., Backer, J. M., Siddle, K., and White, M. F. (1991). The insulin receptor functions normally in Chinese hamster ovary cells after truncation of the C terminus. *J. Biol. Chem.* **266,** 10616–10623.

Mynarcik, D. C., Yu, G. Q., and Whittaker, J. (1996). Alanine-scanning mutagenesis of a C-terminal ligand binding domain of the insulin receptor alpha subunit. *J. Biol. Chem.* **271,** 2439–2442.

Mynarcik, D. C., Williams, P. F., Schaffer, L., Yu, G. Q., and Whittaker, J. (1997). Analog binding properties of insulin receptor mutants. Identification of amino acids interacting with the COOH terminus of the B-chain of the insulin molecule. *J. Biol. Chem.* **272,** 2077–2081.

O'Brien, R. M., Soos, M. A., and Siddle, K. (1986). Monoclonal antibodies for the human insulin receptor stimulate intrinsic receptor-kinase activity. *Biochem. Soc. Trans.* **14,** 1021–1023.

Pandini, G., Medico, E., Conte, E., Sciacca, L., Vigneri, R., and Belfiore, A. (2003). Differential gene expression induced by insulin and insulin-like growth factor-II through the insulin receptor isoform A. *J. Biol. Chem.* **278,** 42178–42189.

Parpal, S., Karlsson, M., Thorn, H., and Strålfors, P. (2001). Cholesterol depletion disrupts caveolae and insulin receptor signaling for metabolic control via insulin receptor substrate-1, but not for mitogen-activated protein kinase control. *J. Biol. Chem.* **276,** 9670–9678.

Pierce, K. L., Maudsley, S., Daaka, Y., Luttrell, L. M., and Lefkowitz, R. J. (2000). Role of endocytosis in the activation of the extracellular signal-regulated kinase cascade by sequestering and nonsequestering G protein-coupled receptors. *Proc. Natl. Acad. Sci. USA* **97,** 1489–1494.

Pillutla, R. C., Hsiao, K. C., Beasley, J. R., Brandt, J., Østergaard, S., Hansen, P. H., Spetzler, J. C., Danielsen, G. M., Andersen, A. S., Brissette, R. E., Lennick, M., Fletcher, P. W., *et al.* (2002). Peptides identify the critical hotspots involved in the biological activation of the insulin receptor. *J. Biol. Chem.* **277,** 22590–22594.

Pol, A., Calvo, M., and Enrich, C. (1998). Isolated endosomes from quiescent rat liver contain the signal transduction machinery. Differential distribution of activated Raf-1 and Mek in the endocytic compartment. *FEBS Lett.* **441,** 34–38.

Pullen, R. A., Lindsay, D. G., Wood, S. P., Tickle, I. J., Blundell, T. L., Wollmer, A., Krail, G., Brandenburg, D., Zahn, H., Gliemann, J., and Gammeltoft, S. (1976). Receptor-binding region of insulin. *Nature* **259,** 369–373.

Rakatzi, I., Ramrath, S., Ledwig, D., Dransfeld, O., Bartels, T., Seipke, G., and Eckel, J. (2003). A novel insulin analog with unique properties: LysB3,GluB29 insulin induces prominent activation of insulin receptor substrate 2, but marginal phosphorylation of insulin receptor substrate 1. *Diabetes* **52,** 2227–2238.

Reddy, C. C., Niyogi, S. K., Wells, A., Wiley, H. S., and Lauffenburger, D. A. (1996). Engineering epidermal growth factor for enhanced mitogenic potency. *Nat. Biotechnol.* **14,** 1696–1699.

Rizzo, M. A., Shome, K., Watkins, S. C., and Romero, G. (2000). The recruitment of Raf-1 to membranes is mediated by direct interaction with phosphatidic acid and is independent of association with Ras. *J. Biol. Chem.* **275,** 23911–23918.

Saltiel, A. R., and Pessin, J. E. (2003). Insulin signaling in microdomains of the plasma membrane. *Traffic* **4,** 711–716.

Sawka-Verhelle, D., Baron, V., Mothe, I., Filloux, C., White, M. F., and Van-Obberghen, E. (1997a). Tyr624 and Tyr628 in insulin receptor substrate-2 mediate its association with the insulin receptor. *J. Biol. Chem.* **272,** 16414–16420.

Sawka-Verhelle, D., Filloux, C., Tartare-Deckert, S., Mothe, I., and Van-Obberghen, E. (1997b). Identification of Stat 5B as a substrate of the insulin receptor. *Eur. J. Biochem.* **250,** 411–417.

Sawka-Verhelle, D., Tartare-Deckert, S., White, M. F., and Van-Obberghen, E. (1996). Insulin receptor substrate-2 binds to the insulin receptor through its phosphotyrosine-binding domain and through a newly identified domain comprising amino acids 591–786. *J. Biol. Chem.* **271,** 5980–5983.

Schäffer, L. (1994). A model for insulin binding to the insulin receptor. *Eur. J. Biochem.* **221,** 1127–1132.

Schäffer, L., Brissette, R. E., Spetzler, J. C., Pillutla, R. C., Østergaard, S., Lennick, M., Brandt, J., Fletcher, P. W., Danielsen, G. M., Hsiao, K. C., Andersen, A. S., Dedova, O., *et al.* (2003). Assembly of high-affinity insulin receptor agonists and antagonists from peptide building blocks. *Proc. Natl. Acad. Sci. USA* **100,** 4435–4439.

Schmitz-Peiffer, C. (2002). Protein kinase C and lipid-induced insulin resistance in skeletal muscle. *Ann. N. Y. Acad. Sci.* **967,** 146–157.

Sciacca, L., Mineo, R., Pandini, G., Murabito, A., Vigneri, R., and Belfiore, A. (2002). In IGF-I receptor-deficient leiomyosarcoma cells autocrine IGF-II induces cell invasion and protection from apoptosis via the insulin receptor isoform A. *Oncogene* **21,** 8240–8250.
Seino, S., and Bell, G. I. (1989). Alternative splicing of human insulin receptor messenger RNA. *Biochem. Biophys. Res. Commun.* **159,** 312–316.
Seino, S., Seino, M., Nishi, S., and Bell, G. I. (1989). Structure of the human insulin receptor gene and characterization of its promoter. *Proc. Natl. Acad. Sci. USA* **86,** 114–118.
Siddiqui, N. I. (2007). Insulin analogues: New dimension of management of diabetes mellitus. *Mymensingh Med. J.* **16,** 117–121.
Soos, M., Taylor, R., and Siddle, K. (1986a). Insulin-inhibitory and insulin-like efffects of monoclonal antibodies for the human insulin receptor. *Biochem. Soc. Trans.* **14,** 317–318.
Soos, M. A., Siddle, K., Baron, M. D., Heward, J. M., Luzio, J. P., Bellatin, J., and Lennox, E. S. (1986b). Monoclonal antibodies reacting with multiple epitopes on the human insulin receptor. *Biochem. J.* **235,** 199–208.
Sorkin, A., Eriksson, A., Heldin, C. H., Westermark, B., and Claesson-Welsh, L. (1993). Pool of ligand-bound platelet-derived growth factor beta-receptors remain activated and tyrosine phosphorylated after internalization. *J. Cell Physiol.* **156,** 373–382.
Sorkin, A., McClure, M., Huang, F., and Carter, R. (2000). Interaction of EGF receptor and grb2 in living cells visualized by fluorescence resonance energy transfer (FRET) microscopy. *Curr. Biol.* **10,** 1395–1398.
Sparrow, L. G., Gorman, J. J., Strike, P. M., Robinson, C. P., McKern, N. M., Epa, V. C., and Ward, C. W. (2007). The location and characterisation of the O-linked glycans of the human insulin receptor. *Proteins Struct. Funct. Genet.* **66,** 261–265.
Sparrow, L. G., Lawrence, M. C., Gorman, J. J., Strike, P. M., Robinson, C. P., McKern, N. M., and Ward, C. W. (2008). N-linked glycans of the human insulin receptor and their distribution over the crystal structure. *Proteins* **71,** 426–439.
Stein, E. G., Gustafson, T. A., and Hubbard, S. R. (2001). The BPS domain of Grb10 inhibits the catalytic activity of the insulin and IGF1 receptors. *FEBS Lett.* **493,** 106–111.
Strack, V., Hennige, A. M., Krützfeldt, J., Bossenmaier, B., Klein, H. H., Kellerer, M., Lammers, R., and Häring, H. U. (2000). Serine residues 994 and 1023/25 are important for insulin receptor kinase inhibition by protein kinase C isoforms beta2 and theta. *Diabetologia* **43,** 443–449.
Sun, X. J., Rothenberg, P., Kahn, C. R., Backer, J. M., Araki, E., Wilden, P. A., Cahill, D. A., Goldstein, B. J., and White, M. F. (1991). Structure of the insulin receptor substrate IRS-1 defines a unique signal transduction protein. *Nature* **352,** 73–77.
Sung, C. K. (1992). Insulin receptor signaling through non-tyrosine kinase pathways: Evidence from anti-receptor antibodies and insulin receptor mutants. *J. Cell Biochem.* **48,** 26–32.
Taniguchi, C. M., Emanuelli, B., and Kahn, C. R. (2006). Critical nodes in signalling pathways: Insights into insulin action. *Nat. Rev. Mol. Cell Biol.* **7,** 85–96.
Taylor, R., Soos, M. A., Wells, A., Argyraki, M., and Siddle, K. (1987). Insulin-like and insulin-inhibitory effects of monoclonal antibodies for different epitopes on the human insulin receptor. *Biochem. J.* **242,** 123–129.
Ueki, K., Kondo, T., and Kahn, C. R. (2004). Suppressor of cytokine signaling 1 (SOCS-1) and SOCS-3 cause insulin resistance through inhibition of tyrosine phosphorylation of insulin receptor substrate proteins by discrete mechanisms. *Mol. Cell Biol.* **24,** 5434–5446.
Uhles, S., Moede, T., Leibiger, B., Berggren, P. O., and Leibiger, I. B. (2003). Isoform-specific insulin receptor signaling involves different plasma membrane domains. *J. Cell Biol.* **163,** 1327–1337.

Uhles, S., Moede, T., Leibiger, B., Berggren, P. O., and Leibiger, I. B. (2007). Selective gene activation by spatial segregation of insulin receptor B signaling. *FASEB J.* **21,** 1609–1621.

Ullrich, A., Bell, J. R., Chen, E. Y., Herrera, R., Petruzzelli, L. M., Dull, T. J., Gray, A., Coussens, L., Liao, Y. C., Tsubokawa, M., *et al.* (1985). Human insulin receptor and its relationship to the tyrosine kinase family of oncogenes. *Nature* **313,** 756–761.

Vieira, A. V., Lamaze, C., and Schmid, S. L. (1996). Control of EGF receptor signaling by clathrin-mediated endocytosis. *Science (New York, N. Y.)* **274,** 2086–2089.

Wang, C. C., Goldfine, I. D., Fujita-Yamaguchi, Y., Gattner, H. G., Brandenburg, D., and De-Meyts, P. (1988). Negative and positive site-site interactions, and their modulation by pH, insulin analogs, and monoclonal antibodies, are preserved in the purified insulin receptor. *Proc. Natl. Acad. Sci. USA* **85,** 8400–8404.

Ward, C., Lawrence, M., Streltsov, V., Garrett, T., McKern, N., Lou, M. Z., Lovrecz, G., and Adams, T. (2008). Structural insights into ligand-induced activation of the insulin receptor. *Acta Physiol. (Oxf)* **192,** 3–9.

Ward, C. W., Lawrence, M. C., Streltsov, V. A., Adams, T. E., and McKern, N. M. (2007). The insulin and EGF receptor structures: New insights into ligand-induced receptor activation. *Trends Biochem. Sci.* **32,** 129–137.

White, M. F., Livingston, J. N., Backer, J. M., Lauris, V., Dull, T. J., Ullrich, A., and Kahn, C. R. (1988a). Mutation of the insulin receptor at tyrosine 960 inhibits signal transmission but does not affect its tyrosine kinase activity. *Cell* **54,** 641–649.

White, M. F., Shoelson, S. E., Keutmann, H., and Kahn, C. R. (1988b). A cascade of tyrosine autophosphorylation in the beta-subunit activates the phosphotransferase of the insulin receptor. *J. Biol. Chem.* **263,** 2969–2980.

Whittaker, L., Hao, C., Fu, W., and Whittaker, J., (2008). High-affinity insulin binding: Insulin interacts with two receptor ligand binding sites. *Biochemistry* **47,** 12900–12909.

Wiegmann, K., Schütze, S., Machleidt, T., Witte, D., and Krönke, M. (1994). Functional dichotomy of neutral and acidic sphingomyelinases in tumor necrosis factor signaling. *Cell* **78,** 1005–1015.

Wilden, P. A., Backer, J. M., Kahn, C. R., Cahill, D. A., Schroeder, G. J., and White, M. F. (1990). The insulin receptor with phenylalanine replacing tyrosine-1146 provides evidence for separate signals regulating cellular metabolism and growth. *Proc. Natl. Acad. Sci. USA* **87,** 3358–3362.

Wilden, P. A., Siddle, K., Haring, E., Backer, J. M., White, M. F., and Kahn, C. R. (1992). The role of insulin receptor kinase domain autophosphorylation in receptor-mediated activities Analysis with insulin and anti-receptor antibodies. *J. Biol. Chem.* **267,** 13719–13727.

Williams, P. F., Mynarcik, D. C., Yu, G. Q., and Whittaker, J. (1995). Mapping of an NH2-terminal ligand binding site of the insulin receptor by alanine scanning mutagenesis. *J. Biol. Chem.* **270,** 3012–3016.

Wu, J., Tseng, Y. D., Xu, C., Neubert, T. A., White, M. F., and Hubbard, S. R. (2008). Structural and biochemical characterization of the KRLB region in insulin receptor substrate-2. *Nat. Struct. Mol. Biol.* **15,** 251–258.

Youngren, J. F. (2007). Regulation of insulin receptor function. *Cell. Mol. Life Sci.* **64,** 873–891.

CHAPTER FOUR

c-Abl and Insulin Receptor Signalling

Marco Genua, Giuseppe Pandini, Maria Francesca Cassarino, Rosa Linda Messina, *and* Francesco Frasca

Contents

Department of Internal Medicine, Endocrinology Unit, University of Catania, P.O. Garibaldi Nesima, Via Palermo 636, 95122 Catania, Italy

Vitamins and Hormones, Volume 80
ISSN 0083-6729, DOI: 10.1016/S0083-6729(08)00604-3

Abstract

Insulin Receptor (IR) and IGF-I receptor (IGF-IR) are homolog but display distinct functions: IR is mainly metabolic, while IGF-IR is mitogenic. However, in some conditions like foetal growth, cancer and diabetes, IR may display some non-metabolic effects like proliferation and migration. The molecular mechanisms underlying this 'functional switch of IR' have been attributed to several factors including overexpression of ligands and receptors, predominant IR isoform expression, preferential recruitment of intracellular substrates. Here, we report that c-Abl, a cytoplasmic tyrosine kinase regulating several signal transduction pathways, is involved in this functional switch of IR. Indeed, c-Abl tyrosine kinase is involved in IR signalling as it shares with IR some substrates like Tub and SORBS1 and is activated upon insulin stimulation. Inhibition of c-Abl tyrosine kinase by STI571 attenuates the effect of insulin on Akt/GSK-3β phosphorylation and glycogen synthesis, and at the same time, it enhances the effect of insulin on ERK activation, cell proliferation and migration. This effect of STI571 is specific to c-Abl inhibition, because it does not occur in Abl-null cells and is restored in c-Abl-reconstituted cells. Numerous evidences suggest that focal adhesion kinase (FAK) is involved in mediating this c-Abl effect. First, c-Abl tyrosine kinase activation is concomitant with FAK dephosphorylation in response to insulin, whereas c-Abl inhibition is accompanied by FAK phosphorylation in response to insulin, a response similar to that observed with IGF-I. Second, the c-Abl effects on insulin signalling are not observed in cells devoid of FAK (FAK$^{-/-}$ cells). Taken together these results suggest that c-Abl activation by insulin, via a modification of FAK response, may play an important role in directing mitogenic *versus* metabolic insulin receptor signalling.

I. Introduction

The IGF system is composed by Insulin receptor (IR), IGF-I receptor (IGF-IR) and their ligands (insulin for IR and IGF-I/IGF-II for IGF-IR), which form a complex interplay regulating several important cellular processes (Fig. 4.1). In recent years, it has become evident that the system plays an important role in physiology, like development and metabolic homeostasis, and in disease like diabetes and cancer development (Frasca *et al.*, 2008). Despite the high degree of homology between IR and IGF-IR, the two receptors maintain distinct and separate functions: IR is mainly involved in glucose homeostasis, while IGF-IR in cell proliferation. However, in particular condition like cancer and diabetes, the IR may acquire the ability to stimulate cell proliferation. This functional switch of IR is dependent on several mechanisms, which include receptor/ligand overexpression, predominat IR isoforms expression, formation of IR/IGF-IR hybrid receptors

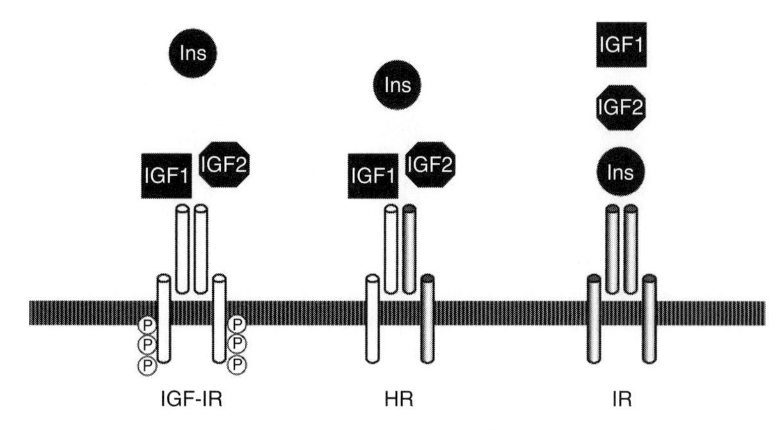

Figure 4.1 Schematic representation of the members of the IGF system. Insulin receptor (IR) binds insulin with high affinity, while IGF-I receptor (IGF-IR) binds IGF-I and IGF-II. In cells and tissues expressing both IR and IGF-IR, IR hemireceptors may heterodimerize with IGF-IR hemireceptors, leading to the formation of hybrid IR/IGF-IRs (HRs), which bind IGF-I and IGF-II with high affinity and insulin with a much lower affinity.

(HRs), different recruitment and activation kinetics of intracellular substrates (Frasca *et al.*, 2008) (Fig. 4.4). Since deregulation of the IGF system is a common event in human disease several investigators have focused their efforts to understand the key steps in IGF system regulation to develop strategies aimed at modulating its specific activities. The focus of this review is the emerging role of c-Abl tyrosine kinase in regulating the metabolic and mitogenic potential of IR. The results described here underline the role of c-Abl as a molecular switch for the IR functions: activation of c-Abl restrains the mitogenic activity and enhances the metabolic effects of IR. This effect of c-Abl is mainly dependent on the inhibition of the activity of focal adhesion kinase (FAK). These results indicate that c-Abl could be considered as molecular targets for novel anti-cancer and anti-diabetes therapies, aimed at down regulating the IR mitogenic effects in cancer and promoting the metabolic ones in diabetes and insulin resistance status.

II. Insulin and IGF-IRs

A. Structure and function

The IR and the IGF-IR are close homologs and belong to the super family of receptor with tyrosine kinase activity (Ebina *et al.*, 1985; Ullrich *et al.*, 1985). Both regulate crucial functions of multi-cellular organism (Drakas *et al.*, 2004). Two distinct genes, which are believed to derive by duplication of a common ancestral one, encode these two receptors.

Homology between IR and IGF-IR ranges from 45–65% in the ligand binding domain to 60–85% in the tyrosine kinase and substrate recruitment domains (Andersen *et al.*, 1995; Mynarcik *et al.*, 1997; Ullrich *et al.*, 1986; Whittaker *et al.*, 2001; Yip *et al.*, 1988). Both the IR and the IGF-IR are tetrameric glycoproteins composed of two extracellular α-and two transmembrane β-subunits linked by disulfide bounds. Each α-subunit, containing the ligand binding sites, is $\sim$130 kDa, whereas each β-subunit, containing the tyrosine kinase domain, is $\sim$90–95 kDa.

In both receptors, the ligand binding sites are predominantly located at a cysteine-rich region, in the extra-cellular α-subunit (Andersen *et al.*, 1995; Whittaker and Whittaker, 2005). Moreover, the C-terminus of the IR α-subunit also plays a role in insulin binding. This region contains 12 amino acidic residues encoded by exon 11 and is present only in the IR isoform B (IR-B) (Moller *et al.*, 1989; Mosthaf *et al.*, 1990), while IR-A is generated by exon 11 exclusion.

The C-terminus of the intracellular β-subunit is also structurally and functionally different between IR and IGF-IR (Faria *et al.*, 1994). Three tyrosines are present in the C-terminus of the IGF-IR, while two tyrosines are present in the corresponding region of the IR.

Deletion experiments have demonstrated that the β-subunit C-terminus is required for the transforming effect of the IGF-IR (Blakesley *et al.*, 1996). This region is, however, dispensable for the anti-apoptotic or the mitogenic effects. Deletion of the C-terminus may actually increase the anti-apoptotic or the mitogenic effect of IGF-IR (Miura *et al.*, 1995; O'Connor *et al.*, 1997). In contrast, the C-terminus of the IR β-subunit lacks the transforming potential while has a role in modulating mitogenesis.

The transmembrane domain has a crucial role in recruiting intracellular mediators (Kato *et al.*, 1994). Two tyrosine residues (Tyr 965 and Tyr 972) are sites of auto phosphorylation in the IR transmembrane region. Both sites are conserved in the IGF-IR (Tyr 938 and Tyr 945). The catalytic domain of the β-subunit is also well conserved in the two receptors. It contains three tyrosines (Tyr 1131, Tyr 1135, Tyr 1136) in the IGF-IR and three tyrosines (Tyr 1158, Tyr 1162, Tyr 1163) in the IR.

B. IR isoforms

The human IR is encoded by a single gene, which is located on chromosome 19 and contains 22 exons. The mature IR exists as two isoforms, IR-A and IR-B, which result from the alternative splicing of the primary transcript (Fig. 4.2). The IR-B differs from IR-A by the inclusion of exon 11, which encodes a 12 amino acid fragment (residues 717–728) of the IR α-subunit. Inclusion of this exon is differently regulated in various tissues and diseases. IR is expressed at high levels in adult muscle, adipose tissue and liver, where it regulates glucose metabolism in response to insulin (Fig. 4.2).

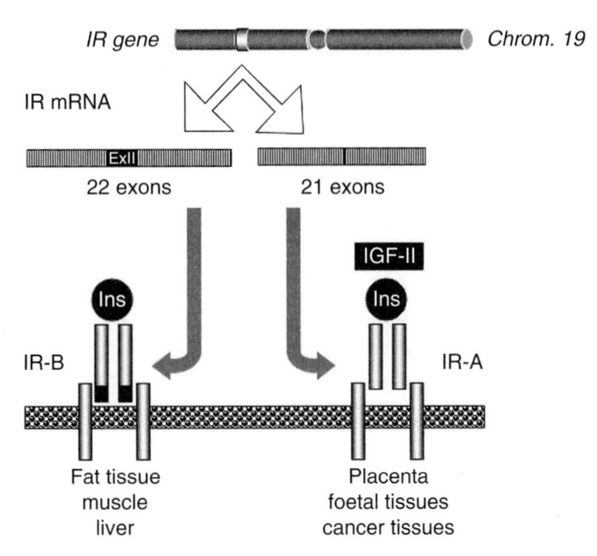

Figure 4.2 Schematic representation of IR splicing, isoform expression and function. The IR gene is located on chromosome 19 and contains 22 exons. Inclusion or exclusion of exon 11 in IR mRNA leads to two different IR isoforms: IR-A (exon 11−) and IR-B (exon 11+). Exon 11 inclusion is regulated by SF2/ASF and MLBNs, while exon 11 exclusion is favoured by hnRNP-A1 and the CELF family of splicing factors. IR-B is mainly expressed in tissue targets of the metabolic effects of insulin. IR-A is expressed in foetal, cancer tissue, CNS and haematopoietic cells. While IR-B binds only insulin, IR-A may bind both insulin and insulin-like growth factor II (IGF-II). In this view, IR-A may mediate some of the IGF-II effects on cell proliferation in cancer and foetal development.

These target tissues of insulin mainly express one of the two IR isoforms, IR-B. The second isoform, IR-A, is present in these tissues at a lower relative abundance (Mosthaf *et al.*, 1990).

IR-A is mainly expressed in foetal tissue and is up-regulated in several diseases including type 2 diabetes, cancer and myotonic dystrophy (Denley *et al.*, 2003). Currently, the mechanisms underlying IR splicing are poorly understood. Experiments with minigenes spanning 10–12 have allowed the identification of the important sequences for the splicing process. Indeed, a 48-nucleotide purine-rich sequence, at the 5′ end of intron 10, causes an increase in exon 11 inclusion (Kosaki *et al.*, 1998), while a 43-nucleotide sequence favouring skipping of exon 11 has been mapped upstream of intron 10. Further minigene analysis indicated that sequences in exon 10, exon 11 and exon 12 are responsible for the splicing process, maybe because they are recognized by specific splicing factors including U1 snRNP, SF1 and U2AF65/35 (Webster *et al.*, 2004).

Overall, these data suggest that IR splicing generates two receptors with different role and functions: while IR-B is the classical receptor for metabolic effects of insulin in muscle, liver and adipose tissues, IR-A is may be a growth promoting and anti-apoptotic receptor under physiological conditions like embryonic development and pathological conditions like cancer.

C. Hybrid receptors (HRs)

Given the close homology of the IR and the IGF-IR, hybrid insulin/IGF-IRs (HRs) exist in cells and tissues co-expressing both receptors. Indeed, heterodimerization between Insulin hemireceptor and IGF-I hemireceptor may occur, leading to the formation of hybrid IR/IGF-IR (Figs. 4.1 and 4.3). Although the biological role of the HRs is not fully clarified, early and functional studies with purified HRs have indicated that they bind IGF-I with an affinity similar to that of typical IGF-IR, while they bind insulin with a much lower affinity (Pandini *et al.*, 1999; Soos *et al.*, 1993) (Fig. 4.1).

In accordance with the high affinity for IGF-I, HRs have been measured as the proportion of total IGF-I binding sites that may be immunoprecipitated with an anti-IR antibody. This method indicated that in normal tissues HRs represent 40–90% of total IGF-I binding sites (Soos *et al.*, 1993). Studies performed in transfected cells have shown that, HR formation occurs by random assembly of hemidimers (Siddle *et al.*, 1994). However, the modulation of the assembly of either homodimeric receptors or HRs by unknown factors cannot be excluded.

Further studies were designed to evaluate HRs with respect to IR-A and IR-B. Experiments performed in transfected cells indicated that both IR-A and IR-B could form hybrids with IGF-IR with the same efficiency (HR-A and HR-B), in close accordance with the random assembly model (Pandini *et al.*, 2007). This observation indicates that the relative abundance of

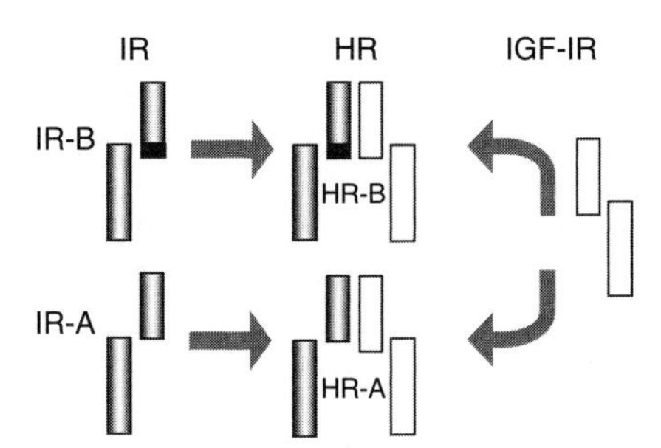

Figure 4.3 Schematic representation of HR-A and HR-B. In cells and tissues expressing either IR isoform A (IR-A) or B (IR-B) and IGF-IR, IR hemireceptors may heterodimerize with IGF-IR hemireceptors, leading to the formation of either hybrids containing IR-A (HR-A) or IR-B (HR-B).

HR-A (containing IR-A) or HR-B (containing IR-B) simply depends on the relative abundance of the two IR isoforms (Fig. 4.3).

III. Metabolic Versus Mitogenic Effect of IR

A. Mitogenic effect of IR: Genetic evidences

Although the predominant effects of IR activation by insulin are metabolic, there is evidence that mitogenic effects may also be elicited.

Genetic evidence obtained in knockout mice has indicated that IR is unnecessary for glucose metabolism in the embryo and that IGF-IR and IR cooperate for an optimal embryonic development (Louvi *et al.*, 1997). Liu and colleague have demonstrate that mice lacking of IR ($IR^{-/-}$ mice) born with near-normal body size but die within few days after birth by diabetic ketoacidosis (Liu *et al.*, 1993). Double mutant mice lacking both IR and IGF-IR are smaller than mice lacking only IGF-IR, thus confirm that IR can also promote foetal growth. The ability of IR to mediate mitogenic signalling, however, remains to be fully elucidated.

B. Possible intracellular mediators involved in the mitogenic effect of IR

The binding of ligand to the α-subunit of IR stimulates the tyrosine kinase activity intrinsic to the β-subunit of the receptor (Ebina *et al.*, 1985a,b; Ullrich *et al.*, 1985, 1986), which, in turn, phosphorylates several immediate substrates including IR substrate proteins (IRS1-4). Upon tyrosine phosphorylation, IRS proteins interact with the p85 regulatory subunit of PI 3-kinase, leading to the activation of the protein kinase Akt to the plasma membrane (Harrington *et al.*, 2005). Akt activation is essential for metabolic effects of insulin, such as glucose uptake, glycogen synthesis, fatty acid synthesis and protein synthesis (Fig. 4.4).

The Akt pathway is also crucial for the gene transcription and apoptosis protection. In addition to p85, phosphorylated IRS-1 and IRS-2 are able to recruit the Grb2/Sos complex that is able to activate MAP kinase (ERK), a key enzyme in cell cycle entry and progression (Ceresa and Pessin, 1998). Moreover, SHC phosphorylation in response to insulin is able to activate the same pathway and is required for sustained MAP kinase activation and a normal mitogenic response to insulin and insulin-like growth factors (Sasaoka and Kobayashi, 2000). It is reasonable to suppose, therefore, that the preferential activation of a specific signalling pathway may be responsible for the prevalence of either metabolic or mitogenic effect of insulin (Fig. 4.4).

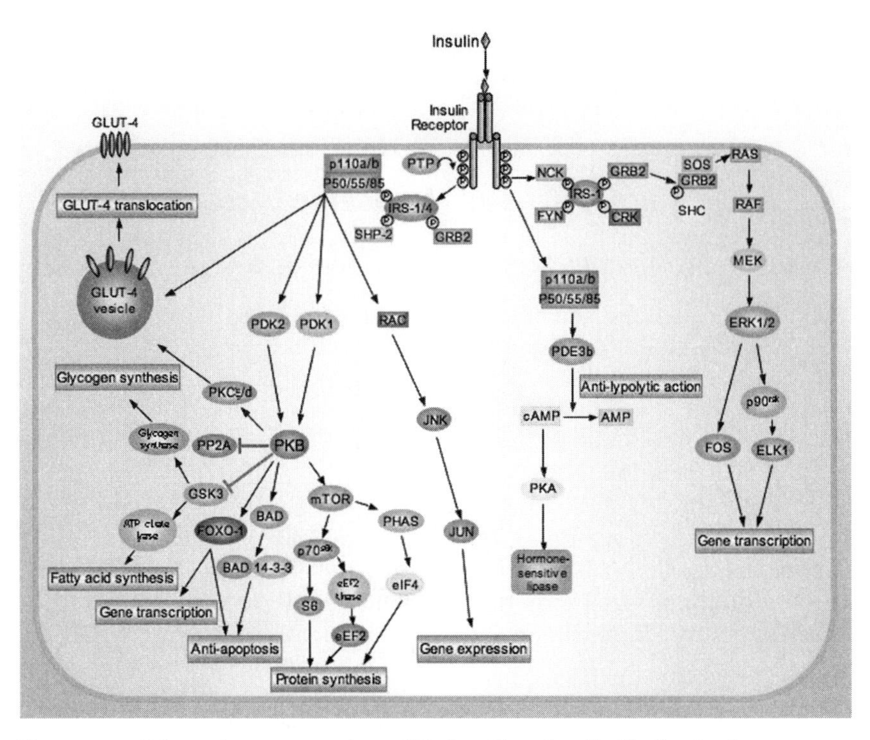

Figure 4.4 Schematic representation of IR signalling. Insulin binding leads to receptor autophosphorylation and phosphorylation of several intracellular substrates including IRS1/4. Phosphorylated IRS-1 recruits Grb2/Sos complex, which triggers the RAS/RAF/MEK/ERK pathway (on the right). This pathway is mainly involved in mediating the mitogenic effect of insulin and insulin like growth factors (IGF-I and IGF-II). Recruitment of p85 on IRS-1 and IRS-2 leads to PI-3 kinase activation, and, as a consequence, Akt pathway activation and Glut4 translocation (on the left). This pathway is mainly involved in mediating the metabolic effects of insulin, including glucose uptake, glycogen and protein synthesis. Moreover, Akt pathway activation is responsible for the anti-apoptotic effect of insulin, IGF-I and IGF-II.

C. IR isoform expression pattern affects the mitogenic/metabolic effect of IR

As mentioned earlier, the human IR exists in two isoforms, IR-A and IR-B, which show some functional differences in respect to insulin binding and signalling. IR-A has a twofold higher affinity than the IR-B, while the IR-B has a more efficient kinase activity (Kellerer *et al.*, 1992; Kosaki *et al.*, 1995; Mosthaf *et al.*, 1990; Vogt *et al.*, 1991).

It is reasonable to hypothesize that the different expression of IR isoform is responsible for the different pathway activation. More recent evidence indicates that IR-A, but not IR-B is activated by IGF-II at high affinity (Frasca *et al.*, 1999). This finding was observed by transfecting either IR-A or IR-B human cDNAs in R-mouse fibroblasts, which are IGF-IR gene deficient by homologous recombination. In the absence of interference of IGF-IR, the binding affinity of IGF-II for IR-A was very high (ED50 = 3.0 + 0.4 nM) similar to that observed with the classical IGF-IR (1.6 + 0.3 nM), (Frasca *et al.*, 1999). Effects of IGF-II and insulin, studied in R-cells transfected with the IR-A, indicated that IGF-II is more effective than insulin in stimulating cell proliferation, while insulin is more potent than IGF-II in stimulating glucose uptake (Frasca *et al.*, 1999). Similar data were also found by other authors (Morrione *et al.*, 1997). When signalling pathways were analyzed in R-/IR-A cells, quantitative and temporal differences in the phosphorylation of intracellular substrates were observed in response to insulin or IGF-II. In particular, both the IRS/PI3K and the Shc/ERK pathways were less intensely and more transiently activated after IGF-II than after insulin stimulation (Frasca *et al.*, 1999).

In SKUT-1 leiomyosarcoma cells, which lack functional IGF-IR, results were even clearer: insulin was more potent that IGF-II in stimulating the PI3K/Akt pathway and in inhibiting cell apoptosis, while IGF-II was more potent than insulin in activating the Shc/ERK pathway and in stimulating cell chemo-invasion (Sciacca *et al.*, 2002). These observations raised the possibility that IGF-II, by acting on IR-A, may be more effective than that can be predicted by its affinity to the receptor.

Post-receptor events differentially activated by insulin and IGF-II via IR-A, were evaluated by micro-array analysis in R-/IR-A cells (Pandini *et al.*, 2004): while 214 transcripts were similarly regulated by insulin and IGF-II, only 45 genes were differentially affected. In more detail, 18 of these differentially regulated genes were exclusively responsive to one of the two ligands (12 to insulin and 6 to IGF-II). These data, showing that IGF-II is more potent than insulin in the induction of certain genes, are in line with previous ones showing that IGF-II was more potent than insulin in stimulating mitogenesis and cell migration. These data provide a molecular basis for understanding the biological role of IR-A in embryonic/foetal growth and the selective biological advantage for malignant cells producing IGF-II and expressing IR-A. The IR-A expression in foetal cells is very important for embryonic development, as it may mediate the growth promoting effect of IGF-II. Analysis of mouse dwarfing phenotypes resulting from targeted mutagenesis of the *IGF1* and *IGF2* genes and the cognate *IR* and *IGFIR* genes (Liu *et al.*, 1993; Louvi *et al.*, 1997) indicate that embryos lacking both IGF-IR and IGF-II are more severely growth-retarded than single IGF-IR knockout mice (Liu *et al.*, 1993; Louvi *et al.*, 1997), suggesting that IGF-II acts also via another receptor, which could be IR-A. Accordingly, the phenotype of double IGF-IR/IGFII knockout mice is similar to that of double IGF-IR/IR knockout mice, thereby suggesting that IR mediates some of the IGF-II effects (Liu *et al.*, 1993; Louvi *et al.*, 1997). Taken together these results suggest that while IR-B is a receptor for the metabolic effects of insulin, IR-A is a receptor involved in mediating the mitogenic effect of IGF-II in embryonic life. Overexpression of IR-A may be responsible for the increased cell proliferation in response to insulin and IGF-II.

D. HRs may contribute to mitogenic effect of insulin

As mentioned earlier, the fine regulation of IR-A/IR-B ratio expression may be rendered more complex by the co-expression of the cognate IGF-IR, which may form HRs with IR. As HR-As are overexpressed and functional in cancer cells, which are often exposed to high levels of autocrine/paracrine IGF-I/II, it is of interest to understand their ligand specificity and signalling capacity. In R-cells transfected with IGF-IR and with either IR-A or IR-B and expressing similar amounts of either HR-A or HR-B, HR-A binds IGF-I, IGF-II and insulin, while HR-B mostly binds IGF-I (Pandini *et al.*, 2002). Remarkably, IGF-I bound HR-A with higher affinity than HR-B. As a consequence, cells expressing HR-A are more sensitive to biological effects of both IGFs and insulin, such as proliferation and migration, as compared to cells transfected with HR-B (Pandini *et al.*, 2002, 2007) (Fig. 4.5).

The intracellular signalling of HRs is not completely elucidated. However, there is evidence suggesting that β-subunit moieties, belonging to

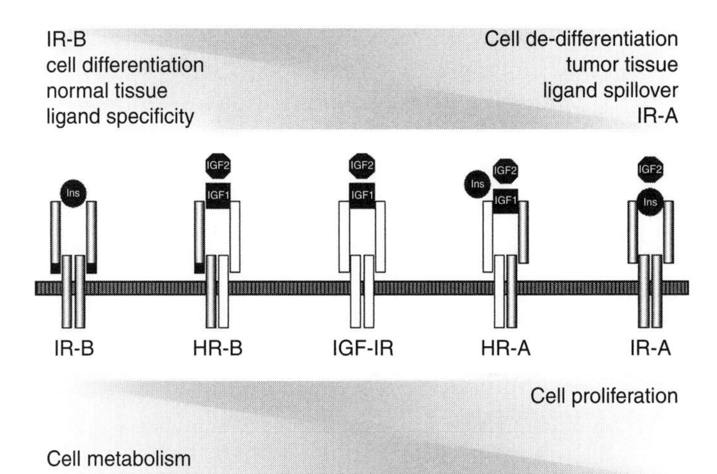

Figure 4.5 Schematic diagram of different ligand and receptor subtypes of the IGF system in physiological and pathological conditions. IR-A and HR-A, which are especially increased in cancer and poorly differentiated tissues, represent receptors with broad ligand specificity and have the final effect of up-regulating the IGF signalling pathway. IR-B and HR-B, which are expressed predominantly in well differentiated tissues, target of the metabolic effects of insulin, restrain ligand specificity and decrease tissue response to mitogenic effect of insulin and insulin-like growth factors.

both IR and IGF-IR, are phosphorylated in HRs. Interestingly, insulin binding to HR-A is able to activate the β-subunit of the IGFIR moiety, and, as a consequence, both IGF-I and insulin are able to induce phosphorylation of Crk-II, a specific substrate of IGF-IR (Pandini *et al.*, 2002). These data indicate, therefore, that insulin binding to HR-A may provide a way for insulin to effectively cross-talk with the signalling capability of IGF-IR as insulin binding affinity for HR-A (ED50 = 4.5 nM insulin) is much higher than its affinity for IGF-IR (ED50 > 1000 nM insulin). These effects may be particularly relevant in foetal and cancer tissues, where IR-A is expressed at high levels and/or in the presence of hyperinsulinemia (obesity, insulin resistance, type 2 diabetes). In contrast, the predominant IR-B expression occurring in normal differentiated cells reduces cell sensitivity to IGFs by causing sequestration of IGF-IR moieties into low affinity HR-B. Therefore IR-A overexpression sensitizes cells to autocrine IGF-II and also to circulating insulin, especially when insulin levels are chronically high. At the same time, it leads to increased formation of HR-A, confers an abnormally high mitogenic:metabolic ratio of various ligands (Fig. 4.5).

Structure, expression and ligand affinity differences of IRs may clarify the ability to induce metabolic or mitogenic effects; moreover IRs have the capacity to recruit different substrates and the understanding in intracellular signalling and into the protein tyrosine kinases (PTKs) involved could better elucidate this ability.

IV. c-Abl Tyrosine Kinase

A. The tyrosine kinase family

PTKs are a large and diverse multigene family. Their principal functions involve the regulation of multicellular aspects of the organism: cell-to-cell signals concerning growth, differentiation, adhesion, motility, and death, are frequently transmitted through tyrosine kinases.

Historically, tyrosine kinases define the prototypical class of human malignancies; moreover, tyrosine kinase genes have also been linked to a wide variety of congenital syndromes (Robinson *et al.*, 2000). Intensive study of this relevant gene family over the past 20 years has produced numerous insights into the structure, regulation and function of these genes and their products. In humans, tyrosine kinases have been demonstrated to play significant roles in the development of many disease states, including diabetes and cancer.

As the sequencing of the human genome is completed by the Human Genome Project, a search for tyrosine kinase coding elements identified 90 unique kinase genes, along with five pseudogenes. Of the 90 tyrosine

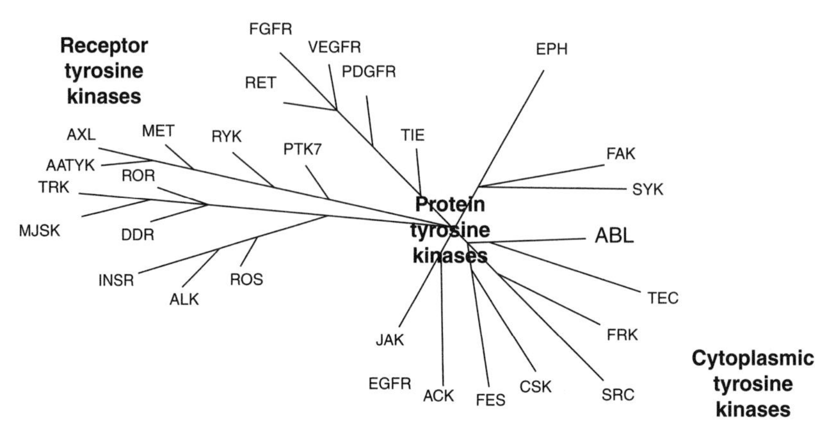

Figure 4.6 Schematic diagram of different human tyrosine kinases. Human tyrosine kinases are classified into cytoplasmic (right) and receptor tyrosine kinases. The branching tree is represented on the basis of sequence homology between the different kinases.

kinases, 58 are receptor type, distribuited into 20 subfamilies. The 32 non-receptor tyrosine kinases can be placed in 10 subfamiles (Robinson *et al.*, 2000) (Fig. 4.6).

The proto-oncoprotein c-Abl is a 140 kDa member of ABL family of non-receptor tyrosine kinases. c-Abl was originally identified as the cellular homolog of the v-Abl oncogene of the Abelson murine leukemia virus.

B. c-Abl structure and localization

c-Abl is the product of ABL1 gene, localized in chromosome 9q34.1. The ABL1 gene encodes two transcript variants that differ into exon1 incorporation. Transcript variant a includes exon 1a, but not exon 1b and give a 1130 amino acidic protein that localized into the nucleus; transcript variant b includes exon 1b, but not exon 1a and contains a different N-terminal. In this protein, longer than a isoform of 19 amino acids, the presence of an N-terminal glycine which could be myristylated indicates the possibility to be directed to plasma membrane (Fig. 4.7A).

The ABL gene family is also represented by ABL2 gene, localized in chromosome 1q24-q25. The gene product is a cytoplasmic tyrosine kinase that is closely related to but distinct from ABL1. ABL2, also know as Abelson-related gene (Arg), encodes three different transcript variants and is expressed in both normal and cancer cells. These three variants give different protein isoforms that presents similarity in tyrosine kinase, SH2 and SH3 domains (Fig. 4.7A).

Some of the functional domains of c-Abl have been characterized (Shaul, 2000). Common features to this family are the myrostoylation site

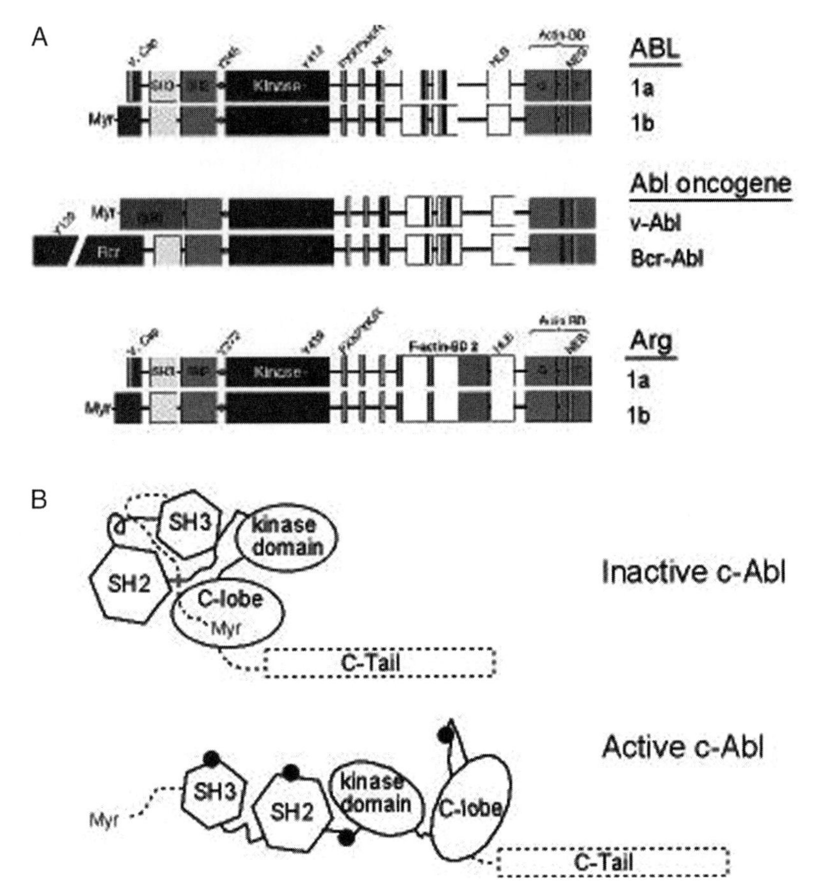

Figure 4.7 c-Abl: structure, function and activation. (A) The extreme N-terminus of the c-Abl protein contains a consensus motif for N-terminal myristoylation (Myr−). The Src homology-3 domain (SH3), Src homology-2 domain (SH2) and the catalytic domain (Kinase) make up the remaining N-terminal half of c-Abl. In the C-terminal half there are four PXXPXK/R sequences, three nuclear localization sequences (NLS), one nuclear export sequence (NES) and three high mobility group-like boxes (HLB). In addition, at the extreme C-terminus there is an actin-binding domain (Actin-BD), which contains a region that mediates binding to monomeric actin (G) and a consensus motif that mediates binding to filamentous actin (F). The tyrosines Y245 and Y412 are regulatory. (B) The crystal structure of the N-terminal region of c-Abl suggests it is folded into an inactive conformation through three intramolecular interactions: (1) Myr-CAP interaction with the C-lobe of the kinase domain, (2) SH3 interaction with the SH2-CAT linker and (3) placement of the activation loop in such a way that hinders substrate entry. The C-terminus and the Cap are depicted with dotted lines, because the position of these regions was not elucidated in the current c-Abl crystal structure. The inactive conformation may involve the C-terminal region. Disruption of the inactive conformation can be achieved by several mechanisms (bottom). For example, the Myr-Cap may be unlatched through membrane binding. SH3 or SH2 ligands may unclamp the kinase domain from the SH3-SH2 regulatory domains.

(found at the N terminus), the tyrosine kinase domain and the Src homology 2 (SH2) and 3 (SH3) domains, both regulating c-Abl activity by mediating discriminate protein–protein interactions (Cohen *et al.*, 1995; Pawson, 1994).

The SH3 domain of c-Abl is ~50 amino-acids in length, and preferentially interacts with proline rich regions containing PxxP motif (Cicchetti *et al.*, 1992; Ren *et al.*, 1993) (Fig. 4.7B). This domain negatively regulates c-Abl and deletion of SH3 turns ABL1 gene into an oncogene; moreover the t(9:22) (Pandini *et al.*, 1999; Whittaker *et al.*, 2001) translocation resulting in the head-to-tail fusion of the BCR and ABL1 genes, produce a constitutively active tyrosine kinase lacking in SH3 domain that is present in many cases of chronic myelogenous leukemia (CML).

The SH2 domain, about 100 amino-acids long, interacts with tyrosine phosphorylated residues (Fig. 4.7B). c-Abl is also characterized by a long C-terminal tail, which is crucial for c-Abl function: it contains three NLSs (nuclear localization signals) and a single NES (nuclear export signal) motif, a putative DNA-binding domain with three high mobility group-like domains and an actin binding domain (Fig. 4.7A). With all of these structural domains, c-Abl is likely to simultaneously participate in many processes by direct protein–protein interactions.

In animal cells, c-Abl is ubiquitous but with different subcellular localization. In fibroblasts, it resides predominantly in the nucleus, while in primary haematopoietic cells and neurons c-Abl is more cytoplasmatic. In sharp contrast, all the transforming Abl variants are exclusively cytoplasmatic. The cellular localization of c-Abl is also controlled by NLSs and NES and confirms its possible involvement in multiple molecular pathways.

C. Mechanisms of c-Abl activation

The Abl SH3 domain participates in a sandwich interaction, binding the SH2-kinase linker, which adopts a polyproline type II helix, between itself and the amino-terminal lobe of the kinase domain (Barila and Superti-Furga, 1998). As for Src-family kinases, the kinase domain of Abl needs to be able to adopt the proper conformation in order for the linker and SH3 domain to 'dock'. Many examples for proteins that contain SH3-domain ligands and are able to activate c-Abl have been studied in detail, including

---

Proteins with SH3 domains that bind to the Abl PXXP motifs may also activate Abl. Phosphorylation of Y245 in the SH2-CAT linker and Y412 in the activation loop may stabilize the active conformation. PDGF-dependent stimulation of c-Abl requires Src and PLC−. Src can phosphorylate Y412 whereas PLC− is proposed to reduce the level of PtdIns(4,5)P_2 to activate c-Abl. Once activated, c-Abl phosphorylates substrate proteins to regulate various F-actin-based processes, such as membrane ruffling, filopodial exploration, neurite extension and cell migration.

c-Cbl, c-Jun, Dok-R, RFX1 and ST5 (Barila *et al.*, 2000; Katan *et al.*, 1997; Majidi *et al.*, 1998; Master *et al.*, 2003; Miyoshi-Akiyama *et al.*, 2001). Unfortunately, many of these studies fail to show that upon mutation of the PxxP motif Abl activation is impaired. In addition, activation *in vitro* with an SH3 domain ligand peptide has not yet been reported, probably due to the low level of affinity of SH3 peptide ligands for their respective binding site. Most SH3 domain-dependent activators are, in turn, also phosphorylated by c-Abl. These phosphorylated residues could bind to the Abl SH2 domain, thereby further activating c-Abl kinase activity (Fig. 4.7B).

SH2 domains bind tyrosine-phosphorylated peptides in a particular sequence specific context, as these domains contain two surface pockets, one that recognizes the phospho-tyrosine and a second one that binds to specific residues that are located downstream of the phospho-tyrosine.

The SH3 domain of Abl is already bound to its respective ligand in autoinhibited Abl. In contrast, the SH2 domain has no ligand engaging the phospho-tyrosine-binding pocket in autoinhibited c-Abl (Fig. 4.7B). This is due to the tight protein–protein interface of the SH2 domain with the carboxy-terminal lobe of the kinase domain, which is stabilized by a network of interlocking hydrogen bonds and partly occludes access to the SH2 domain, as the kinase domain masks the binding sites for the residues that are immediately upstream of the phospho-tyrosine (Fig. 4.7B).

In support of this, and similar to the situation for c-Src, tyrosine-phosphorylated peptides that were derived from c-Abl substrates are capable of activating c-Abl *in vitro*. A concentration-dependent increase in c-Abl activity was observed and required an intact SH2-domain (Hantschel *et al.*, 2003). This unexpected finding offered an explanation for the repeatedly observed activation of c-Abl by high concentration of substrates. Examples found in the literature include Abl interactor 1 (Abi-1), Nck, paxillin, Cbl, Cables and possibly also p73, c-Crk, Ena (enabled) and disabled (Juang and Hoffmann, 1999; Miyoshi-Akiyama *et al.*, 2001; Smith *et al.*, 1999). Furthermore, this may form the basis for the described nuclear tyrosine phosphorylation circuit involving c-Abl, its substrate c-Jun and the MAP-kinase JNK (Barila *et al.*, 2000).

D. c-Abl and tyrosine kinase receptors

SH2 domain-dependent interactions of c-Abl with activated EphB2 and Trk receptor tyrosine kinases (Yano *et al.*, 2000; Yu *et al.*, 2001), as well as the B-cell receptor ancillary protein CD1937 were also described and are likely to reflect the activation mechanism of c-Abl by receptor tyrosine kinases. Moreover, increased c-Abl tyrosine kinase acitivty has been observed upon activation of PDGF-R (platelet-derived growth factor-R) and Met (Frasca *et al.*, 2001; Plattner *et al.*, 1999). However, for most of the examples presented earlier, the phosphorylation site in the substrates/activators has not been mapped and confirmed by mutating the site. Therefore, the precise molecular mechanism awaits further analysis.

E. Abl substrates

Similar to its role in kinases of the Src-family, the Abl SH2 domain also appears to be important for processive phosphorylation as substrates that contain multiple phosphorylation sites are more efficiently phosphorylated than substrates with only single phosphorylation sites (Duyster *et al.*, 1995; Mayer *et al.*, 1995; Pellicena *et al.*, 1998). Although differently interpreted in the past, a 'priming' site may initially be phosphorylated through the basal kinase activity of c-Abl or other kinases and initiate the positive feedback loop described earlier. The substrate specificity of non-receptor tyrosine kinases and the binding specificity of their associated SH2 domain strongly correlate. This means that good Abl substrates are also good binders to the Abl SH2 domain and vice versa. This supports the known concept of co-evolution of SH2-binding and kinase specificities. However, it is important to note though that most Abl phosphorylation sites that are found in physiological substrates do not match the optimal substrate sequence. Interestingly, most of the proteins that were shown to activate c-Abl by binding of a PxxP motif to the Abl SH3 domain also become phosphorylated by Abl and subsequently serve as SH2 domain-dependent activators (Fig. 4.7B). Vice versa, most proteins that act as SH2 domain-dependent activators also contain SH3 domain ligands.

One of the major differences between c-Abl and its oncogenic counterparts is that c-Abl is normally not phosphorylated on tyrosine residues. The activation loop of the Abl kinase domain folds into the active site and prevents binding of both the substrate and ATP. In most kinases, phosphorylation of one or more residues in the activation loop stabilizes it in a conformation that serves as a binding platform for the peptide substrate and facilitates the phosphotransfer reaction (Nolen *et al.*, 2004). In c-Abl, phosphorylation of Tyr412 in the activation loop is coupled to a concomitant increase in catalytic activity and can occur by an autocatalytic mechanism in *trans* (Dorey *et al.*, 2001) (Fig. 4.7B). Furthermore, treatment of cells with PDGF results in activation of c-Abl through measurable Tyr412 phosphorylation by Src-family kinases (Dorey *et al.*, 2001; Plattner *et al.*, 2003, 1999). Besides Tyr412, the second proline of the PXXP motif of the SH2–kinase linker to which the c-Abl SH3 domain binds is replaced with a tyrosine (Tyr245) that can be phosphorylated. Mutation of Tyr245 inhibited the autophosphorylation-induced activation of wild-type c-Abl by 50% (Brasher and Van Etten, 2000; Tanis *et al.*, 2003).

In addition to phosphorylation on Tyr412 and Tyr245, active forms of c-Abl, such as c-Abl that is activated by constitutively active Src, as well as

Bcr-Abl and v-Abl are phosphorylated on more than 20 residues. Interestingly, besides tyrosine residues, many serine and threonine residues are also phosphorylated. For the vast majority of these phosphorylation events, the responsible kinase is unknown. Moreover, the functional consequences of these phosphorylation events on activity, structure and regulation of c-Abl have not yet been studied systematically. In particular, whether a possible phosphorylation event would be able to initiate activation by itself or whether it would merely stabilize an activated conformation of c-Abl is not understood. Mapping of the phosphorylation sites on the structure of autoinhibited c-Abl shows that many of these sites are not easily accessible by a cognate kinase and conformational changes are required to accommodate the phosphate group. This suggests that most phosphorylation events in c-Abl might rather stabilize a conformation associated with activity, rather than being able to initiate activation per se.

V. c-Abl and IR Signalling

A. IR, IGF-IR and focal adhesion kinase (FAK)

As mentioned earlier, the reason of different action of IR and IGF-IR have been extensively studied and depend also on the different recruitment and activation of intracellular substrates including Grb10, 14-3-3, CrkII and FAK (Frasca *et al.*, 2008). FAK is a ubiquitously expressed cytoplasmic tyrosine kinase, involved in integrin and growth factor receptor signalling (Fig. 4.8). It is widely accepted that FAK is a point of convergence in the actions of the extracellular matrix (ECM) and growth factors (Mitra *et al.*, 2005). FAK tyrosine phosphorylation occurs rapidly in response to growth factor stimulation including PDGF, epidermal growth factor and IGF-I. In particular, FAK phosphorylation at tyrosine 925 creates a binding site for the SH2 domain of Grb2, thereby activating the Ras/ERK pathway (Schlaepfer *et al.*, 1994) (Fig. 4.8). Moreover, the activated FAK phosphorylates several adapter proteins including the p130CAS (Harte *et al.*, 1996) and paxillin (Hildebrand *et al.*, 1995), which in turn provide docking sites to the small SH2-containing proteins Crk and Nck (Harte *et al.*, 1996; Ruest *et al.*, 2001; Tachibana *et al.*, 1997) (Fig. 4.8). Similarly to Grb2, also Crk and Nck recruitment results in the activation of the Ras/ERK pathway (Mitra *et al.*, 2005) (Fig. 4.8). Because FAK is able to trigger the Ras pathway in response to several stimuli, it has been proposed that FAK plays a major role in mediating the effect of tyrosine kinase receptors on the ERK activation.

At variance with most growth factors, which stimulate FAK phosphorylation, insulin may cause either FAK phosphorylation or dephosphorylation (Baron *et al.*, 1998; Leventhal *et al.*, 1997; Pillay *et al.*, 1995), depending

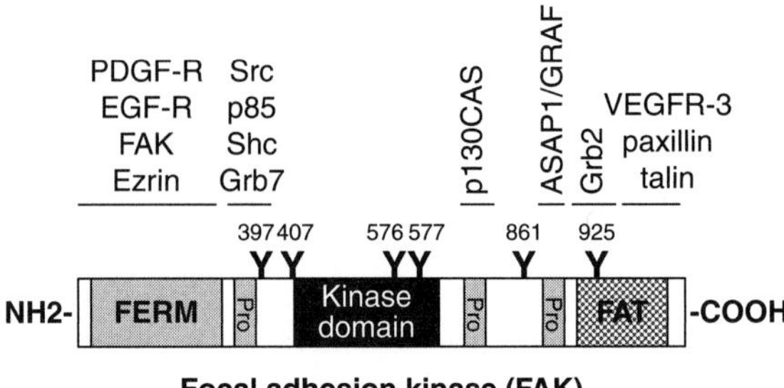

Figure 4.8 Focal adhesion kinase (FAK) structural features and binding partners. The kinase domain of FAK is flanked by the N-terminus that harbours the FERM domain, and by the C-terminus that consists, in addition to proline-rich domains, of the FAT domain. The FERM domain interacts with growth factor receptors, the autophosphorylation site of FAK (tyrosine 397) is required for its activity and is a binding partner for Src, p85, Shc and Grb7. The kinase domain contains two important tyrosines, 576 and 577, whose phosphorylation is necessary for full kinase activity. The C-terminal domain harbours two proline-rich domains that interact with SH3 domain containing proteins like p130CAS, Graf and ASAP1. Furthermore, it contains an important tyrosine, 925, which, after phosphorylation, interacts with Grb2 and a FAT sequence that localizes FAK to the integrins, both directly and indirectly via the scaffolding proteins paxillin, talin and vinculin.

on the cell context and experimental conditions, including ligand concentration, cell adhesion and IR expression level (Baron *et al.*, 1998; Cheung *et al.*, 2000; Huang *et al.*, 2006; Knight *et al.*, 1995; Schaller *et al.*, 1992; Wang *et al.*, 1998). This suggests that the inhibition of the integrin/FAK/ERK signalling might be an important prerequisite in the metabolic, rather than mitogenic, effect of insulin.

B. c-Abl and IR signalling

As mentioned above, c-Abl is a non-receptor tyrosine kinase found in both the cytoplasm and the nucleus activated by cell adhesion (Taagepera *et al.*, 1998) and growth factors like PDGF, HGF and NGF (Frasca *et al.*, 2001; Plattner *et al.*, 1999). c-Abl activation by PDGF results in cell proliferation; cAbl activation by Met has a negative effect reducing cell migration in response to HGF in thyroid cancer cells; c-Abl activation by Trk results in CrkII phosphorylation which drives axonal growth and guidance.

The role of c-Abl in IR pathway is unknown, but it has been shown that c-Abl shares with IR some intracellular substrates including Tub, which is phosphorylated by both IR and c-Abl, and SORBS1, which interact both with IR and c-Abl in insulin-dependent manner (Kapeller *et al.*, 1999; Lin *et al.*, 2001). In particular, mutation in the *tub* gene leads to maturity-onset obesity, insulin resistance, and progressive retinal and cochlear degeneration in mice. *Tub* is a member of a family of genes that encode proteins of unknown function that are conserved across species. The structural analysis of Tub suggests that Tub is an intracellular protein. Additional sequence analysis revealed the presence of tyrosine phosphorylation motifs and Src homology 2 (SH2)-binding sites. Indeed, experiments performed in CHO-IR cells transfected with Tub indicate that Tub is phosphorylated on tyrosine in response to insulin and insulin-like growth factor-1 (Kapeller *et al.*, 1999). Moreover, in PC12 cells, insulin but not EGF induced tyrosine phosphorylation of endogenous Tub. *In vitro*, Tub is phosphorylated by purified IR kinase as well as by Abl and JAK 2 but not by EGF-R and Src kinases. Upon tyrosine phosphorylation, Tub associates selectively with the SH2 domains of Abl, Lck, and the C-terminal SH2 domain of phospholipase Cγ and insulin enhances the association of Tub with endogenous phospholipase Cγ in CHO-IR cells. These data suggest that Tub may function as an adaptor protein linking the IR, and possibly other protein-tyrosine kinases, to c-Abl signalling.

SORBS1 (sorbin and SH3 domain containing 1) display multiple transcripts among different tissues including adipose, liver and skeletal muscle tissues. This gene was mapped to human chromosome 10q23.3–q24.1, which is a candidate region for insulin resistance found in Pima Indians. Experiments performed in human hepatoma Hep3B cells indicated that SORBS1 is partly dissociated from the IR complex and bound to c-Abl protein upon insulin stimulation. This interaction with c-Abl is through the third SH3 domain and a possible conformational change of SORBS1 is induced by insulin stimulation. These data therefore suggest that c-Abl tyrosine kinase via SORBS1 might play a role in the insulin signalling pathway (Lin *et al.*, 2001).

The first solid evidence of the involvement of cAbl in IR signalling is due to the availability of the Abl tyrosine kinase inhibitor STI571 (Imatinib Mesylate, Gleevec^{TM}), currently used in the treatment of CML (Druker, 2003). Indeed, it is well known that while IGF-IR tyrosine kinase is able to interact and phosphorylate CrkII, IR is able to bind CrkII but not to phosphorylate it. However, insulin stimulation in HepG2 cells results in CrkII phosphoryation, which is prevented by the pre-incubation with the Abl inhibitor STI571. These evidences strongly suggest that CrkII phosphorylation in response to insulin stimulation is mediated by c-Abl activation. This hypothesis is confirmed by *in vitro* kinase experiments in HepG2 cells indicating an increased c-Abl tyrosine kinase activity after insulin, but not after IGF-I stimulation (Frasca *et al.*, 2007).

Signal transduction experiments in HepG2 cells indicate that blockade of c-Abl activation by STI571 results in increased ERK activation and decreased Akt activation in response to insulin. As a consequence, inhibition of c-Abl results in increased proliferation and migration and decreased Glycogen Synthesis in response to insulin. This effect of c-Abl inhibition is not observed with IGF-IR signalling and effects. These results are confirmed in cells from double Arg Abl Knockout Mice (*abl*$^{-/-}$ *arg*$^{-/-}$ cells) reconstituted with wild type and Kinase Defective c-Abl. Further signal transduction studies indicated that the effect of c-Abl is not dependent on proximal events of IR signalling, including IRS-1 phosphorylation and recruitment of the small adapter protein p85 and Grb2 but rather on the distal effect on FAK phosphorylation. Indeed, when cells are exposed to insulin and c-Abl is stimulated and FAK is dephosphorylated, activation of AKT/GSK3 pathway and glycogen synthesis may occur. When c-Abl is inhibited, FAK is phosphorylated and the effect of insulin on the ERK pathway activation, cell proliferation and migration is increased. There is evidence, therefore, that when c-Abl is inhibited or absent, there is a change in the response of FAK to insulin stimulation, and insulin stimulates FAK phosphorylation in a manner similar to that of IGF-I. c-Abl inhibition, therefore, selectively modifies IR signalling and effects in such a way that IR signalling will mimic IGF-IR signalling with respect to FAK phosphorylation (Fig. 4.9). c-Abl-dependent changes of FAK phosphorylation are a necessary step for the effect of c-Abl on insulin signalling (Frasca *et al.*, 2007).

Those changes in FAK phosphorylation indicate that c-Abl influences insulin signalling only in the presence of FAK, as demonstrated in cells devoid of FAK (*FAK*$^{-/-}$ cells) in which c-Abl inhibition is without effect on insulin signalling and actions. This hypothesis is in line with previous observations indicating that in cells transfected with a mutant IR in which Tyr 1210 is replaced by Phe, after insulin stimulus there is no evidence of FAK dephosphorylation. Interestingly, in presence of this mutant, insulin is more effective in stimulating ERK rather than phosphatidylinositol 3-kinase when compared with wild-type IR (Van der Zon *et al.*, 1996) (Fig. 4.9). These results confirm the importance of FAK dephosphorylation for activating insulin metabolic effects and for restraining non-metabolic effects.

Since c-Abl is also activated by cell adhesion (Taagepera *et al.*, 1998), it may play a relevant role in the complex interplay between integrin and insulin signalling, that cause cell interaction with ECM, whilst insulin regulates the cell energy metabolism and storage. Indeed, cell detachment from ECM has similar effects of c-Abl inhibition, causing a dramatic change of FAK phosphorylation in response to insulin stimulus (Baron *et al.*, 1998). When cells attach, c-Abl is activated and it causes FAK dephosphorylation in response to insulin, and consequently reduction of migration and proliferation, whilst cell metabolic activity increase. In contrast, when c-Abl activity is down regulated, as it occurs when cells detach, metabolic effects of insulin (like glycogen synthesis) are blunted, whereas effects on cell proliferation and migration are potentiated to an extent similar to that elicited by IGF-I. By this mechanism detached cells may obtain the

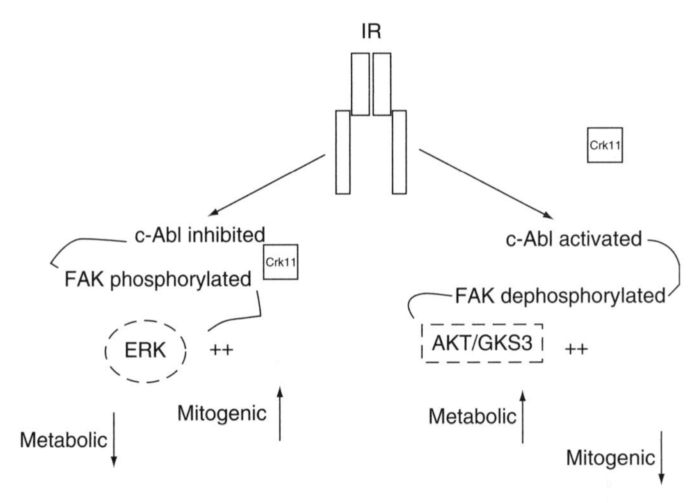

Figure 4.9 Theoretical model of the molecular mechanism underlying the effect of c-Abl on IR signalling. Right, in cells with a functional c-Abl, activated IR activation may recruit the SH2 of CrkII which, in turn may, recruit and activate c-Abl, via SH3 domain. Activated c-Abl phosphorylates CrkII, which refolds into an inactive conformation, incapable to interact with FAK. This molecular sequence results in a decreased level of FAK phosphorylation and, as a consequence, a more restrained effect of insulin focused mainly onto metabolic activities. Left, in cells with an inactive c-Abl, CrkII is not phosphorylated and maintains an active conformation, capable to interact with FAK, thereby enhancing FAK phosphorylation levels. Increased phosphorylation of FAK in response to insulin makes cell signalling similar to that of IGF-I, thereby promoting cell proliferation and migration.

biochemical instruments favouring their re-localization and proliferation. These data may also explain why insulin can cause FAK phosphorylation or dephosphorylation depending on different cell context and experimental conditions (Baron *et al.*, 1998; Cheung *et al.*, 2000; Huang *et al.*, 2006; Knight *et al.*, 1995; Schaller *et al.*, 1992; Wang *et al.*, 1998).

C. Possible role of CrkII in mediating the interaction between c-Abl and FAK

The Crk family of adaptor proteins (c-CrkII, c-CrkI, CrkL) are Src Homology 2 (SH2) and SH3 domain containing proteins that have been implicated in many signalling events of proliferation, differentiation, cell adhesion and cytoskeletal reorganization (Birge *et al.*, 1996; Feller *et al.*, 1998). The role of Crk in the signalling pathways is primarily mediated by the SH2 and the first SH3 domains which form specific interactions with intracellular proteins.

The SH2 domain of Crk binds in the context of phospho Tyr-X-X-Pro, and primarily interact with tyrosine phosphorylated focal adhesion proteins p130CAS and paxillin (Birge and Hanafusa, 1993; Sakai *et al.*, 1994; Songyang *et al.*, 1994). The N-terminal SH3 domain of Crk binds to proline-rich sequences in the context of Pro-X-X-Pro-X-Lys (Knudsen *et al.*, 1994) and interacts with a limited number of cellular proteins including C3G, a guanine nucleotide exchange factor for Rap1 and R-Ras (Mochizuki *et al.*, 2000), DOCK180, a regulator of Rac1 (Hasegawa *et al.*, 1996; Kiyokawa *et al.*, 1998), the hematapoietic progenitor kinase 1 (PRK1), an upstream activator of JNK (Ling *et al.*, 1999) and the tyrosine kinase c-Abl (Feller *et al.*, 1994).

While much has been learned about the function of the SH2 and N-terminal SH3 domains via the identification of specific interacting proteins, the functions of the C-terminal SH3 domain and the SH3 linker regions of Crk are not well understood. Previously, it has been demonstrated that the C-terminal region of Crk II contains negative regulatory elements important for mediating distinct functions associated with the protein (Ogawa *et al.*, 1994).

Mutants in the C-terminal SH3 domain (W276K Crk) or deletions in the entire C-terminal SH3 domain, enhance Abl binding to Crk and increase Tyr phosphorylation. Whilst a disruption in the SH3 linker/ C-terminal SH3 domain boundary results in FAK activation, the tyrosine phosphorylation of focal adhesion proteins p130cas and paxillin, and increased numbers of focal adhesions in fibroblasts. These results are consistent with the proposed negative regulatory role of the Crk C-terminus, and suggest that the C-terminal region of Crk may contribute to the regulation of distinct tyrosine kinase pathways involving Abl and FAK (Fig. 4.9).

VI. Concluding Remarks

The present data regarding the role of c-Abl in IR signalling may be summarized as follows: (1) c-Abl and IR share commons substrates like Tub and SORBS1, (2) c-Abl is activated upon insulin but not upon IGF-I stimulation, (3) c-Abl activation by insulin results in FAK dephosphorylation in response to insulin stimulation, (4) c-Abl activation by insulin stimulation is a pre-requisite to restrain the non-metabolic effect of insulin (Frasca *et al.*, 2007; Kain and Klemke, 2001), (5) c-Abl is convergent point between insulin and adhesion signalling, whose function is to modulate cell responsiveness to either metabolic or non-metabolic effects of insulin, depending on the cell context and adhesion conditions. Taken together, these results identify c-Abl as a potential target for novel anti-cancer and anti-diabetes therapies. It is reasonable to suppose that c-Abl tyrosine kinase activity manipulation may be useful in IR overexpressing tumors or to improve insulin sensitivity in diabetes. These *in vitro* data, however, may be very different or even opposite than *in vivo* results. Indeed, it has been already reported that the c-Abl inhibitor Imanitib Mesylate is able to improve glycemic control in patients affected by CML, and this result is the opposite than expected by the *in vitro* observations (Tsapas *et al.*, 2008). The mechanism(s) underlying this effect is not clear. One hypothesis is that reduction of leukemia cells in those patient may result in the reduction of cytokines like TNFα, which is responsible for the increased insulin resistance via the inhibition of IR signalling (Hotamisligil *et al.*, 1996; Tsapas *et al.*, 2008). Moreover, other reports obtained in mice indicated that c-Abl inhibition results in increased resistance to apoptosis of pancreatic β cells and, as a consequence, improved insulin secretion (Hagerkvist *et al.*, 2006, 2007). These observations underline the concept that the effect of c-Abl inhibition on insulin activities *in vivo* arises from the contribution of different tissue compartments. Moreover, manipulation of either wild type or oncogenic Abl (like in CML) may provide different, even opposite, results. Although results obtained *in vitro* and *in vivo* may be somehow conflicting, the importance of these results rely on the identification of modulators of insulin signalling, which may have potential therapeutic implications.

ACKNOWLEDGMENTS

We thank American Italian Cancer Foundation (AICF) for a fellowship to Giuseppe Pandini. These Studies were also supported by grants from Associazione Italiana Ricerca sul Cancro (AIRC) and Ministero dell'Università e della Ricerca (MIUR).

REFERENCES

Andersen, A. S., Wiberg, F. C., and Kjeldsen, T. (1995). Localization of specific amino acids contributing to insulin specificity of the insulin receptor. *Ann. N. Y. Acad. Sci.* **766,** 466–468.

Barila, D., and Superti-Furga, G. (1998). An intramolecular SH3-domain interaction regulates c-Abl activity. *Nat. Genet.* **18,** 280–282.

Barila, D., Mangano, R., Gonfloni, S., Kretzschmar, J., Moro, M., Bohmann, D., and Superti-Furga, G. (2000). A nuclear tyrosine phosphorylation circuit: c-Jun as an activator and substrate of c-Abl and JNK. *EMBO J.* **19,** 273–281.

Baron, V., Calleja, V., Ferrari, P., Alengrin, F., and Van Obberghen, E. (1998). p125Fak focal adhesion kinase is a substrate for the insulin and insulin-like growth factor-I tyrosine kinase receptors. *J. Biol. Chem.* **273,** 7162–7168.

Birge, R. B., and Hanafusa, H. (1993). Closing in on SH2 specificity. *Science* **262,** 1522–1524.

Birge, R. B., Knudsen, B. S., Besser, D., and Hanafusa, H. (1996). SH2 and SH3-containing adaptor proteins: Redundant or independent mediators of intracellular signal transduction. *Genes Cells* **1,** 595–613.

Blakesley, V. A., Kalebic, T., Helman, L. J., Stannard, B., Faria, T. N., Roberts, C. T. Jr., and LeRoith, D. (1996). Tumorigenic and mitogenic capacities are reduced in transfected fibroblasts expressing mutant insulin-like growth factor (IGF)-I receptors. The role of tyrosine residues 1250, 1251, and 1316 in the carboxy-terminus of the IGF-I receptor. *Endocrinology* **137,** 410–417.

Brasher, B. B., and Van Etten, R. A. (2000). c-Abl has high intrinsic tyrosine kinase activity that is stimulated by mutation of the Src homology 3 domain and by autophosphorylation at two distinct regulatory tyrosines. *J. Biol. Chem.* **275,** 35631–35637.

Ceresa, B. P., and Pessin, J. E. (1998). Insulin regulation of the Ras activation/inactivation cycle. *Mol. Cell. Biochem.* **182,** 23–29.

Cheung, A. T., Wang, J., Ree, D., Kolls, J. K., and Bryer-Ash, M. (2000). Tumor necrosis factor-alpha induces hepatic insulin resistance in obese Zucker (fa/fa) rats via interaction of leukocyte antigen-related tyrosine phosphatase with focal adhesion kinase. *Diabetes* **49,** 810–819.

Cicchetti, P., Mayer, B. J., Thiel, G., and Baltimore, D. (1992). Identification of a protein that binds to the SH3 region of Abl and is similar to Bcr and GAP-rho. *Science* **257,** 803–806.

Cohen, G. B., Ren, R., and Baltimore, D. (1995). Modular binding domains in signal transduction proteins. *Cell* **80,** 237–248.

Denley, A., Wallace, J. C., Cosgrove, L. J., and Forbes, B. E. (2003). The insulin receptor isoform exon 11- (IR-A) in cancer and other diseases: A review. *Horm. Metab. Res.* **35,** 778–785.

Dorey, K., Engen, J. R., Kretzschmar, J., Wilm, M., Neubauer, G., Schindler, T., and Superti-Furga, G. (2001). Phosphorylation and structure-based functional studies reveal a positive and a negative role for the activation loop of the c-Abl tyrosine kinase. *Oncogene* **20,** 8075–8084.

Drakas, R., Tu, X., and Baserga, R. (2004). Control of cell size through phosphorylation of upstream binding factor 1 by nuclear phosphatidylinositol 3-kinase. *Proc. Natl. Acad. Sci. USA* **101,** 9272–9276.

Druker, B. J. (2003). Chronic myeloid leukemia in the imatinib era. *Semin. Hematol.* **40,** 1–3.

Duyster, J., Baskaran, R., and Wang, J. Y. (1995). Src homology 2 domain as a specificity determinant in the c-Abl-mediated tyrosine phosphorylation of the RNA polymerase II carboxyl-terminal repeated domain. *Proc. Natl. Acad. Sci. USA* **92,** 1555–1559.

Ebina, Y., Edery, M., Ellis, L., Standring, D., Beaudoin, J., Roth, R. A., and Rutter, W. J. (1985a). Expression of a functional human insulin receptor from a cloned cDNA in Chinese hamster ovary cells. *Proc. Natl. Acad. Sci. USA* **82,** 8014–8018.

Ebina, Y., Ellis, L., Jarnagin, K., Edery, M., Graf, L., Clauser, E., Ou, J. H., Masiarz, F., Kan, Y. W., Goldfine, I. D., Roth, R. A., and Rutter, W. J. (1985b). The human insulin receptor cDNA: The structural basis for hormone-activated transmembrane signalling. *Cell* **40,** 747–758.

Faria, T. N., Blakesley, V. A., Kato, H., Stannard, B., LeRoith, D., and Roberts, C. T. Jr. (1994). Role of the carboxyl-terminal domains of the insulin and insulin-like growth factor I receptors in receptor function. *J. Biol. Chem.* **269,** 13922–13928.

Feller, S. M., Knudsen, B., and Hanafusa, H. (1994). c-Abl kinase regulates the protein binding activity of c-Crk. *EMBO J.* **13,** 2341–2351.

Feller, S. M., Posern, G., Voss, J., Kardinal, C., Sakkab, D., Zheng, J., and Knudsen, B. S. (1998). Physiological signals and oncogenesis mediated through Crk family adapter proteins. *J. Cell. Physiol.* **177,** 535–552.

Frasca, F., Pandini, G., Scalia, P., Sciacca, L., Mineo, R., Costantino, A., Goldfine, I. D., Belfiore, A., and Vigneri, R. (1999). Insulin receptor isoform A, a newly recognized, high-affinity insulin-like growth factor II receptor in foetal and cancer cells. *Mol. Cell. Biol.* **19,** 3278–3288.

Frasca, F., Vigneri, P., Vella, V., Vigneri, R., and Wang, J. Y. (2001). Tyrosine kinase inhibitor STI571 enhances thyroid cancer cell motile response to hepatocyte growth factor. *Oncogene* **20,** 3845–3856.

Frasca, F., Pandini, G., Malaguarnera, R., Mandarino, A., Messina, R. L., Sciacca, L., Belfiore, A., and Vigneri, R. (2007). Role of c-Abl in directing metabolic versus mitogenic effects in insulin receptor signalling. *J. Biol. Chem.* **282,** 26077–26088.

Frasca, F., Pandini, G., Sciacca, L., Pezzino, V., Squatrito, S., Belfiore, A., and Vigneri, R. (2008). The role of insulin receptors and IGF-I receptors in cancer and other diseases. *Arch. Physiol. Biochem.* **114,** 23–37.

Hagerkvist, R., Makeeva, N., Elliman, S., and Welsh, N. (2006). Imatinib mesylate (Gleevec) protects against streptozotocin-induced diabetes and islet cell death *in vitro*. *Cell Biol. Int.* **30,** 1013–1017.

Hagerkvist, R., Sandler, S., Mokhtari, D., and Welsh, N. (2007). Amelioration of diabetes by imatinib mesylate (Gleevec): Role of beta-cell NF-kappaB activation and antiapoptotic preconditioning. *FASEB J.* **21,** 618–628.

Hantschel, O., Nagar, B., Guettler, S., Kretzschmar, J., Dorey, K., Kuriyan, J., and Superti-Furga, G. (2003). A myristoyl/phosphotyrosine switch regulates c-Abl. *Cell* **112,** 845–857.

Harrington, L. S., Findlay, G. M., and Lamb, R. F. (2005). Restraining PI3K: mTOR signalling goes back to the membrane. *Trends Biochem. Sci.* **30,** 35–42.

Harte, M. T., Hildebrand, J. D., Burnham, M. R., Bouton, A. H., and Parsons, J. T. (1996). p130Cas, a substrate associated with v-Src and v-Crk, localizes to focal adhesions and binds to focal adhesion kinase. *J. Biol. Chem.* **271,** 13649–13655.

Hasegawa, H., Kiyokawa, E., Tanaka, S., Nagashima, K., Gotoh, N., Shibuya, M., Kurata, T., and Matsuda, M. (1996). DOCK180, a major CRK-binding protein, alters cell morphology upon translocation to the cell membrane. *Mol. Cell. Biol.* **16,** 1770–1776.

Hildebrand, J. D., Schaller, M. D., and Parsons, J. T. (1995). Paxillin, a tyrosine phosphorylated focal adhesion-associated protein binds to the carboxyl terminal domain of focal adhesion kinase. *Mol. Biol. Cell* **6,** 637–647.

Hotamisligil, G. S., Peraldi, P., Budavari, A., Ellis, R., White, M. F., and Spiegelman, B. M. (1996). IRS-1-mediated inhibition of insulin receptor tyrosine kinase activity in TNF-alpha- and obesity-induced insulin resistance. *Science* **271,** 665–668.

Huang, D., Khoe, M., Ilic, D., and Bryer-Ash, M. (2006). Reduced expression of focal adhesion kinase disrupts insulin action in skeletal muscle cells. *Endocrinology* **147,** 3333–3343.

Juang, J. L., and Hoffmann, F. M. (1999). *Drosophila* abelson interacting protein (dAbi) is a positive regulator of abelson tyrosine kinase activity. *Oncogene* **18,** 5138–5147.

Kain, K. H., and Klemke, R. L. (2001). Inhibition of cell migration by Abl family tyrosine kinases through uncoupling of Crk-CAS complexes. *J. Biol. Chem.* **276,** 16185–16192.

Kapeller, R., Moriarty, A., Strauss, A., Stubdal, H., Theriault, K., Siebert, E., Chickering, T., Morgenstern, J. P., Tartaglia, L. A., and Lillie, J. (1999). Tyrosine phosphorylation of tub and its association with Src homology 2 domain-containing proteins implicate tub in intracellular signalling by insulin. *J. Biol. Chem.* **274,** 24980–24986.

Katan, Y., Agami, R., and Shaul, Y. (1997). The transcriptional activation and repression domains of RFX1, a context-dependent regulator, can mutually neutralize their activities. *Nucleic Acids Res.* **25,** 3621–3628.

Kato, H., Faria, T. N., Stannard, B., Roberts, C. T. Jr., and LeRoith, D. (1994). Essential role of tyrosine residues 1131, 1135, and 1136 of the insulin-like growth factor-I (IGF-I) receptor in IGF-I action. *Mol. Endocrinol.* **8,** 40–50.

Kellerer, M., Lammers, R., Ermel, B., Tippmer, S., Vogt, B., Obermaier-Kusser, B., Ullrich, A., and Haring, H. U. (1992). Distinct alpha-subunit structures of human insulin receptor A and B variants determine differences in tyrosine kinase activities. *Biochemistry* **31,** 4588–4596.

Kiyokawa, E., Hashimoto, Y., Kobayashi, S., Sugimura, H., Kurata, T., and Matsuda, M. (1998). Activation of Rac1 by a Crk SH3-binding protein, DOCK180. *Genes Dev.* **12,** 3331–3336.

Knight, J. B., Yamauchi, K., and Pessin, J. E. (1995). Divergent insulin and platelet-derived growth factor regulation of focal adhesion kinase (pp125FAK) tyrosine phosphorylation, and rearrangement of actin stress fibers. *J. Biol. Chem.* **270,** 10199–10203.

Knudsen, B. S., Feller, S. M., and Hanafusa, H. (1994). Four proline-rich sequences of the guanine-nucleotide exchange factor C3G bind with unique specificity to the first Src homology 3 domain of Crk. *J. Biol. Chem.* **269,** 32781–32787.

Kosaki, A., Pillay, T. S., Xu, L., and Webster, N. J. (1995). The B isoform of the insulin receptor signals more efficiently than the A isoform in HepG2 cells. *J. Biol. Chem.* **270,** 20816–20823.

Kosaki, A., Nelson, J., and Webster, N. J. (1998). Identification of intron and exon sequences involved in alternative splicing of insulin receptor pre-mRNA. *J. Biol. Chem.* **273,** 10331–10337.

Leventhal, P. S., Shelden, E. A., Kim, B., and Feldman, E. L. (1997). Tyrosine phosphorylation of paxillin and focal adhesion kinase during insulin-like growth factor-I-stimulated lamellipodial advance. *J. Biol. Chem.* **272,** 5214–5218.

Lin, W. H., Huang, C. J., Liu, M. W., Chang, H. M., Chen, Y. J., Tai, T. Y., and Chuang, L. M. (2001). Cloning, mapping, and characterization of the human sorbin and SH3 domain containing 1 (SORBS1) gene: A protein associated with c-Abl during insulin signalling in the hepatoma cell line Hep3B. *Genomics* **74,** 12–20.

Ling, P., Yao, Z., Meyer, C. F., Wang, X. S., Oehrl, W., Feller, S. M., and Tan, T. H. (1999). Interaction of hematopoietic progenitor kinase 1 with adapter proteins Crk and CrkL leads to synergistic activation of c-Jun N-terminal kinase. *Mol. Cell. Biol.* **19,** 1359–1368.

Liu, J. P., Baker, J., Perkins, A. S., Robertson, E. J., and Efstratiadis, A. (1993). Mice carrying null mutations of the genes encoding insulin-like growth factor I (Igf-1) and type 1 IGF receptor (Igf1r). *Cell* **75,** 59–72.

Louvi, A., Accili, D., and Efstratiadis, A. (1997). Growth-promoting interaction of IGF-II with the insulin receptor during mouse embryonic development. *Dev. Biol.* **189,** 33–48.

Majidi, M., Hubbs, A. E., and Lichy, J. H. (1998). Activation of extracellular signal-regulated kinase 2 by a novel Abl-binding protein, ST5. *J. Biol. Chem.* **273,** 16608–16614.

Master, Z., Tran, J., Bishnoi, A., Chen, S. H., Ebos, J. M., Van Slyke, P., Kerbel, R. S., and Dumont, D. J. (2003). Dok-R binds c-Abl and regulates Abl kinase activity and mediates cytoskeletal reorganization. *J. Biol. Chem.* **278,** 30170–30179.

Mayer, B. J., Hirai, H., and Sakai, R. (1995). Evidence that SH2 domains promote processive phosphorylation by protein-tyrosine kinases. *Curr. Biol.* **5,** 296–305.

Mitra, S. K., Hanson, D. A., and Schlaepfer, D. D. (2005). Focal adhesion kinase: In command and control of cell motility. *Nat. Rev. Mol. Cell. Biol.* **6,** 56–68.

Miura, M., Surmacz, E., Burgaud, J. L., and Baserga, R. (1995). Different effects on mitogenesis and transformation of a mutation at tyrosine 1251 of the insulin-like growth factor I receptor. *J. Biol. Chem.* **270,** 22639–22644.

Miyoshi-Akiyama, T., Aleman, L. M., Smith, J. M., Adler, C. E., and Mayer, B. J. (2001). Regulation of Cbl phosphorylation by the Abl tyrosine kinase and the Nck SH2/SH3 adaptor. *Oncogene* **20,** 4058–4069.

Mochizuki, N., Ohba, Y., Kobayashi, S., Otsuka, N., Graybiel, A. M., Tanaka, S., and Matsuda, M. (2000). Crk activation of JNK via C3G and R-Ras. *J. Biol. Chem.* **275,** 12667–12671.

Moller, D. E., Yokota, A., Caro, J. F., and Flier, J. S. (1989). Tissue-specific expression of two alternatively spliced insulin receptor mRNAs in man. *Mol. Endocrinol.* **3,** 1263–1269.

Morrione, A., Valentinis, B., Xu, S. Q., Yumet, G., Louvi, A., Efstratiadis, A., and Baserga, R. (1997). Insulin-like growth factor II stimulates cell proliferation through the insulin receptor. *Proc. Natl. Acad. Sci. USA* **94,** 3777–3782.

Mosthaf, L., Grako, K., Dull, T. J., Coussens, L., Ullrich, A., and McClain, D. A. (1990). Functionally distinct insulin receptors generated by tissue-specific alternative splicing. *EMBO J.* **9,** 2409–2413.

Mynarcik, D. C., Williams, P. F., Schaffer, L., Yu, G. Q., and Whittaker, J. (1997). Identification of common ligand binding determinants of the insulin and insulin-like growth factor 1 receptors. Insights into mechanisms of ligand binding. *J. Biol. Chem.* **272,** 18650–18655.

Nolen, B., Taylor, S., and Ghosh, G. (2004). Regulation of protein kinases; controlling activity through activation segment conformation. *Mol. Cell* **15,** 661–675.

O'Connor, R., Kauffmann-Zeh, A., Liu, Y., Lehar, S., Evan, G. I., Baserga, R., and Blattler, W. A. (1997). Identification of domains of the insulin-like growth factor I receptor that are required for protection from apoptosis. *Mol. Cell. Biol.* **17,** 427–435.

Ogawa, S., Toyoshima, H., Kozutsumi, H., Hagiwara, K., Sakai, R., Tanaka, T., Hirano, N., Mano, H., Yazaki, Y., and Hirai, H. (1994). The C-terminal SH3 domain of the mouse c-Crk protein negatively regulates tyrosine-phosphorylation of Crk associated p130 in rat 3Y1 cells. *Oncogene* **9,** 1669–1678.

Pandini, G., Vigneri, R., Costantino, A., Frasca, F., Ippolito, A., Fujita-Yamaguchi, Y., Siddle, K., Goldfine, I. D., and Belfiore, A. (1999). Insulin and insulin-like growth factor-I (IGF-I) receptor overexpression in breast cancers leads to insulin/IGF-I hybrid receptor overexpression: Evidence for a second mechanism of IGF-I signalling. *Clin. Cancer Res.* **5,** 1935–1944.

Pandini, G., Frasca, F., Mineo, R., Sciacca, L., Vigneri, R., and Belfiore, A. (2002). Insulin/insulin-like growth factor I hybrid receptors have different biological characteristics depending on the insulin receptor isoform involved. *J. Biol. Chem.* **277,** 39684–39695.

Pandini, G., Conte, E., Medico, E., Sciacca, L., Vigneri, R., and Belfiore, A. (2004). Igf-ii binding to insulin receptor isoform a induces a partially different gene expression profile from insulin binding. *Ann. N. Y. Acad. Sci.* **1028,** 450–456.

Pandini, G., Wurch, T., Akla, B., Corvaia, N., Belfiore, A., and Goetsch, L. (2007). Functional responses and *in vivo* antitumour activity of h7C10: A humanised monoclonal antibody with neutralising activity against the insulin-like growth factor-1 (IGF-1) receptor and insulin/IGF-1 hybrid receptors. *Eur. J. Cancer* **43,** 1318–1327.

Pawson, T. (1994). SH2 and SH3 domains in signal transduction. *Adv. Cancer Res.* **64,** 87–110.

Pellicena, P., Stowell, K. R., and Miller, W. T. (1998). Enhanced phosphorylation of Src family kinase substrates containing SH2 domain binding sites. *J. Biol. Chem.* **273,** 15325–15328.

Pillay, T. S., Sasaoka, T., and Olefsky, J. M. (1995). Insulin stimulates the tyrosine dephosphorylation of pp125 focal adhesion kinase. *J. Biol. Chem.* **270,** 991–994.

Plattner, R., Kadlec, L., DeMali, K. A., Kazlauskas, A., and Pendergast, A. M. (1999). c-Abl is activated by growth factors and Src family kinases and has a role in the cellular response to PDGF. *Genes Dev.* **13,** 2400–2411.

Plattner, R., Irvin, B. J., Guo, S., Blackburn, K., Kazlauskas, A., Abraham, R. T., York, J. D., and Pendergast, A. M. (2003). A new link between the c-Abl tyrosine kinase and phosphoinositide signalling through PLC-gamma1. *Nat. Cell Biol.* **5,** 309–319.

Ren, R., Mayer, B. J., Cicchetti, P., and Baltimore, D. (1993). Identification of a ten-amino acid proline-rich SH3 binding site. *Science* **259,** 1157–1161.

Robinson, D. R., Wu, Y. M., and Lin, S. F. (2000). The protein tyrosine kinase family of the human genome. *Oncogene* **19,** 5548–5557.

Ruest, P. J., Shin, N. Y., Polte, T. R., Zhang, X., and Hanks, S. K. (2001). Mechanisms of CAS substrate domain tyrosine phosphorylation by FAK and Src. *Mol. Cell. Biol.* **21,** 7641–7652.

Sakai, R., Iwamatsu, A., Hirano, N., Ogawa, S., Tanaka, T., Mano, H., Yazaki, Y., and Hirai, H. (1994). A novel signalling molecule, p130, forms stable complexes *in vivo* with v-Crk and v-Src in a tyrosine phosphorylation-dependent manner. *EMBO J.* **13,** 3748–3756.

Sasaoka, T., and Kobayashi, M. (2000). The functional significance of Shc in insulin signalling as a substrate of the insulin receptor. *Endocr. J.* **47,** 373–381.

Schaller, M. D., Borgman, C. A., Cobb, B. S., Vines, R. R., Reynolds, A. B., and Parsons, J. T. (1992). pp125FAK a structurally distinctive protein-tyrosine kinase associated with focal adhesions. *Proc. Natl. Acad. Sci. USA* **89,** 5192–5196.

Schlaepfer, D. D., Hanks, S. K., Hunter, T., and van der Geer, P. (1994). Integrin-mediated signal transduction linked to Ras pathway by GRB2 binding to focal adhesion kinase. *Nature* **372,** 786–791.

Sciacca, L., Mineo, R., Pandini, G., Murabito, A., Vigneri, R., and Belfiore, A. (2002). In IGF-I receptor-deficient leiomyosarcoma cells autocrine IGF-II induces cell invasion and protection from apoptosis via the insulin receptor isoform A. *Oncogene* **21,** 8240–8250.

Shaul, Y. (2000). c-Abl: Activation and nuclear targets. *Cell Death Differ.* **7,** 10–16.

Siddle, K., Soos, M. A., Field, C. E., and Nave, B. T. (1994). Hybrid and atypical insulin/insulin-like growth factor I receptors. *Horm. Res.* **41**(Suppl. 2), 56–64discussion 65.

Smith, J. M., Katz, S., and Mayer, B. J. (1999). Activation of the Abl tyrosine kinase *in vivo* by Src homology 3 domains from the Src homology 2/Src homology 3 adaptor Nck. *J. Biol. Chem.* **274,** 27956–27962.

Songyang, Z., Shoelson, S. E., McGlade, J., Olivier, P., Pawson, T., Bustelo, X. R., Barbacid, M., Sabe, H., Hanafusa, H., Yi, T., *et al.* (1994). Specific motifs recognized by the SH2 domains of Csk, 3BP2, fps/fes, GRB-2, HCP, SHC, Syk, and Vav. *Mol. Cell. Biol.* **14,** 2777–2785.

Soos, M. A., Field, C. E., and Siddle, K. (1993a). Purified hybrid insulin/insulin-like growth factor-I receptors bind insulin-like growth factor-I, but not insulin, with high affinity. *Biochem. J.* **290**(Pt. 2), 419–426.

Soos, M. A., Nave, B. T., and Siddle, K. (1993b). Immunological studies of type I IGF receptors and insulin receptors: Characterisation of hybrid and atypical receptor subtypes. *Adv. Exp. Med. Biol.* **343,** 145–157.

Taagepera, S., McDonald, D., Loeb, J. E., Whitaker, L. L., McElroy, A. K., Wang, J. Y., and Hope, T. J. (1998). Nuclear-cytoplasmic shuttling of C-ABL tyrosine kinase. *Proc. Natl. Acad. Sci. USA* **95,** 7457–7462.

Tachibana, K., Urano, T., Fujita, H., Ohashi, Y., Kamiguchi, K., Iwata, S., Hirai, H., and Morimoto, C. (1997). Tyrosine phosphorylation of Crk-associated substrates by focal adhesion kinase. A putative mechanism for the integrin-mediated tyrosine phosphorylation of Crk-associated substrates. *J. Biol. Chem.* **272,** 29083–29090.

Tanis, K. Q., Veach, D., Duewel, H. S., Bornmann, W. G., and Koleske, A. J. (2003). Two distinct phosphorylation pathways have additive effects on Abl family kinase activation. *Mol. Cell. Biol.* **23,** 3884–3896.

Tsapas, A., Vlachaki, E., Sarigianni, M., Klonizakis, F., and Paletas, K. (2008). Restoration of insulin sensitivity following treatment with imatinib mesylate (Gleevec) in nondiabetic patients with chronic myelogenic leukemia (CML). *Leuk. Res.* **32,** 674–675.

Ullrich, A., Bell, J. R., Chen, E. Y., Herrera, R., Petruzzelli, L. M., Dull, T. J., Gray, A., Coussens, L., Liao, Y. C., Tsubokawa, M., Mason, A., Seeburg, P. H., *et al.* (1985). Human insulin receptor and its relationship to the tyrosine kinase family of oncogenes. *Nature* **313,** 756–761.

Ullrich, A., Gray, A., Tam, A. W., Yang-Feng, T., Tsubokawa, M., Collins, C., Henzel, W., Le Bon, T., Kathuria, S., Chen, E., Jacobs, S., *et al.* (1986). Insulin-like growth factor I receptor primary structure: comparison with insulin receptor suggests structural determinants that define functional specificity. *EMBO J.* **5,** 2503–2512.

Van der Zon, G. C., Ouwens, D. M., Dorrestijn, J., and Maassen, J. A. (1996). Replacement of the conserved tyrosine 1210 by phenylalanine in the insulin receptor affects insulin-induced dephosphorylation of focal adhesion kinase but leaves other responses intact. *Biochemistry* **35,** 10377–10382.

Vogt, B., Carrascosa, J. M., Ermel, B., Ullrich, A., and Haring, H. U. (1991). The two isotypes of the human insulin receptor (HIR-A and HIR-B) follow different internalization kinetics. *Biochem. Biophys. Res. Commun.* **177,** 1013–1018.

Wang, Q., Bilan, P. J., and Klip, A. (1998). Opposite effects of insulin on focal adhesion proteins in 3T3-L1 adipocytes and in cells overexpressing the insulin receptor. *Mol. Biol. Cell* **9,** 3057–3069.

Webster, N. J., Evans, L. G., Caples, M., Erker, L., and Chew, S. L. (2004). Assembly of splicing complexes on exon 11 of the human insulin receptor gene does not correlate with splicing efficiency *in vitro*. *BMC Mol. Biol.* **5,** 7.

Whittaker, J., and Whittaker, L. (2005). Characterization of the functional insulin binding epitopes of the full length insulin receptor. *J. Biol. Chem.* **280**(22), 20932–20936.

Whittaker, J., Groth, A. V., Mynarcik, D. C., Pluzek, L., Gadsboll, V. L., and Whittaker, L. J. (2001). Alanine scanning mutagenesis of a type 1 insulin-like growth factor receptor ligand binding site. *J. Biol. Chem.* **276,** 43980–43986.

Yano, H., Cong, F., Birge, R. B., Goff, S. P., and Chao, M. V. (2000). Association of the Abl tyrosine kinase with the Trk nerve growth factor receptor. *J. Neurosci. Res.* **59,** 356–364.

Yip, C. C., Hsu, H., Patel, R. G., Hawley, D. M., Maddux, B. A., and Goldfine, I. D. (1988). Localization of the insulin-binding site to the cysteine-rich region of the insulin receptor alpha-subunit. *Biochem. Biophys. Res. Commun.* **157,** 321–329.

Yu, H. H., Zisch, A. H., Dodelet, V. C., and Pasquale, E. B. (2001). Multiple signalling interactions of Abl and Arg kinases with the EphB2 receptor. *Oncogene* **20,** 3995–4006.

CHAPTER FIVE

CXCL14 and Insulin Action

Takahiko Hara* *and* Yuki Nakayama*,†

Contents

* Stem Cell Project Group, The Tokyo Metropolitan Institute of Medical Science, Tokyo Metropolitan Organization for Medical Research, 3-18-22 Honkomagome, Bunkyo-ku, Tokyo 113-8613, Japan
† Current address: Priority Organization for Innovation and Excellence, Kumamoto University, 2-39-1 Kurokami, Kumamoto City, Kumamoto, 860-8555, Japan

Vitamins and Hormones, Volume 80
ISSN 0083-6729, DOI: 10.1016/S0083-6729(08)00605-5

Abstract

CXCL14 is a member of CXC chemokine family. The physiological roles of CXCL14 and its receptor/signal transduction pathway remain largely unknown. In the human, CXCL14 exhibits chemoattractive activity for activated monocytes and dendritic precursor cells. Recruitment of dendritic precursor cells and inhibition of angiogenesis by CXCL14 suggest that this chemokine has a tumor suppressive function. However, analysis of CXCL14-deficient (CXCL14$^{-/-}$) mice revealed that CXCL14 is dispensable for development and maintenance of tissue macrophages and dendritic cells. CXCL14$^{-/-}$ female mice, but not male mice, weigh significantly less than wild-type mice and are protected from obesity-induced hyperglycemia, hyperinsulinemia, hypoadiponectinemia, and insulin resistance. CXCL14 expression is elevated in white adipose tissue (WAT) of high-fat diet (HFD)-fed obese mice and leptin-system defective mutant mice. Phenotypes of HFD-fed CXCL14$^{-/-}$ female mice indicate that CXCL14 is involved in recruitment of macrophages into WAT, which causes chronic inflammation and contributes to insulin resistance. Transgenic overexpression of CXCL14 in skeletal muscle restores obesity-induced insulin resistance in CXCL14$^{-/-}$ female mice. In addition, CXCL14 attenuates insulin-stimulated glucose uptake in cultured myocytes. Based on these data, it is evident that CXCL14 is a novel regulator of glucose metabolism that acts by recruiting macrophages to WAT and interacting with insulin signaling pathways in skeletal muscle.

I. Introduction

Type-2 diabetes is caused in large part by obesity. In obese individuals, white adipose tissue (WAT) secretes increased amounts of free fatty acids and inflammatory cytokines, including tumor necrosis factor-α (Nguyen *et al.*, 2005; Uysal *et al.*, 1997) and interleukin (IL)-6 (Cai *et al.*, 2005). These factors are correlated with the pathogenesis of insulin resistance and cardiovascular disease which are associated with metabolic syndrome. Recent reports have demonstrated that WAT in obese individuals exhibits increased infiltration of macrophages producing pro-inflammatory cytokines (Weisberg *et al.*, 2003; Xu *et al.*, 2003). Activation of c-Jun N-terminal kinase-1 (Hirosumi *et al.*, 2002) and NF-κB-mediated signaling pathways (Arkan *et al.*, 2005) in macrophages has been implicated in the mechanism of obesity-induced insulin resistance.

A candidate factor important for macrophage recruitment into WAT is CCL2 (also known as monocyte chemoattractant protein-1, or MCP-1). CCL2 is up-regulated in WAT of high-fat diet (HFD)-induced obese mice (Sartipy and Loskutoff, 2003; Takahashi *et al.*, 2003), and glucose metabolism is improved in HFD-fed obese mice lacking CCL2 (Kanda *et al.*, 2006) or its receptor, CCR2 (Weisberg *et al.*, 2006). Furthermore, transgenic mice overexpressing CCL2 mRNA in adipose tissue exhibit enhanced macrophage infiltration into WAT and whole body insulin resistance (Kamei *et al.*, 2006; Kanda *et al.*, 2006). CCL2 directly inhibits insulin-stimulated glucose uptake in cultured adipocytes (Sartipy and Loskutoff, 2003) and myocytes (Kamei *et al.*, 2006), and enhances glucose production in the liver of transgenic mice (Kanda *et al.*, 2006). These results indicate that the CCL2/CCR2 axis contributes to WAT macrophage recruitment and impaired glucose metabolism in obese mice. However, conflicting reports have shown that the number of WAT macrophages and glucose metabolism in CCL2-deficient obese mice are not different from control mice (Chen *et al.*, 2005; Inouye *et al.*, 2007; Kirk *et al.*, 2008). We recently discovered that CXCL14 is an obesity-inducible factor, and that CXCL14-deficient HFD-fed female mice exhibit a reduction in WAT macrophages and improved insulin sensitivity compared to HFD-fed obese control mice (Nara *et al.*, 2007). In this chapter, we describe the basic characteristics of this relatively unstudied chemokine and the physiological roles of CXCL14, as determined by analyses of CXCL14-deficient mice.

II. Basic Properties of CXCL14

A. Classification of chemokines and cloning of CXCL14

Chemokines are small proteins possessing chemotactic activity mainly for leukocytes and lymphocytes. Two major chemokine classes, CC and CXC, have been defined by the spacing of two N-terminal cysteine residues that are either adjacent to each other or separated by one amino acid residue, respectively (Nelson and Krensky, 2001; Zlotnik and Yoshie, 2000). Mammalian CXC chemokines (CXCL1-CXCL16) are further subdivided into two groups, ELR^+ and ELR^- chemokines, based on the presence or absence of a three amino acid sequence Glu-Leu-Arg (the ELR motif) preceding the CXC signature (Table 5.1). ELR^+ chemokines (CXCL1, CXCL5, etc.) are implicated in chemoattraction of neutrophils, whereas most of the ELR^- chemokines (CXCL4, CXCL9, etc.) are involved in lymphocyte chemotaxis. Actions of CXC chemokines extend beyond the immune system. Many CXC chemokines have been identified as pivotal factors regulating tumor invasion and neo-angiogenesis. CXCL1, CXCL5, and CXCL8, which are ELR^+ chemokines, are angiostatic, whereas CXCL4, CXCL9, and CXCL10, which are ELR^- chemokines, are inhibitory for angiogenesis both *in vivo* and *in vitro* (Belperio *et al.*, 2000). Although most ELR^- chemokines are induced by interferon-γ, some are constitutively expressed in the central nervous system.

Table 5.1 Classification of CXC chemokine family members and their receptors

Name	ELR subclass	Other nomenclature	Human chromosome	Receptor(s)
CXCL1	ELR^+	GROα/MGSA-α/KC	4q21.1	CXCR2 > CXCR1
CXCL2	ELR^+	GROβ/MGSA-β/MIP-2α	4q21.1	CXCR2
CXCL3	ELR^+	GROγ/MGSA-γ/MIP-2β	4q21.1	CXCR2
CXCL4	ELR^-	PF4	4q21.1	CXCR3-B
CXCL5	ELR^+	ENA-78	4q21.1	CXCR2
CXCL6	ELR^+	GCP-2/CKα-3	4q21.1	CXCR1, CXCR2
CXCL7	ELR^+	NAP-2/CTAPIII	4q21.1	CXCR2
CXCL8	ELR^+	IL-8/NAP-1	4q21.1	CXCR1, CXCR2
CXCL9	ELR^-	Mig	4q21.1	CXCR3-A
CXCL10	ELR^-	IP-10/CRG-2	4q21.1	CXCR3-A
CXCL11	ELR^-	I-TAC	4q21.1	CXCR3-A
CXCL12	ELR^-	SDF-1α/$\tilde{\beta}$PBSF	10q11.21	CXCR4, CCR7
CXCL13	ELR^-	BCA-1/BLC	4q21.1	CXCR5
CXCL14	ELR^-	BRAK/BMAC/MIP-2γ	5q31.1	Unknown
CXCL15	ELR^+	Lungkine	Not found in human	Unknown
CXCL16	ELR^-	SR-Psox	17p13	CXCR6

CXCL14 (originally designated BRAK, BMAC, or Mip-2γ) was originally cloned as a gene whose expression is down-regulated in human cancer cell lines and tumor specimens (Frederick *et al.*, 2000; Hromas *et al.*, 1999; Sleeman *et al.*, 2000). The deduced 99 amino acid residue precursor of CXCL14 has a 22 amino acid signal peptide that is cleaved to produce a 77 amino acid mature protein (Fig. 5.1). The human precursor CXCL14 contains 12 additional amino acid residues at the N-terminus and a 34 amino acid sequence that serves as a signal peptide (Cao *et al.*, 2000). The calculated molecular weights of human and mouse CXCL14 are 9419 Da and 9437 Da, respectively. CXCL14 is a basic protein (isoelectric point = 10.26) and lacks the ELR sequence preceding the CXC motif, thereby belonging to the ELR^- chemokine subgroup. Studies have shown that CXCL14 is a chemoattractant for activated monocytes (Kurth *et al.*, 2001), monocyte-derived immature dendritic cells (Schaerli *et al.*, 2005;

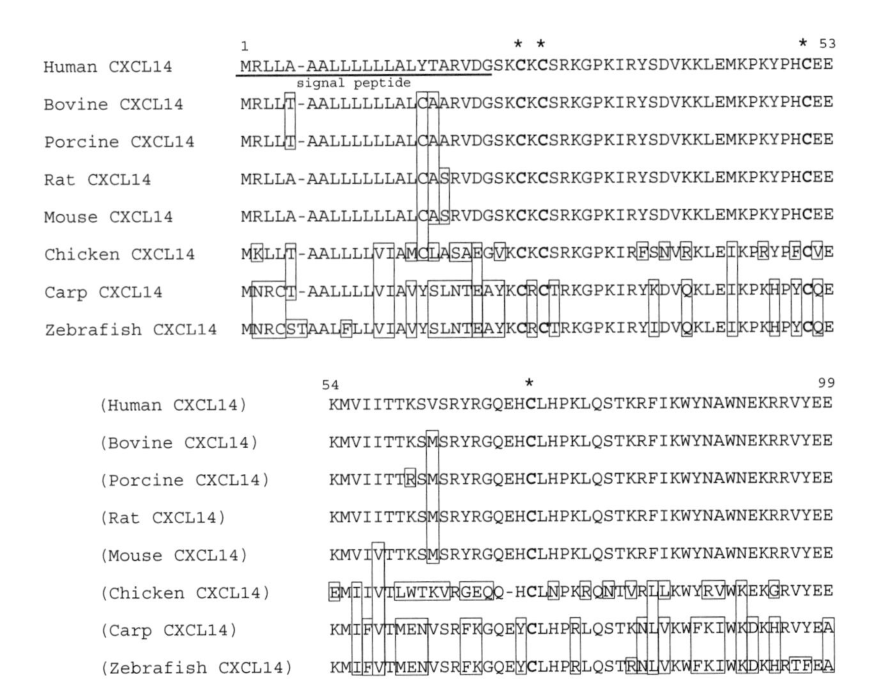

Figure 5.1 Comparison of the amino acid sequence of human CXCL14 with various vertebrate orthologues. The primary amino acid sequence of human CXCL14 is aligned with bovine, porcine, rat, mouse, chicken, carp, and zebrafish CXCL14. The alignment is shown from the N-terminus to C-terminus in two panels. Amino acid residues distinct from human CXCL14 are boxed. Hyphens indicate gaps. Location of a secretory protein signal peptide is underlined. Four cysteine residues conserved in the CXC chemokine family are typed in bold and marked by asterisks. Accession numbers are as follows: human CXCL14, NP_004878; bovine CXCL14, NP_001029582; porcine CXCL14, AY308800; rat CXCL14, NP_001013155; mouse CXCL14, NP_062514; chicken CXCL14, NP_990043; carp CXCL14, CAD59917; zebrafish CXCL14, NP_571702.

Shellenberger *et al.*, 2004; Shurin *et al.*, 2005), epithelial tumor cells (Allinen *et al.*, 2004), and activated natural killer cells (Starnes *et al.*, 2006). Similar to CXCL10, CXCL14 is a potent inhibitor of angiogenesis stimulated by CXCL8, basic fibroblast growth factor, or vascular endothelial growth factor (Shellenberger *et al.*, 2004).

B. Structure of the CXCL14 gene and mRNA expression

The human CXCL14 gene is located on chromosome 5q31.1, while the mouse orthologue is located on chromosome 13. There are 4 exons spanning an ~8.6 kb genomic region in human, or a 7.9 kb genomic region in mouse. The predicted size of the CXCL14 mRNA is 1685 bases in human and 1823 bases in mouse.

In the human, CXCL14 is abundantly expressed in kidney, small intestine, liver, brain, and skeletal muscle, but is barely detectable in lung or ovary (Frederick *et al.*, 2000; Hromas *et al.*, 1999; Sleeman *et al.*, 2000). The CXCL14 transcript is present at high levels in the basal layer of epidermal keratinocytes and squamous epithelium (Frederick *et al.*, 2000; Schaerli *et al.*, 2005). CXCL14 mRNA is produced by lipopolysaccharide-stimulated monocytes and B cells, but not by T cells (Frederick *et al.*, 2000). Monocyte-derived dendritic cells and the human monocytic leukemia cell line THP-1 are also positive for CXCL14 mRNA (Cao *et al.*, 2000). Distinct from other ELR^- chemokines, CXCL14 is not inducible in peripheral blood mononuclear cells by interferon-γ stimulation (Frederick *et al.*, 2000).

The expression pattern of CXCL14 mRNA in the mouse differs from the human. While CXCL14 is not detectable in the small intestine or liver of adult mice, a relatively large amount of CXCL14 mRNA is produced in the lung and ovary (Nara *et al.*, 2007; Sleeman *et al.*, 2000). Further investigation revealed that, in mice, the CXCL14 transcript is produced in both brown adipose tissue (BAT) and WAT, and that the expression levels in BAT and WAT are strikingly elevated in HFD-induced obese mice, leptin-deficient *ob/ob* mice, and leptin receptor-defective *db/db* mice (Nara *et al.*, 2007; Takahashi *et al.*, 2007). Although currently available antibodies against CXCL14 are not sensitive enough to quantify CXCL14 protein levels in WAT or other tissues, an increase in CXCL14 protein in serum upon HFD-feeding has recently been reported (Takahashi *et al.*, 2007). Thus, as with CCL2, CXCL14 is an obesity-induced gene. In the WAT of HFD-fed obese female mice, adipocytes, stromal cells, and infiltrated macrophages contribute to the increase in CXCL14 expression (Nara *et al.*, 2007). Phosphorylation of c-Jun derepresses inflammation-associated gene expression, including CXCL14, in macrophages by removing the nuclear receptor–corepressor complex (Ogawa *et al.*, 2004). Therefore, increased free fatty acids, endoplasmic reticulum stress, and/or reactive oxygen species in WAT of obese mice may induce CXCL14 transcription via the c-Jun N-terminal kinase-1. Additionally, elevated insulin levels in obese mice may contribute to the up-regulation of CXCL14, since administration of insulin to mice results in increased serum CXCL14 concentrations (Takahashi *et al.*, 2007).

Expression of CXCL14 mRNA in brain is remarkably high in both mouse (Nara *et al.*, 2007; Sleeman *et al.*, 2000) and fish (Huising *et al.*, 2004). In addition, substantial amounts of CXCL14 transcript are produced in unfertilized carp eggs and this expression continues through the first 48 h of development (Huising *et al.*, 2004), which is before lymphoid organs develop. As the CXCL14 gene is evolutionally conserved (see the following section), these observations suggest that it may play roles in the central nervous system and in early development.

C. CXCL14 orthologues in other species

CXCL14 orthologues have been identified in nonhuman mammals, birds, and fish. As shown in Fig. 5.1, the primary amino acid sequence of CXCL14 is highly conserved in mammals. Human CXCL14 is 96% identical to bovine CXCL14 (95/99) at the amino acid level. Likewise, the sequence identity between human CXCL14 and the porcine, rat, and mouse proteins is 95% (94/99), 96% (95/99), and 95% (94/99), respectively. In the mature 77 amino acid form, human and mouse CXCL14 differ by only two residues. Therefore, human CXCL14 is able to cross-activate mouse macrophages in chemotaxis assays (Nara *et al.*, 2007). More surprisingly, the amino acid sequence of human CXCL14 is well conserved even in birds and fish, and is 60% (59/99) identical to chicken CXCL14, 58% (57/99) identical to carp CXCL14, and 54% (53/99) identical to zebrafish CXCL14. Reflecting the high degree of conservation of the primary amino acid sequence of CXCL14 in vertebrates, the exon–intron structure of the CXCL14 gene is also well conserved between human and fish (Huising *et al.*, 2004).

D. Receptor and signal transduction

To date, a receptor for CXCL14 has not been identified (Table 5.1). Thus, intracellular signaling events elicited by CXCL14 stimulation have not been fully explored in any species. Human monocytes acquire CXCL14-responsiveness and lose their chemotactic response to CCL2 after prostaglandin E_2 treatment, suggesting that expression of a putative CXCL14 receptor or signal transduction machinery of CXCL14 may be distinct from that of other inflammatory monokines (Kurth *et al.*, 2001). NF-κB is activated by CXCL14 stimulation in both immature dendritic cells and pancreatic cancer cells (Shurin *et al.*, 2005; Wente *et al.*, 2008).

III. Biological Activities of CXCL14

A. Chemotactic activity

CXCL14 is a highly selective chemoattractant for human monocytes that have been pretreated with prostaglandin E_2 or forskolin, agents that activate adenylate cyclase (Kurth *et al.*, 2001). Since a high concentration of CXCL14 (>100 nM) is required for maximum chemotaxis by prostaglandin E_2-treated macrophages, CXCL14 is likely to be a homeostatic chemokine, rather than an inflammatory chemokine, and to regulate the trafficking of tissue macrophages in skin and mucosal tissues.

However, recent studies revealed that human monocyte-derived immature dendritic cells, but not mature dendritic cells, are responsive to CXCL14 in the absence of prostaglandin E_2 treatment. In human, CXCL14 is a chemoattractant for dendritic cell precursors which are induced *in vitro* from $CD14^+$ monocytes and $CD34^+$ hematopoietic progenitor cells, and for immature dendritic cells directly isolated from peripheral blood (Schaerli *et al.*, 2005; Shellenberger *et al.*, 2004; Shurin *et al.*, 2005). As CXCL14 is constitutively expressed in normal epidermis (Schaerli *et al.*, 2005), CXCL14 could play an important role in recruitment of cutaneous dendritic precursor cells, which eventually differentiate into Langerhans cells under steady-state conditions. In *in vitro* experiments, loss of CXCL14 in head and neck squamous cell carcinomas is correlated with loss of chemoattraction of immature dendritic cells, whereas CXCL14-expressing carcinoma cells are able to recruit dendritic cells (Shurin *et al.*, 2005). In addition, CXCL14 up-regulates the expression of dendritic cell maturation markers and enhances the proliferation of allogenic T cells in mixed lymphocyte reactions (Shurin *et al.*, 2005). Therefore, chemoattraction of dendritic cell precursors by CXCL14 would contribute substantially to antitumor immunity.

Other types of cells for which CXCL14 acts as a chemoattractant include activated natural killer cells (Starnes *et al.*, 2006) and breast cancer-derived epithelial tumor cells (Allinen *et al.*, 2004). In the latter case, CXCL14 acts as a paracrine factor to facilitate tumor progression, in contrast to its tumor-suppressive role in other cell types.

All of the above evidence emanates from experiments using human cells. In the mouse, we recently reported that CXCL14 mediates migration of $Mac1^+$ peripheral blood leukocytes (Nara *et al.*, 2007). There is, however, as yet no evidence that CXCL14 acts as a chemoattractant for immature dendritic cells or natural killer cells of mouse origin. Furthermore, a very recent report demonstrated that the total number of macrophages and dendritic cells in the epidermis was not significantly different between CXCL14-deficient mice and wild-type mice (Meuter *et al.*, 2007).

B. Involvement in tumorigenesis

Initial studies have demonstrated that CXCL14 is preferentially expressed in normal tissues in the absence of inflammatory stimulation. In some tumor cell lines and malignant tumor samples, expression of CXCL14 mRNA is diminished (Frederick *et al.*, 2000; Hromas *et al.*, 1999; Sleeman *et al.*, 2000). In several cancer cell lines, CXCL14 protein is degraded via ubiquitin-mediated proteolysis by the 26S proteasome (Peterson *et al.*, 2006) and five amino acid residues (Val^{41}-Ser-Arg-Tyr-Arg^{45}), which are unique to CXCL14 among CXC chemokines, are responsible for this degradation. Thus, expression of CXCL14 is negatively regulated by both transcriptional and posttranslational mechanisms in cancer cells. Since CXCL14 potentially recruits dendritic precursor cells and activated natural killer cells, down-regulation of CXCL14 in malignant tumors may be a strategy by which tumor cells escape from immune surveillance. Strong expression of CXCL14 in inflammatory cells and stromal cells adjacent to squamous cell carcinomas of the tongue implies a tumor-suppressive role for CXCL14 (Frederick *et al.*, 2000). Inhibition of neovascularization by blocking endothelial cell chemotaxis (Shellenberger *et al.*, 2004) could be an additional antitumor function of CXCL14. Consistent with this hypothesis, overexpression of CXCL14 in both a prostate cancer cell line and an oral carcinoma cell line lowered the tumor forming capacity of the cells in nude mice (Ozawa *et al.*, 2006; Schwarze *et al.*, 2005).

However, CXCL14 expression is not always absent in cancer tissues. CXCL14 is, in fact, overexpressed in prostate and pancreatic cancers in human (Schwarze *et al.*, 2005; Wente *et al.*, 2008). Moreover, CXCL14 enhances the invasiveness of breast and pancreatic cancer cell lines (Allinen *et al.*, 2004; Wente *et al.*, 2008). Therefore, it appears that CXCL14 is suppressive for solid tumors, whereas malignant tumor cells may take advantage of the chemotactic activity of CXCL14 to enhance their invasiveness and metastasis. *In vivo* transplantation of tumor cells or carcinogen-induced tumorigenesis experiments using CXCL14-deficient mice could further our understanding of the roles of CXCL14 in the near future.

IV. Novel Functions of CXCL14 Revealed by Knockout Mice

A. General properties of CXCL14-deficient mice

To understand the physiological roles of CXCL14 in mice, we recently established a CXCL14-knockout mouse strain (CXCL14$^{-/-}$), independent of the previous knockout strain from the Moser laboratory. Both of these knockout strains produce a birth ratio of CXCL14$^{-/-}$ mice that is lower than the predicted Mendelian frequency and the surviving CXCL14$^{-/-}$ adult mice weigh $\sim$10–20% less than littermate CXCL14$^{+/-}$ mice or wild-type mice (Meuter *et al.*, 2007; Nara *et al.*, 2007). However, CXCL14$^{-/-}$ mice exhibited no macroscopic abnormalities and both male and female knockout mice are fertile, at least in their genetic background after crossing with C57BL/6 mice up to five generations (Meuter *et al.*, 2007; Takahiko Hara, unpublished data).

Reflecting the lighter body weight, the WAT mass of regular diet (RD)-fed CXCL14$^{-/-}$ mice is significantly reduced, while the weights of the liver and kidney are indistinguishable from littermate CXCL14$^{+/-}$ mice (Nara *et al.*, 2007). The lean phenotype and WAT hypoplasia of CXCL14$^{-/-}$ mice appears to be caused by decreased food intake

(Nara *et al.*, 2007), although more precise investigations are required to interpret this unexpected phenotype of CXCL14$^{-/-}$ mice. The phenotype may be related to the abundant expression of CXCL14 in the central nervous system, but the possibility of a somatotropic defect in the CXCL14$^{-/-}$ mice cannot currently be excluded.

Under HFD feeding conditions, CXCL14$^{-/-}$ mice gain a significant amount of weight and the total amount of visceral fat and subcutaneous fat accumulated by CXCL14$^{-/-}$ mice is indistinguishable from that of HFD-fed littermate CXCL14$^{+/-}$ mice (Nara *et al.*, 2007). Moreover, obesity-associated hypertrophy of adipocytes in WAT occurs in HFD-fed CXCL14$^{-/-}$ mice (Nara *et al.*, 2007). Therefore, CXCL14$^{-/-}$ mice are not defective in adipogenesis or WAT development. After 12 weeks of HFD feeding, livers of CXCL14$^{+/-}$ mice are enlarged and accumulate a large amount of fat. Interestingly, CXCL14$^{-/-}$ mice are partially protected from this HFD-induced hepatic hypertrophy and steatosis (Nara *et al.*, 2007). It remains to be determined whether this phenotype is directly related to the action of CXCL14 or is an indirect effect of the CXCL14-deficiency.

B. Macrophage infiltration into WAT in HFD-induced obese mice

Under RD feeding conditions, the total numbers of macrophages in WAT of CXCL14$^{-/-}$ mice are not significantly different from those of CXCL14$^{+/-}$ mice. In contrast, when mice are fed a HFD for more than 12 weeks, much larger numbers of Mac1^{+} macrophages are recruited to the WAT of CXCL14$^{+/-}$ mice compared to CXCL14$^{-/-}$ mice (Nara *et al.*, 2007). Immunohistochemical analysis also revealed a lower frequency of F4/80^{+} macrophages surrounding mature adipocytes in WAT from HFD-fed CXCL14$^{-/-}$ mice (Nara *et al.*, 2007). As mentioned earlier, expression of CXCL14 mRNA is strongly induced in WAT of HFD-induced obese mice. Taken together with the chemoattractive activity of CXCL14 for peripheral blood macrophages in mice, these results suggest that obesity-induced up-regulation of CXCL14 is responsible for the increase of WAT macrophages in HFD-fed obese mice.

C. Insulin sensitivity in control of blood glucose levels

Macrophage infiltration into visceral WAT is thought to be a critical cause of obesity-induced insulin resistance (Weisberg *et al.*, 2003; Xu *et al.*, 2003). Consistent with this hypothesis, HFD-fed CXCL14$^{-/-}$ mice exhibit a stronger insulin response to lower blood glucose levels when compared to HFD-fed CXCL14$^{+/-}$ mice (Nara *et al.*, 2007). Intriguingly, differences in the *in vivo* insulin sensitivity between CXCL14$^{-/-}$ and CXCL14$^{+/-}$ mice under HFD-feeding conditions are much more pronounced in female mice than in male mice (Nara *et al.*, 2007). Thus, amelioration of HFD-induced insulin resistance in CXCL14$^{-/-}$ mice is largely dependent on gender.

Fasting blood glucose levels of HFD-fed CXCL14$^{-/-}$ mice are significantly lower than those of HFD-fed CXCL14$^{+/-}$ mice. Under RD feeding conditions, however, the insulin sensitivity of CXCL14$^{-/-}$ mice is similar to CXCL14$^{+/-}$ mice.

In the liver, skeletal muscle, and WAT of HFD-fed CXCL14$^{-/-}$ mice, insulin-stimulated phsophorylation of Ser^{473} of Akt is much more robust than in the corresponding organs of HFD-fed CXCL14$^{+/-}$ mice (Nara *et al.*, 2007). Improvement in the insulin response in HFD-fed CXCL14$^{-/-}$ mice is more prominent in skeletal muscle than in liver or WAT.

D. Insulin secretion upon glucose infusion

Serum insulin concentrations of HFD-fed CXCL14$^{-/-}$ female mice are significantly lower than those of HFD-fed CXCL14$^{+/-}$ female mice (Nara *et al.*, 2007), suggesting that CXCL14$^{-/-}$ mice are protected from HFD-induced hyperinsulinemia. However, HFD-fed CXCL14$^{-/-}$ mice are defective in glucose disposal based on the intraperitoneal glucose tolerance test. Serum insulin levels in HFD-fed CXCL14$^{-/-}$ mice are consistently lower than in HFD-fed CXCL14$^{+/-}$ mice even when large amounts of glucose are administered (Nara *et al.*, 2007). CXCL14 deficiency is associated not only with improved insulin sensitivity, but also impaired insulin production, which would result in CXCL14$^{-/-}$ mice being glucose intolerant under HFD-feeding conditions.

Serum insulin levels of CXCL14$^{-/-}$ mice are also lower than CXCL14$^{+/-}$ mice even under RD feeding conditions, implying a positive correlation between the presence of CXCL14 and insulin secretion. The reduced body weight of CXCL14$^{-/-}$ mice may also affect the serum insulin levels. Nevertheless, RD-fed CXCL14$^{-/-}$ mice exhibit a slightly better glucose disposal capacity compared to CXCL14$^{+/-}$ mice in the intraperitoneal glucose tolerance test (Nara *et al.*, 2007).

E. Effect of CXCL14 on serum lipid levels and glucose transporter expression

Serum cholesterol in HFD-fed CXCL14$^{-/-}$ mice is significantly lower than in CXCL14$^{+/-}$ mice, while the concentrations of triglyceride and FFA are unchanged (Nara *et al.*, 2007). The concentration of triglyceride in the liver is decreased in HFD-fed CXCL14$^{-/-}$ mice (Nara *et al.*, 2007), suggesting that the CXCL14-deficiency affects lipid turnover to some extent.

In HFD-fed CXCL14$^{-/-}$ mice, the mRNA levels of glucose transporter-4 (GLUT4) in BAT and skeletal muscle are unaltered compared to HFD-fed CXCL14$^{+/-}$ mice. In contrast, down-regulation of GLUT4 mRNA in WAT occurred in both HFD-fed CXCL14$^{-/-}$ and CXCL14$^{+/-}$ mice (Nara *et al.*, 2007). Therefore, CXCL14 action is independent of obesity-induced down-regulation of GLUT4 in WAT, which is a common and crucial feature of insulin resistance (Shepherd and Kahn, 1999).

F. Effect of CXCL14 on the expression of adipokines and cytokines

Leptin, adiponectin, and retinol-binding protein-4 (RBP4) are key adipose tissue-derived regulators of obesity-associated type 2 diabetes. Serum leptin concentrations are lower in RD-fed CXCL14$^{-/-}$ mice compared to RD-fed CXCL14$^{+/-}$ mice, but are comparable in the two strains under HFD feeding conditions (Nara *et al.*, 2007). Serum leptin levels in these mice reflect the differences in WAT mass in mice fed a RD or a HFD.

In contrast, serum adiponectin levels are significantly higher in HFD-fed CXCL14$^{-/-}$ mice compared to HFD-fed CXCL14$^{+/-}$ mice (Nara *et al.*, 2007), indicating that CXCL14$^{-/-}$ mice are insensitive to HFD-induced hypoadiponectinemia. Moreover, the serum concentration of RBP4 in CXCL14$^{-/-}$ mice is significantly lower than in CXCL14$^{+/-}$ mice under both RD and HFD feeding conditions (Nara *et al.*, 2007). The levels of phosphoenolpyruvate kinase and glucose-6-phosphatase mRNAs in the livers of HFD-fed CXCL14$^{-/-}$ mice are much lower than in the livers of HFD-fed CXCL14$^{+/-}$ mice (Nara *et al.*, 2007). These observations are consistent with previous reports demonstrating that RBP4 enhances gluconeogenesis in the liver (Yang *et al.*, 2005) and that adiponectin suppresses hepatic glucose production (Yamauchi *et al.*, 2002). Presumably, the decrease in RBP4 and the increase in adiponectin in HFD-fed CXCL14$^{-/-}$ mice in part accounts for the protection of CXCL14$^{-/-}$ mice from obesity-induced gluconeogenesis in the liver and obesity-associated impairment of insulin signaling in WAT, liver, and skeletal muscle.

IL-6 and CCL2 are crucial mediators of obesity-induced insulin resistance. HFD feeding up-regulates the expression of IL-6 in WAT of CXCL14$^{+/-}$ mice, but not CXCL14$^{-/-}$ mice (Nara *et al.*, 2007). However, HFD-induced up-regulation of CCL2 occurs in both CXCL14$^{+/-}$ and CXCL14$^{-/-}$ mice (Nara *et al.*, 2007). Therefore, obesity-induced up-regulation of CCL2 in WAT does not compensate for the impaired macrophage infiltration into WAT of HFD-fed CXCL14$^{-/-}$ mice.

G. Contribution of skeletal muscle-derived CXCL14 in induction of insulin resistance

Obesity-associated up-regulation of CXCL14 mRNA occurs in BAT, WAT, and skeletal muscle. We recently generated a skeletal muscle specific CXCL14 transgenic (CXCL14^{Tg}) mouse line, by utilizing the mouse muscle creatine kinase (MCK) promoter/enhancer. Under HFD feeding conditions, insulin sensitivity is significantly blunted in CXCL14$^{-/-Tg}$ mice, similar to HFD-fed CXCL14$^{+/+}$ or CXCL14$^{+/-}$ mice (Nara *et al.*, 2007). Thus, constitutive expression of CXCL14 from skeletal muscle can restore HFD-induced whole body insulin resistance in CXCL14$^{-/-}$ mice. Under RD feeding conditions, however, the insulin response of the double mutant CXCL14$^{-/-Tg}$ mice is indistinguishable from that of CXCL14$^{+/-}$ or CXCL14$^{-/-}$ mice (Nara *et al.*, 2007). Overexpression of CXCL14 itself is not sufficient to induce insulin resistance. Obesity-associated changes such as larger WAT for macrophage infiltration and/or impaired expression of various adipokines would be required to induce whole body insulin resistance in combination with CXCL14.

V. Signal Cross-Talk Between CXCL14 and Insulin

As expression of CXCL14 mRNA increases in skeletal muscle of HFD-fed obese mice, CXCL14 may transduce some signals in myocytes. Such is indeed the case. When differentiated myotubes derived from C2C12 cells are pretreated for 1 h with CXCL14, insulin-stimulated Ser^{473}-phosphorylation of Akt is attenuated by ~40% (Nara *et al.*, 2007). Moreover, a 1 h preincubation with CXCL14 significantly inhibits the insulin-stimulated uptake of 2-deoxyglucose in C2C12-derived myocytes (Nara *et al.*, 2007). CXCL14 also inhibits glucose uptake in 3T3-L1-derived adipocytes, but the effect is marginal when compared to cultured myocytes. A very recent report indicates that overnight exposure of 3T3-L1-derived adipocytes to CXCL14 substantially augments insulin-mediated signal transduction (Takahashi *et al.*, 2007). Thus, CXCL14 may trigger adipocytes to produce secondary factors which coordinately modulate the insulin sensitivity of the adipocytes themselves. In any case, putative CXCL14 receptors must be expressed in myocytes and adipocytes, and CXCL14-mediated signals may cross-talk with the insulin-elicited signal transduction pathways. CCL2 possesses similar insulin-inhibitory activity in cultured adipocytes and myocytes (Kamei *et al.*, 2006; Sartipy and Loskutoff, 2003).

VI. CXCL14 as a Metabolic Regulator

As described, recent studies using CXCL14-knockout mice revealed that CXCL14 is an obesity-associated regulator of glucose metabolism in female mice. As schematically shown in Fig. 5.2, obesity-induced

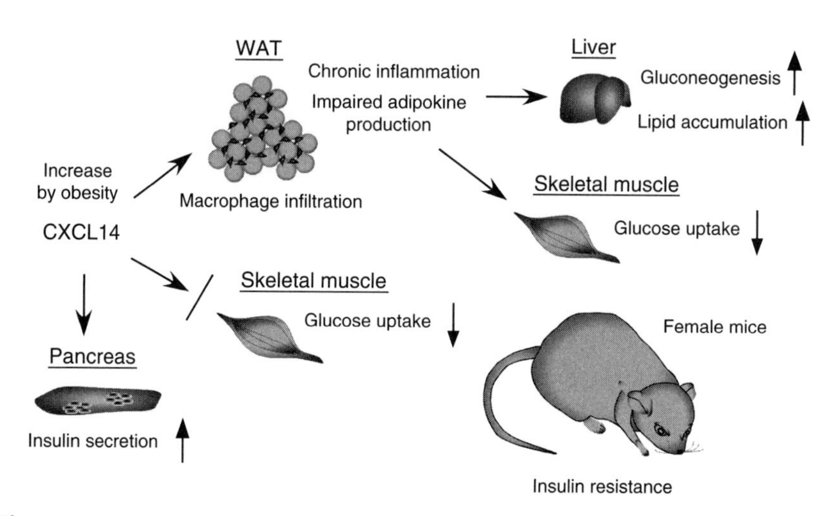

Figure 5.2 Proposed functions of CXCL14 as a metabolic regulator. Based on the phenotypic abnormalities of CXCL14-deficient female mice under HFD feeding conditions, CXCL14 is involved in obesity-induced insulin resistance. Target organs and outcomes of CXCL14 action are shown schematically.

up-regulation of CXCL14 in WAT promotes the infiltration of macrophages into WAT and subsequent inflammatory responses. Enhanced production of CXCL14 in HFD-fed mice alters the expression of adipokines, including adiponectin, RBP4, and IL-6, thereby enhancing gluconeogenesis in the liver and blunting glucose uptake in skeletal muscle. The dramatic increase in macrophages in WAT and a direct CXCL14 action on skeletal muscle are likely to play important roles in this diabetic cascade. CXCL14 indirectly participates in fatty liver formation as well, which may also have a large impact on glucose metabolism.

Previous reports have demonstrated that CCL2 inhibits insulin-stimulated glucose uptake in adipocytes and skeletal muscle cells (Kamei *et al.*, 2006; Sartipy and Loskutoff, 2003). In our recent study, we have shown that CXCL14 attenuates insulin-stimulated glucose uptake in skeletal muscle (Nara *et al.*, 2007). Thus, chemokines play crucial roles in the obesity-induced impairment of glucose metabolism, more than simply recruiting inflammatory cells to the visceral WAT. It will be important to elucidate the molecular basis of the insulin inhibitory signals elicited by CCR2 and/or the putative CXCL14 receptor. In addition, further investigations are necessary to answer why amelioration of obesity-induced whole body insulin resistance occurs only in female CXCL14$^{-/-}$ mice. A similar gender-specific insulin sensitive phenotype has been reported in Toll-like receptor 4 knockout mice.

VII. Conclusions

CXCL14 research began in 1999. Over the last 9 years, our knowledge of the physiological roles of this relatively young member of the CXC chemokines has increased incrementally. In human systems, the chemoattractive activity of CXCL14 to recruit dendritic precursor cells strengthens immunological barriers against malignant cancer cells. The angiostatic activity of CXCL14 may also be part of the tumor-suppressive role of CXCL14. However, these biological activities may not be unique to CXCL14 in mice, as numbers of tissue macrophages and dendritic precursor cells are unchanged in CXCL14-deficient mice. Instead, phenotypes of $CXCL14^{-/-}$ female mice clearly demonstrate that CXCL14 is involved in body weight control, as well as regulation of blood glucose levels by recruitment of WAT macrophages and direct interaction with insulin signaling pathways in skeletal muscle. Elucidation of the CXCL14 receptor structure and signal transduction machinery will be key in understanding the physiological functions of this novel metabolic regulator at the molecular level.

REFERENCES

Allinen, M., Beroukhim, R., Cai, L., Brennan, C., Lahti-Domenici, J., Huang, H., Porter, D., Hu, M., Chin, L., Richardson, A., Schnitt, S., Sellers, W. R., *et al.* (2004). Molecular characterization of the tumor microenvironment in breast cancer. *Cancer Cell* **6,** 17–32.

Arkan, M. C., Hevener, A. L., Greten, F. R., Maeda, S., Li, Z. W., Long, J. M., Wynshaw-Boris, A., Poli, G., Olefsky, J., and Karin, M. (2005). IKK-beta links inflammation to obesity-induced insulin resistance. *Nat. Med.* **11,** 191–198.

Belperio, J. A., Keane, M. P., Arenberg, D. A., Addison, C. L., Ehlert, J. E., Burdick, M. D., and Strieter, R. M. (2000). CXC chemokines in angiogenesis. *J. Leukoc. Biol.* **68,** 1–8.

Cai, D., Yuan, M., Frantz, D. F., Melendez, P. A., Hansen, L., Lee, J., and Shoelson, S. E. (2005). Local and systemic insulin resistance resulting from hepatic activation of IKK-beta and NF-kappaB. *Nat. Med.* **11,** 183–190.

Cao, X., Zhang, W., Wan, T., He, L., Chen, T., Yuan, Z., Ma, S., Yu, Y., and Chen, G. (2000). Molecular cloning and characterization of a novel CXC chemokine macrophage inflammatory protein-2 gamma chemoattractant for human neutrophils and dendritic cells. *J. Immunol.* **165,** 2588–2595.

Chen, A., Mumick, S., Zhang, C., Lamb, J., Dai, H., Weingarth, D., Mudgett, J., Chen, H., MacNeil, D. J., Reitman, M. L., and Qian, S. (2005). Diet induction of monocyte chemoattractant protein-1 and its impact on obesity. *Obes. Res.* **13,** 1311–1320.

Frederick, M. J., Henderson, Y., Xu, X., Deavers, M. T., Sahin, A. A., Wu, H., Lewis, D. E., El-Naggar, A. K., and Clayman, G. L. (2000). *In vivo* expression of the novel CXC chemokine BRAK in normal and cancerous human tissue. *Am. J. Pathol.* **156,** 1937–1950.

Hirosumi, J., Tuncman, G., Chang, L., Gorgun, C. Z., Uysal, K. T., Maeda, K., Karin, M., and Hotamisligil, G. S. (2002). A central role for JNK in obesity and insulin resistance. *Nature* **420,** 333–336.

Hromas, R., Broxmeyer, H. E., Kim, C., Nakshatri, H., Christopherson, K. II,, Azam, M., and Hou, Y. H. (1999). Cloning of BRAK, a novel divergent CXC chemokine preferentially expressed in normal versus malignant cells. *Biochem. Biophys. Res. Commun.* **255,** 703–706.

Huising, M. O., van der Meulen, T., Flik, G., and Verburg-van Kemenade, B. M. (2004). Three novel carp CXC chemokines are expressed early in ontogeny and at nonimmune sites. *Eur. J. Biochem.* **271,** 4094–4106.

Inouye, K. E., Shi, H., Howard, J. K., Daly, C. H., Lord, G. M., Rollins, B. J., and Flier, J. S. (2007). Absence of CC chemokine ligand 2 does not limit obesity-associated infiltration of macrophages into adipose tissue. *Diabetes* **56,** 2242–2250.

Kamei, N., Tobe, K., Suzuki, R., Ohsugi, M., Watanabe, T., Kubota, N., Ohtsuka-Kowatari, N., Kumagai, K., Sakamoto, K., Kobayashi, M., Yamauchi, T., Ueki, K., *et al.* (2006). Overexpression of monocyte chemoattractant protein-1 in adipose tissues causes macrophage recruitment and insulin resistance. *J. Biol. Chem.* **281,** 26602–26614.

Kanda, H., Tateya, S., Tamori, Y., Kotani, K., Hiasa, K. I., Kitazawa, R., Kitazawa, S., Miyachi, H., Maeda, S., Egashira, K., and Kasuga, M. (2006). MCP-1 contributes to macrophage infiltration into adipose tissue, insulin resistance, and hepatic steatosis in obesity. *J. Clin. Invest.* **116,** 1494–1505.

Kirk, E. A., Sagawa, Z. K., McDonald, T. O., O'Brien, K. D., and Heinecke, J. W. (2008). Macrophage chemoattractant protein deficiency fails to restrain macrophage infiltration into adipose tissue. *Diabetes* **57,** 1254–1261.

Kurth, I., Willimann, K., Schaerli, P., Hunziker, T., Clark-Lewis, I., and Moser, B. (2001). Monocyte selectivity and tissue localization suggests a role for breast and kidney-expressed chemokine (BRAK) in macrophage development. *J. Exp. Med.* **194,** 855–861.

Meuter, S., Schaerli, P., Roos, R. S., Brandau, O., Bosl, M. R., von Andrian, U. H., and Moser, B. (2007). Murine CXCL14 is dispensable for dendritic cell function and localization within peripheral tissues. *Mol. Cell. Biol.* **27,** 983–992.

Nara, N., Nakayama, Y., Okamoto, S., Tamura, H., Kiyono, M., Muraoka, M., Tanaka, K., Taya, C., Shitara, H., Ishii, R., Yonekawa, H., Minokoshi, Y., *et al.* (2007). Disruption of CXC motif chemokine ligand-14 in mice ameliorates obesity-induced insulin resistance. *J. Biol. Chem.* **282,** 30794–30803.

Nelson, P. J., and Krensky, A. M. (2001). Chemokines, chemokine receptors, and allograft rejection. *Immunity* **14,** 377–386.

Nguyen, M. T., Satoh, H., Favelyukis, S., Babendure, J. L., Imamura, T., Sbodio, J. I., Zalevsky, J., Dahiyat, B. I., Chi, N. W., and Olefsky, J. M. (2005). JNK and tumor necrosis factor-alpha mediate free fatty acid-induced insulin resistance in 3T3-L1 adipocytes. *J. Biol. Chem.* **280,** 35361–35371.

Ogawa, S., Lozach, J., Jepsen, K., Sawka-Verhelle, D., Perissi, V., Sasik, R., Rose, D. W., Johnson, R. S., Rosenfeld, M. G., and Glass, C. K. (2004). A nuclear receptor corepressor transcriptional checkpoint controlling activator protein 1-dependent gene networks required for macrophage activation. *Proc. Natl. Acad. Sci. USA* **101,** 14461–14466.

Ozawa, S., Kato, Y., Komori, R., Maehata, Y., Kubota, E., and Hata, R. (2006). BRAK/CXCL14 expression suppresses tumor growth *in vivo* in human oral carcinoma cells. *Biochem. Biophys. Res. Commun.* **348,** 406–412.

Peterson, F. C., Thorpe, J. A., Harder, A. G., Volkman, B. F., and Schwarze, S. R. (2006). Structural determinants involved in the regulation of CXCL14/BRAK expression by the 26 S proteasome. *J. Mol. Biol.* **363,** 813–822.

Sartipy, P., and Loskutoff, D. J. (2003). Monocyte chemoattractant protein 1 in obesity and insulin resistance. *Proc. Natl. Acad. Sci. USA* **100,** 7265–7270.

Schaerli, P., Willimann, K., Ebert, L. M., Walz, A., and Moser, B. (2005). Cutaneous CXCL14 targets blood precursors to epidermal niches for langerhans cell differentiation. *Immunity* **23,** 331–342.

Schwarze, S. R., Luo, J., Isaacs, W. B., and Jarrard, D. F. (2005). Modulation of CXCL14 (BRAK) expression in prostate cancer. *Prostate* **64,** 67–74.

Shellenberger, T. D., Wang, M., Gujrati, M., Jayakumar, A., Strieter, R. M., Burdick, M. D., Ioannides, C. G., Efferson, C. L., El-Naggar, A. K., Roberts, D., Clayman, G. L., and Frederick, M. J. (2004). BRAK/CXCL14 is a potent inhibitor of angiogenesis and a chemotactic factor for immature dendritic cells. *Cancer Res.* **64,** 8262–8270.

Shepherd, P. R., and Kahn, B. B. (1999). Glucose transporters and insulin action—implications for insulin resistance and diabetes mellitus. *N. Engl. J. Med.* **341,** 248–257.

Shurin, G. V., Ferris, R., Tourkova, I. L., Perez, L., Lokshin, A., Balkir, L., Collins, B., Chatta, G. S., and Shurin, M. R. (2005). Loss of new chemokine CXCL14 in tumor tissue is associated with low infiltration by dendritic cells (DC), while restoration of human CXCL14 expression in tumor cells causes attraction of DC both *in vitro* and *in vivo*. *J. Immunol.* **174,** 5490–5498.

Sleeman, M. A., Fraser, J. K., Murison, J. G., Kelly, S. L., Prestidge, R. L., Palmer, D. J., Watson, J. D., and Kumble, K. D. (2000). B cell- and monocyte-activating chemokine (BMAC), a novel non-ELR alpha-chemokine. *Int. Immunol.* **12,** 677–689.

Starnes, T., Rasila, K. K., Robertson, M. J., Brahmi, Z., Dahl, R., Christopherson, K., and Hromas, R. (2006). The chemokine CXCL14 (BRAK) stimulates activated NK cell migration: Implications for the downregulation of CXCL14 in malignancy. *Exp. Hematol.* **34,** 1101–1105.

Takahashi, K., Mizuarai, S., Araki, H., Mashiko, S., Ishihara, A., Kanatani, A., Itadani, H., and Kotani, H. (2003). Adiposity elevates plasma MCP-1 levels leading to the increased CD11b-positive monocytes in mice. *J. Biol. Chem.* **278,** 46654–46660.

Takahashi, M., Takahashi, Y., Takahashi, K., Zolotaryov, F. N., Hong, K. S., Iida, K., Okimura, Y., Kaji, H., and Chihara, K. (2007). CXCL14 enhances insulin-dependent glucose uptake in adipocytes and is related to high-fat diet-induced obesity. *Biochem. Biophys. Res. Commun.* **364,** 1037–1042.

Uysal, K. T., Wiesbrock, S. M., Marino, M. W., and Hotamisligil, G. S. (1997). Protection from obesity-induced insulin resistance in mice lacking TNF-alpha function. *Nature* **389,** 610–614.

Weisberg, S. P., McCann, D., Desai, M., Rosenbaum, M., Leibel, R. L., and Ferrante, A. W. Jr., (2003). Obesity is associated with macrophage accumulation in adipose tissue. *J. Clin. Invest.* **112,** 1796–1808.

Weisberg, S. P., Hunter, D., Huber, R., Lemieux, J., Slaymaker, S., Vaddi, K., Charo, I., Leibel, R. L., and Ferrante, A. W. (2006). CCR2 modulates inflammatory and metabolic effects of high-fat feeding. *J. Clin. Invest.* **116,** 115–124.

Wente, M. N., Mayer, C., Gaida, M. M., Michalski, C. W., Giese, T., Bergmann, F., Giese, N. A., Buchler, M. W., and Friess, H. (2008). CXCL14 expression and potential function in pancreatic cancer. *Cancer Lett.* **259,** 209–217.

Xu, H., Barnes, G. T., Yang, Q., Tan, G., Yang, D., Chou, C. J., Sole, J., Nichols, A., Ross, J. S., Tartaglia, L. A., and Chen, H. (2003). Chronic inflammation in fat plays a crucial role in the development of obesity-related insulin resistance. *J. Clin. Invest.* **112,** 1821–1830.

Yamauchi, T., Kamon, J., Minokoshi, Y., Ito, Y., Waki, H., Uchida, S., Yamashita, S., Noda, M., Kita, S., Ueki, K., Eto, K., Akanuma, Y., *et al.* (2002). Adiponectin stimulates glucose utilization and fatty-acid oxidation by activating AMP-activated protein kinase. *Nat. Med.* **8,** 1288–1295.

Yang, Q., Graham, T. E., Mody, N., Preitner, F., Peroni, O. D., Zabolotny, J. M., Kotani, K., Quadro, L., and Kahn, B. B. (2005). Serum retinol binding protein 4 contributes to insulin resistance in obesity and type 2 diabetes. *Nature* **436,** 356–362.

Zlotnik, A., and Yoshie, O. (2000). Chemokines: A new classification system and their role in immunity. *Immunity* **12,** 121–127.

CHAPTER SIX

Crosstalk Between Growth Hormone and Insulin Signaling

Jie Xu* *and* Joseph L. Messina[†,‡]

Contents

Abstract

Growth Hormone (GH) is a major growth-promoting and metabolic regulatory hormone. Interaction of GH with its cell surface GH receptor (GHR) causes activation of the GHR-associated cytoplasmic tyrosine kinase, JAK2, and activation of several signaling pathways, including the STATs, ERK1/2, and PI3K pathways. Insulin is also a key hormone regulating metabolism and growth.

* Department of Medicine, Division of Endocrinology, Diabetes, and Metabolism, University of Alabama at Birmingham, Birmingham, Alabama 35294-0019
† Department of Pathology, Division of Molecular and Cellular Pathology, 1670 University Boulevard, University of Alabama at Birmingham, Birmingham, Alabama 35294-0019
‡ Veterans Affairs Medical Center, Birmingham, Alabama 35233

Vitamins and Hormones, Volume 80
ISSN 0083-6729, DOI: 10.1016/S0083-6729(08)00606-7

Insulin binding to the insulin receptor (IR) results in phosphorylation/activation of the IR, and activates the PI3K/Akt and ERK1/2 pathways. Due to their important roles in growth and metabolism, GH and insulin can functionally interact with each other, regulating cellular metabolism. In addition, recent data suggests that GH and insulin can directly interact by signaling crosstalk. Insulin regulation of GH signaling depends on the duration of exposure to insulin. Transient insulin exposure enhances GH-induced activation of MEK/ERK pathway through post-GHR mechanisms, whereas prolonged insulin exposure inhibits GH-induced signaling at both receptor and postreceptor levels. Chronic excessive GH interferes with insulin's activation of the IR/IRS/PI3K pathway and several proteins are involved in the mechanisms underlying GH-induced insulin resistance.

I. Introduction

Growth hormone (GH) is a 22-kDa polypeptide hormone, synthesized and secreted primarily by somatotrophs of the anterior pituitary. GH is an important regulator of somatic growth and metabolism. It directly stimulates proliferation and differentiation of chondrocyte precursors, which enter the growth plate as proliferative chondrocytes (Le *et al.*, 2001). GH also induces the expression of insulin-like growth factor-1 (IGF-1), which is required for chondrocyte hypertrophy (Isgaard *et al.*, 1986; Le *et al.*, 2001). A lack of GH results in dwarfism, as seen in patients with congenital GH deficiency (Baumann, 2001; Vance and Mauras, 1999) and GH-deficient animal models (Laron, 2005; Okuma, 1984; Sornson *et al.*, 1996). Excessive GH is the cause of acromegaly, with overgrowth of bone and connective tissue, and, if occurring prior to epiphyseal closure, excessive height (Baumann, 2001; Vance and Mauras, 1999). GH transgenic mice exhibit an increase in adult body size ranging from ~30% to nearly 100% in different lines (Bartke *et al.*, 1994; Palmiter *et al.*, 1982; Searle *et al.*, 1992).

In addition to promoting growth, GH also has important metabolic actions, including decreasing fat and increasing lean body mass (Fain *et al.*, 1999; Waxman *et al.*, 1995). New research continues to reveal other potential roles of GH, including regulation of cardiac and immune function, and aging. GH protects cardiac myocytes against apoptosis (Gu *et al.*, 2001), promotes cell cycle progression, and prevents apoptosis of lymphoid cells (Jeay *et al.*, 2002). The data from short-lived GH transgenic mice and long-lived GHR knockout mice suggests that GH reduces lifespan (Brown-Borg *et al.*, 1996; Coschigano *et al.*, 2000, 2003).

Given its physiological roles in growth and metabolism, it is not surprising that GH may functionally interact with other hormones and growth factors, such as insulin. For example, GH can exert both insulin-like and insulin-antagonistic effects on certain target cells and tissues. Acute exposure to GH results in immediate insulin-like effects including an inhibition of lipolysis and stimulation of lipogenesis (Smal *et al.*, 1987). Chronically, GH opposes insulin action, promoting insulin resistance and diabetes (Dominici *et al.*, 1999a; Smith *et al.*, 1997; Thirone *et al.*, 1997). This review focuses on the crosstalk between GH and insulin signaling, particularly the regulation of GH signaling by insulin.

II. GH Signaling

GH exerts its effects by interacting with the GH receptor (GHR), a member of the cytokine receptor superfamily. The GHR is expressed in multiple tissues, including muscle, adipose, heart, kidney, intestine, and bone, with expression being most abundant in the liver. Binding of GH to its receptor results in the rapid phosphorylation/activation of Janus activating kinase 2 (JAK2) and GHR (Argetsinger *et al.*, 1993; Carter-Su *et al.*, 1996). Activation of the GHR/JAK2 complex leads to activation of signal transducers and activators of transcription (STATs), ERK/MAPK Kinase (MEK)/extracellularly regulated kinase (ERK), and phosphatidylinositol 3 kinase (PI3K)/Akt pathways (Bennett *et al.*, 2007b; Carter-Su *et al.*, 1996; Frank *et al.*, 1995; Moller *et al.*, 1992; VanderKuur *et al.*, 1997; Xu *et al.*, 2005, 2006a) (Fig. 6.1).

A. STAT pathway

STAT proteins are recruited to activated JAK2 by binding to the GHR/JAK2 complex through the STAT Src-homology-2 (SH2) domains. STATs are then tyrosine phosphorylated and activated by JAK2. Activated STATs then homo- or heterodimerize through a reciprocal SH2 domain-phosphotyrosine interaction, translocate to the nucleus, bind to their corresponding DNA response elements, and stimulate gene transcription, including IGF-1 (Davey *et al.*, 1999).

GH can utilize STAT1, STAT3, and STAT5 to regulate a variety of genes, but STAT5 is the predominant STAT protein activated by GH (Bergad *et al.*, 1995; Campbell *et al.*, 1995; Choi and Waxman, 2000; Delesque-Touchard *et al.*, 2000; Ji *et al.*, 1999; Lahuna *et al.*, 2000; Meyer *et al.*, 1994; Sotiropoulos *et al.*, 1995; Waxman *et al.*, 1995; Xu *et al.*, 2005). The two isoforms of STAT5 (STAT5a and STAT5b) are encoded by two homologous genes, which are ~90% identical and possess both overlapping and distinct functions in GH signal transduction (Herrington *et al.*, 2000; Shuai, 1999). STAT5b is the major STAT5 isoform activated by GH in the

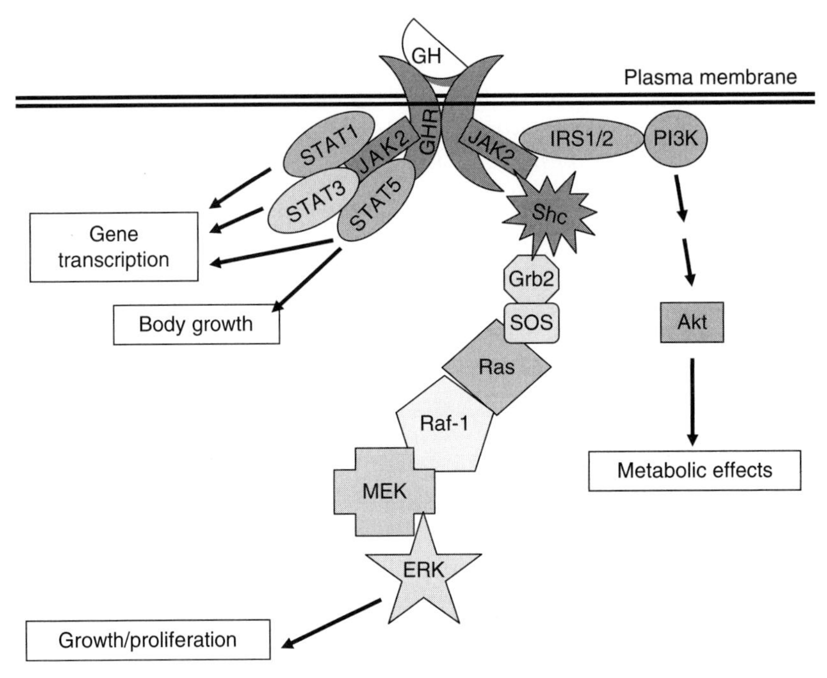

Figure 6.1 The main intracellular signaling pathways activated by GH. Major signaling pathways engaged by the activated GHR/JAK2 complex include the STATs, MEK/ERK, and PI3K/Akt pathways, as described in the text.

liver and is vital for sexually dimorphic hepatic gene expression in response to the male (pulsatile) or the female (continuous) GH secretory pattern (Waxman *et al.*, 1995). STAT5b is also required for GH-stimulated body growth and lipolysis in adipose tissue (Fain *et al.*, 1999; Waxman *et al.*, 1995) and STAT5b is necessary for GH-stimulated IGF-1 gene transcription (Chia *et al.*, 2006; Woelfle *et al.*, 2003a,b). A GH pulse can induce the binding of homo- and heterodimers of STAT1 and STAT3 to the c-sis-inducible element (SIE) of the c-fos promoter, resulting in the regulation of c-fos gene expression (Campbell *et al.*, 1995; Gronowski and Rotwein, 1994; Herrington *et al.*, 2000; Meyer *et al.*, 1994; Ram *et al.*, 1996) and likely many other genes.

B. MEK/ERK pathway

Unlike STAT5 activation which requires the GHR cytoplasmic domain in addition to JAK2 activation, the activation of ERK1/2 by GH only requires the GHR regions responsible for GHR interaction with JAK2 (Frank *et al.*, 1995; Hansen *et al.*, 1996; Moller *et al.*, 1992; Sotiropoulos *et al.*, 1994). GH activates ERK1/2 by both JAK2-dependent and JAK2-independent mechanisms.

The JAK2-dependent mechanism is a cascade involving Shc, Grb2, SOS, Ras, Raf-1, and MEK1/2 (Cobb, 1999; Lewis *et al.*, 1998; VanderKuur *et al.*, 1997; Winston and Hunter, 1995; Xu *et al.*, 2006a). Coupling of GHR to a catalytically active JAK2 results in the recruitment and tyrosine phosphorylation of Shc, which then interacts with the adapter protein Grb2. Preassociated with Grb2 is SOS, a guanine nucleotide exchange factor which is recruited to the cell membrane via Shc/Grb2 association. At the cell membrane, SOS stimulates the formation of active GTP-bound Ras, which then initiates a sequence of phosphorylation events, activating a cascade of protein serine/threonine kinases, including Raf-1 kinase. Activated Raf-1 phosphorylates and activates MEK1/2, which in turn phosphorylates and activates ERK1/2. A JAK2-independent mechanism of ERK1/2 activation has also been reported which involves Src-dependent activation of Ral and phospholipase D (Zhu *et al.*, 2002). GH induces activation of ERK1/2 in several model systems, including primary hepatocytes, rat hepatoma H4IIE cells, and murine 3T3-F442A preadipocyte fibroblasts (Anderson, 1992; Campbell *et al.*, 1992; Moller *et al.*, 1992; Winston and Bertics, 1992; Xu *et al.*, 2006a,b). However, ERK1/2 is not activated in all cells in which GH activates JAK2, such as IM-9 lymphocytes (Love *et al.*, 1998), suggesting that the activation of ERK1/2 by GH is cell-type dependent. GH-activated ERK1/2 induces activation of transcription factors and downstream gene expression, including the c-fos, Egr-1, and Jun-B genes (Herrington *et al.*, 2000; Hodge *et al.*, 1998). ERK1/2 activation also mediates GH-induced cell proliferation and differentiation (Hikida *et al.*, 1995; Hodge *et al.*, 1998; Liang *et al.*, 1999; Nguyen *et al.*, 1996) and plays an important role in cross-talk between GH and other growth factor signaling pathways (Huang *et al.*, 2003; Kim *et al.*, 1999; Liang *et al.*, 1999, 2000; Xu *et al.*, 2006a,b). GH utilizes members of the epidermal growth factor receptor (EGFR) family, such as EGFR and ErbB-2 (Kim *et al.*, 1999; Yamauchi *et al.*, 1997) and GH-stimulated phosphorylation of EGFR can serve as a scaffold to further elicit the activation of ERK1/2 (Yamauchi *et al.*, 1997). GH stimulates the tyrosine dephosphorylation and serine/threonine phosphorylation of ErbB2 via ERK1/2 pathway in 3T3-F442A preadipocytes in which cotreatment with EGF and GH results in a marked decline in DNA synthesis and a substantial decrease of cyclin D1 expression (Kim *et al.*, 1999).

C. PI3K/Akt pathway

GH stimulates tyrosine phosphorylation of IRS-1, IRS-2, and IRS-3 (Argetsinger *et al.*, 1995, 1996; Ridderstrale *et al.*, 1995; Souza *et al.*, 1994; Yamauchi *et al.*, 1998). PI3K is a pivotal effector molecule downstream of IRSs, and GH has been demonstrated to activate or tyrosine phosphorylate PI3K, which is associated with IRS-1 or IRS-2

(White, 1998; Yenush and White, 1997). PI3K plays important roles in many cellular processes, including cell proliferation, cell survival, and cellular metabolism (White, 1998; Yenush and White, 1997).

D. Mutations of GH signal transduction

Cellular GH signaling can be modulated, either augmented or attenuated, by alteration of GHR abundance or availability, or post-GHR signaling molecules. $GHR^{-/-}$ mice are viable and of nearly normal size at birth, but lag behind their normal siblings in postnatal growth and reach ~50% of normal adult body weight (Zhou *et al.*, 1997). GH resistance can be caused by a GHR mutation and over 50 different mutations of the GHR gene have been identified, including nonsense, frameshift, splicing, and missense mutations. These mutations may impair GHR gene expression, GH binding, or possibly GH signaling. A GH variant manifesting normal activation of STAT5, but reduced activation of ERK1/2, was found in a child with short stature (Lewis *et al.*, 2004). A separate mutant GHR, found in a family with markedly short stature, shows an increased GH activation of STAT5, with only minimal activation of the ERK1/2 pathway, compared with wild-type GHR (Metherell *et al.*, 2001). Downstream of the GHR, a missense mutation of the STAT5b gene has been found in a patient with severe postnatal growth failure due to the inability to phosphorylate/activate STAT5b protein (Hwa *et al.*, 2004; Kofoed *et al.*, 2003).

III. Insulin Signaling

Insulin is secreted by the β-cells of the pancreas and exerts profound effects on a variety of cellular processes. Biological responses to insulin include stimulation of glucose uptake and lipid and protein synthesis, inhibition of lipolysis and proteolysis, and modulation of cellular growth and differentiation (Hill and Milner, 1985; Messina, 1999). These diverse effects are mediated by the binding of insulin to cell-surface insulin receptors (IR) which belongs to the family of receptors with intrinsic tyrosine kinase activity. Insulin binding to its receptor initiates the recruitment of intracellular SH2 domain-containing molecules, such as IRSs and Shc, to the activated IR and subsequent tyrosine phosphorylation of IRSs and Shc (Fig. 6.2).

The association between the phosphorylated IRSs and p85 subunit of PI3K leads to activation of p110 catalytic subunit of PI3K (Biddinger and Kahn, 2006; Dominici *et al.*, 2005; Taniguchi *et al.*, 2006). PI3K plays a crucial role in insulin actions, including stimulation of glucose transport, activation of glycogen synthase, and inhibition of hepatic gluconeogenesis (Biddinger and Kahn, 2006; Dominici *et al.*, 2005; Taniguchi *et al.*, 2006).

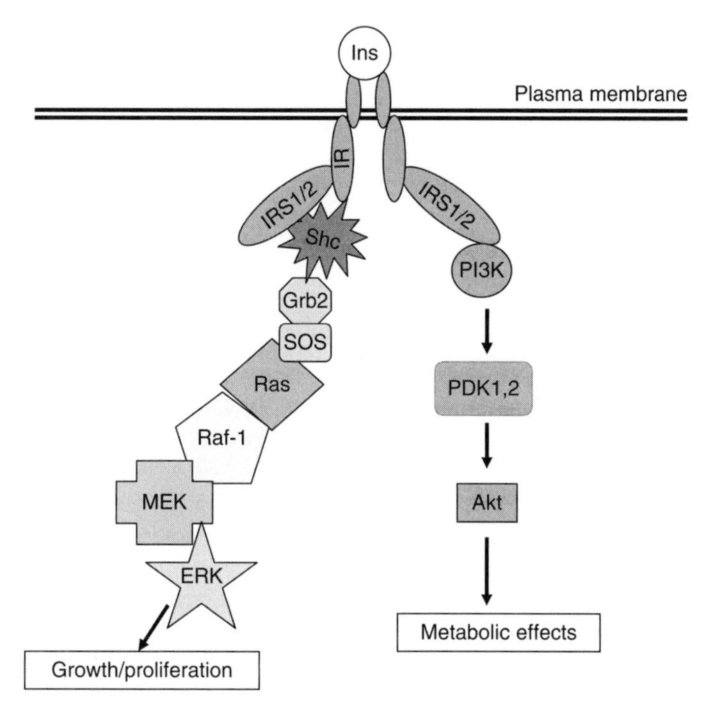

Figure 6.2 The main intracellular signaling pathways activated by GH. Major signaling pathways activated by insulin include the PI3K/Akt and MEK/ERK pathways, as described in the text.

The serine/threonine kinase Akt is one of the major downstream effectors of PI3K and mediates several insulin responses including insulin-induced glucose uptake and glycogen synthase activation (Biddinger and Kahn, 2006; Dominici *et al.*, 2005; Taniguchi *et al.*, 2006). The Ras/Raf-1/MEK/ERK signaling pathway is another major pathway activated by insulin. Both the IRSs and Shc participate in this pathway and activation of this pathway by insulin has a direct role in the promoting growth (Biddinger and Kahn, 2006; Dominici *et al.*, 2005; Taniguchi *et al.*, 2006).

IV. Regulation of GH Signaling by Insulin

A. Enhancement of cellular responsiveness to GH by transient insulin

1. Maintenance of GHR levels by insulin

Physiologic levels of insulin, which transiently increase following a meal, appear to be necessary for normal liver GH responsiveness (Baxter *et al.*, 1980; Bereket *et al.*, 1995; Ji *et al.*, 1999; Menon *et al.*, 1992, 1994).

In Type 1 diabetic patients, insulin deficiency is associated with reduced levels of circulating GH binding protein (GHBP), which is cleaved from cell surface GHR and shed into the circulation (Baxter and Turtle, 1978; Clayton *et al.*, 1994; Hanaire-Broutin *et al.*, 1996a,b; Kratzsch *et al.*, 1996; Mercado *et al.*, 1992, 1995). In rodents treated with streptozotocin, a Type 1 diabetes-inducing compound, liver GHR levels and GH binding decrease, as does plasma IGF-1. Insulin treatment restores GH binding and IGF-1 levels (Baxter *et al.*, 1980; Bornfeldt *et al.*, 1989; Chen *et al.*, 1997; Maes *et al.*, 1983; Menon *et al.*, 1994).

Insulin is known to promote both fetal and postnatal body growth (Messina, 1999). Pancreatectomy of sheep fetus result in reduced growth, which is reversible by insulin treatment (Fowden, 1992). Conversely, infants of hyperglycemic mothers (resulting in fetal hyperinsulinemia) exhibit an increased frequency of macrosomia (Schwartz *et al.*, 1994). The loss of insulin secretion in humans (Type 1 diabetes) is often accompanied by diminished growth in childhood (Hoskins *et al.*, 1985), which can be substantially improved by insulin treatment (Chiarelli *et al.*, 2004; Donaghue *et al.*, 2003). The growth-promoting effects of insulin may be at least partially explained by the fact that it maintains liver GHR levels, but it may also affect GH-induced postreceptor signaling.

2. Enhancement of insulin on GH-induced activation of the MEK/ERK pathway

In rat H4IIE hepatoma cells, insulin pretreatment (10 nM, 4 h or less) selectively enhances GH-induced phosphorylation of MEK1/2 and ERK1/2, but not GH-induced activation of STAT5 and Akt (Xu *et al.*, 2006a). Although insulin pretreatment alters GH-induced formation of the Shc/Grb2/SOS complex, it does not significantly affect GH-induced activation of other signaling intermediates upstream of MEK/ERK, including JAK2, Ras, and Raf-1. Immunofluorescent staining indicates that insulin pretreatment facilitates GH-induced cell membrane translocation of MEK1/2. Insulin pretreatment also increases the amount of MEK associating with its scaffolding protein, KSR. Therefore, short-term insulin treatment selectively enhances GH activation of the MEK/ERK pathway through post-GHR mechanisms, most likely from an increased MEK1/2 association with scaffolding proteins and cell membrane translocation of MEK 1/2 (Xu *et al.*, 2006a) (Fig. 6.3).

3. Insulin-regulated resensitization of GH signaling

Since GH is secreted in a pulsatile pattern *in vivo* (Jansson *et al.*, 1985; Tannenbaum and Martin, 1976), this raised the question of whether insulin affects signaling induced by repeated GH pulses. In rat H4IIE hepatoma cells, repeated GH pulses can reinduce STAT5 phosphorylation following a GH-free recovery period. However, following the first pulse of GH, and

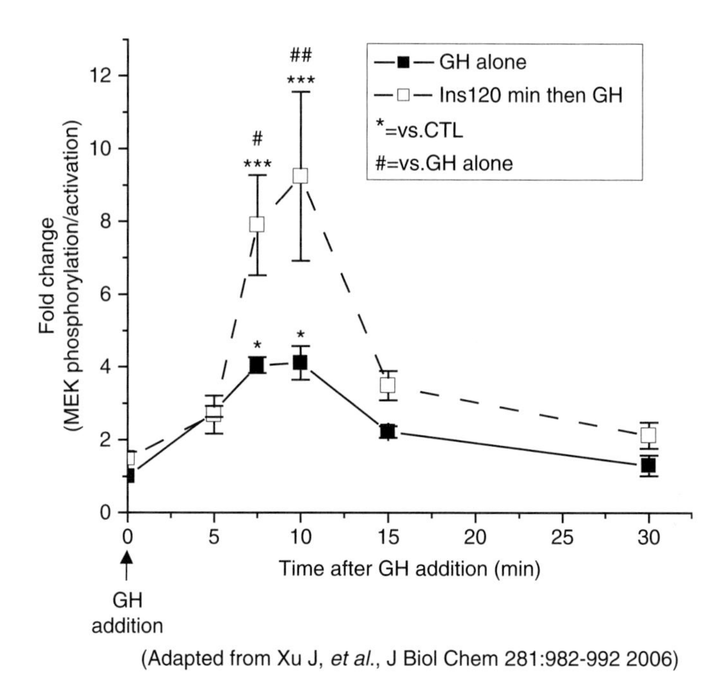

Figure 6.3 Insulin enhances GH-induced MEK1/2 phosphorylation. H4IIE cells were pretreated without or with 10 nM insulin for 120 min, followed by 500 ng/ml GH or vehicle for the indicated times. Western blot analysis was performed with PS-MEK1/2 antibody. Densitometric analysis of autoradiographs from three separate experiments was performed to quantify PS-MEK1/2 levels. The data are expressed as mean ± S.E. The PS-MEK1/2 levels in untreated samples were arbitrarily set to 1 (* or # = $p < 0.05$, ## = $p < 0.01$).

then a GH-free recovery period, activation of the MEK/ERK and PI3k/Akt pathways remain desensitized to GH stimulation by later GH pulses (Ji *et al.*, 2002) (Fig. 6.4A and 6.4B). A second GH application is able to activate the signaling intermediates upstream of MEK/ERK, including JAK2, Ras, and Raf-1 (Xu *et al.*, 2006b). This correlates with recovery of GHR levels that are decreased following the initial GH pulse. However, the recovery of GHR levels is insufficient for GH-induced phosphorylation of MEK1/2 and ERK1/2. Insulin treatment restores the ability of a second GH exposure to induce phosphorylation of MEK1/2 and ERK1/2 without altering GHR levels or GH-induced phosphorylation/activation of JAK2 and Raf-1 (Fig. 6.4B). Insulin increases the amount of MEK associated with the scaffolding protein, KSR. Insulin also significantly increases the ability of GH to stimulate tyrosine phosphorylation of KSR. Previous GH exposure also induces desensitization of STAT1 and STAT3 phosphorylation/activation, but this desensitization is *not* reversed by insulin (Fig. 6.4C).

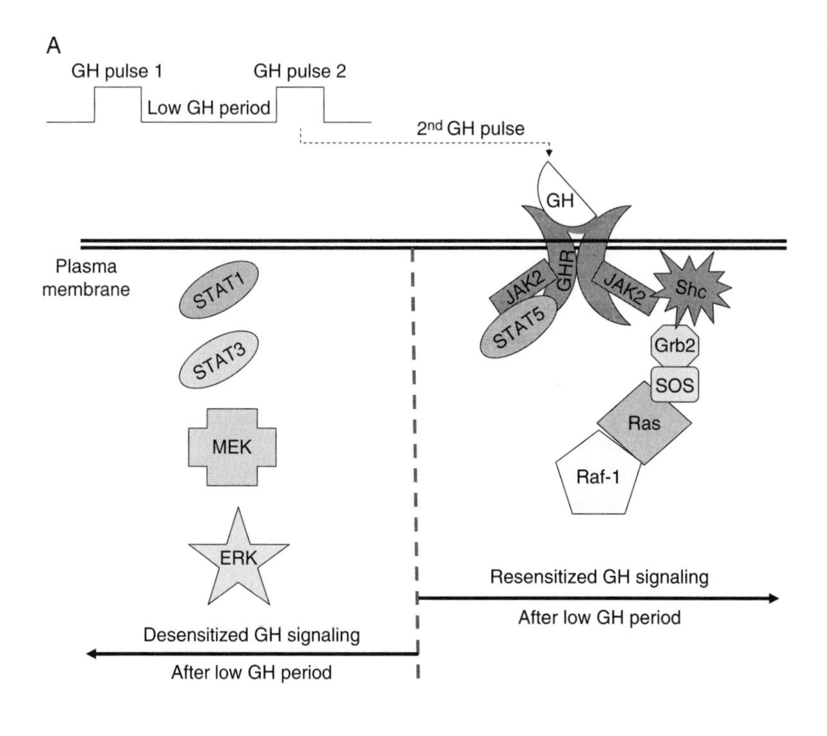

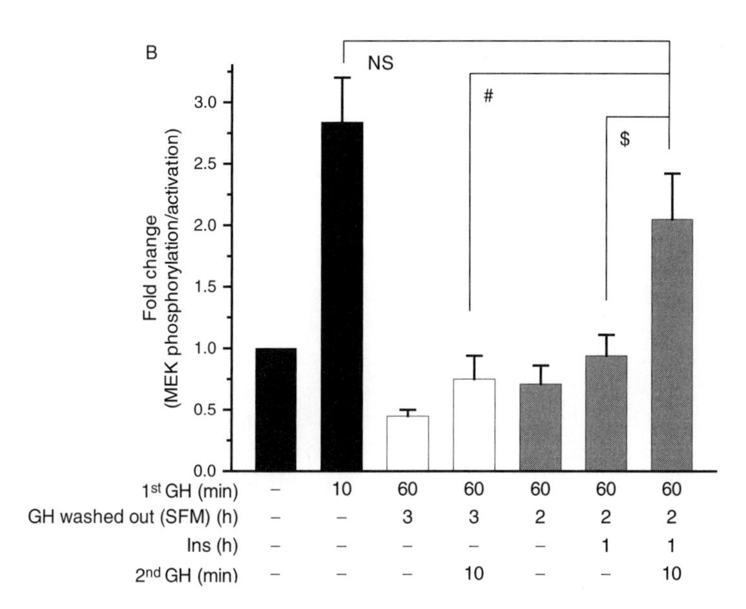

(Adapted from Xu J, *et al.*, J Biol Chem 281:21594-21606 2006)

Figure 6.4 Continued

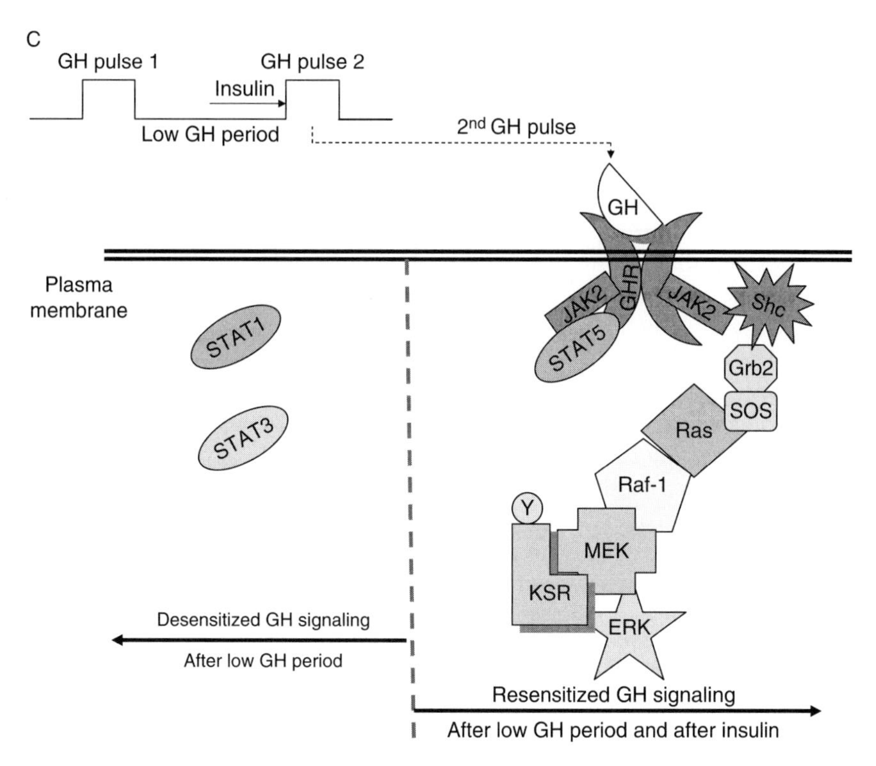

Figure 6.4 Insulin reverses GH-induced homologous desensitization. (A) GH induces desensitization of MEK, ERK, STAT1, and STAT3. Following a previous GH pulse (GH1) plus a GH-free interval, a second GH pulse (GH2) is able to activate the JAK2/Ras/Raf-1 pathway, but not the MEK/ERK pathway. This suggests a disrupted signal transduction from Raf-1 to MEK. The activation of STAT1 and STAT3 is also desensitized to the second GH pulse. The dashed line separates the desensitized MEK, ERK, STAT1, and STAT3 from the resensitized JAK2, Ras, Raf-1, and STAT5. (B) Insulin restores the ability of a second GH treatment to induce MEK1/2 phosphorylation. H4IIE cells were treated with GH1 or vehicle for the indicated time, washed, and incubated in GH-free, serum-free medium for 3 h. They were then treated without or with 10 nM insulin for 1 h and followed by GH2. Western blot analysis was performed with PS-MEK1/2 antibody. Densitometric analysis of autoradiographs from at least three separated experiments was performed to quantify PS-MEK1/2 levels. The data are expressed as mean $\pm$ S.E. The PS-MEK1/2 levels in untreated samples were arbitrarily set to 1 (Ins, insulin; NS, not significantly different; # or \$ $= p < 0.05$). (C) Insulin resensitizes the activation of the MEK/ERK pathway to a second GH pulse without altering the activation of JAK2, STAT5, or Raf-1. This resensitization is selective to the MEK/ERK pathway because the STAT1 and STAT3 remain desensitized. The dashed line separates the desensitized STAT1 and STAT3 from the resensitized GH signaling molecules.

This suggests that insulin-regulated resensitization of GH signaling may be necessary to reset responsiveness to GH after a normal, physiologic pulse of GH and is specific for the MEK/ERK signaling pathway (Xu *et al.*, 2006b).

4. Biological implications of resensitization of GH signaling by insulin

Resensitization of GH activation of the MEK/ERK pathway by insulin treatment may have important implications under physiologic conditions in which insulin levels transiently increase following a meal and drop thereafter. In normal human subjects, basal insulin concentrations in the hepatic portal circulation are approximately 0.2 nM (Blackard and Nelson, 1970) while peak postprandial insulin concentrations reach approximately 3 nM in the hepatic portal circulation (Blackard and Nelson, 1970; Cerasi *et al.*, 1970). In rats, basal hepatic portal insulin concentrations are similar, around 0.5 nM, while postprandial insulin concentrations in the hepatic portal circulation have not been reported (Wei *et al.*, 2004). If the ratio of postprandial to basal insulin in rats is similar to that found in humans, portal postprandial insulin concentrations in rats would be expected to reach ~10 nM, as used in the studies of H4IIE cells (Xu *et al.*, 2006a,b). Thus, normal postprandial concentrations of insulin may be necessary to resensitize liver GH signaling via the MEK/ERK pathway. In patients with conditions associated with deficient insulin secretion, such as malnutrition or Type 1 diabetes, the biological effects of GH on ERK1/2 activation may be weakened or lost.

Proper nutrition plays an important role in prepubertal and pubertal growth. When malnutrition occurs, growth is inhibited (Styne, 2003), probably due to the decreased insulin secretion, and which may result in decreased ERK1/2 activation by GH. Type 1 diabetes is well known to adversely affect linear growth (Chiarelli *et al.*, 2004; Guest, 1953). The decreased growth in children with type 1 diabetes has been substantially improved by daily multiple-dose insulin treatment compared with single-dose insulin treatment (Chiarelli *et al.*, 2004; Donaghue *et al.*, 2003; Guest, 1953). One possible explanation is that following the multiple doses of insulin, as our data suggests, there is an insulin-induced resensitization of GH-inducible ERK signaling, thus allowing a more complete growth response to the multiple GH secretory pulses. The role of GH in promoting growth is thought to be mediated by STAT5 activation. Several recent reports, however, indicate that STAT5 may not be the only pathway involved. A mutant GHR found in a family with markedly short stature shows increased ability to activate STAT5 with only minimal activation of the ERK1/2 pathway, compared with wild-type GHR after GH stimulation (Metherell *et al.*, 2001). Similarly, a GH variant manifesting normal activation of STAT5, but reduced activation of ERK1/2, was found in a child with short stature (Lewis *et al.*, 2004). These reports suggest that ERK1/2 might also play an important role in mediating the growth-promoting effects of GH. The role of insulin in body growth may also involve its effects on the MEK/ERK signaling pathway, and one possible mechanism is via its effects to resensitize GH-induced MEK/ERK signaling. Since insulin does not markedly affect GH-induced activation of STAT5, this effect of GH signaling seems specific to the MEK/ERK pathway. Therefore, when insulin secretion is deficient, the GH-induced and STAT5-mediated actions of GH, such as sexually dimorphic hepatic gene expression, may remain intact, whereas GH-induced and ERK1/2-dependent growth may be inhibited. This would be an efficient biological control that allows differentiation between a GH pulse that would lead to growth or a GH secretory pulse that would result in only the nongrowth effects of GH.

The synergistic effects of GH and insulin in promoting cell proliferation suggest that repeated GH pulse-induced ERK1/2 activation in the presence of insulin may be necessary for liver growth (Messina, 1999; Xu *et al.*, 2006b). GH transgenic mice demonstrate significantly increased growth of internal organs, but the liver in particular is enlarged compared with other organs (Shea *et al.*, 1987; Wanke *et al.*, 1991), and there are life-long high levels of hepatocellular replication in this model (Snibson *et al.*, 1999). GH also plays a critical role in liver regeneration after hepatectomy (Pennisi *et al.*, 2004). It is not clear which pathway (or combination of pathways) is responsible for the action of GH in hepatocyte proliferation, but since the ERK1/2 pathway is frequently associated with cell proliferation in the growth response to many growth factors, the ERK1/2 pathway is a potential candidate pathway for GH-induced liver cell proliferation. If the ERK pathway mediates GH-stimulated liver growth, the presence of insulin may well be necessary to allow multiple GH pulses to activate multiple cycles of ERK activation. Thus, the synergistic effects of GH and insulin may be necessary for proper whole body growth, at least in part because repeated ERK1/2 activation by GH following multiple GH pulses requires insulin resensitization of GH-induced MEK/ERK signaling (Messina, 1999; Xu *et al.*, 2006b).

B. Inhibition of cellular responsiveness to GH by prolonged exposure to insulin

Patients with Type 2 diabetes and peripheral insulin resistance exhibit reduced plasma IGF-1 levels, the mRNA of which is regulated by GH, possibly resulting from a decrease in GH responsiveness following prolonged exposure to high levels of insulin (Kratzsch *et al.*, 1996; Mercado *et al.*, 1992). Continuous insulin treatment of osteoblasts decreases the fraction of cellular GHR presented at the plasma membrane via inhibition of surface translocation of GHR with no effect on the total content of GHR (Leung *et al.*, 1997). In a study of the effects of insulin on the GH/IGF-1 axis during a period of protracted critical illness, 363 patients requiring intensive care for more than 7 days were randomly assigned to conventional or intensive insulin therapy (Mesotten *et al.*, 2004). Conventional insulin therapy maintains glycemia between 180 and 200 mg/dl, whereas intensive insulin therapy can achieve normoglycemia. Intensive insulin therapy resulted in a significant reduction of serum GHBP levels, an index of GHR expression, and reduction of GH-dependent parameters, indicating the presence of GH resistance. These *in vivo* findings suggest that prolonged, and intensive, insulin treatment inhibits GHR expression and induces GH resistance.

In H4IIE cells, prolonged insulin exposure (8 or more hours) significantly inhibits GH-induced tyrosine phosphorylation of STAT5b without any effect on STAT5b protein abundance (Ji *et al.*, 1999). This desensitization of GH-induced JAK2/STAT5b signaling by long-term insulin treatment is due to a corresponding decrease in total cellular GHR abundance and cell surface GH binding (Fig. 6.5). The molecular basis for this negative modulation of GHR abundance by insulin is via inhibition of GHR mRNA expression, dependent on insulin signaling via both the PI3K and MEK/ERK pathways (Bennett *et al.*, 2007a,b). Insulin levels, even in the low physiologic range for the liver, for extended periods of time, may

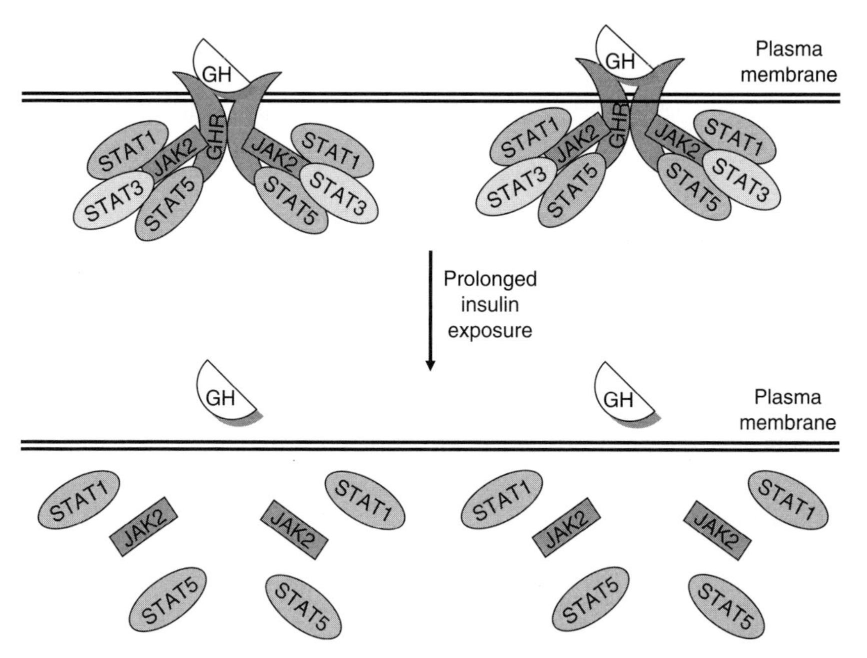

Figure 6.5 Prolonged exposure to insulin inhibits GH-induced activation of the JAK2/STATs pathways, including STAT1, STAT3, and STAT5. Following prolonged exposure to insulin, GHR levels are significantly reduced and activation of all three STATs is inhibited. The inhibitory effects of prolonged insulin exposure on GH-activation of STAT3 also involve a reduction in STAT3 protein abundance.

dramatically reduce expression of both GHR mRNA and protein. Continued presence of insulin is required to maintain the reduction of GHR mRNA and protein expression and there is a significant delay before GHR protein decreases (Bennett *et al.*, 2007a,b). Thus, transient increases in insulin levels are unlikely to significantly reduce GHR levels. Reduced insulin levels result in a recovery of GHR mRNA and then protein (Bennett *et al.*, 2007a,b). Therefore, insulin is a potent regulator of GHR gene expression, and the time and level of exposure to insulin may be critical determinants of insulin's effects on hepatic GHR mRNA. Extended periods of elevated insulin concentrations may cause reduction of GHR expression, while exposure to transient insulin may produce little or no change.

The GH to insulin ratio is thought important to increase lean muscle mass (relatively high GH or GH action) versus fat synthesis (relatively high insulin). Liver GH resistance may develop when a patient exhibits chronic hyperinsulinemia, a condition often observed in patients with obesity and in the early stages of Type 2 diabetes (Ji *et al.*, 1999). Thus, a reduction in GHR, with a concomitant decrease in GH action, due to the hyperinsulinemia, would favor fat deposition and muscle wasting, a characteristic common to obese individuals and many Type 2 diabetic patients. Patients with peripheral insulin resistance and hyperinsulinemia often exhibit abdominal adiposity, a morphology similar to that found in patients or animal models with GH deficiency or a disruption of GH signaling, such as STAT5b knockout mice (Angulo *et al.*, 1996; Bjorntorp, 1997; Laron, 1995; Udy *et al.*, 1997; Zhou *et al.*, 1997).

GH-induced tyrosine phosphorylation of STAT3 and STAT1 is also greatly reduced following prolonged insulin pretreatment in H4IIE cells (Xu *et al.*, 2005). Total STAT5b and STAT1 protein levels are not altered by prolonged insulin treatment. However, prolonged insulin treatment (16 h; 10 or 100 nM) results in a 30–40% reduction of total STAT3 protein, with little change at 0.1 and 1.0 nM insulin. Thus, there is a selective reduction of total STAT3 protein levels by prolonged high concentrations of insulin. Basal tyrosine phosphorylation of STAT3 (PY-STAT3) is also inhibited by prolonged insulin treatment. This reduction can only be partially explained by the decreased STAT3 protein abundance since the inhibition of basal PY-STAT3 by insulin is greater than the reduction of total STAT3 protein levels. Thus, there are at least three mechanisms by which prolonged insulin treatment inhibits STAT3 activation: (1) reduction of GHR, GH binding, and GH signaling; (2) reduction of STAT3 protein levels; and (3) decrease of basal PY-STAT3 (Fig. 6.5).

The inhibitory effect of insulin on total STAT3 protein and basal PY-STAT3 levels is dependent on activation of the MEK/ERK pathway and not the PI3K pathway. Under conditions of insulin resistance, insulin stimulation of the PI3K pathway is significantly reduced, whereas the ERK1/2 pathway remains sensitive to insulin stimulation. If compensatory hyperinsulinemia develops due to decreased metabolic regulation by decreased insulin-induced PI3K pathway signaling, activation of the ERK1/2 pathway may increase (Cusi *et al.*, 2000; Lebovitz, 2001). This suggests that the insulin-regulated reduction of STAT3 protein abundance, and of basal PY-STAT3 levels, may occur even in insulin-resistant states, such as obesity and Type 2 diabetes, which are associated with elevated insulin levels. A study of liver-specific STAT3 knock-out mice revealed that STAT3 is also essential for normal glucose homeostasis by inhibiting gluconeogenic gene expression (Klasing, 1988). If the inhibitory effect of insulin on STAT3 observed in H4IIE cells also occurs *in vivo*, then the impaired STAT3 signaling caused by prolonged hyperinsulinemia may lead to increased gluconeogenesis, further contributing to high blood glucose levels. In parallel, it may also inhibit other IL-6 actions due to decreased STAT3 signaling as well as actions of other cytokines and chemokines that function via STAT3 (Campos and Baumann, 1992; Fujita *et al.*, 2003; Kuwahara *et al.*, 2003; O'Riordain *et al.*, 1995; Thompson *et al.*, 1991; Xu *et al.*, 2005).

Therefore, the effects of insulin on GH-activated signaling pathways depend not only on the concentration of insulin and duration of insulin exposure, but also on whether cells have been previously exposed to GH. Transient insulin treatment does not significantly alter GHR expression, but enhances GH-induced postreceptor signaling via ERK1/2. Previous exposure to GH desensitizes the MEK/ERK pathway to repeated GH pulses, and insulin restores the ability of later GH pulses to activate the MEK/ERK pathway. In both cases, the effect of insulin on GH signaling is selective to the MEK/ERK pathway without affecting the JAK2/STAT pathway. In contrast, the effects of prolonged insulin treatment may occur at two levels: (1) downregulation of GHR mRNA and protein expression, resulting in the inhibition of GH-activated STAT5, STAT3 and STAT1 (receptor mechanism); and (2) reduction of STAT3 protein abundance and basal activation (a postreceptor mechanism).

V. Regulation of Insulin Signaling by Chronic GH

A. Effects of GH on insulin action

Under conditions of previous GH deprivation, the acute effects of GH are insulin-like and GH stimulates glucose and amino acid transport, lipogenesis, and protein synthesis (Davidson, 1987; Dominici *et al.*, 2005; Jorgensen *et al.*, 2004). However, after a lag period of a few hours, the anti-insulin effects of GH arise including hyperglycemia and hyperinsulinemia, and persist chronically if exposure to GH continues (Davidson, 1987;

Dominici *et al.*, 2005; Jorgensen *et al.*, 2004). Transgenic mice overexpressing GH are hyperinsulinemic and insulin resistant (Dominici *et al.*, 1999a,b; Frick *et al.*, 2001; Kopchick *et al.*, 1999; Olsson *et al.*, 2003; Valera *et al.*, 1993). Both GH-deficient Ames dwarf mice and GHR-deficient mice display hypoglycemia and hypoinsulinemia due to heightened insulin sensitivity (Borg *et al.*, 1995; Dominici *et al.*, 2000, 2002; Hauck *et al.*, 2001; Liu *et al.*, 2004). Similar to $GHR^{-/-}$ and Ames dwarfs, transgenic mice overexpressing a GH antagonist (GHa) exhibit elevated insulin sensitivity and decreased blood glucose and serum insulin levels, although their growth retardation is much milder (Yakar *et al.*, 2004). Insulin resistance is often observed in acromegalic patients, and about half of these patients are diagnosed with diabetes mellitus (Arya *et al.*, 1997; Ezzat *et al.*, 1994; Hansen *et al.*, 1986). Chronic GH treatment of GH-deficient patients in physiological or pharmacological doses is also reported to cause diabetes mellitus (Lippe *et al.*, 1981). The incidence of Type 2 diabetes in children receiving GH treatment is sixfold higher than that reported in children not receiving GH (Cutfield *et al.*, 2000). The development of insulin resistance and diabetes under chronic excessive GH conditions may be at least partially due to the ability of GH to interfere with insulin signaling.

B. Inhibition of insulin signaling by chronic GH

Chronic GH excess in rats bearing GH-producing tumors reduces IR levels and increases basal phosphorylation and activity of the IR in liver (Venkatesan and Davidson, 1995), in liver and muscle of GH-transgenic mice (Balbis *et al.*, 1996; Dominici *et al.*, 1998, 1999a,b), and in liver from GH-treated rats (Johansen *et al.*, 2005). IRS-1 and PI3K protein levels decrease in liver of GH-treated rats (Thirone *et al.*, 1997). Insulin-stimulated phosphorylation of the IR, IRS-1/2, IRS-1/p85 association, and PI3K activity are decreased in liver and skeletal muscle from both GH-treated rats and GH-transgenic mice (Dominici *et al.*, 1999a; Smith *et al.*, 1997; Thirone *et al.*, 1997). Studies on rats pretreated with streptozotocin suggest that the effects of GH are direct and not secondary to GH-induced hyperinsulinemia (Smith *et al.*, 1997).

There are several proteins involved in the mechanisms by which GH interferes with insulin signaling. PI3K plays an essential role in insulin actions in metabolism. PI3K is a heterodimer consisting of one p85 regulatory and one p110 catalytic subunit, each of which occurs in multiple isoforms (Ueki *et al.*, 2002). Although the p110-p85 complex of PI3K mediates insulin's activation of Akt, free p85α (not bound to the p110 catalytic subunit of PI3K) is an important negative regulator of PI3K activity (Fruman *et al.*, 2000; Mauvais-Jarvis *et al.*, 2002; Terauchi *et al.*, 1999; Ueki *et al.*, 2000). Optimal signaling through the PI3K pathway depends on balance between p85 and p110 levels. Transgenic mice overexpressing GH exhibit increased abundance of the p85α subunit of PI3K, increased basal association of p85α to IRS-1, and decreased insulin-stimulated PI3K activity in skeletal muscle (Barbour *et al.*, 2004; Dominici *et al.*, 1999a). Also, GH regulates p85α expression and PI3K activity in white adipose tissue (del Rincon *et al.*, 2007). GH-deficient mice demonstrate decreased p85α expression and a concomitant increase in insulin-induced IRS1-associated PI3K activity, whereas mice with GH excess have increased p85α levels and a concomitant suppression of the IRS1-associated PI3K activity in response to insulin (del Rincon *et al.*, 2007).

In rats, GH treatment reduces IRS-1 protein levels and inhibits insulin's activation of the IR/IRS/PI3K pathway. In rat liver and muscle, prolonged GH treatment increases the association of JAK2 with IRS-1, resulting in a reduced pool of IRS-1 available as a substrate for phosphorylation by the activated IR (Johansen *et al.*, 2005). GH also leads to serine phosphorylation of IRS-1. Aspirin pretreatment reverses GH-induced insulin resistance and serine phosphorylation of IRS-1 in liver, muscle, and white adipose tissue, in parallel with a reduction of c-jun N terminal kinase (JNK) activity, possibly indicating that JNK is one of the causes of GH-induced insulin resistance (Prattali *et al.*, 2005).

Suppressor of cytokine signaling (SOCS) proteins, inducible by GH and other cytokines, are also potential mediators of GH-induced insulin resistance. SOCS-1 and SOCS-6 interact with the IR and inhibit insulin-induced phosphorylation of IRS-1 *in vitro*, and activation of ERK1/2 and Akt *in vivo* (Mooney *et al.*, 2001). SOCS-1 and SOCS-3 bind to recombinant and endogenous IRS-1 and IRS-2 and promote their ubiquitination and subsequent degradation. For instance, adenoviral-mediated expression of SOCS-1 in mouse liver dramatically reduces hepatic IRS-1 protein levels and causes glucose intolerance (Rui *et al.*, 2002). Overexpression of SOCS-1 and SOCS-3 in liver causes insulin resistance, whereas inhibition of SOCS-1 and -3 in obese diabetic mice improves insulin sensitivity (Ueki *et al.*, 2004). SOCS-1 deficiency results in sustained tyrosine phosphorylation of IRS-1 in response to insulin, and SOCS-1 deficient mice exhibit a significantly low level of blood glucose. Conversely, the forced expression of SOCS-1 reduces tyrosine phosphorylation of IRS-1. SOCS-1 achieves this inhibition both by binding directly to IRS-1 and by suppressing JAK (Kawazoe *et al.*, 2001). Therefore, SOCS proteins interfere with proper insulin signaling at both receptor and postreceptor levels (IRS-1 and JAK).

VI. Conclusions

It is well established that insulin promotes body growth, whereas an insulin deficit, as in Type 1 diabetes, or insulin resistance adversely affects linear growth. Following conditions of GH deprivation, the transient effects of GH are insulin-like and stimulate glucose and amino acid transport, and lipogenesis. However, high levels of GH causes insulin resistance, in part by reducing IR levels, insulin-induced phosphorylation of the IR and IRS-1, and reduction of insulin-induced PI3K activity, resulting in an increased likelihood of developing Type 2 diabetes. Inhibition of insulin signaling by high levels of GH appears to involve several mechanisms, including increased free p85 and SOCS levels, increased JAK2/IRS association, and increased JNK activation.

Recent observations regarding crosstalk between GH and insulin signaling expand our understanding of the molecular mechanisms of the functional interaction between these two important hormones in metabolism and growth. We hypothesize that transient exposure to insulin, as observed after a meal, is necessary for normal liver GH responsiveness, possibly due to its maintenance of GHR levels, enhancement of GH signaling and recovery of GH-induced GH desensitization. Short-term insulin treatment does not significantly alter GHR expression, but enhances GH-induced postreceptor signaling via ERK1/2. Previous exposure to GH desensitizes the MEK/ERK pathway to GH, but insulin restores the ability of a later GH pulses to induce MEK/ERK phosphorylation. The effect of transient insulin on GH signaling is selective to the MEK/ERK pathway without affecting the JAK2/STAT pathway. In contrast, prolonged exposure to insulin, as observed in patients with insulin resistance and the resultant hyperinsulinemia, downregulates GHR mRNA and protein expression, inhibits GH-activated JAK/STAT signaling, and reduces STAT3 protein abundance. Thus, there are multiple forms of cross-talk between GH and insulin that are context dependent. In pathological conditions, this results in the ability of excess insulin or GH to inhibit the signaling of the other. However, possibly even more important is the normal physiological interaction of insulin and GH to promote growth.

ACKNOWLEDGMENTS

We thank the many trainees and other laboratory personnel for their efforts over the years in the performance of the work cited. In particular, this includes Drs. A. B. Keeton, W. L. Bennett, and S. Ji, as well as J. L. Franklin and D. Y. Venable. We also thank Drs. S. J. Frank, T. L. Clemens, as well as Drs. X. Li, Z. Liu, and R. Guan for their collaboration and contributions.

The studies carried out in the author's laboratory were supported in part by National Institutes of Health Grants DK40456 and DK62071, and a VA Merit Review (to J. L. M.).

REFERENCES

Anderson, N. G. (1992). Growth hormone activates mitogen-activated protein kinase and S6 kinase and promotes intracellular tyrosine phosphorylation in 3T3-F442A preadipocytes. *Biochem. J.* **284,** 649–652.

Angulo, M., Castro-Magana, M., Mazur, B., Canas, J. A., Vitollo, P. M., and Sarrantonio, M. (1996). Growth hormone secretion and effects of growth hormone therapy on growth velocity and weight gain in children with Prader-Willi syndrome. *J. Pediatr. Endocrinol. Metab.* **9,** 393–400.

Argetsinger, L. S., Campbell, G. S., Yang, X., Witthuhn, B. A., Silvennoinen, O., Ihle, J. N., and Carter-Su, C. (1993). Identification of JAK2 as a growth hormone receptor-associated tyrosine kinase. *Cell* **74,** 237–244.

Argetsinger, L. S., Hsu, G. W., Myers, M. G. Jr., Billestrup, N., White, M. F., and Carter-Su, C. (1995). Growth hormone, interferon-gamma, and leukemia inhibitory factor promoted tyrosyl phosphorylation of insulin receptor substrate-1. *J. Biol. Chem.* **270,** 14685–14692.

Argetsinger, L. S., Norstedt, G., Billestrup, N., White, M. F., and Carter-Su, C. (1996). Growth hormone, interferon-gamma, and leukemia inhibitory factor utilize insulin receptor substrate-2 in intracellular signaling. *J. Biol. Chem.* **271,** 29415–29421.

Arya, K. R., Pathare, A. V., Chadda, M., and Menon, P. S. (1997). Diabetes in acromegaly—a study of 34 cases. *J. Indian Med. Assoc.* **95,** 546–547.

Balbis, A., Bartke, A., and Turyn, D. (1996). Overexpression of bovine growth hormone in transgenic mice is associated with changes in hepatic insulin receptors and in their kinase activity. *Life Sci.* **59,** 1363–1371.

Barbour, L. A., Shao, J., Qiao, L., Leitner, W., Anderson, M., Friedman, J. E., and Draznin, B. (2004). Human placental growth hormone increases expression of the p85 regulatory unit of phosphatidylinositol 3-kinase and triggers severe insulin resistance in skeletal muscle. *Endocrinology* **145,** 1144–1150.

Bartke, A., Turyn, D., Aguilar, C. C., Sotelo, A. I., Steger, R. W., Chen, X. Z., and Kopchick, J. J. (1994). Growth hormone (GH) binding and effects of GH analogs in transgenic mice. *Proc. Soc. Exp. Biol. Med.* **206,** 190–194.

Baumann, G. (2001). Growth hormone and its disorders. *In* "Principles and Practice of Endocrinology and Metabolism" (K. L. Becker, Ed.), pp. 129–145. Philadelphia, Lippincott Williams & Wilkins.

Baxter, R. C., and Turtle, J. R. (1978). Regulation of hepatic growth hormone receptors by insulin. *Biochem. Biophys. Res. Commun.* **84,** 350–357.

Baxter, R. C., Bryson, J. M., and Turtle, J. R. (1980). Somatogenic receptors of rat liver: Regulation by insulin. *Endocrinology* **107,** 1176–1181.

Bennett, W. L., Ji, S., and Messina, J. L. (2007a). Insulin regulation of growth hormone receptor gene expression evidence for a transcriptional mechanism of down-regulation in rat hepatoma cells. *Mol. Cell. Endocrinol.* **274,** 53–59.

Bennett, W. L., Keeton, A. B., Ji, S., Xu, J., and Messina, J. L. (2007b). Insulin regulation of growth hormone receptor gene expression: Involvement of both the PI-3 kinase and MEK/ERK signaling pathways. *Endocrine* **32,** 219–226.

Bereket, A., Lang, C. H., Blethen, S. L., Gelato, M. C., Fan, J., Frost, R. A., and Wilson, T. A. (1995). Effect of insulin on the insulin-like growth factor system in children with new-onset insulin-dependent diabetes mellitus. *J. Clin. Endocrinol. Metab.* **80,** 1312–1317.

Bergad, P. L., Shih, H. M., Towle, H. C., Schwarzenberg, S. J., and Berry, S. A. (1995). Growth hormone induction of hepatic serine protease inhibitor 2.1 transcription is mediated by a Stat5-related factor binding synergistically to two gamma-activated sites. *J. Biol. Chem.* **270,** 24903–24910.

Biddinger, S. B., and Kahn, C. R. (2006). From mice to men: Insights into the insulin resistance syndromes. *Annu. Rev. Physiol.* **68,** 123–158.

Bjorntorp, P. (1997). Body fat distribution, insulin resistance, and metabolic diseases. *Nutrition* **13,** 795–803.

Blackard, W. G., and Nelson, N. C. (1970). Portal and peripheral vein immunoreactive insulin concentrations before and after glucose infusion. *Diabetes* **19,** 302–306.

Borg, K. E., Brown-Borg, H. M., and Bartke, A. (1995). Assessment of the primary adrenal cortical and pancreatic hormone basal levels in relation to plasma glucose and age in the unstressed Ames dwarf mouse. *Proc. Soc. Exp. Biol. Med.* **210,** 126–133.

Bornfeldt, K. E., Arnqvist, H. J., Enberg, B., Mathews, L. S., and Norstedt, G. (1989). Regulation of insulin-like growth factor-I and growth hormone receptor gene expression by diabetes and nutritional state in rat tissues. *J. Endocrinol.* **122,** 651–656.

Brown-Borg, H. M., Borg, K. E., Meliska, C. J., and Bartke, A. (1996). Dwarf mice and the ageing process. *Nature* **384,** 33.

Campbell, G. S., Pang, L., Miyasaka, T., Saltiel, A. R., and Carter-Su, C. (1992). Stimulation by growth hormone of MAP kinase activity in 3T3-F442A fibroblasts. *J. Biol. Chem.* **267,** 6074–6080.

Campbell, G. S., Meyer, D. J., Raz, R., Levy, D. E., Schwartz, J., and Carter-Su, C. (1995). Activation of acute phase response factor (APRF)/Stat3 transcription factor by growth hormone. *J. Biol. Chem.* **270,** 3974–3979.

Campos, S. P., and Baumann, H. (1992). Insulin is a prominent modulator of the cytokine-stimulated expression of acute-phase plasma protein genes. *Mol. Cell. Biol.* **12,** 1789–1797.

Carter-Su, C., Schwartz, J., and Smit, L. S. (1996). Molecular mechanism of growth hormone action. *Annu. Rev. Physiol.* **58,** 187–207.

Cerasi, E., Hallberg, D., and Luft, R. (1970). Simultaneous determination of insulin in brachial and portal veins during glucose infusion in normal and prediabetic subjects. *Horm. Metab. Res.* **2,** 302–303.

Chen, N. Y., Chen, W. Y., and Kopchick, J. J. (1997). Liver and kidney growth hormone (GH) receptors are regulated differently in diabetic GH and GH antagonist transgenic mice. *Endocrinology* **138,** 1988–1994.

Chia, D. J., Ono, M., Woelfle, J., Schlesinger-Massart, M., Jiang, H., and Rotwein, P. (2006). Characterization of distinct Stat5b binding sites that mediate growth hormone-stimulated IGF-I gene transcription. *J. Biol. Chem.* **281,** 3190–3197.

Chiarelli, F., Giannini, C., and Mohn, A. (2004). Growth, growth factors, and diabetes. *Eur. J. Endocrinol.* **151**(Suppl. 3), U109–U117.

Choi, H. K., and Waxman, D. J. (2000). Pulsatility of growth hormone (GH) signalling in liver cells: Role of the JAK-STAT5b pathway in GH action. *Growth Horm. IGF Res.* **10** (Suppl. B), S1–S8.

Clayton, K. L., Holly, J. M., Carlsson, L. M., Jones, J., Cheetham, T. D., Taylor, A. M., and Dunger, D. B. (1994). Loss of the normal relationships between growth hormone, growth hormone-binding protein, and insulin-like growth factor-I in adolescents with insulin-dependent diabetes mellitus. *Clin. Endocrinol. (Oxf.)* **41,** 517–524.

Cobb, M. H. (1999). MAP kinase pathways. *Prog. Biophys. Mol. Biol.* **71,** 479–500.

Coschigano, K. T., Clemmons, D., Bellush, L. L., and Kopchick, J. J. (2000). Assessment of growth parameters and life span of GHR/BP gene-disrupted mice. *Endocrinology* **141,** 2608–2613.

Coschigano, K. T., Holland, A. N., Riders, M. E., List, E. O., Flyvbjerg, A., and Kopchick, J. J. (2003). Deletion, but not antagonism, of the mouse growth hormone receptor results in severely decreased body weights, insulin, and insulin-like growth factor I levels and increased life span. *Endocrinology* **144,** 3799–3810.

Cusi, K., Maezono, K., Osman, A., Pendergrass, M., Patti, M. E., Pratipanawatr, T., DeFronzo, R. A., Kahn, C. R., and Mandarino, L. J. (2000). Insulin resistance differentially affects the PI 3-kinase- and MAP kinase-mediated signaling in human muscle. *J. Clin. Invest.* **105,** 311–320.

Cutfield, W. S., Wilton, P., Bennmarker, H., Bertsson-Wikland, K., Chatelain, P., Ranke, M. B., and Price, D. A. (2000). Incidence of diabetes mellitus and impaired glucose tolerance in children and adolescents receiving growth-hormone treatment. *Lancet* **355,** 610–613.

Davey, H. W., Wilkins, R. J., and Waxman, D. J. (1999). STAT5 signaling in sexually dimorphic gene expression and growth patterns. *Am. J. Hum. Genet.* **65,** 959–965.

Davidson, M. B. (1987). Effect of growth hormone on carbohydrate and lipid metabolism. *Endocr. Rev.* **8,** 115–131.

Delesque-Touchard, N., Park, S. H., and Waxman, D. J. (2000). Synergistic action of hepatocyte nuclear factors 3 and 6 on CYP2C12 gene expression and suppression by growth hormone-activated STAT5b. Proposed model for female specific expression of CYP2C12 in adult rat liver. *J. Biol. Chem.* **275,** 34173–34182.

del Rincon, J. P., Iida, K., Gaylinn, B. D., McCurdy, C. E., Leitner, J. W., Barbour, L. A., Kopchick, J. J., Friedman, J. E., Draznin, B., and Thorner, M. O. (2007). Growth hormone regulation of p85alpha expression and phosphoinositide 3-kinase activity in adipose tissue: Mechanism for growth hormone-mediated insulin resistance. *Diabetes* **56,** 1638–1646.

Dominici, F. P., Argentino, D. P., Munoz, M. C., Miquet, J. G., Sotelo, A. I., and Turyn, D. (2005). Influence of the crosstalk between growth hormone and insulin signalling on the modulation of insulin sensitivity. *Growth Horm. IGF Res.* **15,** 324–336.

Dominici, F. P., Arostegui, D. G., Bartke, A., Kopchick, J. J., and Turyn, D. (2000). Compensatory alterations of insulin signal transduction in liver of growth hormone receptor knockout mice. *J. Endocrinol.* **166,** 579–590.

Dominici, F. P., Balbis, A., Bartke, A., and Turyn, D. (1998). Role of hyperinsulinemia on hepatic insulin receptor concentration and autophosphorylation in the presence of high growth hormone levels in transgenic mice overexpressing growth hormone gene. *J. Endocrinol.* **159,** 15–25.

Dominici, F. P., Cifone, D., Bartke, A., and Turyn, D. (1999a). Alterations in the early steps of the insulin-signaling system in skeletal muscle of GH-transgenic mice. *Am. J. Physiol.* **277,** E447–E454.

Dominici, F. P., Cifone, D., Bartke, A., and Turyn, D. (1999b). Loss of sensitivity to insulin at early events of the insulin signaling pathway in the liver of growth hormone-transgenic mice. *J. Endocrinol.* **161,** 383–392.

Dominici, F. P., Hauck, S., Argentino, D. P., Bartke, A., and Turyn, D. (2002). Increased insulin sensitivity and upregulation of insulin receptor, insulin receptor substrate (IRS)-1 and IRS-2 in liver of Ames dwarf mice. *J. Endocrinol.* **173,** 81–94.

Donaghue, K. C., Kordonouri, O., Chan, A., and Silink, M. (2003). Secular trends in growth in diabetes: Are we winning? *Arch. Dis. Child.* **88,** 151–154.

Ezzat, S., Forster, M. J., Berchtold, P., Redelmeier, D. A., Boerlin, V., and Harris, A. G. (1994). Acromegaly. Clinical and biochemical features in 500 patients. *Medicine (Baltimore)* **73,** 233–240.

Fain, J. N., Ihle, J. H., and Bahouth, S. W. (1999). Stimulation of lipolysis but not of leptin release by growth hormone is abolished in adipose tissue from Stat5a and b knockout mice. *Biochem. Biophys. Res. Commun.* **263,** 201–205.

Fowden, A. L. (1992). The role of insulin in fetal growth. *Early Hum. Dev.* **29,** 177–181.

Frank, S. J., Yi, W., Zhao, Y., Goldsmith, J. F., Gilliland, G., Jiang, J., Sakai, I., and Kraft, A. S. (1995). Regions of the JAK2 tyrosine kinase required for coupling to the growth hormone receptor. *J. Biol. Chem.* **270,** 14776–14785.

Frick, F., Bohlooly, Y., Linden, D., Olsson, B., Tornell, J., Eden, S., and Oscarsson, J. (2001). Long-term growth hormone excess induces marked alterations in lipoprotein metabolism in mice. *Am. J. Physiol. Endocrinol. Metab.* **281,** E1230–E1239.

Fruman, D. A., Mauvais-Jarvis, F., Pollard, D. A., Yballe, C. M., Brazil, D., Bronson, R. T., Kahn, C. R., and Cantley, L. C. (2000). Hypoglycaemia, liver necrosis and perinatal death in mice lacking all isoforms of phosphoinositide 3-kinase p85 alpha. *Nat. Genet.* **26,** 379–382.

Fujita, N., Sakamaki, H., Uotani, S., Takahashi, R., Kuwahara, H., Kita, A., Oshima, K., Yamasaki, H., Yamaguchi, Y., and Eguchi, K. (2003). Intracerebroventricular administration of insulin and glucose inhibits the anorectic action of leptin in rats. *Exp. Biol. Med. (Maywood)* **228,** 1156–1161.

Gronowski, A. M., and Rotwein, P. (1994). Rapid changes in nuclear protein tyrosine phosphorylation after growth hormone treatment *in vivo*. Identification of phosphorylated mitogen-activated protein kinase and STAT91. *J. Biol. Chem.* **269,** 7874–7878.

Gu, Y., Zou, Y., Aikawa, R., Hayashi, D., Kudoh, S., Yamauchi, T., Uozumi, H., Zhu, W., Kadowaki, T., Yazaki, Y., and Komuro, I. (2001). Growth hormone signalling and apoptosis in neonatal rat cardiomyocytes. *Mol. Cell. Biochem.* **223,** 35–46.

Guest, G. (1953). The Mauriac syndrome: Dwarfism, hepatomegaly and obesity with juvenile diabetes mellitus. *Diabetes* **2,** 415–417.

Hanaire-Broutin, H., Sallerin-Caute, B., Poncet, M. F., Tauber, M., Bastide, R., Chale, J. J., Rosenfeld, R., and Tauber, J. P. (1996a). Effect of intraperitoneal insulin delivery on growth hormone binding protein, insulin-like growth factor (IGF)-I, and IGF-binding protein-3 in IDDM. *Diabetologia* **39,** 1498–1504.

Hanaire-Broutin, H., Sallerin-Caute, B., Poncet, M. F., Tauber, M., Bastide, R., Rosenfeld, R., and Tauber, J. P. (1996b). Insulin therapy and GH-IGF-I axis disorders in diabetes: Impact of glycaemic control and hepatic insulinization. *Diabetes Metab.* **22,** 245–250.

Hansen, I., Tsalikian, E., Beaufrere, B., Gerich, J., Haymond, M., and Rizza, R. (1986). Insulin resistance in acromegaly: Defects in both hepatic and extrahepatic insulin action. *Am. J. Physiol.* **250,** E269–E273.

Hansen, L. H., Wang, X., Kopchick, J. J., Bouchelouche, P., Nielsen, J. H., Galsgaard, E. D., and Billestrup, N. (1996). Identification of tyrosine residues in the intracellular domain of the growth hormone receptor required for transcriptional signaling and Stat5 activation. *J. Biol. Chem.* **271,** 12669–12673.

Hauck, S. J., Hunter, W. S., Danilovich, N., Kopchick, J. J., and Bartke, A. (2001). Reduced levels of thyroid hormones, insulin, and glucose, and lower body core temperature in the growth hormone receptor/binding protein knockout mouse. *Exp. Biol. Med. (Maywood)* **226,** 552–558.

Herrington, J., Smit, L. S., Schwartz, J., and Carter-Su, C. (2000). The role of STAT proteins in growth hormone signaling. *Oncogene* **19,** 2585–2597.

Hikida, R. S., Knapp, J. R., Chen, W. Y., Gozdanovic, J. A., and Kopchick, J. J. (1995). Effects of bovine growth hormone analogs on mouse skeletal muscle structure. *Growth Dev. Aging* **59,** 121–128.

Hill, D. J., and Milner, R. D. G. (1985). Insulin as a growth factor. *Pediatr. Res.* **19,** 879–886.

Hodge, C., Liao, J. F., Stofega, M., Guan, K. L., Cartersu, C., and Schwartz, J. (1998). Growth hormone stimulates phosphorylation and activation of elk-1and expression of c-fos, egr-1, and junB through activation of extracellular signal-regulated kinases 1 and 2. *J. Biol. Chem.* **273,** 31327–31336.

Hoskins, P. J., Leslie, R. D., and Pyke, D. A. (1985). Height at diagnosis of diabetes in children: A study in identical twins. *Br. Med. J. (Clin. Res. Ed.)* **290,** 278–280.

Huang, Y., Kim, S. O., Jiang, J., and Frank, S. J. (2003). Growth hormone-induced phosphorylation of epidermal growth factor (EGF) receptor in 3T3-F442A cells. Modulation of EGF-induced trafficking and signaling. *J. Biol. Chem.* **278,** 18902–18913.

Hwa, V., Little, B., Kofoed, E. M., and Rosenfeld, R. G. (2004). Transcriptional regulation of insulin-like growth factor-I by interferon-gamma requires STAT-5b. *J. Biol. Chem.* **279,** 2728–2736.

Isgaard, J., Nilsson, A., Lindahl, A., Jansson, J. O., and Isaksson, O. G. (1986). Effects of local administration of GH and IGF-1 on longitudinal bone growth in rats. *Am. J. Physiol.* **250,** E367–E372.

Jansson, J. O., Eden, S., and Isaksson, O. (1985). Sexual dimorphism in the control of growth hormone secretion. *Endocr. Rev.* **6,** 128–150.

Jeay, S., Sonenshein, G. E., Postel-Vinay, M. C., Kelly, P. A., and Baixeras, E. (2002). Growth hormone can act as a cytokine controlling survival and proliferation of immune cells: New insights into signaling pathways. *Mol. Cell. Endocrinol.* **188,** 1–7.

Ji, S., Guan, R., Frank, S. J., and Messina, J. L. (1999). Insulin inhibits growth hormone signaling via the growth hormone receptor/JAK2/STAT5B pathway. *J. Biol. Chem.* **274,** 13434–13442.

Ji, S., Frank, S. J., and Messina, J. L. (2002). Growth hormone-induced differential desensitization of STAT5, ERK, and Akt phosphorylation. *J. Biol. Chem.* **277,** 28384–28393.

Johansen, T., Laurino, C., Barreca, A., and Malmlof, K. (2005). Reduction of adiposity with prolonged growth hormone treatment in old obese rats: Effects on glucose handling and early insulin signaling. *Growth Horm. IGF Res.* **15,** 55–63.

Jorgensen, J. O., Krag, M., Jessen, N., Norrelund, H., Vestergaard, E. T., Moller, N., and Christiansen, J. S. (2004). Growth hormone and glucose homeostasis. *Horm. Res.* **62** (Suppl. 3), 51–55.

Kawazoe, Y., Naka, T., Fujimoto, M., Kohzaki, H., Morita, Y., Narazaki, M., Okumura, K., Saitoh, H., Nakagawa, R., Uchiyama, Y., Akira, S., and Kishimoto, T. (2001). Signal transducer and activator of transcription (STAT)-induced STAT inhibitor 1 (SSI-1)/suppressor of cytokine signaling 1 (SOCS1) inhibits insulin signal transduction pathway through modulating insulin receptor substrate 1 (IRS-1) phosphorylation. *J. Exp. Med.* **193,** 263–269.

Kim, S. O., Houtman, J. C., Jiang, J., Ruppert, J. M., Bertics, P. J., and Frank, S. J. (1999). Growth hormone-induced alteration in ErbB-2 phosphorylation status in 3T3-F442A fibroblasts. *J. Biol. Chem.* **274,** 36015–36024.

Klasing, K. C. (1988). Nutritional aspects of leukocytic cytokines. *J. Nutr.* **118,** 1436–1446.

Kofoed, E. M., Hwa, V., Little, B., Woods, K. A., Buckway, C. K., Tsubaki, J., Pratt, K. L., Bezrodnik, L., Jasper, H., Tepper, A., Heinrich, J. J., and Rosenfeld, R. G. (2003). Growth hormone insensitivity associated with a STAT5b mutation. *N. Engl. J. Med.* **349,** 1139–1147.

Kopchick, J. J., Bellush, L. L., and Coschigano, K. T. (1999). Transgenic models of growth hormone action. *Annu. Rev. Nutr.* **19,** 437–461.

Kratzsch, J., Kellner, K., Zilkens, T., Schmidt-Gayk, H., Selisko, T., and Scholz, G. H. (1996). Growth hormone-binding protein related immunoreactivity is regulated by the degree of insulinopenia in diabetes mellitus. *Clin. Endocrinol. (Oxf.)* **44,** 673–678.

Kuwahara, H., Uotani, S., Abe, T., Degawa-Yamauchi, M., Takahashi, R., Kita, A., Fujita, N., Ohshima, K., Sakamaki, H., Yamasaki, H., Yamaguchi, Y., and Eguchi, K. (2003). Insulin attenuates leptin-induced STAT3 tyrosine-phosphorylation in a hepatoma cell line. *Mol. Cell. Endocrinol.* **205,** 115–120.

Lahuna, O., Rastegar, M., Maiter, D., Thissen, J. P., Lemaigre, F. P., and Rousseau, G. G. (2000). Involvement of STAT5 (signal transducer and activator of transcription 5) and HNF-4 (hepatocyte nuclear factor 4) in the transcriptional control of the hnf6 gene by growth hormone. *Mol. Endocrinol.* **14,** 285–294.

Laron, Z. (1995). Prismatic cases: Laron syndrome (primary growth hormone resistance) from patient to laboratory to patient. *J. Clin. Endocrinol. Metab.* **80,** 1526–1531.

Laron, Z. (2005). Do deficiencies in growth hormone and insulin-like growth factor-1 (IGF-1) shorten or prolong longevity? *Mech. Ageing Dev.* **126,** 305–307.

Le, R. D., Bondy, C., Yakar, S., Liu, J. L., and Butler, A. (2001). The somatomedin hypothesis: 2001. *Endocr. Rev.* **22,** 53–74.

Lebovitz, H. E. (2001). Insulin resistance: Definition and consequences. *Exp. Clin. Endocrinol. Diabetes* **109**(Suppl. 2), S135–S148.

Leung, K. C., Waters, M. J., Markus, I., Baumbach, W. R., and Ho, K. K. (1997). Insulin and insulin-like growth factor-I acutely inhibit surface translocation of growth hormone receptors in osteoblasts: A novel mechanism of growth hormone receptor regulation. *Proc. Natl. Acad. Sci. USA* **94,** 11381–11386.

Lewis, M. D., Horan, M., Millar, D. S., Newsway, V., Easter, T. E., Fryklund, L., Gregory, J. W., Norin, M., Del Valle, C. J., Lopez-Siguero, J. P., Canete, R.,

Lopez-Canti, L. F., *et al.* (2004). A novel dysfunctional growth hormone variant (Ile179Met) exhibits a decreased ability to activate the extracellular signal-regulated kinase pathway. *J. Clin. Endocrinol. Metab.* **89,** 1068–1075.

Lewis, T. S., Shapiro, P. S., and Ahn, N. G. (1998). Signal transduction through MAP kinase cascades. *Adv. Cancer Res.* **74,** 49–139.

Liang, L., Zhou, T., Jiang, J., Pierce, J. H., Gustafson, T. A., and Frank, S. J. (1999). Insulin receptor substrate-1 enhances growth hormone-induced proliferation. *Endocrinology* **140,** 1972–1983.

Liang, L., Jiang, J., and Frank, S. J. (2000). Insulin receptor substrate-1-mediated enhancement of growth hormone-induced mitogen-activated protein kinase activation. *Endocrinology* **141,** 3328–3336.

Lippe, B. M., Kaplan, S. A., Golden, M. P., Hendricks, S. A., and Scott, M. L. (1981). Carbohydrate tolerance and insulin receptor binding in children with hypopituitarism: Response after acute and chronic human growth hormone administration. *J. Clin. Endocrinol. Metab.* **53,** 507–513.

Liu, J. L., Coschigano, K. T., Robertson, K., Lipsett, M., Guo, Y., Kopchick, J. J., Kumar, U., and Liu, Y. L. (2004). Disruption of growth hormone receptor gene causes diminished pancreatic islet size and increased insulin sensitivity in mice. *Am. J. Physiol. Endocrinol. Metab.* **287,** E405–E413.

Love, D. W., Whatmore, A. J., Clayton, P. E., and Silva, C. M. (1998). Growth hormone stimulation of the mitogen-activated protein kinase pathway is cell type specific. *Endocrinology* **139,** 1965–1971.

Maes, M., Ketelslegers, J. M., and Underwood, L. E. (1983). Low plasma somatomedin-C in streptozotocin-induced diabetes mellitus. Correlation with changes in somatogenic and lactogenic liver binding sites. *Diabetes* **32,** 1060–1069.

Mauvais-Jarvis, F., Ueki, K., Fruman, D. A., Hirshman, M. F., Sakamoto, K., Goodyear, L. J., Iannacone, M., Accili, D., Cantley, L. C., and Kahn, C. R. (2002). Reduced expression of the murine p85alpha subunit of phosphoinositide 3-kinase improves insulin signaling and ameliorates diabetes. *J. Clin. Invest.* **109,** 141–149.

Menon, R. K., Arslanian, S., May, B., Cutfield, W. S., and Sperling, M. A. (1992). Diminished growth hormone-binding protein in children with insulin-dependent diabetes mellitus. *J. Clin. Endocrinol. Metab.* **74,** 934–938.

Menon, R. K., Stephan, D. A., Rao, R. H., Shen-Orr, Z., Downs, L. S. Jr., Roberts, C. T. Jr., Leroith, D., and Sperling, M. A. (1994). Tissue-specific regulation of the growth hormone receptor gene in streptozocin-induced diabetes in the rat. *J. Endocrinol.* **142,** 453–462.

Mercado, M., and Baumann, G. (1995). Characteristics of the somatotropic axis in insulin dependent diabetes mellitus. *Arch. Med. Res.* **26,** 101–109.

Mercado, M., Molitch, M. E., and Baumann, G. (1992). Low plasma growth hormone binding protein in IDDM. *Diabetes* **41,** 605–609.

Mesotten, D., Wouters, P. J., Peeters, R. P., Hardman, K. V., Holly, J. M., Baxter, R. C., and Van den Berghe, G. (2004). Regulation of the somatotropic axis by intensive insulin therapy during protracted critical illness. *J. Clin. Endocrinol. Metab.* **89,** 3105–3113.

Messina, J. L. (1999). Insulin as a growth-promoting hormone. *In* "Handbook of Physiology" (J. L. Kostyo, Ed.), Section 7: The Endocrine System. Volume V: Hormonal Control of Growth, pp. 783–811. Oxford University Press, New York.

Metherell, L. A., Akker, S. A., Munroe, P. B., Rose, S. J., Caulfield, M., Savage, M. O., Chew, S. L., and Clark, A. J. (2001). Pseudoexon activation as a novel mechanism for disease resulting in atypical growth-hormone insensitivity. *Am. J. Hum. Genet.* **69,** 641–646.

Meyer, D. J., Campbell, G. S., Cochran, B. H., Argetsinger, L. S., Larner, A. C., Finbloom, D. S., Carter-Su, C., and Schwartz, J. (1994). Growth hormone induces a DNA binding factor related to the interferon-stimulated 91-kDa transcription factor. *J. Biol. Chem.* **269,** 4701–4704.

Moller, C., Hansson, A., Enberg, B., Lobie, P. E., and Norstedt, G. (1992). Growth hormone (GH) induction of tyrosine phosphorylation and activation of mitogen-activated protein kinases in cells transfected with rat GH receptor cDNA. *J. Biol. Chem.* **267,** 23403–23408.

Mooney, R. A., Senn, J., Cameron, S., Inamdar, N., Boivin, L. M., Shang, Y., and Furlanetto, R. W. (2001). Suppressors of cytokine signaling-1 and -6 associate with and inhibit the insulin receptor. A potential mechanism for cytokine-mediated insulin resistance. *J. Biol. Chem.* **276,** 25889–25893.

Nguyen, A. P., Chandorkar, A., and Gupta, C. (1996). The role of growth hormone in fetal mouse reproductive tract differentiation. *Endocrinology* **137,** 3659–3666.

Okuma, S. (1984). Study of growth hormone in spontaneous dwarf rat. *Nippon Naibunpi Gakkai Zasshi* **60,** 1005–1014.

Olsson, B., Bohlooly, Y., Brusehed, O., Isaksson, O. G., Ahren, B., Olofsson, S. O., Oscarsson, J., and Tornell, J. (2003). Bovine growth hormone-transgenic mice have major alterations in hepatic expression of metabolic genes. *Am. J. Physiol. Endocrinol. Metab.* **285,** E504–E511.

O'Riordain, M. G., Ross, J. A., Fearon, K. C. H., Maingay, J., Farouk, M., Garden, O. J., and Carter, D. C. (1995). Insulin and counterregulatory hormones influence acute-phase protein production in human hepatocytes. *Am. J. Physiol. Endocrinol. Metab.* **269,** E323–E330.

Palmiter, R. D., Brinster, R. L., Hammer, R. E., Trumbauer, M. E., Rosenfeld, M. G., Birnberg, N. C., and Evans, R. M. (1982). Dramatic growth of mice that develop from eggs microinjected with metallothionein-growth hormone fusion genes. *Nature* **300,** 611–615.

Pennisi, P. A., Kopchick, J. J., Thorgeirsson, S., Leroith, D., and Yakar, S. (2004). Role of growth hormone (GH) in liver regeneration. *Endocrinology* **145,** 4748–4755.

Prattali, R. R., Barreiro, G. C., Caliseo, C. T., Fugiwara, F. Y., Ueno, M., Prada, P. O., Velloso, L. A., Saad, M. J., and Carvalheira, J. B. (2005). Aspirin inhibits serine phosphorylation of insulin receptor substrate 1 in growth hormone treated animals. *FEBS Lett.* **579,** 3152–3158.

Ram, P. A., Park, S. H., Choi, H. K., and Waxman, D. J. (1996). Growth hormone activation of Stat 1, Stat 3, and Stat 5 in rat liver—differential kinetics of hormone desensitization and growth hormone stimulation of both tyrosine phosphorylation and serine/threonine phosphorylation. *J. Biol. Chem.* **271,** 5929–5940.

Ridderstrale, M., Degerman, E., and Tornqvist, H. (1995). Growth hormone stimulates the tyrosine phosphorylation of the insulin receptor substrate-1 and its association with phosphatidylinositol 3-kinase in primary adipocytes. *J. Biol. Chem.* **270,** 3471–3474.

Rui, L., Yuan, M., Frantz, D., Shoelson, S., and White, M. F. (2002). SOCS-1 and SOCS-3 block insulin signaling by ubiquitin-mediated degradation of IRS1 and IRS2. *J. Biol. Chem.* **277,** 42394–42398.

Schwartz, R., Gruppuso, P. A., Petzold, K., Brambilla, D. H. V., and Teramo, K. A. (1994). Hyperinsulinemia and macrosomia in the fetus of the diabetic mother. *Diabetes Care* **17,** 640–648.

Searle, T. W., Murray, J. D., and Baker, P. J. (1992). Effect of increased production of growth hormone on body composition in mice: Transgenic versus control. *J. Endocrinol.* **132,** 285–291.

Shea, B. T., Hammer, R. E., and Brinster, R. L. (1987). Growth allometry of the organs in giant transgenic mice. *Endocrinology* **121,** 1924–1930.

Shuai, K. (1999). The STAT family of proteins in cytokine signaling. *Prog. Biophys. Mol. Biol.* **71,** 405–422.

Smal, J., Closset, J., Hennen, G., and De Meyts, P. (1987). Receptor binding properties and insulin-like effects of human growth hormone and its 20 kDa-variant in rat adipocytes. *J. Biol. Chem.* **262,** 11071–11079.

Smith, T. R., Elmendorf, J. S., David, T. S., and Turinsky, J. (1997). Growth hormone-induced insulin resistance: Role of the insulin receptor, IRS-1, GLUT-1, and GLUT-4. *Am. J. Physiol. Endocrinol. Metab.* **272,** E1071–E1079.

Snibson, K. J., Bhathal, P. S., Hardy, C. L., Brandon, M. R., and Adams, T. E. (1999). High, persistent hepatocellular proliferation and apoptosis precede hepatocarcinogenesis in growth hormone transgenic mice. *Liver* **19,** 242–252.

Sornson, M. W., Wu, W., Dasen, J. S., Flynn, S. E., Norman, D. J., O'Connell, S. M., Gukovsky, I., Carrière, C., Ryan, A. K., Miller, A. P., Zuo, L., Gleiberman, A. S., *et al.* (1996). Pituitary lineage determination by the *Prophet of Pit-1* homeodomain factor defective in Ames dwarfism. *Nature* **384,** 327–333.

Sotiropoulos, A., Perrot-Applanat, M., Dinerstein, H., Pallier, A., Postel-Vinay, M. C., Finidori, J., and Kelly, P. A. (1994). Distinct cytoplasmic regions of the growth hormone receptor are required for activation of JAK2, mitogen-activated protein kinase, and transcription. *Endocrinology* **135,** 1292–1298.

Sotiropoulos, A., Moutoussamy, S., Binart, N., Kelly, P. A., and Finidori, J. (1995). The membrane proximal region of the cytoplasmic domain of the growth hormone receptor is involved in the activation of Stat 3. *FEBS Lett.* **369,** 169–172.

Souza, S. C., Frick, G. P., Yip, R., Lobo, R. B., Tai, L. R., and Goodman, H. M. (1994). Growth hormone stimulates tyrosine phosphorylation of insulin receptor substrate-1. *J. Biol. Chem.* **269,** 30085–30088.

Styne, D. M. (2003). The regulation of pubertal growth. *Horm. Res.* **60,** 22–26.

Taniguchi, C. M., Emanuelli, B., and Kahn, C. R. (2006). Critical nodes in signalling pathways: Insights into insulin action. *Nat. Rev. Mol. Cell Biol.* **7,** 85–96.

Tannenbaum, G. S., and Martin, J. B. (1976). Evidence for an endogenous ultradian rhythm governing growth hormone secretion in the rat. *Endocrinology* **98,** 562–570.

Terauchi, Y., Tsuji, Y., Satoh, S., Minoura, H., Murakami, K., Okuno, A., Inukai, K., Asano, T., Kaburagi, Y., Ueki, K., Nakajima, H., Hanafusa, T., *et al.* (1999). Increased insulin sensitivity and hypoglycaemia in mice lacking the p85 alpha subunit of phosphoinositide 3-kinase. *Nat. Genet.* **21,** 230–235.

Thirone, A. C., Carvalho, C. R., Brenelli, S. L., Velloso, L. A., and Saad, M. J. (1997). Effect of chronic growth hormone treatment on insulin signal transduction in rat tissues. *Mol. Cell. Endocrinol.* **130,** 33–42.

Thompson, D., Harrison, S. P., Evans, S. W., and Whicher, J. T. (1991). Insulin modulation of acute-phase protein production in a human hepatoma cell line. *Cytokine* **3,** 619–626.

Udy, G. B., Towers, R. P., Snell, R. G., Wilkins, R. J., Park, S. H., Ram, P. A., Waxman, D. J., and Davey, H. W. (1997). Requirement of STAT5b for sexual dimorphism of body growth rates and liver gene expression. *Proc. Natl. Acad. Sci. USA* **94,** 7239–7244.

Ueki, K., Algenstaedt, P., Mauvais-Jarvis, F., and Kahn, C. R. (2000). Positive and negative regulation of phosphoinositide 3-kinase-dependent signaling pathways by three different gene products of the p85alpha regulatory subunit. *Mol. Cell. Biol.* **20,** 8035–8046.

Ueki, K., Fruman, D. A., Brachmann, S. M., Tseng, Y. H., Cantley, L. C., and Kahn, C. R. (2002). Molecular balance between the regulatory and catalytic subunits of phosphoinositide 3-kinase regulates cell signaling and survival. *Mol. Cell. Biol.* **22,** 965–977.

Ueki, K., Kondo, T., Tseng, Y. H., and Kahn, C. R. (2004). Central role of suppressors of cytokine signaling proteins in hepatic steatosis, insulin resistance, and the metabolic syndrome in the mouse. *Proc. Natl. Acad. Sci. USA* **101,** 10422–10427.

Valera, A., Rodriguez-Gil, J. E., Yun, J. S., McGrane, M. M., Hanson, R. W., and Bosch, F. (1993). Glucose metabolism in transgenic mice containing a chimeric P-enolpyruvate carboxykinase/bovine growth hormone gene. *FASEB J.* **7,** 791–800.

Vance, M. L., and Mauras, N. (1999). Growth hormone therapy in adults and children. *N. Engl. J. Med.* **341,** 1206–1216.

VanderKuur, J. A., Butch, E. R., Waters, S. B., Pessin, J. E., Guan, K. L., and Carter-Su, C. (1997). Signaling molecules involved in coupling growth hormone receptor to mitogen-activated protein kinase activation. *Endocrinology* **138,** 4301–4307.

Venkatesan, N., and Davidson, M. B. (1995). Insulin resistance in rats harboring growth hormone-secreting tumors: Decreased receptor number but increased kinase activity in liver. *Metab. Clin. Exp.* **44,** 75–84.

Wanke, R., Hermanns, W., Folger, S., Wolf, E., and Brem, G. (1991). Accelerated growth and visceral lesions in transgenic mice expressing foreign genes of the growth hormone family: An overview. *Pediatr. Nephrol.* **5,** 513–521.

Waxman, D. J., Ram, P. A., Park, S. H., and Choi, H. K. (1995). Intermittent plasma growth hormone triggers tyrosine phosphorylation and nuclear translocation of a liver-expressed, Stat 5-related DNA binding protein. Proposed role as an intracellular regulator of male-specific liver gene transcription. *J. Biol. Chem.* **270,** 13262–13270.

Wei, Y., Bizeau, M. E., and Pagliassotti, M. J. (2004). An acute increase in fructose concentration increases hepatic glucose-6-phosphatase mRNA via mechanisms that are independent of glycogen synthase kinase-3 in rats. *J. Nutr.* **134,** 545–551.

White, M. F. (1998). The IRS-signaling system: A network of docking proteins that mediate insulin and cytokine action. *Recent Prog. Horm. Res.* **53,** 119–138.

Winston, L. A., and Bertics, P. J. (1992). Growth hormone stimulates the tyrosine phosphorylation of 42- and 45-kDa ERK-related proteins. *J. Biol. Chem.* **267,** 4747–4751.

Winston, L. A., and Hunter, T. (1995). JAK2, Ras, and Raf are required for activation of extracellular signal-regulated kinase mitogenactivated protein kinase by growth hormone. *J. Biol. Chem.* **270,** 30837–30840.

Woelfle, J., Billiard, J., and Rotwein, P. (2003a). Acute control of insulin-like growth factor-I gene transcription by growth hormone through Stat5b. *J. Biol. Chem.* **278,** 22696–22702.

Woelfle, J., Chia, D. J., and Rotwein, P. (2003b). Mechanisms of growth hormone action: Identification of conserved Stat5 binding sites that mediate GH-induced insulin-like growth factor-I gene activation. *J. Biol. Chem.* **278,** 51261–51266.

Xu, J., Ji, S., Venable, D. Y., Franklin, J. L., and Messina, J. L. (2005). Prolonged insulin treatment inhibits GH signaling via STAT3 and STAT1. *J. Endocrinol.* **184,** 481–492.

Xu, J., Keeton, A. B., Franklin, J. L., Li, X., Venable, D. Y., Frank, S. J., and Messina, J. L. (2006a). Insulin enhances growth hormone induction of the MEK/ERK signaling pathway. *J. Biol. Chem.* **281,** 982–992.

Xu, J., Liu, Z., Clemens, T. L., and Messina, J. L. (2006b). Insulin reverses growth hormone-induced homologous desensitization. *J. Biol. Chem.* **281,** 21594–21606.

Yakar, S., Setser, J., Zhao, H., Stannard, B., Haluzik, M., Glatt, V., Bouxsein, M. L., Kopchick, J. J., and Leroith, D. (2004). Inhibition of growth hormone action improves insulin sensitivity in liver IGF-1-deficient mice. *J. Clin. Invest.* **113,** 96–105.

Yamauchi, T., Ueki, K., Tobe, K., Tamemoto, H., Sekine, N., Wada, M., Honjo, M., Takahashi, M., Takahashi, T., Hirai, H., Tushima, T., Akanuma, Y., *et al.* (1997). Tyrosine phosphorylation of the EGF receptor by the kinase Jak2 is induced by growth hormone. *Nature* **390,** 91–96.

Yamauchi, T., Kaburagi, Y., Ueki, K., Tsuji, Y., Stark, G. R., Kerr, I. M., Tsushima, T., Akanuma, Y., Komuro, I., Tobe, K., Yazaki, Y., and Kadowaki, T. (1998). Growth hormone and prolactin stimulate tyrosine phosphorylation of insulin receptor substrate-1, -2, and -3, their association with p85 phosphatidylinositol 3-kinase (PI3-kinase), and concomitantly PI3-kinase activation via JAK2 kinase. *J. Biol. Chem.* **273,** 15719–15726.

Yenush, L., and White, M. F. (1997). The IRS-signalling system during insulin and cytokine action. *BioEssays* **19,** 491–500.

Zhou, Y., Xu, B. C., Maheshwari, H. G., He, L., Reed, M., Lozykowski, M., Okada, S., Cataldo, L., Coschigamo, K., Wagner, T. E., Baumann, G., and Kopchick, J. J. (1997).

A mammalian model for Laron syndrome produced by targeted disruption of the mouse growth hormone receptor/binding protein gene (the Laron mouse). *Proc. Natl. Acad. Sci. USA* **94,** 13215–13220.

Zhu, T., Ling, L., and Lobie, P. E. (2002). Identification of a JAK2-independent pathway regulating growth hormone (GH)-stimulated p44/42 mitogen-activated protein kinase activity. GH activation of Ral and phospholipase D is Src-dependent. *J. Biol. Chem.* **277,** 45592–45603.

CHAPTER SEVEN

Intracellular Retention and Insulin-Stimulated Mobilization of GLUT4 Glucose Transporters

Bradley R. Rubin *and* Jonathan S. Bogan

Contents

Abstract

GLUT4 glucose transporters are expressed nearly exclusively in adipose and muscle cells, where they cycle to and from the plasma membrane. In cells not stimulated with insulin, GLUT4 is targeted to specialized GLUT4 storage vesicles (GSVs), which sequester it away from the cell surface. Insulin acts within minutes to mobilize these vesicles, translocating GLUT4 to the plasma membrane to enhance glucose uptake. The mechanisms controlling GSV sequestration and mobilization are poorly understood.

An insulin-regulated aminopeptidase that cotraffics with GLUT4, IRAP, is required for basal GSV retention and insulin-stimulated mobilization. TUG and

Section of Endocrinology and Metabolism, Department of Internal Medicine and Department of Cell Biology, Yale University School of Medicine, New Haven, Connecticut 06520-8020

Vitamins and Hormones, Volume 80
ISSN 0083-6729, DOI: 10.1016/S0083-6729(08)00607-9

Ubc9 bind GLUT4, and likely retain GSVs within unstimulated cells. These proteins may be components of a retention receptor, which sequesters GLUT4 and IRAP away from recycling vesicles. Insulin may then act on this protein complex to liberate GLUT4 and IRAP, discharging GSVs into a recycling pathway for fusion at the cell surface. How GSVs are anchored intracellularly, and how insulin mobilizes these vesicles, are the important topics for ongoing research.

Regulation of GLUT4 trafficking is tissue-specific, perhaps in part because the formation of GSVs requires cell type-specific expression of sortilin. Proteins controlling GSV retention and mobilization can then be more widely expressed. Indeed, GLUT4 likely participates in a general mechanism by which the cell surface delivery of various membrane proteins can be controlled by extracellular stimuli. Finally, it is not known if defects in the formation or intracellular retention of GSVs contribute to human insulin resistance, or play a role in the pathogenesis of type 2 diabetes.

I. Introduction

The regulation of glucose uptake by muscle and adipose tissue is critical to the overall control of carbohydrate metabolism. Insulin is the main hormone that controls the rate of glucose uptake, and it does so by modulating the transport of glucose across the plasma membranes of fat and muscle cells. Transport of glucose into muscle is decreased in humans with type 2 diabetes mellitus, and accounts for muscle insulin resistance (Cline *et al.*, 1999). This compromise in the ability of insulin to stimulate glucose utilization is a key factor in the pathogenesis of type 2 diabetes and the metabolic syndrome (Biddinger and Kahn, 2006; Petersen *et al.*, 2007). Thus, in large measure because of its potential importance for diabetes, the question of how insulin controls glucose transport in fat and muscle cells has been an area of intensive investigation.

The current, molecular understanding of how insulin regulates glucose uptake began with the discovery that insulin redistributes a glucose-transporting activity out of internal membranes and to the plasma membrane (Cushman and Wardzala, 1980; Suzuki and Kono, 1980). Like all cell membranes, the plasma membrane is a lipid bilayer that is impermeable to small, hydrophilic molecules such as glucose. Insulin had earlier been proposed to increase the transport of hexoses across plasma membranes (Levine *et al.*, 1949, 1950). However, it had been anticipated that the transporters would already be present in plasma membranes, and that these were simply activated by insulin. The insight that insulin causes translocation of glucose transporters from an internal reservoir provided an alternative explanation, which could account for the rapidity and reversibility of insulin action. This "translocation hypothesis" has been the basis for much work in this field in the decades since its original formulation.

The insulin-mobilized, glucose-transporting activity was subsequently identified as the GLUT4 glucose transporter (Birnbaum, 1989; Charron *et al.*, 1989; Fukumoto *et al.*, 1989; James *et al.*, 1988, 1989; Kaestner *et al.*, 1989). The fourth-identified member of the GLUT family of facilitative hexose transporters, GLUT4 is expressed nearly exclusively in skeletal and cardiac muscle, and in brown and white adipose tissues. GLUT4 contains 12 transmembrane segments, and cycles among internal membrane compartments and the plasma membrane. In contrast to most proteins that recycle at the cell surface, GLUT4 is efficiently retained within cells that are not stimulated with insulin. This basal, intracellular sequestration of GLUT4 excludes the transporters from the plasma membrane, and restricts glucose uptake. Insulin is then able to act within minutes to increase the fraction of GLUT4 in plasma membranes by up to 30-fold and, correspondingly, to increase glucose uptake. This process is reversible, and upon insulin withdrawal the transporters are again sequestered away from the plasma membrane.

Insulin likely acts by similar mechanisms in both fat and muscle cells. Firm knowledge of whether this is indeed the case awaits molecular characterization of the components that are involved. In skeletal and cardiac muscle, contraction and ischemia also redistribute GLUT4 to the plasma membrane, and these effects are distinct from that of insulin (Chavez *et al.*, 2008; Chen *et al.*, 2008; Fujii *et al.*, 2006; Ishiki and Klip, 2005; Jessen and Goodyear, 2005; Ploug and Ralston, 2002; Ploug *et al.*, 1998; Russell *et al.*, 2004; Taylor *et al.*, 2008; Yang and Holman, 2005; Zhou *et al.*, 2008). Although muscle accounts for a much larger proportion of overall, insulin-regulated glucose uptake, most work to understand how insulin regulates GLUT4-mediated transport has been done using adipocytes. This is because adipocytes, and particularly the cultured 3T3-L1 adipocyte cell line, are a more tractable experimental system. Here, we focus on insulin regulation of GLUT4 in adipocytes.

The ability of insulin to regulate GLUT4 targeting is both cell type-specific and GLUT isoform-specific. Thus, exogenous expression of GLUT4 in fibroblasts or other nonadipose, nonmuscle cells typically does not result in insulin-regulated trafficking or glucose uptake. Likewise, other glucose transporters (e.g., GLUT1) do not exhibit highly insulin-responsive trafficking even when studied in adipose or muscle cells. Finally, whether insulin activates the intrinsic, glucose-transporting activity of GLUT4, in addition to causing its relocalization to the plasma membrane, remains an open question (Kandror, 2003; Ribe *et al.*, 2005; Somwar *et al.*, 2002; Zierler, 1998).

Work to understand how insulin stimulates glucose uptake has focused both on signaling pathways and trafficking mechanisms. Greater progress has been made by working downstream from the insulin receptor, to understand the signaling pathways involved, than by working upstream from GLUT4, to understand the trafficking components. Signaling through the Akt2 kinase to AS160 and other substrates, through atypical protein kinase C, and through the Rho-family GTPase, TC10α, have all been implicated in directing GLUT4 translocation. These signaling pathways may act at different steps in GLUT4's itinerary to control its overall distribution among intracellular compartments. These insulin signaling pathways are well reviewed elsewhere (Cartee and Wojtaszewski, 2007; Chang *et al.*, 2004; Cohen, 2006; Farese *et al.*, 2007; Huang and Czech, 2007; Kanzaki, 2006; Taniguchi *et al.*, 2006).

The translocation hypothesis for insulin action transformed a question that had been framed in terms of organism-level physiology into one that could be approached in terms of cell biology. Concurrently, there has been remarkable progress in understanding the trafficking of membrane proteins (Cai *et al.*, 2007; Mellman and Warren, 2000). Yet despite this progress, the particular mechanisms that regulate GLUT4 targeting have remained elusive. Here, we focus on current understanding of some of these mechanisms, which regulate the distribution of GLUT4 among intracellular, membrane-bound compartments. In particular, we discuss how GLUT4 is targeted and retained in a pool of nonendosomal "GLUT4 storage vesicles (GSVs)" in the absence of insulin, and how insulin mobilizes these vesicles to the plasma membrane to augment glucose uptake.

II. GLUT4 Storage Vesicles (GSVs)

In insulin-responsive cell types, GLUT4 is targeted to unique, intracellular membranes, termed "GSVs" or "Insulin-Responsive Vesicles" (see Fig. 7.1). GSVs are nonendosomal vesicles that accumulate in cells not stimulated with insulin. By sequestering GLUT4 away from the plasma membrane, the efficient targeting of GLUT4 to GSVs restricts basal glucose uptake. Within minutes of insulin addition, GSVs move to the cell surface and fuse, inserting GLUT4 transporters into the plasma membrane to augment glucose uptake. Thus, to understand how insulin stimulates glucose uptake, it is essential to understand how GSVs are formed and retained within cells, and how GLUT4 is so efficiently targeted to these specialized organelles. Put another way, it is the efficient targeting and intracellular retention of GLUT4, in unstimulated cells, that is the key that enables insulin to effect such a dramatic redistribution.

In addition to mobilizing GSVs, insulin regulates other steps in GLUT4 trafficking. Evidence supports the notion that insulin reduces the rate of GLUT4 internalization by about twofold, by modulating the endocytic mechanism that is used (Blot and McGraw, 2006). The more marked action of insulin is to increase the rate of GLUT4 externalization to the plasma

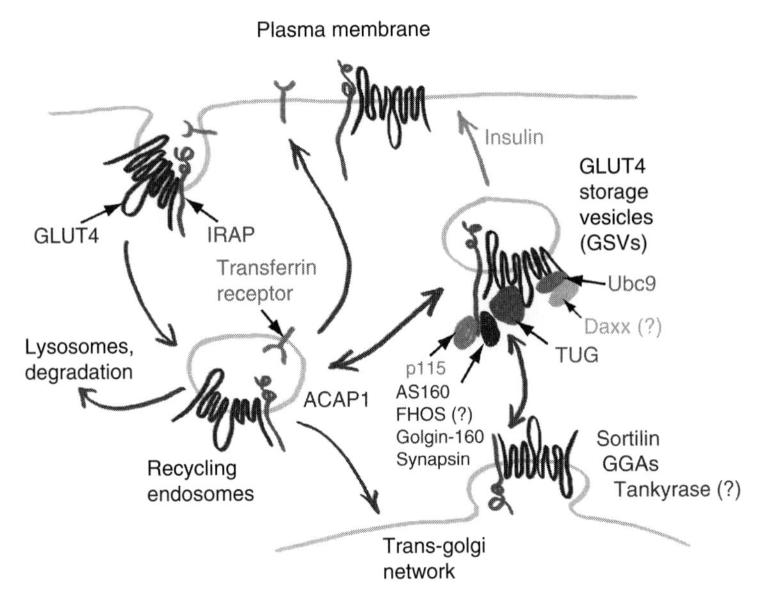

Figure 7.1 A model for GLUT4 glucose transporter trafficking. GLUT4 and IRAP undergo endocytosis at the plasma membrane, and are present together with endosomal proteins such as the transferrin receptor in sorting and recycling endosomes. In contrast to the transferrin receptor, which is returned to the plasma membrane, GLUT4 and IRAP are efficiently targeted to "GLUT4 storage vesicles" (GSVs) in cells not stimulated with insulin. GSVs constitute a specialized compartment that is formed in a cell type-specific manner. To reach GSVs, GLUT4/IRAP may traffic through the trans-Golgi network or directly from endosomes. ACAP1 and sortilin are cargo adaptors that facilitate the incorporation of GLUT4 and IRAP into GSVs, and Golgi-localized, gamma-ear-containing, ARF binding proteins (GGAs) and tankyrase may cooperate in this process. In unstimulated cells, GSVs are retained intracellularly through interactions of GLUT4 and IRAP with a retention receptor complex, which comprises TUG, Ubc9, and other proteins. Insulin is then able to mobilize GSVs for fusion at the plasma membrane by stimulating the disassembly of this complex. Daxx has been shown to bind GLUT4, and may participate stably in the complex or act to stimulate its disassembly. Proteins that may anchor the complex to intracellular structures include the Golgi matrix proteins p115 and Golgin-160, as well as FHOS, AS160, and synapsin. By mobilizing GSVs to the cell surface, insulin markedly increases the number of glucose transporters present in the plasma membrane and augments glucose uptake. Disruption of TUG, Ubc9, sortilin, or IRAP causes GLUT4 not to accumulate in GSVs, and accelerates the movement of GLUT4 to lysosomes as well as back to the plasma membrane.

membrane. In addition to mobilizing GSVs, insulin increases the rate of GSV production from donor membranes, which may be particularly important during sustained insulin exposure (Xu and Kandror, 2002). How insulin stimulates the moblization of GSVs is poorly understood. Insulin may liberate GSVs from an anchoring site(s), or it may redirect their intracellular trafficking, or it may load these vesicles onto motor proteins to target them for delivery to the cell surface (Bose *et al.*, 2002, 2004; Chen *et al.*, 2007; Imamura *et al.*, 2003; Semiz *et al.*, 2003; Yoshizaki *et al.*, 2007). These possible actions are not mutually exclusive. Insulin also activates the exocyst, an octameric protein complex that tethers the vesicles to the plasma membrane and facilitates their docking (Bao *et al.*, 2008; Chen *et al.*, 2007; Ewart *et al.*, 2005; Inoue *et al.*, 2003, 2006). Finally, insulin signals to proteins that regulate SNARE complex formation, and thus controls the actual fusion of GLUT4-containing vesicles with the plasma membrane (Bai *et al.*, 2007; Huang *et al.*, 2007; Jiang *et al.*, 2008; Koumanov *et al.*, 2005; Sano *et al.*, 2005; van Dam *et al.*, 2005; Yamada *et al.*, 2005).

Most data are consistent with the idea that basal intracellular retention of GSVs, and insulin-stimulated mobilization of these vesicles, exerts the greatest effect on overall GLUT4 distribution. Yet it has been difficult to quantify the relative contribution of GLUT4 trafficking through GSVs, as compared to the other insulin-regulated steps noted earlier. Why has this been so? One problem is that there have been few experimental systems to distinguish GSV formation or budding, mobilization, tethering, docking, and fusion at the plasma membrane. The identification of molecules controlling each of these processes, together with the application of new microscopy approaches and the development of cell-free assays, will enable progress in this area (Huang *et al.*, 2007; Jiang *et al.*, 2008; Koumanov *et al.*, 2005; Ramm and James, 2005). Additionally, GLUT4 is distributed among multiple intracellular pools, including GSVs, early and recycling endosomes, and the trans-Golgi network (TGN) (Bogan and Lodish, 1999; Bryant *et al.*, 2002; Slot *et al.*, 1991a,b, 1997). Insulin acts at many (if not all) of these locations, complicating the interpretation of experiments to determine which effects are most responsible for the overall shift in GLUT4 distribution. Finally, GSVs have been defined primarily in negative terms, since other protein markers are excluded from these vesicles (Hashiramoto and James, 2000; Kupriyanova *et al.*, 2002; Lim *et al.*, 2001; Livingstone *et al.*, 1996). Because of these considerations, it has remained uncertain how these vesicles are formed and traffic among intracellular compartments.

Two models have been proposed for how GSVs function in insulin action in 3T3-L1 adipocytes. According to one model, GLUT4 participates in a "dynamic equilibrium" that determines the fraction of total GLUT4 that is present in each location within cells (Karylowski *et al.*, 2004; Martin *et al.*, 2006; Zeigerer *et al.*, 2002). This model posits that the entire pool of GLUT4 traverses the plasma membrane under both basal and insulin-stimulated conditions. Insulin does not alter the size of the cycling pool, but shifts the accumulation of GLUT4 solely by altering the rate constants for externalization and internalization. A second model, termed "static retention," holds that the pools of GLUT4 that recycle in basal and insulin-stimulated cells are different sizes (Coster *et al.*, 2004; Govers *et al.*, 2004). According to this model, a fraction of the total cellular GLUT4 is sequestered in GSVs by a static mechanism in unstimulated cells. This GLUT4 does not recycle at the plasma membrane or exchange with endosomes in the absence of insulin. Insulin acts in a concentration-dependent manner to discharge these transporters into the recycling pathway. Thus, in this model, insulin acts not only by altering trafficking rate constants, but by increasing the amount of GLUT4 that participates in plasma membrane recycling.

Very recently, the finding that insulin releases GLUT4 from a static storage compartment into the recycling pathway has been confirmed (Muretta *et al.*, 2008). Furthermore, the differing observations that led to the two models above may have resulted from differing 3T3-L1 adipocyte culture conditions. To be sure, the static and dynamic models described above may differ by only a single rate: that of GLUT4 mobilization from GSVs in unstimulated cells. If this rate is so low as to be undetectable, then the GLUT4 is effectively sequestered away from the recycling pathway by a static retention mechanism. If this rate is low, but detectable, then the sequestration is less efficient, and dynamic equilibrium best describes the overall regulation of GLUT4 trafficking. The particular cell culture methodologies used may have altered this rate, perhaps by influencing the degree of adipose-specific gene expression.

Cellular differentiation is known to have a large impact on the basal, intracellular sequestration of GLUT4, and upon mobilization of these transporters to the cell surface by insulin. In 3T3-L1 adipocytes, GSVs are formed in a cell type-specific, developmentally regulated manner (Shi and Kandror, 2005). 3T3-L1 cells can be differentiated into cells with characteristics of adipocytes over the course of eight days (Green and Kehinde, 1976). Endogenous GLUT4 is expressed from about the fifth day of this differentiation period, and the fully mature cells demonstrate insulin-stimulated glucose uptake (Garcia de Herreros and Birnbaum, 1989). A mechanism for intracellular sequestration and insulin-stimulated mobilization of GLUT4 exists prior to the onset of GLUT4 expression (Yang *et al.*, 1992). Biochemical methods show that vesicles with properties of GSVs first appear on about the third day of 3T3-L1 differentiation (El-Jack *et al.*, 1999). This developmental stage coincides with enhanced basal, intracellular retention of exogenously expressed GLUT4, and more robust insulin-stimulated translocation (Bogan *et al.*, 2001). Sortilin, a sorting adaptor that binds GLUT4, is first expressed at this time and is necessary and sufficient for the formation of GSVs in 3T3-L1 cells, as discussed below (Shi and Kandror, 2005, 2007). A similar role for sortilin has been proposed in myocytes (Ariga *et al.*, 2008). Earlier kinetic analyses had suggested the presence of a small pool of insulin-mobilized GLUT4 in 3T3-L1 preadipocytes and Chinese Hamster Ovary (CHO) cells, which have limited adipocyte-like characteristics, but not in "undifferentiated" cells such as NIH 3T3 fibroblasts (Bogan *et al.*, 2001; Gros *et al.*, 1999; Lampson *et al.*, 2000). Yet, a distinct population of GSVs is not biochemically detectable in such poor adipocyte models. Thus although 3T3-L1 preadipocytes and CHO cells may contain some components of a mechanism for insulin-responsive GLUT4 trafficking, the mechanism is certainly not well developed in these cell types.

GSVs are present as preformed, 50–80 nM vesicles in unstimulated fat and muscle cells (Xu and Kandror, 2002). In both 3T3-L1 adipocytes and L6 myoblasts, these vesicles are present in a perinuclear location, as well as in more peripheral locations throughout the cytosol (Bryant *et al.*, 2002; Dugani *et al.*, 2008; Piper *et al.*, 1991). GSVs are depleted of endosomal proteins such as the transferrin receptor (TfnR), and contain GLUT4, Insulin-Regulated Aminopeptidase (IRAP), VAMP2, and other proteins in lower abundance (Foster *et al.*, 2006; Kupriyanova *et al.*, 2002; Ramm *et al.*, 2000, 2006; Shi and Kandror, 2005). Particular motifs in GLUT4 are required for insulin responsive trafficking, and presumably function to sort GLUT4 into GSVs, or to permit insulin to mobilize these vesicles (Capilla *et al.*, 2007; Govers *et al.*, 2004; Khan *et al.*, 2004; Palacios *et al.*, 2001; Piper *et al.*, 1993; Song *et al.*, 2008). Interestingly, vesicles that contain IRAP and have properties of GSVs are formed and traffic in an insulin responsive manner at an early stage of 3T3-L1 adipocyte differentiation, before GLUT4 is present (El-Jack *et al.*, 1999). The presence of IRAP seems to be largely sufficient to confer insulin-responsive trafficking in appropriate cell types, as discussed below. These vesicles likely contain GLUT1, a ubiquitously expressed glucose transporter that is not considered to be highly insulin-responsive. In fact, GLUT1 does traffic in a somewhat insulin-responsive manner (Liao *et al.*, 2006). Compared to GLUT4, GLUT1 is not efficiently targeted or retained in GSVs; however, it is likely not entirely excluded from these vesicles.

III. Insulin-Regulated Aminopeptidase (IRAP)

IRAP is the only protein known to cotraffic with GLUT4 throughout its itinerary (Keller, 2003). IRAP contains a single membrane spanning domain, a cytosolic amino terminus, and a large, carboxy-terminal, extracellular region that has aminopeptidase activity. First identified as vp165 or gp160, IRAP is present in insulin-responsive, GLUT4-containing vesicles, and is widely expressed (Kandror *et al.*, 1994; Keller *et al.*, 1995; Mastick *et al.*, 1994). A physiologic substrate of IRAP is vasopressin, which is inactivated by IRAP's aminopeptidase activity (Wallis *et al.*, 2007). IRAP also cleaves oxytocin and angiotensins III and IV *in vitro*, but whether these are physiologic substrates remains unknown. The interaction of angiotensin IV with IRAP present at the surface of neurons has been proposed to function in spatial memory formation (Chai *et al.*, 2004; Stragier *et al.*, 2008).

Why should insulin modulate the degradation of vasopressin, and possibly oxytocin and other peptide hormones? One answer is that the physiologic functions of insulin are still being elucidated, and the significance of these effects for organism physiology remains to be seen. Perhaps more important, the degradation of vasopressin may be accelerated by other stimuli that redistribute GSVs, and thus IRAP, to the cell surface. Muscle contraction and ischemia redistribute IRAP as well as GLUT4, and this may have importance for organism-level physiology. Other hormones, including vasopressin acting on kidney cells and oxytocin acting on endothelial cells, regulate IRAP cell surface targeting and aminopeptidase activity toward extracellular substrates (Masuda *et al.*, 2003; Nakamura *et al.*, 2000). As discussed further below, the action of insulin on GSVs is likely but one example of a fundamental, cellular mechanism that regulates a wide range of physiologic processes.

Characterization of IRAP trafficking has provided essential clues to the mechanism for GSV formation and insulin-stimulated translocation. Initial data suggesting that IRAP and GLUT4 cotraffic throughout their intracellular itinerary have been further supported by more detailed kinetic analyses, biochemical data, and electron microscopy (Carvalho *et al.*, 2004; Garza and Birnbaum, 2000; Kupriyanova *et al.*, 2002; Ramm *et al.*, 2000; Ross *et al.*, 1997; Slot *et al.*, 1991b). No direct interaction of IRAP and GLUT4 has been described. However, both IRAP and GLUT4 bind sortilin, a cell type-specific sorting adaptor that mediates the formation of GSVs (Shi and Kandror, 2005, 2007). To recruit GLUT4 and IRAP to GSVs, sortilin interacts with these proteins in the vesicle lumen. Peptide motifs in the cytosolic portions of GLUT4 and IRAP are also required to direct the trafficking of these proteins to GSVs, as well as for insulin-stimulated mobilization of these vesicles (Johnson *et al.*, 1998, 2001; Lee and Jung, 1997; Song *et al.*, 2008; Waters *et al.*, 1997). Finally, the cytosolic, amino terminal domain of IRAP interacts with several proteins that may be important for its targeting (Chi and Lodish, 2000; Hosaka *et al.*, 2005; Katagiri *et al.*, 2002; Peck *et al.*, 2006; Ramm *et al.*, 2006; Tojo *et al.*, 2003). Of particular note, microinjection of a peptide corresponding to a portion of the cytosolic amino terminus of IRAP caused translocation of GLUT4 to the cell surface in cells not stimulated with insulin (Waters *et al.*, 1997). This result suggests that IRAP binds an intracellular sorting or retention protein in the absence of insulin, and the peptide competed with endogenous IRAP for this binding site. Similar results were obtained with a GLUT4 carboxyl terminal peptide (Lee and Jung, 1997). As described below, a mechanism in which GLUT4 and IRAP bind to proteins that mediate basal intracellular retention has been proposed, and TUG may function as one such retention protein for GLUT4 (Bogan *et al.*, 2003; Yu *et al.*, 2007).

IV. Stability and Trafficking of GLUT4 and IRAP

Studies of IRAP targeting in cells lacking GLUT4, and of GLUT4 targeting in cells lacking IRAP, suggest that both proteins are required for fully insulin-responsive GSV trafficking. Of the two, IRAP appears to play the more essential role, as suggested initially by studies of partially differentiated cultured adipocytes, and more recently by RNAi-mediated depletion of IRAP or GLUT4 (El-Jack *et al.*, 1999; Gross *et al.*, 2004; Yeh *et al.*, 2007). Studies done in mice with genetic manipulation of the genes encoding IRAP and GLUT4 are more definitive. A key to understanding these data is the notion that the stability of GLUT4 protein is secondarily affected by its trafficking.

The main route of GLUT4 degradation is in lysosomes (Shi and Kandror, 2005). Although most GLUT4 that is endocytosed at the plasma membrane is efficiently targeted to GSVs, some GLUT4 accumulates in early endosomes, which may be carried to late endosomes and eventually to lysosomes. Manipulations that restrict the accumulation of GLUT4 in GSVs result in increased traffic through the endosomal system, and increased lysosomal degradation. Thus, expression of sortilin in undifferentiated 3T3-L1 preadipocytes sequesters GLUT4 in GSVs, drawing it out of the endosomal system and slowing its lysosomal degradation (Shi and Kandror, 2005). Similarly, Ubc9 likely regulates GLUT4 abundance indirectly, through effects on targeting, as described further below (Giorgino *et al.*, 2000; Liu *et al.*, 2007). Finally, TUG has been observed to regulate GLUT4 stability, likely as a consequence of effects on sequestration in GSVs (Yu *et al.*, 2007). All three of these proteins bind GLUT4 directly, and likely participate in the targeting or retention of GLUT4 in GSVs of unstimulated cells. By sequestering GLUT4 away from the plasma membrane, these proteins also sequester GLUT4 away from lysosomes and prolong its half-life.

Consistent with this interpretation, insulin stimulation accelerates GLUT4 degradation (Kim *et al.*, 1994; Sargeant and Paquet, 1993). Data from the mid-1990s show that, in unstimulated 3T3-L1 adipocytes, the half-life of GLUT4 protein is ~50 h. By contrast, in the continuous presence of insulin, the half-life of GLUT4 decreases to about 15.5 h. In retrospect, this threefold acceleration in the rate of GLUT4 degradation may result from mobilization of GLUT4 out of GSVs, with consequently enhanced flux of GLUT4 through the endocytic pathway to lysosomes. As described below, data suggest that disruption of TUG, by RNAi or a dominant negative protein fragment, also mobilizes GLUT4 that would otherwise be retained in GSVs of unstimulated cells (Bogan *et al.*, 2003; Yu *et al.*, 2007). TUG disruption causes GLUT4 targeting to the plasma membrane, and this effect is nearly equal to that of insulin. Interestingly, after transient, RNAi-mediated knockdown of TUG, GLUT4 protein abundance is decreased by about 60% in 3T3-L1 adipocytes (Yu *et al.*, 2007). This effect appears to result from accelerated lysosomal degradation of GLUT4, and may underestimate the steady-state effect on GLUT4 abundance. Yet, the main point is that TUG depletion and insulin exposure both accelerate GLUT4 degradation, and these effects are quantitatively similar in 3T3-L1 adipocytes.

The proportion of total cellular GLUT4 that is sequestered in GSVs is greater in primary adipocytes than it is in cultured, 3T3-L1 adipocytes. In 3T3-L1 adipocytes, about half of the intracellular GLUT4 is thought to be in endosomes, and the stability of this pool is likely not affected by disruption of GLUT4 traffic through GSVs. By contrast, most intracellular GLUT4 in primary adipocytes is present in GSVs, and GLUT4 that is not trapped in GSVs may be susceptible to degradation in lysosomes. Therefore, large effects on GLUT4 stability may result from impairment of GSV biogenesis, of GLUT4 targeting to GSVs, or of intracellular retention of GSVs within unstimulated primary adipocytes. In primary cells, the ability of GLUT4 trafficking defects to masquerade as reduced GLUT4 abundance may be quite marked.

In mice with genetic ablation of IRAP, GLUT4 protein abundance is markedly reduced (Keller *et al.*, 2002). The remaining GLUT4 is normally distributed, and moves to the cell surface upon insulin stimulation. However, GLUT4 abundance in muscle ranged 15–43% of that in control mice, and GLUT4 abundance in fat was reduced by about 50%. It is not known if this effect is specific to GLUT4 protein, or if there is a reduction in GLUT4 mRNA. From the discussion above, it is certainly possible that GLUT4 degradation is accelerated due to impaired targeting or trapping of GLUT4 in GSVs.

Conversely, in mice with adipose-specific GLUT4 disruption or overexpression, IRAP abundance is secondarily affected (Carvalho *et al.*, 2004). IRAP protein abundance is reduced by 35% in adipose specific (Cre-lox mediated) knockout of GLUT4, and is increased by 45% in mice with tenfold overexpression of GLUT4. IRAP mRNA abundance is not changed, consistent with the notion that these effects are due to alterations in IRAP protein stability. Similar results were observed in cardiomyocyte-specific GLUT4 knockout mice, in which IRAP abundance is reduced by 70% (Abel *et al.*, 2004). An earlier study of mice with global disruption of the GLUT4 gene did not find any consistent effect on IRAP abundance (Jiang *et al.*, 2001). However, in this study, basal redistribution of IRAP to the plasma membrane was observed. Taken together, these observations suggest that IRAP and GLUT4 rely on each other for efficient sequestration in GSVs, and for insulin-stimulated mobilization from this pool.

V. TUG, an Essential Component of a Retention Receptor for GLUT4

A. Cloning and initial characterization

TUG was identified in the first functional screen for proteins that alter the proportion of GLUT4 at the cell surface (Bogan *et al.*, 2003). This was essentially a genetic approach, which stood in contrast to previous biochemical approaches to identify regulators of GLUT4 trafficking. Briefly, earlier work had conceptualized and validated a GLUT4 reporter protein, which permitted measurement of fluorescence intensities corresponding to cell-surface and total GLUT4 (Bogan *et al.*, 2001). The ratio of fluorescence intensities is proportional to the proportion of GLUT4 at the cell surface, and was measured in individual, living cells using flow cytometry. To identify proteins that alter this ratio, a 3T3-L1 adipocyte cDNA library was expressed using a retrovirus vector (Liu *et al.*, 2000). Flow sorting was then used in an iterative strategy to enrich for particular clones that altered the ratio of surface-to-total GLUT4.

Kinetic analyses had suggested that after insulin addition, the GLUT4 that first appears at the cell surface originates from GSVs (Bogan *et al.*, 2001; Holman *et al.*, 1994; Lee *et al.*, 2000; Yeh *et al.*, 1995). It remains uncertain whether GLUT4 that recycles to the cell surface in the continuous, steady-state presence of insulin traffics through GSVs. If it does, these vesicles are likely mobilized immediately upon their formation, so that no large accumulation is evident. Alternatively, it may be that in the continuous presence of insulin, GLUT4 is never targeted to GSVs and traffics only through endosomes. In practical terms, to bias the screen to detect proteins involved in trafficking of GLUT4 into or out of GSVs, cells were stimulated only briefly with insulin.

The clone initially identified in this screen was a partial-length cDNA that reduced the initial movement of GLUT4 to the cell surface after insulin addition. This partial clone was hypothesized to produce a dominant negative protein fragment, termed TUG UBX-Cter, which was further proposed to inhibit the accumulation of GLUT4 in GSVs. The hypothesis that TUG UBX-Cter blocks the intracellular retention of GLUT4 in GSVs of unstimulated cells has since received substantial support, as reviewed below (Bogan *et al.*, 2003; Yu *et al.*, 2007). Yet, it is unclear whether the initial cloning of the partial-length TUG cDNA resulted from this dominant negative effect to deplete the pool of GSVs. An alternative possibility is that the protein that was produced was not the UBX-Cter fragment, but a much larger portion of full-length TUG, which might have been produced as a fusion with retroviral gag. If so, this protein would likely have been sufficient to bind and retain GLUT4 intracellularly. Therefore the partial-length TUG clone may have emerged from the screen by a positive effect to retain GLUT4 intracellularly. Regardless, the UBX-Cter fragment has proven to be a useful and effective dominant-negative protein.

According to the model proposed, TUG traps GLUT4 that is incorporated into GSVs, and retains this pool of GLUT4 intracellularly in cells not stimulated with insulin. In functional terms, TUG "tethers" GLUT4 to an intracellular structure to prevent GSV movement to the cell surface. Insulin then releases this tether to mobilize the GSVs. Distinct and separable regions of TUG are required to bind GLUT4 and to retain the transporters intracellularly; the latter region presumably binds an as-yet-unidentified intracellular anchoring site. This model also suggests how the UBX-Cter fragment can act as a dominant negative fragment: it may bind and occupy an essential anchoring site, and exclude the interaction of endogenous, full-length TUG at this location. TUG contains an amino terminal domain with similarity to ubiquitin, as well as a ubiquitin-like UBX domain near the carboxyl terminus (Bogan *et al.*, 2003; Tettamanzi *et al.*, 2006). Hence, TUG was named according to its proposed function, as a "*T*ether, containing a *U*BX domain, for *G*LUT4."

The human ortholog of the mouse TUG protein was cloned as part of a fusion, produced by a chromosome translocation, in alveolar soft-part sarcomas and renal cell carcinomas (Heimann *et al.*, 2001; Ladanyi *et al.*, 2001). This protein, termed ASPL or RCC17, is 76% identical to murine TUG. In malignancies, the amino terminus of ASPL is fused to the carboxyl terminus of TFE3, a transcription factor (Argani and Ladanyi, 2005). This causes marked overexpression of TFE3, which in turn upregulates MET, a tyrosine kinase that functions as a receptor for hepatocyte growth factor (Tsuda *et al.*, 2007). Activation of MET signaling contributes to the growth of alveolar soft part sarcomas.

B. Evidence that TUG retains GLUT4 in GSVs

Several lines of evidence support the model described above, in which the long splice variant of TUG functions as an essential component of a retention receptor for GLUT4. First, insulin stimulates the dissociation of a TUG–GLUT4 protein complex in 3T3-L1 adipocytes, and this dissociation precedes movement of GLUT4 out of GSV-containing light microsomes and to the plasma membrane (Bogan *et al.*, 2001, 2003). More important, the number of TUG–GLUT4 complexes that dissociate correlates with the number of GLUT4 transporters that are rapidly mobilized to the cell surface. Both amounts are increased by overexpression of intact TUG, and decreased by expression of the UBX-Cter fragment, in 3T3-L1 adipocytes.

Second, TUG colocalizes with GSVs in unstimulated cells. TUG is present both in the cytosol and in a membrane-associated pool. The membrane-bound TUG may be equivalent to the GLUT4-bound TUG, and cofractionates with GLUT4 in intracellular vesicles with characteristics of GSVs. TUG remains in this membrane fraction after insulin stimulates the movement of GLUT4 to the plasma membrane, consistent with dissociation of a TUG–GLUT4 complex. By confocal microscopy as well, TUG colocalizes with intracellular GLUT4-containing vesicles, and not with TfnR-containing endosomes.

Third, TUG binds with much greater affinity to GLUT4 than to GLUT1, and low affinity binding to GLUT1 can be detected only when both proteins are markedly overexpressed or when recombinant proteins are studied. Grossly, the affinity of TUG for GLUT4 versus GLUT1 parallels the degree to which these transporters are mobilized by insulin (Liao *et al.*, 2006). Manipulation of TUG function has no effect on TfnR trafficking at the cell surface (Bogan *et al.*, 2003). Intriguingly, TUG and GLUT4 interact through two distinct regions in each protein. An amino terminal region of TUG binds directly to the large intracellular loop of GLUT4. A more central region of TUG binds to another region of GLUT4, which may be the cytosolic amino terminus. The use of GLUT4–GLUT1 chimeras has identified these regions as necessary and sufficient for insulin-responsive trafficking of newly synthesized GLUT4 (Khan *et al.*, 2004).

Fourth, RNAi-mediated depletion of TUG results in substantial movement of GLUT4 to plasma membranes of unstimulated 3T3-L1 adipocytes, and correspondingly increased glucose uptake (Yu *et al.*, 2007). In cells with stable disruption of TUG function, achieved by RNAi or expression of the UBX-Cter fragment, insulin stimulated a twofold further increase in glucose uptake. This additional stimulation may result from insulin action to accelerate fusion of GLUT4-containing vesicles at the plasma membrane, or to reduce the rate of GLUT4 endocytosis, or to regulate intrinsic glucose transport activity (Blot and McGraw, 2006; Kandror, 2003; Koumanov *et al.*, 2005). Yet, the data show that the bulk of insulin's ability to control GLUT4 distribution and glucose uptake in 3T3-L1 adipocytes is mediated by a TUG-regulated step.

Fifth, TUG disruption not only mimics the effect of insulin on GLUT4 targeting, but also mimics the effect of insulin to accelerate GLUT4 degradation (Yu *et al.*, 2007). As discussed above, this is likely a secondary consequence of effects on targeting, which lead to an increased rate of GLUT4 movement to lysosomes. This effect is what one would predict for a retention receptor that has a major regulatory role in GLUT4 targeting.

Sixth, intracellular GLUT4 is drawn out of TfnR-containing endosomes when TUG is overexpressed, and overlaps almost entirely with these endosomes when TUG function is disrupted by the dominant negative TUG UBX-Cter fragment (Yu *et al.*, 2007). Although difficult to quantify, these confocal microscopy findings are exactly those predicted by the model outlined above.

Very recently, it was shown that in muscle, similar to adipose, insulin stimulates the dissociation of GLUT4 and TUG (Schertzer *et al.*, 2009). This observation suggests that insulin acts on a TUG-GLUT4 protein complex by similar mechanisms in both of these tissues.

It remains possible that TUG functions primarily as a sorting adaptor that targets GLUT4 to GSVs. Alternatively, TUG may prevent GLUT4 from being targeted to the cell surface, perhaps by controlling its association with molecular motors implicated in insulin-stimulated movement (Bose *et al.*, 2002, 2004; Imamura *et al.*, 2003; Semiz *et al.*, 2003). These possibilities are not mutually exclusive with the proposed role of TUG to bind and retain GLUT4 intracellularly in GSVs until insulin stimulates the mobilization of these vesicles. Indeed, it would not be surprising if these processes were mechanistically coupled, and TUG could well play a role to coordinate them.

The particular domains within TUG suggest some additional possibilities for how this protein may function. TUG contains a ubiquitin-like domain at its amino terminus, and a ubiquitin-like UBX domain near its carboxyl terminus (Bogan *et al.*, 2003; Tettamanzi *et al.*, 2006). Much of the central region of the protein has low sequence complexity, and may be unstructured in the absence of binding partners. TUG is an unstable protein that appears to turn over rapidly. In general, UBX domains mediate the recruitment of p97/VCP, an ATPase that has been implicated in endoplasmic reticulum-associated degradation, membrane fusion, and in the disassembly of protein complexes and proteasomal degradation of component proteins (Dreveny *et al.*, 2004; Schuberth and Buchberger, 2008; Wang *et al.*, 2004). Not surprisingly, TUG binds p97 (Cresswell, Li, and Bogan, unpublished data). However, the functional significance of this interaction remains uncertain, and it is not known if p97 activity is required to disassemble a TUG–GLUT4 protein complex. Recent data suggest that insulin stimulates the destruction of a small fraction of total cellular TUG, which may correspond to that bound to GLUT4 (Cresswell and Bogan, unpublished data). This mechanism of action would fit well with the idea that sequestered GSVs are released into the actively recycling pool in a graded or quantal manner, which scales with the concentration of insulin that is used to stimulate the cells (Govers *et al.*, 2004).

If TUG is a key regulator of GLUT4 trafficking and glucose uptake, then why was it not identified earlier, given the intensity of research in this field? The answer may lie in the approaches that have been used to study GLUT4 trafficking. In general, to identify regulating proteins, one can take a genetic or a biochemical approach. Genetic approaches to understand how insulin regulates GLUT4 trafficking have not previously been possible, since this process occurs exclusively in differentiated, postmitotic, vertebrate tissues (i.e., adipose and muscle). Thus, genetic screens done in yeast or other simple organisms cannot be used directly to study this mechanism.

Similarly, biochemical approaches have been frustrated by the fact that GLUT4 has a complicated topology, including 12 transmembrane domains. This has limited the utility of studies based on protein–protein interactions (e.g., two-hybrid screens). In adipocytes and muscle, although GLUT4 is present and adopts its proper conformation, the identification of interacting proteins is limited by the possibility that binding proteins may be degraded during purification.

With the above in mind, two characteristics of TUG suggest why it would not likely have been identified using biochemical methods. First, TUG binds GLUT4 coordinately through two interacting regions in each protein, and these involve noncontiguous intracellular sequences in GLUT4. Second, TUG is a short-lived protein that is degraded rapidly in lysates. The use of thiol capping reagents, which partially block p97/VCP activity, seem to partially prevent this degradation. Thus, the fact that TUG was isolated in the first functional screen for regulators of GLUT4 trafficking, which was essentially a genetic overexpression screen, is important. This approach provided, for the first time, the possibility of identifying a protein with these properties.

C. A role for Ubc9 and possibly Daxx

Ubc9, the only known conjugating enzyme for the ubiquitin-like modifier, SUMO, binds GLUT4 and controls its abundance in L6 skeletal muscle cells (Giorgino *et al.*, 2000). SUMO is a ubiquitin-like modifier that, like ubiquitin, is covalently attached to target proteins by the action of an enzymatic cascade (Geiss-Friedlander and Melchior, 2007; Kerscher *et al.*, 2006). This cascade typically includes activating (E1), conjugating (E2), and ligating (E3) enzymes; however, SUMO modification (sumoylation) can occur in the absence of an E3 (Bernier-Villamor *et al.*, 2002). Sumoylation is known to modulate protein targeting and stability, and may antagonize the effect of ubiquitin to promote protein degradation. GLUT4 can be detected in sumoylated form (Giorgino *et al.*, 2000; Lalioti *et al.*, 2002). These observations were initially felt to be most consistent with an effect of SUMO-modification to stabilize GLUT4 directly.

More recent data suggest that Ubc9 more likely functions primarily to control GLUT4 targeting. It was noted in the initial publication that the effect of Ubc9 to stabilize GLUT4 might be secondary to an effect on protein trafficking (Giorgino *et al.*, 2000). This fits with the notion that GLUT4 targeting and stability are intimately related, and that the main route of GLUT4 degradation is lysosomal, as discussed above. Furthermore, the fact that a very small fraction of GLUT4 can be detected in sumoylated form does not necessarily indicate that GLUT4 is a physiologic target of sumoylation. SUMO E1, E2, and E3 enzymes are all autosumoylated in yeast (Zhou *et al.*, 2004). It seems possible that

GLUT4 may function as a SUMO E3, which may bring together the E2 (Ubc9) and the target substrate of SUMO modification. If so, then candidate sumoylation targets would include Daxx, TUG, and possibly other proteins.

Recently, it was reported that overexpression of Ubc9 in 3T3-L1 adipocytes promoted the accumulation of GLUT4 in GSVs, whereas RNAi-mediated depletion of Ubc9 caused a selective loss of GLUT4 in GSVs (Liu *et al.*, 2007). These data confirm and extend the earlier studies, which were done in L6 myoblasts, and are consistent with the hypothesis that effects on GLUT4 stability result from effects on GLUT4 sequestration in GSVs. In addition to reducing GLUT4 abundance, Ubc9 knockdown selectively reduces the abundance of IRAP, sortilin, mannose 6-phosphate receptor, and syntaxin-6, suggesting that GSVs may derive from the TGN (see below). In glucose uptake experiments, Ubc9 depletion reduces insulin-stimulated glucose uptake, but had no effect on basal glucose transport. These data are compatible with the finding that heterozygous Ubc9 knockout mice have a mild but significant increase in blood glucose concentrations (Nacerddine *et al.*, 2005). In 3T3-L1 adipocytes, taking into account the ~50% reduction in overall GLUT4 abundance, it is likely that the proportion of GLUT4 in the plasma membranes of unstimulated cells is somewhat increased. This phenotype is similar to that observed when TUG is depleted (Yu *et al.*, 2007). Given that Ubc9 and TUG bind distinct regions of GLUT4, it seems possible that these proteins participate in a common retention receptor complex.

Intriguingly, overexpression of a catalytically inactive form of Ubc9 also targets GLUT4 to GSVs and enhances GLUT4 abundance, similar to overexpression of the wildtype protein and opposite to the effect of Ubc9 knockdown (Liu *et al.*, 2007). Thus, Ubc9-mediated sumoylation is not required for targeting or retention of GLUT4 in GSVs, or in the acute effect of insulin to mobilize these vesicles. This finding is in contrast to those of Giorgino *et al.*, in which stable expression of dominant negative Ubc9 had effects that were opposite to those of the wildtype protein (Giorgino *et al.*, 2000). In the report from Liu *et al.*, Ubc9 was expressed transiently, using an adenovirus to introduce the protein into mature, post-mitotic 3T3-L1 adipocytes. Therefore, one possibility to resolve this apparent discrepancy may be that SUMO conjugation activity is required subsequent to the initial mobilization of GSVs by insulin, perhaps for the return of translocated GLUT4 back to the insulin-responsive, GSV pool.

Daxx binds the GLUT4 carboxyl terminus adjacent to the Ubc9 binding site, and is a known SUMO-binding protein and target of sumoylation (Lalioti *et al.*, 2002; Lin *et al.*, 2006; Ryu *et al.*, 2000). The effect of manipulating Daxx on GLUT4 targeting is not known. However, in contrast to Ubc9, which binds GLUT1 as well as GLUT4, Daxx binds specifically to GLUT4, suggesting some functional role in the trafficking of this isoform (Lalioti *et al.*, 2002). If Daxx is involved in insulin action, then it may be recruited transiently or stably to an insulin-responsive retention receptor complex, or it may facilitate entry of GLUT4 into a GSV pool by assisting in the formation of such a complex.

D. Potential anchoring sites for TUG and GLUT4

Where does TUG retain GLUT4 within unstimulated cells? To what protein(s) does TUG link GLUT4, and what proteins comprise a retention receptor complex? As discussed above, IRAP is likely essential for regulated GSV retention and release. Our preliminary data suggest that IRAP binds and is regulated by TUG, consistent with the idea that it participates together with GLUT4 in a retention receptor complex (Rubin and Bogan, unpublished data). How such a complex may be anchored intracellularly within unstimulated cells is not known. Several recent reports suggest the involvement of Golgi proteins in GLUT4 trafficking, and it may be that TUG links GLUT4, ultimately, to proteins present at the Golgi complex. In particular, Golgi matrix proteins and TGN SNARE proteins may be important to anchor GSVs intracellularly in unstimulated cells.

Golgin-160 has been implicated in the regulated exocytosis of various membrane proteins, including GLUT4 and IRAP (Bundis *et al.*, 2006; Hicks and Machamer, 2005; Hicks *et al.*, 2006; Williams *et al.*, 2006). In some examples, this action may be mediated through its interaction with PIST, a *P*DZ-domain containing protein that *I*nteracts *S*pecifically with *T*C10 (Hicks and Machamer, 2005; Neudauer *et al.*, 2001). RNAi-mediated depletion of Golgin-160 results in insulin-independent translocation of GLUT4 and IRAP to the plasma membrane (Williams *et al.*, 2006). This effect appears to be specific, because there is no gross alteration in Golgi morphology and disruption of Golgin-160 function does not affect proteins that traffic constitutively. The effects of RNAi-mediated depletion of Golgin-160 and of TUG are similar; however, no direct interaction of these proteins has been observed.

Also called FIG, GOPC, and CAL, PIST is widely expressed, binds syntaxin-6 as well as Golgin-160, and is a downstream effector of signaling through the Rho-family GTPase TC10 (Charest *et al.*, 2001; Cheng *et al.*, 2002; Neudauer *et al.*, 2001; Yao *et al.*, 2001). PIST binds and controls trafficking of several proteins that move to the cell surface, including frizzled proteins, ion channels such as CFTR and ClC-3B, and receptors such as AMPA-type glutamate, somatostatin, and β1-adrenergic receptors (Cheng *et al.*, 2002; Gentzsch *et al.*, 2003; He *et al.*, 2004; Wente *et al.*, 2005; Yao *et al.*, 2001; Yue *et al.*, 2002). PIST sequesters CFTR intracellularly, and GTP-bound TC10 enhances CFTR targeting to the cell surface (Cheng *et al.*, 2004, 2005). Interestingly, it is proposed that PIST traps AMPA receptors exiting the TGN, and these can then be released to the cell surface in a regulated manner (Cuadra *et al.*, 2004). A similar model may apply to

GLUT4, which may be trapped upon its exit from the TGN by a complex comprising TUG and other proteins.

Insulin is known to activate TC10, and this pathway is required to translocate GLUT4 to the cell surface (Chiang *et al.*, 2001, 2002; Maffucci *et al.*, 2003). Although conflicting data have been reported, the possibility that TC10 activation stimulates the mobilization of GSVs fits with other data, which show that GLUT4 is mobilized from the perinuclear region by a phosphatidylinositol 3-kinase-independent mechanism (Mitra *et al.*, 2004; Semiz *et al.*, 2003). Thus, PIST is an excellent candidate to transmit the insulin signal from TC10 to Golgin-160 or syntaxin-6, and ultimately to mobilize GSVs by causing dissociation of a GLUT4-TUG protein complex. The hypothesis that these proteins function together to control GSV mobilization remains to be tested.

p115 is another Golgi matrix protein implicated in controlling GLUT4 trafficking, which was identified by affinity purification of IRAP-binding proteins (Hosaka *et al.*, 2005). Overexpression of the amino terminus of p115 inhibits the insulin-stimulated movement of GLUT4 and IRAP to the cell surface, and has minimal effect on the distribution of these proteins to the surface of unstimulated cells. The amino terminal p115 fragment binds IRAP, and presumably displaces the endogenous, full-length p115 from this site. If so, the defect in insulin action may result from dispersal of GSVs in the basal state, so that they are no longer able to be mobilized by insulin. Because the vesicles do not default to the cell surface, it may be that there is another anchoring protein(s) present. Whether p115 interacts and cooperates with other proteins discussed above remains unknown.

In neurons, synapsins are phosphoproteins that coat regulated exocytic vesicles in a "reserve pool," which can be mobilized for exocytosis at the plasma membrane (Sudhof, 2004). Synapsin II is present in adipocytes, colocalizes with perinuclear GLUT4, and has been suggested to participate in intracellular retention of GLUT4 within unstimulated cells (Muretta *et al.*, 2007). Although this effect is specific to GLUT4, and not constitutively recycling proteins, it remains uncertain how synapsin may interact with other proteins implicated in GLUT4 regulation. In particular, it is not known if synapsin II phosphorylation is regulated by insulin, how it is linked to GLUT4-containing membranes, or whether it ties these vesicles to the actin cytoskeleton.

Another protein that is required for full intracellular retention of GLUT4 within unstimulated cells is AS160/TBC1D4 (reviewed in Sakamoto and Holman, 2008). Originally identified as a substrate for insulin-stimulated Akt kinase activity, AS160 contains a domain that activates the hydrolysis of GTP by Rab proteins. Because Rab proteins are well known to direct vesicular traffic, there is much excitement about the idea that AS160 links insulin signaling through Akt to the trafficking of GLUT4-containing vesicles. A related protein, TBC1D1, serves this function in muscle, and may additionally be regulated through AMP-activated protein kinase to facilitate GLUT4 translocation in response to muscle contraction (Sakamoto and Holman, 2008). AS160 has been shown to bind the cytosolic amino terminus of IRAP, and to be required for full intracellular retention of GLUT4 in unstimulated cells (Eguez *et al.*, 2005; Larance *et al.*, 2005; Thong *et al.*, 2007). It is not known if AS160 interacts with IRAP in GSVs or in other locations, nor whether it binds TUG or other proteins in a retention receptor complex. It remains possible that AS160 controls the entry of GLUT4 into GSVs in the basal state, and its GTPase activating activity is required for GLUT4 retention (Eguez *et al.*, 2005). Still another possibility is that AS160 acts at a post-GSV step, perhaps to control docking of GLUT4-containing vesicles at the plasma membrane (Capilla *et al.*, 2007; Jiang *et al.*, 2008; van Dam *et al.*, 2005).

VI. From What Membranes do GSVs Originate?

It remains uncertain whether GSVs form at the TGN, from recycling endosomes, or from both of these compartments. Present data support the hypothesis that newly synthesized GLUT4 can enter GSVs without first traversing the plasma membrane (Liu *et al.*, 2005; Watson *et al.*, 2004). Most GLUT4 that enters GSVs derives from the endocytic pathway, and it remains uncertain if the route taken is from endosomes directly to GSVs, or if this endocytosed GLUT4 passes through the TGN. A few observations suggest the latter possibility. Syntaxin-16, a SNARE protein that is reasonably specific to the TGN, is required for efficient intracellular retention of GLUT4 (Proctor *et al.*, 2006). Intriguingly, syntaxin-16 is a phosphoprotein, which is dephosphorylated by insulin stimulation, suggesting that this step may be a point of GLUT4 regulation (Perera *et al.*, 2003).

Other SNARE proteins that are known to form a complex with syntaxin-16 have also been implicated in GLUT4 trafficking, and may control the entry of GLUT4 into GSVs. In particular, syntaxin-6, Vti1a, and VAMP4 form a complex with syntaxin-16 to mediate endosome to TGN fusion (Kreykenbohm *et al.*, 2002; Mallard *et al.*, 2002; Wang and Tang, 2006; Wendler and Tooze, 2001). These proteins have all been implicated functionally in GLUT4 targeting, and have additionally been observed in purified GSVs using mass spectrometry (Bose *et al.*, 2005; Foster *et al.*, 2006; Perera *et al.*, 2003; Ramm *et al.*, 2006; Shewan *et al.*, 2003; Watson and Pessin, 2008; Williams and Pessin, 2008). As noted above, syntaxin-6 binds PIST and thus may be a target of insulin signaling through TC10 (Charest *et al.*, 2001; Neudauer *et al.*, 2001). Additionally, IRAP that has been enzymatically deglycosylated at the cell surface is endocytosed and resialylated, presumably in the TGN or in *trans* Golgi cisternae (Shewan *et al.*, 2003).

Finally, the distribution of intracellular GLUT4 has been found to overlap with that of TGN markers (Bogan and Lodish, 1999; Martin *et al.*, 2000; Ramm *et al.*, 2000; Slot *et al.*, 1997). Thus, it has been proposed that GLUT4 cycles from endosomes to the TGN. It seems most likely that it traffics directly to GSVs from the TGN; however, it remains possible that GLUT4 cycles back to endosomes before becoming trapped in GSVs.

It has been proposed that, in the absence of insulin, GLUT4 participates in an intracellular cycle among GSVs and the TGN (Bryant *et al.*, 2002). If so, the rate at which GSVs fuse with the TGN may be relatively slow, because GSVs accumulate in unstimulated cells. Together with other proteins, TUG may trap GSVs in unstimulated cells and hold them in a relatively static configuration, as discussed above. Alternatively, present data are entirely compatible with a model in which TUG functions to constrain an intracellular cycle of GSV fusion and budding at the TGN. Insulin might then redirect GSVs, so that they fuse with the plasma membrane rather than the TGN.

Several proteins have been suggested to function in the formation of GSVs from TGN or endosomal "donor membranes." Shi and Kandror used an *in vitro* reconstitution assay to show that membranes contain a developmentally regulated protein that supports the production of GSVs (Shi and Kandror, 2005). This protein was identified as sortilin, an integral membrane protein that serves as a cargo adaptor. Sortilin expression is induced upon differentiation of 3T3-L1 adipocytes, and interacts with both GLUT4 and IRAP through their luminal domains (Shi and Kandror, 2005, 2007). Therefore, this protein recruits both GLUT4 and IRAP to insulin responsive vesicles. Exogenous expression of sortilin in 3T3-L1 preadipocytes is sufficient to cause the formation of insulin-responsive GSVs, and RNAi-mediated knockdown of this protein in mature 3T3-L1 adipocytes inhibits GSV formation and insulin-stimulated glucose uptake. Sortilin has been implicated in GLUT4 trafficking in muscle as well as fat, and additionally promotes myogenesis (Ariga *et al.*, 2008). Tissue-specific expression of sortilin can thus account for the observed, tissue-specific formation of insulin-responsive GSVs, as well as for the observed cotrafficking of GLUT4 and IRAP in these vesicles.

Sortilin is a member of a family of cargo adaptors similar to the yeast carboxypeptidase Y receptor, Vps10p, and is abundant in GSVs and in TGN membranes (Hashiramoto and James, 2000; Lin *et al.*, 1997; Morris *et al.*, 1998). Sortilin and related proteins recruit various cargos and bind Golgi-localized, gamma-ear-containing, ARF binding proteins (GGAs) and AP-1 adaptors, which in turn recruit ARF GTPases and clathrin to initiate the budding of coated vesicles (Nielsen *et al.*, 2001; Takatsu *et al.*, 2001). Sortilin is retrieved to the TGN by the action of retromer, a five-component protein complex that mediates endosome-to-Golgi targeting (Canuel *et al.*, 2008; Seaman, 2004, 2007). The retromer functions together with EHD1/2, which have been implicated in GLUT4 trafficking (Gokool *et al.*, 2007; Guilherme *et al.*, 2004). Finally, GGAs have independently been shown to function in the recruitment of GLUT4 and IRAP to GSVs (Hou *et al.*, 2006; Li and Kandror, 2005; Watson *et al.*, 2004). Thus it is likely that GLUT4-containing GSVs derive from the TGN, and this occurs in a cell type-specific manner requiring sortilin.

It is also possible that GLUT4 and IRAP internalized from the plasma membrane enter GSVs directly from endosomes, without passing through the TGN. In particular, ACAP1 is required to recruit GLUT4 from recycling endosomes to the GSV pool (Li and Kandror, 2005). ACAP1 is a GTPase-activating protein for ARF6, which binds directly to GLUT4 and participates in the formation of clathrin coated vesicles from endosomes. In contrast to TUG, the association of ACAP1 and GLUT4 is not affected by insulin, however, ACAP1 may be a target of signaling through Akt (Dai *et al.*, 2004; Li *et al.*, 2005). RNAi-mediated depletion of ARF6 also markedly inhibited GLUT4 redistribution and glucose uptake after insulin. These observations suggest that GSVs may derive from endosomes as well as from the TGN. Consistent with this assertion, other data show that IRAP internalized from the plasma membrane does not require a GGA-dependent step to enter GSVs (Watson and Pessin, 2008). Thus, ACAP1 and sortilin may both function as cargo adaptors to incorporate GLUT4 and IRAP into GSVs derived from endosomes and the TGN.

Tankyrase is another protein that has been suggested to play a role in the entry of GLUT4 into GSVs. Tankyrase is a poly(ADP-ribose) polymerase (PARP) that binds the cytosolic tail of IRAP through its ankyrin repeat domain (Chi and Lodish, 2000; Sbodio and Chi, 2002). First characterized as a regulator of telomeres, it is abundant in the area of the Golgi complex. Tankyrase is present in two isoforms, which can oligomerize (Sbodio *et al.*, 2002). Through its PARP activity, tankyrase can modify itself, IRAP, and potentially other proteins with polymers of ADP-ribose, a reversible, post-translational modification. Both RNAi-mediated depletion of tankyrase and pharmacologic inhibition of PARP activity alter the distribution GLUT4 and IRAP in unstimulated 3T3-L1 adipocytes, and impair the ability of insulin to stimulate GLUT4 translocation (Yeh *et al.*, 2007). Data are most consistent with a model in which tankyrase is required for the entry of GLUT4 and IRAP into GSVs. Whether tankyrase's PARP activity is required to form or to stabilize a retention-receptor complex, or whether this protein acts at an earlier step in GSV formation, remains unknown.

VII. A General Mechanism for the Regulated Targeting of Membrane Proteins

The molecular mechanisms controlling GLUT4 localization have been the focus of research because of their relevance to type 2 diabetes. In muscle of insulin-resistant individuals, GLUT4 is not effectively mobilized to the cell surface by insulin. Defects in insulin signaling, particularly at the level of Insulin Receptor Substrate (IRS) proteins, are clearly important for this effect, and have been invoked as a main cause of insulin resistance in skeletal muscle (Biddinger and Kahn, 2006; Petersen *et al.*, 2007). Whether altered GLUT4 targeting may also play a role in insulin resistance is not known. Pathophysiologic defects in insulin signaling were uncovered only because physiologic insulin signaling mechanisms were understood well enough to permit examination of these pathways in human subjects and in rodent models (Cohen, 2006; Taniguchi *et al.*, 2006). By contrast, the physiologic mechanisms controlling GLUT4 distribution are not well understood.

Some data suggest that defects in GLUT4 targeting, particularly in unstimulated cells, may contribute to insulin resistance. In humans with diabetes, the distribution of GLUT4 among intracellular compartments is altered compared to control subjects, both in fat and in skeletal muscle, in fasting individuals (Garvey *et al.*, 1998; Maianu *et al.*, 2001). Other recent work used platelet derived growth factor to mimic insulin action without requiring IRS proteins, and found that in various models of insulin resistance, defects in GLUT4 translocation can still be observed (Hoehn *et al.*, 2008). These findings raise the possibility that insulin resistance may result, at least in part, from defects in GLUT4 trafficking in unstimulated cells. If the GLUT4 is not positioned in an insulin-responsive configuration, from which it can be mobilized to the cell surface, then insulin action will be defective. In effect, targets of insulin action, GSVs, are missing. Thus, understanding how GSVs are formed and retained intracellularly will likely have importance for understanding the pathogenesis of insulin resistance and type 2 diabetes.

More broadly, the molecular mechanisms that control GLUT4 trafficking are likely shared among several other transporters, receptors, and ion channels. As discussed above, IRAP is important for this mechanism and is widely expressed, and is likely mobilized to the cell surface by various stimuli in a cell type-dependent manner. Other proteins that undergo regulated translocation include AQP2 water channels in the renal collecting duct, H^-/K^+-pumps in gastric parietal cells, AMPA-type glutamate receptors in neurons, and H^+-pumps in the kidney (Chieregatti and Meldolesi, 2005). Still other proteins that participate in regulated translocation mechanisms, which may or may not be similar to those controlling GLUT4, are EAAC1 (a glutamate transporter), CD36 (which mediates fatty acid uptake), HMIT (a H^+/myo-inositol symporter), ABC transporters (for bile and organic ions), UT-A1 (a urea transporter), and other proteins (Cheng *et al.*, 2005; Gonzalez *et al.*, 2007; Klein *et al.*, 2006; Uldry *et al.*, 2004; van Oort *et al.*, 2008; Wakabayashi *et al.*, 2006).

Of particular relevance are AQP2 water channels, which are mobilized by vasopressin to mediate water resorption in the renal collecting duct. Both AQP2 and H^-/K^+-pumps are targeted to the cell surface in response to stimuli that signal through G-protein coupled receptors and intracellular cAMP. This signaling pathway is quite distinct from that of insulin, which binds a receptor with tyrosine kinase activity. Yet, the downstream proteins that regulate targeting of AQP2 and GLUT4 may be quite similar. IRAP and sortilin are present in renal cells, and IRAP is mobilized to the cell surface by vasopressin (Barile *et al.*, 2005; Masuda *et al.*, 2003). In 3T3-L1 adipocytes, exogenously expressed AQP2 is translocated to the cell surface by agents that increase cAMP (Procino *et al.*, 2005). Although AQP2 did not colocalize with GLUT4 in these studies, the stoichiometry is such that it may yet colocalize with IRAP, and it is targeted to a compartment that is clearly distinct from endosomes.

Intriguingly, cAMP stimulates protein kinase A-mediated phosphorylation of Ser256, a residue in the cytosolic carboxyl terminal tail of AQP2, and phosphorylation of this residue is necessary and sufficient to cause AQP2 translocation (van Balkom *et al.*, 2002). Sequences surrounding this site are similar to the Daxx binding peptide in the GLUT4 carboxyl terminus (Lalioti *et al.*, 2002). The phosphorylated residue in AQP2 corresponds to a glutamic acid in GLUT4, which may mimic a phosphoserine. Thus, it is tempting to speculate that phosphorylation of AQP2 may recruit Daxx (or another protein that binds AQP2 adjacent to Daxx), and that this may trigger the mobilization of AQP2-containing vesicles to the cell surface. GLUT4 may be mobilized by some other input, which may yet function through a similar protein complex in a Daxx-dependent manner to mobilize GSVs to the cell surface. These hypotheses are testable, and may shed light on the generality of the translocation mechanisms that are used.

AQP2 can be detected in urine using mass spectrometry, and exosomes containing AQP2 are secreted into the renal collecting duct (Pisitkun *et al.*, 2004; Takata, 2006). Exosomes are membrane-bound vesicles, derived from the internal membranes of prelysosomal multivesicular bodies, which are extruded into the extracellular space (Lakkaraju and Rodriguez-Boulan, 2008). If GLUT4 and AQP2 participate in similar trafficking mechanisms, then it is possible that a similar phenomenon occurs with GLUT4. Indeed, data suggest that GLUT4 is targeted to lysosomes, particularly when sequestration in GSVs is impaired (Shi and Kandror, 2005; Yu *et al.*, 2007). It is reasonable to speculate that impaired sequestration in GSVs, as may occur in insulin resistance, might result in the extrusion of GLUT4-containing membranes into the extracellular space. If so, this might initiate the monocytic infiltrate and low grade inflammation that are present in adipose tissue in the setting of insulin resistance (Zeyda and Stulnig, 2007). According to this hypothesis, it is not the inflammation that causes insulin resistance (at least, initially), but the insulin resistance that causes the inflammation. Testing this hypothesis may reveal new pathophysiologic mechanisms important in the development of insulin resistance and diabetes.

VIII. Conclusions

A hallmark action of insulin is to stimulate glucose uptake. Since the elucidation of the translocation mechanism in 1980, and the molecular identification of GLUT4 in 1989, much progress has been made to understand the insulin signaling pathways responsible for stimulating glucose uptake. The trafficking mechanisms that regulate GLUT4 remain much less well understood. Here, we have attempted to focus on particular aspects of GLUT4 targeting, and to present a framework with which to understand GSV formation, retention, and release. Insulin acts at other sites in GLUT4's intracellular itinerary, and so the above discussion is by no means a comprehensive review of GLUT4 trafficking. Yet, GSVs are a major site of insulin regulation, and the molecular mechanisms controlling these vesicles are beginning to be uncovered. Therefore, there is reason for optimism that this field will progress more rapidly in the coming years.

Finally, it is becoming clear that mechanisms controlling GLUT4 sequestration and release are likely used by multiple other, physiologically important membrane proteins that are targeted to the cell surface in a regulated manner. Therefore, work on GLUT4 will not only elucidate aspects of how insulin regulates glucose homeostasis, but will likely also lead to new insights in many other areas of physiology. Conversely, insights drawn from AQP2, and from other instances in which translocation is regulated by extracellular cues, may shed light on mechanisms important for glucose homeostasis and insulin resistance. This convergence of research brings the promise of deeper understanding and, eventually, of improved prevention and treatment of type 2 diabetes.

Acknowledgments

We thank James Cresswell, Charisse Orme, Chenfei Yu, Hongjie Li, and Emily Stoops for helpful discussions and comments on the manuscript. Work in the authors' laboratory is supported by grants to J.S.B. from the NIH (DK075772), the American Diabetes Association, and the W. M. Keck Foundation.

References

Abel, E. D., Graveleau, C., Betuing, S., Pham, M., Reay, P. A., Kandror, V., Kupriyanova, T., Xu, Z., and Kandror, K. V. (2004). Regulation of insulin-responsive aminopeptidase expression and targeting in the insulin-responsive vesicle compartment of glucose transporter isoform 4-deficient cardiomyocytes. *Mol. Endocrinol.* **18,** 2491–2501.

Argani, P., and Ladanyi, M. (2005). Translocation carcinomas of the kidney. *Clin. Lab. Med.* **25,** 363–78.

Ariga, M., Nedachi, T., Katagiri, H., and Kanzaki, M. (2008). Functional role of sortilin in myogenesis and development of insulin-responsive glucose transport system in C2C12 myocytes. *J. Biol. Chem.* **283,** 10208–10220.

Bai, L., Wang, Y., Fan, J., Chen, Y., Ji, W., Qu, A., Xu, P., James, D. E., and Xu, T. (2007). Dissecting multiple steps of GLUT4 trafficking and identifying the sites of insulin action. *Cell Metab.* **5,** 47–57.

Bao, Y., Lopez, J. A., James, D. E., and Hunziker, W. (2008). Snapin interacts with the Exo70 subunit of the exocyst and modulates GLUT4 trafficking. *J. Biol. Chem.* **283,** 324–331.

Barile, M., Pisitkun, T., Yu, M. J., Chou, C. L., Verbalis, M. J., Shen, R. F., and Knepper, M. A. (2005). Large scale protein identification in intracellular aquaporin-2 vesicles from renal inner medullary collecting duct. *Mol. Cell. Proteomics* **4,** 1095–1106.

Bernier-Villamor, V., Sampson, D. A., Matunis, M. J., and Lima, C. D. (2002). Structural basis for E2-mediated SUMO conjugation revealed by a complex between ubiquitin-conjugating enzyme Ubc9 and RanGAP1. *Cell* **108,** 345–356.

Biddinger, S. B., and Kahn, C. R. (2006). From mice to men: Insights into the insulin resistance syndromes. *Annu. Rev. Physiol.* **68,** 123–158.

Birnbaum, M. J. (1989). Identification of a novel gene encoding an insulin-responsive glucose transporter protein. *Cell* **57,** 305–315.

Blot, V., and McGraw, T. E. (2006). GLUT4 is internalized by a cholesterol-dependent nystatin-sensitive mechanism inhibited by insulin. *EMBO J.* **25,** 5648–5658.

Bogan, J. S., Hendon, N., McKee, A. E., Tsao, T. S., and Lodish, H. F. (2003). Functional cloning of TUG as a regulator of GLUT4 glucose transporter trafficking. *Nature* **425,** 727–733.

Bogan, J. S., and Lodish, H. F. (1999). Two compartments for insulin-stimulated exocytosis in 3T3-L1 adipocytes defined by endogenous ACRP30 and GLUT4. *J. Cell. Biol.* **146,** 609–620.

Bogan, J. S., McKee, A. E., and Lodish, H. F. (2001). Insulin-responsive compartments containing GLUT4 in 3T3-L1 and CHO cells: Regulation by amino acid concentrations. *Mol. Cell. Biol.* **21,** 4785–4806.

Bose, A., Guilherme, A., Huang, S., Hubbard, A. C., Lane, C. R., Soriano, N. A., and Czech, M. P. (2005). The v-SNARE Vti1a regulates insulin-stimulated glucose transport and Acrp30 secretion in 3T3-L1 adipocytes. *J. Biol. Chem.* **280,** 36946–36951.

Bose, A., Guilherme, A., Robida, S. I., Nicoloro, S. M., Zhou, Q. L., Jiang, Z. Y., Pomerleau, D. P., and Czech, M. P. (2002). Glucose transporter recycling in response to insulin is facilitated by myosin Myo1c. *Nature* **420,** 821–824.

Bose, A., Robida, S., Furcinitti, P. S., Chawla, A., Fogarty, K., Corvera, S., and Czech, M. P. (2004). Unconventional myosin Myo1c promotes membrane fusion in a regulated exocytic pathway. *Mol. Cell Biol.* **24,** 5447–5458.

Bryant, N. J., Govers, R., and James, D. E. (2002). Regulated transport of the glucose transporter GLUT4. *Nat. Rev. Mol. Cell Biol.* **3,** 267–277.

Bundis, F., Neagoe, I., Schwappach, B., and Steinmeyer, K. (2006). Involvement of Golgin-160 in cell surface transport of renal ROMK channel: Co-expression of Golgin-160 increases ROMK currents. *Cell Physiol. Biochem.* **17,** 1–12.

Cai, H., Reinisch, K., and Ferro-Novick, S. (2007). Coats, tethers, Rabs, and SNAREs work together to mediate the intracellular destination of a transport vesicle. *Dev. Cell* **12,** 671–682.

Canuel, M., Lefrancois, S., Zeng, J., and Morales, C. R. (2008). AP-1 and retromer play opposite roles in the trafficking of sortilin between the Golgi apparatus and the lysosomes. *Biochem. Biophys. Res. Commun.* **366,** 724–730.

Capilla, E., Suzuki, N., Pessin, J. E., and Hou, J. C. (2007). The glucose transporter 4 FQQI motif is necessary for Akt substrate of 160-kilodalton-dependent plasma membrane translocation but not Golgi-localized (gamma)-ear-containing Arf-binding protein-dependent entry into the insulin-responsive storage compartment. *Mol. Endocrinol.* **21,** 3087–3099.

Cartee, G. D., and Wojtaszewski, J. F. (2007). Role of Akt substrate of 160 kDa in insulin-stimulated and contraction-stimulated glucose transport. *Appl. Physiol. Nutr. Metab.* **32,** 557–566.

Carvalho, E., Schellhorn, S. E., Zabolotny, J. M., Martin, S., Tozzo, E., Peroni, O. D., Houseknecht, K. L., Mundt, A., James, D. E., and Kahn, B. B. (2004). GLUT4 overexpression or deficiency in adipocytes of transgenic mice alters the composition of GLUT4 vesicles and the subcellular localization of GLUT4 and insulin-responsive aminopeptidase. *J. Biol. Chem.* **279,** 21598–21605.

Chai, S. Y., Fernando, R., Peck, G., Ye, S. Y., Mendelsohn, F. A., Jenkins, T. A., and Albiston, A. L. (2004). The angiotensin IV/AT4 receptor. *Cell Mol. Life Sci.* **61,** 2728–2737.

Chang, L., Chiang, S. H., and Saltiel, A. R. (2004). Insulin signaling and the regulation of glucose transport. *Mol. Med.* **10,** 65–71.

Charest, A., Lane, K., McMahon, K., and Housman, D. E. (2001). Association of a novel PDZ domain-containing peripheral Golgi protein with the Q-SNARE (Q-soluble *N*-ethylmaleimide-sensitive fusion protein (NSF) attachment protein receptor) protein syntaxin 6. *J. Biol. Chem.* **276,** 29456–29465.

Charron, M. J., Brosius, F. C. 3rd., Alper, S. L., and Lodish, H. F. (1989). A glucose transport protein expressed predominately in insulin-responsive tissues. *Proc. Natl. Acad. Sci. USA* **86,** 2535–2539.

Chavez, J. A., Roach, W. G., Keller, S. R., Lane, W. S., and Lienhard, G. E. (2008). Inhibition of GLUT4 translocation by Tbc1d1, a Rab GTPase activating protein abundant in skeletal muscle, is partially relieved by AMPK-activated protein kinase activation. *J. Biol. Chem.* **283,** 9187–9195.

Chen, S., Murphy, J., Toth, R., Campbell, D. G., Morrice, N. A., and Mackintosh, C. (2008). Complementary regulation of TBC1D1 and AS160 by growth factors, insulin and AMPK activators. *Biochem. J.* **409,** 449–459.

Chen, X. W., Leto, D., Chiang, S. H., Wang, Q., and Saltiel, A. R. (2007). Activation of RalA is required for insulin-stimulated Glut4 trafficking to the plasma membrane via the exocyst and the motor protein Myo1c. *Dev. Cell* **13,** 391–404.

Cheng, J., Moyer, B. D., Milewski, M., Loffing, J., Ikeda, M., Mickle, J. E., Cutting, G. R., Li, M., Stanton, B. A., and Guggino, W. B. (2002). A Golgi-associated PDZ domain protein modulates cystic fibrosis transmembrane regulator plasma membrane expression. *J. Biol. Chem.* **277,** 3520–3529.

Cheng, J., Wang, H., and Guggino, W. B. (2004). Modulation of mature cystic fibrosis transmembrane regulator protein by the PDZ domain protein CAL. *J. Biol. Chem.* **279,** 1892–1898.

Cheng, J., Wang, H., and Guggino, W. B. (2005). Regulation of cystic fibrosis transmembrane regulator trafficking and protein expression by a Rho family small GTPase TC10. *J. Biol. Chem.* **280,** 3731–3739.

Chi, N. W., and Lodish, H. F. (2000). Tankyrase is a golgi-associated mitogen-activated protein kinase substrate that interacts with IRAP in GLUT4 vesicles. *J. Biol. Chem.* **275,** 38437–38444.

Chiang, S. H., Baumann, C. A., Kanzaki, M., Thurmond, D. C., Watson, R. T., Neudauer, C. L., Macara, I. G., Pessin, J. E., and Saltiel, A. R. (2001). Insulin-stimulated GLUT4 translocation requires the CAP-dependent activation of TC10. *Nature* **410,** 944–948.

Chiang, S. H., Hou, J. C., Hwang, J., Pessin, J. E., and Saltiel, A. R. (2002). Cloning and functional characterization of related TC10 isoforms, a subfamily of Rho proteins involved in insulin-stimulated glucose transport. *J. Biol. Chem.* **277,** 13067–13073.

Chieregatti, E., and Meldolesi, J. (2005). Regulated exocytosis: New organelles for non-secretory purposes. *Nat. Rev. Mol. Cell Biol.* **6,** 181–187.

Cline, G. W., Petersen, K. F., Krssak, M., Shen, J., Hundal, R. S., Trajanoski, Z., Inzucchi, S., Dresner, A., Rothman, D. L., and Shulman, G. I. (1999). Impaired glucose transport as a cause of decreased insulin-stimulated muscle glycogen synthesis in type 2 diabetes. *N. Engl. J. Med.* **341,** 240–246.

Cohen, P. (2006). The twentieth century struggle to decipher insulin signalling. *Nat. Rev. Mol. Cell. Biol.* **7,** 867–873.

Coster, A. C., Govers, R., and James, D. E. (2004). Insulin stimulates the entry of GLUT4 into the endosomal recycling pathway by a quantal mechanism. *Traffic* **5,** 763–771.

Cuadra, A. E., Kuo, S. H., Kawasaki, Y., Bredt, D. S., and Chetkovich, D. M. (2004). AMPA receptor synaptic targeting regulated by stargazin interactions with the Golgi-resident PDZ protein nPIST. *J. Neurosci.* **24,** 7491–7502.

Cushman, S. W., and Wardzala, L. J. (1980). Potential mechanism of insulin action on glucose transport in the isolated rat adipose cell. Apparent translocation of intracellular transport systems to the plasma membrane. *J. Biol. Chem.* **255,** 4758–4762.

Dai, J., Li, J., Bos, E., Porcionatto, M., Premont, R. T., Bourgoin, S., Peters, P. J., and Hsu, V. W. (2004). ACAP1 promotes endocytic recycling by recognizing recycling sorting signals. *Dev. Cell* **7,** 771–776.

Dreveny, I., Pye, V. E., Beuron, F., Briggs, L. C., Isaacson, R. L., Matthews, S. J., McKeown, C., Yuan, X., Zhang, X., and Freemont, P. S. (2004). p97 and close encounters of every kind: A brief review. *Biochem. Soc. Trans.* **32,** 715–720.

Dugani, C. B., Randhawa, V. K., Cheng, A. W., Patel, N., and Klip, A. (2008). Selective regulation of the perinuclear distribution of glucose transporter 4 (GLUT4) by insulin signals in muscle cells. *Eur. J. Cell Biol.* **87,** 337–351.

Eguez, L., Lee, A., Chavez, J. A., Miinea, C. P., Kane, S., Lienhard, G. E., and McGraw, T. E. (2005). Full intracellular retention of GLUT4 requires AS160 Rab GTPase activating protein. *Cell Metab.* **2,** 263–272.

El-Jack, A. K., Kandror, K. V., and Pilch, P. F. (1999). The formation of an insulin-responsive vesicular cargo compartment is an early event in 3T3-L1 adipocyte differentiation. *Mol. Biol. Cell* **10,** 1581–1594.

Ewart, M. A., Clarke, M., Kane, S., Chamberlain, L. H., and Gould, G. W. (2005). Evidence for a role of the exocyst in insulin-stimulated Glut4 trafficking in 3T3-L1 adipocytes. *J. Biol. Chem.* **280,** 3812–3816.

Farese, R. V., Sajan, M. P., Yang, H., Li, P., Mastorides, S., Gower, W. R. Jr., Nimal, S., Choi, C. S., Kim, S., Shulman, G. I., Kahn, C. R., Braun, U., *et al.* (2007). Muscle-specific knockout of PKC-lambda impairs glucose transport and induces metabolic and diabetic syndromes. *J. Clin. Invest.* **117,** 2289–2301.

Foster, L. J., Rudich, A., Talior, I., Patel, N., Huang, X., Furtado, L. M., Bilan, P. J., Mann, M., and Klip, A. (2006). Insulin-dependent interactions of proteins with GLUT4 revealed through stable isotope labeling by amino acids in cell culture (SILAC). *J. Proteome Res.* **5,** 64–75.

Fujii, N., Jessen, N., and Goodyear, L. J. (2006). AMP-activated protein kinase and the regulation of glucose transport. *Am. J. Physiol. Endocrinol. Metab.* **291,** E867–E877.

Fukumoto, H., Kayano, T., Buse, J. B., Edwards, Y., Pilch, P. F., Bell, G. I., and Seino, S. (1989). Cloning and characterization of the major insulin-responsive glucose transporter expressed in human skeletal muscle and other insulin-responsive tissues. *J. Biol. Chem.* **264,** 7776–7779.

Garcia de Herreros, A., and Birnbaum, M. J. (1989). The regulation by insulin of glucose transporter gene expression in 3T3 adipocytes. *J. Biol. Chem.* **264,** 9885–9890.

Garvey, W. T., Maianu, L., Zhu, J. H., Brechtel-Hook, G., Wallace, P., and Baron, A. D. (1998). Evidence for defects in the trafficking and translocation of GLUT4 glucose transporters in skeletal muscle as a cause of human insulin resistance. *J. Clin. Invest.* **101,** 2377–2386.

Garza, L. A., and Birnbaum, M. J. (2000). Insulin-responsive aminopeptidase trafficking in 3T3-L1 adipocytes. *J. Biol. Chem.* **275,** 2560–2567.

Geiss-Friedlander, R., and Melchior, F. (2007). Concepts in sumoylation: A decade on. *Nat. Rev. Mol. Cell Biol.* **8,** 947–956.

Gentzsch, M., Cui, L., Mengos, A., Chang, X. B., Chen, J. H., and Riordan, J. R. (2003). The PDZ-binding chloride channel ClC-3B localizes to the Golgi and associates with cystic fibrosis transmembrane conductance regulator-interacting PDZ proteins. *J. Biol. Chem.* **278,** 6440–6449.

Giorgino, F., de Robertis, O., Laviola, L., Montrone, C., Perrini, S., McCowen, K. C., and Smith, R. J. (2000). The sentrin-conjugating enzyme mUbc9 interacts with GLUT4 and GLUT1 glucose transporters and regulates transporter levels in skeletal muscle cells. *Proc. Natl. Acad. Sci. USA* **97,** 1125–1130.

Gokool, S., Tattersall, D., and Seaman, M. N. (2007). EHD1 interacts with retromer to stabilize SNX1 tubules and facilitate endosome-to-Golgi retrieval. *Traffic* **8,** 1873–1886.

Gonzalez, M. I., Susarla, B. T., Fournier, K. M., Sheldon, A. L., and Robinson, M. B. (2007). Constitutive endocytosis and recycling of the neuronal glutamate transporter, excitatory amino acid carrier 1. *J. Neurochem.* **103,** 1917–1931.

Govers, R., Coster, A. C., and James, D. E. (2004). Insulin increases cell surface GLUT4 levels by dose dependently discharging GLUT4 into a cell surface recycling pathway. *Mol. Cell Biol.* **24,** 6456–6466.

Green, H., and Kehinde, O. (1976). Spontaneous heritable changes leading to increased adipose conversion in 3T3 cells. *Cell* **7,** 105–113.

Gros, J., Gerhardt, C. C., and Strosberg, A. D. (1999). Expression of human (β)3-adrenergic receptor induces adipocyte-like features in CHO/K1 fibroblasts. *J. Cell Sci.* **112,** 3791–3797.

Gross, D. N., Farmer, S. R., and Pilch, P. F. (2004). Glut4 storage vesicles without Glut4: Transcriptional regulation of insulin-dependent vesicular traffic. *Mol. Cell Biol.* **24,** 7151–7162.

Guilherme, A., Soriano, N. A., Furcinitti, P. S., and Czech, M. P. (2004). Role of EHD1 and EHBP1 in perinuclear sorting and insulin-regulated GLUT4 recycling in 3T3-L1 adipocytes. *J. Biol. Chem.* **279,** 40062–40075.

Hashiramoto, M., and James, D. E. (2000). Characterization of insulin-responsive GLUT4 storage vesicles isolated from 3T3-L1 adipocytes. *Mol. Cell. Biol.* **20,** 416–427.

He, J., Bellini, M., Xu, J., Castleberry, A. M., and Hall, R. A. (2004). Interaction with cystic fibrosis transmembrane conductance regulator-associated ligand (CAL) inhibits β1-adrenergic receptor surface expression. *J. Biol. Chem.* **279,** 50190–50196.

Heimann, P., El Housni, H., Ogur, G., Weterman, M. A., Petty, E. M., and Vassart, G (2001). Fusion of a novel gene, RCC17, to the TFE3 gene in t(X;17)(p11.2;q25.3)-bearing papillary renal cell carcinomas. *Cancer Res.* **61,** 4130–4135.

Hicks, S. W., Horn, T. A., McCaffery, J. M., Zuckerman, D. M., and Machamer, C. E. (2006). Golgin-160 promotes cell surface expression of the β-1 adrenergic receptor. *Traffic* **7,** 1666–1677.

Hicks, S. W., and Machamer, C. E. (2005). Isoform-specific interaction of golgin-160 with the Golgi-associated protein PIST. *J. Biol. Chem.* **280,** 28944–28951.

Hoehn, K. L., Hohnen-Behrens, C., Cederberg, A., Wu, L. E., Turner, N., Yuasa, T., Ebina, Y., and James, D. E. (2008). IRS1-independent defects define major nodes of insulin resistance. *Cell Metab.* **7,** 421–433.

Holman, G. D., Lo Leggio, L., and Cushman, S. W. (1994). Insulin-stimulated GLUT4 glucose transporter recycling. A problem in membrane protein subcellular trafficking through multiple pools. *J. Biol. Chem.* **269,** 17516–17524.

Hosaka, T., Brooks, C. C., Presman, E., Kim, S. K., Zhang, Z., Breen, M., Gross, D. N., Sztul, E., and Pilch, P. F. (2005). p115 Interacts with the GLUT4 vesicle protein, IRAP, and plays a critical role in insulin-stimulated GLUT4 translocation. *Mol. Biol. Cell* **16,** 2882–2890.

Hou, J. C., Suzuki, N., Pessin, J. E., and Watson, R. T. (2006). A specific dileucine motif is required for the GGA-dependent entry of newly synthesized insulin-responsive aminopeptidase into the insulin-responsive compartment. *J. Biol. Chem.* **281,** 33457–33466.

Huang, S., and Czech, M. P. (2007). The GLUT4 glucose transporter. *Cell Metab.* **5,** 237–252.

Huang, S., Lifshitz, L. M., Jones, C., Bellve, K. D., Standley, C., Fonseca, S., Corvera, S., Fogarty, K. E., and Czech, M. P. (2007). Insulin stimulates membrane fusion and GLUT4 accumulation in clathrin coats on adipocyte plasma membranes. *Mol. Cell. Biol.* **27,** 3456–3469.

Imamura, T., Huang, J., Usui, I., Satoh, H., Bever, J., and Olefsky, J. M. (2003). Insulin-induced GLUT4 translocation involves protein kinase C-lambda-mediated functional coupling between Rab4 and the motor protein kinesin. *Mol. Cell Biol.* **23,** 4892–4900.

Inoue, M., Chang, L., Hwang, J., Chiang, S. H., and Saltiel, A. R. (2003). The exocyst complex is required for targeting of Glut4 to the plasma membrane by insulin. *Nature* **422,** 629–633.

Inoue, M., Chiang, S. H., Chang, L., Chen, X. W., and Saltiel, A. R. (2006). Compartmentalization of the exocyst complex in lipid rafts controls Glut4 vesicle tethering. *Mol. Biol. Cell* **17,** 2303–2311.

Ishiki, M., and Klip, A. (2005). Minireview: Recent developments in the regulation of glucose transporter-4 traffic: New signals, locations and partners. *Endocrinology* **146,** 5071–5078.

James, D. E., Brown, R., Navarro, J., and Pilch, P. F. (1988). Insulin-regulatable tissues express a unique insulin-sensitive glucose transport protein. *Nature* **333,** 183–185.

James, D. E., Strube, M., and Mueckler, M. (1989). Molecular cloning and characterization of an insulin-regulatable glucose transporter. *Nature* **338,** 83–87.

Jessen, N., and Goodyear, L. J. (2005). Contraction signaling to glucose transport in skeletal muscle. *J. Appl. Physiol.* **99,** 330–337.

Jiang, H., Li, J., Katz, E. B., and Charron, M. J. (2001). GLUT4 ablation in mice results in redistribution of IRAP to the plasma membrane. *Biochem. Biophys. Res. Commun.* **284,** 519–525.

Jiang, L., Fan, J., Bai, L., Wang, Y., Chen, Y., Yang, L., Chen, L., and Xu, T. (2008). Direct quantification of fusion rate reveals a distal role for AS160 in insulin-stimulated fusion of GLUT4 storage vesicles. *J. Biol. Chem.* **283,** 8508–8516.

Johnson, A. O., Lampson, M. A., and McGraw, T. E. (2001). A di-leucine sequence and a cluster of acidic amino acids are required for dynamic retention in the endosomal recycling compartment of fibroblasts. *Mol. Biol. Cell* **12,** 367–381.

Johnson, A. O., Subtil, A., Petrush, R., Kobylarz, K., Keller, S. R., and McGraw, T. E. (1998). Identification of an insulin-responsive, slow endocytic recycling mechanism in chinese hamster ovary cells. *J. Biol. Chem.* **273,** 17968–17977.

Kaestner, K. H., Christy, R. J., McLenithan, J. C., Braiterman, L. T., Cornelius, P., Pekala, P. H., and Lane, M. D. (1989). Sequence, tissue distribution, and differential expression of mRNA for a putative insulin-responsive glucose transporter in mouse 3T3-L1 adipocytes. *Proc. Natl. Acad. Sci. USA* **86,** 3150–3154.

Kandror, K. V. (2003). A long search for Glut4 activation. *Sci. STKE.* 2003, PE5.

Kandror, K. V., Yu, L., and Pilch, P. F. (1994). The major protein of GLUT4-containing vesicles, gp160, has aminopeptidase activity. *J. Biol. Chem.* **269,** 30777–30780.

Kanzaki, M. (2006). Insulin receptor signals regulating GLUT4 translocation and actin dynamics. *Endocr. J.* **53,** 267–293.

Karylowski, O., Zeigerer, A., Cohen, A., and McGraw, T. E. (2004). GLUT4 is retained by an intracellular cycle of vesicle formation and fusion with endosomes. *Mol. Biol. Cell* **15,** 870–882.

Katagiri, H., Asano, T., Yamada, T., Aoyama, T., Fukushima, Y., Kikuchi, M., Kodama, T., and Oka, Y. (2002). Acyl-coenzyme A dehydrogenases are localized on GLUT4-containing vesicles via association with insulin-regulated aminopeptidase in a manner dependent on its dileucine motif. *Mol. Endocrinol.* **16,** 1049–1059.

Keller, S. R. (2003). The insulin-regulated aminopeptidase: A companion and regulator of GLUT4. *Front Biosci.* **8,** s410–s420.

Keller, S. R., Davis, A. C., and Clairmont, K. B. (2002). Mice deficient in the insulin-regulated membrane aminopeptidase show substantial decreases in glucose transporter GLUT4 levels but maintain normal glucose homeostasis. *J. Biol. Chem.* **277,** 17677–17686.

Keller, S. R., Scott, H. M., Mastick, C. C., Aebersold, R., and Lienhard, G. E. (1995). Cloning and characterization of a novel insulin-regulated membrane aminopeptidase from Glut4 vesicles. *J. Biol. Chem.* **270,** 23612–23618.

Kerscher, O., Felberbaum, R., and Hochstrasser, M. (2006). Modification of proteins by ubiquitin and ubiquitin-like proteins. *Annu. Rev. Cell. Dev. Biol.* **22,** 159–180.

Khan, A. H., Capilla, E., Hou, J. C., Watson, R. T., Smith, J. R., and Pessin, J. E. (2004). Entry of newly synthesized GLUT4 into the insulin-responsive storage compartment is dependent upon both the amino terminus and the large cytoplasmic loop. *J. Biol. Chem.* **279,** 37505–37511.

Kim, S. S., Bae, J. W., and Jung, C. Y. (1994). GLUT-4 degradation rate: Reduction in rat adipocytes in fasting and streptozotocin-induced diabetes. *Am. J. Physiol.* **267,** E132–E139.

Klein, J. D., Frohlich, O., Blount, M. A., Martin, C. F., Smith, T. D., and Sands, J. M. (2006). Vasopressin increases plasma membrane accumulation of urea transporter UT-A1 in rat inner medullary collecting ducts. *J. Am. Soc. Nephrol.* **17,** 2680–2686.

Koumanov, F., Jin, B., Yang, J., and Holman, G. D. (2005). Insulin signaling meets vesicle traffic of GLUT4 at a plasma-membrane-activated fusion step. *Cell Metab.* **2,** 179–189.

Kreykenbohm, V., Wenzel, D., Antonin, W., Atlachkine, V., and von Mollard, G. F. (2002). The SNAREs vti1a and vti1b have distinct localization and SNARE complex partners. *Eur. J. Cell. Biol.* **81,** 273–280.

Kupriyanova, T. A., Kandror, V., and Kandror, K. V. (2002). Isolation and characterization of the two major intracellular Glut4 storage compartments. *J. Biol. Chem.* **277,** 9133–9138.

Ladanyi, M., Lui, M. Y., Antonescu, C. R., Krause-Boehm, A., Meindl, A., Argani, P., Healey, J. H., Ueda, T., Yoshikawa, H., Meloni-Ehrig, A., Sorensen, P. H., Mertens, F., *et al.* (2001). The der(17)t(X;17)(p11;q25) of human alveolar soft part sarcoma fuses the TFE3 transcription factor gene to ASPL, a novel gene at 17q25. *Oncogene.* **20,** 48–57.

Lakkaraju, A., and Rodriguez-Boulan, E. (2008). Itinerant exosomes: Emerging roles in cell and tissue polarity. *Trends Cell Biol.* **18,** 199–209.

Lalioti, V. S., Vergarajauregui, S., Pulido, D., and Sandoval, I. V. (2002). The insulin-sensitive glucose transporter, GLUT4, interacts physically with Daxx. Two proteins with capacity to bind Ubc9 and conjugated to SUMO1. *J. Biol. Chem.* **277,** 19783–19791.

Lampson, M. A., Racz, A., Cushman, S. W., and McGraw, T. E. (2000). Demonstration of insulin-responsive trafficking of GLUT4 and vpTR in fibroblasts. *J. Cell Sci.* **113,** 4065–4076.

Larance, M., Ramm, G., Stockli, J., van Dam, E. M., Winata, S., Wasinger, V., Simpson, F., Graham, M., Junutula, J. R., Guilhaus, M., and James, D. E. (2005). Characterization of the role of the Rab GTPase-activating protein AS160 in insulin-regulated GLUT4 trafficking. *J. Biol. Chem.* **280,** 37803–37813.

Lee, W., and Jung, C. Y. (1997). A synthetic peptide corresponding to the GLUT4 C-terminal cytoplasmic domain causes insulin-like glucose transport stimulation and GLUT4 recruitment in rat adipocytes. *J. Biol. Chem.* **272,** 21427–21431.

Lee, W., Ryu, J., Spangler, R. A., and Jung, C. Y. (2000). Modulation of GLUT4 and GLUT1 recycling by insulin in rat adipocytes: Kinetic analysis based on the involvement of multiple intracellular compartments. *Biochemistry* **39,** 9358–9366.

Levine, R., Goldstein, M., Klein, S., and Huddlestun, B. (1949). The action of insulin on the distribution of galactose in eviscerated nephrectomized dogs. *J. Biol. Chem.* **179,** 985–986.

Levine, R., Goldstein, M. S., Huddlestun, B., and Klein, S. P. (1950). Action of insulin on the 'permeability' of cells to free hexoses, as studied by its effect on the distribution of galactose. *Am. J. Physiol.* **163,** 70–76.

Li, J., Ballif, B. A., Powelka, A. M., Dai, J., Gygi, S. P., and Hsu, V. W. (2005). Phosphorylation of ACAP1 by Akt regulates the stimulation-dependent recycling of integrin beta1 to control cell migration. *Dev. Cell* **9,** 663–673.

Li, L. V., and Kandror, K. V. (2005). Golgi-localized, γ-ear-containing, Arf-binding protein adaptors mediate insulin-responsive trafficking of glucose transporter 4 in 3T3-L1 adipocytes. *Mol. Endocrinol.* **19,** 2145–2153.

Liao, W., Nguyen, M. T., Imamura, T., Singer, O., Verma, I. M., and Olefsky, J. M. (2006). Lentiviral short hairpin ribonucleic acid-mediated knockdown of GLUT4 in 3T3-L1 adipocytes. *Endocrinology* **147,** 2245–2252.

Lim, S. N., Bonzelius, F., Low, S. H., Wille, H., Weimbs, T., and Herman, G. A. (2001). Identification of discrete classes of endosome-derived small vesicles as a major cellular pool for recycling membrane proteins. *Mol. Biol. Cell* **12,** 981–995.

Lin, B. Z., Pilch, P. F., and Kandror, K. V. (1997). Sortilin is a major protein component of Glut4-containing vesicles. *J. Biol. Chem.* **272,** 24145–24147.

Lin, D. Y., Huang, Y. S., Jeng, J. C., Kuo, H. Y., Chang, C. C., Chao, T. T., Ho, C. C., Chen, Y. C., Lin, T. P., Fang, H. I., Hung, C. C., Suen, C. S., *et al.* (2006). Role of SUMO-interacting motif in Daxx SUMO modification, subnuclear localization, and repression of sumoylated transcription factors. *Mol. Cell* **24,** 341–354.

Liu, G., Hou, J. C., Watson, R. T., and Pessin, J. E. (2005). Initial entry of IRAP into the insulin-responsive storage compartment occurs prior to basal or insulin-stimulated plasma membrane recycling. *Am. J. Physiol. Endocrinol. Metab.* **289,** E746–E752.

Liu, L. B., Omata, W., Kojima, I., and Shibata, H. (2007). The SUMO conjugating enzyme Ubc9 is a regulator of GLUT4 turnover and targeting to the insulin-responsive storage compartment in 3T3-L1 adipocytes. *Diabetes* **56,** 1977–1985.

Liu, X., Constantinescu, S. N., Sun, Y., Bogan, J. S., Hirsch, D., Weinberg, R. A., and Lodish, H. F. (2000). Generation of mammalian cells stably expressing multiple genes at predetermined levels. *Anal. Biochem.* **280,** 20–28.

Livingstone, C., James, D. E., Rice, J. E., Hanpeter, D., and Gould, G. W. (1996). Compartment ablation analysis of the insulin-responsive glucose transporter (GLUT4) in 3T3-L1 adipocytes. *Biochem. J.* **315**(Pt. 2), 487–495.

Maffucci, T., Brancaccio, A., Piccolo, E., Stein, R. C., and Falasca, M. (2003). Insulin induces phosphatidylinositol-3-phosphate formation through TC10 activation. *EMBO J.* **22,** 4178–4189.

Maianu, L., Keller, S. R., and Garvey, W. T. (2001). Adipocytes exhibit abnormal subcellular distribution and translocation of vesicles containing glucose transporter 4 and insulin-regulated aminopeptidase in type 2 diabetes mellitus: Implications regarding defects in vesicle trafficking. *J. Clin. Endocrinol. Metab.* **86,** 5450–5456.

Mallard, F., Tang, B. L., Galli, T., Tenza, D., Saint-Pol, A., Yue, X., Antony, C., Hong, W., Goud, B., and Johannes, L. (2002). Early/recycling endosomes-to-TGN transport involves two SNARE complexes and a Rab6 isoform. *J. Cell Biol.* **156,** 653–664.

Martin, O. J., Lee, A., and McGraw, T. E. (2006). GLUT4 distribution between the plasma membrane and the intracellular compartments is maintained by an insulin-modulated bipartite dynamic mechanism. *J. Biol. Chem.* **281,** 484–490.

Martin, S., Millar, C. A., Lyttle, C. T., Meerloo, T., Marsh, B. J., Gould, G. W., and James, D. E. (2000). Effects of insulin on intracellular GLUT4 vesicles in adipocytes: Evidence for a secretory mode of regulation. *J. Cell Sci.* **113**(Pt. 19), 3427–3438.

Mastick, C. C., Aebersold, R., and Lienhard, G. E. (1994). Characterization of a major protein in GLUT4 vesicles. Concentration in the vesicles and insulin-stimulated translocation to the plasma membrane. *J. Biol. Chem.* **269,** 6089–6092.

Masuda, S., Hattori, A., Matsumoto, H., Miyazawa, S., Natori, Y., Mizutani, S., and Tsujimoto, M. (2003). Involvement of the V2 receptor in vasopressin-stimulated translocation of placental leucine aminopeptidase/oxytocinase in renal cells. *Eur. J. Biochem.* **270,** 1988–1994.

Mellman, I., and Warren, G. (2000). The road taken: Past and future foundations of membrane traffic. *Cell* **100,** 99–112.

Mitra, K., Ubarretxena-Belandia, I., Taguchi, T., Warren, G., and Engelman, D. M. (2004). Modulation of the bilayer thickness of exocytic pathway membranes by membrane proteins rather than cholesterol. *Proc. Natl. Acad. Sci. USA* **101,** 4083–4088.

Morris, N. J., Ross, S. A., Lane, W. S., Moestrup, S. K., Petersen, C. M., Keller, S. R., and Lienhard, G. E. (1998). Sortilin is the major 110-kDa protein in GLUT4 vesicles from adipocytes. *J. Biol. Chem.* **273,** 3582–3587.

Muretta, J. M., Romenskaia, I., Cassiday, P. A., and Mastick, C. C. (2007). Expression of a synapsin IIb site 1 phosphorylation mutant in 3T3-L1 adipocytes inhibits basal intracellular retention of Glut4. *J. Cell Sci.* **120,** 1168–1177.

Muretta, J. M., Romenskaia, I., and Mastick, C. C. (2008). Insulin releases Glut4 from static storage compartments into cycling endosomes and increases the rate constant for Glut4 exocytosis. *J. Biol. Chem.* **283,** 311–323.

Nacerddine, K., Lehembre, F., Bhaumik, M., Artus, J., Cohen-Tannoudji, M., Babinet, C., Pandolfi, P. P., and Dejean, A. (2005). The SUMO pathway is essential for nuclear integrity and chromosome segregation in mice. *Dev. Cell* **9,** 769–779.

Nakamura, H., Itakuara, A., Okamura, M., Ito, M., Iwase, A., Nakanishi, Y., Okada, M., Nagasaka, T., and Mizutani, S. (2000). Oxytocin stimulates the translocation of oxytocinase of human vascular endothelial cells via activation of oxytocin receptors. *Endocrinology* **141,** 4481–4485.

Neudauer, C. L., Joberty, G., and Macara, I. G. (2001). PIST: A novel PDZ/coiled-coil domain binding partner for the rho-family GTPase TC10. *Biochem. Biophys. Res. Commun.* **280,** 541–547.

Nielsen, M. S., Madsen, P., Christensen, E. I., Nykjaer, A., Gliemann, J., Kasper, D., Pohlmann, R., and Petersen, C. M. (2001). The sortilin cytoplasmic tail conveys Golgi-endosome transport and binds the VHS domain of the GGA2 sorting protein. *EMBO J.* **20,** 2180–2190.

Palacios, S., Lalioti, V., Martinez-Arca, S., Chattopadhyay, S., and Sandoval, I. V. (2001). Recycling of the insulin-sensitive glucose transporter GLUT4. Access of surface internalized GLUT4 molecules to the perinuclear storage compartment is mediated by the Phe5-Gln6-Gln7-Ile8 motif. *J. Biol. Chem.* **276,** 3371–3383.

Peck, G. R., Ye, S., Pham, V., Fernando, R. N., Macaulay, S. L., Chai, S. Y., and Albiston, A. L. (2006). Interaction of the Akt substrate, AS160, with the glucose transporter 4 vesicle marker protein, insulin-regulated aminopeptidase. *Mol. Endocrinol.* **20,** 2576–2583.

Perera, H. K., Clarke, M., Morris, N. J., Hong, W., Chamberlain, L. H., and Gould, G. W. (2003). Syntaxin 6 regulates Glut4 trafficking in 3T3-L1 adipocytes. *Mol. Biol. Cell* **14,** 2946–2958.

Petersen, K. F., Dufour, S., Savage, D. B., Bilz, S., Solomon, G., Yonemitsu, S., Cline, G. W., Befroy, D., Zemany, L., Kahn, B. B., Papademetris, X., Rothman, D. L., *et al.* (2007). The role of skeletal muscle insulin resistance in the pathogenesis of the metabolic syndrome. *Proc. Natl. Acad. Sci. USA* **104,** 12587–12594.

Piper, R. C., Hess, L. J., and James, D. E. (1991). Differential sorting of two glucose transporters expressed in insulin-sensitive cells. *Am. J. Physiol.* **260,** C570–C580.

Piper, R. C., Tai, C., Kulesza, P., Pang, S., Warnock, D., Baenziger, J., Slot, J. W., Geuze, H. J., Puri, C., and James, D. E. (1993). GLUT-4 NH2 terminus contains a phenylalanine-based targeting motif that regulates intracellular sequestration. *J. Cell Biol.* **121,** 1221–1232.

Pisitkun, T., Shen, R. F., and Knepper, M. A. (2004). Identification and proteomic profiling of exosomes in human urine. *Proc. Natl. Acad. Sci. USA* **101,** 13368–13373.

Ploug, T., and Ralston, E. (2002). Exploring the whereabouts of GLUT4 in skeletal muscle (review). *Mol. Membr. Biol.* **19,** 39–49.

Ploug, T., van Deurs, B., Ai, H., Cushman, S. W., and Ralston, E. (1998). Analysis of GLUT4 distribution in whole skeletal muscle fibers: Identification of distinct storage compartments that are recruited by insulin and muscle contractions. *J. Cell Biol.* **142,** 1429–1446.

Procino, G., Caces, D. B., Valenti, G., and Pessin, J. E. (2005). Adipocytes support cAMP-dependent translocation of Aquaporin 2 (AQP2) from intracellular sites distinct from the insulin-responsive GLUT4 storage compartment. *Am. J. Physiol. Renal Physiol.* **290,** F985–F994.

Proctor, K. M., Miller, S. C., Bryant, N. J., and Gould, G. W. (2006). Syntaxin 16 controls the intracellular sequestration of GLUT4 in 3T3-L1 adipocytes. *Biochem. Biophys. Res. Commun.* **347,** 433–438.

Ramm, G., and James, D. E. (2005). GLUT4 trafficking in a test tube. *Cell Metab.* **2,** 150–152.

Ramm, G., Larance, M., Guilhaus, M., and James, D. E. (2006). A role for 14-3-3 in insulin-stimulated GLUT4 translocation through its interaction with the RabGAP AS160. *J. Biol. Chem.* **281,** 29174–29180.

Ramm, G., Slot, J. W., James, D. E., and Stoorvogel, W. (2000). Insulin recruits GLUT4 from specialized VAMP2-carrying vesicles as well as from the dynamic endosomal/trans-Golgi network in rat adipocytes. *Mol. Biol. Cell* **11,** 4079–4091.

Ribe, D., Yang, J., Patel, S., Koumanov, F., Cushman, S. W., and Holman, G. D. (2005). Endofacial competitive inhibition of glucose transporter-4 intrinsic activity by the mitogen-activated protein kinase inhibitor SB203580. *Endocrinology* **146,** 1713–1717.

Ross, S. A., Herbst, J. J., Keller, S. R., and Lienhard, G. E. (1997). Trafficking kinetics of the insulin-regulated membrane aminopeptidase in 3T3-L1 adipocytes. *Biochem. Biophys. Res. Commun.* **239,** 247–251.

Russell, R. R. 3rd., Li, J., Coven, D. L., Pypaert, M., Zechner, C., Palmeri, M., Giordano, F. J., Mu, J., Birnbaum, M. J., and Young, L. H. (2004). AMP-activated protein kinase mediates ischemic glucose uptake and prevents postischemic cardiac dysfunction, apoptosis, and injury. *J. Clin. Invest.* **114,** 495–503.

Ryu, S. W., Chae, S. K., and Kim, E. (2000). Interaction of Daxx, a Fas binding protein, with sentrin and Ubc9. *Biochem. Biophys. Res. Commun.* **279,** 6–10.

Sakamoto, K., and Holman, G. D. (2008). Emerging role for AS160/TBC1D4 and TBC1D1 in the regulation of GLUT4 traffic. *Am. J. Physiol. Endocrinol. Metab.* **295,** E29–E37.

Sano, H., Kane, S., Sano, E., and Lienhard, G. E. (2005). Synip phosphorylation does not regulate insulin-stimulated GLUT4 translocation. *Biochem. Biophys. Res. Commun.* **332,** 880–884.

Sargeant, R. J., and Paquet, M. R. (1993). Effect of insulin on the rates of synthesis and degradation of GLUT1 and GLUT4 glucose transporters in 3T3-L1 adipocytes. *Biochem. J.* **290**(Pt. 3), 913–919.

Sbodio, J. I., and Chi, N. W. (2002). Identification of a tankyrase-binding motif shared by IRAP, TAB182, and human TRF1 but not mouse TRF1. NuMA contains this RXXPDG motif and is a novel tankyrase partner. *J. Biol. Chem.* **277,** 31887–31892.

Sbodio, J. I., Lodish, H. F., and Chi, N. W. (2002). Tankyrase-2 oligomerizes with tankyrase-1 and binds to both TRF1 (telomere-repeat-binding factor 1) and IRAP (insulin-responsive aminopeptidase). *Biochem. J.* **361,** 451–459.

Schertzer, J. D., Antonescu, C. N., Bilan, P. J., Jain, S., Huang, X., Liu, Z., Bonen, A., and Klip, A. (2009). A transgenic mouse model to study GLUT4myc regulation in skeletal muscle. *Endocrinology*. In press (2008 Dec 12). doi: 10.1210/en.2008-1372.

Schuberth, C., and Buchberger, A. (2008). UBX domain proteins: Major regulators of the AAA ATPase Cdc48/p97. *Cell Mol. Life Sci.* **65,** 2360–2371.

Seaman, M. N. (2004). Cargo-selective endosomal sorting for retrieval to the Golgi requires retromer. *J. Cell Biol.* **165,** 111–122.

Seaman, M. N. (2007). Identification of a novel conserved sorting motif required for retromer-mediated endosome-to-TGN retrieval. *J. Cell Sci.* **120,** 2378–2389.

Semiz, S., Park, J. G., Nicoloro, S. M., Furcinitti, P., Zhang, C., Chawla, A., Leszyk, J., and Czech, M. P. (2003). Conventional kinesin KIF5B mediates insulin-stimulated GLUT4 movements on microtubules. *EMBO J.* **22,** 2387–2399.

Shewan, A. M., van Dam, E. M., Martin, S., Luen, T. B., Hong, W., Bryant, N. J., and James, D. E. (2003). GLUT4 recycles via a trans-Golgi network (TGN) subdomain enriched in Syntaxins 6 and 16 but not TGN38: Involvement of an acidic targeting motif. *Mol. Biol. Cell* **14,** 973–986.

Shi, J., and Kandror, K. V. (2005). Sortilin is essential and sufficient for the formation of Glut4 storage vesicles in 3T3-L1 adipocytes. *Dev. Cell* **9,** 99–108.

Shi, J., and Kandror, K. V. (2007). The luminal Vps10p domain of sortilin plays the predominant role in targeting to insulin-responsive Glut4-containing vesicles. *J. Biol. Chem.* **282,** 9008–9016.

Slot, J. W., Garruti, G., Martin, S., Oorschot, V., Posthuma, G., Kraegen, E. W., Laybutt, R., Thibault, G., and James, D. E. (1997). Glucose transporter (GLUT-4) is targeted to secretory granules in rat atrial cardiomyocytes. *J. Cell Biol.* **137,** 1243–1254.

Slot, J. W., Geuze, H. J., Gigengack, S., James, D. E., and Lienhard, G. E. (1991a). Translocation of the glucose transporter GLUT4 in cardiac myocytes of the rat. *Proc. Natl. Acad. Sci. USA* **88,** 7815–7819.

Slot, J. W., Geuze, H. J., Gigengack, S., Lienhard, G. E., and James, D. E. (1991b). Immuno-localization of the insulin regulatable glucose transporter in brown adipose tissue of the rat. *J. Cell Biol.* **113,** 123–135.

Somwar, R., Koterski, S., Sweeney, G., Sciotti, R., Djuric, S., Berg, C., Trevillyan, J., Scherer, P. E., Rondinone, C. M., and Klip, A. (2002). A dominant-negative p38 MAPK mutant and novel selective inhibitors of p38 MAPK reduce insulin-stimulated glucose uptake in 3T3-L1 adipocytes without affecting GLUT4 translocation. *J. Biol. Chem.* **277,** 50386–50395.

Song, X. M., Hresko, R. C., and Mueckler, M. (2008). Identification of amino acid residues within the carboxyl terminus of the Glut4 glucose transporter that are essential for insulin-stimulated redistribution to the plasma membrane. *J. Biol. Chem.* **283,** 12571–12585.

Stragier, B., De Bundel, D., Sarre, S., Smolders, I., Vauquelin, G., Dupont, A., Michotte, Y., and Vanderheyden, P. (2008). Involvement of insulin-regulated aminopeptidase in the effects of the renin-angiotensin fragment angiotensin IV: A review. *Heart Fail. Rev.* **13,** 321–337.

Sudhof, T. C. (2004). The synaptic vesicle cycle. *Annu. Rev. Neurosci.* **27,** 509–547.

Suzuki, K., and Kono, T. (1980). Evidence that insulin causes translocation of glucose transport activity to the plasma membrane from an intracellular storage site. *Proc. Natl. Acad. Sci. USA* **77,** 2542–2545.

Takata, K. (2006). Aquaporin-2 (AQP2): Its intracellular compartment and trafficking. *Cell Mol. Biol. (Noisy-le-grand)* **52,** 34–39.

Takatsu, H., Katoh, Y., Shiba, Y., and Nakayama, K. (2001). Golgi-localizing, γ-adaptin ear homology domain, ADP-ribosylation factor-binding (GGA) proteins interact with acidic dileucine sequences within the cytoplasmic domains of sorting receptors through their Vps27p/Hrs/STAM (VHS) domains. *J. Biol. Chem.* **276,** 28541–28545.

Taniguchi, C. M., Emanuelli, B., and Kahn, C. R. (2006). Critical nodes in signalling pathways: Insights into insulin action. *Nat. Rev. Mol. Cell. Biol.* **7,** 85–96.

Taylor, E. B., An, D., Kramer, H. F., Yu, H., Fujii, N. L., Roeckl, K. S., Bowles, N., Hirshman, M. F., Xie, J., Feener, E. P., and Goodyear, L. J. (2008). Discovery of TBC1D1 as an insulin-, AICAR-, and contraction-stimulated signaling nexus in mouse skeletal muscle. *J. Biol. Chem* **283,** 9787–9796.

Tettamanzi, M. C., Yu, C., Bogan, J. S., and Hodsdon, M. E. (2006). Solution structure and backbone dynamics of an N-terminal ubiquitin-like domain in the GLUT4-regulating protein, TUG. *Protein Sci.* **15,** 498–508.

Thong, F. S., Bilan, P. J., and Klip, A. (2007). The Rab GTPase-activating protein AS160 integrates Akt, protein kinase C, and AMP-activated protein kinase signals regulating GLUT4 traffic. *Diabetes* **56,** 414–423.

Tojo, H., Kaieda, I., Hattori, H., Katayama, N., Yoshimura, K., Kakimoto, S., Fujisawa, Y., Presman, E., Brooks, C. C., and Pilch, P. F. (2003). The Formin family protein, formin homolog overexpressed in spleen, interacts with the insulin-responsive aminopeptidase and profilin IIa. *Mol. Endocrinol.* **17,** 1216–1229.

Tsuda, M., Davis, I. J., Argani, P., Shukla, N., McGill, G. G., Nagai, M., Saito, T., Lae, M., Fisher, D. E., and Ladanyi, M. (2007). TFE3 fusions activate MET signaling by transcriptional up-regulation, defining another class of tumors as candidates for therapeutic MET inhibition. *Cancer Res.* **67,** 919–929.

Uldry, M., Steiner, P., Zurich, M. G., Beguin, P., Hirling, H., Dolci, W., and Thorens, B. (2004). Regulated exocytosis of an H+/myo-inositol symporter at synapses and growth cones. *EMBO J.* **23,** 531–540.

van Balkom, B. W., Savelkoul, P. J., Markovich, D., Hofman, E., Nielsen, S., van der Sluijs, P., and Deen, P. M. (2002). The role of putative phosphorylation sites in the targeting and shuttling of the aquaporin-2 water channel. *J. Biol. Chem.* **277,** 41473–41479.

van Dam, E. M., Govers, R., and James, D. E. (2005). Akt activation is required at a late stage of insulin-induced GLUT4 translocation to the plasma membrane. *Mol. Endocrinol.* **19,** 1067–1077.

van Oort, M. M., van Doorn, J. M., Bonen, A., Glatz, J. F., van der Horst, D. J., Rodenburg, K. W., and Luiken, J. J. (2008). Insulin-induced translocation of CD36 to the plasma membrane is reversible and shows similarity to that of GLUT4. *Biochim. Biophys. Acta* **1781,** 61–71.

Wakabayashi, Y., Kipp, H., and Arias, I. M. (2006). Transporters on demand: Intracellular reservoirs and cycling of bile canalicular ABC transporters. *J. Biol. Chem.* **281,** 27669–27673.

Wallis, M. G., Lankford, M. F., and Keller, S. R. (2007). Vasopressin is a physiological substrate for the insulin-regulated aminopeptidase IRAP. *Am. J. Physiol. Endocrinol. Metab.* **293,** E1092–E1102.

Wang, Q., Song, C., and Li, C. C. (2004). Molecular perspectives on p97-VCP: Progress in understanding its structure and diverse biological functions. *J. Struct. Biol.* **146,** 44–57.

Wang, Y., and Tang, B. L. (2006). SNAREs in neurons—beyond synaptic vesicle exocytosis. *Mol. Membr. Biol.* **23,** 377–384.

Waters, S. B., D'Auria, M., Martin, S. S., Nguyen, C., Kozma, L. M., and Luskey, K. L. (1997). The amino terminus of insulin-responsive aminopeptidase causes Glut4 translocation in 3T3-L1 adipocytes. *J. Biol. Chem.* **272,** 23323–23327.

Watson, R. T., Khan, A. H., Furukawa, M., Hou, J. C., Li, L., Kanzaki, M., Okada, S., Kandror, K. V., and Pessin, J. E. (2004). Entry of newly synthesized GLUT4 into the insulin-responsive storage compartment is GGA dependent. *EMBO J.* **23,** 2059–2070.

Watson, R. T., and Pessin, J. E. (2008). Recycling of IRAP from the plasma membrane back to the insulin-responsive compartment requires the Q-SNARE syntaxin 6 but not the GGA clathrin adaptors. *J. Cell Sci.* **121,** 1243–1251.

Wendler, F., and Tooze, S. (2001). Syntaxin 6: The promiscuous behaviour of a SNARE protein. *Traffic* **2,** 606–611.

Wente, W., Efanov, A. M., Treinies, I., Zitzer, H., Gromada, J., Richter, D., and Kreienkamp, H. J. (2005). The PDZ/coiled-coil domain containing protein PIST modulates insulin secretion in MIN6 insulinoma cells by interacting with somatostatin receptor subtype 5. *FEBS Lett.* **579,** 6305–6310.

Williams, D., Hicks, S. W., Machamer, C. E., and Pessin, J. E. (2006). Golgin-160 is required for the Golgi membrane sorting of the insulin-responsive glucose transporter GLUT4 in adipocytes. *Mol. Biol. Cell* **17,** 5346–5355.

Williams, D., and Pessin, J. E. (2008). Mapping of R-SNARE function at distinct intracellular GLUT4 trafficking steps in adipocytes. *J. Cell Biol.* **180,** 375–387.

Xu, Z., and Kandror, K. V. (2002). Translocation of small preformed vesicles is responsible for the insulin activation of glucose transport in adipose cells. Evidence from the *in vitro* reconstitution assay. *J. Biol. Chem.* **277,** 47972–47975.

Yamada, E., Okada, S., Saito, T., Ohshima, K., Sato, M., Tsuchiya, T., Uehara, Y., Shimizu, H., and Mori, M. (2005). Akt2 phosphorylates Synip to regulate docking and fusion of GLUT4-containing vesicles. *J. Cell Biol.* **168,** 921–928.

Yang, J., Clark, A. E., Kozka, I. J., Cushman, S. W., and Holman, G. D. (1992). Development of an intracellular pool of glucose transporters in 3T3-L1 cells. *J. Biol. Chem.* **267,** 10393–10399.

Yang, J., and Holman, G. D. (2005). Insulin and contraction stimulate exocytosis, but increased AMP-activated protein kinase activity resulting from oxidative metabolism stress slows endocytosis of GLUT4 in cardiomyocytes. *J. Biol. Chem.* **280,** 4070–4078.

Yao, R., Maeda, T., Takada, S., and Noda, T. (2001). Identification of a PDZ domain containing Golgi protein, GOPC, as an interaction partner of frizzled. *Biochem. Biophys. Res. Commun.* **286,** 771–778.

Yeh, J. I., Verhey, K. J., and Birnbaum, M. J. (1995). Kinetic analysis of glucose transporter trafficking in fibroblasts and adipocytes. *Biochemistry* **34,** 15523–15531.

Yeh, T. Y., Sbodio, J. I., Tsun, Z. Y., Luo, B., and Chi, N. W. (2007). Insulin-stimulated exocytosis of GLUT4 is enhanced by IRAP and its partner tankyrase. *Biochem. J.* **402,** 279–290.

Yoshizaki, T., Imamura, T., Babendure, J. L., Lu, J. C., Sonoda, N., and Olefsky, J. M. (2007). Myosin 5a is an insulin-stimulated Akt2 (protein kinase Bβ) substrate modulating GLUT4 vesicle translocation. *Mol. Cell. Biol.* **27,** 5172–5183.

Yu, C., Cresswell, J., Loffler, M. G., and Bogan, J. S. (2007). The glucose transporter 4-regulating protein TUG is essential for highly insulin-responsive glucose uptake in 3T3-L1 adipocytes. *J. Biol. Chem.* **282,** 7710–7722.

Yue, Z., Horton, A., Bravin, M., DeJager, P. L., Selimi, F., and Heintz, N. (2002). A novel protein complex linking the delta 2 glutamate receptor and autophagy: Implications for neurodegeneration in lurcher mice. *Neuron* **35,** 921–933.

Zeigerer, A., Lampson, M. A., Karylowski, O., Sabatini, D. D., Adesnik, M., Ren, M., and McGraw, T. E. (2002). GLUT4 retention in adipocytes requires two intracellular insulin-regulated transport steps. *Mol. Biol. Cell* **13,** 2421–2435.

Zeyda, M., and Stulnig, T. M. (2007). Adipose tissue macrophages. *Immunol. Lett.* **112,** 61–67.

Zhou, Q. L., Jiang, Z. Y., Holik, J., Chawla, A., Hagan, G. N., Leszyk, J., and Czech, M. P. (2008). Akt substrate TBC1D1 regulates GLUT1 expression through the mTOR pathway in 3T3-L1 adipocytes. *Biochem. J.* **411,** 647–655.

Zhou, W., Ryan, J. J., and Zhou, H. (2004). Global analyses of sumoylated proteins in *Saccharomyces cerevisiae*. Induction of protein sumoylation by cellular stresses. *J. Biol. Chem.* **279,** 32262–32268.

Zierler, K. (1998). Does insulin-induced increase in the amount of plasma membrane GLUTs quantitatively account for insulin-induced increase in glucose uptake? *Diabetologia* **41,** 724–730.

CHAPTER EIGHT

Compartmentalization and Regulation of Insulin Signaling to GLUT4 by the Cytoskeleton

Craig A. Eyster *and* Ann Louise Olson

Contents

Abstract

One of the early events in the development of Type 2 diabetes appears to be an inhibition of insulin-mediated GLUT4 redistribution to the cell surface in tissues that express GLUT4. Understanding this process, and how it begins to breakdown in the development of insulin resistance is quite important as we face treatment and prevention of metabolic diseases. Over the past few years, and increasing number of laboratories have produced compelling data to demonstrate a role for both the actin and microtubule networks in the regulation of insulin-mediated GLUT4 redistribution to the cell surface. In this review, we

Department of Biochemistry and Molecular Biology, Oklahoma University Health Sciences Center, Oklahoma City, Oklahoma 73126

Vitamins and Hormones, Volume 80
ISSN 0083-6729, DOI: 10.1016/S0083-6729(08)00608-0

explore this process from insulin-signal transduction to fusion of GLUT4 membrane vesicles, focusing on studies that have implicated a role for the cytoskeleton. We see from this body of work that both the actin network and the microtubule cytoskeleton play roles as targets of insulin action and effectors of insulin signaling leading to changes in GLUT4 redistribution to the cell surface and insulin-mediated glucose uptake.

I. Introduction

Over the past several years, many laboratories have been interested in roles that cytoskeletal structures play in the redistribution of GLUT4 to the cell surface. In the course of this exploration, several possible interactions between the GLUT4 vesicles and either the actin cytoskeleton or the microtubule network have been revealed. The lion's share of the work has been carried out in cultured monolayers of 3T3-L1 adipocytes or L6 myotubes. It is not clear how cytoskeletal structures regulate cellular processes when the cells are essentially growing in two dimensions compared to a three-dimensional organization in tissues. With this caveat in mind, we can use the mechanistic information obtained from cultured cells as a guide to understand what role the cytoskeleton may play in the regulation of GLUT4 protein in insulin resistance *in vivo*.

It is clear that insulin exerts a major regulatory influence on glucose homeostasis by signaling the redistribution of an internal vesicular compartment containing the GLUT4 facilitative glucose transporter to the plasma membrane (Bryant *et al.*, 2002). Redistribution of GLUT4 increases, by about fivefold, the number of GLUT4 transporters at the plasma membrane, thereby increasing glucose transport into the cell (Yang *et al.*, 1992). Defects in GLUT4 redistribution are thought to contribute to the development of insulin resistance and Type 2 diabetes (Rothman *et al.*, 1995). Furthermore, insulin-resistant glucose transport may contribute to the pathogenesis of metabolic syndrome. The observation supporting this hypothesis is that lean subjects appear to divert dietary glucose away from muscle glycogen synthesis toward hepatic lipogenesis (Petersen *et al.*, 2007). In turn, increased hepatic lipogenesis lead to increased plasma triacylglycerol levels and decreased plasma high-density lipoprotein particles, a lipid profile consistent with atherogenic dyslipidemia of the metabolic syndrome. With these clinical observations in mind, it seems likely that disruption of optimal GLUT4 responses may play a role in development of metabolic diseases. Fully understanding the insulin-signaling and GLUT4 redistribution mechanisms is essential for treating, and, hopefully, preventing the metabolic syndrome and Type 2 diabetes. In this review, we will examine several areas of insulin signaling and GLUT4 regulation with respect to roles that the cytoskeleton plays in this process.

Numerous laboratories have implicated both the actin and the microtubule cytoskeletal networks in insulin-dependent GLUT4 redistribution. Studies examining the role of the actin cytoskeleton in this process in L6 myotubes, 3T3-L1 adipocytes, and rat adipocytes have shown that disruption of the actin cytoskeleton by either cytochalasin D or latrunculin A inhibits insulin-mediated GLUT4 redistribution (Omata *et al.*, 2000; Tsakiridis *et al.*, 1994; Wang *et al.*, 1998). The actin network may function to transduce the insulin signal to GLUT4 vesicles (Khayat *et al.*, 2000; Tsakiridis *et al.*, 1995), or to move GLUT4-containing vesicles in the cortical region of the cell (Bose *et al.*, 2002). The cortical actin network undergoes rearrangement when 3T3-L1 adipocytes are treated with insulin, showing that this structure is a target of insulin action (Kanzaki and Pessin, 2001). Exactly how the actin network is signaled by insulin stimulation for rearrangement has remained a mystery. The small GTPases TC10 and Cdc42, which both interact with the actin network and actin-binding proteins, may be activated by insulin signaling (Kanzaki *et al.*, 2002; Usui *et al.*, 2003). It is possible that these Rho kinase members, or others, transmit the insulin signal to filamentous actin. From these studies, we see that several possible mechanisms have been revealed; however, no single mechanism appears to account for all of the specific roles that the cellular cytoskeleton seems to play in insulin action.

II. Insulin Signaling to GLUT4 Vesicles

A. Insulin signaling through PI3-kinase

Insulin signaling is initiated by the peptide hormone insulin binding to the $\alpha 2$–$\beta 2$-heterotetrameric insulin receptor located in adipose and muscle cell plasma membrane (for review, see Cheatham and Kahn, 1995). Insulin receptor activation causes asymmetric autophosphorylation whereby one β-subunit phosphorylates the other β-subunit in the same receptor leading to the activation of tyrosine kinase activity in the cytoplasmic domain (Cobb *et al.*, 1989; Frattali *et al.*, 1992; Lee *et al.*, 1993; Shoelson *et al.*, 1991). The autophosphorylation cascade leads to multiple phosphorylated tyrosine residues within the cytoplasmic tail of the receptor. These activated tyrosine phosphorylated residues can signal downstream events.

The activated insulin receptor with a full complement of phosphorylated tyrosines can act as a docking site for insulin receptor substrates proteins (IRS 1/2) (White, 1998). These adapter proteins recognize phosphorylated tyrosine residues on the receptor leading to a stable associating between the receptor and the IRS proteins. Stable association leads to tyrosine phosphorylation of the IRS proteins by the insulin receptor β-subunit (Chang *et al.*, 2004). The importance of the IRS2 isoform in insulin-mediated GLUT4 redistribution has been demonstrated by the specific knockout in mice, which develop severe insulin resistance (Withers *et al.*, 1998). IRS1 knockout mice were growth retarded but an increase in insulin secretion prevented development of diabetes (Withers *et al.*, 1998). In addition, biochemical evidence for isolated brown adipocytes has supported the essential role for IRS2 in insulin stimulated GLUT4 redistribution (Fasshauer *et al.*, 2000). While the IRS proteins contain no intrinsic enzyme activity, they are essential scaffolding molecules that bridge insulin receptor activity to downstream targets (Myers *et al.*, 1994).

The tyrosine phosphorylation sites on IRS 1/2 can act as docking site for recruitment of phosphatidylinositol 3-kinase (PI3-kinase) to the cell surface (Backer *et al.*, 1992). The importance of activation of PI3-kinase in insulin stimulated GLUT4 redistribution has been firmly established (Saltiel and Kahn, 2001). Relocation of PI3K to the cell surface allows the enzyme to act on lipid substrates, in particular phosphatidylinositol 4,5-bisphosphate, in the plasma membrane causing the generation of phosphatidylinositol 3,4,5-trisphosphate. Generation of PIP3 patches in the plasma membrane can act as docking sites for pleckstrin homology (PH) domain-containing proteins including protein kinase B also known as Akt (Cheatham and Kahn, 1995).

Akt/PKB activation is crucial for insulin signaling to GLUT4 redistribution (Welsh *et al.*, 2005). Akt is present in three isoforms with Akt1 contributing largely to growth and Akt2 contributing to insulin-mediated GLUT4 redistribution (Zhou *et al.*, 2004). A constitutively active Akt (myr-Akt) redistributes GLUT4 to the plasma membrane in the absence of insulin and the Akt2 knockout mouse develops insulin resistance and Type 2 diabetes (Cho *et al.*, 2001; Kohn *et al.*, 1996). To become activated Akt is recruited to the plasma membrane by its N-terminal PH domain.

Akt activation upon membrane recruitment is dependent on two phosphorylation events: one on threonine 308 and one on serine 473. Both are necessary for maximal Akt enzyme activity (Czech and Corvera, 1999). Threonine 308 phosphorylation is mediated by phosphoinositide-dependent protein kinase-1 which is also recruited to the plasma membrane by interaction with PIP3 (Yamada *et al.*, 2002). The identification or discovery of the serine 473 kinase, also known as PDK2 or the hydrophobic motif kinase, has been an extremely controversial area with many proteins identified as harboring this activity (Dong and Liu, 2005). Evidence has emerged that the mTOR–RICTOR complex mediates serine 473 phosphorylation *in vivo* (Sarbassov *et al.*, 2005). This result has been confirmed in 3T3-L1 adipocytes, a frequently used model for insulin-mediated GLUT4 redistribution (Hresko and Mueckler, 2005). A possible mechanism for activation of mTOR–RICTOR by insulin signaling has not yet been identified.

However, PDK2 activity, in 3T3-L1 adipocytes, has been shown to be associated with the cytoskeleton (Hresko *et al.*, 2003). It has also been shown that serine 473 phosphorylation likely precedes threonine 308 phosphorylation, and is important for maximal activation by PDK1 (Sarbassov *et al.*, 2005; Scheid *et al.*, 2002).

Recently, AS160 (Akt substrate of 160kDa) has been identified as a downstream target of insulin stimulated Akt activity (Kane *et al.*, 2002). Sequence analysis suggested that AS160 has a predicted Rab GTPase-activating protein (GAP) domain. Rabs are known to be importantly involved in vesicle movement and fusion (Zerial and McBride, 2001). AS160 is phosphorylated by Akt on multiple residues and these phosphorylations have been hypothesized to lead to deactivation of the Gap function of the protein. Indeed, AS160 has been shown to be required for insulin-mediated GLUT4 redistribution and when Akt phosphorylation sites on AS160 are mutated, insulin is no longer able to signal GLUT4 redistribution (Sano *et al.*, 2003). Mutations of the GAP domain of AS160 suggest a functional requirement for GAP activity for GLUT4 redistribution (Sano *et al.*, 2003). The search for the downstream Rab or other effector modified by AS160 phosphorylation is ongoing.

It has been difficult to identify the downstream Rab(s) involved in GLUT4 redistribution and those regulated by AS160. Rabs 10, 11, and 14 have been shown to be associated with GLUT4 vesicles and represent possible mediators of GLUT4 trafficking (Larance *et al.*, 2005). Rabs 8A and 14 have been shown to be targets of AS160 in muscle cells and when overexpressed can functionally overcome the block of the constitutively active $AS160_{4P}$ mutant that prevents GLUT4 redistribution (Ishikura *et al.*, 2007). Despite these advances, it still remains unclear whether a specific Rab is involved in insulin stimulated GLUT4 redistribution or if multiple Rabs can or do function in the pathway.

The recently identified AS160 related protein Tbc1d1 has supported the case for inhibition of GAP activity as a functional requirement for GLUT4 redistribution. Tbc1d1 belongs to the same protein family as AS160 and also contains a Rab-Gap domain (Roach *et al.*, 2007). Tbc1d1 blocks GLUT4 redistribution like AS160 when overexpressed and a GAP domain mutant had no effect (Roach *et al.*, 2007). Most strikingly, a mutant of Tbc1d1 has been implicated in increased obesity susceptibility in humans (Stone *et al.*, 2006). This provides an important physiological connection between alleviation of Rab-GAP activity and proper insulin signaling to GLUT4 redistribution that may represent a common mechanism for GLUT4 redistribution. While the exact Rab(s) important for GLUT4 redistribution remains controversial, it seems clear that the mechanism of GAP inactivation by Akt phosphorylation is likely a key component of GLUT4 redistribution.

14-3-3 proteins have also been suggested to be involved in GLUT4 redistribution (Ramm *et al.*, 2006). These proteins are thought to interact with an immense number of cellular processes through their ability to interact with specific discrete phosphoserine and phosphothreonine motifs (Bridges and Moorhead, 2005). Several 14-3-3 isoforms were identified in a search for proteins that interact with AS160 (Ramm *et al.*, 2006). Importantly, the 14-3-3 interaction site with AS160 is suggested to be at threonine 642, a residue phosphorylated by Akt and this correlates with the effects of the $AS160_{T642}$ and $AS160_{4P}$ mutants (Ramm *et al.*, 2006; Sano *et al.*, 2003). Interestingly, a physiologic study of the effects of exercise on insulin sensitivity suggested a partial role for the interaction between AS160 and 14-3-3 proteins during acute exercise (Howlett *et al.*, 2007). 14-3-3 proteins represent another interesting possible family of proteins that may interact in insulin signaling to GLUT4 redistribution downstream of Akt/PKB activation.

While there remain missing links between insulin signaling and GLUT4 redistribution, it seems clear that recent advances suggest connections between insulin signaling through AS160 to components such as Rabs, 14-3-3, and possibly SNARE complex proteins that are likely associated with GLUT4-containing vesicles or fusion sites. This importantly suggests that the interaction between insulin signaling and GLUT4 trafficking could be occurring at the level of docking and or fusion of the vesicles. Also, there has not yet been a firm connection drawn between the central insulin-signaling pathway through PI3-kinase and Akt/PKB and a GLUT4 trafficking step upstream of docking and or fusion.

The cellular cytoskeleton, particularly actin, appears to intersect with the Insulin signaling at several points. Specifically, phosphoinositide metabolism is tied to f-actin assembly at the cell cortex through PIP 4,5-bisphosphate in cellular models of insulin-responsive tissues (Chen *et al.*, 2004; McCarthy *et al.*, 2006). In these model systems, prolonged incubation in the presence of insulin leads to decreased membrane-associated PIP 4,5-bisphosphate, decreased cortical actin and insulin resistance. Interestingly, cortical actin in skeletal muscle of insulin-resistant mouse models is diminished consistent with a role of cortical actin in insulin resistance. Several questions remain to be answered, including how PIP 4,5-bisphosphate is metabolized in insulin-resistant tissues.

It is likely that signal-dependent changes in actin dynamics as described above can feed forward to modify insulin-signaling pathways. Recent work from our laboratory has suggested that the actin network requirement is upstream of Akt activation (Eyster *et al.*, 2005). In this work, we found that disruption of the actin network did not effect receptor activation or phosphorylation of IRS1. Rather, PI3-kinase activity was inhibited suggesting that actin is organizing the insulin-signaling complex at the level of the plasma membrane. Our results with the actin network are consistent with the model put forth by Klip and coworkers whereby insulin-mediated actin reorganization in L6 myotubes coordinates production of PIP3 and activation of Akt/PKB (Patel *et al.*, 2003). The activation of PDK2 activity is another important step in activation that may require the actin cytoskeleton. Mueckler and coworkers have partially purified a cellular fraction with PDK2 activity associated with the membrane cytoskeleton in 3T3-L1 adipocytes (Hresko *et al.*, 2003). Reorganization of the actin network following insulin treatment may consist of the simple formation of a specialized scaffold for assembling molecules involved in the signaling complex (Patel *et al.*, 2003).

To understand how actin might regulate insulin signaling requires understanding molecular details of points of intersection between the insulin-signaling complex, the actin network, and GLUT4 regulation. The unconventional myosin actin motor protein Myo 1C has been suggested to be important for GLUT4 redistribution (Bose *et al.*, 2002, 2004). These studies show that Myo 1C plays a role in GLUT4 vesicle membrane fusion and that overexpression of this motor protein will permit membrane fusion in the presence of PI3-kinase inhibitors, suggesting that Myo 1C is part of the PI3-kinase pathway (Bose *et al.*, 2004). The molecular details of Myo 1C function have remained unclear; however, it is possible that this motor protein could play an important role in the activation of PI3-kinase, Akt/PKB, and the remaining signaling cascade leading to GLUT4 vesicle fusion. It has also been suggested that Akt/PKB associates with filamentous actin and that interaction was enhanced by insulin stimulation (Cenni *et al.*, 2003). The connection between Myo 1C, Akt, and actin provides a possible actin based regulatory system for insulin-mediated GLUT4 redistribution and is consistent with our results demonstrating that actin function is necessary to activate Akt/PKB (Eyster *et al.*, 2005).

B. PI3-kinase-independent signaling pathways

Some evidence suggests that insulin-signaling pathway through PI3-kinase activation and Akt2 may not be the only pathway activated by insulin that can affect GLUT4 redistribution (for recent review, see Chang *et al.*, 2004). These studies have suggested non-PI3-kinase-dependent pathways for activation of the small GTPase TC10 and also the activation of atypical protein kinase C (aPKC) that may be important in insulin stimulated GLUT4 redistribution.

TC10 activation by insulin occurs through activation of Cbl which is associated with the insulin receptor through an interaction with the adapter protein CAP that can bind the tyrosine phosphorylated insulin receptor (Chiang *et al.*, 2001). Insulin receptor tyrosine phosphorylation of Cbl activates the protein causing it to recruit CrkII–C3G complex that can activate TC10 (Chiang *et al.*, 2001). The activation of TC10 has been lined to a variety of downstream effectors and attempts have been made to tie the pathway into GLUT4 trafficking. These include activation of actin comet propulsion of GLUT4-containing vesicles (Kanzaki *et al.*, 2001), activation of cortical actin remodeling through N-WASP (Kanzaki *et al.*, 2002), and most recently activation of the exocyst complex (Inoue *et al.*, 2003), which is important for vesicle delivery in a variety of systems. None of these mechanisms have been substantiated or confirmed and several lines of evidence argue against absolute requirements for this pathway. Constitutively active Akt can signal redistribution of GLUT4 in the absence of insulin, suggesting that the TC10 pathway is either not obligatory or feeds into signaling events upstream of Akt activity (Kohn *et al.*, 1996). Other labs have suggested that RNAi knockdown of CAP, Cbl, and Crk have no effect on insulin-mediated GLUT4 redistribution (Mitra *et al.*, 2004). Evidence for the necessity of this pathway is also not supported by work in transgenic mice lacking components of the pathway (Minami *et al.*, 2003). Taken together the evidence for any involvement of the Cbl/Crk/TC10 and how exactly its activation can impinge on GLUT4 redistribution is still yet to be fully determined.

aPKC has also been suggested to be a downstream target of activation of PI3-kinase activity (Ishiki and Klip, 2005). Two isoforms of aPKC have been suggested to activated by insulin, λ and ζ, and be functionally important for GLUT4 redistribution to the plasma membrane (Bandyopadhyay *et al.*, 1999). How exactly the activation of these aPKCs interacts with the GLUT4 trafficking pathway is not clear. Proposed connections of aPKC pathways in GLUT4 trafficking or the cytoskeleton have not been thoroughly studied (Imamura *et al.*, 2003). While these alternate insulin-signaling pathways continue to be heavily investigated, their requirement for GLUT4 redistribution is yet to be substantiated *in vivo*.

III. GLUT4 Vesicle Membrane Trafficking

A. GLUT4 storage compartment

The pool of GLUT4 vesicles that responds to insulin is biochemically distinct from GLUT4 vesicle populations that redistribute to the plasma membrane under other stimuli such as GTPγS (Millar *et al.*, 2000) or exercise (Kandror *et al.*, 1995; Lund *et al.*, 1995). GLUT4 vesicles were first identified as part of the endosomal compartment by immunolocalization and colocalization with endosomal markers such as the transferrin receptor (Livingstone *et al.*, 1996; Martin *et al.*, 1996; Slot *et al.*, 1991a,b). In adipocytes, the GLUT4 storage compartment forms early in differentiation, even before GLUT4 is synthesized to significant levels (El-Jack *et al.*, 1999). Newly synthesized GLUT4 is targeted directly to the plasma membrane (Al-Hasani *et al.*, 1998) from which it gains entry into the early endosome. The insulin-responsive pool of GLUT4 vesicles forms as a specialized postendocytic compartment that is separate from the recycling endosome pool (Kupriyanova *et al.*, 2002; Lampson *et al.*, 2001; Martin *et al.*, 1996; Zeigerer *et al.*, 2002). GLUT4 storage vesicles (GSVs) are small, approximately 50nm in diameter, and contain few proteins in addition to GLUT4 (Bryant *et al.*, 2002). Insulin-responsive aminopeptidase, GLUT4, and the v-SNARE Vamp-2 are major components of the specialized insulin-responsive pool of GSVs (Hashiramoto and James, 2000). In addition, Rabs 10, 11, and 14 have been suggested to be associated with GSVs along with AS160 in the basal state (Larance *et al.*, 2005). The lack of TR and other recycling endocytic components has established the specific nature of GSVs as a specialized postendocytic compartment that has a higher insulin responsiveness than regular recycling endosomes that can also contain GLUT4 (Hashiramoto and James, 2000).

B. GLUT4 endocytosis and exocytosis

Although the connection between the insulin-signaling pathway and GLUT4 redistribution is unknown, a large amount of work has been done to determine the itinerary of GLUT4 as another approach to determine insulin regulated steps (Bryant *et al.*, 2002). GLUT4 is initially targeted to the cell surface after synthesis where it is recycled through early endosomes into recycling endosomes. In the recycling endosome, GLUT4 is sorted away from typical recycling receptors such as the TR into GSVs. GLUT4 is known to slowly appear at the cell surface in the basal state and that its appearance is increased approximately fivefold in the presence of insulin. Kinetic studies demonstrate that the rate of GLUT4 exocytosis is accelerated by insulin treatment (Czech and Buxton, 1993; Holman *et al.*, 1994; Robinson *et al.*, 1992; Yang and Holman, 1993). Increased exocytosis may result from increased fission of GLUT4 vesicles from an endosomal precursor, stimulation of the rate of movement of vesicles to the plasma membrane, increased activation of docking and/or fusion machinery at the plasma membrane, or a combination of two or more of these processes. There have been many steps in the GLUT4 itinerary that have been hypothesized to be insulin dependent but none of the processes or the molecular mechanisms have been firmly identified as targets of insulin signaling. Current thinking in the field has suggested that insulin may regulate docking steps of GLUT4 vesicles with the plasma membrane, regulation of fusion machinery such as SNAREs, or trafficking of GLUT4 to the cell surface/plasma membrane (Bryant *et al.*, 2002; James, 2005).

The translocation of GSVs from their perinuclear location to the cell surface for fusion was initially hypothesized from work following glucose transport activity in isolated adipocytes (Cushman and Wardzala, 1980; Horuk *et al.*, 1983). These initial studies led to the development of the translocation hypothesis that gained favor in the glucose transporter field for many years. The translocation hypothesis suggested that insulin stimulation would increase the trafficking of GSVs to the cell surface. Studies attempting to link the insulin-signaling cascade to some aspect of this suggested trafficking event have been numerous including aspects of the cytoskeleton or release from a molecular tether (Bogan *et al.*, 2003; Semiz *et al.*, 2003). However, work has also suggested that the GSVs are highly mobile in the absence of insulin and that docking/fusion may be the important regulated steps (Bai *et al.*, 2007; Eyster *et al.*, 2006; Huang *et al.*, 2007; Lizunov *et al.*, 2005). This important shift in understanding that the docking and fusion steps may be the key inputs of insulin signaling may lead to a firm interaction between insulin signaling and GLUT4 trafficking pathways that has been lacking.

C. Dynamic retention versus static retention

Multiple models have arisen suggesting how the insulin-responsive compartment is maintained and how it cycles to the cells surface. A dynamic retention model has been proposed that suggest that GSVs continuously cycle to the plasma membrane (Karylowski *et al.*, 2004; Zeigerer *et al.*, 2002). This model has been supported in other systems where the entire pool of GLUT4 visits the plasma membrane in a relatively short period of time supporting the dynamic retention model (Foster *et al.*, 2001). In addition, this model has been supported by recent TIRFM studies demonstrating constant, rapid movement of GSVs close to the cell surface in both basal and insulin stimulated states (Bai *et al.*, 2007; Huang *et al.*, 2007; Lizunov *et al.*, 2005).

Other models had suggested that GLUT4 retention occurred as a result of tethering in the perinuclear compartment or some other form of static retention that would not allow GLUT4 access to the plasma membrane and thereby, keep GLUT4 sequestered in the absence of stimulus. This model suggests that increasing amounts of insulin would increase the GLUT4 pool available for fusion (Coster *et al.*, 2004). This model was also supported by the suggestion of a tethering protein, TUG, which was suggested to control release of GSVs from their perinuclear storage pool (Bogan *et al.*, 2003). Interestingly, microtubules are required for perinuclear organization of GSV (see SectionIII.D) and it is possible that TUG may play a role in tethering GLUT4 vesicles to microtubules. Recent evidence continues to support a role for TUG in GLUT4 redistribution through a variety of knockdown experiments and evidence that TUG can bind directly to GLUT4 (Yu *et al.*, 2007). While the dynamic retention model has gained much support from recent advance in microscopic visualization of mobility of GLUT4-containing vesicles, reconciling the findings that support these two models of GLUT4 retention will be helpful in establishing the control of insulin-responsive GLUT4 redistribution.

Another important regulated step in GLUT4 trafficking is GLUT4 sorting into GSVs from the recycling endosomal system. Identification of the adaptor protein GGA3 as a major player in this process has been an important step forward in our understanding of GLUT4 sorting (Li and Kandror, 2005; Watson *et al.*, 2004). This mechanism has also been suggested for IRAP, another resident member of GSVs, and its delivery into the insulin-responsive GLUT4 storage compartment (Hou *et al.*, 2006). This important identification of an adaptor protein involved in GLUT4 trafficking may open up addition lines of investigation for understanding GLUT4 trafficking through the endosomal system.

GLUT4 endocytosis is another key event regulating insulin-dependent glucose uptake. GLUT4 has been suggested to endocytosed by both clathrin-dependent and caveolae-dependent mechanisms (Cohen *et al.*, 2003; Volchuk *et al.*, 1998). Insulin has been suggested to inhibit endocytosis of GLUT4 those prolonging the exposure on the cell surface (Huang *et al.*, 2001). Recent work has suggested that GLUT4 is endocytosed by two separate pathways, one involving a cholesterol-dependent pathway and one involving AP-2 and clathrin (Blot and McGraw, 2006). Under basal conditions, the majority GLUT4 is rapidly endocytosed through the cholesterol-dependent pathway (Blot and McGraw, 2006). Under insulin stimulation, the cholesterol-dependent pathway is blocked causing all GLUT4 to be internalized slowly through the AP-2/clathrin pathway (Blot and McGraw, 2006). These studies continue to define how insulin is able to cause a change in GLUT4 endocytosis, prolonging the exposure of GLUT4 on the cell surface.

D. GLUT4 trafficking and microtubules

Several groups, using a variety of pharmacological agents, have shown that disruption of the microtubule network in different ways inhibits insulin-mediated GLUT4 redistribution in 3T3-L1 adipocytes (Emoto *et al.*, 2001; Fletcher *et al.*, 2000; Guilherme *et al.*, 2000; Olson *et al.*, 2001; Patki *et al.*, 2001) and primary rat adipocytes (Liu *et al.*, 2003). Depolymerization of the microtubule network causes a redistribution of the perinuclear intracellular GLUT4 storage compartment and probably the *trans*-Golgi network in 3T3-L1 adipocytes (Guilherme *et al.*, 2000; Molero *et al.*, 2001). Experiments using a dominant-interfering light chain mutant of conventional kinesin (KIF5B) to disrupt insulin-dependent GLUT4 redistribution, and live cell microscopy to observe tracking of GLUT4 vesicles along fluorescently labeled microtubules, strongly support involvement of microtubules in insulin-mediated GLUT4 vesicle movement (Semiz *et al.*, 2003). These data suggest that GLUT4 vesicles may use microtubules as a platform for movement.

In contrast to the work described by Semiz *et al.* (2003), we used fluorescence recovery after photobleaching (FRAP) to test how mobility of a GFP-tagged GLUT4 protein changes in a live cell under insulin action and without or with intact cytoskeletal structures (Eyster *et al.*, 2006). While insulin treatment did not change GLUT4 mobility within the cell, we did find that mobility of GLUT4 required an intact microtubule network, but not an intact actin network (Eyster *et al.*, 2006). Using TIRF microscopy, Lizunov *et al.* (2005) demonstrated that the basal movement of GLUT4 near the plasma membrane occurred along microtubules, and that chemical disruption of the microtubules inhibited movement. The actual fusion of GLUT4 vesicles with plasma membrane is not microtubule dependent since a constitutively active Akt mutant supports GLUT4 redistribution to the cell surface in the absence of an intact or a dynamic microtubule network (Eyster *et al.*, 2005).

E. Intersection of microtubules and actin in GLUT4 regulation

In addition to possibly serving as a platform for GLUT4 vesicle movement, data from our laboratory suggest that microtubule dynamics are required for the regulation of insulin-mediated GLUT4 redistribution (Olson *et al.*, 2003). Treatment with either low doses (3.3μM) or high doses (33μM) of nocodazole causes a similar decrease in GLUT4 redistribution to the plasma membrane in 3T3-L1 adipocytes (Eyster *et al.*, 2006). At low doses of nocodazole only the dynamic properties of microtubules should be affected and this has been substantiated by immunohistochemical staining of microtubules in the presence of various nocodazole concentrations (Olson *et al.*, 2003). These data suggest that microtubule dynamics are also relevant to insulin-mediated GLUT4 redistribution.

Data from our laboratory also suggest that insulin regulates the equilibrium between dimeric and polymeric tubulin (Olson *et al.*, 2003). Treatment of cells with insulin causes an increase in antibody staining for tubulin and an increase in the amount of polymerized tubulin determined by differential detergent extraction. Whether this increase in polymerized tubulin is due to an increased rate of polymerization, stabilization of microtubules, or a combination of both is still unknown. There is no increase in γ-tubulin staining at the microtubule organizing center suggesting that the increase in polymerized tubulin is not due the creation of new microtubules (Olson *et al.*, 2003). This action of insulin on microtubules is dependent upon the actin network (Olson *et al.*, 2003). Treatment with pharmacological inhibitor latrunculin B, which depolymerizes the actin network, causes the insulin-stimulated increase in polymerized tubulin to be inhibited (Olson *et al.*, 2003). The function of actin needed for the increase in polymerized tubulin is yet to be determined. One possibility consistent with our data is that actin is required to stabilize microtubules at the cell periphery. A second possibility is that long-range movements of GLUT4 vesicles within the cell depend on an interaction between microtubules and actin filaments.

Indirect support for such a model comes from the observation that myosin 5a, an actin based motor protein, is an Akt2 substrate and that phosphorylation of myosin 5a mediates an association of this motor protein with actin filaments and GLUT4 vesicles (Yoshizaki *et al.*, 2007). While myosin 5a increases in abundance at the plasma membrane in response to insulin, it has not been shown that myosin 5a or actin fibers actually deliver GLUT4 vesicles to the plasma membrane.

IV. GLUT4 Vesicle Fusion

A. Snare complex assembly

SNARE proteins have been established as the necessary final steps in fusion of vesicles with target membranes (for recent reviews, see Bonifacino and Glick, 2004; Jahn and Scheller, 2006). Every SNARE-mediated fusion event requires the formation of a four helix bundle, one from the vesicle-associated SNARE (v-SNARE) and three from the two or three target membrane SNAREs (t-SNAREs). It is thought that the energy released from the four-helix bundle formation between the SNAREs provides the free energy necessary to overcome the energy barrier necessary to fuse the opposing phospholipid bilayers. Previous work has suggested that the SNARE complex necessary to fuse the GSVs involves the t-SNAREs syntaxin-4 and SNAP-23 and VAMP-2 as the v-SNARE (Omata *et al.*, 2000). Additional studies have continued to support this subset of SNARE proteins as the important mediators of fusion of GSVs with the plasma membrane (Macaulay *et al.*, 1997; Randhawa *et al.*, 2000).

SNARE-mediated fusion events may represent an important control point that may be downstream of insulin signaling. Indeed, this type of control is now well established for synaptic vesicle release in neuronal cells, the classic model for SNARE-mediated fusion (Sudhof, 2004). Consequently, there has been a large interest in fully understanding SNARE-mediated fusion events in insulin-mediated GLUT4 redistribution and the SNARE-binding proteins that are thought to regulate their intrinsic fusion capacity.

Several SNARE-binding proteins have been implicated in GLUT4 redistribution. Synip is one such SNARE-binding protein (Min *et al.*, 1999; Okada *et al.*, 2007). Synip has been suggested to be phosphorylated by Akt2 and this phosphorylation event has been suggested to be important for disassociation from syntaxin-4 thereby allowing it to be free to interact in success fusion competent complexes or with the Vamp-2 on arriving GLUT4 vesicles (Yamada *et al.*, 2005). However, this phosphorylation site on Synip has also been suggested to not essential for GLUT4 redistribution as well (Sano *et al.*, 2005). Therefore, the role of Synip in GLUT4 redistribution remains controversial. Synip remains an intriguing possible connection between insulin signaling, particularly Akt2 activity, and GLUT4 redistribution but further studies will be necessary to confirm this interaction between the pathways.

A SNARE-binding protein of the Sec1p/Munc18 (SM) family, Munc18c, has been suggested to be involved in GLUT4 redistribution (for recent review, see James, 2005). Munc proteins have been suggested to control vesicle fusion via their ability to interact with t-SNAREs, like syntaxin-4, and keep them segregated from fusion competent complexes involving other t-SNAREs, such as SNAP-23 (Jahn and Sudhof, 1999). Munc18c was initially shown to be critical in insulin-mediated GLUT4 redistribution in a variety of cell biological and biochemical assays both in 3T3-L1 adipocytes and in L6 myotubes. Unlike other recently described proteins suggested to be involved in insulin-mediated GLUT4 redistribution, Munc18c heterozygous knockout mice have been developed and show increased susceptibility for insulin resistance when placed on a high-fat diet (Oh *et al.*, 2005). Munc18c has also been suggested to interact with aPKC-ζ, once again attempting to link a kinase activated by insulin signaling to a SNARE effector (Hodgkinson *et al.*, 2005). The evidence for a physiological role of Munc18c in insulin stimulated GLUT4 redistribution is more compelling than that for Synip, yet the molecular mechanism tying insulin signaling to Munc18c interaction with SNARE complex proteins and the importance of this interaction remains to be fully elucidated.

Additional SNARE interacting proteins have been suggested to be involved in insulin-mediated GLUT4 redistribution. Tomosyn, a cytosolic syntaxin-binding protein, is thought to regulate interactions between syntaxins and Muncs (Gerst, 2003). This interaction was suggested to have importance in insulin-mediated GLUT4 redistribution as overexpression of Tomosyn in 3T3-L1 adipocytes inhibited the redistribution of GLUT4 to the plasma membrane (Widberg *et al.*, 2003). However, this study did not utilize an epitope-tagged GLUT4–GFP and only quantified cells that exhibited plasma membrane GFP fluorescence by eye. Tomosyn was also shown to interact with SNARE complexes involving syntaxin-4, SNAP-23, and Munc18c both *in vivo* and *in vitro* (Widberg *et al.*, 2003). This represents another intriguing possible point of regulation that needs to be further developed. However, a suggested role for Tomosyn continues to support a necessary role for Munc18c and for insulin-mediated changes of SNARE complex availability and activity.

SNARE turnover and the availability of free t-SNAREs to complex together to form fusion competent complexes is another possible control point for insulin-mediated GLUT4 fusion. The proportion of free SNAREs is largely mediated by the ATPase NSF and its binding partner α-SNAP (Bonifacino and Glick, 2004; Jahn and Scheller, 2006). The importance of

NSF activity in insulin-mediated GLUT4 redistribution has been suggested using multiple techniques including pulldown of NSF with GSVs and the necessary role of ATPase activity in SNARE complexes purified from rat adipose cells (Mastick and Falick, 1997; Timmers *et al.*, 1996). Recent work has suggest that different levels of NSF activity are necessary for different components of GLUT4 trafficking, consistent with the idea that different SNARE complexes, with different numbers of types and availability of SNARE components, are functioning in both the exocytosis and endocytosis/trafficking of GLUT4 (Chen *et al.*, 2005). It seems clear that NSF activity is important for SNARE turnover in insulin-mediated GLUT4 redistribution; however, a connection between NSF activity and insulin signaling has not been established.

Additional SNARE proteins have been implicated in GLUT4 trafficking at steps other than the final exocytotic event involving syntaxin-4, SNAP-23, and Vamp-2. Syntaxin-6 has been suggested to be involved with retrieval and trafficking of GLUT4 from the plasma membrane and into GSVs (Perera *et al.*, 2003). Syntaxins 6 along with 16 have also been suggested to be important for trafficking of GLUT4 from the *trans*-Golgi network to the insulin-responsive compartment via an important acidic domain in the C-terminal tail of GLUT4 (Shewan *et al.*, 2003). Determination of the SNARE requirements along the GLUT4 retrieval and sorting pathways will provide addition insight into how GLUT4 is segregated and if that process is controlled by insulin stimulation.

B. Cytoskeleton and SNARE proteins regulating GLUT4 vesicle fusion

While a significant body of literature addresses the identification of SNARE proteins responsible for GLUT4 vesicle membrane fusion, much less published data identify the role that the cytoskeleton plays in membrane fusion. Toward this end, Pilch and coworkers have recently shown that a component of the cortical actin network, fodrin, undergoes a dramatic remodeling in response to insulin, much more so than actin itself (Liu *et al.*, 2006). In this study, fodrin interacted with syntaxin-4 and NSF and this interaction was enhanced by insulin (Liu *et al.*, 2006). These findings suggest that the fodrin–actin network may function to organize the GLUT4 vesicle fusion machinery in an insulin-dependent manner. Clearly, this paper brings illustrates a further layer of complexity of the interaction of insulin signaling, the actin network and GLUT4 vesicle fusions. It is of considerable interest to develop a deeper mechanistic understanding of how the actin–fodrin network regulates formation of a competent fusion complex.

V. Conclusions

Much of the work in insulin stimulated GLUT4 redistribution has concentrated in two major areas, attempting to find additional proteins activated by insulin signaling that could control GLUT4 redistribution, and understanding GLUT4 trafficking to better understand possible points of regulation by insulin signaling. Importantly, recent advances in microscopy using TIRFM and advances in our knowledge of GLUT4 trafficking kinetics demonstrates that the relevant insulin action points on GLUT4 redistribution is at the cell surface at the docking and fusion steps (Bai *et al.*, 2007; Huang *et al.*, 2007; Lizunov *et al.*, 2005). This has been an important step forward as it localizes the search for insulin-dependent activities to proteins acting in those mechanisms. The membrane cytoskeleton appears to participate in several levels in the complex physiologic pathway by which insulin signals the redistribution of GLUT4 to the cell surface. Part of the difficulty in studying the role of the cytoskeleton stems from the reliance on pharmacological inhibitors to modulate the cytoskeleton. The use of these compounds, with varying potencies and specificities leads to difficulties in interpretation of experimental details, and controversies within the literature. In the end, it will be important to understand all of the roles that the cytoskeleton plays in GLUT4 regulation to understand how GLUT4 translocation is affected during the development of insulin resistance, and to understand how this process can be targeted to treat and prevent Type 2 diabetes.

ACKNOWLEDGMENTS

This work was supported by a grant from the National Institutes of Health (DK068438). We would like to acknowledge our many colleagues who have contributed important research in this area. Due to space limitations, the inclusion of this entire body of work was not possible.

REFERENCES

Al-Hasani, H., Hinck, C. S., and Cushman, S. W. (1998). Endocytosis of the glucose transporter GLUT4 is mediated by the GTPase dynamin. *J. Biol. Chem.* **273,** 17504–17510.

Backer, J. M., Myers, M. G. Jr., Shoelson, S. E., Chin, D. J., Sun, X. J., Miralpeix, M., Hu, P., Margolis, B., Skolnik, E. Y., Schlessinger, J., *et al.* (1992). Phosphatidylinositol 3′-kinase is activated by association with IRS-1 during insulin stimulation. *EMBO J.* **11,** 3469–3479.

Bai, L., Wang, Y., Fan, J., Chen, Y., Ji, W., Qu, A., Xu, P., James, D. E., and Xu, T. (2007). Dissecting multiple steps of GLUT4 trafficking and identifying the sites of insulin action. *Cell Metab.* **5,** 47–57.

Bandyopadhyay, G., Standaert, M. L., Kikkawa, U., Ono, Y., Moscat, J., and Farese, R. V. (1999). Effects of transiently expressed atypical (zeta, lambda), conventional (alpha, beta) and novel (delta, epsilon) protein kinase C isoforms on insulin-stimulated translocation of epitope-tagged GLUT4 glucose transporters in rat adipocytes: Specific interchangeable effects of protein kinases C-zeta and C-lambda. *Biochem. J.* **337**(Pt. 3), 461–470.

Blot, V., and McGraw, T. E. (2006). GLUT4 is internalized by a cholesterol-dependent nystatin-sensitive mechanism inhibited by insulin. *EMBO J.* **25,** 5648–5658.

Bogan, J. S., Hendon, N., McKee, A. E., Tsao, T. S., and Lodish, H. F. (2003). Functional cloning of TUG as a regulator of GLUT4 glucose transporter trafficking. *Nature* **425,** 727–733.

Bonifacino, J. S., and Glick, B. S. (2004). The mechanisms of vesicle budding and fusion. *Cell* **116,** 153–166.

Bose, A., Guilherme, A., Robida, S. I., Nicoloro, S. M., Zhou, Q. L., Jiang, Z. Y., Pomerleau, D. P., and Czech, M. P. (2002). Glucose transporter recycling in response to insulin is facilitated by myosin Myo1c. *Nature* **420,** 821–824.

Bose, A., Robida, S., Furcinitti, P. S., Chawla, A., Fogarty, K., Corvera, S., and Czech, M. P. (2004). Unconventional myosin Myo1c promotes membrane fusion in a regulated exocytic pathway. *Mol. Cell. Biol.* **24,** 5447–5458.

Bridges, D., and Moorhead, G. B. (2005). 14-3-3 proteins: A number of functions for a numbered protein. *Sci. STKE* **2005,** re10.

Bryant, N. J., Govers, R., and James, D. E. (2002). Regulated transport of the glucose transporter GLUT4. *Nat. Rev. Mol. Cell Biol.* **3,** 267–277.

Cenni, V., Sirri, A., Riccio, M., Lattanzi, G., Santi, S., de Pol, A., Maraldi, N. M., and Marmiroli, S. (2003). Targeting of the Akt/PKB kinase to the actin skeleton. *Cell. Mol. Life Sci.* **60,** 2710–2720.

Chang, L., Chiang, S. H., and Saltiel, A. R. (2004). Insulin signaling and the regulation of glucose transport. *Mol. Med.* **10,** 65–71.

Cheatham, B., and Kahn, C. R. (1995). Insulin action and the insulin signaling network. *Endocr. Rev.* **16,** 117–142.

Chen, G., Raman, P., Bhonagiri, P., Strawbridge, A. B., Pattar, G. R., and Elmendorf, J. S. (2004). Protective effect of phosphatidyl 4,5-bisphosphate against cortical filamentous actin loss and insulin resistance induced by sustained exposure of 3T3-L1 adipocytes to insulin. *J. Biol. Chem.* **279,** 39705–39709.

Chen, X., Matsumoto, H., Hinck, C. S., Al-Hasani, H., St-Denis, J. F., Whiteheart, S. W., and Cushman, S. W. (2005). Demonstration of differential quantitative requirements for NSF among multiple vesicle fusion pathways of GLUT4 using a dominant-negative ATPase-deficient NSF. *Biochem. Biophys. Res. Commun.* **333,** 28–34.

Chiang, S. H., Baumann, C. A., Kanzaki, M., Thurmond, D. C., Watson, R. T., Neudauer, C. L., Macara, I. G., Pessin, J. E., and Saltiel, A. R. (2001). Insulin-stimulated GLUT4 translocation requires the CAP-dependent activation of TC10. *Nature* **410,** 944–948.

Cho, H., Mu, J., Kim, J. K., Thorvaldsen, J. L., Chu, Q., Crenshaw, E. B. III, Kaestner, K. H., Bartolomei, M. S., Shulman, G. I., and Birnbaum, M. J. (2001). Insulin resistance and a diabetes mellitus-like syndrome in mice lacking the protein kinase Akt2 (PKB beta). *Science* **292,** 1728–1731.

Cobb, M. H., Sang, B. C., Gonzalez, R., Goldsmith, E., and Ellis, L. (1989). Autophosphorylation activates the soluble cytoplasmic domain of the insulin receptor in an intermolecular reaction. *J. Biol. Chem.* **264,** 18701–18706.

Cohen, A. W., Combs, T. P., Scherer, P. E., and Lisanti, M. P. (2003). Role of caveolin and caveolae in insulin signaling and diabetes. *Am. J. Physiol. Endocrinol. Metab.* **285,** E1151–E1160.

Coster, A. C., Govers, R., and James, D. E. (2004). Insulin stimulates the entry of GLUT4 into the endosomal recycling pathway by a quantal mechanism. *Traffic* **5,** 763–771.
Cushman, S. W., and Wardzala, L. J. (1980). Potential mechanism of insulin action on glucose transport in the isolated rat adipose cell. Apparent translocation of intracellular transport systems to the plasma membrane. *J. Biol. Chem.* **255,** 4758–4762.
Czech, M. P., and Buxton, J. M. (1993). Insulin action on the internalization of the GLUT4 glucose transporter in isolated rat adipocytes. *J. Biol. Chem.* **268,** 9187–9190.
Czech, M. P., and Corvera, S. (1999). Signaling mechanisms that regulate glucose transport. *J. Biol. Chem.* **274,** 1865–1868.
Dong, L. Q., and Liu, F. (2005). PDK2: The missing piece in the receptor tyrosine kinase signaling pathway puzzle. *Am. J. Physiol. Endocrinol. Metab.* **289,** E187–E196.
El-Jack, A. K., Kandror, K. V., and Pilch, P. F. (1999). The formation of an insulin-responsive vesicular cargo compartment is an early event in 3T3-L1 adipocyte differentiation. *Mol. Biol. Cell* **10,** 1581–1594.
Emoto, M., Langille, S. E., and Czech, M. P. (2001). A role for kinesin in insulin-stimulated GLUT4 glucose transporter translocation in 3T3-L1 adipocytes. *J. Biol. Chem.* **276,** 10677–10682.
Eyster, C. A., Duggins, Q. S., and Olson, A. L. (2005). Expression of constitutively active Akt/protein kinase B signals GLUT4 translocation in the absence of an intact actin cytoskeleton. *J. Biol. Chem.* **280,** 17978–17985.
Eyster, C. A., Duggins, Q. S., Gorbsky, G. J., and Olson, A. L. (2006). Microtubule network is required for insulin signaling through activation of Akt/protein kinase B: Evidence that insulin stimulates vesicle docking/fusion but not intracellular mobility. *J. Biol. Chem.* **281,** 39719–39727.
Fasshauer, M., Klein, J., Ueki, K., Kriauciunas, K. M., Benito, M., White, M. F., and Kahn, C. R. (2000). Essential role of insulin receptor substrate-2 in insulin stimulation of Glut4 translocation and glucose uptake in brown adipocytes. *J. Biol. Chem.* **275,** 25494–25501.
Fletcher, L. M., Welsh, G. I., Oatey, P. B., and Tavare, J. M. (2000). Role for the microtubule cytoskeleton in GLUT4 vesicle trafficking and in the regulation of insulin-stimulated glucose uptake. *Biochem. J.* **352**(Pt. 2), 267–276.
Foster, L. J., Li, D., Randhawa, V. K., and Klip, A. (2001). Insulin accelerates inter-endosomal GLUT4 traffic via phosphatidylinositol 3-kinase and protein kinase B. *J. Biol. Chem.* **276,** 44212–44221.
Frattali, A. L., Treadway, J. L., and Pessin, J. E. (1992). Transmembrane signaling by the human insulin receptor kinase. Relationship between intramolecular beta subunit *trans-* and *cis-*autophosphorylation and substrate kinase activation. *J. Biol. Chem.* **267,** 19521–19528.
Gerst, J. E. (2003). SNARE regulators: Matchmakers and matchbreakers. *Biochim. Biophys. Acta* **1641,** 99–110.
Guilherme, A., Emoto, M., Buxton, J. M., Bose, S., Sabini, R., Theurkauf, W. E., Leszyk, J., and Czech, M. P. (2000). Perinuclear localization and insulin responsiveness of GLUT4 requires cytoskeletal integrity in 3T3-L1 adipocytes. *J. Biol. Chem.* **275,** 38151–38159.
Hashiramoto, M., and James, D. E. (2000). Characterization of insulin-responsive GLUT4 storage vesicles isolated from 3T3-L1 adipocytes. *Mol. Cell. Biol.* **20,** 416–427.
Hodgkinson, C. P., Mander, A., and Sale, G. J. (2005). Protein kinase-zeta interacts with munc18c: Role in GLUT4 trafficking. *Diabetologia* **48,** 1627–1636.
Holman, G. D., Lo Leggio, L., and Cushman, S. W. (1994). Insulin-stimulated GLUT4 glucose transporter recycling. A problem in membrane protein subcellular trafficking through multiple pools. *J. Biol. Chem.* **269,** 17516–17524.

Horuk, R., Rodbell, M., Cushman, S. W., and Wardzala, L. J. (1983). Proposed mechanism of insulin-resistant glucose transport in the isolated guinea pig adipocyte. Small intracellular pool of glucose transporters. *J. Biol. Chem.* **258,** 7425–7429.

Hou, J. C., Suzuki, N., Pessin, J. E., and Watson, R. T. (2006). A specific dileucine motif is required for the GGA-dependent entry of newly synthesized insulin-responsive aminopeptidase into the insulin-responsive compartment. *J. Biol. Chem.* **281,** 33457–33466.

Howlett, K. F., Sakamoto, K., Garnham, A., Cameron-Smith, D., and Hargreaves, M. (2007). Resistance exercise and insulin regulate AS160 and interaction with 14-3-3 in human skeletal muscle. *Diabetes* **56,** 1608–1614.

Hresko, R. C., and Mueckler, M. (2005). mTOR.RICTOR is the Ser473 kinase for Akt/protein kinase B in 3T3-L1 adipocytes. *J. Biol. Chem.* **280,** 40406–40416.

Hresko, R. C., Murata, H., and Mueckler, M. (2003). Phosphoinositide-dependent kinase-2 is a distinct protein kinase enriched in a novel cytoskeletal fraction associated with adipocyte plasma membranes. *J. Biol. Chem.* **278,** 21615–21622.

Huang, J., Imamura, T., and Olefsky, J. M. (2001). Insulin can regulate GLUT4 internalization by signaling to Rab5 and the motor protein dynein. *Proc. Natl. Acad. Sci. USA* **98,** 13084–13089.

Huang, S., Lifshitz, L. M., Jones, C., Bellve, K. D., Standley, C., Fonseca, S., Corvera, S., Fogarty, K. E., and Czech, M. P. (2007). Insulin stimulates membrane fusion and GLUT4 accumulation in clathrin coats on adipocyte plasma membranes. *Mol. Cell. Biol.* **27,** 3456–3469.

Imamura, T., Huang, J., Usui, I., Satoh, H., Bever, J., and Olefsky, J. M. (2003). Insulin-induced GLUT4 translocation involves protein kinase C-lambda-mediated functional coupling between Rab4 and the motor protein kinesin. *Mol. Cell. Biol.* **23,** 4892–4900.

Inoue, M., Chang, L., Hwang, J., Chiang, S. H., and Saltiel, A. R. (2003). The exocyst complex is required for targeting of Glut4 to the plasma membrane by insulin. *Nature* **422,** 629–633.

Ishiki, M., and Klip, A. (2005). Minireview: Recent developments in the regulation of glucose transporter-4 traffic: New signals, locations, and partners. *Endocrinology* **146,** 5071–5078.

Ishikura, S., Bilan, P. J., and Klip, A. (2007). Rabs 8A and 14 are targets of the insulin-regulated Rab-GAP AS160 regulating GLUT4 traffic in muscle cells. *Biochem. Biophys. Res. Commun.* **353,** 1074–1079.

Jahn, R., and Scheller, R. H. (2006). SNAREs—Engines for membrane fusion. *Nat. Rev. Mol. Cell Biol.* **7,** 631–643.

Jahn, R., and Sudhof, T. C. (1999). Membrane fusion and exocytosis. *Annu. Rev. Biochem.* **68,** 863–911.

James, D. E. (2005). MUNC-ing around with insulin action. *J. Clin. Invest.* **115,** 219–221.

Kandror, K. V., Coderre, L., Pushkin, A. V., and Pilch, P. F. (1995). Comparison of glucose-transporter-containing vesicles from rat fat and muscle tissues: Evidence for a unique endosomal compartment. *Biochem. J.* **307**(Pt. 2), 383–390.

Kane, S., Sano, H., Liu, S. C., Asara, J. M., Lane, W. S., Garner, C. C., and Lienhard, G. E. (2002). A method to identify serine kinase substrates. Akt phosphorylates a novel adipocyte protein with a Rab GTPase-activating protein (GAP) domain. *J. Biol. Chem.* **277,** 22115–22118.

Kanzaki, M., and Pessin, J. E. (2001). Insulin-stimulated GLUT4 translocation in adipocytes is dependent upon cortical actin remodeling. *J. Biol. Chem.* **276,** 42436–42444.

Kanzaki, M., Watson, R. T., Khan, A. H., and Pessin, J. E. (2001). Insulin stimulates actin comet tails on intracellular GLUT4-containing compartments in differentiated 3T3L1 adipocytes. *J. Biol. Chem.* **276,** 49331–49336.

Kanzaki, M., Watson, R. T., Hou, J. C., Stamnes, M., Saltiel, A. R., and Pessin, J. E. (2002). Small GTP-binding protein TC10 differentially regulates two distinct populations of filamentous actin in 3T3L1 adipocytes. *Mol. Biol. Cell* **13,** 2334–2346.

Karylowski, O., Zeigerer, A., Cohen, A., and McGraw, T. E. (2004). GLUT4 is retained by an intracellular cycle of vesicle formation and fusion with endosomes. *Mol. Biol. Cell* **15,** 870–882.

Khayat, Z. A., Tong, P., Yaworsky, K., Bloch, R. J., and Klip, A. (2000). Insulin-induced actin filament remodeling colocalizes actin with phosphatidylinositol 3-kinase and GLUT4 in L6 myotubes. *J. Cell Sci.* **113**(Pt. 2), 279–290.

Kohn, A. D., Summers, S. A., Birnbaum, M. J., and Roth, R. A. (1996). Expression of a constitutively active Akt Ser/Thr kinase in 3T3-L1 adipocytes stimulates glucose uptake and glucose transporter 4 translocation. *J. Biol. Chem.* **271,** 31372–31378.

Kupriyanova, T. A., Kandror, V., and Kandror, K. V. (2002). Isolation and characterization of the two major intracellular Glut4 storage compartments. *J. Biol. Chem.* **277,** 9133–9138.

Lampson, M. A., Schmoranzer, J., Zeigerer, A., Simon, S. M., and McGraw, T. E. (2001). Insulin-regulated release from the endosomal recycling compartment is regulated by budding of specialized vesicles. *Mol. Biol. Cell* **12,** 3489–3501.

Larance, M., Ramm, G., Stockli, J., van Dam, E. M., Winata, S., Wasinger, V., Simpson, F., Graham, M., Junutula, J. R., Guilhaus, M., and James, D. E. (2005). Characterization of the role of the Rab GTPase-activating protein AS160 in insulin-regulated GLUT4 trafficking. *J. Biol. Chem.* **280,** 37803–37813.

Lee, J., O'Hare, T., Pilch, P. F., and Shoelson, S. E. (1993). Insulin receptor autophosphorylation occurs asymmetrically. *J. Biol. Chem.* **268,** 4092–4098.

Li, L. V., and Kandror, K. V. (2005). Golgi-localized, gamma-ear-containing, Arf-binding protein adaptors mediate insulin-responsive trafficking of glucose transporter 4 in 3T3-L1 adipocytes. *Mol. Endocrinol.* **19,** 2145–2153.

Liu, L. B., Omata, W., Kojima, I., and Shibata, H. (2003). Insulin recruits GLUT4 from distinct compartments via distinct traffic pathways with differential microtubule dependence in rat adipocytes. *J. Biol. Chem.* **278,** 30157–30169.

Liu, L., Jedrychowski, M. P., Gygi, S. P., and Pilch, P. F. (2006). Role of insulin-dependent cortical fodrin/spectrin remodeling in glucose transporter 4 translocation in rat adipocytes. *Mol. Biol. Cell* **17,** 4249–4256.

Livingstone, C., James, D. E., Rice, J. E., Hanpeter, D., and Gould, G. W. (1996). Compartment ablation analysis of the insulin-responsive glucose transporter (GLUT4) in 3T3-L1 adipocytes. *Biochem. J.* **315**(Pt. 2), 487–495.

Lizunov, V. A., Matsumoto, H., Zimmerberg, J., Cushman, S. W., and Frolov, V. A. (2005). Insulin stimulates the halting, tethering, and fusion of mobile GLUT4 vesicles in rat adipose cells. *J. Cell Biol.* **169,** 481–489.

Lund, S., Holman, G. D., Schmitz, O., and Pedersen, O. (1995). Contraction stimulates translocation of glucose transporter GLUT4 in skeletal muscle through a mechanism distinct from that of insulin. *Proc. Natl. Acad. Sci. USA* **92,** 5817–5821.

Macaulay, S. L., Hewish, D. R., Gough, K. H., Stoichevska, V., MacPherson, S. F., Jagadish, M., and Ward, C. W. (1997). Functional studies in 3T3L1 cells support a role for SNARE proteins in insulin stimulation of GLUT4 translocation. *Biochem. J.* **324**(Pt. 1), 217–224.

Martin, S., Tellam, J., Livingstone, C., Slot, J. W., Gould, G. W., and James, D. E. (1996). The glucose transporter (GLUT-4) and vesicle-associated membrane protein-2 (VAMP-2) are segregated from recycling endosomes in insulin-sensitive cells. *J. Cell Biol.* **134,** 625–635.

Mastick, C. C., and Falick, A. L. (1997). Association of *N*-ethylmaleimide sensitive fusion (NSF) protein and soluble NSF attachment proteins-alpha and -gamma with glucose transporter-4-containing vesicles in primary rat adipocytes. *Endocrinology* **138,** 2391–2397.

McCarthy, A. M., Spisak, K. O., Brozinick, J. T., and Elmendorf, J. S. (2006). Loss of cortical actin filaments in insulin-resistant skeletal muscle cells impairs LGUT4 vesicle trafficking and glucose transport. *Am. J. Physiol. Cell Physiol.* **291,** C860–C868.

Millar, C. A., Meerloo, T., Martin, S., Hickson, G. R., Shimwell, N. J., Wakelam, M. J., James, D. E., and Gould, G. W. (2000). Adipsin and the glucose transporter GLUT4 traffic to the cell surface via independent pathways in adipocytes. *Traffic* **1,** 141–151.

Min, J., Okada, S., Kanzaki, M., Elmendorf, J. S., Coker, K. J., Ceresa, B. P., Syu, L. J., Noda, Y., Saltiel, A. R., and Pessin, J. E. (1999). Synip: A novel insulin-regulated syntaxin 4-binding protein mediating GLUT4 translocation in adipocytes. *Mol. Cell* **3,** 751–760.

Minami, A., Iseki, M., Kishi, K., Wang, M., Ogura, M., Furukawa, N., Hayashi, S., Yamada, M., Obata, T., Takeshita, Y., Nakaya, Y., Bando, Y., Izumi, K., Moodie, S. A., Kajiura, F., Matsumoto, M., Takatsu, K., Takaki, S., and Ebina, Y. (2003). Increased insulin sensitivity and hypoinsulinemia in APS knockout mice. *Diabetes* **52,** 2657–2665.

Mitra, P., Zheng, X., and Czech, M. P. (2004). RNAi-based analysis of CAP, Cbl, and CrkII function in the regulation of GLUT4 by insulin. *J. Biol. Chem.* **279,** 37431–37435.

Molero, J. C., Whitehead, J. P., Meerloo, T., and James, D. E. (2001). Nocodazole inhibits insulin-stimulated glucose transport in 3T3-L1 adipocytes via a microtubule-independent mechanism. *J. Biol. Chem.* **276,** 43829–43835.

Myers, M. G. Jr., Wang, L. M., Sun, X. J., Zhang, Y., Yenush, L., Schlessinger, J., Pierce, J. H., and White, M. F. (1994). Role of IRS-1–GRB-2 complexes in insulin signaling. *Mol. Cell. Biol.* **14,** 3577–3587.

Oh, E., Spurlin, B. A., Pessin, J. E., and Thurmond, D. C. (2005). Munc18c heterozygous knockout mice display increased susceptibility for severe glucose intolerance. *Diabetes* **54,** 638–647.

Okada, S., Ohshima, K., Uehara, Y., Shimizu, H., Hashimoto, K., Yamada, M., and Mori, M. (2007). Synip phosphorylation is required for insulin-stimulated Glut4 translocation. *Biochem. Biophys. Res. Commun.* **356,** 102–106.

Olson, A. L., Trumbly, A. R., and Gibson, G. V. (2001). Insulin-mediated GLUT4 translocation is dependent on the microtubule network. *J. Biol. Chem.* **276,** 10706–10714.

Olson, A. L., Eyster, C. A., Duggins, Q. S., and Knight, J. B. (2003). Insulin promotes formation of polymerized microtubules by a phosphatidylinositol 3-kinase-independent, actin-dependent pathway in 3T3-L1 adipocytes. *Endocrinology* **144,** 5030–5039.

Omata, W., Shibata, H., Li, L., Takata, K., and Kojima, I. (2000). Actin filaments play a critical role in insulin-induced exocytotic recruitment but not in endocytosis of GLUT4 in isolated rat adipocytes. *Biochem. J.* **346**(Pt. 2), 321–328.

Patel, N., Rudich, A., Khayat, Z. A., Garg, R., and Klip, A. (2003). Intracellular segregation of phosphatidylinositol-3,4,5-trisphosphate by insulin-dependent actin remodeling in L6 skeletal muscle cells. *Mol. Cell. Biol.* **23,** 4611–4626.

Patki, V., Buxton, J., Chawla, A., Lifshitz, L., Fogarty, K., Carrington, W., Tuft, R., and Corvera, S. (2001). Insulin action on GLUT4 traffic visualized in single 3T3-L1 adipocytes by using ultra-fast microscopy. *Mol. Biol. Cell* **12,** 129–141.

Perera, H. K., Clarke, M., Morris, N. J., Hong, W., Chamberlain, L. H., and Gould, G. W. (2003). Syntaxin 6 regulates Glut4 trafficking in 3T3-L1 adipocytes. *Mol. Biol. Cell* **14,** 2946–2958.

Petersen, K. F., Dufour, S., Savage, D. B., Bilz, S., Solomon, G., Yonemitsu, S., Cline, G. W., Befroy, D., Zemany, L., Kahn, B. B., Papdemetris, X., Rothman, D. L., and Shulman, G. I. (2007). The role of skeletal muscle insulin resistance in the pathogenesis of the metabolic syndrome. *Proc. Natl. Acad. Sci. USA* **104,** 12587–12594.

Ramm, G., Larance, M., Guilhaus, M., and James, D. E. (2006). A role for 14-3-3 in insulin-stimulated GLUT4 translocation through its interaction with the RabGAP AS160. *J. Biol. Chem.* **281,** 29174–29180.

Randhawa, V. K., Bilan, P. J., Khayat, Z. A., Daneman, N., Liu, Z., Ramlal, T., Volchuk, A., Peng, X. R., Coppola, T., Regazzi, R., Trimble, W. S., and Klip, A. (2000). VAMP2, but not VAMP3/cellubrevin, mediates insulin-dependent incorporation of GLUT4 into the plasma membrane of L6 myoblasts. *Mol. Biol. Cell* **11,** 2403–2417.

Roach, W. G., Chavez, J. A., Miinea, C. P., and Lienhard, G. E. (2007). Substrate specificity and effect on GLUT4 translocation of the Rab GTPase-activating protein Tbc1d1. *Biochem. J.* **403,** 353–358.

Robinson, L. J., Pang, S., Harris, D. S., Heuser, J., and James, D. E. (1992). Translocation of the glucose transporter (GLUT4) to the cell surface in permeabilized 3T3-L1 adipocytes: Effects of ATP insulin, and GTP gamma S and localization of GLUT4 to clathrin lattices. *J. Cell Biol.* **117,** 1181–1196.

Rothman, D. L., Magnusson, I., Cline, G., Gerard, D., Kahn, C. R., Shulman, R. G., and Shulman, G. I. (1995). Decreased muscle glucose transport/phosphorylation is an early defect in the pathogenesis of non-insulin-dependent diabetes mellitus. *Proc. Natl. Acad. Sci. USA* **92,** 983–987.

Saltiel, A. R., and Kahn, C. R. (2001). Insulin signalling and the regulation of glucose and lipid metabolism. *Nature* **414,** 799–806.

Sano, H., Kane, S., Sano, E., Miinea, C. P., Asara, J. M., Lane, W. S., Garner, C. W., and Lienhard, G. E. (2003). Insulin-stimulated phosphorylation of a Rab GTPase-activating protein regulates GLUT4 translocation. *J. Biol. Chem.* **278,** 14599–14602.

Sano, H., Kane, S., Sano, E., and Lienhard, G. E. (2005). Synip phosphorylation does not regulate insulin-stimulated GLUT4 translocation. *Biochem. Biophys. Res. Commun.* **332,** 880–884.

Sarbassov, D. D., Guertin, D. A., Ali, S. M., and Sabatini, D. M. (2005). Phosphorylation and regulation of Akt/PKB by the rictor–mTOR complex. *Science* **307,** 1098–1101.

Scheid, M. P., Marignani, P. A., and Woodgett, J. R. (2002). Multiple phosphoinositide 3-kinase-dependent steps in activation of protein kinase B. *Mol. Cell. Biol.* **22,** 6247–6260.

Semiz, S., Park, J. G., Nicoloro, S. M., Furcinitti, P., Zhang, C., Chawla, A., Leszyk, J., and Czech, M. P. (2003). Conventional kinesin KIF5B mediates insulin-stimulated GLUT4 movements on microtubules. *EMBO J.* **22,** 2387–2399.

Shewan, A. M., van Dam, E. M., Martin, S., Luen, T. B., Hong, W., Bryant, N. J., and James, D. E. (2003). GLUT4 recycles via a *trans*-Golgi network (TGN) subdomain enriched in Syntaxins 6 and 16 but not TGN38: Involvement of an acidic targeting motif. *Mol. Biol. Cell* **14,** 973–986.

Shoelson, S. E., Boni-Schnetzler, M., Pilch, P. F., and Kahn, C. R. (1991). Autophosphorylation within insulin receptor beta-subunits can occur as an intramolecular process. *Biochemistry* **30,** 7740–7746.

Slot, J. W., Geuze, H. J., Gigengack, S., James, D. E., and Lienhard, G. E. (1991a). Translocation of the glucose transporter GLUT4 in cardiac myocytes of the rat. *Proc. Natl. Acad. Sci. USA* **88,** 7815–7819.

Slot, J. W., Geuze, H. J., Gigengack, S., Lienhard, G. E., and James, D. E. (1991b). Immuno-localization of the insulin regulatable glucose transporter in brown adipose tissue of the rat. *J. Cell Biol.* **113,** 123–135.

Stone, S., Abkevich, V., Russell, D. L., Riley, R., Timms, K., Tran, T., Trem, D., Frank, D., Jammulapati, S., Neff, C. D., Iliev, D., Gress, R., *et al.* (2006). TBC1D1 is a candidate for a severe obesity gene and evidence for a gene/gene interaction in obesity predisposition. *Hum. Mol. Genet.* **15,** 2709–2720.

Sudhof, T. C. (2004). The synaptic vesicle cycle. *Annu. Rev. Neurosci.* **27,** 509–547.

Timmers, K. I., Clark, A. E., Omatsu-Kanbe, M., Whiteheart, S. W., Bennett, M. K., Holman, G. D., and Cushman, S. W. (1996). Identification of SNAP receptors in rat adipose cell membrane fractions and in SNARE complexes co-immunoprecipitated with epitope-tagged *N*-ethylmaleimide-sensitive fusion protein. *Biochem. J.* **320**(Pt. 2), 429–436.

Tsakiridis, T., Vranic, M., and Klip, A. (1994). Disassembly of the actin network inhibits insulin-dependent stimulation of glucose transport and prevents recruitment of glucose transporters to the plasma membrane. *J. Biol. Chem.* **269,** 29934–29942.

Tsakiridis, T., Vranic, M., and Klip, A. (1995). Phosphatidylinositol 3-kinase and the actin network are not required for the stimulation of glucose transport caused by mitochondrial uncoupling: Comparison with insulin action. *Biochem. J.* **309**(Pt. 1), 1–5.

Usui, I., Imamura, T., Huang, J., Satoh, H., and Olefsky, J. M. (2003). Cdc42 is a Rho GTPase family member that can mediate insulin signaling to glucose transport in 3T3-L1 adipocytes. *J. Biol. Chem.* **278,** 13765–13774.

Volchuk, A., Narine, S., Foster, L. J., Grabs, D., De Camilli, P., and Klip, A. (1998). Perturbation of dynamin II with an amphiphysin SH3 domain increases GLUT4 glucose transporters at the plasma membrane in 3T3-L1 adipocytes. Dynamin II participates in GLUT4 endocytosis. *J. Biol. Chem.* **273,** 8169–8176.

Wang, Q., Bilan, P. J., Tsakiridis, T., Hinek, A., and Klip, A. (1998). Actin filaments participate in the relocalization of phosphatidylinositol 3-kinase to glucose transporter-containing compartments and in the stimulation of glucose uptake in 3T3-L1 adipocytes. *Biochem. J.* **331**(Pt. 3), 917–928.

Watson, R. T., Khan, A. H., Furukawa, M., Hou, J. C., Li, L., Kanzaki, M., Okada, S., Kandror, K. V., and Pessin, J. E. (2004). Entry of newly synthesized GLUT4 into the insulin-responsive storage compartment is GGA dependent. *EMBO J.* **23,** 2059–2070.

Welsh, G. I., Hers, I., Berwick, D. C., Dell, G., Wherlock, M., Birkin, R., Leney, S., and Tavare, J. M. (2005). Role of protein kinase B in insulin-regulated glucose uptake. *Biochem. Soc. Trans.* **33,** 346–349.

White, M. F. (1998). The IRS-signaling system: A network of docking proteins that mediate insulin and cytokine action. *Recent Prog. Horm. Res.* **53,** 119–138.

Widberg, C. H., Bryant, N. J., Girotti, M., Rea, S., and James, D. E. (2003). Tomosyn interacts with the t-SNAREs syntaxin4 and SNAP23 and plays a role in insulin-stimulated GLUT4 translocation. *J. Biol. Chem.* **278,** 35093–35101.

Withers, D. J., Gutierrez, J. S., Towery, H., Burks, D. J., Ren, J. M., Previs, S., Zhang, Y., Bernal, D., Pons, S., Shulman, G. I., Bonner-Weir, S., and White, M. F. (1998). Disruption of IRS-2 causes type 2 diabetes in mice. *Nature* **391,** 900–904.

Yamada, T., Katagiri, H., Asano, T., Tsuru, M., Inukai, K., Ono, H., Kodama, T., Kikuchi, M., and Oka, Y. (2002). Role of PDK1 in insulin-signaling pathway for glucose metabolism in 3T3-L1 adipocytes. *Am. J. Physiol. Endocrinol. Metab.* **282,** E1385–E1394.

Yamada, E., Okada, S., Saito, T., Ohshima, K., Sato, M., Tsuchiya, T., Uehara, Y., Shimizu, H., and Mori, M. (2005). Akt2 phosphorylates Synip to regulate docking and fusion of GLUT4-containing vesicles. *J. Cell Biol.* **168,** 921–928.

Yang, J., and Holman, G. D. (1993). Comparison of GLUT4 and GLUT1 subcellular trafficking in basal and insulin-stimulated 3T3-L1 cells. *J. Biol. Chem.* **268,** 4600–4603.

Yang, J., Clark, A. E., Harrison, R., Kozka, I. J., and Holman, G. D. (1992). Trafficking of glucose transporters in 3T3-L1 cells. Inhibition of trafficking by phenylarsine oxide implicates a slow dissociation of transporters from trafficking proteins. *Biochem. J.* **281** (Pt. 3), 809–817.

Yoshizaki, T., Imamura, T., Babendure, J. L., Lu, J. C., Sonoda, N., and Olefsky, J. M. (2007). Myosin 5a is an insulin-stimulated Akt2 (Protein Kinase B) substrate modulating GLUT4 vesicle translocation. *Mol. Cell. Biol.* **27,** 5172–5183.

Yu, C., Cresswell, J., Loffler, M. G., and Bogan, J. S. (2007). The glucose transporter 4-regulating protein TUG is essential for highly insulin-responsive glucose uptake in 3T3-L1 adipocytes. *J. Biol. Chem.* **282,** 7710–7722.

Zeigerer, A., Lampson, M. A., Karylowski, O., Sabatini, D. D., Adesnik, M., Ren, M., and McGraw, T. E. (2002). GLUT4 retention in adipocytes requires two intracellular insulin-regulated transport steps. *Mol. Biol. Cell* **13,** 2421–2435.

Zerial, M., and McBride, H. (2001). Rab proteins as membrane organizers. *Nat. Rev. Mol. Cell Biol.* **2,** 107–117.

Zhou, Q. L., Park, J. G., Jiang, Z. Y., Holik, J. J., Mitra, P., Semiz, S., Guilherme, A., Powelka, A. M., Tang, X., Virbasius, J., and Czech, M. P. (2004). Analysis of insulin signalling by RNAi-based gene silencing. *Biochem. Soc. Trans.* **32,** 817–821.

CHAPTER NINE

Nutrient Modulation of Insulin Secretion

Nimbe Torres, Lilia Noriega, *and* Armando R. Tovar

Contents

Abstract

The presence of different nutrients regulates the β-cell response to secrete insulin to maintain glucose in the physiological range and appropriate levels of fuels in different organs and tissues. Glucose is the only nutrient secretagogue capable of promoting alone the release of insulin release. The mechanisms of Insulin secretion are dependent or independent of the closure of ATP-sensitive K^+ channels. In addition, insulin secretion in response to glucose and other nutrients is modulated by several hormones as incretins, glucagon, and leptin. Fatty acids (FAs), amino acids, and keto acids influence secretion as well. The exact mechanism for which nutrients induce insulin secretion is complicated because nutrient signaling shows one of the most complex transduction systems, which exists for the reason that nutrient have to be metabolized. FAs in the absence of glucose induce FA oxidation and insulin secretion in a lesser extent. However, FAs in the presence of glucose produce high concentration of malonyl-CoA that repress FA oxidation and increase the formation of LC-CoA amplifying the insulin release. Long-term exposure to fatty acids and glucose results in glucolipotoxicity and decreases in insulin release. The amino acid pattern produced after the consumption of a dietary protein regulates insulin

Departamento de Fisiología de la Nutrición, Instituto Nacional de Ciencias Médicas y Nutrición Vasco de Quiroga No 15, México DF 14000, Mexico

Vitamins and Hormones, Volume 80
ISSN 0083-6729, DOI: 10.1016/S0083-6729(08)00609-2

secretion by generating anaplerotic substrates that stimulates ATP synthesis or by activating specific signal transduction mediated by mTOR, AMPK, and SIRT4 or modulating the expression of genes involved in insulin secretion. Finally, dietary bioactive compounds such as isoflavones play an important role in the regulation of insulin secretion.

I. Introduction

A feature essential for life of higher vertebrates is their ability to maintain a relatively constant blood glucose concentration since the brain and its neurons are dependent upon a continuous supply of blood-delivered glucose. Glucose is essential as an energy source for all cells, although some of them can utilize others "fuel metabolites" such as amino acids or fatty acids (FAs). There are other factors that modulate blood glucose as a consequence of physiological and metabolic events. These can include (1) the intestinal absorption and transport of nutrients to storage depots such as liver, muscle, and/or adipose tissue; (2) muscular activity; (3) thermogenesis; (4) starvation; (5) pregnancy/lactation; or (6) disease or injury states (Norman and Litwack, 1987).

The endocrine system largely responsible for the maintenance of blood glucose levels and the proper cellular uptake and exchange of the "fuel metabolites" is the islet of Langerhans, which is composed of four major cell types: insulin-secreting β-cells, glucagon-secreting α-cells, somatostatin-releasing δ-cells, and pancreatic polypeptide-producing cells. The main function of insulin is to stimulate anabolic reactions for carbohydrates, proteins, and fats, all of which will have the metabolic consequences of producing a lowered blood glucose level. Glucagon can be thought as an indirect antagonist of insulin. Glucagon stimulates catabolic reactions, which lead ultimately to an elevation of blood glucose levels. Thus, islets of Langerhans are continuously adjusting the relative amounts of glucagon and insulin secreted in response to the continuous perturbations of blood glucose and other fuel metabolites occurring as a consequence of changes in anabolism and catabolism in the various tissues (Norman and Litwack, 1987). Furthermore, the islet of Langerhans can be viewed as a "fuel sensor" which simultaneously integrates the signals of many nutrients and modulators to secrete insulin according to the needs of the organism.

The secretion of insulin by β-cell is modulated by various factors including nutrients, neurotransmitters, or peptide hormones. Glucose is the only nutrient secretagogue capable of promoting alone *in vitro* the release of insulin at concentrations within its physiological range. Nonetheless, many additional nutrients including fatty acids, amino acids and keto acids influence secretion as well. The unique feature of the β-cell is that it possesses a transduction system for nutrient signals which is entirely different from that of neuromodulators or peptide hormones (Prentki *et al.*, 1997). Indeed, nutrients must be metabolized in the cell to cause secretion. By contrast, neuromodulators, such as the potent incretin glucagon-like peptide-1 (GLP-I), influence the secretory process following their interaction with specific cell-surface receptors. The exact mechanism for which nutrients induce insulin secretion is complicated because nutrient signaling shows one of the most complex transduction systems, which exists for the reason that nutrient have to be metabolized. Furthermore, diet composition is the main contributor of changes in plasma nutrient levels, representing an important modulator of insulin secretion. In this review we will focus on the mechanisms by which macronutrients such as carbohydrates, protein, and fat constituents in the diet modify the process of insulin secretion.

II. Overview of Insulin Secretion

Insulin secretion *in vivo* in man and animals is pulsatile. This oscillatory nature of the β-cell derives from the oscillatory metabolism of glucose (Lefebvre *et al.*, 1987). The β-cells of the pancreatic islets store insulin in secretory granules, which are released by exocytosis when the concentration of nutrients increased in the blood. The insulin secretion is regulated by different mechanisms.

A. Mechanisms of insulin secretion

1. K_{ATP}-dependent mechanism

Glucose enters the pancreatic β-cell through the noninsulin-dependent glucose transporter (GLUT2). GLUT2 has an elevated K_m (15–20 mmol/l) and a very high V_{max} that ensures a sustained transport of glucose when blood glucose concentration increases. Once inside the cells, glucose is phosphorylated and metabolized through glycolysis. Pyruvate is formed as the terminal metabolite of glycolysis in the cytoplasm and is transported into the mitochondria where it is converted to acetyl-CoA. Subsequently, Acetyl-CoA comes into the Krebs cycle leading to the nicotinamide adenine dinucleotide (NADH) and flavine adenine dinucleotide ($FADH_2$) production. Glucose metabolism generates adenosine triphosphate (ATP) and the ATP/ADP ratio rises in the cytoplasm. The high ATP/ADP ratio provokes the closure of the ATP-sensitive K^+ channels that leads to membrane depolarization and opening of voltage-sensitive calcium channels. Ca^{2+} influx through these channels raises the intracellular Ca^{2+} concentrations and triggers exocytosis of insulin containing granules (Fig. 9.1) (Haber *et al.*, 2002; Malaisse, 1973; Thorens *et al.*, 1988).

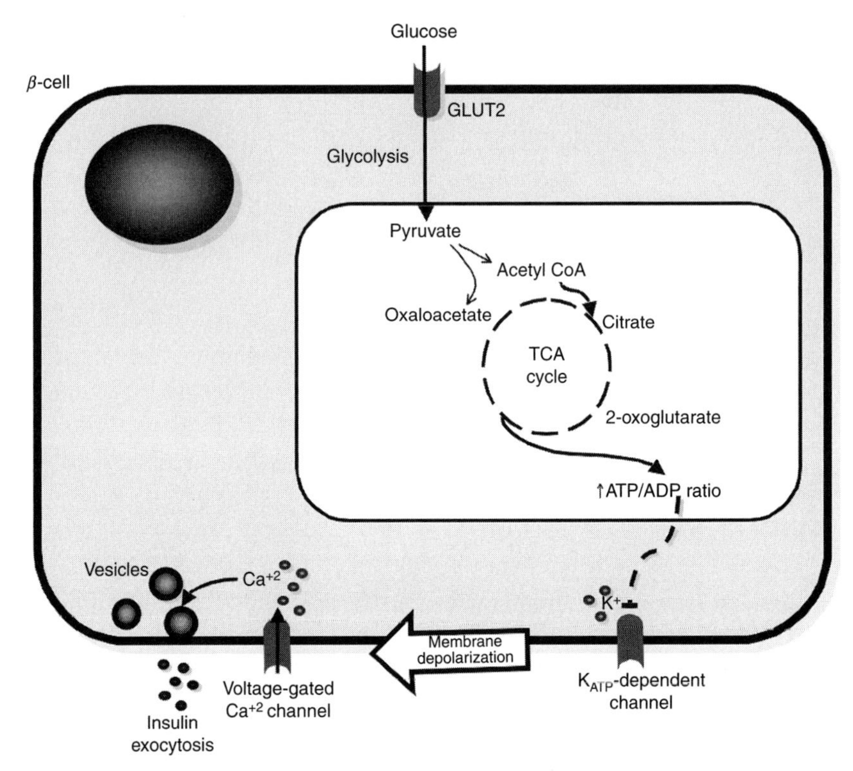

Figure 9.1 Mechanism of glucose-stimulated insulin secretion. Glucose uptake into the β-cell is carried out by the GLUT2 transporter. Intracellular glucose is oxidized to pyruvate by glycolysis, and converted to acetyl-CoA into the mitochondria by the pyruvate dehydrogenase complex. In addition, pyruvate is also converted to oxaloacetate through the anaplerotic reaction catalyzed by the pyruvate carboxylase. Both molecules are the initial substrates for the tricarboxylic acid (TCA) cycle that produces NADH + H^+ and $FADH_2$, which are used by respiratory chain, and oxidative phosphorylation to synthesize ATP. As a result there is an elevation of the ATP/ADP ratio, leading in the closure of the ATP-sensitive potassium voltage channel. Membrane depolarization activates the voltage-gated Ca^{2+} channels, increasing the intracellular concentration of Ca^{2+} which in turn promotes exocytosis of insulin vesicles resulting in insulin secretion.

2. K_{ATP}-independent mechanism

Recent evidence indicates that the mechanism of insulin secretion may be more complex, and that the K_{ATP}/Ca^{2+} pathway does not fully account for the action of nutrient stimuli. It has been demonstrated that glucose and the keto acid of leucine, the α-ketoisocaproate (KIC) stimulate insulin release independently on the K_{ATP} channel (Komatsu *et al.*, 2002). Adenine nucleotides and protein acylation are the best candidates as regulators of insulin secretion through the K_{ATP} channel-independent pathway of glucose signaling (Sato and Henquin, 1998; Straub *et al.*, 2002). The role of cAMP has been the most studied; activators of adenylate cyclase such as forskolin or others hormone as GLP-1, glucose-dependent insulinotropic polypeptide (GIP), and pituitary adenylate cyclase-activating polypeptide (PACAP) increase the levels of intracellular cAMP potentiating insulin release (Yajima *et al.*, 1999) (Fig. 9.2).

B. Hormone-dependent insulin secretion

Insulin secretion in response to glucose and other nutrients can be potentiated or repressed by hormones. The concept of the "enteroinsular axis" was conceived to describe all "intestine factors" that contribute to enhanced

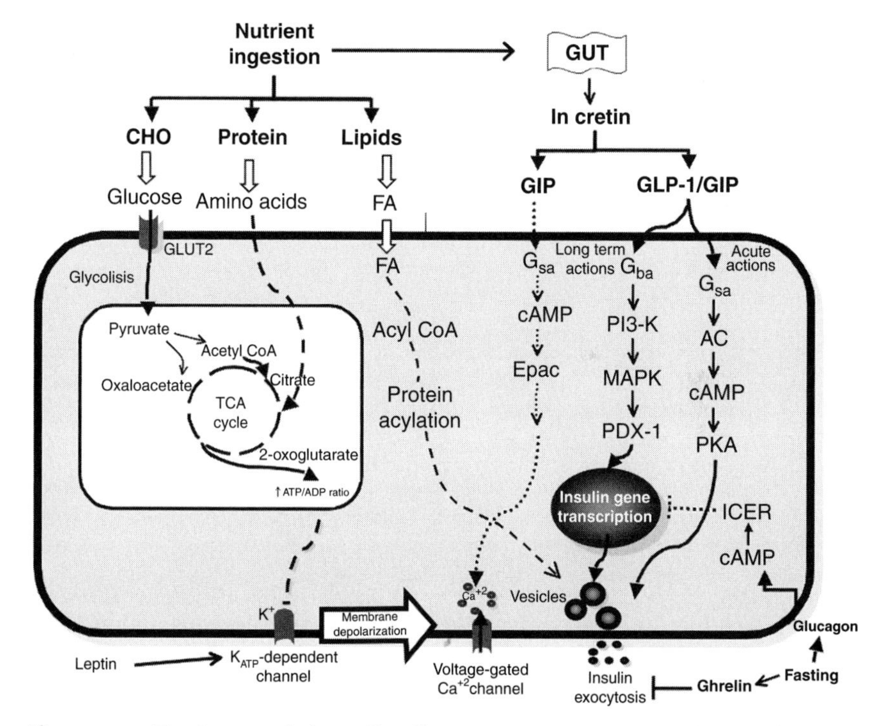

Figure 9.2 Nutrient regulation of insulin secretion. Nutrients regulate insulin secretion by different mechanisms. Glucose stimulates insulin secretion through its oxidation by glycolysis and TCA cycle increasing the ATP/ADP ratio and closure of the ATP-sensitive potassium voltage channel. Fatty acids (FA) generate acyl-CoA-stimulating exocytosis of insulin vesicles by acylation of proteins. Amino acids provide anaplerotic substrates to TCA cycle that also increase ATP/ADP ratio resulting in an increase of insulin secretion. Nutrients may stimulate release of the incretins GIP and GLP-1 from the gut. GIP activates in β-cell voltage-gated Ca^{2+} channel via cAMP. GLP-1 mediates insulin transcription via PI3-K. Glucagon and ghrelin prevent insulin gene transcription and insulin exocytosis, respectively.

insulin secretion following the ingestion of a meal (Unger and Eisentraut, 1969). It was suggested later that the axis comprised nutrient, neural, and hormonal components. An incretin must be released in response to nutrients, particularly carbohydrates, and secondly, it must stimulate insulin secretion at physiological concentrations (Creutzfeldt, 1979). The two main incretin hormones, GLP-1 and GIP, are known for their antihyperglycemic actions. On the contrary, glucagon and ghrelin are secreted in response to low nutrient availability, as in the case of fasting.

1. Glucagon

It is secreted during fasting and suppresses insulin secretion. This hormone is essential for the maintenance of blood glucose levels by stimulation of hepatic glucose output. Glucagon or the PACAP specifically induces the expression of the transcriptional repressor inducible cAMP early repressor (ICER) in pancreatic β-cells, resulting in a repression of the transcriptional expression of the insulin gene (Fig. 9.2). Remarkably, glucagon, GLP-1, and PACAP all stimulate the formation of cAMP to a comparable extent in rat pancreatic islets, but only glucagon activates the expression of ICER and represses insulin gene transcription in β-cells (Fig. 9.2). These findings suggest that the hyperglucagonemia may additionally aggravate the diabetic phenotype via a suppression of insulin gene expression mediated by the transcriptional repressor ICER (Hussain *et al.*, 2000).

2. Glucagon-like peptide-1

It is a major focus of interest because of its strong antidiabetic action, which includes the inhibition of glucagon release. GLP-1 is a potent incretin hormone produced in the L-cells of the distal ileum and colon. In the L-cells, GLP-1 is generated by tissue-specific posttranslational processing of the proglucagon gene (MacDonald *et al.*, 2002). Nutrients, including glucose, fatty acids, and dietary fiber, are all known to upregulate GLP-1 by different mechanisms. One of the effects of GLP-1 is the inhibition of β-cell K_{ATP} channels. The resulting membrane depolarization induced by K_{ATP} channel closure initiates Ca^{2+} influx through voltage-dependent Ca^{2+} channel (VDCC) and triggers the exocytotic release of insulin. This inhibitory effect of GLP-1 on K_{ATP} channels is cAMP/PKA dependent (Fig. 9.2). The physiological consequences of GLP-1 facilitated K_{ATP} channel closure would be to (1) augment the excitability of cells already above the threshold for insulin release and (2) increase the percentage of β-cells actively secreting insulin at glucose concentrations normally subthreshold for the release of insulin. In addition, GLP-1 stimulates insulin gene transcription through a phosphatidylinositol 3-kinase (PI3-K)-dependent pathway via the upregulation of the transcription factor pancreatic duodenal homeobox-1 (PDX-I). Thus, at least two signaling pathways, PI3-K and PKA, can be implicated in the effects of GLP-I on β-cell proliferation. There is evidence that suggests that both GLP-1 and GIP enhance pancreatic β-cell mass by reducing β-cell apoptosis, increasing islet cell proliferation and causing differentiation of cells to a β-cell phenotype (Flatt and Gree, 2006). The proliferative effect of GLP-1 on β-cells may partly be attributed to the stimulation of insulin gene transcription and insulin biosynthesis (Wang *et al.*, 1997).

3. Glucose-dependent insulinotropic polypeptide

It is secreted from K cells in the duodenum. Secretion of this peptide is stimulated by absorbable carbohydrates and by lipids. GIP secretion is therefore greatly increased in response to meal ingestion resulting in 10–20-fold elevations of the plasma concentrations and it is low in the fasted state. The molecular mechanisms whereby GIP potentiates glucose-dependent insulin secretion overlap considerably with those of GLP-1 and include increases in cAMP, inhibition of K_{ATP} channels, increases in intracellular Ca^{2+}, and stimulation of exocytosis. GIP stimulation of insulin secretion is mediated by activation of both cAMP/PKA and cAMP/Epac2, in addition to phospholipase A2 and specific protein kinase signaling pathways. Reduction in Epac2 expression substantially attenuates the effects of GLP-1 on insulin secretion (Drucker, 2006; Siegel and Creutzfeldt, 1985). GIP also upregulates β-cell insulin gene transcription and biosynthesis, as well as the expression of components of β-cell glucose sensors.

4. Ghrelin

It is an acylated 28-amino acid peptide isolated from stomach that inhibits insulin secretion in mice (Reimer *et al.*, 2003) and humans (Broglio *et al.*, 2001). Many studies have demonstrated an inverse relationship between fasting ghrelin and fasting insulin levels in adults and children (Heijboer *et al.*, 2006). The inhibitory effect of ghrelin on insulin secretion was suggested to be due to a tonic regulation of pancreatic β-cells, and prompting inhibition of both insulin and pancreatic somatostatin secretion (Egido *et al.*, 2002). However, more recently, it has been shown that ghrelin is also produced in a new type of islet cell, the ε -cell (Heller *et al.*, 2005), that would affect β-cells via a paracrine mechanism that may require higher local levels of ghrelin than those found in plasma (Egido *et al.*, 2002; Prado *et al.*, 2004; Wierup *et al.*, 2004). In fact, it has been proposed that insulin and ghrelin cells share a common progenitor and that Nkx2•2 and Pax4, two homeodomain proteins, are required to specify or maintain differentiation of the β-cell fate. This finding also suggests that there is a genetic component underlying the balance between insulin and ghrelin in regulating glucose metabolism (Prado *et al.*, 2004).

5. Leptin

It is one of the hormones that affect the amount of insulin that is secreted. Leptin induces β-cell hyperpolarization by opening the K_{ATP} channels on β-cells (Kieffer *et al.*, 1997) inhibiting insulin secretion and insulin mRNA levels in rat isolated pancreatic islets (Pallett *et al.*, 1997). Leptin-induced opening of K_{ATP} channels may be due to the decrease in cytosolic ATP levels in β-cells, as leptin reduces glucose transport and decreases cytosolic ATP levels in β-cells. A vital function of leptin at the cellular level is to protect nonadipose tissues from damage by nonoxidative metabolic products of long-chain FFAs via increased β-oxidation of FFAs and reduction of lipogenesis. The mechanism underlying the opening of K_{ATP} channels by leptin may involve inhibition of tyrosine kinases or activation of tyrosine phosphatases and subsequent tyrosine dephosphorylation (Harvey and Ashford, 1998). Another mechanism by which leptin may open K_{ATP} channels is through lipid metabolism in β-cells. It has been proposed that leptin increases the storage of triglycerides (TGs) in adipocytes to maintain a constant level of intracellular triglycerides in nonadipocytes, such as β-cells (Unger *et al.*, 1999). Culture of isolated islets with leptin depletes the content of triglycerides and increases the oxidation of free fatty acids. This action of leptin appears to be dependent on formation of fatty acyl-CoA. Remarkably, long-chain acyl-CoA esters (LC-CoA), the metabolically active form of free fatty acids, bind to and open K_{ATP} channels in pancreatic β-cells (Larsson *et al.*, 1996). Therefore, it is possible that elevations in LC-CoA within β-cells after exposure to leptin result in activation of K_{ATP} channels and thereby inhibition of insulin secretion.

III. Nutrient Regulation of Insulin Secretion

A. Carbohydrates

1. Regulation of insulin secretion and gene expression by carbohydrate in β-cells

The study of the mechanism for which carbohydrates induce insulin secretion was initiated with glucose in the 1960s, and rapidly led to three key discoveries. First, glucose must be metabolized by β-cells to induce insulin secretion. Second Ca^{2+} has an essential role in insulin secretion and third, pancreatic β-cells are electrically excitable. Glucose stimulation of insulin secretion involves the generation of both triggering and amplifying signals through distinct pathways (Henquin, 2000). The triggering pathway involves the entry of glucose by facilitated diffusion, metabolism of glucose by oxidative glycolysis, rise in the ATP-to-ADP ratio, closure of ATP-sensitive K^+ channels, membrane depolarization, opening of voltage-operated Ca^{2+} channels, Ca^{2+} influx, rise in cytoplasmic free Ca^{2+}, and activation of the exocytotic machinery. The amplifying pathway consists of an increase in efficacy of Ca^{2+} on exocytosis of insulin granules. The amplifying pathway serves to optimize the secretory response and is activated by all metabolized nutrients and requires glucose metabolism.

Also, elevated glucose concentrations stimulate the expression of several genes likely to impact on the differentiated function of β-cells. These includes genes involved in regulating glycolytic flux (Sic2a2 coding for glucose transporter 2, glucokinase), lipogenesis [sterol regulatory element-binding protein (SREBP-1c) (Diraison *et al.*, 2008), fatty acid synthase, acetyl-CoA carboxylase, stearoyl-CoA desaturase, carbohydrate-responsive element-binding protein], and electrical activity (Abcc8 and Kcnj11 coding for the ATP-sensitive potassium channel subunits and the sulfonylurea receptor 1). Chronic exposure of β-cells to elevated glucose concentration also reduces the PPARα gene expression, raises the malonyl-CoA concentration, inhibits fat oxidation, and promotes fatty acid esterification (Roduit *et al.*, 2000). Chronic high glucose treatment in INS-1 cells led to pronounced induction of the ER stress marker genes, BIP and Chop10 enhancing SREBP-1 binding to the human IRS2 promoter suggesting that SREBP-1 activation caused by ER stress is implicated in β-cell glucolipotoxicity (*J. Cell Sci.* 2005 Wang, H.).

2. Insulin secretion by dietary carbohydrates with different glycemic index

With the prevalence of obesity increasing worldwide, the place of carbohydrates in the diet has recently been under closer examination. This has led to the development of methods for analyzing the effects of dietary carbohydrates. Among these methods is the glycemic index (GI), a measure of the effect of food on blood glucose levels, which was initially designed as a method for determining suitable carbohydrates for people with diabetes (Wylie-Rosett *et al.*, 2004). Many high-carbohydrate foods common to Western diets produce a high glycemic response, known as high GI foods, promoting postprandial carbohydrate oxidation at the expense of fat oxidation. These diets produce substantial increases in blood glucose and insulin levels after ingestion. Within a few hours after their consumption, blood sugar levels begin to decline rapidly due to an exaggerated increase in insulin secretion (Bell and Sears, 2003) and a profound state of hunger is created. In contrast, diets based on low-fat, high foods that produce a low glycemic response (low-GI foods) they promote satiety, minimize postprandial insulin secretion, and maintain insulin sensitivity (Brand-Miller *et al.*, 2002). The GI is aimed to identifying foods that stimulate insulin secretion rather than foods that stimulate insulin resistance. Diets high in simple sugars usually have high GI and stimulate an excessive insulin response (Albrink, 1978). This has a clinical significance since the insulin response was highly correlated with the degree of hypertriglyceridemia developed by the patients who received high-carbohydrate formula (Reaven *et al.*, 1967). Besides that, long-term feeding of sucrose increases the secretion and biosynthesis of insulin (Laube *et al.*, 1976). Otherwise, fiber-rich foods generally have a low GI and show several beneficial effects, including lower postprandial glucose and insulin responses, an improved lipid profile, and, possibly, reduced insulin resistance (Riccardi *et al.*, 2008). Fiber consumption predicted lower insulin levels, weight gain, and other CVD risk factors more strongly than did total or saturated fat consumption. The mechanism by which high-fiber diets may protect against obesity and CVD could be by lowering insulin levels (Ludwig *et al.*, 1999). The insulin response to a low-fiber meal was twice as great as that to a high-fiber meal containing an equivalent amount of carbohydrates. The results suggest that carbohydrate-induced hyperlipidemia does not occur if the high-carbohydrate diet is rich in dietary fiber, and furthermore that the insulin-stimulating potential of foods in a very high-carbohydrate diet is a critical determinant of the magnitude of carbohydrate-induced lipemia (Albrink *et al.*, 1979). It has been demonstrated that three types of dietary fibers—sugarcane fiber (SCF), psyllium (PSY), and cellulose (CEL)—attenuate weight gain, enhance insulin sensitivity, and modulate leptin and GLP-1 secretion and gastric ghrelin gene expression (Wang *et al.*, 2007). Also, dietary fiber and protein increase secretion of the anorexigenic and insulinotropic hormone, GLP-1 (Reimer and Russell, 2008). Thus, the effect of fiber on insulin secretion could be an indirect effect resulting of the changes induced in hormones that regulates insulin secretion. Nevertheless, it is necessary further analysis of the mechanism involved in the regulation of dietary fiber on insulin secretion. Other carbohydrate with low GI is fructose, which has been linked with the obesity and cardiovascular disease epidemic (Segal *et al.*, 2007). Fructose is the sweetest naturally occurring monosaccharide (Bantle, 2006). Fructose does not stimulate insulin secretion *in vitro*, probably because the β-cells of the pancreas lack the fructose transporter GLUT5 (Curry, 1989; Sato *et al.*, 1996). Thus, when fructose is given *in vivo* as part of a mixed meal, the increase in glucose and insulin is much smaller than when a similar amount of glucose is given. Moreover, the increase in fructose consumption is associated with no change in ghrelin concentration and low stimulation of leptin secretion, suggesting that fructose mainly as high-fructose corn syrup in beverages, is an important factor in the epidemic of obesity. Fructose ingestion favors *de novo* lipogenesis, which could increase adiposity. In addition, fructose intake could increase overall food intake because of decreased satiety (Bray *et al.*, 2004; Teff *et al.*, 2004). One of the consequences of high concentrations of glucose is the formation of hexosamines that induce insulin resistance, impair insulin secretion, and affect hepatic glucose production. This has led to the hypothesis that cells use hexosamine flux as a glucose- and satiety-sensing pathway. Overexpression of the rate-limiting enzyme glutamine: fructose-6-phosphate aminotransferase (GFA) for the hexosamine synthesis in transgenic animals, produces insulin resistant in skeletal muscle, accumulation of fatty acids in liver, and increase in insulin secretion in the β-cell leading to hyperinsulinemia (McClain, 2002). Thus, regulation of other proteins through the hexosamine pathway contributes to the regulation of insulin secretion (Ohtsubo *et al.*, 2005).

B. Lipids

1. Overview of insulin secretion by lipids

The insulin secretion is influenced, at any given time, by the blood glucose concentration and by the prevalent FA in the circulation. FA and other lipid molecules are important for many cellular functions in the β-cell, including vesicle exocytosis. For the pancreatic β-cell, while the presence of some FAs is essential for glucose-stimulated insulin secretion (GSIS), FAs have enormous capacity to amplify GSIS, which is particularly operative in situations of β-cell compensation for insulin resistance. Recently, the β-cell was considered as lipogenic tissue that uses anaplerosis to synthesize lipids. The predominant lipids in unstimulated or secretagogue-stimulated pancreatic β-cells are cholesterol esters followed by phospholipids and FFAs (MacDonald *et al.*, 2008).

FAs regulate insulin secretion via three interdependent processes, assigned as "trident model" of β-cell signaling (Nolan *et al.*, 2006b). The first two pathways of the model implicate intracellular metabolism of FAs, whereas the third is related to membrane free fatty acid receptor (FFAR) activation.

a. First pathway: Malonyl-CoA/long-chain CoA signaling Pancreatic β-cell is a tissue that uses anaplerosis to synthesize lipids than a tissue that oxidizes lipids for fuel (MacDonald *et al.*, 2008). In pancreatic β-cells, the cytoplasmic FFA are converted to long-chain acyl-CoA (LC-CoA) by acyl-CoA synthase increasing the availability of LC-CoA that is important to stimulate the insulin exocytotic machinery (Nolan *et al.*, 2006a). On basal conditions, the LC-CoA molecules are transported into the mitochondria via carnitine-palmitoyl-transferase 1 (CPT-1), where β-oxidation takes place (Fig. 9.3A). However, high levels of glucose forms acetyl-CoA and subsequently malonyl-CoA that supplies the two-carbon units for fatty acid synthesis inhibiting β-oxidation by CPT-1 and inducing a marked rise in the cytoplasmic content of LC-CoA (Fig. 9.3B). Thus, malonyl-CoA acts by switching β-cell metabolism from FA oxidation to glucose oxidation. LC-CoA is responsible to trigger different mechanisms to release insulin. LC-CoA signaling includes (1) activation of certain types of protein kinase C (PKC) (Yaney *et al.*, 2000) that interact with components of the

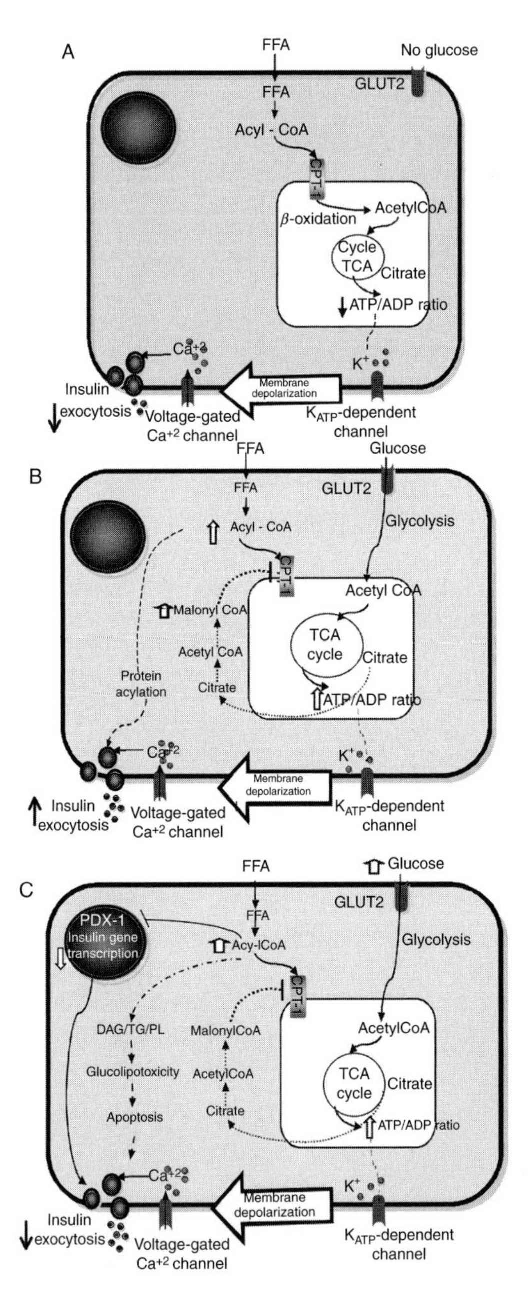

Figure 9.3 Insulin secretion during fasting, acute, and chronic of fatty acids and glucose. (A) During fasting, free fatty acids (FFA) are activated in the β-cell, and then oxidized via β-oxidation. The absence of glucose decreases the production of anaplerotic substrates reducing the ATP/ADP ratio preventing the closure of ATP-sensitive potassium voltage channels reducing insulin exocytosis. (B) Acute exposition of FA and

microtubular/exocytotic machinery (Deeney *et al.*, 1996), (2) modulation of ion channels in direct or indirect way (Ordway *et al.*, 1991), and (3) protein acylation of GTP-binding proteins (Hwang and Rhee, 1999).

a.1. Activation of PKC FA and their metabolites are able to activate PKC signal-transduction pathway involved in GSIS. Increased diacylglycerols (DAG) and Ca^{2+} lead to insulin secretion mediated by PKC that interact with components of the microtubular/exocytotic machinery. Another function purposed for the PKC is the activation of adenylate cyclase and subsequent increase of the intracellular content of cAMP. In addition, elevation of cAMP and protein kinase A (PKA) activity may exert and inhibitory effect on inositol phospholipids turnover and the PKC pathway. The mechanism by which PKA activation augments glucose-stimulated insulin release may involve potentiation of glucose-induced Ca^{2+} influx by either cAMP-dependent protein phosphorylation of the voltage-sensitive Ca^{2+} channels, or by sensitization of the secretory mechanism to Ca^{2+} increasing the efficiency of the secretory pathway by reversible phosphorylation (Lester *et al.*, 2001).

a.2. Modulation of ion channels FA can affect the plasma membrane ionic channels in a direct or indirect way. The direct effect of FA (myristic and arachidonic acids) on ionic channels was demonstrated in K^+ and Ca^{2+} channels. The medium-chain FA (MCFA) are more potent than the shorter- or long-chain FA. The reason for this specificity is probably due to the hydrophobicity of the MCFA, which is not so insoluble in water as the large-chain FA and not so soluble as the short-chain FA. MCFA are then sufficiently hydrophobic to interact with proteic portions of the plasma membrane. The FA can directly interact with the ion channel or are inserted as component of the plasma membrane lipid bilayer. The indirect effects of these compounds are mediated by the products of its metabolism, especially by the ones that can be oxygenated through cyclooxygenase, lipoxygenase, and P-450 cytochrome oxidation ways, as arachidonic acid (AA). AA is major acyl component of glycerolipids, constituting >30% of the glycerolipid fatty acid mass in rodent islets (Haber *et al.*, 2002).

glucose increase the ATP/ADP ratio, membrane depolarization via activation of ATP-sensitive potassium channels resulting in insulin exocytosis, FFA oxidation is reduced due to an increase of malonyl-CoA, thus FFA are activated with CoA, increasing protein acylation resulting in an amplification of the insulin secretion response. (C) After chronic consumption of glucose and FA, there is an excess in the formation of acyl-CoA, resulting in a decrease of insulin gene transcription, an excessive accumulation of lipids that generate glucolipotoxicity and β-cell apoptosis resulting in a severe decrease in insulin secretion.

a.3. Protein acylation of GTP-binding proteins Fatty acylation of proteins seems to be essential for the process of signaling through α-subunits of trimeric GTP-binding proteins (G proteins) and Src family tyrosine kinases. The role of proteins acylation include: directioning proteins to appropriate membranes sites, stabilizing protein–protein interactions, and regulating enzymatic activities in mitochondria (Berthiaume *et al.*, 1994), and binding to nuclear transcriptional factors such as the peroxisome proliferators-activated receptors (PPAR) which influence gene expression in islets (Lemberger *et al.*, 1996).

b. Second pathway: Fatty acids and lipid receptor signaling FFAs are natural ligands for a small group of G protein-coupled transmembrane receptors, including GPR40/FFAR1 selectively expressed in β-cells (Itoh *et al.*, 2003). Activation of GPR40 by FFAs causes an increase in intracellular Ca^{2+} levels, which is believed to be via activation of the Gαθ-phospholipase C pathway with release of Ca^{2+} from the endoplasmic reticulum increasing Ca^{2+} uptake from the extracellular Ca^{2+} pool, which then affects exocytosis. FFA exerts divergent effects on insulin secretion from β-cell: acute exposure to palmitic acid stimulate insulin secretion through GPR40, whereas chronic exposure of saturated fatty acids impairs insulin secretion (Stenneberg *et al.*, 2005) producing accumulation of triglycerides, ceramides, and apoptosis of the β-cell, an effect called lipotoxicity.

c. Third pathway: Triglyceride/FFA cycling In addition to FA oxidation and esterification, lipolysis is the major pathway of intracellular FA partitioning. Lipolysis of intracellular TG refers to the hydrolytic removal of the fatty acyl chains from the glycerol backbone by lipase enzymes. It was demonstrated that elevated glucose increased lipolysis increasing the levels of LC-CoA, DAG, phospholipids, and FFAs. LC-CoA can be used to acylate proteins. DAG not only activate PKC, which is implicated in insulin secretion, but also bind to the C1 domain of the synaptic vesicle priming protein Munc-13, which has recently been shown to be important for normal insulin secretion. TG/FFA will also affect membrane glycerophospholipid metabolism, which could influence secretion via alteration of membrane physicochemical properties. The glycerophospholipids may also have more direct effects. It may be that TG/FFA cycle is a means of targeting the delivery of FFAs, and perhaps specific FFAs such as AA, to a particular subcellular site within the β-cell (Fig. 9.4).

2. The role of fatty acids on insulin secretion

Fasting is associated with a decreased insulin response to glucose. In rats deprived of food for 18–24 h, the ability of the β-cell to secrete insulin in response to a glucose load is fully dependent upon the elevated levels of

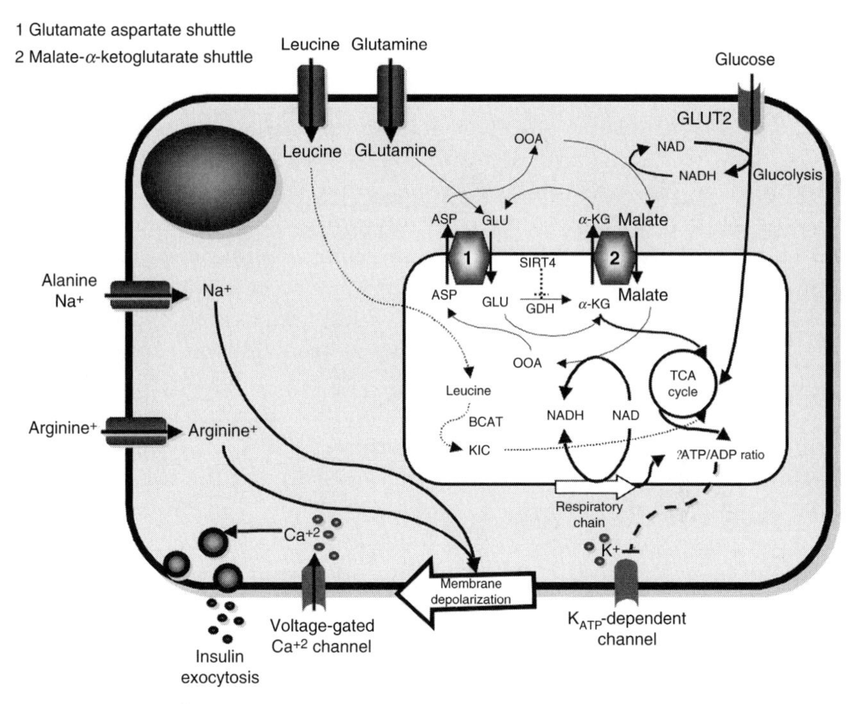

Figure 9.4 Insulin secretion mediated by amino acids. Some amino acids such as alanine and arginine are cotransported with Na^+ into the β-cell provoking a membrane depolarization activating voltage-gated Ca^{2+} channels increasing insulin vesicles exocytosis. Other amino acids such as leucine and glutamine provide anaplerotic substrates to the TCA cycle that increase ATP/ADP ratio increasing insulin secretion. SIRT4 represses the activity of glutamate dehydrogenase through ADP-ribosylation. In this way, SIRT4 downregulates insulin secretion by β-cells in response to amino acids.

circulating FFA characteristic of the fasted state. In fact, circulating FAs are essential for an efficient glucose stimulation of insulin release after prolonged fasting in humans and rats. Two major mechanisms have been postulated to explain the fact that GSIS relies on the circulating levels of FFA in fasted rats. Under starvation, the β-cell is not able to convert glucose into malonyl-CoA, possibly as consequence of the low pyruvate dehydrogenase activity. As a result, there is no suppression of CPT-1 activity, which leads to insufficient rise of the cytosolic acyl-CoA concentration and so the insulin secretory process does not occur (Fig. 9.3A). It has been shown that FA deprivation of islet tissue causes loss of GSIS, a process rapidly reversible by replacement with exogenous FFAs (Boden *et al.*, 1998; Stein *et al.*, 1996).

Studies with isolated islets and insulin-secreting cell lines have shown that exposure to FA can have a myriad of effects on metabolism and gene expression in the islet depending of the time of exposure to the fatty acid, the type of fatty acid and the presence of glucose. Acute exposure (1–3 h) of pancreatic islets to FFA enhances insulin secretion and plays a critical role to modulate the stimulatory effect of glucose on insulin release (Vara *et al.*, 1988). However, long-term exposure (6–24 h) to high FFA concentration can inhibit insulin secretion. Long-term exposure to high FFA concentrations particularly in association with elevated glucose can reduce insulin biosynthesis (Poitout *et al.*, 2006) and secretion (Prentki *et al.*, 2002), and induce cell apoptosis (El-Assaad *et al.*, 2003; Lee *et al.*, 1994; McGarry, 2002; Prentki *et al.*, 2002) (Fig. 9.3C). Rat and human islets exposed to FA for 48 h show increased insulin release at basal glucose concentration (3 mM) and decreased release when glucose concentration is high (27 mM) (Zhou and Grill, 1995). An analogous pattern is evident in long-term experiments in rats fed enriched saturated (SFA) or polyunsaturated diets. Comparison of two diets, one rich in SFAs and the other with the same amount of calories from MUFAs, have shown an improvement in insulin sensitivity in the latter, while β-cell function was unchanged. However, the beneficial impact of fat quality on insulin sensitivity was poor when the fat intake was larger than 37% of total daily intake, and the addition of long-chain ω_3-fatty acids to the diet failed to produce any improvement in insulin sensitivity. We have observed that prolonged exposure of rats to enriched diets with different degree of saturation for two months, indicated that animals fed more saturated fat as coconut oil showed the highest plasma insulin concentrations pM (28.7 $\pm$ 3.9) followed by butter (19.8 $\pm$ 1.5), lard (18.6 $\pm$ 1.8), whereas animals fed mono- or polyunsaturated fatty acids as oleic (12.3 $\pm$ 2.9) or soybean oil (11.6 $\pm$ 2.4) showed the lowest insulin concentrations. Other experiments have shown that exposure to high saturated fat diets showed high insulin responses during hyperglycemic clamps but not by unsaturated fat (Dobbins *et al.*, 2002).

3. Lipids and gene expression in β-cell

Analysis of the relative contribution of elevated glucose and FA in the etiology of impaired β-cell function has led to the suggestion that neither hyperglycemia alone nor elevated circulating FAs alone are necessarily detrimental to β-cell function. Problems arise when both glucose and FA concentrations are concomitantly high. Thus, whereas an elevation in FA without hyperglycemia increase FA oxidation, a combination of excess glucose and FAs is proposed to result in the diversion of FAs from oxidation toward the formation of extramitochondrial lipid-derived signaling molecules, which initially modify insulin secretion but ultimately cause β-cell death, process called glucolipotoxicity (Prentki *et al.*, 2002). Glucolipotoxicity implies biochemical changes that increase malonyl-CoA concentration resulting in altered lipid partitioning (Fig. 9.3C). In addition, there are also changes in the expression of several transcription factors including peroxisome proliferators-activated receptor PPARα and PPARγ, SREBP-1c and PDX, and insulin gene expression (Manco *et al.*, 2004). Both PPARα and PPARγ dose-dependently increased the FA uptake of INS-1E cells. However, the two PPARs directed the incoming FAs along different metabolic pathways. Whereas activation of PPARα induces β-oxidation of the FAs, ectopic expression of PPARγ facilitated lipid accumulation in the β-cell. Under normal physiological conditions, β-cell expression of PPARα exceeds that of PPARγ. However, under hyperglycemic or hyperlipidemic conditions, the picture would change, mimicked by PPARγ overexpression, and lipid accumulation would be facilitated. Chronic exposure of intact islets to high FA enhances endogenous PPARα expression and PPARα-linked genes and elevates insulin secretion at low glucose concentration. However the presence of high glucose concentrations suppresses PPARα expression (Sugden and Holness, 2004). These results reflect which fuel is more dominant under the precise pathophysiological condition studied.

Unsaturated fatty acids such as oleate and linoleate bind to the proliferators-activated receptor-α (PPARα) or (PPARγ) with dissociation constants in the nanomolar range (Lin *et al.*, 1999), whereas saturated fatty acid as palmitate and estearate do not activate PPARα at physiological concentrations.

One gene that is upregulated by PPARα in islets is uncoupling protein-2 (UCP-2). Uncoupled proteins disrupt the electrochemical gradient across the inner mitochondrial membrane, producing heat instead of ATP from respiration. The induction of UCP2 suppresses insulin secretion in normal islets (Tordjman *et al.*, 2002), since UCP2 enhances proton leak in isolated mitochondria from INS-1 cells, suggesting that PPARα-mediated induction of UCP2 uncouples respiration and prevents a rise in ATP despite accelerated fatty acid oxidation and downregulates the ability of β-cells to secrete insulin in response to glucose (Zhang *et al.*, 2001). Prolonged *in vitro* exposure of islets to above physiological concentrations of palmitate increase lipogenic activity resulting in a increase in the neutral lipids diacylglycerol (DAG) and triglycerides content, and decreases the insulin content and impairs insulin gene expression only in the presence of elevated glucose levels (Briaud *et al.*, 2001). This effect seems to be mediated by a decreased insulin gene promoter activity in HIT-T15 cells and a decreased binding of the transcription factor, pancreas–duodenum homeobox-1 (PDX1) and MafA to the insulin gene in islets (Poitout and Robertson, 2007).

Alternatively, the saturation of cellular fatty acids would influence the fluidity of the β-cell membrane. The inhibition of protein acylation in the pancreatic islets impairs nutrient stimulation of insulin release and the acylation of proteins.

The exposure of isolated rat pancreatic islets to palmitate for 48 h, in the presence of excessive glucose concentrations, decreases the expression of the transcription factor IDX-1 as well as the genes transactivated by IDX-1. IDX-1 is a transcription factor expressed in the duodenum and pancreatic beta and delta cells. It is required for embryonic development of the pancreas and transactivates the GLUT2, glucokinase, insulin, and somatostatin gene expression. Other enzyme involved in the oxidation of fatty acids in the β-cell is CPT-1. CPT-1 may serve as a key gene controlling fuel partitioning in β-cells, leading to production of lipids that can downregulate GLUT2 and glucokinase expression.

C. Proteins

1. Effect of dietary proteins on insulin secretion

Although our understanding of fat and carbohydrates as nutrients affecting insulin secretion has greatly increased in the past two decades, the role of proteins and type of protein and amino acids in the regulation of insulin secretion has been less characterized. Ingestion of dietary protein is known to induce both insulin and glucagon secretion. These responses may be affected by the type of protein consumed. Previous studies in our group have demonstrated that consumption of a high quality vegetable protein as soy induces in lesser extent insulin secretion than high quality animal protein as casein, and it was associated with a lower induction of hepatic SREBP-1 reducing lipogenesis (Ascencio *et al.*, 2004). In addition, we have also demonstrated that long-term consumption of soy protein increases glucagon concentration resulting in a low insulin:glucagon ratio. These results suggest that the type of protein modulates insulin secretion even the presence of high-fat content in the diet (Torre-Villavazo *et al.*, 2008). However, it has been questioned whether the effect observed by soy protein is due to its amino acid pattern or the presence of bioactive compounds such as isoflavones associated with the protein. Recently, we demonstrated that the plasma amino acid pattern generated after the consumption of soy protein stimulated in a lesser extent insulin secretion. However, the infusion of genistein one of the main isoflavones in the soy also repressed insulin secretion, indicating that both the amino acid pattern and the isoflavones regulate insulin secretion (Noriega-Lopez *et al.*, 2007). Also, we and others have demonstrated that consumption of fish or vegetable proteins improves insulin sensitivity (Lavigne *et al.*, 2000; Noriega-Lopez *et al.*, 2007).

The effect of dietary protein on insulin secretion seems to depend of both, the quality and quantity of proteins. Diets with low protein content have been associated with a reduction in insulin secretion in lean and genetically obese rats (Tse *et al.*, 1995). Low protein diet reduces the expression of PKAα, PKCα, PLC (Ferreira *et al.*, 2004) and S6K-1 (Filiputti *et al.*, 2008) involved in the insulin secretion pathway resulting in reduced insulin secretion by glucose. On the other hand, high protein diets causes hypersecretion of insulin *in vivo* and *in vitro* (Usami *et al.*, 1982) and increase β-cell responsiveness to glucose. Nevertheless, dietary protein per se does not directly induce changes in insulin secretion, the amino acids and peptides product of the dietary protein degradation seems to be the main responsible of this effect. Proteins in the intestine can be partially degraded to peptides, which have specific amino acid sequences that can express some biological functions. Many proteins such as casein, whey protein, fish meat, wheat, and soybean protein are source of bioactive peptides (Saito, 2008).

2. Mechanism of insulin secretion by dietary amino acids

The absorption rates of individual amino acids are highly dependent on the protein source, therefore, proteins such as casein, soy or cod protein generates different amino acids patterns after its consumption (Lavigne *et al.*, 2001). The proportion and amount of amino acids conforming a specific pattern has shown to produce different response on insulin secretion (Noriega-Lopez *et al.*, 2007). The role of amino acids and their regulatory effects on insulin secretion are complex and involve several pathways in the β-cell (Newsholme *et al.*, 2007b) (Fig. 9.4). However, unlike glucose, individual amino acids do not provoke insulin secretion *in vitro* when added at physiological concentrations. Combinations of amino acids at physiological concentrations or high concentrations of individual amino acids are much more effective. The mechanisms by which amino acids enhance insulin secretion are varied depending on the type of amino acid. *L-Arginine*, a cationic-charged amino acid, enhances insulin secretion by direct depolarization of the plasma membrane at neutral pH (Blachier *et al.*, 1989). *L-Alanine* is cotransported with Na^+ and depolarizes the cell membrane as a consequence of Na^+ transport that induces insulin secretion by activating voltage-dependent calcium channels. The metabolism of alanine resulting in partial oxidation, may initially increase the cellular content of ATP, leading to closure of the ATP-sensitive K^+ channel, depolarization of the plasma membrane, activation of voltage-activated Ca^{2+} channel, Ca^{2+} influx, and insulin exocytosis (Dunne *et al.*, 1990; Henquin and Meissner, 1986). *Aspartate and glutamate* play a key role in NADH shuttles, the latter represents an important mechanism for the maintenance of glucose metabolism. The malate–aspartate shuttle allows the movement of NADH from the cytoplasm to the mitochondria. Into the mitochondria, NADH is oxidized into NAD and the derived electrons are transferred to the electron transport chain, which creates the proton electrochemical gradient driving ATP synthesis with the consequent insulin release. Inhibition of the malate–aspartate shuttle attenuates the secretory response to nutrients, and overexpression of this transporter enhanced glucose-evoked NADH generation, electron transport chain activity and mitochondrial ATP formation (Eto *et al.*, 1999; Rubi *et al.*, 2004). In addition, *L-glutamate* is the most debated amino acid with respect to the possible molecular mechanism of its action on promotion of insulin secretion (Newsholme *et al.*, 2007a). It has been suggested that glutamate may serve as intracellular messenger (Hoy *et al.*, 2002; Maechler *et al.*, 1997) since it could be transported into secretory granules, thereby promoting Ca^{2+}-dependent exocytosis (Bai *et al.*, 2001; Maechler and Wollheim, 1999). Other support for the glutamate hypothesis arise from a study in which β-cells overexpressing L-glutamate decarboxylase showed a reduced glutamate content and reduction in GSIS (Rubi *et al.*, 2001). *L-Glutamine* promotes and maintains the function of the pancreatic β-cells. L-Glutamine is consumed at high rates by β-cells to maintain the high protein turnover for purine and pyrimidine synthesis, mRNA production, and protein synthesis. Although L-glutamine is rapidly taken up and metabolized by islets, it alone does not stimulate insulin secretion or enhance glucose-induced insulin secretion. However, activation of glutamate dehydrogenase (GDH) by addition of leucine enhances insulin secretion by increasing the entry of glutamine carbon into the tricarboxylic acid cycle (Sener and Malaisse, 1980). It is known that glucose inhibits glutaminolysis in β-cells, presumably via GTP-dependent allosteric inhibition of GDH, resulting in accumulation of L-glutamate and thus product-dependent inhibition of glutaminase, and as result blocks leucine-stimulated insulin secretion. Furthermore, glutamine–insulin secretion is a mechanism that offers a pathway of regulation of insulin secretion under conditions that span the range of food availability. Loss of SIRT4, a mitochondrial enzyme that uses NAD to ADP-ribosylate and downregulate GDH activity, increases considerably insulin secretion stimulated by amino acids. A similar effect was observed in β-cells from mice on the dietary regimen of caloric restriction (Ahuja *et al.*, 2007; Haigis *et al.*, 2006). It has been suggested that caloric restriction downregulates SIRT4 activity increasing GDH activity. This increase in GDH activity during caloric restriction potentiates amino acids, particularly glutamine, as insulin secretagogue. Branched chain amino acids (BCAAs) play pivotal roles in hormonal secretion and action as well as in intracellular signaling. When dogs are infused intravenously with individual amino acids, the magnitude of insulin secretion was greatest with tryptophan, leucine, and aspartate, whereas valine, alanine, and histidine had minimal effect on insulin (Rocha *et al.*, 1972). These results provide an example of two amino acids with similar structures (both leucine and valine are BCAAs) but with different physiological effects (Nair and Short, 2005). Interestingly, α-ketoisocaproate, the transamination product of leucine by BCAA transferase, elicited a dramatic insulin secretory response in BTBR islets. This response was associated with an increase in glutamate transamination triggering α-ketoglutarate formation, which is metabolized by the TCA cycle increasing insulin secretion (Rabaglia *et al.*, 2005). These results suggest that the main responsible of insulin secretion stimulated by leucine is the KIC, since depletion of branched chain amino acid transferase, decrease considerably insulin levels in spite of high leucine concentrations (She *et al.*, 2007). Finally, dietary amino acids also influence insulin secretion by indirect mechanisms. For example, dietary amino acids may stimulate incretin release, such as GLP-1, from intestinal l-cells and therefore stimulate insulin secretion (Gameiro *et al.*, 2005; Reimann *et al.*, 2004). In addition, in periods of fasting or starvation, amino acid release from skeletal muscle (primarily L-glutamine and L-alanine) may modulate glucagon release from pancreatic α-cells, which subsequently may influence insulin secretion from β-cells (Chang and Goldberg, 1978). Amino acids may also modulate lipid metabolism; L-glutamine upregulates Acetyl-CoA carboxylase (ACC) responsible for malonyl-CoA synthesis, which in turn stimulate fatty acid synthesis thereby enhancing Ca^{2+}-evoked insulin exocytosis (Corless *et al.*, 2006). ACC is also regulated by phosphorylation via AMP-regulated kinase an enzyme sensitive to amino acid concentration (Xiang *et al.*, 2004).

3. Amino acids and gene expression in β-cells

Amino acids also regulate the expression of genes related to β-cell signal transduction, metabolism and apoptosis (Newsholme *et al.*, 2005). Incubation of BRIN-BD11 β-cells line with L-alanine revealed that a total of 66 genes increased 1.8-fold or greater after 24-h incubation (Cunningham *et al.*, 2005). Otherwise, 10 mmol/l L-glutamine increased insulin secretion rate of BRIN-BD11 β-cell line by 30% compared with 1 mmol/l glutamine, which was associated with upregulation of 148 genes at least 1.8-fold and downregulation of 18 genes. These microarray analyses indicates that glutamine may be required for the optimal *in vivo* and *in vitro* differentiation of pancreas-derived stem cells toward the β-cell phenotype, β-cell endocrine function, and optimal lipid synthesis (Corless *et al.*, 2006). In addition to the effect of individual amino acid on gene expression, an important issue to be analyzed is the effect of combination of several amino acids on β-cell gene expression. The plasma amino acid concentration observed after casein consumption increases PPARγ and GLUT2 expression levels in pancreatic islets with respect to the plasma amino acid concentration after soy protein consumption (Noriega-Lopez) indicating that specific pattern of amino acids modulates de entrance of glucose to β-cell and regulates insulin secretion.

Amino acids also are involved in activating and/or modulating signaling pathways that can affect insulin action at the cellular and molecular levels via mTOR (mammalian target of rampamycin) (Averous *et al.*, 2003) by controlling components of the translational machinery, including initiation and elongation factors. The most effective activator is the single amino acid leucine that at physiological concentrations stimulates p70S6K phosphorylation via the mTOR pathway. This activation contribute to enhanced

β-cell function by stimulating growth-related protein synthesis and proliferation associated with the maintenance of β-cell mass (Xu *et al.*, 2001). Recently, It has been demonstrated that phosphorylation of p70S6K can be blocked by activation of AMPK that might be an important component of the mechanism for nutrient-stimulated mTOR activity (Gleason *et al.*, 2007). Finally, as we mention above, the β-cell expresses components of the insulin-signaling cascade. Insulin and IGFs in concert with the nutrients leucine, glutamine, and glucose have the ability of modulate protein translation through mTOR in β-cells (McDaniel *et al.*, 2002). These combined transduction processes make the β-cell an unique cell to study metabolic and autocrine regulation of mTOR signaling and its effects on insulin secretion.

REFERENCES

Ahuja, N., Schwer, B., Carobbio, S., Waltregny, D., North, B. J., Castronovo, V., Maechler, P., and Verdin, E. (2007). Regulation of insulin secretion by SIRT4, a mitochondrial ADP-ribosyltransferase. *J. Biol. Chem.* **282,** 33583–33592.

Albrink, M. J. (1978). Dietary fiber, plasma insulin, and obesity. *Am. J. Clin. Nutr.* **31,** S277–S279.

Albrink, M. J., Newman, T., and Davidson, P. C. (1979). Effect of high- and low-fiber diets on plasma lipids and insulin. *Am. J. Clin. Nutr.* **32,** 1486–1491.

Ascencio, C., Torres, N., Isoard, F., Gómez-Pérez, F. J., Hernández-Pando, R., and Tovar, A. R. (2004). Soy protein affects serum insulin and hepatic SREBP-1 mRNA and reduces fatty liver in rats. *J. Nutr.* **134,** 522–529.

Averous, J., Bruhat, A., Mordier, S., and Fafournoux, P. (2003). Recent advances in the understanding of amino acid regulation of gene expression. *J. Nutr.* **133,** 2040S–2045S.

Bai, L., Xu, H., Collins, J. F., and Ghishan, F. K. (2001). Molecular and functional analysis of a novel neuronal vesicular glutamate transporter. *J. Biol. Chem.* **276,** 36764–36769.

Bantle, J. P. (2006). Is fructose the optimal low glycemic index sweetener? *Nestle Nutr. Workshop Ser. Clin. Perform. Programme* **11,** 83–91discussion 92–95.

Bell, S. J., and Sears, B. (2003). Low-glycemic-load diets: Impact on obesity and chronic diseases. *Crit. Rev. Food Sci. Nutr.* **43,** 357–377.

Berthiaume, L., Deichaite, I., Peseckis, S., and Resh, M. D. (1994). Regulation of enzymatic activity by active site fatty acylation. A new role for long chain fatty acid acylation of proteins. *J. Biol. Chem.* **269,** 6498–6505.

Blachier, F., Mourtada, A., Sener, A., and Malaisse, W. J. (1989). Stimulus–secretion coupling of arginine-induced insulin release. Uptake of metabolized and nonmetabolized cationic amino acids by pancreatic islets. *Endocrinology* **124,** 134–141.

Boden, G., Chen, X., and Iqbal, N. (1998). Acute lowering of plasma fatty acids lowers basal insulin secretion in diabetic and nondiabetic subjects. *Diabetes* **47,** 1609–1612.

Brand-Miller, J. C., Holt, S. H., Pawlak, D. B., and McMillan, J. (2002). Glycemic index and obesity. *Am. J. Clin. Nutr.* **76,** 281S–285S.

Bray, G. A., Nielsen, S. J., and Popkin, B. M. (2004). Consumption of high-fructose corn syrup in beverages may play a role in the epidemic of obesity. *Am. J. Clin. Nutr.* **79,** 537–543.

Briaud, I., Harmon, J. S., Kelpe, C. L., Segu, V. B. G., and Poitout, V. (2001). Lipotoxicity of the pancreatic b-cell is associated with glucose-dependent esterification of fatty acids into neutral lipids. *Diabetes* **50,** 315–321.

Broglio, F., Arvat, E., Benso, A., Gottero, C., Muccioli, G., Papotti, M., van der Lely, A. J., Deghenghi, R., and Ghigo, E. (2001). Ghrelin, a natural GH secretagogue produced by the stomach, induces hyperglycemia and reduces insulin secretion in humans. *J. Clin. Endocrinol. Metab.* **86,** 5083–5086.

Chang, T. W., and Goldberg, A. L. (1978). The origin of alanine produced in skeletal muscle. *J. Biol. Chem.* **253,** 3677–3684.

Corless, M., Kiely, A., McClenaghan, N. H., Flatt, P. R., and Newsholme, P. (2006). Glutamine regulates expression of key transcription factor, signal transduction, metabolic gene, and protein expression in a clonal pancreatic beta-cell line. *J. Endocrinol.* **190,** 719–727.

Creutzfeldt, W. (1979). The incretin concept today. *Diabetologia* **16,** 75–85.

Cunningham, G. A., McClenaghan, N. H., Flatt, P. R., and Newsholme, P. (2005). L-Alanine induces changes in metabolic and signal transduction gene expression in a clonal rat pancreatic beta-cell line and protects from pro-inflammatory cytokine-induced apoptosis. *Clin. Sci. (Lond.)* **109,** 447–455.

Curry, D. L. (1989). Effects of mannose and fructose on the synthesis and secretion of insulin. *Pancreas* **4,** 2–9.

Deeney, J. T., Cunningham, B. A., Chheda, S., Bokvist, K., Juntti-Berggren, L., Lam, K., Korchak, H. M., Corkey, B. E., and Berggren, P. O. (1996). Reversible Ca^{2+}-dependent translocation of protein kinase C and glucose-induced insulin release. *J. Biol. Chem.* **271,** 18154–18160.

Diraison, F., Ravier, M. A., Richards, S. K., Smith, R. M., Shimano, H., and Rutter, G. A. (2008). SREBP-1 is required for the induction by glucose of pancreatic beta-cell genes involved in glucose sensing. *J. Lipid Res.* **49,** 814–822.

Dobbins, R. L., Szczepaniak, L. S., Myhill, J., Tamura, Y., Uchino, H., Giacca, A., and McGaary, J. D. (2002). The composition of dietary fat directly influences glucose-stimulated insulin secretion in rats. *Diabetes* **51,** 1825–1833.

Drucker, D. J. (2006). The biology of incretin hormones. *Cell Metab.* **3,** 153–165.

Dunne, M. J., Yule, D. I., Gallacher, D. V., and Petersen, O. H. (1990). Effects of alanine on insulin-secreting cells: Patch-clamp and single cell intracellular Ca^{2+} measurements. *Biochim. Biophys. Acta* **1055,** 157–164.

Egido, E. M., Rodriguez-Gallardo, J., Silvestre, R. A., and Marco, J. (2002). Inhibitory effect of ghrelin on insulin and pancreatic somatostatin secretion. *Eur. J. Endocrinol.* **146,** 241–244.

El-Assaad, W., Buteau, J., Peyot, M. L., Nolan, C., Roduit, R., Hardy, S., Joly, E., Dbaibo, G., Rosenberg, L., and Prentki, M. (2003). Saturated fatty acids synergize with elevated glucose to cause pancreatic beta-cell death. *Endocrinology* **144,** 4154–4163.

Eto, K., Tsubamoto, Y., Terauchi, Y., Sugiyama, T., Kishimoto, T., Takahashi, N., Yamauchi, N., Kubota, N., Murayama, S., Aizawa, T., Akanuma, Y., Aizawa, S., Kasai, H., Yazaki, Y., and Kadowaki, T. (1999). Role of NADH shuttle system in glucose-induced activation of mitochondrial metabolism and insulin secretion. *Science* **283,** 981–985.

Ferreira, F., Barbosa, H. C., Stoppiglia, L. F., Delghingaro-Augusto, V., Pereira, E. A., Boschero, A. C., and Carneiro, E. M. (2004). Decreased insulin secretion in islets from rats fed a low protein diet is associated with a reduced PKAalpha expression. *J. Nutr.* **134,** 63–67.

Filiputti, E., Ferreira, F., Souza, K. L., Stoppiglia, L. F., Arantes, V. C., Boschero, A. C., and Carneiro, E. M. (2008). Impaired insulin secretion and decreased expression of the nutritionally responsive ribosomal kinase protein S6K-1 in pancreatic islets from malnourished rats. *Life Sci.* **82,** 542–548.

Flatt, P. R., and Gree, B. D. (2006). Nutrient regulation. *Biochem. Soc. Trans.* **34**(Pt. 5), 774–778.

Gameiro, A., Reimann, F., Habib, A. M., O'Malley, D., Williams, L., Simpson, A. K., and Gribble, F. M. (2005). The neurotransmitters glycine and GABA stimulate glucagon-like peptide-1 release from the GLUTag cell line. *J. Physiol.* **569,** 761–772.
Gleason, C. E., Lu, D., Witters, L. A., Newgard, C. B., and Birnbaum, M. J. (2007). The role of AMPK and mTOR in nutrient sensing in pancreatic beta-cells. *J. Biol. Chem.* **282,** 10341–10351.
Haber, E. P., Ximenes, H. M. A., Procopio, J., Carvalho, C. R. O., Curi, R., and Carpinelli, A. R. (2002). Pleiotropic effects of fatty acids on pancreatic beta cell. *J. Cell Physiol.* **194,** 1–12.
Haigis, M. C., Mostoslavsky, R., Haigis, K. M., Fahie, K., Christodoulou, D. C., Murphy, A. J., Valenzuela, D. M., Yancopoulos, G. D., Karow, M., Blander, G., Wolberger, C., Prolla, T. A., Weindruch, R., Alt, F. W., and Guarente, L. (2006). SIRT4 inhibits glutamate dehydrogenase and opposes the effects of calorie restriction in pancreatic beta cells. *Cell* **126,** 941–954.
Harvey, J., and Ashford, M. L. (1998). Role of tyrosine phosphorylation in leptin activation of ATP-sensitive K^+ channels in the rat insulinoma cell line CRI-G1. *J. Physiol.* **510**(Pt. 1), 47–61.
Heijboer, A. C., Pijl, H., Van den Hoek, A. M., Havekes, L. M., Romijn, J. A., and Corssmit, E. P. (2006). Gut-brain axis: Regulation of glucose metabolism. *J. Neuroendocrinol.* **18,** 883–894.
Heller, R. S., Jenny, M., Collombat, P., Mansouri, A., Tomasetto, C., Madsen, O. D., Mellitzer, G., Gradwohl, G., and Serup, P. (2005). Genetic determinants of pancreatic epsilon-cell development. *Dev. Biol.* **286,** 217–224.
Henquin, J. C. (2000). Triggering and amplifying pathways of regulation of insulin secretion by glucose. *Diabetes* **49,** 1751–1760.
Henquin, J. C., and Meissner, H. P. (1986). Cyclic adenosine monophosphate differently affects the response of mouse pancreatic beta-cells to various amino acids. *J. Physiol.* **381,** 77–93.
Hoy, M., Maechler, P., Efanov, A. M., Wollheim, C. B., Berggren, P. O., and Gromada, J. (2002). Increase in cellular glutamate levels stimulates exocytosis in pancreatic beta-cells. *FEBS Lett.* **531,** 199–203.
Hussain, M. A., Daniel, P. B., and Habener, J. F. (2000). Glucagon stimulates expression of the inducible cAMP early repressor and suppresses insulin gene expression in pancreatic beta-cells. *Diabetes* **49,** 1681–1690.
Hwang, D., and Rhee, S. H. (1999). Receptor-mediated signaling pathways: Potential targets of modulation by dietary fatty acids. *Am. J. Clin. Nutr.* **70,** 545–556.
Itoh, Y., Kawamaata, Y., Harada, M., Kobayashi, M., Fujii, R., Fukusumi, S., Ogi, K., Hosoya, M., Tanaka, Y., Uejima, H., Tanaka, H., , M., , M., Satoh, R., Okubo, S., Kizawa, H., Komatsu, H., Matsumura, F., Noguchi, Y., Shinohara, T., Hinuma, S., Fujisawa, Y., and Fujino, M. (2003). Free fatty acids regulate insulin secretion from pancreatic b cells through GPR40. *Nature* **422,** 173–176.
Kieffer, T. J., Heller, R. S., Leech, C. A., Holz, G. G., and Habener, J. F. (1997). Leptin suppression of insulin secretion by the activation of ATP-sensitive K^+ channels in pancreatic beta cells. *Diabetes* **46,** 1087–1093.
Komatsu, M., Sato, Y., Yamada, S., Yamauchi, K., Hashizume, K., and Aizawa, T. (2002). Triggering of insulin release by a combination of cAMP signal and nutrients: An ATP-sensitive K^+ channel-independent phenomenon. *Diabetes* **51**(Suppl. 1), S29–S32.
Larsson, O., Deeney, J. T., Branstrom, R., Berggren, P. O., and Corkey, B. E. (1996). Activation of the ATP-sensitive K^+ channel by long chain acyl-CoA. A role in modulation of pancreatic beta-cell glucose sensitivity. *J. Biol. Chem.* **271,** 10623–10626.
Laube, H., Schatz, H., Nierle, C., Fussganger, R., and Pfeiffer, E. F. (1976). Insulin secretion and biosynthesis in sucrose fed rats. *Diabetologia* **12,** 441–446.

Lavigne, C., Marette, A., and Jacques, H. (2000). Cod and soy proteins compared with casein improve glucose tolerance and insulin sensitivity in rats. *Am. J. Physiol. Endocrinol. Metab.* **278,** E491–E500.

Lavigne, C., Tremblay, F., Asselin, G., Jacques, H., and Marette, A. (2001). Prevention of skeletal muscle insulin resistance by dietary cod protein in high fat-fed rats. *Am. J. Physiol. Endocrinol. Metab.* **281,** E62–E71.

Lee, Y., Hirose, H., Ohneda, M., Johnson, J. H., McGarry, J. D., and Unger, R. H. (1994). Beta-cell lipotoxicity in the pathogenesis of non-insulin-dependent diabetes mellitus of obese rats: Impairment in adipocyte-beta-cell relationships. *Proc. Natl. Acad. Sci. USA* **91,** 10878–10882.

Lefebvre, P. J., Paolisso, G., Sccheen, A. J., and Henquin, J. C. (1987). Pulsatility of insulin and glucagon release: Physiological significance and pharmacological implications. *Diabetologia* **30,** 443–452.

Lemberger, T., Desvergne, B., and Wahli, W. (1996). Peroxisome proliferator-activated receptors: A nuclear receptor signaling pathway in lipid physiology. *Annu. Rev. Cell Dev. Biol.* **12,** 335–363.

Lester, L. B., Faux, M. C., Nauert, J. B., and Scott, J. D. (2001). Targeted protein kinase A and PP-2B regulate insulin secretion through reversible phosphorylation. *Endocrinology* **142,** 1218–1227.

Lin, Q., Ruuska, S. E., Shaw, N. S., Dong, D., and Noy, N. (1999). Ligand selectivity of the peroxisome proliferator-activated receptor alpha. *Biochemistry* **38,** 185–190.

Ludwig, D. S., Pereira, M. A., Kroenke, C. H., Hilner, J. E., Van Horn, L., Slattery, M. L., and Jacobs, D. R. Jr., (1999). Dietary fiber, weight gain, and cardiovascular disease risk factors in young adults. *JAMA* **282,** 1539–1546.

MacDonald, P. E., El-Kholy, W., Riedel, M. J., Salapatek, A. M., Light, P. E., and Wheeler, M. B. (2002). The multiple actions of GLP-1 on the process of glucose-stimulated insulin secretion. *Diabetes* **51,** S434–S442.

MacDonald, M. J., Dobrzyn, A., Ntambi, J., and Stoker, S. W. (2008). The role of rapid lipogenesis in insulin secretion: Insulin secretagogues acutely alter lipid composition of INS-1 832/13 cells. *Arch. Biochem. Biophys.* **470,** 153–162.

Maechler, P., and Wollheim, C. B. (1999). Mitochondrial glutamate acts as a messenger in glucose-induced insulin exocytosis. *Nature* **402,** 685–689.

Maechler, P., Kennedy, E. D., Pozzan, T., and Wollheim, C. B. (1997). Mitochondrial activation directly triggers the exocytosis of insulin in permeabilized pancreatic beta-cells. *EMBO J.* **16,** 3833–3841.

Malaisse, W. J. (1973). Insulin secretion: Multifactorial regulation for a single process of release. The Minkowski award lecture delivered on September 7, 1972 before the European Association for the study of Diabetes at Madrid, Spain. *Diabetologia* **9,** 167–173.

Manco, M., Calvani, M., and Mingrone, G. (2004). Effects of dietary fatty acids on insulin sensitivity and secretion. *Diabetes Obes. Metab.* **6,** 402–413.

McClain, D. A. (2002). Hexosamines as mediators of nutrient sensing and regulation in diabetes. *J. Diabetes Complications* **16,** 72–80.

McDaniel, M. L., Marshall, C. A., Pappan, K. L., and Kwon, G. (2002). Metabolic and autocrine regulation of the mammalian target of rapamycin by pancreatic beta-cells. *Diabetes* **51,** 2877–2885.

McGarry, J. D. (2002). Banting lecture 2001: Dysregulation of fatty acid metabolism in the etiology of type 2 diabetes. *Diabetes* **51,** 7–18.

Nair, K. S., and Short, K. R. (2005). Hormonal and signaling role of branched-chain amino acids. *J. Nutr.* **135,** 1547S–1552S.

Newsholme, P., Brennan, L., Rubi, B., and Maechler, P. (2005). New insights into amino acid metabolism, beta-cell function and diabetes. *Clin. Sci. (Lond.)* **108,** 185–194.

Newsholme, P., Bender, K., Kiely, A., and Brennan, L. (2007a). Amino acid metabolism, insulin secretion and diabetes. *Biochem. Soc. Trans.* **35,** 1180–1186.

Newsholme, P., Bender, K., Kiely, A., and Brennan, L. (2007b). Amino acid metabolism, insulin secretion and diabetes. *Biochem. Soc. Trans.* **35**(Pt. 5), 1180–1186.
Nolan, C. J., Leahy, J. L., Delghingaro-Augusto, V., Moibi, J., Soni, K., Peyot, M. L., Fortier, M., Guay, C., Lamontagne, J., Barbeau, A., Przybytkowski, E., Joly, E., Masiello, P., Wang, S., Mitchell, G. A., and Prentki, M. (2006a). Beta cell compensation for insulin resistance in Zucker fatty rats: Increased lipolysis and fatty acid signalling. *Diabetologia* **49,** 2120–2130.
Nolan, C. J., Madiraju, M. S., Delghingaro-Augusto, V., Peyot, M. L., and Prentki, M. (2006b). Fatty acid signaling in the beta-cell and insulin secretion. *Diabetes* **55**(Suppl. 2), S16–S23.
Noriega-Lopez, L., Tovar, A. R., Gonzalez-Granillo, M., Hernandez-Pando, R., Escalante, B., Santillan-Doherty, P., and Torres, N. (2007). Pancreatic insulin secretion in rats fed a soy protein high fat diet depends on the interaction between the amino acid pattern and isoflavones. *J. Biol. Chem.* **282,** 20657–20666.
Norman, A. W., and Litwack, G. (1987). Pancreatic Hormones: Insulin and Glucagon. Harcourt Brace Jovanovich, New York.
Ohtsubo, K., Takamatsu, S., Minowa, M. T., Yoshida, A., Takeuchi, M., and Marth, J. D. (2005). Dietary and genetic control of glucose transporter 2 glycosylation promotes insulin secretion in suppressing diabetes. *Cell* **123,** 1307–1321.
Ordway, R. W., Singer, J. J., and Walsh, J. V. Jr., (1991). Direct regulation of ion channels by fatty acids. *Trends Neurosci.* **14,** 96–100.
Pallett, A. L., Morton, N. M., Cawthorne, M. A., and Emilsson, V. (1997). Leptin inhibits insulin secretion and reduces insulin mRNA levels in rat isolated pancreatic islets. *Biochem. Biophys. Res. Commun.* **238,** 267–270.
Poitout, V., and Robertson, R. P. (2007). Glucolipotoxicity: Fuel excess and beta-cell dysfunction. *Endocrine Rev.* doi:10.1210/er.2007-0023.
Poitout, V., Hagman, D., Stein, R., Artner, I., Robertson, R. P., and Harmon, J. S. (2006). Regulation of the insulin gene by glucose and fatty acids. *J. Nutr.* **136,** 873–876.
Prado, C. L., Pugh-Bernard, A. E., Elghazi, L., Sosa-Pineda, B., and Sussel, L. (2004). Ghrelin cells replace insulin-producing beta cells in two mouse models of pancreas development. *Proc. Natl. Acad. Sci. USA* **101,** 2924–2929.
Prentki, M., Tornheim, K., and Corkey, B. E. (1997). Signal transduction mechanisms in nutrient-induced insulin secretion. *Diabetologia* **40,** S32–S41.
Prentki, M., Joly, E., El-Assaad, W., and Roduit, R. (2002). Malonyl-CoA signaling, lipid partitioning, and glucolipotoxicity: Role in beta-cell adaptation and failure in the etiology of diabetes. *Diabetes* **51**(Suppl. 3), S405–S413.
Rabaglia, M. E., Gray-Keller, M. P., Frey, B. L., Shortreed, M. R., Smith, L. M., and Attie, A. D. (2005). Alpha-Ketoisocaproate-induced hypersecretion of insulin by islets from diabetes-susceptible mice. *Am. J. Physiol. Endocrinol. Metab.* **289,** E218–E224.
Reaven, G. M., Lerner, R. L., Stern, M. P., and Farquhar, J. W. (1967). Role of insulin in endogenous hypertriglyceridemia. *J. Clin. Invest.* **46,** 1756–1767.
Reimann, F., Williams, L., da Silva Xavier, G., Rutter, G. A., and Gribble, F. M. (2004). Glutamine potently stimulates glucagon-like peptide-1 secretion from GLUTag cells. *Diabetologia* **47,** 1592–1601.
Reimer, R. A., and Russell, J. C. (2008). Glucose tolerance, lipids, and GLP-1 secretion in JCR:LA-cp rats fed a high protein fiber diet. *Obesity (Silver Spring)* **16,** 40–46.
Reimer, M. K., Pacini, G., and Ahren, B. (2003). Dose-dependent inhibition by ghrelin of insulin secretion in the mouse. *Endocrinology* **144,** 916–921.
Riccardi, G., Rivellese, A. A., and Giacco, R. (2008). Role of glycemic index and glycemic load in the healthy state, in prediabetes, and in diabetes. *Am. J. Clin. Nutr.* **87,** 269S–274S.
Rocha, D. M., Faloona, G. R., and Unger, R. H. (1972). Glucagon-stimulating activity of 20 amino acids in dogs. *J. Clin. Invest.* **51,** 2346–2351.

Roduit, R., Morin, J., Masse, F., Segall, L., Roche, E., Newgard, C. B., Assimacopoulos-Jeannet, F., and Prentki, M. (2000). Glucose down-regulates the expression of the peroxisome proliferator-activated receptor-alpha gene in the pancreatic beta-cell. *J. Biol. Chem.* **275,** 35799–35806.
Rubi, B., Ishihara, H., Hegardt, F. G., Wollheim, C. B., and Maechler, P. (2001). GAD65-mediated glutamate decarboxylation reduces glucose-stimulated insulin secretion in pancreatic beta cells. *J. Biol. Chem.* **276,** 36391–36396.
Rubi, B., del Arco, A., Bartley, C., Satrustegui, J., and Maechler, P. (2004). The malate–aspartate NADH shuttle member Aralar1 determines glucose metabolic fate, mitochondrial activity, and insulin secretion in beta cells. *J. Biol. Chem.* **279,** 55659–55666.
Saito, T. (2008). Antihypertensive peptides derived from bovine casein and whey proteins. *Adv. Exp. Med. Biol.* **606,** 295–317.
Sato, Y., and Henquin, J. C. (1998). The K^{+}-ATP channel-independent pathway of regulation of insulin secretion by glucose: In search of the underlying mechanism. *Diabetes* **47,** 1713–1721.
Sato, Y., Ito, T., Udaka, N., Kanisawa, M., Noguchi, Y., Cushman, S. W., and Satoh, S. (1996). Immunohistochemical localization of facilitated-diffusion glucose transporters in rat pancreatic islets. *Tissue Cell* **28,** 637–643.
Segal, M. S., Gollub, E., and Johnson, R. J. (2007). Is the fructose index more relevant with regards to cardiovascular disease than the glycemic index? *Eur. J. Nutr.* **46,** 406–417.
Sener, A., and Malaisse, W. J. (1980). L-Leucine and a nonmetabolized analogue activate pancreatic islet glutamate dehydrogenase. *Nature* **288,** 187–189.
She, P., Reid, T. M., Bronson, S. K., Vary, T. C., Hajnal, A., Lynch, C. J., and Hutson, S. M. (2007). Disruption of BCATm in mice leads to increased energy expenditure associated with the activation of a futile protein turnover cycle. *Cell Metab.* **6,** 181–194.
Siegel, E. G., and Creutzfeldt, W. (1985). Stimulation of insulin release in isolated rat islets by GIP in physiological concentrations and its relation to islet cyclic AMP content. *Diabetologia* **28,** 857–861.
Stein, D. T., Esser, V., Stevenson, B. E., Lane, K. E., Whiteside, J. H., Daniels, M. B., Chen, S., and McGarry, J. D. (1996). Essentiality of circulating fatty acids for glucose-stimulated insulin secretion in the fasted rat. *J. Clin. Invest.* **97,** 2728–2735.
Stenneberg, P., Rubins, N., Bartoov-Shifman, R., Walker, M. D., and Edlund, H. (2005). The FFA receptor GPR40 links hyperinsulinemia, hepatic steatosis, and impaired glucose homeostasis in mouse. *Cell Metab.* **1,** 245–258.
Straub, S. G., Yajima, H., Komatsu, M., Aizawa, T., and Sharp, G. W. (2002). The effects of cerulenin, an inhibitor of protein acylation, on the two phases of glucose-stimulated insulin secretion. *Diabetes* **51**(Suppl. 1), S91–S95.
Sugden, M. C., and Holness, M. J. (2004). Potential role of peroxisome proliferator-activated receptor-a in the modulation of glucose-stimulated insulin secretion. *Diabetes* **53,** S71–S81.
Teff, K. L., Elliott, S. S., Tschop, M., Kieffer, T. J., Rader, D., Heiman, M., Townsend, R. R., Keim, N. L., D'Alessio, D., and Havel, P. J. (2004). Dietary fructose reduces circulating insulin and leptin, attenuates postprandial suppression of ghrelin, and increases triglycerides in women. *J. Clin. Endocrinol. Metab.* **89,** 2963–2972.
Thorens, B., Sarkar, H. K., Kaback, H. R., and Lodish, H. F. (1988). Cloning and functional expression in bacteria of a novel glucose transporter present in liver, intestine, kidney, and beta-pancreatic islet cells. *Cell* **55,** 281–290.
Tordjman, K., Standley, K. N., Bernal-Mizrachi, C., Leone, T. C., Coleman, T., Kelly, D. P., and Semenkovich, C. F. (2002). PPARa suppresses insulin secretion and induces UCP2 in insulinoma cells. *J. Lipid Res.* **43,** 936–943.

Torre-Villavazo, I., Tovar, A. R., Ramos-Barragan, V. E., Cerbón-Cervantes, M. A., and Torres, N. (2008). Soy protein ameliorates metabolic abnormalities in liver and adipose tissue of rats fed a high fat diet. *J. Nutr.* **138,** 462–468.

Tse, E. O., Gregoire, F. M., Magrum, L. J., Johnson, P. R., and Stern, J. S. (1995). A low protein diet lowers islet insulin secretion but does not alter hyperinsulinemia in obese Zucker (fa/fa) rats. *J. Nutr.* **125,** 1923–1929.

Unger, R. H., and Eisentraut, A. M. (1969). Entero-insular axis. *Arch. Intern. Med.* **123,** 261–266.

Unger, R. H., Zhou, Y. T., and Orci, L. (1999). Regulation of fatty acid homeostasis in cells: Novel role of leptin. *Proc. Natl. Acad. Sci. USA* **96,** 2327–2332.

Usami, M., Seino, Y., Seino, S., Takemura, J., Nakahara, H., Ikeda, M., and Imura, H. (1982). Effects of high protein diet on insulin and glucagon secretion in normal rats. *J. Nutr.* **112,** 681–685.

Vara, E., Fernández-Martin, O., Garcia, C., and Tamarit-Rodriguez, J. (1988). Palmitate dependence of insulin secretion, 'de novo' phospholipid synthesis and Ca^{2+} turnover in glucose turnover in glucose-stimulated rat islets. *Diabetologia* **31,** 687–693.

Wang, Y., Perfetti, R., Greig, N. H., Holloway, H. W., DeOre, K. A., Montrose-Rafizadeh, C., Elahi, D., and Egan, J. M. (1997). Glucagon-like peptide-1 can reverse the age-related decline in glucose tolerance in rats. *J. Clin. Invest.* **99,** 2883–2889.

Wang, Z. Q., Zuberi, A. R., Zhang, X. H., Macgowan, J., Qin, J., Ye, X., Son, L., Wu, Q., Lian, K., and Cefalu, W. T. (2007). Effects of dietary fibers on weight gain, carbohydrate metabolism, and gastric ghrelin gene expression in mice fed a high-fat diet. *Metabolism* **56,** 1635–1642.

Wierup, N., Yang, S., McEvilly, R. J., Mulder, H., and Sundler, F. (2004). Ghrelin is expressed in a novel endocrine cell type in developing rat islets and inhibits insulin secretion from INS-1 (832/13) cells. *J. Histochem. Cytochem.* **52,** 301–310.

Wylie-Rosett, J., Segal-Isaacson, C. J., and Segal-Isaacson, A. (2004). Carbohydrates and increases in obesity: Does the type of carbohydrate make a difference? *Obes. Res.* **12** (Suppl. 2), 124S–129S.

Xiang, X., Saha, A. K., Wen, R., Ruderman, N. B., and Luo, Z. (2004). AMP-activated protein kinase activators can inhibit the growth of prostate cancer cells by multiple mechanisms. *Biochem. Biophys. Res. Commun.* **321,** 161–167.

Xu, G., Kwon, G., Cruz, W. S., Marshall, C. A., and McDaniel, M. L. (2001). Metabolic regulation by leucine of translation initiation through the mTOR-signaling pathway by pancreatic beta-cells. *Diabetes* **50,** 353–360.

Yajima, H., Komatsu, M., Schermerhorn, T., Aizawa, T., Kaneko, T., Nagai, M., Sharp, G. W., and Hashizume, K. (1999). cAMP enhances insulin secretion by an action on the ATP-sensitive K^+ channel-independent pathway of glucose signaling in rat pancreatic islets. *Diabetes* **48,** 1006–1012.

Yaney, G. C., Korchak, H. M., and Corkey, B. E. (2000). Long-chain acyl CoA regulation of protein kinase C and fatty acid potentiation of glucose-stimulated insulin secretion in clonal beta-cells. *Endocrinology* **141,** 1989–1998.

Zhang, C., Baffy, G., Perret, P., Kraauss, S., Peroni, O., Grujic, D., Hagen, T., CVidal-Puig, A. J., Boss, O., Kim, Y., Zheng, X., Wheeler, M. B., Shulman, G. I., Chan, C. B., and Lowell, B. (2001). Uncoupling protein-2 negatively regulates insulin secretion and is a major link between obesity, b cell dysfunction, and type 2 diabetes. *Cell Metab.* **105,** 745–755.

Zhou, Y. P., and Grill, V. E. (1995). Palmitate-induced beta cell insensitivity to glucose is coupled to decreased pyruvate dehydrogenase activity and enhanced kinase activity in rat pancreatic islets. *Diabetes* **44,** 394–399.

CHAPTER TEN

How Insulin Regulates Glucose Transport in Adipocytes

Joseph M. Muretta* *and* Cynthia Corley Mastick†

Contents

* Department of Biochemistry, Molecular Biology, and Biophysics, University of Minnesota, Minneapolis, Minnesota 55455
† Department of Biochemistry and Molecular Biology, University of Nevada School of Medicine, Reno, Nevada 89557

Vitamins and Hormones, Volume 80
ISSN 0083-6729, DOI: 10.1016/S0083-6729(08)00610-9

Abstract

Insulin stimulates glucose storage and metabolism by the tissues of the body, predominantly liver, muscle and fat. Storage in muscle and fat is controlled to a large extent by the rate of facilitative glucose transport across the plasma membrane of the muscle and fat cells. Insulin controls this transport. Exactly how remains debated. Work presented in this review focuses on the pathways responsible for the regulation of glucose transport by insulin. We present some historical work to show how the prevailing model for regulation of glucose transport by insulin was originally developed, then some more recent data challenging this model. We finish describing a unifying model for the control of glucose transport, and some very recent data illustrating potential molecular machinery underlying this regulation. This review is meant to give an overview of our current understanding of the regulation of glucose transport through the regulation of the trafficking of Glut4, highlighting important questions that remain to be answered. A more detailed treatment of specific aspects of this pathway can be found in several excellent recent reviews (Brozinick *et al.*, 2007; Hou and Pessin, 2007; Huang and Czech, 2007; Larance *et al.*, 2008; Sakamoto and Holman, 2008; Watson and Pessin, 2007; Zaid *et al.*, 2008). One of the main objectives of this review is to discuss the results of the experiments measuring the kinetics of Glut4 movement between subcellular compartments in the context of our emerging model of the Glut4 trafficking pathway.

I. Introduction

A. Glucose transport in fat and muscle

Facilitative glucose uptake into cells is mediated by members of the SLC2 or Glut family of hexose transporters (Uldry and Thorens, 2004). There are currently 13 known isoforms, of which six are known to function as glucose transporters (Gluts 1–4, 8, and 10). These transporters allow the bidirectional passage of hexose molecules across the plasma membrane. The class 1 family members (Gluts 1–4) all share similar structural topology (Fig. 10.1). Each contains 12 transmembrane alpha helixes that form an aqueous pore, accessible only to glucose. The transmembrane helixes are in turn connected sequentially via flexible loops exposed alternately to the cytosolic and exofacial sides of the membrane. The N- and C-termini of the Glut proteins are both cytosolic. Each transporter also contains two large loop domains. The first loop is extracellular and contains a conserved N-linked glycosylation site while loop 6 is intracellular. The flexible N- and C-terminal cytosolic tails of the transporters as well as loop 6 contain conserved signal sequences that specify and regulate the differential

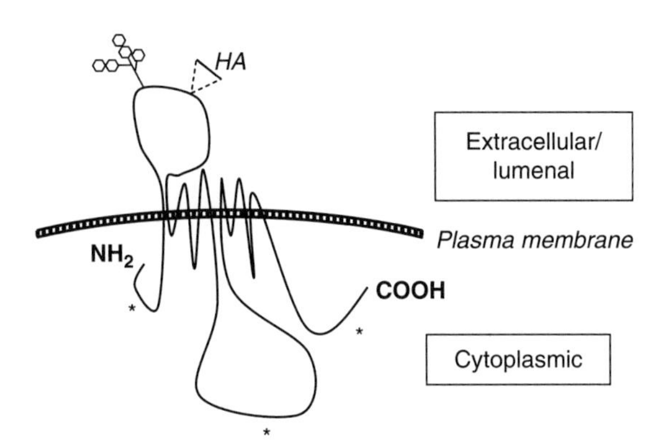

Figure 10.1 Orientation of Gluts (1–4) in membranes. Class I members of the Glut family of facilitative glucose transporters share similar structural topology: twelve transmembrane alpha helixes, connected via flexible loops. The three major domains of the protein exposed to the cytosol, the N- and C-termini and loop 6, contain sequences that specify the differential subcellular targeting of the Gluts (*). The major exofacial domain, loop 1, contains an N-linked glycosylation site which masks extracellular epitopes of the protein. Epitope tags (i.e., HA or myc) have been inserted into this domain to allow detection of Glut4 at the cell surface with antibodies. Note: when the Gluts are localized to vesicles within the cell the exofacial/extracellular domain faces the lumen of the vesicle.

intracellular trafficking of each Glut family member (Blot and McGraw, 2008; Song *et al.*, 2008). For example, the Glut expressed in hepatocytes and pancreatic β-cells (Glut2) is found predominantly at the cell surface, the ubiquitous basal transporter (Glut1) distributes between the plasma membrane and endosomes, and the insulin responsive transporter in adipocytes and muscle (Glut4) is sequestered in specialized intracellular compartments in basal cells, but redistributes after insulin-stimulation.

The expression of the Glut proteins varies greatly between cell types (Uldry and Thorens, 2004). Glut1 is expressed in all cells and serves to maintain basal glucose uptake, Glut2 is expressed primarily in liver and pancreatic beta cells, while Glut3 is expressed in brain. Glut5 is expressed in the intestinal epithelium and testis, Glut6 in spleen, leukocytes and brain. Interestingly, Glut8 is localized to intracellular vesicles where it transports hexose molecules from the lysosome into the cytoplasm following degradation of glycosylated proteins (Augustin *et al.*, 2005). Glut9 is expressed in liver and kidney, Glut10 in liver and pancreas, Glut11 in muscle, and Glut12 in heart and prostate. The major insulin-sensitive glucose transporter is the Glut4 isoform. Glut4 is expressed predominantly in muscle and adipose tissue. Glut4 is also expressed in specific regions of the brain including the hypothalamus where it plays a role in insulin-regulated glucose sensing and metabolic regulation (Leloup *et al.*, 1996).

The rate of glucose uptake in muscle and fat is controlled by several parameters: the concentration of glucose outside vs. inside of the cell, the total number and type of Glut transporters localized to the plasma membrane, and the specific kinetic activities of these transporters. In fat and muscle, hexokinase rapidly converts intracellular glucose to glucose-6-phosphate, which cannot bind to or be transported by the Glut transporters. Conversion of intracellular glucose to glucose-6-phosphate keeps the concentration of free glucose low inside the cell and maintains a concentration gradient which drives the diffusion of glucose into the cell through the facilitative Glut transporters. The kinetic activities and mechanisms of regulation of the different transporters vary. For example, Glut1 is allosterically regulated by ATP, which binds to the cytosolic face of Glut1 (Blodgett *et al.*, 2007). This binding decreases the K_m for glucose. It also decreases the transporters k_{cat} and thus the V_{max} for transport. Thus, ATP binding allows Glut1 to sense the net energy status inside the cell. When ATP levels inside the cell decrease, Glut1 becomes a more efficient transporter. Insulin controls glucose transport in fat and muscle cells by specifically regulating Glut4. Though debated for many years, most data indicate that the kinetic activity of Glut4 is not affected by insulin. Instead, insulin increases the rate of glucose uptake in muscle and fat by increasing the total number of Glut4 transporters localized to the plasma membrane (Fig. 10.2). This increase can be as great as 50-fold in metabolically active brown adipose

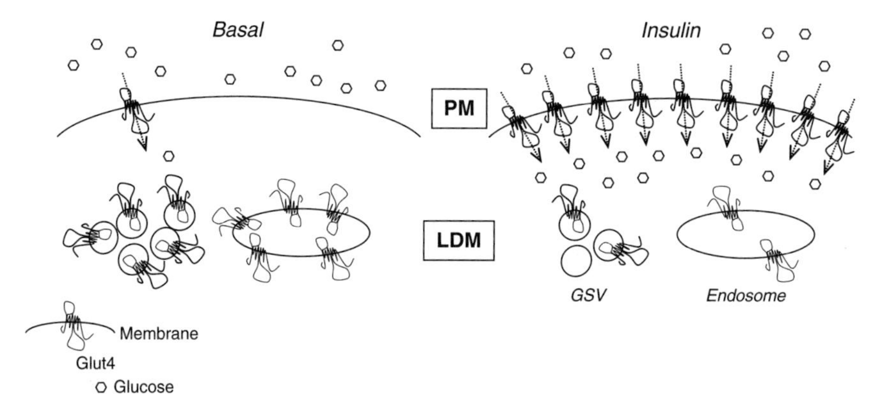

Figure 10.2 The translocation hypothesis. Glucose transport in fat and muscle is largely controlled by regulating the total number of Glut4 molecules inserted into the plasma membrane. In basal cells, the Glut4 is largely sequestered within small vesicular compartments in the cell. These vesicles are highly enriched in the low density microsomes or LDM, a heterogeneous subcellular fraction containing small membrane compartments and cytoskeletal elements. In response to insulin there is a redistribution of Glut4 from the intracellular compartments to the plasma membrane (PM), increasing the rate of glucose transport into the cells.

tissue and can exceed 10-fold in cultured cells (Bryant *et al.*, 2002; Foster and Klip, 2000).

Insulin controls glucose uptake in fat and muscle by controlling the subcellular localization of Glut4. It does this by regulating the membrane trafficking events that move the transporter between successive communicating membrane compartments, rather than through effects on the total amount of Glut4 in the cell. Exercise also causes Glut4 redistribution to the plasma membrane. Genetic and biochemical evidence indicates that in skeletal muscle the same protein machinery is used to control the levels of Glut4 at the plasma membrane in response to exercise and insulin, although the signal transduction pathways activating this machinery are different for these two stimuli (Foster and Klip, 2000; Goodyear and Kahn, 1998). In individuals with type-II diabetes, insulin fails to increase glucose transport and metabolism following feeding. However, exercise still causes Glut4 translocation in the muscle of these patients. This demonstrates that the Glut4 trafficking machinery is intact in type-II diabetic muscle, it is simply not responding to the insulin. Therefore, the defect in Glut4 regulation that occurs is not due to a failure of the Glut4 trafficking machinery but rather a proximal defect in insulin signaling. Because of this, the Glut4 trafficking machinery is an attractive target for the treatment of type-II diabetes. Compounds which act on this machinery could be used to trigger Glut4 redistribution to the plasma membrane and increased glucose disposal despite defective insulin signaling.

Glut4 is regulated by the same protein machinery in muscle and in fat (Bryant *et al.*, 2002; Foster and Klip, 2000). For technical reasons, biochemical analysis of Glut4 traffic in muscle is difficult compared to analysis in adipocytes. It is relatively easy to isolate and culture insulin-responsive primary adipocytes, but much more difficult to obtain and culture isolated muscle. Highly enriched/purified plasma membrane and intracellular compartments containing insulin responsive Glut4 are easily obtained from adipose tissue by differential centrifugation. They are much more difficult to obtain from isolated primary muscle tissue. Cell culture adipocytes (3T3-L1 cells) exhibit high levels of insulin-stimulated glucose transport and metabolism while cell culture models for muscle (L6 myoblasts and myotubes as well as C2C12 myoblasts and myotubes) exhibit smaller insulin-stimulated changes (~2-fold) in glucose transport and metabolism. Because of these issues, adipocytes have been the primary model used for the study of insulin-dependent Glut4 regulation. Two models of adipocyte cells are commonly used: primary adipocytes (isolated from the epididymal fat pads of male rats or mice) and 3T3-L1 cells (differentiated into adipocytes in culture). This review will focus on work that has been done in adipocytes.

II. Historical Perspective

A. The translocation hypothesis and identification of Glut4

The biological affects of insulin on metabolism have been studied since Banting and Best first isolated the hormone in 1921. One of the main affects of insulin action in the body is to increase glucose transport in skeletal muscle and fat. Despite an extensive history, the molecular machinery that controls glucose transport in response to insulin remains elusive. Insulin activates glucose transport in muscle and fat *in vivo*. It also activates transport in isolated muscle strips and adipocytes in cultured cells. Studies by Gliemann and others in the 1970s showed that glucose uptake by cells follows simple Michaelis–Menton kinetics and can be described by the equation $V = k_{cat}E_{tot}S/(K_m + S)$, where V is the initial transport velocity under steady state conditions, k_{cat} is the transport turnover rate, E_{tot} is the total number of transporters participating in transport, S is the concentration of transportable glucose, and K_m is the concentration of glucose required to achieve half maximal transport velocity (Okuno and Gliemann, 1987; Vinten *et al.*, 1976; Wardzala *et al.*, 1978). Insulin increases glucose transport in isolated adipocytes by increasing the V_{max} for transport, with no effect on the transport K_m for glucose. In Michaelis–Menton kinetics, $V_{max} = k_{cat}E_{tot}$. Thus, the change in V_{max} could be due to either an increase in the intrinsic turnover rate of the transporters or in the total number of active transporters localized to the plasma membrane of cells. Later work showed that insulin-stimulated changes in E_{tot} accounted for the increase in transport velocity (Ludvigsen and Jarett, 1980). Two processes could explain this increase. Stimulation could promote conversion of plasma membrane glucose transporters from an inactive to an active form ($E_i \rightarrow E_a$). Alternatively, insulin could increase the total number of active transporters in the membrane by causing a pool of transporters to move from intracellular compartments to the plasma membrane (Fig. 10.2). The later proposal was termed the translocation hypothesis (Cushman and Wardzala, 1980; Suzuki and Kono, 1980).

Several toxins potently inhibit glucose uptake in cells. Among these, cytochalasin-B exhibits significant potency. The Lienhard group used cytochalasin-B to inhibit glucose transport in red blood cells (Gorga and Lienhard, 1981; Zoccoli *et al.*, 1978). These studies showed that cytochalasin-B was a competitive inhibitor of glucose binding to the red blood cell glucose transporter. It therefore directly interacted with the glucose binding site on the transporter. Cytochalasin-B also inhibited glucose transport in muscle and fat. Cushman and Wardzala used ^{3}H-cytochalasin-B binding to measure the number of glucose transporters in subcellular fractions of basal and insulin-stimulated adipocytes (Cushman and Wardzala, 1980). This work demonstrated conclusively that insulin increased glucose transport by causing the transporter to move from low density microsomes (subcellular fractions highly enriched in small intracellular vesicles) to the plasma membrane fraction isolated from primary adipocytes. In a simultaneous report, Suzuki and Kono observed similar phenomena (Suzuki and Kono, 1980). Careful analysis of the kinetics of cytochalasin-B binding to the transport sites suggested that more than one class of glucose transporter existed in muscle and fat (James *et al.*, 1988, 1989). Whether these two classes of sites existed on the same peptide, or represented genetically and/or functionally distinct transport systems could not be determined from this data.

During the 1980s investigators attempted to isolate the insulin-responsive glucose transport system expressed in muscle and adipose tissue. Antibodies raised against the red cell transporter, Glut1, cross reacted with a glucose-inhibited cytochalasin-B binding protein in muscle and primary adipocytes and in the 3T3-L1 cell culture model adipocytes (Lienhard *et al.*, 1982; Schroer *et al.*, 1986). Subcellular fractionation was used to analyze the effect of insulin on the subcellular distribution of Glut1 in adipocytes. Glut1 moved from the low density microsome fraction to the plasma membrane fraction in stimulated cells. However, the analysis of subcellular fractions by western blotting did not agree with cytochalasin-B binding studies (Joost *et al.*, 1988). Insulin increased the cytochalasin-B binding capacity of the adipocyte plasma membrane as much as 15-fold, while stimulation had a much smaller affect (2-fold) on the anti-Glut1 immunoreactivity of the plasma membrane fraction. These discrepancies, together with kinetic analysis of glucose transport and cytochalasin-B binding, supported the hypothesis that muscle and fat expressed an additional unidentified insulin-responsive transporter. This was confirmed by isolation of a monoclonal antibody that recognized a unique, insulin-regulatable glucose transport protein expressed in muscle and fat (James *et al.*, 1988). Using degenerate PCR probes directed toward the known Glut1 isoform, James and Mueckler cloned a previously unidentified transporter from heart and adipose tissue (James *et al.*, 1989). Additional work showed that this protein translocated from the low density microsomes to the plasma membrane in response to insulin and that this translocation could account for the large increase in glucose transport in muscle and fat following insulin stimulation. The newly identified transporter was termed Glut4. These results were supported by work from a number of competing labs (Birnbaum, 1989; Fukumoto *et al.*, 1989; Kaestner *et al.*, 1989; Zorzano *et al.*, 1989).

B. Ultrastructural analysis: The subcellular localization of Glut4 in adipocytes

To further test the translocation hypothesis and characterize the compartments through which Glut4 traffics, a number of labs collaborated with Dr. Jan Slot to analyze the subcellular localization of Glut4 by immunoelectron microscopic analysis of basal and insulin-stimulated cells. Cells were isolated from a number of insulin responsive tissues including brown fat, white fat, skeletal muscle, and cardiac muscle (Haney *et al.*, 1991; Malide *et al.*, 2000; Rodnick *et al.*, 1992; Slot *et al.*, 1991a,b). One of the most comprehensive of these studies was performed in intact brown adipose tissue, where the morphology of the tissue allowed a detailed quantitative analysis of the distribution of Glut4 between distinct intracellular compartments (Slot *et al.*, 1991b). Insulin causes a large increase in glucose transport by brown adipocytes (typically >15-fold, in some reports as great as 50-fold). Immuno-gold labeling with antibodies against Glut4 showed that under basal conditions, approximately 80% of the total Glut4 label was localized to tubulo-vesicular structures distributed in clusters throughout the cytoplasm of the cell and in the *trans*-Golgi region (TGN; summarized in Fig. 10.3). Very little of the Glut4 (<1%) was found in the plasma membrane under basal conditions. Little Glut4 colocalized with endocytosed albumin or with cathepsin (markers for late endosomes/multivesicular bodies and lysosomes). Insulin caused a redistribution of Glut4 from the tubulo-vesicular structures to the plasma membrane. After stimulation, ~40% of the Glut4 was found at the plasma membrane, distributed evenly across the entire plasma membrane as well as in coated pits. Insulin decreased the amount of Glut4 in the tubulo-vesicular compartments by ~50% (from 80% to 37%), in very good agreement with the 50% reduction in Glut4 immunoreactivity seen in the low density microsome cell fraction with insulin stimulation. Therefore, the ultrastructural data strongly supported the translocation hypothesis as originally developed from the cell fractionation data.

Interestingly, insulin also increased the number of transporters localized to early endosomes (from 1.9% to 9.4%). The amount of Glut4 found in late endosomes and lysosomes did not change with insulin stimulation. This suggested that cells regulate the distribution of Glut4 not only between the plasma membrane and the tubulo-vesicular compartments but also its traffic to and from cycling endosomes. Because Glut4 occupies both plasma membrane and endosomes, the investigators proposed that the transporter constitutively cycles between these compartments under both basal and insulin-stimulated conditions. This data was the first indication that Glut4 followed the same trafficking pathway utilized by other cycling proteins, such as the transferrin receptor. This continuous cycling was later verified (Jhun *et al.*, 1992; Satoh *et al.*, 1993).

C. Endosomes and Glut4 trafficking

The transferrin receptor is expressed at low levels in all cells types and at high levels in actively dividing cells. This transmembrane receptor facilitates the receptor mediated endocytosis of the iron carrying protein transferrin (Aisen, 2004). Iron-loaded (holo) transferrin binds with high affinity to its

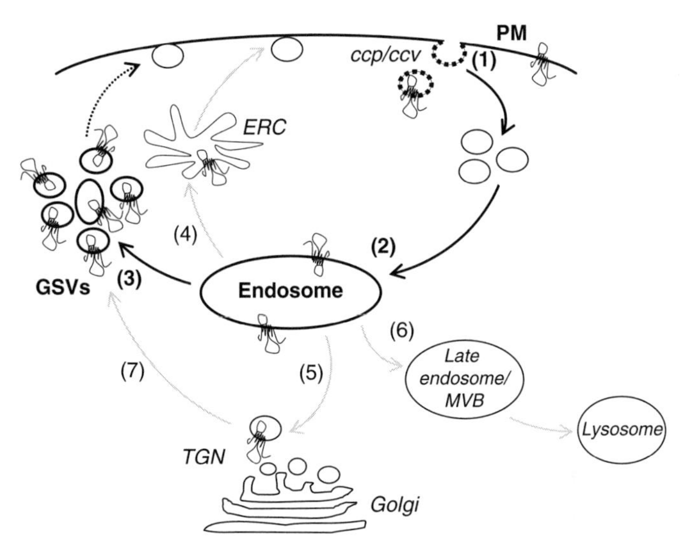

	Basal		Insulin	
PM	0.9	3.9	36.8	48.9
ccp/ccv	1.1		2.7	
Early endosome	1.9		9.4	
Recycling vesicle (ERC)	3.8		3.5	
TV cluster (GSV)	66.5	79.9	31.7	37.4
TGN	13.4		5.7	
Late endo/MVB/lyso	0.8		0.2	
Cytoplasm	11.6		10.0	

Figure 10.3 The trafficking itinerary of Glut4. Glut4 is distributed between many compartments within adipocytes. This suggested the following trafficking itinerary. (1) Glut4 at the plasma membrane (PM) is internalized via clathrin coated pits/vesicles (CCP/CCV). The clathrin dissociates after vesicles are formed. (2) These endocytic vesicles fuse to form endosomes. (3) The Glut4 in endosomes is sorted into specialized storage compartments (GSVs) away from proteins/receptors that (4) constitutively recycle to the plasma membrane through the ECR. The GSVs are made up small tubular/vesicular structures distributed throughout the cytoplasm (TV clusters) and (5) near the *trans*-golgi network (TGN). (6) Very little Glut4 is found in the degradative pathway which includes late endosomes, multivesicular bodies, and lysosomes (late endo/MVB/lyso). (7) Glut4 is also delivered to the GSVs from the golgi during biosynthesis. Dotted line: insulin regulates the trafficking of Glut4 from the GSVs to the plasma membrane. Table: summary of EM data from Slot *et al.* (1991b).

receptors at the plasma membrane at neutral pH. The transferrin/transferrin receptor complex is internalized from the plasma membrane (endocytosed) and delivered to endosomes where the mildly acidic pH (6.0–6.5) promotes release of iron from the carrier protein. The iron-depleted (apo) transferrin remains bound to its receptor at mildly acidic pH. The transferrin/transferrin receptor complex is sorted out of endosomes into exocytic vesicles and recycles back to the plasma membrane. The apo transferrin has low affinity for its receptor at neutral pH, and is released at the plasma membrane, allowing binding of another molecule of holo-transferrin. This mechanism allows for the efficient reutilization of both the receptor and carrier protein for many cycles of iron uptake. Many receptors and other proteins involved in nutrient uptake utilize this trafficking pathway. Due to its unique trafficking itinerary and the fact that it can be easily purified and labeled, transferrin is commonly used as a marker for the early endosomal recycling pathway. The molecular machinery that controls traffic of the transferrin receptor has been extensively investigated and has served as a general model for receptor mediated endocytosis and endosomal recycling.

Insulin causes the transferrin receptor to translocate to the plasma membrane in adipocytes (Tanner and Lienhard, 1987). The magnitude of this translocation is small (2-fold) compared to the effect of stimulation on plasma membrane Glut4 (10–15-fold). The effect of insulin on transferrin receptor traffic reflects a general acceleration of endosomal recycling in stimulated cells. Consistent with this, stimulation causes other resident proteins of the cycling endosomal system to translocate (also by ~2-fold). These include Glut1, the cation-independent mannose-6-phosphate/Insulin-like Growth Factor II receptor, and the α-2-macroglobulin receptor (Corvera *et al.*, 1989; Haney *et al.*, 1991; Kandror and Pilch, 1996; Ko *et al.*, 2001). The differential effects of insulin on general endosomal traffic and Glut4 traffic reflect different aspects of insulin action. The large translocation of Glut4 is the result of both the small effect of insulin on the general endocytic pathway as well as larger effects on the specialized Glut4 trafficking machineries.

Cell surface proteins including Glut4 and the transferrin receptor are endocytosed from the plasma membrane (Fig. 10.3). Glut4 and the transferrin receptor are both internalized via clathrin coated pits into clathrin coated vesicles. However, work by McGraw and colleagues indicates that even at this early step there are differences in the endocytosis of these two proteins (Blot and McGraw, 2006). Shortly after clathrin mediated endocytosis, the clathrin lattice depolymerizes from endocytic vesicles. The endocytosed vesicles then fuse, forming early endosomes. The majority of the transferrin receptor is sorted out of early endosomes and returns to the plasma membrane either directly or after first passing through the endosomal recycling compartment (ECR) (Maxfield and McGraw, 2004). In contrast to the transferrin receptor, the majority of Glut4 is segregated away from the rapidly recycling endosomal pathway into slowly cycling/noncycling specialized intracellular compartments termed Glut4 storage vesicles (GSVs). Trafficking from GSVs to the plasma membrane is highly regulated by insulin. Glut4 is delivered to the GSVs both through recycling from endosomes and directly from the Golgi in the biosynthetic pathway.

Because Glut4 and the transferrin receptor are rapidly separated from each other upon endocytosis, these two proteins have been used to distinguish between effects of insulin on general protein recycling through the endosomal/TGN membrane system and on specialized Glut4 traffic (Eguez *et al.*, 2005; Hashiramoto and James, 2000; Karylowski *et al.*, 2004; Livingstone *et al.*, 1996). It is important to distinguish between these two because, from a therapeutic perspective, the specific effects are the most likely targets for drug intervention in type-II diabetes. General effects on protein recycling in the cells endosomal/TGN system will alter membrane traffic in all cells whereas targeting the Glut4-specific trafficking machinery will only alter traffic of Glut4 in insulin-responsive tissues.

D. Molecular characterization of Glut4-containing compartments

Identification of the molecular machinery that controls Glut4 trafficking has been a long standing aim in the Glut4 field. Using highly specific antibodies that recognize the carboxy-terminal cytoplasmic tail of Glut4, Glut4-containing intracellular membrane compartments were isolated from fractionated cells (Cain *et al.*, 1992; Hanpeter and James, 1995; James and Pilch, 1988; James *et al.*, 1987; Kandror and Pilch, 1994; Larance *et al.*, 2005; Laurie *et al.*, 1993; Mastick *et al.*, 1994; Thoidis *et al.*, 1993). Previous work had shown that the majority of the Glut4 that translocated to the plasma membrane originates from compartments found in the low density microsome fraction of homogenized adipocytes (Wardzala *et al.*, 1978). Therefore, Glut4-containing compartments in the low density microsome fraction were purified by anti-Glut4 immuno-adsorption. In a series of reports, the protein composition of these purified compartments was examined by SDS page, total protein staining, western blotting and peptide sequencing. These studies identified a several proteins that copurified with Glut4 which also translocated to the plasma membrane in response to insulin. Insulin stimulates the translocation of endosomal markers such as the transferrin receptor and the cation-independent mannose-6-phosphate receptor (Corvera *et al.*, 1989; Kandror and Pilch, 1996; Ko *et al.*, 2001). Both the transferrin receptor and the mannose-6-phosphate receptor in the low density microsomes coimmunoprecipitated with Glut4, indicating that Glut4 colocalized to some extent with these endosomal proteins. Importantly, while immuno-adsorption with anti-Glut4 antibody depleted the transferrin receptor from the low density microsome fraction, adsorption with antitransferrin receptor antibody removed only a portion of the Glut4 from the fraction (Livingstone *et al.*, 1996). This indicates that while there is overlap between Glut4 and proteins trafficking in the general endocytic pathway, the majority of Glut4 in the low density microsomes resides in nonendosomal compartments devoid of transferrin receptor. The insulin-responsive pool of Glut4 was found predominantly in these nonendosomal compartments. These nonendosomal specialized compartments were termed the GSVs. Thus, the immunoprecipitation data were in very good agreement with the morphological data.

Several additional proteins were also present in the anti-Glut4 purified fractions. These included a previously unknown protein now termed the Insulin Responsive Amino Peptidase (IRAP) and the synaptic vesicle (SV) protein VAMP2/synaptobrevin (Cain *et al.*, 1992; Keller *et al.*, 1995; Mastick *et al.*, 1994). The biochemical activity of neither protein was known at the time, but both translocated to the plasma membrane in response to insulin. IRAP translocated to the same extent as Glut4, suggesting that it was an additional cargo protein carried in the GSVs. Further, genetic, biochemical and electron microscopy analysis showed that VAMP2 and IRAP were strongly localized to the Glut4-enriched tubulo-vesicular compartments of basal adipocytes and muscle (Martin *et al.*, 1996, 1997; Ramm *et al.*, 2000). The secretory carrier membrane proteins or SCAMPs were also coimmunoprecipitated with Glut4-containing membrane compartments (Laurie *et al.*, 1993). However, unlike VAMP2 and IRAP, the SCAMPs did not translocate following insulin stimulation.

In later work, VAMP2 was determined to be the vesicle SNARE protein required for Glut4 exocytosis (Cheatham *et al.*, 1996; Martin *et al.*, 1998; Tamori *et al.*, 1996; Volchuk *et al.*, 1995). The physiological activity of IRAP has been more difficult to determine. IRAP exhibits a broad range of peptidase activities *in vitro*, cleaving arginine-vasopressin, oxytocin, and the neuro-peptide methionine-Enk. A recent report concluded that vasopressin is the physiologically relevant IRAP substrate (Wallis *et al.*, 2007). IRAP also binds and is inhibited by angiotensin (Lew *et al.*, 2003; Ye *et al.*, 2007). The physiological relevance of these activities and why IRAP translocates with Glut4 in muscle and fat is not known. Interestingly, IRAP is expressed in most cell types including nondifferentiated fibroblasts. IRAP does not translocate following insulin treatment in these cells. However, IRAP plays an important role in the formation and retention of GSVs in fat and muscle. Overexpression of the cytoplasmic tail of IRAP decreases Glut4 retention in the basal state (Waters *et al.*, 1997). In addition, total Glut4 levels are significantly decreased in IRAP knockout animals (Keller *et al.*, 2002), and siRNA-induced knockdown of IRAP impairs Glut4 translocation (Yeh *et al.*, 2007). IRAP binds directly to the insulin-sensitive regulator of Glut4 traffic AS160 (Peck *et al.*, 2006; Ramm *et al.*, 2006). This binding is proposed to help localize AS160 to GSVs. AS160 is phosphorylated by Akt, and is considered to be a major insulin-responsive effector lying downstream of the activation of phosphatidylinositol-3 kinase. AS160 is a Rab GAP and is thought to regulate Rab 10 as well as other potential Rab activities required for Glut4 traffic (Miinea *et al.*, 2005; Sano *et al.*, 2003, 2007). The precise function of the SCAMPs remains unknown, although they clearly play an important role in regulated exocytosis (Castle and Castle, 2005; Liao *et al.*, 2008). They are ubiquitously expressed, and are localized to the TGN and to ECR. However, like Glut4, they are localized to a pathway that is distinct from the constitutive endosomal recycling pathway followed by transferrin. They have been proposed to function in vesicle formation and/or fusion.

E. Kinetic analysis of Glut4 traffic in adipocytes using photo reactive glucose analogues: Development of kinetics models of Glut4 trafficking

Subcellular fractionation experiments showed that Glut4 translocated to the plasma membrane from cell fractions enriched in small intracellular vesicles. Immunoprecipitation experiments showed that the intracellular Glut4 resides in at least two different compartments: endosomes, where the Glut4 is colocalized with general endocytic markers such as the transferrin and cation-independent-mannose-6-phosphate receptors, and specialized nonendosomal compartments (GSVs) where it colocalizes with IRAP and VAMP2. The electron microscopy studies showed that Glut4 translocates from small tubulo-vesicular intracellular compartments to the plasma membrane and endosomes. The localization of Glut4 to endosomes in both basal and insulin-stimulated cells suggested that Glut4 continuously cycles between the plasma membrane and intracellular compartments under both basal and insulin-stimulated conditions. Together these data suggested a simple model for Glut4 trafficking (Fig. 10.3). Glut4 at the plasma membrane is endocytosed and delivered to early endosomes, where it colocalizes with proteins in the general endosomal pathway. While proteins such as the transferrin receptor are rapidly recycled back to the plasma membrane from the endosomes, Glut4 is delivered instead to the specialized GSVs. Insulin regulates the trafficking of Glut4 from the GSVs to the plasma membrane. While most current models for Glut4 trafficking are considerably more complicated than this (i.e., multiple steps at the plasma membrane, more intracellular compartments, cross-talk between compartments; Blot and McGraw, 2006; Larance *et al.*, 2008), they are all modifications of this simple 3-pool model which includes: (1) Glut4 at the cell surface, (2) Glut4 in the general endocytic pathway, and (3) Glut4 trafficking through specialized, insulin-sensitive, nonendosomal compartments (GSVs).

Several mechanisms could explain the translocation of Glut4 (Fig. 10.4). Insulin could increase plasma membrane Glut4 by increasing the rate of transporter exocytosis from the GSVs. Alternatively, insulin could increase surface Glut4 by inhibiting transporter endocytosis from the plasma membrane. Finally, the GSVs might behave like regulated exocytic vesicles such as SV, which are largely segregated from the cycling endosomal pathway in unstimulated cells. In the later case, insulin would trigger release of Glut4 from noncycling GSVs into the cycling plasma membrane/endosomal system, increasing the total amount of cycling Glut4 in both the plasma

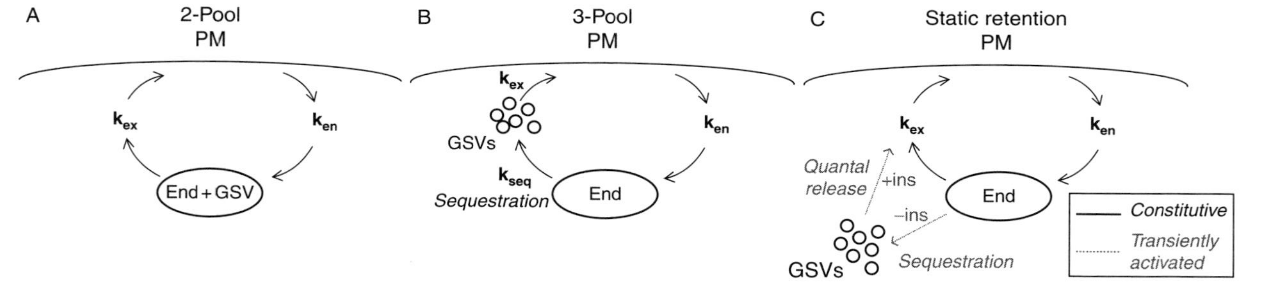

Figure 10.4 Kinetic models of Glut4 trafficking. (A) In the kinetics experiments, Glut4 is distributed into two populations: a pool at the plasma membrane (p) and an intracellular pool (i). The intracellular pool includes Glut4 in the endosomes (end) and GSVs. The rate of exchange of Glut4 between these pools is a function of the rates of internalization (k_{en}) and delivery to the plasma membrane (k_{ex}). The relative partitioning of Glut4 between the two pools, or partition coefficient, equals the ratio of the rate constants ($P = [p]/[i] = k_{ex}/k_{en}$). In all of the kinetics experiments (transition, equilibration, uptake, efflux), the observed relaxation rate constant is the sum of the two rate constants ($k_{obs} = k_{en} + k_{ex}$). (B) The two pool model cannot fully account for the observed behavior of Glut4: the kinetics of the transition from the basal to insulin-stimulated steady state distribution are faster than predicted from the equilibrium kinetics. Holman *et al.* proposed a 3-pool model to account for this discrepancy (Holman *et al.*, 1994). In this model, Glut4 trafficking is rate limited by the rate of fusion of GSVs with the plasma membrane in basal cells, but by the rate of transfer of Glut4 from the endosomes to the GSVs (sequestration, k_{seq}) in insulin-stimulated cells. (C) An alternative model has been proposed (Coster *et al.*, 2004). In static retention, the Glut4 in the GSVs is not actively cycling, and there is only a transient exchange of Glut4 between the cycling Glut4 and the GSVs upon insulin stimulation and insulin withdrawal. The kinetics of Glut4 trafficking and the partition coefficient follows the two pool model. However, the internal pool includes only the Glut4 in the endosomal cycling pathway, not in the GSVs. Quantal release of Glut4 from the GSVs increases the size of the total cycling pool.

membrane and in endosomes. These possibilities could not be distinguished by static methods such as subcellular fractionation and electron microscopy. These methods do not differentiate Glut4 that is actively trafficking from Glut4 that is not cycling. Therefore, to explore these questions further, investigators developed methods to measure the kinetics of Glut4 trafficking between the plasma membrane and intracellular compartments. Two types of kinetics were measured: transition kinetics—the rate at which Glut4 moves from the basal to the insulin steady-state distribution (or insulin to basal distribution after insulin withdrawal), and equilibration kinetics—the rate at which Glut4 in one compartment redistributes and comes to equilibrium with all cycling Glut4. For transition kinetics, the rate of change of total Glut4 in the subcellular fractions (plasma membrane or low density microsomes) with time after perturbation is measured. For the equilibration kinetics, Glut4 in one compartment (plasma membrane) is labeled, and the redistribution of label into all cell fractions measured (as a decrease in plasma membrane label and a reciprocal increase in intracellular labeling).

A number of techniques were used in these studies to monitor the movement of the transporter between intracellular compartments. For the transition kinetics, the rate of glucose transport was used as a measure of insertion or removal of transporters in the plasma membrane. However, this does not distinguish between Glut1 and Glut4. The total amount of Glut4 in the plasma membrane fraction or low density microsomes was also measured directly by western blotting. While this gives specific information about Glut4 translocation, it does not distinguish between cycling Glut4 and Glut4 that is not cycling (i.e., any Glut4 in the biosynthetic or degradative pathways). This can lead to errors in the estimate of the partition coefficient, P ($P = [p]/[i]$, where [p] is the amount of Glut4 in the plasma membrane, and [i] is the concentration of Glut4 in intracellular compartments). This has important consequences for the evaluation of the kinetics data (see below; Muretta *et al.*, 2008). To label the plasma membrane pool of Glut4 in intact live cells, membrane impermeant photoreactive glucose analogues were developed independently by the Holman and Jung labs (Calderhead *et al.*, 1990; Jhun *et al.*, 1992; Satoh *et al.*, 1993; Yang and Holman, 1993). These reagents covalently modify glucose transporters localized at the plasma membrane. Since both Glut4 and Glut1 are labeled, this analysis requires an additional immunoprecipitation step to specifically measure the labeled Glut4. Using these probes, both groups of investigators studied the kinetics of Glut4 exocytosis, endocytosis, and cycling between endosomes and plasma membrane.

Glut4 at the cell surface was covalently labeled at 4 °C with the photoreactive glucose analog (membrane trafficking stops at 4 °C). As expected, there was more Glut4 (10–20 times greater) labeled at 4 °C in the insulin-stimulated cells than in the basal cells. The label was recovered largely in the plasma membrane fraction. The cells were then rapidly warmed to 37 °C and the equilibration kinetics of the redistribution of the labeled Glut4 between the plasma membrane and intracellular compartments measured. After reaching steady state, 12% of the labeled Glut4 remained in the plasma membrane fraction in basal cells, while 88% of the label was found in the low density microsomes ($P = 12/88 = 0.14$; Jhun *et al.*, 1992). In contrast, in insulin-stimulated cells, ~30% remained in the plasma membrane fractions, and ~70% was found in the low density microsomes ($P = 30/70 = 0.42$). Similarly, there was a 3-fold difference in the partition coefficients measured in cells incubated in the continuous presence of insulin or in cells after insulin withdrawal (collagenase treatment; $P = 0.77$ vs. 0.23; Satoh *et al.*, 1993). Glut4 (both labeled and native) was rapidly cleared from the plasma membrane following removal of insulin ($t_{1/2} = 9$ min). Restimulation caused the reappearance of both the labeled and unlabeled Glut4 at the plasma membrane. These studies showed that Glut4 continuously cycles between the plasma membrane and intracellular compartments under both basal and insulin-stimulated conditions, as suggested by the electron microscopy data. Thus, the Glut4 at the plasma membrane is actively cycling and is not simply inserted into the plasma membrane upon insulin stimulation. Furthermore, Glut4 is recycled from the plasma membrane back to insulin-sensitive compartments after insulin withdrawal and reused during successive insulin stimulation events. Consistent with this, delivery of Glut4 to the plasma membrane does not require functional protein synthesis and recovery to the basal state is not blocked by inhibition of protein degradation.

The rate at which Glut4 redistributed together with the measurements of the steady state distribution of Glut4 between the plasma membrane and intracellular compartments were used to determine values for the rate constants for the delivery of Glut4 to the plasma membrane by exocytosis (k_{ex}) and the removal of Glut4 from the plasma membrane by endocytosis (k_{en}). In all of these experiments (transition kinetics, equilibration kinetics, insulin withdrawal experiments) the relaxations follow single first order exponentials. Fitting the data to an exponential association or decay yields a single relaxation rate constant, k_{obs}. This means that within the level of resolution of the experiments, Glut4 trafficking behaves like a simple two phase reversible process, with a single rate limiting step in both exocytosis and endocytosis. It does not mean there is a single step in the pathway, however. Importantly, the rate limiting step for either delivery of Glut4 to the plasma membrane or removal of Glut4 from the plasma membrane may be different in basal and insulin-stimulated cells. Thus, "k_{ex}" and "k_{en}" are not measurements of a specific step in Glut4 trafficking (i.e., fusion of GSVs with the plasma membrane) but rather a description of the rate limiting step of the entire itinerary of the Glut4. Furthermore, the rate limiting step can differ for different measurements of Glut4 translocation. For example, Glut4 vesicles that are docked to the plasma membrane but not fused to the plasma membrane (occluded) would be measured as translocated Glut4 in a cell fractionation experiment, but would be invisible in a cell surface labeling experiment or a glucose transport assay (Holman *et al.*, 1994). All studies to date have referred to these kinetic constants as "k_{ex}" and "k_{en}." However, referring to these processes as endocytosis and exocytosis is inferring a specific mechanism. Because of these complexities, it is probably more appropriate to refer to the rate constants derived from the k_{obs} in the relaxation experiments as the rate constants for delivery of Glut4 to the plasma membrane and removal from the plasma membrane, or as the rate constants for the conversion of Glut4 from a form that is not accessible to surface labeling to a form that can be labeled. These distinctions will become increasingly more important as new assays are developed to measure specific steps in the Glut4 trafficking pathway.

Importantly, both groups explicitly assumed that insulin does not change the total amount of cycling Glut4. Therefore, the effect of insulin had to be entirely accounted for by changes in the intrinsic rate constants for Glut4 endocytosis and exocytosis (k_{en} and k_{ex}). This model is referred to as dynamic exchange or dynamic equilibrium. The data can be fit using a simple two pool model with two rate constants: k_{ex} and k_{en} (Fig. 10.4A). In a two pool model, the partition coefficient for the steady state distribution of Glut4 between the plasma membrane and intracellular compartments equals the ratio of the exocytic and endocytic rate constants ($P = [p]/[i] = k_{ex}/k_{en}$). The observed relaxation rate constant is the sum of the exocytic and endocytic rate constants ($k_{obs} = k_{ex} + k_{en}$). These relationships hold for both the transition and equilibration kinetics experiments. The fold-change in plasma membrane Glut4 equals the ratio of the fold-changes in the exocytic and endocytic rate constants ($\Delta[p] = [p]_{insulin}/[p]_{basal} = \Delta k_{ex}/\Delta k_{en}$). Thus, if insulin increases k_{ex} 3-fold and decreases k_{en} by 1/3, the net result will be a 9-fold increase in plasma membrane Glut4.

Using the simple 2-pool model to analyze the equilibration kinetics, both labs found that insulin increases Glut4 k_{ex} ~3-fold. The effect of insulin on endocytosis was more difficult to determine. While Jung *et al.* found that insulin decreased the rate of Glut4 endocytosis ~3-fold, Satoh *et al.* observed no effect on endocytosis. Whether or not insulin affects the rate constant of Glut4 endocytosis remains controversial (Antonescu *et al.*, 2008; Blot and McGraw, 2006; Muretta *et al.*, 2008). The problem is the difficulty in measuring k_{en}. The exocytic rate constant k_{ex} can be measured independently of k_{en} in a continuous uptake experiment (see Section III.B; Govers *et al.*, 2004; Martin *et al.*, 2006; Muretta *et al.*, 2008). However, all experiments that measure k_{en} are also affected by k_{ex}. Therefore, an accurate estimate of the partition coefficient is essential to determine the relative values of k_{ex} and k_{en}, and any inaccuracies in this measurement will lead to inaccuracies of the estimates of the relative rate constants (Muretta *et al.*, 2008). It is important to note that researchers in the endocytosis field often measure only the early time points in surface equilibration experiments or uptake assays and use the initial slope of the internal to surface ratio ($[i]/[p]$) versus time (in/sur) plot as an estimate of k_{en}. This is an inappropriate use of the technique developed by Wiley and Cunningham to estimate the endocytic rate constant for ligands such as EGF that dissociate from their receptors within cells (Wiley and Cunningham, 1982). The rationale is that the early uptake kinetics will be dominated by endocytosis, since exocytosis has not yet occurred. While this is intuitively satisfying, it is mathematically incorrect, and could lead to incorrect conclusions. The slope of the line is actually a complex function of both k_{ex} and k_{en}, and can be changed significantly through changes in k_{ex} with no change in k_{en}. Thus, it is more appropriate to measure the kinetics of the full relaxation and calculate k_{ex} and k_{en} from the measured k_{obs} and the partition coefficient.

To further explore the effect of insulin on the trafficking of Glut4, the kinetics data were used to build models of the trafficking itinerary of Glut4 (Fig. 10.4) (Holman *et al.*, 1994). The rate of Glut4 exchange between communicating compartments was treated as a first order process such that $V_{(m \to n)} = k_{(m \to n)}[G]_m$ where $V_{(m \to n)}$ is the rate of Glut4 traffic from one compartment (m) to a second compartment (n) and $[G]_m$ is the Glut4 content of compartment m. Two, three, four, and five compartment models were considered. As described above, the simple two pool model (plasma membrane vs. intracellular Glut4) is a good model to use to fit the kinetics data and derive values for "k_{ex}" and "k_{en}." However, a two pool model is inadequate to completely explain the measured behavior of Glut4. Notably, the rate constant measured for the transition of plasma membrane Glut4 levels from the basal to insulin-stimulated states ($t_{1/2} = 2.7$ min) was much faster than the rate constant measured for the equilibration of surface-labeled Glut4 between the plasma membrane and intracellular compartments in insulin-stimulated cells ($t_{1/2} = 10.6$ min). A two pool model predicts that these will be the same. A three compartment model fit the experimental data most succinctly: increasing the number of compartments did not improve the numerical fits of the data. In the three compartment model, Glut4 moves between a plasma membrane pool, a rapidly cycling endosomal pool, and a slow cycling intracellular storage pool. Insulin increases plasma membrane Glut4 content largely by increasing the net rate of Glut4 exocytosis from the slow cycling storage pool.

In Holman's three pool model, the rate of transfer of Glut4 from the GSVs to the plasma membrane equals the rate constant for exocytosis times the amount of Glut4 in the GSVs ($V_{(GSV \to PM)} = k_{ex}[G]_{GSV}$). In basal cells, transfer from GSVs to the plasma membrane is very slow ($k_{ex} = 0.001$ min^{-1}, $t_{1/2} > 10$ h), and rate limiting. This causes an accumulation of Glut4 in GSVs (~97% of the total). Insulin significantly increases k_{ex} (>100-fold). After insulin stimulation, the rate of exocytosis from the GSVs is faster than the rate of transfer from the endosomes to the GSVs (k_{seq}). This causes accumulation of Glut4 in the endosomes, and the transfer of Glut4 from the endosomes to the GSVs (sequestration) becomes rate limiting in insulin-stimulated cells. This allows for a rapid increase in Glut4 at the plasma membrane following insulin stimulation (the rapid transition kinetics are dominated by a large, rapid increase in k_{ex}, releasing the bolus of Glut4 accumulated in the GSVs), with a slower steady state cycling rate (the slower equilibration kinetics are limited by k_{seq}). The kinetic analysis indicates that there are *at least* three steps in the cycling itinerary that control Glut4 trafficking. This does not mean that there are only three compartments, just that there are three steps that are rate limiting. More recent modeling of the movement of Glut4 between subcellular compartments by the Jung lab relied on a four compartment model (Hah *et al.*, 2002). The four compartment model was based on the movement of Glut4 through four different compartments in subcellular fractionation experiments. The affect of insulin on Glut4 traffic in the Jung model is similar to that in the three compartment model: insulin increases the net rate of Glut4 exocytosis from a slow cycling storage pool. This dynamic equilibrium model has been the prevailing view in the field since these kinetics studies were first published.

While dynamic equilibrium can be used to model the observed trafficking behaviors of Glut4, a discrepancy remained. If dynamic equilibrium were the sole determinant of Glut4 distribution in adipose tissue, then k_{ex} would have to be 99-times slower than k_{en} in basal cells to obtain the very efficient sequestration of Glut4 observed ($P_{basal} = 1/99$; Fig. 10.3), and the ratio of k_{ex}/k_{en} would have to increase > 50-fold with insulin ($P_{insulin} = 37/63$). However, the reported estimates of k_{ex} were only 5–15-times slower than k_{en}, and only modest changes (3–9-fold) changes in k_{ex}/k_{en} were observed. k_{ex} was only three times slower in basal cells than in stimulated cells: the dynamic equilibrium model predicts that basal exocytosis of Glut4 would be much slower than was actually observed. Although complicated models of trafficking have been drawn with Glut4 moving bidirectionally between many intracellular compartments, "k_{ex}" will always be dominated by the rate limiting step in any continuous loop, since Glut4 will accumulate behind the slowest step.

An alternative model for the regulation of Glut4 trafficking has been proposed (Fig. 10.4C; Coster *et al.*, 2004; Govers *et al.*, 2004; Martin *et al.*, 2000; Muretta *et al.*, 2008). In this model, there are two separate pools of Glut4 in cells: one that is actively cycling between the plasma membrane and intracellular compartments in both basal and insulin-stimulated cells, and a separate noncycling GSV pool. Insulin increases the total amount of Glut4 in the rapidly cycling membrane pool by releasing it from the GSV pool. Insulin also increases the rate constant of exocytosis of the cycling pool 3-fold. This model of regulation is termed static retention. Because the rate that Glut4 is exocytosed is dependent upon the total amount of Glut4 in the cycling pool as well as k_{ex}, a release of Glut4 from static GSVs would cause the net rate of Glut4 exocytosis to increase. In this model, the fold-change in plasma membrane

Glut4 with insulin stimulation is a function of both the change in size of the cycling pool of Glut4 ($\Delta[\text{Glut4}]_{\text{total}} = ([p] + [i])_{\text{insulin}}/([p] + [i])_{\text{basal}}$) and the relative changes in the exocytic and endocytic rate constants ($\Delta[p] = \Delta[\text{Glut4}] \cdot \Delta k_{ex}/\Delta k_{en}$). Thus, if insulin increases the size of the cycling pool 10-fold, then a modest 3-fold increase in k_{ex} leads to a 30-fold increase in plasma membrane Glut4. In this model, the transporter is also partitioned between three pools: the plasma membrane, endosomes, and GSVs. However, the GSV storage pool does not communicate with the endosomal pool or with the plasma membrane under basal conditions, and the relaxation rate constant for the equilibration experiment is determined by the rate of trafficking from endosomes to the plasma membrane in both basal and insulin-stimulated cells.

Interestingly, the electron microscopy showed that the amount of Glut4 in the rapidly cycling general endosomal pathway (plasma membrane, clathrin coated pits, and early/recycling endosomes) increases 12.5-fold with insulin stimulation (from 3.9% to 48.9% of the total Glut4, summarized in Fig. 10.3; Slot *et al.*, 1991b), consistent with the static retention model. If this 12.5-fold increase in cycling Glut4 is combined with the observed 3-fold increase in k_{ex}, this would result in a 37.5-fold increase in Glut4 at the plasma membrane, which is very close to the 37-fold increase that was observed. Therefore, static retention could account for both the electron microscopy data and the kinetics data, if the tubulo-vesicular Glut4-containing compartments observed by Slot *et al.* are static, and not actively communicating with the plasma membrane/endosomal pathway in basal cells. Whether or not these compartments are in communication with the plasma membrane in basal adipose tissue remains to be determined. Consistent with the static retention model, Martin *et al.* observed that insulin-stimulation lead to a decrease in the total number of Glut4 containing compartments in 3T3-L1 adipocytes, but not to a decrease in the relative amount of Glut4 in each compartment (Martin *et al.*, 2000). This indicates that insulin increases the rate of exocytosis/fusion of Glut4 containing-compartments, and not simply the rate of transfer of Glut4 between compartments. This analysis has not yet been done in primary adipocytes.

Surprisingly, the question of whether the storage compartment of Glut4 is a static or cycling pool could have been addressed using the photo-affinity labeling strategy. It simply was not considered. Failure of the early kinetic studies to make these measurements caused a debate in the Glut4 field which has still not been fully resolved. The distinction is important: in the static retention model, Glut4 vesicles behave like other regulated exocytic compartments such as SV and secretory granules. Consistent with this idea, there is considerable overlap between the proteins required for trafficking of Glut4 and SV (see Section III.E.). In contrast, the dynamic equilibrium model requires a different type of regulatory mechanism. As described below, we believe that both mechanisms play important roles in regulating Glut4 distribution in adipocytes.

III. Current Views and Controversies

A. Models for the regulation of Glut4 trafficking

While the simplest kinetics model for Glut4 trafficking involves only three compartments, our current understanding of the pathway indicates that it is far more complex. Based on a huge literature, which includes kinetics data, subcellular fractionation, microscopy, and mutational analysis, a model for Glut4 trafficking can be drawn with at least 10 discrete steps (Fig. 10.5; modified from Larance *et al.* (2008)). (1) Biogenesis of GSVs—delivery of newly synthesized Glut4 from the ER/Golgi to the storage compartments; (2) Release—increase in the rate of exocytosis of GSVs; (3) Transport—movement of GSVs to the cell periphery; (4) Tethering—low affinity binding of vesicles to the plasma membrane (via the exocyst complex); (5) Docking/fusion—high affinity binding and membrane fusion (via SNAREs); (6) Endocytosis—of Glut4 through clathrin coated pits; (7) Maturation—loss of clathrin from the newly formed endocytic vesicles, gain of Rabs, fusion of small vesicles to form endosomes; (8) Sorting—removal of Glut4 from endosomes for delivery to ECR; (9) Recycling—formation of small exocytic vesicles from the ECR; (10) Sequestration—repackaging of Glut4 into GSVs from the endosomes. These steps involve separate pools of Glut4: the biosynthetic pool (step 1), the fast cycling pool (steps 3–9), and the slowly cycling/noncycling GSV pool (steps 2 and 10). A number of proteins have been identified/suggested to be involved in each of these steps (as indicated in the figure). Defects in any of the indicated proteins (i.e., knockout or expression of mutant forms) will change plasma membrane Glut4 levels. However, they will do so by very different effects on Glut4 and/or general endosomal trafficking. While most work in the field has focused solely on cell surface Glut4, new assays have been developed that can distinguish specific trafficking steps. These assays are currently being used to verify the steps affected by specific proteins, and to identify proteins/processes specifically involved in the regulated trafficking of Glut4 in adipocytes, that distinguish the Glut4 pathway from the general endocytic pathways (recent examples include: Bai *et al.*, 2007; Blot and McGraw, 2008; Huang *et al.*, 2007; Jiang *et al.*, 2008; Koumanov *et al.*, 2005; Watson and Pessin, 2008; Williams and Pessin, 2008).

Insulin regulates Glut4 traffic at multiple steps including: release, translocation, and fusion of GSVs, as well as repackaging of Glut4 into GSVs (Bryant *et al.*, 2002). Fusion requires the SNARE proteins VAMP2, syntaxin 4, and SNAP-23, and is regulated by Munc-18 and Rabs (Foster and Klip, 2000). Internalization requires clathrin and dynamin (Al-Hasani *et al.*, 1998, 2002; Kao *et al.*, 1998; Slot *et al.*, 1991b). The proteins that regulate

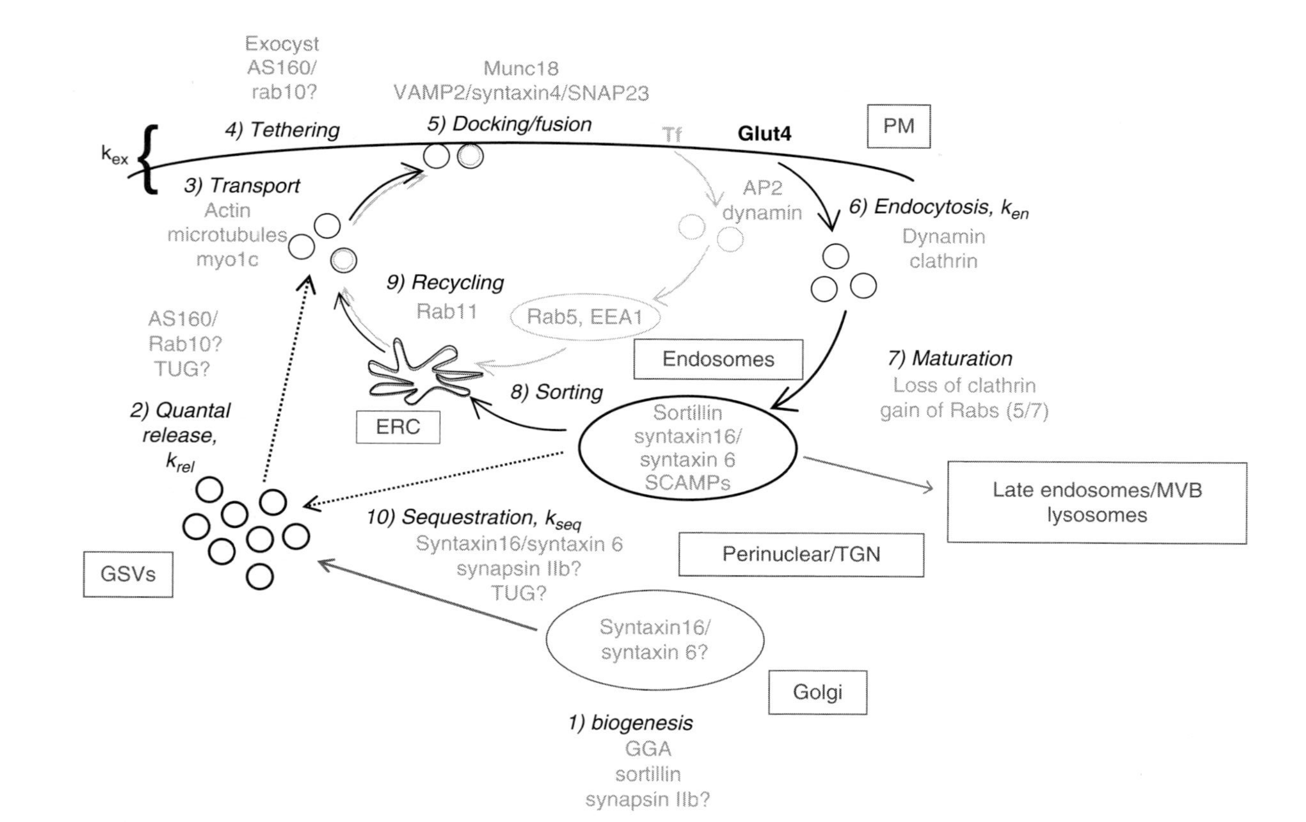

Figure 10.5 Proteins that regulate the trafficking of Glut4. An updated trafficking itinerary, showing static retention. The pool of Glut4 that is cycling follows a modified endocytic pathway (shown in black; the trafficking of the transferrin receptor, Tf, is shown in gray). Cycling Glut4 does not traffic through the GSVs in this model. Dotted lines: insulin stimulates the quantal release of Glut4 from GSVs into the cycling pathway; insulin withdrawal transiently activates the sequestration of Glut4 in endosomes back into GSVs. Proteins that have been implicated in each step are indicated.

GSV translocation, sort Glut4 in endosomes, and deliver Glut4 to GSVs are less well understood, but include AS160, Rabs 10 and 11, syntaxins 6 and 16, synapsin, TUG, sortillin, and GGA (Eguez *et al.*, 2005; Larance *et al.*, 2005; Li and Kandror, 2005; Lin *et al.*, 1997; Morris *et al.*, 1998; Perera *et al.*, 2003; Proctor *et al.*, 2006; Sano *et al.*, 2003, 2007; Shi and Kandror, 2005; Watson *et al.*, 2004; Yu *et al.*, 2007; Zeigerer *et al.*, 2002, 2004).

As described above, exactly how insulin regulates the subcellular distribution of Glut4 remains debated. Two competing models have emerged. In the first, termed dynamic exchange or dynamic equilibrium, Glut4 continually cycles between the plasma membrane, endosomes, and GSVs (Martin *et al.*, 2006). In basal cells, slow exocytosis (release from GSVs) together with fast endocytosis minimizes the amount of Glut4 at the cell surface. Insulin redistributes Glut4 to the plasma membrane by increasing the rate of Glut4 exocytosis (through an increase in k_{ex}), while simultaneously decreasing the rate of Glut4 endocytosis from the plasma membrane (through a decrease in k_{en}). Key features of this model are that all of the Glut4 is actively cycling in both basal and insulin-stimulated cells, and the effect of insulin is completely accounted for by changes in k_{ex} and k_{en}. Although the number of compartments and the connections between compartments varies, kinetically the behavior of Glut4 in this system is analogous to its behavior in the three pool model depicted in Fig. 10.4B.

The second model, termed static retention, differs from dynamic equilibrium in several important respects (Fig. 10.4C). In static retention, a large pool of Glut4 is sequestered in GSVs and does not cycle through the plasma membrane under basal conditions (Coster *et al.*, 2004; Govers *et al.*, 2004). Insulin transiently stimulates the fusion of GSVs with the plasma membrane, releasing quanta of Glut4 into the cycling membrane system (plasma membrane, endosomes, TGN). Upon insulin withdrawal, there is a transient activation of Glut4 trafficking from endosomes back to the GSVs. In this model, the GSVs behave like regulated exocytic vesicles. Insulin increases cell surface Glut4 through an increase in the amount of Glut4 in the actively cycling pool as well as through increases in k_{ex}. The most important feature of the static retention model is that the cycling and static pools of Glut4 remain separate, and the pools only transiently communicate with one another during the basal-to-insulin or insulin-to-basal steady state transitions. Kinetically, the behavior of Glut4 can be described by the two-pool model in this system (Muretta *et al.*, 2008), with many steps potentially contributing the observed rate constants for the delivery of Glut4 to the plasma membrane (k_{ex}) or removal from the plasma membrane (k_{en}). Unlike dynamic equilibrium, the static retention model does not require two different rate limiting steps to account for rapid transition kinetics with slower steady state trafficking.

B. Kinetic analysis of Glut4 traffic in adipocytes using epitope-tagged Glut4 reporter constructs: Dynamic exchange versus static retention

Investigators have tested the role of many proteins in the regulation of Glut4 traffic by RNA knockdown, expression of interfering mutants, and treatment with specific chemical inhibitors. To assess which steps in Glut4 trafficking are affected by interfering treatments, their effects on the kinetics of Glut4 traffic need to be measured. Many of these studies have been performed in the cell culture model 3T3-L1 adipocytes. Though fully differentiated 3T3-L1 cells respond robustly to insulin stimulation by increasing the rate of glucose transport, they do not solely express Glut4. Significant amounts of Glut1 are also expressed. Glut1 is not localized to the insulin-sensitive Glut4 storage compartment. In addition to this, cultures of 3T3-L1 cells contain a significant number of nondifferentiated fibroblasts (Muretta *et al.*, 2008). The fibroblasts do not express Glut4. They do, however, express Glut1. Because of these complexities, glucose transport cannot be used to uniquely monitor insulin-responsive trafficking of Glut4. Synthesis of the photo-affinity probes developed by Holman and Jung is difficult. In addition, the reagents modify other proteins localized to plasma membrane besides Glut4 (including Glut1). Because of this, analysis by photo-affinity labeling requires an affinity purification step using anti-Glut4 antibody. To circumvent these difficulties, Dawson *et al.* developed an epitope-tagged Glut4 reporter construct: HA-Glut4/GFP (Dawson *et al.*, 2001)*. The HA epitope was inserted into the first extracellular loop of the transporter while the GFP was fused to the cytoplasmic carboxy-terminal tail. The HA tag is exposed to the extracellular environment when Glut4 is inserted into the plasma membrane, and can be labeled with anti-HA antibody (Fig. 10.1). This allows direct measurement of the amount of Glut4 localized to the plasma membrane. This labeling is analogous to the photoaffinity labeling of the surface Glut4. However, the labeling is specific for Glut4 and requires no immunopurification step. Neither protein modification affected traffic of the transporter in control experiments. The reporter appears to traffic identically to endogenous Glut4. Therefore, anti-HA immunoreactivity and GFP fluorescence can be used to monitor the subcellular localization and trafficking kinetics of Glut4. Other reporter constructs that have been used include HA-Glut4 (no GFP) and exofacial myc-tagged Glut4.

Two competing labs (the James lab and the McGraw lab) used the anti-HA reporter expressed in differentiated 3T3-L1 adipocytes to examine how Glut4 traffics under basal and insulin-stimulated conditions in these cells

* Although it is highly desirable to be able to analyze the trafficking of the endogenous Glut4, no high affinity antibodies specific for the extracellular epitopes of Glut4 have been reported.

(Govers *et al.*, 2004; Karylowski *et al.*, 2004; Martin *et al.*, 2006). In these experiments, the kinetics of reporter traffic were followed by incubating cells in the continuous presence of anti-HA antibody at 37 °C. With time the antibody accumulates in cells, reaching saturation when the entire pool of cycling Glut4 has passed through the plasma membrane. High concentrations of antibody were used to ensure that binding at the plasma membrane was faster than Glut4 trafficking, and was not rate limiting. Under these conditions, anti-HA binding is essentially instantaneous, and any increase in antibody binding is due to delivery of new epitope-tagged Glut4 to the plasma membrane from the intracellular pools. In one lab (McGraw) anti-HA accumulation was measured using quantitative single cell microscopy. In the other (James) accumulation was followed using a whole plate fluorescence spectroscopy based assay. Antibody binding did not affect reporter trafficking. Both labs showed this by preloading cells expressing the reporter with anti-HA antibody and then monitoring the kinetics of antibody translocation and recycling under basal and insulin-stimulated conditions. In both studies the Glut4 prelabeled with anti-HA antibody behaved the same as the nonlabeled reporter.

As observed in the transition and equilibration kinetic experiments, the kinetics of antibody accumulation can also be described by a single exponential relaxation (Govers *et al.*, 2004; Martin *et al.*, 2006). However, in the uptake experiments the observed relaxation rate constant is only a function of the exocytic rate constant ($k_{obs} = k_{ex}$; at sufficiently high concentrations of antibody). The endocytic rate constant (k_{en}) affects the partition coefficient, but not the relaxation kinetics. Consistent with the earlier kinetic analyses, both groups found that insulin increases k_{ex} by 3–7-fold. The observations in each report differed in several important aspects, however. The McGraw group reported that insulin also inhibited endocytosis, and found no change in the total cycling pool of Glut4 (although it took longer to reach equilibrium, the same maximum amount of antibody was accumulated in basal cells as in insulin-stimulated cells). Their conclusion was that insulin regulates Glut4 cycling through dynamic exchange. In contrast, the James lab found that insulin had little effect on k_{en}, but increased the total cycling Glut4 > 10-fold following stimulation (at steady state there was 10-times more antibody accumulated in stimulated cells than in basal cells). Thus, they concluded that although insulin increased the rate of exocytosis (3-fold), the major effect of stimulation was to release Glut4 from a static noncycling pool. Thus, the Glut4 reporter construct behaved very differently under the experimental conditions used in these two labs. The differences in these experimental results could not be explained by differences in the constructs, the expression systems, or the cell lines used (Martin *et al.*, 2006).

To further explore these discrepancies, we developed a novel flow cytometric assay to monitor Glut4 trafficking (Muretta *et al.*, 2008) based on the assays used by the McGraw and James labs. The flow assay combines the ability of the whole plate assay to analyze thousands of cells (Govers *et al.*, 2004), with the ability of microscopy to examine uptake on a cell-by-cell basis (Martin *et al.*, 2006). Additional advantages include sensitivity of detection, accurate noise correction, and ease of analyzing large numbers of samples. Using this assay we determined that in 3T3-L1 adipocytes, Glut4 is regulated by both static retention and dynamic equilibrium. Importantly, we also found that cell culture conditions are critical: replating after differentiation shifts the regulation from a static mechanism to a dynamic one, explaining in part the differences observed in the previous papers. Replating cells prior to analysis by microscopy is done routinely in many labs. However, the affects of this treatment on Glut4 trafficking and insulin signaling have not been thoroughly characterized. Furthermore, Govers *et al.* showed that the differentiation state of the 3T3-L1 cells is also critically important. The amount of Glut4 sequestered in the static pool increases during differentiation from fibroblasts to adipocytes, shifting the regulation of Glut4 from dynamic equilibrium to static retention (Govers *et al.*, 2004). Most experiments in replated 3T3-L1 cells are done early in differentiation, before large lipid droplets develop; this may also have contributed to the differences in trafficking that were observed. Many labs use the 3T3-L1 tissue culture model because of the relative ease of protein expression/knockdown in these cells compared to primary cells. However, to resolve the discrepancies observed in 3T3-L1 cells, it is critically important that the relative contributions of static retention and dynamic equilibrium in the regulation of Glut4 trafficking be assessed in primary adipocytes, as well as in intact adipose tissue. Several labs are currently working on transgenic animals that can be used for these experiments.

C. Glut4 sequestration is regulated by both static retention and dynamic equilibrium

In intact monolayers of differentiated 3T3-L1 cells, insulin regulates the distribution of Glut4 between the actively cycling pool and the noncycling GSVs (Govers *et al.*, 2004; Muretta *et al.*, 2008). In basal cells, a small fraction of the total Glut4 is cycling between the plasma membrane and endosomes. After maximal insulin, there is a 5–10-fold increase in the maximum amount of anti-HA taken up, showing that 5–10 times more Glut4 is actively cycling. This increase represents the Glut4 that was sequestered in GSVs. The cycling Glut4 distributes between the plasma membrane and endosomes, in a ratio determined by the relative rates of exocytosis and endocytosis (the 2-pool model). Therefore, an increase in the size of the actively cycling pool leads to an increase in cell surface Glut4. In fact, there is a dose-dependent increase in Glut4 at the plasma membrane, due to a dose-dependent increase in the size of the actively cycling pool.

Importantly, even in the continuous presence of insulin, only a fraction of the total Glut4 is labeled at submaximal insulin concentrations. This "quantal release" indicates that there is a transient activation of the GSV release mechanism by insulin, after which the cycling and GSV pools remain segregated and do not mix in the continued presence of insulin. When insulin is withdrawn, the Glut4 is cleared from the cell surface, repackaged into GSVs, and statically retained until the cells are again stimulated with insulin. Thus, insulin withdrawal activates the recycling of Glut4 from the cycling pool to the static GSV pool, again transiently allowing these two populations to mix. Quantal release has important implications for the kinetics modeling. Under these conditions, the rate of Glut4 cycling is determined solely by the rate of cycling through the endocytic pathway. The rate of release of Glut4 from the GSVs after insulin stimulation does not contribute to the observed relaxation at steady state, but does contribute to the rapid translocation kinetics (Muretta *et al.*, 2008).

In addition to increasing the pool size, insulin also increases the rate at which the Glut4 is cycling (k_{ex}) 3–4-fold. No effect on k_{en} was observed. Together with the increase in pool size, this leads to a >15-fold increase in cell surface Glut4. Interestingly, after insulin withdrawal, k_{ex} is even slower in the recovering cells than in control basal cells (Muretta *et al.*, 2008). This causes a rapid clearance of Glut4 from the cell surface (through a rapid shift in the dynamic equilibrium), even though repackaging into the static GSVs is relatively slow. Therefore, the kinetics experiments show that the distribution of Glut4 at the plasma membrane is regulated at least at two different steps: (1) quantal release from the static pool of GSVs and (2) the rate of transit through the cycling pathway, or k_{ex}. Thus both static and dynamic retention mechanisms play important roles. Mathematic modeling indicates that this bipartite regulatory mechanism can accurately account for the observed behaviors of Glut4 in all of the kinetics experiments that have been described (transition, equilibration, and antibody uptake), as well as quantal release (Muretta *et al.*, 2008).

Interestingly, we have identified mutant proteins that differentially affect these two steps in Glut4 trafficking. Expression of a phosphorylation mutant of the SV protein synapsin II (S10A synapsin) in 3T3-L1 adipocytes was insulin-mimetic: it increased cell surface Glut4 3-fold in basal cells (Muretta *et al.*, 2007). Insulin caused a further 3-fold increase in cell surface Glut4 in these cells. S10A synapsin did not affect the distribution of the transferrin receptor. S10A synapsin increased cell surface Glut4 in basal cells by increasing the size of the basal cycling pool of Glut4 by 3-fold (unpublished observation). Insulin further increased plasma membrane Glut4 in S10A synapsin expressing cells by increasing k_{ex} 3-fold. In contrast, expression of a phosphorylation mutant of the regulatory protein AS160 (AS160–4P) decreased Glut4 at the plasma membrane in both basal and insulin-stimulated cells, and inhibited insulin-stimulated Glut4 translocation (Sano *et al.*, 2003;

Zeigerer *et al.*, 2004). AS160 affected Glut4 at the plasma membrane by decreasing k_{ex}, without affecting release of the GSVs. Insulin still increased the size of the cycling pool by 10-fold in cells expressing AS160–4P (unpublished observation), but the amount of Glut4 at the plasma membrane remained low, and the total pool was labeled very slowly, due to the very slow exocytic rate. Therefore, S10A synapsin and AS160–4P differentiate two steps in Glut4 trafficking: (1) release of GSVs from the static pool and (2) an increase in the intrinsic rate of Glut4 exocytosis.

Although many mutant proteins block insulin-stimulated Glut4 exocytosis, S10A synapsin is one of only a small number that inhibit Glut4 retention in basal cells by increasing exocytosis. Others include activated signaling molecules (PI3-kinase and Akt; Eyster *et al.*, 2005; Frevert *et al.*, 1998), overexpression of "cargo" (Glut4 or IRAP), presumably through saturation of the retention mechanism (Carvalho *et al.*, 2004; Waters *et al.*, 1997), shRNA knockdown of AS160 (Eguez *et al.*, 2005; Larance *et al.*, 2005), and expression of mutant forms of Rab10 (Sano *et al.*, 2007), TUG (Yu *et al.*, 2007), or syntaxins 6 and 16 (Perera *et al.*, 2003; Proctor *et al.*, 2006). It is unknown if these proteins affect packaging of Glut4 into the GSVs (from the Golgi or from the endosomes) or the static retention of the GSVs. However, proteins/treatments that specifically increase Glut4 at the plasma membrane without affecting the trafficking of general endocytic markers are likely to be affecting important steps in the specialized Glut4 trafficking pathway. Thus, identification of the molecular machineries controlling these processes is likely to identify novel therapeutic targets.

D. Endosomes and the cycling pool of Glut4 in adipocytes: Glut4 and transferrin have different trafficking kinetics

To fully understand the translocation of Glut4, it is essential to understand the cellular organelles involved in the storage and trafficking of Glut4. Therefore, it is important to determine the itinerary of the cycling Glut4 in basal and insulin-stimulated cells. Insulin causes a translocation of markers of the general endocytic pathway (transferrin receptor, Glut1) through an increase in the rate of exocytosis of these proteins. Thus, one possibility is that all of the cycling Glut4 resides in the constitutively cycling endocytic pathway, and the observed increase in k_{ex} is due to the general increase in endosomal trafficking. In this scenario, quantal release of GSVs would completely account for specialized Glut4 trafficking. However, in our model for Glut4 trafficking we have drawn distinct pathways for cycling Glut4 and transferrin (Fig. 10.5). While there is clearly some overlap between Glut4 and the transferrin receptor (the transferrin receptor coimmunoprecipitates with Glut4), there is accumulating evidence that these markers are largely distinct in cells. For example, by microscopy there is very little overlap seen between Glut4 and the Rab5- and EE1A-positive early endosomal compartments, while there is a high degree of overlap of transferrin with these markers (Cormont *et al.*, 1996; Ramm *et al.*, 2000). However, both Glut4 and transferrin are found in the Rab11-positive recycling endosomes (Zeigerer *et al.*, 2002). Blot and McGraw recently showed that knockdown of the clathrin adaptor protein AP-2 inhibited endocytosis of transferrin, but not Glut4 in adipocytes (Blot and McGraw, 2006). While this shows that Glut4 is internalized through a mechanism that is different than the clathrin-mediated internalization of the transferrin receptor, Glut4 is clearly observed in clathrin coated structures in electron microscopy (Robinson *et al.*, 1992). Furthermore, Glut4 accumulates at the plasma membrane in clusters highly enriched in clathrin (Huang *et al.*, 2007).

Consistent with the idea that Glut4 and transferrin are endocytosed through different pathways, the kinetics of internalization of transferrin and Glut4 are very different. The trafficking kinetics of transferrin have been analyzed in adipocytes using an efflux assay (Tanner and Lienhard, 1987). In these experiments, the cells are loaded with labeled transferrin (either radiolabeled or fluorescently tagged), the cell surface transferrin is removed, and the kinetics of the loss of label measured (transferrin that has cycled through endosomes is readily released once it returns to the plasma membrane). In these experiments, $k_{obs} = k_{ex}$ and $P = k_{ex}/k_{en}$. Using the measured values for k_{ex} (0.111 min^{-1} basal, 0.194 min^{-1} insulin) and P (0.25 basal, 0.5 insulin)†, the endocytic rate constant for transferrin can be estimated. Transferrin is internalized with a halftime of 1.6–2 min ($k_{en} \approx 0.44$ min^{-1} basal, 0.39 min^{-1} insulin). The trafficking kinetics are very similar in 3T3-L1 adipocytes (unpublished observation). In contrast, Glut4 is internalized with a halftime of 8 min ($k_{en} \approx 0.088$ min^{-1} basal, 0.084 min^{-1} insulin) in these same cells (Muretta *et al.*, 2008). Therefore, transferrin is internalized 4–5 times faster than Glut4 in adipocytes, strongly supporting the hypothesis that Glut4 and transferrin are internalized via different mechanisms. The k_{ex} for transferrin (0.12 min^{-1} basal; 0.2 min^{-1} insulin; $t_{1/2} = 3$–6 min) is also 5–10 times faster than the k_{ex} measured for Glut4 in the same cells (0.012 min^{-1} basal, 0.036 min^{-1} insulin; $t_{1/2} = 20$–60 min). These differences in the rates of exocytosis and endocytosis indicate that the cycling pool of Glut4 and transferrin do not simply traffic together through the early endosomal pathway. Thus, although insulin increases the rate of exocytosis from the constitutive early endosomal pathway by ~2-fold, the increase in the rate constant for Glut4 exocytosis is through a distinct mechanism.

Although originally thought to be a single pathway, it was recently shown that receptors that deliver material to the late endosomes for

† About 20% of the transferrin receptor was at the plasma membrane in basal cells and 33% in insulin-stimulated cells.

degradation in the lysosomes (such as the LDL and EGF receptors) are internalized via different clathrin coated pits than the majority of the transferrin (Lakadamyali *et al.*, 2006). By microscopy, all clathrin coated pits contain transferrin, but only a subset also contains LDL and EGF. After internalization, these vesicles are kept largely separate. While the majority of the transferrin (85%) is delivered to large, relatively immobile Rab5- and EE1A-positive early endosomes, the LDL and EGF (and 15% of the transferrin) are delivered to a distinct population of smaller, rapidly moving early endosomes (Rab5- and Rab7-positive) that mature into late endosomes (Rab7-positive). Receptors are recycled from these endosomes, while the ligands are delivered to the lysosomes for degradation. The receptors traffic back to the plasma membrane through the Rab11-positive sorting endosomes, where they are colocalized with transferrin. AP2-knockdown completely inhibited uptake through the transferrin-specific clathrin coated pits, but did not affect the uptake of LDL or EGF (these proteins utilize alternative clathrin adaptor proteins for internalization; Traub, 2003). This segregation requires microtubules; the early and late endosomal proteins were completely colocalized after treatment with nocadozole.

Glut4 colocalizes with markers of the late endosome pathway (the cation-independent mannose-6-phosphate/IGF-II receptor and Rab 7), and sorting endosomes (Rab 11). Based on this colocalization, it is possible that the actively cycling Glut4 in basal and/or insulin-stimulated cells resides within the late endosomal pathway, and is simply being trafficked as cargo in the pathway. Consistent with this, the mannose-6-phosphate receptor and α-2-macroglobulin receptor (LRP) are also translocated to the plasma membrane in response to insulin (Corvera *et al.*, 1989; Kandror and Pilch, 1996; Ko *et al.*, 2001). In the case of LRP, this increase is due to an increase in k_{ex} (Ko *et al.*, 2001). Alternatively, although there is clearly some overlap, the cycling Glut4 pathway may be largely distinct from either the early endosomal or late endosomal pathways. For example, resident SV proteins are recycled in part through an alternative nonendosomal pathway as well as through endosomes (Wenk and De Camilli, 2004). The exact trafficking pathway followed by the cycling pool of Glut4 remains to be determined.

E. The Glut4 storage compartments: GSVs and the reserve pool of SV

Because dynamic equilibrium was the dominant model for the regulation of the translocation of Glut4, it has long been argued that the regulation of Glut4 trafficking must involve a mechanism that is very different than that used by other regulated secretory pathways such as SV or secretory granule exocytosis. However, it is now clear that Glut4 trafficking utilizes protein machinery that is shared by other regulated secretory pathways, including SV exocytosis. The first of these proteins that was identified was VAMP2

(Cain *et al.*, 1992). Since then, others have been found. These include homologues or isoforms of syntaxins, SNAP-23/25, Muncs, and Rabs (Bryant *et al.*, 2002; Foster and Klip, 2000). Isoforms of these proteins are now known to be components of the general membrane trafficking machinery in all cell types. The VAMPs, syntaxins, and SNAP-23/25, along with NSF, α-SNAP and γ-SNAP facilitate membrane fusion, while the Rabs and Muncs play modulatory roles. Like GSVs, fusion of SVs with the plasma membrane requires VAMPs, syntaxins, SNAPs, Muncs, and Rabs (Sudhof, 2004). Like Glut4, resident SV proteins are internalized from the plasma membrane via clathrin coated vesicles and recycled back into SVs in part from endosomes (Bloom *et al.*, 2003; Cameron *et al.*, 1991; Linstedt and Kelly, 1991; Provoda *et al.*, 2000). A number of other proteins involved in SV cycling also play basic roles in vesicle trafficking in all cell types, and are required for Glut4 trafficking. These include synaptotagmin, synaptojanin, synaptophysin/amphiphysin, and dynamin (Sudhof, 2004).

Although at first glance the trafficking of GSVs and SVs seem very different, it is probably more accurate to think of them as similar pathways with specializations unique to each. This is important because the protein machinery regulating SV exocytosis has been so well characterized, and as has been shown repeatedly, most of these proteins (or their homologues) are also important for Glut4 trafficking. Presynaptic terminals contain between 150 and 200 SV on average (Sudhof, 2000, 2004). Like Glut4 vesicles, the SVs are partitioned between an actively cycling/recycling pool of vesicles and a noncycling, reserve pool of vesicles. The cycling pool of SVs can be distinguished from the noncycling vesicles by loading the vesicles with a membrane impermeant fluorescent dye, such as FM1–43 as they cycle through the cell surface (Akbergenova and Bykhovskaia, 2007). Unlike Glut4 vesicles, however, all SV are clustered together near the cytoplasmic surface of the presynaptic plasma membrane. Between 5 and 10 SVs are localized in immediate contact with the membrane. These vesicles, termed the readily releasable pool, are docked with the membrane in a calcium-sensitive, primed, prefusion state. Membrane depolarization causes the concentration of intracellular calcium to rise. This activates the Ca^{2+}-sensitive fusion machinery, triggering extremely rapid (millisecond) exocytosis of the releasable vesicles. The readily releasable pool is rapidly refilled from a small population of recycling vesicles. Under mild stimulation, the vesicles are reformed directly from the plasma membrane or from specialized invaginations of the membrane (Wenk and De Camilli, 2004). Thus, the "active pool" of SVs is very highly regulated at a step just prior to fusion and is controlled through a Ca^{++}-dependent mechanism. In contrast, Glut4 vesicles are not predocked at the plasma membrane. However, there is increasing evidence that Glut4 exocytosis is regulated postdocking at the fusion step (Bai *et al.*, 2007; Huang *et al.*, 2007; Koumanov *et al.*, 2005). The molecular machinery regulating this fusion step in SV trafficking is incompletely understood, but includes the Ca^{++}-sensor synaptotagmin, Muncs, and SNAREs (Martin, 2003; Verhage and Sorensen, 2008; Wojcik and Brose, 2007). Thus, Glut4 trafficking resembles SV trafficking even at this highly specialized step in SV exocytosis, and it is very likely that insulin-sensitive proteins homologous to the Ca^{++}-dependent machinery will be found (Li *et al.*, 2007; Whitehead *et al.*, 2001).

The majority of the SVs in the presynaptic terminal are not actively cycling under normal stimulation conditions, and are not labeled by the membrane impermeant dyes (Akbergenova and Bykhovskaia, 2007). This reserve pool of vesicles resides just behind the releasable/recycling pool, clustered within a dense protein rich matrix (Bloom *et al.*, 2003; Evergren *et al.*, 2007). Under conditions of prolonged or high frequency membrane depolarization, SV recycling cannot keep pace with SV exocytosis. This depletes the vesicles from the releasable/recycling pool and diminishes the response (synaptic depression). When this occurs, the reserve pool is mobilized, resulting in a quantal release of vesicles into the actively cycling pool. This quantal release plays a very important role in synaptic plasticity: it causes an incremental increase in response to subsequent stimuli until the additional SVs are retrieved through recycling. Like GSVs, the reserve SVs are returned to the noncycling reserve pool through recycling through endosomes (Akbergenova and Bykhovskaia, 2007). Thus the trafficking of GSVs and the reserve pool of SV are very similar, and there is likely to be overlap between the proteins that regulated the quantal release of GSVs and SVs.

Genetic and biochemical knock-down experiments in neurons show that the reserve pool of SV is maintained and regulated by the synapsins, a family of abundant peripheral membrane proteins (Bloom *et al.*, 2003; Evergren *et al.*, 2007; Sudhof, 2004). However, exactly where the synapsins function in the SV cycle is not clear (Sudhof, 2004). Significant data in the literature supports the idea that the synapsins specify the recycling of SV proteins from endosomes. Loss of synapsin leads to a loss of the vesicles in the reserve pool and a significant decrease in the total amount of SV protein in neurons (Gaffield and Betz, 2007). Synapsins are associated with tubular extensions of endosomes and with SV in the reserve pool, but are not colocalized with SV proteins in endocytic intermediates derived from the plasma membrane (Bloom *et al.*, 2003; Evergren *et al.*, 2007). This strongly suggests that synapsins are involved in protein delivery from the endosomes to the reserve SVs. Consistent with this idea, synapsin knockout specifically inhibits SV recycling from endosomes, but not through the alternative pathway (Akbergenova and Bykhovskaia, 2007). Synapsin knockout leads to the selective loss of the reserve pool of SV (all SVs in synapsin knockout neurons are in the readily releasable/recycling pool).

Although long considered an exception to the nearly ubiquitous expression of SV proteins in all cell types, synapsin I was recently found in both pancreatic β cells and in liver epithelial cells (Bustos *et al.*, 2001; Krueger *et al.*, 1999; Longuet *et al.*, 2005; Matsumoto *et al.*, 1999), indicating a role in multiple membrane trafficking pathways. Based on the function of synapsin in neurons and the high level of conservation of the membrane traffic machinery, we hypothesized that synapsin might play a role in regulating Glut4 exocytosis. We found that adipocytes express synapsin IIb mRNA and protein (Muretta *et al.*, 2007). Our data show that synapsin IIb plays an important role in Glut4 traffic: expression of a synapsin IIb site 1 phosphorylation mutant (S10A synapsin) interferes with basal Glut4 retention in adipocytes. This loss of retention is due to a failure to package the Glut4 into the static GSVs. Whether the mutant synapsin interferes with Glut4 recycling from endosomes to the GSVs remains to be determined. It is likely that the protein machinery that controls the release of static GSVs into the cycling pool in response to insulin will have additional overlap with the proteins that regulate the release of the reserve pool of SVs. This will need to be tested as new proteins are identified in the SV pathway.

IV. Conclusions and Future Directions

Insulin increases glucose transport into adipocytes 10–20-fold by recruiting the insulin responsive glucose transporter Glut4 from intracellular compartments to the plasma membrane. In basal cells, <5% of the total cellular Glut4 is found in the plasma membrane. After insulin stimulation, 30–50% of the total is found in the plasma membrane. This shift increases the number of transporters at the cell surface, increasing glucose uptake and storage. The large increase in glucose transport is the product of both the efficient sequestration of Glut4 under basal conditions, and the rapid redistribution of Glut4 from intracellular storage compartments to the plasma membrane after insulin stimulation.

In basal cells, Glut4 is sequestered in specialized storage compartments (GSVs). The GSVs behave like regulated secretory vesicles, although their exact nature and composition remain unclear. Insulin triggers the release of Glut4 from GSVs into the actively cycling plasma membrane/endosomal pool. Glut4 at the plasma membrane continuously reenters the cell by endocytosis, after which it passes through endosomes and is cycled back to the plasma membrane. After insulin withdrawal, the cycling Glut4 is sorted back into the static GSVs.

The total amount of Glut4 at the plasma membrane is determined by at least two variables: (1) the total amount of Glut4 that is cycling and (2) the relative rates of Glut4 exocytosis and endocytosis. Insulin regulates both of these variables. In response to insulin there is a dose-dependent increase in the total amount of Glut4 that is actively cycling in adipocytes. This release is quantal: a specific concentration of insulin causes the release of a corresponding number of vesicles from the static GSV pool (as high as 10-fold with maximal insulin). Even with prolonged incubation with insulin, no additional vesicles are released. The released Glut4 molecules continuously cycle between the plasma membrane and intracellular vesicles, through an as yet incompletely characterized endosomal pathway until insulin is withdrawn. After insulin withdrawal, most but not all of the cycling Glut4 is repackaged into static GSVs. A small pool of Glut4 continues to cycle in basal cells. In addition to increasing the total amount of cycling Glut4, insulin also increases the rate of Glut4 cycling through an increase in the rate constant for Glut4 exocytosis, k_{ex} (it is remains controversial whether there is an effect on the rate of Glut4 endocytosis, k_{en}). Upon insulin withdrawal, k_{ex} decreases rapidly, causing a rapid clearance of Glut4 from the plasma membrane. The repackaging of Glut4 into GSVs occurs more slowly. This bipartite control mechanism allows for a very rapid response to changing levels of insulin, as well as very high degree of control over the amount of glucose taken up into adipose tissue. The rapid redistribution of Glut4 between the plasma membrane and intracellular pool is controlled through effects on the intrinsic rate of Glut4 cycling (dynamic equilibrium), while the magnitude of the response is controlled through quantal release of Glut4 from the GSV pool into the cycling pool (static retention).

The quantal nature of Glut4 translocation is a very important aspect of this trafficking. It requires a regulatory mechanism that allows for transient activation of Glut4 release after insulin-stimulation, and transient repackaging of Glut4 back into GSVs after insulin withdrawal. One way to achieve this type of signaling would be to have simultaneous activation of a release mechanism and a feedback inhibition pathway that turns the release mechanism off (an analogy would be the simultaneous expression of cyclins and cyclin-degrading enzymes that is observed during cell cycle progression). The inhibitory pathway would remain active until insulin is withdrawn. Importantly, if this were the type of regulatory mechanism utilized, then at steady state it would be the inhibitory pathway and not the stimulatory pathway that would be active in insulin-stimulated cells. Future work will focus on utilizing the described kinetics assays together with new assays to determine exactly where proteins act in the Glut4 cycle, rather than just determining whether they affect the total amount of Glut4 in the plasma membrane. This information will give us a much more complete picture of Glut4 trafficking and suggest novel ways in which to manipulate this system to regulate glucose transport for the treatment of diabetes.

REFERENCES

Aisen, P. (2004). Transferrin receptor 1. *Int. J. Biochem. Cell Biol.* **36,** 2137–2143.

Akbergenova, Y., and Bykhovskaia, M. (2007). Synapsin maintains the reserve vesicle pool and spatial segregation of the recycling pool in *Drosophila* presynaptic boutons. *Brain Res.* **1178,** 52–64.

Al-Hasani, H., Hinck, C. S., and Cushman, S. W. (1998). Endocytosis of the glucose transporter GLUT4 is mediated by the GTPase dynamin. *J. Biol. Chem.* **273,** 17504–17510.

Al-Hasani, H., Kunamneni, R. K., Dawson, K., Hinck, C. S., Muller-Wieland, D., and Cushman, S. W. (2002). Roles of the N- and C-termini of GLUT4 in endocytosis. *J. Cell Sci.* **115,** 131–140.

Antonescu, C. N., Diaz, M., Femia, G., Planas, J. V., and Klip, A. (2008). Clathrin-dependent and independent endocytosis of glucose transporter 4 (GLUT4) in myoblasts: Regulation by mitochondrial uncoupling. *Traffic* **9,** 1173–1190.

Augustin, R., Riley, J., and Moley, K. H. (2005). GLUT8 contains a [DE]XXXL[LI] sorting motif and localizes to a late endosomal/lysosomal compartment. *Traffic* **6,** 1196–1212.

Bai, L., Wang, Y., Fan, J., Chen, Y., Ji, W., Qu, A., Xu, P., James, D. E., and Xu, T. (2007). Dissecting multiple steps of GLUT4 trafficking and identifying the sites of insulin action. *Cell Metab.* **5,** 47–57.

Birnbaum, M. J. (1989). Identification of a novel gene encoding an insulin-responsive glucose transporter protein. *Cell* **57,** 305–315.

Blodgett, D. M., De Zutter, J. K., Levine, K. B., Karim, P., and Carruthers, A. (2007). Structural basis of GLUT1 inhibition by cytoplasmic ATP. *J. Gen. Physiol.* **130,** 157–168.

Bloom, O., Evergren, E., Tomilin, N., Kjaerulff, O., Low, P., Brodin, L., Pieribone, V. A., Greengard, P., and Shupliakov, O. (2003). Colocalization of synapsin and actin during synaptic vesicle recycling. *J. Cell Biol.* **161,** 737–747.

Blot, V., and McGraw, T. E. (2006). GLUT4 is internalized by a cholesterol-dependent nystatin-sensitive mechanism inhibited by insulin. *EMBO J.* **25,** 5648–5658.

Blot, V., and McGraw, T. E. (2008). Molecular mechanisms controlling GLUT4 intracellular retention. *Mol. Biol. Cell* **19,** 3477–3487.

Brozinick, J. T., Jr., Berkemeier, B. A., Elmendorf, J. S. (2007). "Actin"g on GLUT4: Membrane & cytoskeletal components of insulin action. *Curr. Diabetes Rev.* **3,** 111–122.

Bryant, N. J., Govers, R., and James, D. E. (2002). Regulated transport of the glucose transporter GLUT4. *Nat. Rev. Mol. Cell Biol.* **3,** 267–277.

Bustos, R., Kolen, E. R., Braiterman, L., Baines, A. J., Gorelick, F. S., and Hubbard, A. L. (2001). Synapsin I is expressed in epithelial cells: Localization to a unique *trans*-golgi compartment. *J. Cell Sci.* **114,** 3695–3704.

Cain, C. C., Trimble, W. S., and Lienhard, G. E. (1992). Members of the VAMP family of synaptic vesicle proteins are components of glucose transporter-containing vesicles from rat adipocytes. *J. Biol. Chem.* **267,** 11681–11684.

Calderhead, D. M., Kitagawa, K., Tanner, L. I., Holman, G. D., and Lienhard, G. E. (1990). Insulin regulation of the two glucose transporters in 3T3-L1 adipocytes. *J. Biol. Chem.* **265,** 13801–13808.

Cameron, P. L., Sudhof, T. C., Jahn, R., and De Camilli, P. (1991). Colocalization of synaptophysin with transferrin receptors: Implications for synaptic vesicle biogenesis. *J. Cell Biol.* **115,** 151–164.

Carvalho, E., Schellhorn, S. E., Zabolotny, J. M., Martin, S., Tozzo, E., Peroni, O. D., Houseknecht, K. L., Mundt, A., James, D. E., and Kahn, B. B. (2004). GLUT4 overexpression or deficiency in adipocytes of transgenic mice alters the composition of GLUT4 vesicles and the subcellular localization of GLUT4 and insulin-responsive aminopeptidase. *J. Biol. Chem.* **279,** 21598–21605.

Castle, A., and Castle, D. (2005). Ubiquitously expressed secretory carrier membrane proteins (SCAMPs) 1–4 mark different pathways and exhibit limited constitutive trafficking to and from the cell surface. *J. Cell Sci.* **118,** 3769–3780.

Cheatham, B., Volchuk, A., Kahn, C. R., Wang, L., Rhodes, C. J., and Klip, A. (1996). Insulin-stimulated translocation of GLUT4 glucose transporters requires SNARE-complex proteins. *Proc. Natl. Acad. Sci. USA* **93,** 15169–15173.

Cormont, M., Van Obberghen, E., Zerial, M., and Le Marchand-Brustel, Y. (1996). Insulin induces a change in Rab5 subcellular localization in adipocytes independently of phosphatidylinositol 3-kinase activation. *Endocrinology* **137,** 3408–3415.

Corvera, S., Graver, D. F., and Smith, R. M. (1989). Insulin increases the cell surface concentration of alpha 2-macroglobulin receptors in 3T3-L1 adipocytes. Altered transit of the receptor among intracellular endocytic compartments. *J. Biol. Chem.* **264,** 10133–10138.

Coster, A. C., Govers, R., and James, D. E. (2004). Insulin stimulates the entry of GLUT4 into the endosomal recycling pathway by a quantal mechanism. *Traffic* **5,** 763–771.

Cushman, S. W., and Wardzala, L. J. (1980). Potential mechanism of insulin action on glucose transport in the isolated rat adipose cell. Apparent translocation of intracellular transport systems to the plasma membrane. *J. Biol. Chem.* **255,** 4758–4762.

Dawson, K., Aviles-Hernandez, A., Cushman, S. W., and Malide, D. (2001). Insulin-regulated trafficking of dual-labeled glucose transporter 4 in primary rat adipose cells. *Biochem. Biophys. Res. Commun.* **287,** 445–454.

Eguez, L., Lee, A., Chavez, J. A., Miinea, C. P., Kane, S., Lienhard, G. E., and McGraw, T. E. (2005). Full intracellular retention of GLUT4 requires AS160 Rab GTPase activating protein. *Cell Metab.* **2,** 263–272.

Evergren, E., Benfenati, F., and Shupliakov, O. (2007). The synapsin cycle: A view from the synaptic endocytic zone. *J. Neurosci. Res.* **85,** 2648–2656.

Eyster, C. A., Duggins, Q. S., and Olson, A. L. (2005). Expression of constitutively active Akt/protein kinase B signals GLUT4 translocation in the absence of an intact actin cytoskeleton. *J. Biol. Chem.* **280,** 17978–17985.

Foster, L. J., and Klip, A. (2000). Mechanism and regulation of GLUT-4 vesicle fusion in muscle and fat cells. *Am. J. Physiol. Cell Physiol.* **279,** C877–C890.

Frevert, E. U., Bjorbaek, C., Venable, C. L., Keller, S. R., and Kahn, B. B. (1998). Targeting of constitutively active phosphoinositide 3-kinase to GLUT4-containing vesicles in 3T3-L1 adipocytes. *J. Biol. Chem.* **273,** 25480–25487.

Fukumoto, H., Kayano, T., Buse, J. B., Edwards, Y., Pilch, P. F., Bell, G. I., and Seino, S. (1989). Cloning and characterization of the major insulin-responsive glucose transporter expressed in human skeletal muscle and other insulin-responsive tissues. *J. Biol. Chem.* **264,** 7776–7779.

Gaffield, M. A., and Betz, W. J. (2007). Synaptic vesicle mobility in mouse motor nerve terminals with and without synapsin. *J. Neurosci.* **27,** 13691–13700.

Goodyear, L. J., and Kahn, B. B. (1998). Exercise, glucose transport, and insulin sensitivity. *Annu. Rev. Med.* **49,** 235–261.

Gorga, F. R., and Lienhard, G. E. (1981). Equilibria and kinetics of ligand binding to the human erythrocyte glucose transporter. Evidence for an alternating conformation model for transport. *Biochemistry* **20,** 5108–5113.

Govers, R., Coster, A. C., and James, D. E. (2004). Insulin increases cell surface GLUT4 levels by dose dependently discharging GLUT4 into a cell surface recycling pathway. *Mol. Cell Biol.* **24,** 6456–6466.

Hah, J. S., Ryu, J. W., Lee, W., Kim, B. S., Lachaal, M., Spangler, R. A., and Jung, C. Y. (2002). Transient changes in four GLUT4 compartments in rat adipocytes during the transition, insulin-stimulated to basal: Implications for the GLUT4 trafficking pathway. *Biochemistry* **41,** 14364–14371.

Haney, P. M., Slot, J. W., Piper, R. C., James, D. E., and Mueckler, M. (1991). Intracellular targeting of the insulin-regulatable glucose transporter (GLUT4) is isoform specific and independent of cell type. *J. Cell Biol.* **114,** 689–699.

Hanpeter, D., and James, D. E. (1995). Characterization of the intracellular GLUT-4 compartment. *Mol. Membr. Biol.* **12,** 263–269.

Hashiramoto, M., and James, D. E. (2000). Characterization of insulin-responsive GLUT4 storage vesicles isolated from 3T3-L1 adipocytes. *Mol. Cell Biol.* **20,** 416–427.

Holman, G. D., Lo Leggio, L., and Cushman, S. W. (1994). Insulin-stimulated GLUT4 glucose transporter recycling. A problem in membrane protein subcellular trafficking through multiple pools. *J. Biol. Chem.* **269,** 17516–17524.

Hou, J. C., and Pessin, J. E. (2007). Ins (endocytosis) and outs (exocytosis) of GLUT4 trafficking. *Curr. Opin. Cell Biol.* **19,** 466–473.

Huang, S., and Czech, M. P. (2007). The GLUT4 glucose transporter. *Cell Metab.* **5,** 237–252.

Huang, S., Lifshitz, L. M., Jones, C., Bellve, K. D., Standley, C., Fonseca, S., Corvera, S., Fogarty, K. E., and Czech, M. P. (2007). Insulin stimulates membrane fusion and GLUT4 accumulation in clathrin coats on adipocyte plasma membranes. *Mol. Cell Biol.* **27,** 3456–3469.

James, D. E., and Pilch, P. F. (1988). Fractionation of endocytic vesicles and glucose-transporter-containing vesicles in rat adipocytes. *Biochem. J.* **256,** 725–732.

James, D. E., Lederman, L., and Pilch, P. F. (1987). Purification of insulin-dependent exocytic vesicles containing the glucose transporter. *J. Biol. Chem.* **262,** 11817–11824.

James, D. E., Brown, R., Navarro, J., and Pilch, P. F. (1988). Insulin-regulatable tissues express a unique insulin-sensitive glucose transport protein. *Nature* **333,** 183–185.

James, D. E., Strube, M., and Mueckler, M. (1989). Molecular cloning and characterization of an insulin-regulatable glucose transporter. *Nature* **338,** 83–87.

Jhun, B. H., Rampal, A. L., Liu, H., Lachaal, M., and Jung, C. Y. (1992). Effects of insulin on steady state kinetics of GLUT4 subcellular distribution in rat adipocytes. Evidence of constitutive GLUT4 recycling. *J. Biol. Chem.* **267,** 17710–17715.

Jiang, L., Fan, J., Bai, L., Wang, Y., Chen, Y., Yang, L., Chen, L., and Xu, T. (2008). Direct quantification of fusion rate reveals a distal role for AS160 in insulin-stimulated fusion of GLUT4 storage vesicles. *J. Biol. Chem.* **283,** 8508–8516.

Joost, H. G., Weber, T. M., and Cushman, S. W. (1988). Qualitative and quantitative comparison of glucose transport activity and glucose transporter concentration in plasma membranes from basal and insulin-stimulated rat adipose cells. *Biochem. J.* **249,** 155–161.

Kaestner, K. H., Christy, R. J., McLenithan, J. C., Braiterman, L. T., Cornelius, P., Pekala, P. H., and Lane, M. D. (1989). Sequence, tissue distribution, and differential expression of mRNA for a putative insulin-responsive glucose transporter in mouse 3T3-L1 adipocytes. *Proc. Natl. Acad. Sci. USA* **86,** 3150–3154.

Kandror, K., and Pilch, P. F. (1994). Identification and isolation of glycoproteins that translocate to the cell surface from GLUT4-enriched vesicles in an insulin-dependent fashion. *J. Biol. Chem.* **269,** 138–142.

Kandror, K. V., and Pilch, P. F. (1996). The insulin-like growth factor II/mannose 6-phosphate receptor utilizes the same membrane compartments as GLUT4 for insulin-dependent trafficking to and from the rat adipocyte cell surface. *J. Biol. Chem.* **271,** 21703–21708.

Kao, A. W., Ceresa, B. P., Santeler, S. R., and Pessin, J. E. (1998). Expression of a dominant interfering dynamin mutant in 3T3L1 adipocytes inhibits GLUT4 endocytosis without affecting insulin signaling. *J. Biol. Chem.* **273,** 25450–25457.

Karylowski, O., Zeigerer, A., Cohen, A., and McGraw, T. E. (2004). GLUT4 is retained by an intracellular cycle of vesicle formation and fusion with endosomes. *Mol. Biol. Cell* **15,** 870–882.

Keller, S. R., Scott, H. M., Mastick, C. C., Aebersold, R., and Lienhard, G. E. (1995). Cloning and characterization of a novel insulin-regulated membrane aminopeptidase from Glut4 vesicles. *J. Biol. Chem.* **270,** 23612–23618.

Keller, S. R., Davis, A. C., and Clairmont, K. B. (2002). Mice deficient in the insulin-regulated membrane aminopeptidase show substantial decreases in glucose transporter GLUT4 levels but maintain normal glucose homeostasis. *J. Biol. Chem.* **277,** 17677–17686.

Ko, K. W., Avramoglu, R. K., McLeod, R. S., Vukmirica, J., and Yao, Z. (2001). The insulin-stimulated cell surface presentation of low density lipoprotein receptor-related protein in 3T3-L1 adipocytes is sensitive to phosphatidylinositide 3-kinase inhibition. *Biochemistry* **40,** 752–759.

Koumanov, F., Jin, B., Yang, J., and Holman, G. D. (2005). Insulin signaling meets vesicle traffic of GLUT4 at a plasma-membrane-activated fusion step. *Cell Metab.* **2,** 179–189.

Krueger, K. A., Ings, E. I., Brun, A. M., Landt, M., and Easom, R. A. (1999). Site-specific phosphorylation of synapsin I by Ca2+/calmodulin-dependent protein kinase II in pancreatic betaTC3 cells: Synapsin I is not associated with insulin secretory granules. *Diabetes* **48,** 499–506.

Lakadamyali, M., Rust, M. J., and Zhuang, X. (2006). Ligands for clathrin-mediated endocytosis are differentially sorted into distinct populations of early endosomes. *Cell* **124,** 997–1009.

Larance, M., Ramm, G., Stockli, J., van Dam, E. M., Winata, S., Wasinger, V., Simpson, F., Graham, M., Junutula, J. R., Guilhaus, M., and James, D. E. (2005). Characterization of the role of the Rab GTPase-activating protein AS160 in insulin-regulated GLUT4 trafficking. *J. Biol. Chem.* **280,** 37803–37813.

Larance, M., Ramm, G., and James, D. E. (2008). The GLUT4 code. *Mol. Endocrinol.* **22,** 226–233.

Laurie, S. M., Cain, C. C., Lienhard, G. E., and Castle, J. D. (1993). The glucose transporter GluT4 and secretory carrier membrane proteins (SCAMPs) colocalize in rat adipocytes and partially segregate during insulin stimulation. *J. Biol. Chem.* **268,** 19110–19117.

Leloup, C., Arluison, M., Kassis, N., Lepetit, N., Cartier, N., Ferre, P., and Penicaud, L. (1996). Discrete brain areas express the insulin-responsive glucose transporter GLUT4. *Brain Res. Mol. Brain Res.* **38,** 45–53.

Lew, R. A., Mustafa, T., Ye, S., McDowall, S. G., Chai, S. Y., and Albiston, A. L. (2003). Angiotensin AT4 ligands are potent, competitive inhibitors of insulin regulated aminopeptidase (IRAP). *J. Neurochem.* **86,** 344–350.

Li, L. V., and Kandror, K. V. (2005). Golgi-localized, gamma-ear-containing, Arf-binding protein adaptors mediate insulin-responsive trafficking of glucose transporter 4 in 3T3-L1 adipocytes. *Mol. Endocrinol.* **19,** 2145–2153.

Li, Y., Wang, P., Xu, J., Gorelick, F., Yamazaki, H., Andrews, N., and Desir, G. V. (2007). Regulation of insulin secretion and GLUT4 trafficking by the calcium sensor synaptotagmin VII. *Biochem. Biophys. Res. Commun.* **362,** 658–664.

Liao, H., Zhang, J., Shestopal, S., Szabo, G., Castle, A., and Castle, D. (2008). Nonredundant function of secretory carrier membrane protein isoforms in dense core vesicle exocytosis. *Am. J. Physiol. Cell Physiol.* **294,** C797–C809.

Lienhard, G. E., Kim, H. H., Ransome, K. J., and Gorga, J. C. (1982). Immunological identification of an insulin-responsive glucose transporter. *Biochem. Biophys. Res. Commun.* **105,** 1150–1156.

Lin, B. Z., Pilch, P. F., and Kandror, K. V. (1997). Sortilin is a major protein component of Glut4-containing vesicles. *J. Biol. Chem.* **272,** 24145–24147.

Linstedt, A. D., and Kelly, R. B. (1991). Synaptophysin is sorted from endocytotic markers in neuroendocrine PC12 cells but not transfected fibroblasts. *Neuron* **7,** 309–317.

Livingstone, C., James, D. E., Rice, J. E., Hanpeter, D., and Gould, G. W. (1996). Compartment ablation analysis of the insulin-responsive glucose transporter (GLUT4) in 3T3-L1 adipocytes. *Biochem. J.* **315**(Pt. 2), 487–495.

Longuet, C., Broca, C., Costes, S., Hani el, H., Bataille, D., and Dalle, S. (2005). Extracellularly regulated kinases 1/2 (p44/42 mitogen-activated protein kinases) phosphorylate synapsin I and regulate insulin secretion in the MIN6 beta-cell line and islets of Langerhans. *Endocrinology* **146,** 643–654.

Ludvigsen, C., and Jarett, L. (1980). A comparison of basal and insulin-stimulated glucose transport in rat adipocyte plasma membranes. *Diabetes* **29,** 373–378.

Malide, D., Ramm, G., Cushman, S. W., and Slot, J. W. (2000). Immunoelectron microscopic evidence that GLUT4 translocation explains the stimulation of glucose transport in isolated rat white adipose cells. *J. Cell Sci.* **113**(Pt. 23), 4203–4210.

Martin, L. B., Shewan, A., Millar, C. A., Gould, G. W., and James, D. E. (1998). Vesicle-associated membrane protein 2 plays a specific role in the insulin-dependent trafficking of the facilitative glucose transporter GLUT4 in 3T3-L1 adipocytes. *J. Biol. Chem.* **273,** 1444–1452.

Martin, O. J., Lee, A., and McGraw, T. E. (2006). GLUT4 distribution between the plasma membrane and the intracellular compartments is maintained by an insulin-modulated bipartite dynamic mechanism. *J. Biol. Chem.* **281,** 484–490.

Martin, S., Tellam, J., Livingstone, C., Slot, J. W., Gould, G. W., and James, D. E. (1996). The glucose transporter (GLUT-4) and vesicle-associated membrane protein-2 (VAMP-2) are segregated from recycling endosomes in insulin-sensitive cells. *J. Cell Biol.* **134,** 625–635.

Martin, S., Rice, J. E., Gould, G. W., Keller, S. R., Slot, J. W., and James, D. E. (1997). The glucose transporter GLUT4 and the aminopeptidase vp165 colocalise in tubulo-vesicular elements in adipocytes and cardiomyocytes. *J. Cell Sci.* **110**(Pt. 18), 2281–2291.

Martin, S., Millar, C. A., Lyttle, C. T., Meerloo, T., Marsh, B. J., Gould, G. W., and James, D. E. (2000). Effects of insulin on intracellular GLUT4 vesicles in adipocytes: Evidence for a secretory mode of regulation. *J. Cell Sci.* **113**(Pt. 19), 3427–3438.

Martin, T. F. (2003). Tuning exocytosis for speed: Fast and slow modes. *Biochim. Biophys. Acta* **1641,** 157–165.

Mastick, C. C., Aebersold, R., and Lienhard, G. E. (1994). Characterization of a major protein in GLUT4 vesicles. Concentration in the vesicles and insulin-stimulated translocation to the plasma membrane. *J. Biol. Chem.* **269,** 6089–6092.

Matsumoto, K., Ebihara, K., Yamamoto, H., Tabuchi, H., Fukunaga, K., Yasunami, M., Ohkubo, H., Shichiri, M., and Miyamoto, E. (1999). Cloning from insulinoma cells of synapsin I associated with insulin secretory granules. *J. Biol. Chem.* **274,** 2053–2059.

Maxfield, F. R., and McGraw, T. E. (2004). Endocytic recycling. *Nat. Rev. Mol. Cell Biol.* **5,** 121–132.

Miinea, C. P., Sano, H., Kane, S., Sano, E., Fukuda, M., Peranen, J., Lane, W. S., and Lienhard, G. E. (2005). AS160, the Akt substrate regulating GLUT4 translocation, has a functional Rab GTPase-activating protein domain. *Biochem. J.* **391,** 87–93.

Morris, N. J., Ross, S. A., Lane, W. S., Moestrup, S. K., Petersen, C. M., Keller, S. R., and Lienhard, G. E. (1998). Sortilin is the major 110-kDa protein in GLUT4 vesicles from adipocytes. *J. Biol. Chem.* **273,** 3582–3587.

Muretta, J. M., Romenskaia, I., Cassiday, P. A., and Mastick, C. C. (2007). Expression of a synapsin IIb site 1 phosphorylation mutant in 3T3-L1 adipocytes inhibits basal intracellular retention of Glut4. *J. Cell Sci.* **120,** 1168–1177.

Muretta, J. M., Romenskaia, I., and Mastick, C. C. (2008). Insulin releases glut4 from static storage compartments into cycling endosomes and increases the rate constant for glut4 exocytosis. *J. Biol. Chem.* **283,** 311–323.

Okuno, Y., and Gliemann, J. (1987). Enhancement of glucose transport by insulin at 37 °C in rat adipocytes is accounted for by increased V_{max}. *Diabetologia* **30,** 426–430.

Peck, G. R., Ye, S., Pham, V., Fernando, R. N., Macaulay, S. L., Chai, S. Y., and Albiston, A. L. (2006). Interaction of the Akt substrate, AS160, with the glucose transporter 4 vesicle marker protein, insulin-regulated aminopeptidase. *Mol. Endocrinol.* **20,** 2576–2583.

Perera, H. K., Clarke, M., Morris, N. J., Hong, W., Chamberlain, L. H., and Gould, G. W. (2003). Syntaxin 6 regulates Glut4 trafficking in 3T3-L1 adipocytes. *Mol. Biol. Cell* **14,** 2946–2958.

Proctor, K. M., Miller, S. C., Bryant, N. J., and Gould, G. W. (2006). Syntaxin 16 controls the intracellular sequestration of GLUT4 in 3T3-L1 adipocytes. *Biochem. Biophys. Res. Commun.* **347,** 433–448.

Provoda, C. J., Waring, M. T., and Buckley, K. M. (2000). Evidence for a primary endocytic vesicle involved in synaptic vesicle biogenesis. *J. Biol. Chem.* **275,** 7004–7012.

Ramm, G., Slot, J. W., James, D. E., and Stoorvogel, W. (2000). Insulin recruits GLUT4 from specialized VAMP2-carrying vesicles as well as from the dynamic endosomal/*trans*-golgi network in rat adipocytes. *Mol. Biol. Cell* **11,** 4079–4091.

Ramm, G., Larance, M., Guilhaus, M., and James, D. E. (2006). A role for 14–3–3 in insulin-stimulated GLUT4 translocation through its interaction with the RabGAP AS160. *J. Biol. Chem.* **281,** 29174–29180.

Robinson, L. J., Pang, S., Harris, D. S., Heuser, J., and James, D. E. (1992). Translocation of the glucose transporter (GLUT4) to the cell surface in permeabilized 3T3-L1 adipocytes: Effects of ATP insulin, and GTP gamma S and localization of GLUT4 to clathrin lattices. *J. Cell Biol.* **117,** 1181–1196.

Rodnick, K. J., Slot, J. W., Studelska, D. R., Hanpeter, D. E., Robinson, L. J., Geuze, H. J., and James, D. E. (1992). Immunocytochemical and biochemical studies of GLUT4 in rat skeletal muscle. *J. Biol. Chem.* **267,** 6278–6285.

Sakamoto, K., and Holman, G. D. (2008). Emerging role for AS160/TBC1D4 and TBC1D1 in the regulation of GLUT4 traffic. *Am. J. Physiol. Endocrinol. Metab.* **295,** E29–E37.

Sano, H., Kane, S., Sano, E., Miinea, C. P., Asara, J. M., Lane, W. S., Garner, C. W., and Lienhard, G. E. (2003). Insulin-stimulated phosphorylation of a Rab GTPase-activating protein regulates GLUT4 translocation. *J. Biol. Chem.* **278,** 14599–14602.

Sano, H., Eguez, L., Teruel, M. N., Fukuda, M., Chuang, T. D., Chavez, J. A., Lienhard, G. E., and McGraw, T. E. (2007). Rab10, a target of the AS160 Rab GAP, is required for insulin-stimulated translocation of GLUT4 to the adipocyte plasma membrane. *Cell Metab.* **5,** 293–303.

Satoh, S., Nishimura, H., Clark, A. E., Kozka, I. J., Vannucci, S. J., Simpson, I. A., Quon, M. J., Cushman, S. W., and Holman, G. D. (1993). Use of bismannose photolabel to elucidate insulin-regulated GLUT4 subcellular trafficking kinetics in rat adipose cells. Evidence that exocytosis is a critical site of hormone action. *J. Biol. Chem.* **268,** 17820–17829.

Schroer, D. W., Frost, S. C., Kohanski, R. A., Lane, M. D., and Lienhard, G. E. (1986). Identification and partial purification of the insulin-responsive glucose transporter from 3T3-L1 adipocytes. *Biochim. Biophys. Acta* **885,** 317–326.

Shi, J., and Kandror, K. V. (2005). Sortilin is essential and sufficient for the formation of Glut4 storage vesicles in 3T3-L1 adipocytes. *Dev. Cell* **9,** 99–108.

Slot, J. W., Geuze, H. J., Gigengack, S., James, D. E., and Lienhard, G. E. (1991a). Translocation of the glucose transporter GLUT4 in cardiac myocytes of the rat. *Proc. Natl. Acad. Sci. USA* **88,** 7815–7819.

Slot, J. W., Geuze, H. J., Gigengack, S., Lienhard, G. E., and James, D. E. (1991b). Immuno-localization of the insulin regulatable glucose transporter in brown adipose tissue of the rat. *J. Cell Biol.* **113,** 123–135.

Song, X. M., Hresko, R. C., and Mueckler, M. (2008). Identification of amino acid residues within the C terminus of the Glut4 glucose transporter that are essential for insulin-stimulated redistribution to the plasma membrane. *J. Biol. Chem.* **283,** 12571–12585.

Sudhof, T. C. (2000). The synaptic vesicle cycle revisited. *Neuron* **28,** 317–320.

Sudhof, T. C. (2004). The synaptic vesicle cycle. *Annu. Rev. Neurosci.* **27,** 509–547.

Suzuki, K., and Kono, T. (1980). Evidence that insulin causes translocation of glucose transport activity to the plasma membrane from an intracellular storage site. *Proc. Natl. Acad. Sci. USA* **77,** 2542–2545.

Tamori, Y., Hashiramoto, M., Araki, S., Kamata, Y., Takahashi, M., Kozaki, S., and Kasuga, M. (1996). Cleavage of vesicle-associated membrane protein (VAMP)-2 and cellubrevin on GLUT4-containing vesicles inhibits the translocation of GLUT4 in 3T3-L1 adipocytes. *Biochem. Biophys. Res. Commun.* **220,** 740–745.

Tanner, L. I., and Lienhard, G. E. (1987). Insulin elicits a redistribution of transferrin receptors in 3T3-L1 adipocytes through an increase in the rate constant for receptor externalization. *J. Biol. Chem.* **262,** 8975–8980.

Thoidis, G., Kotliar, N., and Pilch, P. F. (1993). Immunological analysis of GLUT4-enriched vesicles. Identification of novel proteins regulated by insulin and diabetes. *J. Biol. Chem.* **268,** 11691–11696.

Traub, L. M. (2003). Sorting it out: AP-2 and alternate clathrin adaptors in endocytic cargo selection. *J. Cell Biol.* **163,** 203–208.

Uldry, M., and Thorens, B. (2004). The SLC2 family of facilitated hexose and polyol transporters. *Pflugers Arch.* **447,** 480–489.

Verhage, M., and Sorensen, J. B. (2008). Vesicle docking in regulated exocytosis. *Traffic* **9,** 1414–1424.

Vinten, J., Gliemann, J., and Osterlind, K. (1976). Exchange of 3-*O*-methylglucose in isolated fat cells. Concentration dependence and effect of insulin. *J. Biol. Chem.* **251,** 794–800.

Volchuk, A., Sargeant, R., Sumitani, S., Liu, Z., He, L., and Klip, A. (1995). Cellubrevin is a resident protein of insulin-sensitive GLUT4 glucose transporter vesicles in 3T3-L1 adipocytes. *J. Biol. Chem.* **270,** 8233–8240.

Wallis, M. G., Lankford, M. F., and Keller, S. R. (2007). Vasopressin is a physiological substrate for the insulin-regulated aminopeptidase IRAP. *Am. J. Physiol. Endocrinol. Metab.* **293,** E1092–E1102.

Wardzala, L. J., Cushman, S. W., and Salans, L. B. (1978). Mechanism of insulin action on glucose transport in the isolated rat adipose cell. Enhancement of the number of functional transport systems. *J. Biol. Chem.* **253,** 8002–8005.

Waters, S. B., D'Auria, M., Martin, S. S., Nguyen, C., Kozma, L. M., and Luskey, K. L. (1997). The amino terminus of insulin-responsive aminopeptidase causes Glut4 translocation in 3T3-L1 adipocytes. *J. Biol. Chem.* **272,** 23323–23327.

Watson, R. T., and Pessin, J. E. (2007). GLUT4 translocation: The last 200 nanometers. *Cell Signal.* **19,** 2209–2217.

Watson, R. T., and Pessin, J. E. (2008). Recycling of IRAP from the plasma membrane back to the insulin-responsive compartment requires the Q-SNARE syntaxin 6 but not the GGA clathrin adaptors. *J. Cell Sci.* **121,** 1243–1251.

Watson, R. T., Khan, A. H., Furukawa, M., Hou, J. C., Li, L., Kanzaki, M., Okada, S., Kandror, K. V., and Pessin, J. E. (2004). Entry of newly synthesized GLUT4 into the insulin-responsive storage compartment is GGA dependent. *EMBO J.* **23,** 2059–2070.

Wenk, M. R., and De Camilli, P. (2004). Protein–lipid interactions and phosphoinositide metabolism in membrane traffic: Insights from vesicle recycling in nerve terminals. *Proc. Natl. Acad. Sci. USA* **101,** 8262–8269.

Whitehead, J. P., Molero, J. C., Clark, S., Martin, S., Meneilly, G., and James, D. E. (2001). The role of Ca2 + in insulin-stimulated glucose transport in 3T3-L1 cells. *J. Biol. Chem.* **276,** 27816–27824.

Wiley, H. S., and Cunningham, D. D. (1982). The endocytotic rate constant. A cellular parameter for quantitating receptor-mediated endocytosis. *J. Biol. Chem.* **257,** 4222–4229.

Williams, D., and Pessin, J. E. (2008). Mapping of R-SNARE function at distinct intracellular GLUT4 trafficking steps in adipocytes. *J. Cell Biol.* **180,** 375–387.
Wojcik, S. M., and Brose, N. (2007). Regulation of membrane fusion in synaptic excitation-secretion coupling: Speed and accuracy matter. *Neuron* **55,** 11–24.
Yang, J., and Holman, G. D. (1993). Comparison of GLUT4 and GLUT1 subcellular trafficking in basal and insulin-stimulated 3T3-L1 cells. *J. Biol. Chem.* **268,** 4600–4603.
Ye, S., Chai, S. Y., Lew, R. A., and Albiston, A. L. (2007). Insulin-regulated aminopeptidase: Analysis of peptide substrate and inhibitor binding to the catalytic domain. *Biol. Chem.* **388,** 399–403.
Yeh, T. Y., Sbodio, J. I., Tsun, Z. Y., Luo, B., and Chi, N. W. (2007). Insulin-stimulated exocytosis of GLUT4 is enhanced by IRAP and its partner tankyrase. *Biochem. J.* **402,** 279–290.
Yu, C., Cresswell, J., Loffler, M. G., and Bogan, J. S. (2007). The glucose transporter 4-regulating protein TUG is essential for highly insulin-responsive glucose uptake in 3T3-L1 adipocytes. *J. Biol. Chem.* **282,** 7710–7722.
Zaid, H., Antonescu, C. N., Randhawa, V. K., and Klip, A. (2008). Insulin action on glucose transporters through molecular switches, tracks and tethers. *Biochem. J.* **413,** 201–215.
Zeigerer, A., Lampson, M. A., Karylowski, O., Sabatini, D. D., Adesnik, M., Ren, M., and McGraw, T. E. (2002). GLUT4 retention in adipocytes requires two intracellular insulin-regulated transport steps. *Mol. Biol. Cell* **13,** 2421–2435.
Zeigerer, A., McBrayer, M. K., and McGraw, T. E. (2004). Insulin stimulation of GLUT4 exocytosis, but not its inhibition of endocytosis, is dependent on RabGAP AS160. *Mol. Biol. Cell* **15,** 4406–4415.
Zoccoli, M. A., Baldwin, S. A., and Lienhard, G. E. (1978). The monosaccharide transport system of the human erythrocyte. Solubilization and characterization on the basis of cytochalasin B binding. *J. Biol. Chem.* **253,** 6923–6930.
Zorzano, A., Wilkinson, W., Kotliar, N., Thoidis, G., Wadzinkski, B. E., Ruoho, A. E., and Pilch, P. F. (1989). Insulin-regulated glucose uptake in rat adipocytes is mediated by two transporter isoforms present in at least two vesicle populations. *J. Biol. Chem.* **264,** 12358–12363.

CHAPTER ELEVEN

Spatio-Temporal Dynamics of Phosphatidylinositol-3,4,5-Trisphosphate Signalling

Anders Tengholm *and* Olof Idevall-Hagren

Contents

Abstract

Many effects of insulin, insulin-like growth factors and other receptor stimuli are mediated via the phospholipid second messenger phosphatidylinositol-3,4,5-trisphosphate (PIP_3). PIP_3 is formed by the activity of phosphoinositide 3-kinases in the plasma membrane, where it serves to recruit signalling proteins. These proteins coordinate complex events leading to changes in cell metabolism, growth, movement and survival. Over the past decade, new techniques for measurements of PIP_3 in the plasma membrane of individual living cells have markedly improved our understanding of the role of this messenger in a variety of cellular processes. This review summarises the mechanisms involved in formation and degradation of PIP_3 in insulin-responsive cells, how PIP_3 can be measured in individual cells as well as accumulating evidence that the plasma membrane PIP_3 concentration undergoes complex spatio-temporal patterns in many types of cells, with particular emphasis on autocrine insulin-induced PIP_3 oscillations in pancreatic β-cells. © 2009 Elsevier Inc.

Department of Medical Cell Biology, Uppsala University, Biomedical Centre, SE-751 23 Uppsala, Sweden

Vitamins and Hormones, Volume 80
ISSN 0083-6729, DOI: 10.1016/S0083-6729(08)00611-0

I. Introduction

The pleiotropic effects of insulin and IGF-1 are mediated by a number of receptor signalling circuits, many of which involve phosphoinositide 3-kinase (PI3-kinase). Recruitment and activation of this enzyme are necessary for the action of insulin on mitogenesis, glucose transport, hepatic glucose output, and synthesis of glycogen, proteins and triglycerides in insulin target tissues, like liver, muscle and fat (Saltiel and Kahn, 2001). Insulin is released from pancreatic β-cells in response to elevated glucose concentrations. Some of the responses to glucose stimulation of β-cells, such as gene transcription, especially up-regulation of the insulin gene, and insulin feedback effects on insulin secretion are also mediated by PI3-kinase (Leibiger *et al.*, 2002). PI3-kinase catalyses the formation of the lipid second messenger phosphatidylinositol-3,4,5-trisphosphate (PIP_3) in the plasma membrane by phosphorylating the D-3 position of the inositol ring of phosphatidylinositol-4,5-bisphosphate (PtdIns(4,5)P_2) (Vanhaesebroeck *et al.*, 2001) (Fig.11.1). The accumulation of PIP_3 in the plasma membrane serves to recruit signalling proteins with pleckstrin homology (PH) domains to the membrane (Cantley, 2002). Well known examples include the serine–threonine protein kinases Akt (protein kinase B) and phosphoinositide-dependent kinase-1 (PDK1). Akt is activated by phosphorylation mediated by PDK1 (Lawlor and Alessi, 2001) and the rictor-mammalian target of rapamycin kinase complex (Sarbassov *et al.*, 2005). Activated Akt in turn phosphorylates numerous downstream targets involved in the regulation of cell growth, survival, migration and metabolism (reviewed in Manning and Cantley, 2007). Additional proteins that can interact directly with PIP_3 include the Tec family of tyrosine kinases, guanine nucleotide exchange factors for the Rac and Arf families of small GTPases, GTPase-activating proteins, phospholipase Cγ as well as the Grb2-associated binder (Gab) family of scaffold/adaptor proteins (Hawkins *et al.*, 2006).

A major challenge in understanding the versatility of PIP_3 function is to decipher how signalling specificity is obtained. Recent methodological development allowing measurements of PIP_3 in the plasma membrane of individual living cells have markedly improved our understanding of its role in a variety of cellular processes. Real-time microscopy experiments have demonstrated that PIP_3 undergoes rapid turnover, which allows fast and localized changes in the concentration of PIP_3. The generation of spatially and temporally restricted PIP_3 signals may contribute to the specificity in downstream signalling pathways. In the following, we will review the mechanisms involved in regulating PIP_3 turnover in insulin-responsive cells, methods for imaging PIP_3 in individual cells, and recent findings indicating that the PIP_3 concentration shows complex spatio-temporal

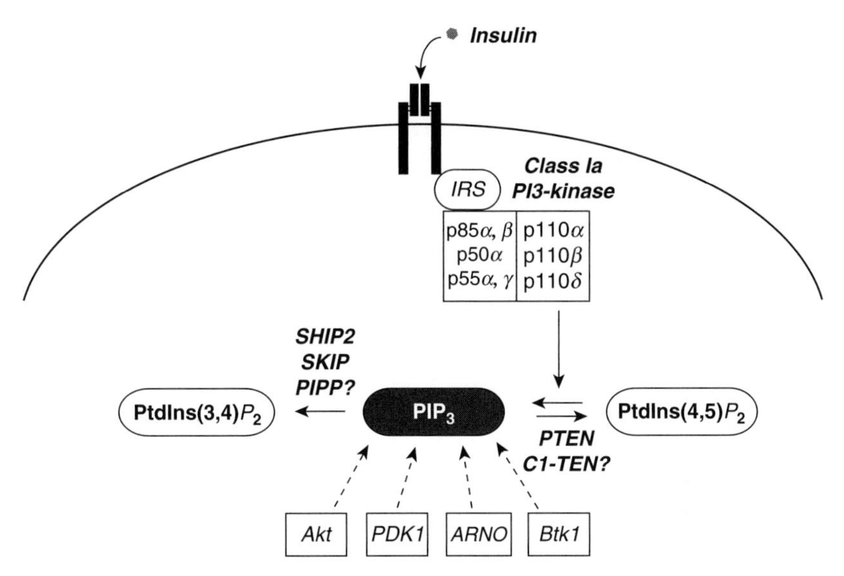

Figure 11.1 Synthesis and degradation of PIP$_3$ in insulin-responsive cells. Activated insulin receptors transmit signals via insulin receptor substrate proteins (IRS) to class Ia PI3-kinase. This type of PI3-kinase is comprised of a regulatory p85, p50 or p55 subunit (five different isoforms) and a catalytic p110 subunit (three isoforms). Class Ia PI3-kinase generates PIP$_3$ by phosphorylation of the D3-position of the inositol ring of phosphatidylinositol-4,5-bisphosphate (PtdIns(4,5)P_2). PIP$_3$ is degraded by the 3′-phosphatase PTEN to PtdIns(4,5)P_2 and by the 5′-phosphatases SHIP2 and SKIP to phosphatidylinositol 3,4-bisphosphate (PtdIns(3,4)P_2). C1-TEN is a putative 3′-phosphatase and PIPP a 5′-phosphatase with yet uncertain function in insulin signalling. PIP$_3$ is recognised by many PH domain containing proteins, such as the protein kinases Akt and PDK1, the Arf6 guanine nucleotide exchange factor ARNO and the tyrosine kinase Btk1.

patterns, including oscillations, in different types of cells. Particular attention will be paid to the role of PIP$_3$ in autocrine insulin receptor activation in insulin-secreting β-cells.

II. Synthesis and Degradation of PIP$_3$

Multiple isoforms of PI3-kinase exist in higher eukaryotes and three main classes have been identified on the basis of their structural features, *in vitro* substrate specificity and mechanism of regulation (Engelman *et al.*, 2006; Hawkins *et al.*, 2006; Vanhaesebroeck *et al.*, 2001). Class I PI3-kinases are well characterised and generate PIP$_3$ in response to tyrosine kinase- or G-protein coupled receptor activation. Class II and III PI3-kinases mainly generate PtdIns(3,4)P_2 and PtdIns(3)P, respectively, but relatively little is known about the regulation and function of these PI3-kinases. The class I PI3-kinases are subdivided into class Ia isoforms, mediating the formation of

PIP_3 in response to insulin, IGF-I and other growth factors, and class Ib, activated by $\beta\gamma$ subunits of heterotrimeric G proteins. The class Ia PI3-kinases are heterodimers comprised of a regulatory p85/50/55 subunit and a catalytic p110 subunit (Fig.11.1). Five major isoforms of the regulatory subunit have been identified (p85α, p55α, p50α, p85β, and p55γ), representing alternatively spliced isoforms encoded from three genes (Engelman *et al.*, 2006; Hawkins *et al.*, 2006). The regulatory subunits contain two SH2-domains that bind to phosphotyrosine residues in activated growth factor receptors or adaptor proteins, such as the insulin receptor substrate (IRS) proteins (Myers and White, 1996), bringing the enzyme in close proximity to its lipid substrate in the plasma membrane. The catalytic subunit of class Ia PI3-kinases occurs in three variants (p110α, β, δ). Inhibitors of class I PI3-kinase or transfection with dominant negative constructs of the enzyme block most metabolic effects of PI3-kinase, including stimulation of glucose transport, glycogen and lipid synthesis (Saltiel and Kahn, 2001; Shepherd *et al.*, 1998). Recent studies using novel isoform-selective chemical inhibitors (Knight *et al.*, 2006) or transgenic mice (Foukas *et al.*, 2006) have revealed that insulin signalling preferentially occurs via the p110α isoform.

PIP_3 signals are terminated via several types of lipid phosphatases. Phosphatase and tensin homologue deleted on chromosome 10 (PTEN) is an inositol lipid phosphatase that dephosphorylates the D3-position of PIP_3 and PtdIns(3,4)P_2 to form PtdIns(4,5)P_2 and PtdIns4P, respectively (Lazar and Saltiel, 2006; Maehama *et al.*, 2001). PTEN is a major regulator of PI3-kinase signalling in many cell types and acts as a tumour suppressor due to antagonism of the antiapoptotic, proliferative and hypertrophic activities of PI3-kinase. PTEN mutations have consequently been found in a variety of human tumours (Maehama *et al.*, 2001). Numerous studies have demonstrated the importance of PTEN for attenuating insulin receptor signalling. For example, overexpression of PTEN in 3T3L1 adipocytes inhibits insulin-induced PtdIns(3,4)P_2 and PIP_3 production, Akt and $p70^{S6K}$ phosphorylation, GLUT4 glucose transporter membrane translocation and glucose uptake (Nakashima *et al.*, 2000; Ono *et al.*, 2001). Downregulation of the phosphatase by small interfering RNAs has been found to enhance Akt activation and glucose uptake in response to insulin (Tang *et al.*, 2005). Moreover, overexpression of catalytically inactive or dominant-negative PTEN mutants also indicated that it is the lipid phosphatase activity of PTEN which is required to downregulate the Akt signalling and glucose uptake in response to insulin (Ono *et al.*, 2001; Wu *et al.*, 1998). Tissue-specific PTEN knockout studies in mice support the conclusion that this lipid phosphatase regulates insulin sensitivity (Butler *et al.*, 2002; Kurlawalla-Martinez *et al.*, 2005; Stiles *et al.*, 2004; Wijesekara *et al.*, 2005). It seems clear that PTEN regulates insulin signalling, but the phosphatase may also be regulated by insulin via changes in subcellular localization and production of reactive oxygen species (Ono *et al.*, 2001; Seo *et al.*, 2005).

The Src-homology 2 containing inositol phosphatases SHIP1 and SHIP2 dephosphorylate the D5-position in PIP_3 to form PtdIns(3,4)P_2 (Astle *et al.*, 2006; Lazar and Saltiel, 2006). This lipid is a second messenger in its own right and may mediate some of the effects of PI3-kinase even independently of PIP_3. Whereas SHIP1 is exclusively expressed in haematopoietic and lymphoid cells, SHIP2 has a wider tissue distribution (Pesesse *et al.*, 1997). Overexpression of SHIP2 in various insulin-responsive cell lines has been reported to markedly reduce insulin-stimulated PIP_3 levels and Akt phosphorylation as well as downstream metabolic signalling (Blero *et al.*, 2001; Ishihara *et al.*, 1999; Lazar and Saltiel, 2006; Sasaoka *et al.*, 2004; Wada *et al.*, 2001). On the contrary, knockdown of SHIP2 in 3T3L1 adipocytes did not affect insulin-stimulated Akt phosphorylation and glucose uptake (Tang *et al.*, 2005). Targeted deletion of SHIP2 in mice increases the sensitivity to insulin in target tissues (Clement *et al.*, 2001). Although the latter finding may have been confounded by concomitant disruption of the adjacent Phox2a gene, another more specific knockout supports the conclusion that SHIP2 is important in regulating the signalling downstream of the insulin receptor (Sleeman *et al.*, 2005). Evidence from animal and human genetic studies suggests that the attenuation of insulin signalling by excessive SHIP2 is involved in insulin resistance in type 2 diabetes (Hori *et al.*, 2002; Kagawa *et al.*, 2005, 2008; Marion *et al.*, 2002).

Skeletal muscle and kidney-enriched inositol phosphatase (SKIP) is a more recently discovered 5′-phosphatase which hydrolyses PIP_3 and PtdIns(4,5)P_2 (Astle *et al.*, 2006). It is ubiquitously expressed with particularly high levels present in skeletal muscle and kidney and has been found to translocate from a perinuclear localization to the plasma membrane upon stimulation with growth factors (Gurung *et al.*, 2003). SKIP has been implicated in metabolic signalling downstream of the insulin receptor. Accordingly, the insulin-induced production of PIP_3 and Akt phosphorylation decreased when SKIP was overexpressed, and increased when endogenous levels of SKIP were reduced by means of antisense oligonucleotides in Chinese hamster ovary cells overexpressing insulin receptors (Ijuin and Takenawa, 2003). Moreover, overexpression of SKIP inhibited insulin-induced glucose uptake and glycogen synthesis in L6 myoblasts (Ijuin and Takenawa, 2003).

A relatively poorly characterised proline-rich inositol polyphosphate 5′-phosphatase (PIPP) has recently been found to hydrolyse PIP_3 to PtdIns (3,4)P_2 (Astle *et al.*, 2006). Overexpression of PIPP in COS cells negatively regulates Akt signalling following EGF stimulation (Ooms *et al.*, 2006). The phosphatase has a wide tissue distribution (Mochizuki and Takenawa, 1999), but little is known about the *in vivo* function, and it is not known if it is involved in insulin or IGF receptor signalling. C1-TEN is a recently discovered putative 3′-phosphatase with predicted structural similarity to PTEN (Hafizi *et al.*, 2005). Cells overexpressing C1-TEN shows impaired

Akt phosphorylation, cell survival, proliferation and migration, although it is not yet known whether these effects are due to deficiency of 3′-phosphorylated lipids.

III. Real-Time Measurements of PIP_3 in Living Cells

In most studies on the regulation of PI3-kinase and lipid phosphatases, phosphoinositide lipids are detected with biochemical methods based on radio-tracer labelling, chromatography or mass spectrometry (Rusten and Stenmark, 2006). Estimation of PIP_3 levels have also been achieved with immunological detection of the lipid (Chen *et al.*, 2002; Yip *et al.*, 2008). These techniques may provide quantitative information about concentrations of specific phosphoinositide isomers, but compartmentalised signals, short-lasting transients, oscillations and other complex time-courses that characterise many signalling systems cannot be studied. Over the past decade, the development of genetically encoded fluorescence biosensors for time-resolved single cell detection of phosphoinositides has greatly improved our understanding of PIP_3 signalling. The green fluorescent protein (GFP) fused to certain PH domains from phosphoinositide binding proteins has become a useful tool for analysis of PIP_3 in living cells (Balla and Varnai, 2002; Halet, 2005). In order to selectively visualise PIP_3, the PH domain must have a binding affinity several orders of magnitude higher than for PtdIns(4,5)P_2, because even in stimulated cells, the concentration of PtdIns(4,5)P_2 is much higher than that of PIP_3. Most PH domains bind membrane phosphoinositides with low affinity and little specificity (Lemmon, 2008), but those from the guanine nucleotide exchange factors general receptor for phosphoinositides-1 (GRP1) and ADP-ribosylation factor nucleotide binding site opener (ARNO) as well as the Bruton's tyrosine kinase-1 (Btk1) all bind PIP_3 with relatively high affinity and specificity and has successfully been used in live cell imaging applications (Varnai *et al.*, 1999; Venkateswarlu *et al.*, 1998; Watton and Downward, 1999). The commonly used PH domain from Akt binds PIP_3 and PtdIns(3,4)P_2 with comparable affinities (Frech *et al.*, 1997; James *et al.*, 1996). At variance with its stereoisomer PtdIns(4,5)P_2, PtdIns(3,4)P_2 occurs in minute amounts in cellular membranes, and the majority of PtdIns(3,4)P_2 is regarded to be formed via 5′-dephosphorylation of PIP_3. Therefore, the Akt PH domain still constitutes a valuable tool for assessing PI3-kinase activity (Kontos *et al.*, 1998). In unstimulated cells with low plasma membrane PIP_3 concentration, the fluorescent PIP_3 reporter constructs are predominantly located in the cytoplasm, but localizes to the membrane when PIP_3 formation is stimulated (Fig.11.2). This translocation can be

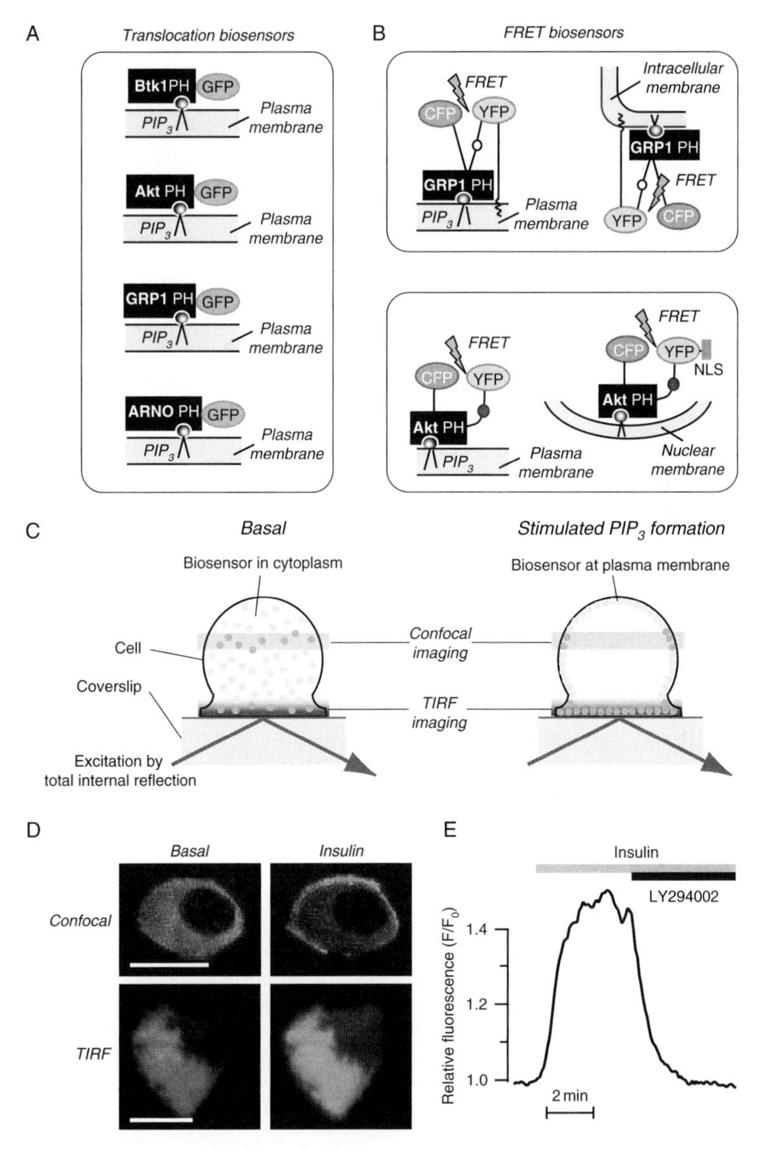

Figure 11.2 Real-time imaging of PIP$_3$ in individual cells. (A) PIP$_3$ translocation biosensors based on the green fluorescent protein (GFP) or its color variants fused to the PH domain from Bruton's tyrosine kinase 1 (Btk1), Akt/protein kinase B, general receptor for phosphoinositides-1 (GRP1) or ADP ribosylation factor nucleotide site opener (ARNO) that bind PIP$_3$ in the plasma membrane. (B) PIP$_3$ biosensors based on intramolecular fluorescence resonance energy transfer (FRET) between cyan and yellow fluorescent protein (CFP and YFP) as a result of a conformational change in the PH domain upon PIP$_3$ binding. In one biosensor based on the PH domain from GRP1 the conformational change is enhanced by membrane targeting to cause a

detected and quantified with fluorescence microscopy. Whereas confocal microscopy has been the predominant method for analysis of living cells, evanescent wave microscopy, also known as total internal reflection fluorescence microscopy, provides advantages for selective visualization of the plasma membrane (Haugh *et al.*, 2000; Idevall-Hagren and Tengholm, 2006; Tengholm and Meyer, 2002; Tengholm *et al.*, 2003).

As an alternative to the PH domain translocation assay, the GRP1-derived PH domain was used in a biosensor that relied on fluorescence resonance energy transfer (FRET) between cyan and yellow fluorescent protein versions (CFP and YFP) fused to each end of the protein such that PH domain binding to PIP_3 would cause increased FRET between the fluorescent proteins (Sato *et al.*, 2003). The biosensor could be targeted to specific membrane compartments to selectively report PIP_3 in the plasma membrane or intracellular membranes. This approach enabled the demonstration that PIP_3 is formed in intracellular membranes following endocyosis of PDGF receptor–ligand complexes (Sato *et al.*, 2003). Using a similar strategy, the PH domain from Akt was fused to a pseudoligand sequence of acidic amino acid residues between CFP and YFP to improve the FRET response of the reporter (Ananthanarayanan *et al.*, 2005). This biosensor was used to compare the PIP_3 responses in the plasma membrane and nucleus following growth factor stimulation. Over the past few years evidence has emerged that PIP_3 is present in the cell nucleus along with components of its biosynthetic machinery, including class I and II PI3-kinases, PTEN and

rotation of rigid linkers around a diglycine hinge engineered within the construct (Sato *et al.*, 2003). This sensor can be targeted to the plasma membrane or to intracellular membranes. Another sensor based on the PH domain from Akt uses a pseudoligand sequence to enhance the FRET response. This reporter offers dual readout of FRET change and translocation in its untargeted form, but it can also be directed to subcellular compartments. For example, using a nuclear localization sequence (NLS), the biosensor can be expressed in the cell nucleus (Ananthanarayanan *et al.*, 2005). (C) Schematic illustration of fluorescent biosensor translocation in response to PIP_3 formation in the plasma membrane. A confocal microscope images an optical section through the cell at any level, whereas a total internal reflection fluorescence (TIRF) microscope visualises a thin volume within ~100 nm from the plasma membrane in the coverslip adhesion region. This part of the cell is selectively excited by an evanescent wave generated by total internal reflection of a laser beam at the coverslip–water interphase. (D) Translocation of a GFP-GRP1 construct expressed in insulin-secreting MIN6 β-cells. Confocal and TIRF microscopy images were acquired before and 5 min after stimulation with 100 nM insulin. In the TIRF microscope, membrane translocation is detected as an overall increase of fluorescence intensity. Scale bars, 10 μm. (E) Time-lapse TIRF recording of the insulin-induced PIP_3 formation in a single MIN6 β-cell expressing a GFP-GRP1 biosensor. Insulin thus triggers biosensor translocation to the membrane (increase of fluorescence), which is counteracted by 100 μM of the PI3-kinase inhibitor LY294002. Fluorescence intensity is expressed in relation to the prestimulatory level (F/F_0). (See Color Insert.)

SHIP2 lipid phosphatases and effectors such as Akt and atypical isoforms of protein kinase C (Deleris *et al.*, 2006; Neri *et al.*, 2002). However, whereas PDGF and IGF-1 triggered marked PIP_3 responses in the plasma membrane of NIH3T3 and HEK293 cells, there were no signs of accumulation of PIP_3 in the nucleus (Ananthanarayanan *et al.*, 2005). Since endogenous proteins compete with the PH-domain-based biosensors for lipid binding, this discrepancy may reflect the presence of PIP_3 pools unavailable to this class of biosensors (Varnai and Balla, 2006). Although FRET-based sensors in principle can be designed for measurements of PIP_3 in cellular compartments other than the plasma membrane, proper detection of FRET is technically difficult and the dynamic range of the signals are typically much smaller than for the translocation-based biosensors.

IV. Spatio-Temporal Patterns of PIP_3 Signals

Real-time measurements of PIP_3 and PtdIns(3,4)P_2 with the Akt PH domain have demonstrated that these lipids have a diffusion coefficient of $\sim 0.5\ \mu m^2/s$ and a half-life of <1 min (Haugh *et al.*, 2000). The combination of rapid turnover and relatively slow diffusion favour generation of rapid changes and steep gradients in signalling lipid concentrations. Although many studies show that PIP_3 localizes throughout the plasma membrane, it may concentrate in highly restricted regions of the membrane or show quite variable time-courses, depending on the cell type and condition (Fig.11.3).

Using the GFP-tagged PH domain from Btk1, PIP_3 was found to be generated in ruffle-like structures of the plasma membrane of fibroblasts stimulated with EGF or PDGF (Varnai *et al.*, 1999). PIP_3 is a crucial messenger for mediating the insulin-stimulated membrane translocation and insertion of GLUT4 glucose transporters in myocytes and adipocytes. The GFP-tagged PH domains from ARNO and GRP1 were used to demonstrate that insulin stimulation of adipocytes is associated with accumulation of PIP_3 in the plasma membrane (Oatey *et al.*, 1999; Venkateswarlu *et al.*, 1998). The latter study showed that PDGF was less efficient than insulin in inducing PIP_3 formation, thereby providing en explanation for the poor ability of PDGF to stimulate glucose uptake (Oatey *et al.*, 1999). Simultaneous evanescent wave microscopy imaging of fluorescence tagged GLUT4 glucose transporter membrane insertion and PIP_3 with the Akt PH domain indicated that the different efficiency of insulin and PDGF to induce glucose uptake largely depended on different time-courses of the lipid signal (Tengholm and Meyer, 2002). Whereas insulin caused a sustained elevation, PDGF only induced a transient elevation of membrane PIP_3.

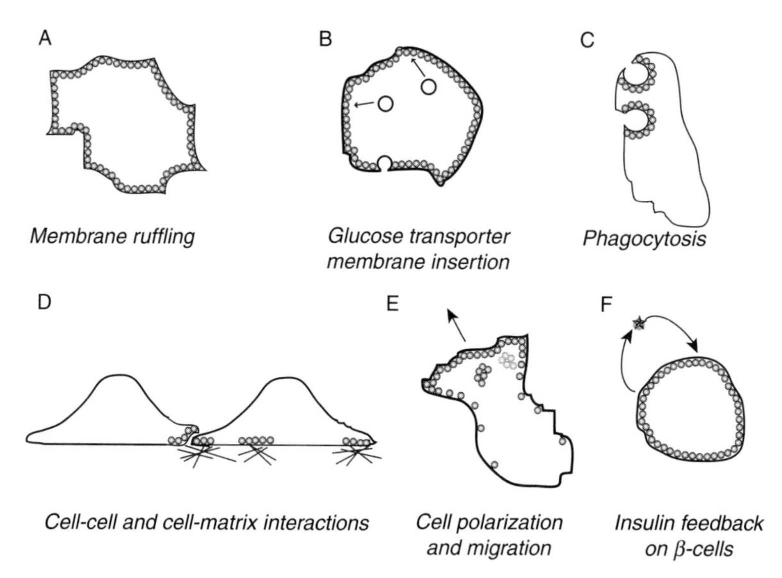

Figure 11.3 Spatio-temporal patterns of PIP_3 signals in different cellular processes. (A) Insulin and growth factor-induced accumulation of PIP_3 in membrane ruffles. (B) Insulin-induced GLUT4 glucose transporter membrane insertion. (C) Local PIP_3 accumulation in the phagocytic cup during phagocytosis. (D) PIP_3 formation at sites of cell–cell and cell–matrix interactions. (E) PIP_3 formation in the leading edge during cell polarization and migration. (F) PIP_3 oscillations as a result of insulin feedback on β-cells.

Focal accumulation of PIP_3 has been found in the phagocytic cup of phagocyting macrophages (Marshall *et al.*, 2001) and neutrophils (Dewitt *et al.*, 2006). PIP_3 or PtdIns(3,4)P_2 restriction at the phagocytic cup were essential for both progression of the cup, phagosome closure and Ca^{2+} signalling (Dewitt *et al.*, 2006). The PI3-kinase-dependent events are important for the cytoskeletal and membrane remodelling that occurs during phagocytosis (Cox *et al.*, 2002).

The generation of PIP_3 seems to be a conserved early event in cell polarization. In epithelial cells, PIP_3 has been found to accumulate at sites of cell–cell and cell–matrix interaction, but not at the apical surface (Watton and Downward, 1999; Yu *et al.*, 2003). Adhesion of epithelial cells to extracellular matrix is known to protect them from apoptosis. The constitutive generation of 3′-phosphoinostides supposedly provide a survival signal via activation of Akt (Watton and Downward, 1999). In a more recent study, application of exogenous PIP_3 caused membrane protrusions, which contained basolateral plasma membrane proteins but excluded apical proteins, suggesting that PIP_3 is directly involved in forming the basolateral plasma membrane in epithelial cells (Gassama-Diagne *et al.*, 2006).

The directional movement of cells towards a source of chemoattractant and the associated change in cell shape is a fundamental cellular process involved in immune responses, wound healing, neuronal patterning and embryogenesis. In migrating cells, 3′-phosphoinositides are primarily formed at the leading edge, creating a steep concentration gradient of the signalling lipids. Such gradients have been directly demonstrated with the GFP-tagged Akt PH domain expressed in neutrophils (Servant *et al.*, 2000; Srinivasan *et al.*, 2003; Wang *et al.*, 2002) and in fibroblasts migrating towards a PDGF source (Haugh *et al.*, 2000). PIP_3 gradients have also been demonstrated in the slime mould *Dictyostelium* migrating in a cAMP gradient (Meili *et al.*, 1999; Parent *et al.*, 1998). The asymmetric distribution of the signalling lipids has been accounted for by PI3-kinase being selectively recruited to the leading membrane, whereas the PTEN phosphatase is simultaneously recruited to regions of the plasma membrane that surrounds the sides and rear of the cell (Funamoto *et al.*, 2002; Iijima and Devreotes, 2002). From studies of primary dendritic cells and fibroblasts it was recently proposed that the global cell polarization is independent of the chemotaxis mechanism (Arrieumerlou and Meyer, 2005). While a PIP_3 gradient across the cell is part of a self-polarization process, chemotaxis is driven by local chemoattractant binding, which results in transient activation of PI3-kinase with concomitant local PIP_3 pulses in the leading edge. Each local PIP_3 pulse was found to be linked to local lamellipod extension and a small turn in the direction of migration (Arrieumerlou and Meyer, 2005).

Cell polarization and migration are key components in embryogenesis. Constitutive PI3-kinase activity with PIP_3 accumulation in junction and apical membranes of blastomeres is necessary for normal preimplantation development by preventing apoptosis (Halet *et al.*, 2008). During vertebrate gastrulation, the accumulation of PIP_3 and PtdIns(3,4)P_2 has strong impact on the velocity of mesendodermal cell migration. Accordingly, the GFP-tagged PH domain from Akt was strongly localized to the plasma membrane of prechordal plate progenitors at sites of cell–cell contact in zebrafish. In cells at the leading edge of the prechordal plate, the PH domain was often asymmetrically distributed from the leading to the rearing edge during process formation and cell elongation (Montero *et al.*, 2003).

V. PIP_3 Oscillations and Autocrine Insulin Signalling in β-Cells

Insulin is secreted from β-cells in the pancreatic islets of Langerhans in response to nutrients, hormones and neural factors. Glucose is a major physiological stimulator of insulin secretion. After rapid transport into the cells via GLUT transporters, the sugar is phosphorylated by glucokinase and further metabolised in the glycolysis and the tricarboxylic acid cycle, which eventually results in an increase of the cytoplasmic ATP/ADP ratio (Henquin, 2000; Maechler and Wollheim, 2001; Prentki and Matschinsky, 1987). This change in adenine nucleotide concentrations causes closure of ATP-sensitive K^+ channels, depolarization of the plasma membrane and activation of voltage-gated Ca^{2+} channels. The resulting influx of Ca^{2+} triggers exocytosis of insulin-containing secretory granules (Ashcroft and Rorsman, 1989). Like many other hormones, insulin is released in a pulsatile fashion with a dominating period of 4–6 min in humans (Pørksen, 2002). This dynamic pattern is due to an extensive synchronization of the secretory activity of individual β-cells within and between islets of Langerhans and is probably important for efficient action of the hormone in target tissues, especially the liver (Gylfe *et al.*, 2000). Interestingly, individual β-cells possess an inherent ability to generate oscillatory signals that stimulate periodic release of insulin. For instance, it has been demonstrated that the cytoplasmic concentrations of Ca^{2+} (Grapengiesser *et al.*, 1988) and cAMP (Dyachok *et al.*, 2006a; Dyachok *et al.*, 2008), the most important second messenger in exocytosis, oscillate in β-cells.

Apart from its effects on the classical target tissues liver, muscle and fat, insulin has also an autocrine function on β-cells. The β-cells thus express insulin and IGF-I receptors (Patel *et al.*, 1982; Van Schravendijk *et al.*, 1987; Verspohl and Ammon, 1980) as well as many of the associated downstream adapter and signalling proteins, including IRS1–4, PI3-kinase and protein kinase B/Akt (Harbeck *et al.*, 1996; Holst *et al.*, 1998; Kulkarni *et al.*, 1999b; Muller *et al.*, 2006; Rothenberg *et al.*, 1995; Velloso *et al.*, 1995; Withers *et al.*, 1998). Activation of insulin receptors by exogenous insulin or glucose-induced insulin secretion have been found to result in insulin receptor and IRS protein phosphorylation, activation of PI3-kinase and formation of PIP_3 (Rothenberg *et al.*, 1995; Velloso *et al.*, 1995). This autocrine activation of insulin receptors has been found to regulate gene transcription (Leibiger *et al.*, 1998; Xu and Rothenberg, 1998), proliferation (Kulkarni *et al.*, 1999a; Withers *et al.*, 1998), glucose metabolism (Borelli *et al.*, 2004; Nunemaker *et al.*, 2004), insulin biosynthesis and secretion (reviewed in Rutter, 1999 and Leibiger *et al.*, 2002).

It is much debated whether insulin stimulates or inhibits its secretion. Reduced insulin secretion in islets or insulinoma cells deficient in insulin receptors (Da Silva Xavier *et al.*, 2004; Kulkarni *et al.*, 1999a), and the observation that insulin or an insulin-mimetic compound stimulates secretion in individual β-cells and islets (Aspinwall *et al.*, 1999; Westerlund *et al.*, 2002) indicate that insulin has a positive feedback effect. In contrast, early measurements of insulin and C-peptide after addition of exogenous insulin *in vivo* (Elahi *et al.*, 1982) or from the perfused pancreas (Iversen and Miles, 1971), provided evidence that insulin inhibits secretion from the β-cells. This concept has later gained support from the observations that insulin secretion from human islets is reduced by insulin receptor activation (Persaud *et al.*, 2002), that Ca^{2+} signalling and insulin secretion is increased in insulinoma cells deficient in insulin receptors (Ohsugi *et al.*, 2005), and that insulin hyperpolarises β-cells in a PI3-kinase dependent manner (Khan *et al.*, 2001). The latter effect is probably mediated by PIP_3 activation of K_{ATP} channels (Harvey *et al.*, 2000; Shyng and Nichols, 1998). Additional support for the conception that negative feedback involves signalling downstream of PIP_3 comes from observations that insulin secretion is enhanced by inhibitors of PI3-kinase (Hagiwara *et al.*, 1995; Zawalich and Zawalich, 2000) or genetic ablation of the PI3-kinase p85 regulatory subunit (Eto *et al.*, 2002). There is still no consensus about insulin feedback effects on insulin secretion and the reasons for the discordant results are still unknown. However, a possible explanation to some of the inconsistencies was recently offered by the observation that low insulin concentrations ($<$100 pM) have a stimulatory effect, whereas high concentrations ($>$250 nM) are inhibitory and intermediate concentrations lack effect on C-peptide secretion from isolated islets (Jimenez-Feltstrom *et al.*, 2004).

The dynamics of autocrine PIP_3 formation in response to insulin secretion was recently investigated in individual mouse insulinoma cells expressing GFP-based PIP_3 biosensors (Idevall-Hagren and Tengholm, 2006). Stimulation of endogenous insulin secretion by depolarizing the plasma membrane with a high concentration of KCl induced concomitant elevation of the cytoplasmic Ca^{2+} concentration and formation PIP_3. There was no PIP_3 response under conditions known to inhibit insulin secretion, such as removal of extracellular Ca^{2+} or activation of adrenergic α_2-receptors with adrenaline. Elevation of the glucose concentration above the threshold for insulin secretion resulted in pronounced oscillations of PIP_3 in the plasma membrane with frequencies similar to those of pulsatile insulin secretion from isolated pancreatic islets or perfused pancreas (Gylfe *et al.*, 2000; Idevall-Hagren and Tengholm, 2006) (Fig.11.4). Simultaneous recordings of the cytoplasmic Ca^{2+} concentration showed that the oscillations of Ca^{2+} were synchronized with and preceded those of PIP_3 by 15–20 s. This delay is consistent with Ca^{2+}-induced exocytosis of insulin resulting in rapid autocrine activation of insulin receptors, since it takes ~15 s from the activation of the receptors to the first detectable translocation of the Akt PH domain (Idevall-Hagren and Tengholm, 2006). The glucose-induced PIP_3 oscillations were completely abolished by insulin receptor antibodies, inhibition of PI3-kinase, omission of extracellular Ca^{2+} or by blocking of voltage-gated Ca^{2+} channels, indicating that the PIP_3 oscillations are generated by autocrine feedback activation of insulin receptors as a result of pulsatile insulin secretion (Dyachok *et al.*, 2008).

The glucose-induced elevation of PIP_3 was higher than that induced by addition of saturating insulin concentrations, suggesting that the sugar may have additional effects on PI3-kinase (Idevall-Hagren and Tengholm,

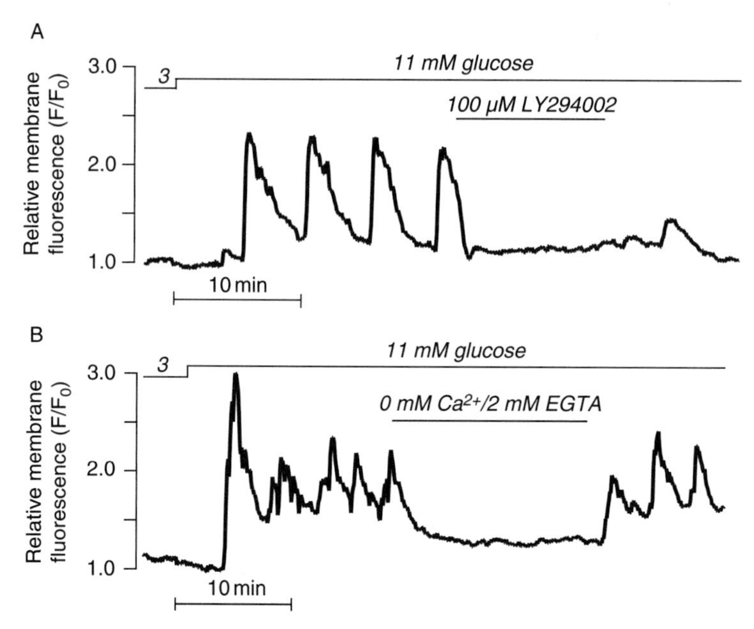

Figure 11.4 Glucose-induced PIP_3 oscillations in insulin-secreting cells. Real-time TIRF microscopy recordings of membrane PIP_3 concentration in individual MIN6 β-cells expressing a GFP-GRP1 biosensor. Elevation of the glucose concentration from 3 to 11 mM triggers after 3–5 min delay a rise of PIP_3 followed by pronounced oscillations from a slightly elevated level. The oscillations were suppressed after inhibition of PI3-kinase with 100 μM LY294002 (A) or by inhibition of glucose-induced insulin secretion by removal of extracellular Ca^{2+} combined with addition of 2 mM EGTA (B).

2006). However, glucose had no effect on PIP_3 under conditions when insulin secretion was inhibited. In contrast, glucose amplified the PIP_3 production induced by a high concentration of insulin under conditions when the insulin receptors were saturated and endogenous insulin secretion inhibited. Since the sugar did not affect the rate of PIP_3 degradation following inhibition of PI3-kinase, it seems that glucose amplification of insulin-induced PIP_3 formation is due to stimulation of PI3-kinase rather than inhibition of lipid phosphatases (Idevall-Hagren and Tengholm, 2006).

VI. Significance of PIP_3 Oscillations

Oscillations is a common feature of many cellular control systems and many second messengers, including Ca^{2+} (Berridge *et al.*, 2000), cAMP (Borodinsky and Spitzer, 2006; Dyachok *et al.*, 2006a), inositol-1,4,5-trisphosphate (Hirose *et al.*, 1999) and diacylglycerol (Codazzi *et al.*, 2001) are known to oscillate. Regulation by oscillations may provide advantages by improving the detection of weak signals and by providing specificity in the regulation of downstream responses, as has been suggested for Ca^{2+} (Cooper *et al.*, 1995; Dolmetsch *et al.*, 1998). Ca^{2+}-sensitive effector proteins have been found to decode the frequency, amplitude and/or duration of the oscillations (De Koninck and Schulman, 1998; Tomida *et al.*, 2003) and it has been found that different patterns of Ca^{2+} oscillations are differently effective in activating various transcription factors (Dolmetsch *et al.*, 1997, 1998). It has been found that oscillations of the cAMP concentration trigger cytoplasmic Ca^{2+} responses, but that the translocation of protein kinase A to the cell nucleus requires sustained elevation of cAMP (Dyachok *et al.*, 2006a,b). It remains to clarify whether oscillations of PIP_3 have similar effects in regulating downstream effectors. As regulators of protein function, phosphoinositides can play both a permissive and a regulatory role, depending on the affinity of the protein for the lipid. For proteins with a high affinity, phosphoinositides probably play a permissive role, whereas proteins with low affinity should be sensitive to both increases and decreases in the lipid concentration. Since PIP_3 regulates cellular events on a time-scale from seconds (electrostatic effects in membranes) over minutes (e.g., cytoskeletal rearrangements) to hours and beyond (gene transcription, cell growth and survival), it is likely that transient elevations of PIP_3 may selectively regulate fast events. Short-lasting local production of PIP_3 should also limit the diffusional spread of the lipid signal. As mentioned above, localised, brief PIP_3 oscillations may be crucial for chemotactic migration in dendritic cells and fibroblasts (Arrieumerlou and Meyer, 2005). The importance of PIP_3 in migration, phagocytosis and cell polarization is at least in part explained by the PIP_3 interaction with PH-domain containing guanine nucleotide exchange factors for the Rho/Rac/Cdc42 and Arf6 families of small GTPases, which in turn are crucial regulators of the actin cytoskeleton (Nobes and Hall, 1995; Song *et al.*, 1998). Arrieumerlou and Meyer (2005) were able to demonstrate that PIP_3 pulses correlated with local lamellipod extension and a small change in the direction of migration. It seems that PIP_3 oscillations triggered by local receptor binding events alone can explain chemotaxis, and that this phenomenon is different from the long-lasting global PIP_3 gradients that are part of the cell polarization process.

Several ion channels are regulated in a PI3-kinase/PIP_3 dependent manner. In some cases the lipid seems to act by inducing translocation and membrane insertion of the channel protein (Viard *et al.*, 2004), but the effect of PIP_3 is more often mediated by direct electrostatic interactions (Gamper and Shapiro, 2007). All major classes of inwardly rectifying K^+ channels are affected by PtdIns(4,5)P_2 and/or PIP_3, and the lipid association with the channels stabilises their open state (Gamper and Shapiro, 2007). For example, PtdIns(4,5)P_2 and PIP_3 interacts with the Kir6.2 subunit of the ATP-sensitive K^+ channel to increase channel activity (Haider *et al.*, 2007). Insulin has been found to hyperpolarise proopiomelanocortin neurons in the hypothalamus via PIP_3 activation of such channels (Plum *et al.*, 2006). A similar mechanism probably explains the PI3-kinase dependent hyperpolarization of pancreatic β-cells by insulin (Khan *et al.*, 2001). The ATP-sensitive K^+ channels are the major determinants of the resting potential in β-cells. Closure of the channels by ATP binding to the regulatory SUR1 subunit causes the membrane depolarization that activates voltage-gated Ca^{2+} channels. Oscillations of PIP_3 due to pulsatile insulin release may consequently be part of a feedback mechanism generating the variations in membrane potential that underlie the pulsatile secretion.

The significance of pulsatile insulin release has mostly been discussed in relation to its effects on the liver (Pørksen, 2002), but insulin oscillations may also be important for the individual β-cell. Due to the pronounced autocrine action on β-cells, the signalling machinery downstream of the insulin receptor can be expected to be strongly and constitutively active, if not desensitised, if the cells were exposed to extended periods of high insulin concentrations. By contrast, pulsatile insulin release is associated with periods with little autocrine stimulation, which enables the cells to sense the background concentrations of circulating insulin and other growth factors of importance for β-cell function. Insulin has been suggested to stimulate β-cell proliferation in states of insulin resistance (Okada *et al.*, 2007). Loss of pulsatile insulin secretion already in early diabetes (Lang *et al.*, 1981; O'Rahilly *et al.*, 1988) may therefore not only contribute to insulin resistance in the extrapancreatic target tissues, but also to impair the compensatory expansion of the β-cell mass.

VII. Concluding Remarks

Numerous studies over the past decade have demonstrated that PIP_3 signals are highly dynamic and that specific spatio-temporal patterns of PIP_3 signals are important for various functions in many types of cells. The pancreatic β-cell constitutes a particularly interesting example, as this cell not only releases insulin, but also expresses many of the components of the insulin receptor signalling machinery and is subject to regulation by insulin at multiple levels. Stimulation of insulin secretion is thus associated with pronounced oscillations of PIP_3 in the β-cell plasma membrane due to autocrine activation of PI3-kinase by pulsatile release of insulin. Important goals for future research include clarification of how different spatio-temporal patterns of PIP_3 signals are generated and how various PIP_3 signalling patterns affect the different downstream effectors. Signalling specificity may be determined not only by the localization of the lipids, but also by the organization of the lipid target molecules in discrete compartments, such as lipid rafts or internalised membranes. Continued studies will combine quantitative measurements of PIP_3 and related signalling lipids with single-cell assays for the localization and activity of their downstream effectors. In this respect, the recent development of fluorescent biosensors reporting the activity of protein kinase B/Akt (Kunkel *et al.*, 2005) and that of the small GTPases Rho, Rac and Cdc42 (Itoh *et al.*, 2002; Yoshizaki *et al.*, 2003) are particularly promising. Likewise, transgenic expression of the biosensors in intact animals (Costello *et al.*, 2002; Nishio *et al.*, 2007), offers possibilities to study PIP_3 signalling in a more physiological setting. These tools will facilitate the exploration of the role of PIP_3 in disease states, such as cancer, diabetes and immunological disorders.

ACKNOWLEDGMENTS

The authors' work is supported by grants from Åke Wiberg's Foundation, the European Foundation for the Study of Diabetes/MSD, the Family Ernfors Foundation, Harald and Greta Jeanssons Foundations, Novo Nordisk Foundation, the Swedish Diabetes Association and the Swedish Research Council.

REFERENCES

Ananthanarayanan, B., Ni, Q., and Zhang, J. (2005). Signal propagation from membrane messengers to nuclear effectors revealed by reporters of phosphoinositide dynamics and Akt activity. *Proc. Natl. Acad. Sci. USA* **102,** 15081–15086.

Arrieumerlou, C., and Meyer, T. (2005). A local coupling model and compass parameter for eukaryotic chemotaxis. *Dev. Cell* **8,** 215–227.

Ashcroft, F. M., and Rorsman, P. (1989). Electrophysiology of the pancreatic β-cell. *Prog. Biophys. Mol. Biol.* **54,** 87–143.

Aspinwall, C. A., Lakey, J. R., and Kennedy, R. T. (1999). Insulin-stimulated insulin secretion in single pancreatic beta cells. *J. Biol. Chem.* **274,** 6360–6365.

Astle, M. V., Seaton, G., Davies, E. M., Fedele, C. G., Rahman, P., Arsala, L., and Mitchell, C. A. (2006). Regulation of phosphoinositide signaling by the inositol polyphosphate 5-phosphatases. *IUBMB Life* **58,** 451–456.

Balla, T., and Varnai, P. (2002). Visualizing cellular phosphoinositide pools with GFP-fused protein-modules. *Sci. STKE* **2002,** PL3.

Berridge, M. J., Lipp, P., and Bootman, M. D. (2000). The versatility and universality of calcium signalling. *Nat. Rev. Mol. Cell. Biol.* **1,** 11–21.

Blero, D., De Smedt, F., Pesesse, X., Paternotte, N., Moreau, C., Payrastre, B., and Erneux, C. (2001). The SH2 domain containing inositol 5-phosphatase SHIP2 controls phosphatidylinositol 3,4,5-trisphosphate levels in CHO-IR cells stimulated by insulin. *Biochem. Biophys. Res. Commun.* **282,** 839–843.

Borelli, M. I., Francini, F., and Gagliardino, J. J. (2004). Autocrine regulation of glucose metabolism in pancreatic islets. *Am. J. Physiol. Endocrinol. Metab.* **286,** E111–E115.

Borodinsky, L. N., and Spitzer, N. C. (2006). Second messenger pas de deux: The coordinated dance between calcium and cAMP. *Sci. STKE* **2006,** pe22.

Butler, M., McKay, R. A., Popoff, I. J., Gaarde, W. A., Witchell, D., Murray, S. F., Dean, N. M., Bhanot, S., and Monia, B. P. (2002). Specific inhibition of PTEN expression reverses hyperglycemia in diabetic mice. *Diabetes* **51,** 1028–1034.

Cantley, L. C. (2002). The phosphoinositide 3-kinase pathway. *Science* **296,** 1655–1657.

Chen, R., Kang, V. H., Chen, J., Shope, J. C., Torabinejad, J., DeWald, D. B., and Prestwich, G. D. (2002). A monoclonal antibody to visualize PtdIns(3,4,5)P_3 in cells. *J. Histochem. Cytochem.* **50,** 697–708.

Clement, S., Krause, U., Desmedt, F., Tanti, J. F., Behrends, J., Pesesse, X., Sasaki, T., Penninger, J., Doherty, M., Malaisse, W., Dumont, J. E., Le Marchand-Brustel, Y., *et al.* (2001). The lipid phosphatase SHIP2 controls insulin sensitivity. *Nature* **409,** 92–97.

Codazzi, F., Teruel, M. N., and Meyer, T. (2001). Control of astrocyte Ca^{2+} oscillations and waves by oscillating translocation and activation of protein kinase C. *Curr. Biol.* **11,** 1089–1097.

Cooper, D. M., Mons, N., and Karpen, J. W. (1995). Adenylyl cyclases and the interaction between calcium and cAMP signalling. *Nature* **374,** 421–424.

Costello, P. S., Gallagher, M., and Cantrell, D. A. (2002). Sustained and dynamic inositol lipid metabolism inside and outside the immunological synapse. *Nat. Immunol.* **3,** 1082–1089.

Cox, D., Berg, J. S., Cammer, M., Chinegwundoh, J. O., Dale, B. M., Cheney, R. E., and Greenberg, S. (2002). Myosin X is a downstream effector of PI(3)K during phagocytosis. *Nat. Cell Biol.* **4,** 469–477.

Da Silva Xavier, G., Qian, Q., Cullen, P. J., and Rutter, G. A. (2004). Distinct roles for insulin and insulin-like growth factor-1 receptors in pancreatic β-cell glucose sensing revealed by RNA silencing. *Biochem. J.* **377,** 149–158.

De Koninck, P., and Schulman, H. (1998). Sensitivity of CaM kinase II to the frequency of Ca^{2+} oscillations. *Science* **279,** 227–230.

Deleris, P., Gayral, S., and Breton-Douillon, M. (2006). Nuclear Ptdlns(3,4,5)P_3 signaling: An ongoing story. *J. Cell. Biochem.* **98,** 469–485.

Dewitt, S., Tian, W., and Hallett, M. B. (2006). Localised PtdIns(3,4,5)P_3 or PtdIns(3,4)P_2 at the phagocytic cup is required for both phagosome closure and Ca^{2+} signalling in HL60 neutrophils. *J. Cell Sci.* **119,** 443–451.

Dolmetsch, R. E., Lewis, R. S., Goodnow, C. C., and Healy, J. I. (1997). Differential activation of transcription factors induced by Ca^{2+} response amplitude and duration. *Nature* **386,** 855–858.

Dolmetsch, R. E., Xu, K., and Lewis, R. S. (1998). Calcium oscillations increase the efficiency and specificity of gene expression. *Nature* **392,** 933–936.

Dyachok, O., Isakov, Y., Sågetorp, J., and Tengholm, A. (2006a). Oscillations of cyclic AMP in hormone-stimulated insulin-secreting β-cells. *Nature* **439,** 349–352.

Dyachok, O., Sågetorp, J., Isakov, Y., and Tengholm, A. (2006b). cAMP oscillations restrict protein kinase A redistribution in insulin-secreting cells. *Biochem. Soc. Trans.* **34,** 498–501.

Dyachok, O., Idevall-Hagren, O., Sågetorp, J., Tian, G., Wuttke, A., Arrieumerlou, C., Akusjärvi, G., Gylfe, E., and Tengholm, A. (2008). Glucose-induced cyclic AMP oscillations regulate pulsatile insulin secretion. *Cell Metab.* **8,** 26–37.

Elahi, D., Nagulesparan, M., Hershcopf, R. J., Muller, D. C., Tobin, J. D., Blix, P. M., Rubenstein, A. H., Unger, R. H., and Andres, R. (1982). Feedback inhibition of insulin secretion by insulin: Relation to the hyperinsulinemia of obesity. *N. Engl. J. Med.* **306,** 1196–1202.

Engelman, J. A., Luo, J., and Cantley, L. C. (2006). The evolution of phosphatidylinositol 3-kinases as regulators of growth and metabolism. *Nat. Rev. Genet.* **7,** 606–619.

Eto, K., Yamashita, T., Tsubamoto, Y., Terauchi, Y., Hirose, K., Kubota, N., Yamashita, S., Taka, J., Satoh, S., Sekihara, H., Tobe, K., Iino, M., *et al.* (2002). Phosphatidylinositol 3-kinase suppresses glucose-stimulated insulin secretion by affecting post-cytosolic [Ca^{2+}] elevation signals. *Diabetes* **51,** 87–97.

Foukas, L. C., Claret, M., Pearce, W., Okkenhaug, K., Meek, S., Peskett, E., Sancho, S., Smith, A. J., Withers, D. J., and Vanhaesebroeck, B. (2006). Critical role for the p110α phosphoinositide-3-OH kinase in growth and metabolic regulation. *Nature* **441,** 366–370.

Frech, M., Andjelkovic, M., Ingley, E., Reddy, K. K., Falck, J. R., and Hemmings, B. A. (1997). High affinity binding of inositol phosphates and phosphoinositides to the pleckstrin homology domain of RAC/protein kinase B and their influence on kinase activity. *J. Biol. Chem.* **272,** 8474–8481.

Funamoto, S., Meili, R., Lee, S., Parry, L., and Firtel, R. A. (2002). Spatial and temporal regulation of 3-phosphoinositides by PI 3-kinase and PTEN mediates chemotaxis. *Cell* **109,** 611–623.

Gamper, N., and Shapiro, M. S. (2007). Regulation of ion transport proteins by membrane phosphoinositides. *Nat. Rev. Neurosci.* **8,** 921–934.

Gassama-Diagne, A., Yu, W., ter Beest, M., Martin-Belmonte, F., Kierbel, A., Engel, J., and Mostov, K. (2006). Phosphatidylinositol-3,4,5-trisphosphate regulates the formation of the basolateral plasma membrane in epithelial cells. *Nat. Cell Biol.* **8,** 963–970.

Grapengiesser, E., Gylfe, E., and Hellman, B. (1988). Glucose-induced oscillations of cytoplasmic Ca^{2+} in the pancreatic β-cell. *Biochem. Biophys. Res. Commun.* **151,** 1299–1304.

Gurung, R., Tan, A., Ooms, L. M., McGrath, M. J., Huysmans, R. D., Munday, A. D., Prescott, M., Whisstock, J. C., and Mitchell, C. A. (2003). Identification of a novel domain in two mammalian inositol-polyphosphate 5-phosphatases that mediates membrane ruffle localization. The inositol 5-phosphatase skip localizes to the endoplasmic reticulum and translocates to membrane ruffles following epidermal growth factor stimulation. *J. Biol. Chem.* **278,** 11376–11385.

Gylfe, E., Ahmed, M., Bergsten, P., Dansk, H., Dyachok, O., Eberhardson, M., Grapengiesser, E., Hellman, B., Lin, J. M., Sundsten, T., Tengholm, A., Vieira, E., *et al.* (2000). Signaling underlying pulsatile insulin secretion. *Ups. J. Med. Sci.* **105,** 35–51.

Hafizi, S., Ibraimi, F., and Dahlbäck, B. (2005). C1-TEN is a negative regulator of the Akt/PKB signal transduction pathway and inhibits cell survival, proliferation, and migration. *FASEB J.* **19,** 971–973.

Hagiwara, S., Sakurai, T., Tashiro, F., Hashimoto, Y., Matsuda, Y., Nonomura, Y., and Miyazaki, J. (1995). An inhibitory role for phosphatidylinositol 3-kinase in insulin secretion from pancreatic B cell line MIN6. *Biochem. Biophys. Res. Commun.* **214,** 51–59.

Haider, S., Tarasov, A. I., Craig, T. J., Sansom, M. S., and Ashcroft, F. M. (2007). Identification of the PIP_2-binding site on Kir6.2 by molecular modelling and functional analysis. *EMBO J.* **26,** 3749–3759.

Halet, G. (2005). Imaging phosphoinositide dynamics using GFP-tagged protein domains. *Biol. Cell* **97,** 501–518.

Halet, G., Viard, P., and Carroll, J. (2008). Constitutive PtdIns(3,4,5)P_3 synthesis promotes the development and survival of early mammalian embryos. *Development* **135,** 425–429.

Harbeck, M. C., Louie, D. C., Howland, J., Wolf, B. A., and Rothenberg, P. L. (1996). Expression of insulin receptor mRNA and insulin receptor substrate 1 in pancreatic islet beta-cells. *Diabetes* **45,** 711–717.

Harvey, J., Hardy, S. C., Irving, A. J., and Ashford, M. L. (2000). Leptin activation of ATP-sensitive $K^+(K_{ATP})$ channels in rat CRI-G1 insulinoma cells involves disruption of the actin cytoskeleton. *J. Physiol.* **527**(Pt. 1), 95–107.

Haugh, J. M., Codazzi, F., Teruel, M., and Meyer, T. (2000). Spatial sensing in fibroblasts mediated by 3′ phosphoinositides. *J. Cell Biol.* **151,** 1269–1280.

Hawkins, P. T., Anderson, K. E., Davidson, K., and Stephens, L. R. (2006). Signalling through Class I PI3Ks in mammalian cells. *Biochem. Soc. Trans.* **34,** 647–662.

Henquin, J. C. (2000). Triggering and amplifying pathways of regulation of insulin secretion by glucose. *Diabetes* **49,** 1751–1760.

Hirose, K., Kadowaki, S., Tanabe, M., Takeshima, H., and Iino, M. (1999). Spatiotemporal dynamics of inositol 1,4,5-trisphosphate that underlies complex Ca^{2+} mobilization patterns. *Science* **284,** 1527–1530.

Holst, L. S., Mulder, H., Manganiello, V., Sundler, F., Ahren, B., Holm, C., and Degerman, E. (1998). Protein kinase B is expressed in pancreatic β cells and activated upon stimulation with insulin-like growth factor I. *Biochem. Biophys. Res. Commun.* **250,** 181–186.

Hori, H., Sasaoka, T., Ishihara, H., Wada, T., Murakami, S., Ishiki, M., and Kobayashi, M. (2002). Association of SH2-containing inositol phosphatase 2 with the insulin resistance of diabetic db/db mice. *Diabetes* **51,** 2387–2394.

Idevall-Hagren, O., and Tengholm, A. (2006). Glucose and insulin synergistically activate PI3-kinase to trigger oscillations of phosphatidylinositol-3,4,5-trisphosphate in β-cells. *J. Biol. Chem.* **281,** 39121–39127.

Iijima, M., and Devreotes, P. (2002). Tumor suppressor PTEN mediates sensing of chemoattractant gradients. *Cell* **109,** 599–610.

Ijuin, T., and Takenawa, T. (2003). SKIP negatively regulates insulin-induced GLUT4 translocation and membrane ruffle formation. *Mol. Cell. Biol.* **23,** 1209–1220.

Ishihara, H., Sasaoka, T., Hori, H., Wada, T., Hirai, H., Haruta, T., Langlois, W. J., and Kobayashi, M. (1999). Molecular cloning of rat SH2-containing inositol phosphatase 2 (SHIP2) and its role in the regulation of insulin signaling. *Biochem. Biophys. Res. Commun.* **260,** 265–272.

Itoh, R. E., Kurokawa, K., Ohba, Y., Yoshizaki, H., Mochizuki, N., and Matsuda, M. (2002). Activation of rac and cdc42 video imaged by fluorescent resonance energy transfer-based single-molecule probes in the membrane of living cells. *Mol. Cell. Biol.* **22,** 6582–6591.

Iversen, J., and Miles, D. W. (1971). Evidence for a feedback inhibition of insulin on insulin secretion in the isolated, perfused canine pancreas. *Diabetes* **20,** 1–9.

James, S. R., Downes, C. P., Gigg, R., Grove, S. J., Holmes, A. B., and Alessi, D. R. (1996). Specific binding of the Akt-1 protein kinase to phosphatidylinositol 3,4,5-trisphosphate without subsequent activation. *Biochem. J.* **315**(Pt. 3), 709–713.

Jimenez-Feltstrom, J., Lundquist, I., Obermuller, S., and Salehi, A. (2004). Insulin feedback actions: Complex effects involving isoforms of islet nitric oxide synthase. *Regul. Pept.* **122,** 109–118.

Kagawa, S., Sasaoka, T., Yaguchi, S., Ishihara, H., Tsuneki, H., Murakami, S., Fukui, K., Wada, T., Kobayashi, S., Kimura, I., and Kobayashi, M. (2005). Impact of SRC homology 2-containing inositol 5′-phosphatase 2 gene polymorphisms detected in a Japanese population on insulin signaling. *J. Clin. Endocrinol. Metab.* **90,** 2911–2919.

Kagawa, S., Soeda, Y., Ishihara, H., Oya, T., Sasahara, M., Yaguchi, S., Oshita, R., Wada, T., Tsuneki, H., and Sasaoka, T. (2008). Impact of transgenic overexpression of SH2-containing inositol 5′-phosphatase 2 on glucose metabolism and insulin signaling in mice. *Endocrinology* **149,** 642–650.

Khan, F. A., Goforth, P. B., Zhang, M., and Satin, L. S. (2001). Insulin activates ATP-sensitive K^+ channels in pancreatic β-cells through a phosphatidylinositol 3-kinase-dependent pathway. *Diabetes* **50,** 2192–2198.

Knight, Z. A., Gonzalez, B., Feldman, M. E., Zunder, E. R., Goldenberg, D. D., Williams, O., Loewith, R., Stokoe, D., Balla, A., Toth, B., Balla, T., Weiss, W. A., *et al.* (2006). A pharmacological map of the PI3-K family defines a role for p110α in insulin signaling. *Cell* **125,** 733–747.

Kontos, C. D., Stauffer, T. P., Yang, W. P., York, J. D., Huang, L., Blanar, M. A., Meyer, T., and Peters, K. G. (1998). Tyrosine 1101 of Tie2 is the major site of association of p85 and is required for activation of phosphatidylinositol 3-kinase and Akt. *Mol. Cell. Biol.* **18,** 4131–4140.

Kulkarni, R. N., Bruning, J. C., Winnay, J. N., Postic, C., Magnuson, M. A., and Kahn, C. R. (1999a). Tissue-specific knockout of the insulin receptor in pancreatic β cells creates an insulin secretory defect similar to that in type 2 diabetes. *Cell* **96,** 329–339.

Kulkarni, R. N., Winnay, J. N., Daniels, M., Bruning, J. C., Flier, S. N., Hanahan, D., and Kahn, C. R. (1999b). Altered function of insulin receptor substrate-1-deficient mouse islets and cultured β-cell lines. *J. Clin. Invest.* **104,** R69–R75.

Kunkel, M. T., Ni, Q., Tsien, R. Y., Zhang, J., and Newton, A. C. (2005). Spatio-temporal dynamics of protein kinase B/Akt signaling revealed by a genetically encoded fluorescent reporter. *J. Biol. Chem.* **280,** 5581–5587.

Kurlawalla-Martinez, C., Stiles, B., Wang, Y., Devaskar, S. U., Kahn, B. B., and Wu, H. (2005). Insulin hypersensitivity and resistance to streptozotocin-induced diabetes in mice lacking PTEN in adipose tissue. *Mol. Cell. Biol.* **25,** 2498–2510.

Lang, D. A., Matthews, D. R., Burnett, M., and Turner, R. C. (1981). Brief, irregular oscillations of basal plasma insulin and glucose concentrations in diabetic man. *Diabetes* **30,** 435–439.

Lawlor, M. A., and Alessi, D. R. (2001). PKB/Akt: A key mediator of cell proliferation, survival and insulin responses? *J. Cell Sci.* **114,** 2903–2910.

Lazar, D. F., and Saltiel, A. R. (2006). Lipid phosphatases as drug discovery targets for type 2 diabetes. *Nat. Rev. Drug Discov.* **5,** 333–342.

Leibiger, I. B., Leibiger, B., Moede, T., and Berggren, P. O. (1998). Exocytosis of insulin promotes insulin gene transcription via the insulin receptor/PI-3 kinase/p70 s6 kinase and CaM kinase pathways. *Mol. Cell* **1,** 933–938.

Leibiger, I. B., Leibiger, B., and Berggren, P. O. (2002). Insulin feedback action on pancreatic β-cell function. *FEBS Lett.* **532,** 1–6.

Lemmon, M. A. (2008). Membrane recognition by phospholipid-binding domains. *Nat. Rev. Mol. Cell. Biol.* **9,** 99–111.

Maechler, P., and Wollheim, C. B. (2001). Mitochondrial function in normal and diabetic β-cells. *Nature* **414,** 807–812.

Maehama, T., Taylor, G. S., and Dixon, J. E. (2001). PTEN and myotubularin: Novel phosphoinositide phosphatases. *Annu. Rev. Biochem.* **70,** 247–279.

Manning, B. D., and Cantley, L. C. (2007). AKT/PKB signaling: Navigating downstream. *Cell* **129,** 1261–1274.

Marion, E., Kaisaki, P. J., Pouillon, V., Gueydan, C., Levy, J. C., Bodson, A., Krzentowski, G., Daubresse, J. C., Mockel, J., Behrends, J., Servais, G., Szpirer, C., *et al.* (2002). The gene INPPL1, encoding the lipid phosphatase SHIP2, is a candidate for type 2 diabetes in rat and man. *Diabetes* **51,** 2012–2017.

Marshall, J. G., Booth, J. W., Stambolic, V., Mak, T., Balla, T., Schreiber, A. D., Meyer, T., and Grinstein, S. (2001). Restricted accumulation of phosphatidylinositol 3-kinase products in a plasmalemmal subdomain during Fc gamma receptor-mediated phagocytosis. *J. Cell Biol.* **153,** 1369–1380.

Meili, R., Ellsworth, C., Lee, S., Reddy, T. B., Ma, H., and Firtel, R. A. (1999). Chemoattractant-mediated transient activation and membrane localization of Akt/PKB is required for efficient chemotaxis to cAMP in *Dictyostelium*. *EMBO J.* **18,** 2092–2105.

Mochizuki, Y., and Takenawa, T. (1999). Novel inositol polyphosphate 5-phosphatase localizes at membrane ruffles. *J. Biol. Chem.* **274,** 36790–36795.

Montero, J. A., Kilian, B., Chan, J., Bayliss, P. E., and Heisenberg, C. P. (2003). Phosphoinositide 3-kinase is required for process outgrowth and cell polarization of gastrulating mesendodermal cells. *Curr. Biol.* **13,** 1279–1289.

Muller, D., Huang, G. C., Amiel, S., Jones, P. M., and Persaud, S. J. (2006). Identification of insulin signaling elements in human β-cells: Autocrine regulation of insulin gene expression. *Diabetes* **55,** 2835–2842.

Myers, M. G. Jr., and White, M. F. (1996). Insulin signal transduction and the IRS proteins. *Annu. Rev. Pharmacol. Toxicol.* **36,** 615–658.

Nakashima, N., Sharma, P. M., Imamura, T., Bookstein, R., and Olefsky, J. M. (2000). The tumor suppressor PTEN negatively regulates insulin signaling in 3T3-L1 adipocytes. *J. Biol. Chem.* **275,** 12889–12895.

Neri, L. M., Borgatti, P., Capitani, S., and Martelli, A. M. (2002). The nuclear phosphoinositide 3-kinase/AKT pathway: A new second messenger system. *Biochim. Biophys. Acta* **1584,** 73–80.

Nishio, M., Watanabe, K., Sasaki, J., Taya, C., Takasuga, S., Iizuka, R., Balla, T., Yamazaki, M., Watanabe, H., Itoh, R., Kuroda, S., Horie, Y., *et al.* (2007). Control of cell polarity and motility by the PtdIns(3,4,5)P_3 phosphatase SHIP1. *Nat. Cell Biol.* **9,** 36–44.

Nobes, C. D., and Hall, A. (1995). Rho, Rac, and Cdc42 GTPases regulate the assembly of multimolecular focal complexes associated with actin stress fibers, lamellipodia, and filopodia. *Cell* **81,** 53–62.

Nunemaker, C. S., Zhang, M., and Satin, L. S. (2004). Insulin feedback alters mitochondrial activity through an ATP-sensitive K^+ channel-dependent pathway in mouse islets and β-cells. *Diabetes* **53,** 1765–1772.

Oatey, P. B., Venkateswarlu, K., Williams, A. G., Fletcher, L. M., Foulstone, E. J., Cullen, P. J., and Tavare, J. M. (1999). Confocal imaging of the subcellular distribution of phosphatidylinositol 3,4,5-trisphosphate in insulin- and PDGF-stimulated 3T3-L1 adipocytes. *Biochem. J.* **344**(Pt. 2), 511–518.

Ohsugi, M., Cras-Meneur, C., Zhou, Y., Bernal-Mizrachi, E., Johnson, J. D., Luciani, D. S., Polonsky, K. S., and Permutt, M. A. (2005). Reduced expression of the insulin receptor in mouse insulinoma (MIN6) cells reveals multiple roles of insulin signaling in gene expression, proliferation, insulin content, and secretion. *J. Biol. Chem.* **280,** 4992–5003.

Okada, T., Liew, C. W., Hu, J., Hinault, C., Michael, M. D., Krtzfeldt, J., Yin, C., Holzenberger, M., Stoffel, M., and Kulkarni, R. N. (2007). Insulin receptors in β-cells are critical for islet compensatory growth response to insulin resistance. *Proc. Natl. Acad. Sci. USA* **104,** 8977–8982.

Ono, H., Katagiri, H., Funaki, M., Anai, M., Inukai, K., Fukushima, Y., Sakoda, H., Ogihara, T., Onishi, Y., Fujishiro, M., Kikuchi, M., Oka, Y., *et al.* (2001). Regulation of phosphoinositide metabolism, Akt phosphorylation, and glucose transport by PTEN (phosphatase and tensin homolog deleted on chromosome 10) in 3T3-L1 adipocytes. *Mol. Endocrinol.* **15,** 1411–1422.

Ooms, L. M., Fedele, C. G., Astle, M. V., Ivetac, I., Cheung, V., Pearson, R. B., Layton, M. J., Forrai, A., Nandurkar, H. H., and Mitchell, C. A. (2006). The inositol polyphosphate 5-phosphatase, PIPP, Is a novel regulator of phosphoinositide 3-kinase-dependent neurite elongation. *Mol. Biol. Cell* **17,** 607–622.

O'Rahilly, S., Turner, R. C., and Matthews, D. R. (1988). Impaired pulsatile secretion of insulin in relatives of patients with non-insulin-dependent diabetes. *N. Engl. J. Med.* **318,** 1225–1230.

Parent, C. A., Blacklock, B. J., Froehlich, W. M., Murphy, D. B., and Devreotes, P. N. (1998). G protein signaling events are activated at the leading edge of chemotactic cells. *Cell* **95,** 81–91.

Patel, Y. C., Amherdt, M., and Orci, L. (1982). Quantitative electron microscopic autoradiography of insulin, glucagon, and somatostatin binding sites on islets. *Science* **217,** 1155–1156.

Persaud, S. J., Asare-Anane, H., and Jones, P. M. (2002). Insulin receptor activation inhibits insulin secretion from human islets of Langerhans. *FEBS Lett.* **510,** 225–228.

Pesesse, X., Deleu, S., De Smedt, F., Drayer, L., and Erneux, C. (1997). Identification of a second SH2-domain-containing protein closely related to the phosphatidylinositol polyphosphate 5-phosphatase SHIP. *Biochem. Biophys. Res. Commun.* **239,** 697–700.

Plum, L., Ma, X., Hampel, B., Balthasar, N., Coppari, R., Munzberg, H., Shanabrough, M., Burdakov, D., Rother, E., Janoschek, R., Alber, J., Belgardt, B. F., *et al.* (2006). Enhanced PIP_3 signaling in POMC neurons causes K_{ATP} channel activation and leads to diet-sensitive obesity. *J. Clin. Invest.* **116,** 1886–1901.

Pørksen, N. (2002). The *in vivo* regulation of pulsatile insulin secretion. *Diabetologia* **45,** 3–20.

Prentki, M., and Matschinsky, F. M. (1987). Ca^{2+}, cAMP, and phospholipid-derived messengers in coupling mechanisms of insulin secretion. *Physiol. Rev.* **67,** 1185–1248.

Rothenberg, P. L., Willison, L. D., Simon, J., and Wolf, B. A. (1995). Glucose-induced insulin receptor tyrosine phosphorylation in insulin-secreting beta-cells. *Diabetes* **44,** 802–809.

Rusten, T. E., and Stenmark, H. (2006). Analyzing phosphoinositides and their interacting proteins. *Nat. Methods* **3,** 251–258.

Rutter, G. A. (1999). Insulin secretion: Feed-forward control of insulin biosynthesis? *Curr. Biol.* **9,** R443–R445.

Saltiel, A. R., and Kahn, C. R. (2001). Insulin signalling and the regulation of glucose and lipid metabolism. *Nature* **414,** 799–806.

Sarbassov, D. D., Guertin, D. A., Ali, S. M., and Sabatini, D. M. (2005). Phosphorylation and regulation of Akt/PKB by the rictor-mTOR complex. *Science* **307,** 1098–1101.

Sasaoka, T., Wada, T., Fukui, K., Murakami, S., Ishihara, H., Suzuki, R., Tobe, K., Kadowaki, T., and Kobayashi, M. (2004). SH2-containing inositol phosphatase 2 predominantly regulates Akt2, and not Akt1, phosphorylation at the plasma membrane in response to insulin in 3T3-L1 adipocytes. *J. Biol. Chem.* **279,** 14835–14843.

Sato, M., Ueda, Y., Takagi, T., and Umezawa, Y. (2003). Production of $PtdInsP_3$ at endomembranes is triggered by receptor endocytosis. *Nat. Cell Biol.* **5,** 1016–1022.

Seo, J. H., Ahn, Y., Lee, S. R., Yeol Yeo, C., and Chung Hur, K. (2005). The major target of the endogenously generated reactive oxygen species in response to insulin stimulation is phosphatase and tensin homolog and not phosphoinositide-3 kinase (PI-3 kinase) in the PI-3 kinase/Akt pathway. *Mol. Biol. Cell* **16,** 348–357.

Servant, G., Weiner, O. D., Herzmark, P., Balla, T., Sedat, J. W., and Bourne, H. R. (2000). Polarization of chemoattractant receptor signaling during neutrophil chemotaxis. *Science* **287,** 1037–1040.

Shepherd, P. R., Withers, D. J., and Siddle, K. (1998). Phosphoinositide 3-kinase: The key switch mechanism in insulin signalling. *Biochem. J.* **333**(Pt. 3), 471–490.

Shyng, S. L., and Nichols, C. G. (1998). Membrane phospholipid control of nucleotide sensitivity of K_{ATP} channels. *Science* **282,** 1138–1141.

Sleeman, M. W., Wortley, K. E., Lai, K. M., Gowen, L. C., Kintner, J., Kline, W. O., Garcia, K., Stitt, T. N., Yancopoulos, G. D., Wiegand, S. J., and Glass, D. J. (2005). Absence of the lipid phosphatase SHIP2 confers resistance to dietary obesity. *Nat. Med.* **11,** 199–205.

Song, J., Khachikian, Z., Radhakrishna, H., and Donaldson, J. G. (1998). Localization of endogenous ARF6 to sites of cortical actin rearrangement and involvement of ARF6 in cell spreading. *J. Cell Sci.* **111**(Pt. 15), 2257–2267.

Srinivasan, S., Wang, F., Glavas, S., Ott, A., Hofmann, F., Aktories, K., Kalman, D., and Bourne, H. R. (2003). Rac and Cdc42 play distinct roles in regulating $PI(3,4,5)P_3$ and polarity during neutrophil chemotaxis. *J. Cell Biol.* **160,** 375–385.

Stiles, B., Wang, Y., Stahl, A., Bassilian, S., Lee, W. P., Kim, Y. J., Sherwin, R., Devaskar, S., Lesche, R., Magnuson, M. A., and Wu, H. (2004). Liver-specific deletion of negative regulator Pten results in fatty liver and insulin hypersensitivity. *Proc. Natl. Acad. Sci. USA* **101,** 2082–2087.

Tang, X., Powelka, A. M., Soriano, N. A., Czech, M. P., and Guilherme, A. (2005). PTEN, but not SHIP2, suppresses insulin signaling through the phosphatidylinositol 3-kinase/Akt pathway in 3T3-L1 adipocytes. *J. Biol. Chem.* **280,** 22523–22529.

Tengholm, A., and Meyer, T. (2002). A PI3-kinase signaling code for insulin-triggered insertion of glucose transporters into the plasma membrane. *Curr. Biol.* **12,** 1871–1876.

Tengholm, A., Teruel, M. N., and Meyer, T. (2003). Single cell imaging of PI3K activity and glucose transporter insertion into the plasma membrane by dual color evanescent wave microscopy. *Sci. STKE* **2003,** PL4.

Tomida, T., Hirose, K., Takizawa, A., Shibasaki, F., and Iino, M. (2003). NFAT functions as a working memory of Ca^{2+} signals in decoding Ca^{2+} oscillation. *EMBO J.* **22,** 3825–3832.

Vanhaesebroeck, B., Leevers, S. J., Ahmadi, K., Timms, J., Katso, R., Driscoll, P. C., Woscholski, R., Parker, P. J., and Waterfield, M. D. (2001). Synthesis and function of 3-phosphorylated inositol lipids. *Annu. Rev. Biochem.* **70,** 535–602.

Van Schravendijk, C. F., Foriers, A., Van den Brande, J. L., and Pipeleers, D. G. (1987). Evidence for the presence of type I insulin-like growth factor receptors on rat pancreatic A and B cells. *Endocrinology* **121,** 1784–1788.

Varnai, P., and Balla, T. (2006). Live cell imaging of phosphoinositide dynamics with fluorescent protein domains. *Biochim. Biophys. Acta* **1761,** 957–967.

Varnai, P., Rother, K. I., and Balla, T. (1999). Phosphatidylinositol 3-kinase-dependent membrane association of the Bruton's tyrosine kinase pleckstrin homology domain visualized in single living cells. *J. Biol. Chem.* **274,** 10983–10989.

Velloso, L. A., Carneiro, E. M., Crepaldi, S. C., Boschero, A. C., and Saad, M. J. (1995). Glucose- and insulin-induced phosphorylation of the insulin receptor and its primary substrates IRS-1 and IRS-2 in rat pancreatic islets. *FEBS Lett.* **377,** 353–357.

Venkateswarlu, K., Oatey, P. B., Tavaré, J. M., and Cullen, P. J. (1998). Insulin-dependent translocation of ARNO to the plasma membrane of adipocytes requires phosphatidylinositol 3-kinase. *Curr. Biol.* **8,** 463–466.

Verspohl, E. J., and Ammon, H. P. (1980). Evidence for presence of insulin receptors in rat islets of Langerhans. *J. Clin. Invest.* **65,** 1230–1237.

Viard, P., Butcher, A. J., Halet, G., Davies, A., Nurnberg, B., Heblich, F., and Dolphin, A. C. (2004). PI3K promotes voltage-dependent calcium channel trafficking to the plasma membrane. *Nat. Neurosci.* **7,** 939–946.

Wada, T., Sasaoka, T., Funaki, M., Hori, H., Murakami, S., Ishiki, M., Haruta, T., Asano, T., Ogawa, W., Ishihara, H., and Kobayashi, M. (2001). Overexpression of SH2-containing inositol phosphatase 2 results in negative regulation of insulin-induced metabolic actions in 3T3-L1 adipocytes via its 5′-phosphatase catalytic activity. *Mol. Cell. Biol.* **21,** 1633–1646.

Wang, F., Herzmark, P., Weiner, O. D., Srinivasan, S., Servant, G., and Bourne, H. R. (2002). Lipid products of PI(3)Ks maintain persistent cell polarity and directed motility in neutrophils. *Nat. Cell Biol.* **4,** 513–518.

Watton, S. J., and Downward, J. (1999). Akt/PKB localisation and 3′ phosphoinositide generation at sites of epithelial cell-matrix and cell-cell interaction. *Curr. Biol.* **9,** 433–436.

Westerlund, J., Wolf, B. A., and Bergsten, P. (2002). Glucose-dependent promotion of insulin release from mouse pancreatic islets by the insulin-mimetic compound L-783,281. *Diabetes* **51**(Suppl. 1), S50–S52.

Wijesekara, N., Konrad, D., Eweida, M., Jefferies, C., Liadis, N., Giacca, A., Crackower, M., Suzuki, A., Mak, T. W., Kahn, C. R., Klip, A., and Woo, M. (2005).

Muscle-specific Pten deletion protects against insulin resistance and diabetes. *Mol. Cell. Biol.* **25,** 1135–1145.
Withers, D. J., Gutierrez, J. S., Towery, H., Burks, D. J., Ren, J. M., Previs, S., Zhang, Y., Bernal, D., Pons, S., Shulman, G. I., Bonner-Weir, S., and White, M. F. (1998). Disruption of IRS-2 causes type 2 diabetes in mice. *Nature* **391,** 900–904.
Wu, X., Senechal, K., Neshat, M. S., Whang, Y. E., and Sawyers, C. L. (1998). The PTEN/MMAC1 tumor suppressor phosphatase functions as a negative regulator of the phosphoinositide 3-kinase/Akt pathway. *Proc. Natl. Acad. Sci. USA* **95,** 15587–15591.
Xu, G. G., and Rothenberg, P. L. (1998). Insulin receptor signaling in the β-cell influences insulin gene expression and insulin content: Evidence for autocrine β-cell regulation. *Diabetes* **47,** 1243–1252.
Yip, S. C., Eddy, R. J., Branch, A. M., Pang, H., Wu, H., Yan, Y., Drees, B. E., Neilsen, P. O., Condeelis, J., and Backer, J. M. (2008). Quantification of PtdIns(3,4,5)P_3 dynamics in EGF-stimulated carcinoma cells: A comparison of PH-domain-mediated methods with immunological methods. *Biochem. J.* **411,** 441–448.
Yoshizaki, H., Ohba, Y., Kurokawa, K., Itoh, R. E., Nakamura, T., Mochizuki, N., Nagashima, K., and Matsuda, M. (2003). Activity of Rho-family GTPases during cell division as visualized with FRET-based probes. *J. Cell Biol.* **162,** 223–232.
Yu, W., O'Brien, L. E., Wang, F., Bourne, H., Mostov, K. E., and Zegers, M. M. (2003). Hepatocyte growth factor switches orientation of polarity and mode of movement during morphogenesis of multicellular epithelial structures. *Mol. Biol. Cell* **14,** 748–763.
Zawalich, W. S., and Zawalich, K. C. (2000). A link between insulin resistance and hyperinsulinemia: Inhibitors of phosphatidylinositol 3-kinase augment glucose-induced insulin secretion from islets of lean, but not obese, rats. *Endocrinology* **141,** 3287–3295.

CHAPTER TWELVE

Serine Kinases of Insulin Receptor Substrate Proteins

Sigalit Boura-Halfon *and* Yehiel Zick

Contents

Abstract

Signaling of insulin and insulin-like growth factor-I (IGF-1) at target tissues is essential for growth, development and for normal homeostasis of glucose, fat, and protein metabolism. Control over this process is therefore tightly regulated. It can be achieved by a negative-feedback control mechanism, whereby downstream components inhibit upstream elements along the insulin and IGF-1 signaling pathway or by signals from other pathways that inhibit insulin/IGF-1 signaling thus leading to insulin/IGF-1 resistance. Phosphorylation of insulin receptor substrates (IRS) proteins on serine residues has emerged as a key step in these control processes both under physiological and pathological

Department of Molecular Cell Biology, Weizmann Institute of Science, Rehovot 76100, Israel

Vitamins and Hormones, Volume 80
ISSN 0083-6729, DOI: 10.1016/S0083-6729(08)00612-2

conditions. The list of IRS kinases is growing rapidly, concomitant with the list of potential Ser/Thr phosphorylation sites in IRS proteins. Here we review a range of conditions that activate IRS kinases to phosphorylate IRS proteins on selected domains. The specificity of this reaction is discussed and its characteristic as an "array" phosphorylation is suggested. Finally, its implications on insulin/IGF-1 signaling, insulin/IGF-1 resistance and diabetes, an emerging epidemic of the twenty-first century are outlined.

I. Introduction

Insulin and IGF-1 resistance are states in which the sensitivity of target cells to respond to ordinary levels of insulin or IGF-1 is reduced. Insulin resistance plays a central role in the development of type-2 diabetes, an emerging epidemic of the twenty-first century. A variety of agents and conditions that induce insulin and IGF-1 resistance activate protein kinases which target elements along the insulin/IGF-1 signaling pathways. Some of these Ser/Thr kinases phosphorylate the insulin receptor substrate (IRS) proteins. Ser/Thr phosphorylation of IRS proteins inhibits their function, and interferes with insulin/IGF-1 signaling in a number of ways, thus leading to the development of an insulin/IGF-1 resistance state.

In this chapter we focus on the key molecular links between Ser/Thr phosphorylation of IRS proteins and the impairment in insulin and IGF-1 signal transduction. We outline the relations between inflammation; stress responses; the activation of IRS kinases; and the induction of insulin/IGF-1 resistance and propose directions for future studies in this field.

II. Insulin and IGF-1 Signaling

Insulin is a major anabolic hormone which promotes proper metabolism, energy balance, and maintenance of normal body weight (Taniguchi *et al.*, 2006). IGF-1 is a mediator of cell growth and differentiation (Baserga, 1999; Butler *et al.*, 1998) and has been implicated in the regulation of cell survival (Kulkarni, 2002; Withers *et al.*, 1999). Its roles include the stimulation of angiogenesis, promotion of re-epithelialization (Bitar and Labbad, 1996), and prevention of cytokine-mediated cell death (Pelosi *et al.*, 2007). These activities are attributed to the antiapoptotic functions of IGF-1 (Baserga, 1999; Butler *et al.*, 1998).

Insulin and IGF-1 actions are mediated by their respective receptors, the insulin and IGF-1 receptor (IR and IGF-1R). IGF-1R and its homologue, the insulin receptor (IR), are composed of two extracellular α-subunits and two transmembrane β-subunits linked by disulfide bonds. The α-subunits contain the ligand-binding domain while the transmembrane β-subunits function as Tyr-specific protein kinases (reviewed in LeRoith, 1995). Three structural regions have been defined within the intracellular part of the β-subunits of IR and IGF-1R. These include the juxtamembrane (JM), the kinase domain, and carboxy-terminal (CT) region (Ebina *et al.*, 1985; Ullrich *et al.*, 1985, 1986).

A. Adaptor proteins for the insulin and IGF-1 receptors

IR and IGF-1R engage with a similar set of intracellular adaptor proteins that mediate their biological activities. Both receptors phosphorylate APS, Shc, Gab-1, and members of the IRS protein family (IRS-1 to IRS-4) to further propagate their metabolic and growth-promoting functions (Saltiel and Kahn, 2001; Youngren, 2007). Still, several proteins are selectively phosphorylated by the insulin or the IGF-1 receptor kinases. IRK selectively phosphorylates Cbl (Ribon and Saltiel, 1997) while the IGF-1R kinase phosphorylates the Crk adaptor protein (Beitner *et al.*, 1996), 14-3-3, Syp, and GTPase-activating protein (GAP) that directly binds to the IGF-1R, but not the IR. While most adaptor proteins interact with the JM domain of the insulin and IGF-1 receptors 14-3-3, Syp, and p85 selectively interact with the CT region (Craparo *et al.*, 1997; Seely *et al.*, 1995).

B. Insulin receptor substrate proteins

IRS-1 and IRS-2 integrate many of the pleiotropic effects of insulin and IGF-1. Although IRS proteins are highly homologous, their relative role in mediating insulin and IGF-1 action is still not fully understood (White, 2002). IRS-1 and IRS-2 appear to function in a distinct manner in the regulation of glucose homeostasis and there is a dynamic relay between IRS-1 and IRS-2 in hepatic insulin signaling during fasting and feeding. IRS-2 mainly functions during fasting and immediately after refeeding, and IRS-1 functions primarily after refeeding (Kubota *et al.*, 2008).

Studies of gene disruption revealed that IRS-1-knockout mice exhibit generalized growth retardation, as well as insulin resistance in peripheral tissues and impaired glucose tolerance (Araki, 1994; Tamemoto, 1994). Still, IRS-2 apparently compensates for the absence of IRS-1 in hepatocytes of IRS-1 null mice and provides the major alternative pathway to phosphatidylinositol 3-kinase (PI3K) activation in skeletal muscle and adipocytes of these animals (Kaburagi *et al.*, 1997; Patti *et al.*, 1995; Yamauchi *et al.*, 1996). Similarly, upregulation of IRS-1 expression and increased IGF-1R signaling were observed in *IRS-2*$^{-/-}$ mice, indicating a compensatory mechanism to overcome IRS-2 deficiency (Freude *et al.*, 2008). IRS-2 null mice develop insulin resistance but have growth defects limited only to certain tissues such as pancreatic β-cells which lead to development of type-2 diabetes (Withers *et al.*, 1998). Similarly, the repression of IRS-2 gene by ATF3, a stress-inducible gene, contributes to pancreatic β-cell apoptosis (Li *et al.*, 2008a).

These findings suggest that IRS-1 and IRS-2 serve complementary, rather than redundant roles in insulin signaling, which could be attributed to selected structural differences, in addition to differences in their tissue distribution and subcellular localization (White, 2002). The role of IRS-3 and IRS-4 in insulin action still needs a better clarification as IRS-3-and IRS-4-knockout mice have only mild defects in growth and metabolism (Sesti *et al.*, 2001).

In terms of structure (Sun *et al.*, 1991, 1995) IRS-1 and IRS-2 contain a conserved pleckstrin homology (PH) domain, located at their amino-terminus, that serves to anchor the IRS proteins to membrane phosphoinositides and helps to localize the IRS proteins in close proximity to their receptors (Voliovitch *et al.*, 1995) (Fig. 12.1). This is evident by the reduced Tyr phosphorylation of IRS-1 whose PH domain has been deleted (Voliovitch *et al.*, 1995). In addition, the PH domain also promotes protein–protein interactions between IRS proteins and other cellular constituents. Such a PH domain-interacting protein, named PHIP, has been identified (Farhang-Fallah *et al.*, 2000, 2002). It selectively binds to the PH domains of IRS-1 and IRS-2 and regulates insulin receptor-stimulated GLUT4 translocation in skeletal muscle cells (Farhang-Fallah *et al.*, 2002). An isoform of PHIP (denoted PHIP1) which represents a novel WD40 repeat-containing protein stimulates IGF-1-dependent and -independent proliferation of β-cells, an event which correlates with transcriptional upregulation of the cyclin D2 promoter and the accumulation of cyclin

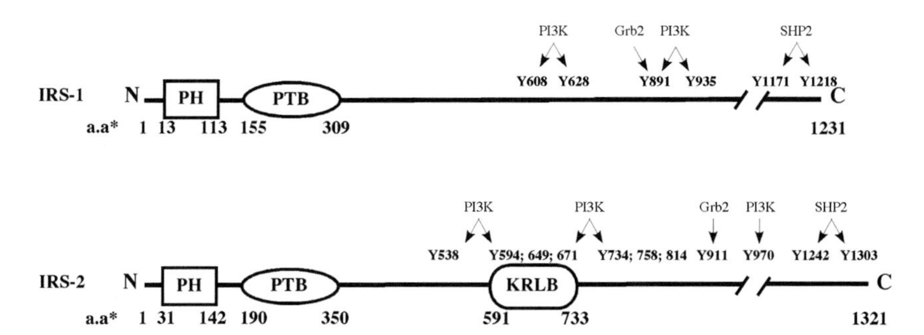

Figure 12.1 Schematic illustration of IRS-1 and IRS-2 structures. Domains organization and the position of tyrosine-phosphorylation sites are illustrated. Upon phosphorylation the latter are recruitment sites for the indicated downstream effectors (* mouse numbering).

D2 protein. These effects can be attributed to a specific role for PHIP1 in the enhancement of IRS-2-dependent signaling responses leading to β-cell growth (Podcheko *et al.*, 2007).

The PH domain of IRS proteins is flanked by a P-Tyr-binding (PTB) domain. The PTB domain, present in a number of signaling molecules (Pawson, 1995), shares 75% sequence identity between IRS-1 and IRS-2 (Sawka *et al.*, 1996) and functions as a binding site to the NPXY motif of the JM region of the insulin and IGF-1 receptors (Eck *et al.*, 1996; Wolf, 1995). The C-terminal region of IRS proteins is poorly conserved. It contains multiple Tyr phosphorylation motifs that serve as docking sites for SH2 domain-containing proteins like the p85α regulatory subunit of PI3K, GRB2, Nck, Fyn, Grb2 and the protein tyrosine phosphatase (PTP) SH-PTP2 that further propagate insulin and IGF-1 signals (Khan and Pessin, 2002; LeRoith and Zick, 2001; Saltiel and Pessin, 2002; Zick, 2001).

IRS proteins contain over seventy potential Ser/Thr phosphorylation sites for kinases like PKA, PKC, AMPK, and MAPK (reviewed in Boura-Halfon and Zick, 2008; Zick, 2001, 2004, 2005). In nontreated cells IRS-1 is strongly phosphorylated on Ser residues and weakly phosphorylated on Tyr residues. Following insulin stimulation, Tyr and Ser phosphorylation of IRS-1 is increased (Capecchi, 1989; Myers *et al.*, 1994). The migration of IRS-1 on SDS-PAGE at an electrophoretic mobility corresponding to 165–185 kDa, compared to its predicted molecular mass of 131 kDa, is consistent with its high basal level of Ser/Thr phosphorylation (Sun *et al.*, 1991). Similar considerations apply to IRS-2.

Phosphorylation of IRS proteins on Ser/Thr residues can induce the dissociation of IRS proteins from the insulin receptor (Liu *et al.*, 2001; Paz *et al.*, 1997); hinder Tyr phosphorylation sites of IRS proteins (Mothe and Van Obberghen, 1996); release the IRS proteins from intracellular complexes that maintain them in close proximity to the receptor (Tirosh *et al.*, 1999); or induce IRS proteins degradation (Pederson *et al.*, 2001). These multiple effects place the spotlight on the Ser/Thr kinases that phosphorylate the IRS protein and on the Ser/Thr sites that are getting phosphorylated.

C. Signaling pathways utilized by the insulin and IGF-1 receptors

Two major signaling pathways are propagated in response to activation of the insulin and IGF-1 receptor kinase: the PI3K and the MAP kinase pathways (Saltiel and Kahn, 2001). The MAP kinase cascade leads to enhanced cell growth, while the PI3K cascade is activated by the IRS proteins to trigger both, the metabolic and growth-promoting functions of insulin and IGF-1.

PI3K, one of the best characterized downstream effector of IRS proteins (Cantley, 2002), associates with Tyr-phosphorylated IRS proteins following insulin stimulation and catalyzes the formation of phosphatidylinositol 3,4,5-trisphosphate, which stimulates phosphoinositide-dependent kinase (PDK1) activity and initiates the activation of its downstream effectors: protein kinase B (PKB, Akt), mTOR, p70S6 kinase (S6K1), as well as the atypical isoforms of PKC (PKCζ/λ), leading to glucose transport, protein and glycogen synthesis (Bandyopadhyay *et al.*, 1997; Wang *et al.*, 1999). Insulin, but not the IGF-1 receptor also activates the Cbl/CAP signaling pathway (Saltiel and Kahn, 2001). The Cbl/CAP signaling cascade mediates glucose transport through activation of the GTP-binding protein TC10 and the recruitment of the CIP4/Gapex-5 complex to the plasma membrane (Lodhi *et al.*, 2007).

The IGF-1 signaling network, similar to the insulin signaling network, involves recruitment and activation of PI3K, PDK1, and Akt. Akt has several substrates such as BAD (Datta *et al.*, 1997; del Peso *et al.*, 1997) and MDM2 (Zhou *et al.*, 2001) that mediate the antiapoptotic action of IGF-1. In addition, the activated Akt translocates to the nucleus where it phosphorylates the forkhead transcription factors, resulting in their removal from the nucleus into the cytoplasm (Salih and Brunet, 2008).

A major function of IGF-1 and insulin is to regulate the output of mTOR signal transduction (Yang and Guan, 2007). The form of mTOR that directly regulates cell growth is TOR complex 1 (TORC1), which contains mTOR in complex with Raptor (regulatory-associated protein of mTOR), mLST8 (also known as GβL) and PRAS40 (proline-rich Akt substrate of 40 kDa, which represses mTOR activity) (Guertin and Sabatini, 2007; Wullschleger *et al.*, 2006). This complex is rapamycin sensitive. TORC2 is a complex of mTOR with GβL, mSIN, and Rictor and is rapamycin insensitive (Lee *et al.*, 2007). It appears that while the TORC2 is involved in cytoskeleton reorganization, TORC1 mediates cell growth regulation in response to insulin (Avruch *et al.*, 2006) and IGF-1 (Feng *et al.*, 2007; Hardie, 2008). Activated mTORC1 phosphorylates two main regulators of mRNA translation and ribosome biogenesis, S6K1 and 4E-BP1, and thus stimulates protein synthesis (Inoki and Guan, 2006).

At least two types of input switch on TORC1: increased availability of amino acids or stimulation by growth factors and hormones like insulin and IGF-1. Availability of amino acids activates small GTP-binding proteins named Rag (Sancak *et al.*, 2008). Rag proteins do not directly stimulate the kinase activity of mTORC1, but promote the intracellular localization of mTOR to a compartment that also contains its activator, the GTP-bound form of the small GTP-binding protein Rheb. Upstream of Rheb is the TSC1:TSC2 heterodimer. TSC2 contains a GTPase-activator protein (GAP) domain that converts Rheb to its inactive Rheb:GDP form. Akt activation in response to insulin or IGF-1 induces phosphorylation of TSC2 (Inoki *et al.*, 2005) and this is thought to inhibit its GAP activity, thus stimulating TORC1. However, Akt also phosphorylates PRAS40 to relieve its inhibitory effect (Sancak *et al.*, 2007). Thus, the Akt pathway may stimulate mTOR via two mechanisms (Hardie, 2008). The AMPK pathway exerts two inhibitory effects on mTOR via phosphorylation of TSC2 and Raptor. Phosphorylation of TSC2 by AMPK stimulates its Rheb–GAP activity (Inoki *et al.*, 2005), while phosphorylation of Raptor induces 14-3-3 binding to raptor and inhibition of mTORC1 activity (Gwinn *et al.*, 2008). RSK-mediated phosphorylation of Raptor positively regulates mTORC1 activity and thus suggests a means by which the Ras/MAPK pathway might promote rapamycin-sensitive signaling independently of the PI3K/Akt pathway (Carriere *et al.*, 2008).

III. Regulation of Insulin and IGF-1 Signaling: Role of Ser/Thr Phosphorylation of IRS Proteins

Control over insulin and IGF-1 signaling can be achieved in an auto regulatory manner, whereby downstream components inhibit upstream elements (negative-feedback control) (Gual *et al.*, 2003; Morino *et al.*, 2006). Alternatively, signals from apparently unrelated pathways can inhibit insulin and IGF-1 signaling. The IRS proteins that are common adaptors of both the insulin and IGF-1 receptors are targets for such feedback control mechanisms that take the form of Ser/Thr phosphorylation (Boura-Halfon and Zick, 2008; Gual *et al.*, 2003; Ishibashi *et al.*, 2001; Liu *et al.*, 2001; Morino *et al.*, 2006; Zick, 2004, 2005). We have previously shown (Herschkovitz *et al.*, 2007; Liu *et al.*, 2001, 2004; Paz, 1999; Paz *et al.*, 1997) that insulin-induced Ser/Thr phosphorylation of IRS proteins dissociates them from IR, prevents their Tyr phosphorylation and inhibits their interactions with downstream effectors. This serves as a physiological feedback control mechanism, utilized by insulin and IGF-1, to turn off their own signaling cascades. However, inducers of insulin and IGF-1 resistance, such as proinflammatory cytokines, take advantage of this mechanism. By inducing the phosphorylation IRS proteins on the same inhibitory Ser/Thr sites they lead to inhibition of IRS proteins function and inhibit the biological activities of insulin and IGF-1 (Zick, 2005). Indeed, increased Ser phosphorylation of IRS proteins has been observed as a result of hyperglycemia (Nakajima *et al.*, 2000); after treatment of cells with activators of PKC (De Fea and Roth, 1997; Griffin *et al.*, 1999; Kellerer *et al.*, 1998); or after incubation of cells with proinflammatory cytokines (Kanety *et al.*, 1995; Liu *et al.*, 2004; Paz *et al.*, 1997; Venters *et al.*, 1999). The above results implicate

IRS kinases, activated by proinflammatory cytokines and other inducers of insulin resistance as potential inhibitors of IRS proteins function, leading to the inhibition of insulin/IGF-1 actions (reviewed in Zick, 2003).

Most studies have focused on IRS-1 as a major target for IRS kinases (Draznin, 2006). However, it is now becoming evident that IRS-2 serve as a target as well (Gurevitch *et al.*, 2008; Sharfi and Eldar-Finkelman, 2007; Solinas *et al.*, 2006). IRS-2 function is also regulated by mechanisms distinct from Ser/Thr phosphorylation. For example, the kinase regulatory loop-binding (KRLB) region (aa 591–733), which is unique to IRS-2, serves as a negative regulatory element to control the extent of Tyr phosphorylation of IRS-2. Hence, inhibition exerted by KRLB domain attenuates insulin signaling and action independent of Ser/Thr phosphorylation (Sawka-Verhelle *et al.*, 1997; Wu *et al.*, 2008).

IRS kinases can be divided into two groups. One includes kinases which are mediators of insulin and IGF-1 signaling. These kinases negatively regulate IRS proteins upon prolong insulin stimulation [e.g., mTOR/S6K1 (Um *et al.*, 2004), MAPK (De Fea and Roth, 1997), and PKCζ (Lee *et al.*, 2008; Liu *et al.*, 2001; Ravichandran *et al.*, 2001b; Sommerfeld *et al.*, 2004)]. The other group consists of kinases that are activated along unrelated pathways to inhibit insulin/IGF-1 action [e.g., GSK-3β (Eldar-Finkelman and Krebs, 1997; Liberman and Eldar-Finkelman, 2005), IKKβ (Gao *et al.*, 2002), c-Jun N-terminus kinase (JNK) (Aguirre *et al.*, 2002; Lee *et al.*, 2003b), mPLK (Kim *et al.*, 2005), and AMPK (Tzatsos and Tsichlis, 2007)].

Of note, several IRS kinases (e.g., S6K1, PKC) are activated both, in response to insulin and inducers of insulin resistance (Morino *et al.*, 2006; Sampson and Cooper, 2006; Um *et al.*, 2006; Zick, 2001). This conclusion is based, for example, upon the fact that insulin as well as several inducers of insulin resistance, such as TNFα, activates PKCζ and its downstream target IKKβ (Lallena *et al.*, 1999), albeit through different mechanisms. Potential TNFα-mediated activation involves the induction of sphingomyelinase (Adam-Klages *et al.*, 1996) and production of ceramide which stimulates PKCζ activity (Muller *et al.*, 1995). Indeed, the inhibitory effects of TNFα on insulin-stimulated Tyr phosphorylation of IRS proteins are mimicked by sphingomyelinase and ceramide analogs (Kanety *et al.*, 1996; Paz *et al.*, 1997). Alternatively, TNFα can induce complex formation between PKCζ, p62, and RIP proteins that serve as adaptors of the TNF receptor and link PKCζ to TNFα signaling (Sanz *et al.*, 1999).

Most information related to the regulation of IRS proteins function is based on studies in cell cultures and in mouse models. Still, accumulating evidence of *in vivo* studies in humans supports the concept that increased Ser/Thr phosphorylation of IRS proteins might turn subjects prone to the development of insulin resistance (Morino *et al.*, 2006). The IRS kinases involved seem to be those already implicated as negative regulators of insulin signaling.

A. Insulin-stimulated Ser/Thr phosphorylation of IRS proteins

Ser/Thr phosphorylation of IRS proteins by insulin-stimulated Ser/Thr kinases serves as a physiological negative-feedback control mechanism that inhibits further Tyr phosphorylation of IRS proteins. The activity of insulin-stimulated IRS kinases is blocked by inhibitors of the PI3K pathway, implicating downstream effectors of PI3K as negative regulators of IRS protein function (Liu *et al.*, 2001; Paz, 1999).

1. PKCζ

A potential IRS kinase is PKCζ, a downstream effector of IRS proteins along the insulin signaling pathway (Standaert *et al.*, 1997). PKCζ is activated in response to insulin to mediate glucose uptake in adipocytes (Bandyopadhyay *et al.*, 2002) and skeletal muscle (Braiman *et al.*, 2001; Liu *et al.*, 2006), downstream of IRS-1 and PI3K (Liu *et al.*, 2006). Subsequent to its activation by insulin and its action as a mediator of insulin action, PKCζ engages in a negative-feedback regulatory process induced by insulin that involves phosphorylation of IRS proteins (Liu *et al.*, 2001; Ravichandran *et al.*, 2001a). This leads to the dissociation of the IR–IRS complexes (Liu *et al.*, 2001); inhibits the ability of IRS proteins to undergo insulin-stimulated Tyr phosphorylation, and as a result, terminates insulin signaling. IRS-1 serves as a substrate for PKCζ *in vitro*, and endogenous IRS-1 forms complexes with PKCζ in an insulin-dependent manner (Ravichandran *et al.*, 2001a).

The timeline of action of PKCζ is still unclear. It is conceivable to assume that PKCζ acts first on its target proteins along the glucose transport machinery to stimulate this process and promote insulin action, before it acts on IRS-1 to dissociate it from the receptor and thus terminate insulin action. The stimulatory roles of aPKCs in insulin action override their inhibitory actions. Support to this conclusion is provided by studies (Farese *et al.*, 2007) where muscle-specific knockout of PKCλ, a postulated mediator for insulin-stimulated glucose transport, was accompanied by systemic insulin resistance; impaired glucose tolerance and islet beta cell hyperplasia, while maintaining intact insulin signaling and actions in muscle, liver, and adipocytes of these mice. Still the molecular mechanism that coordinates these stimulatory/inhibitory functions of atypical PKCs remains to be explored.

2. IKKβ

The above findings suggest that PKCζ can function as an insulin-stimulated IRS kinase, although downstream effectors of PKCζ could also fulfill this role. A potential candidate effector of PKCζ is the Ser/Thr kinase IKKβ. IKKβ is part of the IKK complex that phosphorylates the inhibitor of NF-κB, IκB. This results in the degradation of IκB, allowing for the nuclear translocation and activation of the transcription factor NF-κB (Israel, 2000; Karin, 1999). IKKβ can bind PKCζ both *in vitro* and *in vivo*, serves as an *in vitro* substrate for PKCζ, and is activated by a functional PKCζ (Lallena *et al.*, 1999). IKKβ is activated by insulin in Fao rat hepatoma cells; furthermore, insulin-stimulated Ser phosphorylation of IRS-1 is inhibited by salicylates, implicating IKKβ as an insulin-stimulated IRS kinase (Y. Zick *et al.*, unpublished data). This conclusion is supported by findings that IRS-1 is a direct substrate for IKKβ, activated by stress inducers that phosphorylate IRS-1 on Ser312 (the human homologue of mouse Ser307) (Gao *et al.*, 2002).

3. mTOR and S6K1

A potential candidate to function as an insulin-stimulated IRS kinase downstream of PI3K is the mammalian target of rapamycin (mTOR), which acts as an integrator of nutrient and insulin signals (Fingar and Blenis, 2004; Manning, 2004). mTOR enhances phosphorylation of Ser residues at the COOH-terminus of IRS-1. This phosphorylation inhibits insulin-stimulated Tyr phosphorylation of IRS-1 and its ability to bind PI3K (Li *et al.*, 1999).

Involvement of downstream effectors of PI3K in the negative regulation of insulin signaling gained a strong support in a study implicating S6K1, an effector of PI3K and mTOR, as an IRS kinase. In doing so, S6K1 takes part in the physiological negative-feedback control mechanism induced by insulin to shut off its own action. This pivotal role played by S6K1 is illustrated by the fact that $S6K1^{-/-}$ mice placed on a high-fat diet (HFD) fail to fully autophosphorylate and activate their insulin receptor, still they remain insulin sensitive suggesting that absence of S6K1 facilitates insulin signaling downstream of the insulin receptor (Um *et al.*, 2004). Indeed, whereas wild-type mice placed on HFD are insulin resistant, having increased levels of circulating insulin and a reduced capacity to activate the insulin receptor and its downstream effectors, $S6K1^{-/-}$ mice maintain their capacity to activate downstream effectors like PKB. This suggests that S6K1 elicits a selective inhibitory effect on PKB activation at a point downstream of the insulin receptor. The target of S6K1 seems to be the IRS proteins. Indeed, S6K1 can directly phosphorylate IRS-1 *in vitro* (Harrington *et al.*, 2004) whereas lowering S6K1 expression by siRNA reduces phosphorylation of IRS-1 on selected Ser residues and potentiates insulin-induced PKB phosphorylation (Um *et al.*, 2004). Furthermore, expression of siRNA to S6K1 restores the expression of IRS-1 in $TSC^{-/-}$ mice that is otherwise dampened by S6K1 in these animals (Harrington *et al.*, 2004; Shah *et al.*, 2004). Hence, S6K1 appears to function as an IRS kinase and a central negative regulator of insulin action along the insulin signaling pathway.

4. JNK

JNK, a member of the MAP kinase family of protein kinases and a potential IRS kinase, is activated by insulin by as yet unknown mechanism (Moxham *et al.*, 1996). Recent studies have shown that JNK associates with IRS-1 and phosphorylates it at Ser307 in an insulin-dependent manner (Aguirre *et al.*, 2000, 2002; Lee *et al.*, 2003a). Insulin stimulation of JNK activity requires PI3K and Grb2 signaling. Direct binding of JNK to IRS-1 is not required for JNK activation by insulin, however direct interactions between JNK and IRS-1, which has a JNK-binding motif (Davis, 2000), are required for Ser307 phosphorylation (Lee *et al.*, 2003a). In that respect, JNK resembles IKKβ because both proteins form stable complexes with IRS proteins, and phosphorylate them on Ser307, a process that serves as negative-back control mechanism to terminate insulin signaling.

PKB, mTOR, PKCζ, and IKKβ are downstream effectors of PI3K along the insulin signaling pathway. This suggests that their action should be orchestrated to allow phosphorylation by PKB and sustained activation of IRS-1, prior to the activation of mTOR or PKCζ, the actions of which are expected to terminate insulin signal transduction. Of note, the negative-feedback control mechanism induced by PKCζ (or mTOR) includes a self-attenuation mode, whereby PI3K-mediated activation of mTOR and PKCζ inhibits IRS-1 function; thereby inhibiting further activation of mTOR and PKCζ themselves (Liu *et al.*, 2001; Zick, 2001).

5. Positive regulation of IRS-1 by insulin-mediated Ser phosphorylation

Although Ser phosphorylation mainly inhibits IRS protein function as discussed above, it might play positive roles as well. The first indication linking Ser phosphorylation with improved IRS-1 function (e.g., increased Tyr phosphorylation) was observed when an IRS-1 mutant lacking four potential PKB phosphorylation sites (Ser265, 302, 325, and 358) markedly enhanced the rate of Tyr dephosphorylation implicating one or more of these sites as positive regulator of IRS-1 function (Paz, 1999). In agreement with these findings, Ser302 phosphorylation was shown to be a positive mediator of nutrient availability that promotes mitogenesis and cell growth (Giraud *et al.*, 2004). Using mass spectrometry techniques, Luo *et al.* (2005) could demonstrate that phosphorylation of Ser1223 or Ser629 (human numbering) resulted in increased IRS-1 functions in respond to insulin. Two distinct mechanisms were suggested. Phosphorylation of Ser1223 could reduce the association of IRS-1 with SHP-2, a Tyr phosphatase, thereby increasing IRS-1 Tyr phosphorylation. Phosphorylation of Ser629 was proposed to attenuate the phosphorylation of a second Ser site, Ser636, which was implicated as a negative regulator of insulin signaling and thereby phosphorylation of Ser629 enhanced insulin signaling (Luo *et al.*, 2007).

These findings suggest that upon acute insulin stimulation, IRS-1 is phosphorylated on Tyr residues to propagate insulin signaling, a reaction accompanied by its phosphorylation on "stimulatory" Ser residues which serve as "guardians" of the phosphorylated Tyr residues by inhibiting Tyr phosphatases and/or preventing immediate phosphorylation at "inhibitory" Ser sites. Triggering of the negative-feedback loop and phosphorylation of IRS-1 on "inhibitory" Ser sites then commences with a delayed onset.

B. IRS kinases stimulated by cellular stress and inducers of insulin resistance

Insulin resistance is defined as failure of ordinary levels of insulin to trigger its downstream metabolic actions and is closely associated with obesity and the development of type-2 diabetes (Kahn *et al.*, 2006). Obesity induces insulin resistance because it promotes a state of chronic low-grade inflammation (Hotamisligil *et al.*, 1993; Shoelson *et al.*, 2006; Wellen and Hotamisligil, 2005). This is attributed to release from adipose tissues of free fatty acids (FFAs), hormones (e.g., leptin, adiponectin, ET-1), proinflammatory cytokines (e.g., TNFα, IL-1β, IL-6), and additional products of macrophages that populate adipose tissue in obesity (Fain *et al.*, 2004; Shoelson *et al.*, 2006; Wellen and Hotamisligil, 2005). Increased release of FFAs decreases insulin-mediated glucose transport in skeletal muscle and impairs suppression of glucose production by the liver, a characteristic of insulin resistance (Anderwald and Roden, 2004; Boden and Shulman, 2002; Hirabara *et al.*, 2007). Similarly, proinflammatory cytokines such as TNFα, IL-1β, and IL-6 impair insulin action and promote diabetes (de Luca and Olefsky, 2008).

Several inducers of insulin resistance activate IRS kinases that negatively regulate insulin signaling and action. The list of IRS kinases implicated in the development of insulin resistance is growing rapidly, concomitant with the list of potential Ser/Thr phosphorylation sites in IRS proteins (Draznin, 2006; Zick, 2005).

1. mTOR/S6K1

As already mentioned, mTOR and S6K1 which act as insulin-activated IRS kinases are also activated by inducers of insulin resistance. These observations lead to the obvious question whether inhibition of these kinases improves insulin sensitivity in insulin resistance states. In a short-term *in vivo* study insulin resistance was introduced by infusion of a high concentration of amino acids in humans. Under these conditions, rapamycin, an mTOR inhibitor, improved insulin actions (Krebs *et al.*, 2007). On the contrary, long-term treatment with rapamycin increased rather than decreased the insulin resistance of *Psammomys obesus*, that was induced by high-energy diet (Fraenkel *et al.*, 2008). This could be attributed to the increased activity of stress-response kinases in muscle and islets, subjected to sustained inhibition of mTOR activity. Hence, either sustained activation of the mTOR/S6K1 pathways or their inhibition could result in insulin resistance albeit through different mechanisms.

2. PKCζ and IKKβ

The ability of TNFα to stimulate the activity of PKCζ has already been mentioned (cf. Chapter 3). Another example is IKKβ, an IRS kinase the activity of which is stimulated by FFA, proinflammatory cytokines and other inducers of insulin resistance. IKKβ inhibition by high doses of salicylates or by a reduction in IKKβ gene dose reverses obesity- and diet-induced insulin resistance in animal models (Kim *et al.*, 2001; Yuan *et al.*, 2001), and improves glucose metabolism in type-2 diabetic subjects (Hundal *et al.*, 2002).

Heterozygous deletion of IKKβ ($IKK\beta^{+/-}$) protects against the development of insulin resistance during HFD and in obese $Lep^{ob/ob}$ mice (Kim *et al.*, 2001; Yuan *et al.*, 2001). These findings support a pivotal role for IKKβ in the induction of insulin resistance and diabetes. IKKβ exerts its effects both globally (systemic) and locally (in selected tissues). Mice whose IKKβ was selectively deleted from their myeloid cells preserve their whole-body insulin sensitivity and are protected from insulin resistance induced by HFD (Arkan *et al.*, 2005). Similarly, hepatic expression of the IκBα super-repressor reverses the type-2 diabetes phenotype induced by low-level activation of NF-κB (Cai *et al.*, 2005). In contrast, selective loss of IKKβ in hepatocytes retains insulin sensitivity in liver but not in muscle or fat tissues in response to HFD (Yuan *et al.*, 2001). The reason why expression of IκBα super-repressor but not deletion of IKKβ in hepatocytes, reverses systemic insulin sensitivity of mice on HFD is not clear but could be attributed to different strategies employed to attenuate IKKβ signaling. Nonetheless, IKKβ seems to have a central role in hepatic insulin resistance (Yuan *et al.*, 2001) and in the development of systemic insulin resistance (Cai *et al.*, 2005).

The mechanism by which FFAs activate IKKβ presumably involves an increase in FFA-derived metabolites, such as diacylglycerol and ceramide, which are potent activators of PKCθ (Schmitz-Peiffer *et al.*, 1997) and PKCζ (Muller *et al.*, 1995), both known to activate IKKβ. At the molecular level, inhibition of IKKβ prevents Ser/Thr phosphorylation of IRS proteins induced by HFD, TNFα, or phosphatase inhibitors, whereas it improves insulin-stimulated Tyr phosphorylation of IRS proteins. Hence, IKKβ might serve as a point of convergence, where Ser kinases downstream of insulin signaling and Ser kinases activated by proinflammatory cytokines activate IKKβ to inhibit insulin signaling both under physiological and pathological conditions (Fig. 12.2).

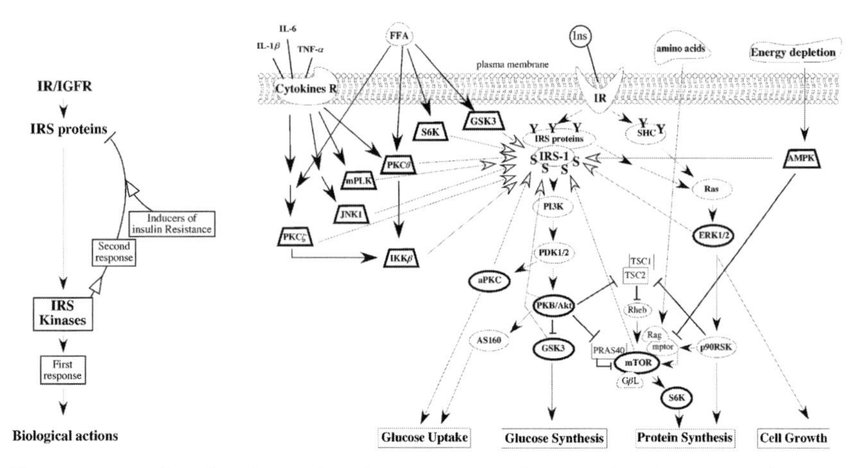

Figure 12.2 Phosphorylation-based negative regulation of IRS proteins by IRS kinases activated by insulin or inducers of insulin resistance. A negative-feedback control mechanism induced by insulin or unrelated pathways is presented schematically (left) or in detail (right). Two major insulin signaling pathways, the MAP kinase and the PI3K cascades, are illustrated. Insulin stimulation activates many IRS kinases (bold ovals) to exert insulin biological actions (filled arrow heads) and subsequently to negatively regulate insulin signaling (open arrow heads) by the phosphorylation of IRS proteins (represented here by IRS-1). Similarly, unrelated pathways activated, for example, upon energy depletion or by inducers of insulin resistance such as cytokines, amino acids, and FFAs interfere with insulin signaling by the activation of IRS kinases (bold trapeze, gray arrows) to phosphorylate the IRS proteins or other intermediate effectors (open arrow head).

3. ERK

A number of MAP kinases were implicated as IRS kinases. A twofold increase in the basal phosphorylation of IRS-1 was observed in muscle biopsies from patients with type-2 diabetes, concomitantly with higher basal activity of ERK1/2. Defects in insulin signaling were evident including reduced PI3K activity, decreased association of PI3K with IRS-1 and reduced Tyr phosphorylation of IRS-1 during insulin stimulation. Inhibition of ERK1/2 by a specific inhibitor inhibited IRS-1 phosphorylation thus implicating ERK1/2 as an IRS kinase associated with the development of insulin resistance and type-2 diabetes (Bouzakri *et al.*, 2003).

4. JNK

JNK is activated by proinflammatory cytokines such as TNFα, and its activity is abnormally elevated in obesity. Conversely, an absence of JNK1 results in decreased adiposity, significantly improved insulin sensitivity and enhanced insulin receptor signaling capacity in two models of mouse obesity (Hirosumi *et al.*, 2002). The inhibitory effects of JNK on insulin signaling can be attributed, at least partially, to its function as an IRS kinase (*vide supra*) (Aguirre *et al.*, 2000, 2002) that phosphorylates Ser307 to uncouple IRS-1 from the insulin receptor.

Palmitic acid induces JNK activation in pancreatic beta cells, resulting in the inhibition of pivotal gene transcription, including insulin. This inhibition results partly due to phosphorylation of IRS-2 at Thr347, which was implicated as a potential inhibitory Thr site (Solinas *et al.*, 2006). Similarly, sequential *in vitro* phosphorylation of IRS-2 on Ser484 and 488 by GSK-3 and JNK, respectively, was suggested to promote hepatic insulin resistance (Sharfi and Eldar-Finkelman, 2007). Hence, both IRS-1 and IRS-2 serve as targets to JNK.

5. PKCθ

Inducers of insulin resistance also activate PKCθ, a novel-type PKC. PKCθ is activated upon increased content of intramuscular long-chain fatty acyl-CoA. An increase in PKCθ activity occurs concomitantly with a decrease in insulin-stimulated Tyr phosphorylation of IRS-1 and a reduction in glucose transport (Griffin *et al.*, 1999; Yu *et al.*, 2002a,b). PKCθ-deficient mice are protected against fat-induced defects in insulin signaling in skeletal muscle, further supporting the role of PKCθ in mediating fatty acid-induced insulin resistance (Kim *et al.*, 2004). The notion that PKCθ plays a role in fatty acid-mediated insulin resistance was challenged by the findings that transgenic mice with muscle-specific expression of a dominant-negative PKCθ have shown age- and obesity-associated glucose intolerance implicating PKCθ as a protective rather than a negative regulator of insulin function (Serra *et al.*, 2003). These discrepant findings may be due to the different technical approaches that were used. For example, a dominant-negative PKCθ could bind (and titrate out) proteins necessary for the proper function of other PKC isoforms. At the molecular level, PKCθ, a known activator of IKKβ and JNK (Kim *et al.*, 2004; Sun *et al.*, 2000), may attenuate insulin signaling directly or via the activation of these IRS kinases. Ser1101 (Li *et al.*, 2004) and Ser307 (Yu *et al.*, 2002a) were suggested as potential target sites of PKCθ in IRS-1.

6. GSK-3

Glycogen synthase kinase-3 (GSK-3) is another kinase, the activity of which is stimulated by inducers of insulin resistance (Cohen and Frame, 2001; Henriksen and Dokken, 2006) and is elevated in diabetic tissues (Eldar-Finkelman *et al.*, 1999). GSK-3α KO mice display enhanced glucose and insulin sensitivity accompanied by reduced fat mass. Hepatic insulin signaling in these mice is increased and so is IRS-1 expression (MacAulay *et al.*, 2007). Similarly, reduction in GSK-3α in human muscle cells increased insulin-stimulated glucose uptake, glycogen synthase activity and IRS-1 expression (Ciaraldi *et al.*, 2007). GSK-3 activity is inhibited upon its phosphorylation by insulin-stimulated PKB (Cross *et al.*, 1995), still insulin-mediated phosphorylation of glycogen synthase were similar in skeletal muscle from both WT and GSK-3α KO mice which may indicate different roles for GSK-3α in mice versus human muscles (Ciaraldi *et al.*, 2007; MacAulay *et al.*, 2007). McManus *et al.* (2005) have generated homozygous knockin mice of constitutive active (CA) GSK-3α, GSK-3β, or GSK-3αβ. Although GSK-3β had a major role in regulating glycogen synthase activity in muscle of CA GSK-3β knockin mice, these animals were not diabetic and their insulin-stimulated PKB activation and glucose uptake were not changed. These findings cast a certain doubt as to the critical role of GSK-3β in the attenuation of insulin signaling (e.g., acting as an IRS kinase) at least in mouse models, although constitutive activation of GSK-3β under normal conditions of regular diet and Low BMI might be necessary but insufficient to promote systemic insulin resistance.

7. AMPK

Members of the AMPK family appear to function as IRS kinases. AMPK which is activated upon energy depletion (glucose deprivation, hypoxia) functions as an IRS kinase. Glucose deprivation, hypoxia, and inhibition of ATP synthesis in the mitochondria stimulate phosphorylation of IRS-1 at Ser794 in a LKB1/AMPK-dependent manner, whereas oxidative stress and 2-deoxyglucose stimulates phosphorylation at this site by a Ca^{2+}/calmodulin-dependent protein kinase beta/AMPK axis (Tzatsos and Tsichlis, 2007). The salt-inducible kinase-2 SIK2, a member of the AMPK family (Horike *et al.*, 2003), phosphorylates IRS-1 at Ser789 (Qiao *et al.*, 2002) at the C-terminal end of IRS-1 and in such a way inhibits IRS-1 function.

IV. The Consequences of Ser Phosphorylation of IRS Proteins

Ser/Thr phosphorylation catalyzed by IRS kinases, similar to other posttranslational modifications, is expected to alter the structure and thereby the function of the IRS proteins. This section outlines the functional consequences of Ser/Thr phosphorylation at different regions of the IRS proteins. As already mentioned, IRS proteins share a similar structure characterized by the presence of a PH domain adjacent to a PTB domain, followed by a variable-length COOH-terminus tail that contains a number of Tyr and Ser phosphorylation sites.

To unveil the importance of phosphorylated Ser/Thr residues of human IRS-1, Yi *et al.* (2007) adopted a mass spectrometry approach. Although this system may suffer some drawbacks, a number of potential Ser/Thr residues involved in the regulation of IRS proteins function could be identified. More than twenty Ser residues of IRS-1 were found to undergo insulin-stimulated phosphorylation in these human muscle biopsies, three of which were newly identified sites: Thr495, Ser527, and Ser1005 (human numbering). These findings validate previous *in vitro* and *in vivo* studies in animal models and suggest that the same strategy could be employed to identify phosphorylated Ser/Thr sites under conditions of insulin resistance, obesity or type-2 diabetes.

A. Ser/Thr phosphorylation at the PH domain

IRS kinases, triggered by inducers of insulin resistance, can phosphorylate sites located within the PH domain of IRS proteins. An example is the mouse Pelle-like kinase (mPLK, homolog of human IL-1 receptor-associated kinase) (Kim *et al.*, 2005). Overexpression of mPLK impairs insulin-stimulated Tyr phosphorylation of IRS-1 and its association with p85α by inducing the phosphorylation of Ser24 (Kim *et al.*, 2005). This Ser residue, located within the PH domain, seems to be critical for IR–IRS-1 complex formation (Farhang-Fallah *et al.*, 2002). Interestingly, conventional and novel PKCs can also phosphorylate this site. Ser24 is a potential phosphorylation site for PKCα (Nawaratne *et al.*, 2006). Accordingly, knockout of PKCα enhances insulin signaling (Leitges *et al.*, 2002). Similarly, insulin-stimulated phosphorylation of Ser24, apparently by PKCδ, diminishes the ability of IRS-1 to bind phosphatidylinositol 4,5-bisphosphate (PIP_2) further supporting the hypothesis that Ser24 is a negative regulatory phosphorylation site in IRS-1 (Greene *et al.*, 2006). Using mass spectrometry analysis, Ser67 (rat numbering), located at the PH domain of IRS-1, was shown to be phosphorylated upon prolonged insulin stimulation. This site as well may be considered as an inhibitory site (Giraud *et al.*, 2007). Consequently, Ser phosphorylation within the PH domain of IRS proteins could account for the development of an insulin resistance state.

B. Ser/Thr phosphorylations at the PTB domain

Protein–protein interactions depend upon spatial matching. Therefore, it is becoming apparent that Ser/Thr phosphorylation of IRS proteins in close proximity to their receptor-binding PTB region affects their interactions with the insulin and IGF-1 receptors. We could show that mutation of seven Ser sites located within or in close proximity to the PTB domain of IRS-1 (Ser265, 302, 325, and 358, S336, 407, and 408) protects it from auto regulatory desensitization induced by insulin and by IRS kinases triggered by inducers of insulin resistance (Herschkovitz *et al.*, 2007; Liu *et al.*, 2004).

Ser307 is located in close proximity to the PTB domain of IRS-1 and is a major "inhibitory" Ser site. Its phosphorylation interferes with the interaction of IR with IRS-1 thus preventing Tyr phosphorylation of IRS-1 (Aguirre *et al.*, 2000). "Novel" PKC isoforms, such as PKCθ, were implicated as mediators of FFA-induced insulin resistance in skeletal muscle, in a process that involves phosphorylation of IRS-1 at Ser307 (Yu *et al.*, 2002b). Ser307 is also phosphorylated by JNK in response to TNFα (Aguirre *et al.*, 2000, 2002) and to palmitic acid. JNK activation in pancreatic β-cells results in the inhibition of gene transcription, including the insulin gene (Solinas *et al.*, 2006). This inhibition is attributed in part to phosphorylation of IRS-1 at Ser307 and of IRS-2 at Thr347 (Solinas *et al.*, 2006). Other kinases phosphorylate Ser307. These include S6K1 (Harrington *et al.*, 2004) and IKKβ (de Alvaro *et al.*, 2004; Gao *et al.*, 2002; Nakamori *et al.*, 2006). In contrast, a reduction in IKKβ levels using specific siRNA fails to prevent TNFα-mediated IRS-1 phosphorylation on Ser312 (human equivalent of mouse Ser307) in primary human skeletal muscle (Austin *et al.*, 2008). This apparent discrepancy can be attributed to the fact that Ser307 is subjected to phosphorylation by a number of IRS kinases in addition to IKKβ. Furthermore, IKKβ itself can phosphorylate sites different from Ser307. For example, mutation of seven Ser sites in IRS-1 (Ser265, 302, 325, 336, 358, 407, and 408) different from Ser307, confers protection from the action of IKKβ implicating additional Ser residue as potential IKKβ-mediated phosphorylation sites (Herschkovitz *et al.*, 2007).

Ser318 of IRS-1 is a potential target for PKCζ, JNK, and kinases along the PI3K/mTOR pathway (Moeschel *et al.*, 2004; Mussig *et al.*, 2005). It is located in close proximity to the PTB domain; therefore, its phosphorylation presumably disrupts the interaction between IR and IRS-1. Phosphorylation of Ser318 is not restricted to insulin stimulation. Elevated plasma levels of leptin, an adipokine produced by adipocytes (Argiles *et al.*, 2005), also stimulates the phosphorylation of Ser318. This downregulates insulin-stimulated Tyr phosphorylation of IRS-1 and glucose uptake (Hennige *et al.*, 2006). A nearby Ser332 of IRS-1 is a potential GSK-3 phosphorylation site with Ser336 being the priming site (Liberman and Eldar-Finkelman, 2005).

Ser/Thr phosphorylation in close proximity to the PTB domain similarly inhibits IRS-2 function. We have recently mutated into Ala, five potential inhibitory Ser sites located proximal to the PTB domain of IRS-2 (Ser303, 343, 362, 381, and 480). Cultured mouse islets expressing the mutated IRS-2 protein, denoted IRS-2^{5A}, but not islets expressing IRS-2^{WT}, were resistant to apoptosis; inhibition of *PDX1* gene transcription and inhibition of glucose-stimulated insulin secretion induced by proinflammatory cytokines. Furthermore, islets expressing IRS-2^{5A} transplanted into islet-deficient diabetic mice, restored the ability of these mice to respond to a glucose load similar to naïve mice, unlike islets expressing IRS-2^{WT} that failed to do so. Thus, elimination of a physiological negative-feedback control mechanism along the insulin signaling pathway that involves Ser/Thr phosphorylation of IRS-2, affords protection against the adverse effects of proinflammatory cytokines and improves β-cell function under stress. IRS-2^{5A} expression in pancreatic islets could therefore be considered a rational treatment against β-cell failure (Gurevitch *et al.*, 2008).

C. Ser/Thr phosphorylation at the C-terminus tail of IRS proteins

Many of the Tyr-phosphorylated consensus motifs (YMXM or YXXM) of IRS proteins are located at their C-terminus tail. Consequently Ser phosphorylation within this area could interrupt with the binding of downstream effectors of the IRS proteins. Indeed, Ser570 of IRS-1, located in the vicinity of the PI3K interaction motif, was shown to be a potential PKCζ phosphorylation site that upon phosphorylation disrupts the IRS-1–p85α complex (Farhang-Fallah *et al.*, 2002). IRS-3 and IRS-4 but not IRS-2 are also substrates for PKCζ at this site (Lee *et al.*, 2008). Potential ERK phosphorylation sites, Ser612, 632, 662, and 731 in IRS-1, located next to Tyr phosphorylation YMXM motifs were shown to be negative regulators for PI3K activity associated with IRS-1 (Mothe and Van Obberghen, 1996). Increased phosphorylation of Ser636 of IRS-1 was observed in myotubes of patients with type-2 diabetes. Inhibition of ERK1/2 with PD98059 reduced this phosphorylation, thereby implicating ERK1/2 in the phosphorylation of Ser636 in human muscle (Bouzakri *et al.*, 2003).

In a recent study, using skeletal muscle biopsies from 11 humans, the mTOR/S6K pathway was shown to negatively modulate glucose metabolism under nutrient abundance (Krebs *et al.*, 2007). In agreement with previous studies, phosphorylation of Ser312 and 636 of IRS-1 was implicated as part of this negative regulation (Krebs *et al.*, 2007; Tremblay *et al.*, 2007).

Ser residues at the C-terminal end of IRS-1 are also targets for phosphorylation by S6K1. S6K1-deficient mice remain sensitive to insulin owing to apparent loss of a negative-feedback loop from S6K1 to IRS-1, which blunts Ser307 and Ser636/639 phosphorylation; sites involved in insulin resistance (Um *et al.*, 2004). Phosphorylation of Ser636/639 mediated by S6K1 inhibits SH2-containing downstream effectors from docking and binding to the P-Tyr residues of IRS-1 located within the YXXM/YMXM motifs at this C-terminal region. Ser1101 was identified as another S6K1 site in IRS-1, the phosphorylation of which is increased upon nutrient overload and obese setting. Phosphorylation of Ser1101 was increased in liver of obese (*db/db*) or wild-type but not of $S6K1^{-/-}$ mice maintained on HFD, implicating S6K1 as the kinase involved (Tremblay *et al.*, 2007).

Additional Ser phosphorylation site at the C-terminus of IRS-1 is Ser789. AMPK phosphorylates IRS-1 on Ser794 (human equivalent of mouse Ser789) which inhibits PI3K/Akt signaling (Tzatsos and Tsichlis,

2007). Ser789 is also a target for the salt-inducible kinase 2 (SIK2, a novel member of the AMPK family) in adipose cells (Horike *et al.*, 2003). The activity and content of SIK2 are elevated in white adipose tissue of *db/db* diabetic mice, suggesting that overexpression of SIK2 induces phosphorylation of Ser789 of IRS-1 and as a result negatively regulates insulin signal transduction (Horike *et al.*, 2003). SIK2 may also function as an AMPK to turn off lipogenesis in low-energy states (Du *et al.*, 2008). Qiao *et al.* (2002), however, claim that unknown Ser/Thr kinase rather than AMPK phosphorylates IRS-1 on Ser789 in insulin-resistant animal model. In contrast, phosphorylation of this site was shown to promote insulin signaling (Jakobsen *et al.*, 2001) in muscle C2C12 cell line. The latter study involved the use of AICAR (5-aminoimidazole-4-carboxamide riboside), a selective inducer of AMPK, and showed correlation between Ser789 phosphorylation of IRS-1 and activation of PI3K. Still, we cannot rule out the possibility that AMPK activates IRS-1 by a different mechanism while the phosphorylation of Ser789 reflects a delayed inhibitory response.

D. Ser/Thr phosphorylation and degradation of IRS proteins

Whereas Ser phosphorylation of IRS proteins is mainly utilized as a short-term mechanism to inhibit insulin signaling, regulated degradation of IRS proteins promotes long-term insulin resistance. Prolonged insulin stimulation substantially reduces IRS-1 and IRS-2 protein levels in multiple cell lines, a process which is blocked by specific inhibitors of the 26S proteasome (Sun *et al.*, 1999). These results suggest that proteasome-mediated degradation, rather than inhibition of transcription and/or translation, determines the cellular content of IRS proteins. In line with this reasoning, the activity of the ubiquitin/proteasome system is elevated in diabetes (Merforth *et al.*, 1999).

Selective inhibitors of PI3K and mTOR have been shown in some, but not all, systems to inhibit insulin-stimulated IRS-1 degradation, consistent with the hypothesis that the degradation of IRS-1 is initiated by Ser/Thr phosphorylation via the PI3K/Akt/mTOR signaling pathway (Egawa *et al.*, 2000; Haruta *et al.*, 2000; Hiratani *et al.*, 2005; Pederson *et al.*, 2001; Smith *et al.*, 1995). Indeed, expression of a constitutively active, membrane-targeted PI3K, $p110^{CAAX}$, induces hyper-Ser/Thr phosphorylation and degradation of IRS-1 (Haruta *et al.*, 2000). Although the nature of the kinases that mediate degradation of IRS protein has not been yet established, phosphorylation of Ser307, in close proximity to the PTB domain, correlates with IRS-1 degradation (Greene *et al.*, 2003). The N-terminal region of IRS-1 including the PH and PTB domains was identified as essential for targeting IRS-1 to the ubiquitin–proteasome degradation pathway (Zhande *et al.*, 2002). Furthermore, insulin-stimulated degradation of a mutant IRS-1 in which Ser307 is replaced with Ala is decreased compared with the wild-type IRS-1 (Greene *et al.*, 2003). These findings suggest that phosphorylation at Ser307 has a dual mechanism of modulating insulin action: uncoupling the interaction of IRS-1 with the insulin receptor as well as targeting of IRS-1 to the degradation pathway (Greene *et al.*, 2003). Phosphorylation of Ser307 might be necessary, but it is insufficient to promote IRS-1 degradation because elimination of an entire Ser/Thr-rich domain of IRS-1, proximal to its PTB domain protects IRS-1 from chronic insulin-induced degradation, although Ser307 remains intact in this deletion mutant (Boura-Halfon *et al.*, 2008). Ser612 is another site whose phosphorylation induces IRS-1 degradation following activation of GRK2 upon chronic endothelin-1 treatment (Ishibashi *et al.*, 2001; Usui *et al.*, 2005).

Other mechanisms exist to induce the degradation of IRS proteins. Many proinflammatory cytokines and inducers of insulin resistance, upregulate suppressors of cytokine signaling (SOCS) proteins. SOCS proteins include eight isoforms that contain an NH2-terminal SH2 domain and a COOH-terminal SOCS box (Yasukawa *et al.*, 2000). SOCS proteins bind via their SH2 domains to activated cytokine receptors or their associated Janus kinases as part of a negative-feedback loop to attenuate cytokine signaling (Naka *et al.*, 2001). They also bind to the elongin BC-containing E3 ubiquitin–ligase complex via the conserved SOCS box (Tyers and Willems, 1999). Disruption of SOCS1 in mice increases insulin sensitivity (Kawazoe *et al.*, 2001), while induction of SOCS1, SOCS3, or SOCS6 inhibit insulin receptor signaling (Emanuelli *et al.*, 2001; Mooney *et al.*, 2001). This process involves IRS proteins degradation because SOCS1/3 promotes the ubiquitination and degradation of both IRS-1 and IRS-2 (Rui *et al.*, 2002). The elongin BC-binding motif in SOCS1 and SOCS3 is required for the ubiquitination and degradation of IRS-1 and IRS-2, revealing an additional mechanism to inhibit insulin action and promote glucose intolerance during infection, inflammation, or metabolic stress. Still, a direct connection between Ser phosphorylation of IRS proteins and their interaction with SOCS proteins remains to be established.

V. Ser Phosphorylation of IRS Proteins as an Array Phenomenon

Data collected so far suggests that an array of Ser/Thr phosphorylation sites of IRS-1 attenuates insulin signaling, rather than phosphorylation of single selected sites (Fig. 12.3). Ser/Thr phosphorylation disrupts at least three interactions of IRS-1: with the plasma membrane; the receptor; or its downstream effectors. Therefore, many potential Ser residues distributed along the IRS proteins could be involved. IRS-1 contains more than 70 Ser residues at potential consensus phosphorylation sites and each signal could

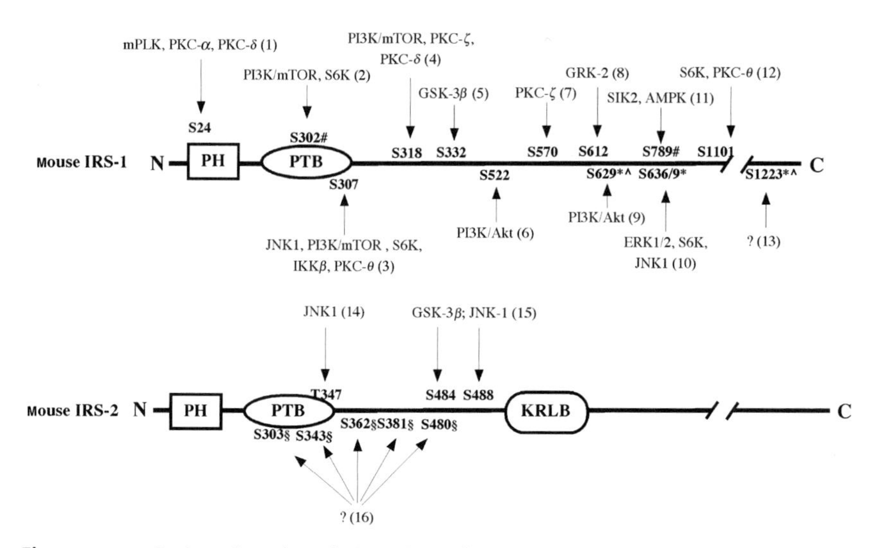

Figure 12.3 **Serine phosphorylation sites of IRS proteins.** Representative list of Ser residues of IRS-1 and IRS-2 which were shown to undergo phosphorylation as a result of various stimuli (see details in the text). Nonmarked sites, sites marked with ^ or # represent, respectively, a suggested negative, positive or negative/positive regulatory phosphorylation site. * human numbering; § sites mutated as a group. List of references: 1. Greene *et al.* (2006), Kim *et al.* (2005), and Nawaratne *et al.* (2006); 2. Giraud *et al.* (2004) and Harrington *et al.* (2004); 3. Aguirre *et al.* (2002), Gao *et al.* (2002), Hiratani *et al.* (2005), Nakamori *et al.* (2006), Solinas *et al.* (2006), Um *et al.* (2004), and Yu *et al.* (2002a); 4. Hennige *et al.* (2006), Moeschel *et al.* (2004), and Mussig *et al.* (2005); 5. Liberman and Eldar-Finkelman (2005); 6. Giraud *et al.* (2007); 7. Sommerfeld *et al.* (2004); 8. Usui *et al.* (2005); 9. Luo *et al.* (2007); 10. Bouzakri *et al.* (2003), Hiratani *et al.* (2005), and Um *et al.* (2004); 11. Horike *et al.* (2003) and Jakobsen *et al.* (2001); 12. Li *et al.* (2004) and Tremblay *et al.* (2007); 13. Luo *et al.* (2005); 14. Solinas *et al.* (2006); 15. Sharfi and Eldar-Finkelman (2007); and 16. Gurevitch *et al.* (2008).

induce the phosphorylation of a different set of residues. Hence, numerous optional phosphorylation combinations are possible (Boura-Halfon and Zick, 2008). To obtain an accurate analysis, all potential sites phosphorylated in response to various stimuli should be examined as opposed to analysis of a small number of residues limited by availability of phospho-specific antibodies. It has already been shown that quite a number of Ser residues undergo phosphorylation upon insulin treatment in human muscle biopsies (Yi *et al.*, 2007) and in $CHO^{IR/IRS}$ cells (Giraud *et al.*, 2007), indicating a robust phosphorylation of an array of Ser sites of IRS-1 in an *in vivo* setting. In both studies, Ser/Thr phosphorylation of IRS-1 was analyzed upon prolonged insulin stimulation (30–120 min) using mass spectrometry and around 20 sites were suggested as potential phosphorylated sites in each study. Of note, these lists were not identical thus suggesting that an even longer list of Ser sites are phosphorylated upon insulin stimulation or upon treatment with inducers of insulin resistance.

The identity of the Ser residues, the phosphorylation of which results in specific alteration in IRS proteins structure and function, is still poorly understood. Ser307 is considered to be involved in IRS-1 degradation (Greene *et al.*, 2003), while we have shown that deletion of a Ser-rich domain of IRS-1 protects it from insulin-stimulated degradation despite the fact that this IRS-1 mutant is still highly phosphorylated on Ser307 (Boura-Halfon *et al.*, 2008). We could also show that mutation of seven Ser sites of IRS-1, different from Ser307, confers upon the mutant protein protection from the inhibitory action of sustained insulin treatment or the inhibitory effects of proinflammatory cytokines (Herschkovitz *et al.*, 2007; Liu *et al.*, 2004). These findings support the notion that phosphorylation of IRS proteins at selected domains (e.g., their PTB domain), rather than phosphorylation of selected sites (e.g., Ser307) is required to inhibit IRS proteins function. Accordingly, mutations of seven Ser residues within a given domain confer stronger protection from IRS kinases than mutations of three or a single residue (Herschkovitz *et al.*, 2007; Liu *et al.*, 2004). This conclusion is supported by a recent study that made use of muscle-specific knockin mice which express a mutant IRS-1 where three serine residues were replaced by alanine (Ser302, 307, and 612) (Morino *et al.*, 2008). The transgenic mice were partially protected from fat-induced insulin resistance.

The concept of a "recognition array" is well established in systems that regulate innate immunity. The detection of an invading pathogen and subsequent activation of antimicrobial responses is often referred to as "pattern recognition," in reference to the proposal that immunity is induced by signals from pattern-recognition receptors (PRRs), which recognize pathogen-associated molecular patterns (PAMPs) (Robinson *et al.*, 2006). PAMPs are conserved groups of molecules that are essential for microbe survival, such as bacterial and fungal cell wall components and viral nucleic acid. In the broadest sense, PRRs comprise any PAMP receptor capable of triggering antimicrobial function (Robinson *et al.*, 2006). Hence, the ability to propagate a biological response might involve recognition of an "array" of specific molecules, rather than binding to a selected molecule located at a given site. Accordingly, mutation of a number of Ser/Thr phosphorylation sites rather than single site-specific mutation sensitizes IRS-1 and protects it from negative-feedback regulation of insulin signaling.

The issue of dominance should also be considered, namely, what happens when an array of "positive" and "negative" Ser sites is mutated. For example, we could show that when such "mixed" mutation is performed the net effect is potentiation of insulin signaling (Herschkovitz *et al.*, 2007; Liu *et al.*, 2004), suggesting the dominance of "inhibitory" Ser sites over the "stimulatory" sites in this particular case. Altogether, these findings indicate the existence of a crosstalk between Ser phosphorylation sites and suggest that the overall phosphorylation pattern dictates IRS-1 functions.

VI. Summary

IRS proteins are major targets for Ser/Thr phosphorylation-based negative regulation that uncouples them from their upstream receptors and their downstream effectors, leading to their degradation and termination of insulin signaling. A positive role for Ser phosphorylation was implicated as well, however data on "stimulatory" Ser sites is less comprehensive and the underlying mechanisms are yet elusive. Ser phosphorylation of IRS proteins involves a number of IRS kinases. Hence, Ser/Thr phosphorylation of IRS proteins represents combinatorial consequences of several kinases, activated by different pathways, acting in concert to phosphorylate multiple sites to generate a rather complicated network (Boura-Halfon and Zick, 2008). This model is supported by studies that show that one stimulus can increase the phosphorylation of many Ser residues of IRS proteins. Secondly, elimination of a number of Ser residues confers upon IRS proteins better protection from inducers of insulin resistance than elimination of single sites. Major questions that remain to be addressed are: which kinases directly phosphorylate IRS proteins, thereby affecting insulin and IGF-1 signaling; which Ser residues are the most critical in regulating IRS function; and are there Ser/Thr-rich domains whose phosphorylation at a number of sites inhibits IRS proteins function.

The spatial and temporal regulatory elements that control this complex phosphorylation network need to be revealed. For example, activation of AMPK that inhibits the activity of mTOR (Gwinn *et al.*, 2008) and thus inhibits insulin and IGF-1 actions, relieves IRS proteins from the inhibitory effects exerted by mTOR on their function. On the other end, AMPK acts as an IRS kinase that inhibits IRS proteins activity (Shargill *et al.*, 1986). Hence, this apparently contradictory actions of AMPK need to be regulated although at the system level there is no contradiction because the dual actions of AMPK enables inhibition of insulin and IGF-1 signaling both at the level of IRS proteins and at the level of activation of mTOR. It is conceivable to assume that Ser kinases activated along the insulin pathway (e.g., S6K1, PKCζ) will be allowed first to execute their action as promoters of insulin signaling, before they induce the phosphorylation of IRS proteins, as part of a negative-feedback mechanisms that will terminate their own activation (Boura-Halfon and Zick, 2008). Then, how these kinases are targeted toward their different substrates (e.g., S6 the "positive" signal vs IRS-1 the "negative" signal) is currently unknown. The issue of "stimulatory" versus "inhibitory" Ser sites of IRS-1 also needs further clarification. Is phosphorylation of "stimulatory" sites precedes that of "inhibitory" sites? and if so, what regulates this process? The issue of "priming" deserves attention. For example, is phosphorylation of IRS proteins at "stimulatory" Ser sites "tags" the protein to further phosphorylation at "inhibitory" Ser sites? The array of Ser/Thr phosphatases that dephosphorylate the different Ser residues and resets the system to its basal state have not been elucidated. It is only conceivable to assume that the activity of these phosphatases is regulated to no lesser extent than the activity of the IRS kinases.

Finally, the impact of other posttranslational modifications on IRS protein functions remains to be addressed. Recent studies demonstrate that insulin enhances the activity of sirtuins (Zhang, 2007), NAD^+-dependent class III histone deacetylases conserved from bacteria to humans (Frye, 2000). Inhibition of SirT1 reduces IGF-1 signaling and increases the resistance of mammalian cells to oxidative stress (Li *et al.*, 2008b). SirT1 increases insulin/IGF-1 signaling in part by deacetylating IRS-2, which promotes its Tyr phosphorylation (Zhang, 2007). These posttranslation modifications in IRS-2 are followed by Ras/ERK activation and sensitization of primary rat neurons to oxidative stress (Li *et al.*, 2008b). Hence, inhibition of Tyr phosphorylation IRS proteins can be achieved either by Ser/Thr phosphorylation or by acetylation, but it still remains to be determined whether these processes are regulated in a concerted fashion or as independent events.

Several strategies could be applied to study the complex regulatory network of IRS proteins function. The fast-developing field of mass spectrometry now enables the identification of arrays of Ser sites of IRS proteins that are subjected to phosphorylation under *in vivo* conditions. Samples of human tissues, processed to isolate their IRS proteins and subject them to such analysis, could be most insightful in defining the extent of phosphorylation of each site under physiological or pathological conditions. Introduction of siRNA technology could help decipher the role of individual kinases, or a kinase combination thereof, in regulating IRS proteins function. Combined with tissue-specific knockout of given kinases, this strategy could enlighten our understanding about the IRS kinases activated under different biological conditions. Then, "knockin" into IRS-1 or IRS-2 null mice of IRS proteins mutated at selected Ser phosphorylation sites, should provide insight into the role of selected Ser sites in the regulation of IRS proteins function.

Addressing these and related questions awaits further studies that will lead us to a better understanding of this complex process. Such research has much clinical relevance and physiological importance as it might direct us toward new potential therapeutic strategies to treat insulin and IGF-1 resistance and their adverse consequences.

REFERENCES

Adam-Klages, S., Adam, D., Wiegmann, K., Struve, S., Kolanus, W., Schneider, M. J., and Kronke, M. (1996). Fan, a novel WD-repeat protein, couples the p55 TNF-receptor to neutral sphingomyelinase. *Cell* **86,** 937–947.

Aguirre, V., Uchida, T., Yenush, L., Davis, R., and White, M. F. (2000). c-Jun NH(2)-terminal kinase promotes insulin resistance during association with insulin receptor substrate-1 and phosphorylation of Ser(307). *J. Biol. Chem.* **275,** 9047–9054.

Aguirre, V., Werner, E. D., Giraud, J., Lee, Y. H., Shoelson, S. E., and White, M. F. (2002). Phosphorylation of Ser307 in insulin receptor substrate-1 blocks interactions with the insulin receptor and inhibits insulin action. *J. Biol. Chem.* **277,** 1531–1537.

Anderwald, C., and Roden, M. (2004). Adipotoxicity and the insulin resistance syndrome. *Pediatr. Endocrinol. Rev.* **1,** 310–319.

Araki, E., Lipes, M. A., Patti, M. E., Bruning, J. C., Haag, B., Johnson, R. S., and Kahn, C. R. (1994). Alternative pathway of insulin signalling in mice with targeted disruption of the IRS-1 gene. *Nature* **372,** 186–190.

Argiles, J. M., Lopez-Soriano, J., Almendro, V., Busquets, S., and Lopez-Soriano, F. J. (2005). Cross-talk between skeletal muscle and adipose tissue: A link with obesity? *Med. Res. Rev.* **25,** 49–65.

Arkan, M. C., Hevener, A. L., Greten, F. R., Maeda, S., Li, Z. W., Long, J. M., Wynshaw-Boris, A., Poli, G., Olefsky, J., and Karin, M. (2005). IKK-beta links inflammation to obesity-induced insulin resistance. *Nat. Med.* **11,** 191–198.

Austin, R. L., Rune, A., Bouzakri, K., Zierath, J. R., and Krook, A. (2008). siRNA-mediated reduction of inhibitor of nuclear factor-kappaB kinase prevents tumor necrosis factor-alpha-induced insulin resistance in human skeletal muscle. *Diabetes* **57,** 2066–2073.

Avruch, J., Hara, K., Lin, Y., Liu, M., Long, X., Ortiz-Vega, S., and Yonezawa, K. (2006). Insulin and amino-acid regulation of mTOR signaling and kinase activity through the Rheb GTPase. *Oncogene* **25,** 6361–6372.

Bandyopadhyay, G., Standaert, M. L., Zhao, L., Yu, B., Avignon, A., Galloway, L., Karnam, P., Moscat, J., and Farese, R. V. (1997). Activation of protein kinase C (alpha, beta, and zeta) by insulin in 3T3/L1 cells. Transfection studies suggest a role for PKC-zeta in glucose transport. *J. Biol. Chem.* **272,** 2551–2558.

Bandyopadhyay, G., Sajan, M. P., Kanoh, Y., Standaert, M. L., Quon, M. J., Lea-Currie, R., Sen, A., and Farese, R. V. (2002). PKC-zeta mediates insulin effects on glucose transport in cultured preadipocyte-derived human adipocytes. *J. Clin. Endocrinol. Metab.* **87,** 716–723.

Baserga, R. (1999). The IGF-I receptor in cancer research. *Exp. Cell Res.* **253,** 1–6.

Beitner, J. D., Blakesley, V. A., Shen, O. Z., Jimenez, M., Stannard, B., Wang, L. M., Pierce, J., and LeRoith, D. (1996). The proto-oncogene product c-Crk associates with insulin receptor substrate-1 and 4PS. Modulation by insulin growth factor-I (IGF) and enhanced IGF-I signaling. *J. Biol. Chem.* **271,** 9287–9290.

Bitar, M. S., and Labbad, Z. N. (1996). Transforming growth factor-beta and insulin-like growth factor-I in relation to diabetes-induced impairment of wound healing. *J. Surg. Res.* **61,** 113–119.

Boden, G., and Shulman, G. I. (2002). Free fatty acids in obesity and type 2 diabetes: Defining their role in the development of insulin resistance and beta-cell dysfunction. *Eur. J. Clin. Invest.* **32**(Suppl. 3), 14–23.

Boura-Halfon, S., and Zick, Y. (2008). Phosphorylation of IRS proteins, insulin action and insulin resistance. *Am. J. Physiol.* (in press).

Boura-Halfon, S., Beck, A., Petrovich, K., Gurevitch, D., Sasson, K., Ronen, D., and Zick, Y. (2008). A novel domain mediates ubiquitination-independent, insulin-induced proteasomal degradation of IRS-1 in: Third Ruseell Berrie D-Cure Symposium (Abstract Book, 2007), pp. 59.

Bouzakri, K., Roques, M., Gual, P., Espinosa, S., Guebre-Egziabher, F., Riou, J. P., Laville, M., Le Marchand-Brustel, Y., Tanti, J. F., and Vidal, H. (2003). Reduced activation of phosphatidylinositol-3 kinase and increased serine 636 phosphorylation of insulin receptor substrate-1 in primary culture of skeletal muscle cells from patients with type 2 diabetes. *Diabetes* **52,** 1319–1325.

Braiman, L., Alt, A., Kuroki, T., Ohba, M., Bak, A., Tennenbaum, T., and Sampson, S. R. (2001). Activation of protein kinase C zeta induces serine phosphorylation of VAMP2 in the GLUT4 compartment and increases glucose transport in skeletal muscle. *Mol. Cell. Biol.* **21,** 7852–7861.

Butler, A. A., Yakar, S., Gewolb, I. H., Karas, M., Okubo, Y., and LeRoith, D. (1998). Insulin-like growth factor-I receptor signal transduction: At the interface between physiology and cell biology. *Comp. Biochem. Physiol. B Biochem. Mol. Biol.* **121,** 19–26.

Cai, D., Yuan, M., Frantz, D. F., Melendez, P. A., Hansen, L., Lee, J., and Shoelson, S. E. (2005). Local and systemic insulin resistance resulting from hepatic activation of IKK-beta and NF-kappaB. *Nat. Med.* **11,** 183–190.

Cantley, L. C. (2002). The phosphoinositide 3-kinase pathway. *Science* **296,** 1655–1657.

Capecchi, M. R. (1989). Altering the genome by homologous recombination. *Science* **244,** 1288–1292.

Carriere, A., Cargnello, M., Julien, L. A., Gao, H., Bonneil, E., Thibault, P., and Roux, P. P. (2008). Oncogenic MAPK signaling stimulates mTORC1 activity by promoting RSK-mediated raptor phosphorylation. *Curr. Biol.* **18,** 1269–1277.

Ciaraldi, T. P., Nikoulina, S. E., Bandukwala, R. A., Carter, L., and Henry, R. R. (2007). Role of glycogen synthase kinase-3 alpha in insulin action in cultured human skeletal muscle cells. *Endocrinology* **148,** 4393–4399.

Cohen, P., and Frame, S. (2001). The renaissance of GSK3. *Nat. Rev. Mol. Cell Biol.* **2,** 769–776.

Craparo, A., Freund, R., and Gustafson, T. A. (1997). 14-3-3 (epsilon) interacts with the insulin-like growth factor I receptor and insulin receptor substrate I in a phosphoserine-dependent manner. *J. Biol. Chem.* **272,** 11663–11669.

Cross, D. A., Alessi, D. R., Cohen, P., Andjelkovich, M., and Hemmings, B. A. (1995). Inhibition of glycogen synthase kinase-3 by insulin mediated by protein kinase B. *Nature* **378,** 785–789.

Datta, S. R., Dudek, H., Tao, X., Masters, S., Fu, H., Gotoh, Y., and Greenberg, M. E. (1997). Akt phosphorylation of BAD couples survival signals to the cell-intrinsic death machinery. *Cell* **91,** 231–241.

Davis, R. J. (2000). Signal transduction by the JNK group of MAP kinases. *Cell* **103,** 239–252.

de Alvaro, C., Teruel, T., Hernandez, R., and Lorenzo, M. (2004). Tumor necrosis factor alpha produces insulin resistance in skeletal muscle by activation of inhibitor kappaB kinase in a p38 MAPK-dependent manner. *J. Biol. Chem.* **279,** 17070–17078.

De Fea, K., and Roth, R. A. (1997). Modulation of insulin receptor substrate-1 tyrosine phosphorylation and function by mitogen-activated protein kinase. *J. Biol. Chem.* **272,** 31400–31406.

de Luca, C., and Olefsky, J. M. (2008). Inflammation and insulin resistance. *FEBS Lett.* **582,** 97–105.

del Peso, L., Gonzalez-Garcia, M., Page, C., Herrera, R., and Nunez, G. (1997). Interleukin-3-induced phosphorylation of BAD through the protein kinase Akt. *Science* **278,** 687–689.

Draznin, B. (2006). Molecular mechanisms of insulin resistance: Serine phosphorylation of insulin receptor substrate-1 and increased expression of p85alpha: The two sides of a coin. *Diabetes* **55,** 2392–2397.

Du, J., Chen, Q., Takemori, H., and Xu, H. (2008). SIK2 can be activated by deprivation of nutrition and it inhibits expression of lipogenic genes in adipocytes. *Obesity (Silver Spring)* **16,** 531–538.

Ebina, Y., Ellis, L., Jarnagin, K., Edery, M., Graf, L., Clauser, E., Ou, J. H., Masiarz, F., Kan, Y., Goldfine, I. D., Roth, R. A., and Rutter, W. J. (1985). The human insulin receptor cDNA: The structural basis for hormone-activated transmembrane signalling. *Cell* **40,** 747–758.

Eck, M. J., Dhe, P. S., Trub, T., Nolte, R. T., and Shoelson, S. E. (1996). Structure of the IRS-1 PTB domain bound to the juxtamembrane region of the insulin receptor. *Cell* **85,** 695–705.

Egawa, K., Nakashima, N., Sharma, P. M., Maegawa, H., Nagai, Y., Kashiwagi, A., Kikkawa, R., and Olefsky, J. M. (2000). Persistent activation of phosphatidylinositol 3-kinase causes insulin resistance due to accelerated insulin-induced insulin receptor substrate-1 degradation in 3T3-L1 adipocytes. *Endocrinology* **141,** 1930–1935.

Eldar-Finkelman, H., and Krebs, E. G. (1997). Phosphorylation of insulin receptor substrate 1 by glycogen synthase kinase 3 impairs insulin action. *Proc. Natl. Acad. Sci. USA* **94,** 9660–9664.

Eldar-Finkelman, H., Schreyer, S. A., Shinohara, M. M., LeBoeuf, R. C., and Krebs, E. G. (1999). Increased glycogen synthase kinase-3 activity in diabetes- and obesity-prone C57BL/6J mice. *Diabetes* **48,** 1662–1666.

Emanuelli, B., Peraldi, P., Filloux, C., Chavey, C., Freidinger, K., Hilton, D. J., Hotamisligil, G. S., and Van Obberghen, E. (2001). SOCS-3 inhibits insulin signaling and is up-regulated in response to tumor necrosis factor-alpha in the adipose tissue of obese mice. *J. Biol. Chem.* **276,** 47944–47949.

Fain, J. N., Madan, A. K., Hiler, M. L., Cheema, P., and Bahouth, S. W. (2004). Comparison of the release of adipokines by adipose tissue, adipose tissue matrix, and adipocytes from visceral and subcutaneous abdominal adipose tissues of obese humans. *Endocrinology* **145,** 2273–2282.

Farese, R. V., Sajan, M. P., Yang, H., Li, P., Mastorides, S., Gower, W. R. J., Nimal, S., Choi, C. S., Kim, S., Shulman, G. I., Kahn, C. R., and Braun, U. M. L. (2007). Muscle-specific knockout of PKC-lambda impairs glucose transport and induces metabolic and diabetic syndromes. *J. Clin. Invest.* **117,** 2289–2301.

Farhang-Fallah, J., Yin, X., Trentin, G., Cheng, A. M., and Rozakis-Adcock, M. (2000). Cloning and characterization of PHIP, a novel insulin receptor substrate-1 pleckstrin homology domain interacting protein. *J. Biol. Chem.* **275,** 40492–40497.

Farhang-Fallah, J., Randhawa, V. K., Nimnual, A., Klip, A., Bar-Sagi, D., and Rozakis-Adcock, M. (2002). The pleckstrin homology (PH) domain-interacting protein couples the insulin receptor substrate 1 PH domain to insulin signaling pathways leading to mitogenesis and GLUT4 translocation. *Mol. Cell. Biol.* **22,** 7325–7336.

Feng, Z., Hu, W., de Stanchina, E., Teresky, A. K., Jin, S., Lowe, S., and Levine, A. J. (2007). The regulation of AMPK beta1, TSC2, and PTEN expression by p53: Stress, cell and tissue specificity, and the role of these gene products in modulating the IGF-1–AKT–mTOR pathways. *Cancer Res.* **67,** 3043–3053.

Fingar, D. C., and Blenis, J. (2004). Target of rapamycin (TOR): An integrator of nutrient and growth factor signals and coordinator of cell growth and cell cycle progression. *Oncogene* **23,** 3151–3171.

Fraenkel, M., Ketzinel-Gilad, M., Ariav, Y., Pappo, O., Karaca, M., Castel, J., Berthault, M. F., Magnan, C., Cerasi, E., Kaiser, N., and Leibowitz, G. (2008). mTOR inhibition by rapamycin prevents beta-cell adaptation to hyperglycemia and exacerbates the metabolic state in type 2 diabetes. *Diabetes* **57,** 945–957.

Freude, S., Leeser, U., Muller, M., Hettich, M. M., Udelhoven, M., Schilbach, K., Tobe, K., Kadowaki, T., Kohler, C., Schroder, H., Krone, W., Bruning, J. C., *et al.* (2008). IRS-2 branch of IGF-1 receptor signaling is essential for appropriate timing of myelination. *J. Neurochem.* **107,** 907–917.

Frye, R. A. (2000). Phylogenetic classification of prokaryotic and eukaryotic Sir2-like proteins. *Biochem. Biophys. Res. Commun.* **273,** 793–798.

Gao, Z., Hwang, D., Bataille, F., Lefevre, M., York, D., Quon, M. J., and Ye, J. (2002). Serine phosphorylation of insulin receptor substrate 1 by inhibitor kappa B kinase complex. *J. Biol. Chem.* **277,** 48115–48121.

Giraud, J., Leshan, R., Lee, Y. H., and White, M. F. (2004). Nutrient-dependent and insulin-stimulated phosphorylation of insulin receptor substrate-1 on serine 302 correlates with increased insulin signaling. *J. Biol. Chem.* **279,** 3447–3454.

Giraud, J., Haas, M., Feener, E. P., Copps, K. D., Dong, X., Dunn, S. L., and White, M. F. (2007). Phosphorylation of Irs1 at SER-522 inhibits insulin signaling. *Mol. Endocrinol.* **21,** 2294–2302.

Greene, M. W., Sakaue, H., Wang, L., Alessi, D. R., and Roth, R. A. (2003). Modulation of insulin-stimulated degradation of human insulin receptor substrate-1 by Serine 312 phosphorylation. *J. Biol. Chem.* **278,** 8199–8211.

Greene, M. W., Ruhoff, M. S., Roth, R. A., Kim, J. A., Quon, M. J., and Krause, J. A. (2006). PKCdelta-mediated IRS-1 Ser24 phosphorylation negatively regulates IRS-1 function. *Biochem. Biophys. Res. Commun.* **349,** 976–986.

Griffin, M. E., Marcucci, M. J., Cline, G. W., Bell, K., Barucci, N., Lee, D., Goodyear, L. J., Kraegen, E. W., White, M. F., and Shulman, G. I. (1999). Free fatty acid-induced insulin resistance is associated with activation of protein kinase C theta and alterations in the insulin signaling cascade. *Diabetes* **48,** 1270–1274.

Gual, P., Gonzalez, T., Gremeaux, T., Barres, R., Le, M., Brustel, Y., and Tanti, J. F. (2003). Hyperosmotic stress inhibits IRS-1 function by distinct mechanisms in 3T3-L1 adipocytes. *J. Biol. Chem.* **278,** 26550–26557.

Guertin, D. A., and Sabatini, D. M. (2007). Defining the role of mTOR in cancer. *Cancer Cell* **12,** 9–22.

Gurevitch, D., Boura-Halfon, S., Issac, R., Alberstein, M., Ronen, D., Lewis, E., and Zick, Y. (2008). Phosphorylation of selected serines at the PTB domain of IRS-2 regulates β cell growth, survival, and insulin secretion in: Keystone Symposia on Islet and Beta Cell Development and Transplantation (2008 Abstract Book) pp. 141.

Gwinn, D. M., Shackelford, D. B., Egan, D. F., Mihaylova, M. M., Mery, A., Vasquez, D. S., Turk, B. E., and Shaw, R. J. (2008). AMPK phosphorylation of raptor mediates a metabolic checkpoint. *Mol. Cell* **30,** 214–226.

Hardie, D. G. (2008). AMPK and raptor: Matching cell growth to energy supply. *Mol. Cell* **30,** 263–265.

Harrington, L. S., Findlay, G. M., Gray, A., Tolkacheva, T., Wigfield, S., Rebholz, H., Barnett, J., Leslie, N. R., Cheng, S., Shepherd, P. R., Gout, I., Downes, C. P., *et al.* (2004). The TSC1–2 tumor suppressor controls insulin-PI3K signaling via regulation of IRS proteins. *J. Cell Biol.* **166,** 213–223.

Haruta, T., Uno, T., Kawahara, J., Takano, A., Egawa, K., Sharma, P. M., Olefsky, J. M., and Kobayashi, M. (2000). A rapamycin-sensitive pathway down-regulates insulin signaling via phosphorylation and proteasomal degradation of insulin receptor substrate-1. *Mol. Endocrinol* **14,** 783–794.

Hennige, A. M., Stefan, N., Kapp, K., Lehmann, R., Weigert, C., Beck, A., Moeschel, K., Mushack, J., Schleicher, E., and Haring, H. U. (2006). Leptin down-regulates insulin action through phosphorylation of serine-318 in insulin receptor substrate 1. *FASEB J.* **20,** 1206–1208.

Henriksen, E. J., and Dokken, B. B. (2006). Role of glycogen synthase kinase-3 in insulin resistance and type 2 diabetes. *Curr. Drug Targets* **7,** 1435–1441.

Herschkovitz, A., Liu, Y. F., Ilan, E., Ronen, D., Boura-Halfon, S., and Zick, Y. (2007). Common inhibitory serine sites phosphorylated by IRS-1 kinases, triggered by insulin and inducers of insulin resistance. *J. Biol. Chem.* **282,** 18018–18027.

Hirabara, S. M., Silveira, L. R., Abdulkader, F., Carvalho, C. R., Procopio, J., and Curi, R. (2007). Time-dependent effects of fatty acids on skeletal muscle metabolism. *J. Cell. Physiol.* **210,** 7–15.

Hiratani, K., Haruta, T., Tani, A., Kawahara, J., Usui, I., and Kobayashi, M. (2005). Roles of mTOR and JNK in serine phosphorylation, translocation, and degradation of IRS-1. *Biochem. Biophys. Res. Commun.* **335,** 836–842.

Hirosumi, J., Tuncman, G., Chang, L., Gorgun, C. Z., Uysal, K. T., Maeda, K., Karin, M., and Hotamisligil, G. S. (2002). A central role for JNK in obesity and insulin resistance. *Nature* **420,** 333–336.

Horike, N., Takemori, H., Katoh, Y., Doi, J., Min, L., Asano, T., Sun, X. J., Yamamoto, H., Kasayama, S., Muraoka, M., Nonaka, Y., and Okamoto, M. (2003). Adipose-specific expression, phosphorylation of Ser794 in insulin receptor substrate-1, and activation in diabetic animals of salt-inducible kinase-2. *J. Biol. Chem.* **278,** 18440–18447.

Hotamisligil, G. S., Shargill, N. S., and Spiegelman, B. M. (1993). Adipose expression of tumor necrosis factor-alpha: Direct role in obesity-linked insulin resistance. *Science* **259,** 87–91.

Hundal, R. S., Petersen, K. F., Mayerson, A. B., Randhawa, P. S., Inzucchi, S., Shoelson, S. E., and Shulman, G. I. (2002). Mechanism by which high-dose aspirin improves glucose metabolism in type 2 diabetes. *J. Clin. Invest.* **109,** 1321–1326.

Inoki, K., and Guan, K. L. (2006). Complexity of the TOR signaling network. *Trends Cell Biol.* **16,** 206–212.

Inoki, K., Corradetti, M. N., and Guan, K. L. (2005). Dysregulation of the TSC–mTOR pathway in human disease. *Nat. Genet.* **37,** 19–24.

Ishibashi, K. I., Imamura, T., Sharma, P. M., Huang, J., Ugi, S., and Olefsky, J. M. (2001). Chronic endothelin-1 treatment leads to heterologous desensitization of insulin signaling in 3T3-L1 adipocytes. *J. Clin. Invest.* **107,** 1193–1202.

Israel, A. (2000). The IKK complex: An integrator of all signals that activate NF-kappaB? *Trends Cell Biol.* **10,** 129–133.

Jakobsen, S. N., Hardie, D. G., Morrice, N., and Tornqvist, H. E. (2001). 5′-AMP-activated protein kinase phosphorylates IRS-1 on Ser-789 in mouse C2C12 myotubes in response to 5-aminoimidazole-4-carboxamide riboside. *J. Biol. Chem.* **276,** 46912–46916.

Kaburagi, Y., Satoh, S., Tamemoto, H., Yamamoto, H. R., Tobe, K., Veki, K., Yamauchi, T., Kono, S. E., Sekihara, H., Aizawa, S., Cushman, S. W., Akanuma, Y., *et al.* (1997). Role of insulin receptor substrate-1 and pp60 in the regulation of insulin-induced glucose transport and GLUT4 translocation in primary adipocytes. *J. Biol. Chem.* **272,** 25839–25844.

Kahn, S. E., Hull, R. L., and Utzschneider, K. M. (2006). Mechanisms linking obesity to insulin resistance and type 2 diabetes. *Nature* **444,** 840–846.

Kanety, H., Feinstein, R., Papa, M. Z., Hemi, R., and Karasik, A. (1995). Tumor necrosis factor alpha-induced phosphorylation of insulin receptor substrate-1 (IRS-1). Possible mechanism for suppression of insulin-stimulated tyrosine phosphorylation of IRS-1. *J. Biol. Chem.* **270,** 23780–23784.

Kanety, H., Hemi, R., Papa, M. Z., and Karasik, A. (1996). Sphingomyelinase and ceramide suppress insulin-induced tyrosine phosphorylation of the insulin receptor substrate-1. *J. Biol. Chem.* **271,** 9895–9897.

Karin, M. (1999). The beginning of the end: IkappaB kinase (IKK) and NF-kappaB activation. *J. Biol. Chem.* **274,** 27339–27342.

Kawazoe, Y., Naka, T., Fujimoto, M., Kohzaki, H., Morita, Y., Narazaki, M., Okumura, K., Saitoh, H., Nakagawa, R., Uchiyama, Y., Akira, S., and Kishimoto, T. (2001). Signal transducer and activator of transcription (STAT)-induced STAT inhibitor 1 (SSI-1)/suppressor of cytokine signaling 1 (SOCS1) inhibits insulin signal transduction pathway through modulating insulin receptor substrate 1 (IRS-1) phosphorylation. *J. Exp. Med.* **193,** 263–269.

Kellerer, M., Mushack, J., Seffer, E., Mischak, H., Ullrich, A., and Haring, H. U. (1998). Protein kinase C isoforms alpha, delta and theta require insulin receptor substrate-1 to inhibit the tyrosine kinase activity of the insulin receptor in human kidney embryonic cells (HEK 293 cells). *Diabetologia* **41,** 833–838.

Khan, A. H., and Pessin, J. E. (2002). Insulin regulation of glucose uptake: A complex interplay of intracellular signalling pathways. *Diabetologia* **45,** 1475–1483.

Kim, J. K., Kim, Y.-J., Filmore, J. J., Chen, Y., Moore, I., Lee, J., Yuan, M., Li, Z. W., Karin, M., Perrer, P., Shoelson, S., and Shulman, G. I. (2001). Prevention of fat-induced insulin resistance by salicylate. *J. Clin. Invest.* **108,** 437–446.

Kim, J. K., Fillmore, J. J., Sunshine, M. J., Albrecht, B., Higashimori, T., Kim, D. W., Liu, Z. X., Soos, T. J., Cline, G. W., O'Brien, W. R., Littman, D. R., and Shulman, G. I. (2004). PKC-theta knockout mice are protected from fat-induced insulin resistance. *J. Clin. Invest.* **114,** 823–827.

Kim, J. A., Yeh, D. C., Ver, M., Li, Y., Carranza, A., Conrads, T. P., Veenstra, T. D., Harrington, M. A., and Quon, M. J. (2005). Phosphorylation of Ser24 in the pleckstrin homology domain of insulin receptor substrate-1 by Mouse Pelle-like kinase/interleukin-1 receptor-associated kinase: Cross-talk between inflammatory signaling and insulin signaling that may contribute to insulin resistance. *J. Biol. Chem.* **280,** 23173–23183.

Krebs, M., Brunmair, B., Brehm, A., Artwohl, M., Szendroedi, J., Nowotny, P., Roth, E., Furnsinn, C., Promintzer, M., Anderwald, C., Bischof, M., and Roden, M. (2007). The Mammalian target of rapamycin pathway regulates nutrient-sensitive glucose uptake in man. *Diabetes* **56,** 1600–1607.

Kubota, N., Kubota, T., Itoh, S., Kumagai, H., Kozono, H., Takamoto, I., Mineyama, T., Ogata, H., Tokuyama, K., Ohsugi, M., Sasako, T., Moroi, M., *et al.* (2008). Dynamic functional relay between insulin receptor substrate 1 and 2 in hepatic insulin signaling during fasting and feeding. *Cell Metab.* **8,** 49–64.

Kulkarni, R. N. (2002). Receptors for insulin and insulin-like growth factor-1 and insulin receptor substrate-1 mediate pathways that regulate islet function. *Biochem. Soc. Trans.* **30,** 317–322.

Lallena, M.-J., Diaz-Meco, M. T., Bren, G., Paya, C. V., and Moscat, J. (1999). Activation of IkappaB kinase beta by protein kinase C isoforms. *Mol. Cell. Biol.* **19,** 2180–2188.

Lee, Y.-H., Giraud, J., Davis, R. J., and White, M. F. (2003a). cJUN N-terminal kinase (JNK) mediates feedback inhibition of the insulin signaling cascade. *J. Biol. Chem.* **278,** 2896–2908.

Lee, Y. H., Giraud, J., Davis, R. J., and White, M. F. (2003b). c-Jun N-terminal kinase (JNK) mediates feedback inhibition of the insulin signaling cascade. *J. Biol. Chem.* **278,** 2896–2902.

Lee, C. H., Inoki, K., and Guan, K. L. (2007). mTOR pathway as a target in tissue hypertrophy. *Annu. Rev. Pharmacol. Toxicol.* **47,** 443–467.

Lee, S., Lynn, E. G., Kim, J. A., and Quon, M. J. (2008). Protein kinase C-ζ phosphorylates insulin receptor substrate-1, -3, and -4, but not -2: Isoform specific determinants of specificity in insulin signaling. *Endocrinology* **149,** 2451–2458.

Leitges, M., Plomann, M., Standaert, M. L., Bandyopadhyay, G., Sajan, M. P., Kanoh, Y., and Farese, R. V. (2002). Knockout of PKC alpha enhances insulin signaling through PI3K. *Mol. Endocrinol.* **16,** 847–858.

LeRoith, D., and Zick, Y. (2001). Recent advances in our understanding of insulin action and insulin resistance. *Diabetes Care* **24,** 588–597.

LeRoith, D., Werner, H., Beitner-Johnson, D., and Roberts, C. T., Jr., (1995). Molecular and cellular aspects of the insulin-like growth factor i receptor. *Endocr. Rev.* **16,** 143–163.

Li, J., DeFea, K., and Roth, R. A. (1999). Modulation of insulin receptor substrate-1 tyrosine phosphorylation by an Akt/phosphatidylinositol 3-kinase pathway. *J. Biol. Chem.* **274,** 9351–9356.

Li, Y., Soos, T. J., Li, X., Wu, J., Degennaro, M., Sun, X., Littman, D. R., Birnbaum, M. J., and Polakiewicz, R. D. (2004). Protein kinase C theta inhibits insulin signaling by phosphorylating IRS1 at Ser(1101). *J. Biol. Chem.* **279,** 45304–45307.

Li, D., Yin, X., Zmuda, E. J., Wolford, C. C., Dong, X., White, M. F., and Hai, T. (2008a). The repression of IRS2 gene by ATF3, a stress-inducible gene, contributes to pancreatic beta-cell apoptosis. *Diabetes* **57,** 635–644.

Li, Y., Xu, W., McBurney, M. W., and Longo, V. D. (2008b). SirT1 inhibition reduces IGF-I/IRS-2/Ras/ERK1/2 signaling and protects neurons. *Cell Metab.* **8,** 38–48.

Liberman, Z., and Eldar-Finkelman, H. (2005). Serine 332 phosphorylation of insulin receptor substrate-1 by glycogen synthase kinase-3 attenuates insulin signaling. *J. Biol. Chem.* **280,** 4422–4428.

Liu, Y. F., Paz, K., Herschkovitz, A., Alt, A., Tennenbaum, T., Sampson, S. R., Ohba, M., Kuroki, T., LeRoith, D., and Zick, Y. (2001). Insulin stimulates PKCzeta-mediated phosphorylation of insulin receptor substrate-1 (IRS-1). A self-attenuated mechanism to negatively regulate the function of IRS proteins. *J. Biol. Chem.* **276,** 14459–14465.

Liu, Y. F., Herschkovitz, A., Boura, H. S., Ronen, D., Paz, K., Leroith, D., and Zick, Y. (2004). Serine phosphorylation proximal to its phosphotyrosine binding domain inhibits insulin receptor substrate 1 function and promotes insulin resistance. *Mol. Cell. Biol.* **24,** 9668–9681.

Liu, L. Z., Zhao, H. L., Zuo, J., Ho, S. K., Chan, J. C., Meng, Y., Fang, F. D., and Tong, P. C. (2006). Protein kinase Czeta mediates insulin-induced glucose transport through actin remodeling in L6 muscle cells. *Mol. Biol. Cell* **17,** 2322–2330.

Lodhi, I. J., Chiang, S. H., Chang, L., Vollenweider, D., Watson, R. T., Inoue, M., Pessin, J. E., and Saltiel, A. R. (2007). Gapex-5, a Rab31 guanine nucleotide exchange factor that regulates Glut4 trafficking in adipocytes. *Cell Metab.* **5,** 59–72.

Luo, M., Reyna, S., Wang, L., Yi, Z., Carroll, C., Dong, L. Q., Langlais, P., Weintraub, S. T., and Mandarino, L. J. (2005). Identification of insulin receptor substrate 1 serine/threonine phosphorylation sites using mass spectrometry analysis: Regulatory role of serine 1223. *Endocrinology* **146,** 4410–4416.

Luo, M., Langlais, P., Yi, Z., Lefort, N., De Filippis, E. A., Hwang, H., Christ-Roberts, C. Y., and Mandarino, L. J. (2007). Phosphorylation of human insulin receptor substrate-1 at serine 629 plays a positive role in insulin signaling. *Endocrinology* **148,** 4895–4905.

MacAulay, K., Doble, B. W., Patel, S., Hansotia, T., Sinclair, E. M., Drucker, D. J., Nagy, A., and Woodgett, J. R. (2007). Glycogen synthase kinase 3alpha-specific regulation of murine hepatic glycogen metabolism. *Cell Metab.* **6,** 329–337.

Manning, B. D. (2004). Balancing Akt with S6K: Implications for both metabolic diseases and tumorigenesis. *J. Cell Biol.* **167,** 399–403.

McManus, E. J., Sakamoto, K., Armit, L. J., Ronaldson, L., Shpiro, N., Marquez, R., and Alessi, D. R. (2005). Role that phosphorylation of GSK3 plays in insulin and Wnt signalling defined by knockin analysis. *EMBO J.* **24,** 1571–1583.

Merforth, S., Osmers, A., and Dahlmann, B. (1999). Alterations of proteasome activities in skeletal muscle tissue of diabetic rats. *Mol. Biol. Rep.* **26,** 83–87.

Moeschel, K., Beck, A., Weigert, C., Lammers, R., Kalbacher, H., Voelter, W., Schleicher, E. D., Haring, H. U., and Lehmann, R. (2004). Protein kinase C-zeta-induced phosphorylation of Ser318 in receptor substrate-1 (IRS-1) attenuates the interaction with the receptor and the tyrosine phosphorylation of IRS-1. *J. Biol. Chem.* **279,** 25157–25163.

Mooney, R. A., Senn, J., Cameron, S., Inamdar, N., Boivin, L. M., Shang, Y., and Furlanetto, R. W. (2001). Suppressors of cytokine signaling-1 and -6 associate with and inhibit the insulin receptor. A potential mechanism for cytokine-mediated insulin resistance. *J. Biol. Chem.* **276,** 25889–25893.

Morino, K., Petersen, K. F., and Shulman, G. I. (2006). Molecular mechanisms of insulin resistance in humans and their potential links with mitochondrial dysfunction. *Diabetes* **55**(Suppl. 2), S9–S15.

Morino, K., Neschen, S., Bilz, S., Sono, S., Tsirigotis, D., Reznick, R. M., Moore, I., Nagai, Y., Samuel, V., Sebastian, D., White, M., Philbrick, W., *et al.* (2008). Muscle specific IRS-1 Ser→Ala transgenic mice are protected from fat-induced insulin resistance in skeletal muscle. *Diabetes* **57,** 2644–2651.

Mothe, I., and Van Obberghen, E. (1996). Phosphorylation of insulin receptor substrate-1 on multiple serine residues, 612, 632, 662, and 731, modulates insulin action. *J. Biol. Chem.* **271,** 11222–11227.

Moxham, C. M., Tabrizchi, A., Davis, R. J., and Malbon, C. C. (1996). Jun N-terminal kinase mediates activation of skeletal muscle glycogen synthase by insulin *in vivo*. *J. Biol. Chem.* **271,** 30765–30773.

Muller, G., Ayoub, M., Storz, P., Rennecke, J., Fabbro, D., and Pfizenmaier, K. (1995). PKC zeta is a molecular switch in signal transduction of TNF-alpha, bifunctionally regulated by ceramide and arachidonic acid. *EMBO J.* **14,** 1961–1969.

Mussig, K., Fiedler, H., Staiger, H., Weigert, C., Lehmann, R., Schleicher, E. D., and Haring, H. U. (2005). Insulin-induced stimulation of JNK and the PI 3-kinase/mTOR pathway leads to phosphorylation of serine 318 of IRS-1 in C2C12 myotubes. *Biochem. Biophys. Res. Commun.* **335,** 819–825.

Myers, M. J., Sun, K. J., and White, M. F. (1994). The IRS-1 signaling system. *Trends Biochem. Sci.* **19,** 289–293.

Naka, T., Tsutsui, H., Fujimoto, M., Kawazoe, Y., Kohzaki, H., Morita, Y., Nakagawa, R., Narazaki, M., Adachi, K., Yoshimoto, T., Nakanishi, K., and Kishimoto, T. (2001). SOCS-1/SSI-1-deficient NKT cells participate in severe hepatitis through dysregulated cross-talk inhibition of IFN-gamma and IL-4 signaling *in vivo*. *Immunity* **14,** 535–545.

Nakajima, K., Yamauchi, K., Shigematsu, S., Ikeo, S., Komatsu, M., Aizawa, T., and Hashizume, K. (2000). Selective attenuation of metabolic branch of insulin receptor down-signaling by high glucose in a hepatoma cell line, HepG2 cells. *J. Biol. Chem.* **275,** 20880–20886.

Nakamori, Y., Emoto, M., Fukuda, N., Taguchi, A., Okuya, S., Tajiri, M., Miyagishi, M., Taira, K., Wada, Y., and Tanizawa, Y. (2006). Myosin motor Myo1c and its receptor NEMO/IKK-gamma promote TNF-alpha-induced serine307 phosphorylation of IRS-1. *J. Cell Biol.* **173,** 665–671.

Nawaratne, R., Gray, A., Jorgensen, C. H., Downes, C. P., Siddle, K., and Sethi, J. K. (2006). Regulation of insulin receptor substrate 1 pleckstrin homology domain by protein kinase C: Role of serine 24 phosphorylation. *Mol. Endocrinol.* **20,** 1838–1852.

Patti, M.-E., Sun, X.-J., Bruening, J. C., Araki, E., Lipes, M. A., White, M. F., and Kahn, C. R. (1995). 4PS/insulin receptor substrate (IRS)-2 is the alternative substrate of the insulin receptor in IRS-1-deficient mice. *J. Biol. Chem.* **270,** 24670–24673.

Pawson, T. (1995). Protein modules and signalling networks. *Nature* **373,** 573–580.

Paz, K., Hemi, R., LeRoith, R., Karasik, A., Elhanany, E., Kanety, H., and Zick, Y. (1997). A Molecular basis for insulin resistance: Elevated serine/threonine phosphorylation of IRS-1 and IRS-2 inhibits their binding to the juxtamembrane region of the insulin receptor and impairs their ability to undergo insulin-induced tyrosine phosphorylation. *J. Biol. Chem.* **272,** 29911–29918.

Paz, K., Yan-Fang, L., Shorer, H., Hemi, R., LeRoith, D., Quon, M., Kanety, H., Seger, R., and Zick, Y. (1999). Phosphorylation of insulin receptor substrate-1 (IRS-1) by PKB positively regulates IRS-1 function. *J. Biol. Chem.* **274,** 28816–28822.

Pederson, T. M., Kramer, D. L., and Rondinone, C. M. (2001). Serine/threonine phosphorylation of IRS-1 triggers its degradation: Possible regulation by tyrosine phosphorylation. *Diabetes* **50,** 24–31.

Pelosi, L., Giacinti, C., Nardis, C., Borsellino, G., Rizzuto, E., Nicoletti, C., Wannenes, F., Battistini, L., Rosenthal, N., Molinaro, M., and Musaro, A. (2007). Local expression of IGF-1 accelerates muscle regeneration by rapidly modulating inflammatory cytokines and chemokines. *FASEB J.* **21,** 1393–1402.

Podcheko, A., Northcott, P., Bikopoulos, G., Lee, A., Bommareddi, S. R., Kushner, J. A., Farhang-Fallah, J., and Rozakis-Adcock, M. (2007). Identification of a WD40 repeat-containing isoform of PHIP as a novel regulator of beta-cell growth and survival. *Mol. Cell. Biol.* **27,** 6484–6496.

Qiao, L. Y., Zhande, R., Jetton, T. L., Zhou, G., and Sun, X. J. (2002). *In vivo* phosphorylation of insulin receptor substrate 1 at serine 789 by a novel serine kinase in insulin-resistant rodents. *J. Biol. Chem.* **277,** 26530–26539.

Ravichandran, L. V., Esposito, D. L., Chen, J., and Quon, M. J. (2001a). PKC-ζ phosphorylates IRS-1 and impairs its ability to activate PI 3-kinase in response to insulin. *J. Biol. Chem.* **276,** 3543–3549.

Ravichandran, L. V., Esposito, D. L., Chen, J., and Quon, M. J. (2001b). Protein kinase C-zeta phosphorylates insulin receptor substrate-1 and impairs its ability to activate phosphatidylinositol 3-kinase in response to insulin. *J. Biol. Chem.* **276,** 3543–3549.

Ribon, V., and Saltiel, A. R. (1997). Insulin stimulates tyrosine phosphorylation of the proto-oncogene product of c-Cbl in 3T3-L1 adipocytes. *Biochem. J.* **324**(Pt. 3), 839–845.

Robinson, M. J., Sancho, D., Slack, E. C., LeibundGut-Landmann, S., and Reis e Sousa, C. (2006). Myeloid C-type lectins in innate immunity. *Nat. Immunol.* **7,** 1258–1265.

Rui, L., Yuan, M., Frantz, D., Shoelson, S., and White, M. F. (2002). SOCS-1 and SOCS-3 block insulin signaling by ubiquitin-mediated degradation of IRS1 and IRS2. *J. Biol. Chem.* **277,** 42394–42398.

Salih, D. A., and Brunet, A. (2008). FoxO transcription factors in the maintenance of cellular homeostasis during aging. *Curr. Opin. Cell Biol.* **20,** 126–136.

Saltiel, A. R., and Kahn, C. R. (2001). Insulin signalling and the regulation of glucose and lipid metabolism. *Nature* **414,** 799–806.

Saltiel, A. R., and Pessin, J. E. (2002). Insulin signaling pathways in time and space. *Trends Cell Biol.* **12,** 65–71.

Sampson, S. R., and Cooper, D. R. (2006). Specific protein kinase C isoforms as transducers and modulators of insulin signaling. *Mol. Genet. Metab.* **89,** 32–47.

Sancak, Y., Thoreen, C. C., Peterson, T. R., Lindquist, R. A., Kang, S. A., Spooner, E., Carr, S. A., and Sabatini, D. M. (2007). PRAS40 is an insulin-regulated inhibitor of the mTORC1 protein kinase. *Mol. Cell* **25,** 903–915.

Sancak, Y., Peterson, T. R., Shaul, Y. D., Lindquist, R. A., Thoreen, C. C., Bar-Peled, L., and Sabatini, D. M. (2008). The Rag GTPases bind raptor and mediate amino acid signaling to mTORC1. *Science* **320,** 1496–1501.

Sanz, L., Sanchez, P., Lallena, M. J., Diaz, M. M., and Moscat, J. (1999). The interaction of p62 with RIP links the atypical PKCs to NF-kappaB activation. *EMBO J.* **18,** 3044–3053.

Sawka, V. D., Tartare, D. S., White, M. F., and Van, O. E. (1996). Insulin receptor substrate-2 binds to the insulin receptor through its phosphotyrosine-binding domain and through a newly identified domain comprising amino acids 591–786. *J. Biol. Chem.* **271,** 5980–5983.

Sawka-Verhelle, D., Baron, V., Mothe, I., Filloux, C., White, M. F., and Van Obberghen, E. (1997). Tyr624 and Tyr628 in insulin receptor substrate-2 mediate its association with the insulin receptor. *J. Biol. Chem.* **272,** 16414–16420.

Schmitz-Peiffer, C., Browne, C. L., Oakes, N. D., Watkinson, A., Chisholm, D. J., Kraegen, E. W., and Biden, T. J. (1997). Alterations in the expression and cellular localization of protein kinase C isozymes epsilon and theta are associated with insulin resistance in skeletal muscle of the high-fat-fed rat. *Diabetes* **46,** 169–178.

Seely, B. L., Reichart, D. R., Staubs, P. A., Jhun, B. H., Hsu, D., Maegawa, H., Milarski, K. L., Saltiel, A. R., and Olefsky, J. M. (1995). Localization of the insulin-like growth factor I receptor binding sites for the SH2 domain proteins p85, Syp, and GTPase activating protein. *J. Biol. Chem.* **270,** 19151–19157.

Serra, C., Federici, M., Buongiorno, A., Senni, M. I., Morelli, S., Segratella, E., Pascuccio, M., Tiveron, C., Mattei, E., Tatangelo, L., Lauro, R., Molinaro, M., *et al.* (2003). Transgenic mice with dominant negative PKC-theta in skeletal muscle: A new model of insulin resistance and obesity. *J. Cell. Physiol.* **196,** 89–97.

Sesti, G., Federici, M., Hribal, M. L., Lauro, D., Sbraccia, P., and Lauro, R. (2001). Defects of the insulin receptor substrate (IRS) system in human metabolic disorders. *FASEB J.* **15,** 2099–2111.

Shah, O. J., Wang, Z., and Hunter, T. (2004). Inappropriate activation of the TSC/Rheb/mTOR/S6K cassette induces IRS1/2 depletion, insulin resistance, and cell survival deficiencies. *Curr. Biol.* **14,** 1650–1656.

Sharfi, H., and Eldar-Finkelman, H. (2007). Sequential phosphorylation of insulin receptor substrate-2 by glycogen synthase kinase-3 and c-Jun NH2-terminal kinase plays a role in hepatic insulin signaling. *Am. J. Physiol. Endocrinol. Metab.* **294,** E307–E315.

Shargill, N. S., Tatoyan, A., el-Refai, M., Pleta, M., and Chan, T. M. (1986). Impaired insulin receptor phosphorylation in skeletal muscle membranes of db/db mice: The use of a novel skeletal muscle plasma membrane preparation to compare insulin binding and stimulation of receptor phosphorylation. *Biochem. Biophys. Res. Commun.* **137,** 286–294.

Shoelson, S. E., Lee, J., and Goldfine, A. B. (2006). Inflammation and insulin resistance. *J. Clin. Invest.* **116,** 1793–1801.

Smith, L. K., Vlahos, C. J., Reddy, K. K., Falck, J. R., and Garner, C. W. (1995). Wortmannin and LY294002 inhibit the insulin-induced down-regulation of IRS-1 in 3T3-L1 adipocytes. *Mol. Cell. Endocrinol.* **113,** 73–81.

Solinas, G., Naugler, W., Galimi, F., Lee, M. S., and Karin, M. (2006). Saturated fatty acids inhibit induction of insulin gene transcription by JNK-mediated phosphorylation of insulin-receptor substrates. *Proc. Natl. Acad. Sci. USA* **103,** 16454–16459.

Sommerfeld, M. R., Metzger, S., Stosik, M., Tennagels, N., and Eckel, J. (2004). *In vitro* phosphorylation of insulin receptor substrate 1 by protein kinase C-zeta: Functional analysis and identification of novel phosphorylation sites. *Biochemistry* **43,** 5888–5901.

Standaert, M. L., Galloway, L., Karnam, P., Bandyopadhyay, G., Moscat, J., and Farese, R. V. (1997). Protein kinase C-zeta as a downstream effector of phosphatidylinositol 3-kinase during insulin stimulation in rat adipocytes. Potential role in glucose transport. *J. Biol. Chem.* **272,** 30075–30082.

Sun, X. J., Rothenberg, P., Kahn, C. R., Backer, J. M., Araki, E., Wilden, P. A., Cahill, D. A., Goldstein, B. J., and White, M. F. (1991). Structure of the insulin receptor substrate IRS-1 defines a unique signal transduction protein. *Nature* **352,** 73–77.

Sun, X. J., Wang, L. M., Zhang, Y., Yenush, L., Myers, M. G., Jr., Glasheen, E., Lane, W. S., Pierce, J. H., and White, M. F. (1995). Role of IRS-2 in insulin and cytokine signalling. *Nature* **377,** 173–177.

Sun, X. J., Goldberg, J. L., Qiao, L. Y., and Mitchell, J. J. (1999). Insulin-induced insulin receptor substrate-1 degradation is mediated by the proteasome degradation pathway. *Diabetes* **48,** 1359–1364.

Sun, Z., Arendt, C. W., Ellmeier, W., Schaeffer, E. M., Sunshine, M. J., Gandhi, L., Annes, J., Petrzilka, D., Kupfer, A., Schwartzberg, P. L., and Littman, D. R. (2000). PKC-theta is required for TCR-induced NF-kappaB activation in mature but not immature T lymphocytes. *Nature* **404,** 402–407.

Tamemoto, H., Kadowaki, T., Tobe, K., Yagi, T., Sakura, H., Hayakawa, T., Terauchi, Y., Ueki, K., Kaburagi, Y., Satoh, S., Sekihara, H., Yoshioka, S., *et al.* (1994). Insulin resistance and growth retardation in mice lacking insulin receptor substrate-1. *Nature* **372,** 182–186.

Taniguchi, C. M., Emanuelli, B., and Kahn, C. R. (2006). Critical nodes in signalling pathways: Insights into insulin action. *Nat. Rev. Mol. Cell Biol.* **7,** 85–96.

Tirosh, A., Potashnik, R., Bashan, N., and Rudich, A. (1999). Oxidative stress disrupts insulin-induced cellular redistribution of insulin receptor substrate-1 and phosphatidylinositol 3-kinase in 3T3-L1 adipocytes. A putative cellular mechanism for impaired protein kinase B activation and GLUT4 translocation. *J. Biol. Chem.* **274,** 10595–10602.

Tremblay, F., Brule, S., Hee Um, S., Li, Y., Masuda, K., Roden, M., Sun, X. J., Krebs, M., Polakiewicz, R. D., Thomas, G., and Marette, A. (2007). Identification of IRS-1 Ser-1101 as a target of S6K1 in nutrient- and obesity-induced insulin resistance. *Proc. Natl. Acad. Sci. USA* **104,** 14056–14061.

Tyers, M., and Willems, A. R. (1999). One ring to rule a superfamily of E3 ubiquitin ligases. *Science* **284,** 603–604.

Tzatsos, A., and Tsichlis, P. N. (2007). Energy depletion inhibits phosphatidylinositol 3-kinase/Akt signaling and induces apoptosis via AMP-activated protein kinase-dependent phosphorylation of IRS-1 at Ser-794. *J. Biol. Chem.* **282,** 18069–18082.

Ullrich, A., Bell, J. R., Chen, E. Y., Herrera, R., Petruzelli, L. M., Dull, T. J., Gray, A., Coussens, L., Liao, Y. C., Tsubokawa, M., Mason, A., Seeburg, P. H., *et al.* (1985). Human insulin receptor and its relationship to the tyrosine kinase family of oncogenes. *Nature* **313,** 756–761.

Ullrich, A., Gray, A., Tam, A., Yang-Feng, W. T., Tsubokawa, M., Collins, C., Henzel, W., Le Bon, T., Kathuria, S., Chen, E., Jacobs, S., Francke, U., *et al.* (1986). Insulin-like growth factor I receptor primary structure: Comparison with insulin receptor suggests structural determinants that define functional specificity. *EMBO J.* **5,** 2503–2512.

Um, S. H., Frigerio, F., Watanabe, M., Picard, F., Joaquin, M., Sticker, M., Fumagalli, S., Allegrini, P. R., Kozma, S. C., Auwerx, J., and Thomas, G. (2004). Absence of S6K1 protects against age- and diet-induced obesity while enhancing insulin sensitivity. *Nature* **431,** 200–205.

Um, S. H., D'Alessio, D., and Thomas, G. (2006). Nutrient overload, insulin resistance, and ribosomal protein S6 kinase 1, S6K1. *Cell Metab.* **3,** 393–402.

Usui, I., Imamura, T., Babendure, J. L., Satoh, H., Lu, J. C., Hupfeld, C. J., and Olefsky, J. M. (2005). G protein-coupled receptor kinase 2 mediates endothelin-1-induced insulin resistance via the inhibition of both Galphaq/11 and insulin receptor substrate-1 pathways in 3T3-L1 adipocytes. *Mol. Endocrinol.* **19,** 2760–2768.

Venters, H. D., Tang, Q., Liu, Q., VanHoy, R. W., Dantzer, R., and Kelley, K. W. (1999). A new mechanism of neurodegeneration: A proinflammatory cytokine inhibits receptor signaling by a survival peptide. *Proc. Natl. Acad. Sci. USA* **96,** 9879–9884.

Voliovitch, H., Schindler, D. G., Hadari, Y. R., Taylor, S. I., Accili, D., and Zick, Y. (1995). Tyrosine phosphorylation of insulin receptor substrate-1 *in vivo* depends upon the presence of its pleckstrin homology region. *J. Biol. Chem.* **270,** 18083–18087.

Wang, Q., Somwar, R., Bilan, P. J., Liu, Z., Jin, J., Woodgett, J. R., and Klip, A. (1999). Protein kinase B/Akt participates in GLUT4 translocation by insulin in L6 myoblasts. *Mol. Cell. Biol.* **19,** 4008–4018.

Wellen, K. E., and Hotamisligil, G. S. (2005). Inflammation, stress, and diabetes. *J. Clin. Invest.* **115,** 1111–1119.

White, M. F. (2002). IRS proteins and the common path to diabetes. *Am. J. Physiol. Endocrinol. Metab.* **283,** E413–E422.

Withers, D. J., Gutierrez, J. S., Towery, H., Burks, D. J., Ren, J. M., Previs, S., Zhang, Y., Bernal, D., Pons, S., Shulman, G. I., Bonner, W. S., and White, M. F. (1998). Disruption of IRS-2 causes type 2 diabetes in mice. *Nature* **391,** 900–904.

Withers, D. J., Burks, D. J., Towery, H. H., Altamuro, S. L., Flint, C. L., and White, M. F. (1999). Irs-2 coordinates Igf-1 receptor-mediated beta-cell development and peripheral insulin signalling. *Nat. Genet.* **23,** 32–40.

Wolf, G., Trüb, T., Ottinger, E., Groninga, L., Lynch, A., White, M. F., Miyazaki, M., Lee, J., and Shoelson, S. E. (1995). PTB domains of IRS-1 and Shc have distinct but overlapping binding specificities. *J. Biol. Chem.* **270,** 27407–27410.

Wu, J., Tseng, Y. D., Xu, C. F., Neubert, T. A., White, M. F., and Hubbard, S. R. (2008). Structural and biochemical characterization of the KRLB region in insulin receptor substrate-2. *Nat. Struct. Mol. Biol.* **15,** 251–258.

Wullschleger, S., Loewith, R., and Hall, M. N. (2006). TOR signaling in growth and metabolism. *Cell* **124,** 471–484.

Yamauchi, T., Tobe, K., Tamemoto, H., Ueki, K., Kaburagi, Y., Yamamoto, H. R., Takahashi, Y., Yoshizawa, F., Aizawa, S., Akanuma, Y., Sonenberg, N., Yazaki, Y., *et al.* (1996). Insulin signalling and insulin actions in the muscles and livers of insulin-resistant, insulin receptor substrate 1-deficient mice. *Mol. Cell. Biol.* **16,** 3074–3084.

Yang, Q., and Guan, K. L. (2007). Expanding mTOR signaling. *Cell Res.* **17,** 666–681.

Yasukawa, H., Sasaki, A., and Yoshimura, A. (2000). Negative regulation of cytokine signaling pathways. *Annu. Rev. Immunol.* **18,** 143–164.

Yi, Z., Langlais, P., De Filippis, E. A., Luo, M., Flynn, C. R., Schroeder, S., Weintraub, S. T., Mapes, R., and Mandarino, L. J. (2007). Global assessment of regulation of phosphorylation of insulin receptor substrate-1 by insulin *in vivo* in human muscle. *Diabetes* **56,** 1508–1516.

Youngren, J. F. (2007). Regulation of insulin receptor function. *Cell. Mol. Life Sci.* **64,** 873–891.

Yu, C., Chen, Y., Cline, G. W., Zhang, D., Zong, H., Wang, Y., Bergeron, R., Kim, J. K., Cushman, S. W., Cooney, G. J., Atcheson, B., White, M. F., *et al.* (2002a). Mechanism by which fatty acids inhibit insulin activation of insulin receptor substrate-1 (IRS-1)-associated phosphatidylinositol 3-kinase activity in muscle. *J. Biol. Chem.* **277,** 50230–50236.

Yu, C., Chen, Y., Zong, H., Wang, Y., Bergeron, R., Kim, J. K., Cline, G. W., Cushman, S. W., Cooney, G. J., Atcheson, B., White, M. F., Kraegen, E. W., *et al.* (2002b). Mechanism by which fatty acids inhibit insulin activation of IRS-1 associated phosphatidylinositol 3-kinase activity in muscle. *J. Biol. Chem.* **277,** 50230–50236.

Yuan, M., Konstantopoulos, N., Lee, J., Hansen, L., Li, Z.-W., Karin, M., and Shoelson, S. E. (2001). Reversal of obesity and diet induced insulin resistance with salicylates or targeted disruption of IKK beta. *Science* **293,** 1673–1677.

Zhande, R., Mitchell, J. J., Wu, J., and Sun, X. J. (2002). Molecular mechanism of insulin-induced degradation of insulin receptor substrate 1. *Mol. Cell. Biol.* **22,** 1016–1026.

Zhang, J. (2007). The direct involvement of SirT1 in insulin-induced insulin receptor substrate-2 tyrosine phosphorylation. *J. Biol. Chem.* **282,** 34356–34364.

Zhou, B. P., Liao, Y., Xia, W., Zou, Y., Spohn, B., and Hung, M. C. (2001). HER-2/neu induces p53 ubiquitination via Akt-mediated MDM2 phosphorylation. *Nat. Cell Biol.* **3,** 973–982.

Zick, Y. (2001). Insulin resistance: A phosphorylation-based uncoupling of insulin signaling. *Trends Cell Biol.* **11,** 437–441.

Zick, Y. (2003). Role of Ser/Thr kinases in the uncoupling of insulin signaling. *Int. J. Obes. Relat. Metab. Disord.* **27**(Suppl. 3), S56–S60.

Zick, Y. (2004). Uncoupling insulin signalling by serine/threonine phosphorylation: A molecular basis for insulin resistance. *Biochem. Soc. Trans.* **32,** 812–816.

Zick, Y. (2005). Ser/Thr phosphorylation of IRS proteins: A molecular basis for insulin resistance. *Sci. STKE 2005.* pe4.

CHAPTER THIRTEEN

Phosphorylation of IRS Proteins: Yin-Yang Regulation of Insulin Signaling

Xiao Jian Sun* *and* Feng Liu†

Contents

Abstract

Growing evidence reveals that insulin signal pathway is not static, but is rather a dynamic, flexible, and fed in by negative (Yin) and positive (Yang) regulation in response to environmental changes. Normal insulin response reflects the balance between Yin and Yang regulation acting upon insulin signaling pathway.

* Department of Medicine, The University of Chicago, Chicago, Illinois 60637
† Departments of Pharmacology and Biochemistry, The University of Texas Health Science Center, San Antonio, Texas 78229

Vitamins and Hormones, Volume 80
ISSN 0083-6729, DOI: 10.1016/S0083-6729(08)00613-4

Conceivably, imbalance between the Yin and Yang results in abnormal insulin sensitivity such as insulin resistance. IRS-proteins are insulin receptor substrates that mediate insulin signaling via multiple tyrosyl phosphorylations. However, they are also substrates for many serine/threonine kinases downstream of other signaling network and become serine phosphorylated in response to various conditions such as inflammation, stress and over nutrients. The serine phosphorylation of IRS-proteins alters the capacities of IRS-proteins to be phosphorylated on tyrosyl, therefore, able to mediate insulin signaling. The unique structure of IRS-proteins render them idea molecules to fulfill the task to sense the environmental cues and integrate them into insulin sensitivity through serine/threonine phosphorylation. This review intends to summarize the role of IRS-proteins in insulin signaling with focuses on the role of Yin and Yang regulation of insulin signaling pathway. Understanding the dynamic of these complicated regulation net work not only provide us a complete picture of what happens in the normal conditions, but also pathaphysiological conditions such as obesity and insulin resistance.

I. Introduction

Insulin, through distinct and well-defined signaling pathways, regulates a variety of important biological events such as glucose homeostasis, protein and lipid metabolism, cell proliferation, and gene expression (Rosen, 1987). Dysregulation of signaling components downstream of the insulin receptor leads to insulin resistance and associated disorders, most notably type-2 diabetes.

Insulin signaling is initiated by the binding of insulin to its plasma membrane receptor, which is composed of two extracellular α-subunits and two transmembrane β-subunits (Kasuga *et al.*, 1982; Ullrich *et al.*, 1985). When the α-subunit of the receptor is bound by insulin, a conformational change is induced, which results in autophosphorylation of tyrosine residues of the β-subunits. The phosphorylated tyrosines function as docking sites for adaptor proteins which may positively or negatively regulate insulin signaling and action; notable examples of such adaptor molecules include the insulin receptor substrate (IRS) proteins, Shc, Grb14, and Grb10 amongst others. During the past three decades, over 100 molecules directly or indirectly involved in the insulin signaling network have been identified (Taniguchi *et al.*, 2006). Among said molecules, IRS-1 and IRS-2 stand out as key regulatory points with respect to insulin signaling. Multiple tyrosine and serine residues in IRS-1 and IRS-2 are targeted by the insulin receptor (IR) tyrosine kinase as well as by many distal serine/threonine protein kinases; this occurs in response to stimulation by various growth factors, stressors, and nutrients. Thus, by serving as a convergence point for diverse extracellular stimuli, phosphorylation of IRS-1/2 provides a mechanistic framework for understanding crosstalk between insulin and other signaling pathways as a means to fine-tune the efficiency of insulin signaling. In the present review, we will discuss recent developments in our understanding of the phosphorylation of the IRS proteins as it pertains to the regulation of insulin signal transduction.

II. Discovery of the IRS Proteins

IRS was first identified by antiphosphotyrosine (αPY) immunoprecipitation as a 185-kDa tyrosyl-phosphorylated protein (pp185) in insulin-stimulated FAO cells (White *et al.*, 1985). Several IRS isoforms were subsequently found in almost all tissues and cells with molecular weights ranging from 60 to 185 kDa. Using antiphosphotyrosine (αPY) affinity chromatography, a protein with a molecular weight of 185 kDa (pp185) was purified from insulin-treated rat liver, and its cDNA subsequently cloned from a rat liver cDNA library (Rothenberg *et al.*, 1991; Sun *et al.*, 1991). The cloned protein was directly tyrosyl-phosphorylated *in vitro* by the purified IR, and underwent insulin-stimulated tyrosine phosphorylation in intact cells coexpressing the IR. As the first of the IRS isoforms to be purified, the protein was thusly termed IRS-1 (Sun *et al.*, 1991, 1992).

Surprisingly, when IRS-1 was depleted by immunoprecipitation using a specific antibody against it, another tyrosine-phosphorylated protein ($pp185^{HMW}$) with an apparent molecular mass slightly larger than that of IRS-1 was identified in insulin-stimulated FAO cells (Miralpeix *et al.*, 1992) as well as in FDC-P1 and FDC-P2 myeloid progenitor cell lines (Wang *et al.*, 1992, 1993), indicating the presence of IRS-1 isoforms. In myeloid progenitor cells, both IL-4 and insulin were found to induce tyrosine phosphorylation of a 185 kDa protein, initially named 4PS (IL-*4*-induced *p*hosphotyrosine *s*ubstrate) (Wang *et al.*, 1992, 1993). The 4PS protein was later purified from insulin-stimulated FDC-P2 cells by affinity chromatography using a GST-fusion protein containing the nSH2 domain of the p85 subunit of PI3-kinase. The cDNA encoding 4PS, which was subsequently cloned from a mouse cDNA library, displayed structural and functional similarities to IRS-1 and was appropriately designated IRS-2 (Sun *et al.*, 1995, 1997). Soon after, the IRS-1 gene was mapped in mouse chromosome 1 and human chromosome 2q36–37, while the IRS-2 gene was mapped in the short arm of mouse chromosome 8 (near the insulin receptor gene) and human 13g34 (Araki *et al.*, 1993; Sun *et al.*, 1997). Several additional IRS isoforms were subsequently identified including IRS-3 (pp60) (Lavan *et al.*, 1997b; Smith-Hall *et al.*, 1997) and IRS-4 (pp160) (Lavan *et al.*, 1997a). These IRS isoforms were found to be located on mouse chromosomes 5 7q21–22 and X q21–22, respectively (Fantin *et al.*, 1999). The functional roles of the IRS isoforms, particularly IRS-1 and IRS-2, were initially thought to be limited to the mediation of insulin signaling; it has since been realized that the IRS proteins also participate in crosstalk with other signaling pathways and may positively or negatively regulate insulin signaling.

In addition to the IRS proteins, the existence of several IRS-related proteins with comparatively less specificity for the insulin receptor tyrosine kinase has been reported. Gab1 (Grb2-associated binder-1) is an adaptor protein that shares amino acid homology and several structural features with IRS-1, and serves as a substrate for both the EGF receptor and the IR (Holgado-Madruga *et al.*, 1996). Other insulin receptor tyrosine kinase substrates include APS (*a*dapter *p*rotein containing PH and *S*H2 domain) (Moodie *et al.*, 1999), Shc (Kaburagi *et al.*, 1995; Sasaoka *et al.*, 1994), c-Cbl (Ribon and Saltiel, 1997), stat5 (Chen *et al.*, 1997), and downstream of kinase proteins (pp62/DOK-1, DOK-2, DOK-3, IRS-5/DOK-4, and IRS-6/DOK-5) (Cai *et al.*, 2003; Cong *et al.*, 1999; Hosomi *et al.*, 1994; Yeh *et al.*, 1996). However, because these proteins also participate significantly in signaling pathways of other (noninsulin) growth factors, and because IRS-1 and IRS-2 are principally involved in insulin signaling, the present review will focus on the latter two molecules.

III. Molecular Structure of the IRS Proteins

The overall amino acid sequence identity between IRS-1 and IRS-2 is only 43%. Three regions within the amino-terminal half of the molecules, though, including the pleckstrin homology (PH) domain, the phosphotyrosine-binding (PTB) domain, and the IRS-homology-3 (IH-3) domain show high sequence homology between the two IRS isoforms (Sun *et al.*, 1995). The significantly tighter homology within the aforementioned regions implies functional importance of these domains in insulin signaling. In addition to said functional domains, IRS proteins contain multiple tyrosine, serine, and threonine phosphorylation sites which are the major structural components that facilitate communication between insulin and other signaling pathways.

A. PH domain

The PH domains of mouse IRS-1 (13–113aa) and mouse IRS-2 (31–142aa) are 62.5% identical and 70% similar in sequence, a likeness significantly greater than that observed between the molecules in their entirety. As mentioned previously, high homology suggests functional significance—

a tenet which applies in this case, as the PH domain is crucial for the localization of IRS-1 and IRS-2 to the activated IR and for their subsequent tyrosine phosphorylation by the insulin receptor tyrosine kinase.

PH domains were first recognized as internal repeats in pleckstrin, a major substrate of PKC in platelets (Musacchio *et al.*, 1993). They are composed of about 100 amino acid residues, and have been found to interact with proteins and phospholipids (Gibson *et al.*, 1994; Macias *et al.*, 1994). The PH domains of IRS-1, -2, and -3 bind to the 3-phosphorylated phosphoinositides with different specificities, pointing toward differential regulation of the cellular location of the IRS isoforms (Razzini *et al.*, 2001). Consistent with this observation, one of the functional roles of the PH domain in IRS proteins is to bring substrates from the cytosol to the plasma membrane, where the IR resides. The PH domain of IRS-1 has also been shown to interact with $G_{\beta\gamma}$-subunits, which would allow crosstalk between IRS proteins and trimeric G proteins (Luttrell *et al.*, 1995; Pitcher *et al.*, 1995).

B. PTB domain

The PTB domains of mouse IRS-1 (aa 155–309) and IRS-2 (aa 191–350) are 75% identical and 85% similar and are the most highly conserved regions among IRS proteins. The PTB domain of IRS-1 binds specifically to the phosphorylated NPXY peptide corresponding to the juxtamembrane region of the IR (O'Neill *et al.*, 1994; Sawka-Verhelle *et al.*, 1996; Wolf *et al.*, 1995); IRS-1 with a truncation deletion in the PTB domain fails to be phosphorylated by the IR in COS and hepatoma cells (O'Neill *et al.*, 1994; Tanaka and Wands, 1996). Thus, a main function of the PTB domain in IRS is to interact with the active insulin receptor tyrosine kinase in response to insulin stimulation, a key step in the tyrosyl phosphorylation of IRS.

In addition to interaction with the IR, the PTB domain of IRS proteins has also been shown to bind to inositol phosphates/phosphoinositides (Takeuchi *et al.*, 1998). Therefore, IRS protein signaling through the PTB domain may be regulated not only by binding to phosphotyrosine residues of the insulin receptor, but also by interaction with inositol phosphates/ phosphoinositides.

C. Other regions

The least well-conserved homologous region (IH-3 domain) between mouse IRS-1 (315–464aa) and IRS-2 (352–541aa) shares a certain degree of similarity with Shc and is involved in the interaction of IRS with the NPXY-motif of the IR (Craparo *et al.*, 1995; Gustafson *et al.*, 1995; He *et al.*, 1995). This region has thus been designated as the SAIN domain for *S*hc *a*nd *I*RS-1 *N*PXY-binding domain. However, the importance of the SAIN domain is questionable based on studies using deletion mutants where IRS-1 was found to still be phosphorylated in SAIN deletion mutants so long as the PTB domain was kept intact (Gustafson *et al.*, 1995; Tanaka and Wands, 1996).

Aside from the SAIN domain, a new interacting region was identified only in IRS-2 (591–786aa), but not in IRS-1 that also interacts with the catalytic domain of the IR. Interaction in this case requires phosphorylation of the IR at Tyr^{1146}, Tyr^{1150}, and $Tyr1^{151}$ (Sawka-Verhelle *et al.*, 1996, 1997). This unique feature of IRS-2 suggests that IRS-1 and IRS-2 have unique functions in transduction of the insulin signal, resulting in diverse cellular effects.

D. Tyrosine phosphorylation sites and the binding of SH2 domain-containing proteins

One of the major and important characteristics of IRS proteins is their rapid tyrosine phosphorylation following insulin stimulation (Sun *et al.*, 1992, 1993). There are 21 and 23 potential tyrosine phosphorylation sites in IRS-1 and IRS-2, respectively, and many of the potential tyrosine phosphorylation sites are conserved between IRS-1 and IRS-2. Eight tyrosine phosphorylation sites in IRS-1 (Tyr^{460}, Try^{608}, Try^{628}, Tyr^{895}, Tyr^{939}, Tyr^{985}, Tyr^{1172}, and Tyr^{1222}) have been confirmed *in vitro* and *in vivo* as IR-mediated phosphorylation sites (Sun *et al.*, 1993); among them, six are well conserved in IRS-2 (Tyr^{649}, Try^{671}, Tyr^{911}, Tyr^{969}, Tyr^{1242}, and Tyr^{1303}) (Sun *et al.*, 1995). A number of these tyrosine phosphorylation sites are surrounded by different sequence motifs, indicating that phosphorylation of these sites plays not only an important, but also a unique role in signaling.

One important outcome of IRS tyrosine phosphorylation is the creation of binding sites for Src homology-2 (SH2) domain-containing proteins (Pawson and Gish, 1992; Songyang *et al.*, 1993). SH2 domains are small protein modules composed of approximately 100 amino acids which are frequently found in signal transduction molecules having distinct biological functions (Koch *et al.*, 1991; Pawson, 1995). The selectivity and specificity of the interaction between an SH2 domain-containing protein and its target depends on both the amino acid sequence surrounding the phosphotyrosine of the target protein and the amino acid sequence of the SH2 domain (Songyang *et al.*, 1993).

A key component of insulin signaling which relies upon IRS tyrosine phosphorylation is the phosphatidylinositol 3-kinase (PI3-K). PI3-K consists of an 85-kDa regulatory subunit (p85) with two SH2 domains at its N-terminus and a 110-kDa subunit (p110) with intrinsic catalytic activity (Dhand *et al.*, 1994a). In response to insulin stimulation, the SH2 domains of the p85 subunit of PI3-K bind with high affinity to the tyrosine-phosphorylated Y^{608}MPM and Y^{939}MNM sequences in IRS-1 (Y^{649}MPM and Y^{969}MNL in IRS-2) (Sun *et al.*, 1993, 1995). The interaction of both SH2 domains of p85 with IRS-1 is critical for full activation of PI3-K (Rordorf-Nikolic *et al.*, 1995). Although the IR has been reported to associate with and activate PI3-kinase, it is generally believed that insulin-stimulated PI3-K

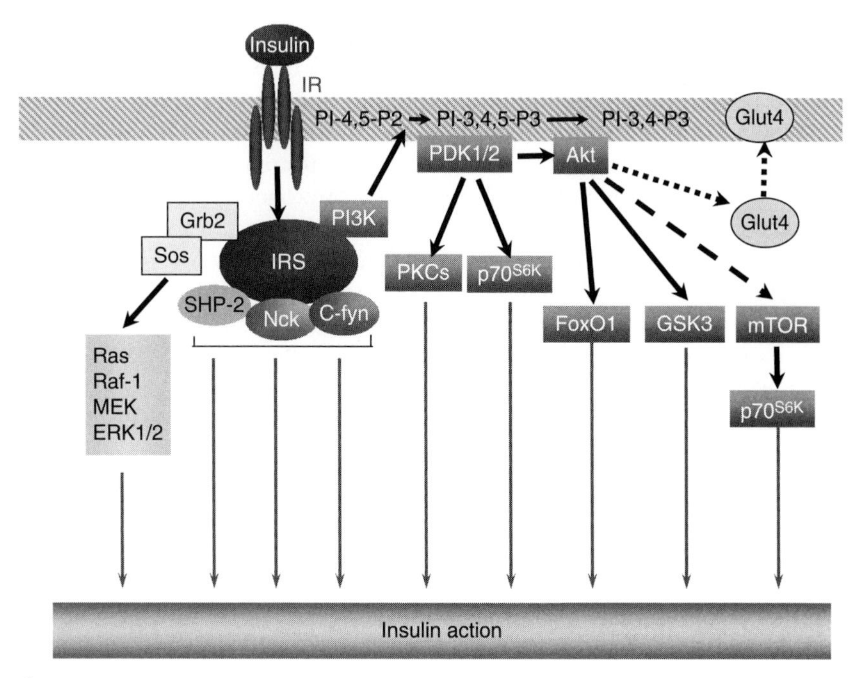

Figure 13.1 IRS-protein signaling pathway. Upon insulin binding, insulin receptor tyrosine kinase becomes activated and phosphorylates IRS proteins on multiple tyrosine residues which serve as binding sites for many downstream signaling molecules.

activation occurs mainly through interaction with IRS proteins (Staubs *et al.*, 1994; Tartare-Deckert *et al.*, 1996; Van Horn *et al.*, 1994).

PI3-K catalyzes the phosphorylation of phosphatidylinositol at the 3′-position and generates 3′-phosphatidylinositol products which recruit PDKs and Akt to the plasma membrane; once properly translocated, Akt is activated through phosphorylation via the PDKs. Activation of Akt results in phosphorylation of downstream molecules including mTOR/$p70^{S6K}$, FoxO1, and GSK-3β (Fig. 13.1) (Dummler and Hemmings, 2007; Fingar and Blenis, 2004). Binding of PI3-K to tyrosyl-phosphorylated IRS proteins, therefore, mediates the PI3-K branch of insulin signaling, which is critical for the majority of insulin's metabolic effects.

In addition to PI3-K, IRS-1 has also been shown to interact with Grb2 in response to insulin stimulation (Baltensperger *et al.*, 1993). Grb2 is a small SH2 and SH3 domain-containing protein that associates with the mammalian homologue of Son of Sevenless (SOS), a guanidine nucleotide exchange factor, to regulate $p21^{ras}$ activity and the MAP kinase pathway (Skolnik *et al.*, 1993). Activation of the MAP kinase cascade is important for insulin action in the nucleus and regulation of gene transcription (Chang and Karin, 2001). The SH2 domain of Grb2 binds to the Y^{895}VNI-motif in IRS-1, which corresponds to Y^{911}INI in IRS-2 (Myers *et al.*, 1994; Sun *et al.*, 1993,

1995). Insulin regulates the p44/42 MAP kinase activation at least partially through the formation of an IRS-1/Grb2/Sos complex (Baltensperger *et al.*, 1993). Specific mutation of this site completely abolishes the binding of Grb2 to IRS-1 and diminishes IRS-1-mediated MAP kinase activation (Myers *et al.*, 1994). In contrast to the PI3-K arm of insulin signaling, the MAP kinase branch is largely responsible for insulin's mitogenic actions (Fig. 13.1).

Although PI3-K and Gtb2 facilitate the bulk of insulin's known effects, other IRS-1/2-binding partners play important regulatory roles with respect to signaling intensity and duration. SHP-2 (also called syp, SHPTP2, PTP1D, PTP2C, and PTPN11) is a protein tyrosine phosphatase that contains two SH2 domains in its N-terminal region (Chan and Feng, 2007; Plutzky *et al.*, 1992; Sugimoto *et al.*, 1994). Insulin stimulates the binding of SHP-2 to tyrosine-phosphorylated IRS-1 at the Y^{1172}IDL- and the Y^{1222}ASI-motifs, which corresponds to the Y^{1242}IAI- and Y^{1303}ASI-motifs in IRS-2, leading to activation of phosphatase activity (Sun *et al.*, 1993). As is the case with PI3-K, the regulation of SHP-2 by IRS-1 also requires that both SH2 domains bind (Eck *et al.*, 1996b; Myers *et al.*, 1998; Pluskey *et al.*, 1995). C-fyn, a member of the Src kinase family, associates with IRS-1 at two conserved tyrosine phosphorylation sites (Y^{895} and Y^{1172}) in response to insulin (Sun *et al.*, 1996). IRS-1 also binds to nck, a small adapter molecule that contains an SH2 domain; the IRS-1-binding site for nck has not yet been assigned (Lee *et al.*, 1993).

E. Serine/threonine phosphorylation sites of IRS proteins and their roles in modulating insulin signaling

In addition to tyrosine phosphorylation, both IRS-1 and IRS-2 contain over 50 consensus phosphorylation sequences for serine/threonine kinases including casein kinase II (CK-II) (S/T-x-x-E/D), cAMP-dependent protein kinase (R/K-R/K-x-S/T), protein kinase C (PKC) (S/T-x-R/K), cdc2 kinase (S/T-P-x-K/R), and MAP kinase (P-x-S/T-P or x-x-S/T-P) (Davis, 1993; Sun *et al.*, 1991). Not surprisingly, IRS proteins are heavily phosphorylated on serine and threonine even in the basal state; additional serine/threonine phosphorylation is induced in response to insulin stimulation, suggesting that IRS serine/threonine phosphorylation plays an important regulatory role in insulin signaling (Sun *et al.*, 1992).

Early studies presented evidence that serine/threonine phosphorylation of IRS-1 attenuates insulin signaling in cultured cells and may be involved in insulin resistance. Treatment of 3T3-L1 adipocytes with okadaic acid, a potent and specific inhibitor of type 1 and 2A protein phosphatase (PP-1 and PP-2A), induces a dose-dependent inhibition of insulin-stimulated IRS-1 tyrosine phosphorylation, PI3-K activation, and glucose uptake without affecting IR tyrosine phosphorylation or kinase activity (Tanti *et al.*, 1991, 1994b). The okadaic acid-induced effects are associated with a reduced electrophoretic mobility of IRS proteins on SDS-PAGE due to serine/threonine phosphorylation of these adaptor proteins. Activation of PKC isoforms by treating cells with TPA also induces serine/threonine phosphorylation of IRS-1, leading to a decrease in insulin-stimulated IRS-1 tyrosine phosphorylation and PI3-K activation (Giorgetti *et al.*, 1993).

It is now well recognized that serine phosphorylation of IRS proteins is associated with altered insulin action and underlies the molecular mechanism of insulin resistance (Gual *et al.*, 2005; Le Marchand-Brustel *et al.*, 2003; White, 2002). Searching for the responsible kinases has led to the identification of many serine/threonine kinases that phosphorylate IRS-1 and modify insulin signaling (Table 13.1 and Fig. 13.2). Serine^{307} in rat and mouse IRS-1 (serine^{312} in human) is the most well-studied phosphorylation site in IRS-1. It was initially identified as a phosphorylation site *in vivo* and *in vitro* for JNK in TNFα-treated cells (Aguirre *et al.*, 2000). Mutation of serine^{307} to alanine eliminates phosphorylation of IRS-1 by JNK and abrogates the inhibitory effect of TNFα on insulin signaling. Indeed, serine^{307} phosphorylation of IRS-1 is dramatically reduced in JNK knockout mice; moreover, these mice are protected from high-fat diet-induced insulin resistance (Hirosumi *et al.*, 2002). Increased serine^{307} phosphorylation of IRS-1 can also be found when insulin resistance is induced by a myriad of other factors including insulin/IGF-1, free fatty acids, amino acids, and various stress signals (Carlson *et al.*, 2004; Gual *et al.*, 2003; Jaeschke and Davis, 2007; Ozcan *et al.*, 2006; Rui *et al.*, 2001a; Solinas *et al.*, 2006). Most of these factors are able to activate JNK, suggesting that JNK negatively regulates insulin signaling through phosphorylation of IRS-1 at serine^{307}. Similarly, serine^{332} (serine^{337} in human) is phosphorylated by GSK-3β, resulting in the inhibition of insulin signaling (Eldar-Finkelman and Krebs, 1997; Liberman and Eldar-Finkelman, 2005). Although both serine^{307} and serine^{332} are well conserved in mouse, rat, and human IRS-1, such conservation is not observed in IRS-2 (Table 13.1).

Serine^{612} and serine^{632} (serine^{616} and serine^{636} in human IRS-1) are typical MAP kinase phosphorylation sites and have been identified by *in vivo* and *in vitro* methods as targets for the MAP kinase (De Fea and Roth, 1997; Yi *et al.*, 2005). Moreover, using liver extracts as a kinase source, phosphorylation of a synthetic IRS-1 peptide containing a MAP kinase site (serine^{612}) was significantly higher in *ob/ob* mice than in lean controls (De Fea and Roth, 1997), implicating increased MAP kinase activity as a contributing mechanism to insulin resistance in this model. Phorbol 12-myristate 13-acetate (PMA) is known to activate MAP kinase through PKC isoforms, resulting in phosphorylation of these phosphorylation sites (serine^{612} and serine^{632}) and insulin resistance. Converting serines to alanines within these sites in IRS-1 prevents PMA-induced insulin resistance (De Fea and Roth, 1997; Li *et al.*, 1999). In contrast to serine^{307}, serines$^{612/632}$ are well conserved between IRS-1 and IRS-2, indicating that IRS-2 may also be a substrate for MAP kinase, and that the latter kinase may

Table 13.1 The list of published serine/threonine phosphorylation sites in IRS-1

IRS-1			mIRS-2			
Mouse	Rat	Human	Conserved in mIRS-2	Enzymes	Effect	References
ps24	ps24	ps24	No	mPLK/hIRAK	Attenuation	Kim *et al.* (2005)
ps24	ps24	ps24	No	PKC	Attenuation	Greene *et al.* (2006) and Nawaratne *et al.* (2006)
pS99	pS99	pS99	No	Casein kinase II		Tanasijevic *et al.* (1993)
pS302	pS302	pS307	High (S343)	mTOR cascade	Enhancing	Giraud *et al.* (2004)
pS302/pS307	pS302/pS307	pS307/pS312	High (S343)/No	JNK	Attenuation	Werner *et al.* (2004)
pS307	pS307	pS312	No	JNK	Attenuation	Aguirre *et al.* (2000)
pS318	pS318	pS323	High (S355)	PKCζ/PKCδ	Attenuation	Moeschel *et al.* (2004) and Weigert *et al.* (2005)
pS332/pS336	pS332/pS336	pS337/pS341	No/high (S381)	GSK-3	Attenuation	Eldar-Finkelman and Krebs (1997) and Liberman and Eldar-Finkelman (2005)
No	pT502	No	No	Casein kinase II		Tanasijevic *et al.* (1993)
pS522	pS522	pS527	High (S573)	Akt1	Attenuation	Giraud *et al.* (2007)
pS612/pS632	pS612/pS632	pS616/pS636	High (S653/S675)	MAP kinase	Attenuation	De Fea and Roth (1997), Li *et al.* (1999), and Yi *et al.* (2005)

pS632/pS635	pS632/pS635	pS636/pS639	High (S675/678)	mTOR	Attenuation	Ozes *et al.* (2001)
pS632	pS632	pS636	High (S675)	MAP kinase	Attenuation	Bouzakri *et al.* (2003)
pS632/pS635	pS632/pS635	pS636/pS639	High (S675/678)	Rho-kinase	Enhancing	Furukawa *et al.* (2005)
pS789	pS789	pS794	Low	AMP kinase	Enhancing	Jakobsen *et al.* (2001)
pS789	pS789	pS794	Low	SIK-2	Attenuation	Horike *et al.* (2003) and Qiao *et al.* (2002)
pS1097	pS1100	pS1101	High (S1138)	PKCθ/p70^{S6K}	Attenuation	Li *et al.* (2004) and Tremblay *et al.* (2007)
pS570/No/1097/1136	pS570/No/1100/1135	pS574/629/1101/1142		PKA		Yi *et al.* (2005)
pS1214	pS1216	pS1223	Low	PKA	Enhancing	Luo *et al.* (2005) and Yi *et al.* (2005)

pS, phosphoserine residues; pT, phosphothreonine residues. Position of phosphorylation sites in mouse, rat, human IRS-1, and mouse IRS-2 are based on the protein sequences published in Sun *et al.* (1995). Highly conserved residues are indicated as "High," low, "Low," and no conserved residues, "No."

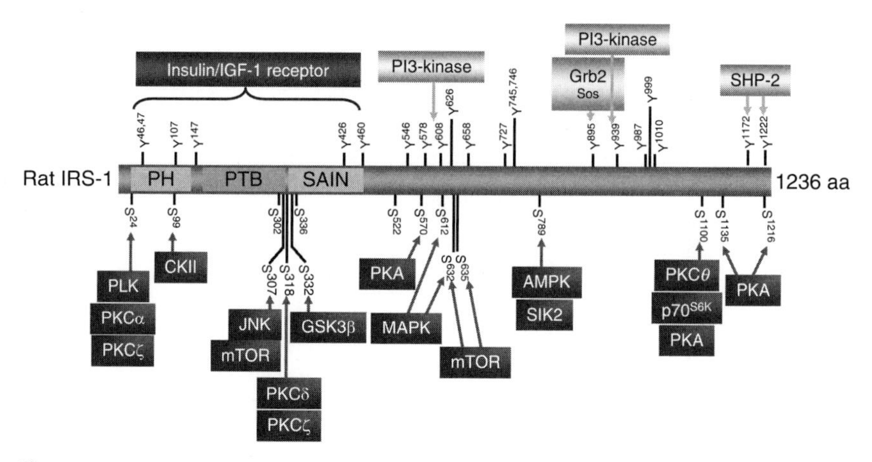

Figure 13.2 Tyrosine and serine phosphorylation of IRS-1. IRS-1 is phosphorylated on tyrosine residues by insulin receptor/IGF-1 receptor tyrosine kinases, eliciting the formation of signaling complexes with multiple downstream proteins (green). IRS-1 is also phosphorylated by many serine/threonine kinases (red) which predetermine the tyrosine phosphorylation events. (See Color Insert.)

negatively regulate insulin signaling via both IRS-1 and IRS-2 through a common mechanism.

Diversity of serine/threonine phosphorylation of IRS proteins is not only reflected by the existence of multiple phosphorylation sites in IRS proteins, but also by the fact that the same site can be phosphorylated by different kinases. Serine^{1100} (serine^{1101} in humans) was initially found to be phosphorylated by PKCθ (Li *et al.*, 2004). However, this site can also be phosphorylated by p70^{S6K}, with resultant attenuation of the insulin signal (Tremblay *et al.*, 2007). This phenomenon is not uncommon; serine^{24} can be phosphorylated by several serine kinases including the mouse pelle-like kinases, PKCα, and PKCδ. In all of these examples, phosphorylation of serine^{24} reduces insulin signaling in cultured cells (Greene *et al.*, 2006; Kim *et al.*, 2005; Nawaratne *et al.*, 2006).

Serine phosphorylation of IRS proteins does not always attenuate insulin signaling. Phosphorylation of serine^{1216} of IRS-1 (serine^{1223} in human IRS-1) has been shown to increase net insulin-stimulated tyrosine phosphorylation (of IRS-1), possibly by decreasing association of the protein tyrosine phosphatase SHP-2 with IRS-1 (Luo *et al.*, 2005). Furthermore, certain sites demonstrate differing associations with insulin resistance, resulting in conflicting conclusions. As an example, following TNFα treatment, phosphorylation of serines$^{632/635}$ of IRS-1 (serines$^{636/639}$ in human IRS-1) increases in accordance with insulin resistance; this phenomenon is refractory to rapamycin, implying mTOR as the responsible kinase (Ozes *et al.*, 2001). Direct phosphorylation by mTOR of serines$^{632/635}$ has not been observed to date, though. It was later found in human skeletal muscle cells that IRS-1 serine^{636} phosphorylation was inversely associated with insulin-stimulated PI3-K activation and that the phosphorylation was sensitive to an inhibitor of MEK, suggesting that serine^{636} might be the target of a MAP kinase (Bouzakri *et al.*, 2003). Recently, serines$^{632/635}$ have been found to be phosphorylated by Rho-kinase *in vitro* and *in vivo* (Furukawa *et al.*, 2005). Contrary to previous results, though, the phosphorylation of serines$^{632/635}$ and the activity of Rho-kinase positively associated with IRS-1 tyrosine phosphorylation and insulin signaling; replacing these serine residues with alanines actually impaired insulin signaling (Furukawa *et al.*, 2005).

Opposing observations with respect to insulin signaling in association with identical phosphorylation events are not limited to the aforementioned sites. Phosphorylation of IRS-1 serine^{302} is drastically increased in insulin-resistant animal models, including *ob/ob* mice and mice fed a high-fat diet (Werner *et al.*, 2004). However, nutrient- or insulin-induced phosphorylation of IRS-1 serine^{302} has been shown to enhance insulin signaling (Giraud *et al.*, 2004); adding to the confusion, PMA-stimulated serine^{302} phosphorylation antagonizes transmission of the insulin signal (Werner *et al.*, 2004). As another example, serine^{789} of IRS-1 is phosphorylated by members of the AMP-activated protein kinase (AMPK) family, including salt-inducible kinases (Horike *et al.*, 2003; Jakobsen *et al.*, 2001). On the one hand, phosphorylation at this site positively correlates with insulin resistance in animals (Horike *et al.*, 2003; Qiao *et al.*, 2002). On the other hand, phosphorylation of the same site positively correlates with sensitivity in C2C12 myotubes (Jakobsen *et al.*, 2001). It is thus apparent that serine phosphorylation of IRS proteins cannot be considered to be a sole or absolute indicator of impaired insulin signaling; the integrative role of the IRS proteins necessitates consideration of other components of the system, both upstream and downstream.

Whether positive or negative in reference to insulin action, it is clear that insulin signaling intermediates (serine/threonine kinases) downstream of IRS-1 and IRS-2 exert a degree of feedback on the originating signaling scaffold. The majority of these serine/threonine kinases are several steps removed from IRS-1/IRS-2; it is possible, though, that regulation occurs immediately downstream of IRS signaling, at the level of PI3-K. PI3-K possesses dual specificity, acting as a phospholipid kinase as well as having serine kinase activity (Carpenter *et al.*, 1993; Dhand *et al.*, 1994b). IRS-1 and IRS-2 form a complex with PI3-K in response to insulin stimulation and are in turn phosphorylated by PI3-K (Freund *et al.*, 1995; Kim *et al.*, 1998; Tanti *et al.*, 1994a). Using membrane-targeted constitutively active PI3-K (p110^{CAAX}), Egawa *et al.* (1999) showed that IRS-1 serine/threonine phosphorylation was increased, suggesting direct phosphorylation of IRS-1 by PI3-K. However, the IRS phosphorylation site for PI3-K has not been identified and we have been unable to observe direct phosphorylation of IRS-1 by PI3-K *in vitro* (X.J. Sun *et al.*, unpublished data). One possibility is that it is not the PI3-K *per se*, but rather a PI3-K-associated kinase that phosphorylates IRS-1/IRS-2 in cells. Indeed, the kinase that phosphorylates IRS-1 in an IRS-1/PI3-K complex is insensitive to wortmannin (a PI3-K inhibitor), implying the involvement of a PI3-K-associated kinase (PAS kinase) in the phosphorylation of IRS-1 (Cengel *et al.*, 1998).

It is now clear that insulin signaling may be modulated by many serine/threonine kinases via phosphorylation of the IRS proteins. Using liver extracts from insulin-resistant animal models as a kinase source, we have detected multiple serine/threonine kinase activities that are able to phosphorylate various regions of IRS-1 and IRS-2 (unpublished data) (Qiao *et al.*, 1999, 2002). It is conceivable that additional IRS kinases will be identified, and identification and characterization of these kinases will shed light on the mechanisms of insulin signaling and insulin resistance.

IV. Biological Function of IRS Proteins in Insulin Action

Insulin has a broad spectrum of biological effects in cells. It promotes growth, induces differentiation, governs metabolism, stimulates protein and DNA synthesis, and regulates gene expression. Recent genetic studies have revealed that these biological functions are associated with the cognate receptor tyrosine kinase (Virkamaki *et al.*, 1999) and are mediated by IRS proteins (White, 2002). Mice lacking IRS-1 exhibit growth retardation and mild systemic insulin resistance secondary to skeletal muscle insulin signaling defects (Araki *et al.*, 1994; Tamemoto *et al.*, 1994). IRS-2-deficient mice, however, develop diabetes as a result of severe hepatic insulin resistance and β-cell failure (Kubota *et al.*, 2000; Simpson *et al.*, 2001; Withers *et al.*, 1998). More detailed physiological studies of IRS-1 and IRS-2 null mice, as well as studies using mice with combined heterozygous null mutations in the insulin receptor and IRS-1 or IRS-2 further support the notion that IRS-2 is of greater importance in mediating insulin signaling in the liver, and IRS-1, in muscle (Kido *et al.*, 2000; Previs *et al.*, 2000). In general, it seems that IRS-1 and IRS-2 are the dominant isoforms; disruption of IRS-3 gene expression does not result in significant phenotypic abnormalities (Liu *et al.*, 1999), while mice lacking IRS-4 show only mild defects in growth, reproduction, and glucose homeostasis (Fantin *et al.*, 2000).

Tissue-specific ablation of IRS-1 or IRS-2 by conditional knockout also supports that IRS-1/2 proteins mediate most, if not all of insulin's actions in insulin-sensitive tissues. Insulin signaling in the liver is essential for glucose homeostasis and liver insulin receptor knockout mice have severe insulin resistance, characterized mainly by the loss of insulin suppression of hepatic glucose production (HGP) (Michael *et al.*, 2000). Ablation of IRS-1 and IRS-2 in liver completely abolishes insulin signaling, and such mice develop hyperglycemia and hyperinsulinemia immediately after birth. The insulin receptor is normal and fully functional in the liver of these animals, suggesting that IRS-1/2 mediate all hepatic insulin signaling (Dong *et al.*, 2006). The critical role of IRS-2 in hepatic insulin action has also been demonstrated in immortalized neonatal hepatocyte cell lines from IRS-2 null mice. In these cells, insulin-induced PI3-kinase activation is decreased by 50%; activation of Akt, GSK-3, Foxo1, and PKC is also blunted, leading to a marked reduction in insulin-stimulated glycogen synthase activity and suppression of gluconeogenic gene expression (Valverde *et al.*, 2003). These defects can be restored by reconstitution of IRS-2(−/−) hepatocytes with adenoviral IRS-2. Consistent with the results from this study, restoration of IRS-2 in the liver of IRS-2 null mice by adenoviral infection enables insulin to suppress HGP to a level similar to that observed in normal mice (Suzuki *et al.*, 2004).

The postprandial maintenance of plasma glucose homeostasis is one of the most important physiological functions of insulin. It is achieved mainly via stimulation of glucose uptake in insulin-sensitive tissues by promoting translocation of the insulin-sensitive glucose transporter (Glut4) to the plasma membrane (Czech, 1995; White, 1996). The IRS-1(−/−) mouse displays a significant decrease in insulin/IGF-1-stimulated glucose uptake *in vivo* and *in vitro* (Araki *et al.*, 1994; Tamemoto *et al.*, 1994; Yamauchi *et al.*, 1996). The remaining insulin/IGF-1 action in the IRS-1(−/−) mouse correlates with tyrosine phosphorylation of IRS-2 which may account for the residual insulin-stimulated glucose uptake in IRS-1(−/−) mice (Araki *et al.*, 1994; Yamauchi *et al.*, 1996).

Although IRS-1 and IRS-2 are highly homologous in sequence, gene knockout studies have revealed unique functions for each molecule, which may be partially due to the tissue-specific expression of these isoforms. To study the unique roles of IRS-1 and IRS-2, siRNA and viruses have been used to either suppress or overexpress these proteins. In differentiated L6 myotubes, a reduction in the levels of IRS-1 or IRS-2 results in different outcomes in insulin response. While insulin-induced ERK phosphorylation is more sensitive to downregulation of IRS-2 than IRS-1 (Thirone *et al.*, 2006), glucose uptake is affected by reduction of IRS-1 but not IRS-2. This confirms that in L6 myotubes, insulin-mediated glucose uptake is controlled mainly by IRS-1 (Thirone *et al.*, 2006).

Recently, insulin action in brain and β-cells has been examined using targeted conditional knockout of the IR. Mice deficient for brain insulin receptor demonstrate increased susceptibility to diet-induced obesity with accompanying insulin resistance (Bruning *et al.*, 2000). Disruption of insulin signaling in β-cells precipitates glucose intolerance and impaired β-cell function and insulin secretion in mice (Kulkarni *et al.*, 1999a). Consistent with these findings, defects in insulin secretion are present in β-cells isolated from IRS-1 null mice (Kulkarni *et al.*, 1999b). Deletion of IRS-2 in β-cells and in the brain leads to obesity, insulin, and leptin resistance; diabetes ultimately develops in these animals (Kubota *et al.*, 2004; Lin *et al.*, 2004). These studies highlight the importance of IRS proteins in mediating insulin signaling in noncanonical insulin-responsive tissues.

V. The Role of IRS Serine Phosphorylation in Mediating the Crosstalk with Other Signaling Pathways

Cellular signaling occurs in the context of an interconnected Web of pathways, as opposed to neat, sequential chains of events. The intensity and breadth of a signaling cascade is widely influenced by inputs from neighboring pathways, with the outcome ultimately governed by environmental stimuli. With respect to insulin, these stimuli include cytokines released in response to cellular trauma (Shoelson *et al.*, 2006; Wellen and Hotamisligil, 2005), circulating nutrients from diet and endogenous production (Argaud *et al.*, 1997; Savage *et al.*, 2007; Summers and Nelson, 2005), free radicals generated as byproducts of metabolism, and endoplasmic reticulum stress occurring secondary to general chemical and physical insults, to name a few (Fridlyand and Philipson, 2006; Wellen and Hotamisligil, 2005). The aforementioned factors initiate signaling networks which crosstalk with insulin signal transducers, often impairing downstream transmission of the insulin directive. Recent studies provide compelling evidence that changes in the serine/threonine phosphorylation status of the IRS proteins is the molecular mechanism by which crosstalk moderates insulin signaling.

A. Cytokines

Proinflammatory cytokines such as TNFα, IL-1, and IL-6 are able to alter insulin sensitivity (He *et al.*, 2006; Hotamisligil, 2003; Hotamisligil *et al.*, 1996; Klover *et al.*, 2003, 2005; Senn *et al.*, 2002). These proinflammatory cytokines activate the NF-κB and JNK signaling pathways. Components from these cascades may subsequently interact with members of the insulin signaling pathway (Cai *et al.*, 2005; Shoelson *et al.*, 2006, 2007). Since obesity and type-2 diabetes commonly associate with a state of chronic inflammation, excessive cytokine-incited signaling has been proposed as a cellular mechanism leading to insulin resistance, particularly in the context of obesity (Hotamisligil, 2003; Shoelson *et al.*, 2006).

Exactly how the aforementioned cytokines communicate with the insulin signaling network remains elusive. However, accumulating genetic and biochemical evidence implies serine phosphorylation of IRS proteins as an interconnecting mechanism. TNFα impairs insulin signaling via activation of JNK, leading to phosphorylation of IRS-1 at serine^{307} (Aguirre *et al.*,

2000; Hotamisligil *et al.*, 1996). Mutation of serine^{307} to alanine prevents phosphorylation of IRS-1 by JNK and abrogates the inhibitory effect of TNFα on insulin-stimulated tyrosine phosphorylation of IRS-1 in cultured cells (Aguirre *et al.*, 2000). Ablating JNK or JIP1 (a scaffold protein necessary for JNK activation), treating animals with JNK-inhibitory peptide, or overexpressing dominant-negative JNK reduces serine^{307} phosphorylation and augments insulin signaling (Hirosumi *et al.*, 2002; Jaeschke *et al.*, 2004; Kaneto *et al.*, 2004; Nakatani *et al.*, 2004).

TNFα-induced insulin resistance has also been linked to the IKKβ pathway (Arkan *et al.*, 2005; Kim *et al.*, 2001; Yuan *et al.*, 2001). Mice lacking IKKβ are protected from high-fat diet-induced insulin resistance with concurrent enhancements in insulin signaling (Arkan *et al.*, 2005). Salicylates, which inhibit IKKβ, offset TNFα-mediated inhibition of insulin-stimulated IRS-1 and IRS-2 tyrosine phosphorylation (Yuan *et al.*, 2001). Improvements in insulin signaling inversely associate with IRS-1 serine^{307} phosphorylation, as TNFα-induced IRS-1 phosphorylation at this site is blunted in cells derived from IKKβ knockout mice or when cells are pretreated with an inhibitor for IKKβ (Gao *et al.*, 2002).

IL-6 and IL-1 may induce serine phosphorylation of IRS-1 and mediate insulin resistance (He *et al.*, 2006; Klover *et al.*, 2003; Rotter *et al.*, 2003). Both IL-6 and IL-1 have been reported to activate IKKβ, JNK, and Erk and phosphorylate IRS-1 at serine^{307} and serine^{612} (Andreozzi *et al.*, 2007; He *et al.*, 2006). The importance of serine^{307} phosphorylation with respect to IL-6 is debatable based on the finding that IL-6 induces serine phosphorylation of IRS-1 at sites other than serine^{307} (Rotter *et al.*, 2003).

In summary, cytokines released in response to inflammatory events may precipitate insulin resistance by binding to their cognate receptors in insulin-sensitive tissues, triggering intracellular signaling pathways which cross-talk with that of insulin. To date, several protein kinases involved in cytokine signaling including JNK, IKKβ, and ERK have been linked to phosphorylation of IRS proteins at distinct sites. These findings provide the framework for a signal-based model detailing the molecular mechanism by which inflammation, a widely observed phenomenon in pathologies with underlying insulin resistance, perturbs insulin function.

B. Nutrients

Nutrient-sensing pathways appear to exert marked influences on insulin signaling. Growing evidence implicates serine phosphorylation of IRS proteins as a central mediating factor in nutrient-induced insulin resistance. The serine/threonine kinase mTOR provides an ideal example for the illustration of this concept. The mTOR/p70^{S6K} pathway is an important nutrient-sensing component with respect to amino acids and glucose, and its perturbation drastically impacts insulin sensitivity (Fisher and White, 2004; Garami *et al.*, 2003; Patti and Kahn, 2004). Incubation of cells with high concentrations of leucine or glucose activates the mTOR/p70^{S6K} pathway, leading to phosphorylation of IRS-1 at serine^{302} and serine^{1100} as well as attenuation of insulin signaling (Giraud *et al.*, 2004; Tremblay *et al.*, 2007). Mouse embryonic fibroblasts (MEFs) lacking the upstream molecules TSC1/TSC2 (inhibitors of mTOR) display enhanced activation of mTOR with corresponding insulin resistance. Serine phosphorylation of IRS-1 at number of sites including serine^{302}, serine^{307}, serine^{612}, and serine$^{632/635}$ is increased in these cells in direct association with insulin resistance (Harrington *et al.*, 2004; Shah and Hunter, 2006). Furthermore, in mice lacking p70^{S6K}, the immediate downstream target of mTOR, there is reduced IRS-1 phosphorylation at serine^{307}, serine$^{632/635}$, and serine^{1100}, with resultant enhancements in insulin signaling and protection from diet-induced obesity (Tremblay *et al.*, 2007; Um *et al.*, 2004). It is possible that mTOR/p70^{S6K}, by serving as a nutrient sensor for amino acids and glucose, may downregulate insulin signaling in an attempt to limit the influx of nutrients during times of sufficiency or excess; the limitation would necessarily involve inhibition of insulin signaling, with serine phosphorylation of the IRS proteins as a potential mechanism.

Certain classes of lipids have been extensively studied with regard to their capacity to antagonize insulin action (Savage *et al.*, 2007). Treating cells with various long-chain fatty acids corresponding to those commonly found in the diet revealed that unsaturated fatty acids have a modest or are without effect with respect to JNK activation. However, exposure to long-chain saturated fatty acids such as palmitate (16:0) or stearate (18:0) cause substantial JNK activation with associated phosphorylation of IRS-1 at serine^{307} and insulin resistance (Jaeschke and Davis, 2007; Solinas *et al.*, 2006). The divergent effects of unsaturated and saturated fatty acids on serine/threonine kinase activation and insulin signaling imply that fatty acid-induced insulin resistance may be mediated by lipid metabolites, particularly those of long-chain saturated fatty acids (e.g., palmitate). As a prime example, ceramide, a member of the sphingolipid family, appears to be a common mediator in both nutrient- and inflammation-induced (e.g., TNFα) insulin resistance (Stratford *et al.*, 2004; Summers, 2006). Palmitate is the direct substrate for serine palmitoyltransferase (SPT), the rate-limiting enzyme for ceramide biosynthesis; stearate, having only two additional carbons, is almost equally ideal a substrate. Unsaturated fatty acids, on the other hand, are not directly capable of being utilized by SPT. Ceramide has been shown to augment IRS-1 phosphorylation at serine^{307} via activation of inflammatory pathways involving JNK, IKK, and NF-κB (Itani *et al.*, 2002; Stratford *et al.*, 2004).

Diacylglycerol (DAG) may also have antagonistic effects on insulin signaling through stimulation of PKC isoforms; unlike ceramide, its direct generation is not limited to saturated fatty acids (Sampson and Cooper, 2006; Schmitz-Peiffer, 2002; Yu *et al.*, 2002). Genetic models have shown that PKCθ and PKCϵ may mediate high-fat diet-induced muscle and liver insulin resistance, respectively (Savage *et al.*, 2007). DAG activates a number of PKC isoforms in a variety of tissues, often resulting in serine phosphorylation of IRS-1 directly and/or indirectly at different sites (De Fea and Roth, 1997; Kim *et al.*, 2004; Liu *et al.*, 2004; Nawaratne *et al.*, 2006; Samuel *et al.*, 2007).

C. Reactive oxygen species and oxidative stress

Reactive oxygen species (ROS) are natural byproducts of metabolism which serve as signaling molecules to regulate many important cellular functions (Lyle and Griendling, 2006; Nakamura *et al.*, 1997). While not normally problematic and indeed essential, ROS can have untoward effects when their production overwhelms the quenching capacity of endogenous antioxidant defenses (Fridlyand and Philipson, 2006). Increased ROS production is associated with obesity and contributes to oxidative stress, inflammation, and insulin resistance by activation of signaling cascades involving JNK, NF-κB, and other kinases (Evans *et al.*, 2002, 2005). ROS may also crosstalk with the insulin signaling pathway via inhibition of phosphatase activities, leading to increased serine/threonine phosphorylation of IRS proteins (Evans *et al.*, 2005). Other free radical classes are known to antagonize insulin signaling as well, but the specific details surrounding their generation and effects are beyond the scope of the present review.

D. Endoplasmic reticulum stress

In obesity, increased inflammatory mediators and lipid accumulation may cause chronic stress at the cellular level, principally affecting the endoplasmic reticulum (ER) (Tsiotra and Tsigos, 2006). The ER of adipocytes is a site for the synthesis, modification, and—together with the Golgi apparatus—secretion of many hormones and cytokines in response to environmental demands (Gregor and Hotamisligil, 2007). Recent studies show that ER stress may participate in the onset of insulin resistance (Gregor and Hotamisligil, 2007; Hotamisligil, 2005; Ozcan *et al.*, 2004). In support of this, ER stress and ROS overproduction lead to JNK-mediated IRS-1 serine^{307} phosphorylation and insulin resistance (Houstis *et al.*, 2006; Ozcan *et al.*, 2004, 2006). Chemical chaperones that reduce ER stress can restore insulin sensitivity in an animal model of type-2 diabetes (Ozcan *et al.*, 2006).

VI. Mechanisms Underlying IRS Serine Phosphorylation-Induced Insulin Resistance

The molecular details connecting serine phosphorylation of IRS proteins to dysfunctional insulin signaling remain largely unknown, but studies from TNFα-induced insulin resistance provide some insight, though.

IRS-1 isolated from TNFα-treated cells has an inhibitory effect on insulin receptor tyrosine kinase activity *in vitro*. This effect is a consequence of phosphorylation of serine residues on the IRS-1 molecule since alkaline phosphatase treatment of IRS-1 abrogates this inhibitory effect (Hotamisligil *et al.*, 1996; Kanety *et al.*, 1995). A similar observation has been reported for IRS-1 isolated from thermally-injured muscle of burn victims (Friedman *et al.*, 1999; Ikezu *et al.*, 1997), as well as IRS-1 that has undergone *in vitro* phosphorylation by GSK-3β (Eldar-Finkelman and Krebs, 1997).

Interaction between the PTB domains of IRS proteins and the juxtamembrane domain of the IR is critical for insulin-induced tyrosyl phosphorylation of IRS proteins (Eck *et al.*, 1996a; Kaburagi *et al.*, 1993; White *et al.*, 1988). TNFα induces phosphorylation of IRS-1 on serine^{307}, which is adjacent to the PTB domain of IRS-1 (Fig. 13.2) (Aguirre *et al.*, 2000). When phosphorylated, the interaction between IRS-1 and the insulin receptor is disrupted, leading to the attenuation of insulin signaling (Aguirre *et al.*, 2002; Nakajima *et al.*, 2000). IRS-1/IRS-2 isolated from cells chronically exposed to TNFα, okadaic acid, or insulin exhibit reduced interaction with the IR *in vitro*; interaction competence can be restored when IRS-1/IRS-2 serine phosphorylation is removed by treatment with alkaline phosphatase (Paz *et al.*, 1997). These data provide strong evidence that serine phosphorylation in response to treatment with factors observed to induce insulin resistance negatively regulates IRS interaction with the IR.

Additional mechanisms by which serine phosphorylation of IRS-1/2 antagonize insulin signaling is provided by data from studies of the 14-3-3 proteins. 14-3-3 is a group of small and highly conserved proteins that binds to a variety of phosphoserine/threonine-containing proteins to regulate their biological functions (Craparo *et al.*, 1997; Freed *et al.*, 1994; Irie *et al.*, 1994; Morrison, 1994; Muslin *et al.*, 1996; Shimizu *et al.*, 1994). The consensus sequence of the 14-3-3-binding site has been proposed to be RSXpSXP (Muslin *et al.*, 1996). In IRS-1 and IRS-2, a highly conserved (RSKS^{265}QP in IRS-1 and RSKS^{303}QS in IRS-2) and a less well-conserved 14-3-3-binding site (KSVS^{637}AP in IRS-1 and RSMS^{429}MP in IRS-2) have been identified, indicating an important role for serine/threonine phosphorylation in regulating the interaction between IRS isoforms and 14-3-3 proteins. Consistent with this, 14-3-3β, ϵ, and ζ isoforms have been shown to interact with IRS-1 in a phosphoserine-dependent manner (Craparo *et al.*, 1997; Kosaki *et al.*, 1998; Ogihara *et al.*, 1997; Xiang *et al.*, 2002), leading to the inhibition of tyrosine phosphorylation of IRS-1 (Kosaki *et al.*, 1998; Ogihara *et al.*, 1997). In addition to 14-3-3, proteins containing "Forkhead-associated" (FHA) domains (a module consisting of 75 amino acids found in protein kinases, phosphatases, and transcription factors) (Durocher *et al.*, 1999; Yaffe and Cantley, 1999) and PDZ domains (a conserved protein interaction module found in transmembrane proteins) (Craven and Bredt, 1998) directly bind to serine-phosphorylated proteins and may have roles in directing intracellular signaling (Barinaga, 1999; Cao *et al.*, 1999; Yaffe and Cantley, 1999). Although the FHA and PDZ modules have not been tested for their binding to serine-phosphorylated IRS proteins, it is reasonable to speculate that serine phosphorylation may create binding sites in IRS isoforms for proteins containing these modules. Such interactions would prohibit binding between IRS-1 and the IR, ultimately leading to impairment of tyrosine phosphorylation of IRS-1 and downstream insulin action.

The absolute quantity of IRS-1/2 plays an important role in the efficacy of insulin signaling. Decreased levels of IRS-1/IRS-2 proteins have been found in insulin-resistant animals and human subjects (Anai *et al.*, 1998a; Bjornholm *et al.*, 1997; Czech *et al.*, 1978; Jiang *et al.*, 1999; Rondinone *et al.*, 1997). Providing a possible explanation for these effects, it is known that ubiquitin–proteasome-mediated degradation controls IRS proteins levels (Lee *et al.*, 2000; Rui *et al.*, 2001b; Sun *et al.*, 1999). Importantly, serine phosphorylation is involved in ubiquitination, which is a critical step in targeting proteins for proteasome degradation (Barinaga, 1999; Yaffe and Cantley, 1999). The conserved tryptophan–tryptophan (WW) domain, WD40 domain, and leucine-rich repeats in some F-box proteins have been found to bind proteins in a serine phosphorylation-dependent manner, and this interaction is required for ubiquitination of targeted proteins (Barinaga, 1999; Kitagawa *et al.*, 1999; Lu *et al.*, 1999; Shirane *et al.*, 1999; Skowyra *et al.*, 1997; Spencer *et al.*, 1999). The WW and WD40 domains often bind to pSer- or pThr-Pro sequences, identified as the PEST-motif (rich in Pro, Glu, Ser, and Thr) (Barinaga, 1999). IRS-1 contains at least 11 such motifs (Smith *et al.*, 1993) and both IRS-1 and IRS-2 have been shown to undergo ubiquitin–proteasome-mediated degradation during chronic insulin or IGF-1 exposure (Lee *et al.*, 2000; Rui *et al.*, 2001b; Sun *et al.*, 1999). It is possible that WW and WD40 domain-containing proteins may interact with serine-phosphorylated IRS-1/IRS-2 during the development of insulin resistance, promoting their degradation through the ubiquitin–proteasome pathway.

Serine phosphorylation may also involve in regulating cellular localization of IRS proteins, an important step in directing insulin signaling down different pathways. Differential distribution of IRS-1 and IRS-2 has been found between the cytosol and the intracellular membrane (IM) (Inoue *et al.*, 1998), high-speed pellet (HSP) (Clark *et al.*, 1998), and low-density microsome (Anai *et al.*, 1998b; Tsuji *et al.*, 2001) fractions. Acute insulin stimulation induces IRS-1 and IRS-2 membrane association (Inoue *et al.*, 1998; Kublaoui *et al.*, 1995). Prolonged insulin treatment promotes the release of IRS-1 and IRS-2 from the membrane faction into the cytosolic fraction (Clark *et al.*, 2002; Inoue *et al.*, 1998). This translocation or redistribution of IRS proteins appears to be controlled by their serine/threonine phosphorylation status (Clark *et al.*, 2002; Inoue *et al.*, 1998). As mentioned previously, 14-3-3 binds to IRS-1 in a serine/threonine phosphorylation-dependent manner (Xiang *et al.*, 2002). This complex is found in the cytosolic but not in the high-speed pellet fraction, suggesting that binding of 14-3-3 may regulate IRS-1 cellular localization (Xiang *et al.*, 2002). Insulin has also been shown to promote IRS-1 nuclear translocation (Chen *et al.*, 2005; Lassak *et al.*, 2002; Prisco *et al.*, 2002; Sun *et al.*, 2003; Trojanek *et al.*, 2006; Wu *et al.*, 2003). Nuclear translocation of IRS-1 is inhibited by serine phosphorylation (of IRS-1) suggesting that serine phosphorylation of IRS-1 may antagonize the ability of insulin to properly regulate gene expression (Lassak *et al.*, 2002).

VII. Conclusion

The insulin signaling pathway is a complex cellular network composed of several nodules and branches spread across multiple cellular compartments (Taniguchi *et al.*, 2006). It has become increasingly clear in recent years that this network is not static, but is rather a dynamic and flexible system. Before activation by insulin, the network is modulated through both negative and positive regulation, constantly communicating with other signaling pathways in response to various environmental cues. Consequently, these modulations predetermine insulin sensitivity. The unique structural features of IRS proteins establish them as ideal molecules to sense these environmental changes and integrate them into the insulin network. Tyrosine phosphorylation of IRS proteins provides a key mechanism to convert IR tyrosine kinase activation into multisignaling arrays. However, the ability of IRS proteins to be tyrosine phosphorylated is preset by the levels of serine/threonine phosphorylation (which is extensive even in the basal state). Thus, serine phosphorylation of IRS proteins functions as a sensing mechanism allowing for real-time adjustment of insulin signaling, and is a critical means of communication with other signaling pathways. Serine/threonine phosphorylation of IRS proteins can therefore lead to dynamic changes in insulin signaling (Fig. 13.3). Serine/threonine phosphorylation of IRS proteins may influence the interaction between IRS proteins and the IR, leading to a global reduction in insulin action in cells. At certain residues, serine/threonine phosphorylation can also interfere with the binding of IRS proteins with unique signaling parterres, thereby altering, in a targeted manner, specific branches in the signaling pathway. The influence of serine phosphorylation of IRS proteins on their cellular distribution may additionally affect insulin sensitivity within different cellular compartment. In conclusion, detailed insight concerning how serine/

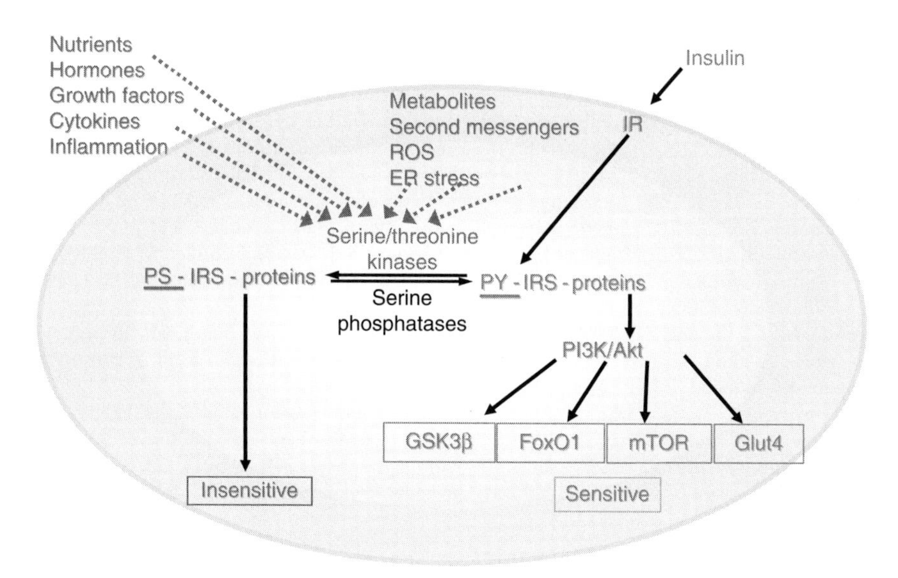

Figure 13.3 Multiple signaling pathways crosstalk with insulin signaling via serine phosphorylation of IRS proteins. Extracellular and intracellular cues determine the level of serine/threonine phosphorylation of IRS proteins, which predetermines the capacity of interaction between IRS proteins with the insulin receptor and downstream components.

threonine phosphorylation of IRS proteins mediates crosstalk will greatly improve our understanding of the molecular mechanisms surrounding the induction, progression, and amelioration of insulin resistance.

REFERENCES

Aguirre, V., Uchida, T., Yenush, L., Davis, R., and White, M. F. (2000). The c-Jun NH(2)-terminal kinase promotes insulin resistance during association with insulin receptor substrate-1 and phosphorylation of Ser(307). *J. Biol. Chem.* **275,** 9047–9054.

Aguirre, V., Werner, E. D., Giraud, J., Lee, Y. H., Shoelson, S. E., and White, M. F. (2002). Phosphorylation of Ser307 in insulin receptor substrate-1 blocks interactions with the insulin receptor and inhibits insulin action. *J. Biol. Chem.* **277,** 1531–1537.

Anai, M., Funaki, M., Ogihara, T., Terasaki, J., Inukai, K., Katagiri, H., Fukushima, Y., Yazaki, Y., Kikuchi, M., Oka, Y., and Asano, T. (1998a). Altered expression levels and impaired steps in the pathway to phosphatidylinositol 3-kinase activation via insulin receptor substrates 1 and 2 in Zucker fatty rats. *Diabetes* **47,** 13–23.

Anai, M., Ono, H., Funaki, M., Fukushima, Y., Inukai, K., Ogihara, T., Sakoda, H., Onishi, Y., Yazaki, Y., Kikuchi, M., Oka, Y., and Asano, T. (1998b). Different subcellular distribution and regulation of expression of insulin receptor substrate (IRS)-3 from those of IRS-1 and IRS-2. *J. Biol. Chem.* **273,** 29686–29692.

Andreozzi, F., Laratta, E., Procopio, C., Hribal, M. L., Sciacqua, A., Perticone, M., Miele, C., Perticone, F., and Sesti, G. (2007). Interleukin-6 impairs the insulin signaling pathway, promoting production of nitric oxide in human umbilical vein endothelial cells. *Mol. Cell. Biol.* **27,** 2372–2383.

Araki, E., Sun, X. J., Haag, B. L., Zhang, Y., Chuang, L. M., Yang-Feng, T., White, M. F., and Kahn, C. R. (1993). Human skeletal muscle insulin receptor substrate-1: Characterization of the cDNA, gene and chromosomal localization. *Diabetes* **42,** 1041–1054.

Araki, E., Lipes, M. A., Patti, M. E., Bruning, J. C., Haag, B., III, Johnson, R. S., and Kahn, C. R. (1994). Alternative pathway of insulin signalling in mice with targeted disruption of the IRS-1 gene. *Nature* **372,** 186–190.

Argaud, D., Kirby, T. L., Newgard, C. B., and Lange, A. J. (1997). Stimulation of glucose-6-phosphatase gene expression by glucose and fructose-2,6-bisphosphate. *J. Biol. Chem.* **272,** 12854–12861.

Arkan, M. C., Hevener, A. L., Greten, F. R., Maeda, S., Li, Z. W., Long, J. M., Wynshaw-Boris, A., Poli, G., Olefsky, J., and Karin, M. (2005). IKK-beta links inflammation to obesity-induced insulin resistance. *Nat. Med.* **11,** 191–198.

Baltensperger, K., Kozma, L. M., Cherniack, A. D., Klarlund, J. K., Chawla, A., Banerjee, U., and Czech, M. P. (1993). Binding of the Ras activator son of sevenless to insulin receptor substrate-1 signaling complexes. *Science* **260,** 1950–1952.

Barinaga, M. (1999). New clues to how proteins link up to run the cell. *Science* **283,** 1247–1249.

Bjornholm, M., Kawano, Y., Lehtihet, M., and Zierath, J. (1997). Insulin receptor substrate-1 phosphorylation and phosphatidylinositol 3-kinase activity in skeletal muscle from NIDDM subjects after *in vivo* insulin stimulation. *Diabetes* **46,** 524–527.

Bouzakri, K., Roques, M., Gual, P., Espinosa, S., Guebre-Egziabher, F., Riou, J. P., Laville, M., Le Marchand-Brustel, Y., Tanti, J. F., and Vidal, H. (2003). Reduced activation of phosphatidylinositol-3 kinase and increased serine 636 phosphorylation of insulin receptor substrate-1 in primary culture of skeletal muscle cells from patients with type 2 diabetes. *Diabetes* **52,** 1319–1325.

Bruning, J. C., Gautam, D., Burks, D. J., Gillette, J., Schubert, M., Orban, P. C., Klein, R., Krone, W., Muller-Wieland, D., and Kahn, C. R. (2000). Role of brain insulin receptor in control of body weight and reproduction. *Science* **289,** 2122–2125.

Cai, D., Dhe-Paganon, S., Melendez, P. A., Lee, J., and Shoelson, S. E. (2003). Two new substrates in insulin signaling, IRS5/DOK4 and IRS6/DOK5. *J. Biol. Chem.* **278,** 25323–25330.

Cai, D., Yuan, M., Frantz, D. F., Melendez, P. A., Hansen, L., Lee, J., and Shoelson, S. E. (2005). Local and systemic insulin resistance resulting from hepatic activation of IKK-beta and NF-kappaB. *Nat. Med.* **11,** 183–190.

Cao, T. T., Deacon, H. W., Reczek, D., Bretscher, A., and von Zastrow, M. (1999). A kinase-regulated PDZ-domain interaction controls endocytic sorting of the beta2-adrenergic receptor. *Nature* **401,** 286–290.

Carlson, C. J., White, M. F., and Rondinone, C. M. (2004). Mammalian target of rapamycin regulates IRS-1 serine 307 phosphorylation. *Biochem. Biophys. Res. Commun.* **316,** 533–539.

Carpenter, C. L., Auger, K. R., Duckworth, B. C., Hou, W. M., Schaffhausen, B., and Cantley, L. C. (1993). A tightly associated serine/threonine protein kinase regulates phosphoinositide 3-kinase activity. *Mol. Cell. Biol.* **13,** 1657–1665.

Cengel, K. A., Kason, R. E., and Freund, G. G. (1998). Phosphatidylinositol 3′-kinase associates with an insulin receptor substrate-1 serine kinase distinct from its intrinsic serine kinase. *Biochem. J.* **335,** 397–404.

Chan, R. J., and Feng, G. S. (2007). PTPN11 is the first identified proto-oncogene that encodes a tyrosine phosphatase. *Blood* **109,** 862–867.

Chang, L., and Karin, M. (2001). Mammalian MAP kinase signalling cascades. *Nature* **410,** 37–40.

Chen, J., Sadowski, H. B., Kohanski, R. A., and Wang, L. H. (1997). Stat5 is a physiological substrate of the insulin receptor. *Proc. Natl. Acad. Sci. USA* **94,** 2295–2300.

Chen, J., Wu, A., Sun, H., Drakas, R., Garofalo, C., Cascio, S., Surmacz, E., and Baserga, R. (2005). Functional significance of type 1 insulin-like growth factor-mediated nuclear translocation of the insulin receptor substrate-1 and beta-catenin. *J. Biol. Chem.* **280,** 29912–29920.

Clark, S. F., Martin, S., Carozzi, A. J., Hill, M. M., and James, D. E. (1998). Intracellular localization of phosphatidylinositide 3-kinase and insulin receptor substrate-1 in adipocytes: Potential involvement of a membrane skeleton. *J. Cell Biol.* **140,** 1211–1225.

Clark, S. F., Molero, J. C., and James, D. E. (2002). Release of insulin receptor substrate proteins from an intracellular complex coincides with the development of insulin resistance. *J. Biol. Chem.* **275**(6), 3819–3826.

Cong, F., Yuan, B., and Goff, S. P. (1999). Characterization of a novel member of the DOK family that binds and modulates Abl signaling. *Mol. Cell. Biol.* **19,** 8314–8325.

Craparo, A., O'Neill, T. J., and Gustafson, T. A. (1995). Non-SH2 domains within the insulin receptor substrate-1 and SHC mediate their phosphotyrosine-dependent interaction with the NPEY motif of the insulin-like growth factor 1 receptor. *J. Biol. Chem.* **270,** 15639–15643.

Craparo, A., Freund, R., and Gustafson, T. A. (1997). 14-3-3 (epsilon) interacts with the insulin-like growth factor I receptor and insulin receptor substrate I in a phosphoserine-dependent manner. *J. Biol. Chem.* **272,** 11663–11669.

Craven, S. E., and Bredt, D. S. (1998). PDZ proteins organize synaptic signaling pathways. *Cell* **93,** 495–498.

Czech, M. P. (1995). Molecular actions of insulin on glucose transport. [Review]. *Annu. Rev. Nutr.* **15,** 441–471.

Czech, M. P., Richardson, D. K., Becker, S. G., Walters, C. G., Gitomer, W., and Heinrich, J. (1978). Insulin response in skeletal muscle and fat cells of the genetically obese Zucker rat. *Metabolism* **27,** 1967–1981.

Davis, R. J. (1993). The mitogen-activated protein kinase signal transduction pathway. *J. Biol. Chem.* **268,** 14553–14556.

De Fea, K., and Roth, R. A. (1997). Protein kinase C modulation of insulin receptor substrate-1 tyrosine phosphorylation requires serine 612. *Biochemistry* **36,** 12939–12947.

Dhand, R., Hara, K., Hiles, I., Bax, B., Gout, I., Panayotou, G., Fry, M. J., Yonezawa, K., Kasuga, M., and Waterfield, M. D. (1994a). PI 3-kinase: Structural and functional analysis of intersubunit interactions. *EMBO J.* **13,** 511–521.

Dhand, R., Hiles, I., Panayotou, G., Roche, S., Fry, M. J., Gout, I., Totty, N. F., Truong, O., Vicendo, P., Yonezawa, K., Kasuga, M., Courtneidge, S. A., *et al.* (1994b). PI-3-kinase is a dual specificity enzyme—Autoregulation by an intrinsic protein-serine kinase activity. *EMBO J.* **13,** 522–533.

Dong, X., Park, S., Lin, X., Copps, K., Yi, X., and White, M. F. (2006). Irs1 and Irs2 signaling is essential for hepatic glucose homeostasis and systemic growth. *J. Clin. Invest.* **116,** 101–114.

Dummler, B., and Hemmings, B. A. (2007). Physiological roles of PKB/Akt isoforms in development and disease. *Biochem. Soc. Trans.* **35,** 231–235.

Durocher, D., Henckel, J., Fersht, A. R., and Jackson, S. P. (1999). The FHA domain is a modular phosphopeptide recognition motif. *Mol. Cell* **4,** 387–394.

Eck, M. J., Dhe-Paganon, S., Trub, T., Nolte, R. T., and Shoelson, S. E. (1996a). Structure of the IRS-1 PTB domain bound to the juxtamembrane region of the insulin receptor. *Cell* **85,** 695–705.

Eck, M. J., Pluskey, S., Trub, T., Harrison, S. C., and Shoelson, S. E. (1996b). Spatial constraints on the recognition of phosphoproteins by the tandem SH2 domains of the phosphatase SH-PTP2. *Nature* **379,** 277–280.

Egawa, K., Sharma, P. M., Nakashima, N., Huang, Y., Huver, E., Boss, G. R., and Olefsky, J. M. (1999). Membrane-targeted phosphatidylinositol 3-kinase mimics insulin actions and induces a state of cellular insulin resistance. *J. Biol. Chem.* **274,** 14306–14314.

Eldar-Finkelman, H., and Krebs, E. G. (1997). Phosphorylation of insulin receptor substrate 1 by glycogen synthase kinase 3 impairs insulin action. *Proc. Natl. Acad. Sci. USA* **94,** 9660–9664.

Evans, J. L., Goldfine, I. D., Maddux, B. A., and Grodsky, G. M. (2002). Oxidative stress and stress-activated signaling pathways: A unifying hypothesis of type 2 diabetes. *Endocr. Rev.* **23,** 599–622.

Evans, J. L., Maddux, B. A., and Goldfine, I. D. (2005). The molecular basis for oxidative stress-induced insulin resistances. *Antioxid. Redox Signal.* **7,** 1040–1052.

Fantin, V. R., Lavan, B. E., Wang, Q., Jenkins, N. A., Gilbert, D. J., Copeland, N. G., Keller, S. R., and Lienhard, G. E. (1999). Cloning, tissue expression, and chromosomal location of the mouse insulin receptor substrate 4 gene. *Endocrinology* **140,** 1329–1337.

Fantin, V. R., Wang, Q., Lienhard, G. E., and Keller, S. R. (2000). Mice lacking insulin receptor substrate 4 exhibit mild defects in growth, reproduction, and glucose homeostasis. *Am. J. Physiol. Endocrinol. Metab.* **278**(1), E127–E133.

Fingar, D. C., and Blenis, J. (2004). Target of rapamycin (TOR): An integrator of nutrient and growth factor signals and coordinator of cell growth and cell cycle progression. *Oncogene* **23,** 3151–3171.

Fisher, T. L., and White, M. F. (2004). Signaling pathways: The benefits of good communication. *Curr. Biol.* **14,** R1005–R1007.

Freed, E., Symons, M., Macdonald, S. G., McCormick, F., and Ruggieri, R. (1994). Binding of 14-3-3 proteins to the protein kinase Raf and effects on its activation. *Science* **265,** 1713–1716.

Freund, G. G., Wittig, J. G., and Mooney, R. A. (1995). The PI3-kinase serine kinase phosphorylates its p85 subunit and IRS-1 in PI3-kinase/IRS-1 complexes. *Biochem. Biophys. Res. Commun.* **206,** 272–278.

Fridlyand, L. E., and Philipson, L. H. (2006). Reactive species and early manifestation of insulin resistance in type 2 diabetes. *Diabetes Obes. Metab.* **8,** 136–145.

Friedman, J. E., Ishizuka, T., Shao, J., Huston, L., Highman, T., and Catalano, P. (1999). Impaired glucose transport and insulin receptor tyrosine phosphorylation in skeletal muscle from obese women with gestational diabetes. *Diabetes* **48,** 1807–1814.

Furukawa, N., Ongusaha, P., Jahng, W. J., Araki, K., Choi, C. S., Kim, H. J., Lee, Y. H., Kaibuchi, K., Kahn, B. B., Masuzaki, H., Kim, J. K., Lee, S. W., *et al.* (2005). Role of Rho-kinase in regulation of insulin action and glucose homeostasis. *Cell Metab.* **2,** 119–129.

Gao, Z., Hwang, D., Bataille, F., Lefevre, M., York, D., Quon, M. J., and Ye, J. (2002). Serine phosphorylation of insulin receptor substrate 1 by inhibitor kappa B kinase complex. *J. Biol. Chem.* **277,** 48115–48121.

Garami, A., Zwartkruis, F. J., Nobukuni, T., Joaquin, M., Roccio, M., Stocker, H., Kozma, S. C., Hafen, E., Bos, J. L., and Thomas, G. (2003). Insulin activation of Rheb, a mediator of mTOR/S6K/4E-BP signaling, is inhibited by TSC1 and 2. *Mol. Cell* **11,** 1457–1466.

Gibson, T. J., Hyvonen, M., Musacchio, A., and Saraste, M. (1994). PH domain: The first anniversary. *Trends Biochem. Sci.* **19,** 349–353.

Giorgetti, S., Ballotti, R., Kowalski-Chauvel, A., Tartare, S., and Van Obberghen, E. (1993). The insulin and insulin-like growth factor-I receptor substrate IRS-1 associates with and activates phosphatidylinositol 3′-kinase *in vitro*. *J. Biol. Chem.* **268,** 7358–7364.

Giraud, J., Leshan, R., Lee, Y. H., and White, M. F. (2004). Nutrient-dependent and insulin-stimulated phosphorylation of insulin receptor substrate-1 on serine 302 correlates with increased insulin signaling. *J. Biol. Chem.* **279,** 3447–3454.

Giraud, J., Haas, M., Feener, E. P., Copps, K. D., Dong, X., Fidyk, N., and White, M. F. (2007). Phosphorylation of irs1 at ser-522 inhibits insulin signaling. *Mol. Endocrinol.* **21**(9), 2294–2302.

Greene, M. W., Ruhoff, M. S., Roth, R. A., Kim, J. A., Quon, M. J., and Krause, J. A. (2006). PKCdelta-mediated IRS-1 Ser24 phosphorylation negatively regulates IRS-1 function. *Biochem. Biophys. Res. Commun.* **349,** 976–986.

Gregor, M. G., and Hotamisligil, G. S. (2007). Adipocyte stress: The endoplasmic reticulum and metabolic disease. *J. Lipid Res.* **48,** 1905–1914.

Gual, P., Gonzalez, T., Gremeaux, T., Barres, R., Le Marchand-Brustel, Y., and Tanti, J. F. (2003). Hyperosmotic stress inhibits insulin receptor substrate-1 function by distinct mechanisms in 3T3-L1 adipocytes. *J. Biol. Chem.* **278,** 26550–26557.

Gual, P., Le Marchand-Brustel, Y., and Tanti, J. F. (2005). Positive and negative regulation of insulin signaling through IRS-1 phosphorylation. *Biochimie* **87,** 99–109.

Gustafson, T. A., He, W., Craparo, A., Schaub, C. D., and O'Neill, T. J. (1995). Phosphotyrosine-dependent interaction of SHC and insulin receptor substrate 1 with the NPEY motif of the insulin receptor via a novel non-SH2 domain. *Mol. Cell. Biol.* **15,** 2500–2508.

Harrington, L. S., Findlay, G. M., Gray, A., Tolkacheva, T., Wigfield, S., Rebholz, H., Barnett, J., Leslie, N. R., Cheng, S., Shepherd, P. R., Gout, I., Downes, C. P., *et al.* (2004). The TSC1-2 tumor suppressor controls insulin-PI3K signaling via regulation of IRS proteins. *J. Cell Biol.* **166,** 213–223.

He, W., O'Neill, T. J., and Gustafson, T. A. (1995). Distinct modes of interaction of SHC and insulin receptor substrate-1 with the insulin receptor NPEY region via non-SH2 domains. *J. Biol. Chem.* **270,** 23258–23262.

He, J., Usui, I., Ishizuka, K., Kanatani, Y., Hiratani, K., Iwata, M., Bukhari, A., Haruta, T., Sasaoka, T., and Kobayashi, M. (2006). Interleukin-1alpha inhibits insulin signaling with phosphorylating insulin receptor substrate-1 on serine residues in 3T3-L1 adipocytes. *Mol. Endocrinol.* **20,** 114–124.

Hirosumi, J., Tuncman, G., Chang, L., Gorgun, C. Z., Uysal, K. T., Maeda, K., Karin, M., and Hotamisligil, G. S. (2002). A central role for JNK in obesity and insulin resistance. *Nature* **420,** 333–336.

Holgado-Madruga, M., Emlet, D. R., Moscatello, D. K., Godwin, A. K., and Wong, A. J. (1996). A Grb2-associated docking protein in EGF- and insulin-receptor signalling. *Nature* **379,** 560–563.

Horike, N., Takemori, H., Katoh, Y., Doi, J., Min, L., Asano, T., Sun, X. J., Yamamoto, H., Kasayama, S., Muraoka, M., Nonaka, Y., and Okamoto, M. (2003). Adipose-specific expression, phosphorylation of Ser794 in insulin receptor substrate-1, and activation in diabetic animals of salt-inducible kinase-2. *J. Biol. Chem.* **278,** 18440–18447.

Hosomi, Y., Shii, K., Ogawa, W., Matsuba, H., Yoshida, M., Okada, Y., Yokono, K., Kasuga, M., Baba, S., and Roth, R. (1994). Characterization of a 60-Kilodalton substrate of the insulin receptor kinase. *J. Biol. Chem.* **269,** 11498–11502.

Hotamisligil, G. S. (2003). Inflammatory pathways and insulin action. *Int. J. Obes. Relat. Metab. Disord.* **27**(Suppl. 3), S53–S55.

Hotamisligil, G. S. (2005). Role of endoplasmic reticulum stress and c-Jun NH2-terminal kinase pathways in inflammation and origin of obesity and diabetes. *Diabetes* **54**(Suppl. 2), S73–S78.

Hotamisligil, G. S., Peraldi, P., Budvari, A., Ellis, R. W., White, M. F., and Spiegelman, B. M. (1996). IRS-1-mediated inhibition of insulin receptor tyrosine kinase activity in TNF-α- and obesity-induced insulin resistance. *Science* **271,** 665–668.

Houstis, N., Rosen, E. D., and Lander, E. S. (2006). Reactive oxygen species have a causal role in multiple forms of insulin resistance. *Nature* **440,** 944–948.

Ikezu, T., Okamoto, T., Yonezawa, K., Tompkins, R. G., and Martyn, J. A. J. (1997). Analysis of thermal injury-induced insulin resistance in rodents. Implication of postreceptor mechanisms [In Process Citation]. *J. Biol. Chem.* **272,** 25289–25295.

Inoue, G., Cheatham, B., Emkey, R., and Kahn, C. R. (1998). Dynamics of insulin signaling in 3T3-L1 adipocytes. Differential compartmentalization and trafficking of insulin receptor substrate (irs)-1 and irs-2 [In Process Citation]. *J. Biol. Chem.* **273,** 11548–11555.

Irie, K., Gotoh, Y., Yashar, B. M., Errede, B., Nishida, E., and Matsumoto, K. (1994). Stimulatory effects of yeast and mammalian 14-3-3 proteins on the Raf protein kinase. *Science* **265,** 1716–1719.

Itani, S. I., Ruderman, N. B., Schmieder, F., and Boden, G. (2002). Lipid-induced insulin resistance in human muscle is associated with changes in diacylglycerol, protein kinase C, and IkappaB-alpha. *Diabetes* **51,** 2005–2011.

Jaeschke, A., and Davis, R. J. (2007). Metabolic stress signaling mediated by mixed-lineage kinases. *Mol. Cell* **27,** 498–508.

Jaeschke, A., Czech, M. P., and Davis, R. J. (2004). An essential role of the JIP1 scaffold protein for JNK activation in adipose tissue. *Genes Dev.* **18,** 1976–1980.

Jakobsen, S. N., Hardie, D. G., Morrice, N., and Tornqvist, H. E. (2001). AMPK phosphorylates IRS-1 on Ser789 in mouse C2C12 myotubes in response to 5-aminoimidazole-4-carboxamide riboside. *J. Biol. Chem.* **276,** 46912–46916.

Jiang, Z. Y., Lin, Y. W., Clemont, A., Feener, E. P., Hein, K. D., Igarashi, M., Yamauchi, T., White, M. F., and King, G. L. (1999). Characterization of selective resistance to insulin signaling in the vasculature of obese zucker (fa/fa) rats. *J. Clin. Invest.* **104,** 447–457.

Kaburagi, Y., Momomura, K., Yamamoto-Honda, R., Tobe, K., Tamori, Y., Sakura, H., Akanuma, Y., Yazaki, Y., and Kadowaki, T. (1993). Site-directed mutagenesis of the juxtamembrane domain of the human insulin receptor. *J. Biol. Chem.* **268,** 16610–16622.

Kaburagi, Y., Yamamoto-Honda, R., Tobe, K., Ueki, K., Yachi, M., Akanuma, Y., Stephens, R. M., Kaplan, D., Yazaki, Y., and Kadowaki, T. (1995). The role of the NPXY motif in the insulin receptor in tyrosine phosphorylation of insulin receptor substrate-1 and Shc. *Endocrinology* **136,** 3437–3443.

Kaneto, H., Nakatani, Y., Miyatsuka, T., Kawamori, D., Matsuoka, T. A., Matsuhisa, M., Kajimoto, Y., Ichijo, H., Yamasaki, Y., and Hori, M. (2004). Possible novel therapy for diabetes with cell-permeable JNK-inhibitory peptide. *Nat. Med.* **10,** 1128–1132.

Kanety, H., Feinstein, R., Papa, M. Z., Hemi, R., and Karasik, A. (1995). Tumor necrosis factor alpha-induced phosphorylation of insulin receptor substrate-1 (IRS-1). Possible mechanism for suppression of insulin-stimulated tyrosine phosphorylation of IRS-1. *J. Biol. Chem.* **270,** 23780–23784.

Kasuga, M., Hedo, J. A., Yamada, K. M., and Kahn, C. R. (1982). The structure of the insulin receptor and its subunits: Evidence for multiple non-reduced forms and a 210K possible proreceptor. *J. Biol. Chem.* **257,** 10392–10399.

Kido, Y., Burks, D. J., Withers, D., Bruning, J. C., Kahn, C. R., White, M. F., and Accili, D. (2000). Tissue-specific insulin resistance in mice with mutations in the insulin receptor, IRS-1, and IRS-2. *J. Clin. Invest.* **105,** 199–205.

Kim, B., Leventhal, P. S., White, M. F., and Feldman, E. L. (1998). Differential regulation of insulin receptor substrate-2 and mitogen-activated protein kinase tyrosine phosphorylation by phosphatidylinositol 3-kinase inhibitors in SH-SY5Y human neuroblastoma cells [In Process Citation]. *Endocrinology* **139,** 4881–4889.

Kim, J. K., Kim, Y. J., Fillmore, J. J., Chen, Y., Moore, I., Lee, J., Yuan, M., Li, Z. W., Karin, M., Perret, P., Shoelson, S. E., and Shulman, G. I. (2001). Prevention of fat-induced insulin resistance by salicylate. *J. Clin. Invest.* **108,** 437–446.

Kim, J. K., Fillmore, J. J., Sunshine, M. J., Albrecht, B., Higashimori, T., Kim, D. W., Liu, Z. X., Soos, T. J., Cline, G. W., O'Brien, W. R., Littman, D. R., and

Shulman, G. I. (2004). PKC-theta knockout mice are protected from fat-induced insulin resistance. *J. Clin. Invest.* **114,** 823–827.
Kim, J. A., Yeh, D. C., Ver, M., Li, Y., Carranza, A., Conrads, T. P., Veenstra, T. D., Harrington, M. A., and Quon, M. J. (2005). Phosphorylation of Ser24 in the pleckstrin homology domain of insulin receptor substrate-1 by Mouse Pelle-like kinase/interleukin-1 receptor-associated kinase: Cross-talk between inflammatory signaling and insulin signaling that may contribute to insulin resistance. *J. Biol. Chem.* **280,** 23173–23183.
Kitagawa, M., Hatakeyama, S., Shirane, M., Matsumoto, M., Ishida, N., Hattori, K., Nakamichi, I., Kikuchi, A., and Nakayama, K. (1999). An F-box protein, FWD1, mediates ubiquitin-dependent proteolysis of beta-catenin. *EMBO J.* **18,** 2401–2410.
Klover, P. J., Zimmers, T. A., Koniaris, L. G., and Mooney, R. A. (2003). Chronic exposure to interleukin-6 causes hepatic insulin resistance in mice. *Diabetes* **52,** 2784–2789.
Klover, P. J., Clementi, A. H., and Mooney, R. A. (2005). Interleukin-6 depletion selectively improves hepatic insulin action in obesity. *Endocrinology* **146,** 3417–3427.
Koch, C. A., Anderson, D. J., Moran, M. F., Ellis, C. A., and Pawson, T. (1991). SH2 and SH3 domains: Elements that control interactions of cytoplasmic signaling proteins. *Science* **252,** 668–674.
Kosaki, A., Yamada, K., Suga, J., Otaka, A., and Kuzuya, H. (1998). 14-3-3beta protein associates with insulin receptor substrate 1 and decreases insulin-stimulated phosphatidylinositol 3′-kinase activity in 3T3L1 adipocytes [In Process Citation]. *J. Biol. Chem.* **273,** 940–944.
Kublaoui, B., Lee, J., and Pilch, P. F. (1995). Dynamics of signaling during insulin-stimulated endocytosis of its receptor in adipocytes. *J. Biol. Chem.* **270,** 59–65.
Kubota, N., Tobe, K., Terauchi, Y., Eto, K., Yamauchi, T., Suzuki, R., Tsubamoto, Y., Komeda, K., Nakano, R., Miki, H., Satoh, S., Sekihara, H., *et al.* (2000). Disruption of insulin receptor substrate 2 causes type 2 diabetes because of liver insulin resistance and lack of compensatory beta-cell hyperplasia. *Diabetes* **49,** 1880–1889.
Kubota, N., Terauchi, Y., Tobe, K., Yano, W., Suzuki, R., Ueki, K., Takamoto, I., Satoh, H., Maki, T., Kubota, T., Moroi, M., Okada-Iwabu, M., *et al.* (2004). Insulin receptor substrate 2 plays a crucial role in beta cells and the hypothalamus. *J. Clin. Invest.* **114,** 917–927.
Kulkarni, R. N., Bruning, J. C., Winnay, J. N., Postic, C., Magnuson, M. A., and Kahn, C. R. (1999a). Tissue-specific knockout of the insulin receptor in pancreatic beta cells creates an insulin secretory defect similar to that in type 2 diabetes. *Cell* **96,** 329–339.
Kulkarni, R. N., Winnay, J. N., Daniels, M., Bruning, J. C., Flier, S. N., Hanahan, D., and Kahn, C. R. (1999b). Altered function of insulin receptor substrate-1-deficient mouse islets and cultured beta-cell lines. *J. Clin. Invest.* **104,** R69–R75.
Lassak, A., Del, V. L., Peruzzi, F., Wang, J. Y., Enam, S., Croul, S., Khalili, K., and Reiss, K. (2002). Insulin receptor substrate 1 translocation to the nucleus by the human JC virus T-antigen. *J. Biol. Chem.* **277,** 17231–17238.
Lavan, B. E., Fantin, V. R., Chang, E. T., Lane, W. S., Keller, S. R., and Lienhard, G. E. (1997a). A novel 160-kDa phosphotyrosine protein in insulin-treated embryonic kidney cells is a new member of the insulin receptor substrate family [In Process Citation]. *J. Biol. Chem.* **272,** 21403–21407.
Lavan, B. E., Lane, W. S., and Lienhard, G. E. (1997b). The 60-kDa phosphotyrosine protein in insulin-treated adipocytes is a new member of the insulin receptor substrate family. *J. Biol. Chem.* **272,** 11439–11443.
Le Marchand-Brustel, Y., Gual, P., Gremeaux, T., Gonzalez, T., Barres, R., and Tanti, J. F. (2003). Fatty acid-induced insulin resistance: Role of insulin receptor substrate 1 serine

phosphorylation in the retroregulation of insulin signalling. *Biochem. Soc. Trans.* **31,** 1152–1156.

Lee, C. H., Li, W., Nishimura, R., Zhou, M., Batzer, A. G., Myers, M. G., White, M. F., Schlessinger, J., and Skolnik, E. Y. (1993). Nck associates with the SH2 domain-docking protein IRS-1 in insulin-stimulated cells. *Proc. Natl. Acad. Sci. USA* **90,** 11713–11717.

Lee, A. V., Gooch, J. L., Oesterreich, S., Guler, R. L., and Yee, D. (2000). Insulin-like growth factor I-induced degradation of insulin receptor substrate 1 is mediated by the 26S proteasome and blocked by phosphatidylinositol 3′-kinase inhibition. *Mol. Cell. Biol.* **20,** 1489–1496.

Li, J., DeFea, K., and Roth, R. A. (1999). Modulation of insulin receptor substrate-1 tyrosine phosphorylation by an Akt/phosphatidylinositol 3-kinase pathway. *J. Biol. Chem.* **274,** 9351–9356.

Li, Y., Soos, T. J., Li, X., Wu, J., Degennaro, M., Sun, X., Littman, D. R., Birnbaum, M. J., and Polakiewicz, R. D. (2004). Protein kinase C Theta inhibits insulin signaling by phosphorylating IRS1 at Ser(1101). *J. Biol. Chem.* **279,** 45304–45307.

Liberman, Z., and Eldar-Finkelman, H. (2005). Serine 332 phosphorylation of insulin receptor substrate-1 by glycogen synthase kinase-3 attenuates insulin signaling. *J. Biol. Chem.* **280,** 4422–4428.

Lin, X., Taguchi, A., Park, S., Kushner, J. A., Li, F., Li, Y., and White, M. F. (2004). Dysregulation of insulin receptor substrate 2 in beta cells and brain causes obesity and diabetes. *J. Clin. Invest.* **114,** 908–916.

Liu, S. C., Wang, Q., Lienhard, G. E., and Keller, S. R. (1999). Insulin receptor substrate 3 is not essential for growth or glucose homeostasis. *J. Biol. Chem.* **274,** 18093–18099.

Liu, Y. F., Herschkovitz, A., Boura-Halfon, S., Ronen, D., Paz, K., LeRoith, D., and Zick, Y. (2004). Serine phosphorylation proximal to its phosphotyrosine binding domain inhibits insulin receptor substrate 1 function and promotes insulin resistance. *Mol. Cell. Biol.* **24,** 9668–9681.

Lu, P. J., Zhou, X. Z., Shen, M., and Lu, K. P. (1999). Function of WW domains as phosphoserine- or phosphothreonine-binding modules. *Science* **283,** 1325–1328.

Luo, M., Reyna, S., Wang, L., Yi, Z., Carroll, C., Dong, L. Q., Langlais, P., Weintraub, S. T., and Mandarino, L. J. (2005). Identification of insulin receptor substrate 1 serine/threonine phosphorylation sites using mass spectrometry analysis: Regulatory role of serine 1223. *Endocrinology* **146,** 4410–4416.

Luttrell, L. M., Hawes, B. E., Touhara, K., van Biesen, T., Kock, W. J., and Lefkowitz, R. J. (1995). Effect of cellular expression of pleckstrin homology domains on Gi-coupled receptor signaling. *J. Biol. Chem.* **270,** 12984–12989.

Lyle, A. N., and Griendling, K. K. (2006). Modulation of vascular smooth muscle signaling by reactive oxygen species. *Physiology (Bethesda)* **21,** 269–280.

Macias, M. J., Musacchio, A., Ponstingl, H., Nilges, M., Saraste, M., and Oschkinat, H. (1994). Structure of the pleckstrin homology domain from beta-spectrin. *Nature* **369,** 675–677.

Michael, M. D., Kulkarni, R. N., Postic, C., Previs, S. F., Shulman, G. I., Magnuson, M. A., and Kahn, C. R. (2000). Loss of insulin signaling in hepatocytes leads to severe insulin resistance and progressive hepatic dysfunction. *Mol. Cell* **6,** 87–97.

Miralpeix, M., Sun, X. J., Backer, J. M., Myers, M. G., Araki, E., and White, M. F. (1992). Insulin stimulates tyrosine phosphorylation of multiple high molecular weight substrates in FAO hepatoma cells. *Biochemistry* **31,** 9031–9039.

Moeschel, K., Beck, A., Weigert, C., Lammers, R., Kalbacher, H., Voelter, W., Schleicher, E. D., Haring, H. U., and Lehmann, R. (2004). Protein kinase C-zeta-induced phosphorylation of Ser318 in insulin receptor substrate-1 (IRS-1) attenuates the interaction with the insulin receptor and the tyrosine phosphorylation of IRS-1. *J. Biol. Chem.* **279,** 25157–25163.

Moodie, S. A., Alleman-Sposeto, J., and Gustafson, T. A. (1999). Identification of the APS protein as a novel insulin receptor substrate [In Process Citation]. *J. Biol. Chem.* **274,** 11186–11193.

Morrison, D. (1994). 14-3-3: Modulators of signaling proteins? [comment]. *Science* **266,** 56–57.

Musacchio, A., Gibson, T., Rice, P., Thompson, J., and Saraste, M. (1993). The PH domain: A common piece in the structural patchwork of signalling proteins. *Trends Biochem. Sci.* **18,** 343.

Muslin, A. J., Tanner, J. W., Allen, P. M., and Shaw, A. S. (1996). Interaction of 14-3-3 with signaling proteins is mediated by the recognition of phosphoserine. *Cell* **84,** 889–897.

Myers, M. G., Jr.,Wang, L. M., Sun, X. J., Zhang, Y., Yenush, L., Schlessinger, J., Pierce, J. H., and White, M. F. (1994). Role of IRS-1–GRB-2 complexes in insulin signaling. *Mol. Cell. Biol.* **14,** 3577–3587.

Myers, M. G., Mendez, R., Shi, P., Pierce, J. H., Rhoads, R., and White, M. F. (1998). The COOH-terminal tyrosine phosphorylation sites on IRS-1 bind SHP-2 and negatively regulate insulin signaling. *J. Biol. Chem.* **273,** 26908–26914.

Nakajima, K., Yamauchi, K., Shigematsu, S., Ikeo, S., Komatsu, M., Aizawa, T., and Hashizume, K. (2000). Selective attenuation of metabolic branch of insulin receptor down-signaling by high glucose in a hepatoma cell line, HepG2 cells. *J. Biol. Chem.* **275,** 20880–20886.

Nakamura, H., Nakamura, K., and Yodoi, J. (1997). Redox regulation of cellular activation. *Annu. Rev. Immunol.* **15,** 351–369.

Nakatani, Y., Kaneto, H., Kawamori, D., Hatazaki, M., Miyatsuka, T., Matsuoka, T. A., Kajimoto, Y., Matsuhisa, M., Yamasaki, Y., and Hori, M. (2004). Modulation of the JNK pathway in liver affects insulin resistance status. *J. Biol. Chem.* **279,** 45803–45809.

Nawaratne, R., Gray, A., Jorgensen, C. H., Downes, C. P., Siddle, K., and Sethi, J. K. (2006). Regulation of insulin receptor substrate 1 pleckstrin homology domain by protein kinase C: Role of serine 24 phosphorylation. *Mol. Endocrinol.* **20,** 1838–1852.

O'Neill, T. J., Craparo, A., and Gustafson, T. A. (1994). Characterization of an interaction between insulin receptor substrate 1 and the insulin receptor by using the two-hybrid system. *Mol. Cell. Biol.* **14,** 6433–6442.

Ogihara, T., Isobe, T., Ichimura, T., Taoka, M., Funaki, M., Sakoda, H., Onishi, Y., Inukai, K., Anai, M., Fukushima, Y., Kikuchi, M., Yazaki, Y., Oka, Y., and Asano, T. (1997). 14-3-3 protein binds to insulin receptor substrate-1, one of the binding sites of which is in the phosphotyrosine binding domain [In Process Citation]. *J. Biol. Chem.* **272,** 25267–25274.

Ozcan, U., Cao, Q., Yilmaz, E., Lee, A. H., Iwakoshi, N. N., Ozdelen, E., Tuncman, G., Gorgun, C., Glimcher, L. H., and Hotamisligil, G. S. (2004). Endoplasmic reticulum stress links obesity, insulin action, and type 2 diabetes. *Science* **306,** 457–461.

Ozcan, U., Yilmaz, E., Ozcan, L., Furuhashi, M., Vaillancourt, E., Smith, R. O., Gorgun, C. Z., and Hotamisligil, G. S. (2006). Chemical chaperones reduce ER stress and restore glucose homeostasis in a mouse model of type 2 diabetes. *Science* **313,** 1137–1140.

Ozes, O. N., Akca, H., Mayo, L. D., Gustin, J. A., Maehama, T., Dixon, J. E., and Donner, D. B. (2001). A phosphatidylinositol 3-kinase/Akt/mTOR pathway mediates and PTEN antagonizes tumor necrosis factor inhibition of insulin signaling through insulin receptor substrate-1. *Proc. Natl. Acad. Sci. USA* **98,** 4640–4645.

Patti, M. E., and Kahn, B. B. (2004). Nutrient sensor links obesity with diabetes risk. *Nat. Med.* **10,** 1049–1050.

Pawson, T. (1995). Protein modules and signalling networks. *Nature* **373,** 573–580.

Pawson, T., and Gish, G. D. (1992). SH2 and SH3 domains: From structure to function. *Cell* **71,** 359–362.

Paz, K., Hemi, R., LeRoith, D., Karasik, A., Elhanany, E., Kanety, H., and Zick, Y. (1997). A molecular basis for insulin resistance. Elevated serine/threonine phosphorylation of IRS-1 and IRS-2 inhibits their binding to the juxtamembrane region of the insulin receptor and impairs their ability to undergo insulin-induced tyrosine phosphorylation. *J. Biol. Chem.* **272,** 29911–29918.

Pitcher, J. A., Touhara, K., Payne, E. S., and Lefkowitz, R. J. (1995). Pleckstrin homology domain-mediated membrane association and activation of the beta-adrenergic receptor kinase requires coordinate interaction with G-beta/gamma subunits and lipid. *J. Biol. Chem.* **270,** 11707–11710.

Pluskey, S., Wandless, T. J., Walsh, C. T., and Shoelson, S. E. (1995). Potent stimulation of SH-PTP2 phosphatase activity by simultaneous occupancy of both SH2 domains. *J. Biol. Chem.* **270,** 2897–2900.

Plutzky, J., Neel, B. G., and Rosenberg, R. D. (1992). Isolation of a src homology 2-containing tyrosine phosphatase. *Proc. Natl. Acad. Sci. USA* **89,** 1123–1127.

Previs, S. F., Withers, D. J., Ren, J. M., White, M. F., and Shulman, G. I. (2000). Contrasting effects of IRS-1 versus IRS-2 gene disruption on carbohydrate and lipid metabolism *in vivo*. *J. Biol. Chem.* **275,** 38990–38994.

Prisco, M., Santini, F., Baffa, R., Liu, M., Drakas, R., Wu, A., and Baserga, R. (2002). Nuclear translocation of insulin receptor substrate-1 by the simian virus 40 T antigen and the activated type 1 insulin-like growth factor receptor. *J. Biol. Chem.* **277,** 32078–32085.

Qiao, L. Y., Goldberg, J. L., Russell, J. C., and Sun, X. J. (1999). Identification of enhanced serine kinase activity in insulin resistance. *J. Biol. Chem.* **274,** 10625–10632.

Qiao, L. Y., Zhande, R., Jetton, T. L., Zhou, G., and Sun, X. J. (2002). *In vivo* phosphorylation of insulin receptor substrate 1 at serine 789 by a novel serine kinase in insulin-resistant rodents. *J. Biol. Chem.* **277,** 26530–26539.

Razzini, G., Ingrosso, A., Brancaccio, A., Sciacchitano, S., Esposito, D. L., and Falasca, M. (2001). Different subcellular localization and phosphoinositides binding of insulin receptor substrate protein pleckstrin homology domains. *Mol. Endocrinol.* **14**(6), 823–836.

Ribon, V., and Saltiel, A. R. (1997). Insulin stimulates tyrosine phosphorylation of the proto-oncogene product of c-Cbl in 3T3-L1 adipocytes. *Biochem. J.* **324**(Pt. 3), 839–845.

Rondinone, C. M., Wang, L. M., Lonnroth, P., Wesslau, C., Pierce, J. H., and Smith, U. (1997). Insulin receptor substrate (IRS) 1 is reduced and IRS-2 is the main docking protein for phosphatidylinositol 3-kinase in adipocytes from subjects with non-insulin-dependent diabetes mellitus. *Proc. Natl. Acad. Sci. USA* **94,** 4171–4175.

Rordorf-Nikolic, T., Van Horn, D. J., Chen, D., White, M. F., and Backer, J. M. (1995). Regulation of phosphatidylinositol 3-kinase by tyrosyl phosphoproteins. Full activation requires occupancy of both SH2 domains in the 85 kDa regulatory subunit. *J. Biol. Chem.* **270,** 3662–3666.

Rosen, O. M. (1987). After insulin binds. *Science* **237,** 1452–1458.

Rothenberg, P. L., Lane, W. S., Karasik, A., Backer, J. M., White, M., and Kahn, C. R. (1991). Purification and partial sequence analysis of pp185, the major cellular substrate of the insulin receptor tyrosine kinase. *J. Biol. Chem.* **266,** 8302–8311.

Rotter, V., Nagaev, I., and Smith, U. (2003). Interleukin-6 (IL-6) induces insulin resistance in 3T3-L1 adipocytes and is, like IL-8 and tumor necrosis factor-alpha, overexpressed in human fat cells from insulin-resistant subjects. *J. Biol. Chem.* **278,** 45777–45784.

Rui, L., Aguirre, V., Kim, J. K., Shulman, G. I., Lee, A., Corbould, A., Dunaif, A., and White, M. F. (2001a). Insulin/IGF-1 and TNF-alpha stimulate phosphorylation of IRS-1 at inhibitory Ser307 via distinct pathways. *J. Clin. Invest.* **107**(2), 181–189.

Rui, L., Fisher, T. L., Thomas, J., and White, M. F. (2001b). Regulation of insulin/IGF-1 signaling by proteasome-mediated degradation of IRS-2. *J. Biol. Chem.* **276,** 40362–40367.

Sampson, S. R., and Cooper, D. R. (2006). Specific protein kinase C isoforms as transducers and modulators of insulin signaling. *Mol. Genet. Metab.* **89,** 32–47.

Samuel, V. T., Liu, Z. X., Wang, A., Beddow, S. A., Geisler, J. G., Kahn, M., Zhang, X. M., Monia, B. P., Bhanot, S., and Shulman, G. I. (2007). Inhibition of protein kinase Cepsilon prevents hepatic insulin resistance in nonalcoholic fatty liver disease. *J. Clin. Invest.* **117,** 739–745.

Sasaoka, T., Rose, D. W., Jhun, B. H., Saltiel, A. R., Draznin, B., and Olefsky, J. M. (1994). Evidence for a functional role of Shc proteins in mitogenic signaling induced by insulin, insulin-like growth factor-1, and epidermal growth factor. *J. Biol. Chem.* **269,** 13689–13694.

Savage, D. B., Petersen, K. F., and Shulman, G. I. (2007). Disordered lipid metabolism and the pathogenesis of insulin resistance. *Physiol. Rev.* **87,** 507–520.

Sawka-Verhelle, D., Tartare-Deckert, S., White, M. F., and Van Obberghen, E. (1996). Insulin receptor substrate-2 binds to the insulin receptor through its phosphotyrosine-binding domain and through a newly identified domain comprising amino acids 591–786. *J. Biol. Chem.* **271,** 5980–5983.

Sawka-Verhelle, D., Baron, V., Mothe, I., Filloux, C., White, M. F., and Van Obberghen, E. (1997). Tyr624 and Tyr628 in insulin receptor substrate-2 mediate its association with the insulin receptor. *J. Biol. Chem.* **272,** 16414–16420.

Schmitz-Peiffer, C. (2002). Protein kinase C and lipid-induced insulin resistance in skeletal muscle. *Ann. N.Y. Acad. Sci.* **967,** 146–157.

Senn, J. J., Klover, P. J., Nowak, I. A., and Mooney, R. A. (2002). Interleukin-6 induces cellular insulin resistance in hepatocytes. *Diabetes* **51,** 3391–3399.

Shah, O. J., and Hunter, T. (2006). Turnover of the active fraction of IRS1 involves raptor–mTOR- and S6K1-dependent serine phosphorylation in cell culture models of tuberous sclerosis. *Mol. Cell. Biol.* **26,** 6425–6434.

Shimizu, K., Kuroda, S., Yamamori, B., Matsuda, S., Kaibuchi, K., Yamauchi, T., Isobe, T., Irie, K., Matsumoto, K., and Takai, Y. (1994). Synergistic activation by Ras and 14-3-3 protein of a mitogen-activated protein kinase kinase kinase named Ras-dependent extracellular signal-regulated kinase kinase stimulator. *J. Biol. Chem.* **269,** 22917–22920.

Shirane, M., Hatakeyama, S., Hattori, K., and Nakayama, K. (1999). Common pathway for the ubiquitination of IkappaBalpha, IkappaBbeta, and IkappaBepsilon mediated by the F-box protein FWD1. *J. Biol. Chem.* **274,** 28169–28174.

Shoelson, S. E., Lee, J., and Goldfine, A. B. (2006). Inflammation and insulin resistance. *J. Clin. Invest.* **116,** 1793–1801.

Shoelson, S. E., Herrero, L., and Naaz, A. (2007). Obesity, inflammation, and insulin resistance. *Gastroenterology* **132,** 2169–2180.

Simpson, L., Li, J., Liaw, D., Hennessy, I., Oliner, J., Christians, F., and Parsons, R. (2001). PTEN expression causes feedback upregulation of insulin receptor substrate 2. *Mol. Cell. Biol.* **21,** 3947–3958.

Skolnik, E. Y., Batzer, A. G., Li, N., Lee, C. H., Lowenstein, E. J., Mohammadi, M., Margolis, B., and Schlessinger, J. (1993). The function of GRB2 in linking the insulin receptor to ras signaling pathways. *Science* **260,** 1953–1955.

Skowyra, D., Craig, K. L., Tyers, M., Elledge, S. J., and Harper, J. W. (1997). F-box proteins are receptors that recruit phosphorylated substrates to the SCF ubiquitin–ligase complex. *Cell* **91,** 209–219.

Smith, L. K., Bradshaw, M., Croall, D. E., and Garner, C. W. (1993). The insulin receptor substrate (IRS-1) is a PEST protein that is susceptible to calpain degradation *in vitro*. *Biochem. Biophys. Res. Commun.* **196,** 767–772.

Smith-Hall, J., Pons, S., Patti, M. E., Burks, D. J., Yenush, L., Sun, X. J., Kahn, C. R., and White, M. F. (1997). The 60 kDa insulin receptor substrate functions like an IRS protein (pp60IRS3) in adipose cells. *Biochemistry* **36,** 8304–8310.

Solinas, G., Naugler, W., Galimi, F., Lee, M. S., and Karin, M. (2006). Saturated fatty acids inhibit induction of insulin gene transcription by JNK-mediated phosphorylation of insulin-receptor substrates. *Proc. Natl. Acad. Sci. USA* **103,** 16454–16459.

Songyang, Z., Shoelson, S. E., Chaudhuri, M., Gish, G. D., Roberts, T., Ratnofsky, S., Lechleider, R. J., Neel, B. G., Birge, R. B., Fajardo, J. E., Chou, M. M., Hanafusa, H., Schaffhausen, B., and Cantley, J. C. (1993). SH2 domains recognize specific phosphopeptide sequences. *Cell* **72,** 767–778.

Spencer, E., Jiang, J., and Chen, Z. J. (1999). Signal-induced ubiquitination of IkappaBalpha by the F-box protein Slimb/beta-TrCP. *Genes Dev.* **13,** 284–294.

Staubs, P. A., Reichart, D. R., Saltiel, A. R., Milarski, K. L., Maegawa, H., Berhanu, P., Olefsky, J. M., and Seely, B. L. (1994). Localization of the insulin receptor binding sites for the SH2 domain proteins p85, syp, and GAP. *J. Biol. Chem.* **269,** 27186–27192.

Stratford, S., Hoehn, K. L., Liu, F., and Summers, S. A. (2004). Regulation of insulin action by ceramide: Dual mechanisms linking ceramide accumulation to the inhibition of Akt/protein kinase B. *J. Biol. Chem.* **279,** 36608–36615.

Sugimoto, S., Lechleider, R. J., Shoelson, S. E., Neel, B. G., and Walsh, C. T. (1994). Expression, purification and characterization of SH2-containing protein tyrosine phosphatase, SH-PTP2. *J. Biol. Chem.* **269,** 13614–13622.

Summers, S. A. (2006). Ceramides in insulin resistance and lipotoxicity. *Prog. Lipid Res.* **45,** 42–72.

Summers, S. A., and Nelson, D. H. (2005). A role for sphingolipids in producing the common features of type 2 diabetes, metabolic syndrome X, and Cushing's syndrome. *Diabetes* **54,** 591–602.

Sun, X. J., Rothenberg, P. L., Kahn, C. R., Backer, J. M., Araki, E., Wilden, P. A., Cahill, D. A., Goldstein, B. J., and White, M. F. (1991). The structure of the insulin receptor substrate IRS-1 defines a unique signal transduction protein. *Nature* **352,** 73–77.

Sun, X. J., Miralpeix, M., Myers, M. G., Glasheen, E. M., Backer, J. M., Kahn, C. R., and White, M. F. (1992). The expression and function of IRS-1 in insulin signal transmission. *J. Biol. Chem.* **267,** 22662–22672.

Sun, X. J., Crimmins, D. L., Myers, M. G., Miralpeix, M., and White, M. F. (1993). Pleiotropic insulin signals are engaged by multisite phosphorylation of IRS-1. *Mol. Cell. Biol.* **13,** 7418–7428.

Sun, X. J., Wang, L. M., Zhang, Y., Yenush, L., Myers, M. G., Glasheen, E., Lane, W. S., Pierce, J. H., and White, M. F. (1995). Role of IRS-2 in insulin and cytokine signalling. *Nature* **377,** 173–177.

Sun, X. J., Pons, S., Asano, T., Myers, M. G., Glasheen, E. M., White, M. F., and Glasheen, E. (1996). The fyn tyrosine kinase binds IRS-1 and forms a distinct signaling complex during insulin stimulation. *J. Biol. Chem.* **271,** 10583–10587.

Sun, X. J., Pons, S., Wang, L.-M., Zhang, Y., Yenush, L., Burks, D., Myers, M. G., Glasheen, E. M., Copeland, N. G., Jenkins, N. A., Pierce, J. H., and White, M. F. (1997). The IRS-2 gene on murine chromosome 8 encodes a unique signaling adapter for insulin and cytokine action. *Mol. Endocrinol.* **11,** 251–262.

Sun, X. J., Goldberg, J. L., Qiao, L. Y., and Mitchell, J. J. (1999). Insulin-induced insulin receptor substrate-1 degradation is mediated by the proteasome degradation pathway. *Diabetes* **48,** 1359–1364.

Sun, H., Tu, X., Prisco, M., Wu, A., Casiburi, I., and Baserga, R. (2003). Insulin-like growth factor I receptor signaling and nuclear translocation of insulin receptor substrates 1 and 2. *Mol. Endocrinol.* **17,** 472–486.

Suzuki, R., Tobe, K., Aoyama, M., Inoue, A., Sakamoto, K., Yamauchi, T., Kamon, J., Kubota, N., Terauchi, Y., Yoshimatsu, H., Matsuhisa, M., Nagasaka, S., Ogata, H., Tokuyama, K., Nagai, R., and Kadowaki, T. (2004). Both insulin signaling defects in the liver and obesity contribute to insulin resistance and cause diabetes in Irs2(−/−) mice. *J. Biol. Chem.* **279,** 25039–25049.

Takeuchi, H., Matsuda, M., Yamamoto, T., Kanematsu, T., Kikkawa, U., Yagisawa, H., Watanabe, Y., and Hirata, M. (1998). PTB domain of insulin receptor substrate-1 binds inositol compounds. *Biochem. J.* **334,** 211–218.

Tamemoto, H., Kadowaki, T., Tobe, K., Yagi, T., Sakura, H., Hayakawa, T., Terauchi, Y., Ueki, K., Kaburagi, Y., Satoh, S., Sekihara, H., Yoshioka, S., Horikoshi, H., Furuta, Y., Ikawa, Y., Kasuga, M., Yazaki, Y., and Aizawa, S. (1994). Insulin resistance and growth retardation in mice lacking insulin receptor substrate-1. *Nature* **372,** 182–186.

Tanaka, S., and Wands, J. R. (1996). A carboxy-terminal truncated insulin receptor substrate-1 dominant negative protein reverses the human hepatocellular carcinoma malignant phenotype. *J. Clin. Invest.* **98,** 2100–2108.

Tanasijevic, M. J., Myers, M. G., Thomas, R. S., Crimmons, D. L., White, M. F., and Sacks, D. B. (1993). Phosphorylation of the insulin receptor substrate IRS-1 by casein kinase II. *J. Biol. Chem.* **268,** 18157–18166.

Taniguchi, C. M., Emanuelli, B., and Kahn, C. R. (2006). Critical nodes in signalling pathways: Insights into insulin action. *Nat. Rev. Mol. Cell Biol.* **7,** 85–96.

Tanti, J. F., Gremeaux, T., Van Obberghen, E., and Le Marchand-Brustel, Y. (1991). Effects of okadaic acid, an inhibitor of protein phosphatases-1 and -2A, on glucose transport and metabolism in skeletal muscle. *J. Biol. Chem.* **266,** 2099–2103.

Tanti, J. F., Gremeaux, T., Van Obberghen, E., and Le Marchand-Brustel, Y. (1994a). Insulin receptor substrate 1 is phosphorylated by the serine kinase activity of phosphatidylinositol 3-kinase. *Biochem. J.* **304,** 17–21.

Tanti, J. F., Gremeaux, T., Van Obberghen, E., and Le Marchand-Brustel, Y. (1994b). Serine/threonine phosphorylation of insulin receptor substrate 1 modulates insulin receptor signaling. *J. Biol. Chem.* **269**(8), 6051–6057.

Tartare-Deckert, S., Murdaca, J., Sawka-Verhelle, D., Holt, K. H., Pessin, J. E., and Van Obberghen, E. (1996). Interaction of the molecular weight 85K regulatory subunit of the phosphatidylinositol 3-kinase with the insulin receptor and the insulin-like growth factor-1 (IGF-I) receptor: Comparative study using the yeast two-hybrid system. *Endocrinology* **137,** 1019–1024.

Thirone, A. C., Huang, C., and Klip, A. (2006). Tissue-specific roles of IRS proteins in insulin signaling and glucose transport. *Trends Endocrinol. Metab.* **17,** 72–78.

Tremblay, F., Brule, S., Hee, U. S., Li, Y., Masuda, K., Roden, M., Sun, X. J., Krebs, M., Polakiewicz, R. D., Thomas, G., and Marette, A. (2007). Identification of IRS-1 Ser-1101 as a target of S6K1 in nutrient- and obesity-induced insulin resistance. *Proc. Natl. Acad. Sci. USA* **104,** 14056–14061.

Trojanek, J., Croul, S., Ho, T., Wang, J. Y., Darbinyan, A., Nowicki, M., Valle, L. D., Skorski, T., Khalili, K., and Reiss, K. (2006). T-antigen of the human polyomavirus JC attenuates faithful DNA repair by forcing nuclear interaction between IRS-1 and Rad51. *J. Cell. Physiol.* **206,** 35–46.

Tsiotra, P. C., and Tsigos, C. (2006). Stress, the endoplasmic reticulum, and insulin resistance. *Ann. N.Y. Acad. Sci.* **1083,** 63–76.

Tsuji, Y., Kaburagi, Y., Terauchi, Y., Satoh, S., Kubota, N., Tamemoto, H., Kraemer, F. B., Sekihara, H., Aizawa, S., Akanuma, Y., Tobe, K., Kimura, S., and Kadowaki, T. (2001). Subcellular localization of insulin receptor substrate family proteins associated with phosphatidylinositol 3-kinase activity and alterations in lipolysis in primary mouse adipocytes from IRS-1 null mice. *Diabetes* **50,** 1455–1463.

Ullrich, A., Bell, J. R., Chen, E. Y., Herrera, R., Petruzzelli, L. M., Dull, T. J., Gray, A., Coussens, L., Liao, Y. C., Tsubokawa, M., Mason, A., Seeburg, P. H., Grunfeld, C., Rosen, O. M., and Ramachandran, J. (1985). Human insulin receptor and its relationship to the tyrosine kinase family of oncogenes. *Nature* **313,** 756–761.

Um, S. H., Frigerio, F., Watanabe, M., Picard, F., Joaquin, M., Sticker, M., Fumagalli, S., Allegrini, P. R., Kozma, S. C., Auwerx, J., and Thomas, G. (2004). Absence of S6K1 protects against age- and diet-induced obesity while enhancing insulin sensitivity. *Nature* **431,** 200–205.

Valverde, A. M., Burks, D. J., Fabregat, I., Fisher, T. L., Carretero, J., White, M. F., and Benito, M. (2003). Molecular mechanisms of insulin resistance in IRS-2-deficient hepatocytes. *Diabetes* **52,** 2239–2248.

Van Horn, D. J., Myers, M. G., and Backer, J. M. (1994). Direct activation of the phosphatidylinositol 3′-kinase by the insulin receptor. *J. Biochem.* **269**(1), 29–32.

Virkamaki, A., Ueki, K., and Kahn, C. R. (1999). Protein–protein interaction in insulin signaling and the molecular mechanisms of insulin resistance. *J. Clin. Invest.* **103,** 931–943.

Wang, L. M., Keegan, A. D., Paul, W. E., Heidaran, M. A., Gutkind, J. S., and Pierce, J. H. (1992). IL-4 activates a distinct signal transduction cascade from IL-3 in factor dependent myeloid cells. *EMBO J.* **11,** 4899–4908.

Wang, L. M., Keegan, A. D., Li, W., Lienhard, G. E., Pacini, S., Gutkind, J. S., Myers, M. G., Sun, X. J., White, M. F., Aaronson, S. A., Paul, W. E., and Pierce, J. H. (1993). Common elements in interleukin 4 and insulin signaling pathways in factor dependent hematopoietic cells. *Proc. Natl. Acad. Sci. USA* **90,** 4032–4036.

Weigert, C., Hennige, A. M., Brischmann, T., Beck, A., Moeschel, K., Schauble, M., Brodbeck, K., Haring, H. U., Schleicher, E. D., and Lehmann, R. (2005). The phosphorylation of Ser318 of insulin receptor substrate 1 is not *per se* inhibitory in skeletal muscle cells but is necessary to trigger the attenuation of the insulin-stimulated signal. *J. Biol. Chem.* **280,** 37393–37399.

Wellen, K. E., and Hotamisligil, G. S. (2005). Inflammation, stress, and diabetes. *J. Clin. Invest.* **115,** 1111–1119.

Werner, E. D., Lee, J., Hansen, L., Yuan, M., and Shoelson, S. E. (2004). Insulin resistance due to phosphorylation of insulin receptor substrate-1 at serine 302. *J. Biol. Chem.* **279,** 35298–35305.

White, M. F. (1996). The IRS-signalling system in insulin and cytokine action. [Review]. *Philos. Trans. R. Soc. Lond. B Biol. Sci.* **351,** 181–189.

White, M. F. (2002). IRS proteins and the common path to diabetes. *Am. J. Physiol. Endocrinol. Metab.* **283,** E413–E422.

White, M. F., Maron, R., and Kahn, C. R. (1985). Insulin rapidly stimulates tyrosine phosphorylation of a Mr 185,000 protein in intact cells. *Nature* **318,** 183–186.

White, M. F., Livingston, J. N., Backer, J. M., Lauris, V., Dull, T. J., Ullrich, A., and Kahn, C. R. (1988). Mutation of the insulin receptor at tyrosine 960 inhibits signal transmission but does not affect its tyrosine kinase activity. *Cell* **54,** 641–649.

Withers, D. J., Gutierrez, J. S., Towery, H., Burks, D. J., Ren, J. M., Previs, S., Zhang, Y., Bernal, D., Pons, S., Shulman, G. I., Bonner-Weir, S., and White, M. F. (1998). Disruption of IRS-2 causes type 2 diabetes in mice. *Nature* **391,** 900–904.

Wolf, G., Trub, T., Ottinger, E., Groninga, L., Lynch, A., White, M. F., Miyazaki, M., Lee, J., and Shoelson, S. E. (1995). The PTB domains of IRS-1 and Shc have distinct but overlapping specificities. *J. Biol. Chem.* **270,** 27407–27410.

Wu, A., Sciacca, L., and Baserga, R. (2003). Nuclear translocation of insulin receptor substrate-1 by the insulin receptor in mouse embryo fibroblasts. *J. Cell. Physiol.* **195,** 453–460.

Xiang, X., Yuan, M., Song, Y., Ruderman, N., Wen, R., and Luo, Z. (2002). 14-3-3 facilitates insulin-stimulated intracellular trafficking of insulin receptor substrate 1. *Mol. Endocrinol.* **16,** 552–562.

Yaffe, M. B., and Cantley, L. C. (1999). Signal transduction. Grabbing phosphoproteins [News] [In Process Citation]. *Nature* **402,** 30–31.

Yamauchi, T., Tobe, K., Tamemoto, H., Ueki, K., Kaburagi, Y., Yamamoto-Honda, R., Takahashi, Y., Yoshizawa, F., Aizawa, S., Akanuma, Y., Sonenberg, N., Yazaki, Y., and Kadowaki, T. (1996). Insulin signalling and insulin actions in the muscles and livers of insulin-resistant, insulin receptor substrate 1-deficient mice. *Mol. Cell. Biol.* **16,** 3074–3084.

Yeh, T. C., Ogawa, W., Danielsen, A. G., and Roth, R. A. (1996). Characterization and cloning of a 58/53-kDa substrate of the insulin receptor tyrosine kinase. *J. Biol. Chem.* **271,** 2921–2928.

Yi, Z., Luo, M., Carroll, C. A., Weintraub, S. T., and Mandarino, L. J. (2005). Identification of phosphorylation sites in insulin receptor substrate-1 by hypothesis-driven high-performance liquid chromatography–electrospray ionization tandem mass spectrometry. *Anal. Chem.* **77,** 5693–5699.

Yu, C., Chen, Y., Cline, G. W., Zhang, D., Zong, H., Wang, Y., Bergeron, R., Kim, J. K., Cushman, S. W., Cooney, G. J., Atcheson, B., White, M. F., Kraegen, E. W., and Shulman, G. I. (2002). Mechanism by which fatty acids inhibit insulin activation of insulin receptor substrate-1 (IRS-1)-associated phosphatidylinositol 3-kinase activity in muscle. *J. Biol. Chem.* **277,** 50230–50236.

Yuan, M., Konstantopoulos, N., Lee, J., Hansen, L., Li, Z. W., Karin, M., and Shoelson, S. E. (2001). Reversal of obesity- and diet-induced insulin resistance with salicylates or targeted disruption of Ikkbeta. *Science* **293,** 1673–1677.

CHAPTER FOURTEEN

IRS-2 and Its Involvement in Diabetes and Aging

Jiandi Zhang *and* Tian-Qiang Sun

Contents

Abstract

Insulin receptor substrate 2 (IRS-2) is a key molecule in insulin signaling pathway, serving as an adaptor protein to insulin receptor to activate downstream kinase cascades, including MAP kinase and PI-3 kinase cascades. While reduced IRS-2 expression is tightly associated with diabetes and insulin resistance in various rodent diabetic models, this gene is also suggested to play a critical role in reproductive system and pancreas development. Recently, IRS-2 is demonstrated to be actively involved in lifespan regulation. In this chapter, we attempt to give a brief review of what we have learned about this molecule in metabolism and growth.

Rigel Pharmaceuticals, Inc., South San Francisco, California 94080

Vitamins and Hormones, Volume 80
ISSN 0083-6729, DOI: 10.1016/S0083-6729(08)00614-6

I. Introduction

Modern society is characterized by comfortable sedative lifestyle, symbolized by various advanced gadgets like computers, automobiles and air conditioners. In the meantime, abundance of food, especially fast food, creates a situation of overnutrition. Consequently, before we realize it, the whole society has paid a hefty price for all this seemingly perfect life, as obesity and various metabolic diseases including diabetes, have prevailed at an unprecedented pace as the result of imbalance of energy homeostasis.

Insulin signaling pathway, where insulin elicits a series of signaling events to maintain steady level of glucose, is clearly a critical component of overall energy homeostasis. This process is initiated by insulin binding to the insulin receptor, leading to the activation of insulin receptor kinase and autophosphorylation of insulin receptor beta subunit. The activated insulin receptor tyrosine kinase phosphorylates a group of adaptor proteins, mainly insulin receptor substrate (IRS) proteins and through these phosphorylated adaptor proteins to activate downstream kinase cascades, including MAPK and PI-3 kinase cascades, and to regulate downstream events at both transcriptional and translational levels (Fig. 14.1). The critical role of insulin signaling pathway in overall energy homeostasis is emphasized by the fact that people carrying homozygous insulin receptor mutations have severe defects in both growth and developments (Taylor *et al.*, 1990). These patients develop mild diabetes with hyperglycemia and hyperinsulinemia. Their endogenous glucose levels are also heavily influenced by food intake. The same observations have also been confirmed in mice, with those mutant mice die after birth with severe growth defects and hyperglycemia (Accili *et al.*, 1996).

As the docking proteins to mediate the effects of insulin receptor activation to downstream kinase cascades, the role of IRS proteins in insulin signaling pathway has also been well recognized, and intensively investigated to fully understand how defects in insulin signaling pathway lead to

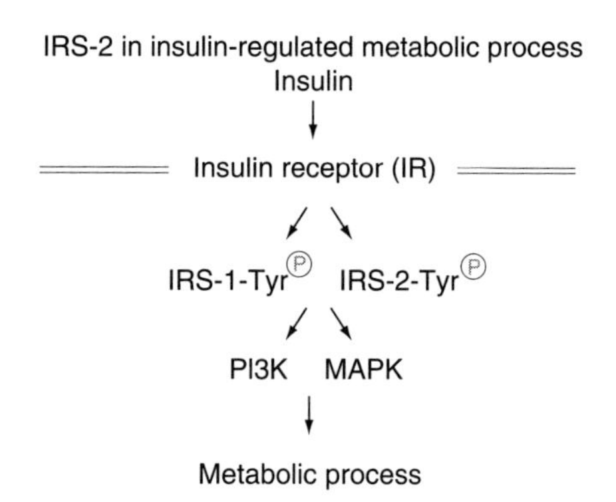

Figure 14.1 Simplified insulin signaling pathway.

dysregulated energy homeostasis. Among four known IRS proteins, IRS-2 protein receives extra attentions, as this protein is the only protein in IRS family where its deletion leads to overt diabetes. In this chapter, the history, the biochemical function, and the involvement of IRS-2 protein in growth and metabolism are reviewed, in addition to its newly identified roles in reproduction and aging.

II. Identification of IRS-2 Protein

Receptor tyrosine phosphorylation is well recognized as the initial step in receptor-mediated signaling pathway. Research on insulin receptor also reveals the presence of tyrosine kinase activity on the alpha subunit of this receptor (Kahn *et al.*, 1996). While in the process of studying the tyrosine kinase activity of insulin receptor, a series of proteins turn out to be heavily tyrosine phosphorylated in addition to the insulin receptor itself. Western blot analysis show that among these tyrosine phosphorylated proteins, a band around 180 kDa always appears side by side with insulin receptor upon insulin treatment (Rothenberg *et al.*, 1991; White *et al.*, 1987). While later studies indicate that two proteins of similar sizes rather than one protein is tyrosine phosphorylated at this position, this observation nonetheless leads to the purification and subsequently cloning of the prototype of IRS family proteins, insulin receptor substrate 1 (IRS-1) protein (Sun *et al.*, 1991). The cloning of IRS-1 protein represents a major step toward understanding insulin signaling pathway.

Consistent with initial observations, IRS-1 protein interacts directly with insulin receptor, and upon tyrosine phosphorylation, transduces insulin signaling to downstream kinase cascades by direct interacting with several SH2-binding proteins, like Grb2 to activate mitogen-activated protein kinase (MAPK) pathway and p85 subunit of PI-3 kinase to activate PI-3 kinase-Akt pathway. Knockout of IRS-1 in mice by several lab confirms the critical role of this gene in insulin signaling pathway, as mice missing this gene show phenotypes reminiscent of insulin receptor knockout mice, including significantly smaller body size, impaired glucose tolerance and impaired insulin-stimulated glucose uptake in muscle (Araki *et al.*, 1994; Tamemoto *et al.*, 1994).

Surprisingly, unlike insulin receptor knockout mice, which are lethal, and severe hyperglycemia, IRS-1 null mice develop only mild insulin resistance, and live significantly longer than the wild type mice, an observation consisting with the putative role of insulin signaling pathway in lifespan regulation. Detailed dissecting of insulin signaling pathway in liver and muscle tissue of IRS-1 knockout mice, on the other hand, show that although IRS-1 protein expression is eliminated, and IRS-1-associated tyrosine phosphorylation is undetectable, there is up-regulation of tyrosine phosphorylation of a similar size protein in IRS-1 mutant mice. Furthermore, tyrosine-phosphorylation-associated PI-3 kinase activity is only reduced slightly in mutant mice (Araki *et al.*, 1994). All these observations indicate the existence of another protein of similar size to IRS-1 to mediate insulin signaling pathway *in vivo*. Of course, this protein is later cloned, and named as IRS-2, as the second member of IRS family proteins (Sun *et al.*, 1995).

III. Basic Structure of IRS Family Proteins

Alignment of IRS-1 and IRS-2 proteins shows that these two proteins share high degree of homology with each other; especially at the N-terminal of the proteins with C-terminal showing certain level of divergence (Sun *et al.*, 1995). These two proteins all contain highly conserved pleckstrin homology (PH) domain and phosphotyrsoine binding (PTB) domain at their N-terminal. Furthermore, there are multiple tyrosine sites on these two proteins, and are being phosphorylated upon insulin treatment. Another common feature of these two proteins is that upon insulin treatment, these two proteins interact with several other adaptor proteins, including Grb2 protein to mediate MAPK cascade and p85 subunit of PI-3 kinase cascade. SHP2 (Src-homology-2 (SH2) domain-containing tyrosine phosphatase-2) is also known to bind to these IRS-2 proteins. Moreover, these two proteins are known to be regulated by several kinases, including p70 S6 kinase, JNK, and AKT (White, 2002). On the other hand, although these two proteins have very similar structures, obviously they not interchangeable, as both cellular and animal studies indicate that these two proteins play divergent roles in insulin signaling pathway and other growth factor signaling pathways. Two other members of this family, IRS-3 and IRS-4 are also cloned based on their structure homology with IRS-1 protein (Lavan *et al.*, 1997a,b). Elimination of either IRS-3 and IRS-4 at whole body level has little effect on overall growth and development, nor does it has any major effect on insulin resistance and glucose homeostasis, indicating a less important role of these two proteins in insulin signaling pathway (Fantin *et al.*, 2000; Liu *et al.*, 1999). In addition, there is no IRS-3 gene in human. Therefore, IRS-1 and IRS-2 are suggested to be the main docking proteins to mediate insulin signaling pathway *in vivo*.

IV. Involvement of IRS Proteins in Other Signaling Pathways

In addition to their critical roles in insulin signaling pathway, IRS proteins are also suggested to be involved in other signaling pathways induced by insulin receptor-like growth factor (IGF-1), epidermal growth factor (EGF), and growth hormone (White, 2002). IRS proteins are also described as a key component of several cytokine signaling pathways, including IL-4, IL-9, IL-13, interferon alpha (Sun *et al.*, 1995). In fact, the role of IRS-2 in cytokine signaling pathway was described well before it is recognized as a key component of insulin signaling pathway, and was initially named 4PS before it is renamed as IRS-2 protein (Fantin *et al.*, 2000; Wang *et al.*, 1993). Like in insulin signaling pathway, the activations of these signaling pathways require tyrosine phosphorylations of their receptors, and subsequently recruiting and tyrosine phosphorylations of IRS proteins to activate downstream kinases cascades, mainly PI-3 kinase pathway. The similar role of IRS-2 protein plays in multiple signaling pathways suggests the existence of conserved signaling pathway to mediate the effects of receptor activation through recruiting docking proteins to activate downstream kinase cascades. Furthermore, it also suggests potential crosstalk among these pathways in growth, differentiation and metabolism.

V. IRS-2 Protein is Well Conserved Across Species

Unlike in mammalian systems, the physiological significance of insulin signaling pathway in lower species including *C. elegans* and *Drosophilae* is still unclear. Nonetheless, the major components of insulin signaling pathway, including insulin receptor, PI-3 kinase and forkhead family proteins, are well conserved in these species. Naturally, as a key component of insulin signaling pathway, IRS homologues have also been identified in these species based on the structural similarity. That is, the presence of PH domain and PTB domain and the putative binding site of the regulatory subunit of PI-3 kinase. These homologues include *IST-1* in *C. elegans* and *Chico* in Drosophila (Bohni *et al.*, 1999; Wolkow *et al.*, 2002). Surprisingly, unlike in mammalian systems with a family of IRS proteins, there is only one IRS protein in each of these two species. The evolvement of IRS proteins from one protein in lower species to four proteins in mammalian system supports a more active and complicate role of IRS proteins in multiple signaling pathways in growth, proliferation and metabolism in mammalian system.

VI. IRS-2, and Its Regulation in Energy Homeostasis

Although IRS proteins have been implied in multiple signaling pathways, their major roles are clearly indicated in metabolism and energy homeostasis. This conclusion is supported by both cellular and animal studies, and is also strengthened by investigations in various rodent diabetic models and in clinical studies.

The cloning of IRS-2 protein by White group in 1995 allows further investigation of the involvement of IRS-2 in insulin signaling pathway (Sun *et al.*, 1991). Human IRS-2 encodes 1339 amino acids, and expresses as a roughly 170 kDa protein. This is a cytoplasmic protein with no expression in nucleus. IRS-2 expression has been documented in liver, heart, adipose tissue, brain, pancreas, muscles, and is clearly regulated by the external signals (Sun *et al.*, 1997; White, 2002). The expression level of this protein is also tightly associated with nutritional statuses at both protein and mRNA levels in liver, with IRS-2 protein up-regulated at fasting state, and down-regulated upon refeeding (Shimomura *et al.*, 2000). As fasting/refeeding best mimics insulin regulation of metabolic processes *in vivo*, this observation implies the possible regulatory effect of insulin on IRS-2 expression. Consistent with this hypothesis, IRS-2 expression level is highly regulated in several rodent diabetic models (Kerouz *et al.*, 1997). In *ob/ob* and *db/db* mice, the widely recognized type II diabetic models with hyperinsulinemia, hyperglycemia and hyperlipidemia, IRS-2 protein level is clearly down-regulated to undetectable level in liver while IRS-1 expression is only slightly down-regulated. This creates a sharp contrast to their expression profile in muscle tissue where both IRS-1 and IRS-2 protein expression levels are significantly down-regulated. The discreet regulation of IRS-1 and IRS-2 expressions at muscle and liver clearly indicate different physiological roles of these two highly homologous proteins in these tissues. In addition, the overall tyrosine phosphorylation levels of these proteins are less affected in liver compared with those in muscle tissues in *ob/ob* mice. Meanwhile, lipodystrophic transgenic Apo-BP-1c mice, another type II diabetes model, also show significant reduced IRS-2 expression level in liver. In contrast, in type I diabetic rodent model of hypoinsulinemia, where beta cells are destroyed with streptocin treatment, IRS-2 mRNA level is significantly increased (Shimomura *et al.*, 2000). All these results support insulin as an active regulator of IRS-2 expression.

Insulin effect on IRS-2 expression is also evaluated directly in several cell lines and in rat primary hepatocytes. Indeed, at cellular level, insulin effectively suppresses IRS-2 expression at both mRNA and protein levels in rat primary hepatocytes (Shimomura *et al.*, 2000; Zhang *et al.*, 2001). In addition, mRNA level of IRS-2 is also subjected to up-regulation by glucocorticoid hormones, as dexmethasone and cAMP significantly increases IRS-2 expression in rat primary hepatocytes. Insulin suppresses IRS-2 mRNA level at transcriptional level rather than though interfering with mRNA stability. Promoter analysis by luciferase assay shows that insulin effect is mediated through a well studied insulin response element (IRE), **TGTTTTG**, at the promoter region of IRS-2, as mutations at this site eliminate insulin effect on IRS-2 mRNA level (Zhang *et al.*, 2001).

This IRE element was first reported in the promoter region of phosphoenoylpyruvate carboxylkinase (PEPCK) by Granner's Group in 1990 (O'Brien *et al.*, 1990), and later is found in the promoter regions of glucose-6-phosphatase (G-6-Pase) and insulin like growth factor binding protein type I (IGFBP-1), two genes well known for their regulation by insulin (O'Brien and Granner, 1996).

Meanwhile, insulin effect on IRS-2 expression at protein level is also investigated in several cell lines (Rui *et al.*, 2001). These cells show decreased IRS-2 protein level within 6 h of insulin treatment, and insulin down-regulates IRS-2 expression at protein level through a proteasome-mediated event, as proteasome inhibitors, like MG132, and lactacystin, interfering with insulin effect on IRS-2 expression at protein level. However, although insulin-induced proteasome-mediated IRS-2 degradation may play a critical role in physiological process, persistent high insulin level in type II diabetic models clearly supports a more aggressive mean of regulation of IRS-2 expression by insulin at mRNA level, and discounts the direct impact of degradation of IRS-2 protein on overall pathphysiological states in type II diabetic patients.

The selective regulation of insulin on IRS-2 expression, rather than that of IRS-1, at least in liver, also supports a dominant role of IRS-2 in the pathphysiological states of type II diabetes. This agrees well with animal studies, where deletion of IRS-1 only leads to mild insulin resistance and growth retardation, yet deletion of IRS-2 leads to overt type II diabetes (Araki *et al.*, 1994; Tamemoto *et al.*, 1994; Withers *et al.*, 1998). However, further studies are required to fully appreciate the divergent roles of IRS-1 and IRS-2 in insulin signaling pathway and overall energy homeostasis.

At least up until now, there are no metabolic diseases has been reported as the direct result of IRS-2 mutation. While this observation itself may discounts the importance of IRS-2 protein in metabolic diseases in human it is equally possible that mutations at IRS-2 protein are associated with severe consequences in human. In the meantime, epidemiological studies suggest that there are some common mutations in IRS-2 coding sequence in general population, including a Glycine 1057Arg mutation, first identified in 1998 (Almind *et al.*, 1999). While no direct associations between this mutation and type II diabetes in the initial study, later studies do suggest complicated interactions among this mutation, obesity and type II diabetes (Le Fur *et al.*, 2002; Stefan *et al.*, 2003, 2004).

VII. Searching for the Regulatory Factor of IRS-2 Transcription

Identification of the well-conserved IRE element at the promoter region of IRS-2 gene suggests that IRS-2, like PEPCK, G-6-Pase, and several other genes, is subjected to insulin regulation through a common protein.

In other word, there exists a common transcriptional factor to mediate insulin effect on the transcription of a group of genes including IRS-2, PEPCK and G-6-Pase. Identification of this factor would provide us much needed information on insulin signaling pathway itself, and the specific role of individual gene in this group in growth, proliferation and metabolism. However, although this factor has been suggested since 1990 (O'Brien *et al.*, 1990), and has been actively pursued in several leading labs, it remains elusive to the scientific world up until now (O'Brien and Granner, 1996).

In the process of searching for this elusive factor, more and more information have been obtained about this protein. It is generally agreed that this factor belongs to a family of forkhead transcriptional factor for several reasons. First, the conserved IRE element in the promoter region of IRS-2 gene fits well with the known binding sequence of forkhead family transcriptional factors, T(G/A)TTT(T/G)(G/T (O'Brien and Granner, 1996). More importantly overexpression of several winged-helix family transcriptional factors, including HNF 3 alpha (foxa1), foxo1a (FKHR), and foxo3a (FKHRL1) can potently activate transcription of constructs carrying this conserved elements (Hall *et al.*, 2000). Consistent with this hypothesis, while it is known that all the forkhead/winged-helix family transcriptional factors are regulated by Akt kinase through PI-3 kinase activation, inhibitors of PI-3 kinase, including wortmannin, and LY294002, also potently inhibit insulin effects on genes with IRE element on their promoter region, including PEPCK, G-6-Pase and IRS-2 gene (Tomizawa *et al.*, 2000; Zhang *et al.*, 2001).

In the past 10 years, several proteins have been suggested as the putative binding factors to this IRE element, including Foxo1a, foxoa1, foxo3a, and C/EBP alpha (Ghosh *et al.*, 2001; Guo *et al.*, 1999; Hall *et al.*, 2000; Unterman *et al.*, 1994). However, none of them are universally recognized. One possible reason is that there is a big family of forkhead transcriptional factors, and these proteins share high homology with each other (Kaestner *et al.*, 2000). These structurally similar proteins greatly complicate the efforts to identify this putative regulatory factor for the IRE element. This is best illustrated in one study where the correlation between the binding of foxo3a with IRE element and its effect on IRE-mediated transcription is studied. While this factor potently bind and activate IRE construct in transfected cells, at least in one case where mutation of one residue in the IRE element known to disrupt IRE-mediated transcription, this factor remain potently activates this mutant IRE-mediated transcription in transfected cells (Hall *et al.*, 2000). Continuing searching of this elusive transcriptional factor may indeed represent another major breakthrough in our understanding of insulin regulation of this family of genes including PEPCK, IGFBP-1 and IRS-2 genes, and provide much needed information on insulin signaling pathway itself. What is more, identification of IRE binding factor is also being complicated by the observations that although IRE element has been identified in both the promoter regions of Glucose-6-phosphatase and PEPCK, there are evidences suggesting that the suppressive effects of insulin on PEPCK and G-6-Pase may be mediated through different mechanisms (O'Brien *et al.*, 2001).

VIII. Phenotype of IRS-2 Null Mice

While cloning of IRS-2 protein provides a lot of information on the role of IRS-2 in insulin signaling pathway, further understanding of the involvement of IRS-2 in insulin signaling pathway and overall energy homeostasis is greatly facilitated by the creation and characterization of the IRS-2 null mice (Kubota *et al.*, 2000; Withers *et al.*, 1998). Elimination of IRS-2 turns out to have profound effects on growth, differentiation, reproduction and metabolism. More importantly, consistent with previous studies, elimination of IRS-2 expression leads to overt type II diabetes in mice.

During the process of evolution, a complicated network has been developed to maintain glucose homeostasis. A main component of this network is the production and secretion of insulin. In type I diabetes, the lack of insulin due to the pancreas failure leads to un-controllable endogenous glucose level and overt diabetes. However, in majority of the population, there is a dynamic relationship between insulin secretion and endogenous glucose level. For example, in IRS-1 null mice, lacking of IRS-1 expression leads to insulin resistance in peripheral tissues, and mild hyperglycemia. To compensate the defect in insulin action in peripheral tissues, the secretion of insulin is significantly increased. As the consequence of persistent production of insulin, the overall pancreas size is also enlarged. The increased production and secretion of insulin thus overcome the defects in insulin action to increase glucose disposal in muscle, and to suppress endogenous glucose production in liver to maintain overall glucose level at physiological level (Araki *et al.*, 1994; Tamemoto *et al.*, 1994). That also explains why in human, there are a fair number of people with insulin resistance, yet these people never reach the stage of type II diabetes.

Therefore, while defects in insulin signaling pathway clearly lead to insulin resistance in peripheral tissues, impaired insulin action itself may not necessary translates into diabetes, as compensating increase of insulin secretion would prevent, or at least delay the development of diabetes. In the meantime, the overall size of pancreas increases significantly to accommodate the increased production of insulin, leading to pancreas hypertrophy. Overt diabetes only show up when insulin production and secretion can not meet the body need to control endogenous glucose level, a situation best illustrated in the IRS-2 null mice (Kubota *et al.*, 2000; Withers *et al.*, 1998).

Compared with wild type mice, IRS-2 null mice show 10% reduced body weight, in contrast to more than 40% reduction of body weight in IRS-1 null mice. Yet, these mice all show severe insulin resistance and develop overt type II diabetes, implying defects in both insulin action and insulin secretion. Indeed, mice of 3 days old show higher level of glucose than wild type, with normal level of insulin, and show hyperglycemia after 3–6 weeks, with threefold increase of blood serum insulin level, until the male mice die of dehydration and hyperosmolar coma (Withers *et al.*, 1998).

Investigation of the insulin signaling pathway shows that deletion of IRS-2 gene in mice is associated with insulin resistance in muscle with impaired glucose disposal, and in liver with inability to suppress endogenous glucose production by insulin. Detailed analysis of insulin signaling pathway show that tyrosine phosphorylation of insulin receptor is not affected by the deletion of IRS-2 gene. However, insulin-induced IRS-1 associated PI-3 kinase activity is significantly reduced in muscle, indicated by the reduced association of p85 subunit of PI-3 kinase with IRS-1. In liver, even at basal state, there is increased association between IRS-1 and p85 subunit of PI-3 kinase, and insulin stimulation can not further increase this association. Presumably, IRS-1 associated PI-3 kinase is already maximally activated at basal state in liver. Elimination of IRS-2 protein expression at whole body level, on the other hand has little impact on insulin signaling pathway in muscle. These observations are in sharp contrast with IRS-1 null mice, where IRS-2 associated tyrosine phosphorylation and PI-3 kinase activity is significantly increased in both liver and muscle. Nonetheless, although these mice have been created for more than 10 years, the molecular basis and physiological significances of these subtle differences between these two mice models remain unclear to the scientific world (Araki *et al.*, 1994; Kubota *et al.*, 2000; Withers *et al.*, 1998).

While elimination of IRS-2 expression is clearly associated with severe insulin resistance at peripheral tissues, the overt diabetic phenotypes commonly observed in male IRS-2 null mice is more likely the direct result of diminished insulin secretion. IRS-2 null mice at 4 weeks already show significantly reduced beta cell mass, and at later stage of development, the impaired insulin secretion in IRS-2 null mice clearly cannot meet the increasing demand in the face of severe insulin resistance in peripheral tissues (Withers *et al.*, 1998). This defect in insulin secretion may be directly associated with total beta cell mass rather than insulin production itself. In IRS-2 null mice, the total mass of beta cells reduces by more than 80%, yet insulin production and secretion are significantly increased in beta cells compared with those of wild type mice. In contrast, in IRS-1 null mice, while the total beta cell mass increases significantly in the presence of insulin resistance, insulin production and secretion from beta cells reduces significantly compared with their wild type counterparts (Kubota *et al.*, 2000). These observations support a critical role of IRS-2 protein in beta cell growth, differentiation and apoptosis, while IRS-1 protein may be directly involved in insulin production and secretion. Nonetheless, IRS-2 null mice with impaired pancreas development and severe insulin resistance, yet minimal growth defects are in sharp contrast with IRS-1 mice with severe growth defect and mild insulin resistance.

The critical role of IRS-2 in beta cell growth, differentiation and apoptosis is further confirmed by one study where IRS-2 expression is restored in pancreas of IRS-2 null mice. These mice remain severely insulin resistant, however, they never develop into type II diabetes (Hennige *et al.*, 2003). On the other hand, by selective deletion of IRS-2 protein in pancreas using Cre-lox system, two research groups independently report significantly reduced IRS-2 expression in pancreas, with severely insulin resistance and hypoinsulinemia (Cantley *et al.*, 2007; Choudhury *et al.*, 2005; Lin *et al.*, 2004). In one report, the deletion of IRS-2 protein in pancreas is directly associated with type II diabetes (Lin *et al.*, 2004). These experiments support the importance of IRS-2 expression in beta cell growth and function. More importantly, they also emphasize the critical role of pancreas in the transition from insulin resistance to diabetes at whole body level.

The role of IRS-2 in beta cell differentiation and proliferation is also suggested by the fact that in wild type mice, high fat feeding leads to significantly increased expression of IRS-2 gene in islet (Terauchi *et al.*, 2007). The expression level of IRS-2 is also sensitive to the overall glucose level, as increased glucose concentration is associated with increased IRS-2 expression in pancreas (Amacker-Francoys *et al.*, 2005; Lingohr *et al.*, 2006). Overexpression of IRS-2 in pancreas has been demonstrated to improve cell growth and proliferation and prevents apoptosis, as reported in several studies (Mohanty *et al.*, 2005). Reduced expression of IRS-2 protein, on the other hand, is clearly associated with increased apoptosis in pancreas (Lingohr *et al.*, 2003). The tight regulation of IRS-2 expression by glucose also explains why under hyperglycemia conditions, there is no corresponding increase of total beta cell mass (beta cell hypertrophy) in IRS-2 null mice.

Exactly how IRS-2 contributes to the development and maintenance of beta cells remains unknown. It has been observed that insulin-like growth factor (IGF-1) signaling pathway is involved in this process, as impaired IGF-1 signaling pathway exacerbates the overall situation in mice with reduced expression level of IRS-2 protein (Withers *et al.*, 1999). In addition, insulin signaling pathway has been demonstrated to be well conserved in beta cells (Muller *et al.*, 2006), and deletion of IRS-2 gene may disrupt insulin signaling pathway through both MAPK and Akt pathways to interfere with beta cell growth, differentiation and apoptosis. In fact, this hypothesis correlates well with the observations that deletion of phosphotyrosine phosphatase 1b (PTP1B) delays the onset of type II diabetes in IRS-2 null mice, possibly through extending insulin action in beta cells (Kushner *et al.*, 2004).

IX. The Role of IRS-2 in Female Reproduction

One unexpected impact of IRS-2 deletion at physiology level is female sterilization, a phenomenon also observed in *Chico* mutant, the *Drosophilae* homologue of IRS family proteins (Burks *et al.*, 2000; Clancy *et al.*, 2001). While both the IRS-1 and IRS-2 male null mice reproduce normally, their female counterparts show reduced fertility, with IRS-2 female mice show severely defects in reproduction when mating with wild type male mice.

Detailed physiological analysis shows that IRS-2 null female mice have small anovulatory ovaries, and reduced follicle number. Another drastic feature of IRS-2 null female mice is that the overall numbers of primary oocytes are significantly lower than those of wild type controls. In addition, these mice have reduced sex steroid hormones and luteinizing hormone. In fact, the reduced sex steroid hormones may lie in the defects in the pituitary glands, as only IRS-2, but not IRS-1, is tyrosine phosphorylated upon insulin treatment in this tissue (Burks *et al.*, 2000).

The fact that male IRS-2 mice behave normally already rules out the possibility that this phenotype is related with insulin resistance. Compared with male mice, female IRS-2 null have less severe phenotype of type II diabetes, including mild insulin resistance and hyperglycemia. In the meantime, female IRS-2 null mice have increased food intake, and store twice more body fat than their age-matched controls, and not surprisingly, these mice have significantly increased leptin level, as much as five times higher than the wild type controls, with hypothalamic leptin resistance. Since *db/db* mice, the leptin receptor mutant mice, are associated with hyperleptinemia, diabetes and female infertility, these observations indicate a potential link between leptin resistance and infertility in female IRS-2 null mice (Burks *et al.*, 2000).

X. The Putative Role of IRS-2 in Aging Process

The classic research in *C. elegans* clearly demonstrates an active role of insulin signaling pathway in lifespan regulation. This pathway has been found to be highly conserved across species including *Drosophilae*, *C. elegans*, and mammals. All the key components of insulin signaling pathway, including Insulin receptor, PI-3 kinase and forkhead family transcriptional factors, all find their counterparts in these animals, and mutations of these counterparts are tightly associated with alteration of overall lifespan in these species. Yet, the physiological significance of this pathway in lower species is still unclear.

The homologue of IRS protein in *C. elegans* and *Drosophilae*, including *ITS-1* in *C. elegans* and *Chico* in *Drosophilae* (Bohni *et al.*, 1999; Wolkow *et al.*, 2002) are also tightly associated with lifespan regulation. While both *ITS-1* and *chico* are identified based on the structural similarity with IRS-1, *chico* mutants also demonstrate phenotypical changes reminiscent of both IRS-1 and IRS-2 null mutants in mice. These flies show dwarfism, typical phenotype of IRS-1 null mice, and impaired oogenesis, a feature commonly observed in IRS-2 null female mice. More importantly, these flies live significantly longer than their wild type counterparts, independent of oxidative stress, suggesting a critical role of IRS proteins in lifespan regulation (Clancy *et al.*, 2001).

While IRS/*Chico* protein as an important regulator of aging process agrees well with the conserved role of insulin signaling pathway in aging process, this theory is further supported by animal studies recently (Selman *et al.*, 2008; Taguchi *et al.*, 2007). Two individual research groups, by studying IRS-1 and IRS-2 heterozygous mice from same origin, clearly demonstrate a critical role of IRS protein in lifespan regulation in mammals. Withers's group, on one hand, demonstrate increased lifespan with both IRS-1 null mice and IRS-1 heterozygous mice compared with their wild type counterpart. However, in this study, the connection between IRS-2 and lifespan regulation can not be recognized, as IRS-2 null mice show significantly shortened lifespan due to complications associated with diabetes, and IRS-2 heterozygous mice show similar lifespan to wild type animals. On the other hand, White's group show that reduction of IRS-2 expression in mice, as in the case of heterozygous mice, significantly extends their overall lifespan, and this regulation is independent from impaired insulin sensitivity in these mice. Although further studies are needed to reconcile these studies concerning the role of IRS-2 in lifespan regulation, these studies clearly suggest a critical role of IRS-1/IRS-2 in aging process.

One putative link between IRS-2 and aging process may lie in the involvement of SirT1 protein in regulation of insulin-induced IRS-2 tyrosine phosphorylation (Fig. 14.2). Calorie restriction, the only physiological mean to modulate lifespan across species, is suggested to be mediated by Sir2 protein, a protein deacetylase. SirT1, the mammalian homologue of Sir2, has been suggested to be involved in various metabolic processes (Pervaiz, 2003; Zhang, 2006). Reduced protein deacetylase activity of SirT1 is suggested to be tightly associated with impaired insulin-induced IRS-2 tyrosine phosphorylation, possibly through prevention of insulin-induced IRS-2 deacetylation (Zhang, 2007). At cellular level, inhibition of SirT1 activity translates into reduced Akt activation, with minimal effect on

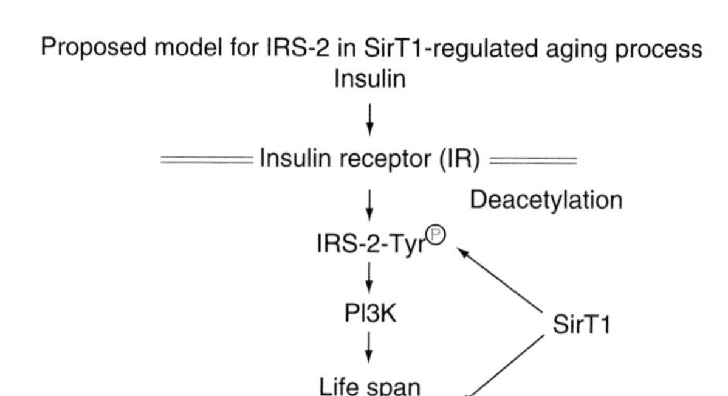

Figure 14.2 Proposed SirT1 involvement in lifespan regulation in mammals. SirT1 is suggested to modulate lifespan in mammals through two pathways. It may regulate lifespan in mammals through deactylation of IRS-2 protein to modulate insulin signaling pathway, in addition to its well-recognized role as the mediator of calorie restriction.

MAPK activation. Further studies show that acetylation of IRS-2 protein interferes with insulin-induced IRS-2 tyrosine phosphorylation, and reduced SirT1 activity prevents insulin-induced deacetylation of IRS-2 protein, and its tyrosine phosphorylation by insulin.

In fact, this connection is hinted in animal studies. Both SirT1 null mice and IRS-2 null mice show impaired female reproduction system due to a common defect in reduced sex hormones and impaired gamete production (McBurney *et al.*, 2003). It is interesting to compare phenotypical changes of these two kinds of mice, and to investigate any other similarity in their pathphysiological states. Nonetheless, although Sir2 is clearly demonstrated in lower species, including *C. elegans*, *Drosophilae*, and yeast, as a positive regulator of overall lifespan, there is little information about the role of SirT1 in lifespan regulation in mammals. IRS-2 as the newly identified substrate of SirT1 would help us better understands the connection between insulin signaling pathway and lifespan regulation, and the putative role of SirT1 protein in this process.

XI. Summary

In the last 20 years, IRSs, especially IRS-1 and IRS-2, have slowly stepped on the center stage of the research on insulin signaling pathway. The importance of IRS-2 protein in growth and metabolism is further emphasized by animal studies where deletion of this gene leads to wide variety of defects in growth, metabolism and reproduction. Severe insulin resistance at peripheral tissues, and defects in insulin secretion in pancreas, two striking features of IRS-2 null mice, are also the trademarks of type II diabetes.

Thus, understanding signaling pathway mediated by IRS-2 may prove critical to our investigation, understanding and future intervention of type II diabetes. Meanwhile, the newly identified link between IRS-2 and SirT1 protein, the putative mediator of calorie restriction, may also provide more information about this protein in metabolism and lifespan regulation. Further investigation in this direction may provide the much needed information on one of the biggest challenge in aging research: How SirT1 modulates lifespan through its modulation of metabolic processes.

Yet, although a great deal of progress has been made concerning the role of IRS-2 protein in growth and metabolism, and it has become a focus point in the fields of diabetes and obesity, there remain many questions unanswered about this protein. To name a few, while this protein clearly mediate insulin effect through interacting with SH2 containing proteins, including p85 subunit of PI-3 kinase and Grb2 proteins, its close relative, the IRS-1 protein, is also found to interact with the same group of proteins. How does it differentiates itself from IRS-1 in insulin signaling pathway is still unclear, as physiological studies strongly against the possibility that these two proteins are redundant in function. In addition, there are multiple tyrosine phosphorylation sites in IRS-2 protein. The physiological significances of these phosphorylations are also unclear. Clearly, the answers to these questions would provide critical information about IRS-2 protein in growth and metabolism, and in various pathphysiological states, and may also provide insight into the molecular basis of diabetes and obesity, one of the toughest challenges facing the mankind in twenty-first century.

REFERENCES

Accili, D., Drago, J., Lee, E. J., Johnson, M. D., Cool, M. H., Salvatore, P., Asico, L. D., Jose, P. A., Taylor, S. I., and Westphal, H. (1996). Early neonatal death in mice homozygous for a null allele of the insulin receptor gene. *Nat. Genet.* **12,** 106–109.

Almind, K., Frederiksen, S. K., Bernal, D., Hansen, T., Ambye, L., Urhammer, S., Ekstrom, C. T., Berglund, L., Reneland, R., Lithell, H., White, M. F., Van Obberghen, E., *et al.* (1999). Search for variants of the gene-promoter and the potential phosphotyrosine encoding sequence of the insulin receptor substrate-2 gene: Evaluation of their relation with alterations in insulin secretion and insulin sensitivity. *Diabetologia* **42,** 1244–1249.

Amacker-Francoys, I., Mohanty, S., Niessen, M., Spinas, G. A., and Trub, T. (2005). The metabolisable hexoses D-glucose and D-mannose enhance the expression of IRS-2 but not of IRS-1 in pancreatic beta-cells. *Exp. Clin. Endocrinol. Diabetes* **113,** 423–429.

Araki, E., Lipes, M. A., Patti, M. E., Bruning, J. C., Haag, B., 3rd, Johnson, R. S., and Kahn, C. R. (1994). Alternative pathway of insulin signalling in mice with targeted disruption of the IRS-1 gene. *Nature* **372,** 186–190.

Bohni, R., Riesgo-Escovar, J., Oldham, S., Brogiolo, W., Stocker, H., Andruss, B. F., Beckingham, K., and Hafen, E. (1999). Autonomous control of cell and organ size by CHICO, a Drosophila homolog of vertebrate IRS1-4. *Cell* **97,** 865–875.

Burks, D. J., Font de Mora, J., Schubert, M., Withers, D. J., Myers, M. G., Towery, H. H., Altamuro, S. L., Flint, C. L., and White, M. F. (2000). IRS-2 pathways integrate female reproduction and energy homeostasis. *Nature* **407,** 377–382.

Cantley, J., Choudhury, A. I., Asare-Anane, H., Selman, C., Lingard, S., Heffron, H., Herrera, P., Persaud, S. J., and Withers, D. J. (2007). Pancreatic deletion of insulin receptor substrate 2 reduces beta and alpha cell mass and impairs glucose homeostasis in mice. *Diabetologia* **50,** 1248–1256.

Choudhury, A. I., Heffron, H., Smith, M. A., Al-Qassab, H., Xu, A. W., Selman, C., Simmgen, M., Clements, M., Claret, M., Maccoll, G., Bedford, D. C., Hisadome, K., *et al.* (2005). The role of insulin receptor substrate 2 in hypothalamic and beta cell function. *J. Clin. Invest.* **115,** 940–950.

Clancy, D. J., Gems, D., Harshman, L. G., Oldham, S., Stocker, H., Hafen, E., Leevers, S. J., and Partridge, L. (2001). Extension of life-span by loss of CHICO, a Drosophila insulin receptor substrate protein. *Science* **292,** 104–106.

Fantin, V. R., Wang, Q., Lienhard, G. E., and Keller, S. R. (2000). Mice lacking insulin receptor substrate 4 exhibit mild defects in growth, reproduction, and glucose homeostasis. *Am. J. Physiol. Endocrinol. Metab.* **278,** E127–133.

Ghosh, A. K., Lacson, R., Liu, P., Cichy, S. B., Danilkovich, A., Guo, S., and Unterman, T. G. (2001). A nucleoprotein complex containing CCAAT/enhancer-binding protein beta interacts with an insulin response sequence in the insulin-like growth factor-binding protein-1 gene and contributes to insulin-regulated gene expression. *J. Biol. Chem.* **276,** 8507–8515.

Guo, S., Rena, G., Cichy, S., He, X., Cohen, P., and Unterman, T. (1999). Phosphorylation of serine 256 by protein kinase B disrupts transactivation by FKHR and mediates effects of insulin on insulin-like growth factor-binding protein-1 promoter activity through a conserved insulin response sequence. *J. Biol. Chem.* **274,** 17184–17192.

Hall, R. K., Yamasaki, T., Kucera, T., Waltner-Law, M., O'Brien, R., and Granner, D. K. (2000). Regulation of phosphoenolpyruvate carboxykinase and insulin-like growth factor-binding protein-1 gene expression by insulin. The role of winged helix/forkhead proteins. *J. Biol. Chem.* **275,** 30169–30175.

Hennige, A. M., Burks, D. J., Ozcan, U., Kulkarni, R. N., Ye, J., Park, S., Schubert, M., Fisher, T. L., Dow, M. A., Leshan, R., Zakaria, M., Mossa-Basha, M., *et al.* (2003). Upregulation of insulin receptor substrate-2 in pancreatic beta cells prevents diabetes. *J. Clin. Invest.* **112,** 1521–1532.

Kaestner, K. H., Knochel, W., and Martinez, D. E. (2000). Unified nomenclature for the winged helix/forkhead transcription factors. *Genes Dev.* **14,** 142–146.

Kahn, C. R., Vicent, D., and Doria, A. (1996). Genetics of non-insulin-dependent (type-II) diabetes mellitus. *Annu. Rev. Med.* **47,** 509–531.

Kerouz, N. J., Horsch, D., Pons, S., and Kahn, C. R. (1997). Differential regulation of insulin receptor substrates-1 and -2 (IRS-1 and IRS-2) and phosphatidylinositol 3-kinase isoforms in liver and muscle of the obese diabetic (ob/ob) mouse. *J. Clin. Invest.* **100,** 3164–3172.

Kubota, N., Tobe, K., Terauchi, Y., Eto, K., Yamauchi, T., Suzuki, R., Tsubamoto, Y., Komeda, K., Nakano, R., Miki, H., Satoh, S., Sekihara, H., *et al.* (2000). Disruption of insulin receptor substrate 2 causes type 2 diabetes because of liver insulin resistance and lack of compensatory beta-cell hyperplasia. *Diabetes* **49,** 1880–1889.

Kushner, J. A., Haj, F. G., Klaman, L. D., Dow, M. A., Kahn, B. B., Neel, B. G., and White, M. F. (2004). Islet-sparing effects of protein tyrosine phosphatase-1b deficiency delays onset of diabetes in IRS2 knockout mice. *Diabetes* **53,** 61–66.

Lavan, B. E., Fantin, V. R., Chang, E. T., Lane, W. S., Keller, S. R., and Lienhard, G. E. (1997a). A novel 160-kDa phosphotyrosine protein in insulin-treated embryonic kidney cells is a new member of the insulin receptor substrate family. *J. Biol. Chem.* **272,** 21403–21407.

Lavan, B. E., Lane, W. S., and Lienhard, G. E. (1997b). The 60-kDa phosphotyrosine protein in insulin-treated adipocytes is a new member of the insulin receptor substrate family. *J. Biol. Chem.* **272,** 11439–11443.

Le Fur, S., Le Stunff, C., and Bougneres, P. (2002). Increased insulin resistance in obese children who have both 972 IRS-1 and 1057 IRS-2 polymorphisms. *Diabetes* **51** (Suppl 3), S304–307.

Lin, X., Taguchi, A., Park, S., Kushner, J. A., Li, F., Li, Y., and White, M. F. (2004). Dysregulation of insulin receptor substrate 2 in beta cells and brain causes obesity and diabetes. *J. Clin. Invest.* **114,** 908–916.

Lingohr, M. K., Dickson, L. M., Wrede, C. E., Briaud, I., McCuaig, J. F., Myers, M. G., Jr., and Rhodes, C. J. (2003). Decreasing IRS-2 expression in pancreatic beta-cells (INS-1) promotes apoptosis, which can be compensated for by introduction of IRS-4 expression. *Mol. Cell Endocrinol.* **209,** 17–31.

Lingohr, M. K., Briaud, I., Dickson, L. M., McCuaig, J. F., Alarcon, C., Wicksteed, B. L., and Rhodes, C. J. (2006). Specific regulation of IRS-2 expression by glucose in rat primary pancreatic islet beta-cells. *J. Biol. Chem.* **281,** 15884–15892.

Liu, S. C., Wang, Q., Lienhard, G. E., and Keller, S. R. (1999). Insulin receptor substrate 3 is not essential for growth or glucose homeostasis. *J. Biol. Chem.* **274,** 18093–18099.

McBurney, M. W., Yang, X., Jardine, K., Hixon, M., Boekelheide, K., Webb, J. R., Lansdorp, P. M., and Lemieux, M. (2003). The mammalian SIR2alpha protein has a role in embryogenesis and gametogenesis. *Mol. Cell Biol.* **23,** 38–54.

Mohanty, S., Spinas, G. A., Maedler, K., Zuellig, R. A., Lehmann, R., Donath, M. Y., Trub, T., and Niessen, M. (2005). Overexpression of IRS2 in isolated pancreatic islets causes proliferation and protects human beta-cells from hyperglycemia-induced apoptosis. *Exp. Cell Res.* **303,** 68–78.

Muller, D., Huang, G. C., Amiel, S., Jones, P. M., and Persaud, S. J. (2006). Identification of insulin signaling elements in human beta-cells: Autocrine regulation of insulin gene expression. *Diabetes* **55,** 2835–2842.

O'Brien, R. M., and Granner, D. K. (1996). Regulation of gene expression by insulin. *Physiol. Rev.* **76,** 1109–1161.

O'Brien, R. M., Lucas, P. C., Forest, C. D., Magnuson, M. A., and Granner, D. K. (1990). Identification of a sequence in the PEPCK gene that mediates a negative effect of insulin on transcription. *Science* **249,** 533–537.

O'Brien, R. M., Streeper, R. S., Ayala, J. E., Stadelmaier, B. T., and Hornbuckle, L. A. (2001). Insulin-regulated gene expression. *Biochem. Soc. Trans.* **29,** 552–558.

Pervaiz, S. (2003). Resveratrol: From grapevines to mammalian biology. *FASEB J.* **17,** 1975–1985.

Rothenberg, P. L., Lane, W. S., Karasik, A., Backer, J., White, M., and Kahn, C. R. (1991). Purification and partial sequence analysis of pp185, the major cellular substrate of the insulin receptor tyrosine kinase. *J. Biol. Chem.* **266,** 8302–8311.

Selman, C., Lingard, S., Choudhury, A. I., Batterham, R. L., Claret, M., Clements, M., Ramadani, F., Okkenhaug, K., Schuster, E., Blanc, E., Piper, M. D., Al-Qassab, H., *et al.* (2008). Evidence for lifespan extension and delayed age-related biomarkers in insulin receptor substrate 1 null mice. *FASEB J.* **22,** 807–818.

Shimomura, I., Matsuda, M., Hammer, R. E., Bashmakov, Y., Brown, M. S., and Goldstein, J. L. (2000). Decreased IRS-2 and increased SREBP-1c lead to mixed insulin resistance and sensitivity in livers of lipodystrophic and ob/ob mice. *Mol. Cell* **6,** 77–86.

Stefan, N., Kovacs, P., Stumvoll, M., Hanson, R. L., Lehn-Stefan, A., Permana, P. A., Baier, L. J., Tataranni, P. A., Silver, K., and Bogardus, C. (2003). Metabolic effects of the Gly1057Asp polymorphism in IRS-2 and interactions with obesity. *Diabetes* **52,** 1544–1550.

Stefan, N., Fritsche, A., Machicao, F., Tschritter, O., Haring, H. U., and Stumvoll, M. (2004). The Gly1057Asp polymorphism in IRS-2 interacts with obesity to affect beta cell function. *Diabetologia* **47,** 759–761.

Sun, X. J., Rothenberg, P., Kahn, C. R., Backer, J. M., Araki, E., Wilden, P. A., Cahill, D. A., Goldstein, B. J., and White, M. F. (1991). Structure of the insulin receptor substrate IRS-1 defines a unique signal transduction protein. *Nature* **352,** 73–77.

Sun, X. J., Wang, L. M., Zhang, Y., Yenush, L., Myers, M. G., Jr., Glasheen, E., Lane, W. S., Pierce, J. H., and White, M. F. (1995). Role of IRS-2 in insulin and cytokine signalling. *Nature* **377,** 173–177.

Sun, X. J., Pons, S., Wang, L. M., Zhang, Y., Yenush, L., Burks, D., Myers, M. G., Jr., Glasheen, E., Copeland, N. G., Jenkins, N. A., Pierce, J. H., and White, M. F. (1997). The IRS-2 gene on murine chromosome 8 encodes a unique signaling adapter for insulin and cytokine action. *Mol. Endocrinol.* **11,** 251–262.

Taguchi, A., Wartschow, L. M., and White, M. F. (2007). Brain IRS2 signaling coordinates life span and nutrient homeostasis. *Science* **317,** 369–372.

Tamemoto, H., Kadowaki, T., Tobe, K., Yagi, T., Sakura, H., Hayakawa, T., Terauchi, Y., Ueki, K., Kaburagi, Y., and Satoh, S. (1994). Insulin resistance and growth retardation in mice lacking insulin receptor substrate-1. *Nature* **372,** 182–186.

Taylor, S. I., Kadowaki, T., Kadowaki, H., Accili, D., Cama, A., and McKeon, C. (1990). Mutations in insulin-receptor gene in insulin-resistant patients. *Diabetes Care* **13,** 257–279.

Terauchi, Y., Takamoto, I., Kubota, N., Matsui, J., Suzuki, R., Komeda, K., Hara, A., Toyoda, Y., Miwa, I., Aizawa, S., Tsutsumi, S., Tsubamoto, Y., *et al.* (2007). Glucokinase and IRS-2 are required for compensatory beta cell hyperplasia in response to high-fat diet-induced insulin resistance. *J. Clin. Invest.* **117,** 246–257.

Tomizawa, M., Kumar, A., Perrot, V., Nakae, J., Accili, D., and Rechler, M. M. (2000). Insulin inhibits the activation of transcription by a C-terminal fragment of the forkhead transcription factor FKHR. A mechanism for insulin inhibition of insulin-like growth factor-binding protein-1 transcription. *J. Biol. Chem.* **275,** 7289–7295.

Unterman, T. G., Fareeduddin, A., Harris, M. A., Goswami, R. G., Porcella, A., Costa, R. H., and Lacson, R. G. (1994). Hepatocyte nuclear factor-3 (HNF-3) binds to the insulin response sequence in the IGF binding protein-1 (IGFBP-1) promoter and enhances promoter function. *Biochem. Biophys. Res. Commun.* **203,** 1835–1841.

Wang, L. M., Keegan, A. D., Li, W., Lienhard, G. E., Pacini, S., Gutkind, J. S., Myers, M. G., Jr., Sun, X. J., White, M. F., Aaronson, S. A., Paul, W. E., and Pierce, J. H. (1993). Common elements in interleukin 4 and insulin signaling pathways in factor-dependent hematopoietic cells. *Proc. Natl. Acad. Sci. USA* **90,** 4032–4036.

White, M. F. (2002). IRS proteins and the common path to diabetes. *Am. J. Physiol. Endocrinol. Metab.* **283,** E413–422.

White, M. F., Stegmann, E. W., Dull, T. J., Ullrich, A., and Kahn, C. R. (1987). Characterization of an endogenous substrate of the insulin receptor in cultured cells. *J. Biol. Chem.* **262,** 9769–9777.

Withers, D. J., Gutierrez, J. S., Towery, H., Burks, D. J., Ren, J. M., Previs, S., Zhang, Y., Bernal, D., Pons, S., Shulman, G. I., Bonner-Weir, S., and White, M. F. (1998). Disruption of IRS-2 causes type 2 diabetes in mice. *Nature* **391,** 900–904.

Withers, D. J., Burks, D. J., Towery, H. H., Altamuro, S. L., Flint, C. L., and White, M. F. (1999). Irs-2 coordinates Igf-1 receptor-mediated beta-cell development and peripheral insulin signalling. *Nat. Genet.* **23,** 32–40.

Wolkow, C. A., Munoz, M. J., Riddle, D. L., and Ruvkun, G. (2002). Insulin receptor substrate and p55 orthologous adaptor proteins function in the *Caenorhabditis elegans* daf-2/insulin-like signaling pathway. *J. Biol. Chem.* **277,** 49591–49597.

Zhang, J. (2006). Resveratrol inhibits insulin responses in a SirT1-independent pathway. *Biochem. J.* **397,** 519–527.

Zhang, J. (2007). The direct involvement of SirT1 in insulin-induced insulin receptor substrate-2 tyrosine phosphorylation. *J. Biol. Chem.* **282,** 34356–34364.

Zhang, J., Ou, J., Bashmakov, Y., Horton, J. D., Brown, M. S., and Goldstein, J. L. (2001). Insulin inhibits transcription of IRS-2 gene in rat liver through an insulin response element (IRE) that resembles IREs of other insulin-repressed genes. *Proc. Natl. Acad. Sci. USA* **98,** 3756–3761.

CHAPTER FIFTEEN

Glucose-Dependent Insulinotropic Polypeptide (Gastric Inhibitory Polypeptide; GIP)

Christopher H. S. McIntosh, Scott Widenmaier, *and* Su-Jin Kim

Contents

Department of Cellular and Physiological Sciences, The Diabetes Research Group, Life Sciences Institute, University of British Columbia, 2350 Health Sciences Mall, Vancouver, BC, Canada V6T 1Z3

Vitamins and Hormones, Volume 80
ISSN 0083-6729, DOI: 10.1016/S0083-6729(08)00615-8

Abstract

Glucose-dependent insulinotropic polypeptide (GIP; gastric inhibitory polypeptide) is a 42 amino acid hormone that is produced by enteroendocrine K-cells and released into the circulation in response to nutrient stimulation. Both GIP and glucagon-like peptide-1 (GLP-1) stimulate insulin secretion in a glucose-dependent manner and are thus classified as incretins. The structure of mammalian GIP is well conserved and both the N-terminus and central region of the molecule are important for biological activity. Following secretion, GIP is metabolized by the endoprotease dipeptidyl peptidase IV (DPP-IV). In addition to its insulinotropic activity, GIP exerts a number of additional actions including promotion of growth and survival of the pancreatic β-cell and stimulation of adipogenesis. The brain, bone, cardiovascular system, and gastrointestinal tract are additional targets of GIP. The GIP receptor is a member of the B-family of G protein-coupled receptors and activation results in the stimulation of adenylyl cyclase and Ca^{2+}-independent phospholipase A_2 and activation of protein kinase (PK) A and PKB. The Mek1/2-Erk1/2 and p38 MAP kinase signaling pathways are among the downstream pathways involved in the regulation of β-cell function. GIP also increases expression of the anti-apoptotic Bcl-2 and decreases expression of the pro-apoptotic Bax, resulting in reduced β-cell death. In adipose tissue, GIP interacts with insulin to increase lipoprotein lipase activity and lipogenesis. There is significant interest in potential clinical applications for GIP analogs and both agonists and antagonists have been developed for preclinical studies.

I. Introduction

The gastrointestinal (GI) tract contains a plethora of bioactive peptides that act as neurotransmitters/neuromodulators, hormones or local regulators of cell function (Chao and Hellmich, 2004; Furness, 2006; Menard, 2004; Rehfeld, 1998; Thomas *et al.*, 2003). Among these are intestinal hormones involved in the regulation of gastric motility and acid secretion. The term "enterogastone" was introduced to describe factors released by dietary fat that cause inhibition of gastric secretion (Kosaka and Lim, 1930), but the definition was later expanded to include substances released by acid and hypertonic solutions (Gregory, 1967). Despite over 75 years of research aimed at identifying physiological enterogastrones, there is still uncertainty as to the major hormones involved in humans. In various species, secretin (Chey and Chang, 2003) and cholecystokinin (CCK) (Fung *et al.*, 1998; Rehfeld, 2004) from the upper small intestine, as well as GLP-1 (Fung *et al.*, 1998), oxyntomodulin, PYY (Fung *et al.*, 1998; Jarrousse *et al.*, 1994) and neurotensin (Zhao and Pothoulakis, 2006), located mainly in the lower bowel, have been shown to inhibit gastric emptying and/or acid secretion. In a series of studies on impure preparations of cholecystokinin (CCK-PZ; cholecystokinin-pancreozymin), Brown and co-workers demonstrated that further purification of CCK-PZ resulted in the removal of a factor that powerfully inhibited acid secretion in dogs (Brown *et al.*, 1969, 1970). The compound responsible was later isolated from porcine intestinal extracts through a heroic series of chromatographic purification steps, and named "gastric inhibitory polypeptide" or "GIP" (Brown, 1971; Brown and Dryburgh, 1971). Purified GIP was shown to inhibit acid and pepsin secretion and motor activity in extrinsically denervated canine gastric pouches (Brown *et al.*, 1989; reviewed in: Brown, 1982; Brown *et al.*, 1989; Pederson, 1994).

In parallel with attempts at identifying the elusive "enterogastrone," studies were ongoing to establish whether the gut played a significant role in regulating the "internal secretion" of the pancreas. Of particular importance, La Barre and co-workers performed a series of investigations on the effects of impure secretin preparations on blood sugar, and introduced the term incrétine (incretin) to describe an upper intestinal factor that reduced blood glucose levels without influencing exocrine pancreatic secretion (La Barre, 1932). Over 30 years later two groups showed that glucose infused via an intragastric (Elrick *et al.*, 1964) or intra-jejunal (McIntyre *et al.*, 1964) route produced a greater insulin response than an equivalent amount administered intravenously, resulting in more efficient glucose disposal. This regulatory link between the gut and the endocrine pancreas was designated the "enteroinsular axis" (Unger and Eisentraut, 1969). Over a number of years, Dupré and co-workers attempted to identify insulin secretagogues in intestinal extracts and observed that similar impure preparations of CCK-PZ to those used by Brown improved glucose tolerance when infused intravenously (Dupré and Beck, 1966; Rabinovitch and Dupré, 1972). The term "incretin" was subsequently reintroduced to describe the hormonal components that stimulated insulin secretion in a glucose-dependent manner (Creutzfeldt, 1979, 2005). It has been estimated that incretins account for 50–65% of the total insulin response, depending on the oral nutrient load (Muscelli *et al.*, 2006; Nauck *et al.*, 1986b).

In collaborative studies between Brown and Dupré it was demonstrated that intravenous GIP stimulated insulin secretion and increased disposal of an intravenous glucose load in normal humans (Brown, 1987; Dupré and

Beck, 1966; Dupré *et al.*, 1973). Since this incretin effect of GIP was considered to be its more important action, an alternative definition of the GIP acronym was introduced: glucose-dependent insulinotropic polypeptide (Brown and Pederson, 1976). The possible existence of a further incretin arose from studies showing that the insulin response to intraduodenal glucose in rats was only partially reduced when GIP antiserum was co-administered (Ebert and Creutzfeldt, 1982; Lauritsen *et al.*, 1981) and that significant insulinotropic activity remained in rat intestinal extracts following immunoaffinity removal of GIP (Ebert *et al.*, 1983). Following discovery of the coding sequence for a glucagon-related peptide in an anglerfish proglucagon cDNA (Lund *et al.*, 1982) a number of groups contributed to the identification of GLP-1(7-36)amide and GLP-1(7-37) as mammalian incretins (Lund, 2005). By convention, they are now jointly called "GLP-1." Although other gastrointestinal hormones, such as secretin and CCK, can stimulate insulin secretion, only GIP and GLP-1 have been shown to act in a glucose-dependent manner in humans at concentrations found during a meal, and they are currently considered the main incretin hormones.

This review focuses fairly specifically on the current state of knowledge regarding the biology of GIP. For detailed reviews of the early literature the following are recommended: Brown (1982), Brown *et al.* (1989), Fehmann *et al.* (1995), Meier *et al.* (2002), Pederson (1994), and Pederson and McIntosh (2004). Many excellent incretin reviews of a more general nature are available, including: Baggio and Drucker (2006, 2007), Creutzfeldt (2005), Drucker (2006, 2007), and Drucker and Nauck (2006).

II. Glucose-Dependent Insulinotropic Polypeptide (GIP)

A. Structure

From initial sequencing of purified porcine GIP, by enzyme digestion and manual dansyl-Edman degradation, it was concluded that the hormone was a 43 amino acid polypeptide (Brown and Dryburgh, 1971). However, when more sophisticated sequencing techniques became available, it was determined that an extra glutamine was erroneously inserted at position 30 and the sequence was therefore corrected (Jörnvall *et al.*, 1981). Peptide or cDNA sequencing, or identification of ortholog sequences in the Ensembl database, has revealed that the GIP molecule is a 42 amino acid polypeptide (GIP_{1-42}) in all species for which full sequence data are available (Fig. 15.1). Human (Moody *et al.*, 1984; Takeda *et al.*, 1987), porcine (Jörnvall *et al.*, 1981), bovine (Carlquist *et al.*, 1984), rat (Higashimoto *et al.*, 1992; Sharma *et al.*, 1992; Tseng *et al.*, 1993), and mouse (Schieldrop *et al.*, 1996) GIP sequences differ by only 1–2 amino acids (Fig. 15.1). However recent

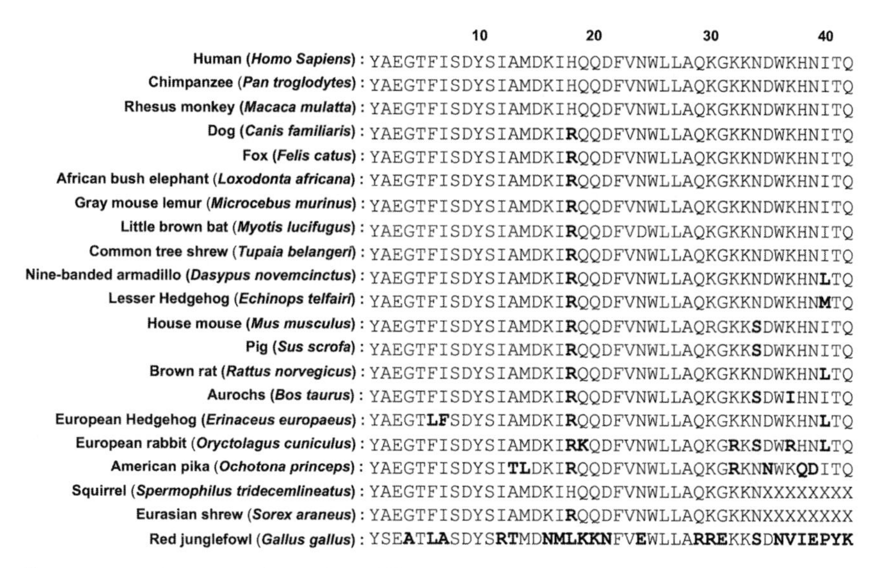

Figure 15.1 *Sequences of GIP from different species.* Sequences originally obtained from peptide sequencing (Pig—Jörnvall *et al.*, 1981; Human—Moody *et al.*, 1984; Aurochs—Carlquist *et al.*, 1984) or cDNA sequencing (Rat—Higashimoto *et al.*, 1992; Mouse—Schieldrop *et al.*, 1996), or identification of ortholog sequences in the Ensembl database.

entries in the Ensembl database have shown that GIP in the European rabbit and American pika differ by six and seven amino acids from the human peptide, respectively. Interestingly, the only nonmammalian GIP sequence reported to date, that of the red junglefowl, exhibits only 48% homology with human GIP. Further comparative studies at both the DNA and protein levels are clearly needed.

GIP belongs to the secretin/glucagon superfamily of peptides that share strong N-terminal regional homology (Sherwood *et al.*, 2000). In humans, identified members of the family include GIP, secretin, glucagon, GLP-1, GLP-2, growth hormone-releasing hormone (GHRH), vasoactive intestinal polypeptide (VIP), peptide histidine methionine (PHM) and pituitary adenylate cyclase-activating polypeptide (PACAP). Ancestral members of the superfamily have been traced back to the invertebrates, with PACAP and glucagon the oldest and most tightly conserved members. It is believed that the superfamily originated by gene or exon duplication, with continued divergence and gene duplications in the vertebrates (Sherwood *et al.*, 2000).

Structure-activity studies on GIP and GIP analogs have identified the N-terminus and central region of the GIP molecule as being critical for biological activity (Hinke *et al.*, 2001, 2003a). Truncated forms of GIP, including GIP_{1-39} (Sandberg *et al.*, 1986) and GIP_{1-30} (Wheeler *et al.*, 1995) retain a high degree of biological activity. However, fairly modest changes to Tyr^1–Ala^2 at the N-terminus can drastically reduce bioactivity

(Hinke *et al.*, 2003a). As discussed later (Section IVA), cleavage by Dipeptidyl peptidase IV (DPP-IV) (Kieffer *et al.*, 1995b; McIntosh, 2008; Mentlein *et al.*, 1993; Pauly *et al.*, 1996) results in a peptide (GIP_{3-42}) lacking insulinotropic activity (Schmidt *et al.*, 1987). The high affinity binding region of GIP_{1-42} resides within the region Phe_6–Lys_{30} (Gelling *et al.*, 1997a) and recently NMR and CD spectroscopy measurements have confirmed predictions (Manhart *et al.*, 2003) that this region exists as an α-helix in solution (Alana *et al.*, 2007). There is a less well defined helix between Lys_{30} and Trp_{36} (Alana *et al.*, 2007). GIP_{6-30} and $GIP_{7-30amide}$ bind to the receptor with high affinity, but act as antagonists (Gelling *et al.*, 1997a; Tseng *et al.*, 1996). GIP_{3-42}, can also act as an antagonist of GIP_{1-42}-induced cAMP production *in vitro* (Hinke *et al.*, 2002). Although similar antagonism was reported to occur with physiological concentrations of GIP_{3-42} in rodents (Gault *et al.*, 2002b), this finding could not be replicated in pigs (Deacon *et al.*, 2006). Both GIP_{1-14} and GIP_{19-30} are capable of receptor binding and activation of adenylyl cyclase (Hinke *et al.*, 2001) and joining the two peptides with linkers that enhance helix formation in the C-terminal (19–30) portion of GIP produces peptides with enhanced *in vitro* activity (Manhart *et al.*, 2003)

B. Localization

Radioimmunoassay (RIA) measurements of tissue extracts demonstrated that GIP is present throughout the small intestine, with highest concentrations in the duodenum (O'Dorisio *et al.*, 1976). Immunocytochemical (ICC) (Buffa *et al.*, 1975; Polak *et al.*, 1973; Usellini *et al.*, 1984; Van Ginneken and Weyns, 2004) and electron microscopic (Buchan *et al.*, 1978) studies on human and dog tissues demonstrated that GIP is produced by enteroendocrine K-cells, mainly located within the crypts and mid-zones of glands in the duodenum and upper jejunum. In the rodent gut, GIP distribution extends through to the ileum (Buchan *et al.*, 1982a), with GIP gene expression decreasing from the upper to lower small intestine (Berghöfer *et al.*, 1997; Tseng *et al.*, 1993). GIP was also reported to be localized in pancreatic α-cells (Alumets *et al.*, 1978; Smith *et al.*, 1977) in the same secretory granules as glucagon (Ahren *et al.*, 1981). However, RIA of tissue extracts failed to identify pancreatic GIP (Bloom and Polak, 1978) and non-specificity of antisera probably explained these results (Buchan *et al.*, 1982b; Larsson and Moody, 1980; Sjölund *et al.*, 1983). GIP mRNA was also reported to be undetectable in pancreata from rat fetuses or pups at any age examined (Tseng *et al.*, 1995). Nevertheless, the question as to whether or not α-cells can produce GIP should probably be reexamined.

GLP-1 has generally been reported to be located mainly in the distal part of the small intestine and colon (Eissele *et al.*, 1992). However, in early ICC studies ~10% of GIP cells in the duodenum of dog, cat, pig, and human also stained with antisera against the proglucagon-derived peptide glicentin (Sjölund *et al.*, 1983). In recent studies, ~60% of GLP-1 cells demonstrated co-localization with GIP in the mid-jejunal to mid-ileal regions of the intestine from several species (Mortensen *et al.*, 2003). Theodorakis and co-workers reported co-staining for GIP and GLP-1 in the human duodenum (Theodorakis *et al.*, 2006). A further point of interest regarding the K-cell is the reported co-localization of xenin and GIP in a subpopulation of cells (Anlauf *et al.*, 2000). Xenin is a 25 amino acid member of the xenopsin/neurotensin/xenin peptide family (Feurle, 1998), that exert effects on both endocrine (Silvestre *et al.*, 2003) and exocrine pancreas, as well as on GI functions (Feurle, 1998) and food intake (Cline *et al.*, 2007). However, circulating xenin in humans appears to be mainly increased during the cephalic phase of feeding (Feurle *et al.*, 2003) and its functional relationship to GIP is therefore unclear.

Only rudimentary details of the lineage development of the K-cell in the intestine are available. Knockout and transgenic mouse studies showed that the transcription factor Pdx1 is critical for postnatal production of intestinal GIP producing cells and appears to influence the frequency at which enteroendocrine progenitor cells are induced to follow the K-cell lineage (Boyer *et al.*, 2006). The two paired box genes, Pax4 and Pax6, are involved at later stages of K-cell development (Larsson *et al.*, 1998). Other sites of GIP gene expression have also been identified, including the submandibular salivary gland (Tseng *et al.*, 1993,1995), stomach (Yeung *et al.*, 1999) and the brain (Nyberg *et al.*, 2005; Sondhi *et al.*, 2006), but there is little known about the tissue specific factors involved in their development.

III. The GIP Gene and Precursor

The structure of the human GIP gene was first elucidated using clones isolated from a genomic DNA library (Inagaki *et al.*, 1989). The gene spans approximately 10 kb and is linked to a gene cluster on chromosome 17q that includes the genes NME1 (non-metastatic cells 1), PPY (pancreatic polypeptide), homeobox B6 (HOXb6), and the nerve growth factor receptor (NGFR) (Anderson *et al.*, 1993; Inagaki *et al.*, 1989; Lewis *et al.*, 1994). The human gene consists of six exons plus five introns, with the precursor, preproGIP, encoded by exons 2–5 and the 5′- untranslated region and 3′- polyadenosine tail in exons 1 and 6, respectively (Inagaki *et al.*, 1989) (Fig. 15.2). The GIP_{1-42} sequence is encoded by exons 3 and 4 (Inagaki *et al.*, 1989) (Fig. 15.2). Subsequently, the rat gene was shown to have an identical exon/intron structure (Higashimoto and Liddle, 1993; Higashimoto *et al.*, 1992). Analyses of human cDNA clones identified a 459 base pair open reading frame that encodes the 153 amino acid

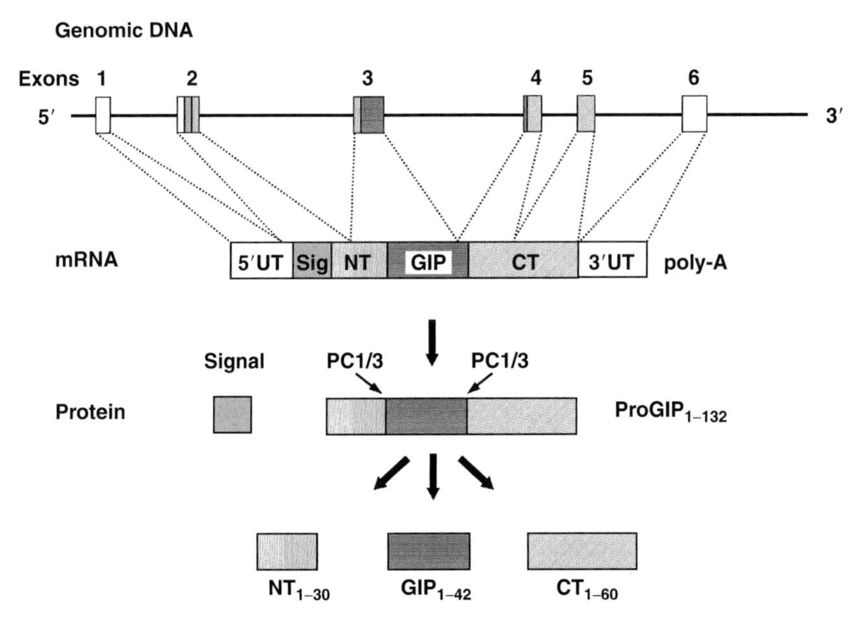

Figure 15.2 Structure of the human GIP Gene, mRNA, precursor protein and peptide products. From data in: Inagaki *et al.* (1989), Takeda *et al.* (1987), Ugleholdt *et al.* (2006).

preproGIP (Takeda *et al.*, 1987), and it was predicted that the precursor consists of a 21 amino acid signal peptide, a 30 residue NH_2-terminal peptide, GIP_{1-42}, and a 60 residue C-terminal peptide (Fig. 15.2). Rodent preproGIP has a similar structure: a 432 base pair open reading frame encoding a 144 amino acid product, with a 21 amino acid signal sequence, a 22 amino acid NH_2-terminal peptide, GIP_{1-42}, and a 59 residue C-terminal peptide (Higashimoto and Liddle, 1993; Higashimoto *et al.*, 1992). The GIP sequence in proGIP is flanked by single arginine residues, sites for cleavage by prohormone convertase (PC) enzymes. GIP and PC1/3 are co-localized in K-cells and PC1/3 has been shown to be sufficient for precursor processing, whereas processing in intestinal extracts from PC1/3 null mice was severely impaired (Ugleholdt *et al.*, 2006). There is a Gly^{31}–Lys^{32}–Lys^{33} sequence in most forms of GIP_{1-42}, and it has been suggested that cellular processing may result in production of an alternative peptide, $GIP_{1-30amide}$, *in vivo* (Rossowski *et al.*, 1992; Schieldrop *et al.*, 1996).

The 5′-upstream region of the human GIP gene contains a TATA motif, consensus Sp1, Ap-1, and Ap-2 sites, as well as two CRE elements, thus suggesting regulation by protein kinases A and C (Inagaki *et al.*, 1989; Someya *et al.*, 1993). Basal promoter activity was found to require the DNA sequence −180 to +14 (relative to the transcriptional initiation site), whereas inducible transcription was primarily mediated by one of the two CRE elements in the regions −164 to −149, and c-Jun was capable of repressing transcription (Someya *et al.*, 1993). Similar regulatory elements have been identified in the rat 5′ promoter region, including TATA and CAAT-like boxes, Ap-1, and Ap-2 sites and three CRE elements (Higashimoto and Liddle, 1993). There is also evidence for a more distal promoter region, with cell-specific expression mediated by a GATA and an ISL-1 motif located between bases −193 and −182 and −156 and −151, respectively (Boylan *et al.*, 1997; Jepeal *et al.*, 2003) and a CATTA region that binds PDX-1 (Jepeal *et al.*, 2005). $PDX^{-/-}$ mice exhibited a 98% reduction in the number of GIP-expressing cells, demonstrating the importance of the transcription factor for K-cell development (Jepeal *et al.*, 2005).

Rat duodenal GIP mRNA transcripts have been detected by day 18 of embryogenesis (Tseng *et al.*, 1995), but it is unclear whether low levels of GIP mRNA remain until birth, with a subsequent postnatal increase (Higashimoto and Liddle, 1994) or if GIP levels are maximal at birth (Tseng *et al.*, 1995). There is also controversy as to whether prolonged (2 day) fasting results in increased (Sharma *et al.*, 1992) or decreased (Higashimoto *et al.*, 1995) intestinal GIP mRNA levels. Despite the reduced GIP mRNA levels observed in the latter study, tissue peptide levels were unchanged, although circulating levels decreased (Higashimoto *et al.*, 1995). Oral administration of either a high glucose meal (Higashimoto *et al.*, 1995) or a 10% glucose solution (Tseng *et al.*, 1994) resulted in increases in both GIP duodenal mRNA and GIP peptide levels, whereas peptone was without effect. Elevating glucose also increased GIP mRNA levels in STC_{6-14} cells (Schieldrop *et al.*, 1996), supporting a key role for glucose in the regulation of GIP gene expression. Increasing dietary fat also caused an increase in intestinal GIP mRNA and circulating GIP levels, but no change in tissue levels of GIP, (Higashimoto *et al.*, 1995). Peptide production in response to both glucose and fat therefore appears to match secretion.

IV. GIP Secretion and Metabolism

Nutrients are the prime stimulators of incretin release. There is an extensive literature on fasting and postprandial plasma levels of GIP (For Reviews of early literature see: (Brown, 1982; Brown *et al.*, 1989). However, many early studies are difficult to assess because of the use of RIAs with antibodies of undefined specificity (Alam and Buchanan, 1993). Antibodies were generally against epitopes of porcine GIP within the region 15–42, where greatest interspecies sequence differences occur, resulting in considerable variability between values from different laboratories (Alam and Buchanan, 1993; Amland *et al.*, 1984). Human fasting immunoreactive (IR-) GIP levels, measured with multiple antisera directed at the mid- and

C-terminal regions of GIP, ranged from 12 to 92 pM, increasing to 35–235 pM postprandially (Jorde *et al.*, 1983b). Multiple IR species of GIP exist in plasma and their differential recognition has been an additional problem (Alam and Buchanan, 1993; Jorde *et al.*, 1983b; Krarup, 1988; Krarup *et al.*, 1985; Pederson, 1994). We now know that circulating IR-GIP consists of high molecular weight forms of unclear origin (Alam and Buchanan, 1993; Krarup, 1988; Ohneda *et al.*, 1984), as well as metabolites, resulting from its degradation. Prior to considering GIP secretion, its metabolism will first be discussed.

A. GIP metabolism

From measurements of canine renal arteriovenous differences in IR-GIP levels, it was suggested that renal extraction occurred (O'Dorisio *et al.*, 1977), a route also important for GLP-1 clearance (Deacon, 2005). Elevated levels of GIP were found in patients with renal failure indicating a similar role for GIP clearance in humans (O'Dorisio *et al.*, 1977). It was later shown that incubation of GIP or GLP-1 in serum or plasma, or with purified forms of the enzyme dipeptidyl peptidase IV (DPP IV; CD26; E.C. 3.4.14.5), resulted in rapid N-terminal truncation by removal of Tyr^1–Ala^2 from $GIP_{1–42}$ and His^1–Ala^2 from GLP-$1_{7–36}$ to form the non-insulinotropic peptides, $GIP_{3–42}$ and GLP-$1_{9–36}$, respectively (Deacon *et al.*, 2000; Kieffer *et al.*, 1995b; Mentlein *et al.*, 1993). N-terminal degradation was also demonstrated with peptides administered intravenously to rodents (Kieffer *et al.*, 1995b; Pauly *et al.*, 1996) or humans (Deacon *et al.*, 2000), as well as with endogenous GIP (Deacon, 2005; Deacon *et al.*, 2000). The IR-GIP levels detected by RIAs using antibodies directed towards the N-terminus of GIP, or the central/C-terminal part of the GIP molecule, have been referred to as "intact" or "active" GIP and "total" GIP, respectively. Using such RIAs, it was shown that the half-lives of $GIP_{1–42}$ and GLP-$1_{7–36}$ were 5–7 and 2–3 min, respectively (Deacon *et al.*, 2000; Deacon *et al.*, 1995; Meier *et al.*, 2004c), and levels did not differ between patients with chronic renal insufficiency and normal subjects (Meier *et al.*, 2004c). By contrast, the concentration of the degradation products, $GIP_{3–42}$ and GLP-$1_{9–36}$, were higher in the patients with renal failure, indicating that renal metabolism and/or uptake are important for final elimination of metabolites (Deacon, 2004). Although hepatic extraction of GIP and GLP-1 was not observed in rats and dogs (Chap *et al.*, 1987; Hanks *et al.*, 1984), the liver was subsequently shown to be a major site of N-terminal degradation of GIP in the pig (Deacon, 2004; Deacon *et al.*, 2001).

B. GIP secretion in response to nutrients

In human studies, total GIP levels increase up to sixfold in response to a meal (Creutzfeldt and Ebert, 1977; Krarup, 1988; Kuzio *et al.*, 1974), with carbohydrates and fats the most potent stimuli for release. Responses of intact GIP and GLP-1 to a meal are proportional to its calorific value (Vilsbøll *et al.*, 2003c) and rate of nutrient delivery into the intestine. Glucose, administered either orally or intraduodenally to humans (Andersen *et al.*, 1978; Creutzfeldt and Ebert, 1977; Krarup *et al.*, 1985; Lavin *et al.*, 1998; Lucey *et al.*, 1984; O'Dorisio *et al.*, 1978), rats (Bryer-Ash *et al.*, 1994), dogs (Greenberg and Pokol-Daniel, 1994; Ohneda *et al.*, 1985) or pigs (Knapper *et al.*, 1995a) results in rapid release of GIP, with levels reaching a peak within 15 to 30 minutes and returning to fasting values with a similar time course to circulating glucose and insulin (3 h). Glucose-stimulated GIP secretion decreases from the duodenum to the ileum (Thomas *et al.*, 1977), in agreement with K-cell cell distribution, and there is a high correlation between the rate of glucose absorption and increases in GIP levels (Wachters-Hagedoorn *et al.*, 2006). Although co-localization of GIP and GLP-1 in the human upper intestine has been demonstrated (Section IIB), the majority of data are consistent with rapidly absorbed nutrients primarily stimulating GIP secretion, whereas starch, and other nutrients requiring metabolism, also release GLP-1 (Holst, 2004). Secretion is also stimulated by galactose and the disaccharides sucrose and maltose, whereas fructose, sorbitol, mannose and lactose are ineffective (Sirinek *et al.*, 1983; Sykes *et al.*, 1980). Rat studies showed that glucose-stimulated GIP secretion involves the sodium-dependent glucose transporter (Sykes *et al.*, 1980) and blockade of glucose absorption reduces GIP secretion (Fushiki *et al.*, 1992).

Triglycerides (TGs) strongly stimulate GIP release in humans (Cleator and Gourlay, 1975; Creutzfeldt and Ebert, 1977; Falko *et al.*, 1975; Krarup, 1988; Krarup *et al.*, 1985), dogs (Pederson *et al.*, 1975) and rats (Ebert *et al.*, 1991; Hampton *et al.*, 1983). The GIP response to oral or intraduodenal fat is slower and more prolonged than with glucose (Krarup *et al.*, 1985; Lucey *et al.*, 1984), partly due to delayed gastric emptying. In most human studies, fat was found to be a more potent GIP secretagogue than isocaloric glucose (Cleator and Gourlay, 1975; Krarup *et al.*, 1985) whereas, surprisingly, in pigs fat was a poor stimulus, although its administration potentiated glucose-induced GIP secretion (Knapper *et al.*, 1995a). Secretion of GIP in response to TGs is dependent upon metabolism, since the presence of bile (Yoshidome *et al.*, 1995) and pancreatic digestive enzymes (Ebert and Creutzfeldt, 1980; Ross and Shaffer, 1981) in the intestinal lumen are required for a response. Long-chain fatty acids (LCFAs) are mainly responsible for stimulating GIP secretion (Ohneda *et al.*, 1984), medium chain FAs are ineffective (Creutzfeldt and Ebert, 1985; Ebert and Creutzfeldt, 1987; O'Dorisio *et al.*, 1976; Ross and Shaffer, 1981). Inhibition of chylomicron formation ablated GIP responses to intraduodenal fat (Creutzfeldt and Ebert, 1985). Although intragastric peptone (protein hydrolyzate) (Wolfe *et al.*, 2000) and intraduodenal lactalbumin hydrolyzate (Lucey *et al.*, 1984) stimulated GIP secretion, responses may have been indirect through stimulation of acid secretion. Duodenal acidification stimulates GIP secretion in rats (Ebert *et al.*, 1979b; LeRoith *et al.*, 1980; Wolfe *et al.*, 2000) and humans (Ebert *et al.*, 1979b) and omeprazole was shown to inhibit responses to peptone administration (Wolfe *et al.*, 2000). However, amino acids are capable of stimulating GIP secretion (Creutzfeldt and Ebert, 1977; O'Dorisio *et al.*, 1976; Schulz *et al.*, 1982; Thomas *et al.*, 1976). Overall responses of GIP to nutrients depend upon the meal size and their rate of absorption from the small intestine (Chaikomin *et al.*, 2005; Schirra *et al.*, 1996). Increasing either carbohydrate (Morgan, 1996) or fat (Murphy *et al.*, 1995) content of a meal results in greater GIP responses and reduced nutrient absorption, as seen in celiac disease, results in lower GIP responses (Besterman *et al.*, 1979)

C. Hormonal and neural regulation of GIP secretion

There appear to be inhibitory systems involved in the regulation of GIP secretion, although the details are unclear. Insulin infusion reduced GIP secretion stimulated by intraduodenal glucose in humans (Sirinek *et al.*, 1978) or oral gavage in rats (Bryer-Ash *et al.*, 1994). Glucose and insulin inhibited GIP responses to oral fat independently (Creutzfeldt *et al.*, 1980) in insulinopenic type I diabetes (T1DM) patients. Interestingly, intravenous infusion of C-peptide also reduced GIP release in response to fat (Dryburgh *et al.*, 1980) and glucagon reduced GIP responses to carbohydrate (Ranganath *et al.*, 1999), findings that warrant further investigation. Further support for feedback inhibition came from studies on DPP-IV inhibitor administration to dogs (Deacon *et al.*, 2002), normal human subjects (El-Ouaghlidi *et al.*, 2007) or type 2 diabetes (T2DM) patients (Herman et al., 2006) in which reductions in incretin levels occurred. It is not clear whether incretin hormones directly inhibited their respective enteroendocrine cells, or whether other mediators, such as insulin or the autonomic nervous system, were responsible (El-Ouaghlidi *et al.*, 2007).

The possibility that GIP secretion was regulated by the autonomic nervous sytem developed from the observation that exaggerated GIP responses to oral glucose developed in patients following a vagotomy and pyloroplasty (Thomford *et al.*, 1974). However, increased rates of gastric emptying and duodenal motility were probably responsible (Cataland *et al.*, 1978; Creutzfeldt and Ebert, 1977). Most studies employing pharmacological inhibition (Flaten, 1981; Nelson *et al.*, 1986), vagal stimulation (Ohneda *et al.*, 1985) or cryogenic blockade of the vagus nerves (Greenberg and Pokol-Daniel, 1994), provided no evidence for a major role of the autonomic nervous system. However, since anti-cholinergic drugs reduce gastric emptying, it has been difficult to completely rule out a minor contribution of the parasympathetic nervous system to early GIP secretory responses (Ahren and Holst, 2001; Chaikomin *et al.*, 2007) and enteric nerves could play a role (Kraenzlin *et al.*, 1985).

D. Cellular mechanisms involved in GIP secretion

Triglyceride-induced GIP release is ablated in cystic fibrosis (Ross and Shaffer, 1981) and inhibition of glucose uptake by phlorizin treatment reduces GIP responses in rats (Creutzfeldt and Ebert, 1977). From these, and other, *in vivo* studies it has been concluded that absorption of nutrients is necessary for GIP secretion (Creutzfeldt and Ebert, 1977). Because of the diffuse nature of the enteroendocrine system, it has proven difficult to isolate cells for studying GIP secretion *in vitro*. Partially purified K-cells were prepared from canine intestinal mucosa (Kieffer *et al.*, 1994) and elevating glucose or gastrin releasing peptide (GRP), or immunoneutralization of somatostatin, all stimulated GIP secretion. A subclone of the intestinal STC-1 endocrine tumor cell line (STC6-14), containing ~30% IR-GIP cells, also responded to glucose (Kieffer *et al.*, 1995a). Both ATP-dependent K^+ channel subunits, SUR1 and Kir6.2, are localized in K- and L-cells in the human small intestine, indicating a similar pathway for glucose-induced responses to that in the pancreatic β-cell (Nielsen *et al.*, 2007). K-cells also contain glucokinase, and it has been suggested that the similar characteristics to β-cells makes the K-cells excellent targets for insulin-gene therapy (Cheung *et al.*, 2000). The G protein-coupled receptor, GPR119, is localized selectively in K- and L-cells (Chu *et al.*, 2008), although its physiological role in this site has not been established. Glucose-induced secretion of GLP-1 has been reported to involve regulation by sweet taste receptors and activation of the taste-associated G protein gustducin (Jang *et al.*, 2007; Rozengurt *et al.*, 2006). It will be of interest to establish whether there is a similar pathway involved in GIP secretion, as well as the mechanisms underlying triglyceride-induced GIP release.

V. The GIP Receptor

A. GIP receptor gene

The human receptor gene is located on chromosome 19q13.3 (Gremlich *et al.*, 1995) and contains 14 exons and 12 introns, with a protein coding region of 12.5 kb (Yamada *et al.*, 1995), whereas that of the rat receptor gene spans ~10.2 kb and contains 13 exons (Boylan *et al.*, 1999). A number of studies have suggested that alternative mRNA splicing results in the production of GIP receptor variants of differing lengths in human (Gremlich *et al.*, 1995; Volz *et al.*, 1995), rat (Boylan *et al.*, 1999) and mouse (Harada *et al.*, 2008) pancreatic islets, mouse adipose tissue and intestine (Harada *et al.*, 2008), endothelial cell lines (Zhong *et al.*, 2000) and adrenocortical tumors (N'Diaye *et al.*, 1999). It is unclear as to the functional significance of most of these splice variants, but a mouse variant that retains intron 8, results in expression of a truncated receptor that acts in a dominant negative manner with respect to wild type receptor cell surface expression (Harada *et al.*, 2008). Interestingly, in mice fed a high fat diet, relative expression of the truncated form is reduced, resulting in increased sensitivity to GIP.

The promoter regions of GIP-R genes have been partially characterized. Both human (Baldacchino *et al.*, 2004, 2005; Lacroix *et al.*, 2004) and rat (Boylan *et al.*, 1999) promoters lack TATA boxes and are GC-rich. Analysis of a 350 bp region upstream of the human GIP-R gene revealed six consensus sequences for Sp1 transcription factors, that can functionally bind Sp1 and Sp3 (Baldacchino *et al.*, 2005). 5′-flanking sequences of the rat gene contain several transcription factor binding motifs, including a cAMP response element (CRE), an octamer binding site, three Sp1 sites and an initiator element. At least one Sp1 binding site is important for transcriptional activity, and distal negative regulatory sequences were suggested to control cell-specific expression (Boylan *et al.*, 1999; Lacroix *et al.*, 2004).

Table 15.1 Size and sequence homology of GIP receptors

Species	Amino acids	Sequence homology to human (%)
Human (*Homo sapiens*)	466	100
Chimpanzee (*Pan troglodytes*)	466	99.4
Rhesus monkey (*Macaca mulatto*)	492	86.2
Gray mouse lemur (*Microcebus murinus*)	446	62.3
Common tree shrew (*Tupaia belangeri*)	465	60.7
Dog (*Canis familiaris*)	466	86.7
Aurochs (*Bos Taurus*)	466	85.6
Little brown bat (*Myotis lucifugus*)	468	46.3
European rabbit (*Oryctolagus cuniculus*)	466	86.5
Fox (*Felis catus*)	388	89.0
American pika (*Ochotona princeps*)	455	86.4
Guinea pig (*Cavia porcellus*)	455	83.6
Lesser Hedgehog (*Echinops telfairi*)	440	83.3
Brown rat (*Rattus norvegicus*)	455	80.1
House mouse (*Mus musculus*)	460	79.5
Eurasian shrew (*Sorex araneus*)	466	72.0

Sequences obtained from cDNAs (rat—Usdin *et al.*, 1993; human—Gremlich *et al.*, 1995) or identification of ortholog sequences in the Ensembl database.

B. GIP receptor structure, activation, desensitization, and internalization

The pancreatic GIP-R is a glycoprotein that was originally identified in insulinoma cell extracts. Cross-linking studies provided an estimated molecular weight of ~59 kDa (Amiranoff *et al.*, 1986; Brown *et al.*, 1981, 1989; Couvineau *et al.*, 1984; McIntosh *et al.*, 1996). GIP-R cDNAs were subsequently cloned from a number of different species (Gremlich *et al.*, 1995; Usdin *et al.*, 1993; Wheeler *et al.*, 1995; Yasuda *et al.*, 1994) and additional sequences have been obtained from the Ensembl database. The GIP-R belongs to the *Secretin*, B-family of the seven transmembrane G-protein-coupled receptor (GPCR) family that includes, among others, the receptors for secretin, glucagon, GLP-1, GLP-2, VIP, GRH, and PACAP (Harmar, 2001; Mayo *et al.*, 2003). Based on data from *Arabidopsis thaliana*, it has been suggested that the "ancient mother-of-all GPCRs" (Perez, 2005) may have been a member of the Secretin or Adhesion family of receptors (Fredriksson and Schiöth, 2005). GIP receptors sequenced to date (Table 15.1) all consist of 440–492 amino acids, apart from the fox receptor (*felix catus*) that has 388 amino acids. The sequences of the chimpanzee (*Pan troglodytes*) and rhesus monkey (*Macaca mulatto*) GIP receptors share 99.4 and 86.2% sequence identity with the human receptor, whereas the most divergent of those sequenced to date is that of the little brown bat (*Myotis lucifugus*) that exhibits only 46.3% homology. The sequences of the rat, human and hamster GIP receptors share 40–47% identity with the GLP-1 and glucagon receptor sequences (Usdin *et al.*, 1993; Yasuda *et al.*, 1994). GIP receptors from non-mammalian species have not been identified or sequenced to date.

The native receptor (Brown *et al.*, 1989; Maletti *et al.*, 1986) and cloned human and rat receptors (Gremlich *et al.*, 1995; Wheeler *et al.*, 1995) expressed in clonal cell lines exhibit K_d values of 0.2–7 nM. Receptor expression studies in primary cell lines (Gremlich *et al.*, 1995; Wheeler *et al.*, 1995) have facilitated detailed analysis of the regions responsible for ligand binding, G-protein coupling, and receptor desensitization and internalization. The amino terminal domain (NT) contains consensus sequences (N-X-S/T) for *N*-glycosylation, supporting the proposal that it is a glycoprotein (Amiranoff *et al.*, 1985) and glycosylation is important for cell surface expression (Lynn, 2003). Chimeric GIP–GLP-1 receptor studies demonstrated that the NT of the GIP receptor constitutes a major part of the ligand-binding domain, and the first transmembrane (TM) domain is important for receptor activation (Gelling *et al.*, 1997b). The crystal structure of a complex of the human GIP receptor extracellular domain bound to GIP_{1-42} has recently been reported (Parthier *et al.*, 2007). GIP binds in an α-helical conformation, with the C-terminal region binding in a surface groove of the receptor, largely through hydrophobic interactions. The N-terminus of GIP remains free to interact with other parts of the receptor.

Thermodynamic binding studies demonstrated that, as expected, both GIP and the receptor undergo conformational changes during binding. Mutation of threonine to proline at position 340 in the sixth TM domain was reported to produce a constitutively active receptor with respect to cAMP production (Tseng and Lin, 1997), but equivalent mutations in the human population have not been identified.

The third intracellular (IC) loop and carboxy-terminal tail (CT) are rich in threonine (Thr) and serine (Ser) residues, which are potential phosphorylation sites (Usdin *et al.*, 1993; Wheeler *et al.*, 1995). Site-directed mutagenesis studies showed that the majority of the GIP receptor CT is not required for signaling, a minimum chain length is required for expression, and sequences within the CT play specific roles in adenylate cyclase coupling, modulating binding affinity and internalization/down-regulation (Tseng and Zhang, 1998a,b; Wheeler *et al.*, 1999).

GIP induces homologous desensitization of the GIP-R (Fehmann and Habener, 1991; Tseng and Zhang, 1998b, 2000). Details of the events involved remain to be clarified, but amino acids Cys_{411} and Ser_{406} in the C-terminal tail of the rat receptor have been implicated in both desensitization and down-regulation, and the G-protein receptor kinase, GRK2, and β-arrestin-1 appear to be involved in the desensitization process (Tseng and Zhang, 1998b, 2000). Interestingly, elevating GIP levels was shown to induce heterologous desensitization of the GLP-1 receptor in INS-1 cells, via a pathway involving $G\alpha_{i2}$ and cAMP/PKA (Rucha and Verspohl, 2005). Chronic elevations of glucose also result in desensitization of the islet GIP-R (Hinke *et al.*, 2000; Lynn *et al.*, 2003), the implications of which are discussed in Section VIII.

C. GIP receptor distribution

Protein and mRNA expression studies have shown that the GIP-R is distributed in multiple rodent tissues, including the pancreatic islet (Moens *et al.*, 1996; Usdin *et al.*, 1993; Wheeler *et al.*, 1995), fat (McIntosh *et al.*, 1999; Yip and Wolfe, 2000), stomach, brain (Nyberg *et al.*, 2005; Usdin *et al.*, 1993), pituitary (Usdin *et al.*, 1993), lung (Usdin *et al.*, 1993; Yasuda *et al.*, 1994), heart and vascular endothelium (Usdin *et al.*, 1993; Zhong *et al.*, 2000), and bone (Bollag *et al.*, 2000). In comparative studies GIP-R mRNA transcripts were detected in rat α- and β-cells, whereas GLP-1R transcripts were restricted to β-cells, implying that effects of GLP-1 on glucagon secretion are indirect (Moens *et al.*, 1996). The rat adrenal cortex has been shown to express GIP-R mRNA (Usdin *et al.*, 1993) and the presence of receptors in isolated rat adrenocortical cells was demonstrated by ^{125}I-GIP binding (Mazzocchi *et al.*, 1999) but the GIP-R does not appear to be expressed in the normal adult human adrenal gland. Expression has been detected in extracts of fetal cortex (Lebrethon *et al.*, 1998), but contamination from endothelial cell mRNA cannot be discounted

(Lacroix *et al.*, 2001). Of clinical interest, was the finding that food-dependent forms of adrenal hyperplasia and Cushing's syndrome exist that exhibit GIP-induced cortisol secretion, due to ectopic receptor expression (Lacroix *et al.*, 1992; N'Diaye *et al.*, 1999; Reznik *et al.*, 1992). Numerous cases have subsequently been reported (N'Diaye *et al.*, 1999; Lacroix *et al.*, 2001) and abundant expression of GIP receptor mRNA has been detected in the adrenocortical adenomas (de Herder *et al.*, 1996). However, cDNA analysis has not revealed any mutations (N'Diaye *et al.*, 1999).

VI. Actions of GIP

Although GIP was originally identified on the basis of its effects on gastric function, its insulinotropic activity and associated metabolic actions are now considered of greater importance. The combined actions of GIP and GLP-1 involve a wide range of physiological responses associated with nutrient intake and disposal. GLP-1 reduces food intake and inhibits gastric emptying, thus promoting optimal delivery of nutrients to the intestine (Drucker, 2006, 2007), whereas GIP contributes to the regulation of glucose uptake (Cheeseman and Tsang, 1996). Both GIP and GLP-1 stimulate insulin biosynthesis and secretion, as well as promoting β-cell growth and survival, while GLP-1 also inhibits glucagon secretion (Drucker, 2006, 2007). Through effects on the cardiovascular system, GLP-1 (Sokos *et al.*, 2006) and GIP (Kogire *et al.*, 1992) optimize delivery of nutrients to the tissues. GIP also plays an important role in promoting triglyceride storage in the adipocyte (Kim *et al.*, 2007b) and GLP-1 may promote hepatic and skeletal muscle glucose storage. Moreover, both incretins are now believed to be involved in bone metabolism (Baggio and Drucker, 2007; Zhong *et al.*, 2007). The physiological importance of these various actions of GIP have been substantiated in studies on GIP-R knockout (GIP-R$^{-/-}$) and double knockout (GIP-R$^{-/-}$/ GLP-1 R$^{-/-}$; DIRKO) mice (Hansotia *et al.*, 2004, 2007; Miyawaki *et al.*, 1999, 2002; Preitner *et al.*, 2004). Unlike GLP-1, which exerts many of its effects indirectly via activation of CNS neurons and through stimulation of afferent vagal nerve fibres (Baggio and Drucker, 2006; Holst and Deacon, 2005; McIntosh, 2008), GIP is believed to mainly act in an endocrine fashion.

A. The endocrine pancreas

Dupré and co-workers (1973) first demonstrated that administration of GIP to normal human subjects resulted in a potentiation of insulin secretion and more rapid disposal of an intravenous glucose load. GIP was subsequently shown to be insulinotropic in the rat (Turner *et al.*, 1974b), baboon (Turner *et al.*, 1974a) and dog (Pederson *et al.*, 1975). From studies on isolated rat islets (Fujimoto *et al.*, 1978; Schauder *et al.*, 1975) and the isolated vascularly perfused rat pancreas (Pederson and Brown, 1976) it was concluded that the effect of GIP was via direct action on the pancreatic islet (Brown and Otte, 1978). Hyperglycemic clamp studies with an oral glucose load (Andersen *et al.*, 1978) or GIP infusion (Elahi *et al.*, 1979) established that GIP was insulinotropic at physiological concentrations and fulfilled all of the criteria (Creutzfeldt, 1979) to establish it as an incretin. The majority of studies have shown that GIP is incapable of stimulating insulin secretion in humans under fasting conditions, there being a glucose threshold for its insulinotropic action (Elahi *et al.*, 1979; Verdonk *et al.*, 1980), although when plasma glucose is clamped at the subject's fasting level (Vilsbøll *et al.*, 2003b), or 1–3 mM above fasting (Elahi *et al.*, 1979; Vilsbøll *et al.*, 2003b), both GIP or GLP-1 stimulate insulin secretion when infused at rates that mimic physiological levels. In the rat this threshold was demonstrated to be ~4.5 mM glucose in the rat pancreas, with maximum potentiation at 16.7 mM glucose (Jia *et al.*, 1995; Pederson and Brown, 1976). GIP potentiated glucose-induced secretion in isolated islets (Siegel and Creutzfeldt, 1985; Szecowka *et al.*, 1982) and a number of different clonal β-cell lines (In 111 (Amiranoff *et al.*, 1984; Maletti *et al.*, 1986), β-TC3 (Kieffer *et al.*, 1993), RIN 1046-38 (Montrose-Rafizadeh *et al.*, 1994), BRIN-BD11 (O'Harte *et al.*, 1998) and INS-1(832/13) (Kim *et al.*, 2005a). There is controversy in the literature over the comparative efficacies of GIP and GLP-1, but both peptides potentiated glucose-induced insulin secretion in the isolated rat pancreas at concentrations as low as 16 pM and exerted equivalent maximal effects (Jia *et al.*, 1995). It has been suggested that GLP-1, but not GIP, regulates fasting glycemia in rodents (Baggio *et al.*, 2000), whereas GIP only potentiates glucose-induced insulin secretion, mainly through potentiation of the early phase of secretion (Lewis *et al.*, 2000). However, in a Danish group of glucose tolerant subjects homozygous for a Glu_{354}/Gln substitution in the GIP receptor, fasting serum C-peptide levels were decreased (Almind *et al.*, 1998), suggesting that GIP may regulate β-cell secretion in the fasting state in humans.

The important contribution of GIP to the incretin response was first shown in immunoneutralization studies in rats (Ebert and Creutzfeldt, 1982; Lauritsen *et al.*, 1981). Studies using immunoadsorption of GIP from intestinal extracts (Ebert *et al.*, 1983), intravenous administration of antibodies against the GIP receptor (Lewis *et al.*, 2000) or GIP receptor antagonist (Gault *et al.*, 2002a; Tseng *et al.*, 1996, 1999) have provided estimates of 50–70% for the contribution of GIP to the overall incretin response. Studies in GIP-R$^{-/-}$ and DIRKO mice have confirmed the importance of GIP action for glucose homeostasis (Hansotia *et al.*, 2004; Miyawaki *et al.*, 1999; Preitner *et al.*, 2004). Both GIP-R$^{-/-}$ and GLP 1-R$^{-/-}$ mice exhibit only small increases in fasting glucose and fairly mild glucose intolerance and reductions in insulin secretion in response to an oral glucose load (Miyawaki *et al.*, 1999; Scrocchi *et al.*, 1996). Responses of GIP-R$^{-/-}$ isolated islets and perfused pancreata to glucose were normal, but enhanced responsiveness to GLP-1 suggests that rodents are capable of compensatory responses to the loss of a single incretin (Pamir *et al.*, 2003). The DIRKO mice show slightly decreased food intake and markedly more disturbed glucose tolerance and insulin secretory responses than single receptor knockout mice (Hansotia *et al.*, 2004; Preitner *et al.*, 2004).

Regulation of β-cell mass involves replication, neogenesis, growth and apoptosis (Rhodes, 2005) and there is growing evidence that GLP-1 and GIP exert positive effects on these processes (Baggio and Drucker, 2007b; Drucker, 2007; Ehses *et al.*, 2003; Trümper *et al.*, 2001, 2002). Although apoptosis plays a central role in the normal development and maintenance of islet mass, increased apoptosis occurs during the development of diabetes. Among the contributing factors are the chronic hyperglycemia and hyperlipidemia associated with T2DM (Rhodes, 2005) and both incretins have been shown to protect rodent islets against glucolipotoxicity-induced apoptosis (Buteau *et al.*, 2004; Kim *et al.*, 2005b). The mechanisms involved are discussed in Section VII.

GIP has also been shown to regulate the secretion of other islet hormones. Unlike GLP-1, which is glucagonostatic in all species studied (Drucker, 2007; Fridolf *et al.*, 1991; Ørskov *et al.*, 1988), GIP stimulates pancreatic glucagon secretion from both rat (Fujimoto *et al.*, 1978; Pederson and Brown, 1978) and dog (Adrian *et al.*, 1978) pancreata, with maximal effects under low glucose conditions. Since GIP potentiates leucine and arginine-induced insulin secretion *in vivo* (Mazzaferri *et al.*, 1983) and *in vitro* with isolated islets (Mazzaferri *et al.*, 1983; Schauder *et al.*, 1977), GIP-induced glucagon release may play an important role in maintaining normoglycemia during high protein meals in these species. There has been controversy over whether GIP plays a similar role in humans since no effect of GIP on glucagon secretion was found in normal individuals (Elahi *et al.*, 1979; Nauck *et al.*, 1993). However, GIP exerted glucagonotropic actions in perfused pancreata from human cadavers (Brunicardi *et al.*, 1990) and in patients with liver cirrhosis and hyperglucagonemia (Dupré *et al.*, 1991). Recently GIP administration was shown to rapidly increase glucagon levels under fasting conditions (Meier *et al.*, 2003a), and a potential physiological role for GIP in the regulation of α-cell secretion needs to be reassessed. Pancreatic secretion of both somatostatin (Ipp *et al.*, 1977) and pancreatic polypeptide (Adrian *et al.*, 1978; Amland *et al.*, 1985) are also stimulated by GIP in some species, although the significance of these effects in humans is unclear.

B. Adipose tissue

The robust responses of GIP to fat ingestion indicate that it probably plays an important role in lipid metabolism (Morgan, 1996). There is strong evidence supporting a role for GIP in triglyceride (TG) disposal *in vivo*.

In dogs, GIP promoted clearance of chylomicron-associated TGs from blood (Wasada *et al.*, 1981). Infusion of GIP also lowered plasma TG responses to intraduodenal fat in rats (Ebert *et al.*, 1991) and immunoneutralization of endogenous GIP resulted in greater plasma TG levels following a fat load (Ebert *et al.*, 1991). In contrast, neither exogenous (Jorde *et al.*, 1984) nor endogenous (Ohneda *et al.*, 1983) GIP influenced the removal of intravenously administered TGs alone, showing that chylomicron transport is necessary. Since GIP normally releases insulin during a meal, and insulin is a major hormonal regulator of lipogenesis, a component of GIP's actions on fat metabolism is indirect. However, a number of studies have shown that GIP plays an important direct role (Beck, 1989; Kim *et al.*, 2007a,b; McIntosh *et al.*, 1999; Morgan, 1996; Yip and Wolfe, 2000).

GIP was shown to stimulate FA synthesis from acetate in adipose tissue explants (Oben *et al.*, 1991), as well as increasing uptake and incorporation of glucose into lipids (Hauner *et al.*, 1988). In humans, dietary fat is more important than *de novo* lipogenesis in the accumulation of adipose-tissue triglycerides (Morgan, 1996) and the hydrolysis of circulating triglycerides and liberation of FFAs for uptake and storage in the adipocyte is primarily mediated by lipoprotein lipase (LPL). GIP enhanced the activity of LPL in cultured preadipocytes and mature adipocytes (Eckel *et al.*, 1979; Kim *et al.*, 2007a,b; Knapper *et al.*, 1995b), including subcutaneous human adipocytes (Kim *et al.*, 2007a) as well as enhancing FA incorporation into adipose tissue (Beck and Max, 1986) via insulin-dependent pathways. A close correlation has been found between GIP responses and plasma levels of post-heparin LPL, a reflection of adipose tissue LPL (Murphy *et al.*, 1995). Therefore, matching the triglyceride load and adipose tissue uptake of FAs may be an important function of GIP (Morgan, 1996).

GIP also stimulates lipolysis (McIntosh *et al.*, 1999; Yip *et al.*, 1998). The significance of a dual lipolytic/lipogenic action of GIP is currently unclear, but GIP may have different functions during fasting from those in the presence of elevated circulating nutrients. The pancreatic β-cell is exquisitely sensitive to circulating FFAs. In their absence, β-cells become insensitive to subsequent glucose stimulation (Dobbins *et al.*, 1998; McGarry, 2002; Stein *et al.*, 1996), whereas chronic elevation of FFA levels in the presence of high glucose is toxic (gluco-lipotoxicity) (Prentki *et al.*, 2002). GIP's lipolytic action on differentiated 3T3-L1 cells is inhibited by insulin (McIntosh *et al.*, 1999) and GIP may play an important role in maintaining circulating FFAs at appropriate levels during fasting, when insulin levels are low, thus priming β-cells for subsequent glucose stimulation. Following nutrient ingestion, GIP stimulates insulin secretion and the elevated insulin levels inhibit GIP's lipolytic action; the two hormones thus combining to stimulate lipogenesis. An alternative interpretation is that the increased FFAs produced in response to GIP-induced lipolysis are re-esterified, thus contributing to lipogenesis (Getty-Kaushik *et al.*, 2006).

GIP may also have long-term effects on lipid metabolism; synthesis of both pancreatic lipase and colipase synthesis were stimulated by GIP (Duan and Erlanson-Albertsson, 1992), and this may partially account for the elevated triglyceride levels observed following triolein ingestion in rats fed a high fat diet (Hampton *et al.*, 1983).

C. Stomach

Although GIP was originally identified on the basis of its ability to inhibit acid secretion in denervated canine Heidenhain gastric pouches (Brown and Pederson, 1970; Brown *et al.*, 1969, 1970; Pederson and Brown, 1972), in studies on dogs with intact autonomic innervation GIP was found to be a much weaker inhibitor of meal-stimulated acid secretion (Soon-Shiong *et al.*, 1979; Yamagishi and Debas, 1980). Similarly, *in vivo* studies showed that physiological levels of GIP exerted no effect on basal or pentagastrin-stimulated acid secretion in humans (Maxwell *et al.*, 1980; Meier *et al.*, 2006, 2004b), although supra-physiological levels produced small gastric inhibitory responses in some studies (Maxwell *et al.*, 1980), but not others (Meier *et al.*, 2006, 2004b). It was demonstrated in perfusion studies that GIP potently stimulates the secretion of somatostatin (SS) from the rat stomach (McIntosh *et al.*, 1981a). Since SS was present in high concentrations in the stomach of a number of species (Arimura *et al.*, 1975; McIntosh, 1985; McIntosh *et al.*, 1978), and had been shown to strongly inhibit gastric acid secretion (Albinus *et al.*, 1975), it was suggested that GIP acted indirectly on the parietal cell through the paracrine action of SS (McIntosh *et al.*, 1981a). This contention was later supported by the observation that GIP-induced inhibition of acid secretion in the gastric fistula rat was inhibited by a selective SS receptor antagonist (Rossowski *et al.*, 1998), There may be species differences in the effect of GIP on gastrin secretion, with a stimulatory action in the rat (Jia *et al.*, 1994; McIntosh *et al.*, 1978; Pederson *et al.*, 1981; Schusdziarra *et al.*, 1983), while inhibiting release in the pig (Holst *et al.*, 1983) and dog (Villar *et al.*, 1976; Wolfe and Reel, 1986; Wolfe *et al.*, 1983).

The autonomic nervous system plays an important role in modulating the gastric effects of GIP. Administration of acetylcholine (McIntosh *et al.*, 1981a) or cholinergic agonists (Soon-Shiong *et al.*, 1984), or stimulation of the vagal (parasympathetic) innervation of the stomach (Holst *et al.*, 1983; McIntosh *et al.*, 1981a,b), inhibited GIP-induced SS secretion. Acetylcholine (McIntosh *et al.*, 1981b) and opioid peptides (McIntosh *et al.*, 1983, 1990) released from intrinsic neurons contribute to the inhibition of SS release. In contrast, splanchnic nerve stimulation at the level of the celiac ganglion (McIntosh *et al.*, 1981b) or infusion of β-adrenergic agonists (Koop *et al.*, 1980) resulted in a stimulation of SS release. These observations suggest that in order for GIP to inhibit acid secretion *in vivo*, parasympathetic tone must be reduced to allow gastric SS secretion. Sympathetic activation of SS release via the enterogastric reflex could then potentiate GIP responsiveness. Experimental demonstration of such modulation during the intestinal phase of digestion is, however, difficult to establish, especially in humans.

D. Other gastrointestinal tissues

Both GIP and GLP-1 receptors are present in several regions of the gastrointestinal system suggesting that they exert a concerted action on the absorption and disposal of nutrients. GIP is expressed in rat submandibular gland (Tseng *et al.*, 1993, 1995) and inhibitory effects of GIP on ductal Na^+ transport in the rabbit mandibular gland have been identified (Denniss and Young, 1978). In humans, fasting salivary GIP levels were reported to be higher than in plasma, with an increase in levels following swallowing (Messenger *et al.*, 2003). GIP may also be a local regulator of mucosal alkalinization in the proximal duodenum (Konturek *et al.*, 1985). Stimulatory effects of GIP on intestinal water and electrolyte transport (Helman and Barbezat, 1977) and glucose uptake (Cheeseman and O'Neill, 1998; Cheeseman and Tsang, 1996) have also been described. GIP-mediated increases in glucose uptake involve both trafficking of the Na^+-dependent glucose transporter 1 (SGLT1) into the brush-border membrane (Cheeseman, 1997) and GLUT2-dependent transport in the basolateral membrane (Cheeseman and O'Neill, 1998; Cheeseman and Tsang, 1996). GIP also potentiated acetylcholine- and CCK-stimulated pancreatic amylase secretion (Mueller *et al.*, 1986, 1987) but inhibited bombesin-stimulated secretion (Rossowski *et al.*, 1992). It is currently unclear whether these effects are through direct actions on the acinar cell or secondary to insulin secretion. GIP administration was shown to inhibit antral and duodenal motor activity in dogs (Castresana *et al.*, 1978) and gastric emptying in rodents (Young *et al.*, 1996). However, such effects may be indirect since GIP is capable of stimulating GLP-1 release from the L-cell in these species (Damholt *et al.*, 1998; Herrmann-Rinke *et al.*, 1995; Roberge and Brubaker, 1993) and GLP-1 is a potent inhibitor of gastric emptying (Baggio and Drucker, 2007b; Drucker, 2007; Young *et al.*, 1996). In humans, where GIP has not been shown to stimulate GLP-1 secretion, it does not inhibit gastric emptying (Meier *et al.*, 2004b).

E. Cardiovascular system

There is considerable interest in the effects of GLP-1 on cardiomyocyte survival and cardiac performance (Baggio and Drucker, 2007b) but, although GIP-R mRNA transcripts have been detected in the heart (Usdin *et al.*, 1993) no major cardiac effects of GIP have been reported. However, GIP may play a physiological role in the regulation of splanchnic blood flow. Intravenous infusion of GIP rapidly increased superior mesenteric arterial blood flow in cats (Fara and Salazar, 1978) and dogs (Kogire *et al.*, 1988). Portal venous blood flow was also increased by GIP (Kogire *et al.*, 1992), whereas both hepatic arterial (Kogire *et al.*, 1992) and pancreatic (Kogire *et al.*, 1988) flow were decreased. In rat studies, although whole pancreas blood flow was decreased by GIP, under hyperglycemic conditions GIP augmented islet blood flow (Svensson *et al.*, 1997). In a comparative study GIP induced endothelin-1 secretion from endothelial cells isolated from the canine hepatic artery and nitric oxide secretion from portal vein endothelial cells (Ding *et al.*, 2004), effects that would be expected to result in vascular changes optimizing delivery of nutrients to the liver during a meal.

F. Bone

During a meal bone metabolism is influenced by acute nutrient-induced effects, for which a number of factors have been suggested as mediators (Clowes *et al.*, 2005). Both GIP-R mRNA and protein are present in normal bone and osteoclast-like (SaSo2 and MG63) cell lines (Bollag *et al.*, 2000, 2001; Zhong *et al.*, 2007). In a series of studies, Isales and co-workers showed that GIP exerts anabolic and proliferative actions in bone. GIP increased alkaline phosphatase activity and collagen type 1 mRNA levels in osteoclasts, responses associated with new bone formation (Bollag *et al.*, 2000). Bone resorption in a fetal bone organ culture system, and resorptive activity of isolated mature osteoclasts was inhibited by GIP (Zhong *et al.*, 2007). GIP therefore also plays a role in the anti-resorptive effects of nutrients. These *in vitro* responses were shown to be physiologically relevant, since GIP increased bone density in ovariectomized rats (Bollag *et al.*, 2001), whereas GIP-R knockout mice demonstrated reduced bone size and mass, and altered bone microarchitecture, biomechanical properties and turnover (Tsukiyama *et al.*, 2006; Xie *et al.*, 2005). By contrast, transgenic mice overexpressing GIP under control of the metallothionein promoter showed increases in markers of bone formation, decreases in bone resorption markers, and increased bone mass (Xie *et al.*, 2007). The significance of these findings for humans is unclear. Although it was reported that GLP-2, but neither GIP nor GLP-1, acutely reduced markers of bone resorption in humans (Henriksen *et al.*, 2003), studies were performed with pharmacological doses of peptides, different administration protocols and lack of placebo controls.

G. Other GIP targets

The GIP-R is expressed in several brain regions, including the cerebral cortex, hippocampus, and olfactory bulb (Usdin *et al.*, 1993) and GIP expression has also been reported (Nyberg *et al.*, 2005; Sondhi *et al.*, 2006).

Only a single reference has alluded to the possibility that GIP may reduce food intake and no details were provided (Lavin *et al.*, 1996). Administration of GIP *in vivo*, or to hippocampal cultures, resulted in increased proliferation of progenitor cells (Nyberg *et al.*, 2005). Adult GIP-$R^{-/-}$ mice had reduced numbers of proliferating cells in the dentate gyrus regions of the hippocampus and GIP-R overexpressing mice exhibited improved sensorimotor coordination and memory (Nyberg *et al.*, 2005). It is therefore possible that GIP plays a role in both neurogenesis and neuron function. Although GIP-R mRNA transcripts have been detected in the pituitary (Usdin *et al.*, 1993), only high concentrations of GIP stimulated ACTH secretion from AtT20 cells or gonadotropins from rat pituitary cells (Nussdorfer *et al.*, 2000). GIP stimulated corticosterone secretion in the rat (Mazzocchi *et al.*, 1999), but no such responses have been reported with normal human adrenal tissue. No GIP-R transcripts have been identified in muscle or liver (Usdin *et al.*, 1993), although GIP was reported to increase muscle glycogen production and glucose uptake and oxidation *in vitro* (O'Harte *et al.*, 1998) and suppress glucagon-dependent hepatic glycogenolysis *in vivo* (Hartmann *et al.*, 1986). However, in humans GIP did have an impact on hepatic insulin action (Service *et al.*, 1990). It is unclear as to whether or not GIP influences insulin clearance in humans (Meier *et al.*, 2003b, 2007; Rudovich *et al.*, 2004), but the suggestion that a GIP-induced reduction in hepatic insulin extraction compensates for impairments in insulin secretion (Rudovich *et al.*, 2004) is of interest.

VII. GIP-Activated Signal-Transduction Pathways

A. The pancreatic islet

Glucose is the primary stimulator of insulin secretion. Following its entry into the β-cell, glucose is metabolized by glycolysis and mitochondrial oxidation. The resulting increase in the ATP/ADP ratio causes closure of ATP-sensitive $K^+(K_{ATP})$ channels, membrane depolarization, activation of voltage-dependent Ca^{2+} channels (VDCCs) and increases in intracellular Ca^{2+}, followed by insulin-granule exocytosis (Ashcroft and Rorsman, 2004) (Fig. 15.3) This is followed by membrane repolarization mediated by voltage-dependent K^+ (K_V) (MacDonald and Wheeler, 2003) and Ca^{2+}-sensitive K (K_{Ca}) channels (MacDonald *et al.*, 2002a). The incretin hormones act by potentiating membrane depolarization and increases in intracellular Ca^{2+} levels, in addition to exerting direct effects on the exocytotic machinery (Ding and Gromada, 1997; Holst and Gromada, 2004; MacDonald *et al.*, 2002a; Wahl *et al.*, 1992).

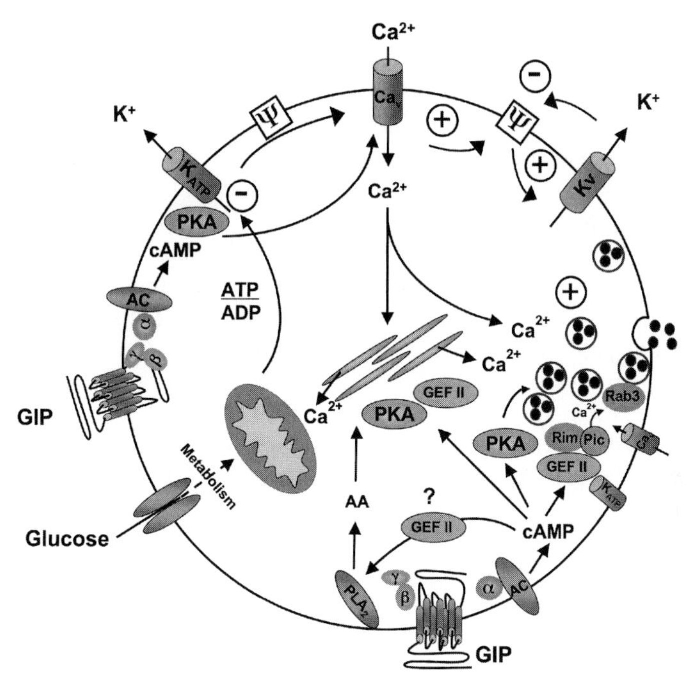

Figure 15.3 Representation of the main signaling pathways by which glucose and GIP are proposed to stimulate insulin secretion. See text for details. cAMP = Cyclic AMP; PKA = Protein kinase A; GEFII = Guanine nucleotide exchange factor II; Pic = Piccolo; Rim2 = Regulating synaptic membrane exocytosis 2; Rab2 (member RAS oncogene family); PLA_2 = Ca^{2+}-independent phospholipase A_2); K_{ATP} = ATP-dependent K^+ channel; Ca_v = Voltage-dependent Ca^{2+} channel; K_v = Voltage-dependent K^+ channel.

GIP has been shown to stimulate adenylate cyclase (AC) in pancreatic cell lines (Amiranoff *et al.*, 1984; Ehses *et al.*, 2001; Lu *et al.*, 1993), isolated pancreatic islets (Siegel and Creutzfeldt, 1985) and GIP-R transfected cells (Wheeler *et al.*, 1995), resulting in localized increases in cAMP (Seino and Shibasaki, 2005). Synergistic interaction between GLP-1 and glucose, resulting in increments in Gsα and Ca^{2+}, has been proposed to activate calmodulin-activated type VIII adenylate cyclase (Delmeire *et al.*, 2003), and a similar interaction is likely for GIP. Increased β-cell cyclic AMP activates both protein kinase A (PKA)-dependent (Ding and Gromada, 1997) and -independent events (Seino and Shibasaki, 2005) involving the cAMP-specific guanine nucleotide exchange factor (GEF) II/exchange protein directly activated by cAMP (EPAC) 2 (Holz, 2004; Seino and Shibasaki, 2005) (Fig. 15.3). In common with GLP-1, PKA-dependent events activated by GIP probably involve anchoring proteins (Fraser *et al.*, 1998). GIP also potentiates α-cell secretion in a PKA-dependent manner (Ding *et al.*, 1997).

There are multiple phosphorylation targets of PKA that are related to its effects on exocytosis. As mentioned, membrane depolarization is dependent upon closure of K_{ATP} channels (Fig. 15.3). PKA-mediated phosphorylation of Ser_{1448} of the SUR1 subunit induces K_{ATP} channel closure via an ADP-dependent mechanism and changes in ADP levels are thought to be central to the glucose-sensitivity of incretin action (Light *et al.*, 2002). Additionally, positive β-cell priming effects of discontinuous exposure to a combination of GIP and GLP-1 have been observed (Delmeire *et al.*, 2004) and GLP-1 was shown to sensitize the β-cell to glucose through a mechanism involving stimulation of mitochondrial ATP synthesis (Tsuboi *et al.*, 2003).

In β-cells and clonal β-cell lines, GIP increased Ca^{2+} influx through voltage-dependent Ca^{2+} (Ca_v) channels and non-selective ion channels, as well as calcium-induced Ca^{2+} release (CICR) from intracellular stores (Lu *et al.*, 1993; Wahl *et al.*, 1992; Wheeler *et al.*, 1995). More detailed studies have been performed on the events involved in GLP-1 modulation of CICR, and intracellular Ca^{2+} release channels (ryanodine receptors, inositol 1,4,5-trisphosphate receptors) appear to respond to simultaneous increases in Ca^{2+} and cAMP, the latter through both PKA- and EPAC-mediated pathways (Holz *et al.*, 2006; Kang *et al.*, 2005). In addition to its effects on membrane depolarization, GIP exerts more direct actions on exocytosis, also via both PKA-dependent (Ding and Gromada, 1997) and -independent (Seino and Shibasaki, 2005) pathways. Protein kinase A phosphorylates proteins that are associated with the exocytotic machinery (Seino and Shibasaki, 2005), whereas a cAMP–GEFII–Rim2 complex plays a central role in PKA-independent actions of cAMP (Holz, 2004; Holz *et al.*, 2006; Kashima *et al.*, 2001; Rucha and Verspohl, 2005) (Fig. 15.3). Under unstimulated conditions GEFII is associated with Kir6.2. Following adenylate cyclase activation, increases in cAMP cause cAMP–GEFII to dissociate from Kir6.2 and induce Ca^{2+}-dependent dimerization of Rim2 and piccolo and interaction with the secretory granule binding protein, Rab3, a core component of the exocytotic system (Kashima *et al.*, 2001; Seino and Shibasaki, 2005) (Fig. 15.3).

Other GIP-activated signaling pathways contribute to the stimulation of insulin secretion. GIP receptors couple functionally to a Group VIA islet Ca^{2+}-independent phospholipase A_2 ($iPLA_2$) both through $G\beta\gamma$ dimer-activated and cAMP-mediated pathways, resulting in increased arachidonic acid production (Ehses *et al.*, 2001) (Fig. 15.3). The importance of such pathways in β-cell function is supported by studies showing that inhibition (Song *et al.*, 2005) or siRNA suppression (Bao *et al.*, 2006a) of $iPLA_2$ in INS-1 β-cells reduced insulin secretion. Additionally, $iPLA_2$ knockout mice exhibit greater glucose intolerance than wild type mice on a high fat diet (Bao *et al.*, 2006b). The downstream effects of $iPLA_2$ activation on insulin secretion have not been completely elucidated. However, a wortmannin-sensitive pathway, probably involving activation of PI3-kinase

γ (Li *et al.*, 2006), has been shown to contribute to GIP-stimulated insulin secretion (Kubota *et al.*, 1997; Straub and Sharp, 1996) and release of Ca^{2+} from intracellular stores by arachidonic acid could also be involved (Fig. 15.3).

Membrane repolarization occurs in response to voltage-dependent K^+ currents that are mediated by K_V and and K_{Ca} channels (MacDonald and Wheeler, 2003; MacDonald *et al.*, 2002a). Two major classes of K_V currents have been identified, based on their biophysical and pharmacological properties: delayed rectifier and A-type currents. Numerous sub-types of K_V channels expressed in β-cells contribute to currents associated with insulin secretion (MacDonald and Wheeler, 2003; MacDonald *et al.*, 2002c). Both GIP and GLP-1 reduce K_V channel currents, thus prolonging β-cell action potentials and potentiating the Ca^{2+} signal. GLP-1 receptor activation reduces K_V channel currents via both PKA and PI3Kinase/ Protein kinase B (PKB/Akt)-mediated effects on delayed rectifier channels, and the $K_V2.1$ channel plays a dominant role (MacDonald and Wheeler, 2003; MacDonald *et al.*, 2002b, 2003). GIP also inhibits delayed rectifier currents in the INS-1 cell line and human β-cells (Choi, Kim et al; Unpublished observations), as well as exerting a novel regulatory action on A-type currents by increasing endocytosis of $K_V1.4$ channels through cAMP/PKA-dependent phosphorylation (Kim *et al.*, 2005a). Interestingly, arachidonic acid has recently been shown to increase the rate of inactivation of $K_V2.1$ channels (Jacobson *et al.*, 2007), raising the possibility that GIP-activation of $iPLA_2$ is also linked to K_V channel regulation.

There is an extensive literature on the effects of GLP-1 receptor agonists on β-cell development, mitogenesis and proliferation (Baggio and Drucker, 2006; Doyle and Egan, 2007; Drucker, 2007; Vahl and D'Alessio, 2003; Wajchenberg, 2007), but very little on the role of GIP. Both GLP-1 and GIP stimulate insulin gene transcription and proinsulin biosynthesis (Drucker, 2007; Drucker *et al.*, 1987; Fehmann and Göke, 1995; Schäfer and Schatz, 1979; Wang *et al.*, 1996), as well as increasing expression of enzymes involved in glucose uptake and metabolism (Doyle and Egan, 2007; Wang *et al.*, 1996). Neither GIP-$R^{-/-}$, GLP-$1R^{-/-}$ nor DIRKO mice demonstrate marked abnormal islet morphology (Pamir *et al.*, 2003; Preitner *et al.*, 2004), indicating that the incretins are not essential for β-cell development, and their major role is probably in islet maintenance and repair. GIP potentiated glucose-induced INS-1 cell proliferation (Ehses *et al.*, 2003; Trümper *et al.*, 2001, 2002) in a comparable fashion to GLP-1 (Ehses *et al.*, 2003) and it activates the MAP kinase or ERK kinase-Extracellular signal-Regulated Kinase 1/2 (MEK1/2-ERK1/2) (Ehses *et al.*, 2002; Kubota *et al.*, 1997; Trümper *et al.*, 2001, 2002) and MKK3/6-p38 (Ehses *et al.*, 2003) mitogen-activated protein kinase (MAPK) modules, as well as the PI3Kinase/PKB pathway (Trümper *et al.*, 2001, 2002) associated with mitogenesis. In INS-1 β-cells and CHO-K1 cells expressing the

GIP-R, GIP activated Mek1/2 and ERK1/2, through phosphorylation (Ehses *et al.*, 2002; Trümper *et al.*, 2001). It was suggested that Rap-1 mediated phosphorylation of Raf-B, but not Src or Ras, was integral to GIP action (Ehses *et al.*, 2002) and a similar pathway has subsequently been shown to be activated by GLP-1 (Doyle and Egan, 2007). GIP-mediated activation of PKB is complex. Under certain experimental conditions, PI3Kinase is involved (Kim *et al.*, 2005b; Trümper *et al.*, 2001). However, an alternative, PI3Kinase-independent pathway may also exist (Widenmaier *et al.*, unpublished observations). Downstream, GIP-activation results in increased phosphorylation of substrates of ERK1/2 (Elk-1; p90 RSK) (Ehses *et al.*, 2002), p38 MAPK (ATF-2) (Ehses *et al.*, 2002) and PKB ($p70^{S6K}$, Foxo1, glycogen synthase kinase (GSK) 3β) (Kim *et al.*, 2005b; Trümper *et al.*, 2001). As with GLP-1 (Baggio and Drucker, 2006), the transcription factor PDX-1 is likely a key factor in GIP-induced β-cell proliferation and proteins involved in cell cycle control, such as cyclin D1, are also involved (Friedrichsen *et al.*, 2006).

Both *in vivo* and *in vitro* studies have shown that GIP and GLP-1 exert powerful prosurvival effects on β-cells (Drucker, 2006; Baggio and Drucker, 2006, 2007). A number of stresses induce apoptosis in β-cells or pancreatic islets, including high glucose + fatty acids (glucolipotoxicity), serum depletion with a low glucose environment or treatment with agents that induce mitochondrial or endoplasmic reticulum (ER) stress (Buteau *et al.*, 2004; Eizirik *et al.*, 2008; Kim *et al.*, 2005b; Lee and Pervaiz, 2007). Under apoptotic conditions, GIP reduces activation of caspase-3 and DNA fragmentation (Ehses *et al.*, 2003; Trümper *et al.*, 2002). These anti-apoptotic effects are associated with activation of the Mek1/2-Erk1/2 module (Ehses *et al.*, 2003; Trümper *et al.*, 2002), although this pathway may make a relatively small contribution to the overall response, and a cAMP-dependent reduction in p38 MAPK phosphorylation may be more important (Ehses *et al.*, 2003). Exendin-4 was recently shown to reduce ER stress-associated β-cell death through a pathway involving PKA-dependent induction of ATF-4 (Yusta *et al.*, 2006), and a similar effect of GIP likely occurs.

Regulation of apoptosis by GIP also occurs at the level of gene expression. GIP stimulates expression of the anti-apoptotic *bcl-2* gene in β-cells through a PKA-mediated pathway involving dephosphorylation of AMP-activated protein kinase (AMPK), increased nuclear entry of cAMP-responsive CREB coactivator 2 (TORC2) and phosphorylation of cAMP response element binding protein (CREB) (Kim *et al.*, 2008) (Fig. 15.4). It is likely that expression of the bcl_{XL} and bcl_W genes are also increased. PKB/Akt is also a major target of GIP (Kim *et al.*, 2005a,b; Trümper *et al.*, 2001, 2002) and its activation induces phosphorylation and nuclear exclusion of the transcription factor Foxo1 (Kim *et al.*, 2005b). Since unphosphorylated Foxo1 is required for expression of the pro-apoptotic protein bax, this results in a reduction in Bax levels (Fig. 15.4). *In vivo*

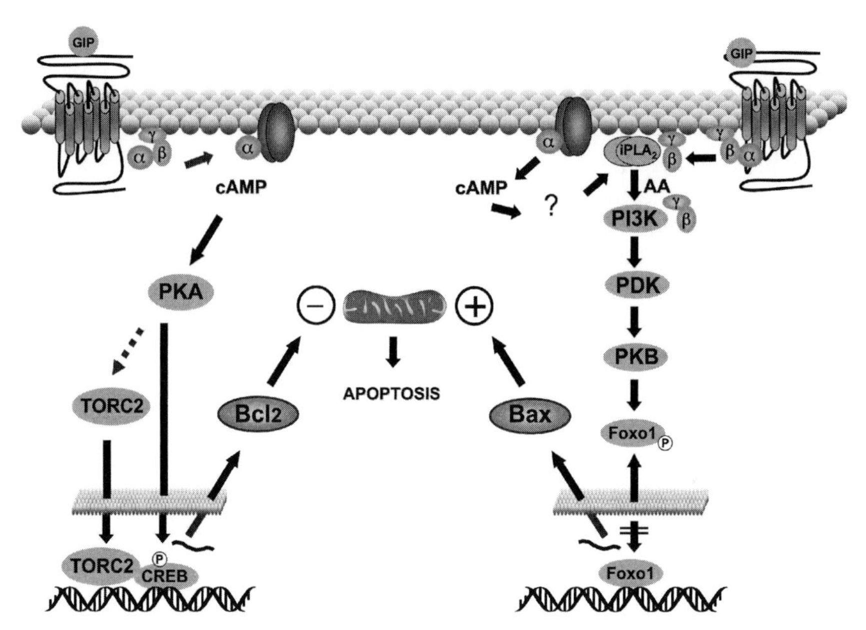

Figure 15.4 Diagram of proposed pathways by which GIP increases expression of Bcl_2 and decreases expression of bax. See test for details. cAMP = Cyclic AMP; PKA = Protein kinase A; TORC2 = cAMP-responsive CREB coactivator 2; CREB = cAMP-response element binding protein; Bcl2 = B-cell leukemia/lymphoma 2; Bax = Bcl2-associated X protein; PI3K = phosphoinositide-3-kinase; Foxo1 = Forkhead Box O1; $iPLA_2$ = Ca^{2+}-independent phospholipase A_2.

administration of GIP over 4 weeks to the VDF Zucker rat resulted in reduced islet apoptosis and similar reciprocal effects on Bcl-2 and Bax to those observed *in vitro* (Kim *et al.*, 2005b). Such protective effects have also been found with GLP-1 in STZ diabetic (Li *et al.*, 2003) and ZDF rats (Farilla *et al.*, 2002), indicating that they are central to the anti-apoptotic effects of the incretins.

B. The adipocyte

There is uncertainty as to the exact role played by GIP in the adipocyte. In keeping with its actions on the β-cell, GIP was shown to stimulate adenylate cyclase (Ebert and Creutzfeldt, 1987; McIntosh *et al.*, 1999; Yip *et al.*, 1998) and lipolysis (Ebert and Creutzfeldt, 1987; Getty-Kaushik *et al.*, 2006; McIntosh *et al.*, 1999; Yip *et al.*, 1998) in differentiated 3T3-L1 cells and rat adipocytes, although the lipolytic activity could not be reproduced in a subsequent study from one of these groups (Song *et al.*, 2007). There appear to be both insulin-independent and -dependent adipocyte actions of GIP (McIntosh *et al.*, 1999), with the latter predominating, and support for an important role in the regulation of adipogenesis/lipogenesis is growing. GIP, through activation of PKB, promoted differentiation of preadipocytes to adipocytes (Song *et al.*, 2007). It also increases delivery of triglycerides to the adipocyte, by stimulating LPL activity (Eckel *et al.*, 1979; Kim *et al.*, 2007a,b; Knapper *et al.*, 1995b), and stimulates lipogenesis by increasing glucose and fatty acid uptake (Beck and Max, 1986; Hauner *et al.*, 1988; Oben *et al.*, 1991). There have been few studies on the signaling mechanisms involved, but GIP has recently been shown to stimulate LPL activity in differentiated 3T3-L1 cells and human adipocytes, through a pathway involving reduction in the activity of AMP Kinase and increased release of the adipocytokine, resistin (Kim *et al.*, 2007b). These effects are insulin-dependent and mediated via signaling through both the SAPK/JNK and p38 MAP Kinase (Kim *et al.*, 2007b). The mechanisms by which GIP increases adipocyte nutrient uptake are still being elucidated, but GLUT-4 cell membrane translocation may be involved (Song *et al.*, 2007).

C. The adrenal cortex

Stimulation of corticosterone secretion from rat adrenocorticol cells by GIP is mediated via activation of adenylyl cylcase and PKA (Mazzocchi *et al.*, 1999). GIP has generally been found to have no effect on normal human adrenocortical cells (Lacroix *et al.*, 1992, 2001). However, GIP increased cAMP production, expression of steroidogenic enzymes and cortisol secretion from adrenal tumor cells (Chabre *et al.*, 1998; Lebrethon *et al.*, 1998). GIP may also have mitogenic effects on tumor cells, since its addition increased [^{3}H]thymidine incorporation and ERK1/2 activity (Chabre *et al.*, 1998). The ability of ectopic adrenal GIP-R expression to induce tumorigenesis was demonstrated by transplanting bovine adrenocortical cells overexpressing the GIP receptor under the kidney capsule of adrenalectomized mice (Mazzuco *et al.*, 2007). Cortisol-secreting adenomatous adrenocortical tissue developed that produced a mild Cushing's syndrome with hyperglycemia (Mazzuco *et al.*, 2007).

D. Other cell types

Studies on other cell types have shown that stimulation of adenylate cyclase is the major pathway by which GIP acts in gastric cell lines (Emami *et al.*, 1986; Gespach *et al.*, 1982) osteoblast-like cells (SaOS2) (Bollag *et al.*, 2000) and endothelial cell lines (Ding *et al.*, 2003; Zhong *et al.*, 2000). However, in a comparative study on endothelial cells isolated from the canine hepatic artery (HAEC) or portal vein (PVEC), GIP induced a Ca^{2+}-dependent endothelin-1 secretion from HAEC and secretion of nitric oxide from PVEC (Ding *et al.*, 2004). These responses would be expected to result in contraction of the hepatic artery and relaxation of the portal vein, thus leading to optimizing delivery of nutrients to the liver during a meal.

VIII. Pathophysiology of GIP

There have been numerous studies targeted at identifying possible cause and effect relationships between altered GIP secretion or responsiveness and disease etiology. As discussed earlier (Section VC), ectopic GIP receptor expression is associated with specific forms of food-dependent hypercortosolism (Lacroix *et al.*, 2001, 2004) and gene array analysis identified adrenal genes whose expression is altered in macronodular hyperplasia (Bourdeau *et al.*, 2004). Changes in GIP secretion associated with gastrointestinal dysfunction (e.g. duodenal ulcer, celiac disease) have generally been concluded to be secondary to the disease (Crockett *et al.*, 1976) and in this section only the possible contribution of GIP to obesity and T2DM etiology, and the potential therapeutic uses of GIP receptor agonists and antagonists, will be considered further.

A. Does GIP contribute to the development of obesity and diabetes?

Fasting levels of total plasma GIP in obese subjects have generally been found to be normal (Jones *et al.*, 1989; Jorde *et al.*, 1983a; Lauritsen *et al.*, 1980) or slightly elevated when compared with normal weight individuals (Creutzfeldt *et al.*, 1978; Ebert *et al.*, 1979a; Elahi *et al.*, 1984; Fukase *et al.*, 1993; Mazzaferri *et al.*, 1985; Theodorakis *et al.*, 2004). However, obese subjects with impaired glucose tolerance or diabetes have been shown to exhibit modestly potentiated GIP responses to a glucose load (Creutzfeldt and Ebert, 1977; Creutzfeldt *et al.*, 1978; Crockett *et al.*, 1976; Jones *et al.*, 1989; Mazzaferri *et al.*, 1985; Theodorakis *et al.*, 2004). In a comprehensive series of studies, Creutzfeldt and co-workers found that obese non-diabetic or diabetic patients exhibited greater increases in total GIP responses to a mixed meal, when compared to normal weight individuals (Creutzfeldt and Ebert, 1977; Creutzfeldt *et al.*, 1978; Willms *et al.*, 1978). In other studies only small increases in total (Osei *et al.*, 1986; Mazzaferri *et al.*, 1985) or intact (Vilsbøll *et al.*, 2001, 2003c) GIP responses have been observed. There are a number of possible explanations for these disparate results. The patient groups and their previous dietary history may be important. Subjects of southern European background demonstrated greater GIP responses to a test meal than those from a northern background (Jackson *et al.*, 2000). Increasing dietary fat content increased subsequent GIP responses to an OGTT (Morgan *et al.*, 1988) and a greater intake of dietary monounsaturated FAs was associated with higher fasting and postprandial GIP levels than with a high saturated FA intake (Morgan, 1996). The test diet itself is also important, with exaggerated GIP responses in obesity only observed in response to a high calorie test meal (≥1000 kcal) (Ebert and Creutzfeldt, 1989). Therefore the patient groups studied may vary considerably as a result of their dietary characteristics, test diet utilized and responsiveness.

From the early secretion studies in obese/diabetic patients the suggestion arose that over-nutrition in Western society chronically produces increased GIP responses. Such hyperGIPemia was predicted to contribute to hyperinsulinemia, with the combined effects of increased GIP and insulin resulting in hyperplasia and/or hypertrophy of adipose tissue (Creutzfeldt *et al.*, 1978; Gault *et al.*, 2003b; Kieffer, 2003; Morgan, 1996). Is there strong evidence to support this suggestion?

Exaggerated GIP responses in obesity/diabetes have been suggested to result from increased gastric emptying (Ebert and Creutzfeldt, 1989), K-cell hyperplasia (Gault *et al.*, 2003b) and reduced inhibitory feedback of FA-induced GIP release by insulin (Creutzfeldt *et al.*, 1978; Ebert *et al.*, 1979a). Surprisingly, there are no published studies on intestinal K-cell number or size in human obesity. Parallels have therefore been sought in obese rodent models, and greatly elevated plasma GIP levels in *ob/ob* and *db/db* obese mice are associated with K-cell hyperplasia (Bailey *et al.*, 1986; Flatt *et al.*, 1983). The elevated GIP responses of obesity were decreased by fasting in both humans (Willms *et al.*, 1978) and *ob/ob* mice (Flatt *et al.*, 1984), suggesting a primary role for nutrients in both species. The contribution of reduced insulin feedback inhibition on the K-cell is controversial, although there is supportive evidence for a direct inhibitory effect of insulin on GIP secretion in humans (Creutzfeldt *et al.*, 1980) and in rats (Bryer-Ash *et al.*, 1994).

A further complication in understanding GIP pathophysiology arose from studies on the relative insulin responses to oral and intravenous glucose in normal subjects and T2DM patients. It was shown that the incretin effect was greatly reduced in the diabetic group (Nauck *et al.*, 1986a) and the decrease has subsequently been attributed to two major defects: reduced circulating levels of intact GLP-1 (Toft-Nielsen *et al.*, 2001; Vilsbøll *et al.*, 2001, 2003c) and defective β-cell responsivenes to GIP (Jones *et al.*, 1987; Krarup *et al.*, 1987; Nauck *et al.*, 1993), whereas responsiveness to GLP-1 is relatively intact (Elahi *et al.*, 1994; Nauck *et al.*, 1986a). The underlying causes of GIP resistance are unknown, but could include reduced receptor cell surface expression resulting from altered gene transcription/translation or receptor down-regulation, faulty GIP-R processing or receptor trafficking, reduced hormone-receptor interaction due to mutations, or defective intracellular signaling (Holst *et al.*, 1997).

Missense mutations have been identified in Japanese (Kubota *et al.*, 1996) and Danish (Almind *et al.*, 1998) populations resulting in substitutions of Gly_{198}/Cys, Glu_{354}/Gln and Ala_{207}/Val. CHO cells expressing the

GIP-R with the Gly_{198}/Cys mutation exhibited elevated EC_{50} values for GIP-induced cAMP production, whereas the other two mutations were without effects on function (Almind *et al.*, 1998; Kubota *et al.*, 1996). However, allelic frequency studies showed no association of the mutations with T2DM. Attempts at identifying whether a genetic component was involved in GIP resistance also included studies on first-degree relatives of type 2 diabetics, but definitive conclusions cannot be drawn from the available data (Meier *et al.*, 2001, 2003c; Nauck *et al.*, 2004). There is therefore no strong support for a mutant form of the GIP-R playing a major role and rodent models have therefore been studied in attempts at understanding potential mechanisms underlying GIP resistance. The VDF rat demonstrates a loss of normal glucose threshold for GIP action (Chan *et al.*, 1985), resulting in potentiation of insulin secretion under fasting conditions and subsequently the obese rats become insulin resistant. Levels of GIP-R mRNA and protein are reduced in both the VDF rat (Lynn *et al.*, 2001) and in a 90% pancreatectomized Sprague Dawley rat model (Xu *et al.*, 2007), both of which exhibit chronic hyperglycemia. Elevating glucose was found to induce down-regulation of the GIP-R *in vivo* (Lynn *et al.*, 2003; Zhou *et al.*, 2007), as well as expression in β-cell lines, by a pathway that involved down-regulation of the transcription factor PPARα (Lynn, 2003). Normalization of glycemia with phlorizin treatment reversed GIP-R down-regulation in VDF (Piteau *et al.*, 2007) and 90% pancreatectomized (Xu *et al.*, 2007) rats, suggesting that hyperglycemia was a major factor involved. However, GIP can also induce β-cell homologous desensitization (Hinke *et al.*, 2000) and down-reulation via receptor ubiqitination and proteosome degradation may also play a role (Zhou *et al.*, 2007). In humans, the defective response to GIP appears to involve largely the late phase of insulin secretion (Vilsbøll *et al.*, 2002) and patients with other forms of diabetes (Vilsbøll *et al.*, 2003a) or impaired glucose tolerance (Muscelli *et al.*, 2006) exhibit similar reductions. While comparative studies of responses to administration of GIP as a bolus or by infusion have been interpreted as favoring a general β-cell defect (Meier *et al.*, 2004a; Vilsbøll *et al.*, 2002), a selective glucose-induced reduction in GIP receptor expression or action cannot be ruled out.

A cause-effect relationship between hyperGIPemia, hyperinsulinemia and insulin resistance in T2DM has also not been established (Theodorakis *et al.*, 2004). Compared to the effects of GIP on the pancreatic islet, we know relatively little about the role of GIP in human adipose tissue, but a lipogenic function would be consistent with its anabolic characteristics and, with moderate nutrient intake, GIP-mediated promotion of triglyceride storage is probably an important component of postprandial nutrient disposal. At present there is no strong evidence supporting a major contribution of GIP to the development of human obesity. However, the potential consequences of increased GIP action have been inferred from studies on homozygous GIP receptor GIP-$R^{-/-}$ (Miyawaki *et al.*, 1999) that were resistant to obesity when fed a high fat diet (Miyawaki *et al.*, 2002) due to preferential utilization of fat as energy substrate during the light phase. Crossbreeding of GIP-$R^{-/-}$ mice with obese $Lep^{ob/+}$ mice greatly reduced the adiposity and insulin resistance in the double homozygous $GIP^{ad-/-}$, Lep^{ob}/Lep^{ob} mice (Miyawaki *et al.*, 2002). Both GIP-$R^{-/-}$ and DIRKO mice exhibited reduced body weight gain and adipose tissue accretion when fed a high fat diet (Hansotia *et al.*, 2007) and 50 week old GIP-$R^{-/-}$ mice raised on a normal diet were reported to have reduced fat deposits and improved insulin sensitivity when compared to wild type mice (Yamada *et al.*, 2007). GIP-$R^{-/-}$ mice are also spontaneously hyperactive (Yamada *et al.*, 2007). Clearly these findings, and the adipocyte actions discussed in Section VIIB, point to an important role for GIP in adipose tissue function and further studies on its role in humans is required.

B. Do GIP analogs have therapeutic potential?

Loss of β-cell mass due to autoimmune responses underlies T1DM. During development of T2DM in humans there is initially an increase in β-cell mass, but there is now strong evidence that loss of β-cells via apoptosis subsequently outstrips the low levels of islet replication and neogenesis resulting in a reduction in β-cell mass (Butler *et al.*, 2003; Sakuraba *et al.*, 2002). Both GLP-1 receptor agonists (Exenatide) and DPP-IV inhibitors (Sitagliptin; Vildagliptin) are currently in use as T2DM therapeutics and they effectively improve glucose tolerance (Baggio and Drucker, 2006; Deacon, 2007; Drucker, 2007; Drucker and Nauck, 2006; McIntosh, 2008; Vahl and D'Alessio, 2003). Levels of both intact GIP and GLP-1 are increased during DPP-IV inhibitor treatment (Deacon *et al.*, 2002; Herman et al., 2006). It is unclear as to the contribution of GIP to the improvements in glucose homeostasis seen during DPP-IV treatment (McIntosh, 2008). However reducing chronic hyperglycemia in diabetic rodents leads to a return of GIP sensitivity (Piteau *et al.*, 2007; Xu *et al.*, 2007) and untreated T2DM patients receiving the sulfonylurea glyburide also demonstrated improved GIP responsiveness (Meneilly *et al.*, 1993). It is therefore possible that, with improved glycemic status, responsiveness to GIP in T2DM patients will return. In view of the demonstrated effects of incretins on β-cell development, proliferation and survival in rodents (Section VIIA) and the beneficial functional effects of treating human islets with GLP-1 prior to transplantation (Farilla *et al.*, 2003) it is therefore understandable that there is significant interest in the potential of incretins for preserving islet mass, and GIP could play a role. Could GIP receptor agonists be useful therapeutics?

The majority of studies on clinical applications of incretin analogs have been targeted at GLP-1, largely due to the β-cell resistance to GIP in

T2DM. Pharmacological levels of GIP analogs also exert significant effects in "GIP-resistant" rodents (Hinke *et al.*, 2002) suggesting that they may also be clinically useful. Since cleavage of GIP by DPP-IV ablates β-cell effects of GIP by removal of the N-terminal dipeptide Tyr^1–Ala^2, several different approaches to rendering GIP resistant to DPP-IV and improving its stability have been taken. Among these are: 1. Chemical modification of the N-terminus, as in *N*-Glucitol-, *N*-acetyl- and *N*-Pyroglutamyl-GIP (Green *et al.*, 2004; O'Harte *et al.*, 1999, 2002). 2. Substitutions at amino acid position 2 with D-Ala (Hinke *et al.*, 2002), Phospho-Ser (Hinke *et al.*, 2004), Gly (Gault *et al.*, 2003a), or at position 3 with Glu or Pro (Gault *et al.*, 2002a). 3. Acetylation with long-chain or short-chain fatty acids (Green *et al.*, 2004; Irwin *et al.*, 2005). 4. Introduction of palmitate residues attached to intrinsic lysine residues at positions 16 or 37 (Irwin *et al.*, 2005). 5. Peptides containing linkers that enhance helix formation in the C-terminal (19–30) portion of GIP (Manhart *et al.*, 2003). Although many modifications have resulted in inactive peptides (Hinke *et al.*, 2003b), DPP-IV resistant forms, including D-Ala^2-GIP (Hinke *et al.*, 2002, 2003b) and *N*-acetyl-GIP (O'Harte *et al.*, 2002) have enhanced biological activity and are effective in improving glucose tolerance in diabetic rodents. Analogs with extended half-lives, resulting from the insertion of albumin-binding palmitate residues may also demonstrate extended actions (Irwin *et al.*, 2005). However, despite their potential, none of these analogs have been shown to improve glucose tolerance in humans.

If a definitive link between elevated GIP responses and the development of obesity and T2DM can be demonstrated then administration of an antagonist may be appropriate. As discussed in Section IIA, GIP_{6-30} and $GIP_{7-30\ amide}$ bind to the receptor with high affinity, but act as antagonists (Gelling *et al.*, 1997a; Tseng *et al.*, 1996), although they have short half-lives *in vivo*. Pro^3-GIP was shown to act as a GIP antagonist both *in vitro* (Gault *et al.*, 2002a) and *in vivo* (Gault *et al.*, 2002a, 2003c). Acute administration of Pro^3-GIP to *ob/ob* mice decreased plasma insulin responses to a glucose load (Gault *et al.*, 2003c) and chronic (11 day) intraperitoneal administration impaired glucose tolerance and insulin sensitivity in normal mice (Irwin *et al.*, 2004). However, similar treatment of *ob/ob* mice improved glucose handling, reduced insulin resistance, and attenuated the development of islet hypertrophy and β-cell hyperplasia (Gault *et al.*, 2005). Administration of the antagonist during development reduced the onset of diabetes in *ob/ob* mice (Irwin *et al.*, 2007). Surprisingly, extending the half-life of Pro^3-GIP by acylation of an internal lysine did not improve its efficacy (Gault *et al.*, 2007). In a recent study, Pro^3-GIP induced weight loss in Swiss mice previously fed a high diet, normalized fasting plasma glucose and reduced circulating triglycerides and hepatic steatosis (McClean *et al.*, 2007). Although these results are intriguing, translation to humans may not be straightforward. The role of GIP in human adipocyte metabolism remains to be defined and there may be important differences between visceral and subcutaneous fat (Rudovich *et al.*, 2007). Additionally, inhibition of the GIP receptor in all body sites could have significant deleterious effects on critical metabolic functions, including bone and the CNS.

ACKNOWLEDGMENTS

Studies from the authors' laboratory referred to in the review were generously supported by funding from the Canadian Institutes of Health Research, Canadian Diabetes Association, and the Canadian Foundation for Innovation.

REFERENCES

Adrian, T. E., Bloom, S. R., Hermansen, K., and Iversen, J. (1978). Pancreatic polypeptide, glucagon and insulin secretion from the isolated perfused canine pancreas. *Diabetologia* **14,** 413–417.

Ahren, B., and Holst, J. J. (2001). The cephalic insulin response to meal ingestion in humans is dependent on both cholinergic and noncholinergic mechanisms and is important for postprandial glycemia. *Diabetes* **50,** 1030–1038.

Ahren, B., Håkanson, R., Lundquist, I., Sjölund, K., and Sundler, F. (1981). GIP-like immunoreactivity in glucagon cells. Interactions between GIP and glucagon on insulin release. *Acta Physiol. Scand.* **112,** 233–242.

Alam, M. J., and Buchanan, K. D. (1993). Conflicting gastric inhibitory polypeptide data: Possible causes. *Diab. Res. Clin. Pract.* **19,** 93–101.

Alana, I., Malthouse, J. P. G., O'Harte, F. P. M., and Hewage, C. M. (2007). The bioactive conformation of glucose-dependent insulinotropic polypeptide by NMR and CD spectroscopy. *Proteins* **68,** 92–99.

Albinus, M., Blair, E. L., Grund, E. R., Reed, J. D., Sanders, D. J., Gomez-Pan, A., Schally, A. V., and Besser, G. M. (1975). The mechanism whereby growth hormone-release inhibiting hormone (somatostatin) inhibits food stimulated gastric acid secretion in the cat. *Agents Actions* **5,** 306–310.

Almind, K., Ambye, L., Urhammer, S. A., Hansen, T., Echwald, S. M., Holst, J. J., Gromada, J., Thorens, B., and Pedersen, O. (1998). Discovery of amino acid variants in the human glucose-dependent insulinotropic polypeptide (GIP) receptor: The impact on the pancreatic beta cell responses and functional expression studies in Chinese hamster fibroblast cells. *Diabetologia* **41,** 1194–1198.

Alumets, J., Håkanson, R., O'Dorisio, T., Sjölund, K., and Sundler, F. (1978). Is GIP a glucagon cell constituent? *Histochemistry* **58,** 253–257.

Amiranoff, B., Vauclin-Jacques, N., and Laburthe, M. (1984). Functional GIP receptors in a hamster pancreatic beta cell line, In 111: Specific binding and biological effects. *Biochem. Biophys. Res. Commun.* **123,** 671–676.

Amiranoff, B., Vauclin-Jacques, N., and Laburthe, M. (1985). Interaction of gastric inhibitory polypeptide (GIP) with the insulin-secreting pancreatic beta cell line, In lll: Characteristics of GIP binding sites. *Life Sci.* **36,** 807–813.

Amiranoff, B., Couvineau, A., Vauclin-Jacques, N., and Laburthe, M. (1986). Gastric inhibitory polypeptide receptor in hamster pancreatic beta cells. Direct cross-linking, solubilization and characterization as a glycoprotein. *Eur. J. Biochem.* **159,** 353–358.

Amland, P. F., Jorde, R., Revhaug, A., Myhre, E. S., Burhol, P. G., and Giercksky, K. E. (1984). Fasting and postprandial GIP values in pigs, rats, dogs, and man measured with five different GIP. antisera. *Scand. J. Gastroenterol.* **19,** 1095–1098.

Amland, P. F., Jorde, R., Aanderud, S., Burhol, P. G., and Giercksky, K. E. (1985). Effects of intravenously infused porcine GIP on serum insulin, plasma C-peptide, and pancreatic polypeptide in non-insulin-dependent diabetes in the fasting state. *Scand. J. Gastroenterol.* **20,** 315–320.

Andersen, D. K., Elahi, D., Brown, J. C., Tobin, J. D., and Andres, R. (1978). Oral glucose augmentation of insulin secretion. Interactions of gastric inhibitory polypeptide with ambient glucose and insulin levels. *J. Clin. Invest.* **62,** 152–161.

Anderson, L. A., Friedman, L., Osborne-Lawrence, S., Lynch, E., Weissenbach, J., Bowcock, A., and King, M. C. (1993). High-density genetic map of the BRCA1 region of chromosome 17q12-q21. *Genomics* **17,** 618–623.

Anlauf, M., Weihe, E., Hartschuh, W., Hamscher, G., and Feurle, G. E. (2000). Localization of xenin-immunoreactive cells in the duodenal mucosa of humans and various mammals. *J. Histochem. Cytochem.* **48,** 1617–1626.

Arimura, A., Sato, H., Dupont, A., Nishi, N., and Schally, A. V. (1975). Somatostatin: Abundance of immunoreactive hormone in rat stomach and pancreas. *Science* **189,** 1007–1009.

Ashcroft, F., and Rorsman, P. (2004). Molecular defects in insulin secretion in type-2 diabetes. *Rev. Endocr. Metabol. Dis.* **4,** 135–142.

Baggio, L. L., and Drucker, D. J. (2006). Therapeutic approaches to preserve islet mass in type 2 diabetes. *Ann. Rev. Med.* **57,** 265–281.

Baggio, L. L., and Drucker, D. J. (2007). Biology of incretins: GLP-1 and GIP. *Gastroenterology* **132,** 2131–2157.

Baggio, L., Kieffer, T. J., and Drucker, D. J. (2000). Glucagon-like peptide-1, but not glucose-dependent insulinotropic peptide, regulates fasting glycemia and nonenteral glucose clearance in mice. *Endocrinology* **141,** 3703–3709.

Bailey, C. J., Flatt, P. R., Kwasowski, P., Powell, C. J., and Marks, V. (1986). Immunoreactive gastric inhibitory polypeptide and K cell hyperplasia in obese hyperglycaemic (ob/ob) mice fed high fat and high carbohydrate cafeteria diets. *Acta Endocrinol.* **112,** 224–229.

Baldacchino, V., Oble, S., Hamet, P., Tremblay, J., Bourdeau, I., and Lacroix, A. (2004). The Sp transcription factor family is involved in the cellular expression of the human GIP-R. gene promoter. *Endocr. Res.* **30,** 805–806.

Baldacchino, V., Oble, S., Decarie, P.-O., Bourdeau, I., Hamet, P., Tremblay, J., and Lacroix, A. (2005). The Sp transcription factors are involved in the cellular expression of the human glucose-dependent insulinotropic polypeptide receptor gene and over-expressed in adrenals of patients with Cushing's syndrome. *J. Mol. Endocrinol.* **35,** 61–71.

Bao, S., Bohrer, A., Ramanadham, S., and Jin, W. S. Z. J. T. (2006a). Effects of stable suppression of Group VIA phospholipase A2 expression on phospholipid content and composition, insulin secretion, and proliferation of INS-1 insulinoma cells. *J. Biol. Chem.* **281,** 187–198.

Bao, S., Song, H., Wohltmann, M., Ramanadham, S., Wu, J., Bohrer, A., and Turk, J. (2006b). Insulin secretory responses and phospholipid composition of pancreatic islets from mice that do not express group VIA phospholipase A2 and the effect of metabolic stress on glucose homeostasis. *J. Biol. Chem.* **281,** 20958–20973.

Beck, B. (1989). Gastric inhibitory polypeptide: A gut hormone with anabolic functions. *J. Mol. Endocrinol.* **2,** 169–174.

Beck, B., and Max, J. P. (1986). Direct metabolic effects of gastric inhibitory polypeptide (GIP): Dissociation at physiological levels of effects on insulin-stimulated fatty acid and glucose incorporation in rat adipose tissue. *Diabetologia* **29,** 68.

Berghöfer, P., Peterson, R. G., Schneider, K., Fehmann, H. C., and Göke, B. (1997). Incretin hormone expression in the gut of diabetic mice and rats. *Metabolism: Clin. Exp.* **46,** 261–267.

Besterman, H. S., Cook, G. C., Sarson, D. L., Christofides, N. D., Bryant, M. G., Gregor, M., and Bloom, S. R. (1979). Gut hormones in tropical malabsorption. *Brit. Med. J.* **2,** 1252–1255.

Bloom, S., and Polak, J. (1978). Gut hormone overview. *In* "Gut Hormones" (S. R. Bloom, Ed.), pp. 3–18. Churchill Livingstone, Edinburgh.

Bollag, R. J., Zhong, Q., Phillips, P., Min, L., Zhong, L., Cameron, R., Mulloy, A. L., Rasmussen, H., Qin, F., Ding, K. H., and Isales, C. M. (2000). Osteoblast-derived cells express functional glucose-dependent insulinotropic peptide receptors. *Endocrinology* **141,** 1228–1235.

Bollag, R. J., Zhong, Q., Ding, K. H., Phillips, P., Zhong, L., Qin, F., Cranford, J., Mulloy, A. L., Cameron, R., and Isales, C. M. (2001). Glucose-dependent insulinotropic peptide is an integrative hormone with osteotropic effects. *Mol. Cell. Endocrinol.* **177,** 35–41.

Bourdeau, I., Antonini, S. R., Lacroix, A., Kirschner, L. S., Matyakhina, L., Lorang, D., Libutti, S. K., and Stratakis, C. A. (2004). Gene array analysis of macronodular adrenal hyperplasia confirms clinical heterogeneity and identifies several candidate genes as molecular mediators. *Oncogene* **23,** 1575–1585.

Boyer, D. F., Fujitani, Y., Gannon, M., Powers, A. C., Stein, R. W., and Wright, C. V. E. (2006). Complementation rescue of Pdx1 null phenotype demonstrates distinct roles of proximal and distal cis-regulatory sequences in pancreatic and duodenal expression. *Dev. Biol.* **298,** 616–631.

Boylan, M. O., Jepeal, L. I., Jarboe, L. A., and Wolfe, M. M. (1997). Cell-specific expression of the glucose-dependent insulinotropic polypeptide gene in a mouse neuroendocrine tumor cell line. *J. Biol. Chem.* **272,** 17438–17443.

Boylan, M. O., Jepeal, L. I., and Wolfe, M. M. (1999). Structure of the rat glucose-dependent insulinotropic polypeptide receptor gene. *Peptides* **20,** 219–228.

Brown, J. C. (1971). A gastric inhibitory polypeptide. I. The amino acid composition and the tryptic peptides. *Can. J. Biochem.* **49,** 255–261.

Brown, J. C. (1982). "Gastric Inhibitory Polypeptide. Monographs on Endocrinology," Vol. 24 Springer-Verlag, New York.

Brown, J. C. (1987). Role of gastric inhibitory polypeptide in regulation of insulin release. *Front. Horm. Res.* **16,** 157–166.

Brown, J. C., and Dryburgh, J. R. (1971). A gastric inhibitory polypeptide. *II. The complete amino acid sequence. Can. J. Biochem.* **49,** 867–872.

Brown, J. C., and Otte, S. C. (1978). Gastrointestinal hormones and the control of insulin secretion. *Diabetes* **27,** 782–787.

Brown, J. C., and Pederson, R. A. (1970). A multiparameter study on the action of preparations containing cholecystokinin-pancreozymin. *Scand. J. Gastroenterol* **5,** 537–541.

Brown, J. C., and Pederson, R. A. (1976). GI hormones and insulin secretion. *Endocrinol.: Proc. 5th. Int. Congr. Endocrinol.* **2,** 568–570.

Brown, J. C., Pederson, R. A., Jorpes, E., and Mutt, V. (1969). Preparation of highly active enterogastrone. *Can. J. Physiol. Pharmacol.* **47,** 113–114.

Brown, J. C., Mutt, V., and Pederson, R. A. (1970). Further purification of a polypeptide demonstrating enterogastrone activity. *J. Physiol.* **209,** 57–64.

Brown, J. C., Dahl, M., Kwauk, S., McIntosh, C. H., Otte, S. C., and Pederson, R. A. (1981). Actions of GIP. *Peptides* **2,** 241–245.

Brown, J. C., Buchan, A. M. J., McIntosh, C. H. S., and Pederson, R. A. (1989). Gastric inhibitory polypeptide. *In* "Handbook of Physiology, Section 6: The Gastrointestinal System" (S. G. Schultz, G. M. Makhlouf, and B. B. Rauner, Eds.), pp. 403–430. American. Physiological Society, Bethesda.

Brunicardi, F. C., Druck, P., Seymour, N. E., Sun, Y. S., Elahi, D., and Andersen, D. K. (1990). Selective neurohormonal interactions in islet cell secretion in the isolated perfused human pancreas. *J. Surg. Res.* **48,** 273–278.

Bryer-Ash, M., Cheung, A., and Pederson, R. A. (1994). Feedback regulation of glucose-dependent insulinotropic polypeptide (GIP) secretion by insulin in conscious rats. *Regul. Pept.* **51,** 101–109.

Buchan, A. M., Polak, J. M., Capella, C., Solcia, E., and Pearse, A. G. (1978). Electronimmunocytochemical evidence for the K cell localization of gastric inhibitory polypeptide (GIP) in man. *Histochemistry* **56,** 37–44.

Buchan, A. M., Ingman-Baker, J., Levy, J., and Brown, J. C. (1982a). A comparison of the ability of serum and monoclonal antibodies to gastric inhibitory polypeptide to detect immunoreactive cells in the gastroenteropancreatic system of mammals and reptiles. *Histochemistry* **76,** 341–349.

Buchan, A. M., Lance, V., and Polak, J. M. (1982b). The endocrine pancreas of Alligator mississippiensis. An. immunocytochemical investigation. *Cell. Tissue Res.* **224,** 117–128.

Buffa, R., Polak, J. M., Pearse, A. G., Solcia, E., Grimelius, L., and Capella, C. (1975). Identification of the intestinal cell storing gastric inhibitory peptide. *Histochemistry* **43,** 249–255.

Butler, A. E., Janson, J., Bonner-Weir, S., Ritzel, R., Rizza, R. A., and Butler, P. C. (2003). Beta-cell deficit and increased beta-cell apoptosis in humans with type 2 diabetes. *Diabetes* **52,** 102–110.

Buteau, J., El-Assaad, W., Rhodes, C. J., Rosenberg, L., Joly, E., and Prentki, M. (2004). Glucagon-like peptide-1 prevents beta cell glucolipotoxicity. *Diabetologia* **47,** 806–815.

Carlquist, M., Maletti, M., Jornvall, H., and Mutt, V. (1984). A novel form of gastric inhibitory polypeptide (GIP) isolated from bovine intestine using a radioreceptor assay. Fragmentation with staphylococcal protease results in GIP1-3 and GIP4-42, fragmentation with enterokinase in GIP1-16 and GIP17-42. *Eur. J. Biochem.* **145,** 573–577.

Castresana, M., Lee, K. Y., Chey, W. Y., and Yajima, H. (1978). Effects of motilin and octapeptide of cholecystokinin on antral and duodenal myoelectric activity in the interdigestive state and during inhibition by secretin and gastric inhibitory polypeptide. *Digestion* **17,** 300–308.

Cataland, S., O'Dorisio, T. M., Crockett, S. E., and Mekhjian, H. S. (1978). Gastric inhibitory polypeptide (GIP) and insulin release in duodenal ulcer patients. *J. Clin. Endocrinol. Metab.* **47,** 615–619.

Chabre, O., Liakos, P., Vivier, J., Bottari, S., Bachelot, I., Chambaz, E. M., Feige, J. J., and Defaye, G. (1998). Gastric inhibitory polypeptide (GIP) stimulates cortisol secretion, cAMP production and DNA synthesis in an adrenal adenoma responsible for food-dependent Cushing's syndrome. *Endocr. Res.* **24,** 851–856.

Chaikomin, R., Doran, S., Jones, K. L., Feinle-Bisset, C., O'Donovan, D., Rayner, C. K., and Horowitz, M. (2005). Initially more rapid small intestinal glucose delivery increases plasma insulin, GIP, and GLP-1 but does not improve overall glycemia in healthy subjects. *Am. J. Physiol. Endocrinol. Metab.* **289,** E504–E507.

Chaikomin, R., Wu, K. L., Doran, S., Jones, K. L., Smout, A. J. P. M., Renooij, W., Holloway, R. H., Meyer, J. H., Horowitz, M., and Rayner, C. K. (2007). Concurrent duodenal manometric and impedance recording to evaluate the effects of hyoscine on motility and flow events, glucose absorption, and incretin release. *Am. J. Physiol. Gastrointest. Liver Physiol.* **292,** G1099–G1104.

Chan, C. B., Pederson, R. A., Buchan, A. M., Tubesing, K. B., and Brown, J. C. (1985). Gastric inhibitory polypeptide and hyperinsulinemia in the Zucker (fa/fa) rat: A developmental study. *Int. J. Obes.* **9,** 137–146.

Chao, C., and Hellmich, M. R. (2004). Bi-directional signaling between gastrointestinal peptide hormone receptors and epidermal growth factor receptor. *Growth Factors* **22,** 261–268.

Chap, Z., O'Dorisio, T. M., Cataland, S., and Field, J. B. (1987). Absence of hepatic extraction of gastric inhibitory polypeptide in conscious dogs. *Dig. Dis. Sci.* **32,** 280–284.

Cheeseman, C. (1997). Upregulation of SGLT-1 transport activity in rat jejunum induced by GLP-1 infusion *in vivo*. *Am. J. Physiol. Regul. Integr. Comp. Physiol.* **273,** R1965–R1971.

Cheeseman, C. I., and O'Neill, D. (1998). Basolateral D-glucose transport activity along the crypt-villus axis in rat jejunum and upregulation induced by gastric inhibitory peptide and glucagon-like peptide-2. *Exp. Physiol.* **83,** 605–616.

Cheeseman, C. I., and Tsang, R. (1996). The effect of GIP. and glucagon-like peptides on intestinal basolateral membrane hexose transport. *Am. J. Physiol.* **271,** G477–G482.

Cheung, A. T., Dayanandan, B., Lewis, J. T., Korbutt, G. S., Rajotte, R. V., Bryer-Ash, M., Boylan, M. O., Wolfe, M. M., and Kieffer, T. J. (2000). Glucose-dependent insulin release from genetically engineered K cells. *Science* **290,** 1959–1962.

Chey, W. Y., and Chang, T.-M. (2003). Secretin, 100 years later. *J. Gastroenterol* **38,** 1025–1035.

Chu, Z.-L., Carroll, C., Alfonso, J., Gutierrez, V., He, H., Lucman, A., Pedraza, M., Mondala, H., Gao, H., Bagnol, D., Chen, R., Jones, R. M., *et al.* (2008). A Role for intestinal endocrine cell-expressed GPR119 in glycemic control by enhancing GLP-1 and GIP release. *Endocrinology* **149,** 2038–2047. doi:10.1210/en.2007-0966.

Cleator, I. G., and Gourlay, R. H. (1975). Release of immunoreactive gastric inhibitory polypeptide (IR-GIP) by oral ingestion of food substances. *Am. J. Surg.* **130,** 128–135.

Cline, M. A., Nandar, W., and Rogers, J. O. (2007). Xenin reduces feed intake by activating the ventromedial hypothalamus and influences gastrointestinal transit rate in chicks. *Behav. Brain Res.* **179,** 28–32.

Clowes, J. A., Khosla, S., and Eastell, R. (2005). Potential role of pancreatic and enteric hormones in regulating bone turnover. *J. Bone Miner. Res.* **20,** 1497–1506.

Couvineau, A., Amiranoff, B., Vauclin-Jacques, N., and Laburthe, M. (1984). The GIP. receptor on pancreatic beta cell tumor: Molecular identification by covalent cross-linking. *Biochem. Biophys. Res. Commun.* **122,** 283–288.

Creutzfeldt, W. (1979). The incretin concept today. *Diabetologia* **16,** 75–85.

Creutzfeldt, W. (2005). The [pre-] history of the incretin concept. *Regul. Pept.* **128,** 87–91.

Creutzfeldt, W., and Ebert, R. (1977). Release of gastric inhibitory polypeptide (GIP. to a test meal under normal and pathological conditions in man. *In* "Diabetes" (J. S. Bajaj, Ed.), pp. 64–75. Excerpta Medica, Amsterdam.

Creutzfeldt, W., and Ebert, R. (1985). New developments in the incretin concept. *Diabetologia* **28,** 565–573.

Creutzfeldt, W., Ebert, R., Willms, B., Frerichs, H., and Brown, J. C. (1978). Gastric inhibitory polypeptide (GIP) and insulin in obesity: Increased response to stimulation and defective feedback control of serum levels. *Diabetologia* **14,** 15–24.

Creutzfeldt, W., Talaulicar, M., Ebert, R., and Willms, B. (1980). Inhibition of gastric inhibitory polypeptide (GIP) release by insulin and glucose in juvenile diabetes. *Diabetes* **29,** 140–145.

Crockett, S. E., Mazzaferri, E. L., and Cataland, S. (1976). Gastric inhibitory polypeptide (GIP) in maturity-onset diabetes mellitus. *Diabetes* **25,** 931–935.

Damholt, A. B., Buchan, A. M., and Kofod, H. (1998). Glucagon-like-peptide-1 secretion from canine L-cells is increased by glucose-dependent-insulinotropic peptide but unaffected by glucose. *Endocrinology* **139,** 2085–2091.

Deacon, C. F. (2004). Circulation and degradation of GIP and GLP-1. *Horm. Metab, Res.* **36,** 761–765.

Deacon, C. F. (2005). What do we know about the secretion and degradation of incretin hormones? *Regul. Pept.* **128,** 117–124.

Deacon, C. F. (2007). Dipeptidyl peptidase 4 inhibition with sitagliptin: A new therapy for type 2 diabetes. *Expert Opin. Inv. Drugs* **16,** 533–545.

Deacon, C. F., Nauck, M. A., Toft-Nielsen, M., Pridal, L., Willms, B., and Holst, J. J. (1995). Both subcutaneously and intravenously administered glucagon-like peptide I are rapidly degraded from. the NH2-terminus in type II diabetic patients and in healthy subjects. *Diabetes* **44,** 1126–1131.

Deacon, C. F., Nauck, M. A., Meier, J., Hucking, K., and Holst, J. J. (2000). Degradation of endogenous and exogenous gastric inhibitory polypeptide in healthy and in type 2 diabetic subjects as revealed using a new assay for the intact peptide. *J. Clin. Endocrinol. Metab.* **85,** 3575–3581.

Deacon, C. F., Danielsen, P., Klarskov, L., Olesen, M., and Holst, J. J. (2001). Dipeptidyl peptidase IV inhibition reduces the degradation and clearance of GIP and potentiates its insulinotropic and antihyperglycemic effects in anesthetized pigs. *Diabetes* **50,** 1588–1597.

Deacon, C. F., Wamberg, S., Bie, P., Hughes, T. E., and Holst, J. J. (2002). Preservation of active incretin hormones by inhibition of dipeptidyl peptidase IV suppresses meal-induced incretin secretion in dogs. *J. Endocrinol.* **172,** 355–362.

Deacon, C. F., Plamboeck, A., Rosenkilde, M. M., de Heer, J., and Holst, J. J. (2006). GIP-(3-42) does not antagonize insulinotropic effects of GIP at physiological concentrations. *Am. J. Physiol. Endocrinol. Metab.* **291,** E468–E475.

de Herder, W. W., Hofland, L. J., Usdin, T. B., de Jong, F. H., Uitterlinden, P., van Koetsveld, P., Mezey, E., Bonner, T. I., Bonjer, H. J., and Lamberts, S. W. (1996). Food-dependent Cushing's syndrome resulting from abundant expression of gastric inhibitory polypeptide receptors in adrenal adenoma cells. *J. Clin. Endocrinol. Metab.* **81,** 3168–3172.

Delmeire, D., Flamez, D., Hinke, S. A., Cali, J. J., Pipeleers, D., and Schuit, F. (2003). Type VIII adenylyl cyclase in rat beta cells: Coincidence signal detector/generator for glucose and GLP-1. *Diabetologia* **46,** 1383–1393.

Delmeire, D., Flamez, D., Moens, K., Hinke, S. A., Van Schravendijk, C., Pipeleers, D., and Schuit, F. (2004). Prior *in vitro* exposure to GLP-1 with or without GIP can influence the subsequent beta cell responsiveness. *Biochem. Pharmacol.* **68,** 33–39.

Denniss, A. R., and Young, J. A. (1978). Modification of salivary duct electrolyte transport in rat and rabbit by physalaemin, VIP, GIP and other enterohormones. *Pflügers Arch. Eur. J. Physiol.* **376,** 73–80.

Ding, W. G., and Gromada, J. (1997). Protein kinase A-dependent stimulation of exocytosis in mouse pancreatic beta-cells by glucose-dependent insulinotropic polypeptide. *Diabetes* **46,** 615–621.

Ding, W. G., Renstrom, E., Rorsman, P., Buschard, K., and Gromada, J. (1997). Glucagon-like peptide I and glucose-dependent insulinotropic polypeptide stimulate $Ca2^{+}$-induced secretion in rat alpha-cells by a protein kinase A-mediated mechanism. *Diabetes* **46,** 792–800.

Ding, K.-H., Zhong, Q., and Isales, C. M. (2003). Glucose-dependent insulinotropic peptide stimulates thymidine incorporation in endothelial cells: Role of endothelin-1. *Am. J. Physiol. Endocrinol. Metab.* **285,** E390–E396.

Ding, K.-H., Zhong, Q., Xu, J., and Isales, C. M. (2004). Glucose-dependent insulinotropic peptide: Differential effects on hepatic artery vs. portal vein endothelial cells. *Am. J. Physiol. Endocrinol. Metab.* **286,** E773–E779.

Dobbins, R. L., Chester, M. W., Stevenson, B. E., Daniels, M. B., Stein, D. T., and McGarry, J. D. (1998). A fatty acid-dependent step is critically important for both glucose- and non-glucose-stimulated insulin secretion. *J. Clin. Invest.* **101,** 2370–2376.

Doyle, M. E., and Egan, J. M. (2007). Mechanisms of action of glucagon-like peptide 1 in the pancreas. *Pharmacol. Ther.* **113,** 546–593.

Drucker, D. J. (2006). The biology of incretin hormones. *Cell. Metab.* **3,** 153–165.

Drucker, D. J. (2007). The role of gut hormones in glucose homeostasis. *J. Clin. Invest.* **117,** 24–32.

Drucker, D. J., and Nauck, M. A. (2006). The incretin system: Glucagon-like peptide-1 receptor agonists and dipeptidyl peptidase-4 inhibitors in type 2 diabetes. *Lancet* **368,** 1696–1705.

Drucker, D., Philippe, J., Mojsov, S., Chick, W., and Habener, J. (1987). Glucagon-like I stimulates insulin gene expression and increases cyclic AMP levels in a rat islet cell line. *Proc. Natl. Acad. Sci. USA* **84,** 3434–3438.

Dryburgh, J. R., Hampton, S. M., and Marks, V. (1980). Endocrine pancreatic control of the release of gastric inhibitory polypeptide. A possible physiological role for C-peptide. *Diabetologia* **19,** 397–401.

Duan, R. D., and Erlanson-Albertsson, C. (1992). Gastric inhibitory polypeptide stimulates pancreatic lipase and colipase synthesis in rats. *Am. J. Physiol.* **262,** G779–G784.

Dupré, J., and Beck, J. C. (1966). Stimulation of release of insulin by an extract of intestinal mucosa. *Diabetes* **15,** 555–559.

Dupré, J., Ross, S. A., Watson, D., and Brown, J. C. (1973). Stimulation of insulin secretion by gastric inhibitory polypeptide in man. *J. Clin. Endocrinol. Metab.* **37,** 826–828.

Dupré, J., Caussignac, Y., McDonald, T. J., and Van Vliet, S. (1991). Stimulation of glucagon secretion by gastric inhibitory polypeptide in patients with hepatic cirrhosis and hyperglucagonemia. *J. Clin. Endocrinol. Metab.* **72,** 125–129.

Ebert, R., and Creutzfeldt, W. (1980). Reversal of impaired GIP and insulin secretion in patients with pancreatogenic steatorrhea following enzyme substitution. *Diabetologia* **19,** 198–204.

Ebert, R., and Creutzfeldt, W. (1982). Influence of gastric inhibitory polypeptide antiserum on glucose-induced insulin secretion in rats. *Endocrinology* **111,** 1601–1606.

Ebert, R., and Creutzfeldt, W. (1987). Metabolic effects of Gastric Inhibitory Polypeptide. *Front. Horm. Res.* **16,** 175–185.

Ebert, R., and Creutzfeldt, W. (1989). Gastric inhibitory polypeptide (GIP) hypersecretion in obesity depends on meal size and is not related to hyperinsulinemia. *Acta Diabetol.* **26,** 1–15.

Ebert, R., Frerichs, H., and Creutzfeldt, W. (1979a). Impaired feedback control of fat induced gastric inhibitory polypeptide (GIP) secretion by insulin in obesity and glucose intolerance. *Eur. J. Clin. Invest.* **9,** 129–135.

Ebert, R., Illmer, K., and Creutzfeldt, W. (1979b). Release of gastric inhibitory polypeptide (GIP) by intraduodenal acidification in rats and humans and abolishment of the incretin effect of acid by GIP-antiserum in rats. *Gastroenterology* **76,** 515–523.

Ebert, R., Unger, H., and Creutzfeldt, W. (1983). Preservation of incretin activity after removal of gastric inhibitory polypeptide (GIP) from rat gut extracts by immunoadsorption. *Diabetologia* **24,** 449–454.

Ebert, R., Nauck, M., and Creutzfeldt, W. (1991). Effect of exogenous or endogenous gastric inhibitory polypeptide (GIP) on plasma triglyceride responses in rats. *Horm. Metabol. Res.* **23,** 517–521.

Eckel, R. H., Fujimoto, W. Y., and Brunzell, J. D. (1979). Gastric inhibitory polypeptide enhanced lipoprotein lipase activity in cultured preadipocytes. *Diabetes* **28,** 1141–1142.

Ehses, J. A., Lee, S. S., Pederson, R. A., and McIntosh, C. H. S. (2001). A new pathway for glucose-dependent insulinotropic polypeptide (GIP) receptor signaling: Evidence for the involvement of phospholipase A2 in GIP-stimulated insulin secretion. *J. Biol. Chem.* **276,** 23667–23673.

Ehses, J. A., Pelech, S. L., Pederson, R. A., and McIntosh, C. H. S. (2002). Glucose-dependent insulinotropic polypeptide activates the Raf-Mek1/2-ERK1/2 module via a cyclic AMP/cAMP-dependent protein kinase/Rap1-mediated pathway. *J. Biol. Chem.* **277,** 37088–37097.

Ehses, J. A., Casilla, V. R., Doty, T., Pospisilik, J. A., Winter, K. D., Demuth, H. U., Pederson, R. A., and McIntosh, C. H. S. (2003). Glucose-dependent insulinotropic polypeptide promotes beta-(INS-1) cell survival via cyclic adenosine monophosphate-mediated caspase-3 inhibition and regulation of p38 mitogen-activated protein kinase. *Endocrinology* **144,** 4433–4445.

Eissele, R., Göke, R., Willemer, S., Harthus, H. P., Vermeer, H., Arnold, R., and Göke, B. (1992). Glucagon-like peptide-1 cells in the gastrointestinal tract and pancreas of rat, pig and man. *Eur. J. Clin. Invest.* **22,** 283–291.

Eizirik, D. L., Cardozo, A. K., and Cnop, M. (2008). The role for endoplasmic reticulum stress in diabetes mellitus. *Endocr. Rev* **29,** 42–61.

Elahi, D., Andersen, D. K., Brown, J. C., Debas, H. T., Hershcopf, R. J., Raizes, G. S., Tobin, J. D., and Andres, R. (1979). Pancreatic alpha- and beta-cell responses to GIP infusion in normal man. *Am. J. Physiol.* **237,** E185–E191.

Elahi, D., Andersen, D. K., Muller, D. C., Tobin, J. D., Brown, J. C., and Andres, R. (1984). The enteric enhancement of glucose-stimulated insulin release. The role of GIP in aging, obesity, and non-insulin-dependent diabetes mellitus. *Diabetes* **33,** 950–957.

Elahi, D., McAloon-Dyke, M., Fukagawa, N. K., Meneilly, G. S., Sclater, A. L., Minaker, K. L., Habener, J. F., and Andersen, D. K. (1994). The insulinotropic actions of glucose-dependent insulinotropic polypeptide (GIP) and glucagon-like peptide-1 (7–37) in normal and diabetic subjects. *Regul. Pept.* **51,** 63–74.

El-Ouaghlidi, A., Rehring, E., Holst, J. J., Schweizer, A., Foley, J., Holmes, D., and Nauck, M. A. (2007). The dipeptidyl peptidase 4 inhibitor vildagliptin does not accentuate glibenclamide-induced hypoglycemia but reduces glucose-induced glucagon-like peptide 1 and gastric inhibitory polypeptide secretion. *J. Clin. Endocrinol. Metab.* **92,** 4165–4171.

Elrick, H., Stimmler, L., Hlad, C. J., and Arai, Y. (1964). Plasma insulin response to oral and intravenous glucose administration. *J. Clin. Endocrinol. Metab.* **24,** 1076–1082.

Emami, S., Chastre, E., Bodere, H., Gespach, C., Bataille, D., and Rosselin, G. (1986). Functional receptors for VIP, GIP, glucagon-29 and -37 in the HGT-1 human gastric cancer cell line. *Peptides* **7,** 121–127.

Falko, J. M., Crockett, S. E., Cataland, S., and Mazzaferri, E. L. (1975). Gastric inhibitory polypeptide (GIP) stimulated by fat ingestion in man. *J. Clin. Endocrinol. Metab.* **41,** 260–265.

Fara, J. W., and Salazar, A. M. (1978). Gastric inhibitory polypeptide increases mesenteric blood flow. *Proc. Soc. Exp. Biol. Med.* **158,** 446–448.

Farilla, L., Hui, H., Bertolotto, C., Kang, E., Bulotta, A., Di Mario, U., and Perfetti, R. (2002). Glucagon-like peptide-1 promotes islet cell growth and inhibits apoptosis in Zucker diabetic rats. *Endocrinology* **143,** 4397–4408.

Farilla, L., Bulotta, A., Hirshberg, B., Li Calzi, S., Khoury, N., Noushmehr, H., Bertolotto, C., Di Mario, U., Harlan, D. M., and Perfetti, R. (2003). Glucagon-like peptide 1 inhibits cell apoptosis and improves glucose responsiveness of freshly isolated human islets. *Endocrinology* **144,** 5149–5158.

Fehmann, H. C., and Göke, B. (1995). Characterization of GIP(1-30) and GIP(1-42) as stimulators of proinsulin gene transcription. *Peptides* **16,** 1149–1152.

Fehmann, H. C., and Habener, J. F. (1991). Homologous desensitization of the insulinotropic glucagon-like peptide-I (7-37) receptor on insulinoma (HIT-T15) cells. *Endocrinology* **128,** 2880–2888.

Fehmann, H. C., Göke, R., and Göke, B. (1995). Cell and molecular biology of the incretin hormones glucagon-like peptide-I and glucose-dependent insulin releasing polypeptide. *Endocr. Rev* **16,** 390–410.

Feurle, G. E. (1998). Xenin-A review. *Peptides* **19,** 609–615.

Feurle, G. E., Ikonomu, S., Partoulas, G., Stoschus, B., and Hamscher, G. (2003). Xenin plasma concentrations during modified sham. feeding and during meals of different composition demonstrated by radioimmunoassay and chromatography. *Regul. Pept.* **111,** 153–159.

Flaten, O. (1981). The effect of adrenoceptor blockade on the release of gastric inhibitory polypeptide after intraduodenal glucose in humans. *Scand. J. Gastroenterol* **16,** 641–645.
Flatt, P. R., Bailey, C. J., Kwasowski, P., Swanston-Flatt, S. K., and Marks, V. (1983). Abnormalities of GIP in spontaneous syndromes of obesity and diabetes in mice. *Diabetes* **32,** 433–435.
Flatt, P. R., Bailey, C. J., Kwasowski, P., Page, T., and Marks, V. (1984). Plasma immunoreactive gastric inhibitory polypeptide in obese hyperglycaemic (ob/ob) mice. *J. Endocrinol.* **101,** 249–256.
Fraser, I. D., Tavalin, S. J., Lester, L. B., Langeberg, L. K., Westphal, A. M., Dean, R. A., Marrion, N. V., and Scott, J. D. (1998). A novel lipid-anchored A-kinase anchoring protein facilitates cAMP-responsive membrane events. *EMBO J.* **17,** 2261–2272.
Fredriksson, R., and Schiöth, H. B. (2005). The repertoire of G-protein-coupled receptors in fully sequenced genomes. *Mol. Pharmacol.* **67,** 1414–1425.
Fridolf, T., Bottcher, G., Sundler, F., and Ahren, B. (1991). GLP-1 and GLP-1(7-36) amide: Influences on basal and stimulated insulin and glucagon secretion in the mouse. *Pancreas* **6,** 208–215.
Friedrichsen, B. N., Neubauer, N., Lee, Y. C., Gram, V. K., Blume, N., Petersen, J. S., Nielsen, J. H., and Moldrup, A. (2006). Stimulation of pancreatic beta-cell replication by incretins involves transcriptional induction of cyclin D1 via multiple signalling pathways. *J. Endocrinol.* **188,** 481–492.
Fujimoto, W. Y., Ensinck, J. W., Merchant, F. W., Williams, R. H., Smith, P. H., and Johnson, D. G. (1978). Stimulation by gastric inhibitory polypeptide of insulin and glucagon secretion by rat islet cultures. *Proc. Soc. Exp. Biol. Med.* **157,** 89–93.
Fukase, N., Igarashi, M., Takahashi, H., Manaka, H., Yamatani, K., Daimon, M., Tominaga, M., and Sasaki, H. (1993). Hypersecretion of truncated glucagon-like peptide-1 and gastric inhibitory polypeptide in obese patients. *Diabet. Med.* **10,** 44–49.
Fung, L. C., Chisholm, C., and Greenberg, G. R. (1998). Glucagon-like peptide-1-(7-36) amide and peptide YY mediate intraduodenal fat-induced inhibition of acid secretion in dogs. *Endocrinology* **139,** 189–194.
Furness, J. (2006). "The Enteric Nervous System." Blackwell Publishing, Oxford.
Fushiki, T., Kojima, A., Imoto, T., Inoue, K., and Sugimoto, E. (1992). An. extract of *Gymnema sylvestre* leaves and purified gymnemic acid inhibits glucose-stimulated gastric inhibitory peptide secretion in rats. *J. Nutr.* **122,** 2367–2373.
Gault, V. A., O'Harte, F. P., Harriott, P., and Flatt, P. R. (2002a). Characterization of the cellular and metabolic effects of a novel enzyme-resistant antagonist of glucose-dependent insulinotropic polypeptide. *Biochem. Biophys. Res. Commun.* **290,** 1420–1426.
Gault, V. A., Parker, J. C., Harriott, P., Flatt, P. R., and O'Harte, F. P. (2002b). Evidence that the major degradation product of glucose-dependent insulinotropic polypeptide, GIP(3-42), is a GIP receptor antagonist *in vivo*. *J. Endocrinol.* **175,** 525–533.
Gault, V. A., Flatt, P. R., Harriott, P., Mooney, M. H., Bailey, C. J., and O'Harte, F. P. (2003a). Improved biological activity of Gly2- and Ser2-substituted analogues of glucose-dependent insulinotrophic polypeptide. *J. Endocrinol.* **176,** 133–141.
Gault, V. A., O'Harte, F. P., and Flatt, P. R. (2003b). Glucose-dependent insulinotropic polypeptide (GIP): Anti-diabetic and anti-obesity potential? *Neuropeptides* **37,** 253–263.
Gault, V. A., O'Harte, F. P. M., Harriott, P., Mooney, M. H., Green, B. D., and Flatt, P. R. (2003c). Effects of the novel (Pro3)GIP antagonist and exendin(9-39)amide on GIP- and GLP-1-induced cyclic AMP generation, insulin secretion and postprandial insulin release in obese diabetic (ob/ob) mice: Evidence that GIP is the major physiological incretin. *Diabetologia* **46,** 222–230.
Gault, V. A., Irwin, N., Green, B. D., McCluskey, J. T., Greer, B., Bailey, C. J., Harriott, P., O'Harte, F. P. M., and Flatt, P. R. (2005). Chemical ablation of gastric inhibitory polypeptide receptor action by daily (Pro3)GIP administration improves

glucose tolerance and ameliorates insulin resistance and abnormalities of islet structure in obesity-related diabetes. *Diabetes* **54,** 2436–2446.

Gault, V. A., Hunter, K., Irwin, N., Greer, B., Green, B. D., Harriott, P., O'Harte, F. P. M., and Flatt, P. R. (2007). Characterisation and glucoregulatory actions of a novel acylated form of the (Pro3)GIP receptor antagonist in type 2 diabetes. *Biol. Chem.* **388,** 173–179.

Gelling, R. W., Coy, D. H., Pederson, R. A., Wheeler, M. B., Hinke, S., Kwan, T., and McIntosh, C. H. S. (1997a). GIP(6-30amide) contains the high affinity binding region of GIP and is a potent inhibitor of GIP1-42 action *in vitro*. *Regul. Pept.* **69,** 151–154.

Gelling, R. W., Wheeler, M. B., Xue, J., Gyomorey, S., Nian, C., Pederson, R. A., and McIntosh, C. H. S. (1997b). Localization of the domains involved in ligand binding and activation of the glucose-dependent insulinotropic polypeptide receptor. *Endocrinology* **138,** 2640–2643.

Gespach, C., Bataille, D., Dutrillaux, M. C., and Rosselin, G. (1982). The interaction of glucagon, gastric inhibitory peptide and somatostatin with cyclic AMP production systems present in rat gastric glands. *Biochim. Biophys. Acta* **720,** 7–16.

Getty-Kaushik, L., Song, D. H., Boylan, M. O., Corkey, B. E., and Wolfe, M. M. (2006). Glucose-dependent insulinotropic polypeptide modulates adipocyte lipolysis and reesterification. *Obesity* **14,** 1124–1131.

Green, B. D., Gault, V. A., O'Harte, F. P., and Flatt, P. R. (2004). Structurally modified analogues of glucagon-like peptide-1 (GLP-1) and glucose-dependent insulinotropic polypeptide (GIP) as future antidiabetic agents. *Curr. Pharm. Des.* **10,** 3651–3662.

Greenberg, G. R., and Pokol-Daniel, S. (1994). Neural modulation of glucose-dependent insulinotropic peptide (GIP) and insulin secretion in conscious dogs. *Pancreas* **9,** 531–535.

Gregory, R. A. (1967). Enterogastrone-a reappraisal of the problem. *In* "Gastric Secretion" (T. K. Shnitka, J. A. L. Gilbert, and R. C. Harrison, Eds.), pp. 467–477. Pergamon Press, New York.

Gremlich, S., Porret, A., Hani, E. H., Cherif, D., Vionnet, N., Froguel, P., and Thorens, B. (1995). Cloning, functional expression, and chromosomal localization of the human pancreatic islet glucose-dependent insulinotropic polypeptide receptor. *Diabetes* **44,** 1202–1208.

Hampton, S. M., Kwasowski, P., Tan, K., Morgan, L. M., and Marks, V. (1983). Effect of pretreatment with a high fat diet on the gastric inhibitory polypeptide and insulin responses to oral triolein and glucose in rats. *Diabetologia* **24,** 278–281.

Hanks, J. B., Andersen, D. K., Wise, J. E., Putnam, W. S., Meyers, W. C., and Jones, R. S. (1984). The hepatic extraction of gastric inhibitory polypeptide and insulin. *Endocrinology* **115,** 1011–1008.

Hansotia, T., Baggio, L. L., Delmeire, D., Hinke, S. A., Yamada, Y., Tsukiyama, K., Seino, Y., Holst, J. J., Schuit, F., and Drucker, D. J. (2004). Double incretin receptor knockout (DIRKO) mice reveal an essential role for the enteroinsular axis in transducing the glucoregulatory actions of DPP-IV inhibitors. *Diabetes* **53,** 1326–1335.

Hansotia, T., Maida, A., Flock, G., Yamada, Y., Tsukiyama, K., Seino, Y., and Drucker, D. J. (2007). Extrapancreatic incretin receptors modulate glucose homeostasis, body weight, and energy expenditure. *J. Clin. Invest.* **117,** 143–152.

Harada, N., Yamada, Y., Tsukiyama, K., Yamada, C., Nakamura, Y., Mukai, E., Hamasaki, A., Liu, X., Toyoda, K., Seino, Y., and Inagaki, N. (2008). A novel GIP receptor splice variant influences GIP sensitivity of pancreatic beta-cells in obese mice. *Am. J. Physiol.– Endocrinol. Metab.* **294,** E61–E68.

Harmar, A. J. (2001). Family-B G-protein-coupled receptors. *Genome Biol.* **2,** 1–10.

Hartmann, H., Ebert, R., and Creutzfeldt, W. (1986). Insulin-dependent inhibition of hepatic glycogenolysis by gastric inhibitory polypeptide (GIP) in perfused rat liver. *Diabetologia* **29,** 112–114.

Hauner, H., Glatting, G., Kaminska, D., and Pfeiffer, E. F. (1988). Effects of gastric inhibitory polypeptide on glucose and lipid metabolism of isolated rat adipocytes. *Ann. Nutr. Metab.* **32,** 282–288.

Helman, C. A., and Barbezat, G. O. (1977). The effect of gastric inhibitory polypeptide on human jejunal water and electrolyte transport. *Gastroenterology* **72,** 376–379.

Henriksen, D. B., Alexandersen, P., Bjarnason, N. H., Vilsboll, T., Hartmann, B., Henriksen, E. E., Byrjalsen, I., Krarup, T., Holst, J. J., and Christiansen, C. (2003). Role of gastrointestinal hormones in postprandial reduction of bone resorption. *J. Bone Miner. Res.* **18,** 2180–2189.

Herman, G. A., Bergman, A., Stevens, C., Kotey, P., Yi, B., Zhao, P., Dietrich, B., Golor, G., Schrodter, A., Keymeulen, B., Lasseter, K. C., Kipnes, M. S., *et al.* (2006). Effect of single oral doses of sitagliptin, a dipeptidyl peptidase-4 inhibitor, on incretin and plasma glucose levels after an oral glucose tolerance test in patients with type 2 diabetes. *J. Clin. Endocrinol. Metab.* **91,** 4612–4619.

Herrmann-Rinke, C., Voge, A., Hess, M., and Goke, B. (1995). Regulation of glucagon-like peptide-1 secretion from rat ileum by neurotransmitters and peptides. *J. Endocrinol.* **147,** 25–31.

Higashimoto, Y., and Liddle, R. A. (1993). Isolation and characterization of the gene encoding rat glucose-dependent insulinotropic peptide. *Biochem. Biophys. Res. Commun.* **193,** 182–190.

Higashimoto, Y., and Liddle, R. A. (1994). Developmental expression of the glucose-dependent insulinotropic polypeptide gene in rat intestine. *Biochem. Biophys. Res. Commun.* **201,** 964–972.

Higashimoto, Y., Simchock, J., and Liddle, R. A. (1992). Molecular cloning of rat glucose-dependent insulinotropic peptide (GIP). *Biochim. Biophys. Acta* **1132,** 72–74.

Higashimoto, Y., Opara, E. C., and Liddle, R. A. (1995). Dietary regulation of glucose-dependent insulinotropic peptide (GIP) gene expression in rat small intestine. *Comp. Biochem. Physiol. C. Pharmacol. Toxicol. Endocrinol.* **110,** 207–214.

Hinke, S. A., Pauly, R. P., Ehses, J., Kerridge, P., Demuth, H. U., McIntosh, C. H., and Pederson, R. A. (2000). Role of glucose in chronic desensitization of isolated rat islets and mouse insulinoma (betaTC-3) cells to glucose-dependent insulinotropic polypeptide. *J. Endocrinol.* **165,** 281–291.

Hinke, S. A., Manhart, S., Pamir, N., Demuth, H., R., W. G., Pederson, R. A., and McIntosh, C. H. (2001). Identification of a bioactive domain in the amino-terminus of glucose-dependent insulinotropic polypeptide (GIP). *Biochim. Biophys. Acta* **1547,** 143–155.

Hinke, S. A., Gelling, R. W., Pederson, R. A., Manhart, S., Nian, C., Demuth, H.-U., and McIntosh, C. H. S. (2002). Dipeptidyl peptidase IV-resistant [D-Ala(2)]glucose-dependent insulinotropic polypeptide (GIP) improves glucose tolerance in normal and obese diabetic rats. *Diabetes* **51,** 652–661.

Hinke, S. A., Gelling, R., Manhart, S., Lynn, F., Pederson, R. A., Kuhn-Wache, K., Rosche, F., Demuth, H.-U., Coy, D., and McIntosh, C. H. S. (2003a). Structure-activity relationships of glucose-dependent insulinotropic polypeptide (GIP). *Biol. Chem.* **384,** 403–407.

Hinke, S. A., Lynn, F., Ehses, J., Pamir, N., Manhart, S., Kuhn-Wache, K., Rosche, F., Demuth, H.-U., Pederson, R. A., and McIntosh, C. H. S. (2003b). Glucose-dependent insulinotropic polypeptide (GIP): Development of DP. IV-resistant analogues with therapeutic potential. *Adv. Exp. Med. Biol.* **524,** 293–301.

Hinke, S. A., Manhart, S., Kuhn-Wache, K., Nian, C., Demuth, H.-U., Pederson, R. A., and McIntosh, C. H. S. (2004). [Ser2]- and [SerP2] incretin analogs: Comparison of dipeptidyl peptidase IV resistance and biological activities *in vitro* and *in vivo*. *J. Biol. Chem.* **279,** 3998–4006.

Holst, J. J. (2004). On the physiology of GIP and GLP-1. *Horm. Metabol. Res.* **36,** 747–754.
Holst, J. J., and Deacon, C. F. (2005). Glucagon-like peptide-1 mediates the therapeutic actions of DPP-IV inhibitors. *Diabetologia* **48,** 612–615.
Holst, J. J., and Gromada, J. (2004). Role of incretin hormones in the regulation of insulin secretion in diabetic and nondiabetic humans. *Am. J. Physiol. Endocrinol. Metab.* **287,** E199–E206.
Holst, J. J., Jensen, S. L., Knuhtsen, S., Nielsen, O. V., and Rehfeld, J. F. (1983). Effect of vagus, gastric inhibitory polypeptide, and HCl on gastrin and somatostatin release from perfused pig antrum. *Am. J. Physiol.* **244,** G515–G522.
Holst, J. J., Gromada, J., and Nauck, M. A. (1997). The pathogenesis of NIDDM involves a defective expression of the GIP. receptor. *Diabetologia* **40,** 984–986.
Holz, G. G. (2004). Epac: A new cAMP-binding protein in support of glucagon-like peptide-1 receptor-mediated signal transduction in the pancreatic beta-cell. *Diabetes* **53,** 5–13.
Holz, G. G., Kang, G., Harbeck, M., Roe, M. W., and Chepurny, O. G. (2006). Cell physiology of cAMP. sensor Epac. *J. Physiol.* **577,** 5–15.
Inagaki, N., Seino, Y., Takeda, J., Yano, H., Yamada, Y., Bell, G. I., Eddy, R. L., Fukushima, Y., Byers, M. G., and Shows, T. B. (1989). Gastric inhibitory polypeptide: Structure and chromosomal localization of the human gene. *Mol. Endocrinol.* **3,** 1014–1021.
Ipp, E., Dobbs, R. E., Harris, V., Arimura, A., Vale, W., and Unger, R. H. (1977). The effects of gastrin, gastric inhibitory polypeptide, secretin, and the octapeptide of cholecystokinin upon immunoreactive somatostatin release by the perfused canine pancreas. *J. Clin. Invest.* **60,** 1216–1219.
Irwin, N., Gault, V. A., Green, B. D., Greer, B., McCluskey, J. T., Harriott, P., O'Harte, F. P., and Flatt, P. R. (2004). Effects of short-term chemical ablation of the GIP receptor on insulin secretion, islet morphology and glucose homeostasis in mice. *Biol. Chem.* **385,** 845–852.
Irwin, N., Green, B. D., Gault, V. A., Greer, B., Harriott, P., Bailey, C. J., Flatt, P. R., and O'Harte, F. P. (2005). Degradation, insulin secretion, and antihyperglycemic actions of two palmitate-derivitized N-terminal pyroglutamyl analogues of glucose-dependent insulinotropic polypeptide. *J. Med. Chem.* **48,** 1244–1250.
Irwin, N., McClean, P. L., O'Harte, F. P. M., Gault, V. A., Harriott, P., and Flatt, P. R. (2007). Early administration of the glucose-dependent insulinotropic polypeptide receptor antagonist (Pro3)GIP prevents the development of diabetes and related metabolic abnormalities associated with genetically inherited obesity in ob/ob mice. *Diabetologia* **50,** 1532–1540.
Jackson, K. G., Zampelas, A., Knapper, J. M., Roche, H. M., Gibney, M. J., Kafatos, A., Gould, B. J., Wright, J. W., and Williams, C. M. (2000). Differences in glucose-dependent insulinotrophic polypeptide hormone and hepatic lipase in subjects of southern and northern Europe: Implications for postprandial lipemia. *Am. J. Clin. Nutr.* **71,** 13–20.
Jacobson, D. A., Weber, C. R., Bao, S., Turk, J., and Philipson, L. H. (2007). Modulation of the pancreatic islet β-cell-delayed rectifier potassium channel Kv2.1 by the polyunsaturated fatty acid arachidonate. *J. Biol. Chem.* **282,** 7442–7449.
Jang, H.-J., Kokrashvili, Z., Theodorakis, M. J., Carlson, O. D., Kim, B. J., Zhou, J., Kim, H. H., Xu, X., Chan, S. L., Juhaszova, M., Bernier, M., Mosinger, B., *et al.* (2007). Gut-expressed gustducin and taste receptors regulate secretion of glucagon-like peptide-1. *Proc. Natl. Acad. Sci. USA* **104,** 15069–15074.
Jarrousse, C., Carles-Bonnet, C., Niel, H., and Bataille, D. (1994). Activity of oxyntomodulin on gastric acid secretion induced by histamine or a meal in the rat. *Peptides* **15,** 1415–1420.

Jepeal, L. I., Boylan, M. O., and Wolfe, M. M. (2003). Cell-specific expression of the glucose-dependent insulinotropic polypeptide gene functions through a GATA and an ISL-1 motif in a mouse neuroendocrine tumor cell line. *Regul. Pept.* **113,** 139–147.

Jepeal, L. I., Fujitani, Y., Boylan, M. O., Wilson, C. N., Wright, C. V., and Wolfe, M. M. (2005). Cell-specific expression of glucose-dependent-insulinotropic polypeptide is regulated by the transcription factor PDX-1. *Endocrinology* **146,** 383–391.

Jia, X., Brown, J. C., Kwok, Y. N., Pederson, R. A., and McIntosh, C. H. S. (1994). Gastric inhibitory polypeptide and glucagon-like peptide-1(7-36) amide exert similar effects on somatostatin secretion but opposite effects on gastrin secretion from the rat stomach. *Can. J. Physiol. Pharmacol.* **72,** 1215–1219.

Jia, X., Brown, J. C., Ma, P., Pederson, R. A., and McIntosh, C. H. S. (1995). Effects of glucose-dependent insulinotropic polypeptide and glucagon-like peptide-I-(7-36) on insulin secretion. *Am. J. Physiol.* **268,** E645–E651.

Jones, I. R., Owens, D. R., Moody, A. J., Luzio, S. D., Morris, T., and Hayes, T. M. (1987). The effects of glucose-dependent insulinotropic polypeptide infused at physiological concentrations in normal subjects and type 2 (non-insulin-dependent) diabetic patients on glucose tolerance and B-cell secretion. *Diabetologia* **30,** 707–712.

Jones, I. R., Owens, D. R., Luzio, S., Williams, S., and Hayes, T. M. (1989). The glucose dependent insulinotropic polypeptide response to oral glucose and mixed meals is increased in patients with type 2 (non-insulin-dependent) diabetes mellitus. *Diabetologia* **32,** 668–677.

Jorde, R., Amland, P. F., Burhol, P. G., Giercksky, K. E., and Ebert, R. (1983a). GIP and insulin responses to a test meal in healthy and obese subjects. *Scand. J. Gastroenterol.* **18,** 1115–1119.

Jorde, R., Burhol, P. G., and Schulz, T. B. (1983b). Fasting and postprandial plasma GIP values in man measured with seven different antisera. *Regul. Pept.* **7,** 87–94.

Jorde, R., Pettersen, J. E., and Burhol, P. G. (1984). Lack of effect of exogenous or endogenous gastric inhibitory polypeptide on the elimination rate of Intralipid in man. *Acta Med. Scand.* **216,** 19–23.

Jörnvall, H., Carlquist, M., Kwauk, S., Otte, S. C., McIntosh, C. H. S., Brown, J. C., and Mutt, V. (1981). Amino acid sequence and heterogeneity of gastric inhibitory polypeptide (GIP). *FEBS. Lett.* **123,** 205–210.

Kang, G., Chepurny, O. G., Rindler, M. J., Collis, L., Chepurny, Z., Li, W. H., Harbeck, M., Roe, M. W., and Holz, G. G. (2005). A cAMP and Ca2+ coincidence detector in support of Ca2+-induced Ca2+ release in mouse pancreatic beta cells. *J. Physiol.* **566,** 173–188.

Kashima, Y., Miki, T., Shibasaki, T., Ozaki, N., Miyazaki, M., Yano, H., and Seino, S. (2001). Critical role of cAMP–GEFII–Rim2 complex in incretin-potentiated insulin secretion. *J. Biol. Chem.* **276,** 46046–46053.

Kieffer, T. J. (2003). GIP or not GIP? That is the question. *Trends Pharm. Sci.* **24,** 110–112.

Kieffer, T. J., Verchere, C. B., Fell, C. D., Huang, Z., Brown, J. C., and Pedersen, R. A. (1993). Glucose-dependent insulinotropic polypeptide stimulated insulin release from a tumor-derived beta-cell line (beta TC3). *Can. J. Physiol. Pharmacol.* **71,** 917–922.

Kieffer, T. J., Buchan, A. M., Barker, H., Brown, J. C., and Pederson, R. A. (1994). Release of gastric inhibitory polypeptide from cultured canine endocrine cells. *Am. J. Physiol.* **267,** E489–E496.

Kieffer, T. J., Huang, Z., McIntosh, C. H. S., Buchan, A. M., Brown, J. C., and Pederson, R. A. (1995a). Gastric inhibitory polypeptide release from a tumor-derived cell line. *Am. J. Physiol.* **269,** E316–E322.

Kieffer, T. J., McIntosh, C. H. S., and Pederson, R. A. (1995b). Degradation of glucose-dependent insulinotropic polypeptide and truncated glucagon-like peptide 1 *in vitro* and *in vivo* by dipeptidyl peptidase IV. *Endocrinology* **136,** 3585–3596.

Kim, S.-J., Choi, W. S., Han, J. S. M., Warnock, G., Fedida, D., and McIntosh, C. H. S. (2005a). A novel mechanism for the suppression of a voltage-gated potassium channel by glucose-dependent insulinotropic polypeptide: Protein kinase A-dependent endocytosis. *J. Biol. Chem.* **280,** 28692–28700.

Kim, S.-J., Winter, K., Nian, C., Tsuneoka, M., Koda, Y., and McIntosh, C. H. S. (2005b). Glucose-dependent insulinotropic polypeptide (GIP) stimulation of pancreatic beta-cell survival is dependent upon phosphatidylinositol 3-kinase (PI3K)/protein kinase B (PKB) signaling, inactivation of the forkhead transcription factor Foxo1, and down-regulation of bax expression. *J. Biol. Chem.* **280,** 22297–22307.

Kim, S.-J., Nian, C., and McIntosh, C. H. S. (2007a). Activation of lipoprotein lipase by glucose-dependent insulinotropic polypeptide in adipocytes. A role for a protein kinase B, LKB1, and AMP-activated protein kinase cascade. *J. Biol. Chem.* **282,** 8557–8567.

Kim, S.-J., Nian, C., and McIntosh, C. H. S. (2007b). Resistin is a key mediator of glucose-dependent insulinotropic polypeptide (GIP) stimulation of lipoprotein lipase (LPL) activity in adipocytes. *J. Biol. Chem.* **282,** 34139–34147.

Kim, S.-J., Nian, C., Widenmaier, S., and McIntosh, C. H. S. (2008). Glucose-dependent insulinotropic polypeptide-mediated up-regulation of beta-cell antiapoptotic Bcl-2 gene expression is coordinated by cyclic AMP (cAMP) response element binding protein (CREB) and cAMP-responsive CREB coactivator 2. *Mol. Cell. Biol.* **28,** 1644–1656.

Knapper, J. M., Heath, A., Fletcher, J. M., Morgan, L. M., and Marks, V. (1995a). GIP and GLP-1(7-36)amide secretion in response to intraduodenal infusions of nutrients in pigs. *Comp. Biochem. Physiol. C. Pharmacol. Toxicol. Endocrinol.* **111,** 445–450.

Knapper, J. M., Puddicombe, S. M., Morgan, L. M., and Fletcher, J. M. (1995b). Investigations into the actions of glucose-dependent insulinotropic polypeptide and glucagon-like peptide-1(7-36)amide on lipoprotein lipase activity in explants of rat adipose tissue. *J. Nutr.* **125,** 183–188.

Kogire, M., Inoue, K., Sumi, S., Doi, R., Takaori, K., Yun, M., Fujii, N., Yajima, H., and Tobe, T. (1988). Effects of synthetic human gastric inhibitory polypeptide on splanchnic circulation in dogs. *Gastroenterology* **95,** 1636–1640.

Kogire, M., Inoue, K., Sumi, S., Doi, R., Yun, M., Kaji, H., and Tobe, T. (1992). Effects of gastric inhibitory polypeptide and glucagon on portal venous and hepatic arterial flow in conscious dogs. *Dig. Dis. Sci.* **37,** 1666–1670.

Konturek, S., Bilski, J., Tasler, J., and Laskiewicz, L. (1985). Gut hormones in stimulation of gastroduodenal alkaline secretion in conscious dogs. *Am. J. Physiol. Gastrointest. Liver Physiol.* **248,** G687–G691.

Koop, H., Behrens, I., McIntosh, C. H. S., Pederson, R. A., Arnold, R., and Creutzfeldt, W. (1980). Adrenergic modulation of gastric somatostatin release in rats. *FEBS. Lett.* **118,** 2.

Kosaka, T., and Lim, R. (1930). Demonstration of the humoral agent in fat inhibition of gastric secretion. *Proc. Soc. Exp. Biol. NY* **27,** 890–891.

Kraenzlin, M. E., Ch'ng, J. L., Mulderry, P. K., Ghatei, M. A., and Bloom, S. R. (1985). Infusion of a novel peptide, calcitonin gene-related peptide (CGRP) in man. Pharmacokinetics and effects on gastric acid secretion and on gastrointestinal hormones. *Regul. Pept.* **10,** 189–197.

Krarup, T. (1988). Immunoreactive gastric inhibitory polypeptide. *Endocr. Rev.* **9,** 122–134.

Krarup, T., Holst, J. J., and Larsen, K. L. (1985). Responses and molecular heterogeneity of IR-GIP after intraduodenal glucose and fat. *Am. J. Physiol.* **249,** E195–E200.

Krarup, T., Saurbrey, N., Moody, A. J., Kuhl, C., and Madsbad, S. (1987). Effect of porcine gastric inhibitory polypeptide on beta-cell function in type I and type II diabetes mellitus. *Metabolism* **36,** 677–682.

Kubota, A., Yamada, Y., Hayami, T., Yasuda, K., Someya, Y., Ihara, Y., Kagimoto, S., Watanabe, R., Taminato, T., Tsuda, K., and Seino, Y. (1996). Identification of two missense mutations in the GIP receptor gene: A functional study and association analysis with NIDDM: No evidence of association with Japanese NIDDM. subjects. *Diabetes* **45,** 1701–1705.

Kubota, A., Yamada, Y., Yasuda, K., Someya, Y., Ihara, Y., Kagimoto, S., Watanabe, R., Kuroe, A., Ishida, H., and Seino, Y. (1997). Gastric inhibitory polypeptide activates MAP kinase through the wortmannin-sensitive and -insensitive pathways. *Biochem. Biophys. Res. Commun.* **235,** 171–175.

Kuzio, M., Dryburgh, J. R., Malloy, K. M., and Brown, J. C. (1974). Radioimmunoassay for gastric inhibitory polypeptide. *Gastroenterology* **66,** 357–364.

La. Barre, J. (1932). Sur les possibilités d'un traitement du diabète par l'incrétine. *Bull. Acad. R. Med. Belg.* **12,** 620–634.

Lacroix, A., Bolte, E., Tremblay, J., Dupre, J., Poitras, P., Fournier, H., Garon, J., Garrel, D., Bayard, F., Taillefer, R., *et al.* (1992). Gastric inhibitory polypeptide-dependent cortisol hypersecretion – A new cause of Cushing's syndrome. *New Engl. J. Med.* **327,** 974–980.

Lacroix, A., Ndiaye, N., Tremblay, J., and Hamet, P. (2001). Ectopic and abnormal hormone receptors in adrenal Cushing's syndrome. *Endocr. Rev* **22,** 75–110.

Lacroix, A., Baldacchino, V., Bourdeau, I., Hamet, P., and Tremblay, J. (2004). Cushing's syndrome variants secondary to aberrant hormone receptors. *Trends Endocrinol. Metab.* **15,** 375–382.

Larsson, L. I., and Moody, A. J. (1980). Glicentin and gastric inhibitory polypeptide immunoreactivity in endocrine cells of the gut and pancreas. *J. Histochem. Cytochem.* **28,** 925–933.

Larsson, L. I., St-Onge, L., Hougaard, D. M., Sosa-Pineda, B., and Gruss, P. (1998). Pax 4 and 6 regulate gastrointestinal endocrine cell development. *Mech. Dev.* **79,** 153–159.

Lauritsen, K. B., Christensen, K. C., and Stokholm, K. H. (1980). Gastric inhibitory polypeptide (GIP) release and incretin effect after oral glucose in obesity and after jejunoileal bypass. *Scand. J. Gastroenterol* **15,** 489–495.

Lauritsen, B., Holst, J. J., and Moody, A. J. (1981). Depression of insulin release by anti-GIP serum after oral glucose in rats. *Scand. J. Gastroenterol* **16,** 417–420.

Lavin, J. H., Wittert, G., Sun, W.-M., Horowutz, M., Morley, J. E., and Read, N. W. (1996). Appetite regulation by carbohydrate; role of blood glucose and gastrointestinal hormones. *Am. J. Physiol.* **271,** E209–E214.

Lavin, J. H., Wittert, G. A., Andrews, J., Yeap, B., Wishart, J. M., Morris, H. A., Morley, J. E., Horowitz, M., and Read, N. W. (1998). Interaction of insulin, glucagon-like peptide 1, gastric inhibitory polypeptide, and appetite in response to intraduodenal carbohydrate. *Am. J. Clin. Nutr.* **68,** 591–598.

Lebrethon, M. C., Avallet, O., Reznik, Y., Archambeaud, F., Combes, J., Usdin, T. B., Narboni, G., Mahoudeau, J., and Saez, J. M. (1998). Food-dependent Cushing's syndrome: Characterization and functional role of gastric inhibitory polypeptide receptor in the adrenals of three patients. *J. Clin. Endocrinol. Metab.* **83,** 4514–4519.

Lee, S. C., and Pervaiz, S. (2007). Apoptosis in the pathophysiology of diabetes mellitus. *Int. J. Biochem. Cell. Biol.* **39,** 497–504.

LeRoith, D., Spitz, I. M., Ebert, R., Liel, Y., Odes, S., and Creutzfeldt, W. (1980). Acid-induced gastric inhibitory polypeptide secretion in man. *J. Clin. Endocrinol. Metab.* **51,** 1385–1389.

Lewis, T. B., Saenz, M., O'Connell, P., and Leach, R. J. (1994). Localization of glucose-dependent insulinotropic polypeptide (GIP) to a gene cluster on chromosome 17q. *Genomics* **19,** 589–591.

Lewis, J. T., Dayanandan, B., Habener, J. F., and Kieffer, T. J. (2000). Glucose-dependent insulinotropic polypeptide confers early phase insulin release to oral glucose in rats: Demonstration by a receptor antagonist. *Endocrinology* **141,** 3710–3716.

Li, Y., Hansotia, T., Yusta, B., Ris, F., Halban, P. A., and Drucker, D. J. (2003). Glucagon-like peptide-1 receptor signaling modulates beta cell apoptosis. *J. Biol. Chem.* **278,** 471–478.

Li, L.-X., MacDonald, P. E., Ahn, D. S., Oudit, G. Y., Backx, P. H., and Brubaker, P. L. (2006). Role of phosphatidylinositol 3-kinasegamma in the beta-cell: Interactions with glucagon-like peptide-1. *Endocrinology* **147,** 3318–3325.

Light, P. E., Manning Fox, J. E., Riedel, M. J., and Wheeler, M. B. (2002). Glucagon-like peptide-1 inhibits pancreatic ATP-sensitive potassium channels via a protein kinase A- and ADP-dependent mechanism. *Mol. Endocrinol.* **16,** 2135–2144.

Lu, M., Wheeler, M. B., Leng, X. H., and Boyd, A. E., 3rd. (1993). The role of the free cytosolic calcium level in beta-cell signal transduction by gastric inhibitory polypeptide and glucagon-like peptide I(7–37). *Endocrinology* **132** 94–100.

Lucey, M. R., Fairclough, P. D., Wass, J. A., Kwasowski, P., Medbak, S., Webb, J., and Rees, L. H. (1984). Response of circulating somatostatin, insulin, gastrin and GIP, to intraduodenal infusion of nutrients in normal man. *Clin. Endocrinol.* **21,** 209–217.

Lund, P. K. (2005). The discovery of glucagon-like peptide 1. *Regul. Pept.* **128,** 93–96.

Lund, P. K., Goodman, R. H., Dee, P. C., and Habener, J. F. (1982). Pancreatic preproglucagon cDNA contains two glucagon-related coding sequences arranged in tandem. *Proc. Natl. Acad. Sci. USA* **79,** 345–349.

Lynn, F. C. (2003). Regulation of GIP receptor expression in the pancreatic beta cell. Ph.D. Thesis, University of British Columbia, Vancouver, B.C., Canada.

Lynn, F. C., Pamir, N., Ng, E. H., McIntosh, C. H. S., Kieffer, T. J., and Pederson, R. A. (2001). Defective glucose-dependent insulinotropic polypeptide receptor expression in diabetic fatty Zucker rats. *Diabetes* **50,** 1004–1011.

Lynn, F. C., Thompson, S. A., Pospisilik, J. A., Ehses, J. A., Hinke, S. A., Pamir, N., McIntosh, C. H. S., and Pederson, R. A. (2003). A novel pathway for regulation of glucose-dependent insulinotropic polypeptide (GIP) receptor expression in beta cells. *FASEB. J.* **17,** 91–93.

MacDonald, P., and Wheeler, M. B. (2003). Voltage-dependent K(+) channels in pancreatic beta cells: Role, regulation and potential as therapeutic targets. *Diabetologia* **46,** 1046–1062.

MacDonald, P. E., El-Kholy, W., Riedel, M. J., Salapatek, A. M. F., Light, P. E., and Wheeler, M. B. (2002a). The multiple actions of GLP-1 on the process of glucose-stimulated insulin secretion. *Diabetes* **51**(Suppl. 3), S434–S442.

MacDonald, P. E., Salapatek, A. M. F., and Wheeler, M. B. (2002b). Glucagon-like peptide-1 receptor activation antagonizes voltage-dependent repolarizing K(+) currents in beta-cells: A possible glucose-dependent insulinotropic mechanism. *Diabetes* **51**(Suppl. 3), S443–S447.

MacDonald, P. E., Sewing, S., Wang, J., Joseph, J. W., Smukler, S. R., Sakellaropoulos, G., Wang, J., Saleh, M. C., Chan, C. B., Tsushima, R. G., Salapatek, A. M. F., and Wheeler, M. B. (2002c). Inhibition of Kv2.1 voltage-dependent K+ channels in pancreatic beta-cells enhances glucose-dependent insulin secretion. *J. Biol. Chem.* **277,** 44938–44945.

MacDonald, P. E., Wang, X., Xia, F., El-kholy, W., Targonsky, E. D., Tsushima, R. G., and Wheeler, M. B. (2003). Antagonism of rat beta-cell voltage-dependent K+ currents by exendin 4 requires dual activation of the cAMP/protein kinase A and phosphatidylinositol 3-kinase signaling pathways. *J. Biol. Chem.* **278,** 52446–52453.

Maletti, M., Carlquist, M., Portha, B., Kergoat, M., Mutt, V., and Rosselin, G. (1986). Structural requirements for gastric inhibitory polypeptide (GIP) receptor binding and stimulation of insulin release. *Peptides* **7,** 75–78.

Manhart, S., Hinke, S. A., McIntosh, C. H., Pederson, R. A., and Demuth, H. U. (2003). Structure-function analysis of a series of novel GIP analogues containing different helical length linkers. *Biochemistry* **42,** 3081–3088.

Maxwell, V., Shulkes, A., Brown, J. C., Solomon, T. E., Walsh, J. H., and Grossman, M. I. (1980). Effect of gastric inhibitory polypeptide on pentagastrin-stimulated acid secretion in man. *Dig. Dis. Sci.* **25,** 113–116.

Mayo, K. E., Miller, L. J., Bataille, D., Dalle, S., Göke, B., Thorens, B., and Drucker, D. J. (2003). International uinion of pharmacology. XXV. The glucagon receptor family. *Pharmacol. Rev.* **55,** 167–194.

Mazzaferri, E. L., Ciofalo, L., Waters, L. A., Starich, G. H., Groshong, J. C., and DePalma, L. (1983). Effects of gastric inhibitory polypeptide on leucine- and arginine-stimulated insulin release. *Am. J. Physiol.* **245,** E114–E120.

Mazzaferri, E. L., Starich, G. H., Lardinois, C. K., and Bowen, G. D. (1985). Gastric inhibitory polypeptide responses to nutrients in Caucasians and American Indians with obesity and noninsulin-dependent diabetes mellitus. *J. Clin. Endocrinol. Metab.* **61,** 313–321.

Mazzocchi, G., Rebuffat, P., Meneghelli, V., Malendowicz, L. K., Tortorella, C., Gottardo, G., and Nussdorfer, G. G. (1999). Gastric inhibitory polypeptide stimulates glucocorticoid secretion in rats, acting through specific receptors coupled with the adenylate cyclase-dependent signaling pathway. *Peptides* **20,** 589–594.

Mazzuco, T. L., Chabre, O., Feige, J. J., and Thomas, M. (2007). Aberrant GPCR expression is a sufficient genetic event to trigger adrenocortical tumorigenesis. *Mol. Cell. Endocrinol.* 23–28**265–266.**

McClean, P. L., Irwin, N., Cassidy, R. S., Holst, J. J., Gault, V. A., and Flatt, P. R. (2007). GIP receptor antagonism reverses obesity, insulin resistance, and associated metabolic disturbances induced in mice by prolonged consumption of high-fat diet. *Am. J. Physiol. Endocrinol. Metab.* **293,** E1746–E1755.

McGarry, J. D. (2002). Dysregulation of fatty acid metabolism in the etiology of type 2 diabetes. *Diabetes* **51,** 7–18.

McIntosh, C. H. S. (1985). Minireview. Gastrointestinal somatostatin: Distribution, secretion and physiological significance. *Life Sci.* **37,** 2043–2058.

McIntosh, C. H. S. (2008). Dipeptidyl peptidase IV inhibitors and diabetes therapy. *Front. Biosci.* **13,** 1753–1773.

McIntosh, C., Arnold, R., Bothe, E., Becker, H., Kobberling, J., and Creutzfeldt, W. (1978). Gastrointestinal somatostatin in man and dog. *Metab.: Clin. Exp.* **27,** 1317–1320.

McIntosh, C. H. S., Pederson, R. A., Koop, H., and Brown, J. C. (1981a). Gastric inhibitory polypeptide stimulated secretion of somatostatinlike immunoreactivity from the stomach: Inhibition by acetylcholine or vagal stimulation. *Can. J. Physiol. Pharmacol.* **59,** 468–472.

McIntosh, C. H. S., Pederson, R. A., Müller, M., and Brown, J. (1981b). Autonomic nervous control of gastric somatostatin secretion from the perfused rat stomach. *Life Sci.* **29,** 1477–1483.

McIntosh, C. H. S., Kwok, Y. N., Mordhorst, T., Nishimura, E., Pederson, R. A., and Brown, J. C. (1983). Enkephalinergic control of somatostatin secretion from the perfused rat stomach. *Can. J. Physiol. Pharmacol.* **61,** 657–663.

McIntosh, C. H. S., Jia, X., and Kowk, Y. N. (1990). Characterization of the opioid receptor type mediating inhibition of rat gastric somatostatin secretion. *Am. J. Physiol.* **259,** G922–G927.

McIntosh, C. H. S., Wheeler, M. B., Gelling, R. W., Brown, J. C., and Pederson, R. A. (1996). GIP receptors and signal-transduction mechanisms. *Acta Physiol. Scand.* **157,** 361–365.

McIntosh, C. H. S., Bremsak, I., Lynn, F. C., Gill, R., Hinke, S. A., Gelling, R., Nian, C., McKnight, G., Jaspers, S., and Pederson, R. A. (1999). Glucose-dependent insulinotropic polypeptide stimulation of lipolysis in differentiated 3T3-L1 cells: Wortmannin-sensitive inhibition by insulin. *Endocrinology* **140,** 398–404.

McIntyre, N., Holdsworth, C., and Turner, D. (1964). New interpretation of oral glucose. *Lancet* **II,** 20–21.

Meier, J. J., Hucking, K., Holst, J. J., Deacon, C. F., Schmiegel, W. H., and Nauck, M. A. (2001). Reduced insulinotropic effect of gastric inhibitory polypeptide in first-degree relatives of patients with type 2 diabetes. *Diabetes* **50,** 2497–2504.

Meier, J. J., Nauck, M. A., Schmidt, W. E., and Gallwitz, B. (2002). Gastric inhibitory polypeptide: The neglected incretin revisited. *Regul. Pept.* **107,** 1–13.

Meier, J. J., Gallwitz, B., Siepmann, N., Holst, J. J., Deacon, C. F., Schmidt, W. E., and Nauck, M. A. (2003a). Gastric inhibitory polypeptide (GIP) dose-dependently stimulates glucagon secretion in healthy human subjects at euglycaemia. *Diabetologia* **46,** 798–801.

Meier, J. J., Gallwitz, B., Siepmann, N., Holst, J. J., Deacon, C. F., Schmidt, W. E., and Nauck, M. A. (2003b). The reduction in hepatic insulin clearance after oral glucose is not mediated by gastric inhibitory polypeptide (GIP). *Regul. Pept.* **113,** 95–100.

Meier, J. J., Nauck, M. A., Siepmann, N., Greulich, M., Holst, J. J., Deacon, C. F., Schmidt, W. E., and Gallwitz, B. (2003c). Similar insulin secretory response to a gastric inhibitory polypeptide bolus injection at euglycemia in first-degree relatives of patients with type 2 diabetes and control subjects. *Metabolism* **52,** 1579–1585.

Meier, J. J., Gallwitz, B., Kask, B., Deacon, C. F., Holst, J. J., Schmidt, W. E., and Nauck, M. A. (2004a). Stimulation of insulin secretion by intravenous bolus injection and continuous infusion of gastric inhibitory polypeptide in patients with type 2 diabetes and healthy control subjects. *Diabetes* **53**(Suppl. 3), S220–S224.

Meier, J. J., Goetze, O., Anstipp, J., Hagemann, D., Holst, J. J., Schmidt, W. E., Gallwitz, B., and Nauck, M. A. (2004b). Gastric inhibitory polypeptide does not inhibit gastric emptying in humans. *Am. J. Physiol. Endocrinol. Metabol* **286,** E621–E625.

Meier, J. J., Nauck, M. A., Kranz, D., Holst, J. J., Deacon, C. F., Gaeckler, D., Schmidt, W. E., and Gallwitz, B. (2004c). Secretion, degradation, and elimination of glucagon-like peptide 1 and gastric inhibitory polypeptide in patients with chronic renal insufficiency and healthy control subjects. *Diabetes* **53,** 654–662.

Meier, J. J., Nauck, M. A., Kask, B., Holst, J. J., Deacon, C. F., Schmidt, W. E., and Gallwitz, B. (2006). Influence of gastric inhibitory polypeptide on pentagastrin-stimulated gastric acid secretion in patients with type 2 diabetes and healthy controls. *World J. Gastroenterol* **12,** 1874–1880.

Meier, J. J., Holst, J. J., Schmidt, W. E., and Nauck, M. A. (2007). Reduction of hepatic insulin clearance after oral glucose ingestion is not mediated by glucagon-like peptide 1 or gastric inhibitory polypeptide in humans. *Am. J. Physiol. Endocrinol. Metab.* **293,** E849–E856.

Menard, D. (2004). Functional development of the human gastrointestinal tract: Hormone- and growth factor-mediated regulatory mechanisms. *Can. J. Gastroenterol.* **18,** 39–44.

Meneilly, G. S., Bryer-Ash, M., and Elahi, D. (1993). The effect of glyburide on beta-cell sensitivity to glucose-dependent insulinotropic polypeptide. *Diab. Care* **16,** 110–114.

Mentlein, R., Gallwitz, B., and Schmidt, W. E. (1993). Dipeptidyl-peptidase IV hydrolyses gastric inhibitory polypeptide, glucagon-like peptide-1(7–36)amide, peptide histidine methionine and is responsible for their degradation in human serum. *Eur. J. Biochem.* **214,** 829–835.

Messenger, B., Clifford, M. N., and Morgan, L. M. (2003). Glucose-dependent insulinotropic polypeptide and insulin-like immunoreactivity in saliva following sham-fed and swallowed meals. *J. Endocrinol.* **177,** 407–412.

Miyawaki, K., Yamada, Y., Yano, H., Niwa, H., Ban, N., Ihara, Y., Kubota, A., Fujimoto, S., Kajikawa, M., Kuroe, A., Tsuda, K., Hashimoto, H., *et al.* (1999). Glucose intolerance caused by a defect in the entero-insular axis: A study in gastric inhibitory polypeptide receptor knockout mice. *Proc. Natl. Acad. Sci. USA* **96,** 14843–14847.

Miyawaki, K., Yamada, Y., Ban, N., Ihara, Y., Tsukiyama, K., Zhou, H., Fujimoto, S., Oku, A., Tsuda, K., Toyokuni, S., Hiai, H., Mizunoya, W., *et al.* (2002). Inhibition of gastric inhibitory polypeptide signaling prevents obesity. *Nat. Med.* **8,** 738–742.
Moens, K., Heimberg, H., Flamez, D., Huypens, P., Quartier, E., Ling, Z., Pipeleers, D., Gremlich, S., Thorens, B., and Schuit, F. (1996). Expression and functional activity of glucagon, glucagon-like peptide I, and glucose-dependent insulinotropic peptide receptors in rat pancreatic islet cells. *Diabetes* **45,** 257–261.
Montrose-Rafizadeh, C., Egan, J. M., and Roth, J. (1994). Incretin hormones regulate glucose-dependent insulin secretion in RIN 1046–38 cells: Mechanisms of action. *Endocrinology* **135,** 589–594.
Moody, A. J., Thim, L., and Valverde, I. (1984). The isolation and sequencing of human gastric inhibitory peptide (GIP). *FEBS. Lett.* **172,** 142–148.
Morgan, L. M. (1996). The metabolic role of GIP: Physiology and pathology. *Biochem. Soc, Trans.* **24,** 585–591.
Morgan, L. M., Tredger, J. A., Hampton, S. M., French, A. P., Peake, J. C., and Marks, V. (1988). The effect of dietary modification and hyperglycaemia on gastric emptying and gastric inhibitory polypeptide (GIP) secretion. *Brit. J. Nutr.* **60,** 29–37.
Mortensen, K., Christensen, L. L., Holst, J. J., and Ørskov, C. (2003). GLP-1 and GIP are colocalized in a subset of endocrine cells in the small intestine. *Regul. Pept.* **114,** 189–196.
Mueller, M. K., Scheck, T., Demol, P., and Goebell, H. (1986). Interaction of acetylcholine and gastric inhibitory polypeptide on endocrine and exocrine rat pancreatic secretion: Augmentation of acetylcholine-induced amylase and volume secretion by the insulinotropic action of gastric inhibitory polypeptide. *Digestion* **33,** 45–52.
Mueller, M. K., Scheck, T., Dreesmann, V., Miodonski, A., and Goebell, H. (1987). GIP potentiates CCK stimulated pancreatic enzyme secretion: Correlation of anatomical structures with the effects of GIP. and CCK on amylase secretion. *Pancreas* **2,** 106–113.
Murphy, M. C., Isherwood, S. G., Sethi, S., Gould, B. J., Wright, J. W., Knapper, J. A., and Williams, C. M. (1995). Postprandial lipid and hormone responses to meals of varying fat contents: Modulatory role of lipoprotein lipase? *Eur. J. Clin. Nutr.* **49,** 578–588.
Muscelli, E., Mari, A., Natali, A., Astiarraga, B. D., Camastra, S., Frascerra, S., Holst, J. J., and Ferrannini, E. (2006). Impact of incretin hormones on beta-cell function in subjects with normal or impaired glucose tolerance. *Am. J. Physiol. Endocrinol. Metab.* **291,** E1144–E1150.
Nauck, M., Stockmann, F., Ebert, R., and Creutzfeldt, W. (1986a). Reduced incretin effect in type 2 (non-insulin-dependent) diabetes. *Diabetologia* **29,** 46–52.
Nauck, M. A., Homberger, E., Siegel, E. G., Allen, R. C., Eaton, R. P., Ebert, R., and Creutzfeldt, W. (1986b). Incretin effects of increasing glucose loads in man calculated from. venous insulin and C-peptide responses. *J. Clin. Endocrinol. Metab.* **63,** 492–498.
Nauck, M. A., Heimesaat, M. M., Orskov, C., Holst, J. J., Ebert, R., and Creutzfeldt, W. (1993). Preserved incretin activity of glucagon-like peptide 1 [7–36 amide] but not of synthetic human gastric inhibitory polypeptide in patients with type-2 diabetes mellitus. *J. Clin. Invest.* **91,** 301–307.
Nauck, M. A., El-Ouaghlidi, A., Gabrys, B., Hucking, K., Holst, J. J., Deacon, C. F., Gallwitz, B., Schmidt, W. E., and Meier, J. J. (2004). Secretion of incretin hormones (GIP and GLP-1) and incretin effect after oral glucose in first-degree relatives of patients with type 2 diabetes. *Regul. Pept.* **122,** 209–217.
N'Diaye, N., Hamet, P., Tremblay, J., Boutin, J. M., Gaboury, L., and Lacroix, A. (1999). Asynchronous development of bilateral nodular adrenal hyperplasia in gastric inhibitory polypeptide-dependent cushing's syndrome. *J. Clin. Endocrinol. Metab.* **84,** 2616–2622.
Nelson, R. L., Go, V. L., McCullough, A. J., Ilstrup, D. M., and Service, F. J. (1986). Lack of a direct effect of the autonomic nervous system on glucose-stimulated gastric inhibitory polypeptide (GIP) secretion in man. *Dig. Dis. Sci.* **31,** 929–935.

Nielsen, L. B., Ploug, K. B., Swift, P., Orskov, C., Jansen-Olesen, I., Chiarelli, F., Holst, J. J., Hougaard, P., Porksen, S., Holl, R., de Beaufort, C., Gammeltoft, S., *et al.* (2007). Co-localisation of the Kir6.2/SUR1 channel complex with glucagon-like peptide-1 and glucose-dependent insulinotrophic polypeptide expression in human ileal cells and implications for glycaemic control in new onset type 1 diabetes. *Eur. J. Endocrinol.* **156,** 663–671.

Nussdorfer, G. G., Bahcelioglu, M., Neri, G., and Malendowicz, L. K. (2000). Secretin, glucagon, gastric inhibitory polypeptide, parathyroid hormone, and related peptides in the regulation of the hypothalamus–pituitary–adrenal axis. *Peptides* **21,** 309–324.

Nyberg, J., Anderson, M. F., Meister, B., Alborn, A. M., Strom, A. K., Brederlau, A., Illerskog, A. C., Nilsson, O., Kieffer, T. J., Hia, M. A., Ricksten, A., and Eriksson, P. S. (2005). Glucose-dependent insulinotropic polypeptide is expressed in adult hippocampus and induces progenitor cell proliferation. *J. Neurosci.* **25,** 1816–1825.

Oben, J., Morgan, L., Fletcher, J., and Marks, V. (1991). Effect of the entero-pancreatic hormones, gastric inhibitory polypeptide and glucagon-like polypeptide-1(7–36) amide, on fatty acid synthesis in explants of rat adipose tissue. *J. Endocrinol.* **130,** 267–272.

O'Dorisio, T. M., Cataland, S., Stevenson, M., and Mazzaferri, E. L. (1976). Gastric inhibitory polypeptide (GIP). Intestinal distribution and stimulation by amino acids and medium-chain triglycerides. *Am. J. Dig. Dis.* **21,** 761–765.

O'Dorisio, T. M., Sirinek, K. R., Mazzaferri, E. L., and Cataland, S. (1977). Renal effects on serum gastric inhibitory polypeptide (GIP). *Metabolism* **26,** 651–656.

O'Dorisio, T. M., Spaeth, J. T., Martin, E. W. Jr., Sirinek, K. R., Thomford, N. R., Mazzaferri, E. L., and Cataland, S. (1978). Mannitol and glucose: Effects on gastric acid secretion and endogenous gastric inhibitory polypeptide (GIP). *Am. J. Dig. Dis.* **23,** 1079–1083.

O'Harte, F. P., Gray, A. M., and Flatt, P. R. (1998). Gastric inhibitory polypeptide and effects of glycation on glucose transport and metabolism in isolated mouse abdominal muscle. *J. Endocrinol.* **156,** 237–243.

O'Harte, F. P., Mooney, M. H., and Flatt, P. R. (1999). NH2-terminally modified gastric inhibitory polypeptide exhibits amino-peptidase resistance and enhanced antihyperglycemic activity. *Diabetes* **48,** 758–765.

O'Harte, F. P., Gault, V. A., Parker, J. C., Harriott, P., Mooney, M. H., Bailey, C. J., and Flatt, P. R. (2002). Improved stability, insulin-releasing activity and antidiabetic potential of two novel N-terminal analogues of gastric inhibitory polypeptide: *N*-acetyl-GIP and pGlu-GIP. *Diabetologia* **45,** 1281–1291.

Ohneda, A., Kobayashi, T., and Nihei, J. (1983). Effect of endogenous gastric inhibitory polypeptide (GIP) on the removal of triacylglycerol in dogs. *Regul. Pept.* **6,** 25–32.

Ohneda, A., Kobayashi, T., and Nihei, J. (1984). Response of gastric inhibitory polypeptide to fat ingestion in normal dogs. *Regul. Pept.* **8,** 123–130.

Ohneda, A., Kobayashi, T., Nihei, J., Imamura, M., Naito, H., and Tsuchiya, T. (1985). Role of vagus nerve in secretion of gastric inhibitory polypeptide in dogs. *Tohoku J. Exp. Med.* **147,** 183–190.

Ørskov, C., Holst, J. J., and Nielsen, O. V. (1988). Effect of truncated glucagon-like peptide-1 [proglucagon-(78–107) amide] on endocrine secretion from pig pancreas, antrum, and nonantral stomach. *Endocrinology* **123,** 2009–2013.

Osei, K., Falko, J. M., O'Dorisio, T. M., Fields, P. G., and Bossetti, B. (1986). Gastric inhibitory polypeptide responses and glucose turnover rates after natural meals in type II diabetic patients. *J. Clin. Endocrinol. Metab.* **62,** 325–330.

Pamir, N., Lynn, F. C., Buchan, A. M., Ehses, J., Hinke, S. A., Pospisilik, J. A., Miyawaki, K., Yamada, Y., Seino, Y., McIntosh, C. H., and Pederson, R. A. (2003). Glucose-dependent insulinotropic polypeptide receptor null mice exhibit compensatory changes in the enteroinsular axis. *Am. J. Physiol. Endocrinol. Metab.* **284,** E931–E939.

Parthier, C., Kleinschmidt, M., Neumann, P., Rudolph, R., Manhart, S., Schlenzig, D., Fanghanel, J., Rahfeld, J. U., Demuth, H. U., and Stubbs, M. T. (2007). Crystal structure of the incretin-bound extracellular domain of a G protein-coupled receptor. *Proc. Natl. Acad. Sci. USA* **104,** 13942–13947.

Pauly, R. P., Rosche, F., Wermann, M., McIntosh, C. H. S., Pederson, R. A., and Demuth, H. U. (1996). Investigation of glucose-dependent insulinotropic polypeptide-(1–42) and glucagon-like peptide-1-(7–36) degradation *in vitro* by dipeptidyl peptidase IV using matrix-assisted laser desorption/ionization-time of flight mass spectrometry. A novel kinetic approach. *J. Biol. Chem.* **271,** 23222–23229.

Pederson, R. A. (1994). Gastric inhibitory polypeptide. *In* "Gut Peptides: Biochemistry and Physiology" (J. H. Walsh and G. J. Dockray, Eds.), pp. 217–259. Raven Press, New York.

Pederson, R. A., and Brown, J. C. (1972). Inhibition of histamine-, pentagastrin-, and insulin-stimulated canine gastric secretion by pure "gastric inhibitory polypeptide". *Gastroenterology* **62,** 393–400.

Pederson, R. A., and Brown, J. C. (1976). The insulinotropic action of gastric inhibitory polypeptide in the perfused isolated rat pancreas. *Endocrinology* **99,** 780–785.

Pederson, R. A., and Brown, J. C. (1978). Interaction of gastric inhibitory polypeptide, glucose, and arginine on insulin and glucagon secretion from the perfused rat pancreas. *Endocrinology* **103,** 610–615.

Pederson, R. A., and McIntosh, C. H. S. (2004). GIP (Gastric Inhibitory Polypeptide). *In* "Encyclopedia of Endocrinology and Endocrine Diseases" (L. Martini, Ed.), pp. 202–207. Elsevier Press, Amsterdam.

Pederson, R. A., Schubert, H. E., and Brown, J. C. (1975). Gastric inhibitory polypeptide. Its physiologic release and insulinotropic action in the dog. *Diabetes* **24,** 1050–1056.

Pederson, R. A., McIntosh, C. H. S., Mueller, M. K., and Brown, J. C. (1981). Absence of a relationship between immunoreactive-gastrin and somatostatin-like immunoreactivity secretion in the perfused rat stomach. *Regul. Pept.* **2,** 53–60.

Perez, D. M. (2005). From. plants to man: The GPCR. "Tree of life". *Mol. Pharmacol.* **67,** 1383–1384.

Piteau, S., Olver, A., Kim, S. J., Winter, K., Pospisilik, J. A., Lynn, F., Manhart, S., Demuth, H. U., Speck, M., Pederson, R. A., and McIntosh, C. H. S. (2007). Reversal of islet GIP receptor down-regulation and resistance to GIP by reducing hyperglycemia in the Zucker rat. *Biochem. Biophys. Res. Commun.* **362,** 1007–1012.

Polak, J. M., Bloom, S. R., Kuzio, M., Brown, J. C., and Pearse, A. G. (1973). Cellular localization of gastric inhibitory polypeptide in the duodenum and jejunum. *In* "Gut Hormones" (S. R. Bloom, Ed.), pp. 284–288. Churchill Livingstone, Edinburgh.

Preitner, F., Ibberson, M., Franklin, I., Binnert, C., Pende, M., Gjinovci, A., Hansotia, T., Drucker, D. J., Wollheim, C., Burcelin, R., and Thorens, B. (2004). Gluco-incretins control insulin secretion at multiple levels as revealed in mice lacking GLP-1 and GIP receptors. *J. Clin. Invest.* **113,** 635–645.

Prentki, M., Joly, E., El-Assaad, W., and Roduit, R. (2002). Malonyl-CoA signaling, lipid partitioning, and glucolipotoxicity. Role in β-cell adaptation and failure in the etiology of diabetes. *Diabetes* **51**(Suppl. 3), S405–S413.

Rabinovitch, A., and Dupré, J. (1972). Insulinotropic and glucagonotropic activities of crude preparations of cholecystokinin-pancreozymin. *Clin. Research* **20,** 945.

Ranganath, L., Schaper, F., Gama, R., Morgan, L., Wright, J., Teale, D., and Marks, V. (1999). Effect of glucagon on carbohydrate-mediated secretion of glucose-dependent insulinotropic polypeptide (GIP) and glucagon-like peptide-1 (7–36 amide) (GLP-1). *Diab. Metab. Res. Rev* **15,** 390–394.

Rehfeld, D. (1998). The new biology of gastrointestinal hormones. *Physiol. Rev* **78,** 1087–1108.

Rehfeld, J. F. (2004). Clinical endocrinology and metabolism. Cholecystokinin. *Best Pract. Res. Clin. Endocrinol. Metab.* **18,** 569–586.

Reznik, Y., Allali-Zerah, V., Chayvialle, J. A., Leroyer, R., Leymarie, P., Travert, G., Lebrethon, M. C., Budi, I., Balliere, A. M., and Mahoudeau, J. (1992). Food-dependent Cushing's syndrome mediated by aberrant adrenal sensitivity to gastric inhibitory polypeptide. *New Engl. J. Med.* **327,** 981–986.

Rhodes, C. (2005). Type 2 diabetes-a matter of β-cell life and death? *Science* **307,** 380–384.

Roberge, J. N., and Brubaker, P. L. (1993). Regulation of intestinal proglucagon-derived peptide secretion by glucose-dependent insulinotropic peptide in a novel enteroendocrine loop. *Endocrinology* **133,** 233–240.

Ross, S. A., and Shaffer, E. A. (1981). The importance of triglyceride hydrolysis for the release of gastric inhibitory polypeptide. *Gastroenterology* **80,** 108–111.

Rossowski, W. J., Zacharia, S., Mungan, Z., Ozmen, V., Ertan, A., Baylor, L. M., Jiang, N. Y., and Coy, D. H. (1992). Reduced gastric acid inhibitory effect of a pGIP(1–30)NH2 fragment with potent pancreatic amylase inhibitory activity. *Regul. Pept.* **39,** 9–17.

Rossowski, W. J., Cheng, B. L., Jiang, N. Y., and Coy, D. H. (1998). Examination of somatostatin involvement in the inhibitory action of GIP, GLP-1, amylin and adrenomedullin on gastric acid release using a new SRIF antagonist analogue. *Brit. J. Pharmacol.* **125,** 1081–1087.

Rozengurt, N., Wu, S. V., Chen, M. C., Huang, C., Sternini, C., and Rozengurt, E. (2006). Colocalization of the alpha-subunit of gustducin with PYY and GLP-1 in L cells of human colon. *Am. J. Physiol. Gastro. Liver Physiol.* **291,** G792–G802.

Rucha, A., and Verspohl, E. J. (2005). Heterologous desensitization of insulin secretion by GIP (glucose-dependent insulinotropic peptide) in INS-1 cells: The significance of Galphai2 and investigations on the mechanism involved. *Cell. Bioch. Funct.* **23,** 205–212.

Rudovich, N. N., Rochlitz, H. J., and Pfeiffer, A. F. (2004). Reduced hepatic insulin extraction in response to gastric inhibitory polypeptide compensates for reduced insulin secretion in normal-weight and normal glucose tolerant first-degree relatives of type 2 diabetic patients. *Diabetes* **53,** 2359–2365.

Rudovich, N., Kaiser, S., Engeli, S., Osterhoff, M., Gogebakan, O., Bluher, M., and Pfeiffer, A. F. H. (2007). GIP receptor mRNA expression in different fat tissue depots in postmenopausal non-diabetic women. *Regul. Pept.* **142,** 138–145.

Sakuraba, H., Mizukami, H., Yagihashi, N., Wada, R., Hanyu, C., and Yagihashi, S. (2002). Reduced beta-cell mass and expression of oxidative stress-related DNA damage in the islet of Japanese Type II. diabetic patients. *Diabetologia* **45,** 85–96.

Sandberg, E., Ahren, B., Tendler, D., Carlquist, M., and Efendi, S. (1986). Potentiation of glucose-induced insulin secretion in the perfused rat pancreas by porcine GIP (gastric inhibitory polypeptide), bovine GIP, and bovine GIP(1–39). *Acta Physiol. Scand.* **127,** 323–326.

Schäfer, R., and Schatz, H. (1979). Stimulation of (Pro-)insulin biosynthesis and release by gastric inhibitory polypeptide in isolated islets of rat pancreas. *Acta Endocrinol.* **91,** 493–500.

Schauder, P., Brown, J. C., Frerichs, H., and Creutzfeldt, W. (1975). Gastric inhibitory polypeptide: Effect on glucose-induced insulin release from isolated rat pancreatic islets *in vitro*. *Diabetologia* **11,** 483–484.

Schauder, P., Schindler, B., Panten, U., Brown, J. C., Frerichs, H., and Creutzfeldt, W. (1977). Insulin release from isolated rat pancreatic islets induced by alpha-ketoisocapronic acid, L-leucine, D-glucose or D-glyceraldehyde: Effect of gastric inhibitory polypeptide or glucagon. *Mol. Cell. Endocrinol.* **7,** 115–123.

Schieldrop, P. J., Gelling, R. W., Elliot, R., Hewitt, J., Kieffer, T. J., McIntosh, C. H. S., and Pederson, R. A. (1996). Isolation of a murine glucose-dependent insulinotropic polypeptide (GIP) cDNA from a tumor cell line (STC6–14) and quantification of glucose-induced increases in GIP mRNA. *Biochim. Biophys. Acta* **1308,** 111–113.

Schirra, J., Katschinski, M., Weidmann, C., Schafer, T., Wank, U., Arnold, R., and Göke, B. (1996). Gastric emptying and release of incretin hormones after glucose ingestion in humans. *J. Clin. Invest.* **97,** 92–103.

Schmidt, W. E., Siegel, E. G., Kummel, H., Gallwitz, B., and Creutzfeldt, W. (1987). Commercially available preparations of porcine glucose-dependent insulinotropic polypeptide (GIP) contain a biologically inactive GIP-fragment and cholecystokinin-33/-39. *Endocrinology* **120,** 835–837.

Schulz, T. B., Jorde, R., and Burhol, P. G. (1982). Gastric inhibitory polypeptide release into the portal vein in response to intraduodenal amino acid loads in anesthetized rats. *Scand. J. Gastroenterol* **17,** 709–713.

Schusdziarra, V., Bender, H., and Pfeiffer, E. F. (1983). Release of bombesin-like immunoreactivity from the isolated perfused rat stomach. *Regul. Pept.* **7,** 21–29.

Scrocchi, L. A., Brown, T. J., McClusky, N., P. L., B., and Drucker, D. J. (1996). Glucose intolerance but normal satiety in mice with a null mutation in the glucagon-like peptide 1 receptor gene. *Nat. Med.* **2,** 1254–1258.

Seino, S., and Shibasaki, T. (2005). PKA-dependent and PKA-independent pathways for cAMP-regulated exocytosis. *Physiol. Rev* **85,** 1303–1342.

Service, F. J., Heiling, V. J., Go, V. L., and Rizza, R. A. (1990). Lack of effect of gastric inhibitory polypeptide on hepatic and extrahepatic insulin action. *J. Clin. Endocrinol. Metab.* **70,** 1398–1402.

Sharma, S. K., Austin, C., Howard, A., Lo, G., Nicholl, C. G., and Legon, S. (1992). Characterization of rat gastric inhibitory peptide cDNA. *J. Mol. Endocrinol.* **9,** 265–272.

Sherwood, N. M., Krueckl, S. L., and McRory, J. E. (2000). The origin and function of the pituitary adenylate cyclase-activating polypeptide (PACAP)/glucagon superfamily. *Endocr. Rev* **21,** 619–670.

Siegel, E. G., and Creutzfeldt, W. (1985). Stimulation of insulin release in isolated rat islets by GIP in physiological concentrations and its relation to islet cyclic AMP content. *Diabetologia* **28,** 857–861.

Silvestre, R. A., Rodriguez-Gallardo, J., Egido, E. M., Hernandez, R., and Marco, J. (2003). Stimulatory effect of xenin-8 on insulin and glucagon secretion in the perfused rat pancreas. *Regul. Pept.* **115,** 25–29.

Sirinek, K. R., Pace, W. G., Crockett, S. E., O'Dorisio, T. M., Mazzaferri, E. L., and Cataland, S. (1978). Insulin-induced attenuation of glucose-stimulated gastric inhibitory polypeptide secretion. *Am. J. Surg.* **135,** 151–155.

Sirinek, K. R., Levine, B. A., O'Dorisio, T. M., and Cataland, S. (1983). Gastric inhibitory polypeptide (GIP) release by actively transported, structurally similar carbohydrates. *Proc. Soc. Exp. Biol. Med.* **173,** 379–385.

Sjölund, K., Ekelund, M., Hakanson, R., Moody, A. J., and Sundler, F. (1983). Gastric inhibitory peptide-like immunoreactivity in glucagon and glicentin cells: Properties and origin. An immunocytochemical study using several antisera. *J. Histochem. Cytochem.* **31,** 811–817.

Smith, P. H., Merchant, F. W., Johnson, D. G., Fujimoto, W. Y., and Williams, R. H. (1977). Immunocytochemical localization of a gastric imhibitory polypeptide-like material within A-cells of the endocrine pancreas. *Am. J. Anat.* **149,** 585–590.

Sokos, G. G., Nikolaidis, L. A., Mankad, S., Elahi, D., and Shannon, R. P. (2006). Glucagon-like peptide-1 infusion improves left ventricular ejection fraction and functional status in patients with chronic heart failure. *J. Card. Fail* **12,** 694–699.

Someya, Y., Inagaki, N., Maekawa, T., Seino, Y., and Ishii, S. (1993). Two 3′,5′-cyclic-adenosine monophosphate response elements in the promoter region of the human gastric inhibitory polypeptide gene. *FEBS. Lett.* **317,** 67–73.

Sondhi, S., Castellano, J. M., Chong, V. Z., Rogoza, R. M., Skoblenick, K. J., Dyck, B. A., Gabriele, J., Thomas, N., Ki, K., Pristupa, Z. B., Singh, A. N., MacCrimmon, D., *et al.* (2006). cDNA array reveals increased expression of glucose-dependent insulinotropic polypeptide following chronic clozapine treatment: Role in atypical antipsychotic drug-induced adverse metabolic effects. *Pharmacogenomics J.* **6,** 131–140.

Song, K., Zhang, X., Zhao, C., Ang, N. T., and Ma, Z. A. (2005). Inhibition of calcium-independent phospholipase A2 Results in insufficient insulin secretion and impaired glucose tolerance. *Mol. Endocrinol.* **19,** 504–515.

Song, D. H., Getty-Kaushik, L., Tseng, E., Simon, J., Corkey, B. E., and Wolfe, M. M. (2007). Glucose-dependent insulinotropic polypeptide enhances adipocyte development and glucose uptake in part through Akt activation. *Gastroenterology* **133,** 1796–1805.

Soon-Shiong, P., Debas, H. T., and Brown, J. C. (1979). The evaluation of gastric inhibitory polypeptide (GIP) as the enterogastrone. *J. Surg. Res.* **26,** 681–686.

Soon-Shiong, P., Debas, H. T., and Brown, J. C. (1984). Bethanechol prevents inhibition of gastric acid secretion by gastric inhibitory polypeptide. *Am. J. Physiol.* **247,** G171–G175.

Stein, D. T., Esser, V., Stevenson, B. E., Lane, K. E., Whiteside, J. H., Daniels, M. B., Chen, S., and McGarry, J. D. (1996). Essentiality of circulating fatty acids for glucose-stimulated insulin secretion in the fasted rat. *J. Clin. Invest.* **97,** 2728–2735.

Straub, S. G., and Sharp, G. W. (1996). Glucose-dependent insulinotropic polypeptide stimulates insulin secretion via increased cyclic AMP and [Ca2+]1 and a wortmannin-sensitive signalling pathway. *Biochem. Biophys. Res. Commun.* **224,** 369–374.

Svensson, A. M., Efendic, S., Ostenson, C. G., and Jansson, L. (1997). Gastric inhibitory polypeptide and splanchnic blood perfusion: Augmentation of the islet blood flow increase in hyperglycemic rats. *Peptides* **18,** 1055–1059.

Sykes, S., Morgan, L. M., English, J., and Marks, V. (1980). Evidence for preferential stimulation of gastric inhibitory polypeptide secretion in the rat by actively transported carbohydrates and their analogues. *J. Endocrinol.* **85,** 201–207.

Szecowka, J., Grill, V., Sandberg, E., and Efendic, S. (1982). Effect of GIP on the secretion of insulin and somatostatin and the accumulation of cyclic AMP *in vitro* in the rat. *Acta Endocrinol.* **99,** 416–421.

Takeda, J., Seino, Y., Tanaka, K., Fukumoto, H., Kayano, T., Takahashi, H., Mitani, T., Kurono, M., Suzuki, T., and Tobe, T. (1987). Sequence of an intestinal cDNA encoding human gastric inhibitory polypeptide precursor. *Proc. Natl. Acad. Sci. USA* **84,** 7005–7008.

Theodorakis, M. J., Carlson, O., Muller, D. C., and Egan, J. M. (2004). Elevated plasma glucose-dependent insulinotropic polypeptide associates with hyperinsulinemia in impaired glucose tolerance. *Diab. Care* **27,** 1692–1698.

Theodorakis, M. J., Carlson, O., Michopoulos, S., Doyle, M. E., Juhaszova, M., Petraki, K., and Egan, J. M. (2006). Human duodenal enteroendocrine cells: Source of both incretin peptides, GLP-1 and GIP. *Am. J. Physiol. Endocrinol. Metab.* **290,** E550–E559.

Thomas, F. B., Mazzaferri, E. L., Crockett, S. E., Mekhjian, H. S., Gruemer, H. D., and Cataland, S. (1976). Stimulation of secretion of gastric inhibitory polypeptide and insulin by intraduodenal amino acid perfusion. *Gastroenterology* **70,** 523–527.

Thomas, F. B., Shook, D. F., O'Dorisio, T. M., Cataland, S., Mekhjian, H. S., Caldwell, J. H., and Mazzaferri, E. L. (1977). Localization of gastric inhibitory polypeptide release by intestinal glucose perfusion in man. *Gastroenterology* **72,** 49–54.

Thomas, R., Hellmich, M., Townsend, C. Jr., and Evers, B. (2003). Role of gastrointestinal hormones in the proliferation of normal and neoplastic tissues. *Endocr. Rev* **24,** 571–599.

Thomford, N. R., Sirinek, K. R., Crockett, S. E., Mazzaferri, E. L., and Cataland, S. (1974). Gastric inhibitory polypeptide. Response to oral glucose after vagotomy and pyloroplasty. *Arch. Surg.* **109,** 177–182.

Toft-Nielsen, M. B., Damholt, M. B., Madsbad, S., Hilsted, L. M., Hughes, T. E., Michelsen, B. K., and Holst, J. J. (2001). Determinants of the impaired secretion of glucagon-like peptide-1 in type 2 diabetic patients. *J. Clin. Endocrinol. Metab.* **86,** 3717–3723.

Trümper, A., Trümper, K., Trusheim, H., Arnold, R., Göke, B., and Hörsch, D. (2001). Glucose-dependent insulinotropic polypeptide is a growth factor for beta (INS-1) cells by pleiotropic signaling. *Mol. Endocrinol.* **15,** 1559–1570.

Trümper, A., Trümper, K., and Hörsch, D. (2002). Mechanisms of mitogenic and anti-apoptotic signaling by glucose-dependent insulinotropic polypeptide in beta(INS-1)-cells. *J. Endocrinol.* **174,** 233–246.

Tseng, C. C., and Lin, L. (1997). A point mutation in the glucose-dependent insulinotropic peptide receptor confers constitutive activity. *Biochem. Biophys. Res. Commun.* **232,** 96–100.

Tseng, C. C., and Zhang, X. Y. (1998a). Role of regulator of G. protein signaling in desensitization of the glucose-dependent insulinotropic peptide receptor. *Endocrinology* **139,** 4470–4475.

Tseng, C. C., and Zhang, X. Y. (1998b). The cysteine of the cytoplasmic tail of glucose-dependent insulinotropic peptide receptor mediates its chronic desensitization and down-regulation. *Mol. Cell Endocrinol.* **139,** 179–186.

Tseng, C. C., and Zhang, X. Y. (2000). Role of G protein-coupled receptor kinases in glucose-dependent insulinotropic polypeptide receptor signaling. *Endocrinology* **141,** 947–952.

Tseng, C. C., Jarboe, L. A., Landau, S. B., Williams, E. K., and Wolfe, M. M. (1993). Glucose-dependent insulinotropic peptide: Structure of the precursor and tissue-specific expression in rat. *Proc. Natl. Acad. Sci. USA* **90,** 1992–1996.

Tseng, C. C., Jarboe, L. A., and Wolfe, M. M. (1994). Regulation of glucose-dependent insulinotropic peptide gene expression by a glucose meal. *Am. J. Physiol.* **266,** G887–G891.

Tseng, C. C., Boylan, M. O., Jarboe, L. A., Williams, E. K., Sunday, M. E., and Wolfe, M. M. (1995). Glucose-dependent insulinotropic peptide (GIP) gene expression in the rat salivary gland. *Mol. Cell. Endocrinol.* **115,** 13–19.

Tseng, C. C., Kieffer, T. J., Jarboe, L. A., Usdin, T. B., and Wolfe, M. M. (1996). Postprandial stimulation of insulin release by glucose-dependent insulinotropic polypeptide (GIP). Effect of a specific glucose-dependent insulinotropic polypeptide receptor antagonist in the rat. *J. Clin. Invest.* **98,** 2440–2445.

Tseng, C. C., Zhang, X. Y., and Wolfe, M. M. (1999). Effect of GIP and GLP-1 antagonists on insulin release in the rat. *Am. J. Physiol.* **276,** E1049–E1054.

Tsuboi, T., da Silva Xavier, G., Holz, G. G., Jouaville, L. S., Thomas, A. P., and Rutter, G. A. (2003). Glucagon-like peptide-1 mobilizes intracellular Ca2+ and stimulates mitochondrial ATP synthesis in pancreatic MIN6 beta-cells. *Biochem. J.* **369,** 287–299.

Tsukiyama, K., Yamada, Y., Yamada, C., Harada, N., Kawasaki, Y., Ogura, M., Bessho, K., Li, M., Amizuka, N., Sato, M., Udagawa, N., Takahashi, N., *et al.* (2006). Gastric inhibitory polypeptide as an endogenous factor promoting new bone formation after food ingestion. *Mol. Endocrinol.* **20,** 1644–1651.

Turner, D. S., Etheridge, L., Jones, J., Marks, V., Meldrum, B., Bloom, S. R., and Brown, J. C. (1974a). The effect of the intestinal polypeptides, IRP and GIP, on insulin release and glucose tolerance in the baboon. *Clin. Endocrinol.* **3,** 489–493.

Turner, D. S., Etheridge, L., Marks, V., Brown, J. C., and Mutt, V. (1974b). Effectiveness of the intestinal polypeptides, IRP, GIP, VIP and motilin on insulin release in the rat. *Diabetologia* **10,** 459–463.

Ugleholdt, R., Poulsen, M. L. H., Holst, P. J., Irminger, J. C., Orskov, C., Pedersen, J., Rosenkilde, M. M., Zhu, X., Steiner, D. F., and Holst, J. J. (2006). Prohormone convertase 1/3 is essential for processing of the glucose-dependent insulinotropic polypeptide precursor. *J. Biol. Chem.* **281,** 11050–11057.

Unger, R., and Eisentraut, A. (1969). Entero-insular axis. *Arch. Intern. Med.* **123,** 261–265.

Usdin, T. B., Mezey, E., Button, D. C., Brownstein, M. J., and Bonner, T. I. (1993). Gastric inhibitory polypeptide receptor, a member of the secretin-vasoactive intestinal peptide receptor family, is widely distributed in peripheral organs and the brain. *Endocrinology* **133,** 2861–2870.

Usellini, L., Capella, C., Solcia, E., Buchan, A. M., and Brown, J. C. (1984). Ultrastructural localization of gastric inhibitory polypeptide (GIP) in a well characterized endocrine cell of canine duodenal mucosa. *Histochemistry* **80,** 85–89.

Vahl, T., and D'Alessio, D. (2003). Enteroinsular signaling: Perspectives on the role of the gastrointestinal hormones glucagon-like peptide 1 and glucose-dependent insulinotropic polypeptide in normal and abnormal glucose metabolism. *Curr. Opin. Clin. Nutr. Metab. Care* **6,** 461–468.

Van Ginneken, C., and Weyns, A. (2004). A stereological evaluation of secretin and gastric inhibitory peptide-containing mucosal cells of the perinatal small intestine of the pig. *J. Anat.* **205,** 267–275.

Verdonk, C. A., Rizza, R. A., Nelson, R. L., Go, V. L., Gerich, J. E., and Service, F. J. (1980). Interaction of fat-stimulated gastric inhibitory polypeptide on pancreatic alpha and beta cell function. *J. Clin. Invest.* **65,** 1119–1125.

Villar, H. V., Fender, H. R., Rayford, P. L., Bloom, S. R., Ramus, N. I., and Thompson, J. C. (1976). Suppression of gastrin release and gastric secretion by gastric inhibitory polypeptide (GIP) and vasoactive intestinal polypeptide (VIP). *Ann. Surg.* **184,** 97–102.

Vilsbøll, T., Krarup, T., Deacon, C. F., Madsbad, S., and Holst, J. J. (2001). Reduced postprandial concentrations of intact biologically active glucagon-like peptide 1 in type 2 diabetic patients. *Diabetes* **50,** 609–613.

Vilsbøll, T., Krarup, T., Madsbad, S., and Holst, J. J. (2002). Defective amplification of the late phase insulin response to glucose by GIP in obese Type II diabetic patients. *Diabetologia* **45,** 1111–1119.

Vilsbøll, T., Knop, F. K., Krarup, T., Johansen, A., Madsbad, S., Larsen, S., Hansen, T., Pedersen, O., and Holst, J. J. (2003a). The pathophysiology of diabetes involves a defective amplification of the late-phase insulin response to glucose by glucose-dependent insulinotropic polypeptide-regardless of etiology and phenotype. *J. Clin. Endocrinol. Metab.* **88,** 4897–4903.

Vilsbøll, T., Krarup, T., Madsbad, S., and Holst, J. J. (2003b). Both GLP-1 and GIP are insulinotropic at basal and postprandial glucose levels and contribute nearly equally to the incretin effect of a meal in healthy subjects. *Regul. Pept.* **114,** 115–121.

Vilsbøll, T., Krarup, T., Sonne, J., Madsbad, S., Vølund, A., Juul, A. G., and Holst, J. J. (2003c). Incretin secretion in relation to meal size and body weight in healthy subjects and people with type 1 and type 2 diabetes mellitus. *J. Clin. Endocrinol. Metab.* **88,** 2706–2713.

Volz, A., Göke, R., Lankat-Buttgereit, B., Fehmann, H. C., Bode, H. P., and Göke, B. (1995). Molecular cloning, functional expression, and signal transduction of the GIP-receptor cloned from a human insulinoma. *FEBS. Lett.* **373,** 23–29.

Wachters-Hagedoorn, R. E., Priebe, M. G., Heimweg, J. A. J., Heiner, A. M., Englyst, K. N., Holst, J. J., Stellaard, F., and Vonk, R. J. (2006). The rate of intestinal glucose absorption is correlated with plasma glucose-dependent insulinotropic polypeptide concentrations in healthy men. *J. Nutr.* **136,** 1511–1516.

Wahl, M. A., Plehn, R. J., Landsbeck, E. A., Verspohl, E. J., and Ammon, H. P. (1992). Are ionic fluxes of pancreatic beta cells a target for gastric inhibitory polypeptide? *Mol. Cell. Endocrinol.* **90,** 117–123.

Wajchenberg, B. L. (2007). Beta-cell failure in diabetes and preservation by clinical treatment. *Endocr. Rev* **28,** 187–218.

Wang, Y., Montrose-Rafizadeh, C., Adams, L., Raygada, M., Nadiv, O., and Egan, J. M. (1996). GIP regulates glucose transporters, hexokinases, and glucose-induced insulin secretion in RIN 1046–38 cells. *Mol. Cell. Endocrinol.* **116,** 81–87.

Wasada, T., McCorkle, K., Harris, V., Kawai, K., Howard, B., and Unger, R. H. (1981). Effect of gastric inhibitory polypeptide on plasma levels of chylomicron triglycerides in dogs. *J. Clin. Invest.* **68,** 1106–1107.

Wheeler, M. B., Gelling, R. W., McIntosh, C. H. S., Georgiou, J., Brown, J. C., and Pederson, R. A. (1995). Functional expression of the rat pancreatic islet glucose-dependent insulinotropic polypeptide receptor: Ligand binding and intracellular signaling properties. *Endocrinology* **136,** 4629–4639.

Wheeler, M. B., Gelling, R. W., Hinke, S. A., Tu, B., Pederson, R. A., Lynn, F., Ehses, J., and McIntosh, C. H. S. (1999). Characterization of the carboxyl-terminal domain of the rat glucose-dependent insulinotropic polypeptide (GIP) receptor. A role for serines 426 and 427 in regulating the rate of internalization. *J. Biol. Chem.* **274,** 24593–24601.

Willms, B., Ebert, R., and Creutzfeldt, W. (1978). Gastric inhibitory polypeptide (GIP) and insulin in obesity: II. Reversal of increased response to stimulation by starvation of food restriction. *Diabetologia* **14,** 379–387.

Wolfe, M. M., and Reel, G. M. (1986). Inhibition of gastrin release by gastric inhibitory peptide mediated by somatostatin. *Am. J. Physiol.* **250,** G331–G335.

Wolfe, M. M., Hocking, M. P., Maico, D. G., and McGuigan, J. E. (1983). Effects of antibodies to gastric inhibitory peptide on gastric acid secretion and gastrin release in the dog. *Gastroenterology* **84,** 941–948.

Wolfe, M. M., Zhao, K. B., Glazier, K. D., Jarboe, L. A., and Tseng, C. C. (2000). Regulation of glucose-dependent insulinotropic polypeptide release by protein in the rat. *Am. J. Physiol. Gastro. Liver Physiol.* **279,** G561–G566.

Xie, D., Cheng, H., Hamrick, M., Zhong, Q., Ding, K.-H., Correa, D., Williams, S., Mulloy, A., Bollag, W., Bollag, R. J., Runner, R. R., McPherson, J. C., *et al.* (2005). Glucose-dependent insulinotropic polypeptide receptor knockout mice have altered bone turnover. *Bone* **37,** 759–769.

Xie, D., Zhong, Q., Ding, K.-H., Cheng, H., Williams, S., Correa, D., Bollag, W. B., Bollag, R. J., Insogna, K., Troiano, N., Coady, C., Hamrick, M., *et al.* (2007). Glucose-dependent insulinotropic peptide-overexpressing transgenic mice have increased bone mass. *Bone* **40,** 1352–1360.

Xu, G., Kaneto, H., Laybutt, D. R., Duvivier-Kali, V. F., Trivedi, N., Suzuma, K., King, G. L., Weir, G. C., and Bonner-Weir, S. (2007). Downregulation of GLP-1 and GIP receptor expression by hyperglycemia: Possible contribution to impaired incretin effects in diabetes. *Diabetes* **56,** 1551–1558.

Yamada, Y., Hayami, T., Nakamura, K., Kaisaki, P. J., Someya, Y., Wang, C. Z., Seino, S., and Seino, Y. (1995). Human gastric inhibitory polypeptide receptor: Cloning of the gene (GIPR) and cDNA. *Genomics* **29,** 773–776.

Yamada, C., Yamada, Y., Tsukiyama, K., Yamada, K., Yamane, S., Harada, N., Miyawaki, K., Seino, Y., and Inagaki, N. (2007). Genetic inactivation of GIP signaling reverses aging-associated insulin resistance through body composition changes. *Biochem. Biophys. Res. Commun.* **364,** 175–180.

Yamagishi, T., and Debas, H. T. (1980). Gastric inhibitory polypeptide (GIP) is not the primary mediator of the enterogastrone action of fat in the dog. *Gastroenterology* **78,** 931–936.

Yasuda, K., Inagaki, N., Yamada, Y., Kubota, A., Seino, S., and Seino, Y. (1994). Hamster gastric inhibitory polypeptide receptor expressed in pancreatic islets and clonal insulin-secreting cells: Its structure and functional properties. *Biochem. Biophys. Res. Commun.* **205,** 1556–1562.

Yeung, C. M., Wong, C. K., Chung, S. K., Chung, S. S., and Chow, B. K. (1999). Glucose-dependent insulinotropic polypeptide gene expression in the stomach: Revealed by a transgenic mouse study, in situ hybridization and immunohistochemical staining. *Mol. Cell. Endocrinol.* **154,** 161–170.

Yip, R. G., and Wolfe, M. M. (2000). GIP biology and fat metabolism. *Life Sci.* **66,** 91–103.

Yip, R. G., Boylan, M. O., Kieffer, T. J., and Wolfe, M. M. (1998). Functional GIP receptors are present on adipocytes. *Endocrinology* **139,** 4004–4007.

Yoshidome, K., Miyata, M., Izukura, M., Mizutani, S., Sakamoto, T., Tominaga, H., and Matsuda, H. (1995). Secretion of gastric inhibitory polypeptide in patients with bile duct obstruction. *Scand. J. Gastroenterol* **30,** 586–589.

Young, A. A., Gedulin, B. R., and Rink, T. J. (1996). Dose-responses for the slowing of gastric emptying in a rodent model by glucagon-like peptide (7–36) NH2, amylin, cholecystokinin, and other possible regulators of nutrient uptake. *Metabolism* **45,** 1–3.

Yusta, B., Baggio, L. L., Estall, J. L., Koehler, J. A., Holland, D. P., Li, H., Pipeleers, D., Ling, Z., and Drucker, D. J. (2006). GLP-1 receptor activation improves beta cell function and survival following induction of endoplasmic reticulum stress. *Cell. Metab.* **4,** 391–406.

Zhao, D., and Pothoulakis, C. (2006). Effects of NT on gastrointestinal motility and secretion, and role in intestinal inflammation. *Peptides* **27,** 2434–2444.

Zhong, Q., Bollag, R. J., Dransfield, D. T., Gasalla-Herraiz, J., Ding, K. H., Min, L., and Isales, C. M. (2000). Glucose-dependent insulinotropic peptide signaling pathways in endothelial cells. *Peptides* **21,** 1427–1432.

Zhong, Q., Itokawa, T., Sridhar, S., Ding, K.-H., Xie, D., Kang, B., Bollag, W. B., Bollag, R. J., Hamrick, M., Insogna, K., and Isales, C. M. (2007). Effects of glucose-dependent insulinotropic peptide on osteoclast function. *Am. J. Physiol. Endocrinol. Metab.* **292,** E543–E548.

Zhou, J., Livak, M. F. A., Bernier, M., Muller, D. C., Carlson, O. D., Elahi, D., Maudsley, S., and Egan, J. M. (2007). Ubiquitination is involved in glucose-mediated downregulation of GIP receptors in islets. *Am. J. Physiol. Endocrinol. Metab.* **293,** E538–E547.

CHAPTER SIXTEEN

Insulin Granule Biogenesis, Trafficking and Exocytosis

June Chunqiu Hou,* Le Min,† *and* Jeffrey E. Pessin‡

Contents

Abstract

It is becoming increasingly apparent that beta cell dysfunction resulting in abnormal insulin secretion is the essential element in the progression of patients from a state of impaired glucose tolerance to frank type 2 diabetes (Del Prato, 2003; Del Prato and Tiengo, 2001). Although extensive studies have examined the molecular, cellular and physiologic mechanisms of insulin granule biogenesis, sorting, and exocytosis the precise mechanisms controlling these processes and their dysregulation in the developed of diabetes remains an area of important investigation. We now know that insulin biogenesis initiates with the synthesis of preproinsulin in rough endoplastic reticulum and conversion of preproinsulin to proinsulin. Proinsulin begins to be packaged in the

* Department of Pharmacological Sciences, Stony Brook University, Stony Brook, NY 11794
† Division of Endocrinology, Diabetes and Hypertension, Brigham and women's hospital, 221 Longwood Ave, Boston, MA 02115
‡ Departments of Medicine and Molecular Pharmacology, Albert Einstein College of Medicine, Bronx, NY 10461

Vitamins and Hormones, Volume 80
ISSN 0083-6729, DOI: 10.1016/S0083-6729(08)00616-X

Trans-Golgi Network and is sorting into immature secretory granules. These immature granules become acidic via ATP-dependent proton pump and proinsulin undergoes proteolytic cleavage resulting the formation of insulin and C-peptide. During the granule maturation process, insulin is crystallized with zinc and calcium in the form of dense-core granules and unwanted cargo and membrane proteins undergo selective retrograde trafficking to either the constitutive trafficking pathway for secretion or to degradative pathways. The newly formed mature dense-core insulin granules populate two different intracellular pools, the readily releasable pools (RRP) and the reserved pool. These two distinct populations are thought to be responsible for the biphasic nature of insulin release in which the RRP granules are associated with the plasma membrane and undergo an acute calcium-dependent release accounting for first phase insulin secretion. In contrast, second phase insulin secretion requires the trafficking of the reserved granule pool to the plasma membrane.

The initial trigger for insulin granule fusion with the plasma membrane is a rise in intracellular calcium and in the case of glucose stimulation results from increased production of ATP, closure of the ATP-sensitive potassium channel and cellular depolarization. In turn, this opens voltage-dependent calcium channels allowing increased influx of extracellular calcium. Calcium is thought to bind to members of the fusion regulatory proteins synaptogamin that functionally repressors the fusion inhibitory protein complexin. Both complexin and synaptogamin interact as well as several other regulatory proteins interact with the core fusion machinery composed of the Q- or t-SNARE proteins syntaxin 1 and SNAP25 in the plasma membrane that assembles with the R- or v-SNARE protein VAMP2 in insulin granules. In this chapter we will review the current progress of insulin granule biogenesis, sorting, trafficking, exocytosis and signaling pathways that comprise the molecular basis of glucose-dependent insulin secretion.

I. Introduction

In the postprandial state, a variety of nutritional factors in circulation including amino acids, fatty acids and glucose serve as insulin secretagogues resulting in the release of insulin that initiates signaling cascades responsible for the suppression hepatic glucose output, increased macromolecular synthesis (glycogen and triglycerides) and stimulation of peripheral tissue (skeletal muscle and adipose tissue) uptake of glucose. In addition, signals in the gastrointestinal track stimulate the release of gut hormones (incretins) in particular glucagon-like peptide-1 (GLP-1) that markedly potentiates glucose-stimulated insulin secretion (Holst, 2007). Defects in the actions of insulin result in a physiologic state of insulin resistance in which relatively higher concentrations of insulin are required to maintain normal glucose homeostasis. However, type II diabetes only ensures when cell insulin secretory properties becomes abnormal and/or the levels of secreted insulin are insufficient to compensate for the increase demand. The dysfunction of insulin secretion and associated hyperglycemia ultimately leads to micro- and macrovascular damage, causing long-term complications including neuropathy, nephropathy, retinopathy and cardiovascular disease that significantly affects quality of life and reduces life expectancy.

As insulin secretion is a unique property of pancreatic beta cells in the Islets of Langerhans, considerable effort has been applied to understand the biology of this cell type. The pancreas is composed of both an exocrine component (responsible for the release of digestive enzymes into the gastrointestinal lumen), and a much smaller endocrine component composed of the islets that are responsible for the regulated release of a variety of hormones into the circulation. Central component of the islet are the insulin secreting beta cells surrounded by a smaller amount of alpha cells (glucagon), delta cells (somatastatin), PP cells (pancreatic polypeptide) (Weir and Bonner-Weir, 1990) and perhaps Ghrelin secreting cells (Volante *et al.*, 2002; Wierup *et al.*, 2002). These cells release these hormones into the portal circulation and following first pass through the liver enter into the systemic circulation. Of all these cell types, the insulin secreting beta cells have been the primary focus of research effort, in part due to the severity of Type 1 diabetes in which cellular autoimmunity results in the destruction of beta cells and the loss of insulin secretion. In addition, Type 2 diabetes typically initiates with peripheral insulin resistance that continues to increase in severity but only progresses to the diabetic state when beta cells are no longer able to compensate for the worsening insulin resistance.

Cumulative studies on insulin secretion have defined several of the basic mechanisms responsible for insulin biogenesis and processing, dense-core granule formation, intracellular sorting and signaling pathways mediating the trafficking and fusion of insulin granules with the plasma membrane. Although this framework has provided significant insight into these processes, there are numerous areas of these signals and molecular pathways that remain to be studied. More importantly, many of the pathophysiological alterations in the coupling between different signaling pathways mediating insulin granule release remain at the forefront of our understanding of beta cell dysfunction that is leading cause of beta cell failure and hence Type II diabetes. In this chapter, we will attempt to review the current progress and our understanding of insulin granule biogenesis, sorting, trafficking and fusion with the plasma membrane resulting in insulin secretion and the mechanisms by which these events are controlled through intracellular signaling and metabolic pathways.

II. Section I

A. Insulin granule biogenesis (Fig. 16.1)

Insulin is initially synthesized as preproinsulin on the rough endoplasmic reticulum (RER) and during co-translational insertion into the lumen is converted to proinsulin by removal of the amino terminal signal sequence (Dodson and Steiner, 1998). The initiation of specific proinsulin sorting probably occurs in RER (Balch *et al.*, 1994; Tooze *et al.*, 1989) as many cargo proteins are concentrated during export from the endoplasmic reticulum (Balch *et al.*, 1994). In addition, zinc and calcium may become packaged with proinsulin in RER lumen en route to the Golgi (Howell *et al.*, 1978). However, the major sorting and packaging events appear to occur in Trans-Golgi Network (TGN).

B. Proinsulin TGN sorting

In the TGN cargo molecules destined for constitutive (unregulated secretion) are packaged into small transport vesicles that either directly traffic to and fuse with the plasma membrane or alternatively enters the

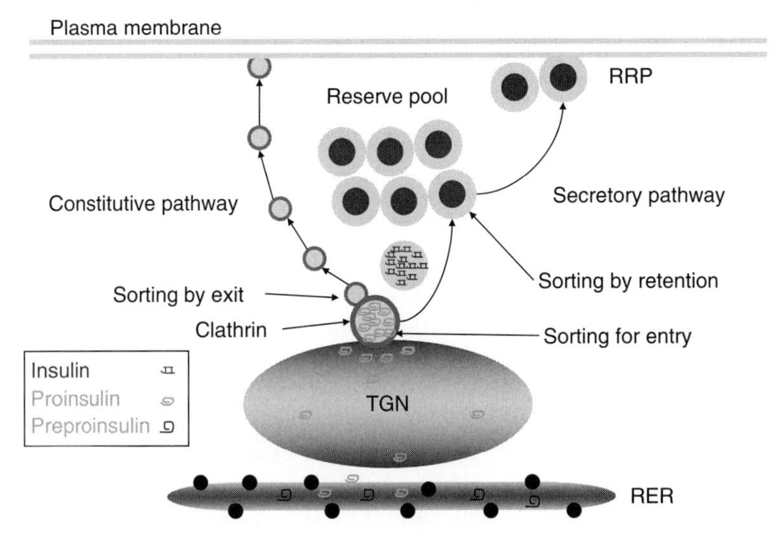

Figure 16.1 *Insulin biogenesis*: Preproinsulin is synthesized in rough endoplasmic reticulum (RER) and is converted to proinsulin. Proinsulin is later transported to Trans-Golgi Network (TGN) and packed into immature granule via sorting for entry. The immature granule loses the clathrin coating and other unwanted components through sorting by exit. In parallel, proinsulin is converted to insulin and condensation/crystallization (sorting by retention) occurs. These processes account for the maturation of the immature insulin granules into a mature insulin secretory granule following TGN exit.

constitutively recycling endosome system and thereby traffic to the plasma membrane. These constitutive endomembrane trafficking pathways allow for rapid movement of membrane phospholipids, integral membrane and secreted proteins from TGN to plasma membrane (Farquhar and Palade, 1981; Kelly, 1985; Palade, 1975). For example, the plasma form of alkaline phosphatase (SEAP) is release in a constitutive manner, and is restricted away from the regulated secretory pathway in beta cells (Molinete *et al.*, 2000b).

In contrast, the regulated secretory pathway involves the concentration and packaging of proinsulin into immature secretory granules, a process termed sorting for entry. The sorting and packaging of proinsulin into immature granule is very dynamic and highly efficient process with the vast majority (over 99%) of proinsulin molecules being correctly sorted to compartments that will directly generate dense-core glucose-stimulated secretory granules (Rhodes and Halban, 1987). Although the exact mechanism of proinsulin sorting in TGN is far from clear, it is thought to occur through interactions between elements of TGN membrane and specific features of the proinsulin molecule. Early studies suggested the presence of a TGN sorting receptor that bound to specific subtypes of cargo proteins allowing for a selective TGN membrane domain that would develop a clatherin coat (Halban and Irminger, 1994; Orci, 1982). The first identified "sorting receptor" was a 25kD Golgi protein but subsequent studies demonstrated that this was not a specific cargo recognizing receptor (Chung *et al.*, 1989; Irminger *et al.*, 1997; Thiele *et al.*, 1997). More recently, carboxypeptidase E (CPE), a prohormone converting enzyme, was suggested to have dual function, both as processing enzyme as well as the sorting receptor in all endocrine cells (Cawley *et al.*, 2003; Dhanvantari *et al.*, 2003). However, CPE loss of function failed to impair proinsulin targeting to the regulated pathway (Natori and Huttner, 1996) and is now thought to serve as a chaperone in the sorting of several prohormones (Kim *et al.*, 2001). Since no specific TGN sorting receptor has been identified, several interesting alternative models have emerged, such as the self-organization of TGN luminal and membrane lipids. In this model, proteins in the lumen of the TGN aggregate in a mildly acidic, high Ca^{2+} concentration environment, and the aggregated proteins directly interact with lipid membrane cholesterol, which in turn leads to reorganization of cholesetrol-rich microdomains. The subsequent budding of these cholesterol-rich microdomains provides the nascent characteristic of immature secretory granules that are destined to enter the regulated secretory pathway.

At present the hunt for specific sorting receptor/mechanisms remains an ongoing effort, however significant progress has been made in the identification of specific features of prohormones that are responsible for secretory pathway trafficking. For example, expression of mutants of the proinsulin at residues 16 (Leu) and 17 (Glu) of A-chain and 13 (Glu) and 17 (Leu) of

B-chain results in proinsulin default to the constitutive pathway in AtT20 or Neuro2a cells (Molinete *et al.*, 2000a). This result suggests that these regions may serve as "sorting domains." Other studies suggested that the sorting domains are located in the insulin B-chain of proinsulin via comparison of various prohormone sequences and molecular modeling techniques (Gorr and Darling, 1995; Kizer and Tropsha, 1991). However, most of these studies did not employ primary β cell or cultured β cell lines, and thus may not necessarily represent the sorting process of proinsulin from TGN to immature granules in β cells *in vivo*. Nevertheless, the current consensus is that proinsulin and other secretory granule proteins have specific sorting information for entry into the secretory granules in TGN.

C. Granule maturation

Although the TGN plays an important role in defining the targeting of proinsulin away from the large amount of bulk flow proteins thereby increasing proinsulin trafficking efficiency (sorting for entry), the immature secretory granule itself probably serves as a critical secondary decision point through the mechanisms termed sorting for exit and sorting by retention (Arvan and Castle, 1992, 1998; Arvan and Halban, 2004). As described above, proinsulin is initially packaged in the immature granules emanating from the TGN. Subsequent granule maturation can be divided into three steps: (i) acidification of the granule lumen, (ii) conversion of proinsulin to insulin and C-peptide through proteolysis by endoproteases PC1/3 and PC2, followed by trimming of the carboxyl termini carboxypeptidase E, and (iii) loss of the coat protein clatherin (Orci, 1982; Orci *et al.*, 1985, 1986). Several studies have shown that mildly acidic immature granule become more acidic as the granule mature (Orci, 1985; Orci *et al.*, 1986). The granule luminal ATP-dependent proton pump serves to produce the acidification of the granule milieu that facilitates the conversion of proinsulin to insulin since both convert enzymes (endoproteases PC1/3 and PC2) display an acidic pH optimum (Davidson *et al.*, 1988). The requirement for initial granule acidification was demonstrated by the pharmacological inhibition of the luminal ATP-dependent pump resulting in an elevation of intra-granular pH and inhibition of proinsulin conversion to insulin (Rhodes *et al.*, 1987). These data are also consistent with proinsulin being initially packaged into immature granule where the conversion to insulin occurs and that these processing events do not occur in the TGN (Huang and Arvan, 1994; Rhodes and Halban, 1987).

Insulin secretory granules exist as large dense-core structures that are easily discernible by electron microscopic as an electron dense interior surrounded by a clear region in a 300–350nm intracellular membrane delineated compartment (Greider *et al.*, 1969; Lange, 1974). The insulin dense-core structures are generated by a calcium and zinc-dependent condensation process (Hill *et al.*, 1991). During the secretory granule maturation, proinsulin converts to insulin by excision of the C-peptide through proteolysis by endoproteases PC1/3 and PC2, followed by trimming C-terminal by carboxypeptidase E. The conversion of $(Zn^{2+})_2(Ca^{2+})(\text{Proinsulin})_6$ to $(Zn^{2+})_2(Ca^{2+})(\text{Insulin})_6$ significantly decreases the solubility of the hexamer leading to crystallization within the lumen of the granule and formation of the dense-core granule (Dunn, 2005).

Several investigators initially suggested that the concentration of insulin initiates during exit from the endoplasmic reticulum and continues as part of the TGN sorting into immature secretory granules (Bauerfeind and Huttner, 1993; Chanat and Huttner, 1991; Sossin *et al.*, 1990; Tooze and Huttner, 1990). More recently, significant evidence has accumulated that the condensation reaction occurs within the secretory granule, rather than in TGN (for review, see Arvan and Castle (1992)). This was based both upon the presence of the prohormone convertase, Zinc transporter and ATP pump being present in insulin secretory granules (Arias *et al.*, 2000; Chimienti *et al.*, 2006; Rhodes *et al.*, 1987). Moreover, proinsulin has very poor avidity to form hexamers and a mutation of proinsulin that is defective for hexamerization tends to associate with constitutively secretory pathway (Carroll *et al.*, 1988; Gross *et al.*, 1989; Kuliawat *et al.*, 2000). On the other hand, other investigators observed that the same mutant was efficiently sorted to secretory granules and was retained in this compartment similar to that of fully processed insulin (Halban and Irminger, 2003). In this regard, guinea-pig insulin does not crystallize but these animals display relatively normal insulin secretion with an appropriate number and concentration of beta cell insulin granules (Arvan and Halban, 2004). Whether or not insulin condensation *per se* is a necessary sorting reaction, insulin granule biogenesis appears to require signals for both sorting by entry at the TGN and subsequent sorting for exit and sorting by retention in the post-TGN immature granules.

This latter process has been demonstrated for several immature granule proteins but is most clearly exemplified by proinsulin packaging into clathrin coated compartments during TGN exit, which is regulated via the AP-1 complex, and the ADP-ribosylation factor Arf-1 (Dittie *et al.*, 1996). Arf-1 is a small GTP binding protein that undergoes successive rounds of GTP binding and GTP hydrolysis to GDP. In the GTP bound state, Arf-1 is in an active conformation and interacts with effectors. However, in the GDP bound state the affinity of Arf-1 for effectors is markedly reduced resulting a dissociation of these interactions (Randazzo and Kahn, 1994).

As indicated above, the immature insulin granules function as a sorting compartment that during the maturation process in which proinsulin is converted to insulin there is a concomitant decrease in clathrin association (Orci *et al.*, 1985, 1986, 1987; Steiner *et al.*, 1987). Insulin and other insoluble granules components are retained within the granule while other non-selective soluble proteins and granule membrane components bud from the immature granule as clathrin-coated transport vesicles. This "sorting for exit" process removes the undesirable components from secretory granules thereby resulting in the maturation of the immature granules to mature granules (Feng and Arvan, 2003). It should be noted that although this post-granular constitutive-like retrieval pathway is generally accepted for insulin granule maturation, it remains controversial in other systems.

III. Section II

A. Insulin granule trafficking

It is generally accepted that there are at least two populations of insulin secretory granules, the readily releasable pool (RRP) that is responsible for the initial (first phase) insulin secretion and a second reserve pool that is responsible for a more prolonged (second phase) insulin secretion (Bratanova-Tochkova *et al.*, 2002; Rorsman and Renstrom, 2003; Rutter, 2001; Straub and Sharp, 2002). The readily releasable granule pool is apparently pre-docked at the cell surface membrane in a complex with SNARE and calcium-regulated proteins that allow for the rapid calcium-dependent fusion of primed insulin granules with the plasma membrane (Daniel *et al.*, 1999). Following this initial rapid first phase of calcium-dependent insulin secretion (1–5 min), a more prolonged (5–60 min) second phase insulin secretion results from the recruitment of reserve granule populations to the same release sites. Although the signaling events and fusion mechanisms have been intensively studied, there remain several unanswered questions with regard to the exact regulatory proteins and specific fusion reactions that define first and second phase secretion.

1. Metabolic signaling pathway leads to insulin secretion (Fig. 16.2)

The β cells coupled insulin secretion to nutrient availability through various pathways leading to calcium mobilization and activation of cAMP-dependent signaling cascades. Although beta cells are highly sensitive to fatty acid, amino acids and muscarinic agonists, glucose has been the most intensively studied metabolite trigger as dysregulation in plasma glucose levels is the hallmark of impaired glucose tolerance (insulin resistance) and the subsequent development of Type 2 Diabetes Mellitus.

The postprandial rise in plasma glucose levels results in increase flux of glucose into β cells via high capacity, low affinity glucose transporter 2 (GLUT2) isoform present in the plasma membrane (Thorens, 1992). The kinetic properties of GLUT2 are such that rate of glucose uptake is

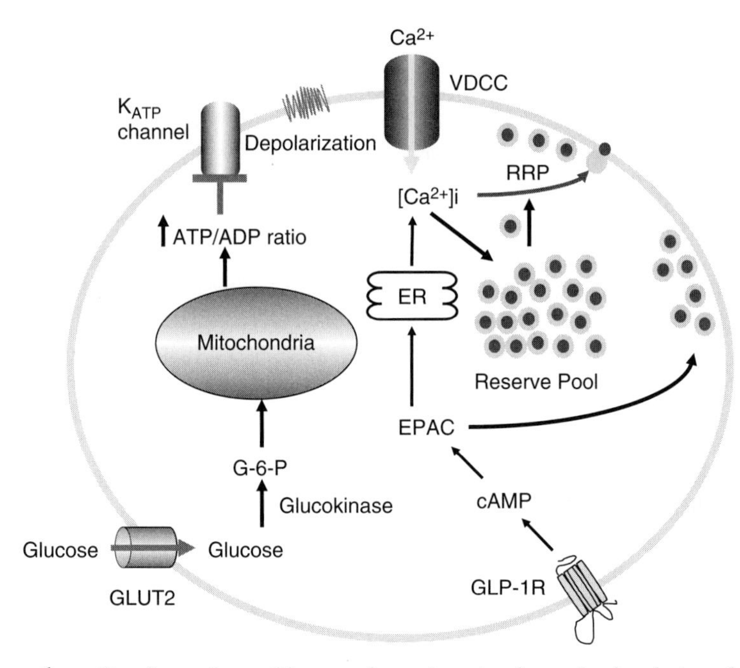

Figure 16.2 *Signaling pathway*: Glucose, the major stimulant, via glycolysis and mitochondrial ATP energy production increases the ATP/ADP ratio that leads to the closure of the ATP-sensitive K^+ (K_{ATP}) channels. The subsequent cellular depolarization activates voltage dependent calcium channels resulting in extracellular Ca^{2+} influx and fusion of insulin granules with the plasma membrane. The incretin hormone GLP-1 acts on its receptor at β cell plasma membrane to activate adenylyl cyclase and increase intracellular cAMP levels. In turn, cAMP binds and activates protein kinase A and EPAC. EPAC then functions to increase intracellular calcium level from intracellular calcium stores in the endoplasmic reticulum and to increase the number of readily releasable pool of insulin granules at the plasma membrane. The combination of these two processes results in a potentiation of insulin secretion.

proportional to the physiologic rise in glucose concentration (Newgard and McGarry, 1995). The essential nature of GLUT2 was demonstrated in GLUT2 null mice that fail to sense ambient glucose concentration but also display defects in pancreatic β cell development (Guillam *et al.*, 1997). Following plasma membrane transport, the cytosolic glucose is converted to glucose-6-phophate by the high K_M glucokinase (hexokinase IV), which is the rate-limiting step in glycolysis (De Vos *et al.*, 1995; Iynedjian, 1993; Newgard and McGarry, 1995). The importance of glucokinase as the "glucose sensor" (Matschinsky, 1990) in insulin secretion has been documented by the finding that the patients with a variety of mutations in glucokinase gene result in impaired insulin secretion and develop a genetic syndrome termed MODY for maturity onset diabetes of the young

(Fajans *et al.*, 2001). The production of glucose-6-phosphate drives an increase in energy production via glycolysis and carbohydrate oxidation most notably through an increase in the cytosolic ATP/ADP ratio (Rorsman *et al.*, 2000). In the pancreatic β cells, ATP-sensitive K^+ (K_{ATP}) channels set the membrane potential and closure of these channels results in depolarization that in turn opens voltage-dependent Ca^{2+} channels (VDCCs) resulting in Ca^{2+} influx. Insulin secretion is directly triggered through Ca^{2+} dependent insulin granule fusion with the plasma membrane as discussed below. However, the subsequent of opening of voltage-dependent K^+ channels limits Ca^{2+} influx and therefore terminates insulin secretion.

In 1995 the β cell K_{ATP} channel along with its regulatory sulphonylurea receptor (SUR) subunits were first cloned from insulinoma cells (Aguilar-Bryan *et al.*, 1995; Inagaki *et al.*, 1992; Sakura *et al.*, 1995). There are four pore forming inward-rectifier K^+ channel subunits (Kir6.2 in β cell) and four regulatory SUR subunits (Aguilar-Bryan and Bryan, 1999; Ashfield *et al.*, 1999). In the basal state, K_{ATP} has a high tendency to open and cause K+ efflux that maintains the membrane potential close to -70mV (Atwater *et al.*, 1979; Dukes and Philipson, 1996). Several modulators are thought to regulate K_{ATP} channels and in particular the Kir6.2 subunit is inhibited by ATP (John *et al.*, 1998; Mikhailov *et al.*, 1998; Tucker *et al.*, 1997) and mutation of the Kir6.2 subunit reverse the effect of ATP on K_{ATP} channel closure. Anti-diabetic sulphonylurea drugs have the similar property to induce the closure of the K_{ATP} channel but this class of drugs interact with SUR1 regulatory subunit (a member of the ATP-binding cassette super-family) to induce cellular depolarization (Aguilar-Bryan *et al.*, 1995; Ashfield *et al.*, 1999; Philipson and Steiner, 1995). Interestingly, the SUR1 subunit can bind to both ATP and MgADP, and binding of MgADP may stabilize the K_{ATP} channel in the open state. Since MgADP can be displaced when the ratio of ATP to ADP increase this allows the Kir6.2 subunit to display ATP inhibition (Reimann *et al.*, 2000). In addition to adenine nucleotides, other metabolic products may also modulate K_{ATP} channels independent of ATP (Ainscow *et al.*, 2002). For example, lipid-derived products such as long chain-coenzyme A derivatives activate mouse and human K_{ATP} channels and hyperpolarize β cell thereby reducing insulin secretion (Branstrom *et al.*, 1997, 2004).

In any case, the closure of ATP-sensitive K^+ channels results in cellular depolarization that activates L-type VDCCs resulting the influx of extracellular Ca^{+2} (Horvath *et al.*, 1998; Keahey *et al.*, 1989; Ligon *et al.*, 1998). The requirement for extracellular Ca^{+2} influx for insulin secretion has been well documented by numerous investigators and includes the use of Ca^{+2} ionophores to induce secretion, blockade of secretion by removal of extracellular Ca^{+2} (Ashby and Speake, 1975; Eddlestone *et al.*, 1995) and directed monitor of cytosolic free Ca^{2+} (Prentki and Wollheim, 1984). Moreover, membrane depolarization by electrophysiologic methods and KCl induced depolarization simulates whereas inhibition of L-type VDCC function inhibits *β* cell insulin secretion (Giugliano *et al.*, 1980). These data demonstrate L-type Ca^{2+} channels are required effectors for glucose-stimulated insulin secretion. In addition, influx of extracellular Ca^{2+} as well as other secretagogues acetylcholine and cholecystokinin can potentiate insulin secretion through the release of intracellular Ca^{2+} stores and activation of protein kinase C (Graves and Hinkle, 2003; Zawalich *et al.*, 1989).

Although the established linked between ATP/ADP ratio as the initiating event resulting in *β* cell depolarization and Ca^2+ mediated insulin secretion has been well documented, the metabolic pathway that generates the ATP responsible for this process has remained unresolved. The first step in glucose metabolism is the generation of glucose-6-phosphate that consumes one mole ATP and the conversion of fructose-6-phosphate to fructose-1,6-bisphosphate consumes another. However, ATP is generated at other steps during the glycolytic process that generates pyruvate, which is transported into the mitochondria to drive the tricarboxylic acid (TCA) cycle. Recent studies using stable isotope (^{13}C NMR) flux analysis, the recycling of pyruvate across the mitochondrial inner membrane was found to directly correlate with glucose-stimulated insulin secretion (Lu *et al.*, 2002). However, previous studies showed that inhibition of pyruvate transport into the mitochondria or reduction in TCA cycling did not prevent glucose-stimulated insulin secretion (Hildyard *et al.*, 2005; MacDonald *et al.*, 2005; Mertz *et al.*, 1996; Zawalich and Zawalich, 1997). Taking together, these data suggest that ATP levels generated by the TCA cycle probably does not contribute to glucose-dependent insulin secretion.

Alternatively, NADH enters the mitochondria via two shuttles (glycerol–phosphate and malate–aspartate shuttles) to produce ATP (Eto *et al.*, 1999; Hedeskov *et al.*, 1987; Malaisse *et al.*, 1979; Pralong *et al.*, 1990). Compared to most other mammalian cells, in *β* cells display high activities of the glycerol–phosphate shuttle and mitochondrial glycerol–phosphate dehydrogenase (mGPDH) suggesting an important role for NADH in *β* cell function (MacDonald *et al.*, 1990). However, inhibition of either shuttle independently fails to inhibit insulin-secretion whereas genetic ablation of both shuttles, completely abolished insulin secretion (Eto *et al.*, 1999). In addition, NADPH is another high energy intermediate in glucose metabolism and high ratios of NADPH to NADH appears to follow secretagogue-stimulated secretion (Ashcroft and Christie, 1979; Hedeskov *et al.*, 1987). Moreover, inhibition of NADPH production was found to decrease insulin secretion (Ammon and Steinke, 1972a,b; MacDonald *et al.*, 1974). Since mitochondrial oxidative metabolism generates nearly 98% of *β* cell ATP, it is likely that NADH and NADPH generated mitochondrial-derived ATP provides the critical regulation of *β* cell K_{ATP} channels (Dukes *et al.*, 1994; Jensen *et al.*, 2006; Ohta *et al.*, 1992).

2. The role of incretin hormones on insulin secretion

In addition to direct activators of insulin secretion, over the past several years a family of gut-derived regulatory hormones has been shown to function as regulators of insulin secretion. After absorption of nutrients, particularly glucose, the intestine endocrine L cells secrete gluco-incretin hormones. GLP-1 and glucose-dependent insulinotropic peptide (GIP, also known as gastric inhibitory polypeptide) are the two main incretin hormones that have been intensively investigated (Drucker *et al.*, 1987; Meier *et al.*, 2002a,b; Mojsov *et al.*, 1987). GLP-1 is produced and secreted in the intestinal L cells by the prohormone convertase PC1/3 mediated posttranslational procession of proglucagon (Dhanvantari *et al.*, 2001). In human, the major biological form of GLP-1 is the GLP-1 (7–36) amide form, another bioactive GLP-1 (7–37) form is also present but at much lower levels. GLP-1 is coupled to a specific G-protein coupled receptor and when activated increases adenylyl cyclase activity and subsequent activation of the cAMP-dependent second messengers pathways such as PKA and EPAC. The acute effect of GLP-1 is to enhance glucose-stimulated insulin secretion and accounts for increase in insulin secretion observed between an oral glucose tolerance test and an intraperitoneal glucose tolerance test. However, GLP-1 also displays long-term effects to promote insulin synthesis, β cell proliferation and neogenesis (Stoffers, 2004). Naturally occurring GLP-1 receptor agonist, exendin-4, now being used to treat type 2 diabetes and significantly improves glucose tolerance, first and second phase insulin secretion along with weight loss (Doyle and Egan, 2007). Physiologically, GLP-1 has a very short half-life in the plasma being inactivated by dipeptidylpeptidase-IV (DDP-IV, also known as CD26) proteolytic cleavage (Hansen *et al.*, 1999). This enzyme presents in blood stream and cell membrane and cleaves GLP-1 (7–36) to the inactive form GLP-1(9–36) (Hansen *et al.*, 1999).

IV. Section III

A. Granule fusion

1. SNARE-dependent fusion

Many models have been proposed to elucidate the mechanism of regulated secretory granule exocytosis. It is now well established that SNARE proteins are the minimal machinery required for *in vivo* membrane fusion (Leabu, 2006). SNARE proteins belong to a superfamily consisting of over 35 proteins that share a common structural SNARE domain (Weimbs *et al.*, 1997). Based upon structural considerations required for the formation of the fusogenic SNAREpin complex, the SNARE proteins have been further classified into R-SNAREs and Q-SNAREs depending on whether the central amino acid in the SNARE domain is an arginine (R-SNARE) or a glutamine (Q-SNARE) residue, respectively (Bock *et al.*, 2001). The subset of SNARE proteins required for plasma membrane fusion in a variety of cell types is restricted to the Q-SNAREs, syntaxin 1–4, SNAP25, SNAP23, and the R-SNAREs (VAMP1–3). Specific interaction between a vesicle-SNARE with the cognate target-SNARE leads to the formation of a SNAREpin complex in which four SNARE motifs assemble into a twisted parallel four-helical bundle (Hong, 2005). It is this helical structure that catalyzes the fusion of vesicles with their target compartments.

In the case of insulin release, several studies have established that the t-SNARE proteins syntaxin 1 (Stx1) and SNAP25 are plasma membrane localized whereas the v-SNARE protein VAMP2 is associated with insulin secretory granules. The functional role for this core SNARE complex is essential for insulin secretion based upon the use of various proteolytic neurotoxins and expression of dominant-interfering and protease resistant mutants (Boyd *et al.*, 1995; Gonelle-Gispert *et al.*, 1999, 2000; Huang *et al.*, 1998; Regazzi *et al.*, 1995, 1996; Sadoul *et al.*, 1995, 1997; Wheeler *et al.*, 1996). However, although the SNARE proteins forms the catalytic entity required for membrane fusion they are not sufficient to account for rapid Ca^{2+} dependent exocytosis in most secretory cells. Thus, in addition to this core SNARE complex, several accessory proteins have also been implicated in the regulation of the granule docking/priming/fusion process. Synaptotagmin is currently the most favorite candidate Ca^{2+}-sensor in synaptic vesicle fusion (Chapman, 2002; Shin *et al.*, 2002; Sugita *et al.*, 2002). Synaptotagmin also comprise a very large group of proteins with most of the family members containing two Ca^{2+} binding sites, termed C2A and C2B. Synaptotagmin I and II are essential for synaptic vesicle exocytosis but do not appear responsible for releasing of large dense-core vesicles (Sugita *et al.*, 2002). In β cells, the Ca^{2+}/phospholipid binding proteins synaptotagmin III, IV, V, VII and VIII and syncollin are insulin granule membrane proteins that have been suggested to be responsible for mediating calcium-dependent secretion. Syncollin has been shown to interact with syntaxin 1 and to dissociate in the presence of physiologic concentrations of calcium that induce membrane fusion (Edwardson *et al.*, 1997). Over expression of syncollin in isolated islets inhibits glucose-stimulated insulin secretion, however, there are no direct data demonstrating a secretagogue-dependent dissociation of syncollin from Stx1 or an increase in VAMP2 association in β cells (Hays *et al.*, 2005; Li *et al.*, 2005). On the other hand, over expression of synaptotagmin III and/or VII increases calcium-mediated insulin secretion whereas blocking antibodies to both synaptotagmin I and II also inhibit calcium-stimulated exocytosis (Gao *et al.*, 2000; Lang *et al.*, 1997). Synaptotagmin peptide inhibitory experiments have suggested that its isoforms,

V, VII and VIII may mediate calcium-regulation of exocytosis whereas siRNA mediated gene silencing indicate that V and IX are responsible (Gut *et al.*, 2001; Iezzi *et al.*, 2004). In addition, synaptotagmin I deficient mice display a prolonged delay and slowed calcium-dependent fusion of large dense-core granules in chromaffin cells (Voets *et al.*, 2001a). Thus, it remains unclear whether a single specific synaptotagmin isoforms is required or that these calcium sensors function in combinatorial manner.

Biochemical and cell biological studies have developed a model for calcium stimulated membrane fusion of primed vesicles assembled at the plasma membrane. Recent studies have indicated that the Q- and R-SNAREs are partially zippered but are prevented from fully assembling into the four-α-helical bundle due to the presence of another protein, termed complexin (Pobbati *et al.*, 2004). Thus complexin appears to function as a repressor of fusion by pausing complete SNAREpin formation. Calcium binding to synaptotagmin blocks the inhibitory function of complexin and thereby allows fusion to rapidly proceed (Tang *et al.*, 2006). Consistent with this model, complexin I and II are both expressed in β cells and siRNA-mediated gene silencing of complexin has been found to inhibit depolarization-stimulated (calcium) mediated insulin secretion (Abderrahmani *et al.*, 2004).

In addition to the potential roles of syncollin, synaptotagmin and complexin as part of the calcium fusion sensor, the voltage-sensitive N- and L-subtype calcium channels directly bind several SNARE proteins. For example, the N-type channels associates with SNAP25 and syntaxin 1 and expression of the interacting domain of the channel inhibited calcium-dependent secretion (Wiser *et al.*, 1999b). Similarly, in pancreatic β cells the L-subtype calcium channel (primarily Ca(V)1.2) can physically associate with syntaxin, synaptotagmin and SNAP25 (Bokvist *et al.*, 1995; Trus *et al.*, 2001; Wiser *et al.*, 1999b; Yang *et al.*, 1999). Interestingly, syntaxin binding appears to modulate the channel kinetics and injection of the channel peptide responsible for SNARE interaction inhibits insulin secretion in response to channel opening but does not alter calcium current properties or the release of insulin in response to photolysis of caged calcium. These data suggest that the calcium channel is part of the fusion machinery directly coupling spatially localized calcium influx to the rapidly releasable granule pool. Similarly the voltage sensitive potassium channel Kv2.1 can also interact with syntaxin 1 (Leung *et al.*, 2003, 2005).

Although not directly part of the calcium regulated secretory process, the syntaxin binding protein Munc18a (n-sec1) plays a critical modulatory role (both positive and negative) in the granule docking/fusion process. In multiple systems, over expression of Munc18 proteins has a marked inhibitory effects to reduce vesicle exocytosis (Houng *et al.*, 2003; Khan *et al.*, 2001; Pevsner *et al.*, 1994b; Schulze *et al.*, 1994; Thurmond *et al.*, 1998, 2000; Verhage *et al.*, 2000; Yang *et al.*, 2000; Zhang *et al.*, 2000).

These data suggest that Munc18 is a negative regulator of syntaxin function and is consistent with the ability of Munc18 to compete for SNAP25 and VAMP2 binding to syntaxin (Araki *et al.*, 1997; Dulubova *et al.*, 2003; Perez-Branguli *et al.*, 2002; Pevsner *et al.*, 1994a; Tamori *et al.*, 1998; Tellam *et al.*, 1997). In addition, structural analysis has revealed that Munc18a maintains syntaxin 1 in a closed conformational state that is inaccessible for SNAP25 and VAMP binding (Bracher and Weissenhorn, 2001; Dulubova *et al.*, 1999; Misura *et al.*, 2000; Yang *et al.*, 2000). Thus, it has been postulated that Munc18a must dissociate and/or undergo a conformational change to allow for the conversion of syntaxin to the open state required for SNAP25 and subsequently VAMP2 binding. Although several studies have suggested that the Munc18 proteins undergo serine/threonine phosphorylation by PKC and Cdk5 these modifications occur very slowly and kinetically cannot account for the initial steps of granule release (Barclay *et al.*, 2003; de Vries *et al.*, 2002; Lilja *et al.*, 2004). Moreover, neither a dissociation nor conformational state change of Munc18a has yet to be detected within the time frame necessary to promote exocytosis.

On the other hand, there is also strong evidence that Munc18a plays a necessary positive role in vesicle membrane exocytosis. For example, using *in vitro* fusion assays, Munc18 was recently shown to accelerate the rate of SNARE-dependent membrane fusion suggesting a positive fusogenic role (Shen *et al.*, 2007; Tareste *et al.*, 2008). Loss of function of Munc18 homologues in yeast, flies and mice completely abrogates the vesicle secretory process (Harrison *et al.*, 1994; Novick *et al.*, 1980; Verhage *et al.*, 2000; Voets *et al.*, 2001b). Although homozygotic Munc18a knockout mice have apparently normal neural development they die at birth and are nearly completely devoid of synaptic transmission (Verhage *et al.*, 2000). Similarly, disruption of the Munc18 interaction with syntaxin through the use of blocking peptides also prevents vesicle fusion events (Dresbach *et al.*, 1998; Thurmond *et al.*, 2000). These data suggest that the Munc18a protein has multiple functions displaying very complex but distinct properties probably in different stages of docking, priming and fusion process. In this regard, studies in Munc18a null chromaffin cells have observed the expected inhibition of calcium-dependent exocytosis with a reduction in the number of docked vesicles. Surprisingly however, the kinetics of the residual release as well as single fusion events was not different from control cells (Voets *et al.*, 2001b). One interpretation of these data is that the relative level of expression of Munc18a is responsible for the docking of secretory vesicles and conversion from the docked to the primed (readily releasable state) but is not involved in the fusion process *per se*. Thus, Munc18a probably is required for the docking of secretory vesicles (positive role) and in its absence the number of docked vesicles is decreased, hence there is decreased secretion. In contrast, over expression would increase the number of docked vesicles but would prevent vesicle priming (formation of the high affinity four helix bundle between syntaxin/SNAP25 and VAMP2), hence reduced secretion. Consistent with this interpretation, Munc18a is expressed in islet cells and also appears to serve as a negative regulator of insulin secretion when over expressed (Zhang *et al.*, 2000). Moreover, Munc18 is decreased along with SNAP25 in islets with impaired insulin secretion from GK rats, a non-obese model of NIDDM (Zhang *et al.*, 2002).

2. Fusion pore opening

There are three general fusion processes that have been proposed. The first is a complete fusion of the granule membrane with the plasma membrane that results in the emptying of the granule contents and complete mixing of the granule membrane contents (membrane proteins and lipids) with the plasma membrane. The second is a kiss and run type mechanisms in which transient pores open between the granule membrane and the plasma membrane allowing for the partial or full release of the granule contents followed by a closure of the membrane fusion pore. A subset of the kiss and run model is a process termed cavicapture in which only selective component of the granule membrane and perhaps granule content undergo exocytosis before the membrane fusion pore closes.

Several approaches have been developed to examine granule fusion and insulin release from pancreatic β cells. Early studies took advantage of cell surface membrane capacitance measurements to examine granule fusion events. Membrane capacitance is proportional to the surface area and therefore an increase in membrane capacitance reflects and increases in the insertion of a granule membrane (Neher, 1998; Neher and Marty, 1982). Since insulin granules are significantly larger (~300–400nm diameter) compared to small transport vesicles (~50 nm diameter) changes in membrane capacitance is sufficiently sensitive to detect single fusion events. These types of analyses demonstrated increases in steady state in β cell membrane capacitance that occurred in a step wise fashion (quantal) suggesting each insulin granule undergoes a full fusion with the plasma membrane (Ma *et al.*, 2004; Orci *et al.*, 1973; Takahashi *et al.*, 2002). Although membrane capacitance is a very powerful and informative tool, it does not distinguish the dynamic interaction between exocytosis and endocytosis but only the change in the steady state membrane surface area. Furthermore it does not discriminate between which membranes are undergoing fusion/endocytosis with the plasma membrane or whether there is an actual release of cargo contents.

To address this latter issue, carbon fiber amperometry provides a direct way to monitor the exocytosis of granule contents following membrane fusion. This technique delivers high voltage that causes oxidation of released substance that generates a current spike and has been successfully used for numerous types of exocytosis events including insulin (Huang *et al.*, 1995). Carbon fiber amperometry studies have indicated that a single secretory granule in a human β cell contains 1.7amol of insulin, equal to 118mmol/l of intra-granular insulin. These studies also demonstrated that pH gradient between the granule lumen and extracelluar fluid is required for insulin release and the dissociation of the Zn-insulin complex limits the rate of hormone release (Aspinwall *et al.*, 1997). As with membrane capacitance measurements, carbon fiber amperometry is a very sensitive measurement of quantal release events. However, this approach only measures content release and does provide information with regard to the mechanism of membrane fusion itself.

More recent advances in fluorescent microscopy have allowed use of various fluorescently tagged reporter proteins to examine the dynamics of membrane fusion. In particular, confocal fluorescent microscopy and total internal reflectance fluorescent (TIRF) microscopy has markedly increased the sensitivity and spatial resolution so that time-lapse visualization can be routinely performed (Ohara-Imaizumi *et al.*, 2002; Steyer *et al.*, 1997). These techniques have primarily utilized chimeric reporter proteins that are fused with various colored fluorescent protein tags, such as green fluorescent protein (GFP), red fluorescent protein (RFP), cyan fluorescent protein (CFP). Although these approaches allow for dynamic visualization, the spatial resolution remains limited to the wavelength of visible light (~220nm) and the introduction of the relatively large fluorescent reporter protein may in of itself alter the properties of the trafficking and fusion events observed.

Several studies have reported that individual insulin granules independently form a plasma membrane fusion and release their intralumenal cargo followed by the flattening and mixing of the membrane components with the plasma membrane (Takahashi *et al.*, 2002, 2004). In this regard, others have suggested that the release of the insulin-containing dense-core granule occurs *en mass* consistent with the formation of a plasma membrane pore that fully expands to encompass the granule membrane proteins and lipids (Ma *et al.*, 2004; Orci *et al.*, 1973). However, in addition to individual insulin granule fusion events, it has also been suggested that insulin granules secretion may occur through a compound exocytosis mechanism whereby already fused granules undergo additional fusion events with other granules, termed sequential exocytosis (Bokvist *et al.*, 2000; Leung *et al.*, 2002; Orci and Malaisse, 1980). These classic models of exocytosis involves complete fusion of the vesicular and plasma membranes, followed by retrieval of membrane at a different site (Ohara-Imaizumi *et al.*, 2002).

However, other studies have suggested that a transient fusion pore can occur that results in either non-selective (kiss and run) or selective (cavicapture) release of granule contents (Tsuboi and Rutter, 2003; Tsuboi *et al.*, 2004). In particular, it was reported that the insulin granule containing membrane protein phogrin (phosphatase-like protein on the granules of insulin containing cells) is not transferred to the plasma membrane following granule secretion (Tsuboi *et al.*, 2004). Furthermore, using TIRF microscopy, dynamin was observed to frequently accumulate at the sites of β cell granule release whereas typical markers of clathrin-mediated endocytosis (clathrin or epsin) were generally absent from these release domains. This study also examined the trafficking/release of single granule events that appeared to be unaffected.

Further studies revealed that a neuropeptide Y-Venus fusion protein, a highly fluorescent and pH-insensitive variant of GFP, could also be released from vesicles without full membrane fusion (Nagai *et al.*, 2002). TIRF microscopy demonstrated that the granule membrane remained intact at cell surface while the release of cargo proteins from the same pool of granules frequently occurred (Rutter and Tsuboi, 2004). In another cell type, PC12 (a pheochromocytoma cel line), Taraaska *et al.* observed the partial release of the cargo protein plasminogen activator (tPA), whereas phogrin again remained on the membrane of the same vesicle. Taking together, these findings proposed a unique mechanism, namely, kiss and run fusion model to elucidate insulin granule exocytosis events. The granule content is released via the fusion pore that transiently and reversibly opens during the exocytosis. In this model, the fusion pore can close even before the release of granule content and without mixing of the granule membrane with the plasma membrane (Galli and Haucke, 2001; Jarousse and Kelly, 2001). In fact, it was estimated that 90% of the release events occurs via a selective kiss and run mechanism with only 10% accounted for by complete fusion (Obermuller *et al.*, 2005).

3. Membrane retrieval

At present it is difficult to reconcile the apparent different results obtained by capacitance measurements versus that observed by fluorescent microscopy. The increase in capacitance must necessarily occur as a result of a steady-state increase in membrane surface area. In contrast, a kiss and run mechanism necessarily results in transient fusion pore openings that will not contribute to a stable increase in membrane capacitance. If a kiss and run mechanism was predominantly responsible for insulin secretion there would not be a requirement for membrane retrieval (endocytosis) as there will be little contribution to an increase in membrane surface area. However, if complete fusion occurs the subsequent increase in plasma membrane mass must be balanced by an equivalent amount of membrane retrieval.

Capacitance measurement in chromaffin cells demonstrated that two kinetically distinct endocytosis events occur after exocytosis, one with a rapid rate (time constant 10s), which requires dynamin 1 binding to the endocytotic vesicles that is calcium dependent but independent of clathrin coats (Eliasson *et al.*, 1996). The second is a slow endocytosis process (time constant 100s) that involves both clathrin and dyanmin 2 but does not require any change in calcium concentration (Gopel *et al.*, 1999).

The rapid endocytosis process appears to be more important following weak stimulation, whereas the slow type is responsible for intense stimulation that resulted in large exocytotic increases in membrane surface area. Tsuboi *et al.* using a DsRed-VAMP2 chimera and endocytosis with dynamin 1 EGFP showed that only 20% of exocytotic events appear to involve in endocytosis in PC12 cells (Tsuboi *et al.*, 2002). However, these earlier studies did not investigate the recruitment of classic endocytotic pathway proteins such as clathrin and other proteins including the adaptor protein Epsin, which binds to adaptor protein-2 and results in formation of clathrin coats around vesicles (Barg *et al.*, 2001). The interaction between the Epsin N-terminal domain and phosphatidylinositol-4,5-bisphosphate initiates the binding of Epsin to the plasma membrane that generates a signal for endocytosis (Wiser *et al.*, 1999a). The insertion of Epsin into the membrane appears to cause membrane curvature (Curry *et al.*, 1968), which in turn facilitates vesicle formation and recapture.

Whether or not full fusion or a kiss and run type mechanism occurs, recent studies have suggested a functional coupling between insulin secretion (exocytosis) and endocytosis. As indicated above, dynamin is a GTPase protein that is required for the severing of the plasma membrane invagination that are typically clathrin coated (coated pits) that will form the endocytotic vesicle (Hill *et al.*, 2001). Inhibition of dynamin function results in an accumulation of coated pits with elongated necks that are unable to undergo scission effectively blocking endocytosis (Damke *et al.*, 1994). Expression of a dominant-interfering dynamin mutant or siRNA knockdown of dynamin 2 in β cells was reported to inhibit depolarization stimulated insulin secretion (Min *et al.*, 2007). Interestingly, the very initial secretion was relatively normal whereas the remaining first and second phase secretion was nearly completely inhibited. Similar findings using a blocking dynamin antibody have also been observed catecholamine release in chromaffin cells (Elhamdani *et al.*, 2001). These data suggest that secretory cells have a mechanism that couples the rate and/or extent of membrane exocytosis with that of endocytosis.

V. Section IV

A. Biphasic insulin secretion

The majority of studies previously described have focused on the initial insulin granule fusion and release process. However, there are at least two populations of insulin secretory granules, the RRP that is responsible for the initial (first phase) insulin secretion and a second reserve pool that is responsible for a more prolonged (second phase) insulin secretion (Bratanova-Tochkova *et al.*, 2002; Rorsman and Renstrom, 2003; Rutter,

2001; Straub and Sharp, 2002). The readily releasable granule pool is apparently pre-docked at the cell surface complex with syntaxin 1 (Stx1), SNAP25 and VAMP2 together with one of the calcium-regulated protein, synaptotagmin (Daniel *et al.*, 1999). Following the initial rapid first phase calcium-dependent insulin secretion, second phase secretion results from the recruitment of the reserve granules to the same release sites that are also dependent on t- and v-SNARE interaction. Interestingly, a recent study has also implicated Stx4 in facilitating biphasic insulin secretion (Spurlin and Thurmond, 2006). The exact mechanism involved, however, is not well characterized and specific details of the fusion process responsible for first and second phase insulin secretion is far from clear.

1. Characteristic feature of insulin granule biphasic insulin secretion

The RRP has been estimated to account for 1–5% of the granules, that is 20–100 granules depending on the conditions of the experiments (Gromada *et al.*, 1999; Renstrom *et al.*, 1997), which can undergo exocytosis following stimulation without any further modification. It is estimated that about 40 granules per beta cell that undergo exocytosis during the first-phase insulin secretion (Rorsman *et al.*, 2000). Thus, the majority of the first-phase insulin secretion could be attributable to exocytosis of RRP granules (Rorsman *et al.*, 2000). Since the first-phase insulin secretion is only transient and once RRP has been depleted, the reserve pool granules start to process exocytosis at a much slower rate to supply new granules for the second-phase insulin secretion. However, recruiting insulin granules from reserve pool that consists of 95–99% of total insulin granules requires a series of ATP, Ca^{2+}, time, and temperature dependent reactions. This process is also referred to as mobilization or priming in order to obtain release competence. Moreover, the capacitance measurements demonstrate exocytosis of RRP granules does not require ATP hydrolysis, in contrast refilling the RRP pool is highly ATP dependent (Eliasson *et al.*, 1997). Interestingly, activation by cAMP elevation and the incretin hormone GLP-1 facilitate granule mobilization and increase the size of RRP in both mouse β cells and mouse islets (Eddlestone *et al.*, 1985; Renstrom *et al.*, 1997). The mobilization involves the interaction between t-SNARE and v-SNARE, the formation of SNARE complexes (Rettig and Neher, 2002; Xu *et al.*, 1999) and additional maturation reactions leading to an increased probability of release.

As mentioned above, the SNARE proteins play very important role in the granule membrane fusion. More importantly SNARE proteins are the security to ensure Ca^{2+} is restricted to the plasma membrane area in close contact with the secretory granules. The L-type Ca^{2+} channel (L-loop) binds to syntaxin, SNAP 25 and synaptotagmin, this in turn will secure the Ca^{2+} channel to the secretory granule. This unique arrangement will allow

RRP granules expose to high level of Ca^{2+} to ensure high efficient exocytosis (Rorsman *et al.*, 2000). Since the first detailed ultrastructural analysis of pancreatic β cells were published (Dean, 1973), electron microscopy become a very powerful tool to study the functional and ultrastructural correlates of insulin secretion. Using electron microscope, Olofesson *et al.* reported that mouse beta cells contain 10,000 granules per cell (Olofsson *et al.*, 2002), with approximately 5% of the granules pre-docked close to membrane and another 2000 granules located less than one granule distance from the plasma membrane. This study used very high extracellular K^+ (depolarization) to stimulate first-phase insulin secretion in the absence of glucose. This increased insulin secretion is associated with 30% reduction of pre-docked insulin granules. Under these conditions, a second treatment with K^+ failed to further increase insulin secretion. In contrast, following K+ depolarization glucose was still able to enhance insulin secretion (Olofsson *et al.*, 2002). Data from electron microscopy revealed that the depletion of docked granule pool is much more than the defined RRP by capacitance measurements. One possibility for this discrepancy is the presence of another pool of granules named "nearly RRP" and these types of granules can maintain the competence of RRP that is responsible for repopulating the RRP in an energy (ATP) dependent process (Henquin *et al.*, 2002). Thus, a current model is that following an initial burst by the RRP the nearly RRP granules account for the total amount quantity of insulin release during first phase secretion. Subsequently, second phase secretion requires the mobilization of the reserve pool to replenish the RRP, which requires energy (ATP) and calcium (Rorsman and Renstrom, 2003).

As previously indicated, the reserve pool represents more than 95% of the secretory insulin granules that are mobilize to and replete the RRP to maintain sustained (second phase) insulin release in respond to secretagogue stimulation. (Rorsman *et al.*, 2000). Cyclic AMP stimulated and Ca^{2+} dependent insulin secretion was associated with the enhancement of insulin release from RRP and the acceleration of refilling of this pool from reserve pool (Renstrom *et al.*, 1997). Several studies have also suggested the existence of two granule population including fast-moving granules and random but restrict movement granules in reserve pool (Ivarsson *et al.*, 2004; Pouli *et al.*, 1998; Varadi *et al.*, 2003). Recent image analysis data examining the tracking of individual granule motions support this model and further indicate that the fasting-moving granules accounting for less than 10% granules are required to replete the RRP (Hao *et al.*, 2005). More recently, the small GTPase RalA has been shown to be involved in the regulation of the RRP mobilization from the reserve pool. Depletion of RalA in islet and cultured INS beta cells blocked insulin secretion accompanied with the reduction in the subsequent mobilization and exocytosis of the reserve pool of granules (Lopez *et al.*, 2008).

In summary, we discussed the insulin biogenesis, insulin granule sorting, maturation, distribution, signaling pathway and exocytosis. Sorting for entry, sorting by exit and sorting by retention are currently proposed models to elucidate insulin granule sorting. During the sorting process, proinsulin is converted to insulin followed by condensation/crystallization while immature proinsulin granule becomes mature secretory granule and is distributed in two different pools, namely, readily release pool and reserve pool. The existence of two pools of insulin secretory granule appears to account for the biphasic secretion pattern. It has been well documented that elevation of intracellular Ca^{2+} play critical role in insulin secretion, facilitating the SNARE complex regulated membrane fusion and enhancing mobilization of the granule from reserve pool to replete RRP. Glucose, the major stimulant, via glycolysis increases the ATP/ADP ratio that lead to close the ATP-sensitive K^+ (K_{ATP}) channels and closure of these channels results in depolarization and subsequent Ca^{2+} influx. The research on the incretin hormones has provided a unique way to treat Type 2 Diabetes Mellitus. Exenatide, an analog of human GLP1 and incretin mimetics, has been successfully used clinically. GLP-1 binds its receptor on β cell plasma membrane via cyclic AMP signal pathway to promote insulin synthesis, β cell proliferation and neogenesis. Exocytosis occurs when the granule membrane fuses with plasma membrane that lead release of the granule components. The role of SNARE complex in insulin exocytosis has been well established and regulator proteins such as syncollin, synaptotagmin and complexin have been proposed as brakes between the t-SNARE and v-SNARE complex. These proteins act as calcium sensors that trigger the fusion event when bind with calcium currently, there are several working models full fusion, Kiss and run and cavicapture that can account for insulin granule fusion. Further studies are still necessary to determine whether one of these models, a combination or an alternative model can fully describe the dynamics of beta cell insulin secretion.

REFERENCES

Abderrahmani, A., Niederhauser, G., Plaisance, V., Roehrich, M. E., Lenain, V., Coppola, T., Regazzi, R., and Waeber, G. (2004). Complexin I regulates glucose-induced secretion in pancreatic beta-cells. *J. Cell Sci.* **117,** 2239–2247.

Aguilar-Bryan, L., and Bryan, J. (1999). Molecular biology of adenosine triphosphate-sensitive potassium channels. *Endocr. Rev.* **20,** 101–135.

Aguilar-Bryan, L., Nichols, C. G., Wechsler, S. W., Clement, J. P. t., Boyd, A. E., 3rd, Gonzalez, G., Herrera-Sosa, H., Nguy, K., Bryan, J., and Nelson, D. A. (1995). Cloning of the beta cell high-affinity sulfonylurea receptor: A regulator of insulin secretion. *Science* **268,** 423–426.

Ainscow, E. K., Mirshamsi, S., Tang, T., Ashford, M. L., and Rutter, G. A. (2002). Dynamic imaging of free cytosolic ATP concentration during fuel sensing by rat hypothalamic neurones: Evidence for ATP-independent control of ATP-sensitive K(+) channels. *J. Physiol.* **544,** 429–445.

Ammon, H. P., and Steinke, J. (1972a). 6-Amnionicotinamide (6-AN) as a diabetogenic agent. *In vitro* and *in vivo* studies in the rat. *Diabetes* **21,** 143–148.

Ammon, H. P., and Steinke, J. (1972b). Effect of 6-aminonicotinamide on insulin release and C-14 glucose oxidation by isolated pancreatic rat islets: Difference between glucose, tolbutamide and aminophylline. *Endocrinology* **91,** 33–38.

Araki, S., Tamori, Y., Kawanishi, M., Shinoda, H., Masugi, J., Mori, H., Niki, T., Okazawa, H., Kubota, T., and Kasuga, M. (1997). Inhibition of the binding of SNAP-23 to syntaxin 4 by Munc18c. *Biochem. Biophys. Res. Commun.* **234,** 257–262.

Arias, A. E., Velez-Granell, C. S., Mayer, G., and Bendayan, M. (2000). Colocalization of chaperone Cpn60, proinsulin and convertase PC1 within immature secretory granules of insulin-secreting cells suggests a role for Cpn60 in insulin processing. *J. Cell Sci.* **113**(Pt 11), 2075–2083.

Arvan, P., and Castle, D. (1992). Protein sorting and secretion granule formation in regulated secretory cells. *Trends Cell Biol.* **2,** 327–331.

Arvan, P., and Castle, D. (1998). Sorting and storage during secretory granule biogenesis: Looking backward and looking forward. *Biochem. J.* **332**(Pt 3), 593–610.

Arvan, P., and Halban, P. A. (2004). Sorting ourselves out: Seeking consensus on trafficking in the beta-cell. *Traffic* **5,** 53–61.

Ashby, J. P., and Speake, R. N. (1975). Insulin and glucagon secretion from isolated islets of Langerhans. The effects of calcium ionophores. *Biochem. J.* **150,** 89–96.

Ashcroft, S. J., and Christie, M. R. (1979). Effects of glucose on the cytosolic ration of reduced/oxidized nicotinamide-adenine dinucleotide phosphate in rat islets of Langerhans. *Biochem. J.* **184,** 697–700.

Ashfield, R., Gribble, F. M., Ashcroft, S. J., and Ashcroft, F. M. (1999). Identification of the high-affinity tolbutamide site on the SUR1 subunit of the K(ATP) channel. *Diabetes* **48,** 1341–1347.

Aspinwall, C. A., Brooks, S. A., Kennedy, R. T., and Lakey, J. R. (1997). Effects of intravesicular H+ and extracellular H+ and Zn2+ on insulin secretion in pancreatic beta cells. *J. Biol. Chem.* **272,** 31308–31314.

Atwater, I., Dawson, C. M., Ribalet, B., and Rojas, E. (1979). Potassium permeability activated by intracellular calcium ion concentration in the pancreatic beta-cell. *J. Physiol.* **288,** 575–588.

Balch, W. E., McCaffery, J. M., Plutner, H., and Farquhar, M. G. (1994). Vesicular stomatitis virus glycoprotein is sorted and concentrated during export from the endoplasmic reticulum. *Cell* **76,** 841–852.

Barclay, J. W., Craig, T. J., Fisher, R. J., Ciufo, L. F., Evans, G. J., Morgan, A., and Burgoyne, R. D. (2003). Phosphorylation of Munc18 by protein kinase C regulates the kinetics of exocytosis. *J. Biol. Chem.* **278,** 10538–10545.

Barg, S., Ma, X., Eliasson, L., Galvanovskis, J., Gopel, S. O., Obermuller, S., Platzer, J., Renstrom, E., Trus, M., Atlas, D., Striessnig, J., and Rorsman, P. (2001). Fast exocytosis with few Ca(2+) channels in insulin-secreting mouse pancreatic B cells. *Biophys. J.* **81,** 3308–3323.

Bauerfeind, R., and Huttner, W. B. (1993). Biogenesis of constitutive secretory vesicles, secretory granules and synaptic vesicles. *Curr. Opin. Cell Biol.* **5,** 628–635.

Bock, J. B., Matern, H. T., Peden, A. A., and Scheller, R. H. (2001). A genomic perspective on membrane compartment organization. *Nature* **409,** 839–841.

Bokvist, K., Eliasson, L., Ammala, C., Renstrom, E., and Rorsman, P. (1995). Co-localization of L-type Ca2+ channels and insulin-containing secretory granules and its significance for the initiation of exocytosis in mouse pancreatic B-cells. *EMBO J.* **14,** 50–57.

Bokvist, K., Holmqvist, M., Gromada, J., and Rorsman, P. (2000). Compound exocytosis in voltage-clamped mouse pancreatic beta-cells revealed by carbon fibre amperometry. *Pflugers Arch.* **439,** 634–645.

Boyd, R. S., Duggan, M. J., Shone, C. C., and Foster, K. A. (1995). The effect of botulinum neurotoxins on the release of insulin from the insulinoma cell lines HIT-15 and RINm5F. *J. Biol. Chem.* **270,** 18216–18218.

Bracher, A., and Weissenhorn, W. (2001). Crystal structures of neuronal squid Sec1 implicate inter-domain hinge movement in the release of t-SNAREs. *J. Mol. Biol.* **306,** 7–13.

Branstrom, R., Corkey, B. E., Berggren, P. O., and Larsson, O. (1997). Evidence for a unique long chain acyl-CoA ester binding site on the ATP-regulated potassium channel in mouse pancreatic beta cells. *J. Biol. Chem.* **272,** 17390–17394.

Branstrom, R., Aspinwall, C. A., Valimaki, S., Ostensson, C. G., Tibell, A., Eckhard, M., Brandhorst, H., Corkey, B. E., Berggren, P. O., and Larsson, O. (2004). Long-chain CoA esters activate human pancreatic beta-cell KATP channels: Potential role in Type 2 diabetes. *Diabetologia* **47,** 277–283.

Bratanova-Tochkova, T. K., Cheng, H., Daniel, S., Gunawardana, S., Liu, Y. J., Mulvaney-Musa, J., Schermerhorn, T., Straub, S. G., Yajima, H., and Sharp, G. W. (2002). Triggering and augmentation mechanisms, granule pools, and biphasic insulin secretion. *Diabetes* **51,** S83–S90.

Carroll, R. J., Hammer, R. E., Chan, S. J., Swift, H. H., Rubenstein, A. H., and Steiner, D. F. (1988). A mutant human proinsulin is secreted from islets of Langerhans in increased amounts via an unregulated pathway. *Proc. Natl. Acad. Sci. USA* **85,** 8943–8947.

Cawley, N. X., Rodriguez, Y. M., Maldonado, A., and Loh, Y. P. (2003). Trafficking of mutant carboxypeptidase E to secretory granules in a beta-cell line derived from Cpe (fat)/Cpe(fat) mice. *Endocrinology* **144,** 292–298.

Chanat, E., and Huttner, W. B. (1991). Milieu-induced, selective aggregation of regulated secretory proteins in the trans-Golgi network. *J. Cell Biol.* **115,** 1505–1519.

Chapman, E. R. (2002). Synaptotagmin: A Ca(2+) sensor that triggers exocytosis? *Nat. Rev. Mol. Cell Biol.* **3,** 498–508.

Chimienti, F., Devergnas, S., Pattou, F., Schuit, F., Garcia-Cuenca, R., Vandewalle, B., Kerr-Conte, J., Van Lommel, L., Grunwald, D., Favier, A., and Seve, M. (2006). *In vivo* expression and functional characterization of the zinc transporter ZnT8 in glucose-induced insulin secretion. *J. Cell Sci.* **119,** 4199–4206.

Chung, K. N., Walter, P., Aponte, G. W., and Moore, H. P. (1989). Molecular sorting in the secretory pathway. *Science* **243,** 192–197.

Curry, D. L., Bennett, L. L., and Grodsky, G. M. (1968). Dynamics of insulin secretion by the perfused rat pancreas. *Endocrinology* **83,** 572–584.

Damke, H., Baba, T., Warnock, D. E., and Schmid, S. L. (1994). Induction of mutant dynamin specifically blocks endocytic coated vesicle formation. *J. Cell Biol.* **127,** 915–934.

Daniel, S., Noda, M., Straub, S. G., and Sharp, G. W. (1999). Identification of the docked granule pool responsible for the first phase of glucose-stimulated insulin secretion. *Diabetes* **48,** 1686–1690.

Davidson, H. W., Rhodes, C. J., and Hutton, J. C. (1988). Intraorganellar calcium and pH control proinsulin cleavage in the pancreatic beta cell via two distinct site-specific endopeptidases. *Nature* **333,** 93–96.

Dean, P. M. (1973). Ultrastructural morphometry of the pancreatic-cell. *Diabetologia* **9,** 115–119.

Del Prato, S. (2003). Loss of early insulin secretion leads to postprandial hyperglycaemia. *Diabetologia* **46**(Suppl 1)), M2–8.

Del Prato, S., and Tiengo, A. (2001). The importance of first-phase insulin secretion: Implications for the therapy of type 2 diabetes mellitus. *Diabetes Metab. Res. Rev.* **17,** 164–174.

De Vos, A., Heimberg, H., Quartier, E., Huypens, P., Bouwens, L., Pipeleers, D., and Schuit, F. (1995). Human and rat beta cells differ in glucose transporter but not in glucokinase gene expression. *J. Clin. Invest.* **96,** 2489–2495.

de Vries, K. J., Geijtenbeek, A., Brian, E. C., de Graan, P. N., Ghijsen, W. E., and Verhage, M. (2002). Dynamics of munc18–1 phosphorylation/dephosphorylation in rat brain nerve terminals. *Eur. J. Neurosci.* **12,** 385–390.

Dhanvantari, S., Izzo, A., Jansen, E., and Brubaker, P. L. (2001). Coregulation of glucagon-like peptide-1 synthesis with proglucagon and prohormone convertase 1 gene expression in enteroendocrine GLUTag cells. *Endocrinology* **142,** 37–42.

Dhanvantari, S., Shen, F. S., Adams, T., Snell, C. R., Zhang, C., Mackin, R. B., Morris, S. J., and Loh, Y. P. (2003). Disruption of a receptor-mediated mechanism for intracellular sorting of proinsulin in familial hyperproinsulinemia. *Mol. Endocrinol.* **17,** 1856–1867.

Dittie, A. S., Hajibagheri, N., and Tooze, S. A. (1996). The AP-1 adaptor complex binds to immature secretory granules from PC12 cells, and is regulated by ADP-ribosylation factor. *J. Cell Biol.* **132,** 523–536.

Dodson, G., and Steiner, D. (1998). The role of assembly in insulin's biosynthesis. *Curr. Opin. Struct. Biol.* **8,** 189–194.

Doyle, M. E., and Egan, J. M. (2007). Mechanisms of action of glucagon-like peptide 1 in the pancreas. *Pharmacol. Ther.* **113,** 546–593.

Dresbach, T., Burns, M. E., O'Connor, V., DeBello, W. M., Betz, H., and Augustine, G. J. (1998). A neuronal Sec1 homolog regulates neurotransmitter release at the squid giant synapse. *J. Neurosci.* **18,** 2923–2932.

Drucker, D. J., Philippe, J., Mojsov, S., Chick, W. L., and Habener, J. F. (1987). Glucagon-like peptide I stimulates insulin gene expression and increases cyclic AMP levels in a rat islet cell line. *Proc. Natl. Acad. Sci. USA* **84,** 3434–3438.

Dukes, I. D., and Philipson, L. H. (1996). K+ channels: generating excitement in pancreatic beta-cells. *Diabetes* **45,** 845–853.

Dukes, I. D., McIntyre, M. S., Mertz, R. J., Philipson, L. H., Roe, M. W., Spencer, B., and Worley, J. F. (1994). Dependence on NADH produced during glycolysis for beta-cell glucose signaling. *J. Biol. Chem.* **269,** 10979–10982.

Dulubova, I., Sugita, S., Hill, S., Hosaka, M., Fernandez, I., Sudhof, T. C., and Rizo, J. (1999). A conformational switch in syntaxin during exocytosis: role of munc18. *EMBO J.* **18,** 4372–4382.

Dulubova, I., Yamaguchi, T., Arac, D., Li, H., Huryeva, I., Min, S. W., Rizo, J., and Sudhof, T. C. (2003). Convergence and divergence in the mechanism of SNARE binding by Sec1/Munc18-like proteins. *Proc. Natl. Acad. Sci. USA* **100,** 32–37.

Dunn, M. F. (2005). Zinc-ligand interactions modulate assembly and stability of the insulin hexamer – A review. *Biometals* **18,** 295–303.

Eddlestone, G. T., Oldham, S. B., Lipson, L. G., Premdas, F. H., and Beigelman, P. M. (1985). Electrical activity, cAMP concentration, and insulin release in mouse islets of Langerhans. *Am. J. Physiol.* **248,** C145–153.

Eddlestone, G. T., Komatsu, M., Shen, L., and Sharp, G. W. (1995). Mastoparan increases the intracellular free calcium concentration in two insulin-secreting cell lines by inhibition of ATP-sensitive potassium channels. *Mol. Pharmacol.* **47,** 787–797.

Edwardson, J. M., An, S., and Jahn, R. (1997). The secretory granule protein syncollin binds to syntaxin in a Ca2(+)-sensitive manner. *Cell* **90,** 325–333.

Elhamdani, A., Palfrey, H. C., and Artalejo, C. R. (2001). Quantal size is dependent on stimulation frequency and calcium entry in calf chromaffin cells. *Neuron* **31,** 819–830.

Eliasson, L., Proks, P., Ammala, C., Ashcroft, F. M., Bokvist, K., Renstrom, E., Rorsman, P., and Smith, P. A. (1996). Endocytosis of secretory granules in mouse pancreatic beta-cells evoked by transient elevation of cytosolic calcium. *J. Physiol.* **493** (Pt 3), 755–767.

Eliasson, L., Renstrom, E., Ding, W. G., Proks, P., and Rorsman, P. (1997). Rapid ATP-dependent priming of secretory granules precedes Ca(2+)-induced exocytosis in mouse pancreatic B-cells. *J. Physiol.* **503**(Pt 2), 399–412.

Eto, K., Tsubamoto, Y., Terauchi, Y., Sugiyama, T., Kishimoto, T., Takahashi, N., Yamauchi, N., Kubota, N., Murayama, S., Aizawa, T., Akanuma, Y., Aizawa, S., *et al.* (1999). Role of NADH shuttle system in glucose-induced activation of mitochondrial metabolism and insulin secretion. *Science* **283,** 981–985.

Fajans, S. S., Bell, G. I., and Polonsky, K. S. (2001). Molecular mechanisms and clinical pathophysiology of maturity-onset diabetes of the young. *N. Engl. J. Med.* **345,** 971–980.

Farquhar, M. G., and Palade, G. E. (1981). The Golgi apparatus (complex)-(1954–1981)-from artifact to center stage. *J. Cell Biol.* **91,** 77s–103s.

Feng, L., and Arvan, P. (2003). The trafficking of alpha 1-antitrypsin, a post-Golgi secretory pathway marker, in INS-1 pancreatic beta cells. *J. Biol. Chem.* **278,** 31486–31494.

Galli, T., and Haucke, V. (2001). Cycling of synaptic vesicles: How far? How fast. *Sci. STKE* 2001 RE1.

Gao, Z., Reavey-Cantwell, J., Young, R. A., Jegier, P., and Wolf, B. A. (2000). Synaptotagmin III/VII isoforms mediate Ca2+-induced insulin secretion in pancreatic islet beta - cells. *J. Biol. Chem.* **275,** 36079–36085.

Giugliano, D., Torella, R., Cacciapuoti, F., Gentile, S., Verza, M., and Varricchio, M. (1980). Impairment of insulin secretion in man by nifedipine. *Eur. J. Clin. Pharmacol.* **18,** 395–398.

Gonelle-Gispert, C., Halban, P. A., Niemann, H., Palmer, M., Catsicas, S., and Sadoul, K. (1999). SNAP-25a and -25b isoforms are both expressed in insulin-secreting cells and can function in insulin secretion. *Biochem. J.* **339,** 159–165.

Gonelle-Gispert, C., Molinete, M., Halban, P. A., and Sadoul, K. (2000). Membrane localization and biological activity of SNAP-25 cysteine mutants in insulin-secreting cells. *J. Cell. Sci.* **113,** 3197–3205.

Gopel, S., Kanno, T., Barg, S., Galvanovskis, J., and Rorsman, P. (1999). Voltage-gated and resting membrane currents recorded from B-cells in intact mouse pancreatic islets. *J. Physiol.* **521**(Pt 3), 717–728.

Gorr, S. U., and Darling, D. S. (1995). An N-terminal hydrophobic peak is the sorting signal of regulated secretory proteins. *FEBS. Lett.* **361,** 8–12.

Graves, T. K., and Hinkle, P. M. (2003). Ca(2+)-induced Ca(2+) release in the pancreatic beta-cell: Direct evidence of endoplasmic reticulum Ca(2+) release. *Endocrinology* **144,** 3565–3574.

Greider, M. H., Howell, S. L., and Lacy, P. E. (1969). Isolation and properties of secretory granules from rat islets of Langerhans. II. Ultrastructure of the beta granule. *J. Cell. Biol.* **41,** 162–166.

Gromada, J., Hoy, M., Renstrom, E., Bokvist, K., Eliasson, L., Gopel, S., and Rorsman, P. (1999). CaM kinase II-dependent mobilization of secretory granules underlies acetylcholine-induced stimulation of exocytosis in mouse pancreatic B-cells. *J. Physiol.* **518**(Pt 3), 745–759.

Gross, D. J., Halban, P. A., Kahn, C. R., Weir, G. C., and Villa-Komaroff, L. (1989). Partial diversion of a mutant proinsulin (B10 aspartic acid) from the regulated to the constitutive secretory pathway in transfected AtT-20 cells. *Proc. Natl. Acad. Sci. USA* **86,** 4107–4111.

Guillam, M. T., Hummler, E., Schaerer, E., Yeh, J. I., Birnbaum, M. J., Beermann, F., Schmidt, A., Deriaz, N., and Thorens, B. (1997). Early diabetes and abnormal postnatal pancreatic islet development in mice lacking Glut-2. *Nat. Genet.* **17,** 327–330.

Gut, A., Kiraly, C. E., Fukuda, M., Mikoshiba, K., Wollheim, C. B., and Lang, J. (2001). Expression and localisation of synaptotagmin isoforms in endocrine beta-cells: Their function in insulin exocytosis. *J. Cell Sci.* **114,** 1709–1716.

Halban, P. A., and Irminger, J. C. (1994). Sorting and processing of secretory proteins. *Biochem. J.* **299**(Pt 1), 1–18.

Halban, P. A., and Irminger, J. C. (2003). Mutant proinsulin that cannot be converted is secreted efficiently from primary rat beta-cells via the regulated pathway. *Mol. Biol. Cell* **14,** 1195–1203.

Hansen, L., Deacon, C. F., Orskov, C., and Holst, J. J. (1999). Glucagon-like peptide-1-(7–36) amide is transformed to glucagon-like peptide-1-(9–36)amide by dipeptidyl peptidase IV in the capillaries supplying the L cells of the porcine intestine. *Endocrinology* **140,** 5356–5363.

Hao, M., Li, X., Rizzo, M. A., Rocheleau, J. V., Dawant, B. M., and Piston, D. W. (2005). Regulation of two insulin granule populations within the reserve pool by distinct calcium sources. *J. Cell Sci.* **118,** 5873–5884.

Harrison, S. D., Broadie, K., van de Goor, J., and Rubin, G. M. (1994). Mutations in the Drosophila Rop gene suggest a function in general secretion and synaptic transmission. *Neuron* **13,** 555–566.

Hays, L. B., Wicksteed, B., Wang, Y., McCuaig, J. F., Philipson, L. H., Edwardson, J. M., and Rhodes, C. J. (2005). Intragranular targeting of syncollin, but not a syncollin GFP chimera, inhibits regulated insulin exocytosis in pancreatic beta-cells. *J. Endocrinol.* **185,** 57–67.

Hedeskov, C. J., Capito, K., and Thams, P. (1987). Cytosolic ratios of free [NADPH]/[NADP+] and [NADH]/[NAD+] in mouse pancreatic islets, and nutrient-induced insulin secretion. *Biochem. J.* **241,** 161–167.

Henquin, J. C., Ishiyama, N., Nenquin, M., Ravier, M. A., and Jonas, J. C. (2002). Signals and pools underlying biphasic insulin secretion. *Diabetes* **51**(Suppl 1)), S60–67.

Hildyard, J. C., Ammala, C., Dukes, I. D., Thomson, S. A., and Halestrap, A. P. (2005). Identification and characterisation of a new class of highly specific and potent inhibitors of the mitochondrial pyruvate carrier. *Biochim. Biophys. Acta* **1707,** 221–230.

Hill, C. P., Dauter, Z., Dodson, E. J., Dodson, G. G., and Dunn, M. F. (1991). X-ray structure of an unusual Ca2+ site and the roles of Zn2+ and Ca2+ in the assembly, stability, and storage of the insulin hexamer. *Biochemistry* **30,** 917–924.

Hill, E., van Der Kaay, J., Downes, C. P., and Smythe, E. (2001). The role of dynamin and its binding partners in coated pit invagination and scission. *J. Cell Biol.* **152,** 309–323.

Holst, J. J. (2007). The physiology of glucagon-like peptide 1. *Physiol. Rev.* **87,** 1409–1439.

Hong, W. (2005). SNAREs and traffic. *Biochim. Biophys. Acta* **1744,** 493–517.

Horvath, A., Szabadkai, G., Varnai, P., Aranyi, T., Wollheim, C. B., Spat, A., and Enyedi, P. (1998). Voltage dependent calcium channels in adrenal glomerulosa cells and in insulin producing cells. *Cell Calcium.* **23,** 33–42.

Houng, A., Polgar, J., and Reed, G. L. (2003). Munc18-syntaxin complexes and exocytosis in human platelets. *J. Biol. Chem.* **278,** 19627–19633.

Howell, S. L., Tyhurst, M., Duvefelt, H., Andersson, A., and Hellerstrom, C. (1978). Role of zinc and calcium in the formation and storage of insulin in the pancreatic beta-cell. *Cell Tissue Res.* **188,** 107–118.

Huang, X. F., and Arvan, P. (1994). Formation of the insulin-containing secretory granule core occurs within immature beta-granules. *J. Biol. Chem.* **269,** 20838–20844.

Huang, L., Shen, H., Atkinson, M. A., and Kennedy, R. T. (1995). Detection of exocytosis at individual pancreatic beta cells by amperometry at a chemically modified microelectrode. *Proc. Natl. Acad. Sci. USA* **92,** 9608–9612.

Huang, X., Wheeler, M. B., Kang, Y. H., Sheu, L., Lukacs, G. L., Trimble, W. S., and Gaisano, H. Y. (1998). Truncated SNAP-25 (1–197), like botulinum neurotoxin A, can inhibit insulin secretion from HIT-T15 insulinoma cells. *Mol. Endocrinol.* **12,** 1060–1070.

Iezzi, M., Kouri, G., Fukuda, M., and Wollheim, C. B. (2004). Synaptotagmin V and IX isoforms control Ca2+ -dependent insulin exocytosis. *J. Cell. Sci.* **117,** 3119–3127.

Inagaki, N., Yasuda, K., Inoue, G., Okamoto, Y., Yano, H., Someya, Y., Ohmoto, Y., Deguchi, K., Imagawa, K., Imura, H., *et al.* (1992). Glucose as regulator of glucose transport activity and glucose-transporter mRNA in hamster beta-cell line. *Diabetes* **41,** 592–597.

Irminger, J. C., Verchere, C. B., Meyer, K., and Halban, P. A. (1997). Proinsulin targeting to the regulated pathway is not impaired in carboxypeptidase E-deficient Cpefat/Cpefat mice. *J. Biol. Chem.* **272,** 27532–27534.

Ivarsson, R., Obermuller, S., Rutter, G. A., Galvanovskis, J., and Renstrom, E. (2004). Temperature-sensitive random insulin granule diffusion is a prerequisite for recruiting granules for release. *Traffic* **5,** 750–762.

Iynedjian, P. B. (1993). Mammalian glucokinase and its gene. *Biochem. J.* **293**(Pt 1), 1–13.

Jarousse, N., and Kelly, R. B. (2001). Endocytotic mechanisms in synapses. *Curr. Opin. Cell Biol.* **13,** 461–469.

Jensen, M. V., Joseph, J. W., Ilkayeva, O., Burgess, S., Lu, D., Ronnebaum, S. M., Odegaard, M., Becker, T. C., Sherry, A. D., and Newgard, C. B. (2006). Compensatory responses to pyruvate carboxylase suppression in islet beta-cells. Preservation of glucose-stimulated insulin secretion. *J. Biol. Chem.* **281,** 22342–22351.

John, S. A., Monck, J. R., Weiss, J. N., and Ribalet, B. (1998). The sulphonylurea receptor SUR1 regulates ATP-sensitive mouse Kir6.2 K+ channels linked to the green fluorescent protein in human embryonic kidney cells (HEK 293). *J. Physiol.* **510**(Pt 2), 333–345.

Keahey, H. H., Rajan, A. S., Boyd, A. E., and Kunze, D. L. (1989). Characterization of voltage-dependent Ca2+ channels in beta-cell line. *Diabetes* **38,** 188–193.

Kelly, R. B. (1985). Pathways of protein secretion in eukaryotes. *Science* **230,** 25–32.

Khan, A. H., Thurmond, D. C., Yang, C., Ceresa, B. P., Sigmund, C. D., and Pessin, J. E. (2001). Munc18c regulates insulin-stimulated glut4 translocation to the transverse tubules in skeletal muscle. *J. Biol. Chem.* **276,** 4063–4069.

Kim, T., Tao-Cheng, J. H., Eiden, L. E., and Loh, Y. P. (2001). Chromogranin A, an "on/off" switch controlling dense-core secretory granule biogenesis. *Cell* **106,** 499–509.

Kizer, J. S., and Tropsha, A. (1991). A motif found in propeptides and prohormones that may target them to secretory vesicles. *Biochem. Biophys. Res. Commun.* **174,** 586–592.

Kuliawat, R., Prabakaran, D., and Arvan, P. (2000). Proinsulin endoproteolysis confers enhanced targeting of processed insulin to the regulated secretory pathway. *Mol. Biol. Cell* **11,** 1959–1972.

Lang, J. (1999). Molecular mechanisms and regulation of insulin exocytosis as a paradigm of endocrine secretion. *Eur. J. Biochem.* **259,** 3–17.

Lang, J., Fukuda, M., Zhang, H., Mikoshiba, K., and Wollheim, C. B. (1997). The first C2 domain of synaptotagmin is required for exocytosis of insulin from pancreatic beta-cells: Action of synaptotagmin at low micromolar calcium. *EMBO J.* **16,** 5837–5846.

Lange, R. H. (1974). Crystalline islet B-granules in the grass snake (Natrix natrix (L.)): Tilting experiments in the electron microscope. *J. Ultrastruct. Res.* **46,** 301–307.

Leabu, M. (2006). Membrane fusion in cells: Molecular machinery and mechanisms. *J. Cell Mol. Med.* **10,** 423–427.

Leung, Y. M., Sheu, L., Kwan, E., Wang, G., Tsushima, R., and Gaisano, H. (2002). Visualization of sequential exocytosis in rat pancreatic islet beta cells. *Biochem. Biophys. Res. Commun.* **292,** 980–986.

Leung, Y. M., Kang, Y., Gao, X., Xia, F., Xie, H., Sheu, L., Tsuk, S., Lotan, I., Tsushima, R. G., and Gaisano, H. Y. (2003). Syntaxin 1A binds to the cytoplasmic C terminus of Kv2.1 to regulate channel gating and trafficking. *J. Biol. Chem.* **278,** 17532–17538.

Leung, Y. M., Kang, Y., Xia, F., Sheu, L., Gao, X., Xie, H., Tsushima, R. G., and Gaisano, H. Y. (2005). Open form of syntaxin-1A is a more potent inhibitor than wild-type syntaxin-1A of Kv2.1 channels. *Biochem. J.* **387,** 195–202.

Li, J., Luo, R., Hooi, S. C., Ruga, P., Zhang, J., Meda, P., and Li, G. (2005). Ectopic expression of syncollin in INS-1 beta-cells sorts it into granules and impairs regulated secretion. *Biochemistry* **44,** 4365–4374.

Ligon, B., Boyd, A. E., and Dunlap, K. (1998). Class A calcium channel variants in pancreatic islets and their role in insulin secretion. *J. Biol. Chem.* **273,** 13905–13911.

Lilja, L., Johansson, J. U., Gromada, J., Mandic, S. A., Fried, G., Berggren, P. O., and Bark, C. (2004). Cyclin-dependent kinase 5 associated with p39 promotes Munc18-1 phosphorylation and Ca(2+)-dependent exocytosis. *J. Biol. Chem.* **279,** 29534–29541.

Lopez, J. A., Kwan, E. P., Xie, L., He, Y., James, D. E., and Gaisano, H. Y. (2008). The RalA GTPase is a central regulator of insulin exocytosis from pancreatic islet beta-cells. *J. Biol. Chem* **283,** 17939–17945.

Lu, D., Mulder, H., Zhao, P., Burgess, S. C., Jensen, M. V., Kamzolova, S., Newgard, C. B., and Sherry, A. D. (2002). 13C NMR isotopomer analysis reveals a connection between pyruvate cycling and glucose-stimulated insulin secretion (GSIS). *Proc. Natl. Acad. Sci. USA* **99,** 2708–2713.

Ma, L., Bindokas, V. P., Kuznetsov, A., Rhodes, C., Hays, L., Edwardson, J. M., Ueda, K., Steiner, D. F., and Philipson, L. H. (2004). Direct imaging shows that insulin granule exocytosis occurs by complete vesicle fusion. *Proc. Natl. Acad. Sci. USA* **101,** 9266–9271.

MacDonald, M. J., Ammon, H. P., Patel, T., and Steinke, J. (1974). Failure of 6-aminonicotinamide to inhibit the potentiating effect of leucine and arginine on glucose-induced insulin release in vitrol. *Diabetologia* **10,** 761–765.

MacDonald, M. J., Warner, T. F., and Mertz, R. J. (1990). High activity of mitochondrial glycerol phosphate dehydrogenase in insulinomas and carcinoid and other tumors of the amine precursor uptake decarboxylation system. *Cancer Res.* **50,** 7203–7205.

MacDonald, P. E., Joseph, J. W., and Rorsman, P. (2005). Glucose-sensing mechanisms in pancreatic beta-cells. *Philos. Trans. Roy. Soc. Lond. B Biol. Sci.* **360,** 2211–2225.

Malaisse, W. J., Sener, A., Herchuelz, A., and Hutton, J. C. (1979). Insulin release: The fuel hypothesis. *Metabolism* **28,** 373–386.

Matschinsky, F. M. (1990). Glucokinase as glucose sensor and metabolic signal generator in pancreatic beta-cells and hepatocytes. *Diabetes* **39,** 647–652.

Meier, J. J., Gallwitz, B., Schmidt, W. E., and Nauck, M. A. (2002a). Glucagon-like peptide 1 as a regulator of food intake and body weight: Therapeutic perspectives. *Eur. J. Pharmacol.* **440,** 269–279.

Meier, J. J., Nauck, M. A., Schmidt, W. E., and Gallwitz, B. (2002b). Gastric inhibitory polypeptide: the neglected incretin revisited. *Regul. Pept.* **107,** 1–13.

Mertz, R. J., Worley, J. F., Spencer, B., Johnson, J. H., and Dukes, I. D. (1996). Activation of stimulus-secretion coupling in pancreatic beta-cells by specific products of glucose metabolism. Evidence for privileged signaling by glycolysis. *J. Biol. Chem.* **271,** 4838–4845.

Mikhailov, M. V., Proks, P., Ashcroft, F. M., and Ashcroft, S. J. (1998). Expression of functionally active ATP-sensitive K-channels in insect cells using baculovirus. *FEBS Lett.* **429,** 390–394.

Min, L., Leung, Y. M., Tomas, A., Watson, R. T., Gaisano, H. Y., Halban, P. A., Pessin, J. E., and Hou, J. C. (2007). Dynamin is functionally coupled to insulin granule exocytosis. *J. Biol. Chem.* **282,** 33530–33536.

Misura, K. M., Scheller, R. H., and Weis, W. I. (2000). Three-dimensional structure of the neuronal-Sec1-syntaxin 1a complex. *Nature* **404,** 355–362.

Mojsov, S., Weir, G. C., and Habener, J. F. (1987). Insulinotropin: Glucagon-like peptide I (7–37) co-encoded in the glucagon gene is a potent stimulator of insulin release in the perfused rat pancreas. *J. Clin. Invest.* **79,** 616–619.

Molinete, M., Irminger, J. C., Tooze, S. A., and Halban, P. A. (2000a). Trafficking/sorting and granule biogenesis in the beta-cell. *Semin. Cell Dev. Biol.* **11,** 243–251.

Molinete, M., Lilla, V., Jain, R., Joyce, P. B., Gorr, S. U., Ravazzola, M., and Halban, P. A. (2000b). Trafficking of non-regulated secretory proteins in insulin secreting (INS-1) cells. *Diabetologia* **43,** 1157–1164.

Nagai, T., Ibata, K., Park, E. S., Kubota, M., Mikoshiba, K., and Miyawaki, A. (2002). A variant of yellow fluorescent protein with fast and efficient maturation for cell-biological applications. *Nat. Biotechnol.* **20,** 87–90.

Natori, S., and Huttner, W. B. (1996). Chromogranin B (secretogranin I) promotes sorting to the regulated secretory pathway of processing intermediates derived from a peptide hormone precursor. *Proc. Natl. Acad. Sci. USA* **93,** 4431–4436.

Neher, E. (1998). Vesicle pools and Ca2+ microdomains: New tools for understanding their roles in neurotransmitter release. *Neuron* **20,** 389–399.

Neher, E., and Marty, A. (1982). Discrete changes of cell membrane capacitance observed under conditions of enhanced secretion in bovine adrenal chromaffin cells. *Proc. Natl. Acad. Sci. USA* **79,** 6712–6716.

Newgard, C. B., and McGarry, J. D. (1995). Metabolic coupling factors in pancreatic beta-cell signal transduction. *Annu. Rev. Biochem.* **64,** 689–719.

Novick, P., Field, C., and Schekman, R. (1980). Identification of 23 complementation groups required for post-translational events in the yeast secretory pathway. *Cell* **21,** 205–215.

Obermuller, S., Lindqvist, A., Karanauskaite, J., Galvanovskis, J., Rorsman, P., and Barg, S. (2005). Selective nucleotide-release from dense-core granules in insulin-secreting cells. *J. Cell Sci.* **118,** 4271–4282.

Ohara-Imaizumi, M., Nakamichi, Y., Tanaka, T., Katsuta, H., Ishida, H., and Nagamatsu, S. (2002). Monitoring of exocytosis and endocytosis of insulin secretory granules in the pancreatic beta-cell line MIN6 using pH-sensitive green fluorescent protein (pHluorin) and confocal laser microscopy. *Biochem. J.* **363,** 73–80.

Ohta, M., Nelson, D., Wilson, J. M., Meglasson, M. D., and Erecinska, M. (1992). Relationships between energy level and insulin secretion in isolated rat islets of Langerhans. Manipulation of [ATP]/[ADP][Pi] by 2-deoxy-d-glucose. *Biochem. Pharmacol.* **43,** 1859–1864.

Olofsson, C. S., Gopel, S. O., Barg, S., Galvanovskis, J., Ma, X., Salehi, A., Rorsman, P., and Eliasson, L. (2002). Fast insulin secretion reflects exocytosis of docked granules in mouse pancreatic B-cells. *Pflugers Arch.* **444,** 43–51.

Orci, L. (1982). Macro- and micro-domains in the endocrine pancreas. *Diabetes* **31,** 538–565.

Orci, L. (1985). The insulin factory: a tour of the plant surroundings and a visit to the assembly line. The Minkowski lecture 1973 revisited. *Diabetologia* **28,** 528–546.

Orci, L., Amherdt, M., Malaisse-Lagae, F., Rouiller, C., and Renold, A. E. (1973). Insulin release by emiocytosis: Demonstration with freeze-etching technique. *Science* **179,** 82–84.

Orci, L., and Malaisse, W. (1980). Hypothesis: Single and chain release of insulin secretory granules is related to anionic transport at exocytotic sites. *Diabetes* **29,** 943–944.

Orci, L., Ravazzola, M., Amherdt, M., Madsen, O., Perrelet, A., Vassalli, J. D., and Anderson, R. G. (1986). Conversion of proinsulin to insulin occurs coordinately with acidification of maturing secretory vesicles. *J. Cell Biol.* **103,** 2273–2281.

Orci, L., Ravazzola, M., Amherdt, M., Madsen, O., Vassalli, J. D., and Perrelet, A. (1985). Direct identification of prohormone conversion site in insulin-secreting cells. *Cell* **42,** 671–681.

Orci, L., Ravazzola, M., Storch, M. J., Anderson, R. G., Vassalli, J. D., and Perrelet, A. (1987). Proteolytic maturation of insulin is a post-Golgi event which occurs in acidifying clathrin-coated secretory vesicles. *Cell* **49,** 865–868.

Palade, G. (1975). Intracellular aspects of the process of protein synthesis. *Science* **189,** 347–358.

Perez-Branguli, F., Muhaisen, and Blasi, A. J. (2002). Munc 18a binding to syntaxin 1A and 1B isoforms defines its localization at the plasma membrane and blocks SNARE assembly in a three-hybrid system assay. *Mol. Cell Neurosci.* **20,** 169–180.

Pevsner, J., Hsu, S. -C., Braun, J. E. A., Calakos, N., Ting, A. E., Bennett, M. K., and Scheller, R. H. (1994a). Specificity and regulation of a synaptic vesicle docking complex. *Neuron* **13,** 353–361.

Pevsner, J., Hsu, S. -C., and Scheller, R. H. (1994b). N-Sec1: A neural-specific syntaxin-binding protein. *Proc. Natl. Acad. Sci. USA* **91,** 1445–1449.

Philipson, L. H., and Steiner, D. F. (1995). Pas de deux or more: The sulfonylurea receptor and K+ channels. *Science* **268,** 372–373.

Pobbati, A. V., Razeto, A., Boddener, M., Becker, S., and Fasshauer, D. (2004). Structural basis for the inhibitory role of tomosyn in exocytosis. *J. Biol. Chem.* **279,** 47192–47200.

Pouli, A. E., Emmanouilidou, E., Zhao, C., Wasmeier, C., Hutton, J. C., and Rutter, G. A. (1998). Secretory-granule dynamics visualized *in vivo* with a phogrin-green fluorescent protein chimaera. *Biochem. J.* **333**(Pt 1), 193–199.

Pralong, W. F., Bartley, C., and Wollheim, C. B. (1990). Single islet beta-cell stimulation by nutrients: Relationship between pyridine nucleotides, cytosolic Ca2+ and secretion. *EMBO J.* **9,** 53–60.

Prentki, M., and Wollheim, C. B. (1984). Cytosolic free Ca2+ in insulin secreting cells and its regulation by isolated organelles. *Experientia* **40,** 1052–1060.

Randazzo, P. A., and Kahn, R. A. (1994). GTP hydrolysis by ADP-ribosylation factor is dependent on both an ADP-ribosylation factor GTPase-activating protein and acid phospholipids. *J. Biol. Chem.* **269,** 10758–10763.

Regazzi, R., Wollheim, C. B., Lang, J., Theler, J. M., Rossetto, O., Montecucco, C., Sadoul, K., Weller, U., Palmer, M., and Thorens, B. (1995). VAMP-2 and cellubrevin are expressed in pancreatic beta-cells and are essential for Ca(2+)-but not for GTP gamma S-induced insulin secretion. *EMBO J.* **14,** 2723–2730.

Regazzi, R., Sadoul, K., Meda, P., Kelly, R. B., Halban, P. A., and Wollheim, C. B. (1996). Mutational analysis of VAMP domains implicated in Ca2+-induced insulin exocytosis. *EMBO J.* **15,** 6951–6959.

Reimann, F., Gribble, F. M., and Ashcroft, F. M. (2000). Differential response of K(ATP) channels containing SUR2A or SUR2B subunits to nucleotides and pinacidil. *Mol. Pharmacol.* **58,** 1318–1325.

Renstrom, E., Eliasson, L., and Rorsman, P. (1997). Protein kinase A-dependent and -independent stimulation of exocytosis by cAMP in mouse pancreatic B-cells. *J. Physiol.* **502**(Pt 1), 105–118.

Rettig, J., and Neher, E. (2002). Emerging roles of presynaptic proteins in Ca++-triggered exocytosis. *Science* **298,** 781–785.

Rhodes, C. J., and Halban, P. A. (1987). Newly synthesized proinsulin/insulin and stored insulin are released from pancreatic B cells predominantly via a regulated, rather than a constitutive, pathway. *J. Cell Biol.* **105,** 145–153.

Rhodes, C. J., Lucas, C. A., Mutkoski, R. L., Orci, L., and Halban, P. A. (1987). Stimulation by ATP of proinsulin to insulin conversion in isolated rat pancreatic islet secretory granules. Association with the ATP-dependent proton pump. *J. Biol. Chem.* **262,** 10712–10717.

Rorsman, P., and Renstrom, E. (2003). Insulin granule dynamics in pancreatic beta cells. *Diabetologia* **46,** 1029–1045.

Rorsman, P., Eliasson, L., Renstrom, E., Gromada, J., Barg, S., and Gopel, S. (2000). The cell physiology of biphasic insulin secretion. *News Physiol. Sci.* **15,** 72–77.

Rutter, G. A. (2001). Nutrient-secretion coupling in the pancreatic islet beta-cell: Recent advances. *Mol. Aspects Med.* **22,** 247–284.

Rutter, G. A., and Tsuboi, T. (2004). Kiss and run exocytosis of dense core secretory vesicles. *Neuroreport* **15,** 79–81.

Sadoul, K., Lang, J., Montecucco, C., Weller, U., Regazzi, R., Catsicas, S., Wollheim, C. B., and Halban, P. A. (1995). SNAP-25 is expressed in islets of Langerhans and is involved in insulin release. *J. Cell Biol.* **128,** 1019–1028.

Sadoul, K., Berger, A., Niemann, H., Weller, U., Roche, P. A., Klip, A., Trimble, W. S., Regazzi, R., Catsicas, S., and Halban, P. A. (1997). SNAP-23 is not cleaved by botulinum neurotoxin E and can replace SNAP-25 in the process of insulin secretion. *J. Biol. Chem.* **272,** 33023–33027.

Sakura, H., Ammala, C., Smith, P. A., Gribble, F. M., and Ashcroft, F. M. (1995). Cloning and functional expression of the cDNA encoding a novel ATP-sensitive potassium channel subunit expressed in pancreatic beta-cells, brain, heart and skeletal muscle. *FEBS Lett.* **377,** 338–344.

Schulze, K. L., Littleton, J. T., Salzberg, A., Halachmi, N., Stern, M., Lev, Z., and Bellen, H. J. (1994). rop, a Drosophila homolog of yeast Sec1 and vertebrate n-Sec1/Munc-18 proteins, is a negative regulator of neurotransmitter release *in vivo*. *Neuron* **13,** 1099–1108.

Shen, J., Tareste, D. C., Paumet, F., Rothman, J. E., and Melia, T. J. (2007). Selective activation of cognate SNAREpins by Sec1/Munc18 proteins. *Cell* **128,** 183–195.

Shin, O. H., Rizo, J., and Sudhof, T. C. (2002). Synaptotagmin function in dense core vesicle exocytosis studied in cracked PC12 cells. *Nat. Neurosci.* **5,** 649–656.

Sossin, W. S., Fisher, J. M., and Scheller, R. H. (1990). Sorting within the regulated secretory pathway occurs in the trans-Golgi network. *J. Cell Biol.* **110,** 1–12.

Spurlin, B. A., and Thurmond, D. C. (2006). Syntaxin 4 facilitates biphasic glucose-stimulated insulin secretion from pancreatic beta cells. *Mol. Endocrinol.* **20,** 183–193.

Steiner, D. F., Michael, J., Houghten, R., Mathieu, M., Gardner, P. R., Ravazzola, M., and Orci, L. (1987). Use of a synthetic peptide antigen to generate antisera reactive with a proteolytic processing site in native human proinsulin: Demonstration of cleavage within clathrin-coated (pro)secretory vesicles. *Proc. Natl. Acad. Sci. USA* **84,** 6184–6188.

Steyer, J. A., Horstmann, H., and Almers, W. (1997). Transport, docking and exocytosis of single secretory granules in live chromaffin cells. *Nature* **388,** 474–478.

Stoffers, D. A. (2004). The development of beta-cell mass: Recent progress and potential role of GLP-1. *Horm. Metab. Res.* **36,** 811–821.

Straub, S. G., and Sharp, G. W. (2002). Glucose-stimulated signaling pathways in biphasic insulin secretion. *Diabetes Metab. Res. Rev.* **18,** 451–463.

Sugita, S., Shin, O. H., Han, W., Lao, Y., and Sudhof, T. C. (2002). Synaptotagmins form a hierarchy of exocytotic Ca(2+) sensors with distinct Ca(2+) affinities. *EMBO J.* **21,** 270–280.

Takahashi, N., Kishimoto, T., Nemoto, T., Kadowaki, T., and Kasai, H. (2002). Fusion pore dynamics and insulin granule exocytosis in the pancreatic islet. *Science* **297,** 1349–1352.

Takahashi, N., Hatakeyama, H., Okado, H., Miwa, A., Kishimoto, T., Kojima, T., Abe, T., and Kasai, H. (2004). Sequential exocytosis of insulin granules is associated with redistribution of SNAP25. *J. Cell Biol.* **165,** 255–262.

Tamori, Y., Kawanishi, M., Niki, T., Shinoda, H., Araki, S., Okazawa, H., and Kasuga, M. (1998). Inhibition of insulin-induced GLUT4 translocation by Munc18c through interaction with syntaxin4 in 3T3-L1 adipocytes. *J. Biol. Chem.* **273,** 19740–19746.

Tang, J., Maximov, A., Shin, O. H., Dai, H., Rizo, J., and Sudhof, T. C. (2006). A complexin/synaptotagmin 1 switch controls fast synaptic vesicle exocytosis. *Cell* **126,** 1175–1187.

Tareste, D., Shen, J., Melia, T. J., and Rothman, J. E. (2008). SNAREpin/Munc18 promotes adhesion and fusion of large vesicles to giant membranes. *Proc. Natl. Acad. Sci. USA* **105,** 2380–2385.

Tellam, J. T., Macaulay, S. L., McIntosh, S., Hewish, D. R., Ward, C. W., and James, D. E. (1997). Characterization of Munc-18c and syntaxin-4 in 3T3-L1 adipocytes. Putative role in insulin-dependent movement of GLUT-4. *J. Biol. Chem.* **272,** 6179–6186.

Thiele, C., Gerdes, H. H., and Huttner, W. B. (1997). Protein secretion: Puzzling receptors. *Curr. Biol.* **7,** R496–500.

Thorens, B. (1992). Molecular and cellular physiology of GLUT-2, a high-Km facilitated diffusion glucose transporter. *Int. Rev. Cytol.* **137,** 209–238.

Thurmond, D. C., Ceresa, B. P., Okada, S., Elmendorf, J. S., Coker, K., and Pessin, J. E. (1998). Regulation of insulin-stimulated GLUT4 translocation by munc18c in 3T3L1 adipocytes. *J. Biol. Chem.* **273,** 33876–33883.

Thurmond, D. C., Kanzaki, M., Khan, A. H., and Pessin, J. E. (2000). Munc18c function is rquired for insulin-stimulated plasma membrane fusion of GLUT4 and insulin-responsive amino peptidase storage vesicles. *Mol. Cell. Biol.* **20,** 379–388.

Tooze, S. A., and Huttner, W. B. (1990). Cell-free protein sorting to the regulated and constitutive secretory pathways. *Cell* **60,** 837–847.

Tooze, J., Kern, H. F., Fuller, S. D., and Howell, K. E. (1989). Condensation-sorting events in the rough endoplasmic reticulum of exocrine pancreatic cells. *J. Cell Biol.* **109,** 35–50.

Trus, M., Wiser, O., Goodnough, M. C., and Atlas, D. (2001). The transmembrane domain of syntaxin 1A negatively regulates voltage-sensitive Ca(2+) channels. *Neuroscience* **104,** 599–607.

Tsuboi, T., and Rutter, G. A. (2003). Multiple forms of "kiss-and-run" exocytosis revealed by evanescent wave microscopy. *Curr. Biol.* **13,** 563–567.

Tsuboi, T., Terakawa, S., Scalettar, B. A., Fantus, C., Roder, J., and Jeromin, A. (2002). Sweeping model of dynamin activity. Visualization of coupling between exocytosis and endocytosis under an evanescent wave microscope with green fluorescent proteins. *J. Biol. Chem.* **277,** 15957–15961.

Tsuboi, T., McMahon, H. T., and Rutter, G. A. (2004). Mechanisms of dense core vesicle recapture following "kiss and run" ("cavicapture") exocytosis in insulin-secreting cells. *J. Biol. Chem.* **279,** 47115–47124.

Tucker, S. J., Gribble, F. M., Zhao, C., Trapp, S., and Ashcroft, F. M. (1997). Truncation of Kir6.2 produces ATP-sensitive K+ channels in the absence of the sulphonylurea receptor. *Nature* **387,** 179–183.

Varadi, A., Tsuboi, T., Johnson-Cadwell, L. I., Allan, V. J., and Rutter, G. A. (2003). Kinesin I and cytoplasmic dynein orchestrate glucose-stimulated insulin-containing vesicle movements in clonal MIN6 beta-cells. *Biochem. Biophys. Res. Commun.* **311,** 272–282.

Verhage, M., Maia, A. S., Plomp, J. J., Brussaard, A. B., Heeroma, J. H., Vermeer, H., Toonen, R. F., Hammer, R. E., van den Berg, T. K., Missler, M., Geuze, H. J., and Sudhof, T. C. (2000). Synaptic assembly of the brain in the absence of neurotransmitter secretion. *Science* **287,** 864–869.

Voets, T., Moser, T., Lund, P. E., Chow, R. H., Geppert, M., Sudhof, T. C., and Neher, E. (2001a). Intracellular calcium dependence of large dense-core vesicle exocytosis in the absence of synaptotagmin I. *Proc. Natl. Acad. Sci. USA* **98,** 11680–11685.

Voets, T., Toonen, R. F., Brian, E. C., de Wit, H., Moser, T., Rettig, J., Sudhof, T. C., Neher, E., and Verhage, M. (2001b). Munc18-1 promotes large dense-core vesicle docking. *Neuron* **31,** 581–591.

Volante, M., Allia, E., Gugliotta, P., Funaro, A., Broglio, F., Deghenghi, R., Muccioli, G., Ghigo, E., and Papotti, M. (2002). Expression of ghrelin and of the GH secretagogue receptor by pancreatic islet cells and related endocrine tumors. *J. Clin. Endocrinol. Metab.* **87,** 1300–1308.

Weimbs, T., Low, S. H., Chapin, S. J., Mostov, K. E., Bucher, P., and Hofmann, K. (1997). A conserved domain is present in different families of vesicular fusion proteins: A new superfamily. *Proc. Natl. Acad. Sci. USA* **94,** 3046–3051.

Weir, G. C., and Bonner-Weir, S. (1990). Islets of Langerhans: The puzzle of intraislet interactions and their relevance to diabetes. *J. Clin. Invest.* **85,** 983–987.
Wheeler, M. B., Sheu, L., Ghai, M., Bouquillon, A., Grondin, G., Weller, U., Beaudoin, A. R., Bennett, M. K., Trimble, W. S., and Gaisano, H. Y. (1996). Characterization of SNARE protein expression in beta cell lines and pancreatic islets. *Endocrinology* **137,** 1340–1348.
Wierup, N., Svensson, H., Mulder, H., and Sundler, F. (2002). The ghrelin cell: A novel developmentally regulated islet cell in the human pancreas. *Regul. Pept.* **107,** 63–69.
Wiser, O., Trus, M., Hernandez, A., Renstrom, E., Barg, S., Rorsman, P., and Atlas, D. (1999a). The voltage sensitive Lc-type Ca2+ channel is functionally coupled to the exocytotic machinery. *Proc. Natl. Acad. Sci. USA* **96,** 248–253.
Wiser, O., Trus, M., Hernandez, A., Renstrom, E., Barg, S., Rorsman, P., and Atlas, D. (1999b). The voltage sensitive Lc-type Ca2+ channel is functionally coupled to the exocytotic machinery. *Proc. Natl. Acad. Sci. USA* **96,** 248–253.
Xu, T., Rammner, B., Margittai, M., Artalejo, A. R., Neher, E., and Jahn, R. (1999). Inhibition of SNARE complex assembly differentially affects kinetic components of exocytosis. *Cell* **99,** 713–722.
Yang, S. N., Larsson, O., Branstrom, R., Bertorello, A. M., Leibiger, B., Leibiger, I. B., Moede, T., Kohler, M., Meister, B., and Berggren, P. O. (1999). Syntaxin 1 interacts with the L(D) subtype of voltage-gated Ca(2+) channels in pancreatic beta cells. *Proc. Natl. Acad. Sci. USA* **96,** 10164–10169.
Yang, B., Steegmaier, M., Gonzalez, L. C. Jr., and Scheller, R. H. (2000). nSec1 binds a closed conformation of syntaxin1A. *J. Cell Biol.* **148,** 247–252.
Zawalich, W. S., and Zawalich, K. C. (1997). Influence of pyruvic acid methyl ester on rat pancreatic islets. Effects on insulin secretion, phosphoinositide hydrolysis, and sensitization of the beta cell. *J. Biol. Chem.* **272,** 3527–3531.
Zawalich, W. S., Zawalich, K. C., and Rasmussen, H. (1989). Interactions between cholinergic agonists and enteric factors in the regulation of insulin secretion from isolated perifused rat islets. *Acta Endocrinol. (Copenh)* **120,** 702–707.
Zhang, W., Efanov, A., Yang, S. N., Fried, G., Kolare, S., Brown, H., Zaitsev, S., Berggren, P. O., and Meister, B. (2000). Munc-18 associates with syntaxin and serves as a negative regulator of exocytosis in the pancreatic beta -cell. *J. Biol. Chem.* **275,** 41521–41527.
Zhang, W., Khan, A., Ostenson, C. G., Berggren, P. O., Efendic, S., and Meister, B. (2002). Down-regulated expression of exocytotic proteins in pancreatic islets of diabetic GK rats. *Biochem. Biophys. Res. Commun.* **291,** 1038–1044.

CHAPTER SEVENTEEN

Glucose, Regulator of Survival and Phenotype of Pancreatic Beta Cells

Geert A. Martens *and* Daniel Pipeleers

Contents

Diabetes Research Center, Brussels Free University-VUB, Laarbeeklaan 103, 1090 Brussels, Belgium

Vitamins and Hormones, Volume 80
ISSN 0083-6729, DOI: 10.1016/S0083-6729(08)00617-1

Abstract

The key role of glucose in regulating insulin release by the pancreatic beta cell population is not only dependent on acute stimulus-secretion coupling mechanisms but also on more long-term influences on beta cell survival and phenotype. Glucose serves as a major survival factor for beta cells via at least three actions: it prevents an oxidative redox state, it suppresses a mitochondrial apoptotic program that is triggered at reduced mitochondrial metabolic activity and it induces genes needed for the cellular responsiveness to glucose and to growth factors. Glucose-regulated pathways may link protein synthetic and proliferative activities, making glucose a permissive factor for beta cell proliferation, in check with metabolic needs. Conditions of inadequate glucose metabolism in beta cells are not only leading to deregulation of acute secretory responses but should also be considered as causes for increased apoptosis and reduced formation of beta cells, and loss of their normal differentiated state.

ABBREVIATIONS

FACS	fluorescence activated cell sorting
GCK	glucokinase
GLUT2	glucose transporter 2 (SLC2A2)
GSIS	glucose-stimulated insulin secretion
GSK3β	glycogen synthase kinase 3β
mTOR	mammalian target of rapamycin
PDX1	pancreatic duodenal homeobox protein 1
PKA	protein kinase A
PKB	protein kinase B
PP1	protein phosphatase 1
ROS	reactive oxygen species

I. SCOPE

Pancreatic beta cells sense minute-to-minute variations in blood sugars, amino acids, fatty acids and ketone bodies. Under physiological conditions, peaks of postprandial glucose (~10 mM) and amino acids alternate, during fasting, with phases of lower glucose (~5 mM) and amino acid concentrations and subtle rises in fatty acid and ketone bodies. The mechanisms through which nutrients acutely regulate insulin secretion have been studied since many years which has led to a progressively better understanding (Fahien and Macdonald, 2002; Henquin, 2000; Henquin *et al.*, 2006; Wiederkehr and Wollheim, 2006). More recent work has also examined to which extent nutrients influence the development and maintenance of the functional beta cell mass—defined here as the integration of the number of beta cells and their individual insulin synthetic and secretory capacity. In the present chapter we will review the role of glucose as a survival and differentiation factor for the beta cells; this will require a distinction between the nutrient's acute effects that rapidly regulate hormone release and its longer-term actions. Our main emphasis will be on the healthy beta cell. When possible we will try to explore potential relevance for beta cell dysfunction in diabetes.

II. Beta Cell Handling of Glucose: Metabolic Specializations to Ensure Low-Affinity/High Capacity Glucose Sensing

Insulin synthesis and secretion by beta cells are acutely (minute–hours) regulated by glucose. Their individual level of activation is determined by the cells' capacity to metabolize glucose, mainly through mitochondrial oxidation (Antinozzi *et al.*, 2002; Kiekens *et al.*, 1992; Wiederkehr and Wollheim, 2006). The underlying mechanisms are outlined in comprehensive reviews (Deeney *et al.*, 2000; Henquin *et al.*, 2006; Macdonald *et al.*, 2005; Newgard, 2002; Schuit *et al.*, 2001; Wollheim and Maechler, 2002). The present section only highlights their major sites, with emphasis on metabolic specializations for glucose sensing by beta cells.

A. Abundant glucose uptake coupled to low affinity/high capacity glucose phosphorylation

The beta cell is equipped with abundant glucose transport systems, GLUT2 in the rodent, and GLUT1-3 in man (De Vos *et al.*, 1995) that mediate the rapid equilibrium of intra-and extracellular glucose concentration so that cellular glucose uptake never becomes limiting for metabolic flux. The metabolic control over the beta cells' functional state is exerted at the level of glucokinase (GCK). Through its low-affinity (high K_m) and high-capacity kinetic properties for glucose phosphorylation, it serves as the gatekeeper for the beta cells' metabolic flux. It is therefore also referred as the main glucose sensor in beta cells (Matschinsky, 2002; Zelent *et al.*, 2005). Both activating and loss-of-function GCK mutations have been reported in man, resulting in, respectively, an excessive and defective insulin secretion (Gloyn *et al.*, 2005; Matschinsky, 1996). Rodent beta cells are shown to be strictly dependent on this low-affinity hexokinase for their glucose metabolism (Schuit *et al.*, 1999), so that they become progressively glucose-deprived at levels under 5 mM glucose (Martens, 2008; Martens and Van de Casteele, 2007). Although it is still unclear whether this is also the case in human beta cells, we believe this characteristic is critical for understanding the chronic glucose regulation of the beta cell functional state. We, and others, have shown that the insulin synthetic and secretory capacity of an individual beta cell depends on its GCK expression and activity level (Heimberg *et al.*, 1993; Jetton and Magnuson, 1992; Jorns *et al.*, 1999; Kiekens *et al.*, 1992; Van Schravendijk *et al.*, 1992), which is subject to chronic regulation *in vivo* (Jorns *et al.*, 1999). GCK expression physiologically varies significantly between individual beta cells, establishing a molecular signature for various functional states that the cells can acquire *in vivo* (Jetton and Magnuson, 1992; Ling *et al.*, 2006; Luther *et al.*, 2006). A down-regulation of GCK expression can herald a loss of the beta cells' glucose responsive function, which could be an early phase of their dysfunction in the natural course of diabetes (Kim *et al.*, 2005; Otani *et al.*, 2004).

The GCK in beta cells is additionally regulated by its binding to another glycolytic enzyme, 6-phosphofructo-2-kinase/fructose-2,6-bisphosphatase (PFK-2/FBPase-2), which is, as GCK, abundantly expressed in beta cells. Their mutual interaction has been shown to accelerate flux through the glucose sensor, and represents a level of regulation that requires further investigation (Massa *et al.*, 2004).

B. Active aerobic oxidation and oxidative phosphorylation

While the GLUT/GCK system makes the beta cell's glucose metabolic rate proportionate to the ambient glucose concentration, the mitochondria will breakdown most glucose through aerobic oxidation (Schuit *et al.*, 1997), generating mitochondrial NADH and $FADH_2$ along rates that closely reflect the sigmoid kinetics of GCK (Bennett *et al.*, 1996; Martens *et al.*, 2005, 2006). The high glucose oxidation rates in the mitochondria of the beta cells can be maintained through a highly active oxidative phosphorylation which prevents even at excessively high glucose levels—the development of a state 4 respiration that is characterized by limited availability of ADP substrate for the F_0F_1ATPase and slow down of mitochondrial H^+-gradient discharge (Martens *et al.*, 2005). This property might be explained by the ability of glucose-metabolizing beta cells to match their increased ADP phosphorylation by a proportionate consumption of the generated ATP in various endergonic processes such as hormone synthesis, lipid anabolism, and exocytosis with ATP-consuming Ca^{2+} pumping. This would be consistent with the linear increase in the ATP/ADP ratio up to 10 mM glucose and its leveling off in the 10–20 mM glucose range (Detimary *et al.*, 1998), where insulin synthesis is no longer increasing but insulin secretion and lipid anabolism further augment (Schuit *et al.*, 1991). Another mechanism that might prevent respiratory slow down in beta cells is their relatively high rate of H^+ dissipation via uncoupling proteins such as UCP2 (Affourtit and Brand, 2006).

C. Anaplerotic/cataplerotic signaling cycles

The beta cell's Krebs cycle is also specialized as a distribution system in which a substantial fraction of imported glucose carbon is exported in anaplerotic/cataplerotic metabolic cycles that can generate signals for beta cell activation. To this end, the beta cell has a high expression of pyruvate carboxylase (Macdonald, 1995; Schuit *et al.*, 1997) through which pyruvate can enter the Krebs cycle as oxaloacetate when the flux through pyruvate dehyrogenase complex is saturated by accumulation of NADH or acetyl-coA. In hepatocytes, pyruvate carboxylase is essential for gluconeogenesis from pyruvate, but beta cells reportedly have no active glucose production capacity (Matschinsky, 1996) despite their abundant expression of the mitochondrial isoform of PEPCK (our unpublished observation), of which the role requires further study. The anaplerotic reaction through pyruvate carboxylase instead fuels cycles via malic enzyme (pyruvate/malate cycle) that generate NADPH, which could function as additional coupling factor in glucose-stimulated insulin secretion (GSIS) (Fransson *et al.*, 2006; Lu *et al.*, 2002; Schuit *et al.*, 1997). Another cataplerotic cycle involves citrate export and via ATP citrate lyase, conversion of glucose-derived acetyl-CoA to malonyl-CoA for cytosolic fatty acid synthesis, generating acyl-coA molecules with signaling potential (Flamez *et al.*, 2002; Roduit *et al.*, 2004). The precise role of these cycles in GSIS is not yet clarified (Joseph *et al.*, 2007).

D. Active lipid synthetic and signaling pathways

Beta cells are also highly active in the synthesis of cholesterol and derived complex lipids. Intermediates have been proposed as possibly synergistic coupling factors for GSIS, for example, NADPH. These anabolic pathways could also function as partly "futile cycles" that "sink" excess glucose carbon (Roche *et al.*, 1998). Also, their adequate function appears essential to preserve glucose metabolism, and thus glucose sensing by the beta cells (Ronnebaum *et al.*, 2008). Among potential additional sites for glucose regulation in beta cells, we have proposed l-3-hydroxyacyl-coA dehydrogenase (HADH, formerly HADHSC). HADH, a key enzyme of the mitochondrial beta-oxidation chain, has a high expression in human and rodent beta cells, which is disproportionate to that of other enzymes of this pathway (Martens *et al.*, 2007) and to the rather low capacity of the beta cell to oxidize exogenous fatty acids (Hellemans *et al.*, 2007). HADH seems involved in glucose sensing as indicated by observations in human loss-of-function mutants and by *in vitro* studies on rodent beta cells (Martens *et al.*, 2007) but the mechanism is not yet known.

In conclusion, beta cells are metabolically activated by glucose flux through glucokinase. Most glucose is catabolized through glycolysis and further aerobically oxidized to generate ATP, the prime signaling factor that couples glucose metabolism to biosynthetic and secretory activation. Part of the glucose carbon also enters into other metabolic cycles that generate additional signals for GSIS and/or that fulfill an anabolic function of potential importance for beta cell functions. Activation of ATP-generating reactions by glucose is functionally coupled to a proportionate activation of ATP-consuming reactions, so that mitochondrial homeostasis is maintained. Healthy beta cells thus couple high metabolic activity to activation of endergonic endocrine functions.

III. Glucose as Regulator of the Differentiated Beta Cell Phenotype

A. Glucose regulation of the beta cell's protein synthesis

Glucose concentration-dependently stimulates protein biosynthesis by beta cells with a predominant effect on their (pro)insulin synthesis (Kiekens *et al.*, 1992; Permutt and Kipnis, 1972). This control is mediated through a concerted activation of various kinases such as mammalian target of rapamycin (mTOR), protein kinase B (PKB/Akt), protein kinase A (PKA), protein kinase C (PKC) and mitogen-activated protein kinases (MAPKs)/extracellular signal related kinase (ERK) kinases (MEKs) (Kwon *et al.*, 2004; Proud, 2004).

The first signals that couple activated glucose metabolism to kinase activation (Summarized in Fig. 17.1) are:

(i) A sustained increase in the beta cell's free cytosolic Ca^{2+} as induced by K^+_{ATP} channel closure and opening of voltage-gated Ca^{2+} channels. Calmodulin-dependent kinases transmit this signal to a cascade coupling PKA—MEK—ERK-1/2 (Briaud *et al.*, 2003), protein phosphatase-1 (Vander Mierde *et al.*, 2007) and possibly also involving PKB and mTOR activity (Wang *et al.*, 2008);
(ii) Increased ATP/ADP-activating p70/S6 kinase ($p70^{S6K}$) via "ATP sensor" mTOR (Briaud *et al.*, 2003)

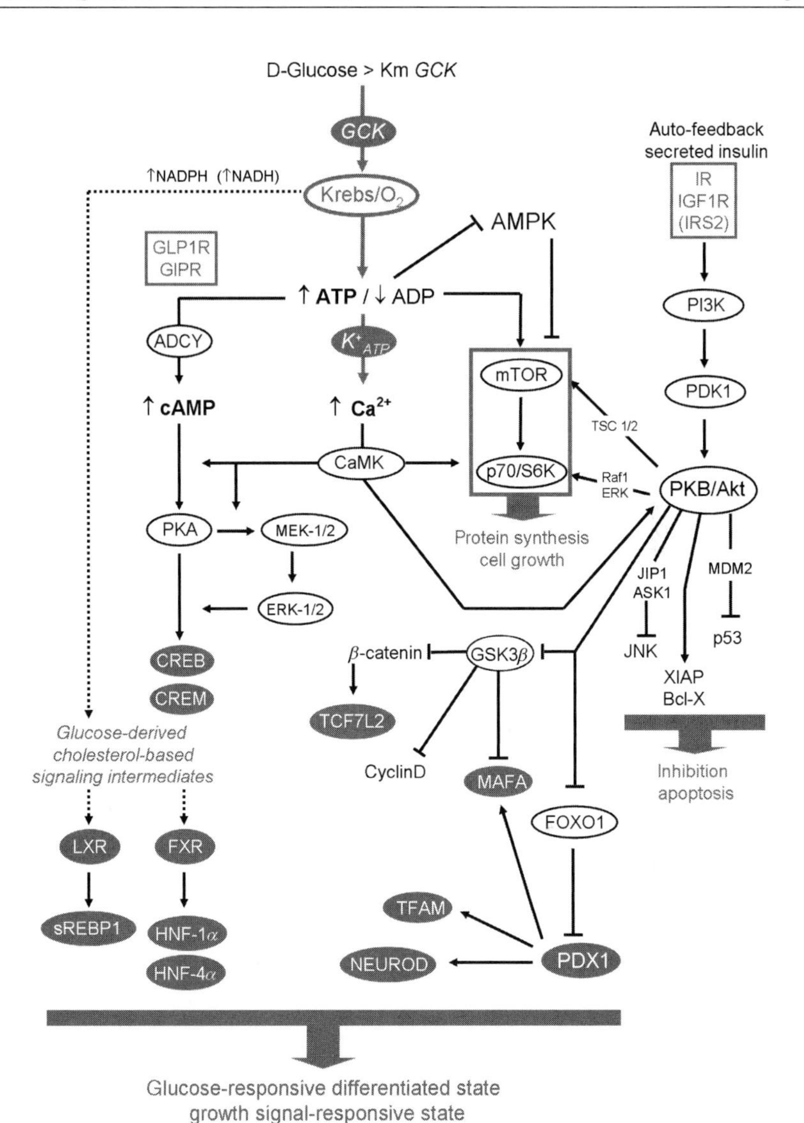

Figure 17.1 The glucose-stimulated beta cell. When glucose is raised above the K_m of glucokinase (GCK), large amounts of ADP are converted to ATP. This will inhibit the AMP-dependent kinase (AMPK) and activate ATP-sensor mTOR and S6-kinase (S6K/ p70), leading to activation of protein (insulin) synthesis and beta cell growth. Closure of ADP/ATP-sensing K^+ channels (K^+_{ATP}) raises cytoplasmic Ca^{2+}, which, through calmodulin-dependent kinase (CaMK) transmits the glucose signaling to the protein kinase A (PKA) and MEK/ERK relays, and to protein kinase B (PKB/Akt). By providing ATP substrate for adenylate cyclase (ADCY), glucose also potentiates beta cell stimulation by incretins through GLP-1 and GIP receptors, and facilitates activation of cyclic AMP (cAMP)-responsive transcription factors CREM and CREB. PKB also acts as integrator for various growth factors, including autofeedback by insulin;

(iii) A glucose-mediated potentiation of the beta cell's cAMP level (Prentki and Matschinsky, 1987)—boosting PKA activation. Glucose could directly augment cAMP production, by providing ATP substrate for the adenylate cyclase; more likely however, glucose stimulation only potentiates the accumulation of incretin-induced cAMP (Schuit and Pipeleers, 1985). The result is a metabolically controlled activation of the beta cells' mRNA translation machinery at various levels: phosphorylation of ribosomal protein S6 (rpS6) by $p70^{S6K}$ and of 4E-binding protein 1 (4E-BP1), associated to a dephosphorylation of eukaryotic elongation factor EF2 (eEF2) and initiation factor 2 alpha (eIF2α). These regulations have mainly been studied in short-term (hours) incubations.

Glucose also chronically (days) activates the translation machinery and protein biosynthesis in a concentration-dependent manner (Ling and Pipeleers, 1996; Ling *et al.*, 1996); while mTOR inhibition by rapamycin can decrease (de)phosphorylation of the translation machinery, this does not clearly affect the level of protein synthetic activation. Chronic glucose-regulation of translation and protein synthesis might thus be more directly dependent on energy state (ATP/ADP) of the cells, or be regulated by as yet unidentified glucose metabolites and signaling pathways (Wang and Ling, personal communication), or via additional effects of glucose on the beta cells' gene transcription, increasing the maximal protein synthetic capacity. In this aspect, it has been shown that sustained glucose stimulation increases the level of insulin mRNA that is available for translation, via increased insulin gene transcription (Greenman *et al.*, 2005), stabilization (Knoch *et al.*, 2004) or via recruitment of insulin mRNA to more active pools of ribosomes (Wicksteed *et al.*, 2001). Glucose also causes a coordinate induction of various genes involved in the beta cells' specialized function, with increased expression of genes involved in hormone synthesis and processing in the endoplasmic reticulum (Martens, 2008; Schuit *et al.*, 2002). Along the same line, we also found that the level of protein synthesis is not only

the kinase not only stimulates protein synthesis and cell growth through tuberous sclerosis (TSC) 1 and 2 and RAF/ERK proteins, but also directly inhibits apoptosis at multiple levels and causes nuclear activation of PDX1 and inactivation of GSK3β. PDX1 directly activates several transcription factors that are important for the differentiated beta cell phenotype (such as NeuroD, TFAM, and MAFA); this goes in part also via inactivation of GSK3β, which has the potential of being a negative regulator of beta cell function and proliferation. Possibly, glucose carbon is converted to complex sterol-based lipids that could fulfill signaling roles through nuclear sensors such as LXR and FXR that control activity of SREBP1 and HNF-factors, another set of factors important for proper metabolic beta cell function. For abbreviations that represent official gene names, the reader is referred to corresponding paragraph of the manuscript.

dependent on the available insulin mRNA, or on the level of translation activation, but is also controlled by the intrinsic glucose responsiveness of the beta cell, which is subject to chronic regulation both *in vitro* and *in vivo* (Martens *et al.*, 2006).

B. Glucose chronically maintains glucose-responsive state of beta cells

The specialized functions of many cell types in the body are regulated by their inductive signals, and these processes are mediated through altered gene transcription in a cell-type specific manner. Examples are physical and memory training. Physical activity enhances the expression of skeletal muscle-specific proteins, thereby mediating their adaptive hypertrophy (Fluck, 2006). Memory formation via long-term potentiation in neurons is abolished when gene transcription (Barondes and Jarvik, 1964) or protein synthesis (Flexner *et al.*, 1962) are blocked. Similarly, we and others have shown that the specialized function or "differentiated state"—defined here as the global set of proteins that are needed for proper glucose sensing by beta cells—is chronically regulated by the quintessential beta cell stimulus, glucose.

As compared to a basal state (3 mM), 10 mM glucose-exposed purified rat beta cells after 24 h show increased expression of various gene clusters involved in energy metabolism and glucose sensing (e.g., GCK, GLUT2, HADH), cataplerosis of glucose carbon into lipid synthetic pathways (ACLY, IDI1), the regulated insulin biosynthetic/secretory pathway (CHGA, SYP, SNAP25), membrane transport, intracellular signaling, gene transcription, growth (receptors) (PRLR, GHR, CCKAR, GNAS, CCND1), incretin effect (GIPR) and protein synthesis/degradation (PCSK1) (Schuit *et al.*, 2002) (Fig. 17.2). This effect is maintained after 3 days, in 10 versus 5 mM glucose-stimulated beta cells (Martens, 2008) (Fig. 17.2) with significant overlap between the genes induced after 1 and 3 days. This glucose regulation is also found, at least in part, in the mouse MIN6 cell line that shows both acute (10 vs. 1 mM glucose, 4 h) (Glauser *et al.*, 2007) and chronic (25 vs. 5 mM, 24 h) glucose-responsive gene expression. It has physiological relevance also *in vivo*, as it can reveal how the beta cells' phenotype is comprehensively regulated by the repetitive cycles of food intake and fasting (Hinke *et al.*, 2004).

The chronic glucose-regulation of the beta cells' gene expression is complex, and the number of regulators involved is ever growing (Fig. 17.1). As outlined above, glucose can rapidly and concentration-dependently (in)activate a multitude of kinases/phosphatases (e.g., PKA, PI3K/Akt, PP1, GSK3-beta) that consequently activate interacting cascades of transcription factors driving beta cell-specific gene expression. In addition, the nutrient can also control stability of specific mRNAs, increasing their availability for translation, or control the level of microRNAs with

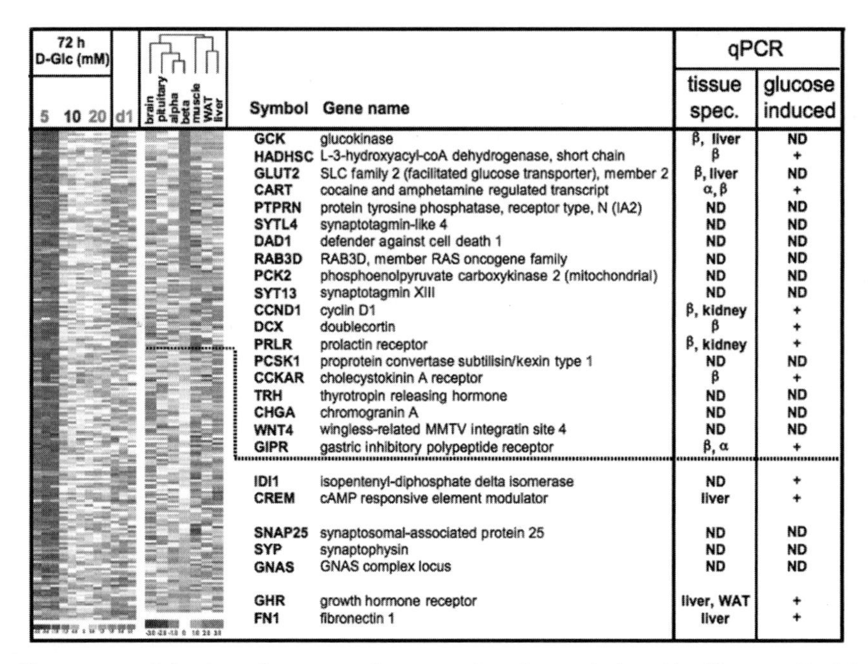

Symbol	Gene name	qPCR tissue spec.	qPCR glucose induced
GCK	glucokinase	β, liver	ND
HADHSC	L-3-hydroxyacyl-coA dehydrogenase, short chain	β	+
GLUT2	SLC family 2 (facilitated glucose transporter), member 2	β, liver	ND
CART	cocaine and amphetamine regulated transcript	α, β	+
PTPRN	protein tyrosine phosphatase, receptor type, N (IA2)	ND	ND
SYTL4	synaptotagmin-like 4	ND	ND
DAD1	defender against cell death 1	ND	ND
RAB3D	RAB3D, member RAS oncogene family	ND	ND
PCK2	phosphoenolpyruvate carboxykinase 2 (mitochondrial)	ND	ND
SYT13	synaptotagmin XIII	ND	ND
CCND1	cyclin D1	β, kidney	+
DCX	doublecortin	β	+
PRLR	prolactin receptor	β, kidney	+
PCSK1	proprotein convertase subtilisin/kexin type 1	ND	ND
CCKAR	cholecystokinin A receptor	β	+
TRH	thyrotropin releasing hormone	ND	ND
CHGA	chromogranin A	ND	ND
WNT4	wingless-related MMTV integratin site 4	ND	ND
GIPR	gastric inhibitory polypeptide receptor	β, α	+
IDI1	isopentenyl-diphosphate delta isomerase	ND	+
CREM	cAMP responsive element modulator	liver	+
SNAP25	synaptosomal-associated protein 25	ND	ND
SYP	synaptophysin	ND	ND
GNAS	GNAS complex locus	ND	ND
GHR	growth hormone receptor	liver, WAT	+
FN1	fibronectin 1	liver	+

Figure 17.2 Selection of representative transcripts that are induced by 10 versus 5 mM glucose in cultured beta cells. The left heat map shows relative mRNA expression level in freshly isolated (day 1, d1) beta cells, and in cells cultured for 3 days at 5, 10, or 20 mM glucose. The right heat map shows corresponding relative mRNA expression in various rat tissue/cell types; genes are ordered by decreasing beta cell-abundance/specificity. The two columns on the right indicate eventual confirmation by Taqman qPCR of tissue-specific mRNA expression (left column) or glucose-induction (right column); (ND, not done; +, confirmed). (See Color Insert.)

regulatory potential. A final and more elusive level of phenotypic regulation is posttranslational modification of proteins (Andrali *et al.*, 2007), affecting their cellular localization, stability, and function.

C. Transcription factors that coregulate glucose-responsive beta cell phenotype

In the following paragraph we will try give a succinct and by no means exhaustive overview of the expanding set of transcription factors involved in the glucose control of beta cell phenotype. These nuclear factors often work in concert with other stimuli, such as fatty acids or cAMP and thus function as global integrators of the metabolic state (Fig. 17.1). The sterol regulatory element binding protein-1 (*SREBP1)* controls lipogenic pathways, and in beta cells has been shown to mediate, both *in vivo* and *in vitro*, the beta cells' adaptation to increased glucose concentrations (30 vs. 8 mM);

SREBP1 in beta cells not only directly regulates lipid synthetic pathways, but also key beta cell genes involved in glucose sensing (e.g., GCK, GLUT2), regulated secretion (K^{+}_{ATP} channel subunits), and maintenance of beta cell differentiation (pancreatic duodenal homeobox 1, PDX1) (Diraison *et al.*, 2008). SREBP1 is under control of liver X receptor beta (*LXR-beta*), explaining the similarities between SREBP1 and LXR-beta knockout beta cells (Zitzer *et al.*, 2006).

PDX1 itself was also shown to regulate mitochondrial nutrient responsiveness in the beta cell, through its activation of mitochondrial transcription factor A (*TFAM*), a nuclear factor that specifically controls mitochondrial genes, and that when defective causes mitochondrial diabetes (Silva *et al.*, 2000). PDX1 can be kinetically activated through PI3K/Akt-mediated phosphorylation, leading to its translocation to the nucleus. Glucose can thus directly activate PDX1, or indirectly via an autofeedback loop involving binding of secreted insulin on the insulin receptor (da Silva *et al.*, 2004; Leibiger *et al.*, 1998), thereby providing a positive feedback loop with *in vivo* relevance (Okada *et al.*, 2007). PDX1 cooperates with other beta cell-specific transcription factors, such as Beta-2 and MafA.

MafA is a potent activator of insulin gene transcription (Kataoka *et al.*, 2002). Its levels are increased by glucose through the nutrient's ability to inhibit glycogen synthase kinase-3 beta (GSK3-beta), leading to decreased MafA phosphorylation and preventing its proteasomal degradation (Han *et al.*, 2007).

The intrinsic link between glucose and the cAMP-increasing incretins (GLP-1, GIP) to maintain the beta cells' "glucose competence" is also found at the transcriptional level. This is molecularly reflected by the cAMP-responsive element modulator/binding protein (*CREB/CREM*) transcription factor family, that operates downstream of PKA and ERK-1/2. These factors coregulate the chronic beta cell adaptation to elevated glucose and fatty acids (e.g., CREM (Zhou *et al.*, 2003)). At the same time they are also vital for the antiapoptotic/mitogenic actions of glucose and incretins (GLP-1, GIP, glucagon), (Costes *et al.*, 2006; Hussain *et al.*, 2006), for example, by up-regulating IRS2, thereby stimulating PKB activation by IGF-1 or insulin (Jhala *et al.*, 2003).

Another important set of regulators are the forkhead proteins, in particular *FOXO1*, which regulate rapid adaptations of beta cell function to high glucose, insulin and growth factors (IGF-1, EGF, GLP-1, GIP), and, in the long term, also mediate adaptation of beta cell mass *in vivo* through activation of Akt/PKB (Reviewed in Glauser and Schlegel, 2007; Kitamura and Ido, 2007). Naturally occurring mutations in humans resulting in maturity-onset diabetes of the young, have also highlighted the critical role of other forkhead proteins, namely the hepatocyte nuclear factor (HNF) family (Hansen *et al.*, 2002). Hepatocyte nuclear factors regulate beta cell development (*HNF1-alpha*), but also glucose-responsive beta cell function, and adaptive growth *in vivo* (*HNF1-alpha en HNF4-alpha*) (Wang *et al.*, 2000). *Forkhead box A2 protein* (*FOXA2, HNF3-beta*) directs beta cells towards their characteristic low-affinity glucose sensing while maintaining their K^{+}_{ATP}-dependent coupling mechanism; at the same time, it appears to limit their functional sensitivity to amino acids (Lantz *et al.*, 2004; Wang *et al.*, 2002).

D. Posttranscriptional regulation by glucose

Another major level of glucose-control is at the posttranscriptional level, by regulation of mRNA stability. This mechanism is best documented for insulin mRNA which was shown to be stabilized by binding of the poly-pyrimidine tract binding protein (*PTB*), also named hnRNP I, to its 3′-untranslated region (Fred *et al.*, 2006). In rat islets, this stabilization is stimulated by glucose, and by hypoxia, probably in an mTOR-dependent manner. Since PTB participates in specialized gene expression in other tissues (e.g., iNOS, VEGF), it might play a role in the concerted regulation of other beta cell-specific genes.

Another emerging class of posttranscriptional regulators are the micro-RNAs; in islet beta cells, abundant expression of at least two such molecules was shown: miR-375, which negatively regulates GSIS at a distal step of stimulus-exocytosis coupling (Poy *et al.*, 2004), and more recently, miR-124a, which negatively regulates FOXA2 and PDX1, thereby possibly modifying beta cellular nutrient responsiveness (Baroukh *et al.*, 2007). More studies are needed to examine if/how glucose can regulate expression and activity of these microRNAs, and whether these regulations are clinically relevant.

IV. Glucose Regulation of Beta Cell Number

A. Glucose as survival factor for beta cells

Our laboratory previously developed fluorescence-activated cell sorting (FACS) protocols that allow beta cell purification from rodents up to 95% insulin-expressing cells (Van De Winkel and Pipeleers, 1983). This enabled us to study the glucose requirements of the beta cell without the confounding effects of other islet cells. Studies on these pure beta cell preparations clearly established glucose as beta cell survival factor (Hoorens *et al.*, 1996; Ling *et al.*, 1994). Purified rodent beta cells survive best at glucose concentration around 10 mM (Ling *et al.*, 1994). When chronically (days) exposed to lower glucose concentrations they lose part of their differentiated gene expression (cfr. infra) and finally die by apoptosis (Hoorens *et al.*, 1996; Martens, 2008).

Along the same line, glucose can also be considered as proliferation factor for beta cells. Until recently, the beta cell was considered by most as essentially postmitotic, and their levels of proliferation—measured by nucleotide incorporation into DNA—are quite low (~0.1%) both *in vivo* (Dor *et al.*, 2004; Georgia and Bhushan, 2004; Teta *et al.*, 2005) and *in vitro* (Weinberg *et al.*, 2007). Few studies actually reported glucose effects on the proliferation of primary beta cell cultures, obtained from rodents and/or humans (Maedler *et al.*, 2006; Parnaud *et al.*, 2008) Moreover, these studies typically found lower or absent beta cell division in basal (up to 5 mM) glucose concentrations, and increased division at glucose >10 mM glucose. The paucity of studies on primary cells contrasts with the numerous, mechanistic reports on glucose-induced proliferation of various kinds of insulinoma cell lines (e.g., rat INS1 and RIN, mouse MIN6) that are characterized by a cell cycling process that is, at least in part, disconnected from normal growth signaling cascades.

B. Possible parallel between glucose control of beta cell proliferation and insulin synthesis

Remarkably, the same kinases and glucose-derived metabolic signals that control glucose-regulated protein synthesis in primary beta cells, have been implied as mediators of glucose-stimulated proliferation of insulinoma cells (Fig. 17.1). This would suggest a mechanistic association between both processes with relevance for the regulation of the functional beta cell mass. Glucose could act as a direct mitogen for beta cells through its activation of PKA, PKB-p70S6K, and ERK-1/2 (Hugl *et al.*, 1998; Maedler *et al.*, 2006), but at the same time also enhance the beta cells' responsiveness to growth factors that signal through these cascades. This has been shown for insulin-like growth factor-1 (IGF-1) (Hugl *et al.*, 1998; Lingohr *et al.*, 2006), growth hormone (GH) (Cousin *et al.*, 1999) and GLP-1 (Kwon *et al.*, 2004). Glucose supports beta cell activation and growth by GLP-1 through its cAMP- and Ca^{2+}-mediated priming of PKA; PKA subsequently coactivates mTOR (Briaud *et al.*, 2003; Kwon *et al.*, 2004), and signals to the nucleus by phosphorylation of transcription factor CREB, in a process that is further stimulated by activation of the MEK-1 to ERK-1/2 relay (Briaud *et al.*, 2003; Costes *et al.*, 2006). Similarly, glucose augments the proliferative response to IGF-1 via a direct action on PKB-p70S6K and ERK-1/2, which are downstream of the IGF-1 receptor/IRS2 (Hugl *et al.*, 1998). In addition, both glucose and cAMP can directly augment expression of IRS2, and thus support beta cell proliferation (Jhala *et al.*, 2003; Mohanty *et al.*, 2005; Schuppin *et al.*, 1998). The potency of GH to stimulate beta cell proliferation (Brelje *et al.*, 1993) is also dependent on the beta cells' sensing. In this case, however, the cytokine's signaling through JAK2/STAT5 is not directly boosted by glucose (Cousin *et al.*, 1999).

C. Glucose, candidate facilitator or permissive factor for beta cell proliferation

Above observations illustrate that glucose can operate as permissive factor for beta cell growth, presumably via a combined activation of the kinases involved in protein synthesis and by providing a minimal threshold of energetic state. This predicts that a minimal glucose-stimulation will also be required to disclose the stimulatory actions of established beta cell mitogens such as prolactin and placental lactogen (Brelje *et al.*, 1993; Nielsen, 1982), and others such as epidermal growth factor, gastrin, platelet-derived growth factor and hepatocyte growth factor. Glucose's effects on proliferation of human and rodent beta cells appear to decrease with age of the animal (Maedler *et al.*, 2006; Swenne, 1983; Swenne and Andersson, 1984). The aging beta cell also becomes less responsive to GH- and PRL-induced growth (Brelje and Sorenson, 1991); it is not clear whether the latter is only due to progressive down-regulation of the receptors involved, or is partly explained by an underlying age-dependent decrease in glucose sensing (Borg *et al.*, 1995). In this context, we found that the mRNA expression of the prolactin (PRLR), growth hormone (GHR), cholecystokinine A (CCKAR), and glucose-dependent insulinotropic polypeptide (GIPR) receptors are all increased by glucose stimulation in rodent beta cells, suggesting that appropriate glucose recognition has important effects on the beta cells' growth factor responsiveness by regulating their receptors (Fig. 17.2). The ability of the beta cell mass to expand in response to glucose stimulation, is also determined by genetic background (Nesher *et al.*, 1999; Swenne and Andersson, 1984) and this correlates with the susceptibility of the beta cell to high glucose-induced deregulation (Leibowitz *et al.*, 2001; Svensson *et al.*, 1993). Similarly, in humans, the genetic background, such as variants in the TCF7L2 regulatory sequence, could affect the adaptive increase in functional beta cell mass in conditions of increased metabolic needs such as in pregnancy or obesity (Lingohr *et al.*, 2002; Sladek *et al.*, 2007). On basis of the presently available *in vitro* evidence, we suggest that glucose serves as a permissive factor for beta cell proliferation rather than a true proliferation factor. When the glucose concentration is lowered below a critical threshold, the beta cell's response to growth/proliferation-factors is blunted, and an apoptotic suicide program is activated (Hoorens *et al.*, 1996; Martens and Van de Casteele, 2007), as discussed below.

V. Beta Cell Handling of Threatening High and Low Glucose Levels

A. Beta cell adaptations to low glucose

1. Acute physiological regulation: Built-in metabolic safety mechanisms that prevent inappropriate beta cell activation

As opposed to the chronically devastating complications induced by hyperglycemia, hypoglycemia can be rapidly lethal: it can instantly cause dysfunction, even death of cells that are highly glucose-dependent such as neurons, leading to life-threatening coma. The normal pancreas contains an amount of insulin that could kill its host within minutes if inappropriately released. That this horror scenario does not happen can be seen as a sign for most efficient safety mechanisms that acutely shut off insulin release when metabolically inappropriate, such as when blood glucose levels reach lower physiologic limits (Quintens *et al.*, 2008). The following mechanisms might be relevant for this function:

(i) The lack of monocarboxylate transporters (MCT) that mediate lactate uptake (Otonkoski *et al.*, 2007): blood lactate levels are typically increased during strenuous exercise when anaerobic glucose metabolism in skeletal muscle produces lactate that is transported to the liver for reduction to pyruvate (Cori cycle). Through their lack of MCT, beta cells will not metabolize circulating lactate and thus not respond by releasing insulin, thereby maintaining the glucose supply for muscle activity.

(ii) The beta cells' relative insensitivity to ketone bodies thus avoiding a sustained insulin secretion during fasting (Biden and Taylor, 1983); it is conceivable that acetoacetate and β-hydroxy-butyrate are deviated from mitochondrial oxidation, for example, by using them in fatty acid synthesis through the action of acetyl-coA synthetase (AACS), an enzyme that is abundantly expressed in beta cells (our unpublished data).

(iii) The lack of a high-affinity hexokinase (Schuit *et al.*, 1999) so that glucose signaling is prevented at low glycemia; the forkhead transcription factor FOXA2 might well be a crucial regulator by suppressing the high affinity hexokinases in beta cells and thus directing the cells towards low-affinity glucose sensing, (Wang *et al.*, 2002).

(iv) The $K^{+}{}_{ATP}$-dependent coupling mechanism of GSIS, considered as a major site in the beta cells' glucose sensing. Nutrients are known to stimulate insulin secretion, proportionate to their tendency to stimulation of the beta cell's ATP/ADP ratio. The high energetic efficiency of the Krebs cycle thus explains why mitochondrial nutrients are among the most efficient secretagogues: for example, glucose, glyceraldehyde, leucine, and succinate (Wiederkehr and Wollheim, 2006). The resultant decrease in the ADP/ATP ratio is molecularly measured by the K^+_{ATP} channels thus leading to cellular depolarization up to a threshold that opens the voltage-gated Ca^{2+} channels and triggers exocytosis (Ashcroft *et al.*, 1984). It is remarkable that the glucose concentration-dependent effects on K^+ efflux are most pronounced between 0 and 6 mM; between 6 and 20 mM, glucose exerts only a moderate effect (Carpinelli and Malaisse, 1981). The 6–20 mM range induces however the most robust effect on insulin secretion, via both K^+_{ATP}-dependent and -independent signaling (Henquin, 2000). In our opinion, the K^+_{ATP}-regulated pathway should also be seen as a safety valve that prevents insulin secretion at blood glucose below 5 mM. Evidently, the kinetics of the ADP/ATP sensing system are ultimately controlled by the amount of phosphorylated ADP, which, in case of a glucose stimulation, will be determined by the glucokinase activity of the beta cells.

2. Chronic physiological regulation: Down-regulation of glucose sensing functions by sustained absence of the glucose stimulus

Glucose is known to induce genes that allow nutrient sensing and subsequent insulin release (Fig. 17.2). Phrased otherwise, when exposed to low glucose levels beta cell down-regulate genes that are of minor importance under this condition. When beta cells are only viewed as insulin-secreting cells, this could be seen as a dedifferentiation while it is in our opinion inherent to the adaptive capacity of differentiated beta cells. In fact, the low-glucose exposed cells seem to up-regulate a set of genes that may help their adaptation to a low level of the glucose survival factor and with the appearance of apoptotic signals (Fig. 17.3). One of the early stress genes is the oncogene/proapoptotic *c-Myc*, which, in concert with other transcription factors, has the ability to reprogram the cells, and which, by itself, can initiate beta cell apoptosis (Van de Casteele *et al.*, 2003). Using gene chips to follow the shift in beta cell transcriptome by low glucose culture we found that beta cells develop several stress-protective pathways (Fig. 17.3) (Martens, 2008): (1) integrated endoplasmic reticulum stress response and/or unfolded protein response (e.g., beta cell abundant genes *DDIT3*, *HERPUD1*, *DNAJB1*, *ATF3*); (2) cell-cycle arrest (*c-Myc*, *CREG1*, *CDKN1A-p21* ^{cip1}) and circadian rhythm (*DBP*, *NRD1D2*); (3) response to DNA damage and DNA repair (*MGMT*, *POLA2*, *CHES1*) with induction of p53-target par excellence CDKN1A-p21 ^{cip1}; (4) response to oxidative stress (*NFE2L2*, *JUN*, *GPX1*, and *catalase*); and (5) free-iron chelation (*TFR*, *LCN2*), as well as other defenses against nutrient deprivation (*TRIB3*, *ATF3*); (6) potential stabilizers of the mitochondrial function (*Bcl-2*) and/or inhibitors of caspase action (*XIAP*).

These data are *in se* insufficient to demonstrate activation of the actual unfolded protein response in glucose-deprived beta cells, but are at least indicative for activation of the integrated stress response, as described by

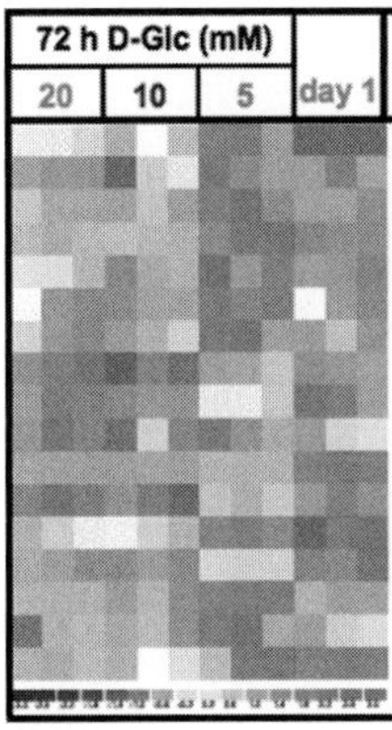

72 h D-Glc (mM)						
20	10	5	day 1	Symbol	Gene name	qPCR
				DBP	D site albumin promotor binding protein	ND
				NR1D2	nuclear receptor subfamily 1, group D, member 2 (Rev-ERB beta)	ND
				CREG1	cellular repressor of E1A-stimulated genes	ND
				MGMT	O-6-methylguanine-DNA methyltransferase	+
				CDKN1A	cyclin-dependent kinase inhibitor 1A (p21, cip1)	+
				POLA2	polymerase (DNA directed), alpha 2	ND
				CHES1	checkpoint suppressor 1	ND
				DDIT3	DNA-damage inducible transcript 3 (GADD153, CHOP)	+
				HERPUD1	HCY-inducible, ER stress-inducible, ubiquitin-like domain member 1	ND
				DNAJB1	DnaJ (Hsp40) homolog, subfamily B, member 1	ND
				ATF3	activating transcription factor 3	ND
				MYC	myelocytomatosis viral oncogene homolog, avian (c-Myc)	+
				NFE2L2	nuclear factor, erythroid derived 2, like 2 (NRF2)	+
				JUN	Jun oncogene	+
				TRIB3	tribbles homolog 3 (Drosophila)	+
				TFR	transferrin receptor	ND
				LCN2	lipocalin 2	ND

Figure 17.3 Selection of transcripts that are induced in 5 versus 10 mM glucose in cultured beta cells. This selection emphasizes relevant stress-protective/pro or anti-apoptotic transcripts and is thus not representative for the whole low glucose-transcriptome. The heat map shows relative expression levels in freshly isolated (day 1) and cells cultured at indicated glucose concentration. The right column indicates eventual confirmation by qPCR of the glucose-regulation. (See Color Insert.)

Jonas *et al.* (Elouil *et al.*, 2007). Simultaneous activation of a DNA-repair program, cell cycle arrest and integrated stress response in low glucose-cultured beta cells, sketches the phenotype of a cell in distress that develops adaptation and/or protection mechanisms against damaging influences (Summarized in Fig. 17.4). The beta cells' adaptation mechanisms to low glucose-stress also disclose potential ways through which beta cell grafts may cope with the hypoxemic/hypoglycemic conditions in the early posttransplant period. Moreover, ER stress could have a critical role in the integration of metabolic and survival responses through the mTOR-mediated regulation of insulin receptor signaling (Ozcan *et al.*, 2008). This way, decreased glucose recognition by beta cells could, apart from initiating a beta cell-intrinsic ROS-mediated apoptotic pathway (cfr. Infra), also decrease the cells responsiveness to growth signals via the insulin receptor.

3. Chronic pathological adaptation: Glucose deprivations triggers an intrinsic mitochondrial apoptotic program

In FACS-purified rodent beta cells, glucose concentration-dependently suppresses programmed cell death (Hoorens *et al.*, 1996). The concentration-response curve of this protection resembles the glucokinase-mediated effects, with markedly less cells dying at glucose levels above 5 mM. It is thus another consequence of the beta cells' dependency on low-affinity glucose phosphorylation. In other cell types, glucose deprivation also leads to cell death however at much lower (~1 mM) concentration, that is, below the K_m of their high affinity hexokinases 1–3 (Lee *et al.*, 1998).

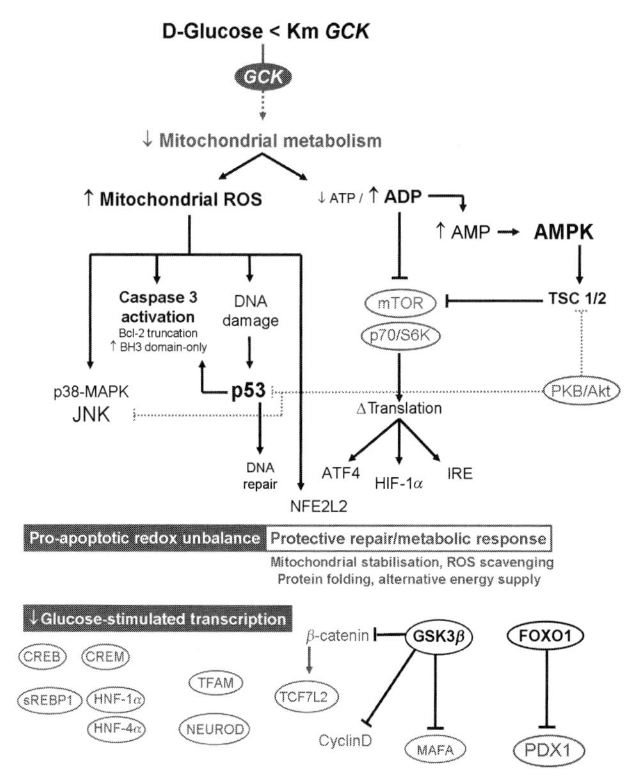

Figure 17.4 The glucose-deprived beta cell. At glucose concentrations below the K_m of GCK, mitochondrial metabolism is decreased below a critical threshold. This causes redox imbalance with accumulation of reactive oxygen species (ROS) that cause DNA damage (p53 activation), activation of p38 and JNK-stress pathways and finally result in instability of the mitochondrial membrane and activation of a caspase 3-dependent apoptotic program. The decreased ATP formation decreases activity of mTOR and PKB and, via adenylate kinase-mediated ATP regeneration from ADP, also increased activity of AMP-dependent kinase. Combined it results in shut-down of ATP-consuming protein synthesis which further increases the cells' vulnerability for the oxidative apoptotic program. The final result (transient shut-down of beta cell function, loss of normal differentiation or apoptosis) will depend on the balance between these proapoptotic signals, and activation of defense systems such mediated by regulators such as NFE2L2, ATF4, genes with IRE sequences and HIF1α that scavenge ROS or repair cellular damage. At the same time, lack of the glucose stimulus causes decreased activity of key beta cell transcription factors, paving the way for loss of beta cell differentiated state.

Low glucose-induced beta cell death results from an activation of the intrinsic mitochondrial apoptotic program, with nuclear fragmentation and caspase-3 activation as final executor (Kefas *et al.*, 2003). It is preceded by a progressive decrease of beta cell-specific functions, with decreased rates of protein synthesis, and blunted GSIS. After 3 days culture at 5 mM glucose, rat beta cell preparations did not exhibit a higher percent dead cells than 10 mM glucose cultured preparations, but their insulin synthetic and secretory capacity was decreased (Ling *et al.*, 1996; Martens, 2008); they did already show proapoptotic features, such as 28S-ribosomal RNA cleavage. Apoptosis could be prevented by (1) addition of mitochondrial nutrients such as leucine, succinate, or pyruvate—indicating that insufficient mitochondrial catabolism is involved in the apoptotic process (Martens *et al.*, 2005; Van de Casteele *et al.*, 2003) and (2) stabilizing the mitochondrial membrane via overexpression of Bcl-2 (Cai *et al.*, 2007). The decreased ATP/ADP level in glucose-deprived cells was molecularly sensed by the AMP-dependent protein kinase. While short-term AMPK activation might protect cells through shutdown of endergonic synthetic (protein, lipid) pathways and preservation of energy for their survival (Rutter *et al.*, 2003), we found that its sustained activation further accelerated the decline of beta cell function and survival (Kefas *et al.*, 2003). The death-inducing actions of sustained AMPK activation could be prevented by sustained activation of PKB/Akt (Cai *et al.*, 2008).

4. Oxidative stress as early trigger of apoptosis in glucose deprived beta cells

Glucose stimulation of metabolically responsive beta cells results in rapid formation and accumulation of NAD(P)H and reduced mitochondrial flavins ($FADH_2$, $FMNH_2$); this results in a shift of endogenous fluorescence that can be measured in individual beta cells by flow cytometry (Bennett *et al.*, 1996; Martens *et al.*, 2005; Van De Winkel and Pipeleers, 1983). At the population level, the mean NAD(P)H level again follows sigmoid kinetics in the 0–20 mM glucose range with K_m around 7.5 mM (Kiekens *et al.*, 1992). The accumulation of $FADH_2$ and $FMNH_2$ becomes prominent above 5 mM glucose, and indicates the threshold at which mitochondrial glucose breakdown in the Krebs cycle becomes prominent (Martens *et al.*, 2005), resulting in reduced redox state of the mitochondria. As discussed, metabolic glucose responsiveness of individual beta cells varies greatly as function of their level of glucokinase expression and activity (Heimberg *et al.*, 1993).

A high flux through the respiratory chain, in glucose-stimulated cells, is generally thought to increase their net formation of reactive oxygen species (ROS) (Brownlee, 2003; Robertson, 2004); this mechanism has been implied to explain accelerated aging/dysfunction of diabetic beta cells, which are thought to be chronically over-stimulated by hyperglycemia. To directly investigate such mechanism we used flow cytometry to simultaneously and real-time assess glucose metabolic rates via measurement of NAD(P)H, $FADH_2$, $FMNH_2$- and superoxide- and peroxide-formations (Martens *et al.*, 2005). It was thus found that glucose suppresses instead of accentuates mitochondrial ROS formation. This effect appeared specific for beta cells as it was not seen in other islet cell types. Beta cells produced progressively more superoxide en H_2O_2 when glucose was lowered under

5 mM, with electron transport complex I as most likely site of formation. The ability of glucose to suppress superoxide production in individual cells correlated with their metabolic responsiveness, with less ROS produced in the beta cell subpopulation that showed a high-glucose-responsiveness in mitochondrial activity. ROS in glucose-deprived beta cells was suppressed by other mitochondrial nutrients—confirming mitochondrial nutrient shortage as cause. The accentuated superoxide formation in beta cells with lower mitochondrial metabolic activity was detectable within minutes, and persisted for days. ROS scavenging in low glucose-cultured cells prevented beta cell apoptosis. This showed that the ROS accumulation was an early but persisting signal that pushed the cells towards apoptosis (Martens *et al.*, 2005), involving destabilization of the mitochondrial membrane that could be prevented by Bcl-2 (Cai *et al.*, 2007).

5. Chronic pathological adaptation: Abnormal functional beta cell mass by intrauterine and early postnatal nutrient restriction

Intrauterine and early postnatal growth retardation can cause insulin resistance and defective beta cell mass later in life. Early epidemiological work (Hales *et al.*, 1991) fitted with findings from animal experiments that showed impaired development of a normal functional beta cell mass after intrauterine or early postnatal food restriction (Swenne *et al.*, 1987, 1992), and vice versa, that early postnatal overfeeding by small litters caused beta cell mass expansion. It has now become clear that nutrient deprivation in critical developmental periods (late gestation, early life) can disturb the development of a normal beta cell mass with full adaptive potential through a deregulation of beta cell neogenesis, replication, and differentiation. The contribution of each of these main branches depends on the nature (e.g., insulin-induced hypoglycemia in utero, uteroplacental artery ligation, isocaloric low protein diet) and time-window of the nutrient deprivation (Dumortier *et al.*, 2007). Studies in rodents and sheep lead to the following view: specific glucose deprivation of the developing fetus interferes with the genesis of beta cells from their precursors; it appears to decrease the number of PDX1- and Ngn3-expressing cells, leading to a lower development of beta cells (Dumortier *et al.*, 2007). Under this condition, fetal glucocorticoid levels are elevated, which can be one factor responsible for the lower beta cell differentiation. Depending on the duration/severity of the glucose deprivation, the number, differentiation and/or function of the formed beta cells is/are affected (Rozance *et al.*, 2006). It is conceivable that the glucose deprivation is expressed in a particular subset of beta cells, or that it reprograms the cells into a phenotype with reduced responsiveness to proliferative factors, leading to failures in beta cell mass adaptation, for example, during pregnancy, or more rapid age-induced decline (Blondeau *et al.*, 1999). On the other hand, proliferation of beta cells appeared unaffected by deprivation. This process can be experimentally reproduced—at least in part—in laboratory models with normal placental blood flow, but low calorie/energy diets (Garofano *et al.*, 1998). It was thus shown that a 50% glucose deprivation resulted in a decreased ATP formation, particularly in beta cells. Glucose deprivation may thus primarily interfere with the birth of beta cells from progenitors, and their subsequent maturation/differentiation into a normal glucose-responsive phenotype but not with the subsequent first wave of proliferation of newly formed beta cells.

Opposed to this, are the effects of selective protein/amino acid deprivation: an isocaloric low-protein diet in utero and/or in the neonatal period, appears not to affect beta cell genesis, but clearly and specifically disturbs the proliferation of the young beta cells (Blondeau *et al.*, 1999; Dahri *et al.*, 1991; Petrik *et al.*, 1999; Swenne *et al.*, 1992). Protein malnutrition decreases the beta cell's production of IGF-2, increases their susceptibility for apoptosis and appears to block the beta cells into a prolonged G1-cell cycle phase. These effects might be direct, or through a negative effect on islet blood vessels.

Nutrient restriction *in vivo* invariably activates counter-regulatory responses, complicating the mechanistic exploration of glucose's specific trophic effect on developing beta cells. Guillemain *et al.* recently examined such specific glucose effect on pancreatic endocrine development, using mouse embryonic pancreas explants (Guillemain *et al.*, 2007). In explants cultured with abundant amino acids but no glucose, the development and/or proliferation of PDX1- and Neurogenin 3 (Ngn3)-expressing progenitors was normal, but their subsequent differentiation to glucagon- and, even more outspokenly, insulin-expressing cells was abolished. Lack of glucose interfered with expression of NeuroD1, a transcription factor that is pivotal for beta cell development and operates downstream of Ngn3. It is unlikely that this glucose effect is mediated via feedback signaling through insulin or IGF-1 receptors; while the latter clearly play a role in postnatal beta cell function/differentiation (Kulkarni *et al.*, 2002), their absence in double or single knockouts does not affect beta cell formation (Okada *et al.*, 2007; Ueki *et al.*, 2006). Moreover, the effect of glucose on beta cell formation in explants was clearly concentration-dependent in the 0–10 mM range, and strikingly, appeared to be sigmoid with major acceleration from 5 mM glucose on. This observation not only suggests that glucokinase plays a dominant role in developing beta cells. It also indicates intercellular differences in glucose sensitivity, from the moment beta cells are born, thus further supporting the concept of functional heterogeneity in the beta cell population (Pipeleers, 1992; Pipeleers *et al.*, 1994).

B. Beta cell adaptations to chronic high glucose: Glucotoxicity or not?

Type 2 diabetes is usually characterized by a slowly progressive deterioration of endogenous insulin control on glucose homeostasis. This deficit could be partly explained by an inadequate adaptation of the beta cell mass to the increasing metabolic demands in conditions of reduced insulin sensitivity. They have also been attributed to a vicious cycle in which moderate but sustained elevations of glycemia over-stimulate the beta cells, leading with time to their dysfunction and death (Reviewed in Butler *et al.*, 2007; Weir and Bonner-Weir, 2004). In the latter context, sustained hyperglycemia has been considered as a betacell-toxic factor; and therefore coined as ''glucotoxicity'' (Brownlee, 2005; Robertson *et al.*, 2004). The term and its significance are currently still under debate.

1. Excessive metabolic activation and degranulation in high glucose responsive beta cells

In a series of *in vitro* studies on rat and human beta cells we did not notice a glucotoxicity following prolonged exposure to high glucose levels, that is, up to 10 days of culture. As discussed in a previous section, the percent dead cells was not increased (Hoorens *et al.*, 1996; Ling *et al.*, 1994; Martens *et al.*, 2005). In functional terms, changes were noticed but they rather appeared adaptations towards an increased state of activity (Ling and Pipeleers, 1996; Ling *et al.*, 1996). It is in this context to be noticed that the glucose concentration-dependency of insulin synthesis and secretion differ: biosynthesis appears linearly proportionate to the glucose oxidation range, up to an oxidation rate that is measured around 10–12 mM glucose—the physiological upper limit of postprandial glycemia (Martens *et al.*, 2006; Schuit *et al.*, 1991). Further increasing glucose concentration up to 20 mM can further stimulate oxidation, and recruits cells into a state of sustained metabolic activation with high NADH (Ling *et al.*, 1996; Martens *et al.*, 2005) but maximal glucose inducible-insulin synthesis levels off. Insulin secretion under this condition however, is further amplified presumably via K+-ATP channel-independent signaling (Henquin, 2000) and the GSIS curve is left-shifted. This imbalance results in progressive glucose-induced loss of cellular insulin content (degranulation), which becomes detectable after 3 days (Martens, 2008) and becomes more marked with time; after 5–7 days, the degranulation of 20 mM glucose-exposed rodent and human beta cells is so severe that their absolute insulin output decreases, despite intrinsic over-activation of their glucose-signaling, reflected by increased fractional secretion rates (Ling and Pipeleers, 1996; Ling *et al.*, 1996). Such sustained high glucose-exposure did not cause glucotoxicity, but resulted in a functional state characterized by (1) maximal metabolic recruitment of the beta cell population with also activation of the lower-glucose responsive cells; (2) left-shift of the GSIS curve with also increased basal insulin synthesis and secretion; and (3) insulin degranulation, reducing net GSIS despite further activation of their intrinsic glucose signaling (fractional secretion). We examined whether such hyper-stimulated state affects gene expression; remarkably we found that, after 3 days, 20 and 10 mM glucose-cultured beta cells showed no significant differences in their mRNA expression patterns (illustrated in Figs.17.2–17.3) (Martens, 2008). The short (3 days) but excessive and sustained elevation of glucose levels thus results in secretory patterns that reflect a functional adaptation and are not the result of an altered gene expression. It can of course not be excluded that exposure for longer than 10 days will influence gene expression and impair beta cell survival and function. It is also clear that extrapolation of *in vitro* findings to the *in vivo* situation should only serve as a working hypothesis, as a guide to further *in vivo* studies.

2. Apoptosis of high-glucose unresponsive beta cells

There is increasing evidence that the beta cell mass decreases during longstanding hyperglycemia. We do not exclude that chronically elevated metabolic rates can be damaging for normal beta cells, but believe the evidence is still insufficient. Besides, we propose that the possibility should also be considered of beta cell apoptosis occurring in cells with inadequate metabolic rates despite the hyperglycemic condition. In the latter case, beta cell death would thus resemble that of glucose deprivation; it could be initiated by an underlying deficit of mitochondrial glucose metabolism, and involve increased formation of oxygen radicals as early signal in the apoptotic execution program. This view is not in contradiction with a number of observations.

In the Weir concept of diabetic progression (Weir and Bonner-Weir, 2004), the initial preclinical phase of disease is followed by phases of "stable adaptation" and then "unstable adaptation," clinically characterized by (1) loss of acute GSIS; (2) moderate but sustained increase in fasting blood glucose levels (from <5.6 to ~ 6.4 mmol/L) and (3) appearance of progressively escalating glucose intolerance. In a rudimentary rodent model for this disease stage, namely rats after partial pancreatectomy, beta cells in stable or unstable adaptation showed prominent signs of dedifferentiation, with enhanced expression of stress- and apoptosis-associated genes (c-myc, NF-κB). These residual beta cells also expressed metabolic enzymes that are normally suppressed in healthy beta cells, such as lactate dehydrogenase, glucose-6-phosphatase, fructose-1,6-bisphosphatase and high-affinity hexokinases. Such modulation of beta cell's metabolic constellation could potentially interfere with normal glucose sensing. It could, for example, divert glucose away from mitochondrial oxidation and anaplerotic reactions, or, in the case of high-affinity hexokinases, interfere with the normal low-affinity hexokinase (GCK) activity.

A review of the literature further supports the view that long standing (weeks–months) hyperglycemia can result in decreased rather than increased glucose metabolic rates in beta cells. Sustained hyperglycemia has been shown to suppress the expression of GCK, GLUT2, and pyruvate carboxylase (Jonas *et al.*, 1999; Kim *et al.*, 2007). Decreased GLUT2 expression or GCK activity hampers glucose metabolism in beta cells

(Guillam *et al.*, 2000; Matschinsky, 2002). The same tendency is observed in beta cells from Zucker Diabetic Fatty rats. Loss of functional leptin receptors in this model causes multiple abnormalities in both lipid and glucose metabolism, resulting in lipid accumulation, associated with decreased GLUT2 and GK expression. The latter was associated with decreased glycolytic and mitochondrial glucose metabolic rate (Ohneda *et al.*, 1994), reversed upon reintroduction of the defective gene (Wang *et al.*, 1998). Deficiency of glucose oxidation was also detected in islets isolated from spontaneously diabetic Goto-Kakizaki rats (Giroix *et al.*, 1993), as well as in beta cells isolated from adult rats that became glucose-intolerant later in life after partial beta cell ablation using streptozotocin in the neonatal period (Giroix *et al.*, 1990). A decreased glucose oxidation rate was also found in rodent, but not human, islets transplanted into a hyperglycemic recipient (Jansson *et al.*, 1995; Korsgren *et al.*, 1990). Direct comparisons of glucose metabolic rate in islets obtained from type 2 diabetic and healthy control humans are, to our knowledge, not available. Yet, Gunton *et al.* (Gunton *et al.*, 2005) recently showed that the expression of several key glycolytic enzymes was markedly suppressed in islets from type 2 diabetic human patients: mRNA levels of GCK, glucose-6-phosphoisomerase, phosphofructokinase, and phosphoglucomutase were up to 5-fold lower in diabetic islets, indicating a beta cell-intrinsic, severe impairment of the glucose-sensing machinery, with possibly also lower mitochondrial glucose metabolism.

ACKNOWLEDGMENTS

Our work has been supported by the Research Foundation Flanders (Grant FWO-G.0183.05, G.0400.07), by the Inter-University Poles of Attraction Program (IUAP P5/17) from the Belgian Science Policy, by the European Commission Sixth Framework Programme (FP6. 512145). We thank the staff of the Diabetes Research Center for their collaboration and expert assistance.

REFERENCES

Affourtit, C., and Brand, M. D. (2006). Stronger control of ATP/ADP by proton leak in pancreatic beta-cells than skeletal muscle mitochondria. *Biochem. J.* **393,** 151–159.

Andrali, S. S., Qian, Q., and Ozcan, S. (2007). Glucose mediates the translocation of NeuroD1 by *O*-linked glycosylation. *J. Biol. Chem.* **282,** 15589–15596.

Antinozzi, P. A., Ishihara, H., Newgard, C. B., and Wollheim, C. B. (2002). Mitochondrial metabolism sets the maximal limit of fuel-stimulated insulin secretion in a model pancreatic beta cell: A survey of four fuel secretagogues. *J. Biol. Chem.* **277,** 11746–11755.

Ashcroft, F. M., Harrison, D. E., and Ashcroft, S. J. (1984). Glucose induces closure of single potassium channels in isolated rat pancreatic beta-cells. *Nature* **312,** 446–448.

Barondes, S. H., and Jarvik, M. E. (1964). The influence of actinomycin-D on brain RNA synthesis and on memory. *J. Neurochem.* **11,** 187–195.

Baroukh, N., Ravier, M. A., Loder, M. K., Hill, E. V., Bounacer, A., Scharfmann, R., Rutter, G. A., and Van, O. E. (2007). MicroRNA-124a regulates Foxa2 expression and intracellular signaling in pancreatic beta-cell lines. *J. Biol. Chem.* **282,** 19575–19588.

Bennett, B. D., Jetton, T. L., Ying, G., Magnuson, M. A., and Piston, D. W. (1996). Quantitative subcellular imaging of glucose metabolism within intact pancreatic islets. *J. Biol. Chem.* **271,** 3647–3651.

Biden, T. J., and Taylor, K. W. (1983). Effects of ketone bodies on insulin release and islet-cell metabolism in the rat. *Biochem. J.* **212,** 371–377.

Blondeau, B., Garofano, A., Czernichow, P., and Breant, B. (1999). Age-dependent inability of the endocrine pancreas to adapt to pregnancy: A long-term consequence of perinatal malnutrition in the rat. *Endocrinology* **140,** 4208–4213.

Borg, L. A., Dahl, N., and Swenne, I. (1995). Age-dependent differences in insulin secretion and intracellular handling of insulin in isolated pancreatic islets of the rat. *Diabete Metab.* **21,** 408–414.

Brelje, T. C., and Sorenson, R. L. (1991). Role of prolactin versus growth hormone on islet B-cell proliferation *in vitro*: Implications for pregnancy. *Endocrinology* **128,** 45–57.

Brelje, T. C., Scharp, D. W., Lacy, P. E., Ogren, L., Talamantes, F., Robertson, M., Friesen, H. G., and Sorenson, R. L. (1993). Effect of homologous placental lactogens, prolactins, and growth hormones on islet B-cell division and insulin secretion in rat, mouse, and human islets: Implication for placental lactogen regulation of islet function during pregnancy. *Endocrinology* **132,** 879–887.

Briaud, I., Lingohr, M. K., Dickson, L. M., Wrede, C. E., and Rhodes, C. J. (2003). Differential activation mechanisms of Erk-1/2 and p70(S6K) by glucose in pancreatic beta-cells. *Diabetes* **52,** 974–983.

Brownlee, M. (2003). A radical explanation for glucose-induced beta cell dysfunction. *J. Clin. Invest.* **112,** 1788–1790.

Brownlee, M. (2005). The pathobiology of diabetic complications: A unifying mechanism. *Diabetes* **54,** 1615–1625.

Butler, P. C., Meier, J. J., Butler, A. E., and Bhushan, A. (2007). The replication of beta cells in normal physiology, in disease and for therapy. *Nat. Clin. Pract. Endocrinol. Metab.* **3,** 758–768.

Cai, Y., Martens, G. A., Hinke, S. A., Heimberg, H., Pipeleers, D., and Van de Casteele, M. (2007). Increased oxygen radical formation and mitochondrial dysfunction mediate beta cell apoptosis under conditions of AMP-activated protein kinase stimulation. *Free Radic. Biol. Med.* **42,** 64–78.

Cai, Y., Wang, Q., Ling, Z., Pipeleers, D., McDermott, P., Pende, M., Heimberg, H., and Van de, C. M. (2008). Akt activation protects pancreatic beta cells from AMPK-mediated death through stimulation of mTOR. *Biochem. Pharmacol.* **75,** 1981–1993.

Carpinelli, A. R., and Malaisse, W. J. (1981). Regulation of 86Rb outflow from pancreatic islets: The dual effect of nutrient secretagogues. *J. Physiol.* **315,** 143–156.

Costes, S., Broca, C., Bertrand, G., Lajoix, A. D., Bataille, D., Bockaert, J., and Dalle, S. (2006). ERK1/2 control phosphorylation and protein level of cAMP-responsive element-binding protein: A key role in glucose-mediated pancreatic beta-cell survival. *Diabetes* **55,** 2220–2230.

Cousin, S. P., Hugl, S. R., Myers, M. G. Jr., White, M. F., Reifel-Miller, A., and Rhodes, C. J. (1999). Stimulation of pancreatic beta-cell proliferation by growth hormone is glucose-dependent: Signal transduction via janus kinase 2 (JAK2)/signal transducer and activator of transcription 5 (STAT5) with no crosstalk to insulin receptor substrate-mediated mitogenic signalling. *Biochem. J.* **344**(Pt. 3), 649–658.

Dahri, S., Snoeck, A., Reusens-Billen, B., Remacle, C., and Hoet, J. J. (1991). Islet function in offspring of mothers on low-protein diet during gestation. *Diabetes* **40**(Suppl. 2), 115–120.

da Silva, X., Qian, Q., Cullen, P. J., and Rutter, G. A. (2004). Distinct roles for insulin and insulin-like growth factor-1 receptors in pancreatic beta-cell glucose sensing revealed by RNA silencing. *Biochem. J.* **377,** 149–158.

Deeney, J. T., Prentki, M., and Corkey, B. E. (2000). Metabolic control of beta-cell function. *Semin. Cell Dev. Biol.* **11,** 267–275.

Detimary, P., Dejonghe, S., Ling, Z., Pipeleers, D., Schuit, F., and Henquin, J. C. (1998). The changes in adenine nucleotides measured in glucose-stimulated rodent islets occur in beta cells but not in alpha cells and are also observed in human islets. *J. Biol. Chem.* **273,** 33905–33908.

De Vos, A., Heimberg, H., Quartier, E., Huypens, P., Bouwens, L., Pipeleers, D., and Schuit, F. (1995). Human and rat beta cells differ in glucose transporter but not in glucokinase gene expression. *J. Clin. Invest.* **96,** 2489–2495.

Diraison, F., Ravier, M. A., Richards, S. K., Smith, R. M., Shimano, H., and Rutter, G. A. (2008). SREBP1 is required for the induction by glucose of pancreatic beta-cell genes involved in glucose sensing. *J. Lipid Res.* **49,** 814–822.

Dor, Y., Brown, J., Martinez, O. I., and Melton, D. A. (2004). Adult pancreatic beta-cells are formed by self-duplication rather than stem-cell differentiation. *Nature* **429,** 41–46.

Dumortier, O., Blondeau, B., Duvillie, B., Reusens, B., Breant, B., and Remacle, C. (2007). Different mechanisms operating during different critical time-windows reduce rat fetal beta cell mass due to a maternal low-protein or low-energy diet. *Diabetologia* **50,** 2495–2503.

Elouil, H., Bensellam, M., Guiot, Y., Vander Mierde, D., Pascal, S. M., Schuit, F. C., and Jonas, J. C. (2007). Acute nutrient regulation of the unfolded protein response and integrated stress response in cultured rat pancreatic islets. *Diabetologia* **50,** 1442–1452.

Fahien, L. A., and Macdonald, M. J. (2002). The succinate mechanism of insulin release. *Diabetes* **51,** 2669–2676.

Flamez, D., Berger, V., Kruhoffer, M., Orntoft, T., Pipeleers, D., and Schuit, F. C. (2002). Critical role for cataplerosis via citrate in glucose-regulated insulin release. *Diabetes* **51,** 2018–2024.

Flexner, J. B., Flexner, L. B., Stellar, E., De La, H. A. B. A., and Roberts, R. B. (1962). Inhibition of protein synthesis in brain and learning and memory following puromycin. *J. Neurochem.* **9,** 595–605.

Fluck, M. (2006). Functional, structural and molecular plasticity of mammalian skeletal muscle in response to exercise stimuli. *J. Exp. Biol.* **209,** 2239–2248.

Fransson, U., Rosengren, A. H., Schuit, F. C., Renstrom, E., and Mulder, H. (2006). Anaplerosis via pyruvate carboxylase is required for the fuel-induced rise in the ATP: ADP ratio in rat pancreatic islets. *Diabetologia* **49,** 1578–1586.

Fred, R. G., Tillmar, L., and Welsh, N. (2006). The role of PTB in insulin mRNA stability control. *Curr. Diabetes Rev.* **2,** 363–366.

Garofano, A., Czernichow, P., and Breant, B. (1998). Beta-cell mass and proliferation following late fetal and early postnatal malnutrition in the rat. *Diabetologia* **41,** 1114–1120.

Georgia, S., and Bhushan, A. (2004). Beta cell replication is the primary mechanism for maintaining postnatal beta cell mass. *J. Clin. Invest.* **114,** 963–968.

Giroix, M. H., Sener, A., Bailbe, D., Portha, B., and Malaisse, W. J. (1990). Impairment of the mitochondrial oxidative response to d-glucose in pancreatic islets from adult rats injected with streptozotocin during the neonatal period. *Diabetologia* **33,** 654–660.

Giroix, M. H., Vesco, L., and Portha, B. (1993). Functional and metabolic perturbations in isolated pancreatic islets from the GK rat, a genetic model of noninsulin-dependent diabetes. *Endocrinology* **132,** 815–822.

Glauser, D. A., and Schlegel, W. (2007). The emerging role of FOXO transcription factors in pancreatic beta cells. *J. Endocrinol.* **193,** 195–207.

Glauser, D. A., Brun, T., Gauthier, B. R., and Schlegel, W. (2007). Transcriptional response of pancreatic beta cells to metabolic stimulation: Large scale identification of immediate-early and secondary response genes. *BMC Mol. Biol.* **8,** 54.

Gloyn, A. L., Odili, S., Zelent, D., Buettger, C., Castleden, H. A., Steele, A. M., Stride, A., Shiota, C., Magnuson, M. A., Lorini, R., d'Annunzio, G., Stanley, C. A., *et al.* (2005). Insights into the structure and regulation of glucokinase from a novel mutation (V62M), which causes maturity-onset diabetes of the young. *J. Biol. Chem.* **280,** 14105–14113.

Greenman, I. C., Gomez, E., Moore, C. E., and Herbert, T. P. (2005). The selective recruitment of mRNA to the ER and an increase in initiation are important for glucose-stimulated proinsulin synthesis in pancreatic beta-cells. *Biochem. J.* **391,** 291–300.

Guillam, M. T., Dupraz, P., and Thorens, B. (2000). Glucose uptake, utilization, and signaling in GLUT2-null islets. *Diabetes* **49,** 1485–1491.

Guillemain, G., Filhoulaud, G., Da Silva-Xavier, G., Rutter, G. A., and Scharfmann, R. (2007). Glucose is necessary for embryonic pancreatic endocrine cell differentiation. *J. Biol. Chem.* **282,** 15228–15237.

Gunton, J. E., Kulkarni, R. N., Yim, S., Okada, T., Hawthorne, W. J., Tseng, Y. H., Roberson, R. S., Ricordi, C., O'Connell, P. J., Gonzalez, F. J., and Kahn, C. R. (2005). Loss of ARNT/HIF1beta mediates altered gene expression and pancreatic-islet dysfunction in human type 2 diabetes. *Cell* **122,** 337–349.

Hales, C. N., Barker, D. J., Clark, P. M., Cox, L. J., Fall, C., Osmond, C., and Winter, P. D. (1991). Fetal and infant growth and impaired glucose tolerance at age 64. *BMJ* **303,** 1019–1022.

Han, S. I., Aramata, S., Yasuda, K., and Kataoka, K. (2007). MafA stability in pancreatic beta cells is regulated by glucose and is dependent on its constitutive phosphorylation at multiple sites by glycogen synthase kinase 3. *Mol. Cell. Biol.* **27,** 6593–6605.

Hansen, S. K., Parrizas, M., Jensen, M. L., Pruhova, S., Ek, J., Boj, S. F., Johansen, A., Maestro, M. A., Rivera, F., Eiberg, H., Andel, M., Lebl, J., *et al.* (2002). Genetic evidence that HNF-1alpha-dependent transcriptional control of HNF-4alpha is essential for human pancreatic beta cell function. *J. Clin. Invest.* **110,** 827–833.

Heimberg, H., De Vos, A., Vandercammen, A., Van Schaftingen, E., Pipeleers, D., and Schuit, F. (1993). Heterogeneity in glucose sensitivity among pancreatic beta-cells is correlated to differences in glucose phosphorylation rather than glucose transport. *EMBO J.* **12,** 2873–2879.

Hellemans, K., Kerckhofs, K., Hannaert, J. C., Martens, G., Van Veldhoven, P. P., and Pipeleers, D. (2007). Peroxisome proliferator-activated receptor alpha-retinoid X receptor agonists induce beta-cell protection against palmitate toxicity. *FEBS J.* **274,** 6094–6105.

Henquin, J. C. (2000). Triggering and amplifying pathways of regulation of insulin secretion by glucose. *Diabetes* **49,** 1751–1760.

Henquin, J. C., Dufrane, D., and Nenquin, M. (2006). Nutrient control of insulin secretion in isolated normal human islets. *Diabetes* **55,** 3470–3477.

Hinke, S. A., Hellemans, K., and Schuit, F. C. (2004). Plasticity of the beta cell insulin secretory competence: Preparing the pancreatic beta cell for the next meal. *J. Physiol.* **558,** 369–380.

Hoorens, A., Van de Casteele, M., Kloppel, G., and Pipeleers, D. (1996). Glucose promotes survival of rat pancreatic beta cells by activating synthesis of proteins which suppress a constitutive apoptotic program. *J. Clin. Invest.* **98,** 1568–1574.

Hugl, S. R., White, M. F., and Rhodes, C. J. (1998). Insulin-like growth factor I (IGF-I)-stimulated pancreatic beta-cell growth is glucose-dependent. Synergistic activation of insulin receptor substrate-mediated signal transduction pathways by glucose and IGF-I in INS-1 cells. *J. Biol. Chem.* **273,** 17771–17779.

Hussain, M. A., Porras, D. L., Rowe, M. H., West, J. R., Song, W. J., Schreiber, W. E., and Wondisford, F. E. (2006). Increased pancreatic beta-cell proliferation mediated by CREB binding protein gene activation. *Mol. Cell. Biol.* **26,** 7747–7759.

Jansson, L., Eizirik, D. L., Pipeleers, D. G., Borg, L. A., Hellerstrom, C., and Andersson, A. (1995). Impairment of glucose-induced insulin secretion in human pancreatic islets transplanted to diabetic nude mice. *J. Clin. Invest.* **96,** 721–726.

Jetton, T. L., and Magnuson, M. A. (1992). Heterogeneous expression of glucokinase among pancreatic beta cells. *Proc. Natl. Acad. Sci. USA* **89,** 2619–2623.

Jhala, U. S., Canettieri, G., Screaton, R. A., Kulkarni, R. N., Krajewski, S., Reed, J., Walker, J., Lin, X., White, M., and Montminy, M. (2003). cAMP promotes pancreatic beta-cell survival via CREB-mediated induction of IRS2. *Genes Dev.* **17,** 1575–1580.

Jonas, J. C., Sharma, A., Hasenkamp, W., Ilkova, H., Patane, G., Laybutt, R., Bonner-Weir, S., and Weir, G. C. (1999). Chronic hyperglycemia triggers loss of pancreatic beta cell differentiation in an animal model of diabetes. *J. Biol. Chem.* **274,** 14112–14121.

Jorns, A., Tiedge, M., and Lenzen, S. (1999). Nutrient-dependent distribution of insulin and glucokinase immunoreactivities in rat pancreatic beta cells. *Virchows Arch.* **434,** 75–82.

Joseph, J. W., Odegaard, M. L., Ronnebaum, S. M., Burgess, S. C., Muehlbauer, J., Sherry, A. D., and Newgard, C. B. (2007). Normal flux through ATP-citrate lyase or fatty acid synthase is not required for glucose-stimulated insulin secretion. *J. Biol. Chem.* **282,** 31592–31600.

Kataoka, K., Han, S. I., Shioda, S., Hirai, M., Nishizawa, M., and Handa, H. (2002). MafA is a glucose-regulated and pancreatic beta-cell-specific transcriptional activator for the insulin gene. *J. Biol. Chem.* **277,** 49903–49910.

Kefas, B. A., Cai, Y., Ling, Z., Heimberg, H., Hue, L., Pipeleers, D., and Van de Casteele, M. (2003). AMP-activated protein kinase can induce apoptosis of insulin-producing MIN6 cells through stimulation of c-Jun-N-terminal kinase. *J. Mol. Endocrinol.* **30,** 151–161.

Kiekens, R., In 't Veld, P., Mahler, T., Schuit, F., Van De Winkel, M., and Pipeleers, D. (1992). Differences in glucose recognition by individual rat pancreatic B cells are associated with intercellular differences in glucose-induced biosynthetic activity. *J. Clin. Invest.* **89,** 117–125.

Kim, W. H., Lee, J. W., Suh, Y. H., Hong, S. H., Choi, J. S., Lim, J. H., Song, J. H., Gao, B., and Jung, M. H. (2005). Exposure to chronic high glucose induces beta-cell apoptosis through decreased interaction of glucokinase with mitochondria: Downregulation of glucokinase in pancreatic beta-cells. *Diabetes* **54,** 2602–2611.

Kim, W. H., Lee, J. W., Suh, Y. H., Lee, H. J., Lee, S. H., Oh, Y. K., Gao, B., and Jung, M. H. (2007). AICAR potentiates ROS production induced by chronic high glucose: Roles of AMPK in pancreatic beta-cell apoptosis. *Cell Signal.* **19,** 791–805.

Kitamura, T., and Ido, K. Y. (2007). Role of FoxO proteins in pancreatic beta cells. *Endocr. J.* **54,** 507–515.

Knoch, K. P., Bergert, H., Borgonovo, B., Saeger, H. D., Altkruger, A., Verkade, P., and Solimena, M. (2004). Polypyrimidine tract-binding protein promotes insulin secretory granule biogenesis. *Nat. Cell Biol.* **6,** 207–214.

Korsgren, O., Jansson, L., Sandler, S., and Andersson, A. (1990). Hyperglycemia-induced B cell toxicity. The fate of pancreatic islets transplanted into diabetic mice is dependent on their genetic background. *J. Clin. Invest.* **86,** 2161–2168.

Kulkarni, R. N., Holzenberger, M., Shih, D. Q., Ozcan, U., Stoffel, M., Magnuson, M. A., and Kahn, C. R. (2002). Beta-cell-specific deletion of the Igf1 receptor leads to hyperinsulinemia and glucose intolerance but does not alter beta-cell mass. *Nat. Genet.* **31,** 111–115.

Kwon, G., Marshall, C. A., Pappan, K. L., Remedi, M. S., and McDaniel, M. L. (2004). Signaling elements involved in the metabolic regulation of mTOR by nutrients, incretins, and growth factors in islets. *Diabetes* **53**(Suppl. 3), 225–232.

Lantz, K. A., Vatamaniuk, M. Z., Brestelli, J. E., Friedman, J. R., Matschinsky, F. M., and Kaestner, K. H. (2004). Foxa2 regulates multiple pathways of insulin secretion. *J. Clin. Invest.* **114,** 512–520.

Lee, Y. J., Galoforo, S. S., Berns, C. M., Chen, J. C., Davis, B. H., Sim, J. E., Corry, P. M., and Spitz, D. R. (1998). Glucose deprivation-induced cytotoxicity and alterations in mitogen-activated protein kinase activation are mediated by oxidative stress in multidrug-resistant human breast carcinoma cells. *J. Biol. Chem.* **273,** 5294–5299.

Leibiger, B., Moede, T., Schwarz, T., Brown, G. R., Kohler, M., Leibiger, I. B., and Berggren, P. O. (1998). Short-term regulation of insulin gene transcription by glucose. *Proc. Natl. Acad. Sci. USA* **95,** 9307–9312.

Leibowitz, G., Yuli, M., Donath, M. Y., Nesher, R., Melloul, D., Cerasi, E., Gross, D. J., and Kaiser, N. (2001). Beta-cell glucotoxicity in the *Psammomys obesus* model of type 2 diabetes. *Diabetes* **50**(Suppl. 1), S113–S117.

Ling, Z., and Pipeleers, D. G. (1996). Prolonged exposure of human beta cells to elevated glucose levels results in sustained cellular activation leading to a loss of glucose regulation. *J. Clin. Invest.* **98,** 2805–2812.

Ling, Z., Hannaert, J. C., and Pipeleers, D. (1994). Effect of nutrients, hormones and serum on survival of rat islet beta cells in culture. *Diabetologia* **37,** 15–21.

Ling, Z., Kiekens, R., Mahler, T., Schuit, F. C., Pipeleers-Marichal, M., Sener, A., Kloppel, G., Malaisse, W. J., and Pipeleers, D. G. (1996). Effects of chronically elevated glucose levels on the functional properties of rat pancreatic beta-cells. *Diabetes* **45,** 1774–1782.

Ling, Z., Wang, Q., Stange, G., in't Veld, P., and Pipeleers, D. (2006). Glibenclamide treatment recruits beta-cell subpopulation into elevated and sustained basal insulin synthetic activity. *Diabetes* **55,** 78–85.

Lingohr, M. K., Buettner, R., and Rhodes, C. J. (2002). Pancreatic beta-cell growth and survival—A role in obesity-linked type 2 diabetes? *Trends Mol. Med.* **8,** 375–384.

Lingohr, M. K., Briaud, I., Dickson, L. M., McCuaig, J. F., Alarcon, C., Wicksteed, B. L., and Rhodes, C. J. (2006). Specific regulation of IRS-2 expression by glucose in rat primary pancreatic islet beta-cells. *J. Biol. Chem.* **281,** 15884–15892.

Lu, D., Mulder, H., Zhao, P., Burgess, S. C., Jensen, M. V., Kamzolova, S., Newgard, C. B., and Sherry, A. D. (2002). 13C NMR isotopomer analysis reveals a connection between pyruvate cycling and glucose-stimulated insulin secretion (GSIS). *Proc. Natl. Acad. Sci. USA* **99,** 2708–2713.

Luther, M. J., Hauge-Evans, A., Souza, K. L., Jorns, A., Lenzen, S., Persaud, S. J., and Jones, P. M. (2006). MIN6 beta-cell-beta-cell interactions influence insulin secretory responses to nutrients and non-nutrients. *Biochem. Biophys. Res. Commun.* **343,** 99–104.

Macdonald, M. J. (1995). Feasibility of a mitochondrial pyruvate malate shuttle in pancreatic islets. Further implication of cytosolic NADPH in insulin secretion. *J. Biol. Chem.* **270,** 20051–20058.

Macdonald, M. J., Fahien, L. A., Brown, L. J., Hasan, N. M., Buss, J. D., and Kendrick, M. A. (2005). Perspective: Emerging evidence for signaling roles of mitochondrial anaplerotic products in insulin secretion. *Am. J. Physiol. Endocrinol. Metab.* **288,** E1–E15.

Maedler, K., Schumann, D. M., Schulthess, F., Oberholzer, J., Bosco, D., Berney, T., and Donath, M. Y. (2006). Aging correlates with decreased beta-cell proliferative capacity and enhanced sensitivity to apoptosis: A potential role for Fas and pancreatic duodenal homeobox-1. *Diabetes* **55,** 2455–2462.

Martens, G. A. (2008). "Glucose and the beta cell: Interplay between the specialized phenotype of insulin-producing beta cells and their glucose sensing." ISBN 978-90-5487-476-8. VUB Press. 19–41. Ref Type: Thesis/Dissertation.

Martens, G. A., and Van de Casteele, M. (2007). Glycemic control of apoptosis in the pancreatic beta cell: Danger of extremes? *Antioxid. Redox Signal.* **9,** 309–317.

Martens, G. A., Cai, Y., Hinke, S., Stange, G., Van de Casteele, M., and Pipeleers, D. (2005). Glucose suppresses superoxide generation in metabolically responsive pancreatic beta cells. *J. Biol. Chem.* **280,** 20389–20396.

Martens, G. A., Wang, Q., Kerckhofs, K., Stange, G., Ling, Z., and Pipeleers, D. (2006). Metabolic activation of glucose low-responsive beta-cells by glyceraldehyde correlates with their biosynthetic activation in lower glucose concentration range but not at high glucose. *Endocrinology* **147,** 5196–5204.

Martens, G. A., Vervoort, A., Van de Casteele, M., Stange, G., Hellemans, K., Van Thi, H. V., Schuit, F., and Pipeleers, D. (2007). Specificity in beta cell expression of L-3-hydroxyacyl-CoA dehydrogenase, short chain, and potential role in down-regulating insulin release. *J. Biol. Chem.* **282,** 21134–21144.

Massa, L., Baltrusch, S., Okar, D. A., Lange, A. J., Lenzen, S., and Tiedge, M. (2004). Interaction of 6-phosphofructo-2-kinase/fructose-2,6-bisphosphatase (PFK-2/FBPase-2) with glucokinase activates glucose phosphorylation and glucose metabolism in insulin-producing cells. *Diabetes* **53,** 1020–1029.

Matschinsky, F. M. (1996). Banting lecture 1995. A lesson in metabolic regulation inspired by the glucokinase glucose sensor paradigm. *Diabetes* **45,** 223–241.

Matschinsky, F. M. (2002). Regulation of pancreatic beta-cell glucokinase: From basics to therapeutics. *Diabetes* **51**(Suppl. 3), S394–S404.

Mohanty, S., Spinas, G. A., Maedler, K., Zuellig, R. A., Lehmann, R., Donath, M. Y., Trub, T., and Niessen, M. (2005). Overexpression of IRS2 in isolated pancreatic islets causes proliferation and protects human beta-cells from hyperglycemia-induced apoptosis. *Exp. Cell Res.* **303,** 68–78.

Nesher, R., Gross, D. J., Donath, M. Y., Cerasi, E., and Kaiser, N. (1999). Interaction between genetic and dietary factors determines beta-cell function in *Psammomys obesus*, an animal model of type 2 diabetes. *Diabetes* **48,** 731–737.

Newgard, C. B. (2002). While tinkering with the beta-cell. . .Metabolic regulatory mechanisms and new therapeutic strategies: American Diabetes Association Lilly Lecture, 2001. *Diabetes* **51,** 3141–3150.

Nielsen, J. H. (1982). Effects of growth hormone, prolactin, and placental lactogen on insulin content and release, and deoxyribonucleic acid synthesis in cultured pancreatic islets. *Endocrinology* **110,** 600–606.

Ohneda, M., Johnson, J. H., Lee, Y. H., Nagasawa, Y., and Unger, R. H. (1994). Post-GLUT-2 defects in beta-cells of non-insulin-dependent diabetic obese rats. *Am. J. Physiol.* **267,** E968–E974.

Okada, T., Liew, C. W., Hu, J., Hinault, C., Michael, M. D., Krtzfeldt, J., Yin, C., Holzenberger, M., Stoffel, M., and Kulkarni, R. N. (2007). Insulin receptors in beta-cells are critical for islet compensatory growth response to insulin resistance. *Proc. Natl. Acad. Sci. USA* **104,** 8977–8982.

Otani, K., Kulkarni, R. N., Baldwin, A. C., Krutzfeldt, J., Ueki, K., Stoffel, M., Kahn, C. R., and Polonsky, K. S. (2004). Reduced beta-cell mass and altered glucose sensing impair insulin-secretory function in betaIRKO mice. *Am. J. Physiol. Endocrinol. Metab.* **286,** E41–E49.

Otonkoski, T., Jiao, H., Kaminen-Ahola, N., Tapia-Paez, I., Ullah, M. S., Parton, L. E., Schuit, F., Quintens, R., Sipila, I., Mayatepek, E., Meissner, T., Halestrap, A. P., *et al.* (2007). Physical exercise-induced hypoglycemia caused by failed silencing of monocarboxylate transporter 1 in pancreatic beta cells. *Am. J. Hum. Genet.* **81,** 467–474.

Ozcan, U., Ozcan, L., Yilmaz, E., Duvel, K., Sahin, M., Manning, B. D., and Hotamisligil, G. S. (2008). Loss of the tuberous sclerosis complex tumor suppressors triggers the unfolded protein response to regulate insulin signaling and apoptosis. *Mol. Cell* **29,** 541–551.

Parnaud, G., Bosco, D., Berney, T., Pattou, F., Kerr-Conte, J., Donath, M. Y., Bruun, C., Mandrup-Poulsen, T., Billestrup, N., and Halban, P. A. (2008). Proliferation of sorted human and rat beta cells. *Diabetologia* **51,** 91–100.

Permutt, M. A., and Kipnis, D. M. (1972). Insulin biosynthesis. I. On the mechanism of glucose stimulation. *J. Biol. Chem.* **247,** 1194–1199.

Petrik, J., Reusens, B., Arany, E., Remacle, C., Coelho, C., Hoet, J. J., and Hill, D. J. (1999). A low protein diet alters the balance of islet cell replication and apoptosis in the fetal and neonatal rat and is associated with a reduced pancreatic expression of insulin-like growth factor-II. *Endocrinology* **140,** 4861–4873.

Pipeleers, D., Kiekens, R., Ling, Z., Wilikens, A., and Schuit, F. (1994). Physiologic relevance of heterogeneity in the pancreatic beta-cell population. *Diabetologia* **37**(Suppl. 2), S57–S64.

Pipeleers, D. G. (1992). Heterogeneity in pancreatic beta-cell population. *Diabetes* **41,** 777–781.

Poy, M. N., Eliasson, L., Krutzfeldt, J., Kuwajima, S., Ma, X., Macdonald, P. E., Pfeffer, S., Tuschl, T., Rajewsky, N., Rorsman, P., and Stoffel, M. (2004). A pancreatic islet-specific microRNA regulates insulin secretion. *Nature* **432,** 226–230.

Prentki, M., and Matschinsky, F. M. (1987). Ca2+, cAMP, and phospholipid-derived messengers in coupling mechanisms of insulin secretion. *Physiol. Rev.* **67,** 1185–1248.

Proud, C. G. (2004). Role of mTOR signalling in the control of translation initiation and elongation by nutrients. *Curr. Top. Microbiol. Immunol.* **279,** 215–244.

Quintens, R., Hendrickx, N., Lemaire, K., and Schuit, F. (2008). Why expression of some genes is disallowed in beta-cells. *Biochem. Soc. Trans.* **36,** 300–305.

Robertson, R. P. (2004). Chronic oxidative stress as a central mechanism for glucose toxicity in pancreatic islet beta cells in diabetes. *J. Biol. Chem.* **279,** 42351–42354.

Robertson, R. P., Harmon, J., Tran, P. O., and Poitout, V. (2004). Beta-cell glucose toxicity, lipotoxicity, and chronic oxidative stress in type 2 diabetes. *Diabetes* **53**(Suppl. 1), S119–S124.

Roche, E., Farfari, S., Witters, L. A., Assimacopoulos-Jeannet, F., Thumelin, S., Brun, T., Corkey, B. E., Saha, A. K., and Prentki, M. (1998). Long-term exposure of beta-INS cells to high glucose concentrations increases anaplerosis, lipogenesis, and lipogenic gene expression. *Diabetes* **47,** 1086–1094.

Roduit, R., Nolan, C., Alarcon, C., Moore, P., Barbeau, A., Delghingaro-Augusto, V., Przybykowski, E., Morin, J., Masse, F., Massie, B., Ruderman, N., Rhodes, C., *et al.* (2004). A role for the malonyl-CoA/long-chain acyl-CoA pathway of lipid signaling in the regulation of insulin secretion in response to both fuel and nonfuel stimuli. *Diabetes* **53,** 1007–1019.

Ronnebaum, S. M., Joseph, J. W., Ilkayeva, O., Burgess, S. C., Lu, D., Becker, T. C., Sherry, A. D., and Newgard, C. B. (2008). Chronic suppression of acetyl-CoA carboxylase 1 in {beta}-cells impairs insulin secretion via inhibition of glucose rather than lipid metabolism. *J. Biol. Chem.* **283,** 14248–14256.

Rozance, P. J., Limesand, S. W., and Hay, W. W. Jr. (2006). Decreased nutrient-stimulated insulin secretion in chronically hypoglycemic late-gestation fetal sheep is due to an intrinsic islet defect. *Am. J. Physiol. Endocrinol. Metab.* **291,** E404–E411.

Rutter, G. A., da, S. X., and Leclerc, I. (2003). Roles of 5′ -AMP-activated protein kinase (AMPK) in mammalian glucose homoeostasis. *Biochem. J.* **375,** 1–16.

Schuit, F., De Vos, A., Farfari, S., Moens, K., Pipeleers, D., Brun, T., and Prentki, M. (1997). Metabolic fate of glucose in purified islet cells. Glucose-regulated anaplerosis in beta cells. *J. Biol. Chem.* **272,** 18572–18579.

Schuit, F., Moens, K., Heimberg, H., and Pipeleers, D. (1999). Cellular origin of hexokinase in pancreatic islets. *J. Biol. Chem.* **274,** 32803–32809.

Schuit, F., Flamez, D., De Vos, A., and Pipeleers, D. (2002). Glucose-regulated gene expression maintaining the glucose-responsive state of beta-cells. *Diabetes* **51**(Suppl. 3), S326–S332.

Schuit, F. C., and Pipeleers, D. G. (1985). Regulation of adenosine 3′,5′-monophosphate levels in the pancreatic B cell. *Endocrinology* **117,** 834–840.

Schuit, F. C., Huypens, P., Heimberg, H., and Pipeleers, D. G. (2001). Glucose sensing in pancreatic beta-cells: A model for the study of other glucose-regulated cells in gut, pancreas, and hypothalamus. *Diabetes* **50,** 1–11.

Schuit, F. C., Kiekens, R., and Pipeleers, D. G. (1991). Measuring the balance between insulin synthesis and insulin release. *Biochem. Biophys. Res. Commun.* **178,** 1182–1187.

Schuppin, G. T., Pons, S., Hugl, S., Aiello, L. P., King, G. L., White, M., and Rhodes, C. J. (1998). A specific increased expression of insulin receptor substrate 2 in pancreatic beta-cell lines is involved in mediating serum-stimulated beta-cell growth. *Diabetes* **47,** 1074–1085.

Silva, J. P., Kohler, M., Graff, C., Oldfors, A., Magnuson, M. A., Berggren, P. O., and Larsson, N. G. (2000). Impaired insulin secretion and beta-cell loss in tissue-specific knockout mice with mitochondrial diabetes. *Nat. Genet.* **26,** 336–340.

Sladek, R., Rocheleau, G., Rung, J., Dina, C., Shen, L., Serre, D., Boutin, P., Vincent, D., Belisle, A., Hadjadj, S., Balkau, B., Heude, B., *et al.* (2007). A genome-wide association study identifies novel risk loci for type 2 diabetes. *Nature* **445,** 881–885.

Svensson, C., Sandler, S., and Hellerstrom, C. (1993). Lack of long-term beta-cell glucotoxicity *in vitro* in pancreatic islets isolated from two mouse strains (C57BL/6J; C57BL/KsJ) with different sensitivities of the beta-cells to hyperglycaemia *in vivo*. *J. Endocrinol.* **136,** 289–296.

Swenne, I. (1983). Effects of aging on the regenerative capacity of the pancreatic B-cell of the rat. *Diabetes* **32,** 14–19.

Swenne, I., and Andersson, A. (1984). Effect of genetic background on the capacity for islet cell replication in mice. *Diabetologia* **27,** 464–467.

Swenne, I., Crace, C. J., and Milner, R. D. (1987). Persistent impairment of insulin secretory response to glucose in adult rats after limited period of protein-calorie malnutrition early in life. *Diabetes* **36,** 454–458.

Swenne, I., Borg, L. A., Crace, C. J., and Schnell, L. A. (1992). Persistent reduction of pancreatic beta-cell mass after a limited period of protein-energy malnutrition in the young rat. *Diabetologia* **35,** 939–945.

Teta, M., Long, S. Y., Wartschow, L. M., Rankin, M. M., and Kushner, J. A. (2005). Very slow turnover of beta-cells in aged adult mice. *Diabetes* **54,** 2557–2567.

Ueki, K., Okada, T., Hu, J., Liew, C. W., Assmann, A., Dahlgren, G. M., Peters, J. L., Shackman, J. G., Zhang, M., Artner, I., Satin, L. S., Stein, R., *et al.* (2006). Total insulin and IGF-I resistance in pancreatic beta cells causes overt diabetes. *Nat. Genet.* **38,** 583–588.

Van de Casteele, M., Kefas, B. A., Cai, Y., Heimberg, H., Scott, D. K., Henquin, J. C., Pipeleers, D., and Jonas, J. C. (2003). Prolonged culture in low glucose induces apoptosis of rat pancreatic beta-cells through induction of c-myc. *Biochem. Biophys. Res. Commun.* **312,** 937–944.

Vander Mierde, D., Scheuner, D., Quintens, R., Patel, R., Song, B., Tsukamoto, K., Beullens, M., Kaufman, R. J., Bollen, M., and Schuit, F. C. (2007). Glucose activates a protein phosphatase-1-mediated signaling pathway to enhance overall translation in pancreatic beta-cells. *Endocrinology* **148,** 609–617.

Van De Winkel, M., and Pipeleers, D. (1983). Autofluorescence-activated cell sorting of pancreatic islet cells: Purification of insulin-containing B-cells according to glucose-induced changes in cellular redox state. *Biochem. Biophys. Res. Commun.* **114,** 835–842.

Van Schravendijk, C. F., Kiekens, R., and Pipeleers, D. G. (1992). Pancreatic beta cell heterogeneity in glucose-induced insulin secretion. *J. Biol. Chem.* **267,** 21344–21348.

Wang, H., Maechler, P., Antinozzi, P. A., Hagenfeldt, K. A., and Wollheim, C. B. (2000). Hepatocyte nuclear factor 4alpha regulates the expression of pancreatic beta-cell genes implicated in glucose metabolism and nutrient-induced insulin secretion. *J. Biol. Chem.* **275,** 35953–35959.

Wang, H., Gauthier, B. R., Hagenfeldt-Johansson, K. A., Iezzi, M., and Wollheim, C. B. (2002). Foxa2 (HNF3beta) controls multiple genes implicated in metabolism-secretion coupling of glucose-induced insulin release. *J. Biol. Chem.* **277,** 17564–17570.

Wang, M. Y., Koyama, K., Shimabukuro, M., Mangelsdorf, D., Newgard, C. B., and Unger, R. H. (1998). Overexpression of leptin receptors in pancreatic islets of Zucker diabetic fatty rats restores GLUT-2, glucokinase, and glucose-stimulated insulin secretion. *Proc. Natl. Acad. Sci. USA* **95,** 11921–11926.

Wang, Q., Heimberg, H., Pipeleers, D., and Ling, Z. (2008). Glibenclamide activates translation in rat pancreatic beta cells through calcium-dependent mTOR, PKA and MEK signalling pathways. *Diabetologia* **51,** 1202–1212.

Weinberg, N., Ouziel-Yahalom, L., Knoller, S., Efrat, S., and Dor, Y. (2007). Lineage tracing evidence for *in vitro* dedifferentiation but rare proliferation of mouse pancreatic beta-cells. *Diabetes* **56,** 1299–1304.

Weir, G. C., and Bonner-Weir, S. (2004). Five stages of evolving beta-cell dysfunction during progression to diabetes. *Diabetes* **53**(Suppl. 3), S16–S21.

Wicksteed, B., Herbert, T. P., Alarcon, C., Lingohr, M. K., Moss, L. G., and Rhodes, C. J. (2001). Cooperativity between the preproinsulin mRNA untranslated regions is necessary for glucose-stimulated translation. *J. Biol. Chem.* **276,** 22553–22558.

Wiederkehr, A., and Wollheim, C. B. (2006). Minireview: Implication of mitochondria in insulin secretion and action. *Endocrinology* **147,** 2643–2649.

Wollheim, C. B., and Maechler, P. (2002). Beta-cell mitochondria and insulin secretion: Messenger role of nucleotides and metabolites. *Diabetes* **51**(Suppl. 1), S37–S42.

Zelent, D., Najafi, H., Odili, S., Buettger, C., Weik-Collins, H., Li, C., Doliba, N., Grimsby, J., and Matschinsky, F. M. (2005). Glucokinase and glucose homeostasis: Proven concepts and new ideas. *Biochem. Soc. Trans.* **33,** 306–310.

Zhou, Y. P., Marlen, K., Palma, J. F., Schweitzer, A., Reilly, L., Gregoire, F. M., Xu, G. G., Blume, J. E., and Johnson, J. D. (2003). Overexpression of repressive cAMP response element modulators in high glucose and fatty acid-treated rat islets. A common mechanism for glucose toxicity and lipotoxicity? *J. Biol. Chem.* **278,** 51316–51323.

Zitzer, H., Wente, W., Brenner, M. B., Sewing, S., Buschard, K., Gromada, J., and Efanov, A. M. (2006). Sterol regulatory element-binding protein 1 mediates liver X receptor-beta-induced increases in insulin secretion and insulin messenger ribonucleic acid levels. *Endocrinology* **147,** 3898–3905.

CHAPTER EIGHTEEN

Matrix Metalloproteinases, T Cell Homing and β-Cell Mass in Type 1 Diabetes

Alexei Y. Savinov *and* Alex Y. Strongin

Contents

Abstract

The pathogenesis of type 1 diabetes begins with the activation of autoimmune T killer cells and is followed by their homing into the pancreatic islets. After penetrating the pancreatic islets, T cells directly contact and destroy insulin-producing β cells. This review provides an overview of the dynamic interactions which link T cell membrane type-1 matrix metalloproteinase (MT1-MMP) and the signaling adhesion CD44 receptor with T cell transendothelial migration and the subsequent homing of the transmigrated cells to the pancreatic islets. MT1-MMP regulates the functionality of CD44 in diabetogenic T cells. By regulating the functionality of T cell CD44, MT1-MMP mediates the transition of T cell

Burnham Institute for Medical Research, Inflammatory and Infectious Disease Center, La Jolla, CA 92037

Vitamins and Hormones, Volume 80
ISSN 0083-6729, DOI: 10.1016/S0083-6729(08)00618-3

adhesion to endothelial cells to the transendothelial migration of T cells, thus, controlling the rate at which T cells home into the pancreatic islets. As a result, the T cell MT1-MMP-CD44 axis controls the severity of the disease. Inhibition of MT1-MMP proteolysis of CD44 using highly specific and potent synthetic inhibitors, which have been clinically tested in cancer patients, reduces the rate of transendothelial migration and the homing of T cells. Result is a decrease in the net diabetogenic efficiency of T cells and a restoration of β cell mass and insulin production in NOD mice. The latter is a reliable and widely used model of type I diabetes in humans. Overall, existing experimental evidence suggests that there is a sound mechanistic rationale for clinical trials of the inhibitors of T cell MT1-MMP in human type 1 diabetes patients.

I. Matrix Metalloproteinases and Their Natural Protein Inhibitors

A. MMPs

Historically, interstitial collagenase (MMP-1) was the first identified member of the now extensive matrix metalloproteinase (MMP) family. MMP-1 was initially discovered in the course of studying collagen remodeling during the metamorphosis of a tadpole into a frog and much later the presence of this enzyme was confirmed in humans (Gross and Lapiere, 1962; Stocker and Bode, 1995). Because collagens, especially type I collagen, represent the major structural proteins of all tissues and serve as the main barrier to migrating cells, more than three decades ago an innovative hypothesis was postulated and this hypothesis has now been proven correct. According to this hypothesis, collagenolytic enzymes including MMP-1 play pivotal roles in multiple physiologic and, especially pathologic processes, which involve both extensive and aberrant collagenolysis. Recent scientific discoveries have vastly expanded our knowledge of MMPs' structures and functions. These discoveries directly implicate a number of the individual MMPs, including MMP-1, in multiple diseases of the cardiovascular, pulmonary, renal, endocrine, gastrointestinal, musculoskeletal, visual, and hematopoietic systems in humans.

MMPs belong to a zinc endopeptidase, metzincin superfamily (Gomis-Ruth, 2003). This superfamily is distinguished from other proteinases by the presence of a strictly conserved HEXXHXXGXX(H/D) histidine sequence motif. This motif exhibits three histidine residues that chelate the active site zinc and also a canonical methionine residue which is the C-terminal to the conserved histidine sequence. The canonical methionine is part of a tight 1,4-beta-turn that loops the polypeptide chain beneath the catalytic zinc ion, forming a hydrophobic floor to the zinc ion binding site. The metzincin family is normally divided into four subfamilies: seralysins, astacins, adamalysins [ADAMs (proteins with a disintegrin and a metalloproteinase domain) and ADAM-TS (ADAM with thrombospondin-like motif)] and MMPs. Although our knowledge of MMP biology is rapidly expanding, we do not as yet understand precisely how these enzymes regulate various cellular functions.

The human MMP family is comprised of 24 currently known zinc-containing enzymes which share several common functional domains. MMPs are often referred to by a descriptive name such as gelatinases (MMP-2 and MMP-9) and collagenases [MMP-1, MMP-8, MMP-13, MMP-14/membrane type-1 matrix metalloproteinase (MT1-MMP) and, some what conclusively, MMP-18] and this name is generally based on a preferred substrate. Collagenases are the only known mammalian enzymes capable of degrading triple-helical fibrillar collagen into distinctive 3/4 and 1/4 fragments. An additional and widely accepted MMP numbering system based on the order of discovery is also in use (Fig. 18.1) (Egeblad and Werb, 2002; Nagase and Woessner, 1999).

In general, MMPs may be described as multifunctional enzymes capable of cleaving the extracellular matrix components (collagens, laminin, fibronectin, vitronectin, aggrecan, enactin, versican, perlecan, tenascin, elastin, and many others), growth factors, cytokines and cell surface-associated adhesion and signaling receptors. Because of their high degrading activity and potentially disastrous effect on the cell microenvironment, cellular MMPs are expressed in small amounts, and their cellular localization and activity are tightly controlled, either positively or negatively, at both the transcriptional and the posttranscriptional levels by cytokines, including interleukins (IL-1, IL-4, and IL-6), growth factors (epidermal growth factor, hepatocyte growth factor and transforming growth factor-β), and tumor necrosis factor-α) (Sternlicht and Werb, 2001; Zucker *et al.*, 2003). In a feedback loop, some of these regulatory factors themselves are proteolytically activated or inactivated by the individual MMPs (McQuibban *et al.*, 2000).

MMPs are synthesized as latent zymogens. The active site zinc of the MMP catalytic domain is coordinated with the three histidines of the active site and with the cysteine of the "cysteine switch" motif of the N-terminal prodomain (Van Wart and Birkedal-Hansen, 1990). To become functionally potent proteinases, the zymogens of MMPs require proteolytic activation. In the process of this activation, the N-terminal inhibitory prodomain is removed and the catalytic site of the emerging enzyme becomes liberated and exposed. The activation of MMPs may occur both intracellularly and extracellularly (Murphy *et al.*, 1999; Pei and Weiss, 1995). MMPs including MMP-11, MMP-28, and several MT-MMPs) with the furin cleavage motif RXK/RR in their propeptide sequence are normally activated in the trans-Golgi network by serine proteinases such as furin and certain additional members of the proprotein convertase family. Activation of MMPs which

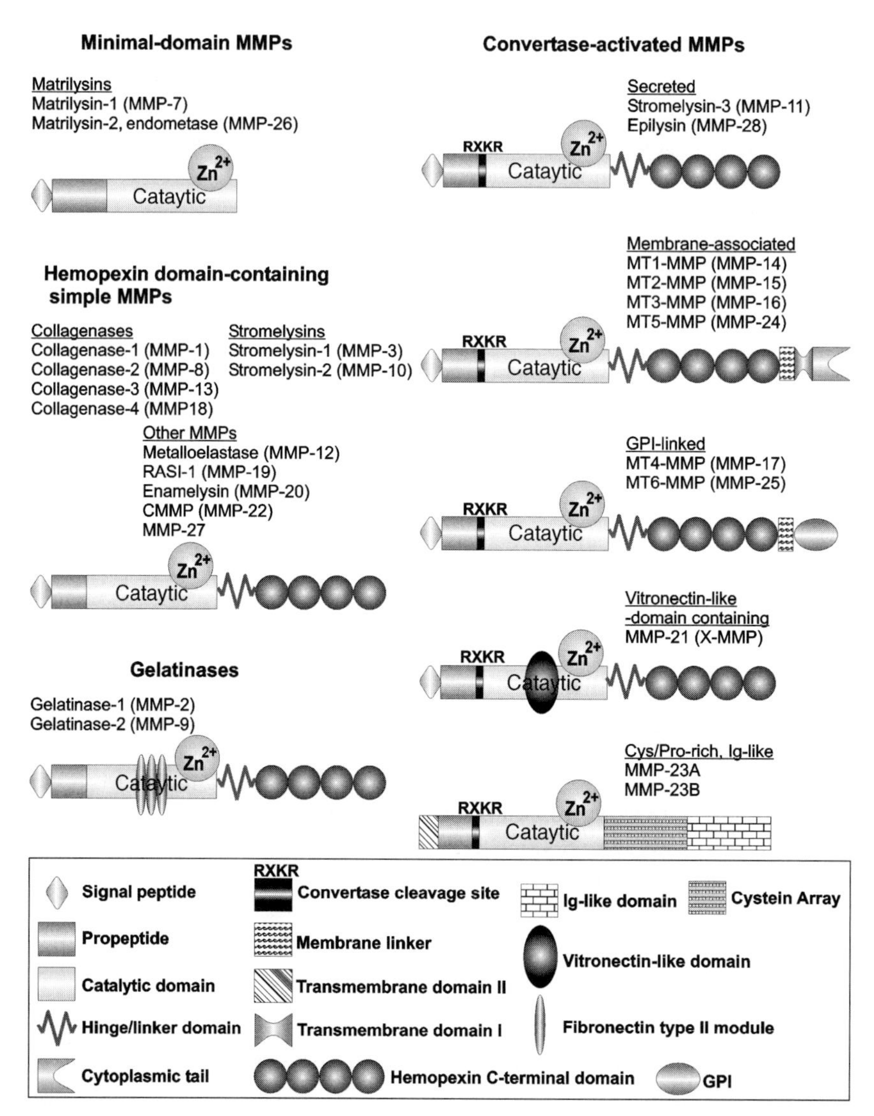

Figure 18.1 *Domain structure of MMPs.* The structure of MMPs is made up of the following homologous domains: (1) a signal peptide; directs MMPs to the secretory or plasma membrane insertion pathway; (2) a prodomain; confers latency to MMPs, (3) a catalytic domain with the active site zinc atom; (4) a hemopexin-like domain; in coordination with the catalytic domain controls the interactions with substrates, and (5) a flexible hinge region; links the catalytic and the hemopexin domain and provides each domain mobility relative to the other. The membrane-type MMPs contain an additional transmembrane domain and a short cytoplasmic tail domain (MMP-14, MMP-15, MMP-16, and MMP-24) or a glycosylphosphatidyl inositol linkage, which attaches MMP-17 and MMP-25 to the cell surface. MMP-2 and MMP-9 contain fibronectin-like type II repeats which assist in collagen substrate binding. A hemopexin domain is absent in MMP-7 and MMP-26. RXKR, furin cleavage motif. Date of

are secreted in the extracellular milieu is frequently mediated by serine proteases, including plasmin, by the membrane type MMPs (e.g., activation of the latent soluble MMP-2 proenzyme by MT1-MMP) or by other pre-existing active MMPs (e.g., activation of the latent soluble enzymes of MMP-1 and MMP-9 by the soluble MMP-3 enzyme).

With the exception of the activated MMP-7 and MMP-26 enzymes, which are represented by the catalytic domain alone, all other MMPs have a C-terminal hemopexin-like domain. This domain regulates the activity and the specificity of the catalytic domain function. The hemopexin domain is separated from the catalytic domain by a flexible hinge region. Membrane-tethered MMPs are distinguished from soluble MMPs by an additional transmembrane domain and a short cytoplasmic tail (MMP-14/MT1-MMP, MMP-15/MT2-MMP, MMP-16/MT3-MMP, and MMP-24/MT5-MMP). In contrast to these four MT-MMPs, MMP-17/MT4-MMP, and MMP-25/MT6-MMP are attached to the cell membrane via a glycosylphosphatidyl inositol (GPI) anchor (Fig. 18.1) Historically, MMPs were initially characterized by their extensive ability to degrade extracellular matrix proteins including collagens, laminin, fibronectin, vitronectin, aggrecan, enactin, tenascin, elastin, and proteoglycans. More recently, it has been recognized that MMPs cleave in addition to the extracellular matrix components, many other protein types including cytokines and cell adhesion signaling receptors.

Because the individual MMPs have overlapping substrate cleavage preferences, MMP knockouts and inactivating mutations in individual MMP genes in mice, with the exception of MT1-MMP, do not elicit an easily recognized phenotype and are non-lethal, at least up to the first few weeks after birth, suggesting functional redundancy among MMP family members. MT1-MMP knockout, however, has a profound effect: MT1-MMP null mice develop dwarfism, extensive bone malformations and die before adulthood, thus supporting an important role of MT1-MMP in collagen type I turnover (Holmbeck *et al.*, 1999, 2003, 2004). Mice lacking both MMP-2 and MT1-MMP die immediately after birth of respiratory failure, abnormal blood vessels, and immature muscle fibers reminiscent of central core disease (Oh *et al.*, 2004).

---

discovery of MMP is shown in parentheses. MMP-1/interstitial collagenase (1986); MMP-2/gelatinase A (1988); MMP-3/stromelysin-1 (1985); MMP-7/matrilysin (1988); MMP-8/collagenase-2 (1990); MMP-9/gelatinase B (1989); MMP-10/stromelysin-2 (1988); MMP-11/stromelysin-3 (1990); MMP-12/metalloelastase (1992); MMP-13/collagenase-3 (1994); MMP-14/MT1-MMP (1994); MMP-15/MT2-MMP (1995); MMP-16/MT3-MMP (1999); MMP-17/MT4-MMP (1996); MMP-18/collagenase-4 (1996); MMP-19/RASI-1 (1996); MMP-20/enamelysin (1997); MMP-21/XMMP (1998); MMPP-22/CMMP (1998); MMP-23 (1999); MMP-24/MT5-MMP (1999); MMP-25/MT6-MMP (1999); MMP-26/endometase (2001); MMP-27 (2000); MMP-28/epilysin (2001).

B. Tissie inhibitors of matrix metalloproteinases

Once activated, MMPs are normally inhibited by tissue inhibitors of metalloproteinases (TIMPs). Four individual species of TIMPs are known in humans (TIMP-1, -2, -3, and -4) (Nagase *et al.*, 2006) (Fig. 18.2). MMP/TIMP balance is believed to be a major factor in the regulation of the net proteolytic activity of the individual activated MMPs. Structurally, TIMPs contain two domains. The inhibitory N-terminal domain binds non-covalently and stoichiometrically to the active site of the active mature MMPs, blocking access of substrates to the catalytic site. The C-terminal domain of TIMP-1 and TIMP-2 binds to the hemopexin domain of the proenzymes of MMP-9 and MMP-2, respectively. The latter binding is essential for the cell surface activation of MMP-2 by MMP-14/MT1-MMP.

In this well-characterized unconventional activation mechanism, MMP-14/MT1-MMP on the cell surface acts as a receptor for TIMP-2. TIMP-2 binds via its N-terminal domain to the active site of MT1-MMP. This binary complex then acts as a receptor for the MMP-2 proenzyme, with the TIMP-2 C-terminal domain binding to the C-terminal hemopexin domain of MMP-2 and with the formation of a trimolecular MT1-MMP-TIMP-2-MMP-2 complex.

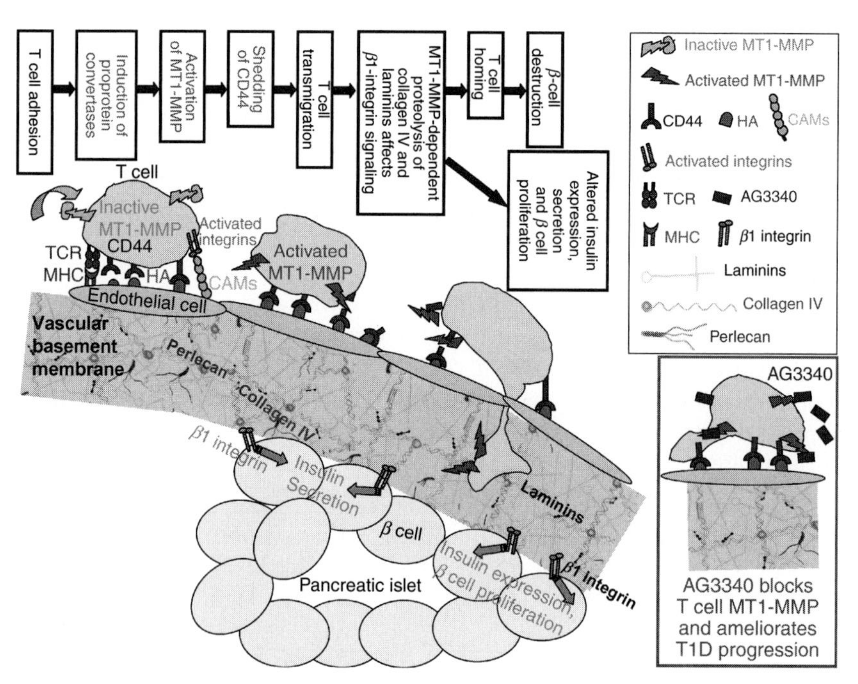

Figure 18.2 *Schematic representation of the role of T cell MT1-MMP in diabetogenesis.* AG3340, a hydroxamate class inhibitor of matrix metalloproteinases; CAMs, cell adhesion molecules; HA, hyaluronan; T1D, type 1 diabetes; TCR, T cell receptor; MHC, major histocompatibility complex.

A TIMP-2-free, second MT1-MMP molecule which is close to the trimolecular complex then cleaves the N-terminal propeptide of the MMP-2 proenzyme, generating an intermediate species. Further proteolysis of the MMP-2 propeptide through an autocatalytic mechanism generates the fully active enzyme of MMP-2 which is then released from the complex.

II. T Cell Membrane Type-1 Matrix Metalloproteinase

A. Structure and function of MT1-MMP

MT1-MMP, a prototypic membrane-type MMP, is distinguished from soluble MMPs by a C-terminal transmembrane domain and a cytoplasmic tail (Egeblad and Werb, 2002). In the human genome, MT1-MMP is encoded by a single copy gene located on chromosome 14. MT1-MMP is widely expressed and its presence has been documented in multiple cell types. Because both the expression and the activity of MT1-MMP are elevated in tumor cells and because high levels of MT1-MMP directly correlate with enhanced cell migration, this proteinase is generally considered pro-invasive and pro-tumorigenic.

Because the prodomain part of MT1-MMP has the furin-cleavage motif, furin is believed to be an essential component of the activation pathway that results in the generation of the active, mature cellular MT1-MMP (Pei and Weiss, 1995; Yana and Weiss, 2000). MT1-MMP was originally thought to exhibit a single function as a membrane activator of soluble MMPs, including MMP-2 (Sato *et al.*, 1994; Strongin *et al.*, 1995) and MMP-13 (Knauper *et al.*, 2002). Recent data, however, has provided evidence that, in addition, MT1-MMP degrades multiple components of the extracellular matrix and a number of cell adhesion and signaling receptors (Strongin, 2006). MT1-MMP is regulated both as a proteinase and as a membrane protein at the transcriptional and posttranscriptional levels by multifaceted, tightly controlled and well-coordinated mechanisms. These multidimensional mechanisms regulate MT1-MMP spatially and temporally, and they are essential not only for the proper functioning of MT1-MMP alone but also for the performance of the normal multiple biological functions of the entire cell. The regulatory mechanisms which control the functional activity of MT1-MMP include control of the extent of activation of the MT1-MMP proenzyme by furin, the level of inhibition of MT1-MMP by TIMPs and self-proteolytic inactivation, a homophilic complex formation involving the hemopexin domain and the cytoplasmic tail, the efficiency of trafficking of MT1-MMP through the cell compartment to the plasma membrane, the rate of the internalization of MT1-MMP into the transient compartment inside the cells and, lastly, the extent of the recycling of MT1-MMP back to the plasma membrane (Itoh and Seiki, 2006; Seiki, 2003; Strongin, 2006).

Internalization via clathrin-coated pits and also through caveolae is also recognized as a important mechanism to regulate MT1-MMP activity (Galvez *et al.*, 2004; Jiang *et al.*, 2001; Labrecque *et al.*, 2004; Rozanov *et al.*, 2004). The "up/down" switch may have been built into the peptide sequence of the MT1-MMP's cytoplasmic tail to regulate the recruitment to the plasma membrane and to target the protease to the invasive front in migrating cells. Transient changes in subcellular compartmentalization of MT1-MMP, which occur in its trafficking and internalization pathways, are, probably, the underlying mechanisms which specifically control the functions of MT1-MMP in malignant cells (Wang *et al.*, 2004a,b).

TIMP-2, TIMP-3, and TIMP-4 are highly potent inhibitors of MT1-MMP. TIMP-1, however, is a very poor inhibitor of this proteinase (Nagase and Woessner, 1999; Nagase *et al.*, 2006; Will *et al.*, 1996). Current evidence suggests that the activity of cellular MT1-MMP is short-lived and that the half-life of active, mature MT1-MMP attached to the plasma membrane is approximately 1 h (Deryugina *et al.*, 2004; Wang *et al.*, 2004b). During this time period, active MT1-MMP is either inactivated by TIMPs or autolytically degraded or internalized with only a subsequent partial recycling (Osenkowski *et al.*, 2004). Because MT1-MMP, in addition to its role in matrix degradation, is directly involved in the cleavage of cell surface receptors, this short-lived proteinase, exerts a long-lasting effect on cell behavior and functions in cancer cells as the main mediator of proteolytic events on the cell surface. Our data and the results of others show that the proteolysis of CD44, integrins, tissue transglutaminase, the low density lipoprotein receptor-related protein (LRP1), E-cadherin and related cell-associated adhesion signaling receptors is the important role of MT1-MMP (Strongin, 2006). By cleaving these receptors, the short-lived MT1-MMP has a long lasting effect on the cell microenvironment and cell behavior. In addition, MT1-MMP, as opposed to the soluble MMPs, is ideally positioned to regulate pericellular proteolysis and the functionality of the neighboring cell receptors, including CD44 (Seiki, 2003).

B. Adhesion and signaling CD44 receptor

The transmembrane, cell adhesion signaling receptor CD44 is the principal receptor for hyaluronan (Turley *et al.*, 2002). CD44 plays a critical role in cell functions, including adhesion, migration, invasion, and survival (Cichy and Pure, 2003). The C-end cytoplasmic tail interacts with ezrin, radixin and moesin and links CD44 to the actin cytoskeleton. CD44 also interacts with the Rho-family GTPases and induces the rearrangement of the cytoskeleton (Turley *et al.*, 2002). The binding of CD44 to hyaluronan stimulates the downstream signaling pathways and leads to the activation of protein kinase Cα (Slevin *et al.*, 2002). Antigenic stimulation of T cells induces the cell surface expression of the highly active form of CD44 which binds hyaluronan with high affinity (Lesley *et al.*, 1993). The presence of high concentrations of hyaluronan in the microcapillaries is consistent with the importance of CD44 in T cell trafficking (Aruffo *et al.*, 1990; Bennett *et al.*, 1995; DeGrendele *et al.*, 1997; Estess *et al.*, 1999; Mohamadzadeh *et al.*, 1998). CD44 is a potent adhesion receptor and it facilitates T cell adhesion on the endothelium and the subsequent transmigration events (Avigdor *et al.*, 2004; Savinov and Strongin, 2007; Savinov *et al.*, 2005, 2006, 2007; Weiss *et al.*, 2000).

CD44 is a marker of activated T cells (DeGrendele *et al.*, 1997; Estess *et al.*, 1998). CD44 is heavily glycosylated (Katoh *et al.*, 1995). Glycosylation regulates the oligomerization and the movement of CD44 across the plasma membrane. CD44 is a target of MT1-MMP proteolysis in tumor cells (Kajita *et al.*, 2001; Mori *et al.*, 2002; Murai *et al.*, 2004; Nakamura *et al.*, 2004; Suenaga *et al.*, 2005). There are three cleavage sites in the ectodomain stem region of CD44 (Mori *et al.*, 2002; Nakamura *et al.*, 2004). MT1-MMP cleaves the $SGG^{192}\downarrow Y^{193}IF$ sequence in the CD44 molecule while two additional sites ($SGG^{233}\downarrow S^{234}HT$ and $HGS^{249}\downarrow Q^{250}EG$) are cleaved either by MT1-MMP or ADAM/ADAM-TS proteases or by both. The inactivation of CD44 functionality by MT1-MMP proteolysis stimulates the migration of cancer cells. Conversely, the inactivation of MT1-MMP by TIMPs (excluding TIMP-1 which is a poor inhibitor MT1-MMP) or synthetic inhibitors protect cell surface-associated CD44 from MT1-MMP proteolysis and these events correlate with migration recess.

C. Synthetic antagonists of MT1-MMP

In many cancer types, MMPs including MT1-MMP are up-regulated and because these proteinases are considered pro-invasive and tumorigenic, considerable effort has been devoted to the development of synthetic MMP inhibitors. This activity has led to the design of a variety of compounds including hydroxamate-based inhibitors the ligand hydroxamate group of which coordinates to the catalytic zinc ion thus rendering MMPs inactive. Hydroxamate inhibitors have showed great promise in animal models of cancer and certain inhibitors have been tested in cancer patients.

Several hydroxamate-based inhibitors, clinically tested in cancer patients, including GM6001 and AG3340, are potent against MT1-MMP. The hydroxamate AG3340 (trade name Prinomastat; 2-{[(hydroxyamino) methyl]-5,6-dimethyl-4-(4-pyridin-4-yloxyphenyl)sulfonyl-morpholine-3-thione} (Agouron-Pfizer) inhibits MT1-MMP with a K_i of 40 pM (Shalinsky *et al.*, 1999; Zucker *et al.*, 2000). The K_i values of AG3340 against MMP-2, MMP-3, and MMP-13 are approximately 100, 300, and 200 pM, respectively. Other individual MMPs are significantly less sensitive to AG3340 inhibition (e.g., the K_i values for MMP-1 and MMP-7 are 10 and 55 nM, respectively). AG3340/Prinomastat was used as an oral angiogenic drug in phase I–III clinical trials in humans with advanced non-small cell lung cancer and prostate cancer (Hande *et al.*, 2004). The trials were halted because of the drug's lack of effectiveness in patients with late-stage disease (Cappuzzo *et al.*, 2003) but there was no question about patient safety. The hydroxamate peptidomimetic inhibitor GM6001 (Galardin; Ilomastat; *N*-[2R)-2-(hydroxamidocarbonylmethyl)-4-methylpentanoyl]-L-tryptophan methylamide) with a K_i value against MT1-MMP also in the sub-nanomolar range was tested on ulcer patents.

A recently designed thiirane inhibitor SB-3CT [(4-phenoxyphenylsulfonyl) butane-1,2-dithiol] exhibits a dithiolate moiety that chelates the active site zinc (Celenza *et al.*, 2008; Lee *et al.*, 2007; Sodek *et al.*, 2007). SB-3CT was specifically designed to target MMP-2/MMP-9 gelatinases, not MT1-MMP. Whereas SB-3CT is an effective and selective gelatinase inhibitor, it either does not inhibit or it poorly inhibits other MMPs and the closely related metalloproteinase TACE (tumor necrosis factor-α converting enzyme) (Ikejiri *et al.*, 2005a,b). The efficacy of SB3-CT was demonstrated in several distinct models where the functional activity of gelatinases was relevant to the disease. Thus, intraperitoneal treatment with SB-3CT (50 mg/kg) inhibited intraosseous growth of human PC3 cells within the marrow of human fetal femur fragments previously implanted in SCID mice (Bonfil *et al.*, 2006). In a transient focal cerebral ischemia model of stroke in mice, MMP-9 contributes directly to neuron apoptosis and brain damage (Gu *et al.*, 2005) by degrading the extracellular matrix laminin. SB-3CT blocks MMP-9 activity, including MMP-9-mediated laminin cleavage, thus rescuing neurons from apoptosis.

A natural compound epigallocatechin gallate (EGCG), a major catechin of green tea, also exhibits inhibitory, albeit largely non-specific, effects on MMP (Annabi *et al.*, 2002; Chiang *et al.*, 2006; Dell'Aica *et al.*, 2002; El Bedoui *et al.*, 2005; Kim *et al.*, 2004; Lee *et al.*, 2005; Song *et al.*, 2004; Vayalil and Katiyar, 2004).

III. Rodent Model of Human Type 1 Diabetes

A. Type I diabetes

Type 1 diabetes (T1D) is a major debilitating human disease with an early childhood onset. Lymphocyte infiltration into the islets of Langerhans is a hallmark of T1D in humans and in rodent models. The pathogenesis of T1D begins with the activation of autoimmune T killer cells which then home into the pancreatic islets. After penetrating the pancreatic islets, T cells directly contact and destroy insulin-producing β cells. Autoreactive IS

$CD8^+$ T killer cells are specific for islet-derived insulin antigen (Savinov *et al.*, 2003a,b; Verdaguer *et al.*, 1997; Wong *et al.*, 1996). This T cell subpopulation is the key player in the destruction of the pancreatic β cells. The transmigration of IS-$CD8^+$ T killer cells from the bloodstream through the pancreatic endothelium and into the islets of Langerhans is essential for the T cell-β cell contact and the subsequent destruction of β cells. The specific molecular mechanisms governing these processes are not, as yet, completely understood (Springer, 1994; Weber, 2003; Worthylake and Burridge, 2001) and, therefore, therapeutic interventions leading to β cell regeneration and the reversal of established T1D are exceedingly limited.

The transendothelial migration of T cells consists of the following basic steps: initial adhesion followed by tethering and rolling, arrest of rolling that translates into firm adhesion, extravasation and migration through the endothelial cell barrier (Butcher and Picker, 1996; Butcher *et al.*, 1999). CD44 and other adhesion receptors including selectins, cadherins, immunoglobulin superfamily cell adhesion molecules (CAMs, and, specifically, VCAM, ICAM-1, and ICAM-2) and integrins, which are expressed in both T cells and endothelial cells, contribute to the adhesion of T cells to the endothelium (Constantin *et al.*, 2000; Stein *et al.*, 2000; Yang *et al.*, 1996). Initial adhesion and slow rolling of T cells on endothelial cells is largely mediated by L-selectin, ligands for tissue-specific P and E selectins, and integrins. The slow rolling of T cells on the endothelial surface is followed by the activation of T cell integrins and firm adhesion. The interactions of T cell CD44 with its abundant endothelial ligand (hyaluronan) are essential for firm adhesion (Butcher *et al.*, 1999; Nandi *et al.*, 2004; Sallusto *et al.*, 2000; Weber, 2003). Thus, neutralizing antibodies to CD44 protect NOD mice from diabetes (Weiss *et al.*, 2000). Chemokines, expressed in a tissue-specific manner, mediate homing specificity by activating cognate G protein-coupled receptors on the T cells (Campbell and Butcher, 2000; Savinov *et al.*, 2003b). G protein-coupled chemokine receptors participate in integrin activation (Campbell *et al.*, 1998; Stein *et al.*, 2000; von Andrian and Mackay, 2000). The transition from firm adhesion to extravasation and diapedesis, which involves the cytoskeletal rearrangement of endothelial cells and the proteolytic degradation of the endothelial cell junctions and sub-endothelial basement membrane, is not yet completely understood. It is clear, however, that the enhancement of firm adhesion will decrease transmigration and diapedesis. Weak adhesion and the inability of T cells to establish firm adhesion will also result in diminished transmigration and diapedesis.

B. NOD mice and T1D

Mice of the NOD inbred strain develop a spontaneous disease closely resembling human T1D, and are widely and successfully used as a model of T1D (Gallegos and Bevan, 2004). $CD8^+$ T lymphocytes are involved in diabetogenesis in NOD mice. NOD mice lacking $CD8^+$ T cells do not develop diabetes (Serreze *et al.*, 1994). Prior work has demonstrated that the IS-$CD8^+$ T cell clone itself is highly diabetogenic, and, after injection, reliably causes the onset of diabetes in NOD mice in a matter of 5–7 days (Savinov *et al.*, 2003b; Wong *et al.*, 1999). After injection into animals, IS-$CD8^+$ T cells recapitulate the natural development of T1D in humans. These cells efficiently adhere to the endothelium and then transmigrate through the pancreatic endothelium. Transmigrated IS-$CD8^+$ T cells destroy β cells and cause diabetes in NOD mice. IS-$CD8^+$ T cells are a convenient, reliable and cost-effective tool to study the regulation of tissue-specific homing of diabetogenic T cells. The IS-$CD8^+$ T cells/NOD mice adoptive transfer model mimics the effector stage of T1D in humans. We believe that inhibitory approaches that use the IS-$CD8^+$ T cells/NOD mice model will greatly increase our understanding of T1D in humans.

IV. T Cell MT1-MMP and CD44 in T1D

A. The MT1-MMP/CD44 axis

Recently, we demonstrated that MT1-MMP dynamically regulates the functionality of the cell surface-associated signaling and adhesion receptor CD44 in diabetogenic IS-$CD8^+$ T cells (Savinov and Strongin, 2007; Savinov *et al.*, 2005, 2006, 2007). The importance of the MT1-MMP-CD44 axis in T1D has been identified both in diabetes transfer model NOD mice and in freshly diabetic NOD mice. By regulating the functionality of CD44, MT1-MMP mediates the transition of T cell adhesion to endothelial cells to the transmigration of T cells which controls the rate of homing of T cells into the pancreatic islets and thus controls the severity of the disease.

Shedding of cellular CD44 by external recombinant MT1-MMP reduces adhesion to the pancreatic endothelium and, as a result, reduces the level of transendothelial migration and the homing of IS-$CD8^+$ T cells into the islets. To establish the role of endogenous T cell MT1-MMP in the proteolysis of CD44 and in the adhesion, the transmigration and the homing of T cells into the pancreatic islets, we determined that endogenous MT1-MMP was latent in non-adherent IS-$CD8^+$ cells, while adhesion of IS-$CD8^+$ cells induced the activation of MT1-MMP, the cleavage of CD44 and the stimulation of T cell transmigration. According to our RT-PCR analysis of the gene expression in the adherent and non-adherent cells, three proprotein convertase family proteinases (furin, PC7 and PACE4) are up-regulated after the adhesion of IS-$CD8^+$ T cells to gelatin-coated plastic. PC7 and, especially, furin and PACE4, are potent in the proteolytic processing and the activation of the proenzyme of MT1-MMP in *in vitro* tests (Remacle *et al.*, 2006). Accordingly, we believe that these three proprotein convertases are largely responsible for the activation of MT1-MMP in the adherent T cells, thus leading to the MT1-MMP-dependent proteolysis of T cell CD44. We have also established that CD44 is the primary receptor involved in the adhesion of IS-$CD8^+$ T cells to the pancreatic endothelium. Thus, inhibition of the CD44 function by the anti-CD44 function-blocking monoclonal antibody led to a 75–80% decrease of IS-$CD8^+$ T cell adhesion leading, in turn, to a decrease of transmigration and homing. We have also demonstrated that MT1-MMP plays the primary role in shedding of T cell CD44. Thus, we have established that MT1-MMP is responsible for 90% of the shedding of CD44 observed in the adherent IS-$CD8^+$ T cells while the combined action of proteinases distinct from MT1-MMP is responsible for only 10% of the shedding.

In agreement with the primary role of MT1-MMP in the shedding of CD44, both AG3340 and GM6001 blocked the functional activity of MT1-MMP and the proteolysis of CD44 in the adherent IS-$CD8^+$ T cells. These events led to the aberrantly extended rather than the temporal adhesion of IS-$CD8^+$ T cells on the pancreatic endothelium. The continuous immobilization on the pancreatic endothelium impeded the transmigration efficiency of the diabetogenic T cells into the pancreas. Overall, the inhibition of MT1-MMP proteolysis of CD44 by either GM6001 or AG3340 reduced the diabetogenic efficiency of T cells by immobilizing the adherent diabetogenic, cytotoxic, T cells on the vascular endothelium, thus preventing T cell homing into the islets. The diminished transmigration led to decreased levels of the homing of IS-$CD8^+$ T cells and this event delayed the transferred diabetes onset in NOD mice which received AG3340 (Savinov and Strongin, 2007; Savinov *et al.*, 2005, 2006, 2007).

B. The specific role of T cell MT1-MMP in T1D

It was shown previously in the model of the onset of type 2 diabetes that MMP-2, MMP-12 and MT1-MMP were up-regulated in diabetic male and high-fat-fed female Zucker diabetic fatty rats as compared to their non-diabetic lean counterparts (Zhou *et al.*, 2005). PD166793 [(*S*)-2-(4′-bromo-biphenyl-4-sulfonylamino)-3-methyl-butyric acid] (a broad-range inhibitor with EC_{50} values of 6100, 47, 12, 7200, 7900, 8, and 240 nM against MMP-1, MMP-2, MMP-3, MMP-7, MMP-9, MMP-13, and MT1-MMP, respectively) (O'Brien *et al.*, 2000; Peterson *et al.*, 2001) preserved β cell mass, presumably, by decreasing the turnover of islet extracellular matrix molecules. The study suggested that pancreatic MMPs play a role in the maintenance of β cell mass. We, in turn, chose to focus our studies on the T cell MT1-MMP instead of on MMPs from the pancreata.

To validate the specific role of T cell MT1-MMP as well as to elucidate the potential significance of other MMPs in the NOD model of T1D, we used AG3340 and two additional inhibitors, EGCG and SB-3CT.

While both EGCG and SB-3CT are poor inhibitors of MT1-MMP, they are capable of targeting MMPs distinct from MT1-MMP. As a result of our cell-based assays and *in vivo* tests employing both a diabetes transfer model in NOD mice and *in vivo* visualization of IS-CD8$^+$ T cells pre-labeled with a fluorescence dye, didodecyl-tetramethylindocarbocyanine perchlorate, and then injected in NOD mice only AG3340, the antagonist of MT1-MMP, delivered clinically-relevant effects. In contrast, EGCG and SB-3CT, were without effect in our studies. Because of the wide-range specificity of the MMP inhibitors, a simultaneous assessment of AG3340, SB-3CT, and EGCG permitted us to conclude that only T cell MT1-MMP plays a significant role in T1D. The combined effect of all other MMPs, including MMP-2 and MMP-9, both of which are efficiently inhibited by SB-3CT, is far less important. We conclude that only MT1-MMP antagonists such as AG3340, as opposed to other broad-range inhibitors of MMPs, are efficient in delaying T1D transfer to NOD mice. We believe that these results confirm the functional importance of the MT1-MMP-CD44 axis in mediating the efficiency of transendothelial migration and the homing of diabetogenic T cells to the pancreatic islets.

C. Potential clinical relevance of targeting T cell MT1-MMP in T1D

Consistent with our biochemical model of the MT1-MMP-CD44 interactions, low dosages of AG3340 (1–5 mg/kg) injected jointly with insulin specifically inhibit T cell intra-islet transmigration, restore β cell functionality, increase insulin-producing β cell mass and alleviate the severity of TID in acutely diabetic NOD mice. As a result, acutely diabetic NOD mice do not require insulin injections for survival for a significant 30-day time period thus providing a promising clue to finding the means to reverse IDDM in humans. The extensive morphometric analyses and the measurements of both the C peptide blood levels and the proinsulin mRNA levels in the islets confirmed the highly beneficial effects of the inhibitor. Diabetes transfer experiments suggest that the inhibitor specifically represses the T cell transmigration and homing processes as opposed to causing general immunosuppression. According to relevant recent publications (Hao *et al.*, 2006) and in agreement with our observations, endothelial precursor stem cells instead of the β-cells themselves are the source of the regenerated, functional, β-cells.

To prove that insulin-producing β-cells were regenerated, NOD mice were allowed to develop T1D. Diseased mice then received insulin alone or insulin jointly with AG3340 for 40 days. Insulin injections were then suspended. Mice which received insulin after the onset of the disease became hyperglycemic in a matter of 2–3 days and were then sacrificed. In contrast, mice which received insulin jointly with the inhibitor restored the pool of insulin-producing β-cells. When insulin injections were cancelled, this β-cell pool was sufficient for the survival of these mice which continued to be normoglycemic/mildly hyperglycemic for several weeks without the use of external insulin.

In contrast to the exocrine pancreatic cells, β cells do not deposit a conventional basement membrane. Instead, they rely on the endothelial cells, which form capillaries with a vascular basement membrane next to the β cells (Nikolova *et al.*, 2006; Yoshitomi and Zaret, 2004). Endothelial basement membranes are rich in laminins, collagen type IV and fibronectin. It is noteworthy that the islet capsule represents an alternative, nonvascular source of laminin and collagen. Because of the spheroid shape of the islets, a large number of β cells is not in contact with the islet capsule, and, therefore, β cells are normally in contact with the vascular basement membrane (Olsson and Carlsson, 2006). Signals from the vascular basement membrane regulate the expression of insulin and control the proliferation of β cells and their progenitors (Duvillie *et al.*, 2002; Lammert *et al.*, 2001, 2003). Specifically, laminin-411 ($\alpha4\beta1\gamma1$) and laminin-511($\alpha5\beta1\gamma1$) are the major laminin species which are present in the vascular basement membrane (Hallmann *et al.*, 2005; Sixt *et al.*, 2001). Plating on laminin-111, -411, and -511 but not on collagen type I, collagen type IV or fibronectin, significantly increases the expression of both the *Ins1* and *Ins2* genes and stimulates the proliferation of murine β cells. Soluble laminins exhibit only a fraction of this effect, suggesting that the interaction of β cells with the laminin matrix is required for the stimulation of these processes (Nikolova *et al.*, 2006). The effects of laminins require the presence of $\beta1$ integrin in β cells. This integrin is predominantly represented in β cells by the $\alpha6\beta1$ integrin, which is a laminin receptor, and by the $\alpha1\beta1$ integrin, which is a collagen receptor (Kaido *et al.*, 2004). The function-blocking $\beta1$ antibody reduces the proliferative cell response and the expression of insulin in β cells plated on laminin-111. Furthermore, collagen type IV, which is normally secreted by the islet endothelial cells, interacts with β cell integrin $\alpha1\beta1$ and this interaction also stimulates insulin secretion by β cells (Kaido *et al.*, 2004). We suggest that the inhibitors of MT1-MMP rescue the laminin-stimulated $\beta1$ integrin signaling in β cells and that this event additionally contributes to survival, rejuvenation and insulin production by β cells (Sixt *et al.*, 2001; Jin *et al.*, 2007).

In summary, we are now confident that AG3340 provides diabetes protection by effectively controlling islet-destructive autoimmunity and stimulating the functional recovery of insulin-producing β-cells and the regeneration of the pancreatic islets, thus providing a sound mechanistic rationale for clinical trials of the inhibitors of MT1-MMP, including AG3340, in T1D in humans. Overall, our current and prior findings (Savinov and Strongin, 2007; Savinov *et al.*, 2005, 2006, 2007) provide a working hypothesis for the novel, antidiabetic, application of the existing inhibitors of MT1-MMP (Fig. 18.2). Our data suggest that inhibition of T cell MT1-MMP is a key to the design of novel and effective therapies of T1D. Because the inhibitor excess to MMPs localized in the poorly angiogenic tumors is limited relative to the cell-surface protease T cell protease, it is highly unlikely that the bioavailability of the T cell MT1-MMP targeting drugs will pose a problem in T1D treatment. Based on the results we obtained with NOD mice, we hypothesize that the pharmacological inhibition of MT1-MMP by specific antagonists, including AG3340, will diminish the homing of T killer cells into the islets and that this event will stimulate the regeneration of insulin-producing β cells leading to a favorable outcome for T1D patients (Chong *et al.*, 2006).

ACKNOWLEDGMENT

This work was supported by National Institutes of Health Grants CA83017, CA77470 and DK071956 (to Strongin).

REFERENCES

Annabi, B., Lachambre, M. P., Bousquet-Gagnon, N., Page, M., Gingras, D., and Beliveau, R. (2002). Green tea polyphenol (−)-epigallocatechin 3-gallate inhibits MMP-2 secretion and MT1-MMP-driven migration in glioblastoma cells. *Biochim. Biophys. Acta* **1542,** 209–220.

Aruffo, A., Stamenkovic, I., Melnick, M., Underhill, C. B., and Seed, B. (1990). CD44 is the principal cell surface receptor for hyaluronate. *Cell* **61,** 1303–1313.

Avigdor, A., Goichberg, P., Shivtiel, S., Dar, A., Peled, A., Samira, S., Kollet, O., Hershkoviz, R., Alon, R., Hardan, I., Ben-Hur, H., Naor, D., *et al.* (2004). CD44 and hyaluronic acid cooperate with SDF-1 in the trafficking of human CD34+ stem/progenitor cells to bone marrow. *Blood* **103,** 2981–2989.

Bennett, K. L., Modrell, B., Greenfield, B., Bartolazzi, A., Stamenkovic, I., Peach, R., Jackson, D. G., Spring, F., and Aruffo, A. (1995). Regulation of CD44 binding to hyaluronan by glycosylation of variably spliced exons. *J. Cell Biol.* **131,** 1623–1633.

Bonfil, R. D., Sabbota, A., Nabha, S., Bernardo, M. M., Dong, Z., Meng, H., Yamamoto, H., Chinni, S. R., Lim, I. T., Chang, M., Filetti, L. C., Mobashery, S., *et al.* (2006). Inhibition of human prostate cancer growth, osteolysis and angiogenesis in a bone metastasis model by a novel mechanism-based selective gelatinase inhibitor. *Int. J. Cancer* **118,** 2721–2726.

Butcher, E. C., and Picker, L. J. (1996). Lymphocyte homing and homeostasis. *Science* **272,** 60–66.

Butcher, E. C., Williams, M., Youngman, K., Rott, L., and Briskin, M. (1999). Lymphocyte trafficking and regional immunity. *Adv. Immunol.* **72,** 209–253.

Campbell, J. J., and Butcher, E. C. (2000). Chemokines in tissue-specific and microenvironment-specific lymphocyte homing. *Curr. Opin. Immunol.* **12,** 336–341.

Campbell, J. J., Hedrick, J., Zlotnik, A., Siani, M. A., Thompson, D. A., and Butcher, E. C. (1998). Chemokines and the arrest of lymphocytes rolling under flow conditions. *Science* **279,** 381–384.

Cappuzzo, F., Bartolini, S., and Crino, L. (2003). Emerging drugs for non-small cell lung cancer. *Expert Opin. Emerg. Drugs* **8,** 179–192.

Celenza, G., Villegas-Estrada, A., Lee, M., Boggess, B., Forbes, C., Wolter, W. R., Suckow, M. A., Mobashery, S., and Chang, M. (2008). Metabolism of (4-phenoxyphenylsulfonyl) methylthiirane, a selective gelatinase inhibitor. *Chem. Biol. Drug. Des.* **71,** 187–196.

Chiang, W. C., Wong, Y. K., Lin, S. C., Chang, K. W., and Liu, C. J. (2006). Increase of MMP-13 expression in multi-stage oral carcinogenesis and epigallocatechin-3-gallate suppress MMP-13 expression. *Oral. Dis.* **12,** 27–33.

Chong, A. S., Shen, J., Tao, J., Yin, D., Kuznetsov, A., Hara, M., and Philipson, L. H. (2006). Reversal of diabetes in non-obese diabetic mice without spleen cell-derived beta cell regeneration. *Science* **311,** 1774–1775.

Cichy, J., and Pure, E. (2003). The liberation of CD44. *J. Cell Biol.* **161,** 839–843.

Constantin, G., Majeed, M., Giagulli, C., Piccio, L., Kim, J. Y., Butcher, E. C., and Laudanna, C. (2000). Chemokines trigger immediate beta2 integrin affinity and mobility changes: Differential regulation and roles in lymphocyte arrest under flow. *Immunity* **13,** 759–769.

DeGrendele, H. C., Estess, P., and Siegelman, M. H. (1997). Requirement for CD44 in activated T cell extravasation into an inflammatory site. *Science* **278,** 672–675.

Dell'Aica, I., Dona, M., Sartor, L., Pezzato, E., and Garbisa, S. (2002). (−)Epigallocatechin-3-gallate directly inhibits MT1-MMP activity, leading to accumulation of nonactivated MMP-2 at the cell surface. *Lab. Invest.* **82,** 1685–1693.

Deryugina, E. I., Ratnikov, B. I., Yu, Q., Baciu, P. C., Rozanov, D. V., and Strongin, A. Y. (2004). Prointegrin maturation follows rapid trafficking and processing of MT1-MMP in Furin-Negative Colon Carcinoma LoVo Cells. *Traffic* **5,** 627–641.

Duvillie, B., Currie, C., Chrones, T., Bucchini, D., Jami, J., Joshi, R. L., and Hill, D. J. (2002). Increased islet cell proliferation, decreased apoptosis, and greater vascularization leading to beta-cell hyperplasia in mutant mice lacking insulin. *Endocrinology* **143,** 1530–1537.

Egeblad, M., and Werb, Z. (2002). New functions for the matrix metalloproteinases in cancer progression. *Nat. Rev. Cancer* **2,** 161–174.

El Bedoui, J., Oak, M. H., Anglard, P., and Schini-Kerth, V. B. (2005). Catechins prevent vascular smooth muscle cell invasion by inhibiting MT1-MMP activity and MMP-2 expression. *Cardiovasc. Res.* **67,** 317–325.

Estess, P., DeGrendele, H. C., Pascual, V., and Siegelman, M. H. (1998). Functional activation of lymphocyte CD44 in peripheral blood is a marker of autoimmune disease activity. *J. Clin. Invest.* **102,** 1173–1182.

Estess, P., Nandi, A., Mohamadzadeh, M., and Siegelman, M. H. (1999). Interleukin 15 induces endothelial hyaluronan expression *in vitro* and promotes activated T cell extravasation through a CD44-dependent pathway *in vivo*. *J. Exp. Med.* **190,** 9–19.

Gallegos, A. M., and Bevan, M. J. (2004). Driven to autoimmunity: The nod mouse. *Cell* **117,** 149–151.

Galvez, B. G., Matias-Roman, S., Yanez-Mo, M., Vicente-Manzanares, M., Sanchez-Madrid, F., and Arroyo, A. G. (2004). Caveolae are a novel pathway for membrane-type 1 matrix metalloproteinase traffic in human endothelial cells. *Mol. Biol. Cell* **15,** 678–687.

Gomis-Ruth, F. X. (2003). Structural aspects of the metzincin clan of metalloendopeptidases. *Mol. Biotechnol.* **24,** 157–202.

Gross, J., and Lapiere, C. M. (1962). Collagenolytic activity in amphibian tissues: A tissue culture assay. *Proc. Natl. Acad. Sci. USA* **48,** 1014–1022.

Gu, Z., Cui, J., Brown, S., Fridman, R., Mobashery, S., Strongin, A. Y., and Lipton, S. A. (2005). A highly specific inhibitor of matrix metalloproteinase-9 rescues laminin from proteolysis and neurons from apoptosis in transient focal cerebral ischemia. *J. Neurosci.* **25,** 6401–6408.

Hallmann, R., Horn, N., Selg, M., Wendler, O., Pausch, F., and Sorokin, L. M. (2005). Expression and function of laminins in the embryonic and mature vasculature. *Physiol. Rev.* **85,** 979–1000.

Hande, K. R., Collier, M., Paradiso, L., Stuart-Smith, J., Dixon, M., Clendeninn, N., Yeun, G., Alberti, D., Binger, K., and Wilding, G. (2004). Phase I and pharmacokinetic study of prinomastat, a matrix metalloprotease inhibitor. *Clin. Cancer Res.* **10,** 909–915.

Hao, E., Tyrberg, B., Itkin-Ansari, P., Lakey, J. R., Geron, I., Monosov, E. Z., Barcova, M., Mercola, M., and Levine, F. (2006). Beta-cell differentiation from nonendocrine epithelial cells of the adult human pancreas. *Nat. Med.* **12,** 310–316.

Holmbeck, K., Bianco, P., Caterina, J., Yamada, S., Kromer, M., Kuznetsov, S. A., Mankani, M., Robey, P. G., Poole, A. R., Pidoux, I., Ward, J. M., and Birkedal-Hansen, H. (1999). MT1-MMP-deficient mice develop dwarfism, osteopenia, arthritis, and connective tissue disease due to inadequate collagen turnover. *Cell* **99,** 81–92.

Holmbeck, K., Bianco, P., Chrysovergis, K., Yamada, S., and Birkedal-Hansen, H. (2003). MT1-MMP-dependent, apoptotic remodeling of unmineralized cartilage: A critical process in skeletal growth. *J. Cell Biol.* **163,** 661–671.

Holmbeck, K., Bianco, P., Yamada, S., and Birkedal-Hansen, H. (2004). MT1-MMP: A tethered collagenase. *J. Cell. Physiol.* **200,** 11–19.

Ikejiri, M., Bernardo, M. M., Bonfil, R. D., Toth, M., Chang, M., Fridman, R., and Mobashery, S. (2005a). Potent mechanism-based inhibitors for matrix metalloproteinases. *J. Biol. Chem.* **280,** 33992–34002.

Ikejiri, M., Bernardo, M. M., Meroueh, S. O., Brown, S., Chang, M., Fridman, R., and Mobashery, S. (2005b). Design, synthesis, and evaluation of a mechanism-based inhibitor for gelatinase A. *J. Org. Chem.* **70,** 5709–5712.

Itoh, Y., and Seiki, M. (2006). MT1-MMP: A potent modifier of pericellular microenvironment. *J. Cell. Physiol.* **206,** 1–8.

Jiang, A., Lehti, K., Wang, X., Weiss, S. J., Keski-Oja, J., and Pei, D. (2001). Regulation of membrane-type matrix metalloproteinase 1 activity by dynamin-mediated endocytosis. *Proc. Natl. Acad. Sci. USA* **98,** 13693–13698.

Jin, E. J., Choi, Y. A., Kyun Park, E., Bang, O. S., and Kang, S. S. (2007). MMP-2 functions as a negative regulator of chondrogenic cell condensation via down-regulation of the FAK-integrin beta1 interaction. *Dev. Biol.* **308,** 474–484.

Kaido, T., Yebra, M., Cirulli, V., and Montgomery, A. M. (2004). Regulation of human beta-cell adhesion, motility, and insulin secretion by collagen IV and its receptor alpha1-beta1. *J. Biol. Chem.* **279,** 53762–53769.

Kajita, M., Itoh, Y., Chiba, T., Mori, H., Okada, A., Kinoh, H., and Seiki, M. (2001). Membrane-type 1 matrix metalloproteinase cleaves CD44 and promotes cell migration. *J. Cell Biol.* **153,** 893–904.

Katoh, S., Zheng, Z., Oritani, K., Shimozato, T., and Kincade, P. W. (1995). Glycosylation of CD44 negatively regulates its recognition of hyaluronan. *J. Exp. Med.* **182,** 419–429.

Kim, H. S., Kim, M. H., Jeong, M., Hwang, Y. S., Lim, S. H., Shin, B. A., Ahn, B. W., and Jung, Y. D. (2004). EGCG blocks tumor promoter-induced MMP-9 expression via suppression of MAPK and AP-1 activation in human gastric AGS cells. *Anticancer Res.* **24,** 747–753.

Knauper, V., Bailey, L., Worley, J. R., Soloway, P., Patterson, M. L., and Murphy, G. (2002). Cellular activation of proMMP-13 by MT1-MMP depends on the C-terminal domain of MMP-13. *FEBS Lett.* **532,** 127–130.

Labrecque, L., Nyalendo, C., Langlois, S., Durocher, Y., Roghi, C., Murphy, G., Gingras, D., and Beliveau, R. (2004). Src-mediated tyrosine phosphorylation of

caveolin-1 induces its association with membrane type 1 matrix metalloproteinase. *J. Biol. Chem.* **279,** 52132–52140.
Lammert, E., Cleaver, O., and Melton, D. (2001). Induction of pancreatic differentiation by signals from blood vessels. *Science* **294,** 564–567.
Lammert, E., Gu, G., McLaughlin, M., Brown, D., Brekken, R., Murtaugh, L. C., Gerber, H. P., Ferrara, N., and Melton, D. A. (2003). Role of VEGF-A in vascularization of pancreatic islets. *Curr. Biol.* **13,** 1070–1074.
Lee, J. H., Chung, J. H., and Cho, K. H. (2005). The effects of epigallocatechin-3-gallate on extracellular matrix metabolism. *J. Dermatol. Sci.* **40,** 195–204.
Lee, M., Hesek, D., Shi, Q., Noll, B. C., Fisher, J. F., Chang, M., and Mobashery, S. (2007). Conformational analyses of thiirane-based gelatinase inhibitors. *Bioorg. Med. Chem. Lett* **18,** 3064–6067.
Lesley, J., Hyman, R., and Kincade, P. W. (1993). CD44 and its interaction with extracellular matrix. *Adv. Immunol.* **54,** 271–335.
McQuibban, G. A., Gong, J. H., Tam, E. M., McCulloch, C. A., Clark-Lewis, I., and Overall, C. M. (2000). Inflammation dampened by gelatinase A cleavage of monocyte chemoattractant protein-3. *Science* **289,** 1202–1206.
Mohamadzadeh, M., DeGrendele, H., Arizpe, H., Estess, P., and Siegelman, M. (1998). Proinflammatory stimuli regulate endothelial hyaluronan expression and CD44/HA-dependent primary adhesion. *J. Clin. Invest.* **101,** 97–108.
Mori, H., Tomari, T., Koshikawa, N., Kajita, M., Itoh, Y., Sato, H., Tojo, H., Yana, I., and Seiki, M. (2002). CD44 directs membrane-type 1 matrix metalloproteinase to lamellipodia by associating with its hemopexin-like domain. *EMBO J.* **21,** 3949–3959.
Murai, S., Umemiya, T., Seiki, M., and Harigaya, K. (2004). Expression and localization of membrane-type-1 matrix metalloproteinase, CD 44, and laminin-5gamma2 chain during colorectal carcinoma tumor progression. *Virchows Arch.* **445,** 271–278.
Murphy, G., Stanton, H., Cowell, S., Butler, G., Knauper, V., Atkinson, S., and Gavrilovic, J. (1999). Mechanisms for pro matrix metalloproteinase activation. *APMIS* **107,** 38–44.
Nagase, H., and Woessner, J. F. Jr. (1999). Matrix metalloproteinases. *J. Biol. Chem.* **274,** 21491–21494.
Nagase, H., Visse, R., and Murphy, G. (2006). Structure and function of matrix metalloproteinases and TIMPs. *Cardiovasc. Res.* **69,** 562–573.
Nakamura, H., Suenaga, N., Taniwaki, K., Matsuki, H., Yonezawa, K., Fujii, M., Okada, Y., and Seiki, M. (2004). Constitutive and induced CD44 shedding by ADAM-like proteases and membrane-type 1 matrix metalloproteinase. *Cancer Res.* **64,** 876–882.
Nandi, A., Estess, P., and Siegelman, M. (2004). Bimolecular complex between rolling and firm adhesion receptors required for cell arrest; CD44 association with VLA-4 in T cell extravasation. *Immunity* **20,** 455–465.
Nikolova, G., Jabs, N., Konstantinova, I., Domogatskaya, A., Tryggvason, K., Sorokin, L., Fassler, R., Gu, G., Gerber, H. P., Ferrara, N., Melton, D. A., and Lammert, E. (2006). The vascular basement membrane: A niche for insulin gene expression and Beta cell proliferation. *Dev. Cell* **10,** 397–405.
O'Brien, P. M., Ortwine, D. F., Pavlovsky, A. G., Picard, J. A., Sliskovic, D. R., Roth, B. D., Dyer, R. D., Johnson, L. L., Man, C. F., and Hallak, H. (2000). Structure-activity relationships and pharmacokinetic analysis for a series of potent, systemically available biphenylsulfonamide matrix metalloproteinase inhibitors. *J. Med. Chem.* **43,** 156–166.
Oh, J., Takahashi, R., Adachi, E., Kondo, S., Kuratomi, S., Noma, A., Alexander, D. B., Motoda, H., Okada, A., Seiki, M., Itoh, T., Itohara, S., *et al.* (2004). Mutations in two

matrix metalloproteinase genes, MMP-2 and MT1-MMP, are synthetic lethal in mice. *Oncogene* **23,** 5041–5048.

Olsson, R., and Carlsson, P. O. (2006). The pancreatic islet endothelial cell: Emerging roles in islet function and disease. *Int. J. Biochem. Cell Biol.* **38,** 492–497.

Osenkowski, P., Toth, M., and Fridman, R. (2004). Processing, shedding, and endocytosis of membrane type 1-matrix metalloproteinase (MT1-MMP). *J. Cell. Physiol.* **200,** 2–10.

Pei, D., and Weiss, S. J. (1995). Furin-dependent intracellular activation of the human stromelysin-3 zymogen. *Nature* **375,** 244–247.

Peterson, J. T., Hallak, H., Johnson, L., Li, H., O'Brien, P. M., Sliskovic, D. R., Bocan, T. M., Coker, M. L., Etoh, T., and Spinale, F. G. (2001). Matrix metalloproteinase inhibition attenuates left ventricular remodeling and dysfunction in a rat model of progressive heart failure. *Circulation* **103,** 2303–2309.

Remacle, A. G., Rozanov, D. V., Fugere, M., Day, R., and Strongin, A. Y. (2006). Furin regulates the intracellular activation and the uptake rate of cell surface-associated MT1-MMP. *Oncogene* **25,** 5648–5655.

Rozanov, D. V., Deryugina, E. I., Monosov, E. Z., Marchenko, N. D., and Strongin, A. Y. (2004). Aberrant, persistent inclusion into lipid rafts limits the tumorigenic function of membrane type-1 matrix metalloproteinase in malignant cells. *Exp. Cell Res.* **293,** 81–95.

Sallusto, F., Mackay, C. R., and Lanzavecchia, A. (2000). The role of chemokine receptors in primary, effector, and memory immune responses. *Annu. Rev. Immunol.* **18,** 593–620.

Sato, H., Takino, T., Okada, Y., Cao, J., Shinagawa, A., Yamamoto, E., and Seiki, M. (1994). A matrix metalloproteinase expressed on the surface of invasive tumour cells. *Nature* **370,** 61–65.

Savinov, A. Y., and Strongin, A. Y. (2007). Defining the roles of T cell membrane proteinase and CD44 in type 1 diabetes. *IUBMB Life* **59,** 6–13.

Savinov, A. Y., Tcherepanov, A., Green, E. A., Flavell, R. A., and Chervonsky, A. V. (2003a). Contribution of Fas to diabetes development. *Proc. Natl. Acad. Sci. USA* **100,** 628–632.

Savinov, A. Y., Wong, F. S., Stonebraker, A. C., and Chervonsky, A. V. (2003b). Presentation of antigen by endothelial cells and chemoattraction are required for homing of insulin-specific CD8+ T cells. *J. Exp. Med.* **197,** 643–656.

Savinov, A. Y., Rozanov, D. V., Golubkov, V. S., Wong, F. S., and Strongin, A. Y. (2005). Inhibition of membrane type-1 matrix metalloproteinase by cancer drugs interferes with the homing of diabetogenic T cells into the pancreas. *J. Biol. Chem.* **280,** 27755–27758.

Savinov, A. Y., Rozanov, D. V., and Strongin, A. Y. (2006). Mechanistic insights into targeting T cell membrane proteinase to promote islet beta-cell rejuvenation in type 1 diabetes. *FASEB J.* **20,** 1793–1801.

Savinov, A. Y., Rozanov, D. V., and Strongin, A. Y. (2007). Specific inhibition of autoimmune T cell transmigration contributes to beta cell functionality and insulin synthesis in non-obese diabetic (NOD) mice. *J. Biol. Chem.* **282,** 32106–32111.

Seiki, M. (2003). Membrane-type 1 matrix metalloproteinase: A key enzyme for tumor invasion. *Cancer Lett.* **194,** 1–11.

Serreze, D. V., Leiter, E. H., Christianson, G. J., Greiner, D., and Roopenian, D. C. (1994). Major histocompatibility complex class I-deficient NOD-B2mnull mice are diabetes and insulitis resistant. *Diabetes* **43,** 505–509.

Shalinsky, D. R., Brekken, J., Zou, H., McDermott, C. D., Forsyth, P., Edwards, D., Margosiak, S., Bender, S., Truitt, G., Wood, A., Varki, N. M., and Appelt, K. (1999). Broad antitumor and antiangiogenic activities of AG3340, a potent and selective MMP inhibitor undergoing advanced oncology clinical trials. *Ann. NY Acad. Sci.* **878,** 236–270.

Sixt, M., Engelhardt, B., Pausch, F., Hallmann, R., Wendler, O., and Sorokin, L. M. (2001). Endothelial cell laminin isoforms, laminins 8 and 10, play decisive roles in T cell recruitment across the blood-brain barrier in experimental autoimmune encephalomyelitis. *J. Cell Biol.* **153,** 933–946.

Slevin, M., Kumar, S., and Gaffney, J. (2002). Angiogenic oligosaccharides of hyaluronan induce multiple signaling pathways affecting vascular endothelial cell mitogenic and wound healing responses. *J. Biol. Chem.* **277,** 41046–41059.

Sodek, K. L., Ringuette, M. J., and Brown, T. J. (2007). MT1-MMP is the critical determinant of matrix degradation and invasion by ovarian cancer cells. *Br. J. Cancer* **97,** 358–367.

Song, X. Z., Xia, J. P., and Bi, Z. G. (2004). Effects of (−)-epigallocatechin-3-gallate on expression of matrix metalloproteinase-1 and tissue inhibitor of metalloproteinase-1 in fibroblasts irradiated with ultraviolet A. *Chin. Med. J.* **117,** 1838–1841.

Springer, T. A. (1994). Traffic signals for lymphocyte recirculation and leukocyte emigration: The multistep paradigm. *Cell* **76,** 301–314.

Stein, J. V., Rot, A., Luo, Y., Narasimhaswamy, M., Nakano, H., Gunn, M. D., Matsuzawa, A., Quackenbush, E. J., Dorf, M. E., and von Andrian, U. H. (2000). The CC chemokine thymus-derived chemotactic agent 4 (TCA-4, secondary lymphoid tissue chemokine, 6Ckine, exodus-2) triggers lymphocyte function-associated antigen 1-mediated arrest of rolling T lymphocytes in peripheral lymph node high endothelial venules. *J. Exp. Med.* **191,** 61–76.

Sternlicht, M. D., and Werb, Z. (2001). How matrix metalloproteinases regulate cell behavior. *Annu. Rev. Cell Dev. Biol.* **17,** 463–516.

Stocker, W., and Bode, W. (1995). Structural features of a superfamily of zinc-endopeptidases: The metzincins. *Curr. Opin. Struct. Biol.* **5,** 383–390.

Strongin, A. Y. (2006). Mislocalization and unconventional functions of cellular MMPs in cancer. *Cance.r Metastasis Rev.* **25,** 87–98.

Strongin, A. Y., Collier, I., Bannikov, G., Marmer, B. L., Grant, G. A., and Goldberg, G. I. (1995). Mechanism of cell surface activation of 72-kDa type IV collagenase. Isolation of the activated form of the membrane metalloprotease. *J. Biol. Chem.* **270,** 5331–5338.

Suenaga, N., Mori, H., Itoh, Y., and Seiki, M. (2005). CD44 binding through the hemopexin-like domain is critical for its shedding by membrane-type 1 matrix metalloproteinase. *Oncogene* **24,** 859–868.

Turley, E. A., Noble, P. W., and Bourguignon, L. Y. (2002). Signaling properties of hyaluronan receptors. *J. Biol. Chem.* **277,** 4589–4592.

Van Wart, H. E., and Birkedal-Hansen, H. (1990). The cysteine switch: A principle of regulation of metalloproteinase activity with potential applicability to the entire matrix metalloproteinase gene family. *Proc. Natl. Acad. Sci. USA* **87,** 5578–5582.

Vayalil, P. K., and Katiyar, S. K. (2004). Treatment of epigallocatechin-3-gallate inhibits matrix metalloproteinases-2 and -9 via inhibition of activation of mitogen-activated protein kinases, c-jun and NF-kappaB in human prostate carcinoma DU-145 cells. *Prostate* **59,** 33–42.

Verdaguer, J., Schmidt, D., Amrani, A., Anderson, B., Averill, N., and Santamaria, P. (1997). Spontaneous autoimmune diabetes in monoclonal T cell nonobese diabetic mice. *J. Exp. Med.* **186,** 1663–1676.

von Andrian, U. H., and Mackay, C. R. (2000). T-cell function and migration. Two sides of the same coin. *N. Engl. J. Med.* **343,** 1020–1034.

Wang, P., Wang, X., and Pei, D. (2004a). Mint-3 regulates the retrieval of the internalized membrane-type matrix metalloproteinase, MT5-MMP, to the plasma membrane by binding to its carboxyl end motif EWV. *J. Biol. Chem.* **279,** 20461–20470.

Wang, X., Ma, D., Keski-Oja, J., and Pei, D. (2004b). Co-recycling of MT1-MMP and MT3-MMP Through the trans-Golgi Network: Identification of DKV582 as a Recycling Signal. *J. Biol. Chem.* **279,** 9331–9336.

Weber, C. (2003). Novel mechanistic concepts for the control of leukocyte transmigration: Specialization of integrins, chemokines, and junctional molecules. *J. Mol. Med.* **81,** 4–19.

Weiss, L., Slavin, S., Reich, S., Cohen, P., Shuster, S., Stern, R., Kaganovsky, E., Okon, E., Rubinstein, A. M., and Naor, D. (2000). Induction of resistance to diabetes in non-obese diabetic mice by targeting CD44 with a specific monoclonal antibody. *Proc. Natl. Acad. Sci. USA* **97,** 285–290.

Will, H., Atkinson, S. J., Butler, G. S., Smith, B., and Murphy, G. (1996). The soluble catalytic domain of membrane type 1 matrix metalloproteinase cleaves the propeptide of progelatinase A and initiates autoproteolytic activation. Regulation by TIMP-2 and TIMP-3. *J. Biol. Chem.* **271,** 17119–17123.

Wong, F. S., Visintin, I., Wen, L., Flavell, R. A., and Janeway, C. A. Jr. (1996). CD8 T cell clones from young nonobese diabetic (NOD) islets can transfer rapid onset of diabetes in NOD mice in the absence of CD4 cells. *J. Exp. Med.* **183,** 67–76.

Wong, F. S., Karttunen, J., Dumont, C., Wen, L., Visintin, I., Pilip, I. M., Shastri, N., Pamer, E. G., and Janeway, C. A. Jr. (1999). Identification of an MHC class I-restricted autoantigen in type 1 diabetes by screening an organ-specific cDNA library. *Nat. Med.* **5,** 1026–1031.

Worthylake, R. A., and Burridge, K. (2001). Leukocyte transendothelial migration: Orchestrating the underlying molecular machinery. *Curr. Opin. Cell Biol.* **13,** 569–577.

Yana, I., and Weiss, S. J. (2000). Regulation of membrane type-1 matrix metalloproteinase activation by proprotein convertases. *Mol. Biol. Cell* **11,** 2387–2401.

Yang, X. D., Michie, S. A., Mebius, R. E., Tisch, R., Weissman, I., and McDevitt, H. O. (1996). The role of cell adhesion molecules in the development of IDDM: Implications for pathogenesis and therapy. *Diabetes* **45,** 705–710.

Yoshitomi, H., and Zaret, K. S. (2004). Endothelial cell interactions initiate dorsal pancreas development by selectively inducing the transcription factor Ptf1a. *Development* **131,** 807–817.

Zhou, Y. P., Madjidi, A., Wilson, M. E., Nothhelfer, D. A., Johnson, J. H., Palma, J. F., Schweitzer, A., Burant, C., Blume, J. E., and Johnson, J. D. (2005). Matrix metalloproteinases contribute to insulin insufficiency in Zucker diabetic fatty rats. *Diabetes* **54,** 2612–2619.

Zucker, S., Cao, J., and Chen, W. T. (2000). Critical appraisal of the use of matrix metalloproteinase inhibitors in cancer treatment. *Oncogene* **19,** 6642–6650.

Zucker, S., Pei, D., Cao, J., and Lopez-Otin, C. (2003). Membrane type-matrix metalloproteinases (MT-MMP). *Curr. Top. Dev. Biol.* **54,** 1–74.

CHAPTER NINETEEN

Role of Wnt Signaling in the Development of Type 2 Diabetes

Michael Bordonaro

Contents

Abstract

Type 2 diabetes is characterized by insulin resistance, insulin deficiency, and hyperglycemia. Susceptibility to type 2 diabetes has been linked to Wnt signaling, which plays an important role in intestinal tumorigenesis. Carriers of variants of the transcription factor 7-like 2 gene, an important component of the Wnt pathway, are at enhanced risk for developing type 2 diabetes. The modulation of proglucagon expression by Wnt activity may partially explain the link between Wnt signaling and diabetes, and one of the transcriptional

The Commonwealth Medical College, Department of Basic Sciences, Scranton, Pennsylvania 18510

Vitamins and Hormones, Volume 80
ISSN 0083-6729, DOI: 10.1016/S0083-6729(08)00619-5

and processing products of the proglucagon gene, the glucagon-like peptide-1 (GLP-1), exhibits a wide variety of antidiabetogenic activities. GLP-1 stimulates Wnt signaling in pancreatic beta cells, enhancing cell proliferation; thus, positive feedback between GLP-1 and Wnt signaling may result in increased proliferation, and suppressed apoptosis, of pancreatic cells. Since beta-cell protection is a potential treatment for type 2 diabetes, stimulation of Wnt activity may represent a valid therapeutic approach.

I. Introduction

Type 2 diabetes is a metabolic disorder that is characterized by insulin resistance, insulin deficiency, and hyperglycemia. Type 2 diabetes is clinically manifested when the decreased levels of insulin secretion are unable to meet the demands of impaired insulin sensitivity, likely due to diminished pancreatic beta-cell numbers and/or deficient beta-cell function (Shu *et al.*, 2008 and references therein). Diabetes afflicts ~150 million persons worldwide and that number is expected to double by 2025 (Zimmet *et al.*, 2001). Due to increasing obesity, type 2 diabetes is becoming more prevalent among adolescents and young adults; in addition, there is significant heritability for the development of this disorder (Grant *et al.*, 2006; Helgason *et al.*, 2007; Zimmet *et al.*, 2001). The increasing health cost incurred to society by this disorder necessitates the development of new therapies.

The possibility for new treatment methodologies is suggested by reports that have established a link between type 2 diabetes and Wnt signaling. The breakthrough in establishing this link came in 2006, when deCODE Genetics identified an association between variants of the transcription factor 7-like 2 (*TCF7L2*) gene, which encodes the transcription factor Tcf4, an important component of Wnt signaling, and increased risk for the development of type 2 diabetes (Grant *et al.*, 2006). Subsequent studies demonstrated that Wnt signaling activity was linked to the production of the incretin hormone GLP-1 in intestinal L cells as well as to proliferation of pancreatic beta cells (discussed below). Wnt activity is associated with insulin/incretin deficiency, as opposed to directly influencing insulin resistance *per se;* nevertheless, the association between Wnt factor variants and risk for type 2 diabetes suggests that Wnt signaling defects are partially responsible for the diabetogenic phenotype. A more thorough understanding of the possible relationship between Wnt activity and the development of type 2 diabetes may lead to novel approaches for the treatment of this disorder.

II. Wnt Signaling

A. Wnt signaling pathway

The Wnt signaling pathway (Fig.19.1) is integrated into many fundamental cellular functions, including proliferation and maintenance of stem cells (e.g., the intestinal stem cell compartment), development and organogenesis; when deregulated through mutation Wnt activity promotes cancer, particularly colorectal cancer (Kinzler and Vogelstein, 1996; Korinek *et al.*, 1997). Wnt signaling is normally induced by the binding of Wnt ligands to their transmembrane frizzled and LRP5/6 cell surface receptors; acting through an association between these receptors and downstream effectors, the Wnt signal disrupts a degradation complex containing APC, Axin, glycogen synthase kinase-3 beta (GSK-3β), beta-catenin, and other proteins. This degradation complex normally phosphorylates beta-catenin and targets it for proteosomal degradation; however, subsequent to Wnt-mediated disruption of the complex, GSK-3β activity is inhibited (Cook *et al.*, 1996; Korinek *et al.*, 1997; Woodgett, 1994), resulting in accumulation of active, dephosphorylated beta-catenin (Korinek *et al.*, 1997; Su *et al.*, 1993). Dephosphorylated beta-catenin interacts with Tcf/Lef DNA binding proteins, Tcf4 being the predominant Tcf/Lef factor in intestinal cells (Korinek *et al.*, 1997). Beta-catenin-Tcf (BCT) transcriptional complexes are detected by their ability to drive transcription from Tcf site-containing promoter constructs (Korinek *et al.*, 1997), and, upon binding to Tcf sites in DNA, activate transcription of Wnt target genes, including genes whose products influence decisions of cell proliferation, differentiation, and apoptosis.

B. Background: Wnt signaling pathway modulation and colorectal cancer

Constitutively activated canonical Wnt signaling is believed to promote cell proliferation and tumorigenesis in the colon. Most human colorectal cancers are initiated by mutations in the Wnt signaling pathway, predominantly by mutation in the adenomatous polyposis coli gene (*APC*, which encodes the APC protein involved in the beta-catenin degradation complex), or, less frequently, in the beta-catenin gene itself. The result of these mutations is constitutive activation of Wnt signaling, leading to the initiation of colorectal neoplasia (Kinzler and Vogelstein, 1996; Korinek *et al.*, 1997).

Our group has shown that sodium butyrate (NaB) and other inhibitors of histone deacetylases (HDACis) hyper-activate Wnt transcriptional activity in CRC cells, partly by initiating an increase in the levels of Ser-37/Thr-41-dephosphorylated (active) beta-catenin, an effect which occurs at

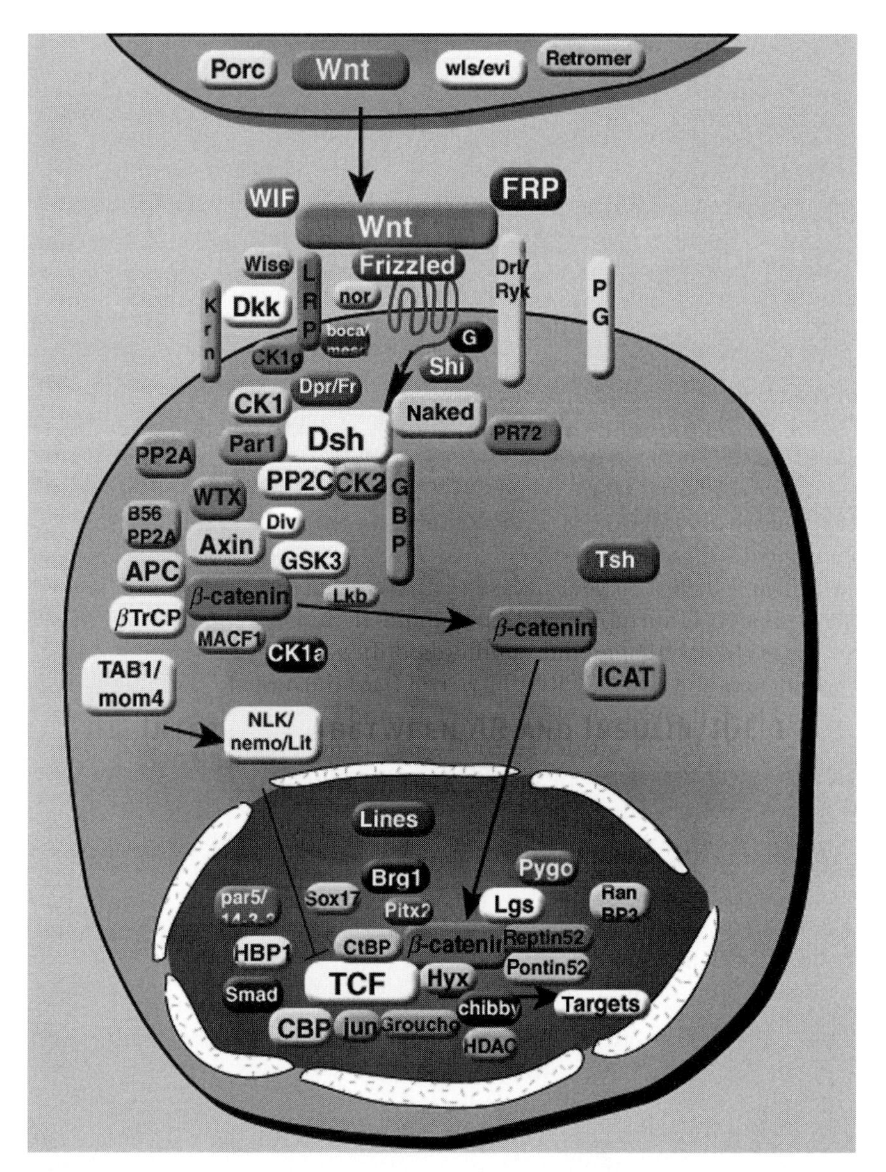

Figure 19.1 Schematic of Wnt signaling. In the absence of Wnt signaling, free intracellular beta-catenin is bound within complexes containing APC, Axin, GSK-3β, and other proteins; beta-catenin is phosphorylated and degraded through the proteosomal pathway. Wnt signaling is normally activated by the interactions of Wnt ligands with their receptors at the plasma membrane. This interaction transmits a signal which results in the disruption of the beta-catenin degradation complex, leading to the accumulation of dephosphorylated, active beta-catenin. The nuclear accumulation of active beta-catenin allows for the formation of a transcriptional complex between beta-catenin and the DNA-binding Tcf factors. In the absence of beta-catenin, Tcf factors

the level of Wnt ligand–receptor interactions at the plasma membrane (Bordonaro *et al.*, 2007). This hyperactivation of Wnt activity in cells that already possess deregulated Wnt signaling induces apoptosis (Bordonaro *et al.*, 2007). Although this finding was unexpected, in that increased Wnt activity is usually associated with enhanced proliferation, the observation was subsequently confirmed in a more comprehensive study, which established a linear relationship between the induction of Wnt transcriptional activity and the levels of apoptosis and the inhibition of clonal growth in ten human CRC cell lines treated with NaB (Lazarova *et al.*, 2004).

The causative relationship between induction of Wnt activity and the level of apoptosis was analyzed in NaB treated CRC cells expressing dominant negative Tcf4, an amino terminal truncated form of Tcf4 which does not bind beta-catenin and which inhibits the activity of endogenous BCT complexes. Dominant negative Tcf4 suppressed the induction of Wnt transcriptional activity by NaB and resulted in reduced levels of apoptosis, suggesting that NaB induced apoptosis of CRC cells is partially dependent upon the hyperactivation of Wnt signaling (Lazarova *et al.*, 2004). In addition to butyrate, other structurally unrelated HDACis, including trichostatin A, suberoylanilide hydroxamic acid, and MS275 also hyperactivate Wnt activity in CRC cells at concentrations that result in apoptotic levels similar to those induced by NaB (Bordonaro *et al.*, 2007).

While the majority of published studies associate Wnt activity with proliferation, several *in vivo* studies support the relationship that we have observed between up-regulated Wnt activity and apoptosis in CRC cells (Lazarova *et al.*, 2004) including: (1) homozygous mutation in the *Drosophila* homolog of *APC* results in neuronal cell apoptosis in the *Drosophila* retina (Ahmed *et al.*, 1998); (2) expression of stabile, amino-terminally truncated beta-catenin results in 3–4-fold higher apoptotic levels in the intestinal villi of transgenic mice (Romagnolo *et al.*, 1999); (3) conditional targeting of *Apc*, the mouse version of *APC*, in murine neural crest cells results in massive apoptosis of cephalic and cardiac neural crest cells at 11.5 days

associate with transcriptional corepressors and therefore downregulate expression of genes with Tcf binding sites. When complexed with beta-catenin, Tcf factors lose their association with transcriptional repressors; beta-catenin contains a transcriptional transactivation domain, and the beta-catenin-Tcf complex associates with transcriptional coactivators. Therefore, Wnt activation results in the expression of target genes containing Tcf binding sites. Mutations in the Wnt signaling pathway, most often in the *APC* and beta-catenin genes, result in constitutive activation of Wnt signaling, which can be independent of the requirement for upstream Wnt ligand–receptor interactions (however, these interactions may still occur, and further upregulate Wnt activity). The deregulated Wnt activity can promote colonic tumorigenesis, although hyperactivation of this pathway induces apoptosis in colorectal cancer cell lines. This figure was reproduced with permission from Dr. R. Nusse from http://www.stanford.edu/~rnusse/pathways/cell2.html. (See Color Insert.)

postcoitum (Hasegawa *et al.*, 2002); and (4) expression of constitutively active beta-catenin in 129/Sv cells of chimeric mice results in apoptosis of these cells (Wong *et al.*, 2002). These findings contrast with the reports that decreased Wnt activity, produced by expression of wild-type APC in $APC^{-/-}$ CRC cells induces apoptosis (Groden *et al.*, 1995; Morin *et al.*, 1996), probably by downregulation of survivin (Zhang *et al.*, 2001), and that increased levels of beta-catenin protect cells from suspension-induced apoptosis (Orford *et al.*, 1999). These contradictory findings can be reconciled by the fact that different levels of Wnt activity are achieved in different experimental systems. Thus, Wong *et al.* (2002) proposed that cells exposed to high levels of Wnt activity undergo apoptosis; whereas, cells exposed to moderate levels of Wnt activity maintain a proliferative state, and cells exposed to low levels of Wnt activity undergo differentiation (terminal differentiation usually being followed by apoptosis).

The "just right hypothesis" for CRC formation is based upon a similar concept that *APC* mutations that result in moderate levels of Wnt signaling are optimal for tumor formation and growth; whereas, *APC* mutations that lead to relatively high levels of Wnt signaling are not selected, most probably due to apoptosis of cells with such mutations (Albuquerque *et al.*, 2002). This model is derived from the observation that in colorectal adenomas of patients with familial adenomatous polyposis there is a selection for *APC* mutants that retain the ability to downregulate Wnt activity. This suggests that complete inactivation of APC leading to high Wnt activity levels is not advantageous for tumor formation, most likely because cells with relatively high Wnt activity levels are eliminated by apoptosis. These findings in CRC cells suggest the existence of a gradient, or continuum, of Wnt activity, which is relevant to the relationship between levels of Wnt activity and development of the diabetogenic phenotype.

C. The Wnt signaling continuum, Tcf factors, and type 2 diabetes

Based upon our findings (Bordonaro *et al.*, 2007; Lazarova *et al.*, 2004), we postulate that both relatively high and relatively low levels of Wnt transcriptional activity lead to CRC cell apoptosis. Therefore, Wnt activity can be viewed and analyzed as a gradient, within which absence of detectable Wnt signaling (such as in cells at the top of the colonic crypt) results in terminal differentiation and apoptosis, relatively low levels of signaling (such as in the stem cell compartment of the colonic crypt) lead to controlled self-renewal, moderate levels of signaling (such as in CRC cells) promote proliferation, and relatively high levels of Wnt activity (such as in CRC cells treated with HDACis) lead to enhanced apoptosis.

This is relevant to the relationship between Wnt activity and type 2 diabetes because it is important to conceptualize Wnt activity as a continuum

(Fig.19.2), rather than as a binary "on/off" switch; varying levels of Wnt activity inputs can result in different physiological outputs in a cell type- and tissue type-dependent manner. Other agents besides HDACis are capable of stimulating Wnt activity; for example, lithium works downstream of the Wnt ligand–receptor interaction by inhibiting GSK-3β activity, thus repressing the phosphorylation of beta-catenin and allowing for the accumulation of active beta-catenin that can complex with Tcf factors (Stambolic *et al.*, 1996). Therefore, lithium and similar agents are capable of shifting Wnt signaling along the continuum of activity, with potential physiological consequences for disorders such as type 2 diabetes and cancer.

Another issue crucial for a complete understanding of the relationship between Wnt activity and type 2 diabetes is the dual role played by Tcf factors in the transcriptional control of relevant target genes. Tcf factors are high mobility box containing, DNA-binding transcription factors, and Tcf4 (encoded by the *TCF7L2* gene) is the member of this family most relevant to both type 2 diabetes and colorectal cancer.

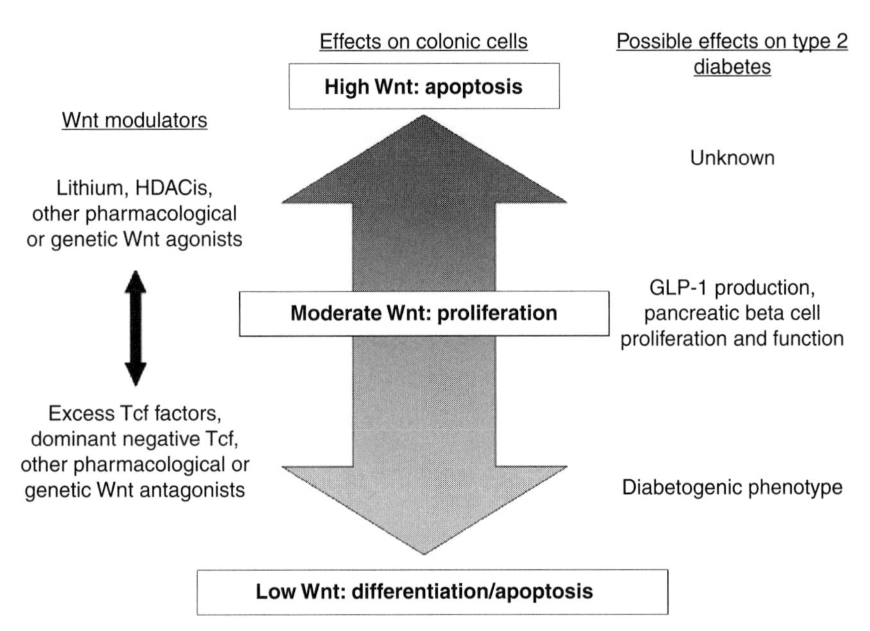

Figure 19.2 Continuum of Wnt signaling activity. Center, range of Wnt signaling activity and physiological consequences of that activity in colonic cells. Right, proposed effects of differential levels of Wnt signaling on the process of diabetogenesis. While the effects of very high levels of Wnt activity on diabetogenesis cannot be predicted at this time, moderate levels of Wnt activity are likely to promote GLP-1 production in intestinal cells, as well as promote pancreatic beta-cell proliferation and function. Repressed Wnt activity, possibly resulting from *TCF7L2* variants, is likely to promote the diabetogenic phenotype. Left, methodologies for up- or downregulating Wnt activity.

In the absence of Wnt signaling and, hence, in the absence of beta-catenin-Tcf complex formation, Tcf factors function as transcriptional repressors of Wnt target genes (Bienz, 1998; Roose and Clevers, 1999). Tcf factors bind to corepressors such as Groucho and C-terminal binding protein 1 (CtBP1), which recruit additional transcriptional corepressors such as histone deacetylases to the promoter/enhancer regions of Wnt target genes (Cavallo *et al.*, 1998; Chen *et al.* 1999; Valenta *et al.*, 2003). The role of histone deacetylases in repressing Wnt target genes suggests that HDACis may stimulate Wnt activity in part by derepressing Tcf-mediated inhibition of Wnt-targeted promoters, in addition to the known role of HDACis in elevating levels of active beta-catenin (Bordonaro *et al.*, 2007).

The importance of Tcf factors as the most downstream fundamental elements of Wnt signaling is underscored by the ability of dominant negative forms of Tcf4 to repress Wnt activity resulting from constitutive activation of the pathway (Kolligs *et al.*, 1999; Korinek *et al.*, 1997) as well as from butyrate-stimulated Wnt activity (Bordonaro *et al.*, 2007; Lazarova *et al.*, 2004). Further, and directly relevant to the relationship between Wnt signaling and type 2 diabetes, dominant negative Tcf can repress both endogenous and lithium-stimulated Wnt-mediated expression of the proglucagon gene in intestinal L cells (Yi *et al.*, 2005).

Finally, we note that the balance between the levels of Tcf and of beta-catenin in CRC cells influences Wnt signaling, with an excess of Tcf factors, uncomplexed with beta-catenin, repressing the ability of HDACis to up-regulate Wnt transcriptional activity (Bordonaro *et al.*, 2007). Therefore, altered Tcf expression is likely to modulate Wnt activity in pancreatic cells (Cauchi and Froguel, 2008), influencing cell physiology and the development of the diabetogenic phenotype (reviewed in Jin and Liu, 2008).

III. *TCF7L2* Variants and Type 2 Diabetes Risk

A. Association of TCG7L2 variants with the risk of type 2 diabetes

In 1999, an association between an area of chromosome 10q and type 2 diabetes was reported for Mexican-Americans (Duggirala *et al.*, 1999). A year later, the *TCF7L2* gene was mapped to chromosome 10q25.3 (Duval *et al.*, 2000), and, in 2003, the same region of chromosome 10 was associated with type 2 diabetes (Reynisdottir *et al.*, 2003). In 2006, deCODE Genetics (Grant e*t al.*, 2006) published data demonstrating an association between variants of the *TCF7L2* gene, which encodes Tcf4, and risk for type 2 diabetes. This group identified and typed 228 microsatellite markers at the chromosome 10q locus in 1185 Icelandic individuals with type 2 diabetes, as well as 931 unrelated population controls. A tetranucleotide repeat, DG10S478, located within intron 3 of the *TCF7L2* gene was associated with risk for type 2 diabetes, and this association was conformed in two additional population cohorts, Danish and European-American. Five single nucleotide polymorphisms (SNPs) most strongly correlated to DG10S478 were identified as rs12255372, rs7903146, rs7901695, rs11196205, and rs7895340, and all five SNPs exhibited association to type 2 diabetes. For the composite allele "X" of the microsatellite marker DG10S478, relative enhanced risk for type 2 diabetes for the population cohorts was 1.45-fold for heterozygous carriers and 2.41-fold for homozygous carriers. Thus, individuals that are heterozygous or homozygous for variants in *TCF7L2* gene are at enhanced risk for the development of type-2 diabetes (Grant *et al.*, 2006; Helgason *et al.*, 2007). This finding has been replicated in several populations (Cauchi and Froguel, 2008; Cauchi *et al.*, 2007; Grant *et al.*, 2006; Helgason *et al.*, 2007); in Europeans, *TCF7L2* has been found to be the most important gene locus linked to a predisposition to type 2 diabetes (Cauchi and Froguel, 2008; Goodarzi and Rotter, 2007; Helgason *et al.*, 2007). With respect to the ancestral T allele of the SNP rs7903146, relative risk for type 2 diabetes is highest in homozygous individuals (TT), particularly when compared to homozygous noncarriers (CC).

B. Possible mechanisms of action of *TCF7L2* variants on the diabetogenic phenotype

It has been reported that the rs7903146 *TCF7L2* polymorphism enhances the expression of Tcf4 in pancreatic beta cells (Cauchi and Froguel, 2008) and that variants of the *TCF7L2* gene may influence the development of type 2 diabetes by modulating the levels of the Wnt target gene glucagon-like peptide 1 (GLP-1) (Yi *et al.*, 2005, 2008). Elevated Tcf4 expression due to specific allelic variants may suppress the expression of GLP-1 and the GLP-induced insulin secretion (Cauchi and Froguel, 2008; Lyssenko *et al.*, 2007). These reports are consistent with our findings that overexpression of Tcf factors inhibits Wnt activity in CRC cells (Bordonaro *et al.*, 2007); therefore, it is likely that excess Tcf4 reduces expression of GLP-1 through suppression of Wnt signaling. However, knockdown of Tcf4 levels with siRNA reduces insulin gene expression and insulin secretion in pancreatic beta cells (Loder *et al.*, 2008), suggesting that a certain level of Tcf4 and, hence, of Wnt signaling, is required for proper pancreatic beta-cell function (Loder *et al.*, 2008). Thus, Wnt activity can be inhibited by both an excess of Tcf4 (Bordonaro *et al.*, 2007) and by low levels of Tcf4; the resulting downregulation of Wnt activity through either mechanism can promote a diabetogenic phenotype. This is an important point that must be stressed; effects of Wnt activity on cellular physiology, whether in type 2 diabetes,

CRC, development, or other processes is unlikely to be mediated by simple binary "on/off" switch. Rather, it is more likely that particular levels of Wnt signaling (Fig.19.2) activate different subsets of Wnt responsive gene promoters that contain binding sites characterized by varying affinities for Tcf factors.

In addition, we have reported that an imbalance of Tcf and beta-catenin in CRC cells, caused by excess Tcf, can repress Wnt activity; this mechanism mediates the development of butyrate resistance in CRC cells exposed to increasing levels of that agent over time (Bordonaro *et al.*, 2007). Therefore, excess production of Tcf4, which may result from altered *TCF7L2* RNA processing, may inhibit Wnt activity in intestinal L cells and pancreatic beta cells (see below) resulting in a diabetogenic phenotype. On the other hand, it has been reported that overexpression of Tcf4 protected pancreatic islet cells from glucose and cytokine-induced apoptosis and impaired function (Shu *et al.*, 2008), which is inconsistent with the idea that excess Tcf4 is necessarily prodiabetogenic in the context of pancreatic cells. In addition, it was reported that carriers of the *TCF7L2* TT risk alleles have 5-fold higher *TCF7L2* mRNA in their pancreatic cells, but that this upregulation of Tcf4 is positively, not negatively, associated with insulin gene expression (although it is negatively associated with insulin release that is stimulated by glucose) (Lyssenko *et al.*, 2007). However, these studies did not determine the levels of active beta-catenin and Wnt activity in the context of glucose and cytokine challenge of the pancreatic cells, nor were the levels of active beta-catenin measured in the islet cells of individuals homozygous for the TT risk alleles. Excess Tcf factors are repressive only if uncomplexed with beta-catenin; for example, we have partially reversed the repressive effects of excess Tcf factors on Wnt activity through lithium-induced upregulation of active beta-catenin (Bordonaro *et al.*, 2007). Therefore, in the context of high levels of beta-catenin, increased expression of Tcf would be expected to result in activation of Wnt signaling and enhanced pancreatic cell viability. In the absence of data on the levels of beta-catenin and of Wnt activity in the relevant cells, it is difficult to interpret the reported association between overexpressed Tcf4 and pancreatic cell protection (Shu *et al.*, 2008).

It is also possible that *TCF7L2* variants exert some of their prodiabetogenic effects by altering Wnt-mediated gene expression in incretin secreting intestinal cells, and that the effects of excess Tcf levels on Wnt activity and type 2 diabetes is tissue-type specific. This is supported by the report that Tcf4 is expressed at high levels in intestinal cells, but not in pancreatic islets (Yi *et al.*, 2005). Furthermore, activators of Wnt activity were able to stimulate proglucagon expression (and, hence, GLP-1 production) from intestinal, but not pancreatic, cells (Yi *et al.*, 2005), further suggesting the importance of the intestinal endocrine cells in mediating Wnt-related effects on diabetogenesis. Findings in the literature based upon *in vitro* studies (see below) suggest that Wnt activity is important in pancreatic beta-cell proliferation, thus possibly linking Wnt signaling and type 2 diabetes through pancreatic cells; however, it is likely that the effects of Wnt signaling in the pancreas is subsequent to, and dependent on, Wnt activity in the intestine.

C. Contrary views and counter-explanations

Another research group (Schafer *et al.*, 2007) reported that nondiabetic carriers of the *TCF7L2* risk alleles do not exhibit defects in measured levels of systemic GLP-1, which is not consistent with a defect in GLP-1 production (e.g., proglucagon expression). On the other hand, this group did observe reduced GLP-1 stimulated insulin secretion in the studied individuals. Does this effectively argue against a role for *TCF7L2* variants in controlling GLP-1 production in the context of diabetes?

The authors note three major caveats to their findings; first, they measured total systemic GLP-1 and may have therefore missed more subtle alterations in active GLP-1 secretion, second, systemic GLP-1 levels may not reflect local concentration of GLP-1, and third, it is possible that TCF7L2-mediated defects in GLP-1 production may be predominantly confined to the brain (Schafer *et al.*, 2007). In addition, it is possible that the effects of aberrant Wnt signaling on GLP-1 production are manifested at particular crucial stages in the diabetogenic process, which were missed by Schafer and colleagues. For example, it has been suggested that the "prediabetic" status may be characterized by compensatory mechanisms that allow individuals with defective GLP-1 production to maintain "normal" systemic levels of GLP-1 in response to glucose challenge; these mechanisms may be lost later in the disease process, in which (possibly Wnt-mediated) defects in GLP-1 production would become clinically relevant (Jin and Liu, 2008). It is important to note in this regard that the study by Schafer and colleagues involved nondiabetic carriers of the risk allele. Therefore, if, (1) changes in GLP-1 production are more evident in diabetic individuals and (2) a predominant role of altered GLP-1 levels in the diabetogenic process involves downregulated stimulation of Wnt activity and proliferation in pancreatic beta cells (see below), then the further decrease in beta-cell volume observed in diabetic individuals (Bonora, 2008) is significant. One may hypothesize that a Wnt-mediated decline in GLP-1 levels in diabetic individuals, as opposed to prediabetic risk allele carriers, would result in the decreased beta-cell mass that is observed at that stage of the disease (Bonora, 2008). Therefore, it is possible that at specific points in the development of type 2 diabetes defects of Wnt signaling may act on the pancreas indirectly through primary effects on intestinal production of GLP-1. Further, the possible influence of Wnt-mediated GLP-1 production in the brain is a heretofore unexplored area that requires further study.

At this point it is again useful to stress the point that the effects of Wnt signaling activity lie along a continuum (Fig.19.2), and that modulation of cell physiology may be quite different at the "extremes" of Wnt signaling compared to moderate levels of this activity. Therefore, some of the ostensibly contradictory findings with respect to the role of *TCF7L2* variants on gene expression and beta-cell function may be due to variation in the levels of Wnt activity. This variation in Wnt activity may result from cell-type specific effects, differences in experimental methodologies, as well as differential effects of specific *TCF7L2* variants on Wnt activity. Whenever possible, it would be useful to ascertain the levels of Wnt activity in the studied cells; this can be easily achieved *in vitro* through reporter assays, and, *in vivo*, through measurement of the levels of active beta-catenin, if the relevant tissue samples can be obtained.

It is also plausible that T*CF7L2* variants act to influence the risk of type 2 diabetes through mechanisms which have not yet been identified and which involve pathways other than Wnt signaling. However, it can be noted that, in addition to *TCF7L2*, variation in the *WNT5B* gene also correlates to enhanced susceptibility to type-2 diabetes (Salpea *et al.*, 2008), and overexpression of Wnt5b in preadipocyte cells suppresses Wnt activity (Kazazawa *et al.*, 2005). The finding that a second Wnt-related gene is associated with type 2 diabetes risk strengthens the hypothesis that Wnt activity is the link between *TCF7L2* variation and diabetes.

In summary, the findings to date on the possible direct role of Tcf-mediated diabetogenic alterations in Wnt-target gene expression have yielded complex, and sometimes, contradictory findings that are, nonetheless, generally promising. More straightforward have been the *in vitro* studies exploring the functional relationship between Wnt signaling, proglucagon expression, and beta-cell proliferation, which are discussed the following section.

IV. Functional Relationship Between Wnt Signaling and Type 2 Diabetes *In Vitro*

A. Wnt signaling, proglucagon expression, and GLP-1 production

GLP-1, a peptide hormone produced by alternative processing of the prohormonal precursor proglucagon (1 and references therein), is released from intestinal enteroendocrine cells after feeding. GLP-1 promotes glucose-dependent insulin secretion, reduces blood glucose levels, enhances peripheral insulin sensitivity, and induces satiety (Liu and Habener, 2008; Yi *et al.*, 2008, and references therein). Additional antidiabetogenic activities of GLP-1 are the stimulation of proliferation of insulin-producing pancreatic cells and the inhibition of apoptosis of these cells (Liu and Habener, 2008 and references therein).

Proglucagon is a Wnt target gene and the activation of proglucagon expression by Wnt signaling takes place at the transcriptional level and is cell type specific (Ni *et al.*, 2003; Yi *et al.*, 2005, 2008). BCT complexes bind to the proglucagon gene promoter in intestinal, proglucagon producing, GLUTag cells in culture; stimulation of these cells with lithium chloride (LiCl), an agent that activates Wnt activity by inhibiting the degradation of beta-catenin, has been shown to result in increased proglucagon promoter activity and mRNA expression (Ni *et al.*, 2003; Yi *et al.*, 2005, 2008). The *in vivo* significance of these findings remains to be determined, given the controversy with respect to the role of *TCF7L2* variants in mediating *in vivo* GLP-1 production (Jin and Liu, 2008; Schafer *et al.*, 2007).

B. Wnt signaling and pancreatic beta cells

Pancreatic beta-cell dysfunction is fundamental in the development of type 2 diabetes; beta-cell volume is reduced in prediabetic, and, to a greater degree, diabetic individuals, and this is primarily due to enhanced apoptosis of these cells (Bonora, 2008). Beta-cell protection is therefore a key preventive and therapeutic aim in combating type 2 diabetes. GLP-1 and its agonist exendin-4 (Exd4), recently utilized in the treatment of type 2 diabetes, stimulate Wnt activity in pancreatic beta cells *in vitro* in a manner independent of GSK-3β but dependent upon beta-catenin, Tcf4, and the activation of AKT and protein kinase A (Liu and Habener, 2008). In the absence of modulators of Wnt signaling, the basal Wnt activity of pancreatic beta cells is dependent upon the inactivation of GSK-3β and AKT activity, but is independent of protein kinase A (Liu and Habener, 2008). Basal and Exd-4 induced *in vitro* pancreatic cell proliferation is dependent upon Wnt activity, which can be blocked by siRNAs to beta-catenin or by expression of a dominant negative form of Tcf4 (Liu and Habener, 2008). Possible mediators of the effects of GLP-1 and Exd4 on Wnt-stimulated pancreatic cell proliferation are the Wnt target genes cyclin D1 and c-Myc, known modulators of cell proliferation, which are upregulated by Exd4 (Liu and Habener, 2008).

V. Conclusions and Future Directions

It is uncertain exactly how *TCF7L2* variants influence the risk for type 2 diabetes. Based upon studies with CRC cells (Bordonaro *et al.*, 2007), it is tempting to speculate that the known effects of the TT risk allele in enhancing levels of Tcf4 results in inhibition of Wnt activity through the repressive function of excess Tcf factors uncomplexed with beta-catenin (Fig.19.3). This repressed Wnt activity would tie into the diabetogenic phenotype possibly through decreased pancreatic beta-cell mass and/or through inhibited incretin expression/secretion. However, the situation in pancreatic cells is complex, with some reports indicating excess Tcf4 enhancing pancreatic cell viability and other reports associating increased levels of Tcf4 with greater levels of insulin expression (Lyssenko *et al.*, 2007; Shu *et al.*, 2008). However, consistent with the findings outlined above on the effects of Wnt signaling on proglucagon expression in intestinal cells, the effects of *TCF7L2* variants on type 2 diabetes may in fact be initiated in intestinal cells rather than solely in the pancreas. Of course, it is possible that the *TCF7L2* risk alleles are in linkage disequilibrium to the actual (other) gene(s) responsible for enhancing the risk of type 2 diabetes, and these other genes may be unrelated to Wnt activity.

However, based on the preponderance of the current evidence, it is likely that Wnt activity in intestinal cells drives the expression of GLP-1, which in turn may activate Wnt signaling in pancreatic beta cells and

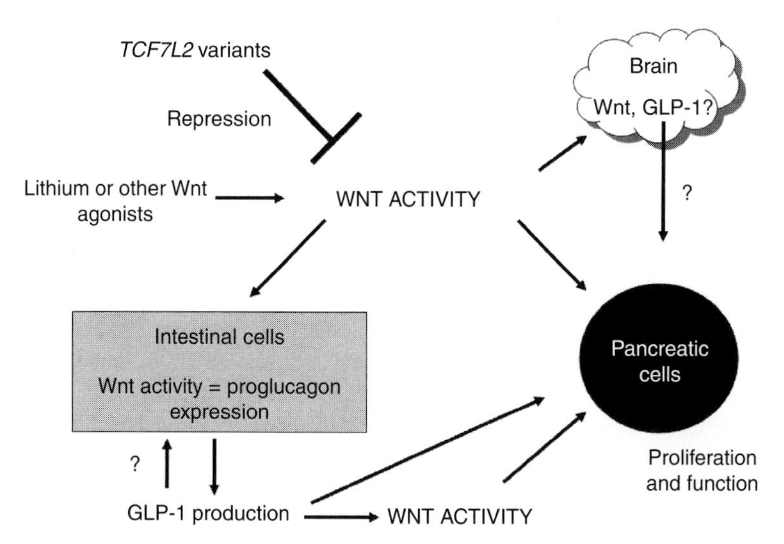

Figure 19.3 Hypothesis concerning the association of Wnt signaling activity and type 2 diabetes. Wnt activity stimulates proglucagon expression and GLP-1 production in the intestine and, possibly, the brain; in addition, Wnt activity is important for pancreatic cell proliferation and function. GLP-1 is proposed to stimulate Wnt activity in pancreatic cells (and, possibly, feeds back to stimulate Wnt activity in the intestine). Lithium and other Wnt agonists promote Wnt activity, while *TCF7L2* variants are proposed to repress Wnt activity, possibly through the production of excess Tcf factors. This repression of Wnt activity likely contributes to the diabetogenic phenotype by inhibiting GLP-1 production and interfering with normal pancreatic cell proliferation and function.

stimulate the proliferation of these cells (Liu and Habener, 2008). The interaction between GLP-1 and the Wnt pathway (Fig.19.3) may influence pancreatic beta-cell protection via two inter-related mechanisms. First, enhanced Wnt activity in intestinal cells may increase levels of GLP-1, which directly stimulates the proliferation of, and blocks the apoptosis of, pancreatic beta cells (Liu and Habener, 2008). Second, GLP-1 may stimulate Wnt activity in pancreatic cells, which also promotes proliferation (Liu and Habener, 2008). The importance of Wnt signaling to pancreatic beta-cell proliferation was underscored by a study showing that Wnt ligand, or activated beta-catenin, can induce pancreatic beta-cell proliferation, insulin production and glucose handling *in vitro* and *in vivo* (Rulifson *et al.*, 2007). It is also possible that GLP-1 positively feeds back to upregulate Wnt activity in intestinal cells. Thus, another processing product of proglucagon, glucagon-like peptide 2 (GLP-2), has been shown to stimulate nuclear beta-catenin levels in mouse intestinal cells via an insulin-like growth factor-1 (IGF-1) dependent pathway (Dube *et al.*, 2008), suggesting that positive feedback from GLP-2 could be an additional mechanism whereby basal Wnt activity in intestinal cells is amplified to enhance GLP-1 production. The possibility of positive feedback between Wnt signaling and GLP-1 is consistent with the findings that Wnt signaling is both an inducer and an effector of GLP-1 and GLP-1 signaling (Gustafson and Smith, 2008).

It has been proposed that insulin resistance found in type-2 diabetes may trigger a compensatory mechanism whereby Wnt activity is induced, resulting in increased levels of GLP-1; enhanced Wnt activity, to moderate levels promoting proliferation (Fig.19.2), may therefore partially explain the association between type-2 diabetes and increased risk for CRC (Jin, 2008). However, lithium, alone or in conjunction with GLP-1/Exd-4, has the potential to result in sustained expression of GLP-1 through enhanced Wnt signaling in normal intestinal enteroendocrine cells, which would obviate the requirement for a "compensatory" mechanism (Jin, 2008). Given the side-effects of lithium treatment, it may be beneficial to develop alternate Wnt agonists capable to upregulating Wnt activity to a similar or greater degree.

Wnt activity, maintained by tumor necrosis factor α (TNF α), can inhibit adipogenesis, which may contribute to insulin resistance (Hammarstedt *et al.*, 2007); however, the net effects of moderate Wnt signaling are likely to be anti-diabetogenic through the mechanisms described in this review. Importantly, it was shown that treatment of preadipocyte cells with pioglitazone suppressed beta-catenin levels during the differentiation process and antagonized the Wnt-mediated inhibition of adipogenesis induced by TNFα.This suggests that treatment with pioglitazone may be used to promote adipogenesis concomitant with treatment with agents that stimulate Wnt activity. This potential treatment regimen could allow for normal adipogenesis while at the same time enhancing Wnt-GLP-1 pathway crosstalk in normal intestinal and pancreatic cells, promoting pancreatic cell proliferation. Therefore, therapeutic applications of Wnt signaling modulation for treatment of type 2 diabetes may require different approaches for various target tissues.

It will be necessary to confirm that the *TCF7L2* variants associated with enhanced risk of type 2 diabetes mediate their effects primarily through the *TCF7L2* gene product, Tcf4, rather than being in linkage disequilibrium with other as-of-yet unidentified alleles responsible for influencing disease risk. The mechanism by which *TCF7L2* intronic sequence variation alters *TCF7L2* gene expression needs to be identified, and a comprehensive study of the effects of increased Tcf levels in pancreatic and intestinal cells, with respect to Wnt signaling, gene expression, and secretion of the relevant gene products, needs to be undertaken. Such a study would necessarily include a formal demonstration of the link between *TCF7L2* gene variants, Tcf4 protein levels, proglucagon gene expression, and the levels and function of GLP-1. The possibility that the *TCF7L2* variants may act to alter GLP-1 levels specifically in the brain (Schafer *et al.*, 2007) also needs to be considered. Further *in vitro* cell culture work on the effects of Wnt signaling levels on intestinal and pancreatic cell gene expression, gene function, and cellular physiology must be followed by *in vivo* studies utilizing mouse models of diabetogenesis, in order to evaluate the therapeutic possibilities of Wnt-targeted interventions for type 2 diabetes. It is necessary to demonstrate that putative Wnt-mediated increased GLP-1 production in the intestine can lead to enhanced Wnt activity in pancreatic beta cells, and that this increase in Wnt signaling is beneficial for beta-cell proliferation and function. It would also be useful to explore the possibility that defective Wnt activity may influence insulin sensitivity, in addition to its possible role in the production of incretin hormones, which was the major focus of this review. Given the complexity of the findings to date, sorting through these issues will take considerable effort; however, further investigation into Wnt pathway-mediated effects on the development of type 2 diabetes may result in novel preventive and therapeutic methodologies for this disorder.

REFERENCES

Ahmed, Y., Hayashi, S., Levine, A., and Wieschaus, E. (1998). Regulation of Armadillo by a *Drosophila* APC inhibits neuronal apoptosis during retinal development. *Cell* **93,** 1171–1182.

Albuquerque, C., Breukel, C., van der Luijt, R., Fidalgo, P., Lage, P., Slors, F. G. M., Leitao, C. N., Fodde, R., and Smits, R. (2002). The just-right signaling model: APC somatic mutations are selected based on a special level of activation of the beta-catenin signaling cascade. *Hum. Mol. Genet.* **11,** 1549–1560.

Bienz, M. (1998). TCF: Transcriptional activator or repressor? *Curr. Opin. Cell Biol.* **10,** 366–372.

Bonora, E. (2008). Protection of pancreatic beta-cells: Is it feasible? *Nutr. Metab. Cardiovasc. Dis.* **18,** 74–83.

Bordonaro, M., Lazarova, D. L., and Sartorelli, A. C. (2007). The activation of beta-catenin by Wnt signaling mediates the effects of histone deacetylase inhibitors. *Exp. Cell Res.* **313,** 1652–1666.

Cauchi, S., and Froguel, P. (2008). TCF7L2 genetic defect and type 2 diabetes. *Curr. Diab. Rep.* **8,** 149–155.

Cauchi, S., El Achhab, Y., Choquet, H., Dina, C., Krempler, F., Weigasser, R., Nejjari, C., Patsch, W., Chikri, M., Meyre, D., and Froguel, P. (2007). TCF7L2 is reproducibly associated with type 2 diabetes in various ethnic groups: A global meta-analysis. *J. Mol. Med.* **8,** 777–782.

Cavallo, R. A., Cox, R. T., Moline, M. M., Roose, J., Polevoy, G. A., Clevers, H., Peifer, M., and Bejsovec, A. (1998). *Drosophila* Tcf and Groucho interact to repress wingless signalling activity. *Nature* **395,** 604–608.

Chen, G., Fernandez, J., Mische, S., and Courey, A. J. (1999). A functional interaction between the histone deacetylase Rpd3 and the corepressor groucho in *Drosophila* development. *Genes Dev.* **13,** 2218–2230.

Cook, D., Fry, M., Hughes, K., Sumathipala, R., Woodgett, J., and Dale, T. (1996). Wingless inactivates glycogen synthase kinase-3 via an intracellular signaling pathway which involves a protein kinase C. *EMBO J.* **15,** 4526–4536.

Dube, P. E., Rowland, K. J., and Brubaker, P. L. (2008). Glucagon-like peptide-2 activates beta-catenin signaling in the mouse intestinal crypt: Role of insulin-like growth factor-I. *Endocrinology* **149,** 291–301.

Duggirala, R., Blangero, J., Almasy, L., Dyer, T. D., Williams, K. L., Leach, R. J., O'Connell, P., and Stern, M. P. (1999). Linkage of type 2 diabetes mellitus and of age at onset to a genetic location on chromosome 10q in Mexican Americans. *Am. J. Hum. Genet.* **64,** 1127–1140.

Duval, A., Busson-Leconiat, M., Berger, R., and Hamelin, R. (2000). Assignment of the TCF-4 gene (TCF7L2) to human chromosome band 10q25.3. *Cytogenet. Cell Genet.* **88,** 264–265.

Goodarzi, M. O., and Rotter, J. I. (2007). Testing the gene or testing a variant? The case of TCF7L2. *Diabetes* **56,** 2417–2419.

Grant, S. F., Thorleifsson, G., Reynisdottir, I., Benediktsson, R., Manolescu, A., Sainz, J., Helga son, A., Stefansson, H., Emilsson, V., Helgadottir, A., Styrkarsdottir, U., Magnusson, K. P., *et al.* (2006). Variant of transcription factor 7-like 2 (TCF7L2) gene confers risk of type 2 diabetes. *Nat. Genet.* **38,** 320–323.

Groden, J., Joslyn, G., Samowitz, W., Jones, D., Bhattacharyya, N., Spirio, L., Thliveris, A., Robertson, M., Egan, S., Meuth, M., and White, R. (1995). Response of colon cancer cell lines to the introduction of APC, a colon-specific tumor suppressor gene. *Cancer Res.* **55,** 1531–1539.

Gustafson, B., and Smith, U. (2008). WNT signaling is both an inducer and effector of glucagons-like-peptide-1. *Diabetologia* **10,** 1768–1770.

Hammarstedt, A., Isakson, P., Gustafson, B., and Smith, U. (2007). Wnt-signaling is maintained and adipogenesis inhibited by TNFα but not MCP-1 and resistin. *Biochem. Biophys. Res. Comm.* **357,** 700–706.

Hasegawa, S., Sato, T., Akazawa, H., Okada, H., Maeno, A., Ito, M., Sugitani, Y., Shibata, H., Miyazaki, J., Katsuki, M., Yamauchi, Y., Yamamura, Ki. K., *et al.* (2002). Apoptosis in neural crest cells by functional loss of APC tumor suppressor gene. *Proc. Natl. Acad. Sci. USA* **99,** 297–302.

Helgason, A., Palsson, S., Thorleifsson, G., Grant, S. F., Emilsson, V., Gunnarsdottir, S., Adeyemo, A., Chen, Y., Chen, G., Reynisdottir, I., Benediktsson, R., Hinney, A., *et al.* (2007). Refining the impact of TCF7L2 variants on type 2 diabetes and adaptive evolution. *Nat. Genet.* **39,** 218–222.

Jin, T. (2008). Why diabetes patients are more prone to the development of colon cancer. *Med. Hypotheses* **71,** 241–244.
Jin, T., and Liu, L. (2008). The Wnt signaling pathway effector TCF7L2 and type II diabetes mellitus. *Mol. Endocrinol.* Epub ahead of print.
Kazazawa, A., Tsukada, S., Kamiyama, M., Yanagimoto, T., Nakajima, M., and Maeda, S. (2005). Wnt5b partially inhibits canonical Wnt/β-catenin signaling pathway and promotes adipogenesis in 3Y3-L1 preadipocytes. *Biochem. Biophys. Res. Commun.* **330,** 505–510.
Kinzler, K. W., and Vogelstein, B. (1996). Lessons from hereditary colorectal cancer. *Cell* **87,** 159–170.
Kolligs, F. T., Hu, G., Dang, C. V., and Fearon, E. R. (1999). Neoplastic transformation of RK3E by mutant beta-catenin requires deregulation of Tcf/Lef transcription but not activation of c-myc expression. *Mol. Cell. Biol.* **19,** 5696–5706.
Korinek, V., Barker, N., Morin, P. J., Van Wichen, D., De Weger, R., Kinzler, K. W., Vogelstein, B., and Clevers, H. (1997). Constitutive transcriptional activation by a beta-catenin-Tcf complex in $APC^{-/-}$ colon carcinoma. *Science* **275,** 1784–1787.
Lazarova, D. L., Bordonaro, M., Carbone, R., and Sartorelli, A. C. (2004). Linear relationship between WNT activity levels and apoptosis in colorectal carcinoma cells exposed to butyrate. *Int. J. Cancer* **11,** 523–531.
Liu, Z., and Habener, J. F. (2008). Glucagon-like peptide-1 activation of TCF7L2-dependent Wnt signaling enhances pancreatic beta-cell proliferation. *J. Biol. Chem.* **283,** 8723–8735.
Loder, M. K., Xavier Gda, S., McDonald, A., and Rutter, G. A. (2008). TCF7L2 controls insulin gene expression and insulin secretion in mature pancreatic beta-cells. *Biochem. Soc. Trans.* **36,** 357–359.
Lyssenko, V., Lupi, R., Marchetti, P., Del Guerra, S., Orho-Melander, M., Aimgren, P., Sjogren, M., Ling, C., Eriksson, K. F., Lethagen, A. L., Mancarella, R., Berglund, G., *et al.* (2007). Mechanisms by which common variants in the TCF7L2 gene increases risk of type 2 diabetes. *J. Clin. Invest.* **117,** 2155–2163.
Morin, P. J., Vogelstein, B., and Kinzler, K. W. (1996). Apoptosis and APC in colorectal tumorigenesis. *Proc. Natl. Acad. Sci. USA* **93,** 7950–7954.
Ni, Z., Anini, Y., Fang, X., Mills, G., Brubaker, P. L., and Jin, T. (2003). Transcriptional activation of the proglucagon gene by lithium and beta-catenin in intestinal endocrine L cells. *J. Biol. Chem.* **278,** 1380–1387.
Orford, K., Orford, C. C., and Byers, S. W. (1999). Exogenous expression of β-catenin regulates contact inhibition, anchorage-independent growth, anoikis, and radiation-induced cell cycle arrest. *J. Cell Biol.* **14,** 855–868.
Reynisdottir, I., Thorleifsson, G., Benediktsson, R., Sigurdsson, G., Emilsson, V., Einarsdottir, A. S., Hjorleifsdottir, E. E., Orlygsdottir, G. T., Bjornsdottir, G. T., Saemundsdottir, J., Halldorsson, S., Hrafnkelsdottir, S., *et al.* (2003). Localization of a susceptibility gene for type 2 diabetes to chromosome 5q34-q35.2. *Am. J. Hum. Genet.* **73,** 323–335.
Romagnolo, B., Berrebi, D., Saadi-Keddoucci, S., Porteu, A., Pichard, A.-I., Peuchmaur, M., Vandewalle, A., Kahn, A., and Perret, C. (1999). Intestinal dysplasia and adenoma in transgenic mice after overexpression of an activated beta-catenin. *Cancer Res.* **59,** 3875–3879.
Roose, J., and Clevers, H. (1999). Tcf transcription factors: Molecular switches in carcinogenesis. *Biochim. Biophys. Acta* **87456,** M23–M27.
Rulifson, I. C., Karnik, S. K., Heiser, P. W., ten Berge, D., Chen, H., Gu, X., Taketo, M. M., Nusse, R., Hebrok, M., and Kim, S. K. (2007). Wnt signaling regulates pancreatic beta cell proliferation. *Proc. Natl. Acad. Sci. USA* **104,** 6247–6252.

Salpea, K. D., Gable, D. R., Cooper, J. A., Stephens, J. W., Hurel, S. J., Ireland, H. A., Feher, M. D., Godsland, I. F., and Humphries, S. E. (2008). The effect of WNT5B IVS3C>G on the susceptibility to type 2 diabetes in UK caucasian subjects. *Nutr. Metab. Cardiovasc. Dis.* Epub ahead of print.

Schafer, S. A., Tschritter, O., Machicao, F., Thamer, C., Stefan, N., Gallwitz, B., Holst, J. J., Dekker, J. M. T., Hart, L. M., Nijpels, G., van Haeften, T. W., Haring, H. U., *et al.* (2007). Impaired glucagon-like peptide-1-induced insulin secretion in carriers of transcription factor 7-like2 (TCF7L2) gene polymorphisms. *Diabetologia* **50,** 2443–2450.

Shu, L., Sauter, N. S., Schulthess, F. T., Matveyenko, A. V., Oberholzer, J., and Maedler, K. (2008). Transcription factor 7-like2 regulates beta-cell survival and function in human pancreatic islets. *Diabetes* **57,** 645–653.

Stambolic, V., Ruel, L., and Woodgett, J. R. (1996). Lithium inhibits glycogen synthase kinase-3 activity and mimics wingless signalling in intact cells. *Curr. Biol.* **6,** 1664–1668.

Su, L.-K., Vogelstein, B., and Kinzler, K. W. (1993). Association of the APC tumor suppresser protein with catenins. *Science* **262,** 1734–1737.

Valenta, T., Lukas, J., and Korinek, V. (2003). HMG box transcription factor TCF-4's interaction with CtBP1 controls the expression of the Wnt target axin2/conductin in human embryonic kidney cells. *Nucleic Acids Res.* **31,** 2369–2380.

Wong, M. H., Huelsken, J., Birchmeier, W., and Gordon, J. I. (2002). Selection of multipotent stem cells during morphogenesis of small intestinal crypts of Lieberkühn is perturbed by β-catenin signaling. *J. Biol. Chem.* **277,** 15843–15850.

Woodgett, J. (1994). Regulation and functions of the glycogen synthase kinase-3 subfamily. *Semin. Cancer Biol.* **5,** 269–275.

Yi, F., Brubaker, P. L., and Jin, T. (2005). TCF-4 mediates cell type-specific regulation of proglucagon gene expression by β-catenin and glycogen synthase kinase-3β. *J. Biol. Chem.* **280,** 1457–1464.

Yi, F., Sun, J., Lim, G. E., Fantus, I. G., Brubaker, P. L., and Jin, T. (2008). Crosstalk between the insulin and Wnt signaling pathways: Evidence from intestinal endocrine cells. *Endocrinology* **149,** 2341–2352.

Zhang, T., Otevrel, T., Gao, Z., Ehrlich, S. M., Fields, J. Z., and Boman, B. M. (2001). Evidence that APC regulates survivin expression: A possible mechanism contributing to the stem cell origin of colon cancer. *Cancer Res.* **61,** 8664–8667.

Zimmet, P., Alberti, K. G. M. M., and Shaw, J. (2001). Global and societal implications of the diabetes epidemic. *Nature* **414,** 782–787.

CHAPTER TWENTY

Retinal Insulin Receptor Signaling in Hyperosmotic Stress

Raju V. S. Rajala, Ivana Ivanovic, *and* Ashok Kumar Dilly

Contents

Departments of Ophthalmology and Cell Biology, and Dean A. McGee Eye Institute, University of Oklahoma Health Sciences Center, 608 Stanton L. Young Boulevard, Oklahoma City, Oklahoma 73104

Vitamins and Hormones, Volume 80
ISSN 0083-6729, DOI: 10.1016/S0083-6729(08)00620-1

Abstract

In the diabetic eye, the increased accumulation of sorbitol in the retina has been implicated in the pathogenesis of diabetic retinopathy (DR). Neurodegeneration is an important component of DR as demonstrated by increased neural apoptosis in the retina during experimental and human diabetes. Insulin receptor (IR) activation has been shown to rescue retinal neurons from apoptosis through a phosphoinositide 3-kinase and protein kinase B (Akt) survival cascade. In this study, we examined the IR signaling in sorbitol-induced hyperosmotic stressed retinas.

Abbreviations

IR	insulin receptor
PI3K	phosphoinositide 3-kinase
IRβ	IR beta subunit
IGF 1R	insulin-like growth factor-1 receptor
SDS-PAGE	sodium dodecyl sulfate polyacrylamide gel electrophoresis
ROS	rod outer segments
IPs	immunoprecipitates
Grb2	growth factor receptor-bound protein 2
Gab-1	Grb2-assoicated binder 1

I. Introduction

Sorbitol is a sugar substitute often used in diet foods (including diet drinks and ice cream) and sugar-free chewing gums. It also occurs naturally in many stone fruits and berries from trees of the genus Sorbus. Ingesting large amounts of sorbtiol can lead to some abdominal pain, gas, and mild to severe diarrhea. Sorbitol can also aggravate irritable bowl syndrome and fructose malabsorption. Sorbitol can be used as a nonstimulant laxative as either an oral suspension or suppository. The drug works by drawing water into the large intestine, thereby stimulating bowl movements (Lederle, 1995). Even in the absence of dietary sorbitol, cells can produce sorbitol naturally. When too much sorbitol is produced inside the cells, it can cause damage (Lorenzi, 2007). Diabetic retinopathy (DR) and neuropathy may be related to excess sorbitol in the cells of the eyes and nerves (Asnaghi *et al.*, 2003; Gabbay, 1973a; Lorenzi, 2007). The source of this sorbitol in diabetes is excess glucose, which goes through the polyol pathway.

The polyol pathway of glucose metabolism is active when the intercellular glucose levels are elevated in the cell (Gabbay, 1973b). Aldose reductase (AR), the first and the rate limiting enzyme in the pathway reduces glucose to sorbitol using NADPH as a cofactor (Lorenzi, 2007). Sorbitol is then metabolized to fructose by sorbitol dehydrogenase (SDH) that used NAD^+ as cofactor (Lorenzi, 2007). Sorbitol is an alcohol that is polyhydroxylated, and strongly hydrophilic and does not diffuse readily through cell membranes and accumulates intracellularly with possible osmotic consequences (Gabbay, 1973b). The fructose produced by the polyol pathway can get phosphorylated to fructose 3-phosphate (Szwergold *et al.*, 1990), which can be further broken down to 3-deoxyglucosone, and both these compounds can be very powerful glycosylating agents that can result in the formation of advanced glycation end products (AGEs) (Szwergold *et al.*, 1990). The usage of NADPH by AR may result in less cofactor becoming available to glutathione reductase, which is critical for the maintenance of the intracellular pools of reduced glutathione (GSH) (Lorenzi, 2007). This reduces the capability of cells to respond to oxidative stress (Barnett *et al.*, 1986). Compensatory increased activity of the glucose monophosphate shunt, the principle supplier of cellular NADPH may occur (Barnett *et al.*, 1986). The usage of NAD by SDH leads to an increased ratio of NADH/ NAD^+ which has been termed "pseudohypoxia" and linked to a multitude of metabolic and signaling changes known to alter cell function (Williamson *et al.*, 1993). Excess NADH serves as a substrate for NADH oxidase and this could be a mechanism for the generation of intracellular oxidant species (Lorenzi, 2007). Thus activation of polyol pathway, by altering the intracellular homeostasis, generating AGEs, and exposing cells to oxidant stress due to decreased antioxidant defense mechanism and generation of oxidant species can initiate several mechanisms of cellular damage.

Accumulation of sorbitol and fructose and the generation or enhancement of oxidative stress has been reported in the whole retina of diabetic animals (Dagher *et al.*, 2004; Gabbay, 1975; Lorenzi, 2007). The retinas of experimentally derived diabetic rats show increased lipid peroxidation (Obrosova *et al.*, 2003), increased nitrotyrosine formation (Obrosova *et al.*, 2005), and depletion of antioxidant enzymes (Obrosova *et al.*, 2003). These abnormalities are prevented by drugs that inhibit AR (Dahlin *et al.*, 1987; Lorenzi, 2007; Narayanan, 1993; Obrosova *et al.*, 2003; Tomlinson *et al.*, 1992, 1994).

Retinas from diabetic patients with retinopathy show more abundant AR immunoreactivity in ganglion cells, nerve fibers, and Muller cells than retinas from nondiabetic individuals (Vinores *et al.*, 1988). It has also been shown that human retinas from nondiabetic eye donors exposed to high glucose levels in organ cultures accumulate sorbitol to the same extent as similarly incubated retinas of nondiabetic rats (Dagher *et al.*, 2004). Retinal ganglion cells, Muller glia, vascular pericytes, and endothelial cells are endowed with AR in all species (Dagher *et al.*, 2004) and these cells are known to be damaged in diabetes (Lorenzi and Gerhardinger, 2001). The retinal vessels of diabetic rats treated with sorbinill, an AR inhibitor for the 9 months duration of diabetes, showed prevention of early complement activation, decreased levels of complement inhibitors, microvascular cell apoptosis, and acellular capillaries (Dagher *et al.*, 2004). Based on the data from the animal models, there is evidence for the concept that polyol pathway activation is a sufficient mechanism for the retinal abnormalities induced by diabetes in rats.

Neurodegeneration is an important component of DR as demonstrated by increased neural apoptosis in the retina during experimental and human diabetes (Barber *et al.*, 1998). IR activation has been shown to rescue retinal neurons from apoptosis through a phosphoinositide 3-kinase and protein kinase B (Akt) survival cascade. A significant decrease of retinal IR kinase activity has been reported after 4 weeks of hyperglycemia in STZ treated rats (Reiter *et al.*, 2006). Hyperosmotic-stress responses interact with the insulin signaling pathways at several levels (Ouwens *et al.*, 2001). Sorbitol has been previously shown to induce the tyrosine phosphorylation of IR (Ouwens *et al.*, 2001).

In the present study, we examined the retinal insulin receptor (IR) signaling in sorbitol-treated retinas *ex vivo* and show that sorbitol activates both the IR and IGF-1R tyrosine kinases, which results in activation of the receptor's direct downstream targets. This receptor activation leads to the activation of PI3K and Akt survival pathway in the retina. With the advent of phospho-site-specific antibody microarry, we observed either increased or decreased phosphorylation of several tyrosine, serine/threonine kinases, and cytoskeletal proteins which are downstream effector molecules of IR and IGF-1R signaling pathways.

II. Experimental Procedures

A. Materials

Human insulin R (rDNA origin) was obtained from Eli Lilly & Co. (Indianapolis, Indiana). The actin antibody was obtained from Affinity BioReagents (Golden, Colorado). Polyclonal anti-IRß, polyclonal anti-Cbl, and monoclonal anti-PY-99 antibodies were obtained from Santa Cruz Biotechnology (Santa Cruz, California). Polyclonal anti-Gab1 antibody was obtained from Upstate Biotechnology (Lake Placid, New York). Sorbitol, BAPTA, and SB 203580 were obtained from Sigma (St. Louis, Missouri). LY294002, anti-pAkt (S473), anti-Akt, anti-p38, and anti-phospho-p38 antibodies were obtained from Cell Signaling (Beverly, Massachusetts). Genestin, HNMP3, PP1, PP2, and PP3 were obtained from Calbiochem (San Diego, California). Actin antibody was obtained from Affinity BioReagents (Golden, Colorado). All other reagents were of analytical grade and from Sigma (St. Louis, Missouri).

B. Animals

All animal work was done in strict accordance with the NIH Guide for the Care and Use of Laboratory Animals and the Association for Research in Vision and Ophthalmology on the Use of Animals in Vision Research. All protocols were approved by the IACUC at the University of Oklahoma Health Sciences Center and the Dean McGee Eye Institute. In all experiments, rats were killed by asphyxiation with carbon dioxide before the retinas were removed.

C. Retinal organ cultures

Retinal organ cultures were carried out as previously described (Rajala *et al.*, 2004, 2007). Retinas were removed from Sprague-Dawley albino rats that were born and raised in dim cyclic light (5 lux; 12 h ON:12 h OFF), and incubated for either 5 min (insulin) or 30 min (sorbitol) at 37 °C in Dulbecco's modified Eagle's (DMEM) medium (Gibco BRL) in the presence of either insulin or sorbitol. Control cultures were carried out in the absence of additives. At the indicated times, retinas were snap-frozen in liquid nitrogen and stored at –80 °C until analyzed. The retinas were lysed in lysis buffer [1% NP 40, 20 mM HEPES (pH 7.4), and 2 mM EDTA] containing phosphatase inhibitors (100 mM NaF, 10 mM Na_4P_2O7, 1 mM $NaVO_3$, and 1 mM molybdate) and protease inhibitors (10 μM leupeptin, 10 μg/ml aprotinin, and 1 mM PMSF), and kept on ice for 10 min followed by centrifugation at 4 °C for 20 min.

D. Preparation of rat rod outer segments

Retinas in culture were stimulated with either 1 μM insulin or 1 M sorbitol for 30 min at 37 °C. After treatment, the rod outer segments (ROS) were prepared using a discontinuous sucrose gradient as previously described (Rajala *et al.*, 2002). Retinas were homogenized in 4.0 ml of ice-cold 47% sucrose solution containing buffer A (100 mM NaCl, 1 mM EDTA,

1 mM $NaVO_3$, 1 mM PMSF, and 10 mM Tris–HCl (pH 7.4)). Retinal homogenates were transferred to 15-ml centrifuge tubes and sequentially overlaid with 3.0 ml of 42%, 3.0 ml of 37%, and 4.0 ml of 32% sucrose dissolved in buffer A. The gradients were spun in a swinging bucket rotor at 82,000*g* for 1 h at 4 °C. The 32/37% interfacial sucrose band, containing the ROS membranes, was harvested and diluted with 10 mM Tris–HCl (pH 7.4) containing 100 mM NaCl and 1 mM EDTA. The band solution was then centrifuged at 27,000*g* for 30 min at 4 °C. The ROS pellets were resuspended in 10 mM Tris–HCl (pH 7.4) containing 100 mM NaCl and 1 mM EDTA, and stored at –20 °C. The non-ROS band designated Band II (37/42%) was also saved for comparison with the ROS fraction. All protein concentrations were determined by BCA reagent (Pierce, Rockford, IL) following the manufacturer's instructions.

E. Cloning, expression, and purification of protein tyrosine phosphatase 1B

Retinal PTP1B was obtained by PCR after reverse transcribing mouse retinal RNA and using PTP1B primers (sense: GAA TTC ATG GAG ATG GAG AAG GAG TTC GAG; antisense: GTC GAC TCA GTG AAA ACA CAC CCG GTA GC). The PCR product was verified by DNA sequencing, digested with EcoR1 and Sal1, and cloned into GST fusion vector, pGEX-4T1. Site-directed mutagenesis was carried out according to the method described earlier (Rajala *et al.*, 2004). Mutant PTP1B-D181A was amplified using primers, (sense: ACC ACA TGG CCT GCC TTT GGA GTC CCC; antisense: GGG GAC TCC AAA GGC AGG CCA TGT GGT). The PCR products were cloned into TOPO vector (Invitrogen) and both the WT and mutant sequences were verified by DNA sequencing. The WT and mutant cDNAs were later excised from the sequencing vector as EcoRI/SalI fragments and cloned into GST fusion vector, pGEX-4T1. An overnight culture of *E. coli* BL21 (DE3) (pGEX-PTP1Bor pGEX-PTP1B-D181A) was diluted 1:10 with 100 μg/ml ampicillin per milliliter, grown for 1 h at 37 °C, and induced for another hour by addition of IPTG to 1 mM. Bacteria were sonicated three times for 20 s each time in lysis buffer containing 10 mM imidazole-HCl (pH 7.2), 1 mM EDTA, 100 mM NaCl, 1 mM dithiothreitol, and 1% Triton X-100. Lysates were clarified by centrifugation, and the supernatants were incubated with 500 μl of 50% glutathione-coupled beads (Amersham Pharmacia) for 30 min at 4 °C. The GST–PTP1B fusion proteins were washed in lysis buffer and eluted twice with 1 ml of 5 mM reduced glutathione (Sigma) in phosphatase buffer (20 mM Tris (pH 7.4), 5% glycerol, 0.05% Trion X-100, 2.5 mM $MgCl_2$, aprotinin (2 μg/ml), and leupeptin (5 μg/ml)). Glycerol was added to a final concentration of 33% (vol/vol), and aliquots of enzyme were stored at -20 °C.

F. PI3-kinase assay

Enzyme assays were carried out as previously described (Rajala *et al.*, 2007). Briefly, assays were performed directly on either IRβ or Cbl immunoprecipitates of retinal lysates prepared from sorbitol treated or untreated lysates in 50 μl of reaction mixture containing 0.2 mg/ml PI-4,5-P_2, 50 μM ATP, 10 μCi [γ^{32}P]ATP, 5 mM $MgCl_2$, and 10 mM HEPES buffer (pH 7.5). The reactions were carried out for 30 min at room temperature and stopped by the addition of 100 μl of 1 N HCl followed by 200 μl of chloroform/methanol (1/1, v/v). Lipids were extracted and resolved on oxalate-coated TLC plates (silica gel 60) with a solvent system of 2-propanol/2 M acetic acid (65/35, v/v). The plates were coated in 1% (w/v) potassium oxalate in 50% (v/v) methanol and then baked in an oven at 100 °C for 1 h prior to use. TLC plates were exposed to X-ray film overnight at –70 °C and radioactive lipids were scraped and quantified by liquid scintillation counting.

G. Phosphatase activity assay

The sorbitol-treated or untreated retinas were lysed in buffer containing 10 mM imidazole-HCl (pH7.2), 1 mM EDTA, 100 mM NaCl, 1 mM dithiothreitol, and 1% Triton X-100. The assays were performed (Takai and Mieskes, 1991) directly on retinal lysates in 80 μl of reaction mixture containing assay buffer (25 mM HEPES (pH 7.2), 50 mM NaCl, 5 mM dithiothritol, 2.5 mM EDTA), 5 μl of 5% BSA and either test or positive control sample. The reactions were preincubated at 37 °C for 15 min, followed by the addition of 120 μl of pNPP (1.5 mg/ml) and incubated for 5–15 min at 37 °C. The reactions were stopped by the addition of 20 μl of 13% K_2HPO_4 and the absorbance read at 405 nm. One unit of enzyme activity was equivalent to 1 nmol PNPP hydrolyzed per minute and the extinction coefficient for pNPP at $A_{405} = 1.78 \times 10^4\ M^{-1}\ cm^{-1}$.

H. Dephosphorylation of tyrosine phosphorylated proteins by PTP1B *in vitro*

Sorbitol treated or untreated retinal proteins were incubated in the presence of catalytically active or inactive PTP1B for 30 min at 30 °C. At the end of the incubation period, the reaction products were subjected to Western blot analysis with anti-PY99 antibody.

I. Antibody microarry

Control and sorbitol-treated retinas in culture were lysed in lysis buffer (1% NP 40, 20 mM HEPES (pH 7.4), 2 mM EDTA, and 1 mM dithiothreitol) containing phosphatase inhibitors (100 mM NaF, 10 mM Na_4P_2O7, 1 mM

$NaVO_3$, and 1 mM molybdate) and protease inhibitors (10 μM leupeptin, 10 μg/ml aprotinin, and 1 mM PMSF), and kept on ice for 10 min followed by centrifugation at 4 °C for 20 min. The protein samples were then subjected to screening using antibody microarry containing 377 pan-specific and 273 phospho-site-specific antibodies (Kinexus Services, Vancouver, Canada).

J. Immunoprecipitation

Retinal lysates were prepared as previously described (Li *et al.*, 2007). Insoluble material was removed by centrifugation at 17,000*g* for 20 min at 4 °C, and the solubilized proteins were precleared by incubation with 40 μl of protein A-Sepharose for 1 h at 4 °C with mixing. The supernatant was incubated with primary antibodies overnight at 4 °C and subsequently with 40 μl of protein A-Sepharose for 2 h at 4 °C. Following centrifugation at 17,000*g* for 1 min at 4 °C, immune complexes were washed three times with ice-cold wash buffer (50 mM HEPES (pH 7.4) 118 mM NaCl, 100 mM NaF, 2 mM $NaVO_3$, 0.1% (w/v) SDS, and 1% (v/v) Triton X-100). The immunoprecipitates were either subjected to Western blotting analysis or used to measure the PI3K activity.

K. SDS-PAGE and Western blotting

Proteins were resolved by 10% SDS-PAGE and transferred onto nitrocellulose membranes. The blots were washed twice for 10 min with TTBS (20 mM Tris–HCl (pH 7.4), 100 mM NaCl, and 0.1% Tween-20) and blocked with either 5% bovine serum albumin or nonfat dry milk powder (Bio-Rad) in TTBS for 1 h at room temperature. Blots were then incubated with anti-PY99 (1:1000), anti-Cbl (1:1000), anti-pAkt (1:1000), anti-Akt (1:1000), or anti-Akt1 (1:1000) or anti-Akt2 (1:1000) or anti-Akt3 (1:1000) or anti-IRß or anti-IGF 1R (1:1000) or anti-actin (1:1000) antibodies overnight at 4 °C. Following primary antibody incubations, immunoblots were incubated with HRP-linked secondary antibodies (either anti-rabbit or anti-mouse) and developed by ECL according to the manufacturer's instructions.

III. Results

A. Sorbitol-induced activation of p38MAP kinase

The p38MAP kinase is known to be activated in stress (Cheng *et al.*, 2002). To determine if sorbitol induces the activation of p38MAP kinase, we subjected the sorbitol-treated retinal proteins to Western blot analysis with anti-phospho-p38 and total p38 antibodies. The results indicate a gradual increase in p38 phosphorylation between 0.2 and 1.0 M sorbitol compared to control (Fig. 20.1A). The total p38 levels were unchanged in the presence of varying concentrations of sorbitol (Fig. 20.1B). These results suggested that sorbitiol induces the activation of p38MAP kinase and the organotypic cultures were mimicking the *in vivo* stress condition.

B. Sorbitol-induced activation of insulin- and insulin-like growth factor-1 receptor

To determine the sorbitol-induced activation of IR and IGF-1 receptors, we immunoprecipitated retinal lysates from control and sorbitol-treated organotypic cultures with anti-IRß (Fig. 20.2B) and anti-IGF 1R (Fig. 20.2E) antibodies followed by Western blot analysis with anti-PY99 antibody. The results indicated the activation of IR (Fig. 20.2A and C) and IGF 1R (Fig. 20.2D and F) in response to sorbitol-induced stress.

C. Sorbitol-induced activation of Akt

To determine if sorbitol-induced the activation of Akt, we incubated rat retinas in organotypic cultures for 30 min in the presence of varying concentrations of sorbitol (0, 0.2, 0.5, 1.0, 2.0, and 3.0 M). At the end of the incubation, the retinas were lysed and subjected to Western blot analysis with anti-pAkt (S473) and anti-Akt antibodies. The results indicated a gradual increase in Akt phosphorylation from 1.0 to 3.0 M sorbitol (Fig. 20.3A). The total Akt levels did not change in response to sorbitol treatment (Fig. 20.3B). These results suggested that sorbitol induced the activation of Akt.

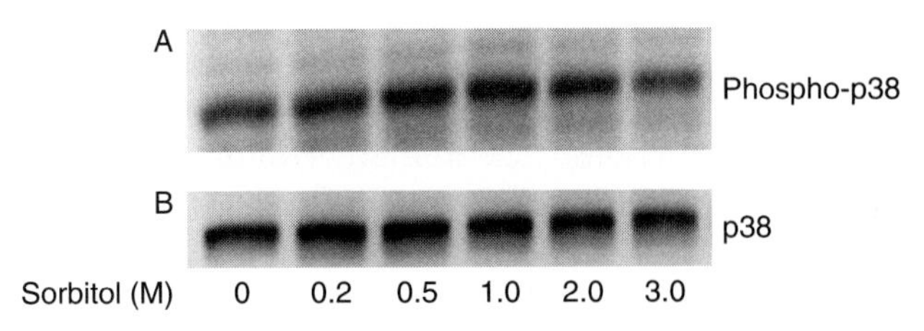

Figure 20.1 Concentration dependent Sorbitol-induced activation of p38 MAP kinase. Various concentrations of sorbitol treated retinal samples were subjected to Western blot analysis with anti-phospho-p38MAP kinase (A) and anti-p38 MAP kinase (B) antibodies.

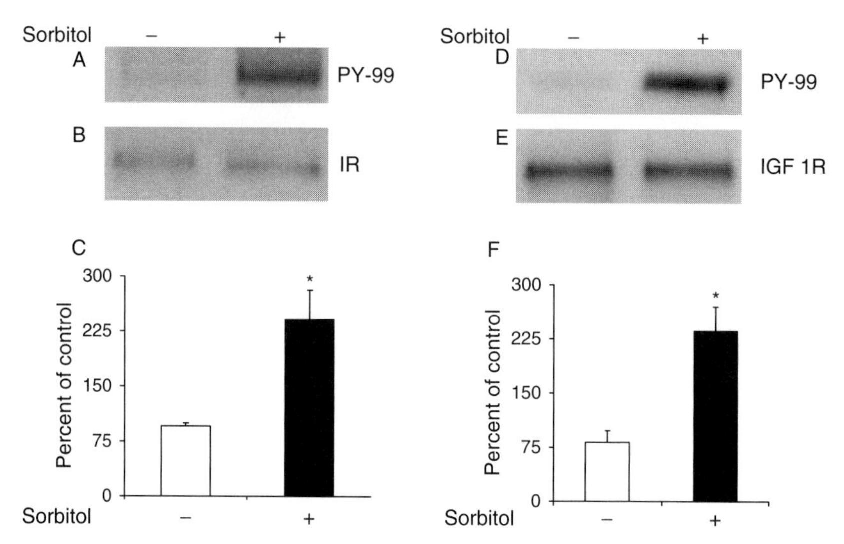

Figure 20.2 Sorbitol-induced activation of IR and IGF-1R. Retinal proteins from control and 1.0 M sorbitol-treated organotypic cultures were immunoprecipitated with anti-IRß antibody followed by Western blot analysis with anti-PY99 antibody (A). The blots were stripped and reprobed with anti-IRß antibody to ensure equal amounts of IR in each immunoprecipitate (B). Densities were calculated from the immunoblots and the results are expressed as percentage of PY99/IR. Absence of sorbitol was taken as 100 percent (C). Data mean ± SD, $n = 3$, $^{*}p < 0.001$. Sorbitol-induced activation of IGF 1R. Retinal proteins from the control and 1.0 M sorbitol-treated organotypic cultures were immunoprecipitated with anti-IGF 1R antibody followed by Western blot analysis with anti-PY99 antibody (D). The blots were stripped and reprobed with anti-IGF 1R antibody to ensure equal amount of IGF 1R in each immunoprecipitate (E). Densities were calculated from the immunoblots and the results are expressed as percentage of PY99/IR. Absence of sorbitol was taken as 100 percent (F). Data mean ± SD, $n = 3$, $^{*}p < 0.001$.

D. Sorbitol-induced activation of Akt2

Akt exist in three isoforms, all of which are expressed in the retina (Reiter *et al.*, 2003). To determine if sorbitol stress activated a specific Akt isoform, we immunoprecipitated retinal lysates from nonstimulated control, insulin-stimulated, and the sorbitol-treated organotypic cultures with anti-Akt1, anti-Akt2, and anti-Akt3 antibodies. The immune complexes were subjected to Western blot analysis with anti-pAkt antibody. The results indicated that insulin activated the Akt1 and the Akt3 isoforms, but not Akt2 (Fig. 20.3C). Sorbitol treatment, however, resulted in the activation of all three isoforms of Akt (Fig. 20.3D). These results suggested that the activation of Akt2 isoform may be stress-dependent.

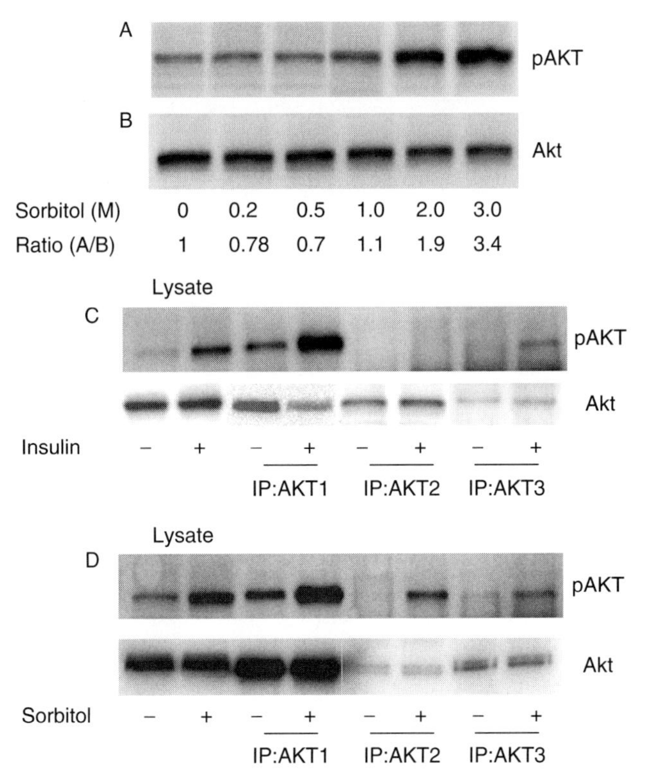

Figure 20.3 Sorbitol-induced activation of Akt isoforms. Various concentrations of sorbitol treated retinal samples were subjected to Western blot analysis with anti-pAkt (S473) (A) and anti-Akt (B) antibodies. Retinas were incubated in culture and treated with either insulin (1 μM) or sorbitol (3 M). Retinas were lysed and the proteins were subjected to immunoprecipitation with anti-Akt1, anti-Akt2 and Anti-Akt3 antibodies followed by Western blot analysis with anti-pAkt (S473) (C and D) antibody. Input, 30 μg of retina lysates stimulated with either presence or absence of insulin or sorbitol.

E. Sorbitol induced activation of IR associated PI3K activity

1. Stress-induced activation of PI3K

We have previously reported the activation of PI3K through tyrosine phosphorylated IR in the retina (Rajala *et al.*, 2007). To determine whether the activation of PI3K is regulated through IR, we have immunoprecipitated IR from retinal lysates that were prepared from non stimulated control and sorbitol-treated (0–3.0 M) organotypic cultures, and measured the PI3K activity. The results indicated an increased PI3K activity with IR from sorbitol-treated retinas (Fig. 20.4). These results suggested that sorbitol-induced activation of PI3K occurs via activation of the IR.

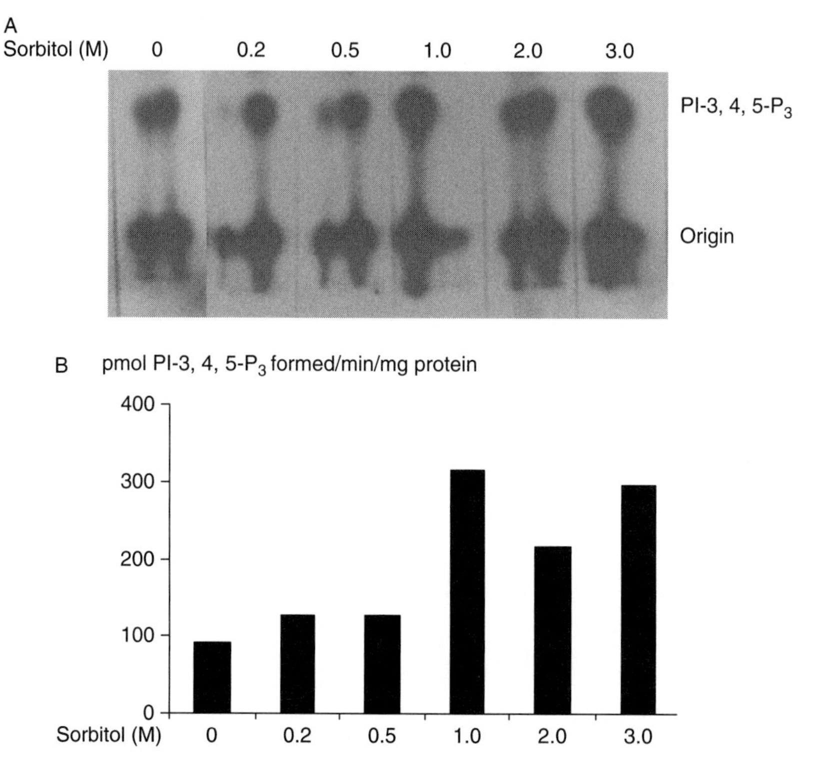

Figure 20.4 Sorbitol-induced activation insulin receptor associated PI3K activity. Retinas were cultured in DMEM and treated with various concentrations of sorbitol (0–3 M) for 30 min at 37°C. TLC autoradiogram of PI3K activity measured in anti-IRβ immunoprecipitates of retinas using PI-4,5-P_2 and [γ^{32}P]ATP as substrates (A). The radioactive spots of PI-3,4,5-P_3 were scraped from the TLC plate and counted (B).

2. PI3K-independent activation of Akt

To determine whether sorbitol also induced the activation of Akt independent of PI3K, we incubated the retinas in the presence of the PI3K inhibitor LY294002 for 30 min before sorbitol or insulin treatment. Retinal proteins were subjected to Western blot analysis with anti-pAkt and total Akt antibodies. The results indicated that LY294002 failed to inhibit the sorbitol-induced activation of Akt, but LY294002 inhibited the insulin-induced activation of Akt (Fig. 20.5A–C), suggesting that Akt activation in hyperosomotic stress can also occur without PI3K activation.

3. p38MAP kinase activation is independent of PI3K activation

To determine if PI3K activation regulates the p38MAP kinase pathway, we preincubated the retinas in the presence of either the PI3K inhibitor LY294002 or the MAP kinase inhibitor SB203580 followed by sorbitol

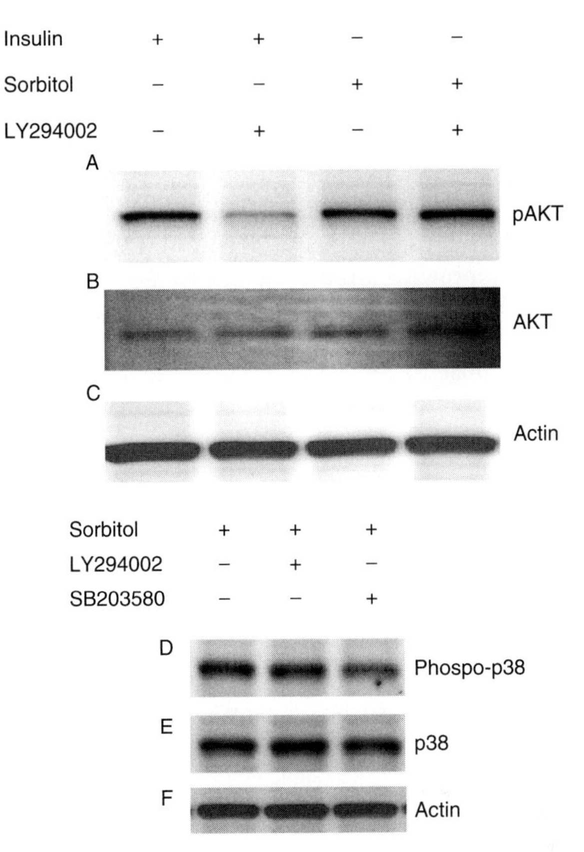

Figure 20.5 PI3K–independent activation of Akt. Rat retinas were pre-incubated in DMEM medium with or without the PI3K inhibitor LY294002 (50 μM) for 30 min prior to either 1μM insulin or 3 M sorbitol. Thirty micrograms of retina lysate were subjected to Western blot analysis with anti-phospho-Akt (Ser 473) (A). The blot was stripped and reprobed with anti-Akt (B) and anti-actin (C) antibodies. PI3K activation is independent of p38 MAP kinase activation. Rat retinas were pre-incubated in DMEM medium with or without the PI3K inhibitor LY294002 (50 μM) or MAP kinase inhibitor SB203580 (50 μM) for 30 min prior to the treatment of either 1μM insulin or 3 M sorbitol. Thirty micrograms of retina lysates were subjected to Western blot analysis with anti-phospho-p38 MAP kinase(D). The blot was stripped and reprobed with anti-p38 MAP kinase (E) and anti-actin (F) antibodies.

treatment. The retinal proteins were subjected to Western blot analysis with anti-phospho-p38 and anti-p38 antibodies. The results indicated that phosphorylation of p38 may be inhibited by the MAP kinase inhibitor SB203580 (Fig. 20.5), whereas the PI3K inhibitor LY294002 failed to inhibit the sorbitol-induced activation of p38 MAP kinase (Fig. 20.5). These results suggested that MAP kinase activation is independent of PI3K activation during hyperosomotic stress.

F. Role of calcium in sorbitol-induced Akt activation

It has been shown that hyperosomotic stress evokes an increase in the cytosolic calcium concentration and triggers calcium signaling (Marchenko and Sage, 2000; Pritchard *et al.*, 2002). To determine whether the sorbitol-induced activation of Akt is calcium dependent, we pre-incubated the retinas in organotypic cultures in the presence of the calcium specific chelator BAPTA followed by 1.0 M sorbitol treatment. The retinas were lysed and the proteins were subjected to Western blot analysis with anti-pAkt, anti-Akt, and anti-actin antibodies. The results indicated that BAPTA failed to reduce the activation of Akt and that the levels of Akt activation were similar to the levels seen in the sorbitol treatment (Fig. 20.6A). The blot was stripped and reprobed with total Akt (Fig. 20.6B) and actin (Fig. 20.6C) to ensure that equal amounts of protein were loaded. These results suggested that under our experimental conditions, calcium has no role in the sorbitol-induced activation of Akt.

G. Tyrosine kinase-induced activation of Akt

To determine the pathway by which Akt undergoes activation in sorbitol stress, we have incubated the retinas in organotypic cultures in the presence of tyrosine kinase inhibitors such as genestin (inhibitor of tyrosine kinases), HNMP3 (inhibitor of IR kinase activity), PP1 (src kinase inhibitor), PP2 (src kinase inhibitor), and PP3 (a negative control for the src family protein tyrosine kinase inhibitor PP2) followed by sorbitol treatment. Retinas were lysed, and the proteins were subjected to Western blot analysis with anti-pAkt, anti-Akt, and anti-actin antibodies. The results indicated that genestin

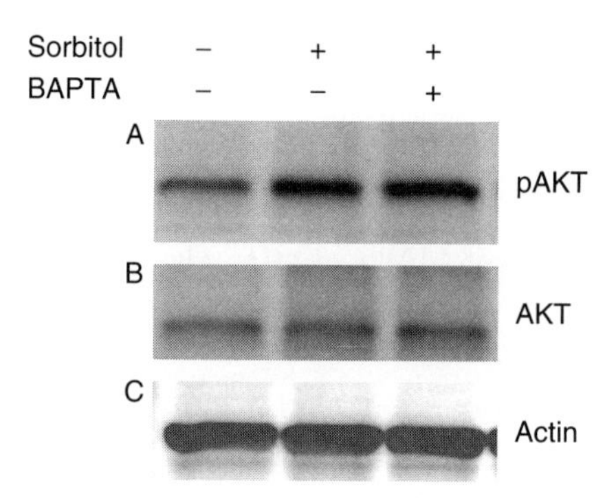

Figure 20.6 Sorbitol-induced activation of Akt is calcium-independent. Rat retinas were pre-incubated in DMEM medium with or without 15 μM BAPTA for 30 min prior to the treatment of 1 M sorbitol for additional 30 min. Thirty micrograms of retina lysates were subjected to Western blot analysis with anti-pAkt (S473) (A) anti-Akt (B), and anti-actin (C) antibodies.

blocked the activation of Akt, but all other tyrosine kinase and nonreceptor tyrosine kinase inhibitors failed to block the activation of Akt (Fig. 20.7A). These results suggested that stress-induced activation of Akt is not under the regulation of nonreceptor src family tyrosine kinase(s). Collectively these experiments suggested that the activation of Akt in osmotic stress could be through tyrosine kinase activation.

H. Cholesterol depletion results in the sorbitol-induced activation of Akt

Localization of IR in caveolae of adipocyte plasma membrane has been reported and cholesterol depletion attenuates IR signaling (Gustavsson *et al.*, 1999). To examine the role of lipid rafts on Akt activation in sorbitol induced stress, we stimulated retinas in organ cultures with either insulin or sorbitol in the presence or absence of cholesterol-sequestering agent, methyl-β-cyclodextrin (MCD), a treatment that disrupts cholesterol-rich detergent-resistant membranes. MCD treatment resulted in a dramatic increase in sorbitol-induced activation of Akt compared to retinas treated with sorbitol in the absence of MCD (Fig. 20.8). Insulin effect on the activation of Akt is significantly lower than either sorbitol or sorbitol in the presence of MCD (Fig. 20.8). These results clearly suggested that disruption of lipid rafts in sorbitol induced stress, which resulted in the activation of Akt.

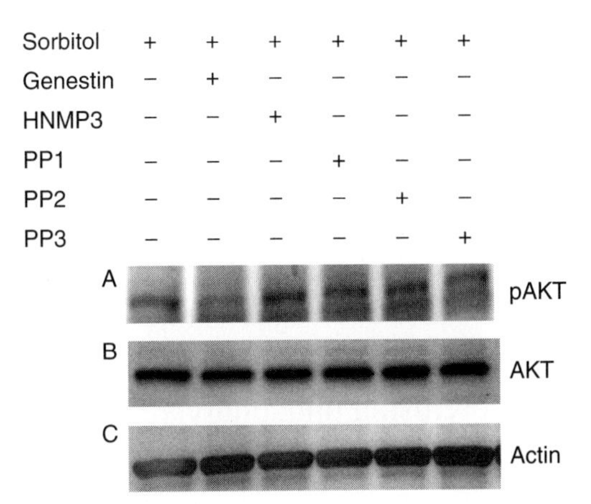

Figure 20.7 Tyrosine kinase induced–activation of Akt. Rat retinas were pre-incubated in DMEM medium with or without 100 μM genestin or HNMP3 or PP1 or PP2 or PP3 for 30 min prior to the treatment of 1 M sorbitol for 30 min. Thirty micrograms of retina lysates were subjected to Western blot analysis with anti-pAkt (S473) (A) anti-Akt (B) and anti-actin (C) antibodies.

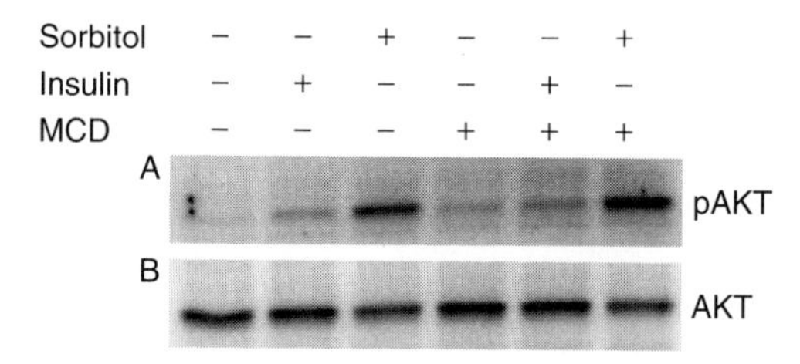

Figure 20.8 Cholesterol-depletion results in the activation of Akt. Rat retinas were incubated in DMEM medium with or without MCD prior to the treatment of 1 M sorbitol for 30 min. Thirty micrograms of retina lysates were subjected to Western blot analysis with anti-pAkt (S473) (A) and anti-Akt (B) antibodies.

I. Sorbitol-induced activation of protein phosphatase (PA) activity

To determine whether sorbitol induced the activation of PA, we did phosphatase assays, using *p*-Nitrophenyl Phosphate (pNPP) as the substrate. The results indicated a significant increase in the PA activity in 3.0 M treated retinas compared to untreated retinas (Fig. 20.9A). The results suggested that hyperosmotic stress increased the activation of PA activity. The total PA activity we measured did not differentiate serine/threonine phosphatase activity from protein tyrosine phosphatase activity (PTP). To differentiate between the two activities, we measured PA activity in the presence of sodium vanadate to inhibit the PTPase activity. The results indicated a significant increase in serine/threonine PA activity in sorbitol treated retinas compared to untreated retinas (Fig. 20.9A). Purified PTP1B was used as control for the sodium vanadate experiment. The results indicated the complete inhibition of PTP1B activity in the presence of sodium vanadate (Fig. 20.9A). The protein samples used for the PA activity were used to examine the phosphorylation of Akt and the results indicated an increased phosphorylation of Akt in 3.0 M sorbitol treated retinas (Fig. 20.9B). The blot were then stripped and reprobed with total Akt to ensure equal amounts of protein were loaded (Fig. 20.9B). Collectively, these results suggested that the observed activation of Akt in hyperosmotic stress is not due to the inhibition of phosphatase activity by sorbitol.

J. Sorbitol-induced tyrosine phosphorylation of several retinal proteins

The PI3K-independent activation of Akt in the sorbitol-treated retinas prompted us to investigate the pathway by which Akt undergoes activation. Retinal proteins were either stressed with sorbitol or stimulated with insulin and subjected to Western blot analysis with the anti-PY99 antibody. The results indicated a significantly increased level of tyrosine phosphorylation in

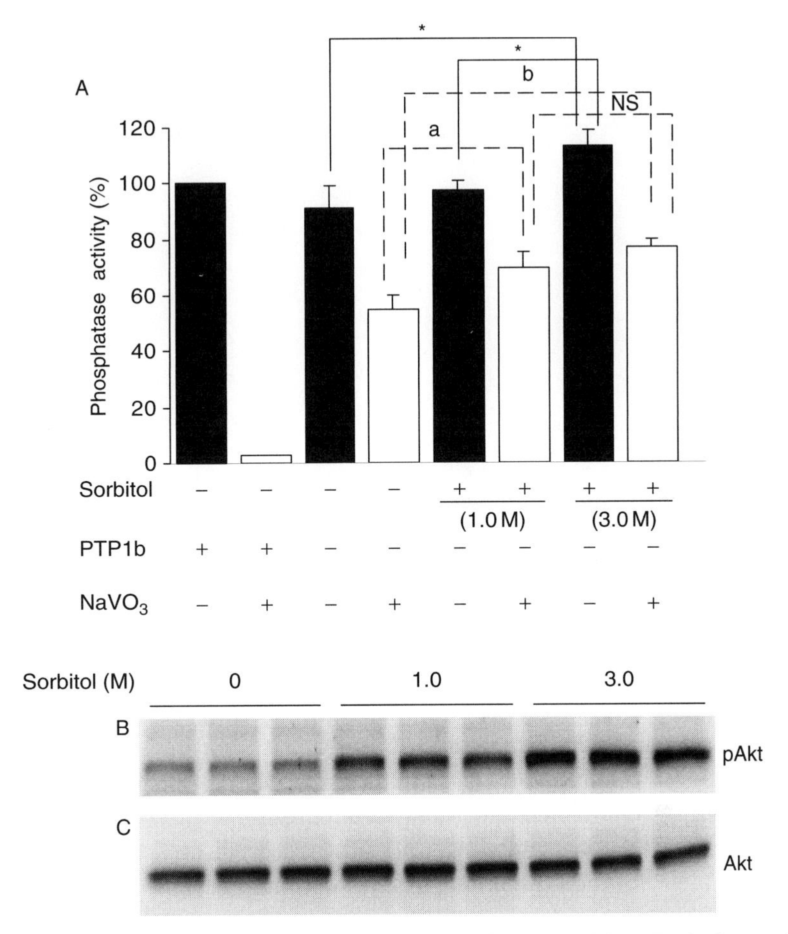

Figure 20.9 Sorbitol-induced activation of phosphatase activity. Retinal proteins from control and sorbitol-treated (1.0 or 3.0 M) or untreated organotypic cultures were subjected to either Western blotting analysis with anti-pAkt (B) and anti-Akt (C) antibodies or measured the phosphatase activity using pNPP as substrate (A). Open bars represent the activity in the presence of 1 mM sodium vanadate. Purified GST-PTP1B (5 μg) was used as positive control. Data mean $\pm$ SD, $n = 3$, $^{*}p < 0.001$.

the retinal proteins in the 1.0 and 2.0 M sorbitol treatment compared to the insulin-stimulated retinas (Fig. 20.10A). We observed the tyrosine phosphorylation of several retinal proteins, with apparent molecular weights of 170, 130, 115, 79, 70, and 41 kDa.

To determine whether the stress-induced tyrosine phosphorylated proteins are localized to the ROS or other retinal membranes, we probed the ROS and the non-ROS fractions with the anti-PY99 antibody. The results indicated a much greater tyrosine phosphorylation in the non-ROS fraction

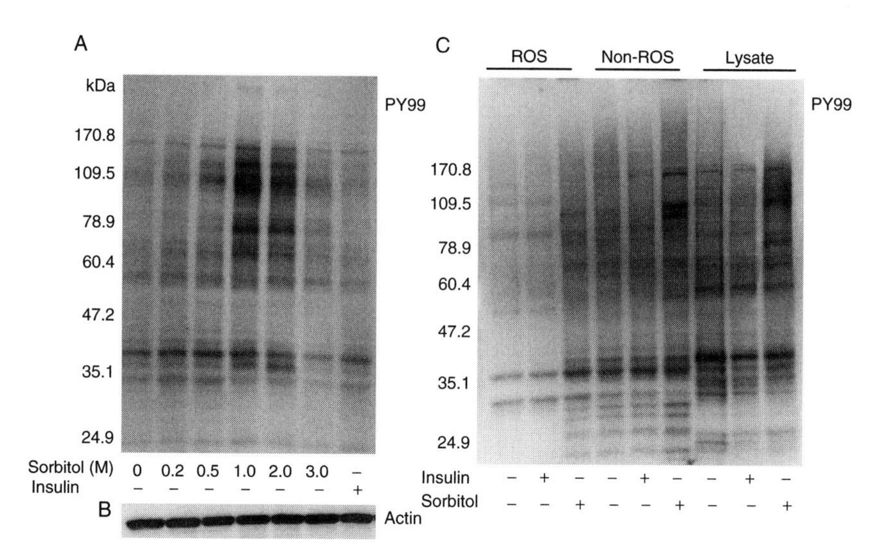

Figure 20.10 Sorbitol-induced tyrosine phosphorylation of several retinal proteins. Retinas were cultured in DMEM in the presence or the absence of either 1 μM insulin or various concentrations of sorbitol for 30 min at 37°C. Thirty micrograms of retinal proteins were subjected to Western blot analysis with anti-PY 99 antibody (A). The blot was reprobed with the anti-actin antibody to ensure equal amount of protein in each lane (B). ROS and non-ROS proteins were subjected to Western blot analysis with the anti-PY99 antibody (C).

compared to the ROS (Fig. 20.10C). These results also suggested that stress response induces a significant tyrosine phosphorylation in inner segment and other retinal cell membranes.

To determine the global changes in the phosphorylation (tyrosine and serine/threonine) of retinal proteins, we examined the phosphorylation state of retinal proteins by antibody microarray. Of 273 phospho-site-specific antibodies 33 proteins were found to exhibit either increased or decreased phosphorylation (data not shown). These proteins include serine/threonine and tyrosine kinases and mainly proteins involved in the cytoskeletal organization. These results clearly suggest that hyperosmotic stress-induces the activation of several protein kinases which may in turn regulate the cytoskeletal reorganization.

K. Sorbitol-induced tyrosine phosphorylated proteins are the substrates of PTP1B *in vitro*

Sorbitol treated retinal proteins were subjected to *in vitro* phosphatase assays in the presence of either catalytically active or inactive PTP1B enzyme. After incubation, the reaction products were subjected to SDS-PAGE followed by Western blot analysis with anti-PY99 antibody. The results indicated that PTP1B dephosphorylates the major 130 and 115 kDa tyrosine phosphorylated proteins in sorbitol-induced stress (Fig. 20.11).

L. Interaction of retinal Cbl with p85 subunit of PI3K under hyperosmotic conditions

To determine the physical interaction between Cbl and p85 subunit of PI3K, we subjected sorbitol-treated or untreated retinal lysates to GST-pull down experiments with GST-p85 full length fusion protein followed by

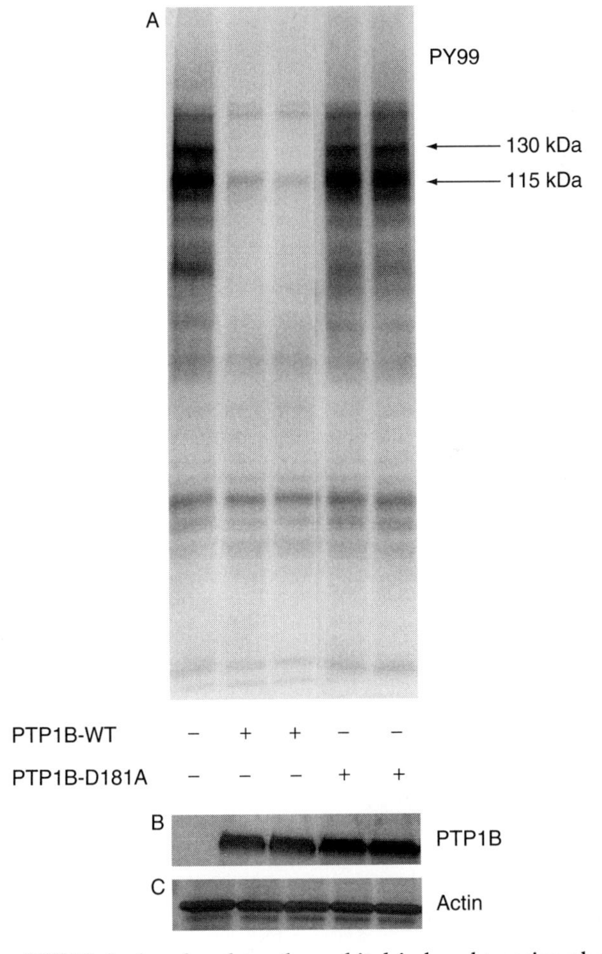

Figure 20.11 PTP1B dephosphorylates the sorbitol-induced tyrosine phosphorylated proteins *in vitro*. Sorbitol-treated retinal proteins were incubated with either wild type PTP1B or catalytically inactive PTP1B (D181A) for 15 min at 37°C. At the end of the reaction, proteins were subjected to SDS-PAGE followed by Western blot analysis with anti-PY99 antibody (A). The blot was reprobed with anti-GST to detect PTP1B (B) and anti-actin (C) antibodies.

Western blot analysis with anti-Cbl antibody. The results showed the binding of Cbl to p85 subunit of PI3K (Fig. 20.12A). To further determine whether the binding is phosphorylation dependent, we immunoprecipitated Cbl from sorbitol-treated or untreated retinal lysates followed by either Western blot analysis with anti-PY 99 antibody (Fig. 20.12B) or directly measured the Cbl associated PI3K activity (Fig. 20.12C). The results clearly indicated that sorbitol-induced the phopshorylation of Cbl (Fig. 20.12B) and that the phosphorylated Cbl was able to associate with the PI3K activity (Fig. 20.12C). In addition to being an adaptor protein, Cbl has been characterized as an E3 ubiquitin ligase that interacts with PI3K and mediates its ubiquitination and proteasome degradation (Fang *et al.*, 2001). This interaction is phosphorylation-independent and it is established through binding of the SH3 domain of p85 with the proline-rich region of Cbl (Fang *et al.*, 2001). In order to identify the ubiquitination state of p85

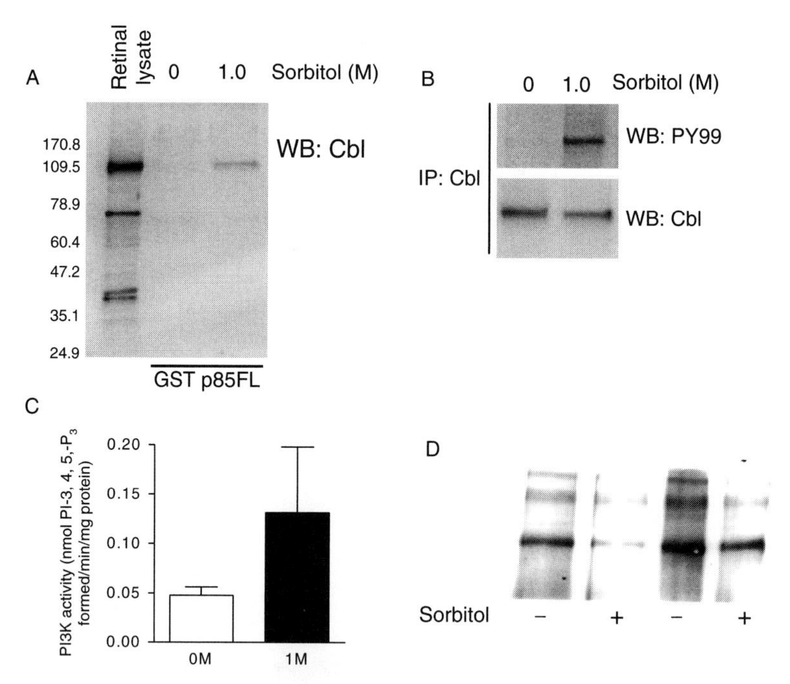

Figure 20.12 Interaction of Cbl with p85 subunit of PI3K. Sorbitol (1.0 M) treated or untreated retinal lysates were subjected to GST-pull down assay with GST-p85 full-length fusion protein followed by Western blot analysis with anti-Cbl antibody (A). Sorbitol treated and untreated retinal lysates were subjected to immunoprecipitation with anti-Cbl antibody followed by either Western blot analysis with anti-PY99 antibody (B) or measured the Cbl associated PI3K activity (C). Sorbitol treated and untreated retinal lysates were immunoprecipitated with anti-ubiquitin antibody followed by Western blot analysis with anti-p85 subunit of PI3K antibody (D).

in sorbitol treated and untreated retinas, immunoblots of anti-Ub IPs were probed with anti-p85 antibody. The results indicated a decrease in p85 polyubiquitination in sorbitol-treated retinas compared to the untreated ones (Fig. 20.12D). This finding suggests that, phosphorylation of Cbl during hyperosmotic stress may act as a "switch" changing the function of Cbl from a E3 ubiquitin ligase to an adaptor function and thereby regulate the activity of PI3K.

IV. Discussion

Our studies clearly indicate that both Akt and MAP kinase pathway are activated in response to sorbitol stress. It has been shown previously that MAPKs acts as glucose transducers for diabetic complications (Tomlinson, 1999). The sorbitol pathway, nonenzymatic glycation of proteins and increased oxidative stress are known to activate protein kinase C which is an effective activator of MAPKs (Tomlinson, 1999). These kinases phosphorylate transcription factors, which in turn alter the balance of gene expression and promote the development of diabetic nephropathy, retinopathy and neuropathy (Tomlinson, 1999). The normal retinal IR exhibits high constitutive activity that is reduced in diabetes (Reiter *et al.*, 2006). The diabetic rat retina further shows loss of PI3K, Akt1 and Akt-3, mTOR and p70S6K activities, and increased GSK3β activity (Reiter and Gardner, 2003). Elevated levels of sorbitol have been shown to be implicated in the pathogenesis of DR (Asnaghi *et al.*, 2003; Dagher *et al.*, 2004; Lorenzi, 2007; Lorenzi and Gerhardinger, 2001; Mizutani *et al.*, 1998). The rate limiting step in the pathway, AR which reduces the glucose to sorbitol is the major therapeutic target for DR (Chandra *et al.*, 2002; Dahlin *et al.*, 1987; Dvornik *et al.*, 1973; Kinoshita *et al.*, 1979; Lorenzi, 2007; Obrosova *et al.*, 2003, 2005; Tomlinson *et al.*, 1992). In this study, like insulin, sorbitol was found to induce tyrosine phosphorylation of IR and IGF-1R. It was reported previously that insertion of IR into the plasma membrane is necessary for sorbitol-induced IR activation (Ouwens *et al.*, 2001). Consistent with these observations, we reported that IRs in ROS of retinas are localized to plasma membrane (Rajala *et al.*, 2007). Further studies, however, are required to understand how the IR kinase activity is being reduced in diabetes.

The organotypic culture system has been successfully used to study protein phosphorylation and provides access to the retina for the addition of exogenous modulators of cellular function(Rajala *et al.*, 2004) We used this system to study the activation of IR, IGF-1, PI3K, Akt, and MAP kinase activation under hyperosmotic stress. Inactivation and dephosphorylation of Akt have been reported during hyperosmotic stress (Meier *et al.*, 1998)

These studies were done in cell culture with 0.5 M sorbitol and at this concentration, sorbitol activated the MAP kinase but not Akt kinase pathway (Meier *et al.*, 1998) Another independent study also failed to demonstrate the activation of Akt in 3T3L1 adipocytes under hyperosmotic stress using 0.6 M sorbitol (Chen *et al.*, 1997) In agreement with these studies, we did not observe activation of Akt at 0.5 M sorbitol in our *ex vivo* retinal organ cultures. However, Akt was activated between 1.0 and 3.0 M sorbitol with the maximum being at 3.0 M. These concentrations are higher than those reported previously (Chen *et al.*, 1997; Meier *et al.*, 1998), but provide the same degree of activation we found in *vivo* light stress (Li *et al.*, 2007), suggesting that sorbitol-induced hyperosmotic stress may mimic the *in vivo* light-stress model. Hyperosmotic-stress has been related to the pathogenesis of retinal detachment in DR (Marmor *et al.*, 1980; Ola *et al.*, 2006; Quintyn and Brasseur, 2004) and the sorbitol-induced *ex vivo* cultures may be useful to study the molecular signaling pathway(s) involved in DR.

In our study, we observed that 3.0 M sorbitol induced the activation of Akt. Inactivation and dephosphorylation of Akt have been reported during hyperosmotic stress (Meier *et al.*, 1998). Our observation of Akt activation suggests that 3.0 M sorbitol could inhibit the protein phosphatase(s) which results in the activation of Akt. To address whether sorbitol could activate Akt signaling or inhibition of phosphatase(s) which indirectly block the dephosphorylation of Akt, we measured the phosphatase activity in hyperosmotic stress. Our results indicate a significant increase in the total phosphatase activity in hyperosomotic stress. To distinguish the activity of serine/threonine phosphatase activity from PTP, we used sodium vanadate to inhibit the PTPase activity and the results still indicate a significant increase in serine/threonine phosphatase activity. These studies clearly suggest that the observed activation of Akt in hyperosmotic stress is not due to the inhibition of phosphatase(s).

In the present study, in response to sorbitol, we have observed increased activation of PI3K through IR activation. In some neuronal cell types, such as cerebellar granular neurons (59) and PC-12 cells (60), receptor activation of PI3K has been shown to protect these cells from stress-induced neurodegeneration. Further, IR activation has been shown to rescue retinal neurons from apoptosis through a phosphoinositide 3-kinase (PI3K) cascade (Barber *et al.*, 1998, 2001). We have previously reported that under physiological conditions, light-induced the tyrosine phosphorylation of retinal IR which leads to the activation of PI3K (Rajala *et al.*, 2002). The earlier studies along with the results from the present study clearly suggests that sorbitol also induces the activation of PI3K associated with the tyrosine phosphorylated IR.

In hyperosomotic stress, we have observed both PI3K-dependent and independent activation of Akt. The PI3K inhibitor LY294002 was found to inhibit the insulin-induced activation of Akt but the same inhibitor failed to inhibit the sorbitol-induced activation of Akt. Activation of PI3K/Akt pathway contributes to cell survival (Datta *et al.*, 1999), however, it has been shown that dopamine induced the PI3K-indepndent activation of Akt in striated neurons (Brami-Cherrier *et al.*, 2002). These results suggest that sorbitol-induced activation of Akt might also be regulated through PI3K-independent mechanism. Consistent with this hypothesis, we have observed an increased tyrosine phosphorylation of several proteins in the retina under hyperosmotic stress. Further, HNMP3, an inhibitor of IR kinase activity has no effect on sorbitol-induced Akt activation. However, genestin, a global tyrosine kinase inhibitor was found to block the sorbitol-induced activation of Akt. It appears from this data that sorbitol-induced Akt activation is mediated through a receptor tyrosine kinase(s), but not IR. The observed Akt activation is not mediated through Src family nonreceptor tyrosine kinases due to the fact that inhibitors of this family failed to block the sorbitol-induced Akt activation. Increased protein tyrosine phosphorylation in the retina after ischemic-reperfusion injury and genestin, ameliorates retinal degeneration after ischemia-reperfusion injury in rats (Hayashi *et al.*, 1996, 1997). These results further support our findings that Akt activation could be trigged in response to stress and retinal injury.

The ability of osmotic shock to directly stimulate tyrosine phosphorylation events was confirmed by phosphotyrosine immunoblotting. Several discrete tyrosine-phosphorylated proteins in the range of 115–170 kDa and 41–79 kDa were clearly induced by osmotic shock treatment. Previous studies have also reported the activation of tyrosine phosphorylation in response to hyperosmotic stress (Chen *et al.*, 1997; Hresko and Mueckler, 2000; Janez *et al.*, 2000). We have probed the phosphotyrosine immunoblots with known tyrosine phosphorylated proteins such as PYK2, FAK, Na^{+} K^{+} ATPase, EGFR, JAK2, TYK2 PDE6beta, HSP90, cSrc, and Grb2-associated binder 1 (Gab-1). Under our experimental conditions, except Gab1 (data not shown), all other proteins were not tyrosine phosphorylated in sorbitol-induced stress conditions. The Grb2 associated binder Gab-1 is shown to be tyrosine phosphorylated following sorbitol stimulation (Janez *et al.*, 2000). We found that Gab1 is rapidly tyrosine phosphorylated immediately after sorbitol treatment and after 5 min, the phosphorylation was greatly diminished (data not shown). We have also demonstrated in this study that genestin completely blocks the Akt activation; which suggests that Akt activation may be signaled through Gab1. Gab-1 is phosphorylated on tyrosine after stimulation with insulin and several growth factors (Janez *et al.*, 2000; Rocchi *et al.*, 1998). It possesses 16 potential phosphotyrosine sites, some of which could serve as binding sites for SH2 domains of the p85 regulatory subunit of PI3K, Grb2, phospholipase C-γ, Nck, and SHP-2 (Gual *et al.*, 2000; Liu and Rohrschneider, 2002; Rocchi *et al.*, 1998; Schlessinger and Lemmon, 2003). It has been shown previously that Gab-1 mediates the neurite outgrowth, DNA synthesis, and survival in PC12 cells (Korhonen *et al.*, 1999). Further, overexpression of Gab-1 has shown to inhibit apoptosis in PC12 cells (Holgado-Madruga *et al.*, 1997). Therefore, it is tempting to speculate that decreased phosphorylation of Gab1 could be a contributory factor for DR.

In this study, we report the specific activation of Akt2 in response to hyperosmotic stress. Consistent with this hypothesis is our earlier finding that Akt2 knockout mice exhibit a greater sensitivity to light-induced retinal degeneration, but not Akt1 knockout mice (Li *et al.*, 2007). Further, mice lacking Akt2 have defects in glucose metabolism that ultimately lead to hyperglycemia and hyperinsulinemia (Cho *et al.*, 2001; Garofalo *et al.*, 2003). These studies clearly suggest the importance of Akt2 in both light-induced retinal degeneration and sorbitol-induced stress.

In the current study, we observed that sorbitol-induced the tyrosine phosphorylation of Cbl and this activation leads to the binding of p85 subunit of PI3K, as we observed an increased Cbl associated PI3K activity in *ex vivo* retinal organ cultures. Although a small fraction of Cbl is constitutively phosphorylated and associated with PI3K, it appears that a more significant fraction of Cbl could potentially negatively regulate p85 by polyubiquitination and degradation as demonstrated in our sorbitol untreated control retinas.

Cbl is a ubiquitously expressed cytosolic protein characterized as both, an adaptor protein and an E3 ubiquitin ligase. Cbl protein has a tyrosine kinase binding (TKB) domain, proline-rich region and five tyrosine phosphorylation sites; Cbl is able to bind to protein tyrosine kinases (PTK), SH3 and SH2 domain containing proteins, respectively (Meisner *et al.*, 1995; Swaminathan and Tsygankov, 2006; Swaminathan *et al.*, 2007; Meisner, 1995; Fukazawa, 1995). Cbl also contains the RING finger domain and a ubiquitin-associated domain which allow for ubiquitination and proteasomal degradation of activated PTK (Levkowitz *et al.*, 1999; Meisner *et al.*, 1997). Stress, extracellular stimuli, growth factors, and hormones stimulate tyrosine phosphorylation of Cbl. Under certain cellular conditions phosphorylated Cbl acts as an adaptor protein and binds to SH2 domain-containing proteins such as Vav guanine nucleotide exchange factor, p85 subunit of PI3K, and Crk proteins to propagate downstream signaling (Feshchenko *et al.*, 1998; Miyake *et al.*, 1997). However, previous findings showed that tyrosine phosphorylation of both Cbl and EGFR tyrosine kinase were necessary for binding, ubiquitination, and degradation of an activated receptor (Levkowitz *et al.*, 1999).

Interaction between p85 subunit of PI3K and Cbl is phosphorylation-dependent and independent. Phosphorylation of Tyr-731 of Cbl is essential for its binding to SH2 domains of p85 (Feshchenko *et al.*, 1998; Swaminathan *et al.*, 2007). Tyrosine phosphorylation significantly increase p85-Cbl binding in membrane fractions when compared to the cytosolic fractions of v-Abl-transformed 3T3 fibroblasts and thus further facilitating

PI3K/Rho downstream cytoskeletal effects (Swaminathan *et al.*, 2007). Another study has shown that the two SH2 domains of p85 were dispensable for p85 and Cbl interaction, subsequent ubiquitination and proteasome degradation therefore suggesting phosphorylation-independent binding (Fang *et al.*, 2001). The same study proved that in Jurkat T cell line, the SH3 domain of p85 and the proline-rich region of Cbl were necessary for the binding of the two proteins and further ubiquitination of p85 by the RING finger domain of Cbl.

Diabetic model studies have shown that chronic and increased sorbitol secretion by the cells establishes a hyperosmotic environment fatal to cell survival and viability. The elevated erythrocyte sorbitol levels were found in diabetic patients with developing DR, a leading cause of blindness (Reddy *et al.*, 2008). Both insulin and sorbitol-induced hyperosmotic shock increased intrinsic tyrosine kinase activity of IR followed by transient tyrosine phosphorylation of insulin-receptor substrate-1 (IRS-1) and Cbl (Ahmed *et al.*, 2000; Rajala and Anderson, 2001). Tyrosine phosphorylated IRS-1 and Cbl harbor signaling proteins involved in activation of PI3K/Akt survival pathway and F-actin polymerization, respectively (Strawbridge *et al.*, 2006). Previous studies have shown that type 2 diabetes, obesity, metabolic syndrome X, and age related insulin resistance were characterized by elevated levels of vasoactive peptide endothelin-1 (ET-1) (Ferri *et al.*, 1995, 1997; Sayama *et al.*, 1999). ET-1 peptide-induced defects in lipid membrane, impaired PI3K signaling, reduced tyrosine phosphorylation of Cbl, and F-actin polymerization as well as GLUT4 trafficking when added to insulin- or sorbitol-incubation conditions; thus, ET-1 challenges and overwrites protective role of Cbl (Strawbridge *et al.*, 2006). Taken together, previous and our current findings suggest that under normal conditions Cbl negatively regulates p85 whereby it potentially minimizes the available PI3K pool utilized for basal signaling by RTK and non-RTK. In our *in vitro* hyperosmolarity stress model, Cbl becomes tyrosine phosphorylated and takes on a role of an adaptor protein aiding to maximize and recruit PI3K necessary for execution of downstream survival pathways. In addition, our laboratory findings show an increase IR tyrosine activity and PI3K/Akt survival signals under the same conditions. Further studies will help us elucidate the function of Cbl as a neuroprotector involved in maintaining cell integrity and viability under the compromising hyperosmotic conditions characteristic of diabetic models.

ACKNOWLEDGMENTS

The project described was supported by grants from the National Eye Institute (R01EY016507 and R01EY00871) and the National Center for Research Resources (P20RR17703). The content is solely the responsibility of the authors and does not necessarily represent the official views of the National Center for Research Resources, the National Eye Institute, or the National Institutes of Health. The authors thank Dr. Yogita Kanan for reading this manuscript.

REFERENCES

Ahmed, Z., Smith, B. J., and Pillay, T. S. (2000). The APS adapter protein couples the insulin receptor to the phosphorylation of c-Cbl and facilitates ligand-stimulated ubiquitination of the insulin receptor. *FEBS Lett.* **475,** 31–34.

Asnaghi, V., Gerhardinger, C., Hoehn, T., Adeboje, A., and Lorenzi, M. (2003). A role for the polyol pathway in the early neuroretinal apoptosis and glial changes induced by diabetes in the rat. *Diabetes* **52,** 506–511.

Barber, A. J., Lieth, E., Khin, S. A., Antonetti, D. A., Buchanan, A. G., and Gardner, T. W. (1998). Neural apoptosis in the retina during experimental and human diabetes. Early onset and effect of insulin. *J. Clin. Invest.* **102,** 783–791.

Barber, A. J., Nakamura, M., Wolpert, E. B., Reiter, C. E., Seigel, G. M., Antonetti, D. A., and Gardner, T. W. (2001). Insulin rescues retinal neurons from apoptosis by a phosphatidylinositol 3-kinase/Akt-mediated mechanism that reduces the activation of caspase-3. *J. Biol. Chem.* **276,** 32814–32821.

Barnett, P. A., Gonzalez, R. G., Chylack, L. T. Jr., and Cheng, H. M. (1986). The effect of oxidation on sorbitol pathway kinetics. *Diabetes* **35,** 426–432.

Brami-Cherrier, K., Valjent, E., Garcia, M., Pages, C., Hipskind, R. A., and Caboche, J. (2002). Dopamine induces a PI3-kinase-independent activation of Akt in striatal neurons: A new route to cAMP response element-binding protein phosphorylation. *J. Neurosci.* **22,** 8911–8921.

Chandra, D., Jackson, E. B., Ramana, K. V., Kelley, R., Srivastava, S. K., and Bhatnagar, A. (2002). Nitric oxide prevents aldose reductase activation and sorbitol accumulation during diabetes. *Diabetes* **51,** 3095–3101.

Chen, D., Elmendorf, J. S., Olson, A. L., Li, X., Earp, H. S., and Pessin, J. E. (1997). Osmotic shock stimulates GLUT4 translocation in 3T3L1 adipocytes by a novel tyrosine kinase pathway. *J. Biol. Chem.* **272,** 27401–27410.

Cheng, H., Kartenbeck, J., Kabsch, K., Mao, X., Marques, M., and Alonso, A. (2002). Stress kinase p38 mediates EGFR transactivation by hyperosmolar concentrations of sorbitol. *J. Cell Physiol.* **192,** 234–243.

Cho, H., Mu, J., Kim, J. K., Thorvaldsen, J. L., Chu, Q., Crenshaw, E. B., 3rd, Kaestner, K. H., Bartolomei, M. S., Shulman, G. I., and Birnbaum, M. J. (2001). Insulin resistance and a diabetes mellitus-like syndrome in mice lacking the protein kinase Akt2 (PKB beta). *Science* **292,** 1728-1731.

Dagher, Z., Park, Y. S., Asnaghi, V., Hoehn, T., Gerhardinger, C., and Lorenzi, M. (2004). Studies of rat and human retinas predict a role for the polyol pathway in human diabetic retinopathy. *Diabetes* **53,** 2404–2411.

Dahlin, L. B., Archer, D. R., and McLean, W. G. (1987). Treatment with an aldose reductase inhibitor can reduce the susceptibility of fast axonal transport following nerve compression in the streptozotocin-diabetic rat. *Diabetologia* **30,** 414–418.

Datta, S. R., Brunet, A., and Greenberg, M. E. (1999). Cellular survival: A play in three Akts. *Genes Dev.* **13,** 2905–2927.

Dvornik, E., Simard-Duquesne, N., Krami, M., Sestanj, K., Gabbay, K. H., Kinoshita, J. H., Varma, S. D., and Merola, L. O. (1973). Polyol accumulation in galactosemic and diabetic rats: Control by an aldose reductase inhibitor. *Science* **182,** 1146–1148.

Fang, D., Wang, H. Y., Fang, N., Altman, Y., Elly, C., and Liu, Y. C. (2001). Cbl-b, a RING-type E3 ubiquitin ligase, targets phosphatidylinositol 3-kinase for ubiquitination in T cells. *J. Biol. Chem.* **276,** 4872–4878.

Ferri, C., Bellini, C., Desideri, G., Di Francesco, L., Baldoncini, R., Santucci, A., and De Mattia, G. (1995). Plasma endothelin-1 levels in obese hypertensive and normotensive men. *Diabetes* **44,** 431–436.

Ferri, C., Bellini, C., Desideri, G., Baldoncini, R., Properzi, G., Santucci, A., and De Mattia, G. (1997). Circulating endothelin-1 levels in obese patients with the metabolic syndrome. *Exp. Clin. Endocrinol. Diabetes* **105**(Suppl. 2), 38–40.

Feshchenko, E. A., Langdon, W. Y., and Tsygankov, A. Y. (1998). Fyn, Yes, and Syk phosphorylation sites in c-Cbl map to the same tyrosine residues that become phosphorylated in activated T cells. *J. Biol. Chem.* **273,** 8323–8331.

Fukazawa, T., Reedquist, K. A., Trub, T., Soltoff, S., Panchamoorthy, G., Druker, B., Cantley, L., Shoelson, S. E., and Band, H. (1995). The SH3 domain-binding T cell tyrosyl phosphoprotein p120. Demonstration of its identity with the c-cbl protooncogene product and *in vivo* complexes with Fyn, Grb2, and phosphatidylinositol 3-kinase. *J.Biol.Chem.* **270,** 19141-19150.

Gabbay, K. H. (1973a). Role of sorbitol pathway in neuropathy. *Adv. Metab. Disord.* **2,** 417–432.

Gabbay, K. H. (1973b). The sorbitol pathway and the complications of diabetes. *N. Engl. J. Med.* **288,** 831–836.

Gabbay, K. H. (1975). Hyperglycemia, polyol metabolism, and complications of diabetes mellitus. *Annu. Rev. Med.* **26,** 521–536.

Garofalo, R. S, Orena, S. J., Rafidi, K., Torchia, A. J., Stock, J. L., Hildebrandt, A. L., Coskran, T., Black, S. C., Brees, D. J., Wicks, J. R., McNeish. J. D., and Coleman, K. G. (2003). Severe diabetes, age-dependent loss of adipose tissue, and mild growth deficiency in mice lacking Akt2/PKBβ. *J. Clin. Invest.* **112,** 197-208.

Gual, P., Giordano, S., Williams, T. A., Rocchi, S., Van Obberghen, E., and Comoglio, P. M. (2000). Sustained recruitment of phospholipase C-gamma to Gab1 is required for HGF-induced branching tubulogenesis. *Oncogene* **19,** 1509–1518.

Gustavsson, J., Parpal, S., Karlsson, M., Ramsing, C., Thorn, H., Borg, M., Lindroth, M., Peterson, K. H., Magnusson, K. E., and Stralfors, P. (1999). Localization of the insulin receptor in caveolae of adipocyte plasma membrane. *FASEB J.* **13,** 1961–1971.

Hayashi, A, Koroma, B.M, Imai, K, and de Juan, E., Jr. (1996). Increase of protein tyrosine phosphorylation in rat retina after ischemia-reperfusion injury. *Invest. Ophthalmol. Vis. Sci.* **37,** 2146–2156.

Hayashi, A., Weinberger, A. W., Kim, H. C., and de Juan, E., Jr. (1997). Genistein, a protein tyrosine kinase inhibitor, ameliorates retinal degeneration after ischemia-reperfusion injury in rat. *Invest. Ophthalmol. Vis. Sci.* **38,** 1193–1202.

Holgado-Madruga, M., Moscatello, D. K., Emlet, D. R., Dieterich, R., and Wong, A. J. (1997). Grb2-associated binder-1 mediates phosphatidylinositol 3-kinase activation and the promotion of cell survival by nerve growth factor. *Proc. Natl. Acad. Sci. USA* **94,** 12419–12424.

Hresko, R. C., and Mueckler, M. (2000). A novel 68-kDa adipocyte protein phosphorylated on tyrosine in response to insulin and osmotic shock. *J. Biol. Chem.* **275,** 18114–18120.

Janez, A., Worrall, D. S., Imamura, T., Sharma, P. M., and Olefsky, J. M. (2000). The osmotic shock-induced glucose transport pathway in 3T3-L1 adipocytes is mediated by gab-1 and requires Gab-1-associated phosphatidylinositol 3-kinase activity for full activation. *J. Biol. Chem.* **275,** 26870–26876.

Kinoshita, J. H., Fukushi, S., Kador, P., and Merola, L. O. (1979). Aldose reductase in diabetic complications of the eye. *Metabolism* **28,** 462–469.

Korhonen, J. M., Said, F. A., Wong, A. J., and Kaplan, D. R. (1999). Gab1 mediates neurite outgrowth, DNA synthesis, and survival in PC12 cells. *J. Biol. Chem.* **274,** 37307–37314.
Lederle, F. A. (1995). Epidemiology of constipation in elderly patients. Drug utilisation and cost-containment strategies. *Drugs Aging* **6,** 465–469.
Levkowitz, G., Waterman, H., Ettenberg, S. A., Katz, M., Tsygankov, A. Y., Alroy, I., Lavi, S., Iwai, K., Reiss, Y., Ciechanover, A., Lipkowitz, S., and Yarden, Y. (1999). Ubiquitin ligase activity and tyrosine phosphorylation underlie suppression of growth factor signaling by c-Cbl/Sli-1. *Mol. Cell* **4,** 1029–1040.
Li, G., Anderson, R. E., Tomita, H., Adler, R., Liu, X., Zack, D. J., and Rajala, R. V. (2007). Nonredundant role of Akt2 for neuroprotection of rod photoreceptor cells from light-induced cell death. *J. Neurosci.* **27,** 203–211.
Liu, Y., and Rohrschneider, L. R. (2002). The gift of Gab. *FEBS Lett.* **515,** 1–7.
Lorenzi, M. (2007). The polyol pathway as a mechanism for diabetic retinopathy: Attractive, elusive, and resilient. *Exp. Diabetes Res.* **2007,** 61038.
Lorenzi, M., and Gerhardinger, C. (2001). Early cellular and molecular changes induced by diabetes in the retina. *Diabetologia* **44,** 791–804.
Marchenko, S. M., and Sage, S. O. (2000). Hyperosmotic but not hyposmotic stress evokes a rise in cytosolic Ca2+ concentration in endothelium of intact rat aorta. *Exp. Physiol.* **85,** 151–157.
Marmor, M. F., Martin, L. J., and Tharpe, S. (1980). Osmotically induced retinal detachment in the rabbit and primate. Electron miscoscopy of the pigment epithelium. *Invest. Ophthalmol. Vis. Sci.* **19,** 1016–1029.
Meier, R., Thelen, M., and Hemmings, B. A. (1998). Inactivation and dephosphorylation of protein kinase Balpha (PKBalpha) promoted by hyperosmotic stress. *EMBO J.* **17,** 7294–7303.
Meisner, H., Conway, B. R., Hartley, D., and Czech, M. P. (1995). Interactions of Cbl with Grb2 and phosphatidylinositol 3′-kinase in activated Jurkat cells. *Mol. Cell. Biol.* **15,** 3571–3578.
Meisner, H., Daga, A., Buxton, J., Fernandez, B., Chawla, A., Banerjee, U., and Czech, M. P. (1997). Interactions of *Drosophila* Cbl with epidermal growth factor receptors and role of Cbl in R7 photoreceptor cell development. *Mol. Cell. Biol.* **17,** 2217–2225.
Miyake, S., Lupher, M. L. Jr., Andoniou, C. E., Lill, N. L., Ota, S., Douillard, P., Rao, N., and Band, H. (1997). The Cbl protooncogene product: From an enigmatic oncogene to center stage of signal transduction. *Crit. Rev. Oncog.* **8,** 189–218.
Mizutani, M., Gerhardinger, C., and Lorenzi, M. (1998). Muller cell changes in human diabetic retinopathy. *Diabetes* **47,** 445–449.
Narayanan, S. (1993). Aldose reductase and its inhibition in the control of diabetic complications. *Ann. Clin. Lab. Sci.* **23,** 148–158.
Obrosova, I. G., Minchenko, A. G., Vasupuram, R., White, L., Abatan, O. I., Kumagai, A. K., Frank, R. N., and Stevens, M. J. (2003). Aldose reductase inhibitor fidarestat prevents retinal oxidative stress and vascular endothelial growth factor overexpression in streptozotocin-diabetic rats. *Diabetes* **52,** 864–871.
Obrosova, I. G., Pacher, P., Szabo, C., Zsengeller, Z., Hirooka, H., Stevens, M. J., and Yorek, M. A. (2005). Aldose reductase inhibition counteracts oxidative-nitrosative stress and poly(ADP-ribose) polymerase activation in tissue sites for diabetes complications. *Diabetes* **54,** 234–242.
Ola, M. S., Berkich, D. A., Xu, Y., King, M. T., Gardner, T. W., Simpson, I., and LaNoue, K. F. (2006). Analysis of glucose metabolism in diabetic rat retinas. *Am. J. Physiol. Endocrinol. Metab.* **290,** E1057–E1067.

Ouwens, D. M., Gomes de Mesquita, D. S., Dekker, J., and Maassen, J. A. (2001). Hyperosmotic stress activates the insulin receptor in CHO cells. *Biochim. Biophys. Acta* **1540,** 97–106.

Pritchard, S., Erickson, G. R., and Guilak, F. (2002). Hyperosmotically induced volume change and calcium signaling in intervertebral disk cells: The role of the actin cytoskeleton. *Biophys. J.* **83,** 2502–2510.

Quintyn, J. C., and Brasseur, G. (2004). Subretinal fluid in primary rhegmatogenous retinal detachment: Physiopathology and composition. *Surv. Ophthalmol.* **49,** 96–108.

Rajala, A., Anderson, R. E., Ma, J. X., Lem, J., Al Ubaidi, M. R., and Rajala, R. V. (2007). G-protein-coupled receptor rhodopsin regulates the phosphorylation of retinal insulin receptor. *J. Biol. Chem.* **282,** 9865–9873.

Rajala, R. V., and Anderson, R. E. (2001). Interaction of the insulin receptor beta-subunit with phosphatidylinositol 3-kinase in bovine ROS. *Invest. Ophthalmol. Vis. Sci.* **42,** 3110–3117.

Rajala, R. V., McClellan, M. E., Ash, J. D., and Anderson, R. E. (2002). *In vivo* regulation of phosphoinositide 3-kinase in retina through light-induced tyrosine phosphorylation of the insulin receptor beta-subunit. *J. Biol. Chem.* **277,** 43319–43326.

Rajala, R. V., McClellan, M. E., Chan, M. D., Tsiokas, L., and Anderson, R. E. (2004). Interaction of the retinal insulin receptor beta-subunit with the P85 subunit of Phosphoinositide 3-Kinase. *Biochemistry* **43,** 5637–5650.

Reddy, G. B., Satyanarayana, A., Balakrishna, N., Ayyagari, R., Padma, M., Viswanath, K., and Petrash, J. M. (2008). Erythrocyte aldose reductase activity and sorbitol levels in diabetic retinopathy. *Mol. Vis.* **14,** 593–601.

Reiter, C. E., and Gardner, T. W. (2003). Functions of insulin and insulin receptor signaling in retina: Possible implications for diabetic retinopathy. *Prog. Retin. Eye Res.* **22,** 545–562.

Reiter, C. E., Sandirasegarane, L., Wolpert, E. B., Klinger, M., Simpson, I. A., Barber, A. J., Antonetti, D. A., Kester, M., and Gardner, T. W. (2003). Characterization of insulin signaling in rat retina *in vivo* and *ex vivo*. *Am. J. Physiol. Endocrinol. Metab.* **285,** E763–E774.

Reiter, C. E., Wu, X., Sandirasegarane, L., Nakamura, M., Gilbert, K. A., Singh, R. S., Fort, P. E., Antonetti, D. A., and Gardner, T. W. (2006). Diabetes reduces basal retinal insulin receptor signaling: Reversal with systemic and local insulin. *Diabetes* **55,** 1148–1156.

Rocchi, S., Tartare-Deckert, S., Murdaca, J., Holgado-Madruga, M., Wong, A. J., and Van Obberghen, E. (1998). Determination of Gab1 (Grb2-associated binder-1) interaction with insulin receptor-signaling molecules. *Mol. Endocrinol.* **12,** 914–923.

Sayama, H., Nakamura, Y., Saito, N., and Konoshita, M. (1999). Does the plasma endothelin-1 concentration reflect atherosclerosis in the elderly? *Gerontology* **45,** 312–316.

Schlessinger, J., and Lemmon, M. A. (2003). SH2 and PTB domains in tyrosine kinase signaling. *Sci. STKE* **2003,** RE12.

Strawbridge, A. B., Elmendorf, J. S., and Mather, K. J. (2006). Interactions of endothelin and insulin: Expanding parameters of insulin resistance. *Curr. Diabetes Rev.* **2,** 317–327.

Swaminathan, G., and Tsygankov, A. Y. (2006). The Cbl family proteins: Ring leaders in regulation of cell signaling. *J. Cell Physiol.* **209,** 21–43.

Swaminathan, G., Feshchenko, E. A., and Tsygankov, A. Y. (2007). c-Cbl-facilitated cytoskeletal effects in v-Abl-transformed fibroblasts are regulated by membrane association of c-Cbl. *Oncogene* **26,** 4095–4105.

Szwergold, B. S., Kappler, F., and Brown, T. R. (1990). Identification of fructose 3-phosphate in the lens of diabetic rats. *Science* **247,** 451–454.

Takai, A., and Mieskes, G. (1991). Inhibitory effect of okadaic acid on the *p*-nitrophenyl phosphate phosphatase activity of protein phosphatases. *Biochem. J.* **275**(Pt. 1), 233–239.

Tomlinson, D. R. (1999). Mitogen-activated protein kinases as glucose transducers for diabetic complications. *Diabetologia* **42,** 1271–1281.

Tomlinson, D. R., Willars, G. B., and Carrington, A. L. (1992). Aldose reductase inhibitors and diabetic complications. *Pharmacol. Ther.* **54,** 151–194.

Tomlinson, D. R., Stevens, E. J., and Diemel, L. T. (1994). Aldose reductase inhibitors and their potential for the treatment of diabetic complications. *Trends Pharmacol. Sci.* **15,** 293–297.

Vinores, S. A., Campochiaro, P. A., Williams, E. H., May, E. E., Green, W. R., and Sorenson, R. L. (1988). Aldose reductase expression in human diabetic retina and retinal pigment epithelium. *Diabetes* **37,** 1658–1664.

Williamson, J. R., Chang, K., Frangos, M., Hasan, K. S., Ido, Y., Kawamura, T., Nyengaard, J. R., van den, E. M., Kilo, C., and Tilton, R. G. (1993). Hyperglycemic pseudohypoxia and diabetic complications. *Diabetes* **42,** 801–813.

CHAPTER TWENTY-ONE

Interleukin-6 and Insulin Resistance

Jeong-Ho Kim, Rebecca A. Bachmann, *and* Jie Chen

Contents

Abstract

Chronic low-grade inflammation has been well recognized as a key feature of obesity that is correlated with insulin resistance and type 2 diabetes. Among the adipose-secreted factors (adipokines), the inflammatory regulator interleukin-6 (IL-6) has emerged as one of the potential mediators that link obesity-derived chronic inflammation with insulin resistance. Adipose tissue contributes to up to 35% of circulating IL-6, the systemic effects of which have been best demonstrated in the liver, where a STAT3—SOCS-3 pathway mediates IL-6 impairment of insulin actions. However, this cytokine displays pleiotropic functions in a tissue-specific and physiological context-dependent manner. In contrast to its role in liver, IL-6 is believed to be beneficial for insulin-regulated glucose metabolism in muscle. Furthermore, the effects of the cytokine are seemingly influenced by whether it is present acutely or chronically; the latter is the setting associated with insulin resistance. Herein we review the *in vivo* and *in vitro* studies that have examined the role of IL-6 in insulin signaling and glucose metabolism in the insulin target tissues: liver, adipose, and skeletal muscle.

Department of Cell and Developmental Biology, University of Illinois at Urbana-Champaign, 601 S. Goodwin Ave. B107, Urbana, IL 61801

Vitamins and Hormones, Volume 80
ISSN 0083-6729, DOI: 10.1016/S0083-6729(08)00621-3

I. Introduction

Insulin resistance is the failure to respond to normal concentrations of circulating insulin. Clinically, insulin resistance is indicated by a plasma glucose concentration between 100 and 125 mg/dL (milligrams of glucose per deciliter of blood) following a 12-hour fast, according to the United States National Institute of Diabetes and Digestive and Kidney Disease (NIDDK, 2005). Insulin resistance is an important facet of several clinical disorders, including type 2 diabetes, obesity, dyslipidemia, and hypertension (Reaven, 1988). Furthermore, GM Reaven recognized two decades ago that these diseases tend to cluster as comorbidities (Reaven, 1988) forming the increasingly more-recognized, but ever-evolving metabolic syndrome (Batsis *et al.*, 2007; Expert-Panel, 2001).

Chronic low-grade inflammation has been shown to be a key feature of obesity and type 2 diabetes. Several obesity-derived circulating factors with systemic effects—such as tumor necrosis factor (TNF)-α, interleukin (IL)-6, IL-1, interferon-γ and various other adipose-secreted factors (otherwise known as adipokines)—have been correlated with the development of insulin resistance. Since the first molecular link between inflammation and obesity, TNF-α, was identified (Hotamisligil *et al.*, 1993), many studies have revealed a role of TNF-α in the induction of insulin resistance and type 2 diabetes (Kern *et al.*, 2001; Uysal *et al.*, 1997; Ventre *et al.*, 1997). The general consensus is that elevated levels of TNF-α can cause insulin resistance (Hotamisligil, 1999).

Similarly, a growing body of evidence has established IL-6 as an important player in metabolic disease states, such as diabetes. This pleiotropic cytokine is known to regulate a variety of functions in different cells and tissues, including the proliferation and differentiation of hematopoietic cells, acute phase response induction in liver cells, and inflammation at sites of tissue injury (Van Snick, 1990). In addition, more recent research has linked IL-6 to the regulation of glucose and lipid metabolism, the former through pathways related to insulin action. The circulating levels of IL-6 have been correlated with adiposity and type 2 diabetes (Bastard *et al.*, 2002; Kern *et al.*, 2001; Pradhan *et al.*, 2001). *In vitro* evidence also supports the link between elevated IL-6 levels and the induction of insulin resistance in hepatic cells (Senn *et al.*, 2002). Still, the role of IL-6 within other tissues, such as muscle and adipose, is controversial or unclear (Carey and Febbraio, 2004). In this review, we consider the accepted role of IL-6 in insulin resistance within the liver, as well as the less straightforward IL-6 effects on peripheral tissues. First, we will begin with a basic primer on insulin signaling and then proceed to discuss the key role of chronic inflammation in perpetuating insulin resistance. A true comprehension of the molecular changes involved in insulin resistance requires a basic review of the normal metabolic pathways that underscore the catabolism and anabolism of glucose, beginning with the tissues that regulate these processes.

II. Insulin Signaling and Insulin Resistance

A. Physiological insulin signaling

Insulin is the anabolic hormone regulating glucose homeostasis, with additional functions including the regulation of protein synthesis, cell growth and differentiation. This hormone is secreted by the beta cells of the islets of Langerhans in the pancreas, in response to increased levels of circulating glucose and amino acids after a meal. In turn, increased levels of insulin stimulate glucose uptake in the periphery primarily by muscle and adipose tissues, and reduce hepatic glucose output through decreased gluconeogenesis (the generation of glucose from nonsugar carbon substrates) and glycogenolysis (the catabolism of glycogen), and increased glycogenesis (the synthesis of glycogen). At the molecular level, binding of insulin to the transmembrane insulin receptor (IR), composed of two α and two β subunits, activates the intrinsic tyrosine kinase activity of the receptor, which results in receptor autophosphorylation and formation of binding sites for the recruitment of immediate effectors, of which insulin receptor substrates (IRSs) are the major targets and signal transducers for IR. Subsequent phosphorylation of IRS by IR initiates the assembly and activation of downstream signaling pathways (White, 2003).

Skeletal muscle glucose transport accounts for ~75% of whole body insulin-stimulated glucose uptake, mediated through the translocation of the glucose transporter (GLUT) 4 to the plasma membrane (Bjornholm and Zierath, 2005; Huang and Czech, 2007). Towards this end, insulin activates the linear signaling cascade IR/IRS/phosphatidylinositol-3-kinase (PI3K)/Akt (Fig. 21.1). Many lines of evidence have suggested that PI3K activation of Akt and atypical protein kinase C (aPKC) is essential for GLUT4 translocation in skeletal muscle as well as adipocytes (Bandyopadhyay *et al.*, 1997; Cong *et al.*, 1997; Kohn *et al.*, 1996; Wang *et al.*, 1999). However, the regulatory mechanism has not been fully elucidated.

In addition, a second pathway is required for insulin-stimulated glucose transport in fat and muscle cells (Baumann *et al.*, 2000). This signaling involves the proto-oncogene Cbl, which is recruited to the IR through its association with the C-terminal SH3 domain of the adaptor protein

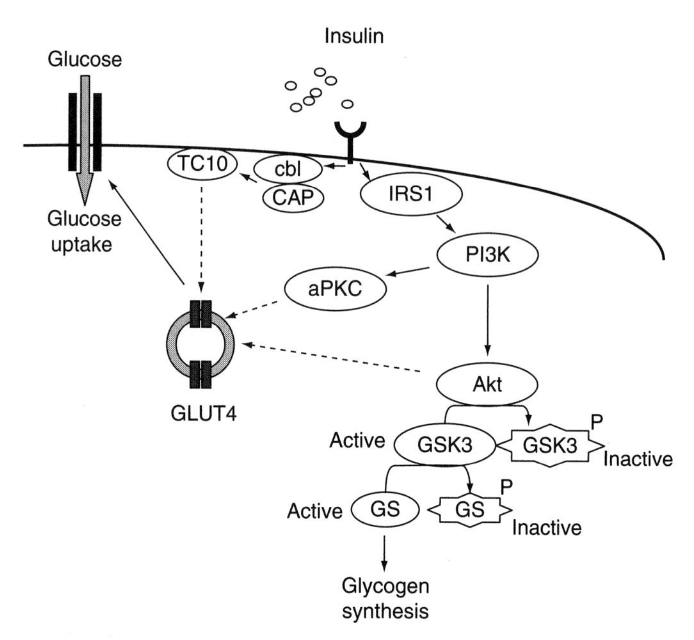

Figure 21.1 Insulin signaling pathways in the regulation of glucose uptake and glycogen synthesis. See text for details.

CAP. Activated IR directly induces tyrosine phosphorylation of Cbl, leading to the activation of the small G protein TC10 (Chiang *et al.*, 2001), which then activates effector proteins, such as TCGAP (Chiang *et al.*, 2003) and/or the exocyst-Snapin complex (Bao *et al.*, 2008; Inoue *et al.*, 2003), to allow GLUT4 translocation (Chiang *et al.*, 2001). The Cbl/CAP/TC10 pathway is required for insulin-induced glucose uptake in parallel with the activation of the PI3K/Akt pathway (Fig. 21.1).

In the liver, gluconeogenesis is regulated by insulin through signal-induced inhibition of the activity of key gluconeogenic enzymes: phosphoenolpyruvate carboxykinase and glucose-6-phosphatase (O'Brien and Granner, 1996). In addition, the synthesis of glycogen—the primary glucose storage macromolecule—is induced upon insulin stimulation through the induction of glycogen synthase activity following feeding. Insulin activates the IR/IRS/PI3K/Akt signaling cascade, leading to the phosphorylation and inactivation of glycogen synthase kinase (GSK) 3 (Cross *et al.*, 1995). Hence, the target of GSK3, glycogen synthase, is freed of inhibitory phosphorylation, and glycogen synthesis is induced upon insulin stimulation (Fig. 21.1). Still, the effects of insulin and ultimately the transport and production of glucose can be further modulated by other physiological and pathological conditions, such as chronic inflammation.

B. Chronic inflammation and insulin resistance

Inflammation is the body's protective reaction to an injury or infection, and the acute phase of inflammation normally promotes recovery back to good health. However, certain conditions can drive the body's defense system awry. For example, hypertension can lead to endothelial dysfunction, which in turn can cause oxidative damage, resulting in a chronic low-grade inflammatory state that ultimately leads to systemic detrimental effects (Hartge *et al.*, 2007). In fact, chronic low-grade inflammation, often linked to oxidative stress, is a common feature of obesity, insulin resistance and type 2 diabetes (Arkan *et al.*, 2005; Houstis *et al.*, 2006). Studies in obese and insulin resistant models have demonstrated a clear link between the chronic activation of inflammatory signaling pathways and insulin insensitivity. Inflammatory cytokines, such as TNF-α, IL-1, and IL-6, are overexpressed in adipose and other tissues in obese mouse models and in overweight humans (Wellen and Hotamisligil, 2005). These inflammatory cytokines can impair insulin signaling and thereby, contribute to the perpetuation of metabolic dysfunction.

An important feature of inflammation is the infiltration of inflamed tissues by immune cells, such as macrophages. A high density of macrophages within adipose tissue has been reported in both obese mice and humans (Weisberg *et al.*, 2003; Xu *et al.*, 2003). The percentage of macrophages in adipose tissue is positively correlated with adiposity and adipocyte size. Also, macrophages in adipose tissue contribute significantly to elevated levels of inflammatory cytokines, including TNF-α and IL-6 (Weisberg *et al.*, 2003). It has thus been proposed that the chronic inflammation associated with obesity-linked insulin resistance is initiated in adipose tissue (Xu *et al.*, 2003).

The removal of several key molecules involved in chronic inflammation and/or insulin signaling can enhance insulin sensitivity, as demonstrated by studies with knockout mice. TNF-α-deficient obese mice are protected from obesity-related insulin resistance in muscle and fat tissues (Uysal *et al.*, 1997). In addition, mice with targeted deletion of signaling molecules induced by inflammatory cytokines, JNK1$^{-/-}$ (Hirosumi *et al.*, 2002), adipose tissue-specific JNK1 knockout (Sabio *et al.*, 2008), or myeloid cell-specific IKK-b knockout (Arkan *et al.*, 2005), are protected from diet-induced insulin resistance. Yet, data from mice deficient in IL-6 has yielded conflicting results (discussed in the later text) (Di Gregorio *et al.*, 2004; Wallenius *et al.*, 2002), highlighting the diversity of the roles played by the adipokine in glucose regulation in various tissues and the complexity of the crosstalk between the liver, muscle, adipose tissue, and the brain. Moreover, the effects of IL-6 are best considered in a tissue specific manner and will be discussed thusly in the subsequent "IL-6 and insulin resistance" section.

C. Molecular mechanisms of insulin resistance

Cellular insulin resistance refers to modifications in the signaling response to the binding of insulin to its receptor. More specifically, this involves changes in the activity of the kinases responsible for insulin signal propagation, concomitant phosphorylation changes, and ultimately differences in target gene expression and/or functional changes (Virkamaki *et al.*, 1999; White, 2002, 2003). Several molecular mechanisms of insulin resistance have been proposed. Serine phosphorylation of IRS appears to be one of the major mechanisms (Fig. 21.2). One possibility is that serine phosphorylation induces a conformational change in the functional domain of IRS-1. For instance, the phosphorylation of Ser307 of IRS-1 (Ser312 in human IRS-1), located at the end of the phosphotyrosine-binding (PTB) domain, has been shown to block the interaction between IRS-1 and IR (Aguirre *et al.*, 2002). In addition, several studies have shown that phosphorylation of serine residues could induce IRS-1 breakdown through the proteosome degradation pathway (Haruta *et al.*, 2000; Pederson *et al.*, 2001; Sun *et al.*, 1999). Many Ser/Thr kinases have been proposed to phosphorylate various serine residues of IRS-1, such as extracellular signal-regulated kinase (ERK) (Bouzakri *et al.*, 2003), c-Jun NH_2-terminal kinase (JNK) (Aguirre *et al.*, 2000), inhibitor kB kinase (IKK) (Gao *et al.*, 2002), ribosomal S6 kinase 1 (S6K1) (Um *et al.*, 2006),

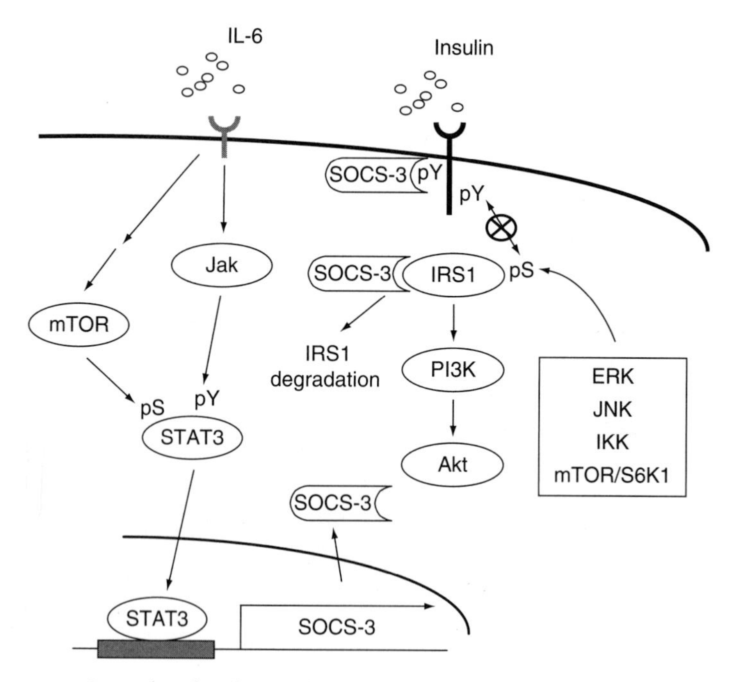

Figure 21.2 Several molecular mechanisms of insulin resistance. See text for details.

and the mammalian target of rapamycin (mTOR) (Tzatsos and Kandror, 2006). Furthermore, serine phosphorylation of IRS-1 mediated through the mTOR-S6K1 pathway has been shown to result in subcellular redistribution of IRS-1, thereby potentially reducing insulin signaling through an additional mechanism (Shah and Hunter, 2006; Takano *et al.*, 2001).

Suppressor of cytokine signaling proteins (SOCSs) also play an important role in the pathogenesis of insulin resistance by integrating the signaling of insulin with cytokine signaling (Krebs and Hilton, 2003). In particular, the expression of SOCS-3 is enhanced by various inflammatory cytokines and hormones, including IL-6 and leptin (Bjorbaek *et al.*, 1999; Emanuelli *et al.*, 2001; Rieusset *et al.*, 2004; Shi *et al.*, 2004). Overexpression of SOCS-3 inhibited insulin-induced glycogen synthase activity in myotubes and glucose uptake in adipocytes (Ueki *et al.*, 2004), while hepatocyte-specific SOCS-3 deletion improved insulin sensitivity in the liver (Torisu *et al.*, 2007). The activation of JNK1 in adipose can trigger insulin resistance in liver, likely through a mechanism involving increased secretion of IL-6 by adipose tissue leading to the elevation of SOCS–3 expression by the liver (Sabio *et al.*, 2008). Mechanistically, SOCS proteins inhibit insulin-induced signaling by directly interfering with IR activation, blocking IRS activation, or inducing IRS degradation (Howard and Flier, 2006; Fig. 21.2).

III. IL-6 and Insulin Resistance

A. IL-6 and the IL-6 signaling pathway

IL-6 regulates a wide range of biological activities not only within the immune system, but also in other systems and for other physiological events. Indeed, IL-6 is secreted by immune cells, adipose tissues (Mohamed-Ali *et al.*, 1997), and skeletal muscle (Keller *et al.*, 2003), while adipocytes, muscle, hepatocytes, and the brain can be targets of IL-6 outside of the immune system.

A brief summary of IL-6 signaling is provided below; the reader is directed to earlier reviews (Heinrich *et al.*, 2003; Van Snick, 1990) for comprehensive discussion of IL-6 structure, signaling, and biological activities. The human IL-6 protein consists of 212 amino acids with a signal peptide of 27 amino acids. Its molecular weight ranges between 21 and 28 kDa, depending on the amount of post-translational modification(s), such as glycosylation and phosphorylation. The IL-6 receptor (IL-6R) is composed of the IL-6 binding α chain and the common signal transducing β chain—gp130, the latter shared by the IL-6 type family of cytokines, such as IL-11, leukemia inhibitory factor (LIF), oncostatin M (OSM), ciliary neurotrophic factor (CNTF), etc.

IL-6 is produced by various types of lymphoid and nonlymphoid cells, such as T-cells, B-cells, keratinocytes, and several types of tumor cells, although the predominant sites of IL-6 production *in vivo* were originally believed to be monocytes/macrophages, fibroblasts, and vascular endothelial cells (Akira *et al.*, 1993). More recent research has revealed that the adipose tissue in fact contributes ~15–35% of the body's basal circulating IL-6 (Mohamed-Ali *et al.*, 1997), and subsequent new studies on this cytokine have been redirected towards assessing its role in mediating metabolic processes.

IL-6 activates two major signaling pathways: the janus kinase/signal transducer and activator of transcription (Jak/STAT) pathway and the mitogen-activated protein kinase (MAPK) pathway (Heinrich *et al.*, 2003). Both pathways can be activated by IL-6 binding to IL-6R and formation of the IL-6/IL-6R/gp130 complex. The cytoplasmic domain of gp130 has no intrinsic kinase activity, instead the oligomerized receptor activates associated tyrosine kinases of the Jak family (Schindler and Darnell, 1995). Activated Jaks phosphorylate several tyrosine residues of gp130 to form the binding sites for downstream effectors with Src-homology 2 (SH2) domains, including STAT transcription factors and SH2-domain containing tyrosine phosphatase (SHP) 2, the latter triggering the MAPK pathway.

IL-6 stimulation can induce the activation of 2 out of 7 members of the STAT family through gp130: STAT1 and STAT3 (Heinrich *et al.*, 2003). Beyond the phosphorylation of tyrosine residues that are responsible for the homo- or heterodimerization of STAT (such as tyrosine 705 of STAT3) (Calo *et al.*, 2003), serine phosphorylation is required for maximal STAT3 activation (Wen *et al.*, 1995). Several protein kinases have been reported to serine phosphorylate STAT3 in response to various stimuli under different cellular contexts, including PKCδ (Jain *et al.*, 1999), JNK (Lim and Cao, 1999), ERK (Chung *et al.*, 1997), MEK kinase 1 (MEKK1) (Lim and Cao, 2001), and mTOR (Kim *et al.*, 2008; Rajan *et al.*, 2003; Yokogami *et al.*, 2000). As a negative feedback control, STATs induce the expression of SOCS proteins, which are characterized by their ability to down-regulate cytokine signaling. SOCS proteins bind directly via their SH2 domains to tyrosine-phosphorylated Jak or activated cytokine receptors to suppress cytokine signaling (Krebs and Hilton, 2000). Indeed, liver or macrophage-specific deletions of SOCS-3 led to prolonged STAT1 and STAT3 activation following IL-6 stimulation (Croker *et al.*, 2003). In addition, as discussed in section I.C above, SOCS proteins have emerged as a major class of inhibitors of insulin signaling.

SH2 domain-containing tyrosine phosphatase (SHP2) is another mediator of IL-6 signaling, following its recruitment to phosphorylated gp130, and phosphorylation by Jak kinase. Tyrosine-phosphorylated SHP2 then interacts with the adaptor protein Grb2, which in turn activates the SOS-Ras-Raf-MAPK cascade (Heinrich *et al.*, 2003).

Although there has been significant progress towards elucidating the intracellular signaling pathways involved in the regulatory mechanisms behind glucose metabolism, the full complexity and interplay between signals has yet to be completely understood. Further research will potentially decrypt other players and connections between the pathways, and lend important insight towards explaining some of the seemingly conflicting data on IL-6 regulation and its breakdown during insulin resistance, which is discussed below.

B. IL-6 in insulin resistance

Many studies have established a correlation between alterations in IL-6 and the development of insulin resistance. Large studies of human populations have demonstrated that elevated plasma IL-6 levels positively correlate with obesity and insulin resistance, and predict the development of type 2 diabetes (Hu *et al.*, 2004; Pradhan *et al.*, 2001; Vozarova *et al.*, 2001), especially in combination with increased levels of IL-1β (Spranger *et al.*, 2003). Authors of a controlled observational study reported higher levels of IL-6 in a group of obese diabetic women than within a prediabetic cohort (Kocak *et al.*, 2007), while a study of morbidly obese patients given gastric surgery demonstrated a significant shift from diabetic or impaired glucose tolerance to normal glucose tolerance along with a significant decrease in serum IL-6 levels (Kopp *et al.*, 2003). Similarly, a small study of obese women and healthy control subjects noted enhanced insulin sensitivity along with weight loss, and observed that significantly lower levels of subcutaneous adipose tissue correlated with reduced serum IL-6 concentrations (Bastard *et al.*, 2000). Furthermore, fasting blood glucose is increased with IL-6 administration in healthy individuals (Tsigos *et al.*, 1997).

Still, these associations do not necessarily imply that IL-6 is a causative factor behind insulin resistance; furthermore, there have seemingly been discrepancies among reports linking IL-6 levels to insulin resistance. As of yet, human genetic studies have not identified a substantial connection between the common polymorphisms in the *IL-6* gene and the circulating IL-6 levels or the risk of developing type 2 diabetes (Qi *et al.*, 2006). Many studies have investigated the role of the C-174G polymorphism in the promoter of *IL-6*. Conflicting data support and oppose the possibility of lowered IL-6 expression due to the promoter polymorphism (Hoene and Weigert, 2008), while a study of healthy humans noted a correlation between reduced insulin sensitivity and the C-174G polymorphism (Kubaszek *et al.*, 2003), although the authors did not provide mechanistic evidence (e.g. changed IL-6 expression levels) of why this might be so. Considerations of the correlation between C-174G and insulin resistance, type 2 diabetes, or weight gain have yielded conflicting results, as well (Hoene and Weigert, 2008).

Systemic deletion of *IL-6* in mice has also not led to clear insights into the role of IL-6 in insulin sensitivity, but has further confirmed the pleiotropic nature of this cytokine. The IL-6 knockout mice develop mature-onset diabetes (Wallenius *et al.*, 2002), an unexpected finding that revealed a new role for IL-6 in the brain. The disruption in energy regulation in these mice could be rescued by low dose infusion of IL-6 into the brain, but not by peripheral IL-6 treatment (Wallenius *et al.*, 2002). Also, a study of healthy and obese humans found a negative correlation between IL-6 levels in the cerebrospinal fluid and total body weight/total body fat, suggesting that insufficient IL-6 levels in the central nervous system may promote the development of obesity (Stenlof *et al.*, 2003). Still, these results must be interpreted with caution, as an independent study of IL-6 knockout mice found no evidence of a higher risk of obesity, nor obvious alterations in metabolism (Di Gregorio *et al.*, 2004). The authors of the second study suggested that the discrepant results could be due to subtle genetic differences introduced during the creation of the two mice strains (Di Gregorio *et al.*, 2004).

Certainly, the presence of other regulatory factors, such as the anti-inflammatory cytokine IL-10, could mute the insulin resistance effects of IL-6, as has been observed in mice (Kim *et al.*, 2004). In addition, there likely are molecular-level differences inherent to the different insulin target tissues that lead to the variations in IL-6 impact on insulin regulation. Still the studies of adipocytes and skeletal muscle have provided particularly conflicting results; the effects of IL-6 have been demonstrated to be beneficial or detrimental, depending on the conditions of the cell culture or the experimental model, or the human population considered and their activity level. In contrast, the results of similar experiments within the liver have produced more consistent results, suggesting that this cytokine has a negative impact on insulin actions in the liver. Since differential effects among tissues involved in insulin resistance have been observed when levels of IL-6 are modified, liver, skeletal muscle, and adipose tissues will each be considered separately.

1. Liver

The liver regulates whole body energy homeostasis, in part through its responsiveness to insulin. Hence, liver dysfunction can have important systemic consequences. In particular, the liver plays a pivotal role in the development of diseases related to glucose metabolic dysfunction (Leclercq *et al.*, 2007). While mice with tissue specific IR deletions in either skeletal muscle or adipose tissue respectively display unchanged glucose homeostasis (Bruning *et al.*, 1998; Kim *et al.*, 2000) or improved insulin sensitivity as a result of protection against obesity (Bluher *et al.*, 2002), IR deletion in hepatic cells yield animals with hyperinsulinemia and severe insulin resistance (Michael *et al.*, 2000). In addition, localized liver inflammation in transgenic mice induced by hepatocyte-specific expression of a constitutively active IKK-β was associated with an insulin resistant phenotype in both liver and skeletal muscle (Cai *et al.*, 2005). As molecular changes within the liver are critical to insulin resistance, studies of hepatocytes are of particular importance.

For hepatocytes, both *in vitro* evidence and *in vivo* observations support a model wherein elevated levels of IL-6 inhibit insulin signaling and lead to insulin resistance. Chronic exposure to IL-6 selectively induces insulin resistance in the liver (Klover *et al.*, 2003), while systemic depletion of IL-6 improves hepatic insulin action in an obese mouse model (Klover *et al.*, 2005). Consistent with a role of IL-6 in hepatic insulin resistance, the hepatocyte-specific expression of a constitutively active IKK-β in mice is associated with increased IL-6 production and insulin resistance (Cai *et al.*, 2005). *In vitro* studies using both mouse primary hepatocytes and human hepatocarcinoma cells have demonstrated that IL-6 causes insulin resistance by suppressing tyrosine phosphorylation of IRS-1 (Senn *et al.*, 2002), mediated through SOCS-3 (Senn *et al.*, 2003; Fig. 21.2). A recent study conducted with similar cell systems further defined the mechanistic link between IL-6 stimulation and insulin resistance in the liver: SOCS-3 expression, and subsequent impairment of insulin signaling, is controlled by mTOR regulation of the phosphorylation of the transcription factor STAT3 (Kim *et al.*, 2008) (Fig. 21.2). Rapamycin, the specific inhibitor of mTOR, ameliorates IL-6 induced insulin resistance in hepatocytes (Kim *et al.*, 2008). It is noteworthy that the mTOR involvement in this particular case is distinct from the well-established contribution of mTOR signaling to insulin resistance through S6K1 phosphorylation of IRS (Um *et al.*, 2006). Overall, this further understanding of the pathways underlying insulin resistance suggests potential new therapeutic avenues.

However, the function of IL-6 may not always be inhibitory for insulin actions in the liver. It has been reported that insulin acting within the brain stimulates hepatic secretion of IL-6, which subsequently acts in an autocrine or paracrine manner to activate STAT3 and suppress hepatic glucose production (Inoue *et al.*, 2006). Mice with a liver-specific STAT3 knockout or systemic deletion of IL-6 display impaired regulation of gluconeogenesis (Inoue *et al.*, 2006). The seemingly contradictory effects of IL-6 signaling in the liver remain to be fully explained, but one could envision that the level and duration of IL-6 exposure might partly dictate the signaling and physiological outcome.

2. Skeletal muscle

Skeletal muscle is the major site of insulin-induced glucose disposal, where IL-6 apparently plays a role distinct from its function in the liver. Although very little IL-6 is found in resting muscle, it is well established that contracting muscle secrete IL-6 leading to a 100-fold increase of plasma IL-6 levels (Pedersen *et al.*, 2004). The increased IL-6 expression is correlated with increased glucose uptake during exercise, although other factors associated with muscle contraction may be required for the effect of IL-6 on glucose homeostasis (Pedersen *et al.*, 2004). An animal study highlighted the importance of the contribution of IL-6 to muscle function during exercise, as IL-6 knockout mice exhibited reduced endurance and energy expenditure (Faldt *et al.*, 2004). Exercise-induced nitric oxide has been implicated as a proximal mediator involved in the control of gene expression, including IL-6 mRNA (Steensberg *et al.*, 2007). A study in healthy humans suggested that IL-6 infusion increases glucose disposal (Carey *et al.*, 2006), while experiments with L6 myotube cultures demonstrated that acute IL-6 treatment increased GLUT4 translocation in an AMP-activated protein kinase (AMPK)-dependent manner (Carey *et al.*, 2006). The requirement for IL-6 in exercise-induced activation of AMPK has also been observed in mice (Kelly *et al.*, 2004). Another investigation with IL-6 stimulated skeletal muscle cells in culture found rapid recruitment of IRS-1 to the IL-6R complex, and activation of IRS-1/Akt signaling was confirmed in the muscle tissue of IL-6 treated mice (Weigert *et al.*, 2006). Notably, there was a differential response in the liver of the IL-6 treated mice, where IRS-1 Ser307 phosphorylation and a 10-fold increase of SOCS-3 expression (as opposed to only a 2-fold increase in muscle) were observed (Weigert *et al.*, 2006), both inhibiting insulin signaling. The authors suggested that acute exposure to IL-6 may have short-term beneficial effects on insulin signaling within skeletal muscle (Weigert *et al.*, 2006). This contrasts with the setting of insulin resistance, which involves low-level chronic IL-6 stimulation. Moreover, the aforementioned studies conducted with healthy resting or exercising muscle may have a very different environment than is present in an insulin resistant individual's muscle.

Overall, studies of diabetic patients have indicated that peripheral insulin resistance is primarily caused by a reduction in glucose uptake by skeletal muscle, mediated through the insulin-inducible GLUT4 glucose transporter (Leclercq *et al.*, 2007). An exercise study noted increased insulin sensitivity following short-term exercise among diabetic participants, however, the insulin sensitivity of this population remained ~60% lower than healthy participants throughout the study and insulin resistance among the diabetic patients was not completely reversed through exercise (Bruce *et al.*, 2004). Indeed, studies of glucose transport and IL-6 during exercise have particularly difficult results to interpret, as exercise, independent of insulin, induces GLUT4 translocation and subsequent glucose uptake (Wallberg-Henriksson *et al.*, 1988).

3. Adipose tissue

Significant attention has focused on the role of adipose tissue in the pathogenesis of insulin resistance, due to changes in the tissue's metabolism and adipokine production, as well as its ongoing inflammatory state. In particular, visceral adipose tissue contributes factors, such as adipokines and reactive oxygen species, that impair insulin signaling in the liver and skeletal muscle. Adipose-derived IL-6 enters circulation and plays a systemic role in modulating insulin actions; under basal conditions up to 35% of systemic IL-6 originates from adipose tissue, secreted both by adipocytes and macrophages located therein (Fried *et al.*, 1998; Mohamed-Ali *et al.*, 1997; Weisberg *et al.*, 2003). Although excess visceral fat has long been associated with chronic inflammation, its direct contribution to the secretion of inflammatory adipokines has only recently been demonstrated by the finding that visceral fat may be a major source of circulating IL-6 (Fontana *et al.*, 2007). Secretions from visceral fat can have a direct impact on the liver through the portal vein, the site through which fat secretions enter circulation and which flows directly to the liver with ~80% of the total liver blood supply (Fontana *et al.*, 2007).

Adipose IL-6 content is inversely correlated to insulin responsiveness *in vivo*, as well as insulin-stimulated glucose transport in isolated adipocytes (Bastard *et al.*, 2002). Seemingly, IL-6 may have differential effects on the constitutive and insulin-regulated glucose transporters, GLUT1 and GLUT4, respectively. While modest enhancement of basal glucose transport mediated by GLUT1 was found in 3T3-L1 adipocytes incubated with IL-6 (Stouthard *et al.*, 1996), other studies in the same type of cells demonstrated an inhibitory effect of chronic IL-6 treatment on insulin signaling and insulin action, accompanied by a reduction in the expression of IRS-1 and GLUT4 genes (Lagathu *et al.*, 2003; Rotter *et al.*, 2003). Certainly, the study with long-term IL-6 treatment more closely mirrors the physiological setting of chronic inflammation. Notably, the proinflammatory cytokine IL-1β induces a similar differential effect on glucose uptake in adipocytes—basal transport through GLUT1 is slightly enhanced, while insulin-stimulated transport via GLUT4 is reduced (Jager *et al.*, 2007).

SOCS proteins, especially SOCS-3, play an important role in the regulation of insulin signaling within adiopocytes (Emanuelli *et al.*, 2000; Shi *et al.*, 2004), and IL-6 has been shown to induce the expression of SOCS-3 in 3T3-L1 adipocytes (Fasshauer *et al.*, 2004; Lagathu *et al.*, 2003). Another potential mediator of the inhibitory effect of IL-6 on insulin action in adipocytes is adiponectin, the expression of which is inhibited by IL-6 in 3T3-L1 cells (Fasshauer *et al.*, 2003).

IV. Conclusions

Although certainly a more recent discovery than its function in regulating inflammation, IL-6's role as a regulator of glucose metabolism is well established. Numerous studies have demonstrated a mechanistic link between IL-6 signaling and the insulin pathways, including the identification of factors, such as SOCS-3, which inhibit insulin signaling at different levels. On the other hand, while there has been correlative data suggesting that IL-6 has insulin-sensitizing effects, no detailed mechanism has been elucidated.

As molecular changes within the liver are critical to insulin resistance, studies of hepatocytes are of particular importance. These studies have unequivocally demonstrated that elevated levels of IL-6 in liver cells, especially in a state of chronic inflammation, can promote insulin resistance and thereby contribute to the development of diabetes, and potentially the metabolic syndrome. The adipose tissue, especially visceral fat, has been identified as the main source of circulating IL-6 that is elevated in obesity, contributing to the chronic inflammatory state. Adipose-produced IL-6 has systemic effects, impairing insulin signaling and action in the liver, in particular. While it is not clear whether IL-6 has significant autocrine effects in adipose tissue *in vivo*, *in vitro* studies have suggested that adipocytes are a target of IL-6 in the negative regulation of insulin signaling and glucose metabolism. Contrary to its role in liver and adipose tissue, IL-6 is believed to be beneficial for insulin-regulated glucose metabolism in muscle.

The available evidence has made clear the importance of considering the pleiotropic role of IL-6 in a tissue-specific and physiological state-dependent manner. It was established in 1988 that exercise, independent from insulin, affects glucose transport in muscle. Thus, conclusions regarding the effects of IL-6 on insulin sensitivity, especially in muscle, cannot be made without noting the setting of the study, such as distinguishing between exercise and at-rest muscle. Notably, most of the studies that have reported sensitizing effects of IL-6 on insulin action were conducted with exercising participants, which are potentially not the most representative of insulin resistant and diabetic populations. The source of IL-6 and the duration of IL-6 exposure also likely contribute to the physiological or pathological outcome, as chronically elevated IL-6 levels, derived from adipose, have been relatively well correlated with insulin resistance states, whereas surges of IL-6 in skeletal muscle induced by exercise facilitate insulin actions.

The sum of the studies on IL-6 and insulin sensitivity highlights the diversity of the roles played by this cytokine in the regulation of glucose metabolism in multiple tissues, as well as the complexity of the crosstalk between the liver, muscle, adipose tissue, and the brain. Further research needs to be directed at understanding the tissue level interplay in order to develop a true picture of the organism level biology of IL-6 and its impact on insulin resistance.

ACKNOWLEDGMENT

Work from the authors' laboratory was supported by funding from the National Institute of Health and the American Diabetes Association.

REFERENCES

Aguirre, V., Uchida, T., Yenush, L., Davis, R., and White, M. F. (2000). The c-Jun NH(2)-terminal kinase promotes insulin resistance during association with insulin receptor substrate-1 and phosphorylation of Ser(307). *J. Biol. Chem.* **275,** 9047–9054.

Aguirre, V., Werner, E. D., Giraud, J., Lee, Y. H., Shoelson, S. E., and White, M. F. (2002). Phosphorylation of Ser307 in insulin receptor substrate-1 blocks interactions with the insulin receptor and inhibits insulin action. *J. Biol. Chem.* **277,** 1531–1537. Epub 2001 Oct 17).

Akira, S., Taga, T., and Kishimoto, T. (1993). Interleukin-6 in biology and medicine. *Adv. Immunol.* **54,** 1–78.

Arkan, M. C., Hevener, A. L., Greten, F. R., Maeda, S., Li, Z. W., Long, J. M., Wynshaw-Boris, A., Poli, G., Olefsky, J., and Karin, M. (2005). IKK-beta links inflammation to obesity-induced insulin resistance. *Nat. Med.* **11,** 191–198.

Bandyopadhyay, G., Standaert, M. L., Zhao, L., Yu, B., Avignon, A., Galloway, L., Karnam, P., Moscat, J., and Farese, R. V. (1997). Activation of protein kinase C (alpha, beta, and zeta) by insulin in 3T3/L1 cells. Transfection studies suggest a role for PKC-zeta in glucose transport. *J. Biol. Chem.* **272,** 2551–2558.

Bao, Y., Lopez, J. A., James, D. E., and Hunziker, W. (2008). Snapin interacts with the Exo70 subunit of the exocyst and modulates GLUT4 trafficking. *J. Biol. Chem.* **283,** 324–331.

Bastard, J. P., Jardel, C., Bruckert, E., Blondy, P., Capeau, J., Laville, M., Vidal, H., and Hainque, B. (2000). Elevated levels of interleukin 6 are reduced in serum and subcutaneous adipose tissue of obese women after weight loss. *J. Clin. Endocrinol. Metab.* **85,** 3338–3342.

Bastard, J. P., Maachi, M., Van Nhieu, J. T., Jardel, C., Bruckert, E., Grimaldi, A., Robert, J. J., Capeau, J., and Hainque, B. (2002). Adipose tissue IL-6 content correlates with resistance to insulin activation of glucose uptake both in vivo and in vitro. *J. Clin. Endocrinol. Metab.* **87,** 2084–2089.

Batsis, J. A., Nieto-Martinez, R. E., and Lopez-Jimenez, F. (2007). Metabolic syndrome: From global epidemiology to individualized medicine. *Clin. Pharmacol. Ther.* **82,** 509–524.

Baumann, C. A., Ribon, V., Kanzaki, M., Thurmond, D. C., Mora, S., Shigematsu, S., Bickel, P. E., Pessin, J. E., and Saltiel, A. R. (2000). CAP defines a second signalling pathway required for insulin-stimulated glucose transport. *Nature* **407,** 202–207.

Bjorbaek, C., El-Haschimi, K., Frantz, J. D., and Flier, J. S. (1999). The role of SOCS-3 in leptin signaling and leptin resistance. *J. Biol. Chem.* **274,** 30059–30065.

Bjornholm, M., and Zierath, J. R. (2005). Insulin signal transduction in human skeletal muscle: Identifying the defects in Type II diabetes. *Biochem. Soc. Trans.* **33,** 354–357.

Bluher, M., Michael, M. D., Peroni, O. D., Ueki, K., Carter, N., Kahn, B. B., and Kahn, C. R. (2002). Adipose tissue selective insulin receptor knockout protects against obesity and obesity-related glucose intolerance. *Dev. Cell* **3,** 25–38.

Bouzakri, K., Roques, M., Gual, P., Espinosa, S., Guebre-Egziabher, F., Riou, J. P., Laville, M., Le Marchand-Brustel, Y., Tanti, J. F., and Vidal, H. (2003). Reduced activation of phosphatidylinositol-3 kinase and increased serine 636 phosphorylation of insulin receptor substrate-1 in primary culture of skeletal muscle cells from patients with type 2 diabetes. *Diabetes* **52,** 1319–1325.

Bruce, C. R., Kriketos, A. D., Cooney, G. J., and Hawley, J. A. (2004). Disassociation of muscle triglyceride content and insulin sensitivity after exercise training in patients with Type 2 diabetes. *Diabetologia* **47,** 23–30.

Bruning, J. C., Michael, M. D., Winnay, J. N., Hayashi, T., Horsch, D., Accili, D., Goodyear, L. J., and Kahn, C. R. (1998). A muscle-specific insulin receptor knockout exhibits features of the metabolic syndrome of NIDDM without altering glucose tolerance. *Mol. Cell* **2,** 559–569.

Cai, D., Yuan, M., Frantz, D. F., Melendez, P. A., Hansen, L., Lee, J., and Shoelson, S. E. (2005). Local and systemic insulin resistance resulting from hepatic activation of IKK-beta and NF-kappaB. *Nat. Med.* **11,** 183–190.

Calo, V., Migliavacca, M., Bazan, V., Macaluso, M., Buscemi, M., Gebbia, N., and Russo, A. (2003). STAT proteins: From normal control of cellular events to tumorigenesis. *J. Cell. Physiol.* **197,** 157–168.

Carey, A. L., and Febbraio, M. A. (2004). Interleukin-6 and insulin sensitivity: Friend or foe? *Diabetologia* **47,** 1135–1142. Epub 2004 Jul 7.

Carey, A. L., Steinberg, G. R., Macaulay, S. L., Thomas, W. G., Holmes, A. G., Ramm, G., Prelovsek, O., Hohnen-Behrens, C., Watt, M. J., James, D. E., Kemp, B. E., Pedersen, B. K., *et al.* (2006). Interleukin-6 increases insulin-stimulated glucose disposal in humans and glucose uptake and fatty acid oxidation in vitro via AMP-activated protein kinase. *Diabetes* **55,** 2688–2697.

Chiang, S. H., Baumann, C. A., Kanzaki, M., Thurmond, D. C., Watson, R. T., Neudauer, C. L., Macara, I. G., Pessin, J. E., and Saltiel, A. R. (2001). Insulin-stimulated GLUT4 translocation requires the CAP-dependent activation of TC10. *Nature* **410,** 944–948.

Chiang, S. H., Hwang, J., Legendre, M., Zhang, M., Kimura, A., and Saltiel, A. R. (2003). TCGAP, a multidomain Rho GTPase-activating protein involved in insulin-stimulated glucose transport. *EMBO J.* **22,** 2679–2691.

Chung, J., Uchida, E., Grammer, T. C., and Blenis, J. (1997). STAT3 serine phosphorylation by ERK-dependent and -independent pathways negatively modulates its tyrosine phosphorylation. *Mol. Cell Biol.* **17,** 6508–6516.

Cong, L. N., Chen, H., Li, Y., Zhou, L., McGibbon, M. A., Taylor, S. I., and Quon, M. J. (1997). Physiological role of Akt in insulin-stimulated translocation of GLUT4 in transfected rat adipose cells. *Mol. Endocrinol.* **11,** 1881–1890.

Croker, B. A., Krebs, D. L., Zhang, J. G., Wormald, S., Willson, T. A., Stanley, E. G., Robb, L., Greenhalgh, C. J., Forster, I., Clausen, B. E., Nicola, N. A., Metcalf, D., *et al.* (2003). SOCS3 negatively regulates IL-6 signaling *in vivo*. *Nat. Immunol.* **4,** 540–545.

Cross, D. A., Alessi, D. R., Cohen, P., Andjelkovich, M., and Hemmings, B. A. (1995). Inhibition of glycogen synthase kinase-3 by insulin mediated by protein kinase B. *Nature* **378,** 785–789.

Di Gregorio, G. B., Hensley, L., Lu, T., Ranganathan, G., and Kern, P. A. (2004). Lipid and carbohydrate metabolism in mice with a targeted mutation in the IL-6 gene: Absence of development of age-related obesity. *Am. J. Physiol. Endocrinol. Metab.* **287,** E182–E187.

Emanuelli, B., Peraldi, P., Filloux, C., Sawka-Verhelle, D., Hilton, D., and Van Obberghen, E. (2000). SOCS-3 is an insulin-induced negative regulator of insulin signaling. *J. Biol. Chem.* **275,** 15985–15991.

Emanuelli, B., Peraldi, P., Filloux, C., Chavey, C., Freidinger, K., Hilton, D. J., Hotamisligil, G. S., and Van Obberghen, E. (2001). SOCS-3 inhibits insulin signaling and is up-regulated in response to tumor necrosis factor-alpha in the adipose tissue of obese mice. *J. Biol. Chem.* **276,** 47944–47949.

Expert-Panel (2001). Executive Summary of The Third Report of The National Cholesterol Education Program (NCEP) Expert Panel on Detection, Evaluation, And Treatment of High Blood Cholesterol In Adults (Adult Treatment Panel III). *Jama* **285,** 2486–2497.

Faldt, J., Wernstedt, I., Fitzgerald, S. M., Wallenius, K., Bergstrom, G., and Jansson, J. O. (2004). Reduced exercise endurance in interleukin-6-deficient mice. *Endocrinology* **145,** 2680–2686.

Fasshauer, M., Kralisch, S., Klier, M., Lossner, U., Bluher, M., Klein, J., and Paschke, R. (2003). Adiponectin gene expression and secretion is inhibited by interleukin-6 in 3T3-L1 adipocytes. *Biochem. Biophys. Res. Commun.* **301,** 1045–1050.

Fasshauer, M., Kralisch, S., Klier, M., Lossner, U., Bluher, M., Klein, J., and Paschke, R. (2004). Insulin resistance-inducing cytokines differentially regulate SOCS mRNA expression via growth factor- and Jak/Stat-signaling pathways in 3T3-L1 adipocytes. *J. Endocrinol.* **181,** 129–138.

Fontana, L., Eagon, J. C., Trujillo, M. E., Scherer, P. E., and Klein, S. (2007). Visceral fat adipokine secretion is associated with systemic inflammation in obese humans. *Diabetes* **56,** 1010–1013.

Fried, S. K., Bunkin, D. A., and Greenberg, A. S. (1998). Omental and subcutaneous adipose tissues of obese subjects release interleukin-6: Depot difference and regulation by glucocorticoid. *J. Clin. Endocrinol. Metab.* **83,** 847–850.

Gao, Z., Hwang, D., Bataille, F., Lefevre, M., York, D., Quon, M. J., and Ye, J. (2002). Serine phosphorylation of insulin receptor substrate 1 by inhibitor kappa B kinase complex. *J. Biol. Chem.* **277,** 48115–48121.

Hartge, M. M., Unger, T., and Kintscher, U. (2007). The endothelium and vascular inflammation in diabetes. *Diab. Vasc. Dis. Res.* **4,** 84–88.

Haruta, T., Uno, T., Kawahara, J., Takano, A., Egawa, K., Sharma, P. M., Olefsky, J. M., and Kobayashi, M. (2000). A rapamycin-sensitive pathway down-regulates insulin signaling via phosphorylation and proteasomal degradation of insulin receptor substrate-1. *Mol. Endocrinol.* **14,** 783–794.

Heinrich, P. C., Behrmann, I., Haan, S., Hermanns, H. M., Muller-Newen, G., and Schaper, F. (2003). Principles of interleukin (IL)-6-type cytokine signalling and its regulation. *Biochem. J.* **374,** 1–20.

Hirosumi, J., Tuncman, G., Chang, L., Gorgun, C. Z., Uysal, K. T., Maeda, K., Karin, M., and Hotamisligil, G. S. (2002). A central role for JNK in obesity and insulin resistance. *Nature* **420,** 333–336.

Hoene, M., and Weigert, C. (2008). The role of interleukin-6 in insulin resistance, body fat distribution and energy balance. *Obes. Rev.* **9,** 20–29.

Hotamisligil, G. S. (1999). Mechanisms of TNF-alpha-induced insulin resistance. *Exp. Clin. Endocrinol. Diabetes* **107,** 119–125.

Hotamisligil, G. S., Shargill, N. S., and Spiegelman, B. M. (1993). Adipose expression of tumor necrosis factor-alpha: Direct role in obesity-linked insulin resistance. *Science* **259,** 87–91.

Houstis, N., Rosen, E. D., and Lander, E. S. (2006). Reactive oxygen species have a causal role in multiple forms of insulin resistance. *Nature* **440,** 944–948.

Howard, J. K., and Flier, J. S. (2006). Attenuation of leptin and insulin signaling by SOCS proteins. *Trends Endocrinol. Metab.* **17,** 365–371.

Hu, F. B., Meigs, J. B., Li, T. Y., Rifai, N., and Manson, J. E. (2004). Inflammatory markers and risk of developing type 2 diabetes in women. *Diabetes* **53,** 693–700.

Huang, S., and Czech, M. P. (2007). The GLUT4 glucose transporter. *Cell Metab.* **5,** 237–252.

Inoue, M., Chang, L., Hwang, J., Chiang, S. H., and Saltiel, A. R. (2003). The exocyst complex is required for targeting of Glut4 to the plasma membrane by insulin. *Nature* **422,** 629–633.

Inoue, H., Ogawa, W., Asakawa, A., Okamoto, Y., Nishizawa, A., Matsumoto, M., Teshigawara, K., Matsuki, Y., Watanabe, E., Hiramatsu, R., Notohara, K., Katayose, K., *et al.* (2006). Role of hepatic STAT3 in brain-insulin action on hepatic glucose production. *Cell. Metab.* **3,** 267–275.

Jager, J., Gremeaux, T., Cormont, M., Le Marchand-Brustel, Y., and Tanti, J. F. (2007). Interleukin-1beta-induced insulin resistance in adipocytes through down-regulation of insulin receptor substrate-1 expression. *Endocrinology* **148,** 241–251.

Jain, N., Zhang, T., Kee, W. H., Li, W., and Cao, X. (1999). Protein kinase C delta associates with and phosphorylates Stat3 in an interleukin-6-dependent manner. *J. Biol. Chem.* **274,** 24392–24400.

Keller, P., Keller, C., Carey, A. L., Jauffred, S., Fischer, C. P., Steensberg, A., and Pedersen, B. K. (2003). Interleukin-6 production by contracting human skeletal muscle: Autocrine regulation by IL-6. *Biochem. Biophys. Res. Commun.* **310,** 550–554.

Kelly, M., Keller, C., Avilucea, P. R., Keller, P., Luo, Z., Xiang, X., Giralt, M., Hidalgo, J., Saha, A. K., Pedersen, B. K., and Ruderman, N. B. (2004). AMPK activity is diminished in tissues of IL-6 knockout mice: The effect of exercise. *Biochem. Biophys. Res. Commun.* **320,** 449–454.

Kern, P. A., Ranganathan, S., Li, C., Wood, L., and Ranganathan, G. (2001). Adipose tissue tumor necrosis factor and interleukin-6 expression in human obesity and insulin resistance. *Am. J. Physiol. Endocrinol. Metab.* **280,** E745–E751.

Kim, J. K., Michael, M. D., Previs, S. F., Peroni, O. D., Mauvais-Jarvis, F., Neschen, S., Kahn, B. B., Kahn, C. R., and Shulman, G. I. (2000). Redistribution of substrates to adipose tissue promotes obesity in mice with selective insulin resistance in muscle. *J. Clin. Invest.* **105,** 1791–1797.

Kim, H. J., Higashimori, T., Park, S. Y., Choi, H., Dong, J., Kim, Y. J., Noh, H. L., Cho, Y. R., Cline, G., Kim, Y. B., and Kim, J. K. (2004). Differential effects of interleukin-6 and -10 on skeletal muscle and liver insulin action in vivo. *Diabetes* **53,** 1060–1067.

Kim, J. H., Kim, J. E., Liu, H. Y., Cao, W., and Chen, J. (2008). Regulation of interleukin-6-induced hepatic insulin resistance by mammalian target of rapamycin through the STAT3-SOCS3 pathway. *J. Biol. Chem.* **283,** 708–715.

Klover, P. J., Zimmers, T. A., Koniaris, L. G., and Mooney, R. A. (2003). Chronic exposure to interleukin-6 causes hepatic insulin resistance in mice. *Diabetes* **52,** 2784–2789.

Klover, P. J., Clementi, A. H., and Mooney, R. A. (2005). Interleukin-6 depletion selectively improves hepatic insulin action in obesity. *Endocrinology* **146,** 3417–3427.

Kocak, H., Oner-Iyidogan, Y., Gurdol, F., Oner, P., Suzme, R., Esin, D., and Issever, H. (2007). Advanced oxidation protein products in obese women: Its relation to insulin resistance and resistin. *Clin. Exp. Med.* **7,** 173–178.

Kohn, A. D., Summers, S. A., Birnbaum, M. J., and Roth, R. A. (1996). Expression of a constitutively active Akt Ser/Thr kinase in 3T3-L1 adipocytes stimulates glucose uptake and glucose transporter 4 translocation. *J. Biol. Chem.* **271,** 31372–31378.

Kopp, H. P., Kopp, C. W., Festa, A., Krzyzanowska, K., Kriwanek, S., Minar, E., Roka, R., and Schernthaner, G. (2003). Impact of weight loss on inflammatory proteins and their association with the insulin resistance syndrome in morbidly obese patients. *Arterioscler. Thromb. Vasc. Biol.* **23,** 1042–1047.

Krebs, D. L., and Hilton, D. J. (2000). SOCS: Physiological suppressors of cytokine signaling. *J. Cell Sci.* **113**(Pt. 16), 2813–2819.

Krebs, D. L., and Hilton, D. J. (2003). A new role for SOCS in insulin action. Suppressor of cytokine signaling. *Sci. STKE* **2003,** PE6.

Kubaszek, A., Pihlajamaki, J., Punnonen, K., Karhapaa, P., Vauhkonen, I., and Laakso, M. (2003). The C-174G promoter polymorphism of the IL-6 gene affects energy expenditure and insulin sensitivity. *Diabetes* **52,** 558–561.

Lagathu, C., Bastard, J. P., Auclair, M., Maachi, M., Capeau, J., and Caron, M. (2003). Chronic interleukin-6 (IL-6) treatment increased IL-6 secretion and induced insulin resistance in adipocyte: Prevention by rosiglitazone. *Biochem. Biophys. Res. Commun.* **311,** 372–379.

Leclercq, I. A., Da Silva Morais, A., Schroyen, B., Van Hul, N., and Geerts, A. (2007). Insulin resistance in hepatocytes and sinusoidal liver cells: Mechanisms and consequences. *J. Hepatol.* **47,** 142–156.

Lim, C. P., and Cao, X. (1999). Serine phosphorylation and negative regulation of Stat3 by JNK. *J. Biol. Chem.* **274,** 31055–31061.
Lim, C. P., and Cao, X. (2001). Regulation of Stat3 activation by MEK kinase 1. *J. Biol. Chem.* **276,** 21004–21011.
Michael, M. D., Kulkarni, R. N., Postic, C., Previs, S. F., Shulman, G. I., Magnuson, M. A., and Kahn, C. R. (2000). Loss of insulin signaling in hepatocytes leads to severe insulin resistance and progressive hepatic dysfunction. *Mol. Cell* **6,** 87–97.
Mohamed-Ali, V., Goodrick, S., Rawesh, A., Katz, D. R., Miles, J. M., Yudkin, J. S., Klein, S., and Coppack, S. W. (1997). Subcutaneous adipose tissue releases interleukin-6, but not tumor necrosis factor-alpha, in vivo. *J. Clin. Endocrinol. Metab.* **82,** 4196–4200.
NIDDK (2005). National Diabetes Statistics fact sheet: General information and national estimates on diabetes in the United States, 2005 N. I. o. H. U.S. Department of Health and Human Services, Ed, Bethesda, MD.
O'Brien, R. M., and Granner, D. K. (1996). Regulation of gene expression by insulin. *Physiol. Rev.* **76,** 1109–1161.
Pederson, T. M., Kramer, D. L., and Rondinone, C. M. (2001). Serine/threonine phosphorylation of IRS-1 triggers its degradation: Possible regulation by tyrosine phosphorylation. *Diabetes* **50,** 24–31.
Pedersen, B. K., Steensberg, A., Fischer, C., Keller, C., Keller, P., Plomgaard, P., Wolsk-Petersen, E., and Febbraio, M. (2004). The metabolic role of IL-6 produced during exercise: Is IL-6 an exercise factor? *Proc. Nutr. Soc.* **63,** 263–267.
Pradhan, A. D., Manson, J. E., Rifai, N., Buring, J. E., and Ridker, P. M. (2001). C-reactive protein, interleukin 6, and risk of developing type 2 diabetes mellitus. *JAMA* **286,** 327–334.
Qi, L., van Dam, R. M., Meigs, J. B., Manson, J. E., Hunter, D., and Hu, F. B. (2006). Genetic variation in IL6 gene and type 2 diabetes: Tagging-SNP haplotype analysis in large-scale case-control study and meta-analysis. *Hum. Mol. Genet.* **15,** 1914–1920.
Rajan, P., Panchision, D. M., Newell, L. F., and McKay, R. D. (2003). BMPs signal alternately through a SMAD or FRAP-STAT pathway to regulate fate choice in CNS stem cells. *J. Cell Biol.* **161,** 911–921.
Reaven, G. M. (1988). Banting lecture 1988. Role of insulin resistance in human disease. *Diabetes.* **37,** 1595–1607.
Rieusset, J., Bouzakri, K., Chevillotte, E., Ricard, N., Jacquet, D., Bastard, J. P., Laville, M., and Vidal, H. (2004). Suppressor of cytokine signaling 3 expression and insulin resistance in skeletal muscle of obese and type 2 diabetic patients. *Diabetes* **53,** 2232–2241.
Rotter, V., Nagaev, I., and Smith, U. (2003). Interleukin-6 (IL-6) induces insulin resistance in 3T3-L1 adipocytes and is, like IL-8 and tumor necrosis factor-alpha, overexpressed in human fat cells from insulin-resistant subjects. *J. Biol. Chem.* **278,** 45777–45784.
Sabio, G., Das, M., Mora, A., Zhang, Z., Jun, J. Y., Ko, H. J., Barrett, T., Kim, J. K., and Davis, R. J. (2008). A stress signaling pathway in adipose tissue regulates hepatic insulin resistance. *Science* **322,** 1539–1543.
Schindler, C., and Darnell, J. E. Jr., (1995). Transcriptional responses to polypeptide ligands: The JAK-STAT pathway. *Annu. Rev. Biochem.* **64,** 621–651.
Senn, J. J., Klover, P. J., Nowak, I. A., and Mooney, R. A. (2002). Interleukin-6 induces cellular insulin resistance in hepatocytes. *Diabetes* **51,** 3391–3399.
Senn, J. J., Klover, P. J., Nowak, I. A., Zimmers, T. A., Koniaris, L. G., Furlanetto, R. W., and Mooney, R. A. (2003). Suppressor of cytokine signaling-3 (SOCS-3), a potential mediator of interleukin-6-dependent insulin resistance in hepatocytes. *J. Biol. Chem.* **278,** 13740–13746.
Shah, O. J., and Hunter, T. (2006). Turnover of the active fraction of IRS1 involves raptor-mTOR- and S6K1-dependent serine phosphorylation in cell culture models of tuberous sclerosis. *Mol. Cell Biol.* **26,** 6425–6434.

Shi, H., Tzameli, I., Bjorbaek, C., and Flier, J. S. (2004). Suppressor of cytokine signaling 3 is a physiological regulator of adipocyte insulin signaling. *J. Biol. Chem.* **279,** 34733–34740.

Spranger, J., Kroke, A., Mohlig, M., Hoffmann, K., Bergmann, M. M., Ristow, M., Boeing, H., and Pfeiffer, A. F. (2003). Inflammatory cytokines and the risk to develop type 2 diabetes: Results of the prospective population-based European Prospective Investigation into Cancer and Nutrition (EPIC)-Potsdam Study. *Diabetes* **52,** 812–817.

Steensberg, A., Keller, C., Hillig, T., Frosig, C., Wojtaszewski, J. F., Pedersen, B. K., Pilegaard, H., and Sander, M. (2007). Nitric oxide production is a proximal signaling event controlling exercise-induced mRNA expression in human skeletal muscle. *FASEB J.* **21,** 2683–2694.

Stenlof, K., Wernstedt, I., Fjallman, T., Wallenius, V., Wallenius, K., and Jansson, J. O. (2003). Interleukin-6 levels in the central nervous system are negatively correlated with fat mass in overweight/obese subjects. *J. Clin. Endocrinol. Metab.* **88,** 4379–4383.

Stouthard, J. M., Oude Elferink, R. P., and Sauerwein, H. P. (1996). Interleukin-6 enhances glucose transport in 3T3-L1 adipocytes. *Biochem. Biophys. Res. Commun.* **220,** 241–245.

Sun, X. J., Goldberg, J. L., Qiao, L. Y., and Mitchell, J. J. (1999). Insulin-induced insulin receptor substrate-1 degradation is mediated by the proteasome degradation pathway. *Diabetes* **48,** 1359–1364.

Takano, A., Usui, I., Haruta, T., Kawahara, J., Uno, T., Iwata, M., and Kobayashi, M. (2001). Mammalian target of rapamycin pathway regulates insulin signaling via subcellular redistribution of insulin receptor substrate 1 and integrates nutritional signals and metabolic signals of insulin. *Mol. Cell Biol.* **21,** 5050–5062.

Torisu, T., Sato, N., Yoshiga, D., Kobayashi, T., Yoshioka, T., Mori, H., Iida, M., and Yoshimura, A. (2007). The dual function of hepatic SOCS3 in insulin resistance *in vivo*. *Genes Cells* **12,** 143–154.

Tsigos, C., Papanicolaou, D. A., Kyrou, I., Defensor, R., Mitsiadis, C. S., and Chrousos, G. P. (1997). Dose-dependent effects of recombinant human interleukin-6 on glucose regulation. *J. Clin. Endocrinol. Metab.* **82,** 4167–4170.

Tzatsos, A., and Kandror, K. V. (2006). Nutrients Suppress Phosphatidylinositol 3-Kinase/ Akt Signaling via Raptor-Dependent mTOR-Mediated Insulin Receptor Substrate 1 Phosphorylation. *Mol. Cell Biol.* **26,** 63–76.

Ueki, K., Kondo, T., and Kahn, C. R. (2004). Suppressor of cytokine signaling 1 (SOCS-1) and SOCS-3 cause insulin resistance through inhibition of tyrosine phosphorylation of insulin receptor substrate proteins by discrete mechanisms. *Mol. Cell Biol.* **24,** 5434–5446.

Um, S. H., D'Alessio, D., and Thomas, G. (2006). Nutrient overload, insulin resistance, and ribosomal protein S6 kinase 1, S6K1. *Cell Metab.* **3,** 393–402.

Uysal, K. T., Wiesbrock, S. M., Marino, M. W., and Hotamisligil, G. S. (1997). Protection from obesity-induced insulin resistance in mice lacking TNF-alpha function. *Nature* **389,** 610–614.

Van Snick, J. (1990). Interleukin-6: An overview. *Annu. Rev. Immunol.* **8,** 253–278.

Ventre, J., Doebber, T., Wu, M., MacNaul, K., Stevens, K., Pasparakis, M., Kollias, G., and Moller, D. E. (1997). Targeted disruption of the tumor necrosis factor-alpha gene: Metabolic consequences in obese and nonobese mice. *Diabetes* **46,** 1526–1531.

Virkamaki, A., Ueki, K., and Kahn, C. R. (1999). Protein-protein interaction in insulin signaling and the molecular mechanisms of insulin resistance. *J. Clin. Invest.* **103,** 931–943.

Vozarova, B., Weyer, C., Hanson, K., Tataranni, P. A., Bogardus, C., and Pratley, R. E. (2001). Circulating interleukin-6 in relation to adiposity, insulin action, and insulin secretion. *Obes. Res.* **9,** 414–417.

Wallberg-Henriksson, H., Constable, S. H., Young, D. A., and Holloszy, J. O. (1988). Glucose transport into rat skeletal muscle: Interaction between exercise and insulin. *J. Appl. Physiol.* **65,** 909–913.

Wallenius, V., Wallenius, K., Ahren, B., Rudling, M., Carlsten, H., Dickson, S. L., Ohlsson, C., and Jansson, J. O. (2002). Interleukin-6-deficient mice develop mature-onset obesity. *Nat. Med.* **8,** 75–79.

Wang, Q., Somwar, R., Bilan, P. J., Liu, Z., Jin, J., Woodgett, J. R., and Klip, A. (1999). Protein kinase B/Akt participates in GLUT4 translocation by insulin in L6 myoblasts. *Mol. Cell Biol.* **19,** 4008–4018.

Weigert, C., Hennige, A. M., Lehmann, R., Brodbeck, K., Baumgartner, F., Schauble, M., Haring, H. U., and Schleicher, E. D. (2006). Direct cross-talk of interleukin-6 and insulin signal transduction via insulin receptor substrate-1 in skeletal muscle cells. *J. Biol. Chem.* **281,** 7060–7067.

Weisberg, S. P., McCann, D., Desai, M., Rosenbaum, M., Leibel, R. L., and Ferrante, A. W. Jr., (2003). Obesity is associated with macrophage accumulation in adipose tissue. *J. Clin. Invest.* **112,** 1796–1808.

Wellen, K. E., and Hotamisligil, G. S. (2005). Inflammation, stress, and diabetes. *J. Clin. Invest.* **115,** 1111–1119.

Wen, Z., Zhong, Z., and Darnell, J. E. Jr., (1995). Maximal activation of transcription by Stat1 and Stat3 requires both tyrosine and serine phosphorylation. *Cell* **82,** 241–250.

White, M. F. (2002). IRS proteins and the common path to diabetes. *Am. J. Physiol. Endocrinol. Metab.* **283,** E413–E422.

White, M. F. (2003). Insulin signaling in health and disease. *Science* **302,** 1710–1711.

Xu, H., Barnes, G. T., Yang, Q., Tan, G., Yang, D., Chou, C. J., Sole, J., Nichols, A., Ross, J. S., Tartaglia, L. A., and Chen, H. (2003). Chronic inflammation in fat plays a crucial role in the development of obesity-related insulin resistance. *J. Clin. Invest.* **112,** 1821–1830.

Yokogami, K., Wakisaka, S., Avruch, J., and Reeves, S. A. (2000). Serine phosphorylation and maximal activation of STAT3 during CNTF signaling is mediated by the rapamycin target mTOR. *Curr. Biol.* **10,** 47–50.

CHAPTER TWENTY-TWO

Structure, Function, and Regulation of Insulin-Degrading Enzyme

Raymond E. Hulse,* Luis A. Ralat,† *and* Wei-Jen Tang*,†

Contents

Abstract

The short half-life of insulin in the human body (4–6 min) prompted the search and discovery of insulin-degrading enzyme (IDE), a 110-kDa metalloprotease that can rapidly degrade insulin into inactive fragments. Genetic and biochemical evidence accumulated in the last sixty years has implicated IDE as an important physiological contributor in the maintenance of insulin levels. Recent structural and biochemical analyses reveal the molecular basis of how IDE uses size and charge distribution of the catalytic chamber and structural flexibility of substrates to selectively recognize and degrade insulin, as well as the regulatory mechanisms of this enzyme. These studies provide a path for potential therapeutics in the control of insulin metabolism by the degradation of insulin.

* Committee on Neurobiology, The University of Chicago, Chicago, Illinois 60637
† Ben-May Department for Cancer Research, The University of Chicago, Chicago, Illinois 60637

Vitamins and Hormones, Volume 80
ISSN 0083-6729, DOI: 10.1016/S0083-6729(08)00622-5

I. Introduction

The ability of insulin-degrading enzyme (IDE) to degrade insulin was reported nearly sixty years ago (Mirsky and Broh-kahn, 1949; Mirsky *et al.*, 1950). Despite decades of research, the role of IDE in the degradation of insulin, as well as the cellular location of this process, remains controversial (Authier *et al.*, 1996; Hersh, 2006). Yet multiple lines of evidence have accumulated, which support a role for IDE as an important protease involved in insulin's catabolism (Duckworth *et al.*, 1998). IDE has an exceptionally high affinity for insulin (~0.1 μM) and can cleave insulin into multiple inactive fragments (Chesneau and Rosner, 2000; Duckworth *et al.*, 1998; Grasso *et al.*, 2007). Several examples of *in vitro* evidence include over-expression studies of IDE in cell lines, which show increased insulin degradation, studies of internalized insulin cross-linking to IDE, as well as injection of monoclonal antibodies to prevent the action of IDE on insulin (Hari *et al.*, 1987; Kuo *et al.*, 1991; Perlman *et al.*, 1993; Shii and Roth, 1986).

Recent evidence has strengthened the physiological relevance of this protease. A decrease in insulin degradation and associated hyperinsulinemia was observed in IDE knockout mice (Farris *et al.*, 2003). In addition, reducing the levels of human IDE in HepG2 cell-line cultured cells using silencing RNA inhibited insulin degradation by up to 76% (Fawcett *et al.*, 2007). The genetic variability of the IDE gene has also been examined in humans and rodents. For example, a large-scale human genetic analysis reveals the association of a single nucleotide polymorphism with type 2 diabetes (Sladek *et al.*, 2007). Additional studies in a different population also revealed a single nucleotide polymorphism in the IDE gene with evidence of hyperinsulinemia as compared to the control group (Marlowe *et al.*, 2006). Further genetic evidence of diabetic susceptibility due to polymorphisms of the IDE gene have been found in the Goto-kakizaki (GK) rat, a widely used rodent model of diabetes (Fakhrai-Rad *et al.*, 2000). Yet, clear genetic linkage between IDE dysfunction and type 2 diabetes remains in dispute (Florez *et al.*, 2006; Groves *et al.*, 2003; Gu *et al.*, 2004; Karamohamed *et al.*, 2003). Importantly, IDE is not the sole enzyme responsible for insulin degradation: cathepsin D has also been shown to participate in the lysosomal degradation of insulin (Authier *et al.*, 2002).

IDE (EC 3.4.24.56, insulysin, or insulinase) is an evolutionarily conserved 110-kDa zinc metalloprotease. It has been described principally as a cytosolic enzyme but is also found in multiple cellular compartments including endosomes, peroxisomes, mitochondria, the cell surface, and in secreted form (Authier *et al.*, 1995; Leissring *et al.*, 2004; Qiu and Folstein, 2006; Qiu *et al.*, 1998; Sudoh *et al.*, 2002; Vekrellis *et al.*, 2000). Its enzymatic activity is optimal at a physiologically relevant pH range

(6.0–8.5) and is sensitive to the metalloprotease inhibitor 1,10-phenanthroline but not to other nonmetalloprotease inhibitors (Duckworth *et al.*, 1998). Although IDE has a high affinity for insulin, it has been implicated in the degradation of other amyloidogenic peptides (Duckworth *et al.*, 1998; Farris *et al.*, 2003; Kurochkin, 2001; Shen *et al.*, 2006). Notably, glucagon, a peptide also implicated in glucose metabolism, is degraded by IDE. The recent structural solution of IDE with insulin B chain, glucagon, and amylin (Shen *et al.*, 2006) and the substrate-free conformation (Im *et al.*, 2007), has increased understanding of the mechanisms of how IDE interacts with insulin and facilitates its degradation.

These multiple lines of evidence along with recent structural analyses all implicate this highly conserved and ubiquitous metalloprotease in the complex metabolic cycle of insulin. Understanding how IDE itself is regulated has become a vital area of research and improves understanding of how IDE degrades insulin. We will focus on the molecular structure and regulatory mechanisms of IDE and how these contribute to insulin metabolism.

II. Structure of IDE

The structural solution of IDE reveals important information about how it associates and degrades substrates. In its monomeric form, IDE is made of two roughly equal sized domains (~55 kDa); IDE-N and IDE-C that are connected by a 28 amino-acid residue loop (Fig. 22.1). When the two halves come together, a crypt is formed to enclose its substrates and the formation of crypt prevents entry or escape of the substrates. Hence, IDE belongs to an emerging protease family referred to as cryptidases, that is, crypt forming peptidases (Malito *et al.*, 2008a). The crypt has a volume of ~15,700 Å^3, which excludes peptides larger than ~70 amino acids (Malito *et al.*, 2008a). The two halves, when closed, share a high surface area (11,496 Å^2) and possess good shape complementarity (Shen *et al.*, 2006). The catalytic site is located inside the crypt in IDE-N (Fig. 22.1). This site is rich in charged, polar, and hydrophobic patches (Shen *et al.*, 2006), which facilitate interaction with both IDE-C and substrates. IDE-N also possesses an exosite, which is ~30 Å away from the catalytic site. Here, the N-termini of ligands are tethered. This site may play a role in the positioning of insulin for cleavage (Im *et al.*, 2007; Shen *et al.*, 2006).

A. Substrate recognition of IDE

The presence of disulfide bonds is a notable feature of certain substrates of IDE (Fig. 22.2). For example, insulin has three disulfide bonds. Interestingly, the cleavage of insulin by IDE does not require the reduction and

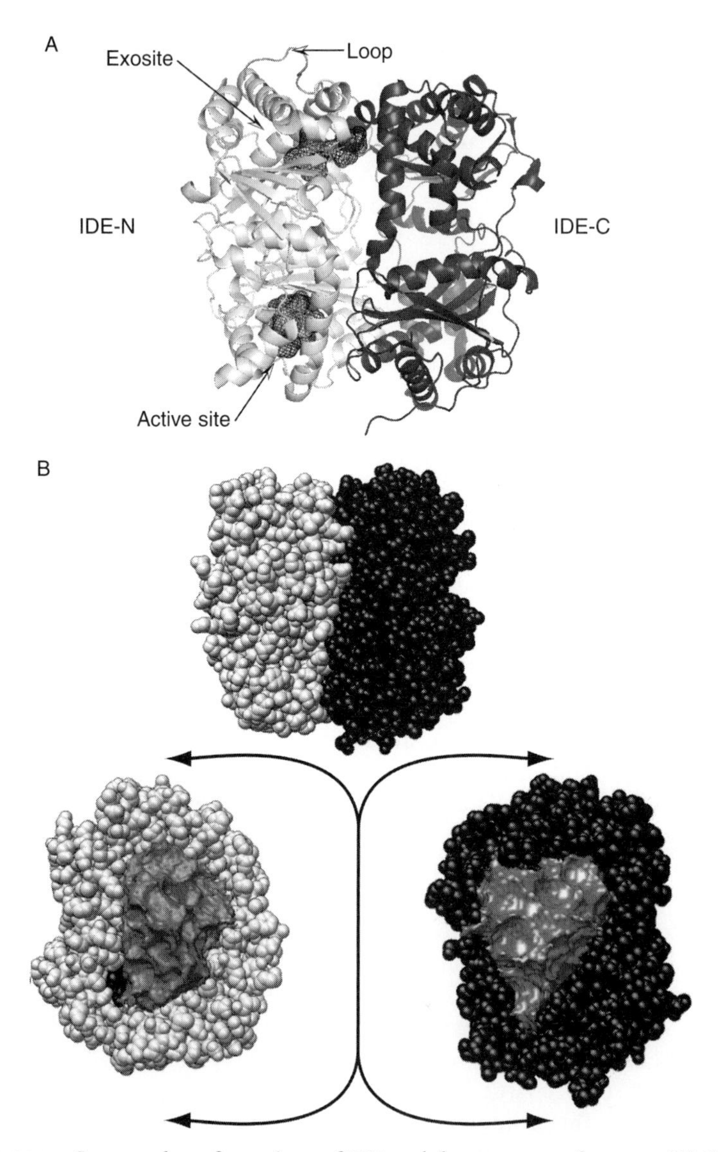

Figure 22.1 Structural conformations of IDE and the presence of a crypt. (A) IDE is a 110,000 Da metalloprotease comprised of two roughly equal-sized domains; IDE-N and IDE-C. IDE-N is the location of several regions that are important in the interaction of substrate to IDE including the exosite and the active site. The active site is also the location of the catalytic zinc ion. The presence of a loop joins the two domains together. (B) When closed, IDE forms a crypt with a volume of 15,700 Å^3 (Malito *et al.*, 2008a) which restricts amyloidogenic peptides to less than 70 amino acids in length. In the closed state, entrance to the catalytic and active site are occluded and substrates may not enter or leave. Structure PDB 2JG4.

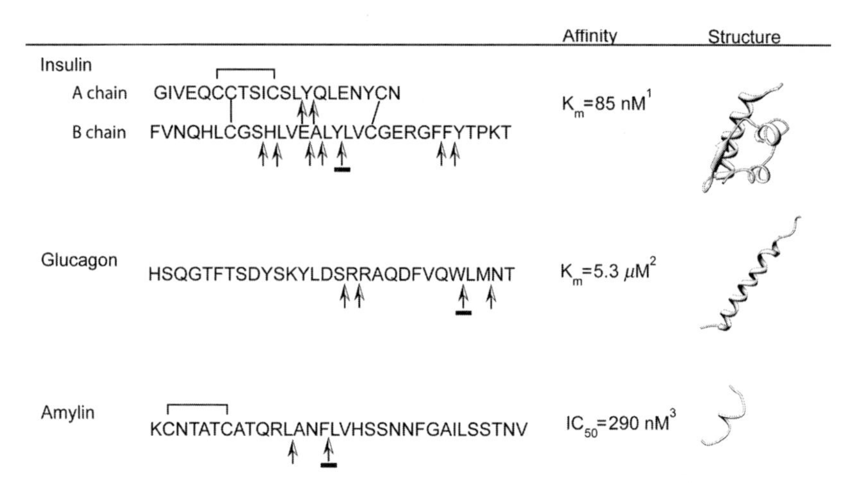

Figure 22.2 Sequence, substrate cleavage and structure of IDE substrates involved in glucose metabolism. [1]Duckworth *et al.*, 1998, [2]Duckworth and Kitabchi 1974, [3]Bennett *et al.*, 2000. PBDs; Full chain insulin 1ZNI, glucagon 1GCN, and partial amylin (residues SNNFGAILSS) 1KUW. Arrows indicate sites of cleavage (Grasso *et al.*, 2007, Shen *et al.*, 2006), underlined arrows correspond to P1–P1′ sites of substrate verified in structures 2G56 (Zn^{2+} free IDE bound to insulin B chain), 2G48 (amylin bound to IDE), and 2G49 (glucagon bound to IDE). Disulfide bonds are indicated by lines joining indicated cysteine residues.

breakdown of these disulfide bonds. Several other substrates of IDE (e.g., amylin, insulin growth factor-II, and tumor growth factor-α) also have disulfide bonds, which could influence their interactions with the catalytic chamber of IDE. Yet not all of IDE's substrates possess disulfide bonds (e.g., glucagon, amyloid-β (Aβ)). Another unique feature in insulin's degradation by IDE is the generation of multiple inactivated insulin fragments. While IDE binds insulin with the high affinity ($\sim$0.1 μM) and the catalytic chamber of IDE only has one reacting center, IDE cuts multiple sites on both A and B chains of IDE (Fig. 22.2; Shen *et al.*, 2006). These cleavage sites are nonrandom, with only 2 and 7 major sites in chain A and B, respectively. The specific products resulted from the cleavages are nondeterministic (or stochastic).

The structural solution of IDE in complex with the insulin B-chain, glucagon, and amylin begins to provide insight into how IDE-substrate interaction and cleavage mechanisms occur. All three substrates do not exceed 70 residues and fit entirely within the crypt switching from an α-helix to a β-strand conformation upon binding to IDE (Shen *et al.*, 2006). This conformational change allows the substrate to interact noncovalently with two areas of IDE-N; the catalytic site and the exosite. Once trapped, the N-terminus of the substrate is anchored to the exosite of the enzyme by interacting with a β-sheet of IDE-N (Shen *et al.*, 2006). The exosite represents an interesting region of IDE, which is highly conserved and may have a regulatory function contributing to substrate specificity and catalysis (Im *et al.*, 2007; Shen *et al.*, 2006). This interaction serves as a molecular tether, allowing the proper positioning of the C-terminal end of the substrate to the catalytic site where cleavage occurs. Alternatively, small peptides (e.g., kinins) have been shown to bind to the exosite of IDE (Malito *et al.*, 2008b). This observation supports evidence that bradykinin activates IDE (Song *et al.*, 2003) and suggests that the exosite may serve as a site of regulation (Malito *et al.*, 2008b).

In addition to the exosite and the catalytic site of IDE, other regions of the enzyme are required for its functioning. IDE-N by itself possesses a mere fraction of catalytic activity yet when IDE-C is added (as an individual component with no activity), $\sim$30% return of activity occurs, suggesting the need for the noncatalytic IDE-C portion (Li *et al.*, 2006). A tyrosine residue (Y831), located in this region, and the primarily positive inner surface of IDE-C may facilitate further positioning of the ligand to the active site by electrostatic tethering. Additional interactions from domain 4 of IDE-C occur with the ligand, including hydrogen bonding from an arginine residue (R824). In both cases, mutation of these residues substantially decreased the specific activity of IDE (Shen *et al.*, 2006).

In order to digest insulin, IDE uses its catalytic chamber (crypt) to engulf insulin entirely. Since the volume of insulin is $\sim$12,000 Å^3 it can be fully accommodated (Malito *et al.*, 2008a). Because insulin is an acidic peptide hormone, its highly negative charged surface could complement the positive charge surface found on the catalytic chamber of IDE-C (Im *et al.*, 2007; Shen *et al.*, 2006). Only a few selective regions of insulin A and B chains are cleaved by IDE (Fig. 22.2). Future structural study will be needed to elucidate how IDE engulfs insulin and selectively targets these cleavage sites.

B. Structural evidence for regulatory mechanisms of IDE from the substrate-free conformation

The structural solution of catalytically active IDE in a substrate-free conformation, along with other factors, such as surface and charge complementarity suggest that this enzyme may normally exist in a closed state (Fig. 22.3). When closed, no path to the crypt is seen; thus ligands have no apparent access to the exosite and catalytic site. It has been hypothesized that the closed conformation of IDE is a rate-limiting step for activity since substrates cannot access the catalytic site. Thus, it is important to elucidate factors that stabilize the closed state since they may also regulate activity of the protease. The observation, by mass spectrometry, of multiple cleavage sites on substrates of IDE, including insulin, suggests that during

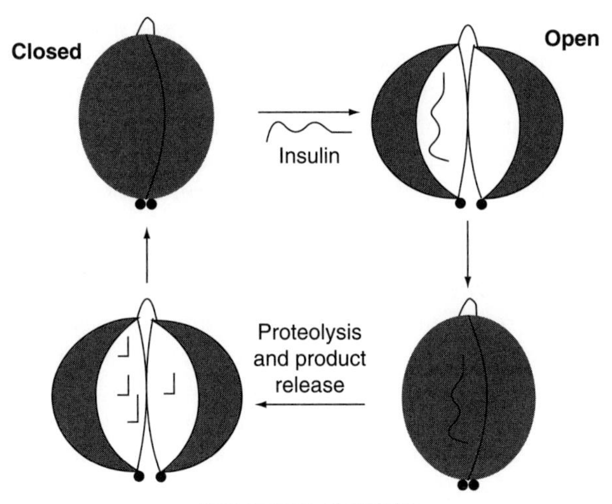

Figure 22.3 Regulators of IDE. Multiple regulatory mechanisms control the ability of IDE to degrade insulin.

entrapment, multiple cuts are made. It is not clear if one catalytic cycle (i.e., open and close) is sufficient to execute all cuts. However, in the open state, substrates and the by-products of degradation would be available to freely diffuse in and out of the crypt.

Several factors are thought to play a role in the regulation of the catalytic cycle. Both the high surface complementarity of IDE-N and IDE-C and the charge characteristics of the interior surfaces may facilitate IDE to be in a closed state when inactive (Im *et al.*, 2007). Our previous structures revealed extensive interactions between IDE-N and IDE-C that are predicted to act like a "latch," which stabilize the closed conformation. Consistent with this model, mutations that destabilize the interactions between IDE-N and IDE-C were found to significantly increase the catalytic rate of IDE.

Consideration of the known structures of IDE and related homologs suggest that the closed state, in the absence of substrate, is unstable. For example, crystal structures of human IDE and a homolog from *Arabidopsis thaliana*, presequence peptidase (*At*PreP), both solved in their closed conformation, also contained substrates bound within their catalytic chambers (Johnson *et al.*, 2006; Im *et al.*, 2007). Furthermore, the structure of the distantly related *Escherichia coli* pitrilysin, which does not have an associated substrate, is in the open conformation (Maskos, 2004). However, Im *et al.* (2007), through the structural characterization of catalytically active IDE, revealed that the natural configuration of the active site is in the closed state, suggesting that a 'molecular switch'' yet to be determined, is responsible for the conformational change between the open and closed state of IDE.

Additional mechanisms of regulation of IDE have been proposed and can be interpreted in the context of the current structural models. A change in oligomeric state, with a dimeric form of IDE showing greater activity versus the tetrameric form has been observed (Song *et al.*, 2003). Such interactions may occur noncovalently. This raises the importance of understanding physiological levels of IDE and how these levels may alter the activity in smaller compartments, such as the peroxisome. Furthermore, the abundance of cysteine residues may play a role in the regulation of IDE via oxidative or possibly nitrosative processes (Malito *et al.*, 2008b). The relative positions of the side chains of cysteine residues for IDE confirm that several residues show exposed thiol groups on the surface or inside the crypt. Indeed, the ability to regulate IDE and *At*PreP, a related cryptidase, has been demonstrated by introducing cysteines to promote disulfide bond formation in oxidizing environments thus locking the cryptidase and prevent substrate access (Johnson *et al.*, 2006; Shen *et al.*, 2006).

III. The Regulation of IDE Activity

Since IDE regulates insulin levels, mechanisms of the regulation of IDE itself have captured the interest of researchers. This work has resulted in the identification of several regulatory mechanisms of IDE function (Fig. 22.3). This regulation can be observed from the molecular to the organ level. It is important to note that the regulation of IDE may not only affect insulin degradation, but also the degradation of all the other peptide substrates.

At the genetic level, the expression of IDE, insulin, and IGF-1 receptor was demonstrated to be highest in the kidney and liver of rats, indicating an anatomical correlation of IDE in the catabolism of insulin and IGF (Bondy *et al.*, 1994). Additionally, the enzymatic activity of IDE from the homogenates of various rat tissues has been classified in a decreasing order with their insulin degrading activity: liver > pancreas > kidney > testis > adrenal gland > spleen > ovary > lung > heart > muscle > brain > fat (Duckworth and Kitabchi, 1974). The highest enzymatic activity found in the liver and kidney is consistent with the hypothesis that IDE plays a crucial role in the catabolism of insulin. Interestingly, while IDE is an important contributor to insulin homeostasis, insulin also contributes to the maintenance of IDE levels. Treatment of primary hippocampal neurons with insulin resulted in an increase of IDE protein levels by ~25%, possibly through a feedback mechanism (Zhao *et al.*, 2004). These findings suggest that genetic regulation may account for the differences in expression of IDE among tissues. While the factors that regulate IDE gene expression have not been identified, small molecules like retinoic acid and synthetic retinoic acid analogs have been indirectly implicated in the regulation of IDE expression (Melino *et al.*, 1996). This suggests the possibility of additional small molecules that may also regulate IDE gene expression and thus insulin levels. A thorough genetic analysis of IDE gene expression and its implications on insulin metabolism is necessary for better understanding of insulin-regulated diabetes-related diseases (Jee *et al.*, 2007) and the development of gene therapy as a means of modulating the activity of IDE.

The discovery that IDE is an allosteric enzyme has introduced exciting and new details about its mechanism of regulation. It has been demonstrated that in solution, IDE exists in an equilibrium of dimers and tetramers, with the dimer having higher activity (Song *et al.*, 2003). Upon substrate binding a 'heterodimer' of IDE (one wild-type and one mutant subunit) activated the adjacent subunit, restoring the activity of the second subunit (Song *et al.*, 2003). This may occur via conformational changes. This finding was followed by the identification of certain small peptide substrates, such as dynorphin B, that can boost the proteolytic activity of IDE towards the cleavage of Aβ, while at the same time inhibiting the hydrolysis of insulin (Song *et al.*, 2003).

In addition to certain peptide substrates, the activity of IDE is also influenced by other factors. For instance, calcium-depleted muscle tissue has decreased ability to degrade insulin and reduced IDE activity. However, the addition of calcium to muscle returns insulin degradation (Ryan *et al.*, 1985). The catalytic activity of IDE *in vitro* was also observed to be inhibited by free long chain fatty acids and acyl-CoA (Hamel *et al.*, 2003). This suggests that elevated intracellular long-chain fatty acid concentrations may act directly on IDE to decrease insulin metabolism. This type of regulatory mechanism may explain the correlation between hyperinsulinemia and insulin resistance with elevated fatty acids and obesity (Hamel *et al.*, 2003).

Insulin degradation by IDE is also affected by ATP through the triphosphate moiety (Camberos *et al.*, 2001; Song *et al.*, 2004). A complete inhibition of insulin degradation was observed with the addition of ATP to purified IDE with the inhibitory effect being greater with ATP than with ADP and AMP. ATP also had the ability to shift the oligomeric equilibrium to a monomer suggesting that IDE contains an allosteric site yet to be determined. Thus, the energy status of the cell may serve as feedback inhibition for insulin hydrolysis. In addition to ATP serving as an allosteric regulator by promoting the transition from tetrameric to dimeric forms of

IDE, ATP might facilitate the transition from the closed to the open state (Im *et al.*, 2007). Im *et al.* (2007) studied the activating effect of ATP and found that its effects are reduced in IDE mutants containing mutations that destabilize the closed state. Furthermore, they provided evidence, through biophysical studies, that ATP induces substantial *intramolecular* conformational changes within IDE. Thus, there is direct evidence that the regulatory mechanism of ATP extends to also facilitating the transition from the closed state to the open state (Im *et al.*, 2007; Fig. 22.3).

The regulation of IDE extends to the cellular compartmentalization and tissue distribution. The subcellular localization of IDE is dependent on the specific cell type. For the most part, the concentration of IDE is highest in the cytosol, (~95%) and a minor percentage (1–5%) is contained in endosomes, peroxisomes, mitochondria, cell surface, and the extracellular milieu (Authier *et al.*, 1995; Duckworth *et al.*, 1998; Leissring *et al.*, 2004; Qiu *et al.*, 1998; Vekrellis *et al.*, 2000). Although IDE has been found in many cellular locations, the exact compartment responsible for insulin hydrolysis is still a controversial issue (Authier *et al.*, 1996; Hersh, 2006). Given the cellular distribution of IDE, insulin hydrolysis may occur intracellularly (either in the cytosol or endosome), at the cell surface, or through the action of secreted IDE. Furthermore, degradation of insulin may occur in different cellular compartments depending on the cell type. For example, primary microglia cells have been shown to secrete IDE while hippocampal neurons possess membrane associated forms (Mentlein *et al.*, 1998; Qiu *et al.*, 1998). Further studies will be necessary to conclusively pinpoint the site of insulin degradation in the cell.

Oxidative and nitrosative stress were recently proposed as other modulators of IDE activity (Malito *et al.*, 2008b; Shinall *et al.*, 2005). IDE has two features that can make the enzyme sensitive to oxidation and nitrosylation. (1) It has a metal on or near the active site which could be particularly sensitive to oxidation, resulting in loss of its catalytic function depending on the extent of the oxidative damage (Stocker and Keaney, 2004). (2) The abundance of cysteine residues that are susceptible to reaction with free radicals.

A variety of human diseases, such as Alzheimer's disease, diabetes, and cardiovascular disease, have been associated with oxidative stress (Stocker and Keaney, 2004) with the modification of several enzymes as a contributing factor. Some examples of proteins that are targeted by oxidative stress include glutamine synthetase, mitochondrial aconitase, adenine nucleotide translocase, carbonic anhydrase III, and calcineurin. IDE has been shown to be vulnerable to treatment with physiologically relevant oxidation and nitrosylation agents *in vitro* (Malito *et al.*, 2008b; Shinall *et al.*, 2005). Yet while insulin degradation was not directly assessed, it is thought that inactivation would extend to insulin cleavage (Shinall *et al.*, 2005). Since it has been demonstrated that oxidative stress can alter IDE activity, it is important to elucidate the molecular and structural basis of IDE oxidation and identify whether this modification can occur *in vivo*. Together, this may provide the knowledge to spare IDE activity in the presence of oxidative stress.

IV. Conclusion

Considerable evidence implicates IDE in the degradation of insulin. While the affinity of this interaction is exceptionally high, the ability to degrade other peptides and an observed gradient of insulin degrading ability based on tissue emphasizes the importance of understanding how this protease is regulated. Structural and biochemical evidence emphasize the importance of several factors including conformational state, presence of ATP, as well as the oligomeric state (Fig. 22.3). These factors, in addition to the presence of a crypt, which engulfs peptides, may all play a role in IDE's ability to trap several different peptides and degrade them thoroughly.

Further structural analysis paired with strategic mutation and biochemical analysis of the regulatory mechanisms of IDE's ability to trap full insulin will extend the knowledge of the nature of these two protein's interactions. Such experiments will increase our ability to understand and influence the degradation of substrates preferentially and so allowing the exploration of the therapeutic potential of IDE. These studies will also determine the most effective strategy for up-or down- regulation and the physiological consequences of increased or decreased proteolytic activity.

ACKNOWLEDGMENTS

NIH R01-GM81539 supported this work for W-J T and LAR, and NIH T32-GM07839 for REH.

REFERENCES

Authier, F., Bergeron, J., Ou, W., Rachubinski, R., Posner, B., and Walton, P. (1995). Degradation of the cleaved leader peptide f thiolase by a peroxisomal proteinase. *Proc. Natl. Acad. Sci.* **92,** 5.

Authier, F., Cameron, P. H., and Taupin, V. (1996). Association of insulin-degrading enzyme with a 70 kDa cytosolic protein in hepatoma cells. *Biochem. J.* **319**(Pt. 1), 149–158.

Authier, F., Metioui, M., Fabrega, S., Kouach, M., and Briand, G. (2002). Endosomal proteolysis of internalized insulin at the C-terminal region of the B chain by cathepsin D. *J. Biol. Chem.* **277,** 9437–9446.

Bennett, R. G., Duckworth, W. C., and Hamel, F. G. (2000). Degradation of amylin by insulin-degrading enzyme. *J. Biol. Chem.* **275,** 36621–36625.

Bondy, C. A., Zhou, J., Chin, E., Reinhardt, R. R., Ding, L., and Roth, R. A. (1994). Cellular distribution of insulin-degrading enzyme gene expression. Comparison with insulin and insulin-like growth factor receptors. *J. Clin. Invest.* **93,** 966–973.

Camberos, M. C., Perez, A. A., Udrisar, D. P., Wanderley, M. I., and Cresto, J. C. (2001). ATP inhibits insulin-degrading enzyme activity. *Exp. Biol. Med. (Maywood)* **226,** 334–341.

Chesneau, V., and Rosner, M. (2000). Functional human insulin-degrading enzyme can be expressed in bacteria. *Protein Expr. Purif.* **19,** 91–98.

Duckworth, W. C., Bennett, R. G., and Hamel, F. G. (1998). Insulin degradation: Progress and potential. *Endocr. Rev.* **19,** 608–624.

Duckworth, W. C., and Kitabchi, A. E. (1974). Insulin and glucagon degradation by the same enzyme. *Diabetes* **23,** 536–543.

Fakhrai-Rad, H., Nikoshkov, A., Kamel, A., Fernström, M., Zierath, J. R., Norgren, S., Luthman, H., and Galli, J. (2000). Insulin-degrading enzyme identified as a candidate diabetes susceptibility gene in GK rats. *Hum. Mol. Genet.* **9,** 2149–2158.

Farris, W., Mansourian, S., Chang, Y., Lindsley, L., Eckman, E. A., Frosch, M. P., Eckman, C. B., Tanzi, R. E., Selkoe, D., and Guenette, S. (2003). Insulin-degrading enzyme regulates the levels of insulin, amyloid beta-protein, and the beta-amyloid precursor protein intracellular domain *in vivo*. *Proc. Natl. Acad. Sci. USA* **100,** 4162–4167.

Fawcett, J., Permana, P. A., Levy, J. L., and Duckworth, W. C. (2007). Regulation of protein degradation by insulin-degrading enzyme: Analysis by small interfering RNA-mediated gene silencing. *Arch. Biochem. Biophys.* **468,** 128–133.

Florez, J. C., Wiltshire, S., Agapakis, C. M., Burtt, N. P., de Bakker, P. I., Almgren, P., Bengtsson Boström, K., Tuomi, T., Gaudet, D., Daly, M. J., Hirschhorn, J. N., McCarthy, M. I., *et al.* (2006). High-density haplotype structure and association testing of the insulin-degrading enzyme (IDE) gene with type 2 diabetes in 4,206 people. *Diabetes* **55,** 128–135.

Grasso, G., Rizzarelli, E., and Spoto, G. (2007). AP/MALDI-MS complete characterization of the proteolytic fragments produced by the interaction of insulin degrading enzyme with bovine insulin. *J. Mass Spectrom.* **42,** 1590–1598.

Groves, C. J., Wiltshire, S., Smedley, D., Owen, K. R., Frayling, T. M., Walker, M., Hitman, G. A., Levy, J. C., O'Rahilly, S., Menzel, S., Hattersley, A. T., and McCarthy, M. I. (2003). Association and haplotype analysis of the insulin-degrading enzyme (IDE) gene, a strong positional and biological candidate for type 2 diabetes susceptibility. *Diabetes* **52,** 1300–1305.

Gu, H. F., Efendic, S., Nordman, S., Ostenson, C. G., Brismar, K., Brookes, A. J., and Prince, J. A. (2004). Quantitative trait loci near the insulin-degrading enzyme (IDE) gene contribute to variation in plasma insulin levels. *Diabetes* **53,** 2137–2142.

Hamel, F. G., Upward, J. L., and Bennett, R. G. (2003). *In vitro* inhibition of insulin-degrading enzyme by long-chain fatty acids and their coenzyme A thioesters. *Endocrinology* **144,** 2404–2408.

Hari, J., Shii, K., and Roth, R. A. (1987). *In vivo* association of [125I]-insulin with a cytosolic insulin-degrading enzyme: Detection by covalent cross-linking and immunoprecipitation with a monoclonal antibody. *Endocrinology* **120,** 829–831.

Hersh, L. (2006). The insulysin (insulin degrading enzyme) enigma. *Cell. Mol. Life Sci.* **63,** 2432–2434.

Im, H., Manolopoulou, M., Malito, E., Shen, Y., Zhao, J., Neant-Fery, M., Sun, C., Meredith, S., Sisodia, S., Leissring, M., and Tang, W. (2007). Structure of substrate-free human insulin-degrading enzyme (IDE) and Biophysical analysis of ATP-induced conformational switch of IDE. *J.Biol. Chem.* **282,** 25453–25463.

Jee, S., Hwang, D., Seo, S., Kim, Y., Kim, C., Kim, B., Shim, S., Lee, S., Sin, J., Bae, C., Lee, B., Jang, M., *et al.* (2007). Microarray analysis of insulin-regulated gene expression in the liver: The use of transgenic mice co-expressing insulin-siRNA and human IDE as an animal model. *Int. J. Mol. Med.* **20,** 829–835.

Johnson, K., Bhushan, S., Ståhl, A., Hallberg, B., Frohn, A., Glaser, E., and Eneqvist, T. (2006). The closed structure of presequence protease PreP forms a unique 10 000 Å3 chamber for proteolysis. *EMBO J.* **25,** 1977–1986.

Karamohamed, S., Demissie, S., Volcjak, J., Liu, C., Heard-Costa, N., Liu, J., Shoemaker, C. M., Panhuysen, C. I., Meigs, J. B., Wilson, P., Atwood, L. D., Cupples, L. A., *et al.* (2003). Polymorphisms in the insulin-degrading enzyme gene are associated with type 2 diabetes in men from the NHLBI Framingham heart study. *Diabetes* **52,** 1562–1567.

Kuo, W. L., Gehm, B. D., and Rosner, M. R. (1991). Regulation of insulin degradation: Expression of an evolutionarily conserved insulin-degrading enzyme increases degradation via an intracellular pathway. *Mol. Endocrinol.* **5,** 1467–1476.

Kurochkin, I. V. (2001). Insulin-degrading enzyme: Embarking on amyloid destruction. *Trends Biochem. Sci.* **26,** 421–425.

Leissring, M. A., Farris, W., Wu, X., Christodoulou, D. C., Haigis, M. C., Guarente, L., and Selkoe, D. J. (2004). Alternative translation initiation generates a novel isoform of insulin-degrading enzyme targeted to mitochondria. *Biochem. J.* **383,** 439–446.

Li, P., Kuo, W. L., Yousef, M., Rosner, M. R., and Tang, W. (2006). The C-terminal domain of human insulin degrading enzyme is required for dimerization and substrate recognition. *Biochem. and Biophys. Res. Commun.* **343,** 1032–1037.

Malito, E., Hulse, R. E., and Tang, W. J. (2008a). Amyloid β-degrading cryptidases: Insulin degrading enzyme, presequence peptidase, and neprilysin. *Cell. Mol. Life Sci.* **65,** 2574–2585.

Malito, E., Ralat, L. A., Manolopoulou, M., Tsay, J. L., Wadlington, N. L., and Tang, W. J. (2008b). Molecular bases for the recognition of the short peptide substrate and cysteine-directed modifications of human insulin-degrading enzyme. *Biochemistry* **47,** 12822–12834.

Marlowe, L., Peila, R., Benke, K. S., Hardy, J., White, L. R., Launer, L. J., and Myers, A. (2006). Insulin-degrading enzyme haplotypes affect insulin levels but not dementia risk. *Neurodegener Dis.* **3,** 320–326.

Maskos, K. (2004). Handbook of Metalloproteins. New York, Wiley.

Melino, G., Draoui, M., Bernardini, S., Bellincampi, L., Reichert, U., and Cohen, P. (1996). Regulation by retinoic acid of insulin-degrading enzyme and of a related endoprotease in human neuroblastoma cell lines. *Cell Growth Differ.* **7,** 787–796.

Mentlein, R., Ludwig, R., and Martensen, I. (1998). Proteolytic degradation of Alzheimer's disease amyloid beta-peptide by a metalloproteinase from microglia cells. *J. Neurochem.* **70,** 721–726.

Mirsky, I. A., and Broh-kahn, R. H. (1949). The inactivation of insulin by tissue extracts; the distribution and properties of insulin inactivating extracts. *Arch. Biochem.* **20,** 1–9.

Mirsky, I. A., Simkin, B., and Broh-Kahn, R. H. (1950). The inactivation of insulin by tissue extracts. VI. The existence, distribution and properties of an insulinase inhibitor. *Arch. Biochem.* **28,** 415–423.

Perlman, R. K., Gehm, B. D., Kuo, W. L., and Rosner, M. R. (1993). Functional analysis of conserved residues in the active site of insulin-degrading enzyme. *J. Biol. Chem.* **268,** 21538–21544.

Qiu, W. Q., and Folstein, M. F. (2006). Insulin, insulin-degrading enzyme, and amyloid-beta peptide in Alzheimer's disease: Review and hypothesis. *Neurobiol. Aging* **27,** 190–198.

Qiu, W. Q., Walsh, D. M., Ye, Z., Vekrellis, K., Zhang, J., Podlisny, M. B., Rosner, M. R., Safavi, A., Hersh, L. B., and Selkoe, D. J. (1998). Insulin-degrading enzyme regulates extracellular levels of amyloid beta-protein by degradation. *J. Biol. Chem.* **273,** 32730–32738.

Ryan, M. P., Gifford, J. D., Solomon, S. S., and Duckworth, W. C. (1985). The calcium dependence of insulin degradation by rat skeletal muscle. *Endocrinology* **117,** 1693–1698.

Shen, Y., Joachimiak, A., Rich Rosner, M., and Tang, W. (2006). Structures of human insulin-degrading enzyme reveal a new substrate recognition mechanism. *Nature* **443,** 870–874.

Shii, K., and Roth, R. A. (1986). Inhibition of insulin degradation by hepatoma cells after microinjection of monoclonal antibodies to a specific cytosolic protease. *Proc. Natl. Acad. Sci. USA* **83,** 4147–4151.

Shinall, H., Song, E. S., and Hersh, L. B. (2005). Susceptibility of amyloid beta peptide degrading enzymes to oxidative damage: A potential Alzheimer's disease spiral. *Biochemistry* **44,** 15345–15350.

Sladek, R., Rocheleau, G., Rung, J., Dina, C., Shen, L., Serre, D., Boutin, P., Vincent, D., Belisle, A., Hadjadj, S., Balkau, B., Heude, B., *et al.* (2007). A genome-wide association study identifies novel risk loci for type 2 diabetes. *Nature* **445,** 881–885.

Song, E. S., Juliano, M. A., Juliano, L., and Hersh, L. B. (2003). Substrate activation of insulin-degrading enzyme (insulysin). A potential target for drug development. *J. Biol. Chem.* **278,** 49789–49794.

Song, E. S., Juliano, M. A., Juliano, L., Fried, M. G., Wagner, S. L., and Hersh, L. B. (2004). ATP effects on insulin-degrading enzyme are mediated primarily through its triphosphate moiety. *J. Biol. Chem.* **279,** 54216–54220.

Stocker, R., and Keaney, J. F. Jr. (2004). Role of oxidative modifications in atherosclerosis. *Physiol. Rev.* **84,** 1381–1478.

Sudoh, S., Frosch, M. P., and Wolf, B. (2002). Differential effects of protease involved in intracellular degradation of amyloid-beta protein between detergent soluble and insolube pools in CHO-695 cells. *Biochemistry* **41,** 9.

Vekrellis, K., Ye, Z., Qiu, W. Q., Walsh, D., Hartley, D., Chesneau, V., Rosner, M. R., and Selkoe, D. J. (2000). Neurons regulate extracellular levels of amyloid beta-protein via proteolysis by insulin-degrading enzyme. *J. Neurosci.* **20,** 1657–1665.

Zhao, L., Teter, B., Morihara, T., Lim, G. P., Ambegaokar, S. S., Ubeda, O. J., Frautschy, S. A., and Cole, G. M. (2004). Insulin-degrading enzyme as a downstream target of insulin receptor signaling cascade: Implications for Alzheimer's disease intervention. *J. Neurosci.* **24,** 11120–11126.

CHAPTER TWENTY-THREE

Modification of Androgen Receptor Function by IGF-1 Signaling: Implications in the Mechanism of Refractory Prostate Carcinoma

Toshihiko Yanase *and* WuQiang Fan

Contents

Abstract

The androgen–androgen receptor (AR) system plays important roles in a variety of biological processes, including prostate cancer (PC) development and progression. Insulin and Insulin-like growth factor-1 (IGF-1) signaling negatively regulate a member of the forkhead box-containing protein O subfamily (FoxO), Foxo-1, and associated biological functions. IGF-1 can potentiate androgen signaling through AR activation. Foxo-1, phosphorylated and inactivated by

Department of Medicine and Bioregulatory Science, Graduate School of Medical Science, Kyushu University, Maidashi 3-1-1, Higashi-ku, Fukuoka 812-8582, Japan

Vitamins and Hormones, Volume 80
ISSN 0083-6729, DOI: 10.1016/S0083-6729(08)00623-7

phosphatidylinositol-3-kinase (PI3K)/Akt kinase induced by IGF-1 or insulin, suppresses ligand-mediated AR transactivation. Foxo-1 reduces expression of androgen-induced AR target genes and suppresses *in vitro* growth of PC cells. These inhibitory effects of Foxo-1 are attenuated by IGF-1, but enhanced when it was rendered Akt-non-phosphorylatable. Foxo-1 directly interacts with the C-terminus of AR in a ligand-dependent manner, and disrupts ligand-induced AR subnuclear compartmentalization. Foxo-1 is recruited by liganded AR to the chromatin of the AR target gene promoter, while IGF-1 or insulin abolishes the Foxo-1 occupancy on the promoter. Liganded AR stimulates IGF-1 receptor expression, suggesting the presence of local positive feedback between IGF-1 and AR signaling in PC cells, presumably resulting in higher IGF-1 signaling tension and further enhancing the functions of the receptor itself. Thus, Foxo-1 is a novel corepressor for AR and IGF-1/insulin signaling may confer stimulatory effects on AR by attenuating Foxo-1 inhibition. Positive feedback between the growth factor and androgen in the local cellular environment may play important roles in AR transactivation regulation in several clinical situations including refractory PC.

Abbreviations

AF-1 domain	activation function-1 domain
AR	androgen receptor
BRCA1	breast cancer susceptibility gene 1
CBP	CREB (cyclic AMP-response element) binding protein
DBD	DNA binding domain
DRIP	VDR (vitamin D receptor) interacting protein ER, estrogen receptor
GFP	green fluorescence protein
GR	glucocorticoid receptor
HAT	histone acetyltransferase
HDAC	histone deacetylation enzyme
LBD	ligand binding domain
MR	mineralocorticoid receptor
NcoR	nuclear receptor corepressor
PCR	polymerase chain reaction
PI3K	phosphatidylinositol-3-kinase
SF-1	steroidogenic factor 1
SMRT	silencing mediator of retinoid and thyroid hormone receptor
SRA	steroid receptor RNA activator

SRC-1	steroid receptor coativator-1
TIF-2	transcriptional intermediary factor-2
TRAP	TR-associated protein
TRRAP	transformation/transcription domain-associated protein
YFP	yellow fluorescence protein.

I. Androgen Receptor Signaling

Steroid hormone receptors such as the androgen receptor (AR), estrogen receptor (ER), and glucocorticoid receptor (GR) are ligand-dependent transcription factors that belong to the nuclear receptor superfamily (Mangelsdorf *et al.*, 1995). Nuclear receptors bind to their cognate response elements in the promoter region of target genes to regulate expression of the gene (McKenna *et al.*, 1999; Yanase *et al.*, 2004). Nuclear receptors share a common structure, which consists of a transcription active domain, a DNA-binding domain and a ligand-binding domain. There are two transcription activation domains; the activation function-1 (AF-1) domain in the N-terminal region and the activation function-2 (AF-2) domain in the C-terminal region. While the AF-2 domain is relatively conserved among nuclear receptors, the AF-1 domain differs widely (Fig. 23.1A). The transactivation function of the AF-1 is ligand-independent and autonomous, while that of the AF-2 is ligand-dependent. When a ligand binds to a receptor, the receptor changes its structure, translocates from the cytoplasm to the nucleus and then binds to the promoter region of the target gene. Coregulator proteins bind to the nuclear receptors and modulate the transcriptional activity of the nuclear receptors in a promoter- and cell-specific manner (Fig. 23.1B). There are two types of coregulators; coactivator proteins, which activate transcription, and corepressor proteins, which repress transcription. CBP/p300, p160 family (SRC-1, TIF-2 etc), and DRIP/TRAP for example have been reported to be typical coactivators, while NCoR and SMRT are typical corepressors. Although most coregulators are AF-2 binding proteins, p300/CBP and SRC-1 interact with both AF-1 and AF-2. In addition, several AF-1 binding coregulators such as BRCA1, SRA cyclin E, and ANT-1(Zhao *et al.*, 2002) have also been identified. Interaction of the AF-1 and AF-2 domains is important for exerting the full nuclear receptor transactivation capacity (Saitoh *et al.*, 2002). Recently, it was found that, rather than operating individually, these coregulators operate together by forming an enormous protein complex. For example, a complex composed mainly of CBP/p300 and the p160 system, has a histone acetyltransferase (HAT) activity that acetylates a basic

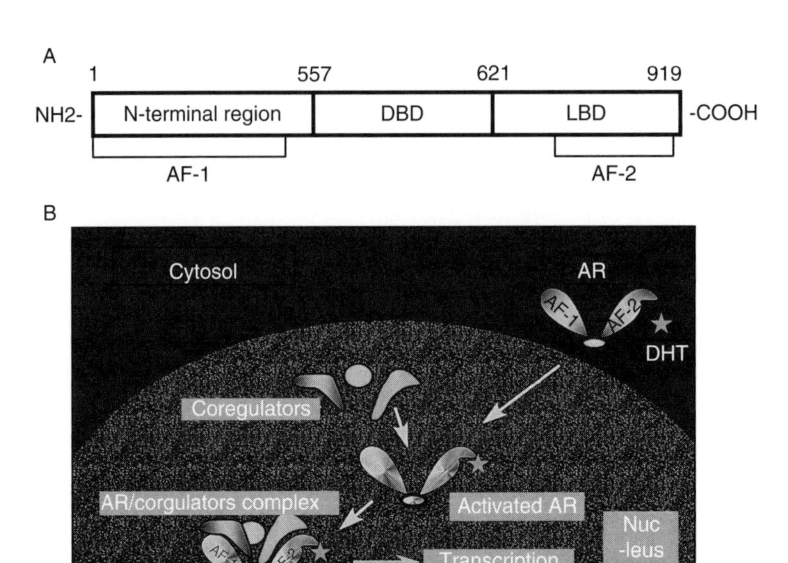

Figure 23.1 (A) Structure of the androgen receptor (AR). (B) AF-1, activation function-1 domain; AF-2, activation function-2 domain; DBD, DNA-binding domain; LBD, ligand-binding domain. Schematic representation of androgen receptor (AR) transactivation. Ligand (DHT)-bound AR translocates from the cytosol to the nucleus and recruits coregulators. AR/coregulators complex transactivates the target gene through binding to the androgen responsive element (ARE).

amino acid of the histone protein and alleviates the chromatin structure. This makes it easier for transcription factors to be recruited on the DNA and thus promotes transcription. However, complexes such as the DRIP/TRAP complex do not have HAT activity. In contrast, the corepressor SMRT/NCoR complex represses transcription activity by coupling with the nuclear receptor that is unbound to the ligand, and recruits the histone deacetylation enzyme (HDAC), which has an opposing effect to HAT on the promoter. It is thus speculated that once the ligand binds to the nuclear receptor, the corepressor complex dissociates from the receptor and the coactivators are recruited onto the promoter.

Ligand-induced subnuclear compartmentalization (foci formation) is closely correlated with the transcriptional activation function of nuclear receptors (Tomura *et al.*, 2001). 5α-dihydrotestosterone (DHT) induces 250–400 fine and well-separated subnuclear AR foci, and the process is accompanied by recruitment of coactivators such as SRC-1, TIF-2, and CBP, which was found to be one of the factors essential for AR foci formation (Saitoh *et al.*, 2002). AR bound to antiandrogens, such as hydroxyflutamide, do not form foci (Tomura *et al.*, 2001); and cofactors that repress AR function readily disrupt DHT-induced receptor foci formation (Chen *et al.*, 2005; Tao *et al.*, 2006). Recent studies of GR (Fletcher *et al.*, 2002; McNally *et al.*, 2000), ER (Stenoien *et al.*, 2001), and SF-1 (Fan *et al.*, 2004) have revealed that receptors undergo continuous and rapid exchange between chromatin regulatory elements and the nucleoplasm compartment when the ligand is constantly available, a process also known as a hit-and-run model (Stenoien *et al.*, 2001).

II. IGF Signaling and Foxo-1

The insulin-like growth factor (IGF) signaling system plays a key role in the growth and development of many normal tissues and regulates overall growth of organisms. The IGF-1 receptor (IGF-1R) is a receptor tyrosine kinase that serves as a key positive regulator of the IGF-1 system (Samani *et al.*, 2007). In response to the stimulatory ligands IGF-1 and IGF-2, IGF-1R signaling results in both proliferative and antiapoptotic effects. In turn, the IGF binding proteins (IGFBPs) and IGFBP proteases play a key role in regulating ligand bioavailability. IGFBPs in extracellular fluid modulate interactions between IGF ligands and IGF receptors; their affinity for IGF-1 and IGF-2 is comparable to that of IGF-1R. Under different physiological conditions, IGFBPs can either increase or decrease IGF signaling, since IGFBPs can prolong the half-lives of IGFs but also can compete with receptors for free IGF-1 and IGF-2. IGF-1R, which is a transmembrane tyrosine kinase composed of two α and two β subunits, is structurally closely related to the insulin receptor, and is widely expressed in human tissues. Binding of the ligands induces conformational changes in the IGF-1R and activation of its intrinsic tyrosine kinase activity. Once phosphorylated, the intracellular portion of the receptor serves as a docking site for several receptor substrates, including insulin receptor substrate (IRS)-1 to IRS-4 and SRC homology and collagen (Shc). These substrates then initiate the phosphorylation cascades that transmit the IGF-1R signal. Phosphorylated IRS-1 can activate phosphatidylinositol-3-kinase (PI3K), leading to activation of protein kinase B (Akt) and phosphorylation of Foxo-1. In addition, recruitment of growth factor receptor-bound protein-2 by phosphorylated IRS-1 or Shc leads to recruitment of Ras and activation of the Raf-1/mitogen-activated protein kinase (MAPK)–extracellular signal-related kinase (ERK) kinase/ERK pathway and downstream nuclear factors, to induce cellular proliferation (Samani *et al.*, 2007).

Considering the above overall signaling pathways, IGF-1 negatively regulates a member of the forkhead box-containing protein O subfamily (FoxO), Foxo-1, and thus its associated biological functions. Foxo-1, also known as FKHR, together with two additional homologs, FKHRL1 and AFX, belongs to the Foxo subfamily of the forkhead transcription factor family, which includes a large array of transcription factors that are characterized by the presence of a conserved 110-amino acid winged helix DNA-binding domain (DBD) (Kops *et al.*, 1999). Foxo subfamily members play important roles in cell cycle regulation and apoptosis, as well as metabolic homeostasis (Nakae *et al.*, 2000). PI3K/Akt signaling, activated by both liganded IGF-1R and insulin receptor phosphorylates each of the Foxo-1 homologs at three different Ser/Thr residues (Nakae *et al.*, 2000). The phosphorylated Foxo-1 proteins are inactive and are exported from the nucleus and subsequently sequestered in the cytoplasm, where they interact with the 14-3-3 protein. The FoxO family is conserved across many species. In mammals, Insulin receptor/IGF-1R–PI3K–Akt signaling inhibits transcription by FoxO-1, FoxO-3a, and FoxO-4. These proteins also possess a forkhead DNA binding domain consisting of ~110 amino acids and a transactivation domain in the C terminus. FoxOs bind to consensus FoxO binding sites [T(G/A)TTTT(G/T)] in the promoter region of their target genes and activate gene expression (Kops *et al.*, 1999).

III. Interaction between AR and Insulin/IGF-1 Signaling

A. Background

Prostate cancer (PC) is the most common malignancy in men worldwide and the second leading cause of cancer-related mortality in the United States (Greenlee *et al.*, 2000). Since over 70% of PCs rely on androgen stimulation for growth, this forms the basis for androgen ablation therapy, which is initially effective but invariably results in treatment resistance after a period of time (Santos *et al.*, 2004). The disease is then referred to as androgen-independent PC, which is usually fatal. Recent loss-of-function studies have revealed that AR still plays a key role in hormone-refractory progression of PC (Chen *et al.*, 2004; Zegarra-Moro *et al.*, 2002). An adaptation of AR signaling may occur to enable functionality under low or absent androgen levels (Buchanan *et al.*, 2001). Among the various mechanisms suggested by which AR may be reactivated in a low androgen environment (Grossmann *et al.*, 2001), growth factor signaling, particularly by IGF-1, is reportedly of significant importance (Burfeind *et al.*, 1996; Nickerson *et al.*, 2001; Pollak *et al.*, 1998; Wolk *et al.*, 1998). Meta-regression analysis revealed that high IGF-1 serum levels are correlated with an increased risk of PC (Renehan *et al.*, 2004), while IGF-1 enhances AR transactivation under very low or absent androgen levels (Culig *et al.*, 1994), and promotes

PC cell proliferation (Burfeind *et al.*, 1996). Recent studies have also revealed that high serum insulin levels are associated with an increased incidence of PC (Amling, 2005; Hsing *et al.*, 2003), although there is a lack of mechanistic studies implicating insulin signaling in AR function regulation.

B. *In vitro* interaction between AR and IGF-1 signaling

Foxo-1 protein is an Akt kinase target and is negatively regulated by phosphorylation in an insulin- and/or IGF-1-dependent manner, with resultant nuclear exportation and cytoplasm sequestration. Therefore, IGF-1 or insulin are expected to attenuate the inhibitory effects of Foxo-1 on AR. Foxo-1 suppresses DHT-dependent AR transactivation that is normally observed by exogenously expressed AR in COS7 cells or using endogenous AR in LNCaP human PC cells (Fan *et al.*, 2007). Promoter activity of a native AR target gene, Prostate specific antigen (PSA), was also suppressed by Foxo-1 in both COS7 and DU145 human PC cells. Thus, the inhibitory effect of Foxo-1 on AR transactivation is not artificial. DHT-mediated endogenous PSA expression at the mRNA or protein level in LNCaP cells was actually suppressed by stably expressing Foxo-1 compared with those levels by the stable expressions of the control plasmids (Fig. 23.2A). Furthermore, Foxo-1 overexpression could reduce proliferation of LNCaP cells at either the basal level or after stimulation with DHT (Fig. 23.2B). The inhibitory effect by Foxo-1 was relatively specific to AR and ERβ but not ERα. As LNCap cells lack functional PTEN, which has dephosphorylation activity, Akt is constitutively active in this cell line. When the impaired PI3K/Akt signaling in LNCAP cells was rescued by cotransfection of PTEN, both wild-type Foxo-1 and a constitutively active mutant, Foxo1-3A, which is a nonphosphorylatable mutant containing all three Akt target residues mutated to alanine, were very effective in suppressing DHT-induced AR transactivation. Importantly, IGF-1 rescued Foxo-1 suppression over AR, when wild-type Foxo-1 was transfected. However, this effect of IGF-1 was not observed when a nonphosphorylated type of mutant, Foxo1-3A, was transfected (Fig. 23.3). These results clearly indicate that IGF-1 rescues Foxo-1 suppression of AR through inactivation of Foxo-1 by phosphorylation. The same results were also observed in the PTEN-intact DU145 PC cells. Interestingly, although less potent, insulin, another PI3K/Akt stimulator, also rescued Foxo-1-mediated suppression of AR transactivation in the PTEN-expressing LNCaP cells. Cotransfection of Foxo-1 with AR in LNCaP did not decrease the AR protein levels amount either in the presence or absence of DHT, but were rather increased in the presence of DHT, suggesting that Foxo1 suppresses the specific activity of the AR (Fan *et al.*, 2007).

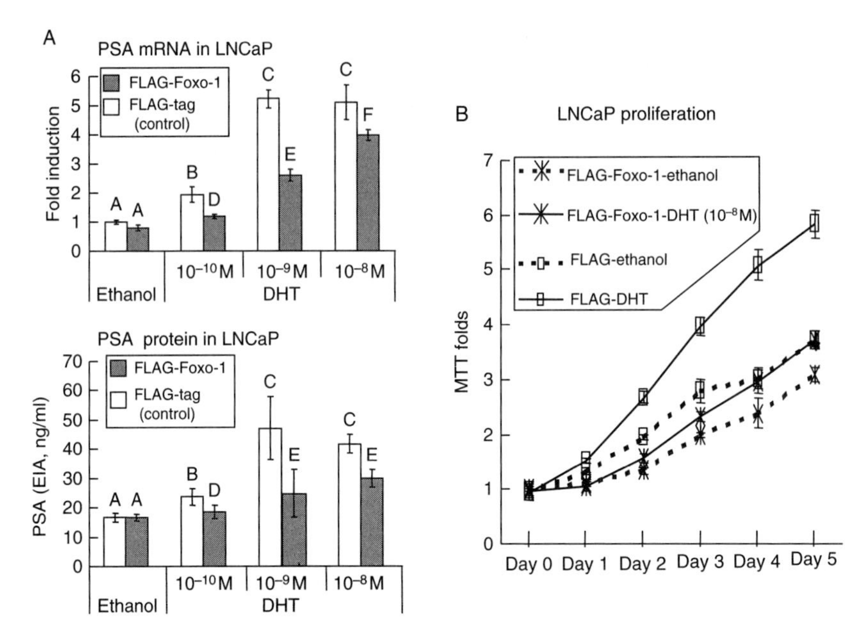

Figure 23.2 Foxo-1 suppresses DHT-mediated AR transactivation and inhibits proliferation of LNCaP cells (from ref. 32). LNCaP cells stably expressing FLAG-Foxo-1 or the FLAG tag were exposed to increasing concentrations of DHT as indicated. Endogenous expression of PSA mRNA and PSA protein secreted into the culture medium were assayed by real-time time PCR and EIA, respectively (A). LNCaP cells stably expressing either FLAG-Foxo1 or the control FLAG tag were grown in the presence of 10^{-8} M DHT or the solvent (ethanol), and cell proliferation was dynamically evaluated using a cell proliferation assay kit. Data are presented as means ± SD (A) or means ± SE (B). Letters above the bars show statistical groups (ANOVA, $P < 0.05$).

C. Interaction between AR and Foxo1: Structural basis and regulation

Direct physical interaction between Foxo-1 and AR has also been shown by immunoprecipitation, mammalian one-hybrid and two-hybrid assays, and ChIP assays (Fan *et al.*, 2007). Immunoprecipitation assays revealed that FLAG-Foxo-1 bands were detected in the anti-AR antibody precipitated immune complexes from cells treated with DHT. The presence of IGF-1 or insulin did not abolish the interaction although the signal intensity was relatively decreased. Based on these findings, Foxo1 may repress DHT-dependent AR transactivation by direct protein–protein interaction. Mammalian two-hybrid assays indicated that the AR C-terminus (amino acids 615–919) but not the AR N-terminus (amino acids 1–660) interacts with Foxo-1 in a hormone-dependent manner. The interaction is AR agonist (DHT)-dependent, and relies on the C-terminus of the

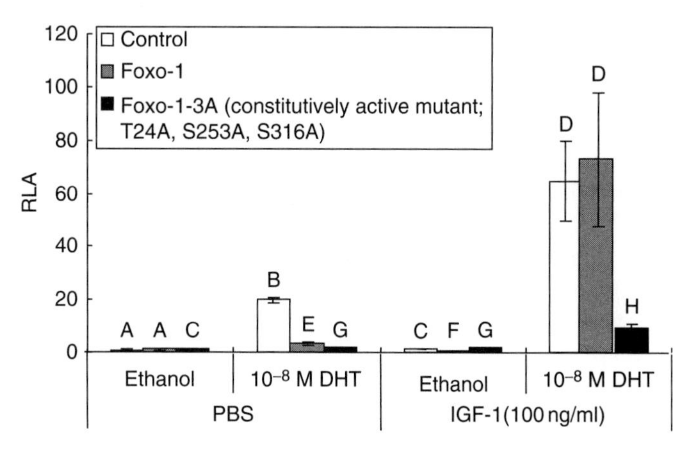

Figure 23.3 IGF-1 rescues Foxo-1 suppression over AR in PTEN-expressing LNCap cells. LNCaP cells growing in 24-well plates were cotransfected with 100 ng of FLAG-Foxo1, FLAG-Foxo1-3A (Akt-non-phosphorylatable constitutively active mutant) or an equimolar amount of pcDNA-FLAG and 400 ng of PSA-Luc, 5 ng of pRL-CMV and 40 ng of PTEN and 40 ng of pCMV-hAR. 48 h after transfection, the cells were treated with 10^{-8} M DHT and/or 100 ng/ml IGF1 in serum-free medium for 24 h before being lysed for luciferase assays. Letters above the bars show statistical groups (ANOVA, $P < 0.05$). The ascending order of the RLA is: F, A, C, G, E, H, B, and D. RLA: Relative luciferase activity.

receptor, which is functionally ligand-dependent. The androgen-dependent Foxo-1-AR C-terminus interaction has also been recently observed by another research group although the authors also reported a ligand-independent interaction between N-terminal A/B region of the receptor and Foxo-1 (Li *et al.*, 2003).

Nuclear receptor cofactors bind to AR via peptide motifs called NR boxes with the consensus sequence of LXXLL (L, leucine; X, any amino acid) (McInerney *et al.*, 1998). Foxo1 contains in its N-terminus an LXXLL peptide motif, which is conserved from mice (^{459}LKELL^{463}) to humans (^{452}LKELL^{466}). However, the role, if any, of this motif in mediating the interaction remains unclear. The interaction is likely weakened, but not completely abolished, by IGF-1. In Chip assays, liganded AR also recruited Foxo-1 to the chromatin of AR target gene PSA promoters (3 ARE), where Foxo-1, in turn, interfered with AR–DNA interactions. IGF-1 or insulin, however, reduced the DNA occupancy by Foxo-1, but consistently enhanced that by AR. Foxo-1 and AR thus form a ligand-dependent complex on chromatin. Furthermore, IGF-1 and insulin are able to remove Foxo1 from the chromatinized DNA. These IGF-1/insulin regulations likely rely on Akt-mediated phosphorylation of Foxo1, since Foxo1-3A inhibited the AR–DNA interaction. Thus, IGF-1, and perhaps also insulin, appears to be able to regulate the Foxo-1--AR interaction via PI3K/Akt phosphorylation of Foxo-1. Via direct interactions between Foxo-1 and AR, ligand-bound AR was found to able to inhibit DNA binding, as well as the transactivation activity of Foxo-1, and impaired the ability of Foxo1 to induce prostate cell apoptosis and cell cycle arrest (Li *et al.*, 2003). Liganded AR was also reportedly able to induce proteolysis of Foxo-1 protein and, therefore, ameliorates Foxo1-related apoptosis of PC cells (Huang *et al.*, 2004). These studies, in combination with our present finding, suggest that AR and Foxo-1 are mutually suppressive in terms of functional regulation.

D. Subcellular interaction and compartmentalization of AR and Foxo-1

DHT-dependent AR transactivation can be visualized by the nuclear foci formation of AR using confocal microscopic analysis (Tomura *et al.*, 2001). Therefore, the subcellular interaction between Foxo-1 and AR in living COS7 cells has been examined using this system. YFP-Foxo-1 was predominantly located in the nucleus in a homogenous manner. Importantly, IGF1 caused complete cytoplasmic fluorescence YFP-Foxo-1 (Fig. 23.4A). As expected, GFP–AR showed diffuse fluorescence in the cytosol in the absence of the ligand; however, DHT induced complete nuclear translocation and typical nuclear foci formation (Fig. 23.4B). However, coexistence of YFP–Foxo-1 and GFP–AR caused impaired AR subnuclear organization. As a result, in cells treated with DHT alone, GFP–AR showed incomplete nuclear translocation and impaired subnuclear foci formation, and the distribution was diffuse in both the nucleus and cytoplasm (Fig. 23.4C). However, treatment with IGF-1 significantly improved subnuclear compartmentalization to the original foci formation of AR (Fig. 23.4D). These results also support the concept that IGF-1 rescues Foxo-1-mediated suppression of AR transactivation activity. Intracellular signaling may also target dynamic cofactors, and eventually regulate receptor-mediated transcriptional events. Foxo1 may compete with coactivators, such as CBP, which is essential for AR foci formation (Saitoh *et al.*, 2002), for interaction with the receptor and, thereby, inhibit the formation of the transcriptionally active complexes. Importantly, our quantitative study has revealed that IGF-1 rescued, at least partially, the Foxo1-, but not the Akt un-phosphorylatable constitutively active mutant Foxo1-3A-, disrupted AR subnuclear compartmentalization. Therefore, Foxo1 interferes with ligand-bound AR subnuclear compartmentalization, an effect that is regulated by IGF-1-PI3K-Akt signaling.

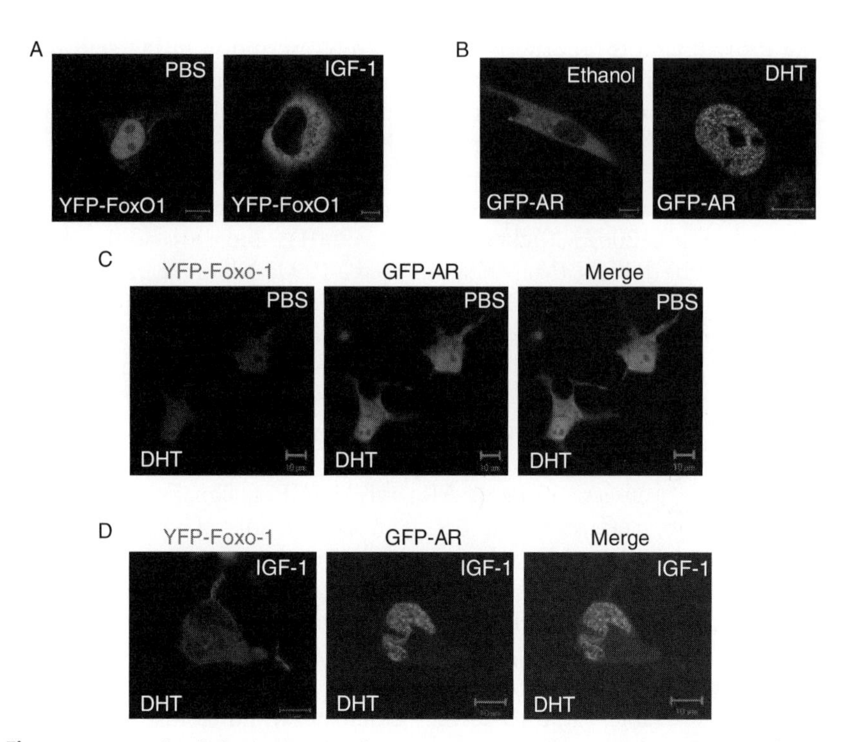

Figure 23.4 Subcellular interaction between Foxo1 and AR in living COS7 cells. The expression of chimeric fluorescent proteins was observed in living COS7 cells using a Zeiss LSM 510 META laser confocal microscope. COS7 cells were transfected with 0.5 μg/dish of EYFP-Foxo1, and then treated with PBS or 100 ng/ml IGF-1 (A). COS7 cells expressing EGFP-AR were exposed to ethanol or DHT (10^{-8} M) (B), respectively. COS7 cells coexpressing EGFP-AR and EYFP-Foxo1 were exposed to DHT alone (C) or DHT+IGF-1 (D). G Scale bars, 10 μm. Phosphorylated nuclear YFP-FOX-1 with IGF-1 exported to the cytoplasm (A). DHT activated GFP-AR and translocated it from cytosol to nucleus to form fine foci in the nucleus (B). By the coexpression of YFP-Foxo-1, formation of fine foci of GFP-AR by DHT was disrupted, showing a diffuse distribution pattern in the nucleus (C). However, the presence of IGF-1 and DHT showed more complete nuclear translocation of GFP-AR, and significantly improved subnuclear compartmentalization (D). (See Color Insert.)

IV. Clinical Implications of Interactions between IGF-1 Signaling and AR

A. Refractory PC

IGF-1 is capable of reactivating AR in low (or absent) androgen environments, and, therefore, represents one of the multiple mechanisms by which PC cells are proposed to progress to the androgen-insensitive stage (Grossmann *et al.*, 2001). However, the mechanisms by which IGF-I and other growth factors regulate AR-mediated transcription in PC cells is not well clarified. IGF-1 exerts its biological role via binding to its receptor, IGF-1R, a cell-surface tyrosine kinase signaling molecule. One of the most important of these pathways is the PI3K/Akt pathway. Activation of the pathway by IGF-1, as well as by other molecules, has been implicated in PC. Specifically, an increase of IGF-1R expression (Grzmil *et al.*, 2004; Hellawell *et al.*, 2002; Nickerson *et al.*, 2001), and loss of the tumor suppressor gene PTEN (McMenamin *et al.*, 1999; Xin *et al.*, 2006) result in increased activity of the PI3K/Akt pathway, and greatly contribute to PC tumor progression. Foxo1, which is endogenously expressed in PC cells, is functionally inhibited by phosphorylation at S^{253}, S^{316}, and T^{24} in response to IGF-1 or insulin through PI3K/Akt kinase (Nakae *et al.*, 2000). Overexpression of Foxo1 in PC cells leads to apoptosis (Modur *et al.*, 2002). From our results (Fan *et al.*, 2007), Foxo-1 directly interacts with and suppresses the transactivation of AR, while IGF-1 signaling ameliorates the suppression, and results in AR gain-of-function. Thus, considering the reported mechanism by which beta-catenin may mediate the stimulatory effect of IGF-1 signaling on AR (Verras *et al.*, 2005), Foxo1 may provide an alternative explanation. Since insulin, similarly to IGF-1, induces PI3K-Akt signaling and the resultant modification of Foxo-1, and also enhances AR transactivation in PTEN expressing LNCaP cells, it is thus of considerable value to further elucidate whether insulin signaling enhances AR function via a similar mechanism. Indeed, it is noteworthy that the syndrome of insulin resistance, which is characterized by an increased insulin level and abdominal obesity, is also associated with an increased incidence of PC and development of a more aggressive form of the disease (Amling *et al.*, 2005; Samani *et al.*, 2007), although the underlying mechanism is currently unclear. Nevertheless, the association between the risk of PC and the serum insulin level is a controversial topic (Stattin *et al.*, 2003). In relation to this, in contrast to systemic circulating growth factors, there is also potential importance of local bioavailability of growth factors and their crosstalk with hormones. Importantly, the expression of IGF-1R was upregulated in LNCaP cells in the presence of DHT. Of note, DHT treatment for 24 h dose-dependently increased IGF-1R mRNA levels in LNCaP cells, and 10^{-8} M DHT increase mRNA levels by 17-fold (Fig. 23.5A; Fan *et al.*, 2007). Ligand-bound AR, which is functionally enhanced by IGF-1-PI3K-Akt signaling-mediated dissociation of the repressor Foxo1, in turn, stimulated the expression of IGF-1R and presumably results in increased tension of IGF-1 signaling, thereby leading to further functional augmentation of the receptor itself (Fig. 23.5B). Consistently, androgen withdrawal leads to a decrease of the PI3K-Akt pathway activity (Baron *et al.*, 2004; Castoria *et al.*, 2004). Thus, a mutually stimulatory feedback circuit (androgen-AR increases IGF-1 signaling tension, while IGF-1 signaling enhances AR transactivation), which works in an

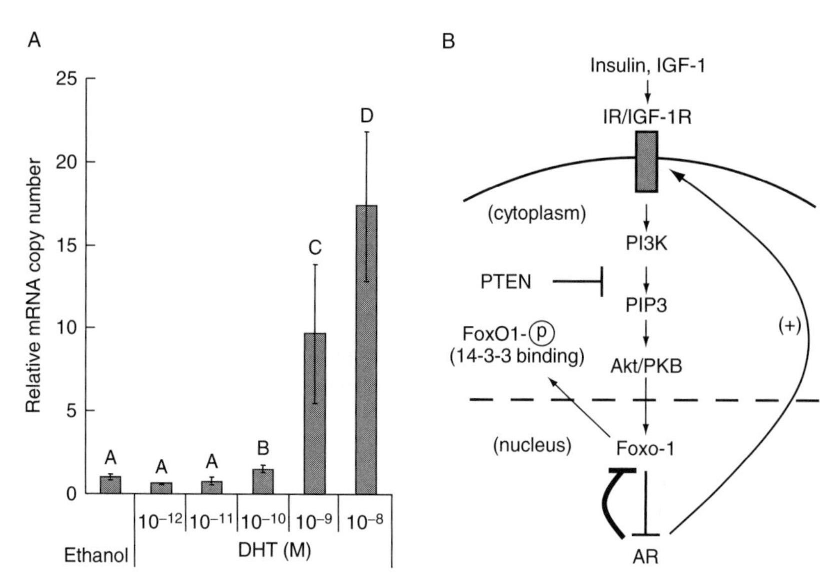

Figure 23.5 (A) DHT upregulates IGF1R mRNA expression. LNCaP cells were serum-starved and then incubated with increasing concentrations of DHT for 24 h The relative mRNA amounts of IGF-1R mRNA to GAPDH mRNA are shown. Triplicate results are expressed as means ± SD. Letters above the bars show statistical groups (ANOVA, $P < 0.05$). (B) Proposed model for the crosstalk between IGF1/insulin signaling and AR. The local bioavailability of IGF-1 (and/or insulin) can be delivered from circulation or produced locally. Following ligand binding to IGF-1R or the insulin receptor, the tyrosine activity is activated, which stimulates signaling pathways through intracellular networks that regulate cell proliferation and cell survival. Akt, or a similar PIP3-dependent kinase, translocates to the nucleus and phosphorylates S253, S316, and T24 of Foxo1. Phosphorylated Foxo1 is exported to the cytoplasm, where it has the potential to bind the 14-3-3 protein and become retained in the cytoplasm. Foxo1 interacts with the C-terminal region of AR in the presence of DHT and interferes with the ligand-induced subnuclear compartmentalization of AR, as well as the liganded receptor-target gene promoter interaction, thereby shutting down AR transactivation. Modification of Foxo1 by the PI3K-Akt system, which is activated by IGF-1 and/or insulin, weakens the Foxo-1-AR interaction and ameliorates the inhibitory effects of Foxo1 on AR. Ligand-bound AR, in turn, stimulates the expression of IGF-1R, resulting in enhanced IGF-1 signaling tension. In parallel, liganded AR may also strengthen the net outcome of the IGF-1-PI3K-Akt-Foxo1 axis by inducing bio-degradation and/or direct functional inhibition of Foxo1, which results in further amelioration of Foxo-1-related apoptosis of PC cells. These mutually inhibitive interactions between AR and Foxo1 in the context of a positive feedback circuit between AR and IGF-1 signaling that works in an autocrine/intracrine manner in the local cellular environment may play important roles in modulation of AR function and prostate cell apoptosis, as well as PC etiology and progression.

autocrine/intracrine manner in the local cellular environment of PC, may play important roles in upregulating AR function and PC progression. Besides the above-mentioned IGF-1R upregulation, the inhibition of Foxo1 provides an additional pathway by which liganded AR may enhance the IGF-1/insulin-PI3K-Akt-Foxo-1 axis. The mutual inhibition between Foxo-1 and AR in the context of IGF-1 signaling may play important roles in AR function and prostate cell apoptosis, as well as the etiology and progression (Fig. 23.5B). Pharmacological strategies that reduce IGF-1/insulin signaling, combined with anti-androgen therapies, may have clinical benefits in combating this disease.

B. Other diseases

A positive modification of IGF-1 over AR function also suggests some possibilities of this pathway in the mechanism or pathogenesis of several clinical diseases other than PC. For example, although the etiology of idiopathic hirsutism is unclear, a small number of patients show an increased 5α-reductase activity, which converts testosterone to DHT (Lakryc *et al.*, 2003). Another plausible explanation for the mechanism of idiopathic hirsutism could be the increased androgen sensitivity in the patients, which might be induced by functional modification of AR by growth factors or cytokines. Thus, it will be interesting to test the local production of IGF-1 or expression of IGF-1R in skin fibroblasts of the patients. In a similar clinical setting, the hirsutism observed in patients with polycystic ovary syndrome (PCOS), who also have hyperinsulinemia as a result of insulin resistance, might be explained not only by androgen excess (Dunaif *et al.*, 1997) but also by the changes of AR activity. However, the latter possibility has yet to be tested.

Some of the physiological actions of the growth hormone (GH)-IGF-1 system and androgens are very similar and overlap. For example, many studies have shown that both hormones act to stimulate muscle strength, increase bone mineral density, and decrease abdominal fat accumulation (Bhasin *et al.*, 1998; Clayton *et al.*, 2007; Gibney *et al.*, 2007; Kawano *et al.*, 2003). Thus, the changes of both hormones associated with aging, namely a decline of IGF-1 (somatopause) and decline of androgen or its action (andropause) may be coordinately linked with each other, thus leading to the various common symptoms or phenomenon observed in aged males. In this sense, GH therapy in adult GH-deficient patients might provide a therapy not only for GH replacement but also for the activation of AR function. To prove this hypothesis, further research is needed to evaluate tissue-specificity in the interaction between GH-IGF-1 and AR.

V. Conclusion

Foxo1, which is endogenously expressed in PC cells, could interact with AR via the C-terminus of the receptor in a ligand-dependent manner, and suppresses ligand-induced AR transactivation. Foxo-1 impairs AR signaling by interfering with ligand-induced AR nuclear translocation and subnuclear compartmentalization, as well as receptor-target gene promoter interactions. Furthermore, IGF-1/insulin-PI3K/Akt pathway-induced phosphorylation of Foxo1 ameliorated the suppression. Intriguingly, liganded AR stimulated IGF-1R expression, suggesting the presence of a local positive feedback loop between IGF-1 and AR signaling in PC cells. In other words, an auto-regulation process may occur, since the upregulated IGF-1R is expected to enhance IGF-1-PI3K-Akt signaling, thereby resulting in a further gain-of-function of AR via phosphorylating the repressive Foxo1.

ACKNOWLEDGMENTS

We would like to thank to Dr. Hajime Nawata for continuous support in this research field. We acknowledge the support of the Health and Labour Sciences Research Grant and a grant from the Ministry of Education, Sports, Science and Culture (No.16086207: Molecular mechanism of sex differentiation).

REFERENCES

Amling, C. T. (2005). Relationship between obesity and prostate cancer. *Curr. Opin. Urol.* **15,** 167–171.

Baron, S., Manin, M., Beaudoin, C., Leotoing, L., Communal, Y., Veyssiere, G., and Morel, L. (2004). Androgen receptor mediates non-genomic activation of phosphatidylinositol 3-OH kinase in androgen-sensitive epithelial cells. *J. Biol. Chem.* **279,** 14579–14586.

Bhasin, S., Bagatell, C. J., Bremner, W. J., Plymate, S. R., Tenover, J. L., Korenman, S. G., and Nieschlag, E. (1998). Issues in testosterone replacement in older men. *J. Clin. Endocrinol. Metab.* **83,** 435–448.

Buchanan, G., Irvine, R. A., Coetzee, G. A., and Tilley, W. D. (2001). Contribution of the androgen receptor to prostate cancer predisposition and progression. *Cancer Metastasis Rev.* **20,** 207–223.

Burfeind, P., Chernicky, C. L., Rininsland, F., and Ilan, J. (1996). Antisense RNA to the type I insulin-like growth factor receptor suppresses tumor growth and prevents invasion by rat prostate cancer cells *in vivo*. *Proc. Natl. Acad. Sci. USA* **93,** 7263–7268.

Castoria, G., Lombardi, M., Barone, M. V., Bilancio, A., Di Domenico, M., De Falco, A., Varricchio, L., Bottero, D., Nanayakkara, M., Migliaccio, A., and Auricchio, F. (2004). Rapid signalling pathway activation by androgens in epithelial and stromal cells. *Steroids* **69,** 517–522.

Chen, C. D., Welsbie, D. S., Tran, C., Baek, S. H., Chen, R., Vessella, R., Rosenfeld, M. G., and Sawyers, C. L. (2004). Molecular determinants of resistance to antiandrogen therapy. *Nat. Med.* **10,** 33–39.

Chen, G., Nomura, M., Morinaga, H., Matsubara, E., Okabe, T., Goto, K., Yanase, T., Zheng, H., Lu, J., and Nawata, H. (2005). Modulation of androgen receptor transactivation by FoxH1. A newly identified androgen receptor corepressor. *J. Biol. Chem.* **280,** 36355–36363.

Clayton, P., Gleeson, H., Monson, J., Popovic, V., Shalet, S. M., and Christiansen, J. S. (2007). Growth hormone replacement throughout life: Insights into age-related responses to treatment. *Growth Horm. IGF Res.* **17,** 369–382.

Culig, Z., Hobisch, A., Cronauer, M. V., Radmayr, C., Trapman, J., Hittmair, A., Bartsch, G., and Klocker, H. (1994). Androgen receptor activation in prostatic tumor cell lines by insulin-like growth factor-I, keratinocyte growth factor, and epidermal growth factor. *Cancer Res.* **54,** 5474–5478.

Dunaif, A. (1997). Insulin resistance and the polycystic ovary syndrome: Mechanism and implications for pathogenesis. *Endocr. Rev.* **18,** 774–800.

Fan, W., Yanase, T., Wu, Y., Kawate, H., Saitoh, M., Oba, K., Nomura, M., Okabe, T., Goto, K., Yanagisawa, J., Kato, S., Takayanagi, R., *et al.* (2004). Protein kinase A potentiates adrenal 4 binding protein/steroidogenic factor 1 transactivation by reintegrating the subcellular dynamic interactions of the nuclear receptor with its cofactors, general control nonderepressed-5/transformation/transcription domain-associated protein, and suppressor, dosage-sensitive sex reversal-1: A laser confocal imaging study in living KGN cells. *Mol. Endocrinol.* **18,** 127–141.

Fan, W., Yanase, T., Morinaga, H., Okabe, T., Nomura, M., Daitoku, H., Fukamizu, A., Kato, S., Takayanagi, R., and Nawata, H. (2007). IGF1/insulin signaling activates androgen signaling through direct interactions of Foxo1 with androgen receptor. *J. Biol. Chem.* **282,** 7329–7338.

Fletcher, T. M., Xiao, N., Mautino, G., Baumann, C. T., Wolford, R., Warren, B. S., and Hager, G. L. (2002). ATP-dependent mobilization of the glucocorticoid receptor during chromatin remodeling. *Mol. Cell. Biol.* **22,** 3255–3263.

Gibney, J., Healy, M. L., and Sönksen, P. H. (2007). The growth hormone/insulin-like growth factor-I axis in exercise and sport. *Endocr. Rev.* **28,** 603–624.

Greenlee, R. T., Murray, T., Bolden, S., and Wingo, P. A. (2000). Cancer statistics CA Cancer. *J. Clin.* **50,** 7–33.

Grossmann, M. E., Huang, H., and Tindall, D. J. (2001). Androgen receptor signaling in androgen-refractory prostate cancer. *J. Natl. Cancer. Inst.* **93,** 1687–1697.

Grzmil, M., Hemmerlein, B., Thelen, P., Schweyer, S., and Burfeind, P. (2004). Blockade of the type I IGF receptor expression in human prostate cancer cells inhibits proliferation and invasion, up-regulates IGF binding protein-3, and suppresses MMP-2 expression. *J. Pathol.* **202,** 50–59.

Hellawell, G. O., Turner, G. D., Davies, D. R., Poulsom, R., Brewster, S. F., and Macaulay, V. M. (2002). Expression of the type 1 insulin-like growth factor receptor is up-regulated in primary prostate cancer and commonly persists in metastatic disease. *Cancer Res.* **62,** 2942–2950.

Huang, H., Muddiman, D. C., and Tindall, D. J. (2004). Androgens negatively regulate forkhead transcription factor FKHR (FOXO1) through a proteolytic mechanism in prostate cancer cells. *J. Biol. Chem.* **279,** 13866–13877.

Hsing, A. W., Gao, Y. T., Chua, S. Jr., Deng, J., and Stanczyk, F. Z. (2003). Insulin resistance and prostate cancer risk. *J. Natl. Cancer Inst.* **95,** 67–71.

Kawano, H., Sato, T., Yamada, T., Sekine, K., Watanabe, T., Nakamura, T., Fukuda, T., Yoshimura, K., Yoshizawa, T., Aihara, K., Yamamoto, Y., Nakamichi, Y., *et al.* (2003). Suppressive function of androgen receptor in bone resorption. *Proc. Natl. Acad. Sci. USA* **100,** 9416–9421.

Kops, G. J., and Burgering, B. M. (1999). Forkhead transcription factors: New insights into protein kinase B (c-akt) signaling. *J. Mol. Med.* **77,** 656–665.

Lakryc, E. M., Motta, E. L., Soares, J. M. Jr., Haidar, M. A., de Lima, G. R., and Baracat, E. C. (2003). The benefits of finasteride for hirsute women with polycystic ovary syndrome or idiopathic hirsutism. *Gynecol. Endocrinol.* **17,** 57–63.

Li, P., Lee, H., Guo, S., Unterman, T. G., Jenster, G., and Bai, W. (2003). AKT-independent protection of prostate cancer cells from apoptosis mediated through complex formation between the androgen receptor and FKHR. *Mol. Cell. Biol.* **23,** 104–118.

Mangelsdorf, D. J., Thummel, C., Beato, M., Herrlich, P., Schutz, G., Umesono, K., Blumberg, B., Kastner, P., Mark, M., Chambon, P., and Evans, R. M. (1995). The nuclear receptor superfamily: The second decade. *Cell* **83,** 835–839.

McInerney, E. M., Rose, D. W., Flynn, S. E., Westin, S., Mullen, T. M., Krones, A., Inostroza, J., Torchia, J., Nolte, R. T., Assa-Munt, N., Milburn, M. V., Glass, C. K., *et al.* (1998). Determinants of coactivator LXXLL motif specificity in nuclear receptor transcriptional activation. *Genes Dev.* **12,** 3357–3368.

McKenna, N. J., Lanz, R. B., and O'Malley, B. W. (1999). Nuclear receptor coregulators: Cellular and molecular biology. *Endcr. Rev.* **20,** 321–344.

McNally, J. G., Muller, W. G., Walker, D., Wolford, R., and Hager, G. L. (2000). The glucocorticoid receptor: Rapid exchange with regulatory sites in living cells. *Science* **287,** 1262–1265.

McMenamin, M. E., Soung, P., Perera, S., Kaplan, I., Loda, M., and Sellers, W. R. (1999). Loss of PTEN expression in paraffin-embedded primary prostate cancer correlates with high Gleason score and advanced stage. *Cancer Res.* **59,** 4291–4296.

Modur, V., Nagarajan, R., Evers, B. M., and Milbrandt, J. (2002). FOXO proteins regulate tumor necrosis factor-related apoptosis inducing ligand expression. Implications for PTEN mutation in prostate cancer. *J. Biol. Chem.* **277,** 47928–47937.

Nakae, J., Barr, V., and Accili, D. (2000). Differential regulation of gene expression by insulin and IGF-1 receptors correlates with phosphorylation of a single amino acid residue in the forkhead transcription factor FKHR. *EMBO J.* **19,** 989–996.

Nickerson, T., Chang, F., Lorimer, D., Smeekens, S. P., Sawyers, C. L., and Pollak, M. (2001). *In vivo* progression of LAPC-9 and LNCaP prostate cancer models to androgen independence is associated with increased expression of insulin-like growth factor I (IGF-I) and IGF-I receptor (IGF-IR). *Cancer Res.* **61,** 6276–6280.

Pollak, M., Beamer, W., and Zhang, J. C. (1998). Insulin-like growth factors and prostate cancer. *Cancer Metastasis Rev.* **17,** 383–390.

Renehan, A. G., Zwahlen, M., Minder, C., O'Dwyer, S. T., Shalet, S. M., and Egger, M. (2004). Insulin-like growth factor (IGF)-I, IGF binding protein-3, and cancer risk: Systematic review and meta-regression analysis. *Lancet* **363,** 1346–1353.

Saitoh, M., Takayanagi, R., Goto, K., Fukamizu, A., Tomura, A., Yanase, T., and Nawata, H. (2002). The presence of the amino- and carboxy-terminal domains in androgen receptor is essential for the completion of a transcriptionally active form with coactivators and intranuclear compartmentalization common to the steroid hormone receptors: A three-dimensional imaging study. *Mol. Endocrinol.* **16,** 694–706.

Samani, A. A., Yakar, S., LeRoith, D., and Brodt, P. (2007). The role of the IGF system in cancer growth and metastasis: Overview and recent insights. *Endocr. Rev.* **28,** 20–47.

Santos, A. F., Huang, H., and Tindall, D. J. (2004). The androgen receptor: A potential target for therapy of prostate cancer. *Steroids* **69,** 79–85.

Stattin, P., and Kaaks, R. (2003). Insulin resistance and prostate cancer risk. *J. Natl. Cancer. Inst.* **95,** 1086–1087.

Stenoien, D. L., Patel, K., Mancini, M. G., Dutertre, M., Smith, C. L., O'Malley, B. W., and Mancini, M. A. (2001). FRAP reveals that mobility of oestrogen receptor-α is ligand- and proteosome-dependent. *Nat. Cell. Biol.* **3,** 15–23.

Tao, R. H., Kawate, H., Wu, Y., Ohnaka, K., Ishizuka, M., Inoue, A., Hagiwara, H., and Takayanagi, R. (2006). Testicular zinc finger protein recruits histone deacetylase 2 and suppresses the transactivation function and intranuclear foci formation of agonist-bound androgen receptor competitively with TIF2. *Mol. Cell. Endocrinol.* **247,** 150–165.

Tomura, A., Goto, K., Morinaga, H., Nomura, M., Okabe, T., Yanase, T., Takayanagi, R., and Nawata, H. (2001). The subnuclear three-dimensional image analysis of androgen receptor fused to green fluorescence protein. *J. Biol. Chem.* **276,** 28395–28401.

Verras, M., and Sun, Z. (2005). Beta-catenin is involved in insulin-like growth factor 1-mediated transactivation of the androgen receptor. *Mol. Endocrinol.* **19,** 391–398.

Wolk, A., Mantzoros, C. S., Andersson, S. O., Bergström, R., Signorello, L. B., Lagiou, P., Adami, H. O., and Trichopoulos, D. (1998). Insulin-like growth factor 1 and prostate cancer risk: A population-based, case-control study. *J. Natl. Cancer Inst.* **90,** 911–915.

Xin, L., Teitell, M. A., Lawson, D. A., Kwon, A., Mellinghoff, I. K., and Witte, O. N. (2006). Progression of prostate cancer by synergy of AKT with genotropic and non-genotropic actions of the androgen receptor. *Proc. Natl. Acad. Sci. USA* **103,** 7789–7794.

Yanase, T., Adachi, M., Takayanagi, R., and Nawata, H. (2004). Coregulator-related disease. *Int. Med.* **43,** 368–373.

Zegarra-Moro, O. L., Schmidt, L. J., Huang, H., and Tindall, D. J. (2002). Disruption of androgen receptor function inhibits proliferation of androgen-refractory prostate cancer cells. *Cancer Res.* **62,** 1008–1013.

Zhao, Y., Goto, K., Saitoh, M., Yanase, T., Nomura, M., Okabe, T., Takayanagi, R., and Nawata, H. (2002). Activation function-1 domain of androgen receptor contributes to the interaction between subnuclear splicing factor compartment and nuclear receptor compartment; Identification of the p102 U5 snRNP binding protein as a coactivator for the receptor. *J. Biol. Chem.* **277,** 3031–3039.

CHAPTER TWENTY-FOUR

Insulin-Like Growth Factor-2/Mannose-6 Phosphate Receptors

Hesham M. El-Shewy* *and* Louis M. Luttrell*,†,‡

Contents

Abstract

The insulin-like growth factor type 2/mannose-6-phosphate (IGF-2/M6P) receptor is a multifunctional single transmembrane glycoprotein that is known to regulate diverse biological functions. It is composed of a large extracytoplasmic domain, a single transmembrane region and a short cytoplasmic tail that lacks intrinsic catalytic activity. The receptor cycles continuously between intracellular compartments and the plasma membrane, and at steady state is predominantly localized in the *trans*-Golgi network and endosomal compartments, and to a lesser extent on the cell surface. The receptor binds IGF-2 with higher affinity than IGF-1 and does not bind insulin. It interacts, via distinct sites, with

* Department of Medicine, Medical University of South Carolina, Charleston, South Carolina 29425
† Research Service of the Ralph H. Johnson Veterans Affairs Medical Center, Charleston, South Carolina 29401
‡ Department of Biochemistry and Molecular Biology, Medical University of South Carolina, Charleston, South Carolina 29425

Vitamins and Hormones, Volume 80
ISSN 0083-6729, DOI: 10.1016/S0083-6729(08)00624-9

lysosomal enzymes and a variety of other M6P-containing ligands. IGF-2/M6P receptors perform diverse cellular functions related to lysosome biogenesis and the regulation of growth and development. It regulates extracellular IGF-2 concentrations, modulating signaling through the growth-stimulatory IGF-1 receptor pathway. It appears to mediate the uptake and processing of M6P-containing cytokines and peptide hormones, such as transforming growth factor-β, leukemia inhibitory factor, and proliferin. Some data suggest that the IGF-2/M6P receptor also functions in signal transduction by transactivating G protein-coupled sphingosine 1-phosphate receptors. Genetic evidence clearly supports a role for IGF-2/M6P receptors in organ development and growth, and recent data indicate that it may play an important role in tumor progression.

I. Introduction

The insulin-like growth factor (IGF) system consists of two signaling peptides, IGF-1 and IGF-2; six IGF binding proteins that regulate local IGF concentration; and two cell surface receptors, the IGF-1 receptor, a member of the insulin receptor family of receptor tyrosine kinases, and the IGF-1/M6P receptor, a single membrane-spanning glycoprotein that lacks intrinsic catalytic activity.

IGF was first identified in 1957 by Salmon and Daughaday and named sulfation factor because of its ability to stimulate the incorporation of radiolabled sulfate into rat cartilage (Salmon and Daughaday, 1957). Sulfation factor was induced by growth hormone, but was not, itself, growth hormone. It was also found to stimulate RNA, DNA, and collagen synthesis and to mimic the effects of insulin (Daughaday and Mariz, 1962; Hall and Uthne, 1971; Salmon and DuVall, 1970). In the 1960s, serum components with insulin-like activity that was not neutralized by anti-insulin antibodies were identified and described as nonsuppressible insulin-like activity (Bürgi *et al.*, 1966; Froesch *et al.*, 1963, 1966). These small peptides were later renamed somatomedins to reflect their general mediation of growth hormone effects on somatic growth (Daughaday *et al.*, 1972). Based on their sequence and their structural and functional homology with insulin, they were ultimately named IGF-1 and IGF-2 (Rinderknecht and Humbel, 1978a,b; Zapf *et al.*, 1978). IGF-1 and IGF-2 are single chain polypeptides that share 62% sequence homology with proinsulin. Both contain A and B domains that are homologous to the A and B chains of insulin. Unlike insulin, IGF peptides do not undergo posttranslational proteolysis, but remain linked in the mature peptides by C domains analogous to the C peptide of insulin (Adams *et al.*, 2000; Dupont and Holzenberger, 2003; Hawkes and Kar, 2004; LeRoith and Roberts, 2003). In addition, both IGF-1 and IGF-2 contain an additional short D domain that is not present in insulin. The unprocessed IGF-1 and IGF-2 prohormones also contain a C terminal E peptide that is cleaved in the Golgi apparatus during secretion. Unlike many peptide hormones, IGFs are not produced and stored within the cells of specific tissues. They arise from almost all cells in the body, and their circulating concentrations are about 100 times higher than most other peptide hormones. IGFs are essential for embryonic and postnatal growth and development. While growth hormone-dependent production of IGF-1 by the liver is the main growth regulator in infants and children, during fetal development IGFs control growth directly, with growth hormone playing little or no role. IGFs are expressed in fetal tissue from zygote formation and implantation until just before birth. In rodents and humans, IGF-2 expression is more extensive in fetal tissue than IGF-1 from mid to late gestation (Randhawa and Cohen, 2005).

IGF bioavailability is determined by its production rate, clearance, and affinity for members of a family of high affinity IGF binding proteins (IGFBP1-6) and low affinity IGFBP-related peptides (IGFBP-rP1-4). These proteins affect the biological half-life of circulating and tissue IGFs, transporting them in the circulation, delivering them to their target cells, and modulating their interaction with receptors (Denley *et al.*, 2005; Duan and Xu, 2005). Binding of IGFs to the binding protein complex decreases IGF activity. IGF stimulation increases the rate synthesis of some binding proteins, changing their levels and providing feedback regulation. Adding to the complexity, the interaction between IGFs and their binding proteins is regulated by proteases that cleave IGFBPs into fragments with lower affinity, allowing for increased levels of bioactive IGFs. The resulting change in the ratio of IGF to IGFBP modulates IGF/IGFBP/IGF receptor interactions and may play a role in normal and abnormal tissue proliferation (Bunn and Fowlkes, 2003; Jones and Clemmons, 1995; Lelbach *et al.*, 2005; Vasylyeva and Ferry, 2007). These proteases fall into three groups. Kallikrein-like serine proteases, which cleave IGFBP-3, include prostate-specific antigen, gamma nerve growth factor, and plasmin. Thrombin, another serine protease, cleaves IGFBP-5 at physiologically relevant concentrations. The second major group, the cathepsins, are a family of intracellular proteases that are activated under acidic conditions and may play a role in certain physiologic and pathologic processes such as neoplastic infiltration. The third group, matrix metalloproteases, compromise a family of peptide hydrolases that function in tissue remodeling by degrading extracellular matrix components such as collagen and proteoglycans.

IGF-1 and IGF-2 exert their biological effects by binding to two structurally distinct plasma membrane receptors (Fig. 24.1). The type 1

IGF receptor is heterotetrameric transmembrane tyrosine kinase receptor that is structurally and functionally related to the insulin receptor. It is composed of two extracellular α subunits containing the ligand-binding domain and two transmembrane β subunits possessing intrinsic ligand-stimulated tyrosine activity (Baserga *et al.*, 1997; Grønborg *et al.*, 1993; LeRoith *et al.*, 1995). Ligand binding to α subunits induces conformational changes and activates the tyrosine kinase of the intrinsic β subunit receptor, which in turn triggers a cascade of tyrosine phosphorylation of various intracellular adapter proteins (AP) involved in signal transduction. IGF-1 and IGF-2 also bind with lower affinity to the insulin receptor, although biologically-relevant activation of the insulin receptor by IGFs probably occurs only in pathophysiologic states characterized by excessive IGF production. Unlike the IGF-1 and insulin receptors, the type 2 IGF receptor (IGF-2/M6P) receptor lacks intrinsic catalytic activity. Instead, it is a type 1 single transmembrane glycoprotein composed of a large extracytoplasmic domain, a single transmembrane region, and a short carboxy-terminal cytoplasmic tail (Ghosh *et al.*, 2003; Hassan, 2003; Kornfeld, 1992).

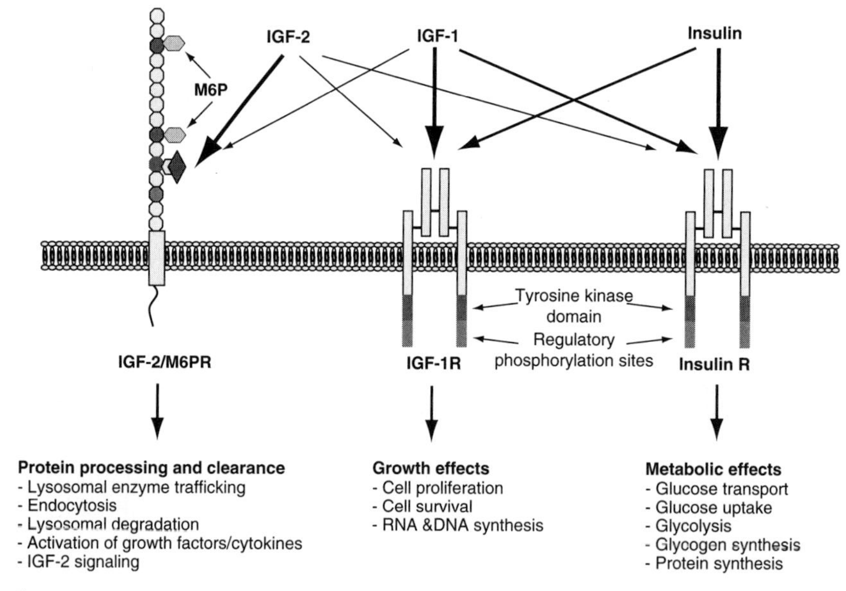

Figure 24.1 Structure of the insulin, IGF-1 and IGF-2/M6P receptors. The IGF-1 and insulin receptors are receptor tyrosine kinases that share high structural homology. Each possesses a heterotetrameric structure composed of two α and two β subunits joined by disulfide bonds. The IGF-2/M6P receptor is a transmembrane glycoprotein consisting of a large extracellular domain, a single transmembrane domain, and a short cytoplasmic tail. The relative binding affinity of IGF-1, IGF-2 and insulin for each receptor differs as indicated by the arrow thickness.

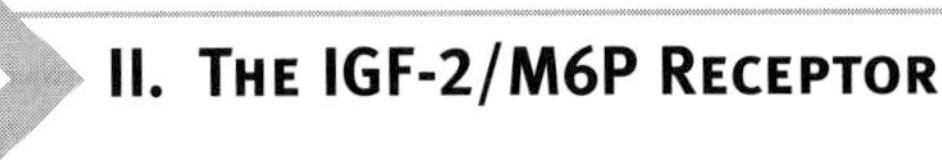

II. The IGF-2/M6P Receptor

In the 1980s, photoaffinity cross-linking studies identified a 250 kDa species that served as a receptor for IGF-2. Meanwhile, a similar sized receptor that bound mannose 6-phosphate (M6P) residues on lysosomal hydrolases was identified and named the cation-independent mannose 6-phosphate receptor (CIMPR) (Morgan *et al.*, 1987; Polychronakos and Piscina, 1988; Polychronakos *et al.*, 1990). The molecular cloning of the CIMPR and IGF-2 receptor revealed a common receptor which is now known as IGF-2/M6P receptor (Ludwig *et al.*, 1994; Szebenyi and Rotwein, 1994).

A. Structure of the IGF-2/M6P receptor

The IGF-2/M6P receptor, also known as cytokine receptor CD222, is structurally distinct from the IGF-1 and insulin receptors. It has higher binding affinity for IGF-2 than for IGF-1 and does not bind insulin. The receptor is a type 1 integral membrane protein and P-type lectin receptor consisting of a large N-terminal extracytoplasmic domain of 2264–2269 residues, a 23-residue transmembrane region, and a short C-terminal cytoplasmic domain of 163–164 residues (Fig. 24.1). The characteristic feature of the extracellular domain is that it is composed of 15 contiguous repeats of approximately 147 amino acids that display 14–38% amino acid sequence identity. Each repeat contains eight conserved cysteines that form intramolecular disulfide bonds that are necessary for proper receptor folding (Braulke, 1999; Dahms and Hancock, 2002; Jones and Clemmons, 1995; Kornfeld, 1992). Nineteen potential *N*-glycosylation sites are distributed throughout the extracytoplasmic domain (Lobel *et al.*, 1987), of which at least two are utilized in forming the mature receptor of 275–300 kDa. Other posttranslational modifications, such as phosphorylation and palmitoylation have also been reported (Dahms and Hancock, 2002; Westcott and Rome, 1988).

The IGF-2/M6P receptor binds IGF-2 and M6P-containing ligands at distinct sites. Repeat 11 comprises the core IGF-2 binding site, whereas repeats 3, 5 and 9 bind proteins bearing M6P moieties (Braulke, 1999; Brown *et al.*, 2002; LeRoith *et al.*, 1995; Reddy *et al.*, 2004; Schmidt *et al.*, 1995). Recent crystallographic studies, including high resolution structures of repeat 11 and of a receptor fragment consisting of repeats 1–3 of the human IGF-2/M6P receptor, have provided detailed insights into the structural features of the receptor (Brown *et al.*, 2002, 2008; Leksa *et al.*, 2002). These structures, along with sequence alignments, suggest that all 15 repeats share a similar topology, consisting of a flattened barrel formed by nine *β*-strands. Analysis of the 1–3 triple-repeat crystal indicates a structure in which repeat 3 sits on the top of repeats 1 and 2, prompting Olson and colleagues to propose that the IGF-2/M6P receptor forms distinct structural units for every three repeats of the extracellular domain, producing five tri-repeat units that stack in back-to-front manner (Olson *et al.*, 2004). In this model, the IGF-2 binding site was proposed to reside on the opposite face of the structure relative to the M6P binding sites.

The cytoplasmic domain of the receptor contains motifs that are important for protein interactions that are required for receptor trafficking and phosphorylation. The single tyrosine-based internalization motif, YSKV, mediates interactions with the clathrin-associated APs, AP-1 and AP-2, and is involved in targeting receptors on the plasma membrane to clathrin-coated vesicles (Dahms and Hancock, 2002; Kornfeld, 1992; Le Borgne and Hoflack, 1998; Pearse and Robinson, 1990). In addition, there are four regions that are potential substrates for protein kinases, including protein kinase C (PKC), cAMP-dependent protein kinase, and casein kinases 1 and II (Dahms and Hancock, 2002; Hawkes and Kar, 2004; Kornfeld, 1992). Additional phosphorylation of threonine and tyrosine residues in the cytosolic tail of the IGF-2/M6PR has been reported (Corvera *et al.*, 1986, 1988; Sahagian and Neufeld, 1983; Zhang *et al.*, 1997a).

B. Soluble form of the IGF-2/M6P receptor

An IGF-2 binding component that was considerably larger than the IGF binding proteins was first observed in fetal serum (Gelato *et al.*, 1989). It was later identified as a truncated form of the IGF-2/M6P receptor, missing the cytoplasmic domain, using antibodies raised against synthetic peptides from the cytoplasmic or extracellular domains (Causin *et al.*, 1988; MacDonald *et al.*, 1989). It has also been found in bovine serum and in the serum, urine, and amniotic fluid of rats and humans (Dahms and Hancock, 2002). The soluble form of the receptor present in the circulation and in medium conditioned by cultured cells is derived from proteolytic cleavage of the extracellular domain of the IGF-2/M6P receptor at the cell surface (Clairmont and Czech, 1991; Scott *et al.*, 1996). It retains its ability to bind IGF-2 (Bobek *et al.*, 1991; Scott and Weiss, 2000) and may therefore be able to modulate IGF-2 activity in a fashion similar to IGF binding proteins as well as representing a pathway for receptor degradation. The soluble form of the IGF-2/M6P receptor is able to inhibit biological responses to IGF-2 in cell culture, such as DNA synthesis and cell proliferation in BRL-3A and mouse 3T3 fibroblast cells and hepatocytes (Scott and Baxter, 1996; Scott *et al.*, 1996). However, it is not clear whether or not the soluble receptor plays an analogous regulatory role *in vivo*.

C. IGF-2/M6P receptor gene structure

The genomic structure of the IGF-2/M6P receptor has been analyzed for the human and the mouse. The murine IGF-2/M6P receptor maps to the centromeric third of chromosome 17 and contains 48 exons spanning 130 kB (Laureys *et al.*, 1988; Szebenyi and Rotwein, 1994). Its human counterpart, also containing 48 exons, spans 136 kB in region 6q25-q27 on the long arm of chromosome 6 (Killian and Jirtle, 1999). Interestingly, the exon boundaries of the IGF-2/M6P receptor do not correspond to its functional or structural domains. Exons 1–46 encode the extracellular region of the receptor with each of its 15 repeats encoded by portions of three to five separate exons (Killian and Jirtle, 1999; Szebenyi and Rotwein, 1994). A 54-bp enhancer, comprised of two E-box motifs and putative binding sites for the transcription factors Sp1 and NGF-1A, has been identified within the 266-bp promoter region (Liu *et al.*, 1995). In the mouse, IGF-2/M6P receptor gene expression is maternally imprinted in peripheral tissues, but for most humans expression is biallelic (Kalscheuer *et al.*, 1993). Like IGF-2, the IGF-2/M6P receptor is expressed at highest levels at day 16–20 of fetal development, most abundantly in the heart, and declines to somewhat lower levels during the postnatal period in rat and mouse tissues (Kornfeld, 1992).

D. Trafficking of IGF-2/M6P receptors

At steady state, ~90% of the IGF-2/M6P receptors localize to the *trans*-Golgi network (TGN) and endosomal compartments, with the remainder on the plasma membrane. The receptor recycles continuously between two cellular pools. Although behaving as a monomer in detergent solutions, the receptor on the cell surface tends to exist as a dimer (Hassan, 2003; York *et al.*, 1999). Dimerization apparently enhances binding affinity for ligands that are multivalent for M6P residues, which bridge adjacent M6P-binding motifs, and alters the kinetics of the receptor internalization from cell surface (Byrd and MacDonald, 2000; Byrd *et al.*, 2000).

Several factors have been shown to modulate recycling and intracellular routing of the IGF-2/M6P receptor. In human fibroblasts, exposure to IGF-1, IGF-2, and epidermal growth factor (EGF) causes a rapid and transient redistribution of the receptor from intracellular pools to the cell surface, resulting in a 2–3-fold increase in the binding and uptake of exogenous lysosomal enzymes (Braulke *et al.*, 1989, 1990; Damke *et al.*, 1992). Insulin also causes a marked redistribution of receptors from internal membranes to the cell surface in rat adipocytes and H-35 hepatoma cells. This effect is associated with an overall decrease in phosphorylation of the receptors present in the plasma membrane (Oka *et al.*, 1984; Oppenheimer *et al.*, 1983). Glucose has been shown to increase IGF-2 binding to the

IGF-2/M6P receptor following an increase in cell surface expression of the receptor in two insulin-secreting cell lines (Zhang *et al.*, 1997a). Extracellular addition of the M6P-containing lysosomal enzyme, β-glucoronidase, has the opposite effect, promoting receptor dimerization and increasing the rate of receptor internalization (York *et al.*, 1999). Conversely, IGF-2/M6P receptor internalization is inhibited by some major histocompatability complex class I-derived peptides in insulin-stimulated rat adipose cells (Stagsted *et al.*, 1993).

The level of IGF-2/M6P receptors expressed on the membrane may also be regulated by phosphorylation (Kiess *et al.*, 1994). PKC-mediated serine phosphorylation or okadaic acid inhibition of serine phosphatases increases the proportion of receptors on the plasma membrane (Braulke and Mieskes, 1992; Hu *et al.*, 1990; Zhang *et al.*, 1997a). The cytoplasmic domain of the bovine IGF-2/M6P receptor contains three potential serine phosphorylation sites; Ser 19, Ser 85, and Ser 156. The latter two reside in casein kinase-like motifs (Corvera *et al.*, 1988; Méresse *et al.*, 1990; Rosorius *et al.*, 1993). Phosphorylation of these sites correlates with TGN and clathrin-coated vesicle localization, suggesting a role for phosphorylation/dephosphorylation in modulating interactions with trafficking proteins and influencing receptor distribution (Méresse and Hoflack, 1993). On the other hand, disruption of these three potential phosphorylation sites by mutagenesis had no detectable effect on the sorting of lysosomal enzymes by the bovine or murine IGF-2/M6P receptor (Chen *et al.*, 1997).

E. IGF-2/M6P receptor ligands

The multifunctional nature of the IGF-2/M6PR is evident in the fact that its extracytoplasmic domain harbors two distinct carbohydrate recognition sites along with its single IGF binding site (Tong *et al.*, 1989; Westlund *et al.*, 1991; Morgan *et al.*, 1987; Dahms and Hancock, 2002). In addition to IGF-2, the IGF-2/M6P receptor binds a diverse spectrum of M6P-containing proteins, including lysosomal enzymes, transforming growth factor-β (TGF-β) (Dennis and Rifkin, 1991), leukemia inhibitory factor (LIF) (Blanchard *et al.*, 1999), proliferin (Lee and Nathans, 1988), thyroglobulin (Herzog *et al.*, 1987), renin precursor (Saris *et al.*, 2001), granzymes A and B, CD26, Herpes simplex viral glycoprotein D, Varicella-Zoster viral glycoprotein I (Dahms and Hancock, 2002), and at least one non-M6P containing ligand, retinoic acid (Kang *et al.*, 1997).

1. IGF-2

IGF-2 is a small acidic polypeptide with structural homology to IGF-1 and insulin (Adams *et al.*, 2000; Dupont and Holzenberger, 2003; Hawkes and Kar, 2004; LeRoith and Roberts, 2003). It has a wide range of biological activities and plays key roles in growth and development, cell division, and differentiation (Lelbach, 2005). A distinct IGF-2 binding site is located on repeat 11 of the extracellular domain of the IGF-2/M6P receptor (Dahms *et al.*, 1994; Garmroudi and MacDonald, 1994; Schmidt *et al.*, 1995). Repeat 11 contains two hydrophobic binding sites, the first being a shallow cleft located at the mouth of the barrel that is spatially similar to the hydrophobic sugar-binding pockets for M6P located in repeats 3 and 9, and the second a region that extends along an external flattened surface. The former, which contains Ile1572, is essential for the initial hydrophobic docking of IGF-2, while the latter appears to provide secondary sites that stabilize IGF-2 binding (Zaccheo *et al.*, 2006). Site directed mutagenesis substituting threonine for isoleucine at position 1572 in the N-terminal half of repeat 11 abolishes IGF-2 binding (Garmroudi *et al.*, 1996; Linnell *et al.*, 2001). Recent structural studies of the interactions between IGF-2 and the receptor, confirmed by mutagenesis, demonstrate that Phe19 and Leu53 of IGF-2 lock into this hydrophobic pocket (Brown *et al.*, 2008). Repeat 11 is also sufficient to mediate internalization of IGF-2 as shown using a chimeric minireceptor in which repeat 11 was fused to the transmembrane and cytoplasmic domain of the receptor (Garmroudi *et al.*, 1996). Nonetheless, IGF-2 binding affinity to the repeat 11 minireceptor is ~10-fold lower than its affinity for the holoreceptor. Examination of IGF-2 binding to minireceptors containing repeats 11–12, 11–13, and 11–15 suggest that an affinity enhancing domain is present in repeat 13, although repeat 13 itself does not bind IGF-2. Interestingly, repeat 13 contains a 43 amino acid insertion similar to the type II repeat of fibronectins, which when deleted from the holoreceptor results in decreased IGF-2 binding affinity (Ghosh *et al.*, 2003; Hassan, 2003; Kornfeld, 1992). In the predicted structure, the IGF-2 binding face of repeat 11 lies adjacent to the region of repeat 13 containing the fibronectin type II-like insert, allowing elements within repeat 13 to slow the rate of IGF-2 dissociation. These interactions enhance receptor affinity for IGF-2 by 5- to 10-fold (Brown *et al.*, 2002; Devi *et al.*, 1998; Garmroudi *et al.*, 1996; Linnell *et al.*, 2001; Schmidt *et al.*, 1995).

2. Lysosomal enzymes

Lysosomes and lysosomal proteins are of considerable biomedical importance, and genetic defects in synthesis, sorting or targeting of lysosomal enzymes result in over 40 human diseases (Sleat *et al.*, 2006). In addition, alterations in the lysosomal system have been implicated in more widespread diseases including cancer, Alzheimer's disease, rheumatoid arthritis, and atherosclerosis. To date, about fifty lysosomal hydrolases have been identified (Ni *et al.*, 2006) and the majority of them are targeted to the lysosomes via binding to the IGF-2/M6P receptor. Two high-affinity M6P binding sites capable of binding lysosomal enzymes reside within repeats 1–3 and 7–11, while a third lower-affinity site has recently been identified in repeat 5 (Hawkes *et al.*, 2007; Reddy *et al.*, 2004). Equilibrium dialysis experiments have demonstrated that the receptor binds 2 moles of M6P or 1 mole of β-galactosidase or equivalent lysosomal enzymes through their M6P residues (Dahms and Hancock, 2002; Westlund *et al.*, 1991). Five conserved amino acid residues in repeat 3 and repeat 9 are essential for carbohydrate recognition by the bovine IGF-2/M6P receptor (Dahms and Hancock, 2002; Dahms *et al.*, 1993b; Hancock *et al.*, 2002). The C-terminal M6P binding site in repeat 9 exhibits optimal binding at pH 6.4–6.5, whereas the N-terminal M6P binding site in repeat 3 demonstrates a higher optimal binding pH of 6.9–7.0. Furthermore, the C-terminal site is highly specific for M6P and M6P phosphomonoester, whereas the amino-terminal site binds M6P phosphodiester and M6-sulfate with only slightly lower affinity than M6P (Dahms and Hancock, 2002; Marron-Terada *et al.*, 2000). These findings suggest that the carbohydrate recognition sites of the IGF-2/M6PR not only bind ligands over a relatively broad pH range, but also recognize a great diversity of ligands. Although the receptor binding sites for lysosomal enzymes and IGF-2 are distinct, there is evidence that lysosomal enzymes like β-galactosidase inhibit IGF-2 binding (Kiess *et al.*, 1988) and conversely, that IGF-2 inhibits the binding of β-galactosidase (Kiess *et al.*, 1989b). The reciprocal inhibition of binding of these two classes of ligand probably reflects steric inhibition, leading to the prediction that the presence of extracellular lysosomal enzymes would inhibit the IGF-2/M6P receptor-mediated degradation of IGF-2, resulting in increased activation of IGF-1 receptors, and, conversely, that the concentration of extracellular lysosomal enzymes would be increased in the presence of IGF-2. Indeed, overexpression of IGF-2 in MCF-7 breast cancer cells (De Leon *et al.*, 1996) or HEK293 human embryonic kidney cells (Hoeflich *et al.*, 1995) increases extracellular levels of cathepsin D and β-hexosaminadase.

3. TGF-β

The IGF-2/M6P receptor plays an essential role in the activation of TGF-β, a cytokine that regulates the differentiation and growth of numerous cell types (Dennis and Rifkin, 1991; Ghahary *et al.*, 1999; Villevalois-Cam *et al.*, 2003). Latency-associated peptide (LAP) and TGF-β1 are synthesized as a single propreprotein that is secreted and stored in the extracellular matrix in an inactive form that must be cleaved to produce mature TGF-β. LAP is glycosylated at three N-linked sites and M6P residues are present on at least two of these carbohydrate side chains. Latent TGF-β binds to the IGF-2/M6P receptor at M6P recognition sites (Kovacina *et al.*, 1989; Purchio *et al.*, 1988), after which it is cleaved by extracellular plasmin (Flaumenhaft *et al.*, 1993). While this model of TGF-β activation has been shown to operate in a cell culture model, it is not known to what extent the pathway involving IGF-2/M6P receptors is functional *in vivo*. Although the matrix glycoprotein, thrombospondin-1, has been reported to mediate the activation of

TGF-β (Crawford *et al.*, 1998), several lines of evidence suggest a role for plasmin-mediated activation of latent TGF-β following its binding to the IGF-2/M6P receptors (Dennis and Rifkin, 1991; Ghahary *et al.*, 1999; Villevalois-Cam *et al.*, 2003). Moreover, some data suggest that IGF-2/M6P receptors complex with the urokinase plasminogen activator receptor and directly bind plasminogen, leading to the generation of active plasmin that in turn proteolotically activates receptor-bound latent TGF-β (Godár *et al.*, 1999, 2003; Kreiling *et al.*, 2003).

4. Leukemia inhibitory factor (LIF)

M6P-sensitive LIF binding to the IGF-2/M6P receptor results in rapid internalization and degradation of the cytokine in numerous cell lines, suggesting that the IGF-2/M6P receptor may regulate the availability of LIF *in vivo* (Blanchard *et al.*, 1999).

5. Proliferin

Proliferin is a member of the prolactin-related glycoprotein family. It is synthesized in placental trophoblast giant cells and acts as a paracrine factor to stimulate endothelial cell migration, promoting angiogenesis during the development of fetal tissues (Linzer and Nathans, 1984; Nilsen-Hamilton *et al.*, 1980). Proliferin, also known as mitogen-regulated protein, was first detected in 3T3 mouse fibroblasts *in vitro*. Its expression has also been demonstrated in mouse placenta and in the circulation of the pregnant mouse, where it reaches peak levels by midgestation (Lee and Nathans, 1988; Lee *et al.*, 1988; Linzer and Nathans, 1985). Proliferin binds to the IGF-2/M6P receptor at M6P recognition sites (Lee and Nathans, 1988; Volpert *et al.*, 1996). Although the signaling pathways that are initiated upon proliferin binding to the IGF-2/M6P receptor are poorly understood, proliferin-stimulated mitogen-activated protein (MAP) kinase activation is reportedly mediated via heterotrimeric G proteins (Groskopf *et al.*, 1997). In the uterus, proliferin also mediates cell proliferation by binding to a specific high affinity receptor that is distinct from the IGF-2/M6P receptor and (Nelson *et al.*, 1995).

6. Thyroglobulin

M6P residues are found in thyroglobulin and radiolabled thyroglobulin binds to the IGF-2/M6P receptor (Herzog *et al.*, 1987). Although thyroglobulin can be endocytosed by human skin fibroblasts via the IGF-2/M6P receptor, the pathway for delivery of thyroglobulin to lysosomes in the thyroid gland does not involve the IGF-2/M6P receptor (Lemansky and Herzog, 1992).

7. Prorenin

Renin is a circulating protease that participates in the renin–angiotensin–aldosterone system. It is secreted from the juxtaglomerular cells of the kidney and catalyzes the first step in the activation of angiotensin, thereby playing a pivotal role in the regulation of blood pressure and extracellular fluid volume (Faust *et al.*, 1987; Lacasse *et al.*, 1985). Renin in the blood is predominantly present in the form of its inactive precursor, prorenin. In *Xenopus* oocytes (Faust *et al.*, 1987), rat cardiomyocytes (Saris *et al.*, 2001), and human umbilical vein endothelial cells (van den Eijnden *et al.*, 2001), binding of M6P-containing prorenin to the IGF-2/M6P receptor promotes its internalization and proteolytic conversion from prorenin to renin.

8. Granzyme B

The serine protease, granzyme B, is an important effector molecule in granule-mediated killing by cytotoxic T lymphocytes and natural killer cells (Lord *et al.*, 2003; Roberts *et al.*, 2003; Trapani and Sutton, 2003). In the target cell, granzyme B activates an apoptotic pathway by cleaving key substrate in the cytoplasm. Thus, the key step is its entry into the target cell. Motyka and colleagues have suggested that grazyme B enters cells by binding to M6P-binding sites on the IGF-2/M6P receptor (Motyka *et al.*, 2000). Doubts have been raised, however, since this study used the small 32 kDa free form of granzyme B (Poe *et al.*, 1991) instead of the predominant physiological form of large granzyme B complexed with the proteoglycan serglycin (Galvin *et al.*, 1999; Metkar *et al.*, 2002).

9. Retinoic acid

Retinoic acid, which binds nuclear retinoic acid receptors and plays an essential role in development, cellular metabolism, and the regulation of cell proliferation, also binds the IGF-2/M6P receptor. Retinoic acid has been shown to stimulate IGF-2/M6P receptor-mediated internalization of IGF-2 and to increase lysosomal enzyme sorting. Correlation of growth inhibition and apoptosis in response to retinoic acid with the presence of IGF-2/M6P receptor led to the proposal that the receptor may play an important role in mediating retinoid-induced apoptosis/growth inhibition (Dahms *et al.*, 1989). Although Kang and Leaf failed to identify a specific binding site for retinoic acid on the receptor using photoaffinity labeling, they demonstrated that neither M6P nor IGF-2 inhibited the retinoic acid-IGF-2/M6P receptor interaction, suggesting a distinct binding site (Kang *et al.*, 1997).

III. Functions of the IGF-2/M6P Receptor

The IGF-2/M6P receptor binds diverse ligands and shuttles continuously between the TGN, plasma membrane and endosomes. These properties enable it to participate in the regulation of cellular and physiologic homeostasis by capturing extracellular cargo and transporting it into the cell via clathrin-dependent endocytosis for processing or degradation. As such, it appears to exert both positive and negative effects on signal transduction. Binding to IGF-2/M6P receptors may promote activation of latent TGF-β, granzyme B and renin precursor, while at the same time enhancing the clearance and degradation of IGF-2 and LIF. Although less well understood than its role in controlling the extracellular concentration lysosomal enzymes, growth factors and cytokines, IGF-2/M6P receptors have themselves been implicated in signal transduction, mediating cellular responses to IGF-2 and proliferin.

A. Sorting of lysosomal enzymes

IGF-2/M6P receptors play a general role in the recapture of endogenous, newly synthesized lysosomal enzymes which escape sorting at the TGN (Fig. 24.2). Targeting of nascent lysosomal enzymes to lysosomes occurs via distinct biosynthetic and endocytic pathways (Hille-Rehfeld, 1995). In the major, biosynthetic arm, newly synthesized lysosomal enzymes are transported to lysosomes by first binding to IGF-2/M6P receptors in the TGN, then traveling in coated vesicles to early and late endosomes where, in the acidic environment of late endosomes, acid hydrolases dissociate from the receptor and are incorporated into lysosomes (Dahms and Hancock, 2002; Hawkes and Kar, 2004). Unoccupied receptors then recycle back to the TGN to bind newly synthesized lysosomal enzymes. The second minor pathway utilizes the 10% of the total cellular IGF-2/M6P receptor population that is present on the plasma membrane to bind extracellular lysosomal enzymes that have escaped the intracellular biosynthetic pathway. The IGF-2/M6P receptor cycles continuously among the membrane compartments of both the biosynthetic and endocytic pathways, and binding lysosomal enzymes increases the rate of receptor internalization. Although the exact mechanisms of enzyme transport have not been elucidated, binding of clathrin-associated AP to an acidic-cluster-dileucine (AC-ALL) motif within the cytoplasmic tail of the IGF-2/M6P receptor is essential for efficient clathrin-mediated transport of lysosomal enzymes to endosomal compartments (Böker *et al.*, 1997; Ghosh *et al.*, 2003; Hille-Rehfeld, 1995; Johnson and Kornfeld, 1992). Recent studies suggest that Golgi-localized γ-ear-containing ADP-ribosylation factor-binding (GGA) proteins play a critical role in IGF-2/M6P receptor transport of lysosomal enzymes between the TGN and endosomes (Boman *et al.*, 2000; Ghosh *et al.*, 2003; Hirst, 2001; Takatsu *et al.*, 2001; Zhu *et al.*, 2001). GGAs are monomeric, multidomain, cytoplasmic proteins consisting of four domains; an N-terminal VHS domain, a GAT domain, a connecting hinge segment, and a C-terminal GAE (adaptin ear homology-a subunit of AP-1) domain (Ghosh *et al.*, 2003; Mullins and Bonifacino, 2001). The GAT domain binds

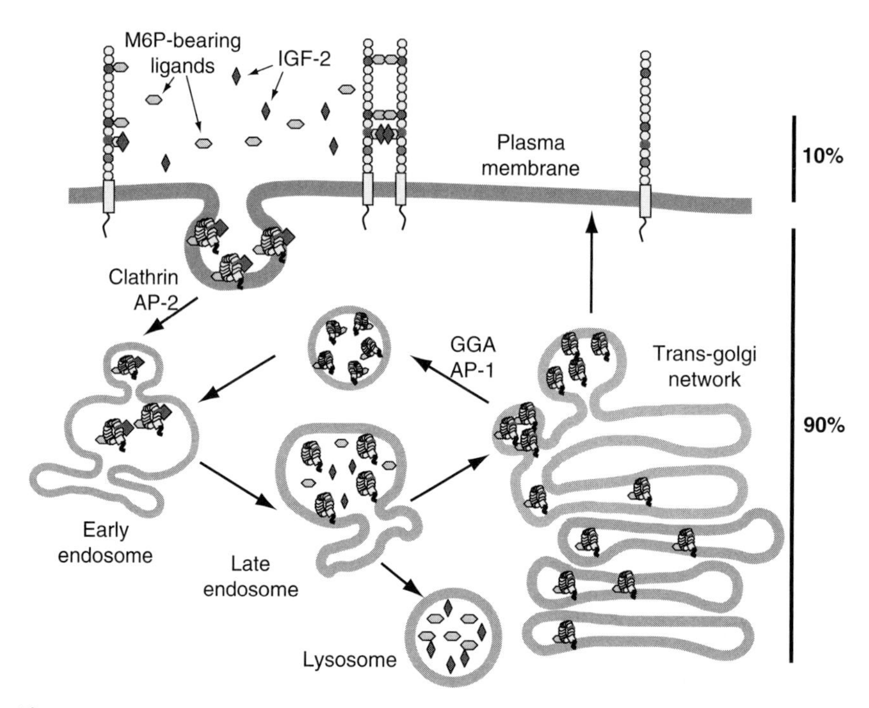

Figure 24.2 Subcellular localization and trafficking of the IGF-2/M6P receptor. Newly synthesized lysosomal enzymes are delivered to lysosomes through binding to IGF-2/M6P receptors in the TGN and clathrin/AP-1/GGA dependent transport to early endosomes. Exposure to the low-PH environment of late endosomes causes lysosomal enzymes to dissociate, after which they are sorted to lysosomes. The receptors are returned to the TGN to repeat this process or trafficked to the cell surface to bind extracellular lysosomal enzymes. Cell surface IGF-1/M6P receptors carrying IGF-2 or M6P-containing ligands undergo clathrin/AP-2 dependent endocytosis, through which they also enter the early-late endosome compartments.

ADP-ribosylation factor-GTP complexes and mediates recruitment of GGAs from the cytosol onto the TGN. The VHS domain interacts with the AC-ALL motif in the cytoplasmic tail of the IGF-2/M6PR. The GAE domain binds proteins that interact with the ear domain of AP1, which play an important role in endosome-to-TGN receptor trafficking. On the other hand, receptor recycling back to TGN from endosomes appears to involve an interaction between the cytoplasmic tail of the receptor and specific tail binding proteins, rather than being clathrin-mediated (Iversen *et al.*, 2001). Two candidate tail binding proteins, phosphofurin acidic cluster sorting protein 1 (PACS-1) and MPR tail interacting protein of 47 kDa (TIP47), have been implicated in receptor recycling (Ghosh *et al.*, 2003; Mullins and Bonifacino, 2001; Orsel *et al.*, 2000). Whereas PACS-1 acts as a connector between the IGF-2/M6P receptor and AP-1 to facilitate recycling of the receptor from early endosomes to the TGN (Hawkes and Kar, 2004),

TIP47 recycles receptors from late endosomes by binding Rab9, a late endosome GTPase (Dahms and Hancock, 2002; Ghosh *et al.*, 2003).

B. Regulation of extracellular IGF-2 levels

IGF-2/M6P receptor mediated internalization and degradation of IGF-2 has been documented in cultured cells, including rat adipocytes (Oka *et al.*, 1985), L6 rat myoblasts (Kiess *et al.*, 1987), rat C6 glial cells (Kiess *et al.*, 1989a), and mouse L cells (Nolan *et al.*, 1990). In L6 cells, for example, the amount of IGF-2 degraded during an 8 h incubation was decreased by 88% by the addition of an IGF-2/M6P receptor blocking antibody (Kiess *et al.*, 1987). However, the most compelling evidence for an important role of the IGF-2/M6P receptor in regulating IGF-2 action comes from gene deletion experiments in mice (Lau *et al.*, 1994; Ludwig *et al.*, 1996; Wang *et al.*, 1994). Disruption of the IGF-2/M6P receptor gene results in prenatal death and fetuses that are 30% larger than normal (Wang *et al.*, 1994). Lau and colleagues found that serum IGF-2 in IGF-2/M6P receptor deficient mice was elevated 2–2.7-fold compared to wild type littermates with no change in IGF-2 mRNA expression in the knockout embryos. (Lau *et al.*, 1994). Others have reported that serum IGF-2 was 4.4-fold higher in the receptor knockout embryos and tissue levels of IGF-2 were also elevated (Ludwig *et al.*, 1996). The hearts of the IGF-2/M6P receptor knockout mice were nearly three times larger than normal and prenatal lethality was attributed to this abnormality (Lau *et al.*, 1994). Histological analysis revealed hyperplasia of the myocardium along with an increase in total DNA content. During normal development, the heart has the highest expression of IGF-2/M6P receptor of any fetal tissue and IGF-2 is also abundant in the heart. Compellingly, the overgrowth phenotype and prenatal lethality of the IGF-2/M6P receptor knockout mouse is reversed by generating IGF-2/M6P receptor knockout mice in an IGF-2 deficient background or by generating mice that were deficient in both IGF-2/M6P receptor and IGF-1 receptors (Ludwig *et al.*, 1996). These studies provide direct evidence that IGF-2/M6P receptors regulate IGF-2 signaling *in vivo*, most likely by regulating extracellular IGF-2 through endocytic clearance. In the absence of IGF-2/M6P receptors, increased local IGF-2 concentrations produce a proliferative and hypertrophic response that is mediated by IGF-2 acting on the IGF-1 receptor.

C. IGF-2/M6P receptors as signal transducers

Unlike its function as a clearance receptor, the role of the IGF-2/M6P receptor in IGF-2 signaling remains controversial and poorly understood. Because the IGF-2/M6P receptor lacks intrinsic catalytic activity, most of the biological effects of IGF-2 have been attributed to activation of the

IGF-1 receptor (Dahms and Hancock, 2002) and insulin receptor isoform A (Frasca *et al.*, 1999). Nonetheless, a growing body of evidence suggests that some of the metabolic actions of IGF-2 are mediated by binding to the IGF-2/M6P receptor. Most of the early evidence implicating IGF-2/M6P receptors in signaling was based on the differences in the relative potency of IGF-2 versus IGF-1, stimulation by IGF-2 analogues which recognize the IGF-2/M6P receptor but not the IGF-1 receptor, and the use of various IGF-2/M6P receptor antibodies to block or mimic responses to IGF-2. Biological responses reportedly mediated by the IGF-2/M6P receptor include calcium influx in mouse embryo fibroblasts (Nishimoto *et al.*, 1987) and rat calvarial osteoblasts (Martinez *et al.*, 1995), increased protein phosphorylation (Hammerman and Gavin, 1984) and alkalinization in proximal renal tubule cells (Mellas *et al.*, 1986), stimulation of Na^+/H^+ exchange and inositol triphosphate production in canine kidney cells (Rogers and Hammerman, 1988), increased amino acid uptake in muscle cells (Schimizu *et al.*, 1986), increased glycogen synthesis in hepatoma cells (Hari *et al.*, 1987), proteoglycan synthesis in human chrondosarcoma cells (Takigawa *et al.*, 1997), calcium mobilization in rabbit articular chondrocytes (Poiraudeau *et al.*, 1997), cell motility in rhabdomyosarcoma cells (El-Badry *et al.*, 1990; Minniti *et al.*, 1992), aromatase activity in placenta cytotrophobalsts (Nestler, 1990), migration of human extravillous trophoblasts (McKinnon *et al.*, 2001), cholinergic neuron function in the central nervous system (Hawkes *et al.*, 2006), and insulin exocytosis in pancreatic beta cells (Zhang *et al.*, 1997b). Studies using receptor blocking antibodies have suggested that IGF-2 stimulated calcium influx and DNA synthesis and induced proliferation of EGF primed BALB/c 3T3 cells through IGF-2/M6P receptor (Kojima *et al.*, 1988). Similarly, thymidine incorporation into DNA in undifferentiated mouse embryonic limb buds (Bhaumick and Bala, 1987), cell proliferation of K562 erythroleukemia cells (Tally *et al.*, 1987), and increased gene expression in spermatocytes (Tsuruta *et al.*, 2000) are reportedly IGF-2/M6P receptor-dependent.

For the most part, the pathways that connect the IGF-2/M6P receptor to these biologic responses have not been elucidated. In several cases, responses attributed to the IGF-2/M6P receptor are *Bordetella pertussis* toxin-sensitive, implicating heterotrimeric Gi/o proteins. Putative G protein-dependent signals transmitted by the IGF-2/M6P receptor include PKC-dependent phosphorylation of intracellular proteins (Zhang *et al.*, 1997a), stimulation of MAP kinases and inhibition of adenylate cyclase activity (El-Shewy *et al.*, 2006, 2007; Hawkes *et al.*, 2007; Ikezu *et al.*, 1995; McKinnon *et al.*, 2001; Nishimoto, 1993; Nishimoto *et al.*, 1987). Until recently, the question of how IGF-2 binding to the receptor activates G proteins has been a controversial subject. In a series of papers, Nishimoto and colleagues studied the molecular basis of their original observation that pertussis toxin inhibited IGF-2 stimulated calcium influx in 3T3 mouse embryo fibroblasts (Ikezu *et al.*, 1995; Kojima *et al.*, 1988; Murayama *et al.*, 1990; Okamoto and Nishimoto 1991; Takahashi *et al.*, 1993). Ultimately, they reported that a segment of the cytoplasmic domain of the IGF-2/M6P receptor interacts directly with Gαi to stimulate GTP exchange. However, Körner and colleagues in a followup study failed to confirm these observations and were unable to demonstrate IGF-2/M6P receptor-mediated GTPγS loading of Gαi in reconstituted phospholipid vesicles (Körner *et al.*, 1995). Recent studies performed using HEK293 cells have proposed a distinct mechanism for IGF-2 mediated activation of G protein signaling (Fig. 24.3). El-Shewy and colleagues reported that IGF-2 promotes rapid membrane recruitment and activation of sphingosine kinase, leading to production of extracellular sphingosine 1-phosphate (S1P), the ligand for G protein-coupled S1P receptors (El-Shewy *et al.*, 2006). Using RNA interference, they showed that downregulation of endogenous IGF-2/M6P receptor expression in the continued presence of the IGF-1 receptor was sufficient to eliminate most of the IGF-2 response, directly implicating IGF-2/M6P receptor in the process (El-Shewy *et al.*, 2007). This triple membrane spanning model of sphingosine kinase-dependent S1P receptor transactivation may have general applicability, accounting for reports that a subset of responses to activation of tyrosine kinase and cytokine receptors

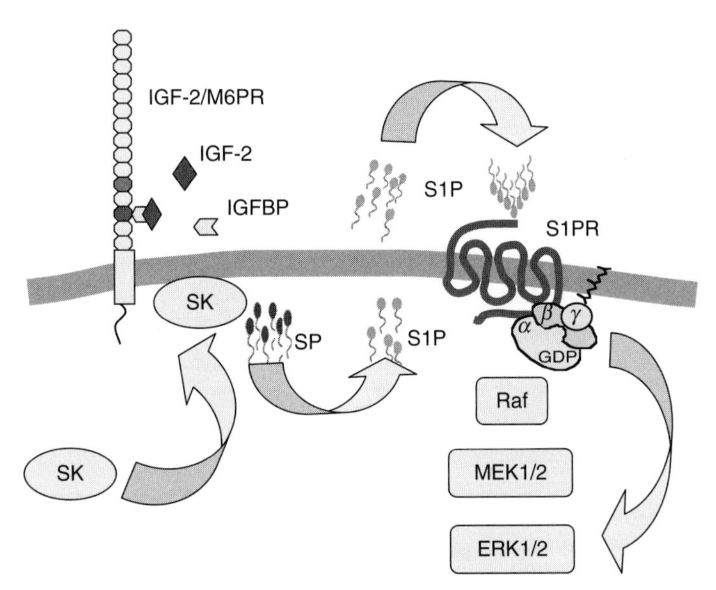

Figure 24.3 IGF-2 induced transactivation of G protein-coupled receptors through sphingosine kinase activation. IGF-2 binding to the IGF-2/M6PR promotes rapid membrane translocation and activation of sphingosine kinase (SK), resulting in production and extracellular release of sphingosine 1-phosphate (S1P) which acts as a ligand for the G protein-coupled S1P group of receptors.

are G protein-dependent (Hobson *et al.*, 2001). Some evidence suggests that M6P containing ligands may also signal in this manner by binding the IGF-2/M6P receptor. In endothelial cells, the chemotactic response to proliferin is both pertussis toxin-sensitive and MAP kinase-dependent (Groskopf *et al.*, 1997). At this point it is unclear whether these responses involve sphingosine kinase activation.

D. Anti-apoptotic effects and role in tumor suppression

It has been suggested that the IGF-2/M6P receptor functions as a tumor suppressor because of its ability to modulate local levels of the mitogen IGF-2 (O'Dell and Day, 1998; Osipo *et al.*, 2001; Wang *et al.*, 1997), to facilitate the activation of TGF-β1 (Dennis and Rifkin, 1991; Ghahary *et al.*, 1999; Yang *et al.*, 2000), modulate circulating LIF levels (Blanchard *et al.*, 1999), and regulate the targeting of lysosomal enzymes to lysosomes (Hille-Rehfeld, 1995; Kornfeld 1992; Le Borgne and Hoflack, 1998). This is supported by evidence that IGF-2 is overexpressed in a number of human cancers (Toretsky and Helman, 1996) and that TGF-β1 inhibits growth of most epithelial cells (Massagué and Wotton, 2000). Furthermore, over-expression of the IGF-2/M6P receptor inhibits cell growth both *in vitro* and *in vivo* (O'Gorman *et al.*, 2002), whereas loss of receptor function is associated with cancer progression (DaCosta *et al.*, 2000; Oates *et al.*, 1998; Osipo *et al.*, 2001). Loss of heterozygosity (LOH) at the IGF-2/M6P receptor gene locus on 6q26-27 has been demonstrated in a number of tumor types, including hepatocellular carcinoma (De Souza *et al.*, 1995a,b; Yamada *et al.*, 1997), breast cancer (Hankins *et al.*, 1996), ovarian cancer (Rey *et al.*, 2000), adrenocortical tumors (Leboulleux *et al.*, 2001), lung carcinoma (Kong *et al.*, 2000), and head and neck tumors (Byrd *et al.*, 1999). In some of these cases, mutations were found in the IGF-2 and M6P binding domains of the remaining allele (De Souza *et al.*, 1995b; Devi *et al.*, 1999; Gemma *et al.*, 2000; Hankins *et al.*, 1996; Yamada *et al.*, 1997). These include single-base deletions in the poly-G region of repeat 9, a target of microsatellite instability in replication/repair error-positive tumors. Microsatellites are oligonucleotide repeat sequences present throughout the human genome, and microsatellite instability is characteristic of disruption of the DNA mismatch repair system. While most microsatellite instability occurs in noncoding DNA, it has been described for the TGF-β1 type II receptor (Markowitz *et al.*, 1995) and IGF-2/M6P receptor (Souza *et al.*, 1996). Microsatellite instability within the IGF-2/M6PR coding region has been reported in gastric, colorectal and endometrial cancers (Ouyang *et al.*, 1997; Souza *et al.*, 1996), generating a frameshift that results in synthesis of a truncated receptor that lacks the transmembrane domain and is presumably secreted as a soluble protein or degraded (Byrd *et al.*, 1999; Devi *et al.*, 1999; Ghosh *et al.*, 2003).

IV. Conclusions

The IGF-2/M6P receptor is a multifunctional protein involved in the uptake, processing, and intracellular trafficking of diverse ligands. It removes IGF-2 from the extracellular space, controlling extracellular ligand concentrations, and modulating signaling via the growth-stimulatory IGF-1 receptor tyrosine kinase. It performs diverse functions related to its binding of M6P containing ligands, participating in the trafficking of lysosomal enzymes and the uptake/processing of cytokines. It also appears to mediate certain effects of IGF-2 directly, including the transactivation of G protein-coupled receptors, which may contribute to IGF-2 effects on cell proliferation and chemotaxis. How such diverse functions came to be incorporated in a single receptor is a matter of speculation. Interestingly, opossum and kangaroo IGF-2/M6P receptors exhibit low affinity for IGF-2 (Dahms *et al.*, 1993a; Yandell *et al.*, 1999), and no significant IGF-2 binding is observed for the IGF-2/M6P receptor from platypus, chicken or frog (Clairmont and Czech, 1989), due to significant differences in the N-terminal portion of domain 11 compared to the mammalian protein (Dahms and Hancock, 2002). This suggests that the regulation of IGF-2 signaling by the IGF-2/M6P receptor is a recent evolutionary development confined to viviparous mammals, while the carbohydrate recognition function of the receptor is widely utilized by mammalian as well as nonmammalian species. Whether the biological processes regulated by the IGF-2/M6P receptor result primarily from its role as a clearance receptor or signal transducer, it is evident that it plays critical roles in the regulation of cell growth, both developmentally and in neoplasia.

References

Adams, T. E., Epa, V. C., Garrett, T. P., and Ward, C. W. (2000). Structure and function of the type 1 insulin-like growth factor receptor. *Cell. Mol. Life Sci.* **57,** 1050–1093.

Baserga, R., Hongo, A., Rubin, M., and Valentinis, B. (1997). The IGF-1 receptor in cell growth, transformation and apoptosis. *Biochem. Biophys. Acta* **1332,** F105–F126.

Bhaumick, B., and Bala, R. M. (1987). Receptors for insulin-like growth factors I and II in developing embryonic mouse limb bud. *Biochem. Biophys. Acta* **927,** 117–128.

Blanchard, F., Duplomb, L., Raher, S., Vusio, P., Hoflack, B., Jacques, Y., and Godard, A. (1999). Mannose 6-Phosphate/Insulin-like growth factor II receptor mediates internalization and degradation of leukemia inhibitory factor but not signal transduction. *J. Biol. Chem.* **274,** 24685–24693.

Bobek, G., Scott, C. D., and Baxter, R. C. (1991). Secretion of soluble insulin-like growth factor-II/mannose 6-phosphate receptor by rat tissues in culture. *Endocrinology* **128,** 2204–2206.

Böker, C., von Figura, K., and Hille-Rehfeld, A. (1997). The carboxy-terminal peptides of 46 kDa and 300 kDa mannose 6-phosphate receptors share partial sequence homology and contain information for sorting in the early endosomal pathway. *J. Cell Sci.* **110,** 1023–1032.

Boman, A. L., Zhang, C., Zhu, X., and Kahn, R. A. (2000). A family of ADP-ribosylation factor effectors that can alter membrane transport through the trans-Golgi. *Mol. Biol Cell.* **11,** 1241–1255.

Braulke, T. (1999). Type-2 IGF receptor: A multiple-ligand binding protein. *Horm. Metab. Res.* **31,** 242–246.

Braulke, T., and Mieskes, G. (1992). Role of protein phosphatases in insulin-like growth factor II (IGF II)-stimulated mannose 6-phosphate/IGF II receptor redistribution. *J. Biol. Chem.* **267,** 17347–17353.

Braulke, T., Tippmer, S., Neher, E., and von Figura, K. (1989). Regulation of the mannos 6-phosphate/IGF II receptor expression at the cell surface by mannose 6-phosphate, insulin-like growth factors and epidermal growth factor. *EMBO J.* **8,** 681–686.

Braulke, T., Tippmer, S., Chao, H. J., and vonFigura, K. (1990). Insulin-like growth factors 1 and II stimulate endocytosis but do not affect sorting of lysosomal enzymes in human fibroblasts. *J. Biol. Chem.* **265,** 6650–6655.

Brown, J., Esnouf, R. M., Jones, M. A., Linnell, J., Harlos, K., Hassan, A. B., and Jones, E. Y. (2002). Structure of a functional IGF2R fragment determined from the anomalous scattering of sulfur. *EMBO J.* **21,** 1054–1062.

Brown, J., Delaine, C., Zaccheo, O. J., Siebold, C., Gilbert, R. J., van Boxel, G., Denley, A., Wallace, J. C., Hassan, A. B., Forbes, B. E., and Jones, E. Y. (2008). Structure and functional analysis of the IGF-II/IGF2R interaction. *EMBO J.* **27,** 265–276.

Bunn, R. C., and Fowlkes, J. L. (2003). Insulin-like growth factor binding protein proteolysis. *Trends Endocrinol. Metab.* **14,** 176–181.

Bürgi, H., Müller, W. A., Humbel, R. E., Labhart, A., and Froesch, E. R. (1966). Non-suppressible insulin-like activity of human serum. I. Physicochemical properties, extraction and partial purification. *Biochim. Biophys. Acta* **121,** 349–359.

Byrd, J. C., and MacDonald, R. G. (2000). Mechanisms for high affinity mannose 6-phosphate ligand binding to the insulin-like growth factor II/mannose 6-phosphate receptor. *J. Biol. Chem.* **275,** 18638–18646.

Byrd, J. C., Devi, G. R., de Souza, A. T., Jirtle, R. L., and MacDonald, R. G. (1999). Disruption of ligand binding to the insulin-like growth factor II/mannose 6-phosphate receptor by cancer-associated missense mutations. *J. Biol. Chem.* **274,** 24408–24416.

Byrd, J. C., Park, J. H., Schaffer, B. S., Garmroudi, F., and MacDonald, R. G. (2000). Dimerization of the insulin-like growth factor II/mannose 6-phosphate receptor. *J. Biol. Chem.* **275,** 18647–18656.

Causin, C., Waheed, A., Braulke, T., Junghans, U., Maly, P., Humbel, R. E., and Von Figura, K. (1988). Mannose 6-phosphate/insulin-like growth factor-II binding proteins in human serum and urine. Their relation to the mannose 6-phosphate/insulin-like growth factor II receptor. *Biochem. J.* **252,** 795–799.

Chen, H. J., Yuan, J., and Lobel, P. (1997). Systematic mutational analysis of the cation-independent mannose 6-phosphate/insulin-like growth factor II receptor cytoplasmic domain. An acidic cluster containing a key aspartate is important for function in lysosomal enzyme sorting. *J. Biol. Chem.* **272,** 7003–7012.

Clairmont, K. B., and Czech, M. P. (1989). Chicken and Xenopus mannose 6-phosphate receptors fail to bind insulin-like growth factor II. *J. Biol. Chem.* **264,** 16390–16392.

Clairmont, K. B., and Czech, M. P. (1991). Extracellular release as the major degradative pathway of the insulin-like growth factor II/mannose 6-phosphate receptor. *J. Biol. Chem.* **266,** 12131–12134.

Corvera, S., Whitehead, R., Mottola, C., and Czech, M. P. (1986). The insulin-like growth factor II receptor is phosphorylated by a tyrosine kinase in adipocyte plasma membranes. *J. Biol. Chem.* **261,** 7675–7679.

Corvera, S., Roach, P. J., DePaoli-Roach, A. A., and Czech, M. P. (1988). Insulin action inhibits insulin-like growth factor-II (IGF-II) receptor phosphorylation in H-35 hepatoma cells. IGF-II receptors isolated from insulin-treated cells exhibit enhanced *in vitro* phosphorylation by casein kinase II. *J. Biol. Chem.* **263,** 3116–3122.

Crawford, S. E., Stellmach, V., Murphy-Ullrich, J. E., Ribeiro, S. M., Lawler, J., Hynes, R. O., Boivin, G. P., and Bouck, N. (1998). Thrombospondin-1 is a major activator of TGF-β1 *in vivo. Cell* **93,** 1159–1170.

DaCosta, S. A., Schumaker, L., and Ellis, M. J. (2000). Mannose 6-phosphate/insulin-like growth factor 2 receptor, a bona fide tumor suppressor gene or just a promising candidate? *J. Mammary Gland Biol. Neoplasia* **5,** 85–94.

Dahms, N. M., and Hancock, M. K. (2002). P-type lectins. *Biochim. Biophys. Acta* **1572,** 317–340.

Dahms, N. M., Lobel, P., and Kornfeld, S. (1989). Mannose 6-phosphate receptors and lysosomal enzyme targeting. *J. Biol. Chem.* **264,** 12115–12118.

Dahms, N. M., Brzycki-Wessell, M. A., Ramanujam, K. S., and Seetharam, B. (1993a). Characterization of mannose 6-phosphate receptors (MPRs) from opossum liver: Opossum cation-independent MPR binds insulin-like growth factor-II. *Endocrinology* **133,** 440–446.

Dahms, N. M., Rose, P. A., Molkentin, J. D., Zhang, Y., and Brzycki, M. A. (1993b). The bovine mannose 6-phosphate/insulin-like growth factor II receptor. The role of arginine residues in mannose 6-phosphate binding. *J. Biol. Chem.* **268,** 5457–5463.

Dahms, N. M., Wick, D. A., and Brzycki-Wessell, M. A. (1994). The bovine mannose 6-phosphate/insulin-like growth factor II receptor. Localization of the insulin-like growth factor II binding site to domains 5–11. *J. Biol. Chem.* **269,** 3802–3809.

Damke, H., vonFigura, K., and Braulke, T. (1992). Simultaneous redistribution of mannose 6-phosphate and transferrin receptors by insulin-like growth factors and phorbol ester. *Biochem. J.* **281,** 225–229.

Daughaday, W. H., and Mariz, I. K. (1962). Conversion of proline U-C14 to labeled hydroxyproline by rat cartilage *in vitro*: Effects of hypophysectomy, growth hormone, and cortisol. *J. Lab. Clin. Med.* **59,** 741–752.

Daughaday, W. H., Hall, K., Raben, M. S., Salmon, W. D. Jr., van den Brande, J. L., and van Wyk, J. J. (1972). Somatomedin: Proposed designation for sulphation factor. *Nature* **235,** 107.

De Leon, D. D., Terry, C., Asmerom, Y., and Nissley, P. (1996). Insulin-like growth factor II modulates the routing of cathepsin D in MCF-7 breast cancer cells. *Endocrinology* **137,** 1851–1859.

De Souza, A. T., Hankins, G. R., Washington, M. K., Fine, R. L., Orton, T. C., and Jirtle, R. L. (1995a). Frequent loss of heterozygosity on 6q at the mannose 6-phosphate/insulin-like growth factor II receptor locus in human hepatocellular tumors. *Oncogene* **10,** 1725–1729.

De Souza, A. T., Hankins, G. R., Washington, M. K., Orton, T. C., and Jirtle, R. L. (1995b). M6P/IGF2R gene is mutated in human hepatocellular carcinomas with loss of heterozygosity. *Nat. Genet.* **11,** 447–449.

Denley, A., Cosgrove, L., Booker, G. W., Wallace, J. C., and Forbes, B. E. (2005). Molecular interactions of the IGF system. *Cytokine Growth Factor Rev.* **16,** 421–439.

Dennis, P. A., and Rifkin, D. B. (1991). Cellular activation of latent transforming growth factor beta requires binding to the cation-independent mannose 6-phosphate/insulin-like growth factor type II receptor. *Proc. Natl. Acad. Sci. USA* **88,** 580–584.

Devi, G. R., Byrd, J. C., Slentz, D. H., and MacDonald, R. G. (1998). An insulin-like growth factor II (IGF-II) affinity-enhancing domain localized within extracytoplasmic repeat 13 of the IGF-II/mannose 6-phosphate receptor. *Mol. Endocrinol.* **12,** 1661–1672.

Devi, G. R., De Souza, A. T., Byrd, J. C., Jirtle, R. L., and MacDonald, R. G. (1999). Altered ligand binding by insulin-like growth factor II/mannose 6-phosphate receptors bearing missense mutations in human cancers. *Cancer Res.* **59,** 14–19.

Duan, C., and Xu, Q. (2005). Roles of insulin-like growth factor (IGF) binding proteins in regulating IGF actions. *Gen. Comp. Endocrinol.* **142,** 44–52.

Dupont, J., and Holzenberger, M. (2003). Biology of insulin-like growth factors in development. *Birth Defects Res. C Embryo Today* **69,** 257–271.

El-Badry, O. M., Minniti, C., Kohn, E. C., Houghton, P. J., Daughaday, W. H., and Helman, L. J. (1990). Insulin-like growth factor II acts as an autocrine growth and motility factor in human rhabdomyosarcoma tumors. *Cell Growth Differ.* **1,** 325–331.

El-Shewy, H. M., Johnson, K. R., Lee, M. H., Jaffa, A. A., Obeid, L. M., and Luttrell, L. M. (2006). Insulin-like growth factors mediate heterotrimeric G protein-dependent ERK1/2 activation by transactivating sphingosine 1-phosphate receptors. *J. Biol. Chem.* **281,** 31399–31407.

El-Shewy, H. M., Lee, M. H., Obeid, L. M., Jaffa, A. A., and Luttrell, L. M. (2007). The insulin-like growth factor type 1 and insulin-like growth factor type 2/mannose-6-phosphate receptors independently regulate ERK1/2 activity in HEK293 cells. *J. Biol. Chem.* **282,** 26150–26157.

Faust, P. L., Chirgwin, J. M., and Kornfeld, S. (1987). Renin, a secretory glycoprotein, acquires phosphomannosyl residues. *J. Cell Biol.* **105,** 1947–1955.

Flaumenhaft, R., Kojima, S., Abe, M., and Rifkin, D. B. (1993). Activation of latent transforming growth factor beta. *Adv. Pharmacol.* **24,** 51–76.

Frasca, F., Pandini, G., Scalia, P., Sciacca, L., Mineo, R., Costantino, A., Goldfine, I. D., Belfiore, A., and Vigneri, R. (1999). Insulin receptor isoform A, a newly recognized, high-affinity insulin-like growth factor II receptor in fetal and cancer cells. *Mol. Cell. Biol.* **19,** 3278–3288.

Froesch, E. R., Bürgi, H., Ramseier, E. B., Bally, P., and Labhart, A. (1963). Antibody-suppressible and nonsuppressible insulin-like activities in human serum and their physiologic significance. *J. Clin. Invest.* **42,** 1816–1834.

Froesch, E. R., Müller, W., Bürgi, H., Waldvogel, M., and Labhart, A. (1966). Nonsuppressible insulin-like activity of human serum.II. Biological properties of plasma extracts with nonsuppressible insulin-like activity. *Biochim. Biophys. Acta* **121,** 360–374.

Galvin, J. P., Spaeny-Dekking, L., Wang, B., Seth, P., Hack, C. E., and Froelich, C. J. (1999). Apoptosis induced by granzyme B-glycosaminoglycan complexes: Implications for granule-mediated apoptosis *in vivo*. *J. Immunol.* **162,** 5345–5350.

Garmroudi, F., and MacDonald, R. G. (1994). Localization of the insulin-like growth factor II (IGF-II) binding/cross-linking site of the IGF-II/mannose 6-phosphate receptor to extracellular repeats 10–11. *J. Biol. Chem.* **269,** 26944–26952.

Garmroudi, F., Devi, G., Slentz, D. H., Schaffer, B. S., and MacDonald, R. G. (1996). Truncated forms of the insulin-like growth factor II (IGF-II)/mannose 6-phosphate receptor encompassing the IGF-II binding site: Characterization of a point mutation that abolishes IGF-II binding. *Mol. Endocrinol.* **10,** 642–651.

Gelato, M. C., Rutherford, C., Stark, R. I., and Daniel, S. S. (1989). The insulin-like growth factor II/mannose-6-phosphate receptor is present in fetal and maternal sheep serum. *Endocrinology* **124,** 2935–2943.

Gemma, A., Hosoya, Y., Uematsu, K., Seike, M., Kurimoto, F., Yoshimura, A., Shibuya, M., and Kudoh, S. (2000). Mutation analysis of the gene encoding the human mannose 6-phosphate/insulin-like growth factor 2 receptor (M6P/IGF2R) in human cell lines resistant to growth inhibition by transforming growth factor beta(1) (TGF-beta(1)). *Lung Cancer* **30,** 91–98.

Ghahary, A., Tredget, E. E., Mi, L., and Yang, L. (1999). Cellular response to latent TGF-beta1 is facilitated by insulin-like growth factor-II/mannose-6-phosphate receptors on MS-9 cells. *Exp. Cell Res.* **25,** 111–120.

Ghosh, P., Dahms, N. M., and Kornfeld, S. (2003). Mannose 6-phosphate receptors: New twist in the tale. *Nat. Rev. Mol. Cell Biol.* **4,** 202–212.

Godár, S., Horejsi, V., Weidle, U. H., Binder, B. R., Hansmann, C., and Stockinger, H. (1999). M6P/IGFII-receptor complexes urokinase receptor and plasminogen for activation of transforming growth factor-beta1. *Eur. J. Immunol.* **29,** 1004–1013.

GrØnborg, M., Wulff, B. S., Rasmussen, J. S., Kjeldsen, T., and Gammeltoft, S. (1993). Structure-function relationship of the insulin-like growth factor-I receptor tyrosine kinase. *J. Biol. Chem.* **268,** 23435–234340.

Groskopf, J. C., Syu, L. J., Saltiel, A. R., and Linzer, D. I. (1997). Proliferin induces endothelial chemotaxis through a G protein-coupled, mitogen-activated protein kinase-dependent pathway. *Endocrinology* **138,** 2835–2840.

Hall, K., and Uthne, I. K. (1971). Some biological properties of purified sulfation factor (SF) from human plasma. *Acta Med. Scand.* **190,** 137–143.

Hammerman, M. R., and Gavin, J. R. III. (1984). Binding of insulin-like growth Factor ii and multiplication-stimulating activity-stimulated phosphorylation in basolateral membranes from dog kidney. *J. Biol. Chem.* **259,** 13511–13517.

Hancock, M. K., Haskins, D. J., Sun, G., and Dahms, N. M. (2002). Identification of residues essential for carbohydrate recognition by the insulin-like growth factor II/mannose 6-phosphate receptor. *J. Biol. Chem.* **277,** 11255–11264.

Hankins, G. R., De Souza, A. T., Bentley, R. C., Patel, M. R., Marks, J. R., Iglehart, J. D., and Jirtle, R. L. (1996). M6P/IGF2 receptor: A candidate breast tumor suppressor gene. *Oncogene* **12,** 2003–2009.

Hari, J., Pierce, S. B., Morgan, D. O., Sara, V., Smith, M. C., and Roth, R. A. (1987). The receptor for insulin-like growth factor II mediates an insulin-like response. *EMBO J.* **6,** 3367–3371.

Hassan, A. (2003). Keys to the hidden treasures of the mannose 6-phosphate/insulin-like growth factor 2 receptor. *Am. J. Pathol.* **162,** 3–6.

Hawkes, C., and Kar, S. (2004). The insulin-like growth factor-II/mannose-6-phosphate receptor: Structure, distribution and function in the central nervous system. *Brain Res. Rev.* **444,** 117–140.

Hawkes, C., Jhamandas, J. H., Harris, S. H., Fu, W., MacDonald, R. G., and Kar, S. (2006). Single transmembrane domain insulin-like growth factor-II/mannose6-phosphate receptor regulates central cholinergic function by activating a G-protein-sensitive, protein kinase C-dependent pathway. *J. Neurosci.* **26,** 585–596.

Hawkes, C., Amritraj, A., Macdonald, R. G., Jhamandas, J. H., and Kar, S. (2007). Heterotrimeric G proteins and the single-transmembrane domain IGF-II/M6P receptor: Functional interaction and relevance to cell signaling. *Mol. Neurobiol.* **35,** 329–345.

Herzog, V., Neumüller, W., and Holzmann, B. (1987). Thyroglobulin, the major and obligatory exportable protein of thyroid follicle cells, carries the lysosomal recognition marker mannose-6-phosphate. *EMBO J.* **6,** 555–560.

Hille-Rehfeld, A. (1995). Mannose 6-phosphate receptors in sorting and transport of lysosomal enzymes. *Biochem. Biophys. Acta* **1241,** 177–194.

Hirst, J., Lindsay, M. R., and Robinson, M. S. (2001). GGAs: Roles of the different domains and comparison with AP-1 and clathrin. *Mol. Biol. Cell.* **12,** 3573–3588.

Hobson, J. P., Rosenfeldt, H. M., Barak, L. S., Olivera, A., Poulton, S., Caron, M. G., Milstien, S., and Spiegel, S. (2001). Role of sphingosine 1-phosphate receptor EDG-1 in PDGF-induced cell motility. *Science* **291,** 1800–1803.

Hoeflich, A., Wolf, E., Braulke, T., Koepf, G., Kessler, U., Brem, G., Rascher, W., Blum, W., and Kiess, W. (1995). Does the overexpression of pro-insulin-like growth factor-II in transfected human embryonic kidney fibroblasts increase the secretion of lysosomal enzymes? *Eur. J. Biochem.* **232,** 172–178.

Hu, K. Q., Backer, J. M., Sahagian, G., Feener, E. P., and King, G. L. (1990). Modulation of the insulin growth factor II/mannose 6-phosphate receptor in microvascular endothelial cells by phorbol ester via protein kinase C. *J. Biol. Chem.* **265,** 13864–13870.

Ikezu, T., Okamoto, T., Giambarella, U., Yokota, T., and Nishimoto, I. (1995). *In vivo* coupling of insulin-like growth factor II/mannose 6-phosphate receptor to heterotrimeric G proteins. Distinct roles of cytoplasmic domains and signal sequestration by the receptor. *J. Biol. Chem.* **270,** 29224–29228.

Iversen, T. G., Skretting, G., Llorente, A., Nicoziani, P., van Deurs, B., and Sandvig, K. (2001). Endosome to Golgi transport of ricin is independent of clathrin and of the Rab9- and Rab11-GTPases. *Mol. Biol. Cell* **12,** 2099–2107.

Johnson, K. F., and Kornfeld, S. (1992). The cytoplasmic tail of the mannose 6-phosphate/insulin-like growth factor-II receptor has two signals for lysosomal enzyme sorting in the Golgi. *J. Cell. Biol.* **119,** 249–257.

Jones, J. I., and Clemmons, D. (1995). Insulin-like growth factors and their binding proteins: Biological actions. *Endocr. Rev.* **16,** 3–34.

Kalscheuer, V. M., Mariman, E. C., Schepens, M. T., Rehder, H., and Ropers, H. H. (1993). The insulin-like growth factor type-2 receptor gene is imprinted in the mouse but not in humans. *Nat. Genet.* **5,** 74–78.

Kang, J. X., Li, Y., and Leaf, A. (1997). Mannose-6-phosphate/insulin-like growth factor-II receptor is a receptor for retinoic acid. *Proc. Natl. Acad. Sci. USA* **94,** 13671–13676.

Kiess, W., Haskell, J. F., Lee, L., Greenstein, L. A., Miller, B. E., Aarons, A. L., Rechler, M. M., and Nissley, S. P. (1987). An antibody that blocks insulin-like growth factor (IGF) binding to the type II IGF receptor is neither an agonist nor an inhibitor of IGF-stimulated biologic responses in L6 myoblasts. *J. Biol. Chem.* **262,** 12745–12751.

Kiess, W., Blickenstaff, G., Sklar, M. M., Thomas, C. L., Nissley, S. P., and Sahagian, G. G. (1988). Biochemical evidence that the type II insulin-like growth factor receptor is identical to the cation-independent mannose 6-phosphate receptor. *J. Biol. Chem.* **263,** 9339–9344.

Kiess, W., Thomas, C. L., Greenstein, L. A., Lee, L., Sklar, M. M., Rechler, M. M., Sahagian, G. G., and Nissley, S. P. (1989a). Insulin-like growth factor-II (IGF-II) inhibits both the cellular uptake of beta-galactosidase and the binding of beta-galactosidase to purified IGF-II/mannose 6-phosphate receptor. *J. Biol. Chem.* **264,** 4710–4714.

Kiess, W., Lee, L., Graham, D. E., Greenstein, L., Tseng, L. Y., Rechler, M. M., and Nissley, S. P. (1989b). Rat C6 glial cells synthesize insulin-like growth factor I (IGF-I) and express IGF-I receptors and IGF-II/mannose 6-phosphate receptors. *Endocrinology* **124,** 1727–1736.

Kiess, W., Yang, Y., Kessler, U., and Hoeflich, A. (1994). Insulin-like growth factor II (IGF-II) and the IGF-II/mannose-6-phosphate receptor: The myth continues. *Horm. Res.* **41,** 66–73.

Killian, J. K., and Jirtle, R. (1999). Genomic structure of the human M6P/IGF2 receptor. *Mamm. Genome* **10,** 74–77.

Kojima, I., Nishimoto, I., Iiri, T., Ogata, E., and Rosenfeld, R. (1988). Evidence that type II insulin-like growth factor receptor is coupled to calcium gating system. *Biochem. Biophys. Res. Commun.* **154,** 9–19.

Kong, F. M., Anscher, M., Washington, M. K., Killian, J. K., and Jirtle, R. L. (2000). M6P/IGF2R is mutated in squamous cell carcinoma of the lung. *Oncogene* **19,** 1572–1578.

Körner, C., Nürnberg, B., Uhde, M., and Braulke, T. (1995). Mannose 6-phosphate/insulin-like growth factor II receptor fails to interact with G-proteins. Analysis of mutant cytoplasmic receptor domains. *J. Biol. Chem.* **270,** 287–295.

Kornfeld, S. (1992). Structure and function of the mannose-6-phosphate/insulin-like growth factor II receptors. *Annu. Rev. Biochem.* **61,** 307–330.
Kovacina, K. S., Steele-Perkins, G., Purchio, A. F., Lioubin, M., Miyazono, K., Heldin, C. H., and Roth, R. A. (1989). Interactions of recombinant and platelet transforming growth factor-beta 1 precursor with the insulin-like growth factor II/mannose 6-phosphate receptor. *Biochem. Biophys. Res. Commun.* **160,** 393–403.
Kreiling, J. L., Byrd, J. C., Deisz, R. J., Mizukami, I. F., Todd, R. F. III., and MacDonald, R. G. (2003). Binding of urokinase-type plasminogen activator receptor (uPAR) to the mannose 6-phosphate/insulin-like growth factor II receptor: Contrasting interactions of full-length and soluble forms of uPAR. *J. Biol. Chem.* **278,** 20628–20637.
Lacasse, J., Ballak, M., Mercure, C., Gutkowska, J., Chapeau, C., Foote, S., Menard, J., Corvol, P., Cantin, M., and Genest, J. (1985). Immunocytochemical localization of renin in Juxtaglomerular cells. *J. Histochem. Cytochem.* **33,** 323–332.
Lau, M. M., Stewart, C. E., Liu, Z., Bhatt, H., Rotwein, P., and Stewart, C. L. (1994). Loss of the imprinted IGF2/cation-independent mannose 6-phosphate receptor results in fetal overgrowth and perinatal lethality. *Genes Dev.* **8,** 2953–2963.
Laureys, G., Barton, D. E., Ullrich, A., and Francke, U. (1988). Chromosomal mapping of the gene for the type II insulin-like growth factor receptor/cation-independent mannose 6-phosphate receptor in man and mouse. *Genomics* **3,** 3224–3229.
Le Borgne, R., and Hoflack, B. (1998). Protein transport from the secretory to the endocytic pathway in mammalian cells. *Biochim. Biophys. Acta* **1404,** 195–209.
Leboulleux, S., Gaston, V., Boulle, N., Le Bouc, Y., and Gicquel, C. (2001). Loss of heterozygosity at the mannose 6-phosphate/insulin-like growth factor 2 receptor locus: A frequent but late event in adrenocortical tumorigenesis. *Eur. J. Endocrinol.* **144,** 163–168.
Lee, S. J., and Nathans, D. (1988). Proliferin secreted by cultured cells binds to mannose 6-phosphate receptors. *J. Biol. Chem.* **263,** 3521–3527.
Lee, S. J., Talamantes, F., Wilder, E., Linzer, D. I., and Nathans, D. (1988). Trophoblastic giant cells of the mouse placenta as the site of proliferin synthesis. *Endocrinology* **122,** 1761–1768.
Leksa, V., Godár, S., Cebecauer, M., Hilgert, I., Breuss, J., Weidle, U. H., Horejsí, V., Binder, B. R., and Stockinger, H. (2002). The N terminus of mannose 6-phosphate/insulin-like growth factor 2 receptor in regulation of fibrinolysis and cell migration. *J. Biol. Chem.* **277,** 40575–40582.
Lelbach, A., Muzes, G., and Feher, J. (2005). The insulin-like growth factor system: IGFs, IGF-binding proteins and IGFBP-proteases. *Acta Physiol. Hung.* **92,** 97–107.
Lemansky, P., and Herzog, V. (1992). Endocytosis of thyroglobulin is not mediated by mannose-6-phosphate receptors in thyrocytes. Evidence for low-affinity-binding sites operating in the uptake of thyroglobulin. *Eur. J. Biochem.* **209,** 111–119.
LeRoith, D., and Roberts, C. J. (2003). The insulin-like growth factor system and cancer. *Cancer Lett.* **195,** 127–137.
LeRoith, D., Warner, H., Beitner-Jonson, D., and Roberts, C. T. Jr. (1995). Molecular and cellular aspects of the insulin-like growth factor 1 receptor. *Endocr. Rev.* **16,** 143–163.
Linnell, J., Groeger, G., and Hassan, A. B. (2001). Real time kinetics of insulin-like growth factor II (IGF-II) interaction with the IGF-II/mannose 6-phosphate receptor: The effects of domain 13 and pH. *J. Biol. Chem.* **276,** 23986–23991.
Linzer, D. I., and Nathans, D. (1984). Nucleotide sequence of a growth-related mRNA encoding a member of the prolactin-growth hormone family. *Proc. Natl. Acad. Sci. USA* **81,** 4255–4259.
Linzer, D. I., and Nathans, D. (1985). A new member of the prolactin-growth hormone gene family expressed in mouse placenta. *EMBO J.* **4,** 1419–1423.

Liu, Z., Mittanck, D. W., Kim, S., and Rotwein, P. (1995). Control of insulin-like growth factor-II/mannose 6-phosphate receptor gene transcription by proximal promoter elements. *Mol. Endocrinol.* **9,** 1477–1487.

Lobel, P., Dahms, N. M., Breitmeyer, J., Chirgwin, J. M., and Kornfeld, S. (1987). Cloning of the bovine 215-kDa cation-independent mannose 6-phosphate receptor. *Proc. Natl. Acad. Sci. USA* **84,** 2233–2237.

Lord, S. J., Rajotta, R. V., Korbutt, G. S., and Bleackley, R. C. (2003). Granzyme B: A natural born killer. *Immunol. Rev.* **193,** 31–38.

Ludwig, T., Tenscher, K., Remmler, J., Hoflack, B., and Lobel, P. (1994). Cloning and sequencing of cDNAs encoding the full-length mouse mannose 6-phosphate/insulin-like growth factor II receptor. *Gene* **142,** 311–312.

Ludwig, T., Eggenschwier, J., Fisher, P., D'Ercole, A. J., Davenport, M. L., and Efstratiadis, A. (1996). Mouse mutants lacking the type 2 IGF receptor (IGF2R) are rescued from perinatal lethality in Igf2 and Igf1r null backgrounds. *Dev. Biol.* **177,** 517–535.

MacDonald, R. G., Tepper, M. A., Clairmont, K. B., Perragaux, S. B., and Czech, M. P. (1989). Serum form of the rat insulin-like growth factor II/mannose 6-phosphate receptor is truncated in the carboxyl terminal domain. *J. Biol. Chem.* **264,** 3256–3261.

Markowitz, S., Wang, J., Myeroff, L., Parsons, R., Sun, L., Lutterbaugh, J., Fan, R. S., Zborowska, E., Kinzler, K. W., Vogelstein, B., Brattain, M., and Willson, J. K. V. (1995). Inactivation of the type II TGF-beta receptor in colon cancer cells with microsatellite instability. *Science* **268,** 1336–1338.

Marron-Terada, P. G., Hancock, M., Haskins, D. J., and Dahms, N. M. (2000). Recognition of Dictyostelium discoideum lysosomal enzymes is conferred by the amino-terminal carbohydrate binding site of the insulin-like growth factor II/mannose 6-phosphate receptor. *Biochemistry* **39,** 2243–2253.

Martinez, D. A., ZusciK, M. J., Ishibe, M., Rosier, R. N., Romano, P. R., Cushing, J. E., and Puzas, J. E. (1995). Identification of functional insulin-like growth factor-II/mannose-6-phosphate receptors in isolated bone cells. *J. Cell Biochem.* **59,** 246–257.

Massagué, J., and Wotton, D. (2000). Transcriptional control by the TGF-beta/Smad signaling system. *EMBO J.* **19,** 1745–1754.

McKinnon, T., Chakraborty, C., Gleeson, L. M., Chidiac, P., and Lala, P. K. (2001). Stimulation of human extravillous trophoblast migration by IGF-II is mediated by IGF type 2 receptor involving inhibitory G protein (S) and phosphorylation of MAPK. *J. Clin. Endocrinol. Metab.* **86,** 3665–3674.

Mellas, J., Gavin, J. R. III., and Hammerman, M. R. (1986). Multiplication-stimulating activity-induced alkalinization of canine renal proximal tubular cells. *J. Biol. Chem.* **261,** 14437–14442.

Méresse, S., and Hoflack, B. (1993). Phosphorylation of the cation-independent mannose 6-phosphate receptor is closely associated with its exit from the trans-Golgi network. *J. Cell Biol.* **120,** 67–75.

Méresse, S., Ludwig, T., Frank, R., and Hoflack, B. (1990). Phosphorylation of the cytoplasmic domain of the bovine cation-independent mannose 6-phosphate receptor. Serines 2421 and 2492 are the targets of a casein kinase II associated to the Golgi-derived HAI adaptor complex. *J. Biol. Chem.* **265,** 18833–18842.

Metkar, S. S., Wang, B., Aguilar-Santelises, M., Raja, S. M., Uhlin-Hansen, L., Podack, E., Trapani, J. A., and Froelich, C. J. (2002). Cytotoxic cell granule-mediated apoptosis: Perforin delivers granzyme B-serglycin complexes into target cells without plasma membrane pore formation. *Immunity* **16,** 417–428.

Minniti, C. P., Kohn, E., Grubb, J. H., Sly, W. S., Oh, Y., Müller, H. L., Rosenfeld, R. G., and Helman, L. J. (1992). The insulin-like growth factor II (IGF-II)/mannose

6-phosphate receptor mediates IGF-II-induced motility in human rhabdomyosarcoma cells. *J. Biol. Chem.* **267,** 9000–9004.

Morgan, D. O., Edman, J., Standring, D. N., Fried, V. A., Smith, M. C., Roth, R. A., and Rutter, W. J. (1987). Insulin-like growth factor II receptor as a multifunctional binding protein. *Nature* **329,** 301–307.

Motyka, B., Korbutt, G., Pinkoski, M. J., Heibein, J. A., Caputo, A., Hobman, M., Barry, M., Shostak, I., Sawchuk, T., Holmes, C. F., Gauldie, J., and Bleackley, R. C. (2000). Mannose 6-phosphate/insulin-like growth factor II receptor is a death receptor for granzyme B during cytotoxic T cell-induced apoptosis. *Cell* **103,** 491–500.

Mullins, C., and Bonifacino, J. S. (2001). The molecular machinery for lysosome biogenesis. *Bioessays* **23,** 333–343.

Murayama, Y., Okamoto, T., Ogata, E., Asano, T., Liri, T., Katada, T., Ui, M., Grubb, J. H., Sly, W. S., and Nishimoto, I. (1990). Distinctive regulation of the functional linkage between the human cation-independent mannose 6-phosphate receptor and GTP-binding proteins by insulin-like growth factor II and mannose 6-phosphate. *J. Biol. Chem.* **265,** 17456–17462.

Nelson, J. T., Rosenzweig, N., and Nilsen-Hamilton, M. (1995). Characterization of the mitogen-regulated protein (proliferin) receptor. *Endocrinology* **136,** 283–288.

Nestler, J. E. (1990). Insulin-like growth factor II is a potent inhibitor of the aromatase activity of human placental cytotrophoblasts. *Endocrinology* **127,** 2064–2070.

Ni, X., Canuel, M., and Morales, C. R. (2006). The sorting and trafficking of lysosomal proteins. *Histol. Histopathol.* **21,** 899–913.

Nilsen-Hamilton, M., Shapiro, J. M., Massoglia, S. L., and Hamilton, R. T. (1980). Selective stimulation by mitogens of incorporation of 35S-methionine into a family of proteins released into the medium by 3T3 cells. *Cell* **20,** 19–28.

Nishimoto, I. (1993). The IGF-II receptor system: A G protein-linked mechanism. *Mol. Reprod. Dev.* **35,** 398–406.

Nishimoto, I., Hata, Y., Ogata, E., and Kojima, I. (1987). Insulin-like growth factor II stimulates calcium influx in competent BALB/c 3T3 cells primed with epidermal growth factor. Characteristics of calcium influx and involvement of GTP-binding protein. *J. Biol. Chem.* **262,** 12120–12126.

Nolan, C. M., Kyle, J. W., Watanabe, H., and Sly, W. S. (1990). Binding of insulin-like growth factor II (IGF-II) by human cation-independent mannose 6-phosphate receptor/IGF-II receptor expressed in receptor-deficient mouse L cells. *Cell Regul.* **1,** 197–213.

O'Dell, S. D., and Day, I. (1998). Molecules in focus; Insulin-like growth factor II (IGF-II). *Int. J. Biochem. Cell Biol.* **30,** 767–771.

O'Gorman, D. B., Weiss, J., Hettiaratchi, A., Firth, S. M., and Scott, C. D. (2002). Insulin-like growth factor-II/mannose 6-phosphate receptor overexpression reduces growth of choriocarcinoma cells *in vitro* and *in vivo*. *Endocrinology* **143,** 4287–4294.

Oates, A. J., Schumaker, L., Jenkins, S. B., Pearce, A. A., DaCosta, S. A., Arun, B., and Ellis, M. J. (1998). The mannose 6-phosphate/insulin-like growth factor 2 receptor (M6P/IGF2R), a putative breast tumor suppressor gene. *Breast Cancer Res. Treat.* **47,** 269–281.

Oka, Y., Mottola, C., Oppenheimer, C. L., and Czech, M. P. (1984). Insulin activates the appearance of insulin-like growth factor II receptors on the adipocyte cell surface. *Proc. Natl. Acad. Sci. USA* **81,** 4028–4032.

Oka, Y., Rozek, L. M., and Czech, M. P. (1985). Direct demonstration of rapid insulin-like growth factor II Receptor internalization and recycling in rat adipocytes. Insulin stimulates 125I-insulin-like growth factor II degradation by modulating the IGF-II receptor recycling process. *J. Biol. Chem.* **260,** 9435–9442.

Okamoto, T., and Nishimoto, I. (1991). Analysis of stimulation-G protein subunit coupling by using active insulin-like growth factor II receptor peptide. *Proc. Natl. Acad. Sci. USA* **88,** 8020–8023.

Olson, L. J., Yammani, R. D., Dahms, N. M., and Kim, J. J. (2004). Structure of uPAR, plasminogen, and sugar-binding sites of the 300 kDa mannose 6-phosphate receptor. *EMBO J.* **23,** 2019–2128.
Oppenheimer, C. L., Pessin, J. E., , M. J., Gitomer, W., and Czech, M. P. (1983). Insulin action rapidly modulates the apparent affinity of the insulin-like growth factor II receptor. *J. Biol. Chem.* **258,** 4824–4830.
Orsel, J. G., Sincock, P. M., Krise, J. P., and Pfeffer, S. R. (2000). Recognition of the 300-kDa mannose 6-phosphate receptor cytoplasmic domain by 47-kDa tail-interacting protein. *Proc. Natl. Acad. Sci. USA* **97,** 9047–9051.
Osipo, C., Dorman, S., and Frankfater, A. (2001). Loss of insulin-like growth factor II receptor expression promotes growth in cancer by increasing intracellular signaling from both IGF-I and insulin receptors. *Exp. Cell Res.* **264,** 388–396.
Ouyang, H., Shiwaku, H., Hagiwara, H., Miura, K., Abe, T., Kato, Y., Ohtani, H., Shiiba, K., Souza, R. F., Meltzer, S. J., and Horii, A. (1997). The insulin-like growth factor II receptor gene is mutated in genetically unstable cancers of the endometrium, stomach, and colorectum. *Cancer Res.* **57,** 1851–1854.
Pearse, B. M., and Robinson, M. S. (1990). Clathrin, adaptors, and sorting. *Annu. Rev. Cell Biol.* **6,** 151–171.
Poe, M., Blake, J., Boulton, D. A., Gammon, M., Sigal, N. H., Wu, J. K., and Zweerink, H. J. (1991). Human cytotoxic lymphocyte granzyme B. Its purification from granules and the characterization of substrate and inhibitor specificity. *J. Biol. Chem.* **266,** 98–103.
Poiraudeau, S., Lieberherr, M., Kergosie, N., and Corvol, M. T. (1997). Different mechanisms are involved in intracellular calcium increase by insulin-like growth factors 1 and 2 in articular chondrocytes: Voltage-gated calcium channels, and/or phospholipase C coupled to a pertussis-sensitive G-protein. *J. Biol. Chem.* **64,** 414–422.
Polychronakos, C., and Piscina, R. (1988). Endocytosis of receptor-bound insulin-like growth factor II is enhanced by mannose-6-phosphate in IM9 cells. *Endocrinology* **123,** 2943–2945.
Polychronakos, C., Guyda, H. J., Janthly, U., and Posner, B. I. (1990). Effects of mannose-6-phosphate on receptor-mediated endocytosis of insulin-like growth factor-II. *Endocrinology* **127,** 1861–1866.
Purchio, A. F., Cooper, J. A., Brunner, A. M., Lioubin, M. N., Gentry, L. E., Kovacina, K. S., Roth, R. A., and Marqardt, H. (1988). Identification of mannose 6-phosphate in two asparagin-linked sugar chains of recombinant transforming growth factor-beta 1 precursor. *J. Biol. Chem.* **263,** 14211–14215.
Randhawa, R., and Cohen, P. (2005). The role of the insulin-like growth factor system in prenatal growth. *Mol. Genet. Metab.* **86,** 84–90.
Reddy, S. T., Chai, W., Childs, R. A., Page, J. D., Feizi, T., and Dahms, N. M. (2004). Identification of a low affinity mannose 6-phosphate-binding site in domain 5 of the cation-independent mannose 6-phosphate receptor. *J. Biol. Chem.* **279,** 38658–38667.
Rey, J. M., Theillet, C., Brouillet, J. P., and Rochefort, H. (2000). Stable amino-acid sequence of the mannose-6-phosphate/insulin-like growth-factor-II receptor in ovarian carcinomas with loss of heterozygosity and in breast-cancer cell lines. *Int. J. Cancer* **85,** 466–473.
Rinderknecht, E., and Humbel, R. (1978a). The amino acid sequence of human insulin-like growth factor I and its structural homology with proinsulin. *J. Biol. Chem.* **253,** 2769–2776.
Rinderknecht, E., and Humbel, R. (1978b). Primary structure of human insulin-like growth factor II. *FEBS Lett.* **89,** 283–286.
Roberts, D. L., Goping, I. S., and Bleackley, R. C. (2003). Mitochondria at the heart of the cytotoxic attack. *Biochem. Biophys. Res. Commun.* **304,** 513–518.

Rogers, S. A., and Hammerman, M. R. (1988). Insulin-like growth factor II stimulates production of inositol trisphosphate in proximal tubular basolateral membranes from canine kidney. *Proc. Natl. Acad. Sci. USA* **85,** 4037–4041.

Rosorius, O., Mieskes, G., Issinger, O. G., Körner, C., Schmidt, B., von Figura, K., and Braulke, T. (1993). Characterization of phosphorylation sites in the cytoplasmic domain of the 300 kDa mannose-6-phosphate receptor. *Biochem. J.* **292,** 833–838.

Sahagian, G. G., and Neufeld, E. F. (1983). Biosynthesis and turnover of the mannose 6-phosphate receptor in cultured Chinese hamster ovary cells. *J. Biol. Chem.* **258,** 7121–7128.

Salmon, W. D., and Daughaday, W. (1957). A hormonally controlled serum factor which stimulates sulfate incorporation by cartilage *in vitro*. *J. Lab. Clin. Med.* **49,** 825–836.

Salmon, W. D., and DuVall, M. R. (1970). A serum fraction with "sulfation factor activity" stimulates in vitro incorporation of leucine and sulfate into protein-polysaccharide complexes, uridine into RNA, and thymidine into DNA of costal cartilage from hypophysectomized rats. *Endocrinology* **86,** 721–727.

Saris, J. J., Derkx, F. H., De Bruin, R. J., Dekkers, D. H., Lamers, J. M., Saxena, P. R., Schalekamp, M. A., and Jan Danser, A. H. (2001). High-affinity prorenin binding to cardiac man-6-P/IGF-II receptors precedes proteolytic activation to renin. *Am. J. Physiol. Heart Circ. Physiol.* **280,** H1706–H1715.

Schimizu, M., Webster, C., Morgan, D. O., Blau, M. H., and Roth, R. A. (1986). Insulin and insulin-like growth factor receptors and responses in cultured human muscle cells. *Am. J. Physiol.* **251,** E611–E615.

Schmidt, B., Kiecke-Siemsen, C., Waheed, A., Braulke, T., and von Figura, K. (1995). Localization of the insulin-like growth factor II binding site to amino acids 1508–1566 in repeat 11 of the mannose 6-phosphate/insulin-like growth factor II receptor. *J. Biol. Chem.* **270,** 14975–14982.

Scott, C. D., and Baxter, R. C. (1996). Regulation of soluble insulin-like growth factor-II/mannose 6-phosphate receptor in hepatocytes from intact and regenerating rat liver. *Endocrinology* **137,** 3864–3870.

Scott, C. D., and Weiss, J. (2000). Soluble insulin-like growth factor II/mannose 6-phosphate receptor inhibits DNA synthesis in insulin-like growth factor II sensitive cells. *J. Cell. Physiol.* **182,** 62–68.

Scott, C. D., Ballesteros, M., Madrid, J., and Baxter, R. C. (1996). Soluble insulin-like growth factor-II/mannose 6-P receptor inhibits deoxyribonucleic acid synthesis in cultured rat hepatocytes. *Endocrinology* **173,** 873–878.

Sleat, D. E., Wang, Y., Sohar, I., Lackland, H., Li, Y., Li, H., Zheng, H., and Lobel, P. (2006). Identification and validation of mannose 6-phosphate glycoproteins in human plasma reveal a wide range of lysosomal and non-lysosomal proteins. *Mol. Cell Proteomics* **5,** 942–956.

Souza, R. F., Appel, R., Yin, J., Wang, S., Smolinski, K. N., Abraham, J. M., Zou, T. T., Shi, Y. Q., Lei, J., Cottrell, J., Cymes, K., Biden, K., *et al.* (1996). Microsatellite instability in the insulin-like growth factor II receptor gene in gastrointestinal tumours. *Nat. Genet.* **14,** 255–257.

Stagsted, J., Olsson, L., Holman, G. D., Cushman, S. W., and Satoh, S. (1993). Inhibition of internalization of glucose transporters and IGF-II receptors. Mechanism of action of MHC class I-derived peptides which augment the insulin response in rat adipose cells. *J. Biol. Chem.* **268,** 22809–22813.

Szebenyi, G., and Rotwein, P. (1994). The mouse insulin-like growth factor II/cation-independent mannose 6-phosphate (IGF-II/MPR) receptor gene: Molecular cloning and genomic organization. *Genomics* **19,** 120–129.

Takahashi, K., Murayama, Y., Okamoto, T., Yokota, T., Ikezu, T., Takahashi, S., Giambarella, U., Ogata, E., and Nishimoto, I. (1993). Conversion of G-protein specificity of insulin-like growth factor II/mannose 6-phosphate receptor by exchanging of a short region with beta-adrenergic receptor. *Proc. Natl. Acad. Sci. USA* **90,** 11772–11776.

Takatsu, H., Katoh, Y., Shiba, Y., and Nakayama, K. (2001). Golgi-localizing, gamma-adaptin ear homology domain, ADP-ribosylation factor-binding (GGA) proteins interact with acidic dileucine sequences within the cytoplasmic domains of sorting receptors through their Vps27p/Hrs/STAM (VHS) domains. *J. Biol. Chem.* **267,** 28541–28545.

Takigawa, M., Okawa, T., Pan, H., Aoki, C., Takahashi, K., Zue, J., Suzuki, F., and Kinoshita, A. (1997). Insulin-like growth factors I and II are autocrine factors in stimulating proteoglycan synthesis, a marker of differentiated chondrocytes, acting through their respective receptors on a clonal human chondrosarcoma-derived chondrocyte cell line, HCS-2/8. *Endocrinology* **138,** 4390–4400.

Tally, M., Li, C. H., and Hall, K. (1987). IGF-2 stimulated growth mediated by the somatomedin type 2 receptor. *Biochem. Biophys. Res. Commun.* **148,** 811–816.

Tong, P. Y., Gregory, W., and Kornfeld, S. (1989). Ligand interactions of the cation-independent mannose 6-phosphate receptor. The stoichiometry of mannose 6-phosphate binding. *J. Biol. Chem.* **264,** 7962–7969.

Toretsky, J. A., and Helman, L. J. (1996). Involvement of IGF-II in human cancer. *J. Endocrinol.* **149,** 367–372.

Trapani, J. A., and Sutton, V. R. (2003). Granzyme B: Pro-apoptotic, antiviral and antitumor functions. *Curr. Opin. Immunol.* **15,** 533–543.

Tsuruta, J. K., Eddy, E. M., and O'Brien, D. A. (2000). Insulin-like growth factor-II/cation-independent mannose 6-phosphate receptor mediates paracrine interactions during spermatogonial development. *Biol. Reprod.* **63,** 1006–1013.

van den Eijnden, J. J., Saris, J. J., de Bruin, R. J., de Wit, E., Sluiter, W., Reudelhuber, T. L., Schalekamp, M. A., Derkx, F. H., and Danser, A. H. (2001). Prorenin accumulation and activation in human endithelial cells: Importance of mannose 6-phosphate receptors. *Arterioscler. Thromb. Vasc. Biol.* **21,** 911–916.

Vasylyeva, T. L., and Ferry, R. J. (2007). Novel roles of the IGF-IGFBP axis in etiopathophysiology of diabetic nephropathy. *Diabetes Res. Clin. Pract.* **76,** 177–186.

Villevalois-Cam, L., Rescan, C., Gilot, D., Ezan, F., Loyer, P., Desbuquois, B., Guguen-Guillouzo, C., and Baffet, G. (2003). The hepatocyte is a direct target for transforming-growth factor beta activation via the insulin-like growth factor II/mannose 6-phosphate receptor. *J. Hepatol.* **38,** 156–163.

Volpert, O., Jackson, D., Bouck, N., and Linzer, D. I. (1996). The insulin-like growth factor II/mannose 6-phosphate receptor is required for proliferin-induced angiogenesis. *Endocrinology* **137,** 2871–3876.

Wang, Z. Q., Fung, M. R., Barlow, D. P., and Wagner, E. F. (1994). Regulation of embryonic growth and lysosomal targeting by the imprinted Igf2/Mpr gene. *Nature* **372,** 464–467.

Wang, S., Souza, R. F., Kong, D., Yin, J., Smolinski, K. N., Zou, T. T., Frank, T., Young, J., Flanders, K. C., Sugimura, H., Abraham, J. M., and Meltzer, S. J. (1997). Deficient transforming growth factor-beta1 activation and excessive insulin-like growth factor II (IGFII) expression in IGFII receptor-mutant tumors. *Cancer Res.* **57,** 2543–2546.

Westcott, K. R., and Rome, L. H. (1988). Cation-independent mannose 6-phosphate receptor contains covalently bound fatty acid. *J. Cell. Biochem.* **38,** 23–33.

Westlund, B., Dahms, N. M., and Kornfeld, S. (1991). The bovine mannose 6-phosphate/insulin-like growth factor II receptor. Localization of mannose 6-phosphate binding sites to domains 1–3 and 7–11 of the extracytoplasmic region. *J. Biol. Chem.* **266,** 23233–23239.

Yamada, T., De Souza, A. T., Finkelstein, S., and Jirtle, R. L. (1997). Loss of the gene encoding mannose 6-phosphate/insulin-like growth factor II receptor is an early event in liver carcinogenesis. *Proc. Natl. Acad. Sci. USA* **94,** 10351–10355.

Yandell, C. A., Dubar, A., Wheldrake, J. F., and Upton, Z. (1999). The kangaroo cation-independent mannose 6-phosphate receptor binds insulin-like growth factor II with low affinity. *J. Biol. Chem.* **274,** 27076–27082.

Yang, L., Tredget, E., and Ghahary, A. (2000). Activation of latent transforming growth factor-beta1 is induced by mannose 6-phosphate/insulin-like growth factor-II receptor. *Wound Repair Regen.* **8,** 538–546.

York, S. J., Amerson, L. S., Gregory, W. T., Dahms, N. M., and Kornfeld, S. (1999). The rate of internalization of the mannose 6-phosphate/insulin-like growth factor II receptor is enhanced by multivalent ligand binding. *J. Biol. Chem.* **274,** 1164–1171.

Zaccheo, O. J., Prince, S. N., Miller, D. M., Williams, C., Kemp, C. F., Brown, J., Jones, E. Y., Catto, L. E., Crump, M. P., and Hassan, A. B. (2006). Kinetics of insulin-like growth factor II (IGF-II) interaction with domain 11 of the human IGF-II/mannose 6-phosphate receptor: Function of CD and AB loop solvent-exposed residues. *J. Mol. Biol.* **359,** 403–421.

Zapf, J., Schoenle, E., and Froesch, E. R. (1978). Insulin-like growth factors I and II: Some biological actions and receptor binding characteristics of two purified constituents of nonsuppressible insulin-like activity of human serum. *Eur. J. Biochem.* **87,** 285–289.

Zhang, Q., Berggen, P. O., and Tally, M. (1997a). Glucose increases both the plasma membrane number and phosphorylation of insulin-like growth factor II/mannose 6-phosphate receptors. *J. Biol. Chem.* **272,** 23703–23706.

Zhang, Q., Tally, M., Larsson, O., Kennedy, R. T., Huang, L., Hall, K., and Berggren, P. O. (1997b). Insulin-like growth factor II signaling through the insulin-like growth factor II/mannose-6-phosphate receptor promotes exocytosis in insulin-secreting cells. *Proc. Natl. Acad. Sci. USA* **94,** 6232–6237.

Zhu, Y., Doray, B., Poussu, A., Lehto, V. P., and Kornfeld, S. (2001). Binding of GGA2 to the lysosomal enzyme sorting motif of the mannose 6-phosphate receptor. *Science* **292,** 1663–1665.

CHAPTER TWENTY-FIVE

Interactions of IGF-II with the IGF2R/Cation-Independent Mannose-6-Phosphate Receptor: Mechanism and Biological Outcomes

J. Brown,* E. Y. Jones,* *and* B. E. Forbes†

Contents

Abstract

The cation-independent mannose-6-phosphate/insulin-like growth factor-II receptor (IGF2R) is a membrane-bound glycoprotein consisting of 15 homologous extracellular repeat domains. The major function of this receptor is trafficking of lysosomal enzymes from the *trans*-Golgi network to the endosomes and their subsequent transfer to lysosomes. The IGF2R also plays a major role in binding and regulating the circulating and tissue levels of IGF-II. As this ligand is important for cell growth, survival, and migration, the maintenance of correct IGF-II levels influences its actions in normal growth and development. Deregulation of IGF2R expression has therefore been associated with growth related disease and cancer. This review highlights recent advances in understanding the IGF2R structure and mechanism of interaction with its ligands, in particular IGF-II. Recent mutagenesis studies combined with the crystal structure of domains 11–14 in complex with IGF-II have mapped the sites of interaction and explain how the IGF2R specificity for IGF-II is achieved. The role of domain 13 in high-affinity

* Cancer Research UK Receptor Structure Research Group, Division of Structural Biology, Wellcome Trust Centre for Human Genetics, University of Oxford, Roosevelt Drive, Headington, Oxford OX3 7BN, United Kingdom
† School of Molecular and Biomedical Science, The University of Adelaide, Adelaide 5005, Australia

Vitamins and Hormones, Volume 80
ISSN 0083-6729, DOI: 10.1016/S0083-6729(08)00625-0

IGF-II binding is also revealed. Characterization of ligand:IGF2R interactions is vital for the understanding of the mechanism of IGF2R actions and will allow the development of specific cancer therapies in the future.

I. Introduction

The insulin-like growth factor-II/cation-independent mannose-6-phosphate receptor (IGF2R) is a 300-kDa transmembrane glycoprotein which is structurally and functionally related to the 46-kDa cation-dependent mannose-6-phosphate receptor (CD-MPR). Together they form the p-type lectin family and play important roles in normal cellular function. Both the IGF2R and CD-MPR are involved in the trafficking of lysosomal enzymes between the *trans*-Golgi network, endosomes, and lysosomes (reviewed by Dahms and Hancock, 2002; Ghosh *et al.*, 2003).

While ~90% of these receptors are found within the cell, both are also found at the cell surface. Under the pH conditions at the cell surface (pH 7.4) the IGF2R, but not the CD-MPR, is able to bind many different mannose-6-phosphate-containing ligands including latent transforming growth factor-β (TGF-β) (Dennis and Rifkin, 1991) and granzyme B (Motyka *et al.*, 2000) as well as nonmannose-6-phosphate-containing peptides including urokinase-type plasminogen activator receptor (uPAR) (Godar *et al.*, 1999), retinoic acid (Kang *et al.*, 1997, 1998), glycosylated leukemia inhibitory factor (LIF) (Blanchard *et al.*, 1998, 1999), and insulin-like growth factor-II (IGF-II) (Tong *et al.*, 1988).

IGF-II plays an important role in normal development and growth. It acts via the type-1 IGF tyrosine kinase receptor (IGF1R) to stimulate cellular proliferation, survival, differentiation, and migration. The IGF2R maintains the correct circulating and local levels of IGF-II, thereby regulating IGF-II's actions in these critical functions (Ghosh *et al.*, 2003; Hawkes and Kar, 2004; Scott and Firth, 2004). Knockout of the *IGF2R* gene in mice results in a lethal phenotype. Embryos develop to term but die at birth due to respiratory problems resulting from lung malformation. Somatic overgrowth and skeletal and cardiac muscle abnormalities are also observed (Lau *et al.*, 1994; Wang *et al.*, 1994). *IGF2R* knockouts are rescued when they carry a second mutation inactivating either the *IGF-II* or *IGF1R* genes (Ludwig *et al.*, 1996). In contrast, tissue-specific knockout in skeletal muscle and liver in adult mice has no effect on phenotype, highlighting the importance of the IGF2R in fetal development (Wylie *et al.*, 2000).

As both IGF-II and the IGF2R are involved in embryonic development it is not surprising that both genes are imprinted. Imprinting is believed to play a role in balancing the needs of the developing fetus with the well-being of the mother and this is particularly important in those mammalian species which are multiparous. Mouse IGF2R is expressed from the maternal allele whereas IGF-II is expressed from the paternal allele (Gicquel and Le Bouc, 2006). However, IGF2R is expressed from both alleles in the mouse central nervous system and in the majority of human tissues (Gicquel and Le Bouc, 2006). As there is relatively little competition in humans between the mother and a single fetus there is no need for such stringent regulation, hence the lack of imprinting of the *IGF2R* gene in humans (Monk *et al.*, 2006).

This review focuses on the molecular interactions between the IGF2R and its ligands, and on the underlying mechanism of action of the IGF2R. The recently reported crystal structure of the IGF2R:IGF-II complex is highlighted (Brown *et al.*, 2008). The resulting biological functions are discussed.

II. The Mechanism of the IGF2R: IGF-II Interaction

A. The IGF2R structure

IGF2R is a type-I transmembrane glycoprotein containing a large N-terminal extracellular region, a single membrane spanning region and a small cytoplasmic tail. Sequence analysis suggests that the extracellular region comprises 15 homologous domains (Lobel *et al.*, 1988) with sequences and cysteine distributions similar to the single extracellular domain of CD-MPR (Roberts *et al.*, 1998). It is therefore likely that each IGF2R domain shares similar structural characteristics, a likelihood strengthened with the publication of domains 1–3 of the bovine protein (Olson *et al.*, 2004b) and domains 11–14 of the human protein (Brown *et al.*, 2008). In addition to illustrating fold conservation, the domain 11–14 structure encompasses the domain 13 fibronectin type-II (FnII) insertion. This is the only significant deviation from the 15 extracellular canonical repeats and comprises a 48-residue insert adopting the typical FnII fold. In the compact arrangement of domains 11–13, this insert projects from domain 13 and nestles close to the IGF-II-binding site of domain 11, interfacing with both domains 11 and 12.

The CD-MPR is present as a homodimer in the membrane and there is some evidence for the IGF2R also forming homodimers (discussed further below). The IGF2R is also found in human serum, urine, and amniotic fluid in a soluble form which results from proteolysis that removes the transmembrane and cytoplasmic domains (Xu *et al.*, 1998). The soluble form maintains high affinity for IGF-II and mannose-6-phosphate-containing proteins (Clairmont and Czech, 1991; Scott and Firth, 2004). The levels of soluble

IGF2R are developmentally regulated with levels highest in infants (5.6 nM) and dropping during adolescence to 3.5 nM (Costello *et al.*, 1999). IGF-II in adults is found at a concentration of 50–100 nM. The soluble form of the IGF2R acts as a carrier protein for IGF-II in the circulation, regulating its bioavailability (Zaina and Squire, 1998), but also plays a role in IGF2R degradation (Clairmont and Czech, 1991).

The cytoplasmic domains of both the IGF2R and the CD-MPR contain recognition motifs for specific binding partners which regulate their internalization and sorting (Ghosh *et al.*, 2003). There is no intrinsic kinase activity within these domains and the receptors are consequently regarded as nonsignaling. The biological functions of the IGF2R can therefore be attributed to its interaction with many binding partners, some of which in turn influence the activity of signaling molecules.

1. IGF-II-binding site on IGF2R

While IGF-II shares over 75% sequence identity with insulin-like growth factor-I (IGF-I) and they are structurally very similar, IGF-I is unable to bind to the IGF2R. The IGF-II-binding site lies in domain 11 (Dahms *et al.*, 1994; Schmidt *et al.*, 1995) with high-affinity binding (10^{-10} M) achieved in the presence of domain 13 (Devi *et al.*, 1998). A point mutation in domain 11 at residue 1572 (Thr to Ile) completely abrogates IGF-II binding (Garmroudi *et al.*, 1996; Linnell *et al.*, 2001). Domain 11 was chosen as the first human IGF2R domain to be structurally characterized because of its importance in IGF-II binding (Brown *et al.*, 2002; Schmidt *et al.*, 1995). As expected, its structure closely resembled that of the previously characterized bovine CD-MPR extracellular domain (Roberts *et al.*, 1998) (see Fig. 25.1). The core IGF2R domain structure is a flattened β-barrel made up of nine β-strands forming two crossed β-sheets. At the N-terminus a β-hairpin caps off the β-barrel; for CD-MPR, this region is α-helical. Four disulfide bonds contribute to fold stability; all domains are predicted to have equivalent disulfides, with the exception of domains 5, 7, and 15, which have only three and domain 13 which has an extra two disulfide bonds in the FnII insert. From analysis of the domain 11 structure, loops at one end of the β-barrel were predicted to form the IGF-II-binding region at a locus spatially equivalent to the carbohydrate-binding pocket of CD-MPR (Brown *et al.*, 2002; Roberts *et al.*, 1998) and this was confirmed later by mutagenesis (Zaccheo *et al.*, 2006).

Details of the mode of IGF-II binding by IGF2R were recently illuminated by the structure of a complex of IGF-II bound to an IGF2R fragment comprising domains 11–13 (Fig. 25.2) (Brown *et al.*, 2008). This structure disproved two previously generated molecular models of the complex (Roche *et al.*, 2006; Williams *et al.*, 2007) but showed, as predicted, that domain 11 does indeed form the primary binding site for IGF-II. While the

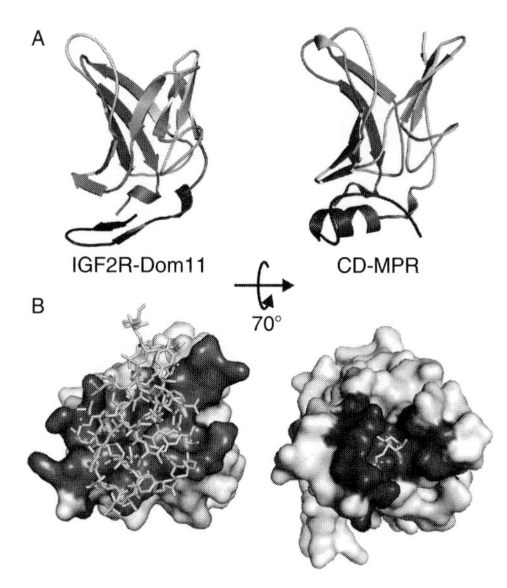

Figure 25.1 Comparison of IGF2R-Dom11 and CD-MPR structures. (A) Ribbon diagrams colored from blue at the N-terminus to red at the C-terminus and based on figures originally published in Brown *et al.* (2002). (B) Surface representations showing residues of the IGF2R domain 11 (left) and CD-MPR (right) undergoing a change in accessible surface area upon ligand binding. IGF-II and mannose-6-phosphate ligands are shown as yellow stick representations bound to their respective receptors. (See Color Insert.)

resolution of the complex structure (4.1 Å) is sufficient to generate a good model of the interaction it is difficult to assign specific side chain contacts between the IGF2R and IGF-II. The structure does confirm that four-domain 11 loops (AB, CD, EF and HI) form the IGF-II-binding site (Brown *et al.*, 2008). Detailed mutagenesis of IGF2R domain 11 residues identified a hydrophobic cluster comprising Tyr1542, Phe1567, and Leu1629 which forms core of the IGF-II-binding site (Zaccheo *et al.*, 2006). The crystal structure reveals that this core surrounds the anchor residue of IGF-II, Phe19 (Delaine *et al.*, 2007).

Surprisingly however, no region of domain 13 (including the fibronectin type-II insert) directly contacts IGF-II. Instead, the enhancing role of the FnII insert appears to be indirect, with the insert maintaining the domain 11 IGF-II-binding site in a high-affinity state prior to ligand binding. In isolated domain 11, a Glu1544Lys mutation in the AB-loop increased affinity for IGF-II by sixfold (Zaccheo *et al.*, 2006). The flexibility of the AB-loop in the isolated domain is probably greater since it does not interface with the domain 13 fibronectin type-II insert. One possible explanation for the enhancement in affinity seen for Glu1544Lys in isolated domain 11 is the introduction of a favorable electrostatic interaction with Asp20 of IGF-II. Domains 11, 12, and 13 thus act in concert to form the

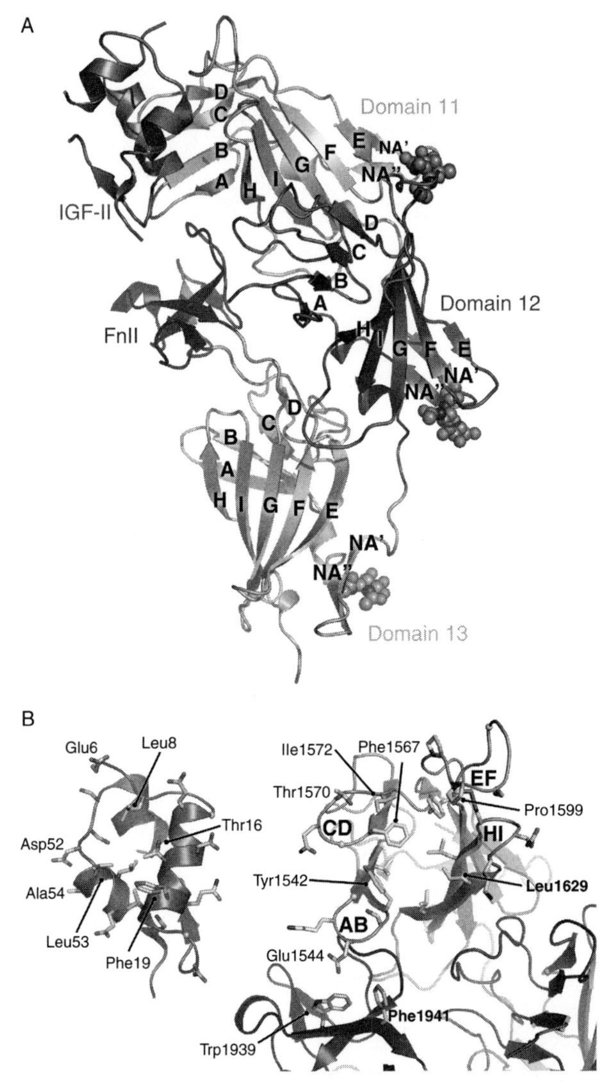

Figure 25.2 The IGF2R/IGF-II interface. (A) Cartoon representation of the complex between IGF2R-Dom11–13 (yellow, red, and blue, respectively) and IGF-II (purple). Glycans are shown as spheres. (B) An open-book view of the complex with side chains undergoing a change in accessible surface area upon complexation shown as sticks. Domains are colored as in (A). Based on figures originally published in Brown *et al.* (2008). (See Color Insert.)

high-affinity IGF-II-binding site with domain 12 acting as a scaffold and potentially playing a role in homodimerization of the IGF2R (discussed later).

2. The IGF2R-binding site on IGF-II

Mutational analyses of IGF-II have identified the residues which play a critical role in IGF2R binding and these include the hydrophobic Leu8, Phe19, and Leu53 along with neighboring residues Glu6, Thr16, and Asp52 (Fig. 25.2) (Brown *et al.*, 2008; Delaine *et al.*, 2007). These residues are buried in the binding interface defined in the crystal structure of the complex (Brown *et al.*, 2008). Interestingly the Phe48, Arg49, and Ser50 motif of IGF-II was originally thought to be a key component in the IGF-II:IGF2R interaction, but is not directly involved in binding to IGF2R (Bach *et al.*, 1993; Brown *et al.*, 2008; Sakano *et al.*, 1991). Other side chains including Ala54 and Leu55 also play a role in IGF2R binding (Brown *et al.*, 2008; Forbes *et al.*, 2001). From sequence comparisons of IGF-I and IGF-II it was previously suggested that specificity of the IGF2R for IGF-II would lie in the regions of most sequence disparity, that is, the C-domain or Ala54 and Leu55 (see Fig. 25.3). However, it turns out that Thr16 (Ala in IGF-I) which is buried in the interface with IGF2R is largely responsible for the different binding affinities of IGF-I and IGF-II for IGF2R (Delaine *et al.*, 2007). Mutation of IGF-II Thr16 to Ala essentially abrogates IGF2R binding, whereas mutation of the equivalent IGF-I Ala13 to Thr introduces the ability to bind the IGF2R, albeit with weak affinity (Delaine *et al.*, 2007).

Interestingly, the IGF2R-binding site on IGF-II involves residues also important for interaction with the N-domain of the IGF-binding proteins (IGFBPs), including residues Glu6 and Phe19 (Sitar *et al.*, 2006). IGFBP-1 to -6 form a family of high-affinity IGF carrier proteins which maintain IGFs in the circulation, deliver them to target tissues, and regulate their availability to bind to the IGF1R and the IGF2R. Nanomolar IGF-binding affinity requires interaction with both the N- and C-terminal domains of the IGFBPs, with the N-domain contributing the most energy to the interaction. It is remarkable that these two high-affinity IGFBP families (the IGFBPs and the IGF2R) have evolved to target the same region of IGF-II to provide tight control of its bioavailability (Brown *et al.*, 2008).

3. Evolution of IGF-II-binding site on IGF2R

Although present across most lineages, the role of IGF2R in controlling IGF-II bioavailability is not universal. Studies have shown that chicken and monotreme IGF2Rs demonstrate no IGF-II binding, whilst marsupial and eutherian IGF2R do (Clairmont and Czech, 1989; Killian *et al.*, 2000; Yandell *et al.*, 1999a). Interestingly, the interaction affinity is lower in marsupials than eutherians, with the affinity of kangaroo IGF2R for kangaroo IGF-II being lower than the equivalent interaction in humans (Yandell *et al.*, 1999a). We now can identify residues in the human IGF2R/IGF-II interface from the crystal structure and look at them in the context of sequence alignments across species including birds, monotremes, marsupials,

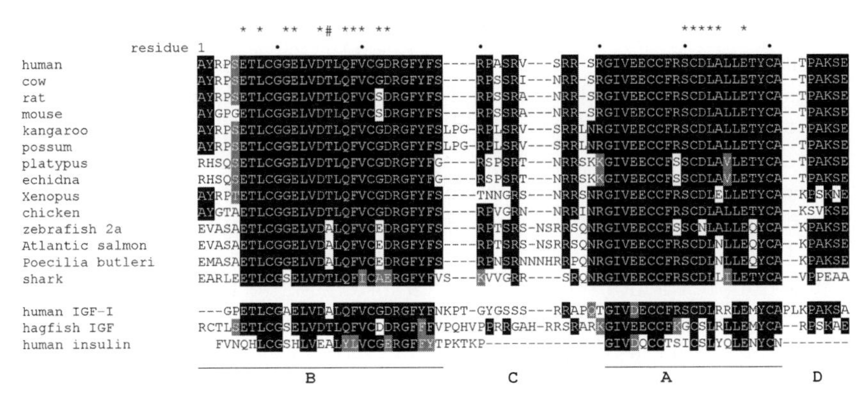

Figure 25.3 Sequence alignment of IGF-II from different species compared with sequences from human IGF-I, insulin, and hagfish IGF. Sequences were aligned in ClustalW and conserved (gray) and identical (black) residues were identified using Boxshade. IGF-II residues buried in the IGF2R:IGF-II interface (a change in accessible surface area of >2 Å^2 upon binding; Brown *et al.*, 2008) are highlighted (*) and Thr16, responsible for the specificity of binding to the IGF2R is marked (#). Every 10th residue of human IGF-II is marked with a (•). The B, C, A, and D domains of IGF-II are highlighted below. Human IGF-I, hagfish IGF, and human insulin sequences are provided to highlight the conservation between the insulin-like peptides. Two live-bearing (shark and *Poecilia butleri*) and one egg-laying (Atlantic salmon) fish are shown to demonstrate that Thr16 is not necessarily associated with placentation in this species. Accession numbers for the IGF-II and insulin-like peptide sequences are human (*Homo sapiens*) NP_00603.1, cow (*Bos taurus*) NP_776512.2, rat (*Rattus norvegicus*) NP_113699.1, mouse (*Mus musculus*) NP_034644.1, kangaroo (Yandell *et al.*, 1999b), possum (*Monodelphis domestica*) ABG22162.1, platypus (*Ornithorhynchus anatinus*) AAT45412.1, echidna (*Tachyglossus aculeatus*) AF339165_1, *Xenopus laevis* NP_001085129.1, chicken NP_001025513.1, zebrafish 2a (*Danio rerio*) XP_001338042.1, Atlantic salmon (*Salmo salar*) ABO36528.1, *Poecilia butleri* ABC60253.1, shark (*Squalus acanthias*) CAA90413.1, human IGF-I AAB21519, hagfish IGF (*Myxine glutinosa*) AAA49265, and human insulin AAA59172.

and eutherians (Fig. 25.4). Such sequence alignments for the IGF2R previously suggested that this evolutionary gain of function is a consequence of only a few key amino acid changes (Killian *et al.*, 2000). We can now verify these key changes; IGF2R residues Tyr1542, 1544, 1569, and 1631 in domain 11 are critical components of the hydrophobic pocket which binds the Phe19 anchor residue of IGF-II, while Glu1544, Gln1569, Ser1600, and Lys1631 also play important roles in IGF-II binding (see Fig. 25.2). These residues are identical in the IGF2R sequence of eutherian species but are divergent in monotremes and chickens. Leu1629 and Ser1600 are also essential for IGF-II binding. Leu1629 is conservatively substituted in monotremes but differs in chickens whereas Ser1600 is different in monotremes but is conservatively substituted in chickens (see Fig. 25.4).

An alignment of IGF-II sequences across species including monotremes, marsupials, eutherians, chicken, and fish is shown in Fig. 25.3. Surprisingly,

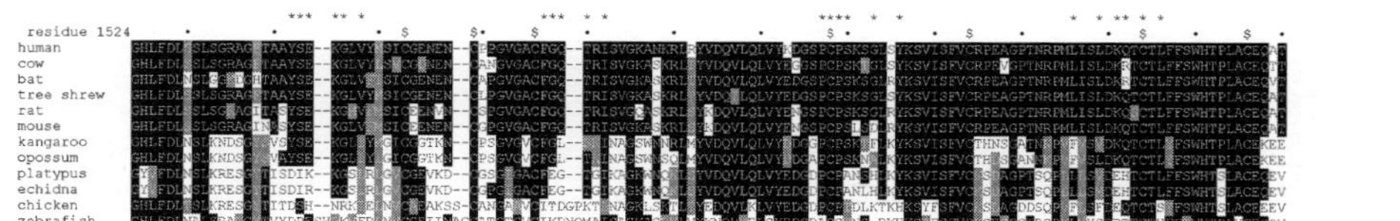

Figure 25.4 Sequence alignment of IGF2R domain 11 residues. Numbering above is based on the human sequence (● = every 10th residue, $ highlights conserved cysteines). Those residues buried in the IGF2R:IGF-II interface (a change in accessible surface area of >2 Å^2 upon binding; Brown *et al.* 2008) are highlighted above (*). Interface residues which are different in both the monotreme and the chicken sequences from the human and are therefore likely to contribute to the lack of IGF-II binding are indicated below (#). Accession numbers for the sequences are human (*Homo sapiens*) NP_000867, cow (*Bos taurus*) P08169, little brown bat (*Myotis lucifugus*) AAK71867.1, common tree shrew (*Tupaia glis*) AAK71869, rat (*Rattus norvegicus*) NP_036888, mouse (*Mus musculus*) NP_034645, kangaroo (*Macropus rufogriseus*) AAK71865, opossum (*Monodelphis domestica*) XP_001371436, platypus (*Ornithorhynchus anatinus*) AAF68173, echidna (*Tachyglossus aculeatus*) AAL23910, chicken (*Gallus gallus*) NP_990301, and zebrafish (*Danio rerio*) NP_001034716.

all residues critical for IGF2R binding including Glu6, Leu8, Phe19, Asp52, and Leu53 are conserved in monotremes and chicken, as is Thr16, the major determinant of the specificity difference between IGF-I and IGF-II. Most fish have Ala at position 16 although sharks have Thr (Fig. 25.3). Therefore, we would predict shark IGF-II to bind to mammalian IGF2R as it has all the critical residues important for this interaction, whereas most fish IGF-IIs would not bind. Conversely, fish IGF2R is not expected to bind mammalian IGF-II as the sequence of zebrafish IGF2R lacks several critical IGF-II-binding residues. Taking the sequence alignments of both the IGF2R and IGF-II into account, it appears that IGF-II has remained unchanged during evolution of the chicken and monotreme species in terms of the residues important for the IGF2R:IGF-II interaction, whereas the IGF2R has gained the ability to bind IGF-II within that period by changing a few key residues within the binding pocket.

4. IGF2R binding of uPAR, plasminogen, TGF-β, and other nonmannose-6-phosphate-containing proteins

Binding sites for uPAR and plasminogen have been located on domain 1 of IGF2R (Kreiling *et al.*, 2003; Leksa *et al.*, 2002; Nykjaer *et al.*, 1998; Olson *et al.*, 2004b). uPAR binds the IGF2R with a stoichiometry of 1:1 with an affinity of 11 μM (Nykjaer *et al.*, 1998) and its binding is not affected by also simultaneously binding IGF-II or latent TGF-β (Godar *et al.*, 1999). The plasminogen interaction can be inhibited by an analogue of lysine (Leksa *et al.*, 2002) and a potential binding site involving three lysine residues has been identified (Olson *et al.*, 2004b). As this site is located on the opposite face of the domains 1–3 structure to the oligosaccharide-binding site, it is possible to simultaneously bind both uPAR/plasminogen and mannose-6-phosphate (Olson *et al.*, 2004b). Latent TGF-β also binds the IGF2R but the location of its binding site has not been defined in detail. Activation of TGF-β is inhibited by mannose-6-phosphate implying that it binds the IGF2R via its mannose-6-phosphate components (Dennis and Rifkin, 1991). A model for the activation of TGF-β has been proposed and involves a complex between the IGF2R and plasminogen, uPAR and TGF-β (Godar *et al.*, 1999). The model suggests that urokinase bound to uPAR converts plasminogen to plasmin. Plasmin then activates latent TGF-β which is also bound to the IGF2R (Dennis and Rifkin, 1991; Godar *et al.*, 1999; Leksa *et al.*, 2005). Active TGF-β then stimulates apoptosis via TGF-β receptors and plays a role in cell migration and invasion (see below).

The binding site for the only other nonmannose-6-phosphate-containing ligand so far shown to bind the IGF2R, retinoic acid, has not yet been defined.

5. IGF2R binding of mannose-6-phosphate-containing proteins

The publication of the crystal structure of domains 1–3 of the bovine IGF2R protein (Olson *et al.*, 2004b) revealed that the core β-barrel/ β-hairpin structure is maintained in these N-terminal domains, with the exception of domain 1 which lacks the β-hairpin (Olson *et al.*, 2004b). Binding of lysosomal enzymes and other mannose-6-phosphate-containing ligands by IGF2R is facilitated by three discrete carbohydrate-binding sites; two high-affinity sites are located in domains 3 and 9 and one low-affinity site in domain 5 (Dahms *et al.*, 1993; Reddy *et al.*, 2004; Westlund *et al.*, 1991). Domain 3 contains a high-affinity carbohydrate-binding site comprising four carbohydrate-binding residues which are conserved in IGF2R domains 5 and 9 and in the CD-MPR (Gln348, Arg391, Glu416, and Tyr421 of the CD-MPR) (Olson *et al.*, 2004a,b; Reddy *et al.*, 2004). A further crucial serine residue (IGF2R Ser386) also participates in a hydrogen-bonding network essential to carbohydrate binding by domain 3 (Olson *et al.*, 2004a). Subtle differences in ligand-binding ability exist between these domains such that domain 3 predominantly binds phosphomonoesters and mannose-6-phosphate at a pH optimum of 6.9–7.0, whereas domain 9 binds phosphodiesters and mannose-6-sulfate at a pH optimum of 6.4–6.5 (Dahms and Hancock, 2002). These domains lack the analogous CD-MPR residue Asp103 which coordinates divalent cations to enhance ligand binding to the CD-MPR (Hancock *et al.*, 2002).

Interestingly, the CD-MPR undergoes considerable structural change upon ligand binding (Olson *et al.*, 2002). There is evidence to suggest that the extracellular region of IGF2R has regions of flexibility (interdomain linkers and protease-susceptible regions), such that the overall structure may also adopt multiple conformations to facilitate binding of diverse mannose-6-phosphate-containing ligands (Olson *et al.*, 2004b). Such flexibility might also contribute to an acidic-release mechanism in lysosomes for stripping IGF2R of these ligands (Olson *et al.*, 2004a). The spatial arrangement of carbohydrate-binding sites coupled with IGF2R flexibility, oligomerization, and glycoprotein heterogeneity is likely to facilitate diverse multivalent interactions between IGF2R and phosphomannosylated ligands.

6. Quaternary structure of IGF2R

Homodimers of IGF2R have been observed to form at the cell surface independent of ligand binding (Byrd *et al.*, 2000; Kreiling *et al.*, 2005), whereas in detergent solutions the most common form is monomeric (York *et al.*, 1999). No clear dimerization regions have been defined, and it appears that multiple interactions occur along the extracellular domains with domain 12 being particularly important (Byrd *et al.*, 2000; Kreiling *et al.*, 2005). Several models for full length IGF2R have been proposed. The first, based on crystal packing data from domain 11, was inconsistent with the domains 1–3 crystal structure (Brown *et al.*, 2002; Olson *et al.*, 2004b). An alternative model based on the domains 1–3 crystal structure was suggested where similar tridomain structural units are repeated through the rest of the extracellular domains. This model, however, proved inconsistent with the domains 11–14 structure which demonstrated that, despite conservation of the core fold, there is no consistency of surface use in interdomain interactions (Brown *et al.*, 2008). Most recently, all of the known IGF2R crystal structures along with biochemical data have been used to suggest a model for IGF2R monomer/dimer (Brown *et al.*, 2008) (Fig. 25.5). The tridomain units equivalent to domains 1–3 are considered as "carbohydrate-binding units" with three copies making up domains 1–9. The domains 11–14 structure, in the form of its crystallographic dimer, forms the basis for an IGF2R dimer. Since isolated bovine domain 5 can dimerize (Reddy *et al.*, 2004), additional dimerization interactions are included between domain 5 from each monomer in the model.

B. The IGF2R function

1. Trafficking of the IGF2R

The major role of the IGF2R is to facilitate the trafficking of lysosomal enzymes between the *trans*-Golgi network, endosomes, and lysosomes (Ghosh *et al.*, 2003). A fraction of newly synthesized lysosomal enzymes escape sorting in the *trans*-Golgi network, are released from the cell, and are recycled back to the lysosomes via the IGF2R (Braulke, 1999; Koster *et al.*,

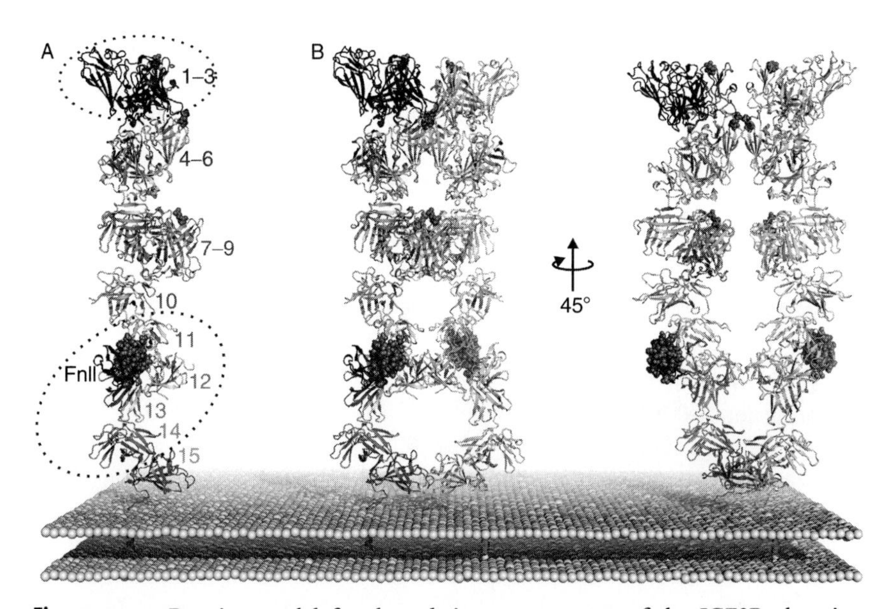

Figure 25.5 Putative model for the relative arrangement of the IGF2R domains. (A) View of an IGF2R monomer, colored from blue at the N-terminus to red at the C-terminus, with FnII in black, IGF-II in magenta and mannose-6-phosphate-binding sites indicated by pink spheres. Dotted ellipses indicate regions of the model for which X-ray crystallography structures have been solved (bovine domains 1–3 and human domains 11–14). (B) Views of a tentative model of an IGF2R dimer, based primarily on crystal packing observations for domains 11–14 and domains 11–13. For each dimer, one monomer is colored as in panel A and the other is colored gray/blue. This figure is from Brown *et al.* (2008).

1994). The IGF2R is also rapidly internalized upon binding of other ligands including IGF-II, granzyme B, uPAR/plasminogen, glycosylated LIF, and retinoids. The recognition motif "YSKV," which binds assembly protein-1 and -2 (AP-1 and -2), is the main mediator of IGF2R internalization and lies in the cytoplasmic tail (Ghosh *et al.*, 2003; Jadot *et al.*, 1992).

Despite their distinct binding sites, mannose-6-phosphate can influence the IGF-II-binding affinity and vice versa (Kiess *et al.*, 1989; MacDonald, 1991). Most of these studies were performed in cell culture and therefore this effect is most likely due to an increase in receptor turnover upon binding of the first ligand which in turn influences the binding of a second ligand. β-glucuronidase can simultaneously bind to more than one IGF2R due to its multivalent nature and receptor crosslinking leads to an increased β-glucuronidase-binding affinity and hence to a higher internalization rate (York *et al.*, 1999). It has been suggested that intermolecular IGF2R binding leads to a structural change in the intracellular internalization signal leading to its optimal recognition via AP-1 and -2 (Ghosh *et al.*, 2003; York *et al.*, 1999).

2. Control of cell growth

The internalization and degradation of IGF-II upon IGF2R binding provides a mechanism to control circulating and tissue levels of this growth factor. IGF-II promotes cell growth, differentiation, and survival acting mainly via the IGF1R or alternatively via the exon 11^- isoform of the insulin receptor (IR-A) (Denley *et al.*, 2003; Scott and Firth, 2004). Thus IGF2R acts as a growth inhibitor by lowering IGF-II bioavailability. IGF-II is particularly important for fetal growth; its overexpression due to a polymorphism disrupting imprinting of the 11p15 region containing the IGF-II gene leads to overgrowth and Beckwith–Wiedemann syndrome (Gicquel and Le Bouc, 2006). A similar overgrowth phenotype is observed in sheep where an imprinting defect of the *Igf2r* gene results in a loss of its expression and a subsequent increase in circulating IGF-II levels (Young *et al.*, 2001). Further studies have shown single nucleotide polymorphisms in the IGF2R are associated with growth control; an increased birth weight is associated with the C.901C>G genotype (results in a Leu252 to Val change) in the Japanese population (Kaku *et al.*, 2007; Killian *et al.*, 2001), and a decreased growth rate in the first 3 years is associated with the Gly1619Arg polymorphism (Petry *et al.*, 2005). Further structure/function analyses may provide a mechanism by which these polymorphisms are affecting growth control. Neither of these residues lies within the IGF-II-binding site but both could be influencing ligand bioavailability by disrupting the overall IGF2R structure.

3. The IGF2R and cancer

The IGF2R gene is regarded as a tumor suppressor gene (Hebert, 2006). It regulates uptake of IGF-II, lysosomal enzymes, glycosylated LIF, and granzyme B (which regulates cytotoxic T cell-induced apoptosis) as well as control of TGF-β activation, all processes requiring tight control to avoid carcinogenesis. Loss of heterozygosity (LOH) of the IGF2R gene and somatic mutations in the remaining allele have been associated with several cancers including primary hepatocellular carcinoma (De Souza *et al.*, 1995; Jang *et al.*, 2008), ovarian cancer (Huang *et al.*, 2006), and prostate cancer (Hu *et al.*, 2006). There is less convincing evidence linking breast cancer with LOH of the *Igf2r* gene but it is frequently mutated in squamous cell carcinoma of the lung (Hebert, 2006). Some of the mutations arising from LOH occur in regions coding for residues within the IGF-II or mannose-6-phosphate-binding sites (Byrd *et al.*, 1999; Devi *et al.*, 1999). In addition, single nucleotide polymorphisms within the *Igf2r* gene associated with growth control have been associated with an increased risk of cancer (Hebert, 2006; Savage *et al.*, 2007) and microsatellite instability within the gene occurs in a large number of gastrointestinal cancers (Souza *et al.*, 1996).

There is substantial experimental evidence to suggest that the IGF2R is a tumor suppressor (Hebert, 2006; Scott and Firth, 2004). For example, overexpression of the IGF2R in breast cancer and choriocarcinoma cells significantly inhibited tumorigenic properties *in vitro* and *in vivo*, leading to fewer tumors and slower growth in nude mice as well as poorer growth and motility *in vitro* (Lee *et al.*, 2003; O'Gorman *et al.*, 2002). For these reasons a modified soluble IGF2R domain 11 is being developed as an IGF-II trap as a potential therapy for IGF-II-producing cancers (Prince *et al.*, 2007).

4. The role of the IGF2R in migration and invasion

Many studies have shown that IGF-II stimulates cell migration and the IGF2R plays a role in this process. Particular interest has been taken in the role of the IGF2R in human extravillous trophoblast (McKinnon *et al.*, 2001) and cancer cell migration and invasion (Gallicchio *et al.*, 2003; Minniti *et al.*, 1992; Scott and Firth, 2004). Using an IGF-II analogue (Tyr27Leu IGF-II) that does not bind the IGF1R, migration of rhabdomyosarcoma cells was stimulated and this was suggested to be an IGF1R-independent effect, possibly via the IGF2R (Gallicchio *et al.*, 2003; Minniti *et al.*, 1992; Scott and Firth, 2004). In each case it is possible that these effects arise by displacement of endogenous IGF-II from the IGF2R which in turn activates the IGF1R or IR-A. Experimentally, breast cancer cell migration can be inhibited by overexpression of the IGF2R, suggesting that IGF2R acts to sequester IGF-II away from the IGF1R or IR-A (Lee *et al.*, 2003).

A role for G protein-coupled receptor signaling associated with the IGF2R has been suggested (Ikezu *et al.*, 1995), particularly in the context of extravillous trophoblast migration (Ikezu *et al.*, 1995; McKinnon *et al.*, 2001). However, others have failed to demonstrate an association of the IGF2R with G protein-coupled receptors (Korner *et al.*, 1995).

Two other mechanisms by which the IGF2R influences cell movement have been identified and are potentially important for trophoblast and cancer cell invasion. Firstly, decreased expression of the IGF2R in cancer may not only lead to increased levels of bioavailable IGF-II (which could act via the IGF1R or the IR-A) but could also deregulate the control of proteolysis at the cell surface, thereby affecting the migration process (Scott and Firth, 2004). Clearance of lysosomal enzymes including the cathepsins influences cell migration and invasion (Mohamed and Sloane, 2006). This is particularly relevant to cancer where increased levels of cathepsins B, D, and L are present in and secreted by many tumor types, which also have decreased IGF2R expression (Hebert, 2006). Cathepsins may be playing a role in degradation of the extracellular matrix, thus assisting tumor cell migration and invasion. Secondly, uPAR/plasminogen activation leads to conversion of latent TGF-β to active TGF-β, an event shown to play a role in motility and chemotaxis (Scott and Firth, 2004). However, there is some contradiction in the literature as to the role of the IGF2R and TGF-β activation in migration as expression of IGF2R in leukocytes decreased plasminogen activation and cell migration (Leivonen and Kahari, 2007; Leksa *et al.*, 2005).

III. Conclusion

Significant advances in our understanding of the IGF2R structure and mechanisms of ligand binding have been made. We now have crystal structures of the mannose-6-phosphate-binding domains 1–3 and domains 11–14 in complex with IGF-II. From these structures and many detailed biochemical analyses a model of the IGF2R quaternary structure has been proposed (Fig. 25.5). Due to the multiligand-binding nature of this receptor it plays important roles in many processes involved in normal growth and development. Deregulation of its expression has been linked to several diseases, and in particular cancer. Complex mechanisms by which the IGF2R regulates these processes have been proposed, whereas perhaps the simple mechanism of receptor internalization and ligand degradation could in fact explain most of IGF2Rs functions. The detailed characterization of the IGF2R:ligand interactions will allow future development of specific tools for dissecting the complex function of the IGF2R. In addition, it will also allow the design of specific modulators of the IGF-II:IGF2R interaction providing new cancer therapies.

REFERENCES

Bach, L. A., Hsieh, S., Sakano, K., Fujiwara, H., Perdue, J. F., and Rechler, M. M. (1993). Binding of mutants of human insulin-like growth factor II to insulin-like growth factor binding proteins 1–6. *J. Biol. Chem.* **268,** 9246–9254.

Blanchard, F., Raher, S., Duplomb, L., Vusio, P., Pitard, V., Taupin, J. L., Moreau, J. F., Hoflack, B., Minvielle, S., Jacques, Y., and Godard, A. (1998). The mannose 6-phosphate/insulin-like growth factor II receptor is a nanomolar affinity receptor for glycosylated human leukemia inhibitory factor. *J. Biol. Chem.* **273,** 20886–20893.

Blanchard, F., Duplomb, L., Raher, S., Vusio, P., Hoflack, B., Jacques, Y., and Godard, A. (1999). Mannose 6-phosphate/insulin-like growth factor II receptor mediates internalization and degradation of leukemia inhibitory factor but not signal transduction. *J. Biol. Chem.* **274,** 24685–24693.

Braulke, T. (1999). Type-2 IGF receptor: A multi-ligand binding protein. *Horm. Metab. Res.* **31,** 242–246.

Brown, J., Esnouf, R. M., Jones, M. A., Linnell, J., Harlos, K., Hassan, A. B., and Jones, E. Y. (2002). Structure of a functional IGF2R fragment determined from the anomalous scattering of sulfur. *EMBO J.* **21,** 1054–1062.

Brown, J., Delaine, C., Zaccheo, O. J., Siebold, C., Gilbert, R. J., van Boxel, G., Denley, A., Wallace, J. C., Hassan, A. B., Forbes, B. E., and Jones, E. Y. (2008). Structure and functional analysis of the IGF-II/IGF2R interaction. *EMBO J.* **27,** 265–276.

Byrd, J. C., Devi, G. R., de Souza, A. T., Jirtle, R. L., and MacDonald, R. G. (1999). Disruption of ligand binding to the insulin-like growth factor II/mannose 6-phosphate receptor by cancer-associated missense mutations. *J. Biol. Chem.* **274,** 24408–24416.

Byrd, J. C., Park, J. H., Schaffer, B. S., Garmroudi, F., and MacDonald, R. G. (2000). Dimerization of the insulin-like growth factor II/mannose 6-phosphate receptor. *J. Biol. Chem.* **275,** 18647–18656.

Clairmont, K. B., and Czech, M. P. (1989). Chicken and Xenopus mannose 6-phosphate receptors fail to bind insulin-like growth factor II. *J. Biol. Chem.* **264,** 16390–16392.

Clairmont, K. B., and Czech, M. P. (1991). Extracellular release as the major degradative pathway of the insulin-like growth factor II/mannose 6-phosphate receptor. *J. Biol. Chem.* **266,** 12131–12134.

Costello, M., Baxter, R. C., and Scott, C. D. (1999). Regulation of soluble insulin-like growth factor II/mannose 6-phosphate receptor in human serum: Measurement by enzyme-linked immunosorbent assay. *J. Clin. Endocrinol. Metab.* **84,** 611–617.

Dahms, N. M., and Hancock, M. K. (2002). P-type lectins. *Biochim. Biophys. Acta* **19,** 2–3.

Dahms, N. M., Rose, P. A., Molkentin, J. D., Zhang, Y., and Brzycki, M. A. (1993). The bovine mannose 6-phosphate/insulin-like growth factor II receptor. The role of arginine residues in mannose 6-phosphate binding. *J. Biol. Chem.* **268,** 5457–5463.

Dahms, N. M., Wick, D. A., and Brzycki-Wessell, M. A. (1994). The bovine mannose 6-phosphate/insulin-like growth factor II receptor. Localization of the insulin-like growth factor II binding site to domains 5–11. *J. Biol. Chem.* **269,** 3802–3809.

De Souza, A. T., Hankins, G. R., Washington, M. K., Orton, T. C., and Jirtle, R. L. (1995). M6P/IGF2R gene is mutated in human hepatocellular carcinomas with loss of heterozygosity. *Nat. Genet.* **11,** 447–449.

Delaine, C., Alvino, C. L., McNeil, K. A., Mulhern, T. D., Gauguin, L., De Meyts, P., Jones, E. Y., Brown, J., Wallace, J. C., and Forbes, B. E. (2007). A novel binding site for the human insulin-like growth factor-II (IGF-II)/mannose 6-phosphate receptor on IGF-II. *J. Biol. Chem.* **282,** 18886–18894.

Denley, A., Wallace, J. C., Cosgrove, L. J., and Forbes, B. E. (2003). The insulin receptor isoform exon 11− (IR-A) in cancer and other diseases: A review. *Horm. Metab. Res.* **35,** 778–785.

Dennis, P. A., and Rifkin, D. B. (1991). Cellular activation of latent transforming growth factor beta requires binding to the cation-independent mannose 6-phosphate/insulin-like growth factor type II receptor. *Proc. Natl. Acad. Sci. USA* **88,** 580–584.

Devi, G. R., Byrd, J. C., Slentz, D. H., and MacDonald, R. G. (1998). An insulin-like growth factor II (IGF-II) affinity-enhancing domain localized within extracytoplasmic repeat 13 of the IGF-II/mannose 6-phosphate receptor. *Mol. Endocrinol.* **12,** 1661–1672.

Devi, G. R., De Souza, A. T., Byrd, J. C., Jirtle, R. L., and MacDonald, R. G. (1999). Altered ligand binding by insulin-like growth factor II/mannose 6-phosphate receptors bearing missense mutations in human cancers. *Cancer Res.* **59,** 4314–4319.

Forbes, B. E., McNeil, K. A., Scott, C. D., Surinya, K. H., Cosgrove, L. J., and Wallace, J. C. (2001). Contribution of residues A54 and L55 of the human insulin-like growth factor-II (IGF-II) A domain to Type 2 IGF receptor binding specificity. *Growth Factors* **19,** 163–173.

Gallicchio, M. A., Kaun, C., Wojta, J., Binder, B., and Bach, L. A. (2003). Urokinase type plasminogen activator receptor is involved in insulin-like growth factor-induced migration of rhabdomyosarcoma cells *in vitro*. *J. Cell. Physiol.* **197,** 131–138.

Garmroudi, F., Devi, G., Slentz, D. H., Schaffer, B. S., and MacDonald, R. G. (1996). Truncated forms of the insulin-like growth factor II (IGF-II)/mannose 6-phosphate receptor encompassing the IGF-II binding site: Characterization of a point mutation that abolishes IGF-II binding. *Mol. Endocrinol.* **10,** 642–651.

Ghosh, P., Dahms, N. M., and Kornfeld, S. (2003). Mannose 6-phosphate receptors: New twists in the tale. *Nat. Rev. Mol. Cell Biol.* **4,** 202–212.

Gicquel, C., and Le Bouc, Y. (2006). Hormonal regulation of fetal growth. *Horm. Res.* **3,** 28–33.

Godar, S., Horejsi, V., Weidle, U. H., Binder, B. R., Hansmann, C., and Stockinger, H. (1999). M6P/IGFII-receptor complexes urokinase receptor and plasminogen for activation of transforming growth factor-beta1. *Eur. J. Immunol.* **29,** 1004–1013.

Hancock, M. K., Haskins, D. J., Sun, G., and Dahms, N. M. (2002). Identification of residues essential for carbohydrate recognition by the insulin-like growth factor II/mannose 6-phosphate receptor. *J. Biol. Chem.* **277,** 11255–11264.

Hawkes, C., and Kar, S. (2004). The insulin-like growth factor-II/mannose-6-phosphate receptor: Structure, distribution and function in the central nervous system. *Brain Res. Brain Res. Rev.* **44,** 117–140.

Hebert, E. (2006). Mannose-6-phosphate/insulin-like growth factor II receptor expression and tumor development. *Biosci. Rep.* **26,** 7–17.

Hu, C. K., McCall, S., Madden, J., Huang, H., Clough, R., Jirtle, R. L., and Anscher, M. S. (2006). Loss of heterozygosity of M6P/IGF2R gene is an early event in the development of prostate cancer. *Prostate Cancer Prostatic Dis.* **9,** 62–67.

Huang, Z., Wen, Y., Shandilya, R., Marks, J. R., Berchuck, A., and Murphy, S. K. (2006). High throughput detection of M6P/IGF2R intronic hypermethylation and LOH in ovarian cancer. *Nucleic Acids Res.* **34,** 555–563.

Ikezu, T., Okamoto, T., Giambarella, U., Yokota, T., and Nishimoto, I. (1995). *In vivo* coupling of insulin-like growth factor II/mannose 6-phosphate receptor to heteromeric G proteins. Distinct roles of cytoplasmic domains and signal sequestration by the receptor. *J. Biol. Chem.* **270,** 29224–29228.

Jadot, M., Canfield, W. M., Gregory, W., and Kornfeld, S. (1992). Characterization of the signal for rapid internalization of the bovine mannose 6-phosphate/insulin-like growth factor-II receptor. *J. Biol. Chem.* **267,** 11069–11077.

Jang, H. S., Kang, K. M., Choi, B. O., Chai, G. Y., Hong, S. C., Ha, W. S., and Jirtle, R. L. (2008). Clinical significance of loss of heterozygosity for M6P/IGF2R in patients with primary hepatocellular carcinoma. *World J. Gastroenterol.* **14,** 1394–1398.

Kaku, K., Osada, H., Seki, K., and Sekiya, S. (2007). Insulin-like growth factor 2 (IGF2) and IGF2 receptor gene variants are associated with fetal growth. *Acta Paediatr.* **96,** 363–367.

Kang, J. X., Li, Y., and Leaf, A. (1997). Mannose-6-phosphate/insulin-like growth factor-II receptor is a receptor for retinoic acid. *Proc. Natl. Acad. Sci. USA* **94,** 13671–13676.

Kang, J. X., Bell, J., Leaf, A., Beard, R. L., and Chandraratna, R. A. (1998). Retinoic acid alters the intracellular trafficking of the mannose-6-phosphate/insulin-like growth factor II receptor and lysosomal enzymes. *Proc. Natl. Acad. Sci. USA* **95,** 13687–13691.

Kiess, W., Thomas, C. L., Greenstein, L. A., Lee, L., Sklar, M. M., Rechler, M. M., Sahagian, G. G., and Nissley, S. P. (1989). Insulin-like growth factor-II (IGF-II) inhibits both the cellular uptake of beta-galactosidase and the binding of beta-galactosidase to purified IGF-II/mannose 6-phosphate receptor. *J. Biol. Chem.* **264,** 4710–4714.

Killian, J. K., Byrd, J. C., Jirtle, J. V., Munday, B. L., Stoskopf, M. K., MacDonald, R. G., and Jirtle, R. L. (2000). M6P/IGF2R imprinting evolution in mammals. *Mol. Cell* **5,** 707–716.

Killian, J. K., Oka, Y., Jang, H. S., Fu, X., Waterland, R. A., Sohda, T., Sakaguchi, S., and Jirtle, R. L. (2001). Mannose 6-phosphate/insulin-like growth factor 2 receptor (M6P/IGF2R) variants in American and Japanese populations. *Hum. Mutat.* **18,** 25–31.

Korner, C., Nurnberg, B., Uhde, M., and Braulke, T. (1995). Mannose 6-phosphate/insulin-like growth factor II receptor fails to interact with G-proteins. Analysis of mutant cytoplasmic receptor domains. *J. Biol. Chem.* **270,** 287–295.

Koster, A., von Figura, K., and Pohlmann, R. (1994). Mistargeting of lysosomal enzymes in M(r) 46,000 mannose 6-phosphate receptor-deficient mice is compensated by carbohydrate-specific endocytotic receptors. *Eur. J. Biochem.* **224,** 685–689.

Kreiling, J. L., Byrd, J. C., Deisz, R. J., Mizukami, I. F., Todd, R. F. III, and MacDonald, R. G. (2003). Binding of urokinase-type plasminogen activator receptor (uPAR) to the mannose 6-phosphate/insulin-like growth factor II receptor: Contrasting interactions of full-length and soluble forms of uPAR. *J. Biol. Chem.* **278,** 20628–20637.

Kreiling, J. L., Byrd, J. C., and MacDonald, R. G. (2005). Domain interactions of the mannose 6-phosphate/insulin-like growth factor II receptor. *J. Biol. Chem.* **280,** 21067–21077.

Lau, M. M., Stewart, C. E., Liu, Z., Bhatt, H., Rotwein, P., and Stewart, C. L. (1994). Loss of the imprinted IGF2/cation-independent mannose 6-phosphate receptor results in fetal overgrowth and perinatal lethality. *Genes Dev.* **8,** 2953–2963.

Lee, J. S., Weiss, J., Martin, J. L., and Scott, C. D. (2003). Increased expression of the mannose 6-phosphate/insulin-like growth factor-II receptor in breast cancer cells alters tumorigenic properties *in vitro* and *in vivo*. *Int. J. Cancer* **107,** 564–570.

Leivonen, S. K., and Kahari, V. M. (2007). Transforming growth factor-beta signaling in cancer invasion and metastasis. *Int. J. Cancer* **121,** 2119–2124.

Leksa, V., Godar, S., Cebecauer, M., Hilgert, I., Breuss, J., Weidle, U. H., Horejsi, V., Binder, B. R., and Stockinger, H. (2002). The N terminus of mannose 6-phosphate/insulin-like growth factor 2 receptor in regulation of fibrinolysis and cell migration. *J. Biol. Chem.* **277,** 40575–40582.

Leksa, V., Godar, S., Schiller, H. B., Fuertbauer, E., Muhammad, A., Slezakova, K., Horejsi, V., Steinlein, P., Weidle, U. H., Binder, B. R., and Stockinger, H. (2005). TGF-beta-induced apoptosis in endothelial cells mediated by M6P/IGFII-R and mini-plasminogen. *J. Cell Sci.* **118,** 4577–4586.

Linnell, J., Groeger, G., and Hassan, A. B. (2001). Real time kinetics of insulin-like growth factor II (IGF-II) interaction with the IGF-II/mannose 6-phosphate receptor: The effects of domain 13 and pH. *J. Biol. Chem.* **276,** 23986–23991.

Lobel, P., Dahms, N. M., and Kornfeld, S. (1988). Cloning and sequence analysis of the cation-independent mannose 6-phosphate receptor. *J. Biol. Chem.* **263,** 2563–2570.

Ludwig, T., Eggenschwiler, J., Fisher, P., D'Ercole, A. J., Davenport, M. L., and Efstratiadis, A. (1996). Mouse mutants lacking the type 2 IGF receptor (IGF2R) are rescued from perinatal lethality in Igf2 and Igf1r null backgrounds. *Dev. Biol.* **177,** 517–535.

MacDonald, R. G. (1991). Mannose-6-phosphate enhances cross-linking efficiency between insulin-like growth factor-II (IGF-II) and IGF-II/mannose-6-phosphate receptors in membranes. *Endocrinology* **128,** 413–421.

McKinnon, T., Chakraborty, C., Gleeson, L. M., Chidiac, P., and Lala, P. K. (2001). Stimulation of human extravillous trophoblast migration by IGF-II is mediated by IGF type 2 receptor involving inhibitory G protein(s) and phosphorylation of MAPK. *J. Clin. Endocrinol. Metab.* **86,** 3665–3674.

Minniti, C. P., Kohn, E. C., Grubb, J. H., Sly, W. S., Oh, Y., Muller, H. L., Rosenfeld, R. G., and Helman, L. J. (1992). The insulin-like growth factor II (IGF-II)/mannose 6-phosphate receptor mediates IGF-II-induced motility in human rhabdomyosarcoma cells. *J. Biol. Chem.* **267,** 9000–9004.

Mohamed, M. M., and Sloane, B. F. (2006). Cysteine cathepsins: Multifunctional enzymes in cancer. *Nat. Rev. Cancer* **6,** 764–775.

Monk, D., Arnaud, P., Apostolidou, S., Hills, F. A., Kelsey, G., Stanier, P., Feil, R., and Moore, G. E. (2006). Limited evolutionary conservation of imprinting in the human placenta. *Proc. Natl. Acad. Sci. USA* **103,** 6623–6628.

Motyka, B., Korbutt, G., Pinkoski, M. J., Heibein, J. A., Caputo, A., Hobman, M., Barry, M., Shostak, I., Sawchuk, T., Holmes, C. F., Gauldie, J., and Bleackley, R. C. (2000). Mannose 6-phosphate/insulin-like growth factor II receptor is a death receptor for granzyme B during cytotoxic T cell-induced apoptosis. *Cell* **103,** 491–500.

Nykjaer, A., Christensen, E. I., Vorum, H., Hager, H., Petersen, C. M., Roigaard, H., Min, H. Y., Vilhardt, F., Moller, L. B., Kornfeld, S., and Gliemann, J. (1998). Mannose 6-phosphate/insulin-like growth factor-II receptor targets the urokinase receptor to lysosomes via a novel binding interaction. *J. Cell Biol.* **141,** 815–828.

O'Gorman, D. B., Weiss, J., Hettiaratchi, A., Firth, S. M., and Scott, C. D. (2002). Insulin-like growth factor-II/mannose 6-phosphate receptor overexpression reduces growth of choriocarcinoma cells *in vitro* and *in vivo*. *Endocrinology* **143,** 4287–4294.

Olson, L. J., Zhang, J., Dahms, N. M., and Kim, J. J. (2002). Twists and turns of the cation-dependent mannose 6-phosphate receptor. Ligand-bound versus ligand-free receptor. *J. Biol. Chem.* **277,** 10156–10161.

Olson, L. J., Dahms, N. M., and Kim, J. J. (2004a). The N-terminal carbohydrate recognition site of the cation-independent mannose 6-phosphate receptor. *J. Biol. Chem.* **279,** 34000–34009.

Olson, L. J., Yammani, R. D., Dahms, N. M., and Kim, J. J. (2004b). Structure of uPAR, plasminogen, and sugar-binding sites of the 300 kDa mannose 6-phosphate receptor. *EMBO J.* **23,** 2019–2028.

Petry, C. J., Ong, K. K., Wingate, D. L., Brown, J., Scott, C. D., Jones, E. Y., Pembrey, M. E., and Dunger, D. B. (2005). Genetic variation in the type 2 insulin-like growth factor receptor gene and disparity in childhood height. *Growth Horm. IGF Res.* **15,** 363–368.

Prince, S. N., Foulstone, E. J., Zaccheo, O. J., Williams, C., and Hassan, A. B. (2007). Functional evaluation of novel soluble insulin-like growth factor (IGF)-II-specific ligand traps based on modified domain 11 of the human IGF2 receptor. *Mol. Cancer Ther.* **6,** 607–617.

Reddy, S. T., Chai, W., Childs, R. A., Page, J. D., Feizi, T., and Dahms, N. M. (2004). Identification of a low affinity mannose 6-phosphate-binding site in domain 5 of the cation-independent mannose 6-phosphate receptor. *J. Biol. Chem.* **279,** 38658–38667.

Roberts, D. L., Weix, D. J., Dahms, N. M., and Kim, J. J. (1998). Molecular basis of lysosomal enzyme recognition: Three-dimensional structure of the cation-dependent mannose 6-phosphate receptor. *Cell* **93,** 639–648.

Roche, P., Brown, J., Denley, A., Forbes, B. E., Wallace, J. C., Jones, E. Y., and Esnouf, R. M. (2006). Computational model for the IGF-II/IGF2r complex that is predictive of mutational and surface plasmon resonance data. *Proteins* **64,** 758–768.

Sakano, K., Enjoh, T., Numata, F., Fujiwara, H., Marumoto, Y., Higashihashi, N., Sato, Y., Perdue, J. F., and Fujita-Yamaguchi, Y. (1991). The design, expression, and characterization of human insulin-like growth factor II (IGF-II) mutants specific for either the IGF-II/cation-independent mannose 6-phosphate receptor or IGF-I receptor. *J. Biol. Chem.* **266,** 20626–20635.

Savage, S. A., Woodson, K., Walk, E., Modi, W., Liao, J., Douglass, C., Hoover, R. N., and Chanock, S. J. (2007). Analysis of genes critical for growth regulation identifies insulin-like growth factor 2 receptor variations with possible functional significance as risk factors for osteosarcoma. *Cancer Epidemiol. Biomarkers Prev.* **16,** 1667–1674.

Schmidt, B., Kiecke-Siemsen, C., Waheed, A., Braulke, T., and von Figura, K. (1995). Localization of the insulin-like growth factor II binding site to amino acids 1508–1566 in repeat 11 of the mannose 6-phosphate/insulin-like growth factor II receptor. *J. Biol. Chem.* **270,** 14975–14982.

Scott, C. D., and Firth, S. M. (2004). The role of the M6P/IGF-II receptor in cancer: Tumor suppression or garbage disposal? *Horm. Metab. Res.* **36,** 261–271.

Sitar, T., Popowicz, G. M., Siwanowicz, I., Huber, R., and Holak, T. A. (2006). Structural basis for the inhibition of insulin-like growth factors by insulin-like growth factor-binding proteins. *Proc. Natl. Acad. Sci. USA* **103,** 13028–13033.

Souza, R. F., Appel, R., Yin, J., Wang, S., Smolinski, K. N., Abraham, J. M., Zou, T. T., Shi, Y. Q., Lei, J., Cottrell, J., Cymes, K., Biden, K., *et al.* (1996). Microsatellite instability in the insulin-like growth factor II receptor gene in gastrointestinal tumours. *Nat. Genet.* **14,** 255–257.

Tong, P. Y., Tollefsen, S. E., and Kornfeld, S. (1988). The cation-independent mannose 6-phosphate receptor binds insulin-like growth factor II. *J. Biol. Chem.* **263,** 2585–2588.

Wang, Z. Q., Fung, M. R., Barlow, D. P., and Wagner, E. F. (1994). Regulation of embryonic growth and lysosomal targeting by the imprinted Igf2/Mpr gene. *Nature* **372,** 464–467.

Westlund, B., Dahms, N. M., and Kornfeld, S. (1991). The bovine mannose 6-phosphate/insulin-like growth factor II receptor. Localization of mannose 6-phosphate binding sites to domains 1–3 and 7–11 of the extracytoplasmic region. *J. Biol. Chem.* **266,** 23233–23239.

Williams, C., Rezgui, D., Prince, S. N., Zaccheo, O. J., Foulstone, E. J., Forbes, B. E., Norton, R. S., Crosby, J., Hassan, A. B., and Crump, M. P. (2007). Structural insights into the interaction of insulin-like growth factor 2 with IGF2R domain 11. *Structure* **15,** 1065–1078.

Wylie, A. A., Murphy, S. K., Orton, T. C., and Jirtle, R. L. (2000). Novel imprinted DLK1/GTL2 domain on human chromosome 14 contains motifs that mimic those implicated in IGF2/H19 regulation. *Genome Res.* **10,** 1711–1718.

Xu, Y., Papageorgiou, A., and Polychronakos, C. (1998). Developmental regulation of the soluble form of insulin-like growth factor-II/mannose 6-phosphate receptor in human serum and amniotic fluid. *J. Clin. Endocrinol. Metab.* **83,** 437–442.

Yandell, C. A., Dunbar, A. J., Wheldrake, J. F., and Upton, Z. (1999a). The kangaroo cation-independent mannose 6-phosphate receptor binds insulin-like growth factor II with low affinity. *J. Biol. Chem.* **274,** 27076–27082.

Yandell, C. A., Francis, G. L., Wheldrake, J. F., and Upton, Z. (1999b). Kangaroo IGF-II is structurally and functionally similar to the human. *J. Endocrinol.* **161,** 445–453.

York, S. J., Arneson, L. S., Gregory, W. T., Dahms, N. M., and Kornfeld, S. (1999). The rate of internalization of the mannose 6-phosphate/insulin-like growth factor II receptor is enhanced by multivalent ligand binding. *J. Biol. Chem.* **274,** 1164–1171.

Young, L. E., Fernandes, K., McEvoy, T. G., Butterwith, S. C., Gutierrez, C. G., Carolan, C., Broadbent, P. J., Robinson, J. J., Wilmut, I., and Sinclair, K. D. (2001). Epigenetic change in IGF2R is associated with fetal overgrowth after sheep embryo culture. *Nat. Genet.* **27,** 153–154.

Zaccheo, O. J., Prince, S. N., Miller, D. M., Williams, C., Kemp, C. F., Brown, J., Jones, E. Y., Catto, L. E., Crump, M. P., and Hassan, A. B. (2006). Kinetics of insulin-like growth factor II (IGF-II) interaction with domain 11 of the human IGF-II/mannose 6-phosphate receptor: Function of CD and AB loop solvent-exposed residues. *J. Mol. Biol.* **359,** 403–421.

Zaina, S., and Squire, S. (1998). The soluble type 2 insulin-like growth factor (IGF-II) receptor reduces organ size by IGF-II-mediated and IGF-II-independent mechanisms. *J. Biol. Chem.* **273,** 28610–28616.

Index

D

J

K

L

O

P

R

W

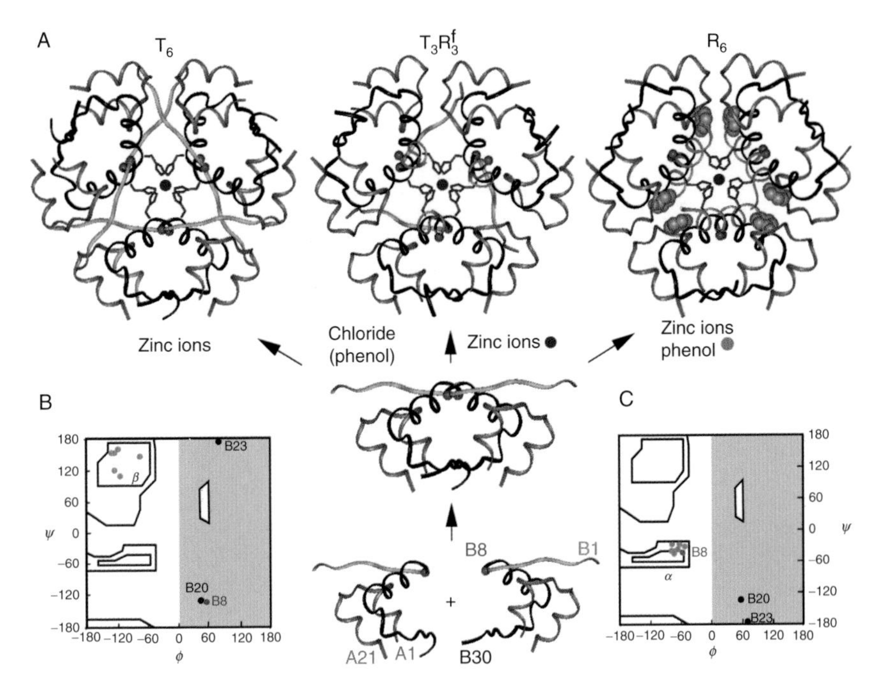

Michael A. Weiss, Figure 2.2 Structural families of insulin hexamers. (A) Left to right, ribbon models of T_6, $T_3R^f_3$, and R_6 zinc insulin hexamers. Central panel indicates pathway of insulin assembly and regulators of conformational reorganization (arrows). The C_α positions of Gly^{B8} are indicated by red balls; the variable secondary structures of the N-terminal segment of the B-chain (residues B1–B7) are shown highlighted in green (extended in T-state) or poweder blue (α-helical in R-state). The B-chain is otherwise shown in black, and A-chain in gray. The side chains of His^{B10} in the metal-ion binding sites are shown in dark blue; zinc ions in magenta; and phenol in burnt amber. (B and C) The TR transition is associated with a conformational change of Gly^{B8} from right to left in the Ramachandran plot. Main-chain dihedral conformations of the three glycines in the B-chain (residues B8, B20, and B23) and residues B2–B8 in a representative T-state (B) or R-state (C) protomer. The conformation of Gly^{B8} is indicated by red; black circles indicate Gly^{B20} and Gly^{B23}. Residues B2–B7 are shown in green (in β-region in T-state) or blue (within α-helical island in R-state).

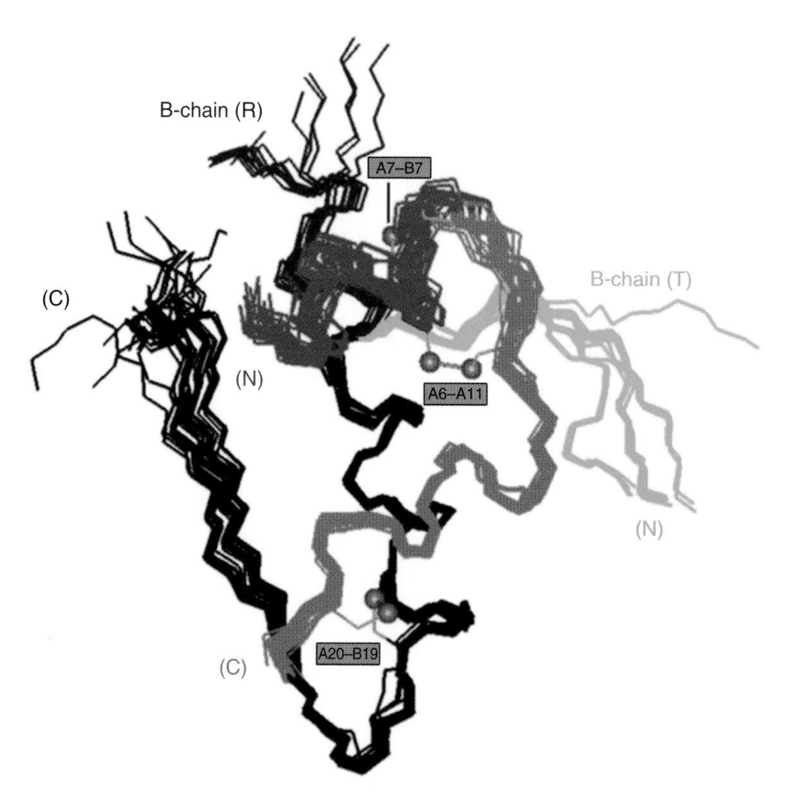

Michael A. Weiss, Figure 2.3 Structural variation among crystallographic protomers. Superposition of crystallographic protomers (15 T states and 15 R states). The structures were aligned according the main-chain atoms of residues B9–B24 and A12–A21. The A1–A8 α-helix in each protomer is shown in red; the variable secondary structure of the N-terminal segment of the B-chain is shown in green (extended in T state; right) or blue (extended α-helix in R state; left). Structures were obtained from the following entries in the Protein Data Bank: (T states), 4INS, 1APH, 1BPH, 1CPH, 1DPH, 1TRZ, 1TYL, 1TYM, 2INS, 1ZNI, 1LPH, 1G7A, 1MSO; (R states), 1EV6, 1ZNJ, 1TRZ, 1ZNI, 1LPH.

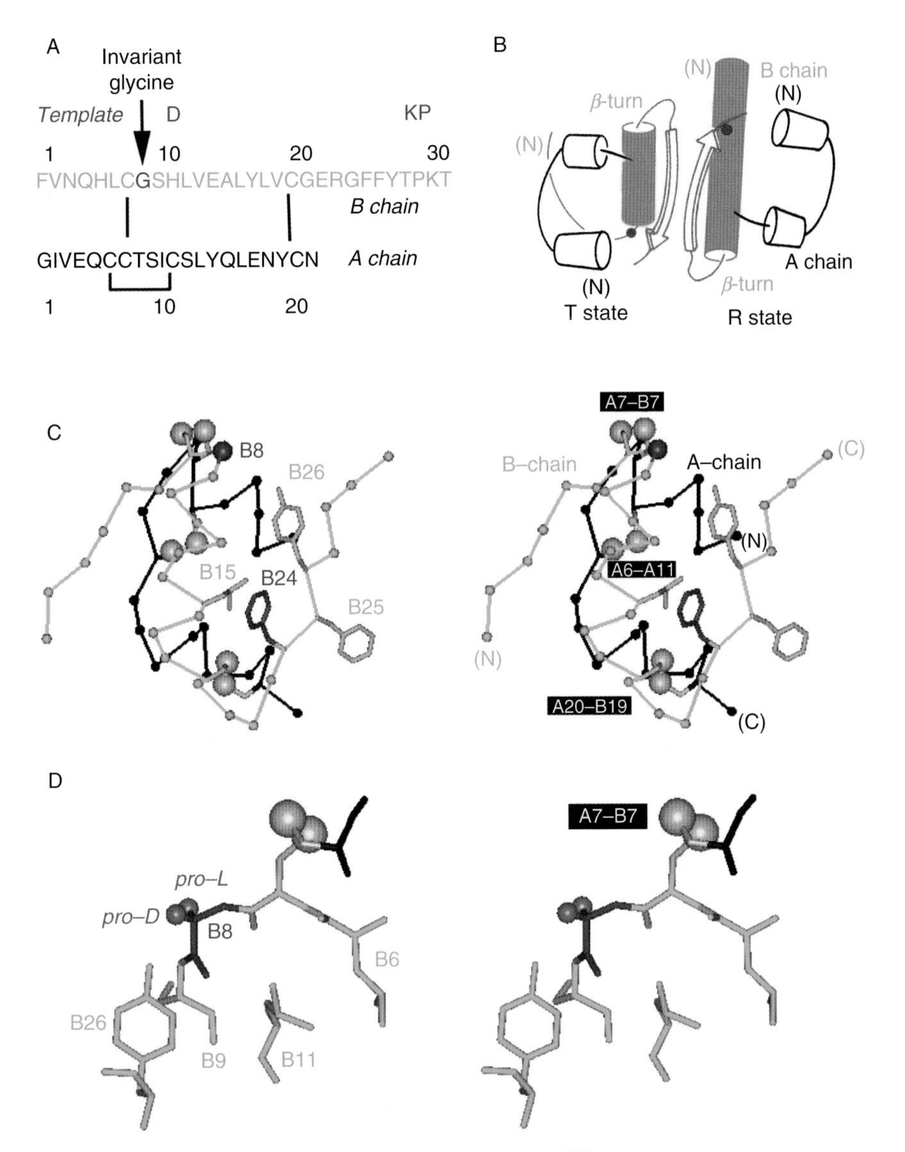

Michael A. Weiss, Figure 2.4 Conformation of Gly^{B8} in a T-state-specific β-turn. (A) Sequence of B chain (top) and A chain (bottom); arrow indicates invariant Gly^{B8} (red). Shown above B chain in magenta are the three substitutions in the monomeric DKP template. (B) Cylinder models of TR dimer based on crystal structure of zinc insulin hexamers (PDB ID: 1TRZ). The T state is at left and R state at right. B-chain α-helices are shown in green; the α-carbons of Gly^{B8} are shown as red circles. Three families of hexamers have been characterized, designated T_6, $T_3R_3^f$, and R_6. The R-state conformation has only been observed within hexamers. (C) Structure of insulin T state (stereo pair) showing positions of selected side chains (labeled at left) relative to Gly^{B8} C_α (red) and disulfide bridges (gold; labeled at right). The B chain is shown in green, and A chain in black. (D) Structure of T-state-specific B7–B10 β-turn (stereo pair). Main chain of Gly^{B8} is shown in red; its pro-L and pro-D H_α atoms are highlighted in blue and magenta, respectively. This figure is reprinted from Hua *et al.* (2006b).

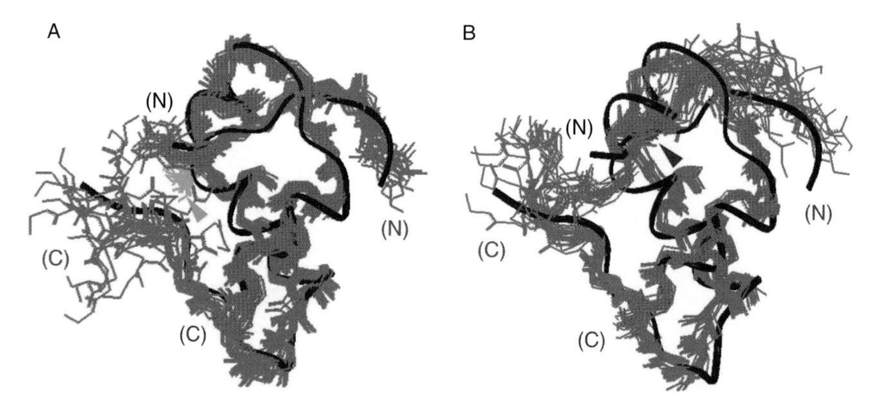

Michael A. Weiss, Figure 2.5 NMR-derived solution structures of D-Ser^{B8} and L-Ser^{B8} analogs of an engineered insulin monomer. Front and back views of (A) D-Ser^{B8}-DKP-insulin and (B) L-Ser^{B8}-DKP-insulin. In each case the A chain is shown in gray, and B chain in blue. The D- and L-^{SerB8} side chains are shown in green and purple, respectively (arrowheads). Ribbons indicate mean structure of DKP-insulin. This figure is reprinted from Hua *et al.* (2006b).

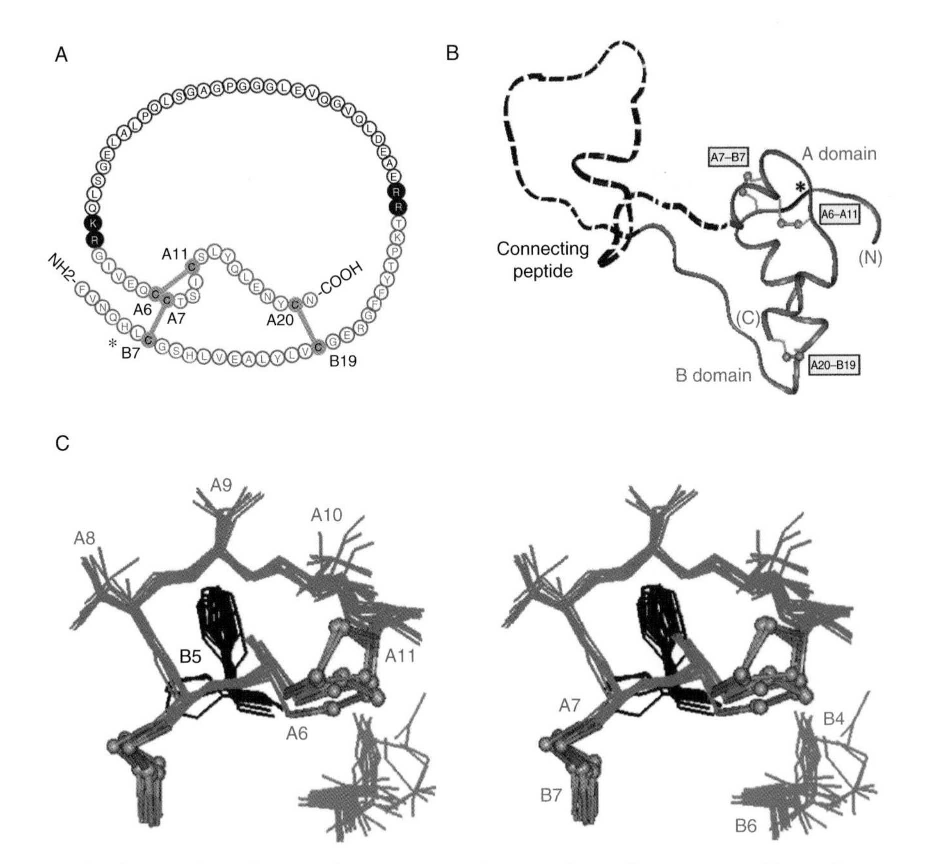

Michael A. Weiss, Figure 2.6 Structure of proinsulin and T-state-specific environment of His^{B5}. (A) Sequence of human proinsulin: insulin moiety is shown in red (A chain) and blue (B chain). The connecting region is shown in black: flanking dibasic cleavage sites (filled circles) and C-peptide (open circles). (B) Structural model of insulin-like moiety and disordered connecting peptide (dashed line). Cystines are labeled in yellow boxes. (C) Structural environment of His^{B5} within A-chain-related crevice. Structures are drawn from T-state coordinates given by PDB code in legend to Figure 2.3. This figure is reprinted from Hua *et al.* (2006a).

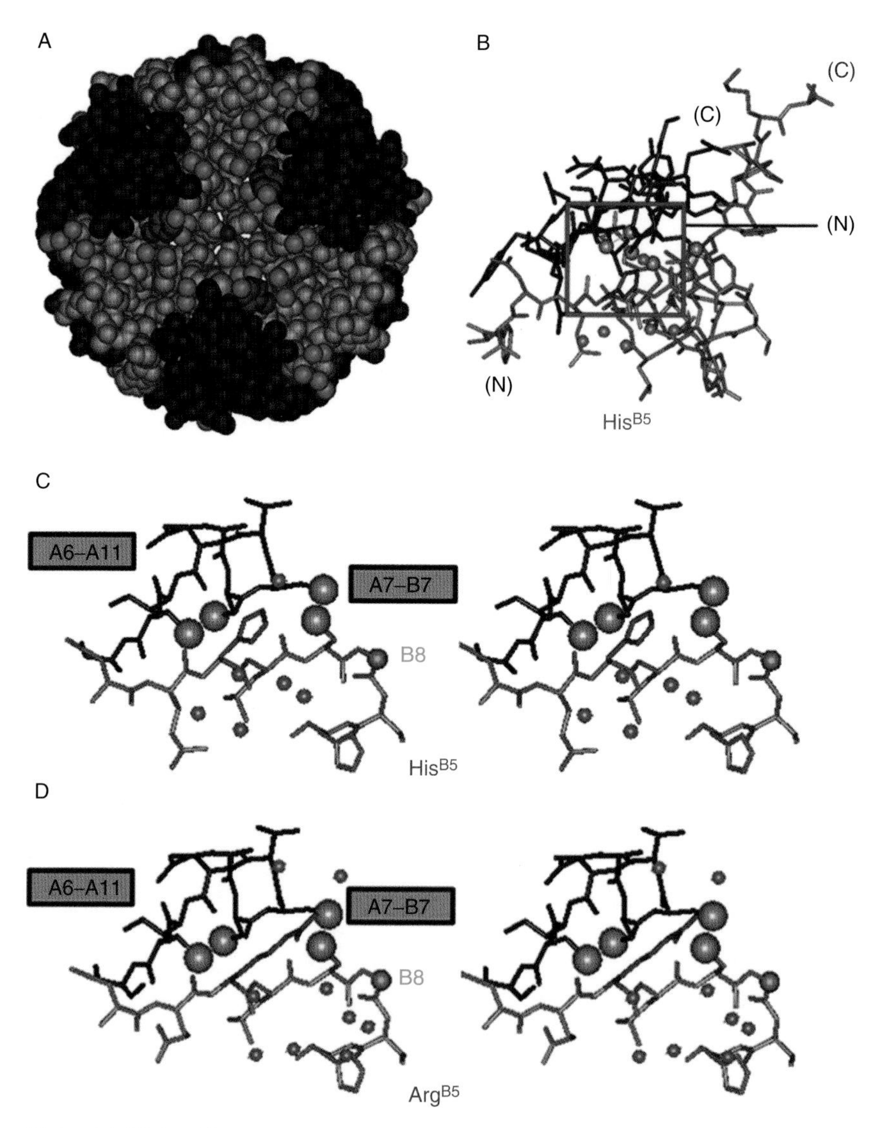

Michael A. Weiss, Figure 2.7 Structural environment of residue B5 in T_6 hexamer and T-state protomers. (A) Spacing filling model of T_6 hexamer showing side chain of His^{B5} (red) lying along protein surface near bound water molecules (blue). (B) Stick model of single T-state protomer with B-chain in black and A-chain in gray. Red box encloses environment of His^{B5} (red) in inter-chain crevice near A7–B7 and A6–A11 disulfide bridges (sulfur atoms shown in gold). (C) Expansion of boxed region in panel B providing stereo view of packing of His^{B5}. (D) Corresponding stereo view of Arg^{B5} (red) in crystallographic T-state protomer. Analogous side-chain NH functions of His^{B5} and Arg^{B5} are near the main-chain of the A-chain. Cystine A7–B7 in each case lies on the protein surface whereas cystine A6–A11 packs within the core of the protomer. Bound water molecules near respective B5-related crevices are shown as blue spheres. This figure is reprinted from Wan *et al.* (2008).

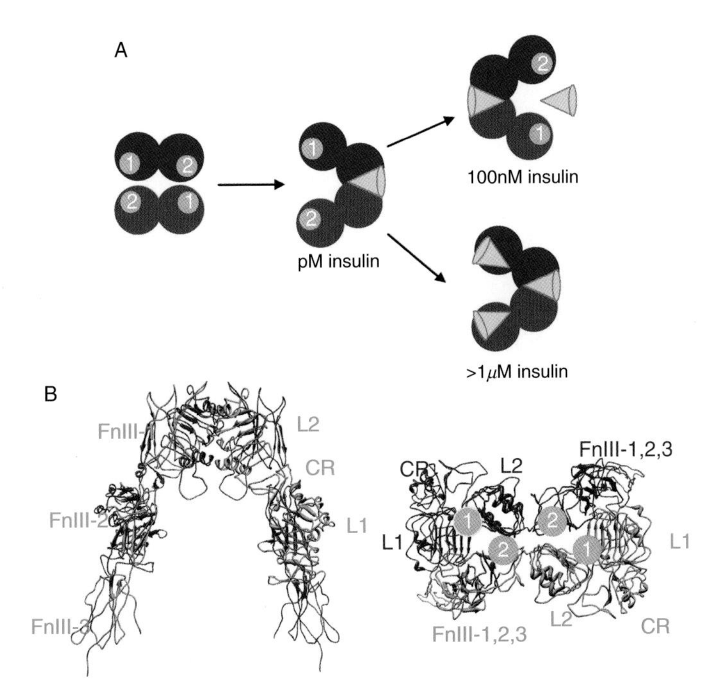

Maja Jensen and Pierre De Meyts, Figure 3.2 Insulin binding to the two binding sites on the IR. Panel A shows the proposed symmetrical model of bivalent-crosslinking insulin-binding mechanism (De Meyts *et al.*, 1994). This model suggests that insulin acts as a bivalent ligand featuring two binding sites on the opposite faces of the molecule that contact two different binding epitopes located on each of the receptor α-subunits. The two α-subunits in each receptor are arranged in an antiparallel fashion, allowing alternative crosslinking of both pairs of binding sites. At low concentrations one insulin molecule binds to a binding site pair (one site on each receptor half) thereby crosslinking the half-receptors (creating a high affinity site). This leaves the second pair of binding sites in a low affinity state. At higher concentrations of insulin (in the range 1–100 nM), insulin binds and crosslink the alternative pair of binding sites. This changes the conformation of the receptor and accelerates the dissociation of the already bound insulin molecule, since it is now monovalently bound. However, at very high insulin concentration (>1 μM), two insulin molecules will occupy the alternative binding sites simultaneously. This will prevent their crosslinking, stabilize the binding of the first bound insulin molecule, and thereby abolish the accelerated dissociation (De Meyts and Whittaker, 2002; De Meyts *et al.*, 1994). Panel B shows the three-dimensional structure of the ectodomain of the IR dimer determined by X-ray crystallography (McKern *et al.*, 2006). The view perpendicular to the ctodomain twofold axis is shown as well as the view along the twofold axis looking towards the cell membrane. One monomer is shown in green and the other in blue. The Individual domains are indicated as well as the insulin binding sites 1 and 2 on the receptor (gray circles). Modeling was carried out using Chimera (pdb file: 2DTG).

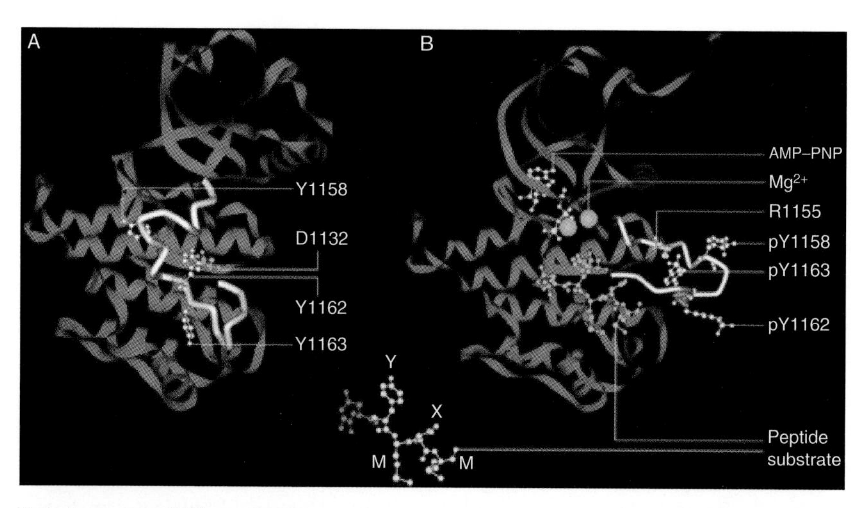

Maja Jensen and Pierre De Meyts, Figure 3.3 Activation of the tyrosine kinase. Panel a shows the inactive IR tyrosine kinase (Hubbard *et al.*, 1994) and panel b the activated kinase (Hubbard, 1997) with a ATP analogue (AMP–PNP), peptide substrate, and magnesium (Mg^{2+}) bound. This figure illustrates the autoinhibition mechanism: The tyrosine Tyr1158 (that is located in the activation loop) is bound in the active site and hydrogen-bonded to a conserved Asp1132 residue in the catalytic loop (panel A) and thereby competes with protein substrates (before autophosphorylation). In the activated state (panel B), the activation loop is tris-phosphorylated (on Tyr1158, Tyr1162, and Tyr1163) and moves out of the active site. Tyr1163 becomes hydrogen-bonded to a conserved Arg1155 residue in the beginning of the activation loop, which stabilizes the repositioned loop. Also shown is the peptide substrate, with the YMXM motif. AMP–PNP, adenylyl imidodiphosphate. The amino acids are numbered based on IR isoform B. From (De Meyts and Whittaker, 2002).

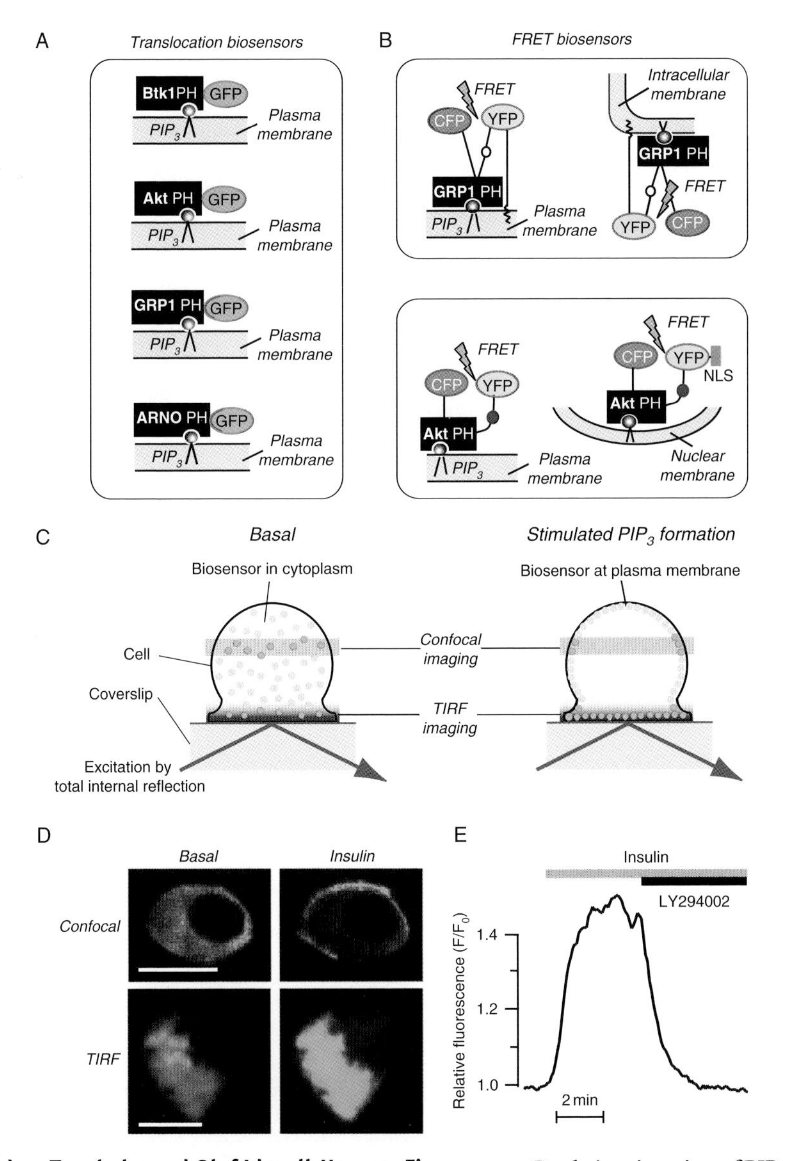

Anders Tengholm and Olof Idevall-Hagren, Figure 11.2 Real-time imaging of PIP_3 in individual cells. (A) PIP_3 translocation biosensors based on the green fluorescent protein (GFP) or its color variants fused to the PH domain from Bruton's tyrosine kinase 1 (Btk1), Akt/protein kinase B, general receptor for phosphoinositides-1 (GRP1) or ADP ribosylation factor nucleotide site opener (ARNO) that bind PIP_3 in the plasma membrane. (B) PIP_3 biosensors based on intramolecular fluorescence resonance energy transfer (FRET) between cyan and yellow fluorescent protein (CFP and YFP) as a result of a conformational change in the PH domain upon PIP_3 binding. In one biosensor based on the PH domain from GRP1 the conformational change is enhanced by membrane targeting to cause a rotation of rigid linkers around a diglycine hinge engineered within the construct (Sato *et al.*, 2003). This sensor can be targeted to the plasma membrane or to intracellular membranes. Another sensor based on the PH domain from Akt uses a pseudoligand sequence to enhance the FRET response. This

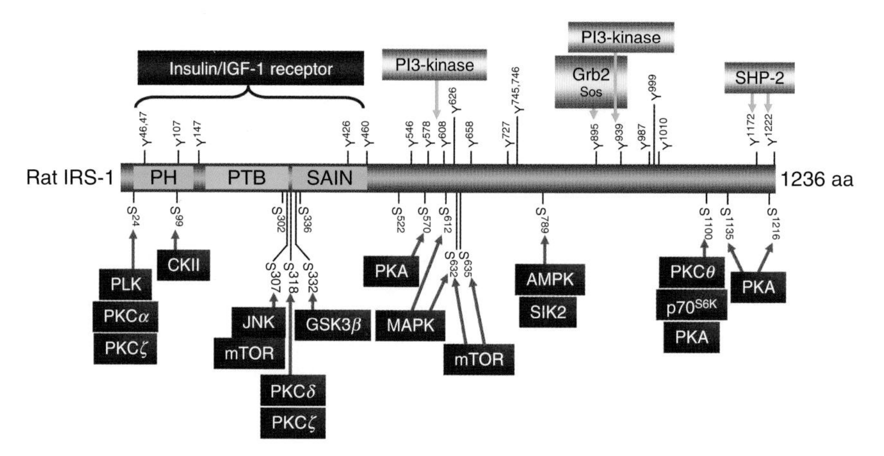

Xiao Jian Sun and Feng Liu, Figure 13.2 Tyrosine and serine phosphorylation of IRS-1. IRS-1 is phosphorylated on tyrosine residues by insulin receptor/IGF-1 receptor tyrosine kinases, eliciting the formation of signaling complexes with multiple downstream proteins (green). IRS-1 is also phosphorylated by many serine/threonine kinases (red) which predetermine the tyrosine phosphorylation events.

---

reporter offers dual readout of FRET change and translocation in its untargeted form, but it can also be directed to subcellular compartments. For example, using a nuclear localization sequence (NLS), the biosensor can be expressed in the cell nucleus (Ananthanarayanan *et al.*, 2005). (C) Schematic illustration of fluorescent biosensor translocation in response to PIP_3 formation in the plasma membrane. A confocal microscope images an optical section through the cell at any level, whereas a total internal reflection fluorescence (TIRF) microscope visualises a thin volume within ~100 nm from the plasma membrane in the coverslip adhesion region. This part of the cell is selectively excited by an evanescent wave generated by total internal reflection of a laser beam at the coverslip–water interphase. (D) Translocation of a GFP-GRP1 construct expressed in insulin-secreting MIN6 β-cells. Confocal and TIRF microscopy images were acquired before and 5 min after stimulation with 100 nM insulin. In the TIRF microscope, membrane translocation is detected as an overall increase of fluorescence intensity. Scale bars, 10 μm. (E) Time-lapse TIRF recording of the insulin-induced PIP_3 formation in a single MIN6 β-cell expressing a GFP-GRP1 biosensor. Insulin thus triggers biosensor translocation to the membrane (increase of fluorescence), which is counteracted by 100 μM of the PI3-kinase inhibitor LY294002. Fluorescence intensity is expressed in relation to the prestimulatory level (F/F_0).

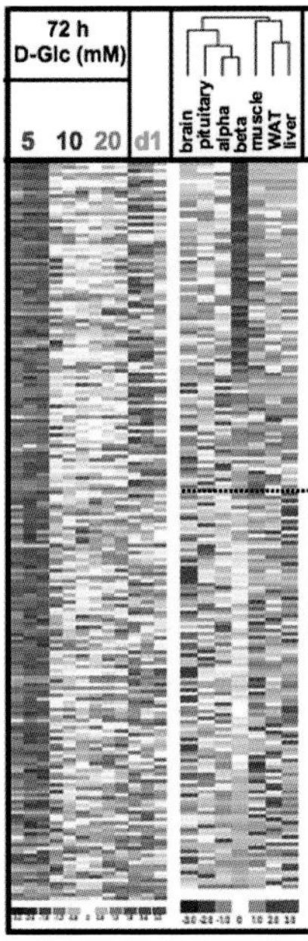

72 h D-Glc (mM): 5, 10, 20, d1	brain, pituitary, alpha, beta, muscle, WAT, liver	Symbol	Gene name	qPCR tissue spec.	qPCR glucose induced
		GCK	glucokinase	β, liver	ND
		HADHSC	L-3-hydroxyacyl-coA dehydrogenase, short chain	β	+
		GLUT2	SLC family 2 (facilitated glucose transporter), member 2	β, liver	ND
		CART	cocaine and amphetamine regulated transcript	α, β	+
		PTPRN	protein tyrosine phosphatase, receptor type, N (IA2)	ND	ND
		SYTL4	synaptotagmin-like 4	ND	ND
		DAD1	defender against cell death 1	ND	ND
		RAB3D	RAB3D, member RAS oncogene family	ND	ND
		PCK2	phosphoenolpyruvate carboxykinase 2 (mitochondrial)	ND	ND
		SYT13	synaptotagmin XIII	ND	ND
		CCND1	cyclin D1	β, kidney	+
		DCX	doublecortin	β	+
		PRLR	prolactin receptor	β, kidney	+
		PCSK1	proprotein convertase subtilisin/kexin type 1	ND	ND
		CCKAR	cholecystokinin A receptor	β	+
		TRH	thyrotropin releasing hormone	ND	ND
		CHGA	chromogranin A	ND	ND
		WNT4	wingless-related MMTV integratin site 4	ND	ND
		GIPR	gastric inhibitory polypeptide receptor	β, α	+
		IDI1	isopentenyl-diphosphate delta isomerase	ND	+
		CREM	cAMP responsive element modulator	liver	+
		SNAP25	synaptosomal-associated protein 25	ND	ND
		SYP	synaptophysin	ND	ND
		GNAS	GNAS complex locus	ND	ND
		GHR	growth hormone receptor	liver, WAT	+
		FN1	fibronectin 1	liver	+

Geert A. Martens and Daniel Pipeleers, Figure 17.2 Selection of representative transcripts that are induced by 10 versus 5 mM glucose in cultured beta cells. The left heat map shows relative mRNA expression level in freshly isolated (day 1, d1) beta cells, and in cells cultured for 3 days at 5, 10, or 20 mM glucose. The right heat map shows corresponding relative mRNA expression in various rat tissue/cell types; genes are ordered by decreasing beta cell-abundance/specificity. The two columns on the right indicate eventual confirmation by Taqman qPCR of tissue-specific mRNA expression (left column) or glucose-induction (right column); (ND, not done; +, confirmed).

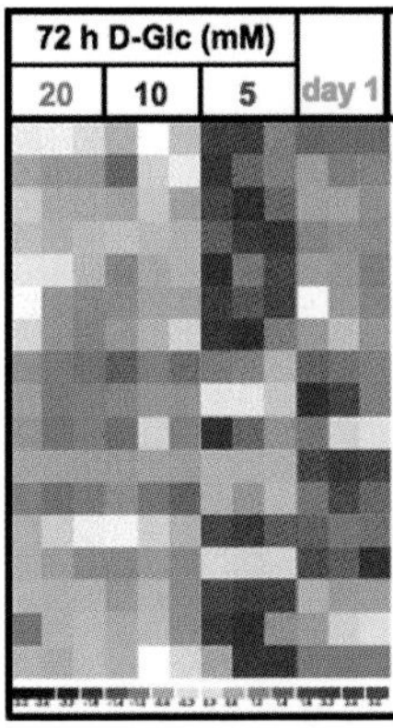

72 h D-Glc (mM): 20, 10, 5, day 1	Symbol	Gene name	qPCR
	DBP	D site albumin promotor binding protein	ND
	NR1D2	nuclear receptor subfamily 1, group D, member 2 (Rev-ERB beta)	ND
	CREG1	cellular repressor of E1A-stimulated genes	ND
	MGMT	O-6-methylguanine-DNA methyltransferase	+
	CDKN1A	cyclin-dependent kinase inhibitor 1A (p21, cip1)	+
	POLA2	polymerase (DNA directed), alpha 2	ND
	CHES1	checkpoint suppressor 1	ND
	DDIT3	DNA-damage inducible transcript 3 (GADD153, CHOP)	+
	HERPUD1	HCY-inducible, ER stress-inducible, ubiquitin-like domain member 1	ND
	DNAJB1	DnaJ (Hsp40) homolog, subfamily B, member 1	ND
	ATF3	activating transcription factor 3	ND
	MYC	myelocytomatosis viral oncogene homolog, avian (c-Myc)	+
	NFE2L2	nuclear factor, erythroid derived 2, like 2 (NRF2)	+
	JUN	Jun oncogene	+
	TRIB3	tribbles homolog 3 (Drosophila)	+
	TFR	transferrin receptor	ND
	LCN2	lipocalin 2	ND

Geert A. Martens and Daniel Pipeleers, Figure 17.3 Selection of transcripts that are induced in 5 versus 10 mM glucose in cultured beta cells. This selection emphasizes relevant stress-protective/pro or antiapoptotic transcripts and is thus not representative for the whole low glucose-transcriptome. The heat map shows relative expression levels in freshly isolated (day 1) and cells cultured at indicated glucose concentration. The right column indicates eventual confirmation by qPCR of the glucose-regulation.

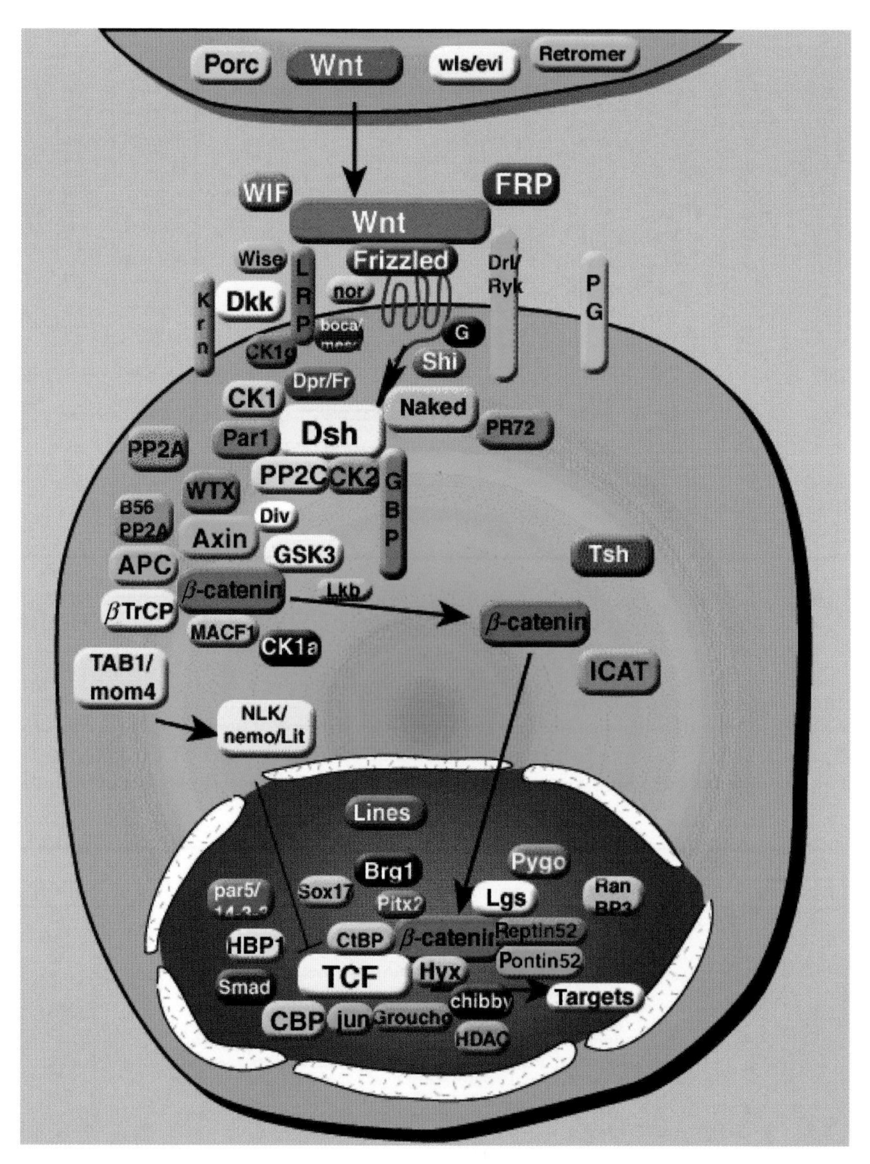

Michael Bordonaro, Figure 19.1 Schematic of Wnt signaling. In the absence of Wnt signaling, free intracellular beta-catenin is bound within complexes containing APC, Axin, GSK-3β, and other proteins; beta-catenin is phosphorylated and degraded through the proteosomal pathway. Wnt signaling is normally activated by the interactions of Wnt ligands with their receptors at the plasma membrane. This interaction transmits a signal which results in the disruption of the beta-catenin degradation complex, leading to the accumulation of dephosphorylated, active beta-catenin. The nuclear accumulation of active beta-catenin allows for the formation of a transcriptional complex between beta-catenin and the DNA-binding Tcf factors. In the absence of beta-catenin, Tcf factors associate with transcriptional corepressors and therefore downregulate expression of genes with Tcf binding sites. When complexed with beta-catenin, Tcf factors lose their association with transcriptional repressors; beta-catenin contains a transcriptional transactivation domain, and the beta-catenin-Tcf complex associates with transcriptional coactivators. Therefore, Wnt activation results in the expression of target genes containing Tcf binding sites. Mutations in the Wnt signaling pathway, most often in the *APC* and beta-catenin genes, result in constitutive activation of Wnt signaling, which can be independent of the requirement for upstream Wnt ligand–receptor interactions (however, these interactions may still occur, and further upregulate Wnt activity). The deregulated Wnt activity can promote colonic tumorigenesis, although hyperactivation of this pathway induces apoptosis in colorectal cancer cell lines. This figure was reproduced with permission from Dr. R. Nusse from http://www.stanford.edu/~rnusse/pathways/cell2.html.

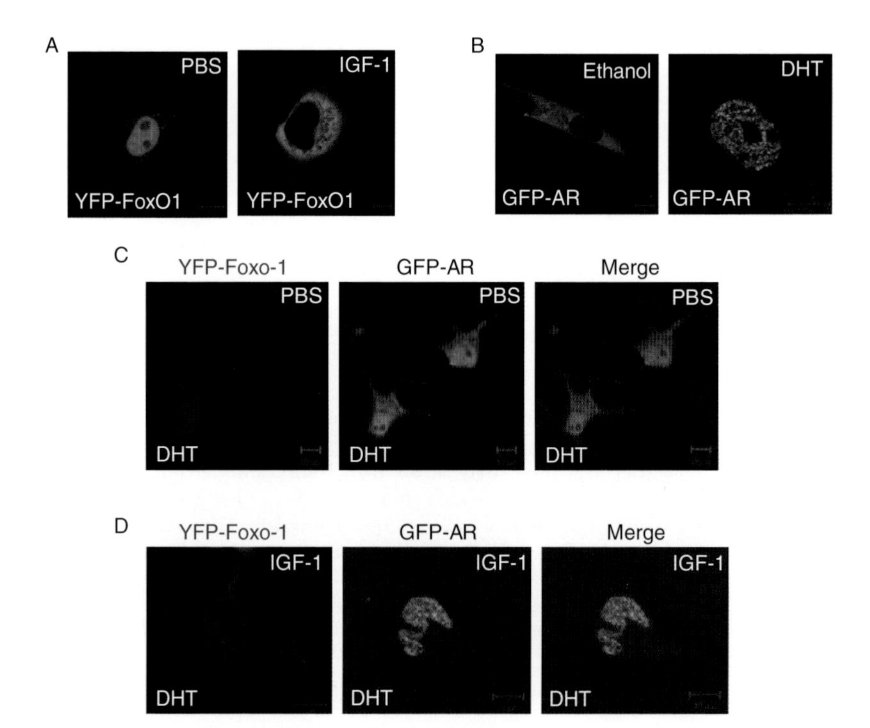

Toshihiko Yanase and WuQiang Fan, Figure 23.4 Subcellular interaction between Foxo1 and AR in living COS7 cells. The expression of chimeric fluorescent proteins was observed in living COS7 cells using a Zeiss LSM 510 META laser confocal microscope. COS7 cells were transfected with 0.5 μg/dish of EYFP-Foxo1, and then treated with PBS or 100 ng/ml IGF-1 (A). COS7 cells expressing EGFP-AR were exposed to ethanol or DHT (10^{-8} M) (B), respectively. COS7 cells coexpressing EGFP-AR and EYFP-Foxo1 were exposed to DHT alone (C) or DHT+IGF-1 (D). G Scale bars, 10 μm. Phosphorylated nuclear YFP-FOX-1 with IGF-1 exported to the cytoplasm (A). DHT activated GFP-AR and translocated it from cytosol to nucleus to form fine foci in the nucleus (B). By the coexpression of YFP-Foxo-1, formation of fine foci of GFP-AR by DHT was disrupted, showing a diffuse distribution pattern in the nucleus (C). However, the presence of IGF-1 and DHT showed more complete nuclear translocation of GFP-AR, and significantly improved subnuclear compartmentalization (D).

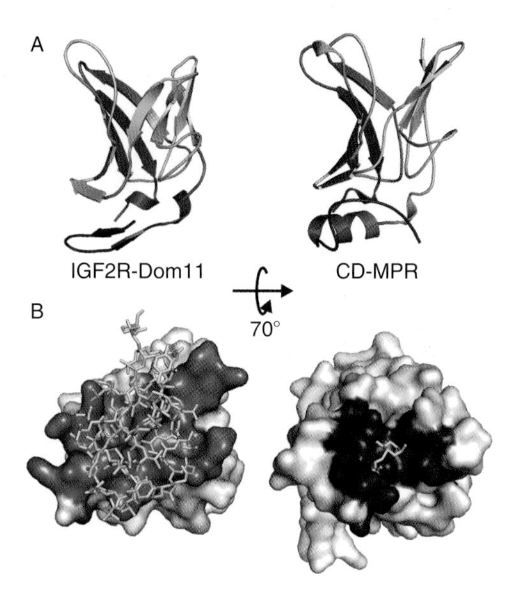

J. Brown *et al.*, Figure 25.1 Comparison of IGF2R-Dom11 and CD-MPR structures. (A) Ribbon diagrams colored from blue at the N-terminus to red at the C-terminus and based on figures originally published in Brown *et al.* (2002). (B) Surface representations showing residues of the IGF2R domain 11 (left) and CD-MPR (right) undergoing a change in accessible surface area upon ligand binding. IGF-II and mannose-6-phosphate ligands are shown as yellow stick representations bound to their respective receptors.

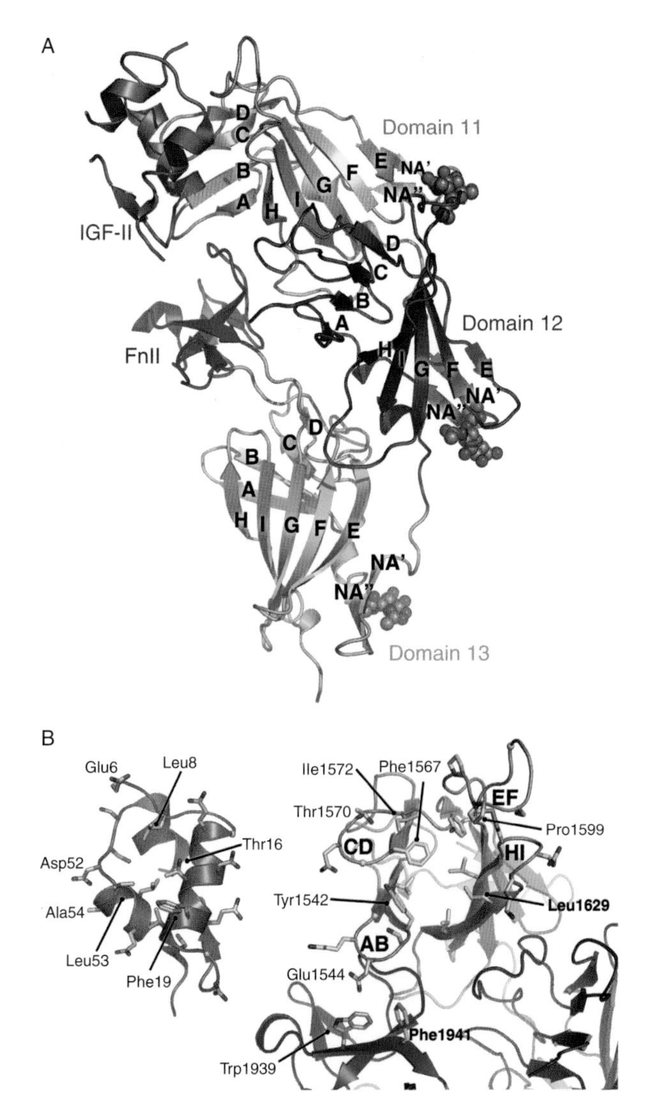

J. Brown *et al.*, Figure 25.2 The IGF2R/IGF-II interface. (A) Cartoon representation of the complex between IGF2R-Dom11–13 (yellow, red, and blue, respectively) and IGF-II (purple). Glycans are shown as spheres. (B) An open-book view of the complex with side chains undergoing a change in accessible surface area upon complexation shown as sticks. Domains are colored as in (A). Based on figures originally published in Brown *et al.* (2008).